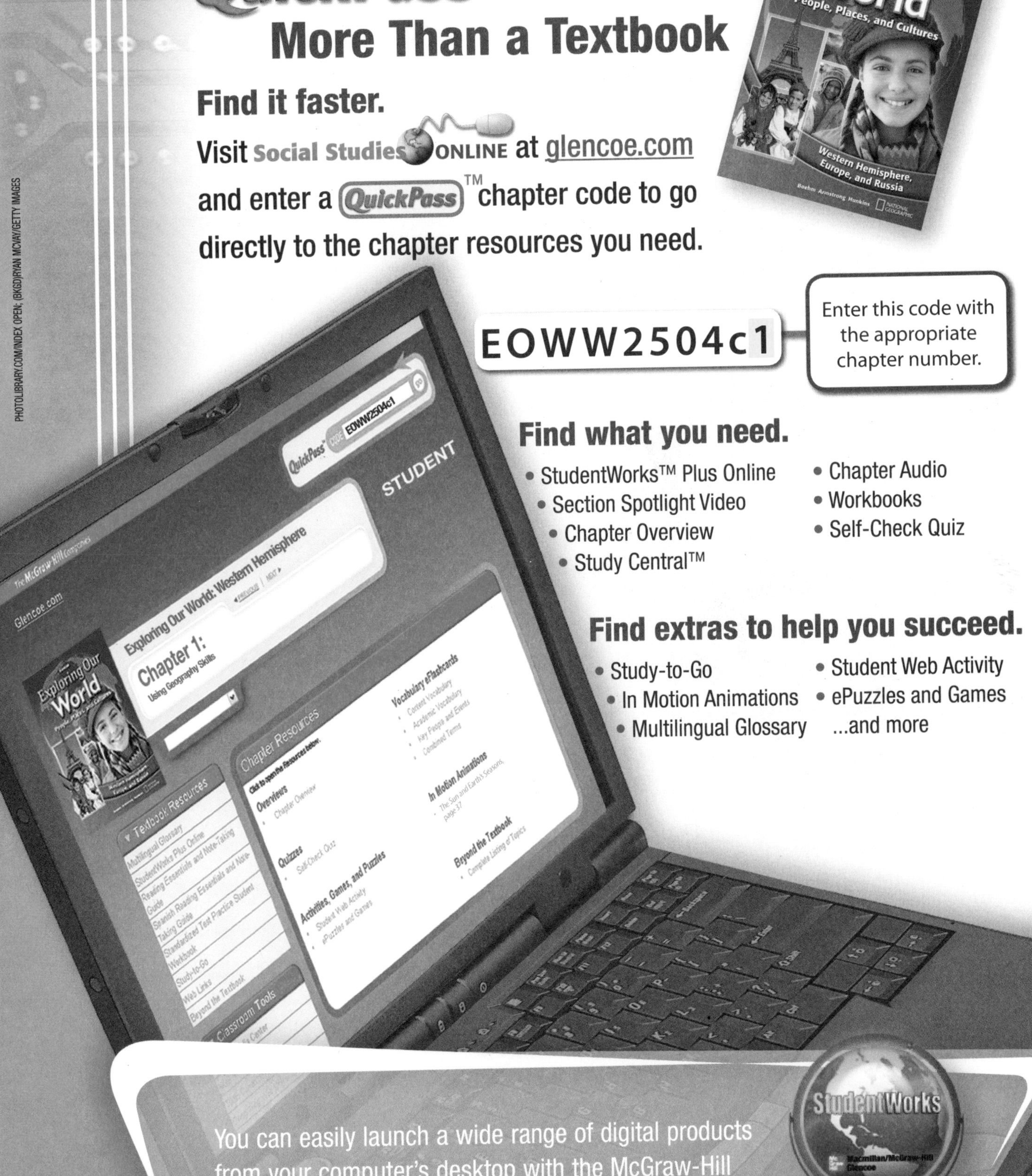

PHOTOLIBRARY.COM/INDEX OPEN; (BKGD)RYAN MCVAY/GETTY IMAGES

Glencoe

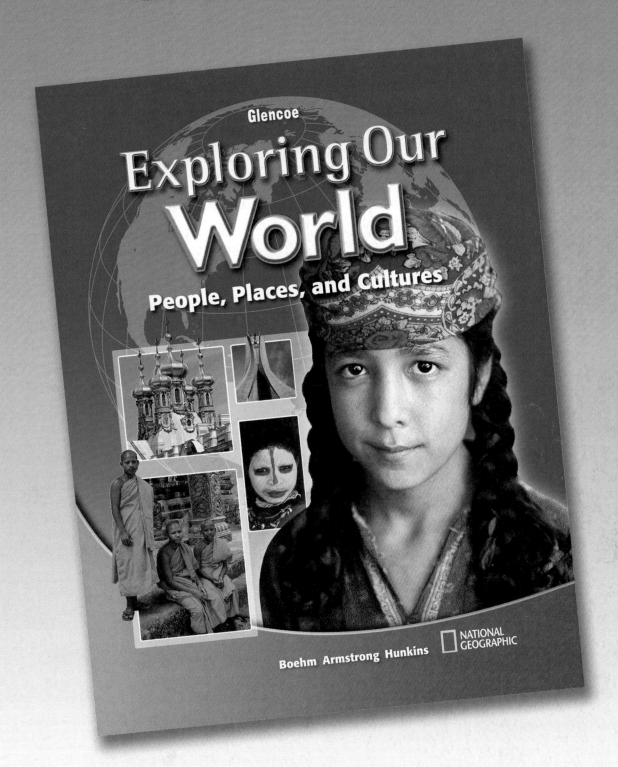

Exploring Our World

People, Places, and Cultures

Boehm Armstrong Hunkins

NATIONAL GEOGRAPHIC

About the Authors

The National Geographic Society, founded in 1888 for the increase and diffusion of geographic knowledge, is the world's largest nonprofit scientific and educational organization. Since its earliest days, the Society has used sophisticated communication technologies, from color photography to holography, to convey geographic knowledge to a worldwide membership. The School Publishing Division supports the Society's mission by developing innovative education programs—ranging from traditional print materials to multimedia programs including CD-ROMs, videos, and software.

Senior Author

Richard G. Boehm

Richard G. Boehm, Ph.D., was one of seven authors of *Geography for Life,* national standards in geography, prepared under Goals 2000: Educate America Act. He was also one of the authors of the *Guidelines for Geographic Education,* in which the Five Themes of Geography were first articulated. Dr. Boehm has received many honors, including "Distinguished Geography Educator" by the National Geographic Society (1990), the "George J. Miller Award" from the National Council for Geographic Education (NCGE) for distinguished service to geographic education (1991), and "Gilbert Grosvenor Honors" in geographic education from the Association of American Geographers (2002). He was President of the NCGE and has twice won the *Journal of Geography* award for best article. He has received the NCGE's "Distinguished Teaching Achievement" award and presently holds the Jesse H. Jones Distinguished Chair in Geographic Education at Texas State University in San Marcos, Texas.

Francis P. Hunkins

Francis P. Hunkins, Ph.D., is Professor of Education at the University of Washington. He began his career as a teacher in Massachusetts. He received his master's degree in education from Boston University and his doctorate from Kent State University with a major in general curriculum and a minor in geography. Dr. Hunkins has written numerous books and articles.

David G. Armstrong

David G. Armstrong, Ph.D., served as Dean of the School of Education at the University of North Carolina at Greensboro. A social studies education specialist with additional advanced training in geography, Dr. Armstrong was educated at Stanford University, University of Montana, and University of Washington.

Dinah Zike

Dinah Zike, M.Ed., is an award-winning author, educator, and inventor known for designing three-dimensional hands-on manipulatives and graphic organizers known as Foldables®. Dinah has developed educational books and materials and is the author of *The Big Book of Books and Activities,* which was awarded Learning Magazine's Teachers' Choice Award. In 2004 Dinah was honored with the CESI Science Advocacy Award. Dinah received her M.Ed. from Texas A&M, College Station, Texas.

RFB&D learning through listening — Students with print disabilities may be eligible to obtain an accessible, audio version of the pupil edition of this textbook. Please call Recording for the Blind & Dyslexic at 1-800-221-4792 for complete information.

The McGraw·Hill Companies

Glencoe

Send all inquiries to:
Glencoe/McGraw-Hill, 8787 Orion Place, Columbus, Ohio 43240-4027

ISBN: 978-0-07-880310-9
MHID: 0-07-880310-1

Printed in the United States of America.
9 QVR/LEH 14 13 12

Academic Consultants

Pratyusha Basu, Ph.D.
Assistant Professor of Geography
University of South Florida
Tampa, Florida

Sari Bennett, Ph.D.
Director, Center for Geographic Education
University of Maryland, Baltimore County
Baltimore, Maryland

David Berger, Ph.D.
Broeklundian Professor of History
Brooklyn College and the Graduate Center
City University of New York
Brooklyn, New York

Dennis Conway, Ph.D.
Professor of Geography
Indiana University
Bloomington, Indiana

Clifford B. Craig, Ph.D.
Professor of Geography and Earth Resources
Utah State University
Logan, Utah

Richard M. Golden, Ph.D.
Professor of History and Director, Jewish Studies Program
University of North Texas
Denton, Texas

Brooks Green, Ph.D.
Associate Professor of Geography
University of Central Arkansas
Conway, Arkansas

Sumit Guha, Ph.D.
Professor of History
Rutgers University
New Brunswick, New Jersey

Gerald T. Hanson, Ph.D.
Professor of Cultural Geography
University of Arkansas at Little Rock
Little Rock, Arkansas

Darrell P. Kruger, Ph.D.
Professor of Geography
Illinois State University
Normal, Illinois

David A. Lanegran, Ph.D.
John S. Holl Professor of Geography
Macalester College
St. Paul, Minnesota

Elizabeth J. Leppman, Ph.D.
Adjunct Professor of Geography
Eastern Kentucky University
Richmond, Kentucky

Catherine M. Lockwood, Ph.D.
Professor of Geography
Chadron State College
Chadron, Nebraska

Farid Mahdavi, Ph.D.
Department of History
San Diego State University
San Diego, California

Paul Nagel, Ph.D.
Coordinator, Louisiana Geography Education Alliance
Northwestern State University
Natchitoches, Louisiana

Bimal Kanti Paul, Ph.D.
Professor of Geography
Kansas State University
Manhattan, Kansas

Joseph P. Stoltman, Ph.D.
Professor of Geography
Western Michigan University
Kalamazoo, Michigan

George W. White, Jr., Ph.D.
Associate Professor of Geography
Frostburg State University
Frostburg, Maryland

Teacher Reviewers

Cynthia L.N. Bloom
Comstock Northeast Middle School
Kalamazoo, Michigan

Ronda Boyd-Roberts
Spring City Middle School
Spring City, Tennessee

Melinda Clay Casey
Truman Middle School
Grand Prairie, Texas

Ann M. Christianson
John Muir Middle School
Wausau, Wisconsin

Cynthia Dabalos
Richview Middle School
Clarksville, Tennessee

Sandra Goldich
Alexandria Middle Magnet School
Alexandria, Louisiana

Julie Hill
Bob Courtway Middle School
Conway, Arkansas

Kathryn Jones
North Reading Middle School
North Reading, Massachusetts

George S. Kuhter
South Tama Middle School
Toledo, Iowa

Clark Norman
Gravelly Hill Middle School
Hillsborough, North Carolina

Robert L. O'Brian
Buffalo Public Schools
Buffalo, New York

Heather Latham Porter
Butler Junior High School
Butler, Pennsylvania

Phyllis M. Quinn
Martin J. Ryerson Middle School
Ringwood, New Jersey

Kim Roberts
University School at East Tennessee State University
Johnson City, Tennessee

Sharon L. Shelerud
Metcalf Junior High School
Burnsville, Minnesota

Sharon Sobierajski
Frederick Law Olmstead Middle School
Buffalo, New York

Melissa L. Younker-Kordish
Calkins Road Middle School
Pittsford, New York

Reading Consultant

ReLeah Lent
National Education Consultant
Alford, Florida

Contents

The Danube River, Hungary

Contents

▼ **Breaking ice on the Arctic Ocean**

West African sandstorm

Contents

Features

Geography & History

Trans-Siberian Railroad train

World Literature

YOU Decide

TIME Features

▲ **Protesting in France**

TIME JOURNAL

Primary Sources

Ed. = Editor	Tr. = Translator	V = Volume

UNIT 4

UNIT 5

UNIT 6

UNIT 7

Maps

NATIONAL GEOGRAPHIC Reference Atlas . . . RA1

UNIT 1

The World

NATIONAL GEOGRAPHIC

The United Kingdom and Ireland

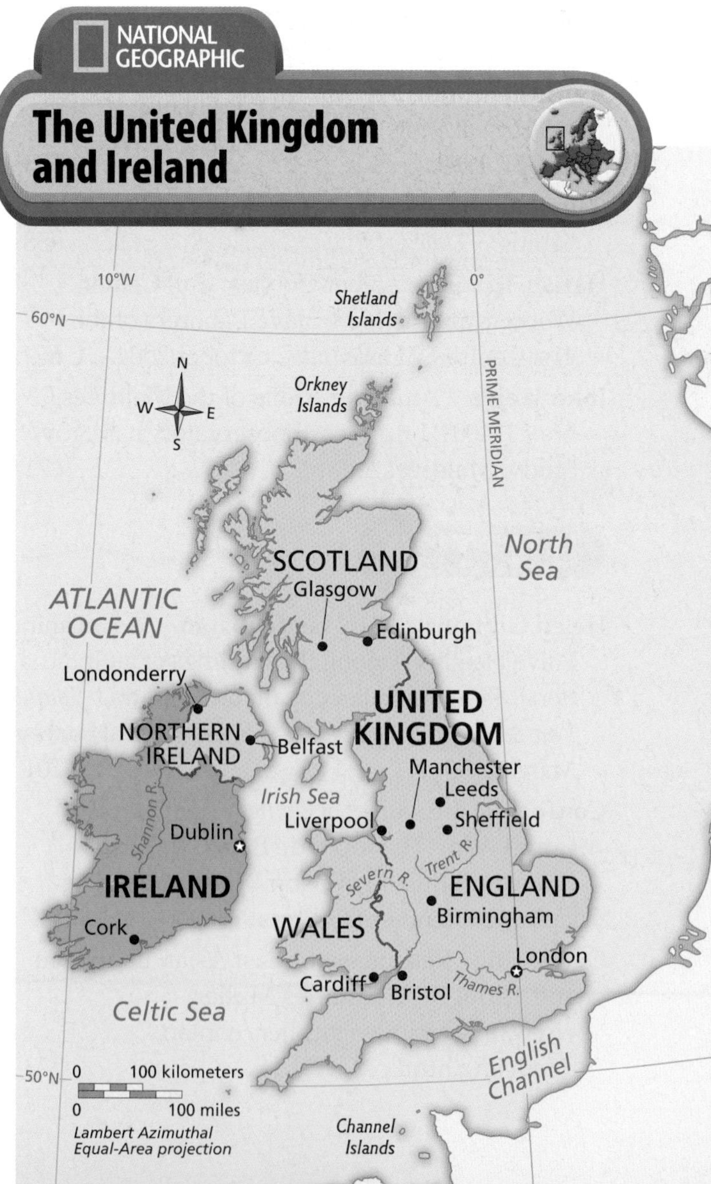

UNIT 2

The United States and Canada

UNIT 3

Latin America

UNIT 4

Europe

UNIT 5

Russia

UNIT 6

North Africa, Southwest Asia, and Central Asia

UNIT 7

Africa South of the Sahara

UNIT 8

South Asia

UNIT 9

East Asia and Southeast Asia

UNIT 10

Australia, Oceania, and Antarctica

NATIONAL GEOGRAPHIC

The Trans-Siberian Railroad

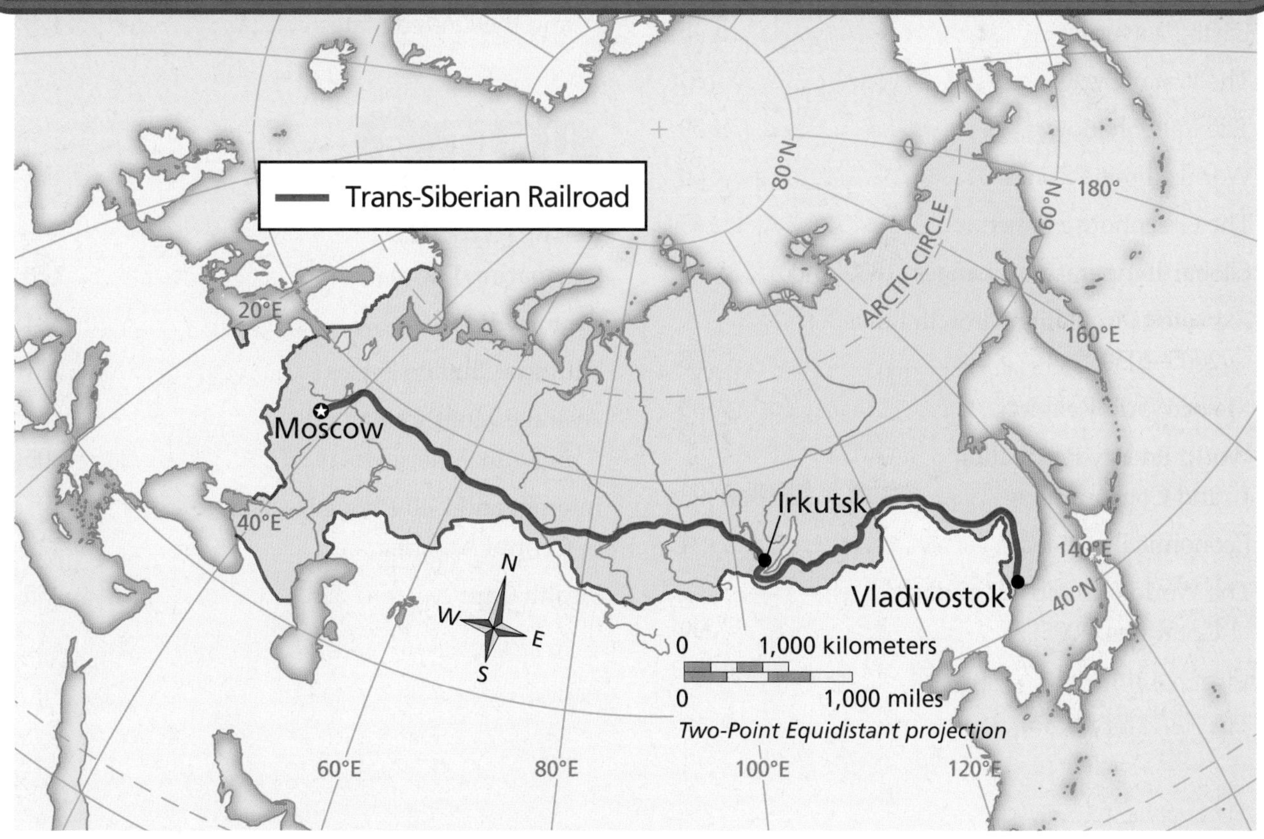

Diagrams, Charts, and Graphs

UNIT 1

The World

UNIT 2

The United States and Canada

UNIT 3

Latin America

UNIT 4

Europe

UNIT 5

Russia

UNIT 6

North Africa, Southwest Asia, and Central Asia

UNIT 7

Africa South of the Sahara

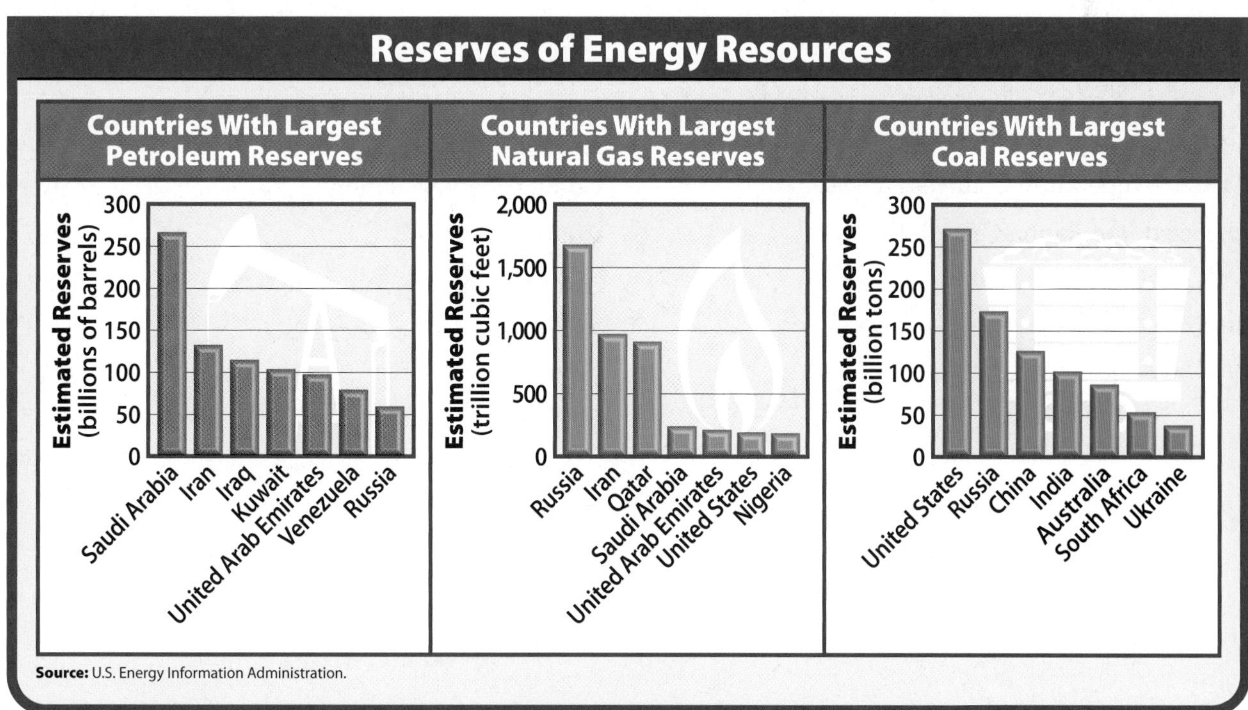

UNIT 8

South Asia

UNIT 9

East Asia and Southeast Asia

UNIT 10

Australia, Oceania, and Antarctica

Scavenger Hunt

Exploring Our World: People, Places, and Cultures contains a wealth of information. The trick is to know where to look to access all the information in the book. If you complete this scavenger hunt exercise with your teacher or parents, you will see how the textbook is organized and how to get the most out of your reading and study time. Let's get started!

1. How many units and how many chapters are in the book?
2. What does Unit 1 cover?
3. Where can you find facts about each country in each unit?
4. In what three places can you learn about the Big Ideas for each section?
5. What does the Foldables Study Organizer at the beginning of Chapter 2 ask you to do?

6. How are the content vocabulary terms throughout your book highlighted in the narrative?
7. Where do you find graphic organizers in your textbook?

8. You want to quickly find all the maps in the book about the world. Where do you look?
9. Where can you practice specific social studies skills in your textbook?
10. Where can you learn about the different types of map projections?

Reference Atlas

ATLAS KEY

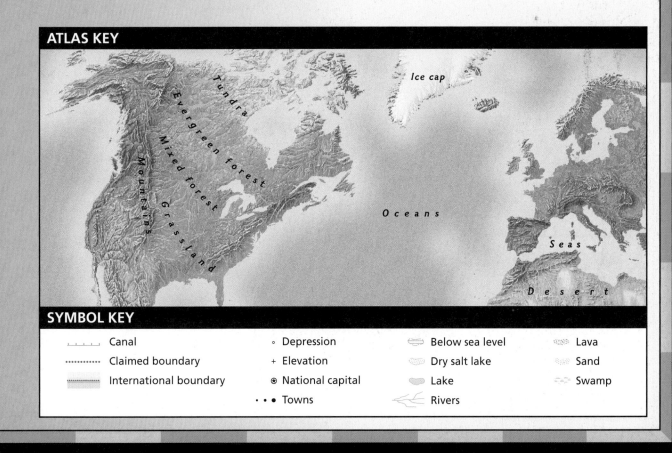

Tundra
Evergreen forest
Mixed forest
Mountains
Grassland
Ice cap
Oceans
Seas
Desert

SYMBOL KEY

Canal	Depression	Below sea level	Lava
Claimed boundary	Elevation	Dry salt lake	Sand
International boundary	National capital	Lake	Swamp
	Towns	Rivers	

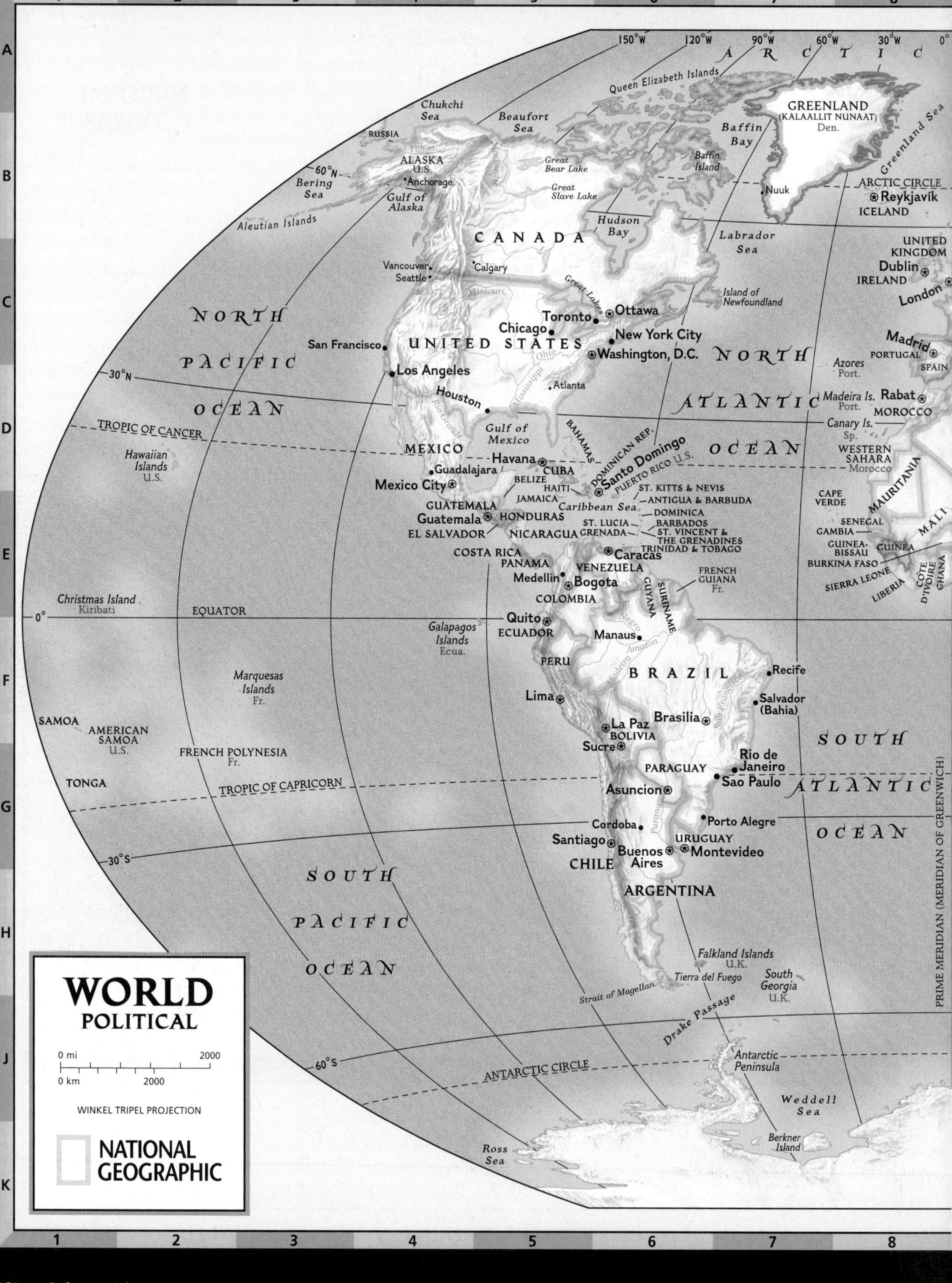

WORLD
POLITICAL
WINKEL TRIPEL PROJECTION
NATIONAL GEOGRAPHIC
0 mi
2000
0 km
2000
ARCTIC
GREENLAND
(KALAALLIT NUNAAT)
Den.
Greenland Sea
Baffin Bay
Baffin Island
Queen Elizabeth Islands
Beaufort Sea
Chukchi Sea
RUSSIA
ALASKA
U.S.
Anchorage
Bering Sea
Gulf of Alaska
Aleutian Islands
Great Bear Lake
Great Slave Lake
Nuuk
ARCTIC CIRCLE
Reykjavík
ICELAND
Hudson Bay
Labrador Sea
CANADA
UNITED KINGDOM
Dublin
IRELAND
London
Vancouver
Seattle
Calgary
Great Lakes
Island of Newfoundland
NORTH PACIFIC OCEAN
Toronto
Ottawa
Chicago
New York City
San Francisco
UNITED STATES
Washington, D.C.
NORTH ATLANTIC OCEAN
Madrid
PORTUGAL
SPAIN
Azores
Port.
Los Angeles
Atlanta
Houston
Madeira Is.
Port.
Rabat
MOROCCO
TROPIC OF CANCER
Gulf of Mexico
BAHAMAS
Canary Is.
Sp.
Hawaiian Islands
U.S.
MEXICO
Havana
CUBA
DOMINICAN REP.
Santo Domingo
PUERTO RICO U.S.
WESTERN SAHARA
Morocco
MAURITANIA
Guadalajara
BELIZE
Mexico City
HAITI
JAMAICA
ST. KITTS & NEVIS
Caribbean Sea
ANTIGUA & BARBUDA
CAPE VERDE
GUATEMALA
Guatemala
HONDURAS
DOMINICA
BARBADOS
SENEGAL
EL SALVADOR
NICARAGUA
ST. LUCIA
GRENADA
ST. VINCENT & THE GRENADINES
GAMBIA
MALI
TRINIDAD & TOBAGO
COSTA RICA
PANAMA
Caracas
VENEZUELA
GUINEA-BISSAU
GUINEA
BURKINA FASO
FRENCH GUIANA
Fr.
SIERRA LEONE
LIBERIA
CÔTE D'IVOIRE
GHANA
Medellín
Bogota
GUYANA
SURINAME
COLOMBIA
Christmas Island
Kiribati
EQUATOR
Quito
ECUADOR
Galapagos Islands
Ecua.
Manaus
Amazon
Negro
Madeira
PERU
Marquesas Islands
Fr.
BRAZIL
Recife
Lima
Salvador (Bahia)
SAMOA
AMERICAN SAMOA
U.S.
Brasilia
La Paz
BOLIVIA
Sucre
FRENCH POLYNESIA
Fr.
SOUTH ATLANTIC OCEAN
Rio de Janeiro
PARAGUAY
Sao Paulo
TONGA
TROPIC OF CAPRICORN
Asuncion
Cordoba
Porto Alegre
Santiago
Buenos Aires
URUGUAY
Montevideo
CHILE
ARGENTINA
SOUTH PACIFIC OCEAN
PRIME MERIDIAN (MERIDIAN OF GREENWICH)
Falkland Islands
U.K.
Tierra del Fuego
South Georgia
U.K.
Strait of Magellan
Drake Passage
Antarctic Peninsula
ANTARCTIC CIRCLE
Weddell Sea
Berkner Island
Ross Sea
150°W
120°W
90°W
60°W
30°W
0°
60°N
30°N
0°
30°S
60°S

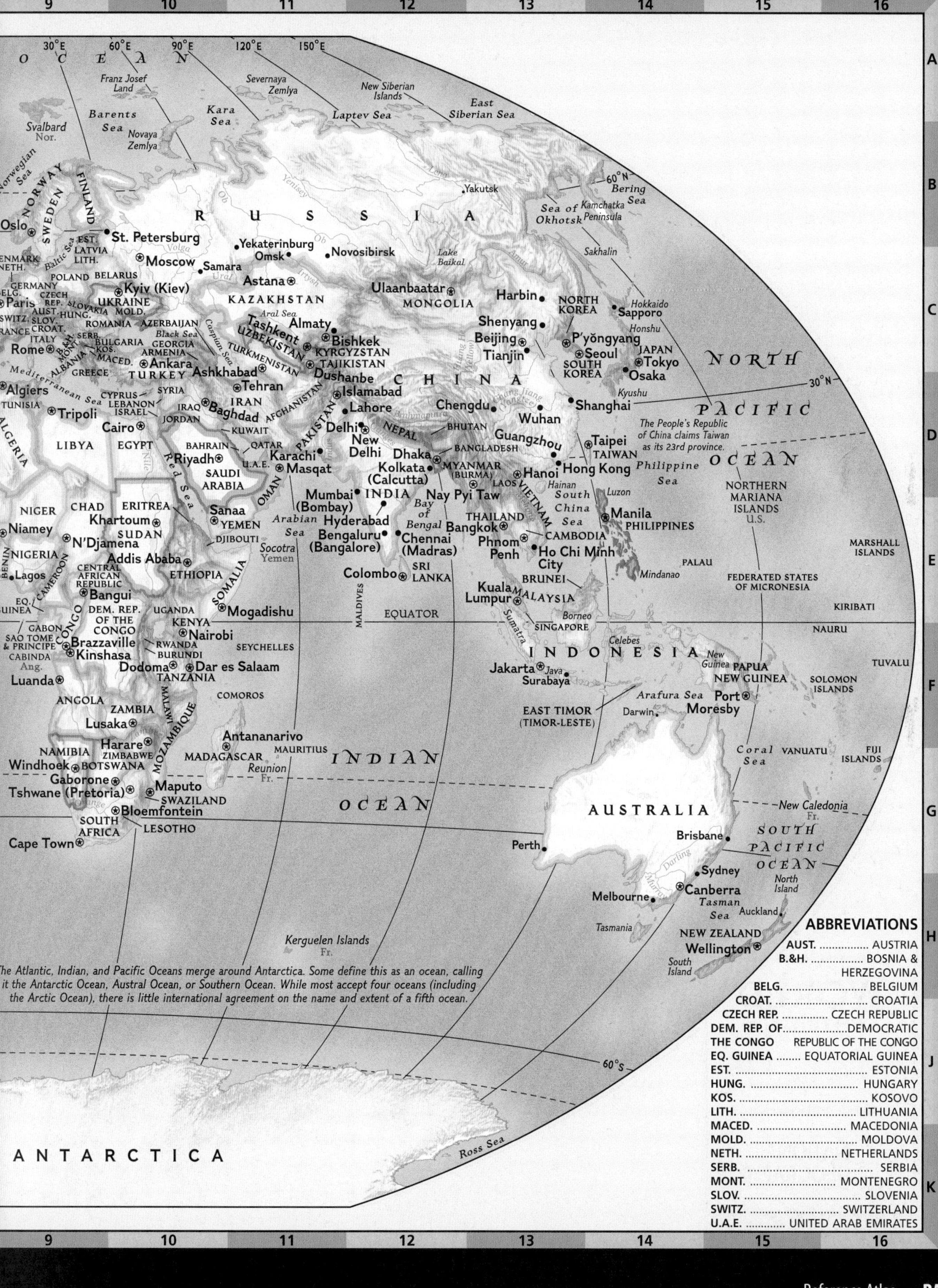

9 10 11 12 13 14 15 16
A B C D E F G H J K
30°E 60°E 90°E 120°E 150°E
O C E A N
Franz Josef Land
Severnaya Zemlya
New Siberian Islands
East Siberian Sea
Laptev Sea
Kara Sea
Barents Sea
Svalbard Nor.
Novaya Zemlya
Norwegian Sea
NORWAY
SWEDEN
FINLAND
R U S S I A
Yakutsk
60°N
Bering Sea
Sea of Okhotsk
Kamchatka Peninsula
Sakhalin
Oslo
St. Petersburg
EST.
LATVIA
LITH.
Baltic Sea
Moscow
Yekaterinburg
Omsk
Novosibirsk
Lake Baikal
Samara
POLAND
BELARUS
GERMANY
CZECH REP.
SLOVAKIA
Kyiv (Kiev)
UKRAINE
MOLD.
Astana
KAZAKHSTAN
Ulaanbaatar
MONGOLIA
Harbin
NORTH KOREA
Hokkaido
Sapporo
Paris
AUST.
HUNG.
SWITZ.
SLOV.
CROAT.
ROMANIA
AZERBAIJAN
Aral Sea
Almaty
Honshu
Shenyang
SERB.
B.&H.
BULGARIA
GEORGIA
Black Sea
Caspian Sea
Tashkent
UZBEKISTAN
Bishkek
KYRGYZSTAN
Beijing
Tianjin
P'yŏngyang
Seoul
SOUTH KOREA
JAPAN
Tokyo
Osaka
Rome
ITALY
MONT.
KOS.
ALBANIA
MACED.
ARMENIA
Ankara
TURKMENISTAN
TAJIKISTAN
Dushanbe
GREECE
TURKEY
Ashkhabad
C H I N A
NORTH
Mediterranean Sea
Algiers
CYPRUS
SYRIA
Tehran
Islamabad
Kyushu
30°N
TUNISIA
LEBANON
ISRAEL
IRAQ
Baghdad
IRAN
AFGHANISTAN
Lahore
Chengdu
Wuhan
Shanghai
Tripoli
Cairo
JORDAN
KUWAIT
PAKISTAN
Delhi
New Delhi
NEPAL
BHUTAN
PACIFIC
The People's Republic of China claims Taiwan as its 23rd province.
ALGERIA
LIBYA
EGYPT
BAHRAIN
QATAR
Karachi
Guangzhou
Taipei
TAIWAN
Riyadh
U.A.E.
Masqat
Dhaka
BANGLADESH
Hong Kong
OCEAN
SAUDI ARABIA
OMAN
Kolkata (Calcutta)
MYANMAR (BURMA)
LAOS
Hanoi
Philippine Sea
Red Sea
Mumbai (Bombay)
INDIA
Nay Pyi Taw
Hainan
VIETNAM
South China Sea
Luzon
NORTHERN MARIANA ISLANDS U.S.
NIGER
CHAD
ERITREA
Khartoum
Sanaa
YEMEN
Arabian Sea
Hyderabad
Bay of Bengal
THAILAND
Bangkok
Manila
PHILIPPINES
Niamey
SUDAN
N'Djamena
DJIBOUTI
Bengaluru (Bangalore)
Chennai (Madras)
Phnom Penh
CAMBODIA
Ho Chi Minh City
MARSHALL ISLANDS
NIGERIA
Addis Ababa
Socotra Yemen
PALAU
CAMEROON
CENTRAL AFRICAN REPUBLIC
ETHIOPIA
Lagos
SOMALIA
Colombo
SRI LANKA
BRUNEI
Mindanao
FEDERATED STATES OF MICRONESIA
Bangui
MALDIVES
Kuala Lumpur
MALAYSIA
EQ. GUINEA
GABON
DEM. REP. OF THE CONGO
UGANDA
KENYA
Mogadishu
EQUATOR
Borneo
KIRIBATI
CONGO
Nairobi
Sumatra
SINGAPORE
NAURU
SAO TOME & PRINCIPE
Brazzaville
Kinshasa
RWANDA
BURUNDI
SEYCHELLES
Celebes
CABINDA Ang.
I N D O N E S I A
New Guinea
Dodoma
Dar es Salaam
TANZANIA
Jakarta
Java
Surabaya
PAPUA NEW GUINEA
TUVALU
Luanda
SOLOMON ISLANDS
ANGOLA
COMOROS
Arafura Sea
Port Moresby
ZAMBIA
MALAWI
EAST TIMOR (TIMOR-LESTE)
Darwin
Lusaka
MOZAMBIQUE
Antananarivo
NAMIBIA
Harare
ZIMBABWE
MAURITIUS
INDIAN
Coral Sea
VANUATU
FIJI ISLANDS
Windhoek
BOTSWANA
MADAGASCAR
Reunion Fr.
Gaborone
Tshwane (Pretoria)
Maputo
SWAZILAND
OCEAN
New Caledonia Fr.
Bloemfontein
AUSTRALIA
SOUTH AFRICA
LESOTHO
SOUTH PACIFIC OCEAN
Brisbane
Cape Town
Perth
Darling
Sydney
Murray
Canberra
North Island
Melbourne
Tasman Sea
Auckland
Tasmania
NEW ZEALAND
Wellington
South Island
Kerguelen Islands Fr.
The Atlantic, Indian, and Pacific Oceans merge around Antarctica. Some define this as an ocean, calling it the Antarctic Ocean, Austral Ocean, or Southern Ocean. While most accept four oceans (including the Arctic Ocean), there is little international agreement on the name and extent of a fifth ocean.
60°S
Ross Sea
ANTARCTICA
ABBREVIATIONS
AUST. .... AUSTRIA
B.&H. .... BOSNIA & HERZEGOVINA
BELG. .... BELGIUM
CROAT. .... CROATIA
CZECH REP. .... CZECH REPUBLIC
DEM. REP. OF THE CONGO .... DEMOCRATIC REPUBLIC OF THE CONGO
EQ. GUINEA .... EQUATORIAL GUINEA
EST. .... ESTONIA
HUNG. .... HUNGARY
KOS. .... KOSOVO
LITH. .... LITHUANIA
MACED. .... MACEDONIA
MOLD. .... MOLDOVA
NETH. .... NETHERLANDS
SERB. .... SERBIA
MONT. .... MONTENEGRO
SLOV. .... SLOVENIA
SWITZ. .... SWITZERLAND
U.A.E. .... UNITED ARAB EMIRATES

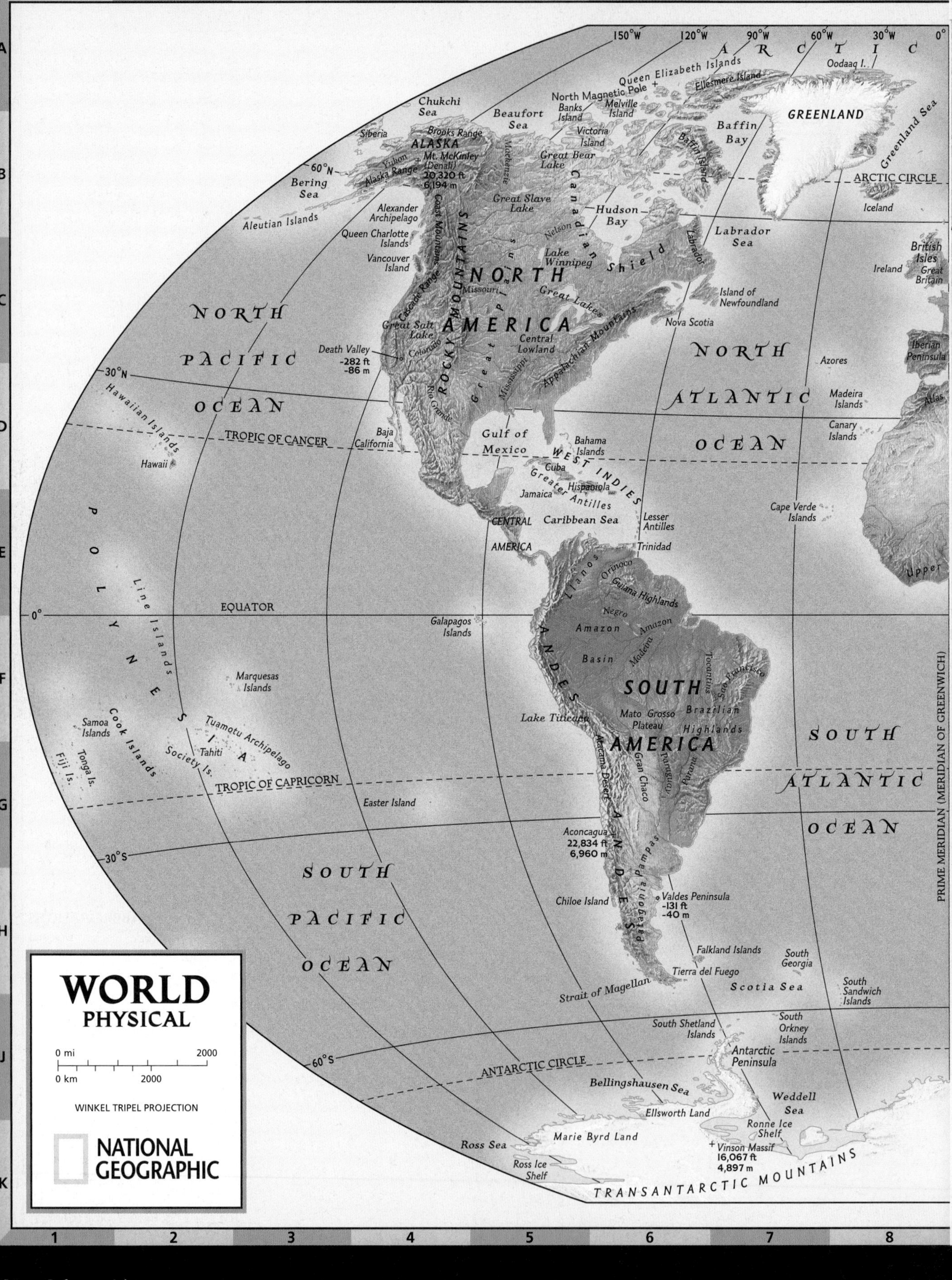
WORLD
PHYSICAL
0 mi 2000
0 km 2000
WINKEL TRIPEL PROJECTION
NATIONAL GEOGRAPHIC
NORTH PACIFIC OCEAN
SOUTH PACIFIC OCEAN
NORTH ATLANTIC OCEAN
SOUTH ATLANTIC OCEAN
ARCTIC
NORTH AMERICA
SOUTH AMERICA
CENTRAL AMERICA
GREENLAND
ALASKA
Mt. McKinley (Denali) 20,320 ft 6,194 m
Death Valley -282 ft -86 m
Aconcagua 22,834 ft 6,960 m
Valdes Peninsula -131 ft -40 m
Vinson Massif 16,067 ft 4,897 m
Queen Elizabeth Islands
North Magnetic Pole
Ellesmere Island
Oodaaq I.
Chukchi Sea
Beaufort Sea
Banks Island
Melville Island
Victoria Island
Baffin Bay
Baffin Island
Greenland Sea
ARCTIC CIRCLE
Siberia
Brooks Range
Yukon
Alaska Range
Bering Sea
Mackenzie
Great Bear Lake
Great Slave Lake
Canadian Shield
Hudson Bay
Iceland
Aleutian Islands
Alexander Archipelago
Queen Charlotte Islands
Vancouver Island
Coast Mountains
Cascade Range
Nelson
Labrador
Labrador Sea
Lake Winnipeg
British Isles
Ireland
Great Britain
ROCKY MOUNTAINS
Missouri
Great Plains
Great Lakes
Island of Newfoundland
Nova Scotia
Great Salt Lake
Colorado
Central Lowland
Appalachian Mountains
Mississippi
Iberian Peninsula
Azores
Madeira Islands
Atlas
Canary Islands
Rio Grande
30°N
Hawaiian Islands
Hawaii
TROPIC OF CANCER
Baja California
Gulf of Mexico
Bahama Islands
WEST INDIES
Cuba
Greater Antilles
Hispaniola
Jamaica
Caribbean Sea
Lesser Antilles
Trinidad
Cape Verde Islands
POLYNESIA
Line Islands
EQUATOR
Galapagos Islands
Llanos
Orinoco
Guiana Highlands
Negro
Amazon
Amazon Basin
Madeira
ANDES
Tocantins
São Francisco
Upper
Marquesas Islands
Lake Titicaca
Mato Grosso Plateau
Brazilian Highlands
Samoa Islands
Cook Islands
Tuamotu Archipelago
Tahiti
Society Is.
Tonga Is.
Fiji Is.
Atacama Desert
Gran Chaco
Paraguay
Paraná
TROPIC OF CAPRICORN
Easter Island
30°S
Pampas
Chiloe Island
Patagonia
Falkland Islands
South Georgia
Tierra del Fuego
Strait of Magellan
Scotia Sea
South Sandwich Islands
South Shetland Islands
South Orkney Islands
Antarctic Peninsula
60°S
60°N
ANTARCTIC CIRCLE
Bellingshausen Sea
Weddell Sea
Ellsworth Land
Ronne Ice Shelf
Marie Byrd Land
Ross Sea
Ross Ice Shelf
TRANSANTARCTIC MOUNTAINS
PRIME MERIDIAN (MERIDIAN OF GREENWICH)
150°W
120°W
90°W
60°W
30°W
0°

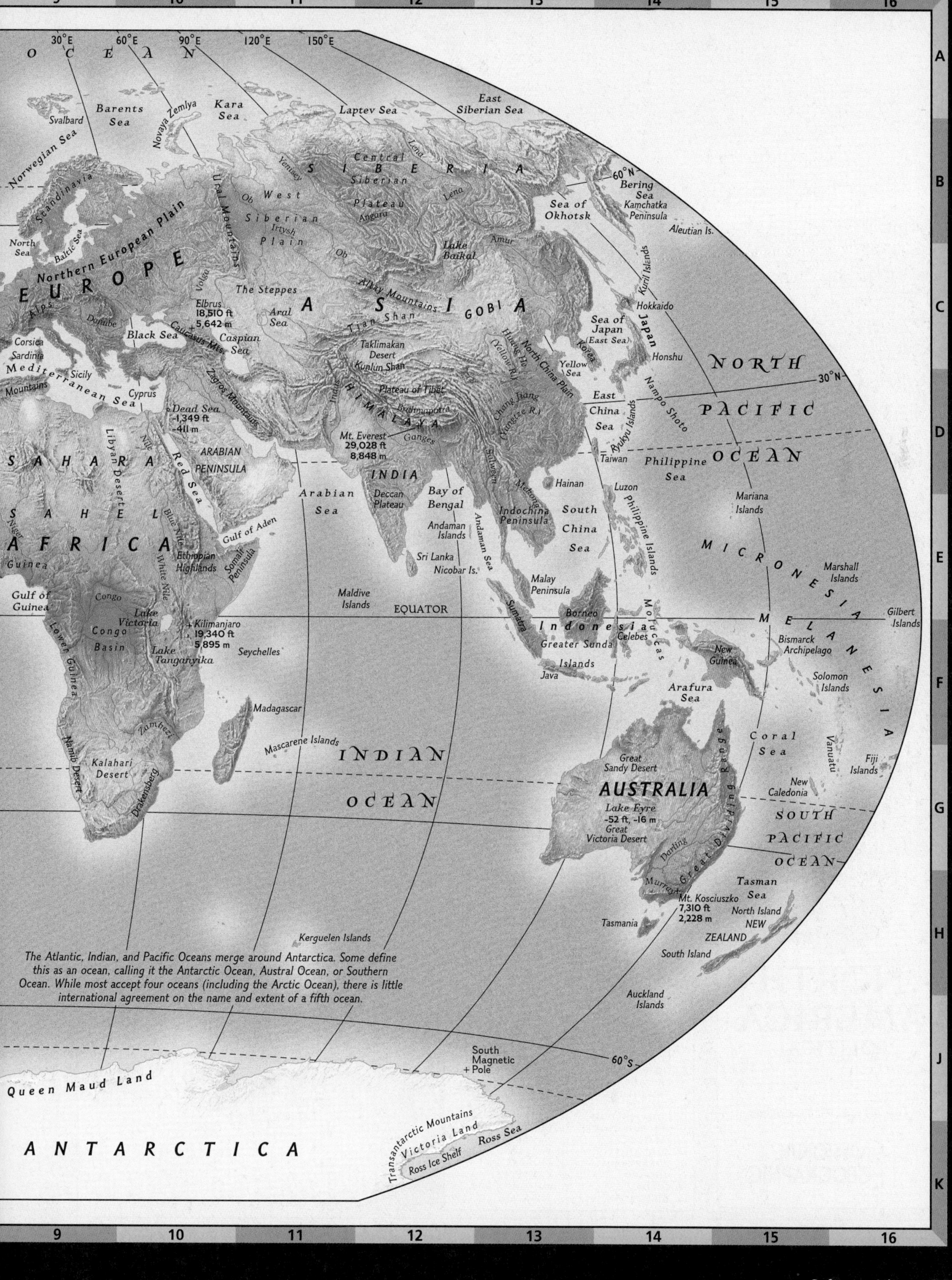

OCEAN
Svalbard
Barents Sea
Novaya Zemlya
Kara Sea
Laptev Sea
East Siberian Sea
Norwegian Sea
Scandinavia
SIBERIA
Central Siberian Plateau
West Siberian Plain
Ural Mountains
Bering Sea
Kamchatka Peninsula
Sea of Okhotsk
Aleutian Is.
North Sea
Baltic Sea
Northern European Plain
EUROPE
Lake Baikal
Kuril Islands
Alps
The Steppes
Elbrus 18,510 ft 5,642 m
Aral Sea
ASIA
Altay Mountains
GOBI
Hokkaido
Japan
Sea of Japan (East Sea)
Honshu
Black Sea
Caucasus Mts.
Caspian Sea
Corsica
Sardinia
Mediterranean Sea
Sicily
Cyprus
Tian Shan
Taklimakan Desert
Kunlun Shan
Plateau of Tibet
HIMALAYA
North China Plain
Korea
Yellow Sea
NORTH PACIFIC OCEAN
Nampo Shoto
East China Sea
Ryukyu Islands
Dead Sea -1,349 ft -411 m
Zagros Mountains
Mt. Everest 29,028 ft 8,848 m
Ganges
Taiwan
Philippine Sea
SAHARA
Libyan Desert
Nile
Red Sea
ARABIAN PENINSULA
INDIA
Hainan
Luzon
SAHEL
Arabian Sea
Deccan Plateau
Bay of Bengal
Indochina Peninsula
South China Sea
Mariana Islands
Gulf of Aden
Andaman Islands
Andaman Sea
Philippine Islands
MICRONESIA
AFRICA
Ethiopian Highlands
Somali Peninsula
Sri Lanka
Nicobar Is.
Marshall Islands
Guinea
Malay Peninsula
Gulf of Guinea
Congo
Lake Victoria
Maldive Islands
EQUATOR
Borneo
Sumatra
Indonesia
Moluccas
Celebes
Gilbert Islands
Kilimanjaro 19,340 ft 5,895 m
Congo Basin
Lake Tanganyika
Seychelles
Greater Sunda Islands
Java
New Guinea
Bismarck Archipelago
MELANESIA
Solomon Islands
Arafura Sea
Madagascar
Mascarene Islands
INDIAN OCEAN
Coral Sea
Vanuatu
Fiji Islands
Kalahari Desert
Namib Desert
Drakensberg
Great Sandy Desert
AUSTRALIA
New Caledonia
Lake Eyre -52 ft, -16 m
Great Victoria Desert
SOUTH PACIFIC OCEAN
Darling
Great Dividing Range
Murray
Tasman Sea
Mt. Kosciuszko 7,310 ft 2,228 m
North Island
NEW ZEALAND
Tasmania
Kerguelen Islands
South Island
The Atlantic, Indian, and Pacific Oceans merge around Antarctica. Some define this as an ocean, calling it the Antarctic Ocean, Austral Ocean, or Southern Ocean. While most accept four oceans (including the Arctic Ocean), there is little international agreement on the name and extent of a fifth ocean.
Auckland Islands
South Magnetic Pole
60°S
Queen Maud Land
Transantarctic Mountains
Victoria Land
Ross Ice Shelf
Ross Sea
ANTARCTICA
30°E
60°E
90°E
120°E
150°E
60°N
30°N

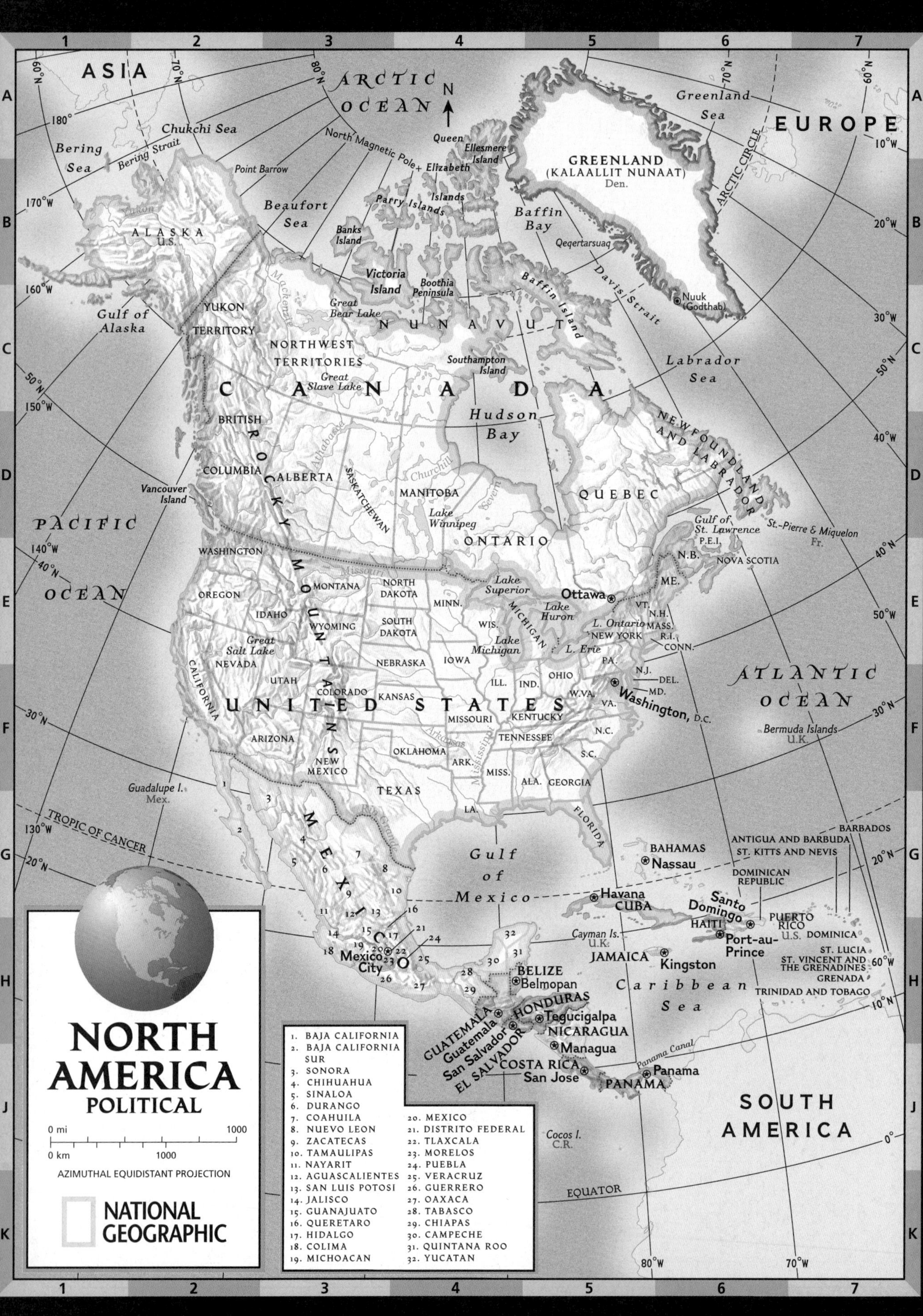

NORTH AMERICA
POLITICAL
0 mi 1000
0 km 1000
AZIMUTHAL EQUIDISTANT PROJECTION
NATIONAL GEOGRAPHIC
1. BAJA CALIFORNIA
2. BAJA CALIFORNIA SUR
3. SONORA
4. CHIHUAHUA
5. SINALOA
6. DURANGO
7. COAHUILA
8. NUEVO LEON
9. ZACATECAS
10. TAMAULIPAS
11. NAYARIT
12. AGUASCALIENTES
13. SAN LUIS POTOSI
14. JALISCO
15. GUANAJUATO
16. QUERETARO
17. HIDALGO
18. COLIMA
19. MICHOACAN
20. MEXICO
21. DISTRITO FEDERAL
22. TLAXCALA
23. MORELOS
24. PUEBLA
25. VERACRUZ
26. GUERRERO
27. OAXACA
28. TABASCO
29. CHIAPAS
30. CAMPECHE
31. QUINTANA ROO
32. YUCATAN
ASIA
ARCTIC OCEAN
N
EUROPE
GREENLAND
(KALAALLIT NUNAAT)
Den.
Greenland Sea
Chukchi Sea
Bering Sea
Bering Strait
Point Barrow
North Magnetic Pole
Queen Elizabeth Islands
Ellesmere Island
Parry Islands
Beaufort Sea
Banks Island
Victoria Island
Boothia Peninsula
Baffin Bay
Qeqertarsuaq
Baffin Island
Davis Strait
Nuuk (Godthab)
ARCTIC CIRCLE
ALASKA
U.S.
Gulf of Alaska
YUKON TERRITORY
NORTHWEST TERRITORIES
Great Bear Lake
Great Slave Lake
NUNAVUT
Southampton Island
Labrador Sea
CANADA
Hudson Bay
BRITISH COLUMBIA
ALBERTA
SASKATCHEWAN
MANITOBA
Lake Winnipeg
ONTARIO
QUEBEC
NEWFOUNDLAND AND LABRADOR
Gulf of St. Lawrence
St.-Pierre & Miquelon
Fr.
P.E.I.
N.B.
NOVA SCOTIA
Vancouver Island
PACIFIC OCEAN
ROCKY MOUNTAINS
WASHINGTON
OREGON
IDAHO
MONTANA
WYOMING
NORTH DAKOTA
SOUTH DAKOTA
MINN.
WIS.
MICHIGAN
Lake Superior
Lake Huron
Lake Michigan
L. Ontario
L. Erie
Ottawa
ME.
VT.
N.H.
MASS.
R.I.
CONN.
NEW YORK
PA.
N.J.
DEL.
MD.
Great Salt Lake
NEVADA
UTAH
CALIFORNIA
COLORADO
NEBRASKA
IOWA
ILL.
IND.
OHIO
W.VA.
VA.
Washington, D.C.
KANSAS
MISSOURI
KENTUCKY
UNITED STATES
ARIZONA
NEW MEXICO
OKLAHOMA
TENNESSEE
N.C.
S.C.
ARK.
MISS.
ALA.
GEORGIA
TEXAS
LA.
FLORIDA
ATLANTIC OCEAN
Bermuda Islands
U.K.
Guadalupe I.
Mex.
TROPIC OF CANCER
MEXICO
Mexico City
Gulf of Mexico
BAHAMAS
Nassau
Havana
CUBA
Cayman Is.
U.K.
JAMAICA
Kingston
HAITI
Port-au-Prince
Santo Domingo
DOMINICAN REPUBLIC
PUERTO RICO
U.S.
ANTIGUA AND BARBUDA
ST. KITTS AND NEVIS
BARBADOS
DOMINICA
ST. LUCIA
ST. VINCENT AND THE GRENADINES
GRENADA
TRINIDAD AND TOBAGO
BELIZE
Belmopan
GUATEMALA
Guatemala
HONDURAS
Tegucigalpa
EL SALVADOR
San Salvador
NICARAGUA
Managua
COSTA RICA
San Jose
PANAMA
Panama
Panama Canal
Caribbean Sea
Cocos I.
C.R.
SOUTH AMERICA
EQUATOR

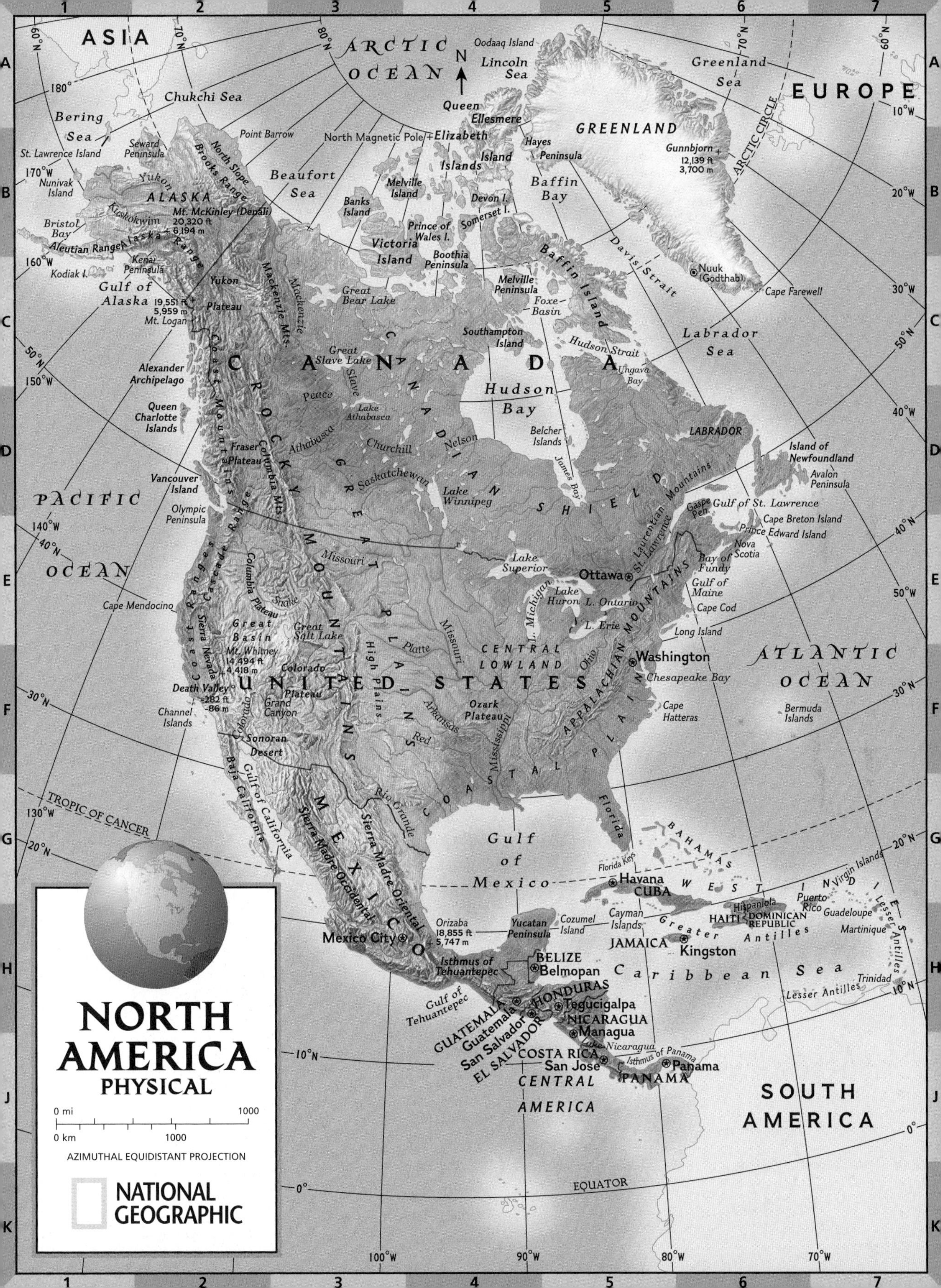
NORTH AMERICA
PHYSICAL
0 mi 1000
0 km 1000
AZIMUTHAL EQUIDISTANT PROJECTION
NATIONAL GEOGRAPHIC
ASIA
EUROPE
SOUTH AMERICA
ARCTIC OCEAN
PACIFIC OCEAN
ATLANTIC OCEAN
Chukchi Sea
Bering Sea
Beaufort Sea
Lincoln Sea
Greenland Sea
Baffin Bay
Labrador Sea
Hudson Bay
Gulf of Alaska
Gulf of Mexico
Caribbean Sea
Gulf of St. Lawrence
Gulf of California
Gulf of Tehuantepec
Gulf of Maine
Bay of Fundy
Chesapeake Bay
ALASKA
CANADA
GREENLAND
LABRADOR
UNITED STATES
MEXICO
CENTRAL AMERICA
CUBA
HAITI
DOMINICAN REPUBLIC
JAMAICA
BELIZE
GUATEMALA
HONDURAS
EL SALVADOR
NICARAGUA
COSTA RICA
PANAMA
BAHAMAS
WEST INDIES
Ottawa
Washington
Mexico City
Havana
Kingston
Belmopan
Guatemala
Tegucigalpa
San Salvador
Managua
San Jose
Panama
Nuuk (Godthab)
ROCKY MOUNTAINS
GREAT PLAINS
CANADIAN SHIELD
CENTRAL LOWLAND
APPALACHIAN MOUNTAINS
COASTAL PLAIN
Laurentian Mountains
Coast Mountains
Cascade Range
Coast Ranges
Sierra Nevada
Sierra Madre Occidental
Sierra Madre Oriental
Brooks Range
Alaska Range
Aleutian Range
Mackenzie Mts.
Columbia Mts.
North Slope
Yukon Plateau
Fraser Plateau
Columbia Plateau
Great Basin
Colorado Plateau
Ozark Plateau
High Plains
Sonoran Desert
Mt. McKinley (Denali) 20,320 ft 6,194 m
19,551 ft 5,959 m Mt. Logan
Mt. Whitney 14,494 ft 4,418 m
Death Valley -282 ft -86 m
Gunnbjorn 12,139 ft 3,700 m
Orizaba 18,855 ft 5,747 m
Great Bear Lake
Great Slave Lake
Lake Athabasca
Lake Winnipeg
Lake Superior
L. Michigan
Lake Huron
L. Ontario
L. Erie
Great Salt Lake
Lake Nicaragua
Yukon
Kuskokwim
Mackenzie
Peace
Slave
Athabasca
Churchill
Nelson
Saskatchewan
Missouri
Snake
Columbia
Platte
Arkansas
Red
Mississippi
Ohio
St. Lawrence
Colorado
Rio Grande
Grand Canyon
Queen Elizabeth Islands
Ellesmere Island
Oodaaq Island
North Magnetic Pole
Point Barrow
Banks Island
Melville Island
Prince of Wales I.
Victoria Island
Devon I.
Somerset I.
Boothia Peninsula
Melville Peninsula
Baffin Island
Foxe Basin
Southampton Island
Hudson Strait
Ungava Bay
Belcher Islands
James Bay
Hayes Peninsula
Davis Strait
Cape Farewell
ARCTIC CIRCLE
Seward Peninsula
St. Lawrence Island
Nunivak Island
Bristol Bay
Kenai Peninsula
Kodiak I.
Alexander Archipelago
Queen Charlotte Islands
Vancouver Island
Olympic Peninsula
Cape Mendocino
Channel Islands
Baja California
Island of Newfoundland
Avalon Peninsula
Gaspe Pen.
Cape Breton Island
Prince Edward Island
Nova Scotia
Cape Cod
Long Island
Cape Hatteras
Bermuda Islands
Florida
Florida Keys
Yucatan Peninsula
Cozumel Island
Cayman Islands
Isthmus of Tehuantepec
Isthmus of Panama
Greater Antilles
Hispaniola
Puerto Rico
Virgin Islands
Guadeloupe
Martinique
Lesser Antilles
Trinidad
TROPIC OF CANCER
EQUATOR

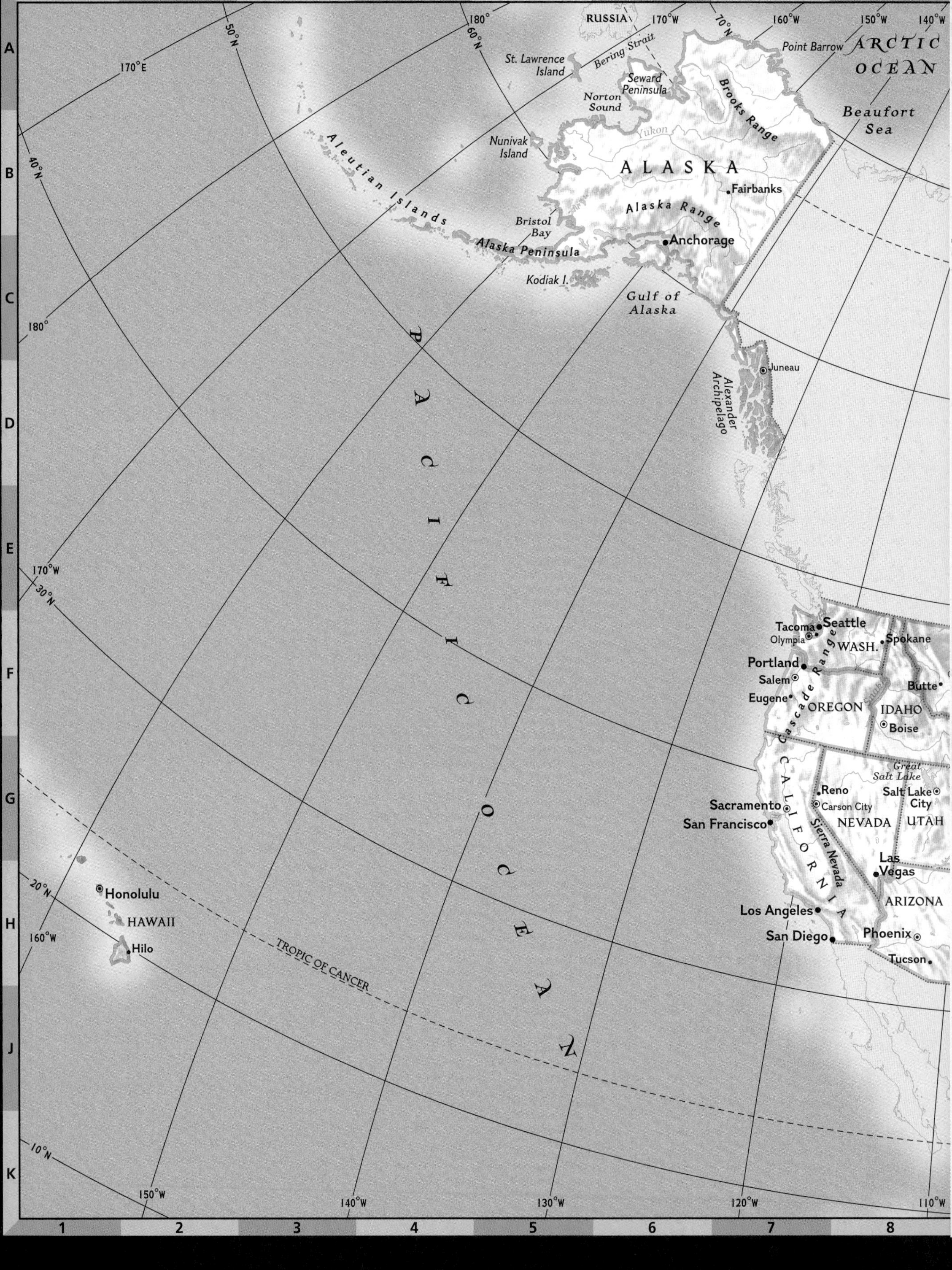
RUSSIA
ARCTIC OCEAN
Beaufort Sea
Point Barrow
Bering Strait
St. Lawrence Island
Seward Peninsula
Norton Sound
Brooks Range
Yukon
Nunivak Island
ALASKA
Fairbanks
Aleutian Islands
Alaska Range
Bristol Bay
Anchorage
Alaska Peninsula
Kodiak I.
Gulf of Alaska
Juneau
Alexander Archipelago
PACIFIC OCEAN
Tacoma
Seattle
Olympia
WASH.
Spokane
Portland
Cascade Range
Salem
Butte
Eugene
OREGON
IDAHO
Boise
CALIFORNIA
Great Salt Lake
Reno
Salt Lake City
Sacramento
Carson City
NEVADA
UTAH
San Francisco
Sierra Nevada
Las Vegas
Honolulu
HAWAII
Hilo
ARIZONA
Los Angeles
San Diego
Phoenix
Tucson
TROPIC OF CANCER

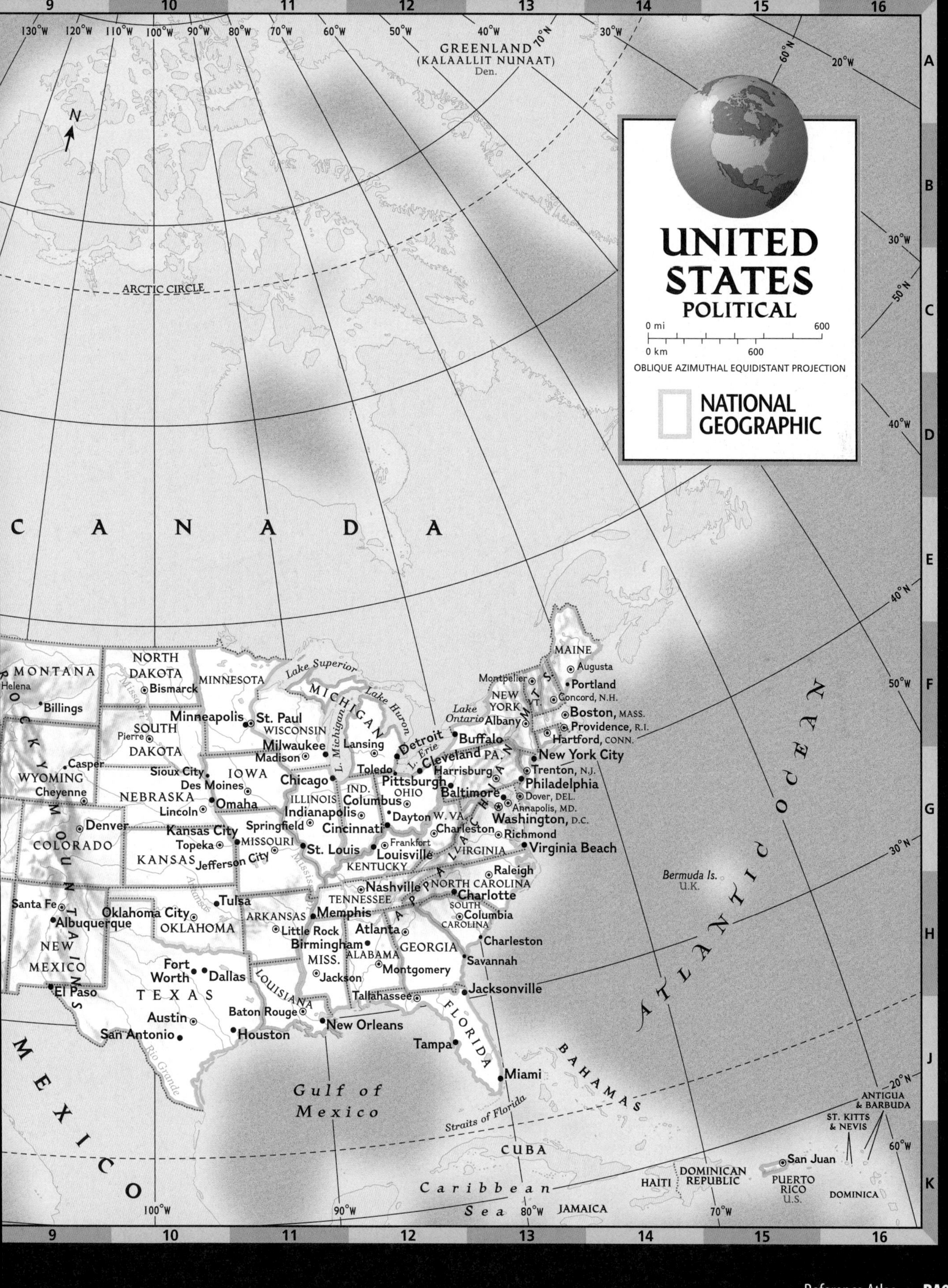
UNITED STATES
POLITICAL
0 mi 600
0 km 600
OBLIQUE AZIMUTHAL EQUIDISTANT PROJECTION
NATIONAL GEOGRAPHIC
GREENLAND
(KALAALLIT NUNAAT)
Den.
ARCTIC CIRCLE
CANADA
MONTANA
NORTH DAKOTA
MINNESOTA
SOUTH DAKOTA
WYOMING
NEBRASKA
IOWA
WISCONSIN
MICHIGAN
ILLINOIS
IND.
OHIO
COLORADO
KANSAS
MISSOURI
KENTUCKY
W. VA.
VIRGINIA
NORTH CAROLINA
TENNESSEE
SOUTH CAROLINA
NEW MEXICO
OKLAHOMA
ARKANSAS
MISS.
ALABAMA
GEORGIA
TEXAS
LOUISIANA
FLORIDA
PA.
NEW YORK
MAINE
VT.
ROCKY MOUNTAINS
APPALACHIAN MTS.
Lake Superior
Lake Huron
L. Michigan
L. Erie
Lake Ontario
Helena
Billings
Bismarck
Minneapolis
St. Paul
Pierre
Casper
Cheyenne
Sioux City
Des Moines
Omaha
Lincoln
Denver
Kansas City
Topeka
Jefferson City
Milwaukee
Madison
Lansing
Chicago
Springfield
Indianapolis
Columbus
Dayton
Cincinnati
St. Louis
Frankfort
Louisville
Detroit
Toledo
Cleveland
Pittsburgh
Buffalo
Harrisburg
Baltimore
Washington, D.C.
Charleston
Richmond
Virginia Beach
Raleigh
Nashville
Charlotte
Columbia
Charleston
Savannah
Atlanta
Birmingham
Memphis
Little Rock
Jackson
Montgomery
Tallahassee
Jacksonville
Tampa
Miami
Tulsa
Oklahoma City
Santa Fe
Albuquerque
El Paso
Fort Worth
Dallas
Austin
San Antonio
Houston
Baton Rouge
New Orleans
Augusta
Portland
Montpelier
Concord, N.H.
Albany
Boston, MASS.
Providence, R.I.
Hartford, CONN.
New York City
Trenton, N.J.
Philadelphia
Dover, DEL.
Annapolis, MD.
Missouri
Mississippi
Arkansas
Rio Grande
MEXICO
Gulf of Mexico
Straits of Florida
BAHAMAS
CUBA
Caribbean Sea
JAMAICA
HAITI
DOMINICAN REPUBLIC
San Juan
PUERTO RICO U.S.
ANTIGUA & BARBUDA
ST. KITTS & NEVIS
DOMINICA
Bermuda Is. U.K.
ATLANTIC OCEAN

Cape Flattery
Mt. Olympus
7,965 ft
2,428 m
Seattle
CASCADE RANGE
COLUMBIA PLATEAU
ROCKY MOUNTAINS
CANADA
Columbia
Blue Mts.
Clearwater Mts.
Bitterroot Range
Salmon River Mts.
Absaroka Range
Bighorn Mts.
Missouri
GREAT PLAINS
Black Hills
COAST RANGES
Great Sandy Desert
Snake
Snake River Plain
Shoshone Falls
Wind River Range
Laramie Mts.
Cape Mendocino
SIERRA NEVADA
Great Salt Lake
Wasatch Range
N. Platte
Sand Hills
Platte
PACIFIC OCEAN
Lake Tahoe
Central Valley
GREAT BASIN
Uinta Mts.
San Francisco
Denver
14,433 ft
4,399 m
Mt. Elbert
High Plains
Mt. Whitney
14,494 ft
4,418 m
Colorado
Lake Powell
Lake Mead
Arkansas
Death Valley
-282 ft, -86 m
San Juan Mts.
Sangre de Cristo Mts.
Point Conception
Mojave Desert
Grand Canyon
Colorado Plateau
Los Angeles
Channel Islands
Salton Sea
San Diego
Phoenix
Sonoran Desert
Rio Grande
Sacramento Mts.
Llano Estacado
Red
Dallas
Brazos
Edwards Plateau
Rio Grande
MEXICO
TROPIC OF CANCER
ARCTIC OCEAN
Point Barrow
Beaufort Sea
Chukchi Sea
North Slope
Brooks Range
RUSSIA
Bering Strait
ARCTIC CIRCLE
Seward Pen.
ALASKA
CANADA
St. Lawrence Island
Yukon
Tanana
Kuskokwim
Alaska Range
Mt. McKinley (Denali)
20,320 ft, 6,194 m
Anchorage
Nunivak Island
Bering Sea
Bristol Bay
Gulf of Alaska
Alexander Archipelago
Alaska Peninsula
Kodiak I.
ALASKA
PACIFIC OCEAN
0 mi 300
0 km 300

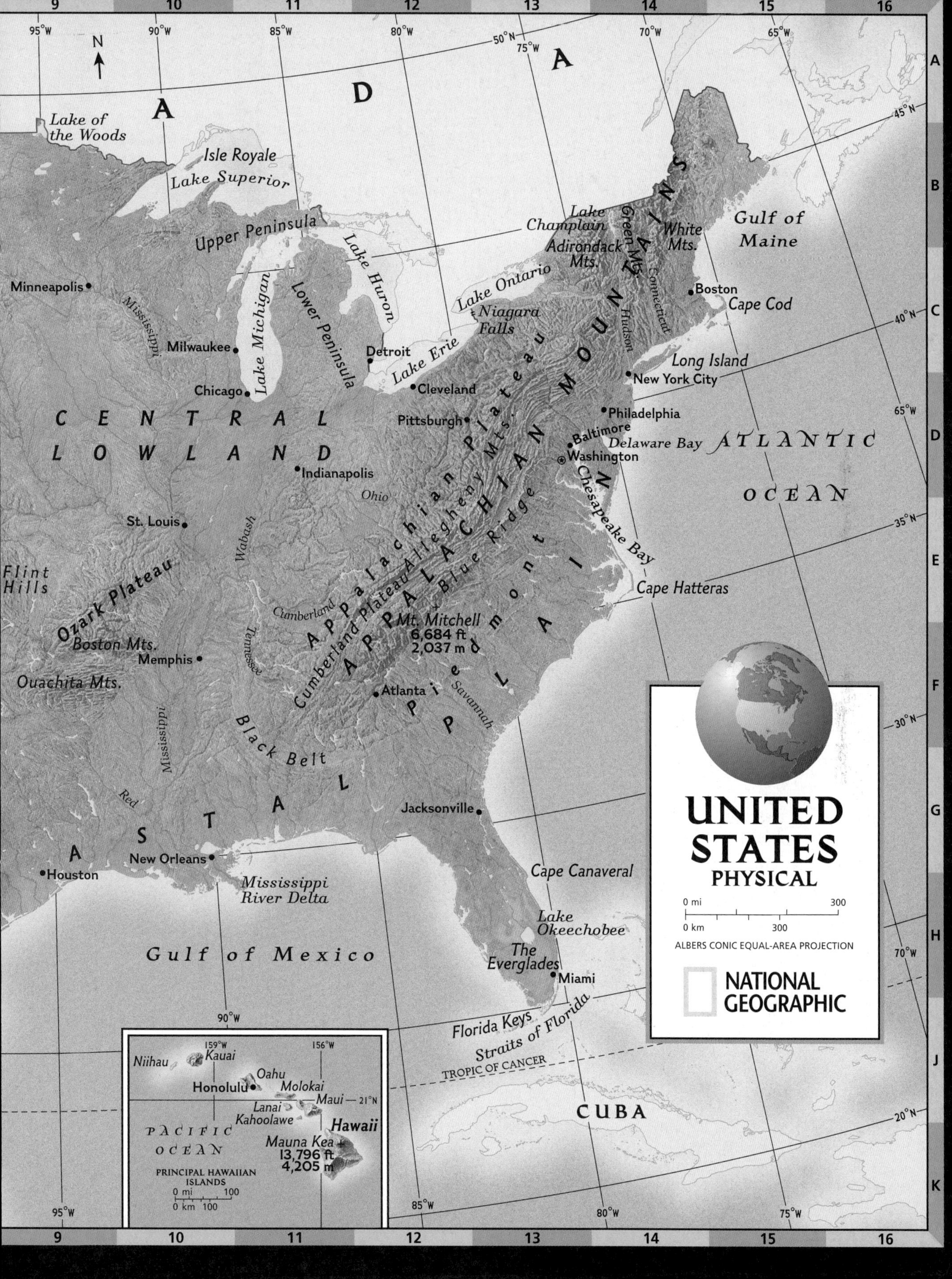
UNITED STATES
PHYSICAL
0 mi 300
0 km 300
ALBERS CONIC EQUAL-AREA PROJECTION
NATIONAL GEOGRAPHIC
CENTRAL LOWLAND
APPALACHIAN MOUNTAINS
Appalachian Plateau
Cumberland Plateau
Allegheny Mts.
Blue Ridge
Piedmont
Mt. Mitchell 6,684 ft 2,037 m
Ozark Plateau
Boston Mts.
Ouachita Mts.
Flint Hills
Black Belt
Lake of the Woods
Isle Royale
Lake Superior
Upper Peninsula
Lower Peninsula
Lake Michigan
Lake Huron
Lake Erie
Lake Ontario
Niagara Falls
Lake Champlain
Adirondack Mts.
Green Mts.
White Mts.
Gulf of Maine
Cape Cod
Long Island
Delaware Bay
Chesapeake Bay
Cape Hatteras
ATLANTIC OCEAN
Cape Canaveral
Lake Okeechobee
The Everglades
Florida Keys
Straits of Florida
TROPIC OF CANCER
CUBA
Gulf of Mexico
Mississippi River Delta
Minneapolis
Milwaukee
Chicago
Detroit
Cleveland
Pittsburgh
Indianapolis
St. Louis
Memphis
Atlanta
Jacksonville
New Orleans
Houston
Miami
Boston
New York City
Philadelphia
Baltimore
Washington
Mississippi
Ohio
Wabash
Cumberland
Tennessee
Red
Savannah
Hudson
Connecticut
PRINCIPAL HAWAIIAN ISLANDS
Niihau
Kauai
Oahu
Honolulu
Molokai
Maui
Lanai
Kahoolawe
Hawaii
Mauna Kea 13,796 ft 4,205 m
PACIFIC OCEAN
0 mi 100
0 km 100

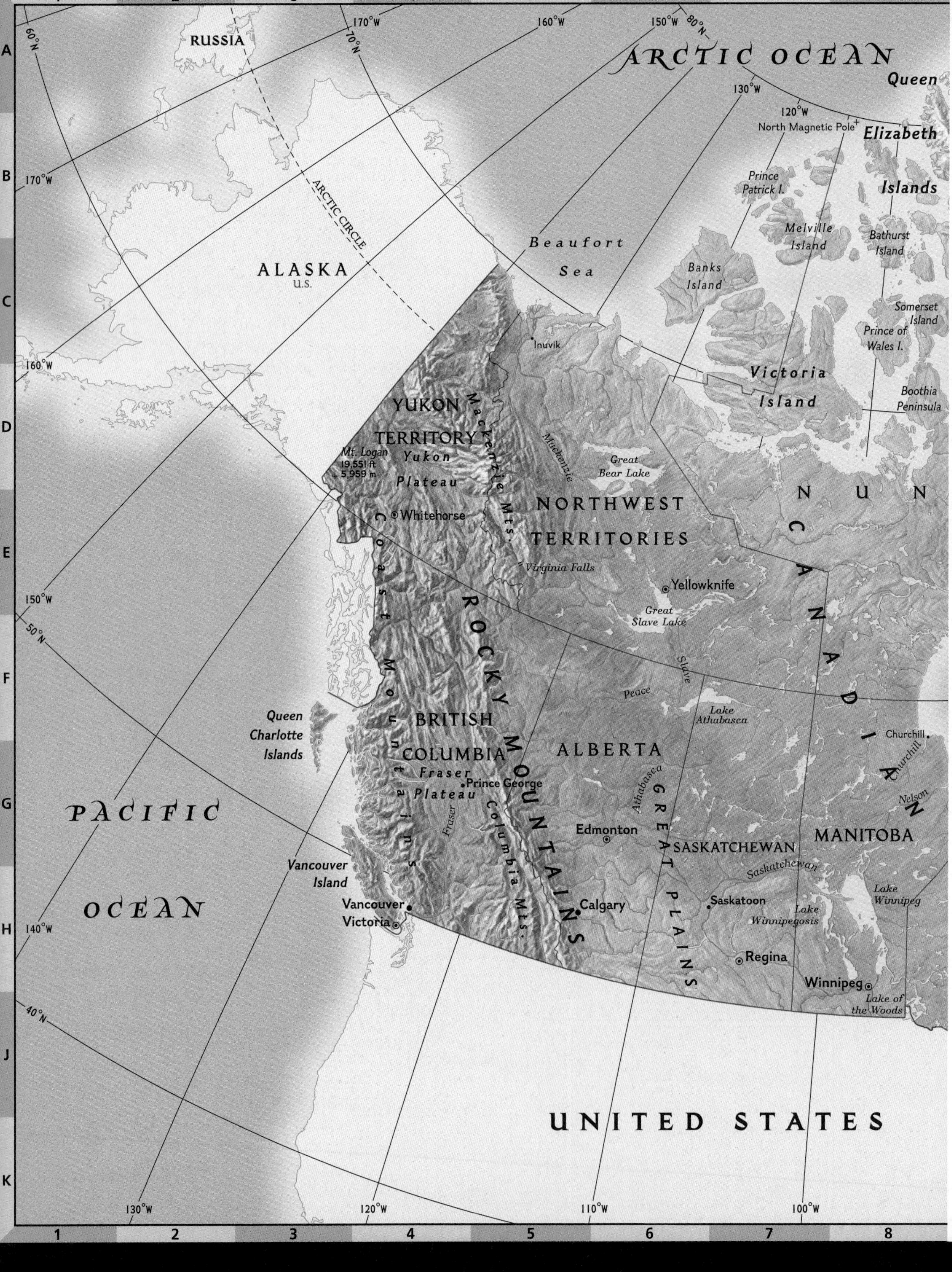

ARCTIC OCEAN
RUSSIA
ALASKA
U.S.
ARCTIC CIRCLE
Beaufort Sea
Queen Elizabeth Islands
North Magnetic Pole
Prince Patrick I.
Melville Island
Bathurst Island
Banks Island
Somerset Island
Prince of Wales I.
Victoria Island
Boothia Peninsula
Inuvik
YUKON TERRITORY
Mackenzie Mts.
Mackenzie
Mt. Logan 19,551 ft 5,959 m
Yukon Plateau
Great Bear Lake
NORTHWEST TERRITORIES
NUNAVUT
Whitehorse
Coast Mountains
Virginia Falls
Yellowknife
Great Slave Lake
CANADIAN
ROCKY MOUNTAINS
Slave
Peace
Lake Athabasca
Queen Charlotte Islands
BRITISH COLUMBIA
ALBERTA
Churchill
Fraser Plateau
Prince George
Athabasca
GREAT PLAINS
Nelson
PACIFIC OCEAN
Fraser
Columbia Mts.
Edmonton
SASKATCHEWAN
MANITOBA
Vancouver Island
Saskatchewan
Lake Winnipeg
Vancouver
Victoria
Calgary
Saskatoon
Lake Winnipegosis
Regina
Winnipeg
Lake of the Woods
UNITED STATES
170°W
160°W
150°W
140°W
130°W
120°W
110°W
100°W
80°N
70°N
60°N
50°N
40°N

CANADA
PHYSICAL/POLITICAL
0 mi 400
0 km 400
AZIMUTHAL EQUIDISTANT PROJECTION
NATIONAL GEOGRAPHIC
GREENLAND
(KALAALLIT NUNAAT)
Den.
ICELAND
Ellesmere Island
Devon Island
Baffin Bay
Baffin Island
Davis Strait
Melville Peninsula
Foxe Basin
Southampton Island
Iqaluit
Hudson Strait
Labrador Sea
Ungava Bay
Hudson Bay
Belcher Islands
James Bay
NEWFOUNDLAND AND LABRADOR
Cartwright
Schefferville
Happy Valley-Goose Bay
Smallwood Reservoir
Churchill Falls
Island of Newfoundland
QUEBEC
Manicouagan Reservoir
Sept-Iles
Anticosti I.
St. John's
Avalon Peninsula
Gulf of St. Lawrence
St.-Pierre & Miquelon
Fr.
Gaspe Pen.
PRINCE EDWARD ISLAND
Cape Breton I.
Charlottetown
NEW BRUNSWICK
NOVA SCOTIA
Fredericton
Saint John
Halifax
Bay of Fundy
ATLANTIC OCEAN
SHIELD
ONTARIO
Lake Nipigon
Chicoutimi
Rouyn-Noranda
Quebec
St. Lawrence
Thunder Bay
Lake Superior
Sudbury
Montreal
Ottawa
Lake Huron
Lake Michigan
Toronto
L. Ontario
Niagara Falls
London
L. Erie
A V U T
80°N
70°N
60°N
50°N
40°N
90°W
80°W
70°W
60°W
50°W
40°W
30°W
20°W
10°W

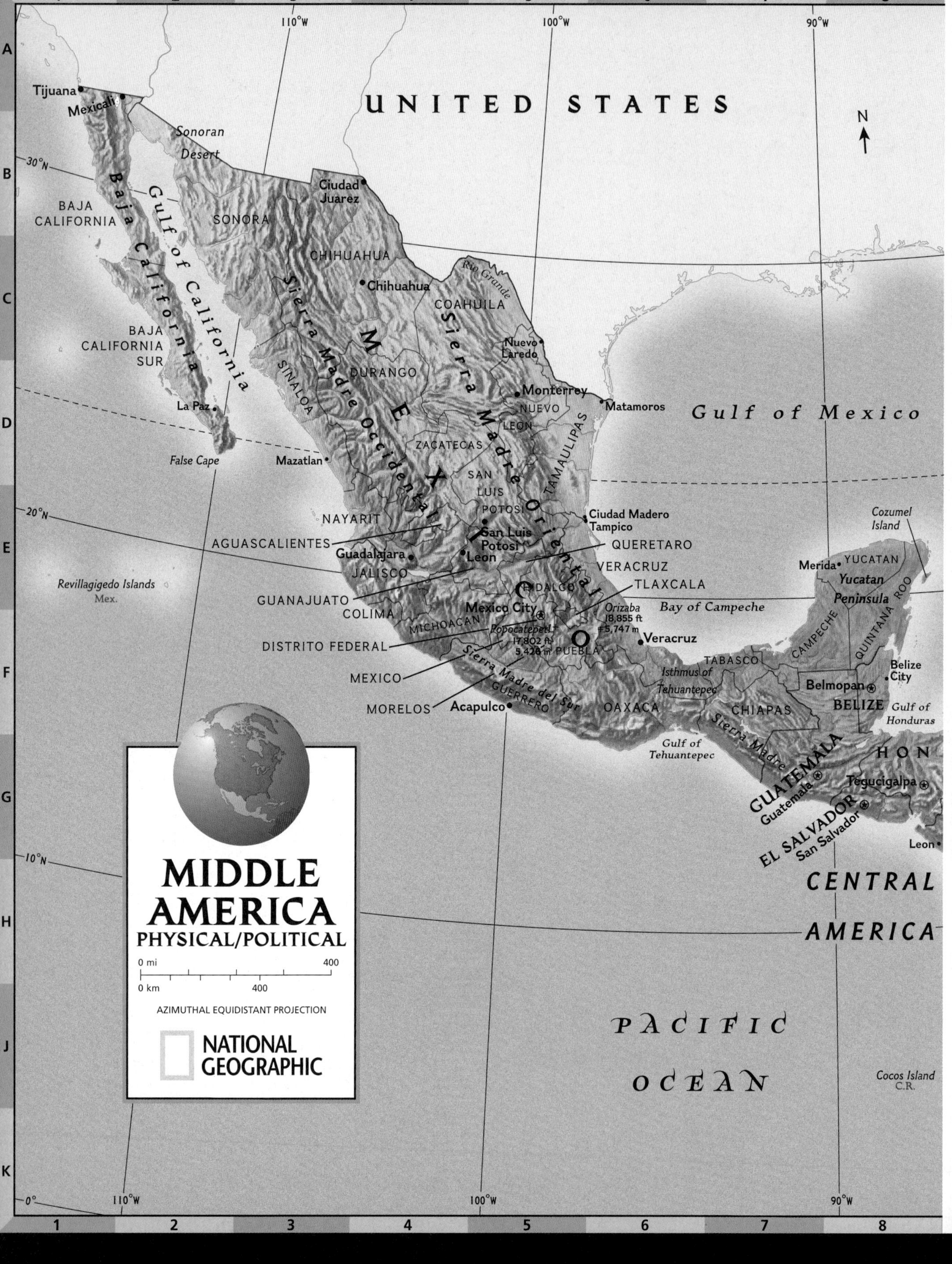
MIDDLE AMERICA
PHYSICAL/POLITICAL
0 mi 400
0 km 400
AZIMUTHAL EQUIDISTANT PROJECTION
NATIONAL GEOGRAPHIC
UNITED STATES
MEXICO
Gulf of Mexico
Bay of Campeche
Gulf of California
Baja California
Sonoran Desert
Sierra Madre Occidental
Sierra Madre Oriental
Sierra Madre del Sur
Sierra Madre
Rio Grande
Yucatan Peninsula
Isthmus of Tehuantepec
Gulf of Tehuantepec
Gulf of Honduras
PACIFIC OCEAN
CENTRAL AMERICA
GUATEMALA
EL SALVADOR
BELIZE
HON
Tijuana
Mexicali
Ciudad Juarez
Chihuahua
Nuevo Laredo
Monterrey
Matamoros
La Paz
Mazatlan
Ciudad Madero
Tampico
San Luis Potosi
Guadalajara
Leon
Mexico City
Veracruz
Acapulco
Merida
Belmopan
Belize City
Guatemala
San Salvador
Tegucigalpa
Leon
BAJA CALIFORNIA
BAJA CALIFORNIA SUR
SONORA
CHIHUAHUA
COAHUILA
SINALOA
DURANGO
NUEVO LEON
TAMAULIPAS
ZACATECAS
SAN LUIS POTOSI
NAYARIT
AGUASCALIENTES
JALISCO
GUANAJUATO
COLIMA
MICHOACAN
DISTRITO FEDERAL
MEXICO
MORELOS
GUERRERO
QUERETARO
VERACRUZ
TLAXCALA
HIDALGO
PUEBLA
OAXACA
TABASCO
CHIAPAS
CAMPECHE
YUCATAN
QUINTANA ROO
Orizaba 18,855 ft 5,747 m
Popocatepetl 17,802 ft 5,426 m
False Cape
Revillagigedo Islands Mex.
Cozumel Island
Cocos Island C.R.
110°W
100°W
90°W
30°N
20°N
10°N
0°
N

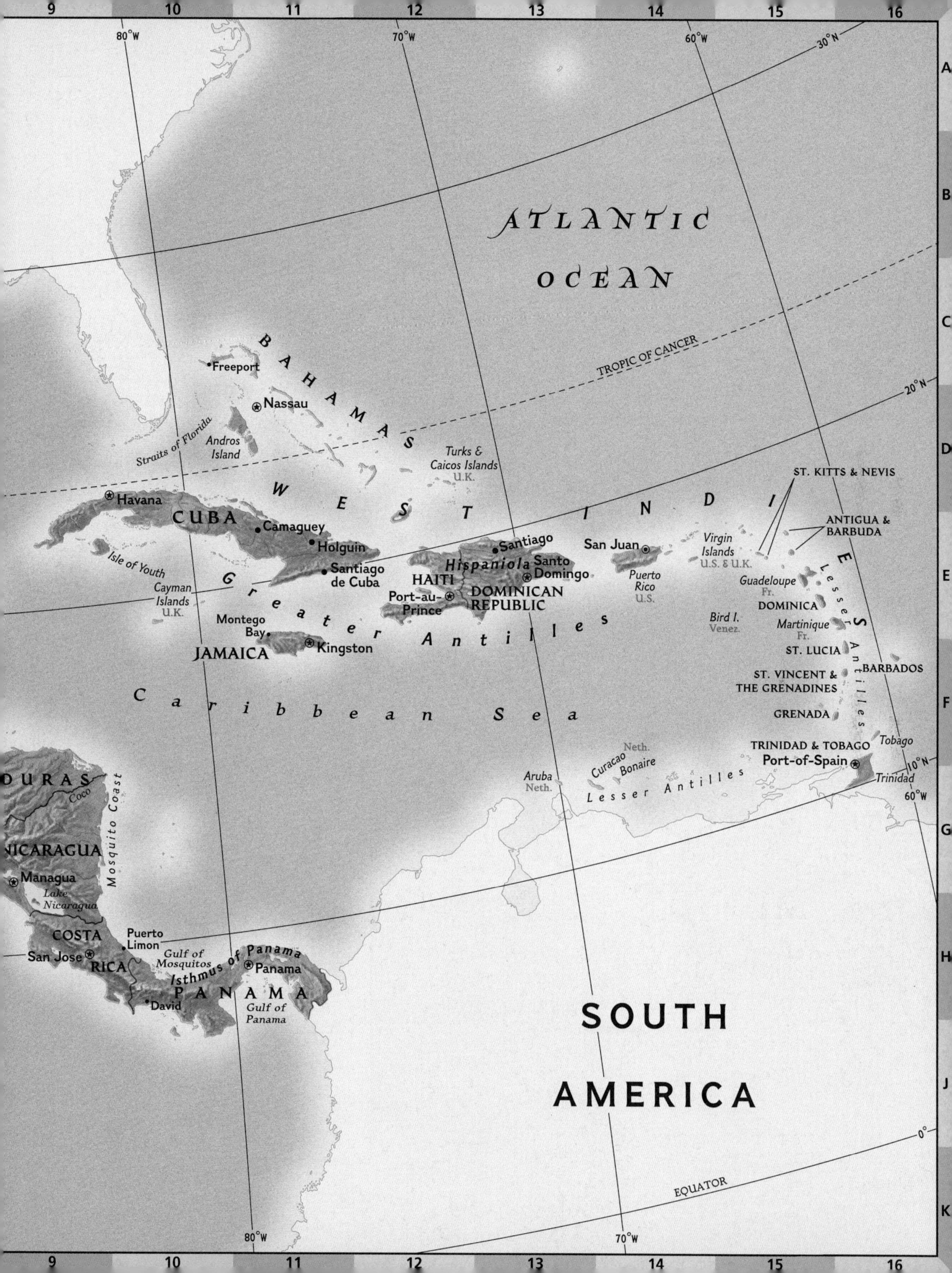
ATLANTIC OCEAN
BAHAMAS
Freeport
Nassau
Andros Island
Straits of Florida
TROPIC OF CANCER
Turks & Caicos Islands
U.K.
WEST INDIES
Havana
CUBA
Camaguey
Holguin
Isle of Youth
Santiago de Cuba
Cayman Islands
U.K.
Greater Antilles
Hispaniola
Santiago
Santo Domingo
HAITI
Port-au-Prince
DOMINICAN REPUBLIC
San Juan
Puerto Rico
U.S.
Virgin Islands
U.S. & U.K.
ST. KITTS & NEVIS
ANTIGUA & BARBUDA
Guadeloupe
Fr.
DOMINICA
Martinique
Fr.
ST. LUCIA
BARBADOS
ST. VINCENT & THE GRENADINES
GRENADA
Lesser Antilles
Bird I.
Venez.
Montego Bay
JAMAICA
Kingston
Caribbean Sea
TRINIDAD & TOBAGO
Port-of-Spain
Tobago
Trinidad
Aruba
Neth.
Curacao
Bonaire
Neth.
Lesser Antilles
HONDURAS
Coco
Mosquito Coast
NICARAGUA
Managua
Lake Nicaragua
COSTA RICA
San Jose
Puerto Limon
Gulf of Mosquitos
Isthmus of Panama
Panama
PANAMA
David
Gulf of Panama
SOUTH AMERICA
EQUATOR
80°W
70°W
60°W
30°N
20°N
10°N
0°
9
10
11
12
13
14
15
16
A
B
C
D
E
F
G
H
J
K

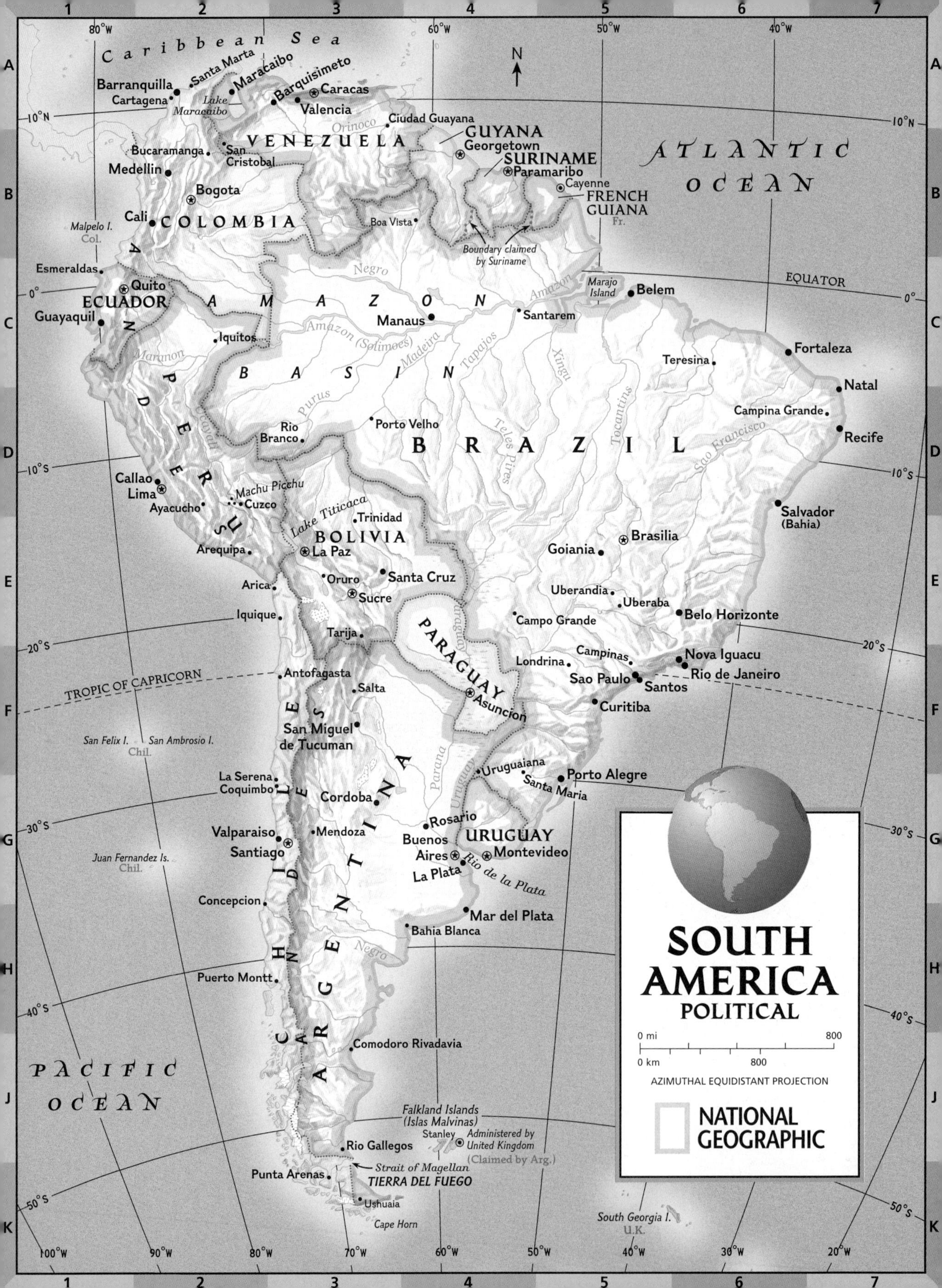

SOUTH AMERICA
POLITICAL
Caribbean Sea
ATLANTIC OCEAN
PACIFIC OCEAN
VENEZUELA
GUYANA
SURINAME
FRENCH GUIANA
Fr.
COLOMBIA
ECUADOR
PERU
BRAZIL
BOLIVIA
PARAGUAY
CHILE
ARGENTINA
URUGUAY
AMAZON BASIN
ANDES
Barranquilla
Cartagena
Santa Marta
Maracaibo
Lake Maracaibo
Barquisimeto
Caracas
Valencia
Ciudad Guayana
Orinoco
Georgetown
Paramaribo
Cayenne
Bucaramanga
San Cristobal
Medellin
Bogota
Cali
Boa Vista
Boundary claimed by Suriname
Malpelo I.
Col.
Esmeraldas
Quito
Guayaquil
Negro
Amazon
Marajo Island
Belem
EQUATOR
Manaus
Santarem
Iquitos
Amazon (Solimoes)
Maranon
Madeira
Tapajos
Xingu
Purus
Fortaleza
Teresina
Natal
Campina Grande
Recife
Tocantins
Sao Francisco
Teles Pires
Ucayali
Porto Velho
Rio Branco
Callao
Lima
Machu Picchu
Cuzco
Ayacucho
Lake Titicaca
Trinidad
La Paz
Salvador (Bahia)
Brasilia
Goiania
Arequipa
Oruro
Santa Cruz
Sucre
Arica
Uberandia
Uberaba
Belo Horizonte
Iquique
Campo Grande
Uruguai
Tarija
Campinas
Nova Iguacu
Rio de Janeiro
Londrina
Sao Paulo
Santos
Antofagasta
TROPIC OF CAPRICORN
Salta
Asuncion
Curitiba
San Miguel de Tucuman
San Felix I.
San Ambrosio I.
Chil.
Parana
Uruguaiana
Santa Maria
Porto Alegre
La Serena
Coquimbo
Cordoba
Uruguay
Rosario
Valparaiso
Mendoza
Santiago
Buenos Aires
Montevideo
La Plata
Rio de la Plata
Juan Fernandez Is.
Chil.
Concepcion
Mar del Plata
Bahia Blanca
Negro
Puerto Montt
Comodoro Rivadavia
Falkland Islands (Islas Malvinas)
Stanley
Administered by United Kingdom
(Claimed by Arg.)
Rio Gallegos
Strait of Magellan
Punta Arenas
TIERRA DEL FUEGO
Ushuaia
Cape Horn
South Georgia I.
U.K.
0 mi 800
0 km 800
AZIMUTHAL EQUIDISTANT PROJECTION
NATIONAL GEOGRAPHIC
N
80°W
60°W
50°W
40°W
10°N
0°
10°S
20°S
30°S
40°S
50°S
100°W
90°W
70°W
30°W
20°W
1 2 3 4 5 6 7
A B C D E F G H J K

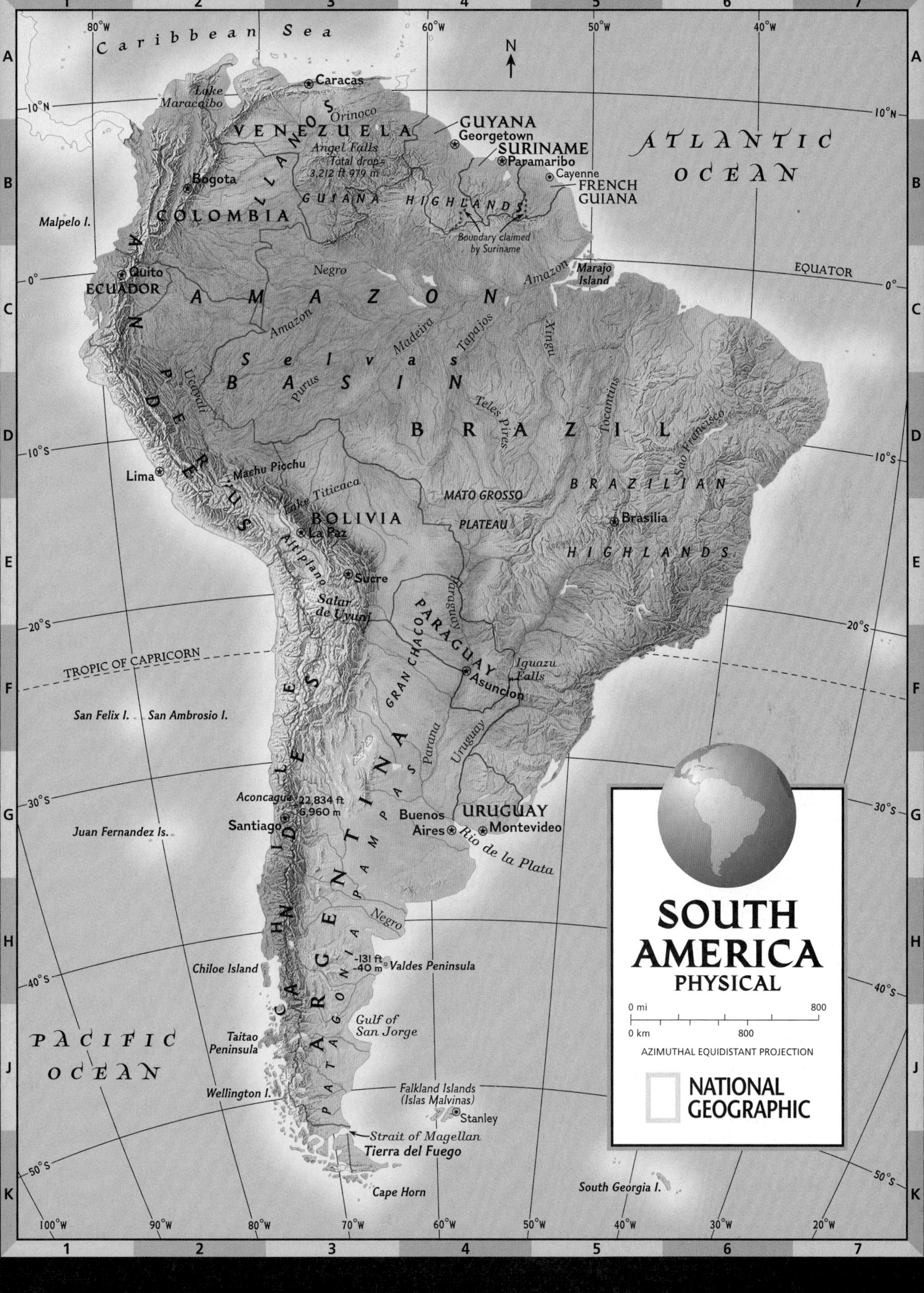
SOUTH AMERICA
PHYSICAL
0 mi 800
0 km 800
AZIMUTHAL EQUIDISTANT PROJECTION
NATIONAL GEOGRAPHIC
Caribbean Sea
ATLANTIC OCEAN
PACIFIC OCEAN
Lake Maracaibo
Caracas
Orinoco
VENEZUELA
LLANOS
GUYANA
Georgetown
SURINAME
Paramaribo
Cayenne
FRENCH GUIANA
Angel Falls
Total drop
3,212 ft 979 m
Bogota
GUIANA HIGHLANDS
Boundary claimed by Suriname
COLOMBIA
Malpelo I.
Quito
ECUADOR
Negro
Amazon
Marajo Island
EQUATOR
AMAZON
Selvas
BASIN
Madeira
Tapajos
Xingu
Purus
Ucayali
Teles Pires
Tocantins
Sao Francisco
BRAZIL
ANDES
Lima
Machu Picchu
Lake Titicaca
MATO GROSSO
PLATEAU
BRAZILIAN
HIGHLANDS
Brasilia
BOLIVIA
La Paz
Altiplano
Sucre
Salar de Uyuni
Paraguay
PARAGUAY
GRAN CHACO
Asuncion
Iguazu Falls
TROPIC OF CAPRICORN
San Felix I.
San Ambrosio I.
Parana
Uruguay
ARGENTINA
PAMPAS
Aconcagua 22,834 ft 6,960 m
Santiago
Juan Fernandez Is.
Buenos Aires
URUGUAY
Montevideo
Rio de la Plata
Negro
Chiloe Island
-131 ft -40 m
Valdes Peninsula
PATAGONIA
Gulf of San Jorge
Taitao Peninsula
Wellington I.
Falkland Islands (Islas Malvinas)
Stanley
Strait of Magellan
Tierra del Fuego
Cape Horn
South Georgia I.

EUROPE
POLITICAL
0 mi 400
0 km 400
AZIMUTHAL EQUIDISTANT PROJECTION
NATIONAL GEOGRAPHIC
ATLANTIC OCEAN
ICELAND
Reykjavík
Akureyri
ARCTIC CIRCLE
Norwegian Sea
PRIME MERIDIAN (MERIDIAN OF GREENWICH)
Faeroe Islands
Den.
Torshavn
Shetland Islands
Lerwick
Orkney Islands
Rockall
U.K.
Isle of Lewis
Inverness
Aberdeen
UNITED KINGDOM
SCOTLAND
Glasgow
Edinburgh
NORTHERN IRELAND
Belfast
IRELAND
Dublin
Cork
Irish Sea
Liverpool
Manchester
WALES
Birmingham
ENGLAND
Cardiff
London
Celtic Sea
Land's End
Southampton
English Channel
North Sea
NORWAY
Tromso
Trondheim
Alesund
Bergen
Stavanger
Oslo
Skagerrak
SWEDEN
Are
Sundsvall
Gulf of
Uppsala
Stockholm
Goteborg
Gotland
Baltic
DENMARK
Arhus
Copenhagen
Malmo
Kiel
Hamburg
Gdansk
Bydgoszcz
POLAND
Lodz
Berlin
GERMANY
NETH.
Amsterdam
The Hague
Brussels
BELGIUM
Bonn
Rhine
LUX.
Frankfurt
Wroclaw
Oder
Prague
CZECH REP.
Bratislava
SLOVAKIA
Munich
Vienna
LIECH.
AUSTRIA
Budapest
HUNGARY
SLOVENIA
Ljubljana
Zagreb
CROATIA
BOSNIA & HERZEGOVINA
Sarajevo
MONTENEGRO
Podgorica
Tiranë
ALBANIA
Le Havre
Brest
Rennes
Paris
Strasbourg
Nantes
FRANCE
La Rochelle
Bay of Biscay
Limoges
Bordeaux
Zurich
Bern
SWITZERLAND
Geneva
Lyon
ALPS
Milan
Turin
Venice
Genoa
MONACO
Nice
Marseille
Toulouse
Donostia-San Sebastian
Pyrenees
ANDORRA
Bilbao
La Coruña
Vigo
Porto
Coimbra
PORTUGAL
Lisbon
Cape St. Vincent
Valladolid
Madrid
Zaragoza
Barcelona
SPAIN
Valencia
Cordoba
Seville
Murcia
Cartagena
Cadiz
Malaga
GIBRALTAR
U.K.
Strait of Gibraltar
Palma
Balearic Islands
Sp.
Corsica
Fr.
Sardinia
It.
Cagliari
SAN MARINO
ITALY
Adriatic Sea
VATICAN CITY
Rome
Naples
Tyrrhenian Sea
Ionian Sea
Palermo
Sicily
Messina
Catania
Valletta
MALTA
Mediterranean
AFRICA
60°N
50°N
40°N
30°N
40°W
30°W
20°W
10°W
0°
10°E
70°N
N
1 2 3 4 5 6 7 8
A B C D E F G H J K

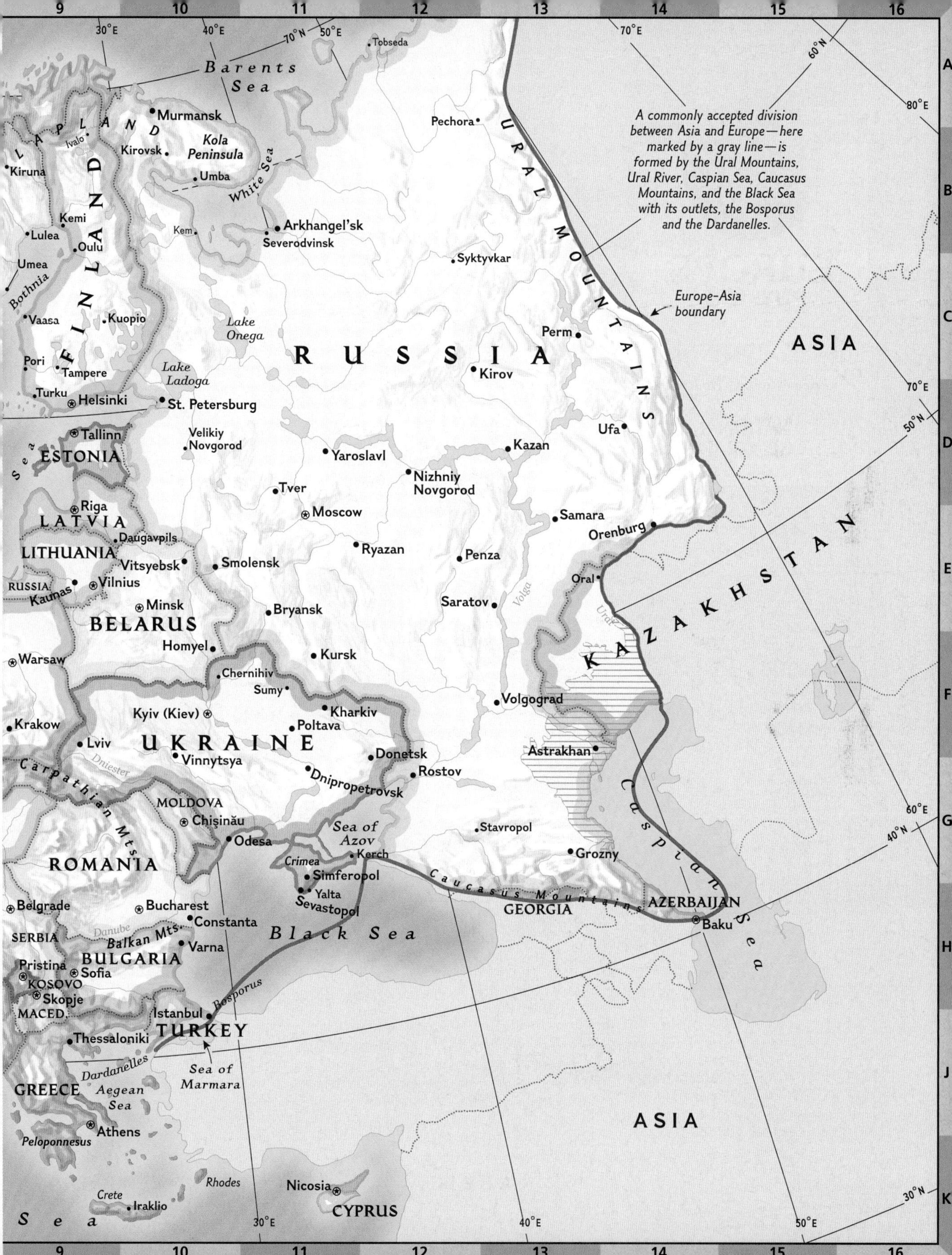

A commonly accepted division between Asia and Europe—here marked by a gray line—is formed by the Ural Mountains, Ural River, Caspian Sea, Caucasus Mountains, and the Black Sea with its outlets, the Bosporus and the Dardanelles.
Europe-Asia boundary
RUSSIA
ASIA
KAZAKHSTAN
URAL MOUNTAINS
Barents Sea
White Sea
Kola Peninsula
LAPLAND
FINLAND
ESTONIA
LATVIA
LITHUANIA
BELARUS
UKRAINE
MOLDOVA
ROMANIA
SERBIA
BULGARIA
KOSOVO
MACED.
GREECE
TURKEY
GEORGIA
AZERBAIJAN
CYPRUS
Black Sea
Caspian Sea
Sea of Azov
Caucasus Mountains
Carpathian Mts.
Balkan Mts.
Aegean Sea
Sea of Marmara
Dardanelles
Bosporus
Crimea
Lake Onega
Lake Ladoga
Moscow
St. Petersburg
Helsinki
Tallinn
Riga
Vilnius
Minsk
Kyiv (Kiev)
Chişinău
Bucharest
Sofia
Warsaw
Athens
Nicosia
Baku
Istanbul
Volgograd
Astrakhan
Kazan
Perm
Ufa
Samara
Orenburg
Saratov
Oral
Murmansk
Arkhangel'sk
Severodvinsk
Pechora
Syktyvkar
Kirov
Yaroslavl
Nizhniy Novgorod
Tver
Ryazan
Penza
Smolensk
Bryansk
Kursk
Kharkiv
Poltava
Donetsk
Rostov
Dnipropetrovsk
Odesa
Kerch
Simferopol
Yalta
Sevastopol
Stavropol
Grozny
Lviv
Vinnytsya
Chernihiv
Sumy
Homyel
Vitsyebsk
Daugavpils
Kaunas
Krakow
Velikiy Novgorod
Constanta
Varna
Belgrade
Pristina
Skopje
Thessaloniki
Peloponnesus
Crete
Iraklio
Rhodes
Kiruna
Lulea
Kemi
Oulu
Umea
Vaasa
Kuopio
Pori
Tampere
Turku
Kirovsk
Umba
Kem
Ivalo
Tobseda
Volga
Ural
Danube
Dniester
Bothnia
RUSSIA
Sea

EUROPE
PHYSICAL
0 mi 400
0 km 400
AZIMUTHEL EQUIDISTANT PROJECTION
NATIONAL GEOGRAPHIC
ATLANTIC OCEAN
Reykjavík
ICELAND
ARCTIC CIRCLE
PRIME MERIDIAN (MERIDIAN OF GREENWICH)
Norwegian Sea
Faeroe Islands
Shetland Islands
Orkney Islands
Outer Hebrides
Highlands
British Isles
Edinburgh
Belfast
UNITED KINGDOM
IRELAND
Dublin
Irish Sea
Great Britain
Cardiff
London
North Sea
English Channel
SCANDINAVIA
NORWAY
SWEDEN
Oslo
Stockholm
Gulf of
Jutland
DENMARK
Copenhagen
Zealand
Baltic
Amsterdam
NETH.
BELGIUM
Brussels
LUX.
Berlin
GERMANY
POLAND
Oder
Elbe
Rhine
NORT
Prague
CZECH REP.
Bratislava
SLOVAKIA
Vienna
Budapest
HUNGARY
Danube
Brittany
Seine
Paris
Loire
FRANCE
Bay of Biscay
Mont Blanc
15,771 ft
4,807 m
Massif Central
Rhone
Bern
SWITZ.
LIECH.
AUSTRIA
ALPS
SLOVENIA
Ljubljana
Zagreb
CROATIA
Drava
Sava
BOSNIA & HERZEGOVINA
Sarajevo
Po
MONACO
Riviera
SAN MARINO
Adriatic Sea
Apennines
Cantabrian Mountains
Pyrenees
Douro
IBERIAN PENINSULA
PORTUGAL
SPAIN
Lisbon
Tagus
Madrid
Ebro
ANDORRA
Corsica
ITALY
VATICAN CITY
Rome
MONTENEGRO
Tiranë
ALBANIA
Sardinia
Tyrrhenian Sea
Balearic Islands
Baetic Mountains
GIBRALTAR
Strait of Gibraltar
Mediterranean
Ionian Sea
Sicily
Etna
10,902 ft
3,323 m
Valletta
MALTA
AFRICA
70°N
60°N
50°N
40°N
30°N
40°W
30°W
20°W
10°W
0°
10°E

Barents Sea
North Cape
Kola Peninsula
White Sea
Pechora
URAL MOUNTAINS
FINLAND
Bothnia
Lake Region
Lake Onega
Lake Ladoga
RUSSIA
Northern Dvina
EUROPEAN PLAIN
Europe-Asia boundary
ASIA
Helsinki
Gulf of Finland
Tallinn
ESTONIA
Kama
Volga
Ural
Riga
LATVIA
Moscow
LITHUANIA
Vilnius
Oka
CENTRAL RUSSIAN UPLAND
Minsk
BELARUS
Dnieper
Don
Warsaw
KAZAKHSTAN
Caspian Depression
Kyiv (Kiev)
UKRAINE
Dniester
Carpathian Mountains
MOLDOVA
Chişinău
Sea of Azov
Crimea
Elbrus 18,510 ft 5,642 m
Caucasus Mountains
ROMANIA
Bucharest
Belgrade
BALKAN PENINSULA
Danube
Black Sea
GEORGIA
AZERBAIJAN
Baku
Caspian Sea
SERBIA
BULGARIA
Balkan Mountains
Sofia
KOSOVO
Skopje
MACED.
Bosporus
TURKEY
Dardanelles
Sea of Marmara
Aegean Sea
GREECE
Athens
Peloponnesus
Rhodes
Crete
Nicosia
CYPRUS
Tisza
Vistula

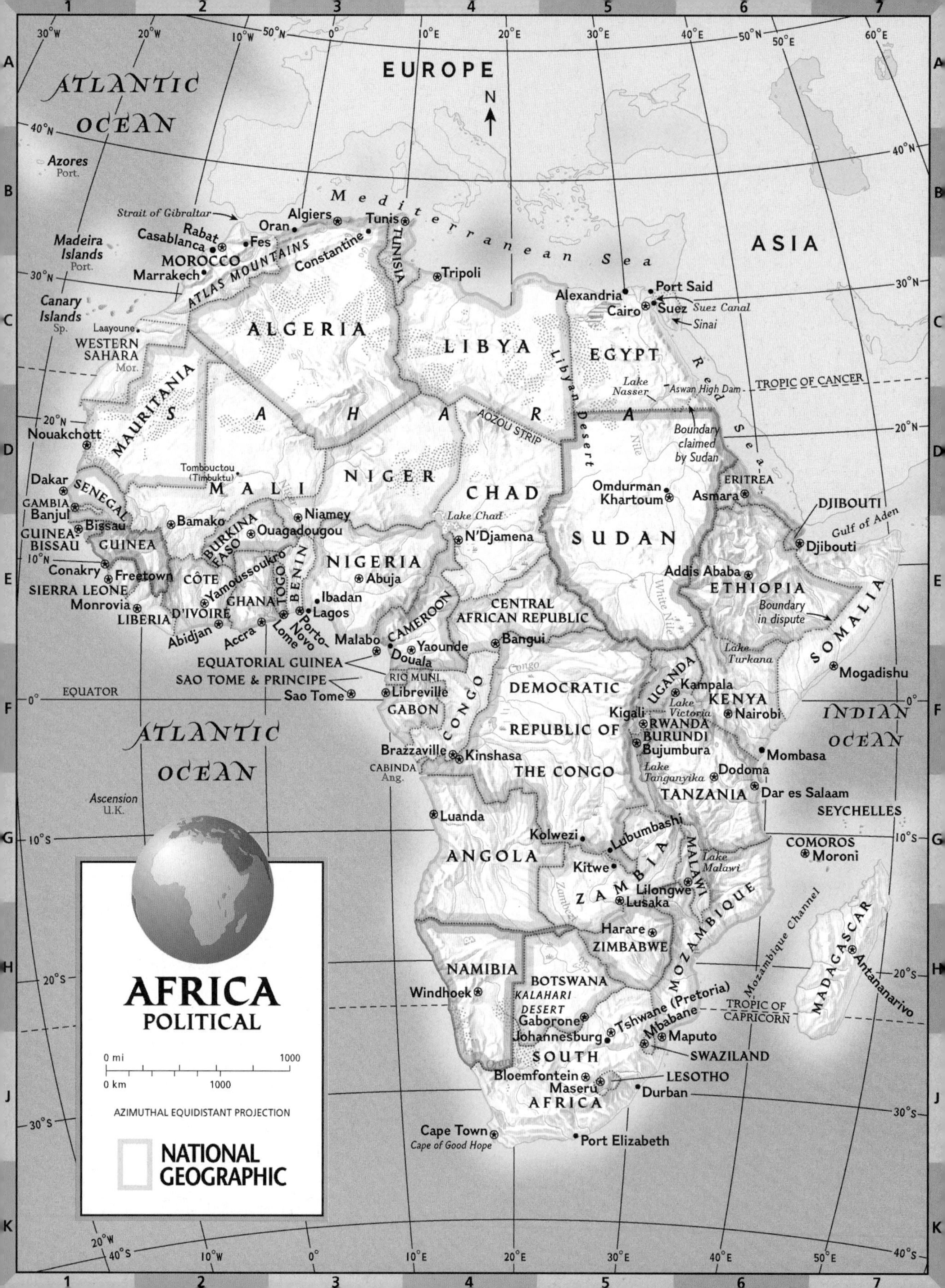

AFRICA
POLITICAL
ATLANTIC OCEAN
EUROPE
ASIA
Mediterranean Sea
Red Sea
INDIAN OCEAN
Gulf of Aden
Mozambique Channel
MOROCCO
ALGERIA
TUNISIA
LIBYA
EGYPT
WESTERN SAHARA
Mor.
MAURITANIA
MALI
NIGER
CHAD
SUDAN
ERITREA
DJIBOUTI
ETHIOPIA
SOMALIA
SENEGAL
GAMBIA
GUINEA-BISSAU
GUINEA
SIERRA LEONE
LIBERIA
CÔTE D'IVOIRE
BURKINA FASO
GHANA
TOGO
BENIN
NIGERIA
CAMEROON
CENTRAL AFRICAN REPUBLIC
EQUATORIAL GUINEA
SAO TOME & PRINCIPE
RIO MUNI
GABON
CONGO
DEMOCRATIC REPUBLIC OF THE CONGO
CABINDA
Ang.
UGANDA
KENYA
RWANDA
BURUNDI
TANZANIA
ANGOLA
ZAMBIA
MALAWI
MOZAMBIQUE
ZIMBABWE
NAMIBIA
BOTSWANA
SOUTH AFRICA
SWAZILAND
LESOTHO
MADAGASCAR
SEYCHELLES
COMOROS
SAHARA
ATLAS MOUNTAINS
Libyan Desert
AOZOU STRIP
KALAHARI DESERT
Azores
Port.
Madeira Islands
Port.
Canary Islands
Sp.
Ascension
U.K.
Strait of Gibraltar
Suez Canal
Sinai
Lake Nasser
Aswan High Dam
Boundary claimed by Sudan
Boundary in dispute
Nile
White Nile
Lake Chad
Lake Turkana
Lake Victoria
Lake Tanganyika
Lake Malawi
Congo
Zambezi
Niger
Senegal
Cape of Good Hope
TROPIC OF CANCER
EQUATOR
TROPIC OF CAPRICORN
Rabat
Casablanca
Fes
Marrakech
Oran
Algiers
Constantine
Tunis
Tripoli
Alexandria
Port Said
Cairo
Suez
Laayoune
Nouakchott
Tombouctou (Timbuktu)
Dakar
Banjul
Bissau
Conakry
Freetown
Monrovia
Bamako
Niamey
Ouagadougou
Yamoussoukro
Abidjan
Accra
Lome
Porto-Novo
Lagos
Ibadan
Abuja
N'Djamena
Omdurman
Khartoum
Asmara
Djibouti
Addis Ababa
Mogadishu
Malabo
Douala
Yaounde
Bangui
Sao Tome
Libreville
Brazzaville
Kinshasa
Kampala
Nairobi
Kigali
Bujumbura
Mombasa
Dodoma
Dar es Salaam
Luanda
Kolwezi
Lubumbashi
Kitwe
Lilongwe
Lusaka
Moroni
Harare
Antananarivo
Windhoek
Gaborone
Tshwane (Pretoria)
Johannesburg
Mbabane
Maputo
Bloemfontein
Maseru
Durban
Cape Town
Port Elizabeth
N
0 mi 1000
0 km 1000
AZIMUTHAL EQUIDISTANT PROJECTION
NATIONAL GEOGRAPHIC

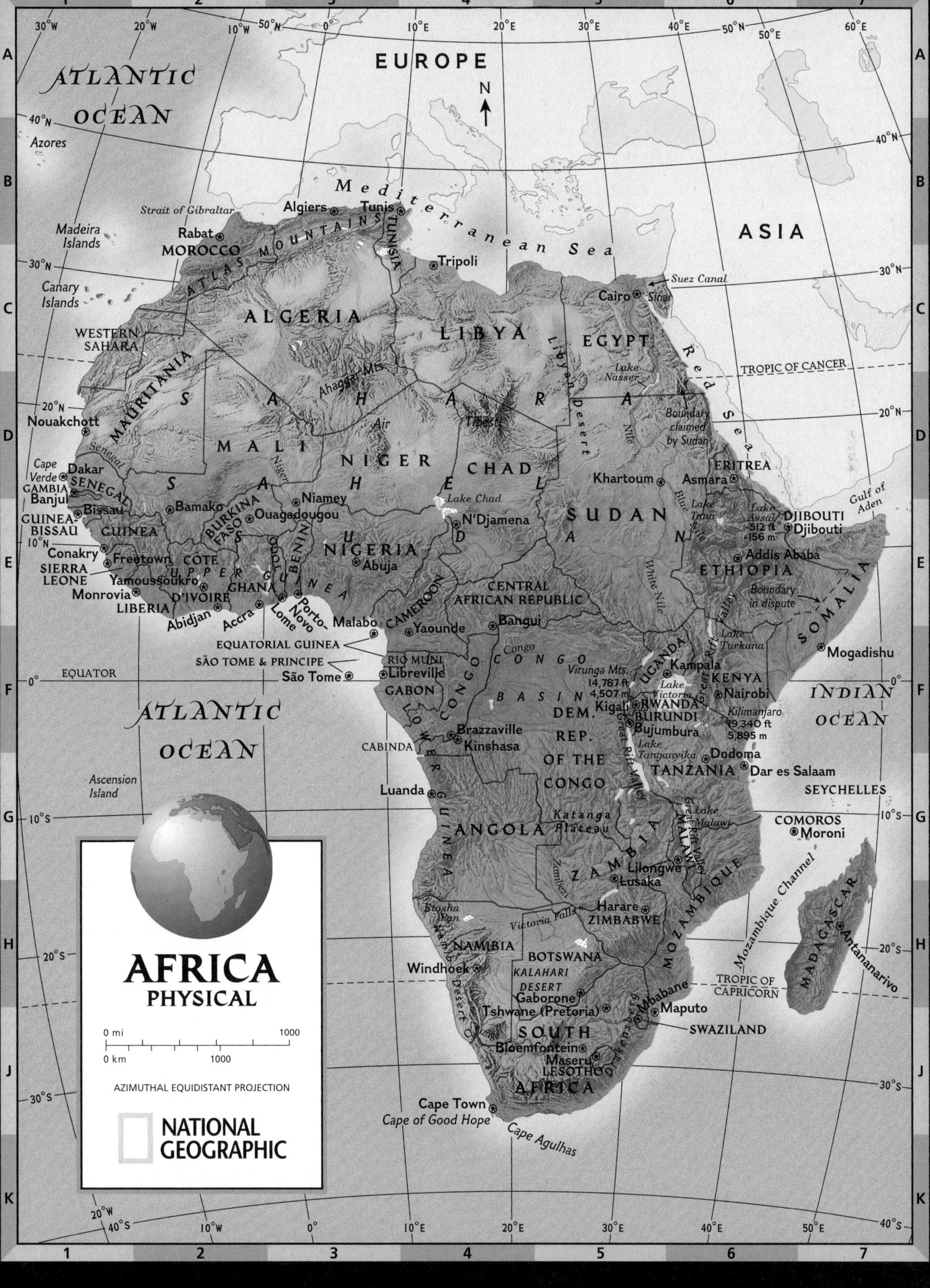
AFRICA
PHYSICAL
EUROPE
ASIA
ATLANTIC OCEAN
INDIAN OCEAN
Mediterranean Sea
Red Sea
ALGERIA
LIBYA
EGYPT
MOROCCO
TUNISIA
WESTERN SAHARA
MAURITANIA
MALI
NIGER
CHAD
SUDAN
NIGERIA
ETHIOPIA
SOMALIA
KENYA
TANZANIA
ANGOLA
NAMIBIA
BOTSWANA
ZAMBIA
ZIMBABWE
MOZAMBIQUE
MADAGASCAR
SOUTH AFRICA
DEM. REP. OF THE CONGO
CONGO
GABON
CAMEROON
CENTRAL AFRICAN REPUBLIC
SAHARA
SAHEL
ATLAS MOUNTAINS
Azores
Madeira Islands
Canary Islands
Strait of Gibraltar
Suez Canal
Sinai
TROPIC OF CANCER
TROPIC OF CAPRICORN
EQUATOR
Cairo
Tripoli
Algiers
Tunis
Rabat
Khartoum
Asmara
Addis Ababa
Mogadishu
Nairobi
Kampala
Kinshasa
Brazzaville
Luanda
Lusaka
Harare
Windhoek
Gaborone
Tshwane (Pretoria)
Cape Town
Cape of Good Hope
Cape Agulhas
Lake Victoria
Lake Chad
Kilimanjaro 19,340 ft 5,895 m
Mozambique Channel
Gulf of Aden
Ascension Island
SEYCHELLES
COMOROS
Moroni
Antananarivo
0 mi 1000
0 km 1000
AZIMUTHAL EQUIDISTANT PROJECTION
NATIONAL GEOGRAPHIC

EUROPE
Black Sea
Sea of Marmara
Istanbul
ANATOLIA
Ankara
TURKEY
Tunis
TUNISIA
Mediterranean Sea
Tripoli
LIBYA
Taurus Mountains
Aleppo
CYPRUS
SYRIA
LEBANON
Damascus
Beirut
ISRAEL
Syrian Desert
Jerusalem
Amman
Alexandria
Cairo
JORDAN
El Giza
Sinai Pen.
See inset below
EGYPT
Nile R.
Hejaz
Aswan High Dam
Red Sea
SAHARA
Boundary claimed by Sudan
SUDAN
AFRICA
Khartoum
10°E
20°E
30°E
40°E
40°N
30°N
Eastern Mediterranean Area
TURKEY
N
Aleppo
CYPRUS
SYRIA
LEBANON
Mediterranean Sea
Beirut
Damascus
Sea of Galilee
Golan Heights
Jordan River
Tel Aviv–Jaffa
West Bank
Suez Canal
Jerusalem
Amman
Gaza Strip
Dead Sea
ISRAEL
JORDAN
El Giza
Cairo
EGYPT
SAUDI ARABIA
Nile River
Gulf of Suez
Gulf of Aqaba
Red Sea
0 mi 100
0 km 100
30°E
30°N

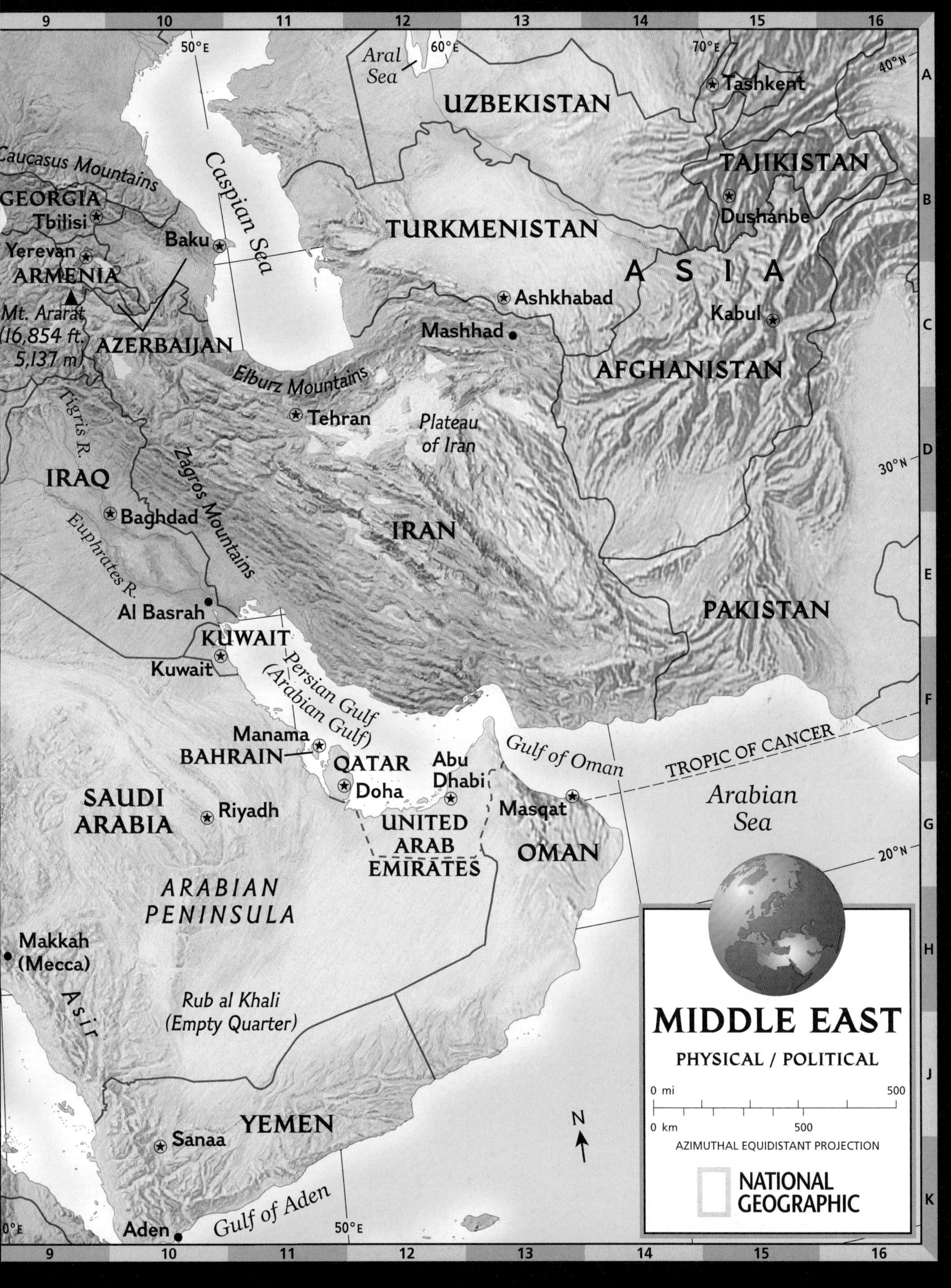

MIDDLE EAST
PHYSICAL / POLITICAL
0 mi
500
0 km
500
AZIMUTHAL EQUIDISTANT PROJECTION
NATIONAL GEOGRAPHIC
UZBEKISTAN
Tashkent
TAJIKISTAN
Dushanbe
TURKMENISTAN
Ashkhabad
ASIA
Kabul
AFGHANISTAN
Mashhad
Aral Sea
Caspian Sea
Caucasus Mountains
GEORGIA
Tbilisi
Baku
Yerevan
ARMENIA
Mt. Ararat (16,854 ft. 5,137 m)
AZERBAIJAN
Elburz Mountains
Tehran
Plateau of Iran
Tigris R.
IRAQ
Baghdad
Zagros Mountains
IRAN
Euphrates R.
Al Basrah
KUWAIT
Kuwait
PAKISTAN
Persian Gulf (Arabian Gulf)
Manama
BAHRAIN
QATAR
Doha
Abu Dhabi
UNITED ARAB EMIRATES
Gulf of Oman
Masqat
TROPIC OF CANCER
Arabian Sea
OMAN
SAUDI ARABIA
Riyadh
ARABIAN PENINSULA
Makkah (Mecca)
Asir
Rub al Khali (Empty Quarter)
YEMEN
Sanaa
Aden
Gulf of Aden
N
50°E
60°E
70°E
40°N
30°N
20°N

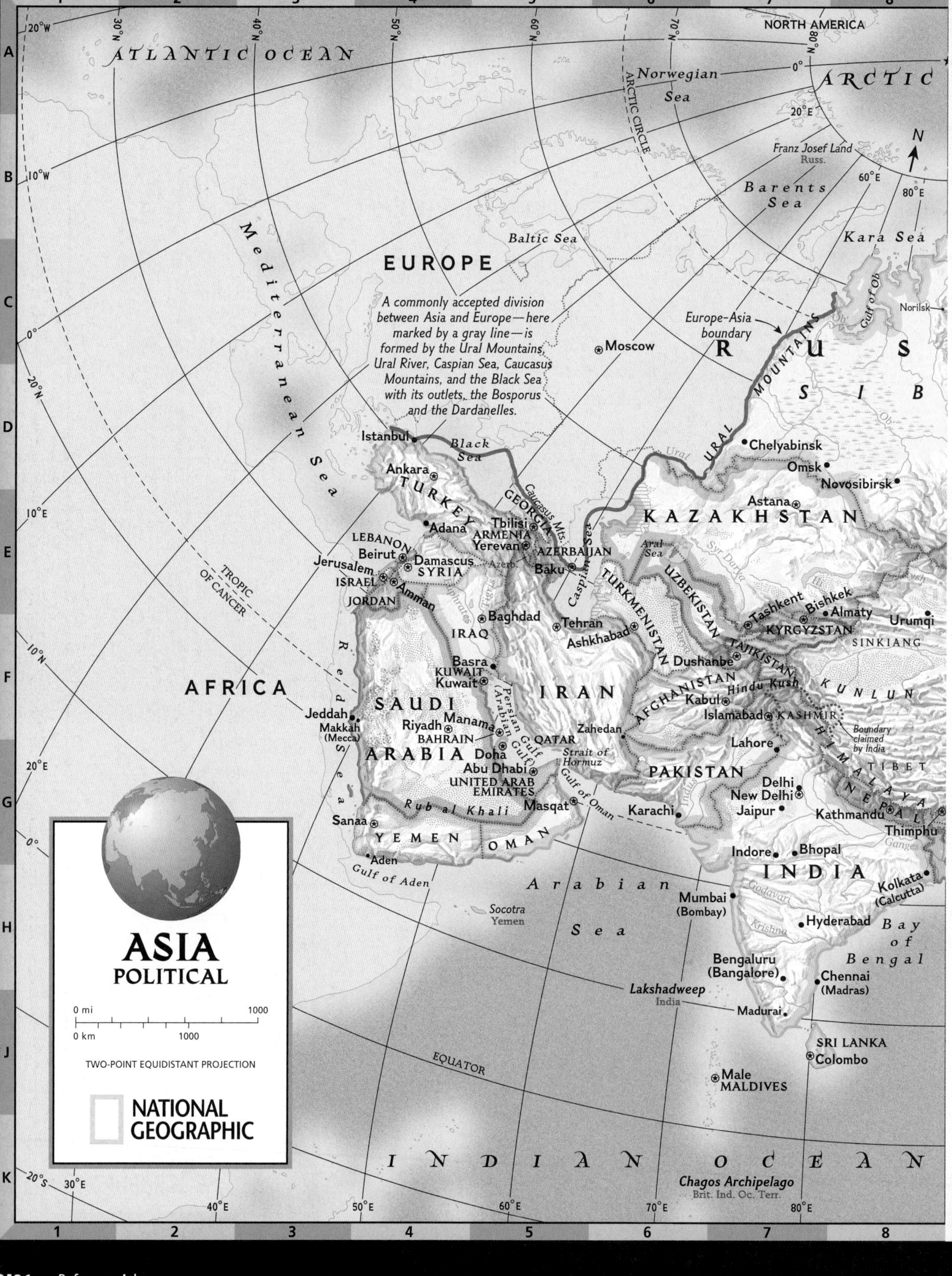

ASIA
POLITICAL
TWO-POINT EQUIDISTANT PROJECTION
NATIONAL GEOGRAPHIC
A commonly accepted division between Asia and Europe—here marked by a gray line—is formed by the Ural Mountains, Ural River, Caspian Sea, Caucasus Mountains, and the Black Sea with its outlets, the Bosporus and the Dardanelles.
ATLANTIC OCEAN
ARCTIC
INDIAN OCEAN
EUROPE
AFRICA
Arabian Sea
Mediterranean Sea
Red Sea
Black Sea
Caspian Sea
Baltic Sea
Norwegian Sea
Barents Sea
Kara Sea
Bay of Bengal
Gulf of Aden
Persian Gulf (Arabian Gulf)
Gulf of Oman
Strait of Hormuz
Europe-Asia boundary
URAL MOUNTAINS
Caucasus Mts.
TURKEY
GEORGIA
ARMENIA
AZERBAIJAN
KAZAKHSTAN
UZBEKISTAN
TURKMENISTAN
KYRGYZSTAN
TAJIKISTAN
AFGHANISTAN
PAKISTAN
IRAN
IRAQ
SYRIA
LEBANON
ISRAEL
JORDAN
KUWAIT
SAUDI ARABIA
BAHRAIN
QATAR
UNITED ARAB EMIRATES
OMAN
YEMEN
INDIA
NEPAL
SRI LANKA
MALDIVES
KASHMIR
TIBET
SINKIANG
Istanbul
Ankara
Adana
Tbilisi
Yerevan
Baku
Beirut
Damascus
Jerusalem
Amman
Baghdad
Basra
Kuwait
Tehran
Ashkhabad
Tashkent
Bishkek
Almaty
Urumqi
Dushanbe
Kabul
Islamabad
Lahore
Zahedan
Karachi
Delhi
New Delhi
Jaipur
Kathmandu
Thimphu
Indore
Bhopal
Kolkata (Calcutta)
Mumbai (Bombay)
Hyderabad
Bengaluru (Bangalore)
Chennai (Madras)
Madurai
Colombo
Male
Riyadh
Manama
Doha
Abu Dhabi
Masqat
Jeddah
Makkah (Mecca)
Sanaa
Aden
Moscow
Chelyabinsk
Omsk
Novosibirsk
Astana
Norilsk
Rub al Khali
Hindu Kush
KUNLUN
HIMALAYA
Boundary claimed by India
Socotra Yemen
Lakshadweep India
Chagos Archipelago Brit. Ind. Oc. Terr.
Franz Josef Land Russ.
NORTH AMERICA
Aral Sea
Syr Darya
Amu Darya
Ural
Ob'
Gulf of Ob
Indus
Ganges
Godavari
Krishna
Euphrates
Tigris
TROPIC OF CANCER
EQUATOR
ARCTIC CIRCLE

NORTH AMERICA
Bering Strait
North Pole
OCEAN
Chukchi Sea
Wrangel I.
Gulf of Anadyr
Anadyr
Bering Sea
East Siberian Sea
North Land
New Siberian Islands
Laptev Sea
Kolyma Range
Commander Is.
Cherskiy Range
Verkhoyanski Mountains
Kamchatka Peninsula
Magadan
Sea of Okhotsk
Yakutsk
SIA
ERIA
Sakhalin
Kuril Islands
Lake Baikal
Hokkaido
Sapporo
Irkutsk
MANCHURIA
Vladivostok
Sea of Japan (East Sea)
JAPAN
Ulaanbaatar
Changchun
Tokyo
MONGOLIA
ALTAY MTS.
Shenyang
NORTH KOREA
Honshu
Marcus I. Jap.
TROPIC OF CANCER
GOBI
P'yŏngyang
Seoul
SOUTH KOREA
Kyoto
Osaka
Beijing
Hiroshima
Shijiazhuang
Qingdao
Yellow Sea
Kyushu
Bonin Is. Jap.
Xuzhou
Volcano Is. Jap.
Lanzhou
Xian
East China Sea
Shanghai
Nanjing
CHINA
Ryukyu Islands
Okinawa
PACIFIC OCEAN
Nanchang
Chengdu
Fuzhou
Taipei
Parece Vela Jap.
Boundary claimed by China
Changsha
TAIWAN
BHUTAN
BANGLADESH
Guiyang
The People's Republic of China claims Taiwan as its 23rd province.
Dhaka
Kunming
Guangzhou
Hong Kong
Macau
Philippine Sea
MYANMAR (BURMA)
Hanoi
Haiphong
Luzon
South China Sea
Hainan
Quezon City
Manila
Samar
Nay Pyi Taw
Vientiane
LAOS
Da Nang
Mindoro
PHILIPPINES
Leyte
Panay
Negros
THAILAND
Bangkok
CAMBODIA
VIETNAM
Palawan
Mindanao
Andaman Islands India
Phnom Penh
Ho Chi Minh City
EQUATOR
Gulf of Thailand
Andaman Sea
Bandar Seri Begawan
BRUNEI
SABAH
Halmahera
Morotai
Biak
Jayapura
New Guinea
Nicobar Islands India
Kuala Lumpur
MALAYSIA
SARAWAK
MALAYSIA
Moluccas
Medan
Kepi
Merauke
Aru Is.
Dolak
SINGAPORE
Borneo
Celebes
Buru
Ceram
INDONESIA
Sumatra
Tanimbar Is.
Jambi
GREATER SUNDA ISLANDS
Mentawai Islands
Dili
EAST TIMOR (TIMOR-LESTE)
Timor Sea
AUSTRALIA
Java Sea
Jakarta
Java
Kupang
80°N
70°N
60°N
50°N
40°N
30°N
20°N
10°N
0°
10°S
20°S
180°
160°E
120°E
100°E
170°W
160°W
170°E
160°E
110°E
130°E
140°E
150°E
9
10
11
12
13
14
15
16
A
B
C
D
E
F
G
H
J
K

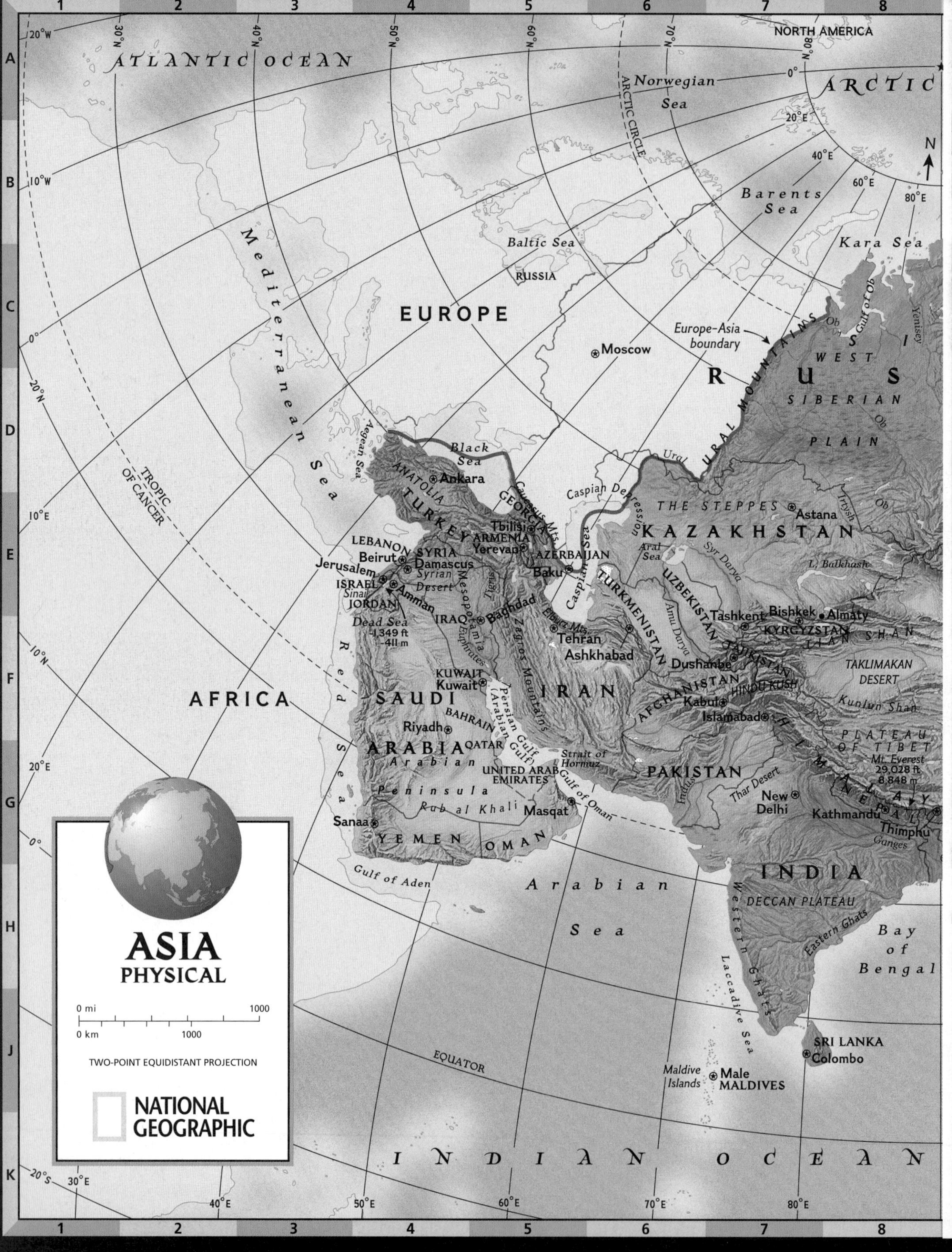
ASIA
PHYSICAL
0 mi
1000
0 km
1000
TWO-POINT EQUIDISTANT PROJECTION
NATIONAL GEOGRAPHIC
NORTH AMERICA
ATLANTIC OCEAN
ARCTIC
Norwegian Sea
ARCTIC CIRCLE
Barents Sea
Kara Sea
Baltic Sea
RUSSIA
EUROPE
Mediterranean Sea
Moscow
Europe-Asia boundary
URAL MOUNTAINS
Gulf of Ob
Ob
Yenisey
WEST SIBERIAN PLAIN
R U S S I
Ural
Caspian Depression
THE STEPPES
Astana
Irtysh
KAZAKHSTAN
Aral Sea
Syr Darya
L. Balkhash
Black Sea
Aegean Sea
ANATOLIA
Ankara
TURKEY
Caucasus Mts.
GEORGIA
Tbilisi
ARMENIA
Yerevan
AZERBAIJAN
Baku
Caspian Sea
LEBANON
Beirut
SYRIA
Damascus
Syrian Desert
Jerusalem
ISRAEL
Sinai
JORDAN
Amman
Dead Sea
-1,349 ft
-411 m
IRAQ
Baghdad
Mesopotamia
Tigris
Euphrates
Zagros Mountains
Elburz Mts.
Tehran
TURKMENISTAN
Ashkhabad
UZBEKISTAN
Amu Darya
Tashkent
Bishkek
Almaty
KYRGYZSTAN
TIAN SHAN
TAJIKISTAN
Dushanbe
TAKLIMAKAN DESERT
AFGHANISTAN
Kabul
HINDU KUSH
Islamabad
Kunlun Shan
PLATEAU OF TIBET
Mt. Everest
29,028 ft
8,848 m
HIMALAYA
KUWAIT
Kuwait
IRAN
Persian Gulf (Arabian Gulf)
BAHRAIN
QATAR
SAUDI ARABIA
Riyadh
Arabian Peninsula
Rub al Khali
UNITED ARAB EMIRATES
Strait of Hormuz
Gulf of Oman
Masqat
OMAN
YEMEN
Sanaa
Red Sea
AFRICA
TROPIC OF CANCER
Gulf of Aden
PAKISTAN
Indus
Thar Desert
New Delhi
NEPAL
Kathmandu
Thimphu
Ganges
INDIA
DECCAN PLATEAU
Western Ghats
Eastern Ghats
Bay of Bengal
Arabian Sea
Laccadive Sea
SRI LANKA
Colombo
Maldive Islands
Male
MALDIVES
EQUATOR
INDIAN OCEAN
N
20°W
10°W
0°
10°E
20°E
30°E
40°E
50°E
60°E
70°E
80°E
20°N
10°N
20°S
30°N
40°N
50°N
60°N
70°N
80°N
A
B
C
D
E
F
G
H
J
K
1
2
3
4
5
6
7
8

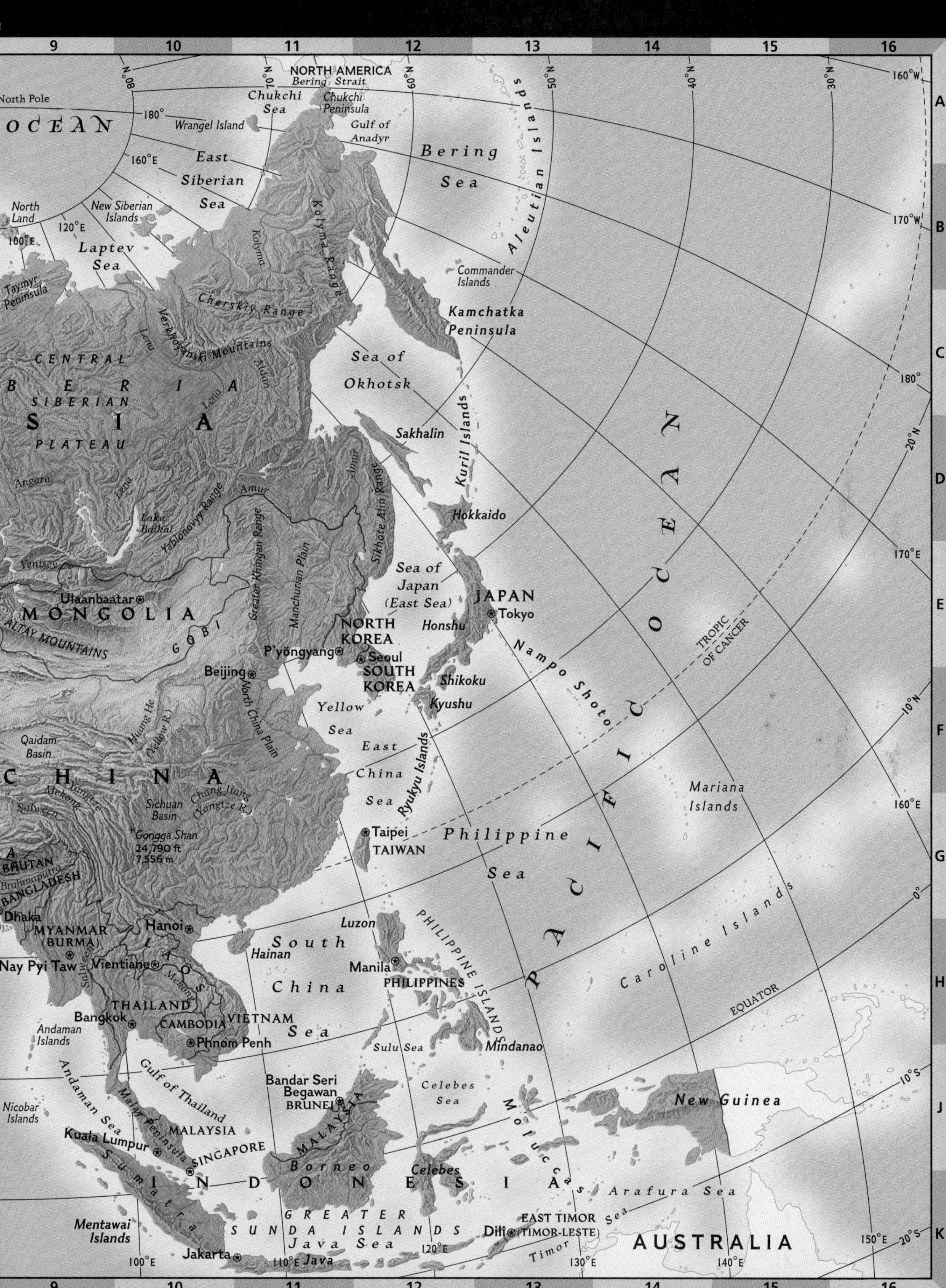

NORTH AMERICA
Bering Strait
North Pole
Chukchi Sea
Chukchi Peninsula
OCEAN
Wrangel Island
Gulf of Anadyr
East Siberian Sea
Bering Sea
Aleutian Islands
North Land
New Siberian Islands
Laptev Sea
Taymyr Peninsula
Kolyma
Kolyma Range
Commander Islands
Cherskiy Range
Verkhoyansk Mountains
Kamchatka Peninsula
CENTRAL SIBERIAN PLATEAU
SIBERIA
ASIA
Lena
Aldan
Sea of Okhotsk
Sakhalin
Kuril Islands
Amur
Sikhote Alin Range
Angara
Lake Baikal
Yablonovyy Range
Hokkaido
Yenisey
Greater Khingan Range
Manchurian Plain
Sea of Japan (East Sea)
JAPAN
Tokyo
Honshu
Ulaanbaatar
MONGOLIA
ALTAY MOUNTAINS
GOBI
NORTH KOREA
P'yŏngyang
Seoul
SOUTH KOREA
Shikoku
Kyushu
Nampo Shoto
Beijing
North China Plain
Huang He (Yellow R.)
Yellow Sea
East China Sea
Ryukyu Islands
Qaidam Basin
CHINA
Mekong
Yangtze
Salween
Sichuan Basin
Chang Jiang (Yangtze R.)
Gongga Shan 24,790 ft 7,556 m
Taipei
TAIWAN
Philippine Sea
PACIFIC OCEAN
TROPIC OF CANCER
Mariana Islands
BHUTAN
Brahmaputra
BANGLADESH
Dhaka
MYANMAR (BURMA)
Nay Pyi Taw
Hanoi
LAOS
Vientiane
Hainan
Luzon
South China Sea
Manila
PHILIPPINES
PHILIPPINE ISLANDS
Caroline Islands
EQUATOR
THAILAND
Bangkok
CAMBODIA
VIETNAM
Phnom Penh
Andaman Islands
Sulu Sea
Mindanao
Andaman Sea
Gulf of Thailand
Malay Peninsula
Nicobar Islands
Bandar Seri Begawan
BRUNEI
MALAYSIA
Celebes Sea
Moluccas
New Guinea
Kuala Lumpur
SINGAPORE
Borneo
Celebes
Sumatra
INDONESIA
Arafura Sea
GREATER SUNDA ISLANDS
Mentawai Islands
Java Sea
Java
Jakarta
EAST TIMOR (TIMOR-LESTE)
Dili
Timor Sea
AUSTRALIA
80°N
70°N
60°N
50°N
40°N
30°N
180°
160°E
120°E
100°E
160°W
170°W
170°E
160°E
150°E
140°E
130°E
120°E
110°E
100°E
20°N
10°N
0°
10°S
20°S
9 10 11 12 13 14 15 16
A B C D E F G H J K

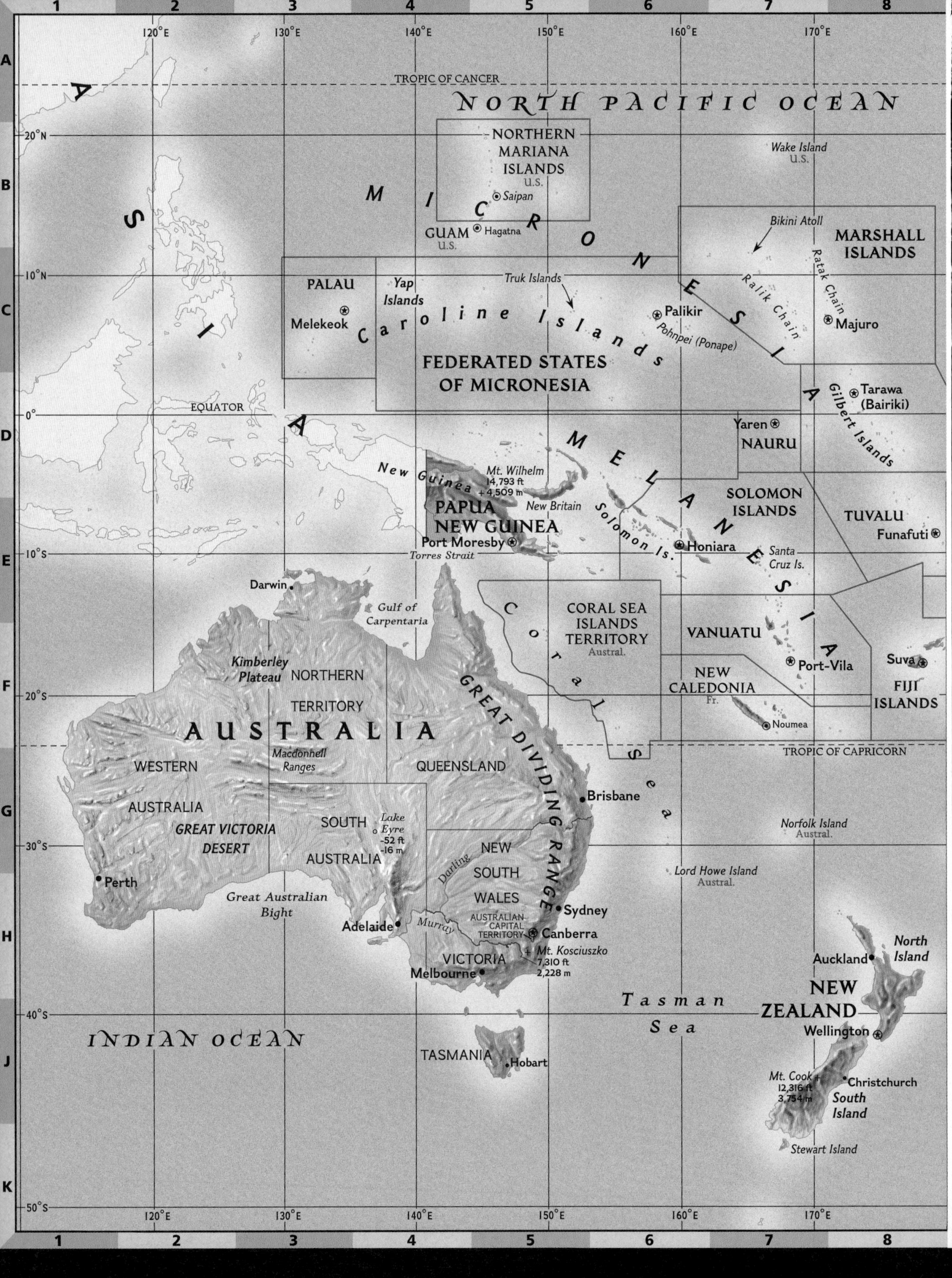

NORTH PACIFIC OCEAN
TROPIC OF CANCER
NORTHERN MARIANA ISLANDS
U.S.
Saipan
GUAM
U.S.
Hagatna
MICRONESIA
Wake Island
U.S.
Bikini Atoll
MARSHALL ISLANDS
Ralik Chain
Ratak Chain
Majuro
PALAU
Melekeok
Yap Islands
Truk Islands
Caroline Islands
Palikir
Pohnpei (Ponape)
FEDERATED STATES OF MICRONESIA
Tarawa (Bairiki)
Gilbert Islands
EQUATOR
ASIA
Yaren
NAURU
MELANESIA
New Guinea
Mt. Wilhelm
14,793 ft
4,509 m
PAPUA NEW GUINEA
New Britain
Solomon Is.
SOLOMON ISLANDS
Honiara
TUVALU
Funafuti
Port Moresby
Torres Strait
Santa Cruz Is.
Darwin
Gulf of Carpentaria
Coral Sea
CORAL SEA ISLANDS TERRITORY
Austral.
VANUATU
Port-Vila
Suva
FIJI ISLANDS
NEW CALEDONIA
Fr.
Noumea
Kimberley Plateau
NORTHERN TERRITORY
AUSTRALIA
GREAT DIVIDING RANGE
TROPIC OF CAPRICORN
Macdonnell Ranges
WESTERN AUSTRALIA
QUEENSLAND
Brisbane
GREAT VICTORIA DESERT
SOUTH AUSTRALIA
Lake Eyre
-52 ft
-16 m
Norfolk Island
Austral.
Perth
NEW SOUTH WALES
Darling
Lord Howe Island
Austral.
Great Australian Bight
Sydney
Adelaide
Murray
AUSTRALIAN CAPITAL TERRITORY
Canberra
Mt. Kosciuszko
7,310 ft
2,228 m
VICTORIA
Melbourne
Auckland
North Island
NEW ZEALAND
Tasman Sea
Wellington
INDIAN OCEAN
TASMANIA
Hobart
Mt. Cook
12,316 ft
3,754 m
Christchurch
South Island
Stewart Island
120°E
130°E
140°E
150°E
160°E
170°E
20°N
10°N
0°
10°S
20°S
30°S
40°S
50°S

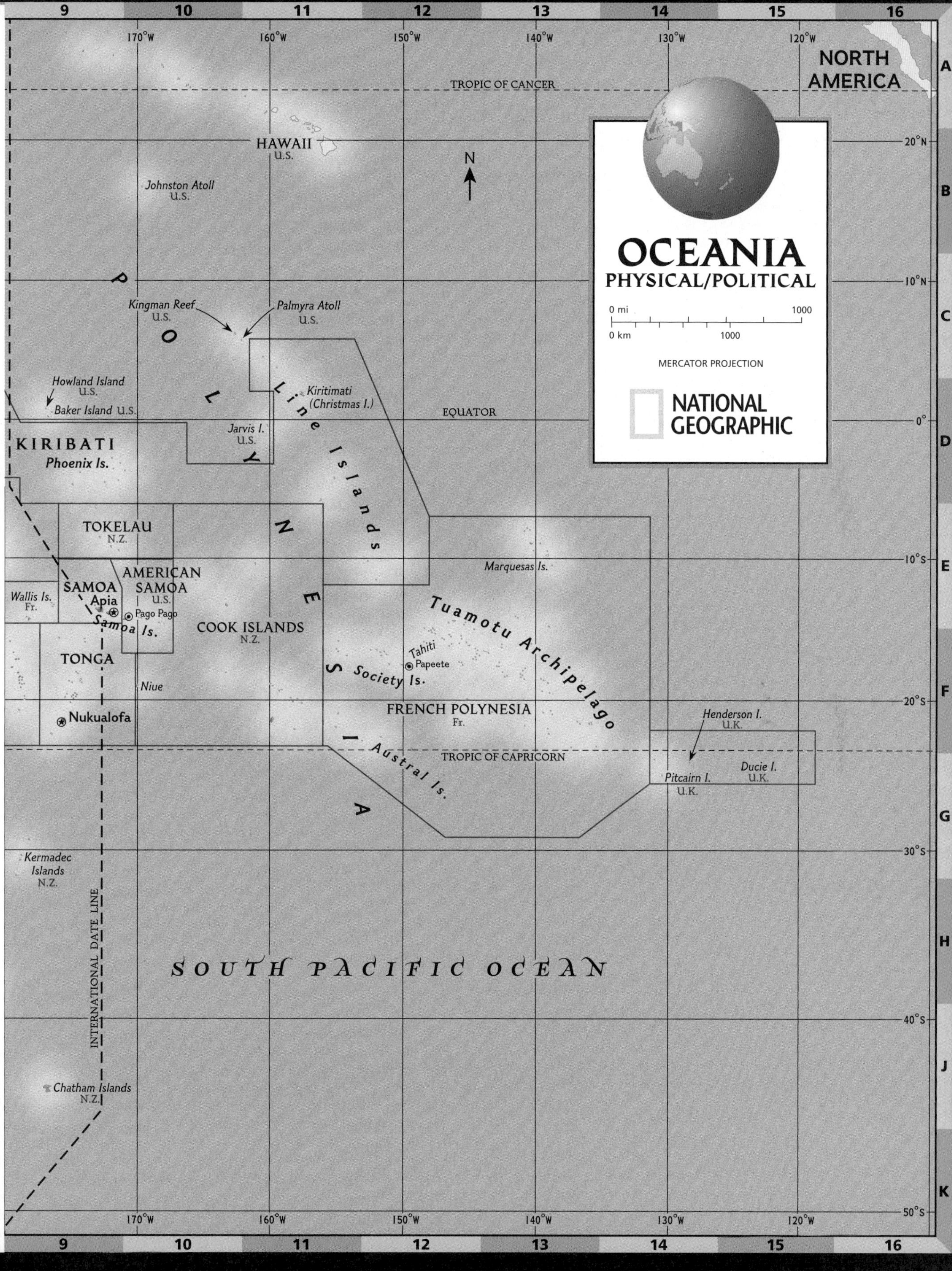
OCEANIA
PHYSICAL/POLITICAL
0 mi 1000
0 km 1000
MERCATOR PROJECTION
NATIONAL GEOGRAPHIC
NORTH AMERICA
TROPIC OF CANCER
HAWAII U.S.
Johnston Atoll U.S.
POLYNESIA
Kingman Reef U.S.
Palmyra Atoll U.S.
Howland Island U.S.
Baker Island U.S.
Kiritimati (Christmas I.)
Line Islands
Jarvis I. U.S.
EQUATOR
KIRIBATI
Phoenix Is.
TOKELAU N.Z.
AMERICAN SAMOA U.S.
SAMOA
Apia
Pago Pago
Samoa Is.
Wallis Is. Fr.
COOK ISLANDS N.Z.
TONGA
Niue
Nukualofa
Marquesas Is.
Tuamotu Archipelago
Tahiti
Papeete
Society Is.
FRENCH POLYNESIA Fr.
Austral Is.
TROPIC OF CAPRICORN
Henderson I. U.K.
Pitcairn I. U.K.
Ducie I. U.K.
Kermadec Islands N.Z.
INTERNATIONAL DATE LINE
SOUTH PACIFIC OCEAN
Chatham Islands N.Z.
170°W
160°W
150°W
140°W
130°W
120°W
20°N
10°N
0°
10°S
20°S
30°S
40°S
50°S
9 10 11 12 13 14 15 16
A B C D E F G H J K

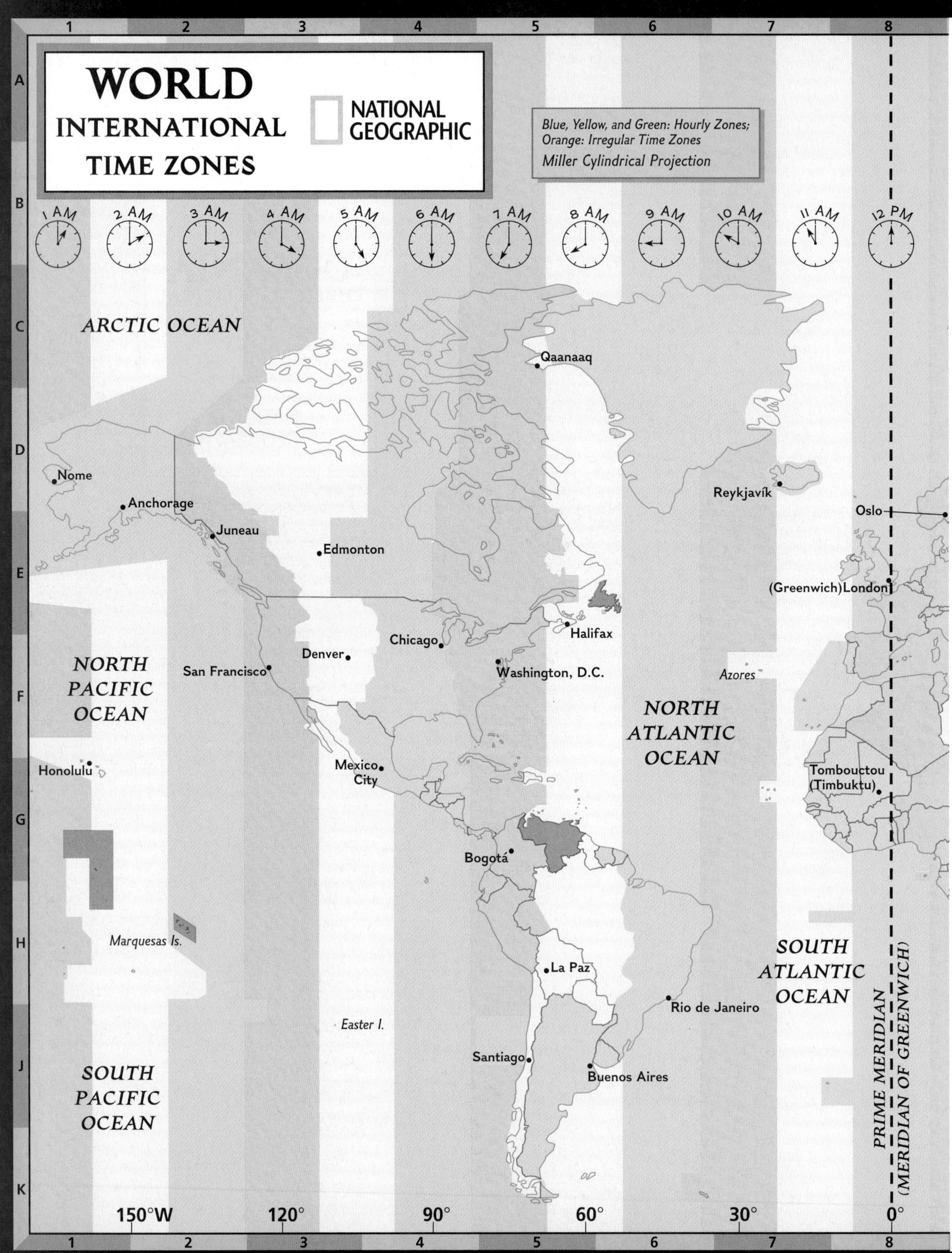
WORLD
INTERNATIONAL
TIME ZONES
NATIONAL GEOGRAPHIC
Blue, Yellow, and Green: Hourly Zones;
Orange: Irregular Time Zones
Miller Cylindrical Projection
1 AM
2 AM
3 AM
4 AM
5 AM
6 AM
7 AM
8 AM
9 AM
10 AM
11 AM
12 PM
ARCTIC OCEAN
Qaanaaq
Nome
Anchorage
Juneau
Edmonton
Reykjavík
Oslo
(Greenwich)London
Halifax
Chicago
Denver
San Francisco
Washington, D.C.
Azores
NORTH PACIFIC OCEAN
NORTH ATLANTIC OCEAN
Honolulu
Mexico City
Tombouctou (Timbuktu)
Bogotá
Marquesas Is.
La Paz
SOUTH ATLANTIC OCEAN
Rio de Janeiro
Easter I.
Santiago
Buenos Aires
SOUTH PACIFIC OCEAN
PRIME MERIDIAN
(MERIDIAN OF GREENWICH)
150°W
120°
90°
60°
30°
0°

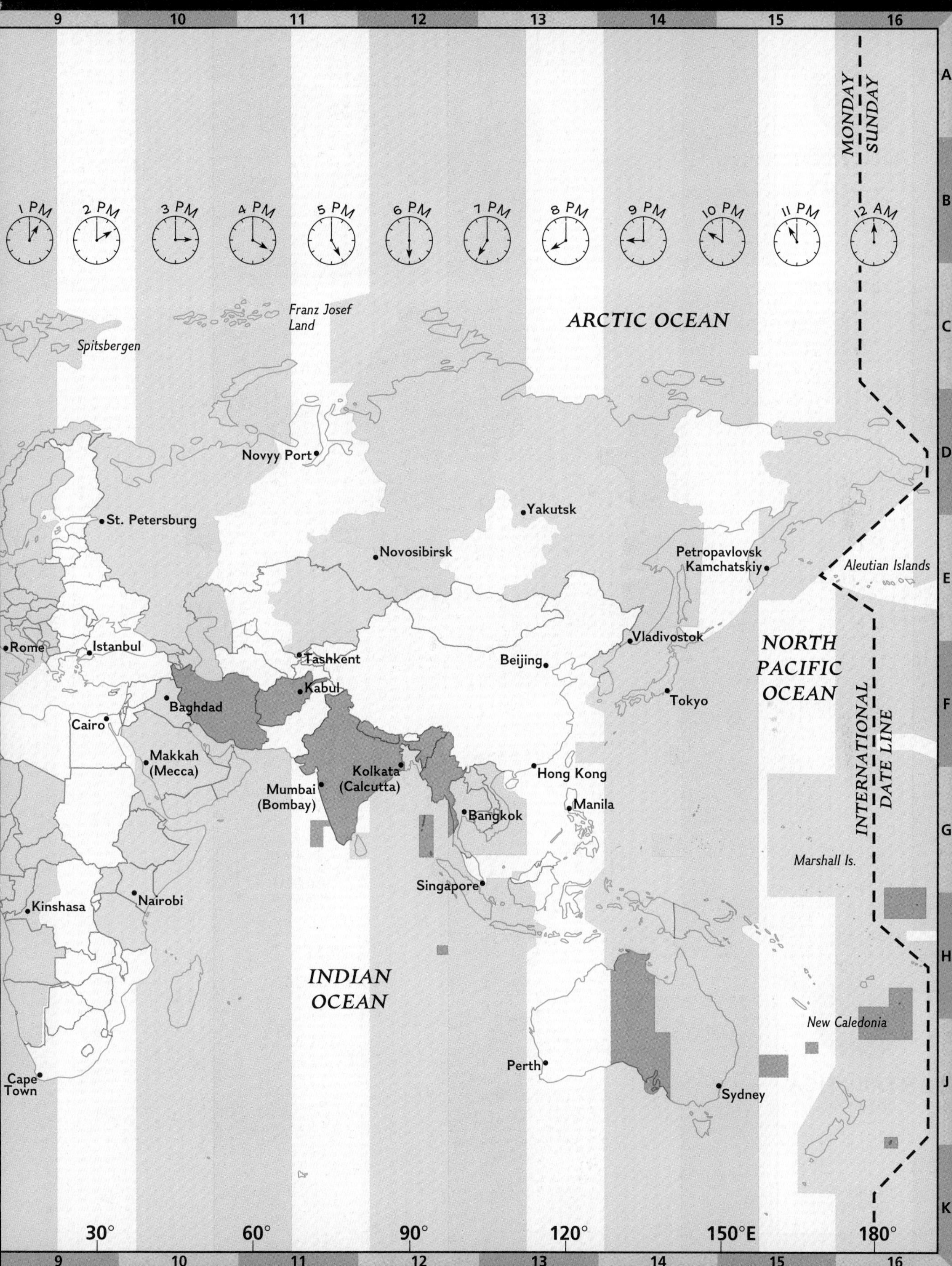

9
10
11
12
13
14
15
16
A
B
C
D
E
F
G
H
J
K
MONDAY
SUNDAY
1 PM
2 PM
3 PM
4 PM
5 PM
6 PM
7 PM
8 PM
9 PM
10 PM
11 PM
12 AM
Franz Josef Land
Spitsbergen
ARCTIC OCEAN
Novyy Port
St. Petersburg
Yakutsk
Novosibirsk
Petropavlovsk Kamchatskiy
Aleutian Islands
Rome
Istanbul
Tashkent
Vladivostok
NORTH PACIFIC OCEAN
Beijing
Kabul
Baghdad
Tokyo
Cairo
Makkah (Mecca)
Kolkata (Calcutta)
Hong Kong
Mumbai (Bombay)
Manila
Bangkok
INTERNATIONAL DATE LINE
Marshall Is.
Singapore
Kinshasa
Nairobi
INDIAN OCEAN
New Caledonia
Perth
Cape Town
Sydney
30°
60°
90°
120°
150°E
180°

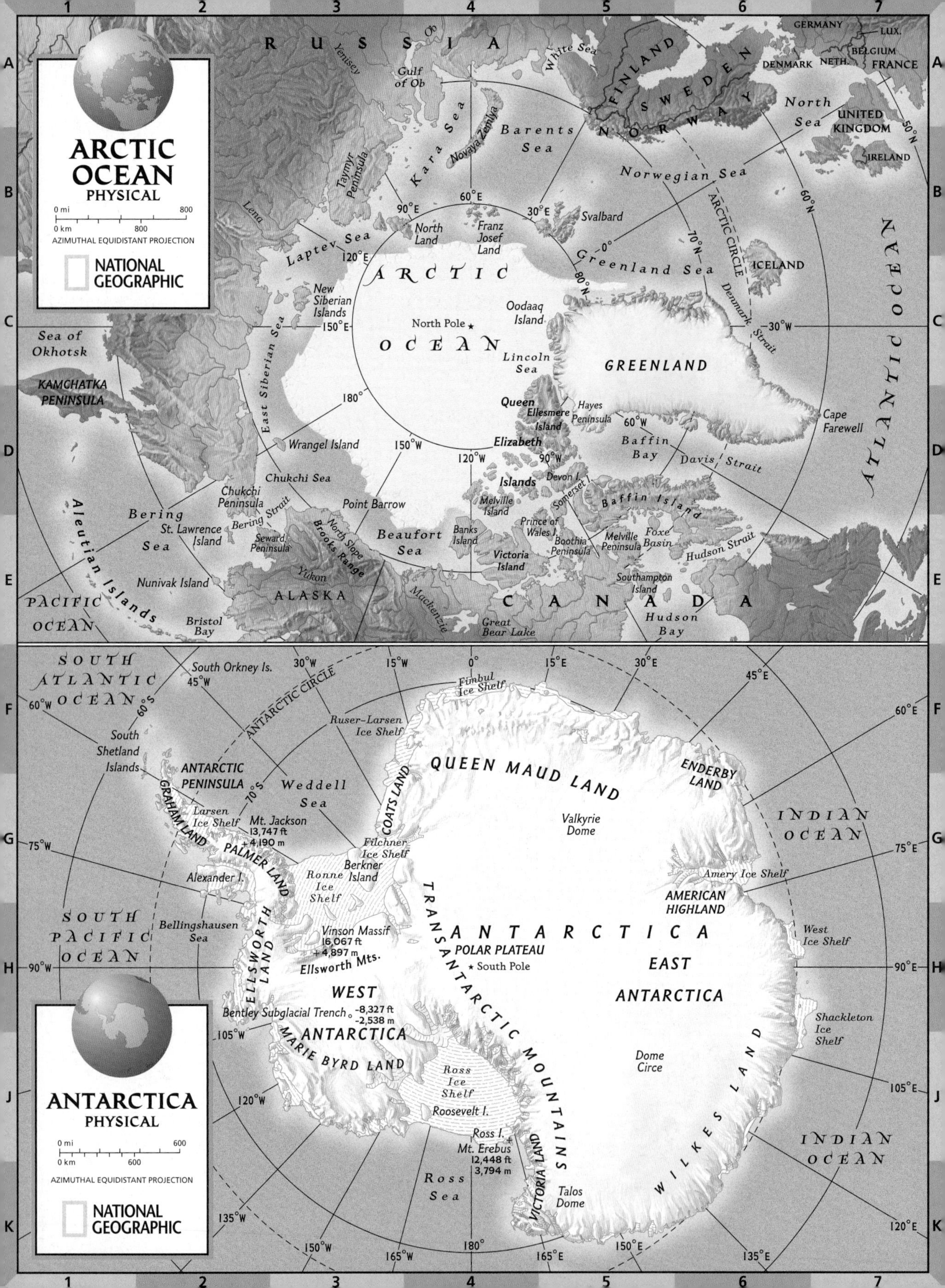

ARCTIC OCEAN
PHYSICAL
AZIMUTHAL EQUIDISTANT PROJECTION
NATIONAL GEOGRAPHIC
RUSSIA
GERMANY
LUX.
BELGIUM
FRANCE
DENMARK
NETH.
FINLAND
SWEDEN
NORWAY
UNITED KINGDOM
IRELAND
North Sea
White Sea
Gulf of Ob
Yenisey
Ob
Kara Sea
Novaya Zemlya
Barents Sea
Norwegian Sea
Taymyr Peninsula
Lena
Laptev Sea
North Land
Franz Josef Land
Svalbard
ARCTIC CIRCLE
ICELAND
Greenland Sea
Denmark Strait
New Siberian Islands
ARCTIC OCEAN
North Pole
Oodaaq Island
Lincoln Sea
GREENLAND
Sea of Okhotsk
KAMCHATKA PENINSULA
East Siberian Sea
Wrangel Island
Queen Elizabeth Islands
Ellesmere Island
Hayes Peninsula
Cape Farewell
Baffin Bay
Davis Strait
ATLANTIC OCEAN
Devon I.
Chukchi Sea
Chukchi Peninsula
Bering Strait
Point Barrow
Melville Island
Somerset I.
Baffin Island
Bering Sea
St. Lawrence Island
Seward Peninsula
North Slope
Brooks Range
Beaufort Sea
Banks Island
Prince of Wales I.
Boothia Peninsula
Melville Peninsula
Foxe Basin
Hudson Strait
Victoria Island
Aleutian Islands
Nunivak Island
Yukon
Southampton Island
PACIFIC OCEAN
ALASKA
Mackenzie
CANADA
Bristol Bay
Great Bear Lake
Hudson Bay
SOUTH ATLANTIC OCEAN
South Orkney Is.
ANTARCTIC CIRCLE
Fimbul Ice Shelf
Ruser-Larsen Ice Shelf
South Shetland Islands
ANTARCTIC PENINSULA
QUEEN MAUD LAND
ENDERBY LAND
Weddell Sea
COATS LAND
GRAHAM LAND
Larsen Ice Shelf
Mt. Jackson 13,747 ft 4,190 m
Valkyrie Dome
INDIAN OCEAN
Filchner Ice Shelf
PALMER LAND
Berkner Island
Amery Ice Shelf
Alexander I.
Ronne Ice Shelf
AMERICAN HIGHLAND
TRANSANTARCTIC MOUNTAINS
SOUTH PACIFIC OCEAN
Bellingshausen Sea
Vinson Massif 16,067 ft 4,897 m
ANTARCTICA
West Ice Shelf
ELLSWORTH LAND
POLAR PLATEAU
Ellsworth Mts.
South Pole
EAST ANTARCTICA
WEST ANTARCTICA
Bentley Subglacial Trench -8,327 ft -2,538 m
Shackleton Ice Shelf
MARIE BYRD LAND
Ross Ice Shelf
Dome Circe
ANTARCTICA
PHYSICAL
AZIMUTHAL EQUIDISTANT PROJECTION
Roosevelt I.
Ross I.
Mt. Erebus 12,448 ft 3,794 m
VICTORIA LAND
WILKES LAND
Ross Sea
Talos Dome
NATIONAL GEOGRAPHIC

A WORLD OF EXTREMES

1. **The largest continent** is Asia with an area of 12,262,691 sq. miles (31,758,898 sq. km).
2. **The smallest continent** is Australia with an area of 2,988,888 sq. miles (7,741,184 sq. km).
3. **The largest country** is Russia with an area of 6,592,819 sq. miles (17,075,322 sq. km).
4. **The smallest country** is Vatican City with an area of 1 sq. mile (2.6 sq. km).
5. **The longest river** is the Nile River with a length of 4,160 miles (6,695 km).
6. **The deepest freshwater lake** is Lake Baikal with a maximum depth of 5,715 feet (1,742 m).
7. **The highest waterfall** is Angel Falls with a height of 3,212 feet (979 m).
8. **The highest mountain** is Mount Everest with a height of 29,028 feet (8,848 m) above sea level.
9. **The largest desert** is the Sahara with an area of 3,500,000 sq. miles (9,065,000 sq. km).

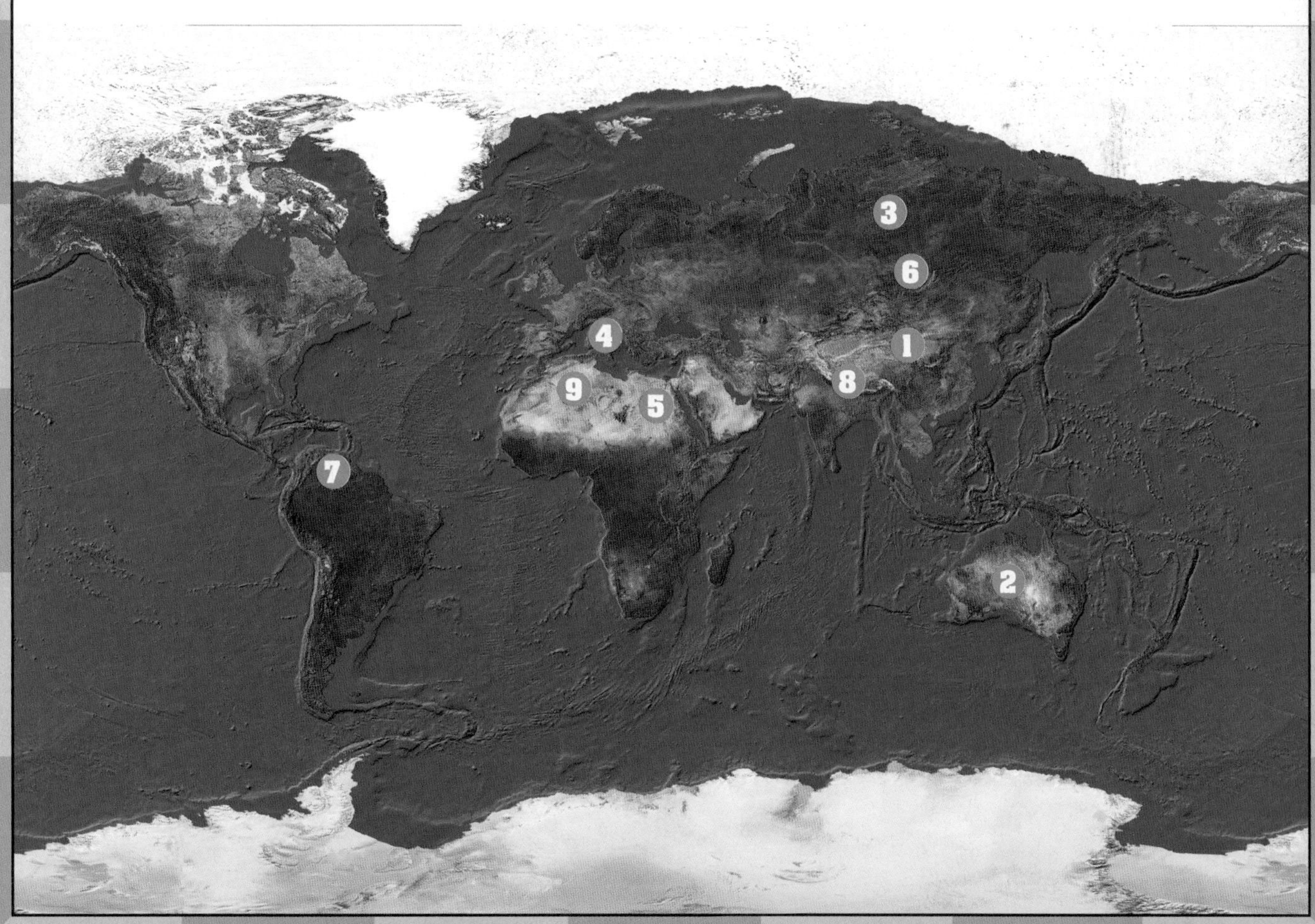

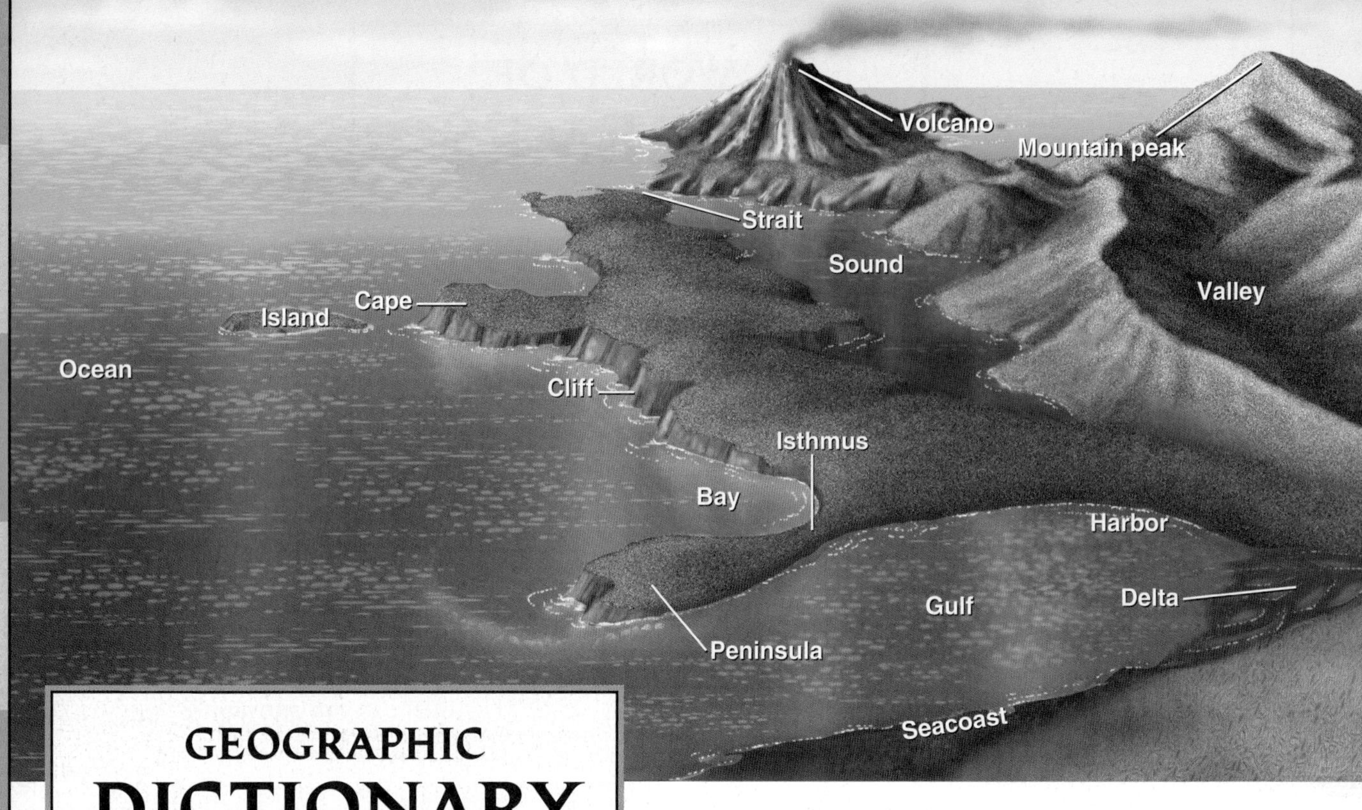

GEOGRAPHIC DICTIONARY

As you read about the world's geography, you will encounter the terms listed below. Many of the terms are pictured in the diagram.

absolute location exact location of a place on the Earth described by global coordinates

basin area of land drained by a given river and its branches; area of land surrounded by lands of higher elevations

bay part of a large body of water that extends into a shoreline, generally smaller than a gulf

canyon deep and narrow valley with steep walls

cape point of land that extends into a river, lake, or ocean

channel wide strait or waterway between two landmasses that lie close to each other; deep part of a river or other waterway

cliff steep, high wall of rock, Earth, or ice

continent one of the seven large landmasses on the Earth

delta flat, low-lying land built up from soil carried downstream by a river and deposited at its mouth

divide stretch of high land that separates river systems

downstream direction in which a river or stream flows from its source to its mouth

elevation height of land above sea level

Equator imaginary line that runs around the Earth halfway between the North and South Poles; used as the starting point to measure degrees of north and south latitude

glacier large, thick body of slowly moving ice

gulf part of a large body of water that extends into a shoreline, generally larger and more deeply indented than a bay

harbor a sheltered place along a shoreline where ships can anchor safely

highland elevated land area such as a hill, mountain, or plateau

hill elevated land with sloping sides and rounded summit; generally smaller than a mountain

island land area, smaller than a continent, completely surrounded by water

isthmus narrow stretch of land connecting two larger land areas

lake a sizable inland body of water

latitude distance north or south of the Equator, measured in degrees

longitude distance east or west of the Prime Meridian, measured in degrees

lowland land, usually level, at a low elevation

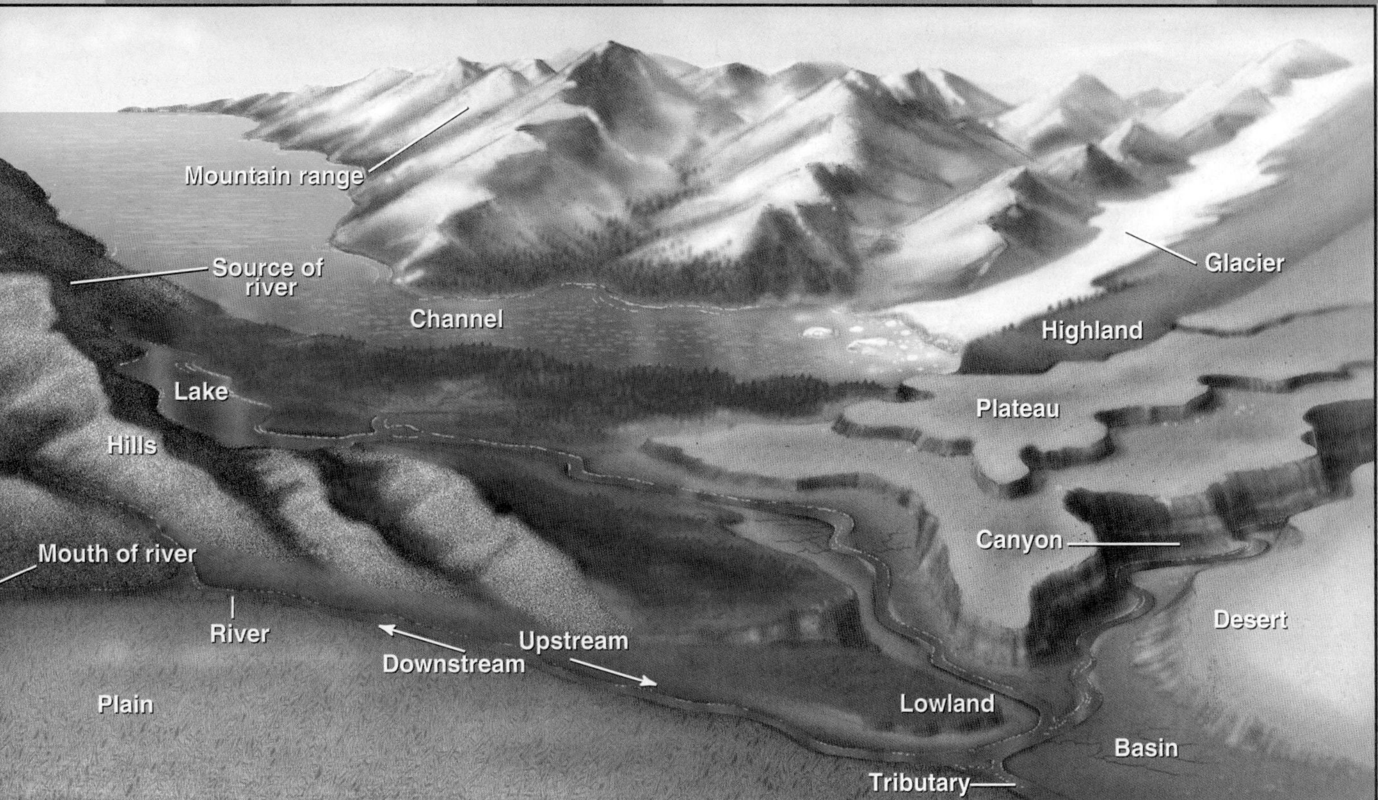

map drawing of the Earth shown on a flat surface

meridian one of many lines on the global grid running from the North Pole to the South Pole; used to measure degrees of longitude

mesa broad, flat-topped landform with steep sides; smaller than a plateau

mountain land with steep sides that rises sharply (1,000 feet or more) from surrounding land; generally larger and more rugged than a hill

mountain peak pointed top of a mountain

mountain range a series of connected mountains

mouth (of a river) place where a stream or river flows into a larger body of water

ocean one of the four major bodies of salt water that surround the continents

ocean current stream of either cold or warm water that moves in a definite direction through an ocean

parallel one of many lines on the global grid that circles the Earth north or south of the Equator; used to measure degrees of latitude

peninsula body of land jutting into a lake or ocean, surrounded on three sides by water

physical feature characteristic of a place occurring naturally, such as a landform, body of water, climate pattern, or resource

plain area of level land, usually at low elevation and often covered with grasses

plateau area of flat or rolling land at a high elevation, about 300 to 3,000 feet (90 to 900 m) high

Prime Meridian line of the global grid running from the North Pole to the South Pole at Greenwich, England; starting point for measuring degrees of east and west longitude

relief changes in elevation over a given area of land

river large natural stream of water that runs through the land

sea large body of water completely or partly surrounded by land

seacoast land lying next to a sea or an ocean

sound broad inland body of water, often between a coastline and one or more islands off the coast

source (of a river) place where a river or stream begins, often in highlands

strait narrow stretch of water joining two larger bodies of water

tributary small river or stream that flows into a large river or stream; a branch of the river

upstream direction opposite the flow of a river; toward the source of a river or stream

valley area of low land usually between hills or mountains

volcano mountain or hill created as liquid rock and ash erupt from inside the Earth

UNIT 1

The World

CHAPTER 1
Using Geography Skills

CHAPTER 2
Earth's Physical Geography

CHAPTER 3
Earth's Human and Cultural Geography

The world at night ▶

NGS ONLINE For more information about the region, see www.nationalgeographic.com/education.

World Atlas

The World
CLIMATE REGIONS

Map Skills

1. **Place** What two climate zones are found in Antarctica?
2. **Regions** In general terms, how would you describe the world climate zones along the Equator?

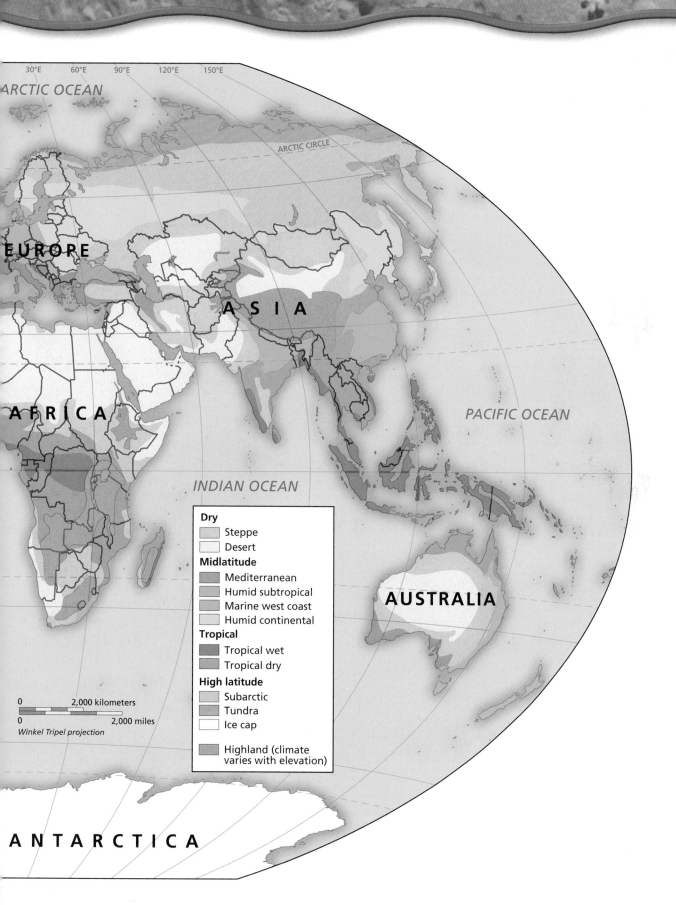

30°E 60°E 90°E 120°E 150°E

ARCTIC OCEAN

ARCTIC CIRCLE

EUROPE

ASIA

AFRICA

PACIFIC OCEAN

INDIAN OCEAN

AUSTRALIA

ANTARCTICA

Dry
- Steppe
- Desert

Midlatitude
- Mediterranean
- Humid subtropical
- Marine west coast
- Humid continental

Tropical
- Tropical wet
- Tropical dry

High latitude
- Subarctic
- Tundra
- Ice cap

- Highland (climate varies with elevation)

0 2,000 kilometers
0 2,000 miles
Winkel Tripel projection

World Atlas

The World
VEGETATION

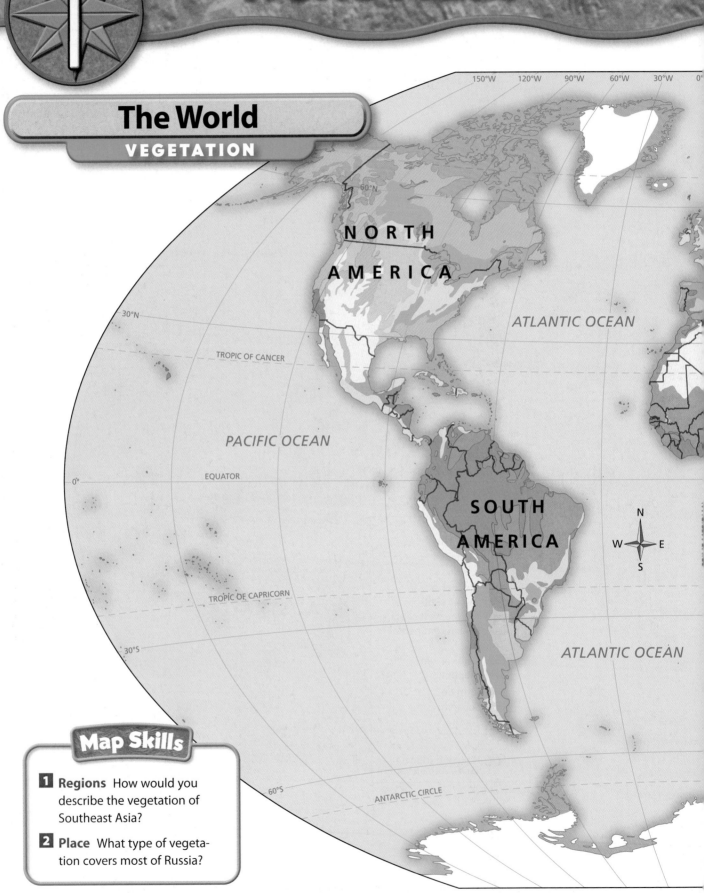

150°W 120°W 90°W 60°W 30°W 0°

60°N

NORTH AMERICA

30°N

ATLANTIC OCEAN

TROPIC OF CANCER

PACIFIC OCEAN

0°

EQUATOR

SOUTH AMERICA

N
W E
S

TROPIC OF CAPRICORN

30°S

ATLANTIC OCEAN

60°S

ANTARCTIC CIRCLE

Map Skills

1 Regions How would you describe the vegetation of Southeast Asia?

2 Place What type of vegetation covers most of Russia?

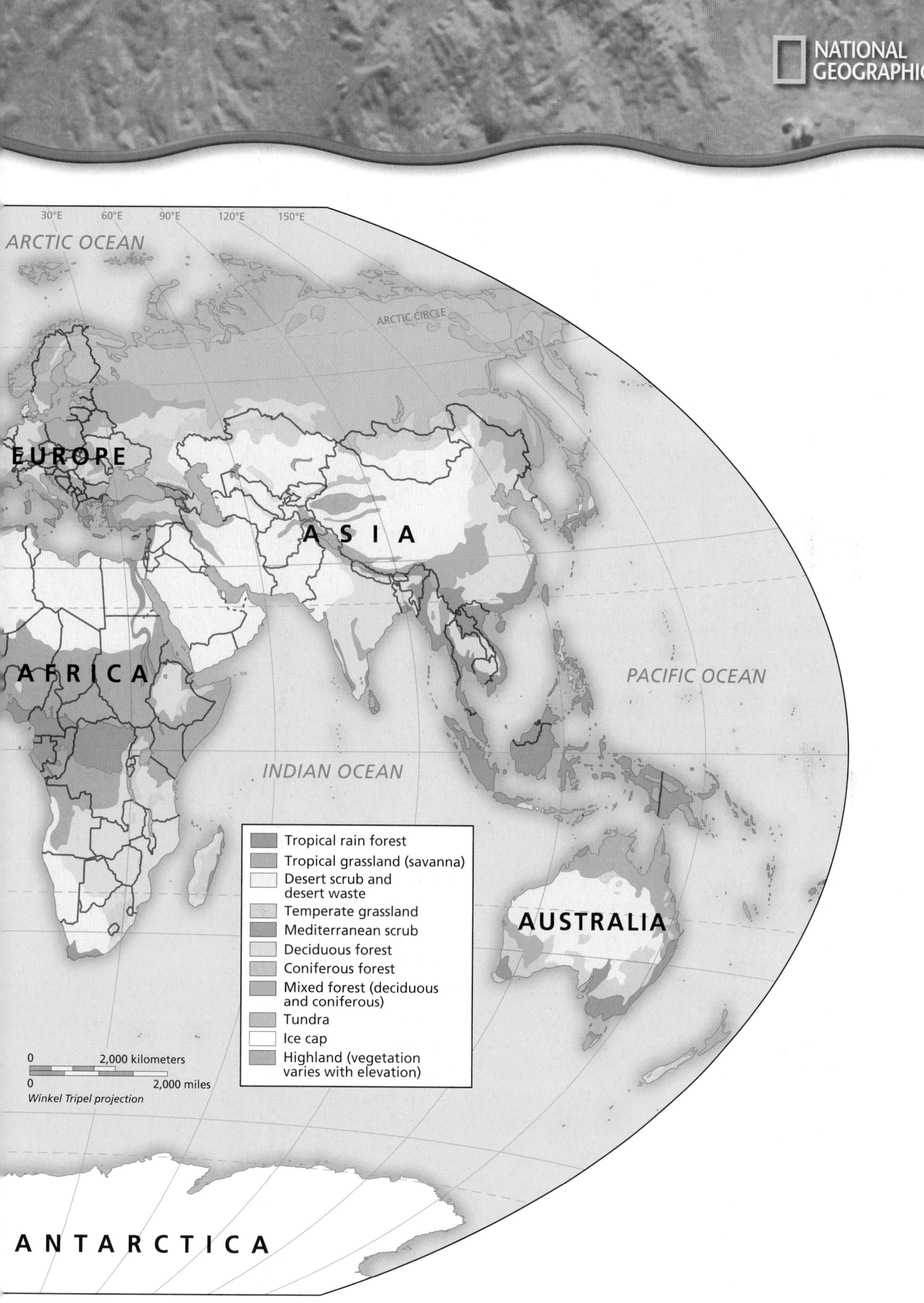
30°E
60°E
90°E
120°E
150°E
ARCTIC OCEAN
ARCTIC CIRCLE
EUROPE
ASIA
AFRICA
PACIFIC OCEAN
INDIAN OCEAN
AUSTRALIA
ANTARCTICA
Tropical rain forest
Tropical grassland (savanna)
Desert scrub and desert waste
Temperate grassland
Mediterranean scrub
Deciduous forest
Coniferous forest
Mixed forest (deciduous and coniferous)
Tundra
Ice cap
Highland (vegetation varies with elevation)
0
2,000 kilometers
0
2,000 miles
Winkel Tripel projection

World Atlas

The World

POPULATION DENSITY

Map Skills

1 **Regions** What areas of the world are the most densely populated?

2 **Place** How are the population patterns of Australia and South America similar? How are they different?

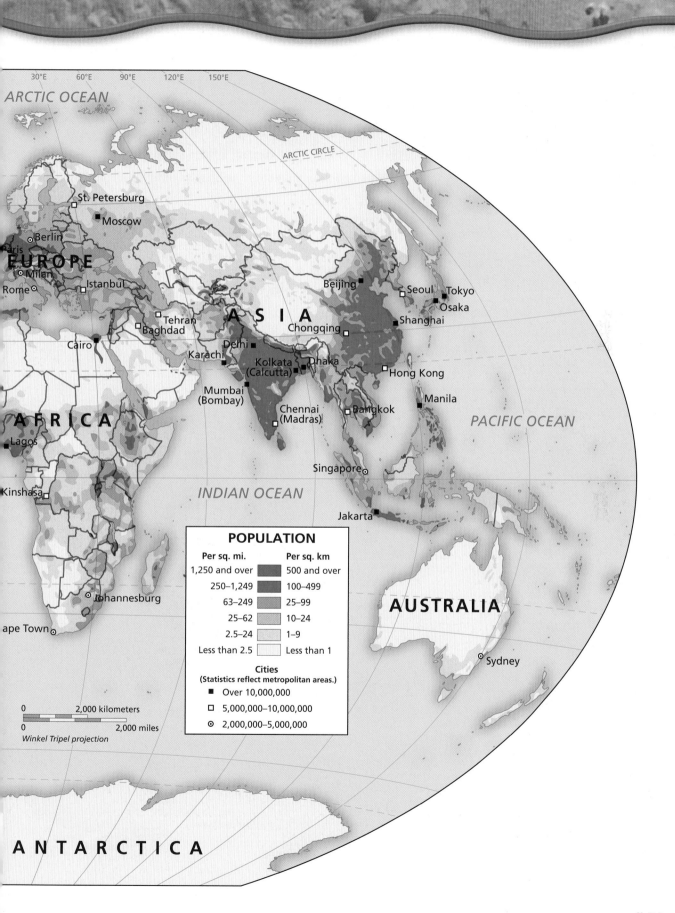

30°E 60°E 90°E 120°E 150°E

ARCTIC OCEAN

ARCTIC CIRCLE

EUROPE

ASIA

AFRICA

AUSTRALIA

ANTARCTICA

PACIFIC OCEAN

INDIAN OCEAN

St. Petersburg
Moscow
Berlin
Paris
Milan
Rome
Istanbul
Tehran
Baghdad
Cairo
Karachi
Delhi
Kolkata (Calcutta)
Dhaka
Mumbai (Bombay)
Chennai (Madras)
Beijing
Seoul
Tokyo
Osaka
Shanghai
Chongqing
Hong Kong
Manila
Bangkok
Singapore
Jakarta
Lagos
Kinshasa
Johannesburg
ape Town
Sydney

POPULATION

Per sq. mi.		Per sq. km
1,250 and over		500 and over
250–1,249		100–499
63–249		25–99
25–62		10–24
2.5–24		1–9
Less than 2.5		Less than 1

Cities
(Statistics reflect metropolitan areas.)
- ■ Over 10,000,000
- □ 5,000,000–10,000,000
- ⊙ 2,000,000–5,000,000

0 2,000 kilometers
0 2,000 miles
Winkel Tripel projection

UNIT 1

World Atlas

The World

RELIGIONS

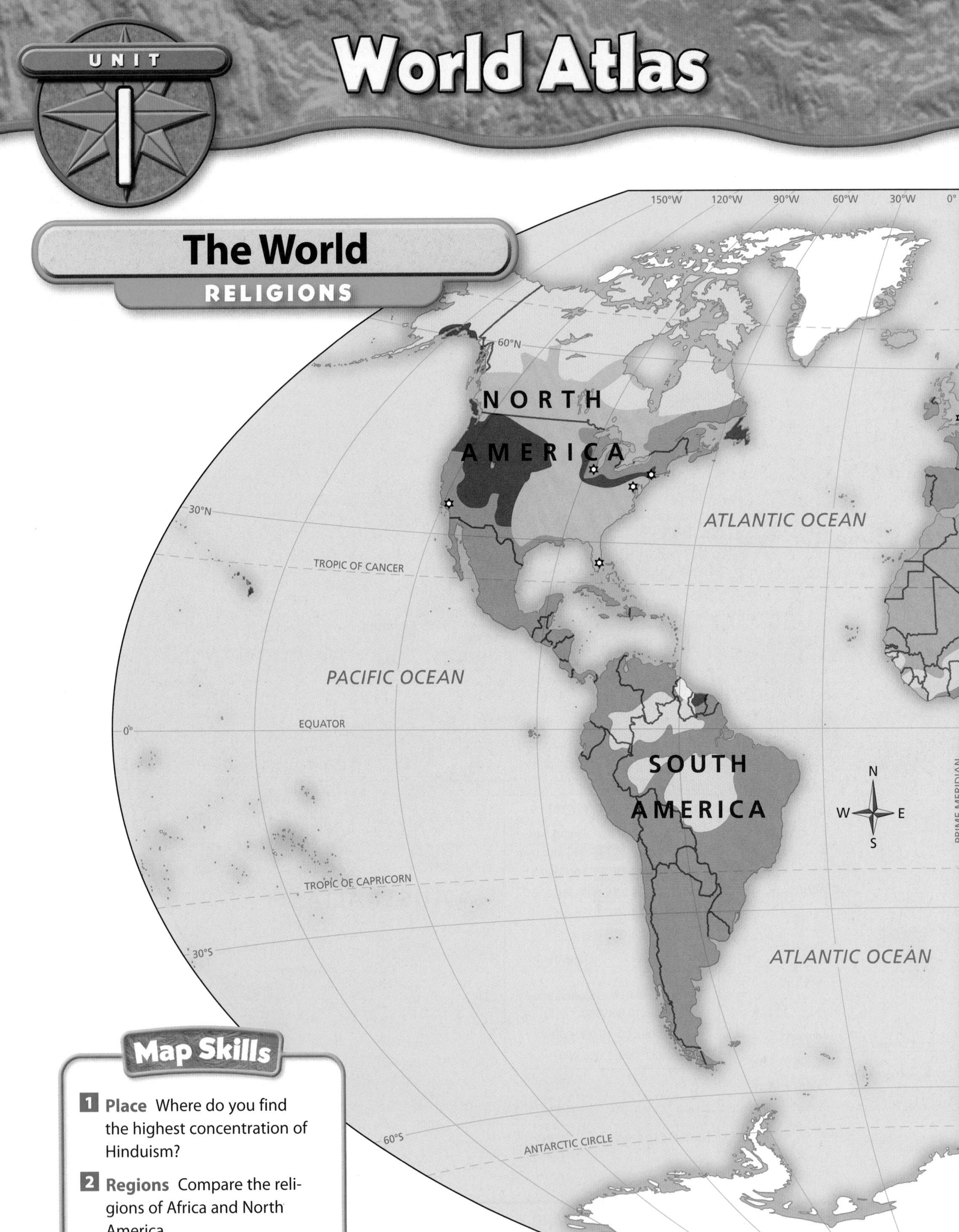

Map Skills

1. **Place** Where do you find the highest concentration of Hinduism?
2. **Regions** Compare the religions of Africa and North America.

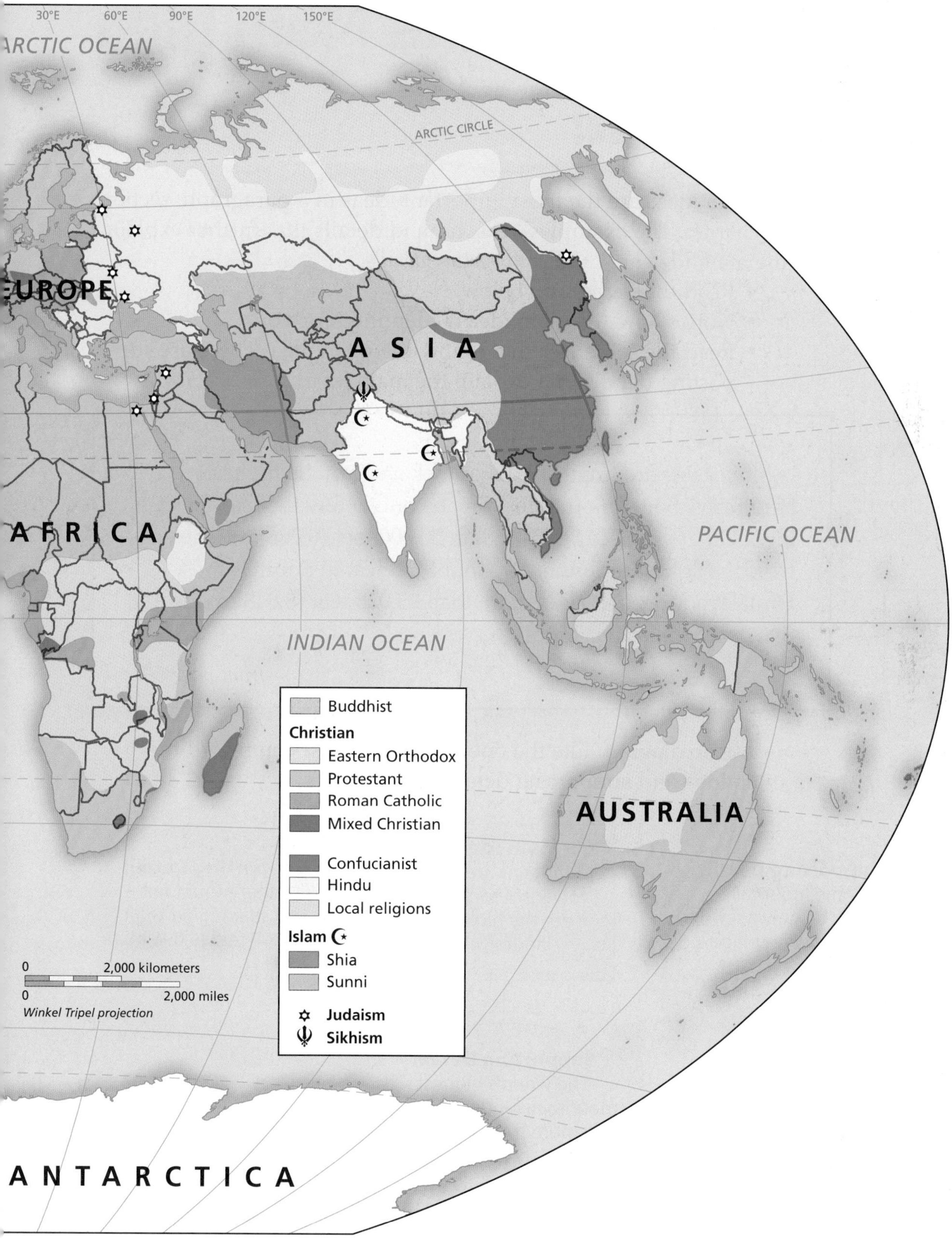

30°E 60°E 90°E 120°E 150°E

ARCTIC OCEAN

ARCTIC CIRCLE

EUROPE

ASIA

AFRICA

PACIFIC OCEAN

INDIAN OCEAN

AUSTRALIA

ANTARCTICA

Buddhist

Christian

Eastern Orthodox

Protestant

Roman Catholic

Mixed Christian

Confucianist

Hindu

Local religions

Islam ☪

Shia

Sunni

✡ **Judaism**

☬ **Sikhism**

0 2,000 kilometers

0 2,000 miles

Winkel Tripel projection

Reading Social Studies

Identifying the Main Idea

Learn It!

Main ideas are the most important ideas in a paragraph, section, or chapter. The examples, reasons, and details that further explain the main idea are called *supporting details.*

- Read the paragraph below.
- Notice how the main idea is identified for you.
- Read the sentences that follow the main idea. These are supporting details that explain the main idea.

Main Idea

Supporting Details

> Mountains are huge towers of rock and are the highest landforms. Some mountains may be only a few thousand feet high. Others can soar higher than 20,000 feet (6,096 m). The world's highest mountain is Mount Everest in South Asia's Himalaya ranges. It rises more than 29,028 feet (8,848 m), nearly five and a half miles high!
>
> *—from page 50*

A web diagram like the one below can help you record the main idea and supporting details.

Main Idea: Mountains are huge towers of rock and are the highest landforms.

Supporting Detail: Some mountains may be only a few thousand feet high.

Supporting Detail: Other mountains can soar higher than 20,000 feet (6,096 m).

Supporting Detail: The world's highest mountain, Mount Everest, rises more than 29,028 feet (8,848 m), nearly five and a half miles high.

Reading Tip

Main ideas often appear in the first sentence, but they can also be found in the middle or at the end of the paragraph.

2 Practice It!

Read the following paragraph from this unit.

- Draw a graphic organizer like the one shown below.
- Write the main idea for the paragraph in the center box.
- Write the supporting details in the ovals surrounding the box.

Remember that you do not need to include every word in the sentence when restating the main idea or supporting details.

Read to Write Activity

Read the main idea for Chapter 3, Section 2, and the paragraphs that follow. Using the main idea as a topic sentence, write a paragraph with supporting details. The supporting details should describe the elements that make up a culture.

> People live on a surprisingly small part of the Earth. Land covers only about 30 percent of the Earth's surface, and only half of this land is usable by humans. Deserts, high mountains, and ice-covered lands cannot support large numbers of people.
>
> —*from page 74*

Sparsely settled Mongolian plain

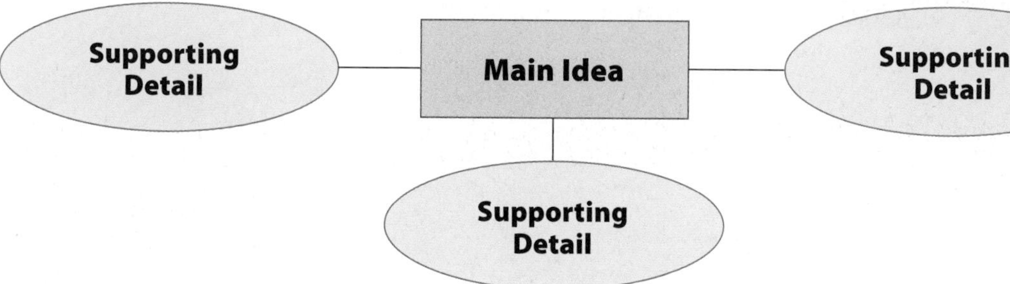

3 Apply It!

Create several web diagrams like the ones found on these pages. As you read Chapters 1, 2, and 3, write the main idea for each section in the center box of the diagram. Write supporting details in ovals surrounding the center box. Use your diagrams to help you study for the chapter assessments.

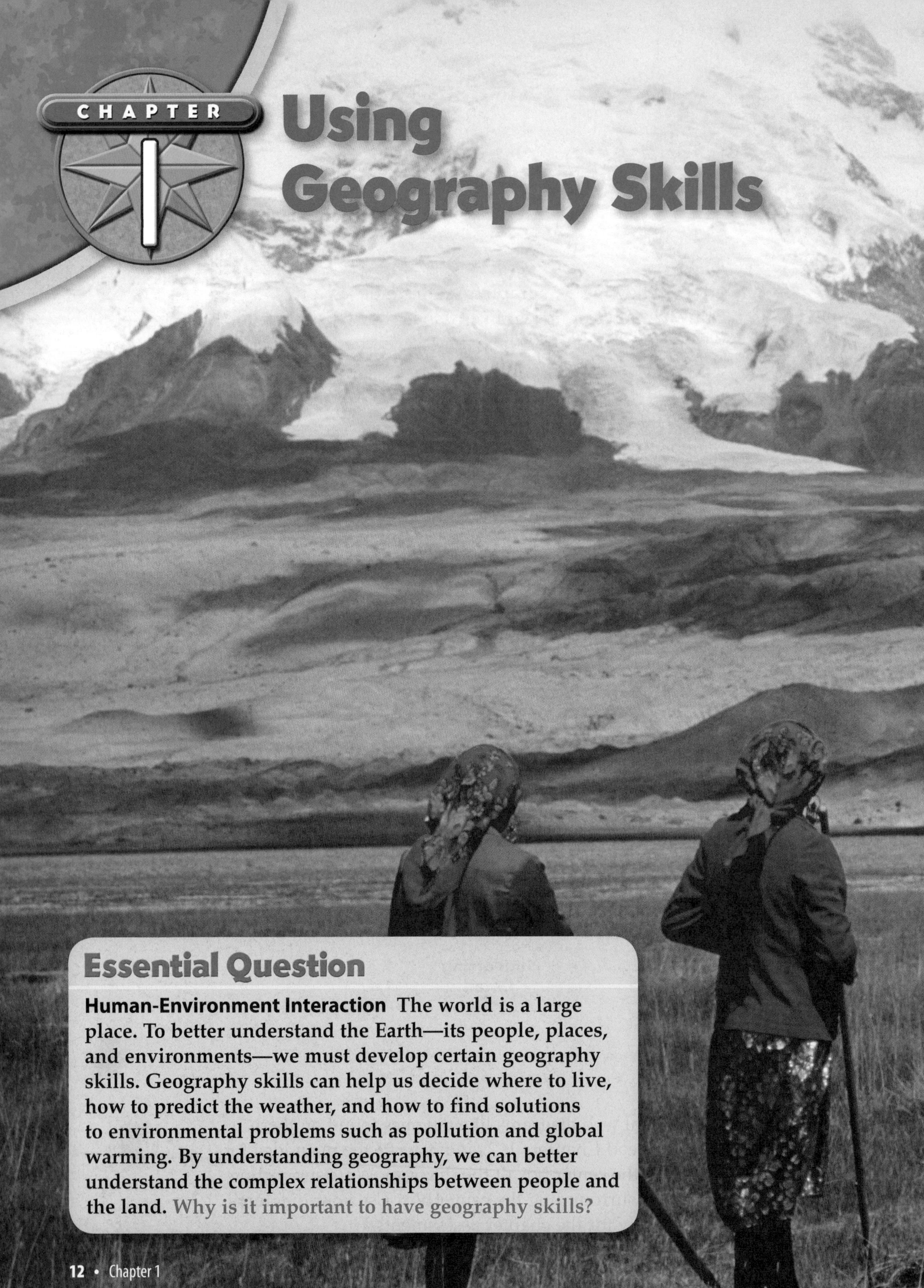

CHAPTER 1 Using Geography Skills

Essential Question

Human-Environment Interaction The world is a large place. To better understand the Earth—its people, places, and environments—we must develop certain geography skills. Geography skills can help us decide where to live, how to predict the weather, and how to find solutions to environmental problems such as pollution and global warming. By understanding geography, we can better understand the complex relationships between people and the land. Why is it important to have geography skills?

Pamir Plateau, China

Section 1: Thinking Like a Geographer

BIG IDEA Geography is used to interpret the past, understand the present, and plan for the future. Geography is the study of the Earth. It is used to analyze the Earth's physical and human features. People can use geographic information to plan, make decisions, and manage resources.

Section 2: The Earth in Space

BIG IDEA Physical processes shape Earth's surface. Earth has different seasons because of the way it tilts and the way it rotates around the sun. The warmth of the sun's rays makes life on Earth possible.

Organizing Information Make this Foldable to help you organize information about the uses of geography and the Earth in space.

Step 1 Fold the sides of an 11x17 sheet of paper to meet in the middle, creating a shutter fold.

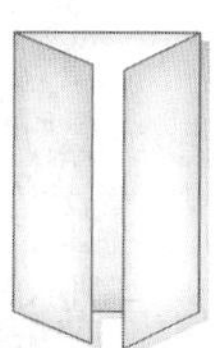

Step 2 Label your Foldable as shown.

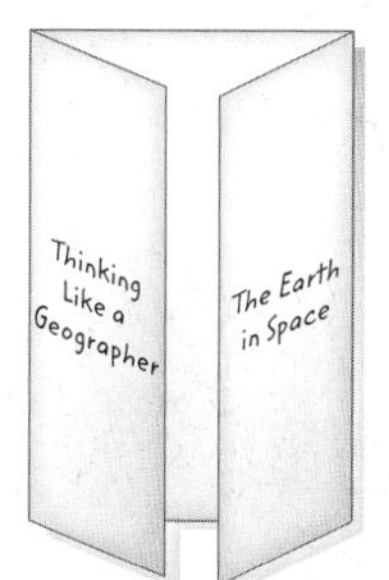

Reading and Writing As you read the chapter, take notes under the appropriate flap of your Foldable. After you have completed your Foldable, use your notes to write a letter encouraging the study of geography by all students.

Visit glencoe.com and enter ***QuickPass***™ code EOW3109c1 for Chapter 1 resources.

Guide to Reading

BIG Idea

Geography is used to interpret the past, understand the present, and plan for the future.

Content Vocabulary

- geography *(p. 15)*
- absolute location *(p. 15)*
- relative location *(p. 15)*
- environment *(p. 15)*
- decade *(p. 16)*
- century *(p. 16)*
- millennium *(p. 16)*
- Global Positioning System (GPS) *(p. 17)*
- Geographic Information Systems (GIS) *(p. 17)*

Academic Vocabulary

- theme *(p. 15)*
- physical *(p. 15)*

Reading Strategy

Identifying Use a chart like the one below to identify two examples for each topic.

Themes of Geography
1.
2.
Types of Geography
1.
2.
Geographer's Tools
1.
2.

Thinking Like a Geographer

Picture This The Italian Research Center, also known as the Pyramid, allows researchers from around the world to study everything from the effects of altitude on humans to the impact of global warming on the Earth. The Pyramid is located at the base of Mount Everest in Nepal. It is completely self-contained and can house up to 20 people. To get to the Pyramid, scientists have to trek through a national park and allow their bodies time to adjust to the extremely high altitude. To learn more about how geographers use information about the world to plan for the future, read Section 1.

▼ Research on Mount Everest

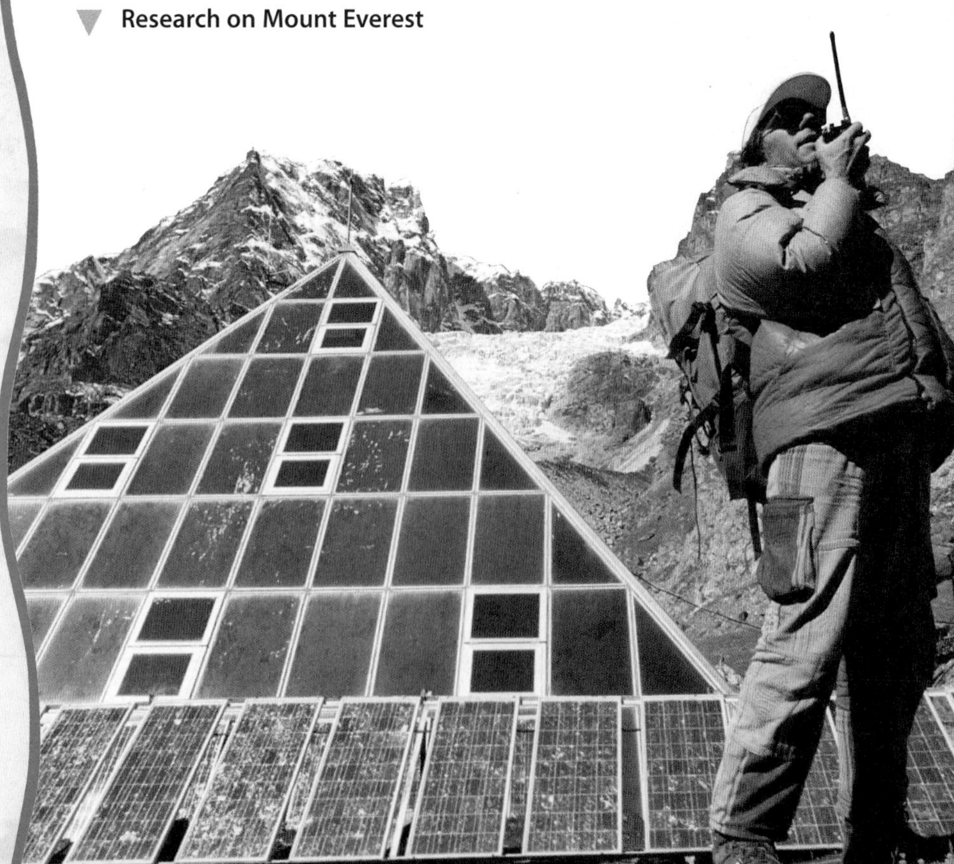

The Five Themes of Geography

Main Idea **Geographers use the Five Themes of Geography to help them study the Earth.**

Geography and You Suppose a teacher tells you to pick a topic for a research paper. How do you organize your ideas? Read to discover how geographers use themes to help them organize ideas about geography.

Geography is the study of the Earth and its people. People who study geography are geographers. Geographers use five **themes,** or topics, to describe places and people. These themes are location, place, human-environment interaction, movement, and regions.

Location

Location is the position of a place on the Earth's surface. Geographers describe location in two ways. **Absolute location** is the exact spot on Earth where a geographic feature, such as a city or mountain, is found. **Relative location** describes where that feature is in relation to the features around it.

Place

Place describes the characteristics of a location that make it unique, or different. A place can be defined by **physical** features, such as landforms, plants, animals, and weather patterns. Other characteristics of a place describe the people who live there—such as what languages they speak.

Human-Environment Interaction

Human-environment interaction describes how people affect their **environment,** or natural surroundings, and how their environment affects them. People affect the environment by using or changing it to meet their needs. Environmental factors that people cannot control, such as temperature and natural disasters, influence how people live.

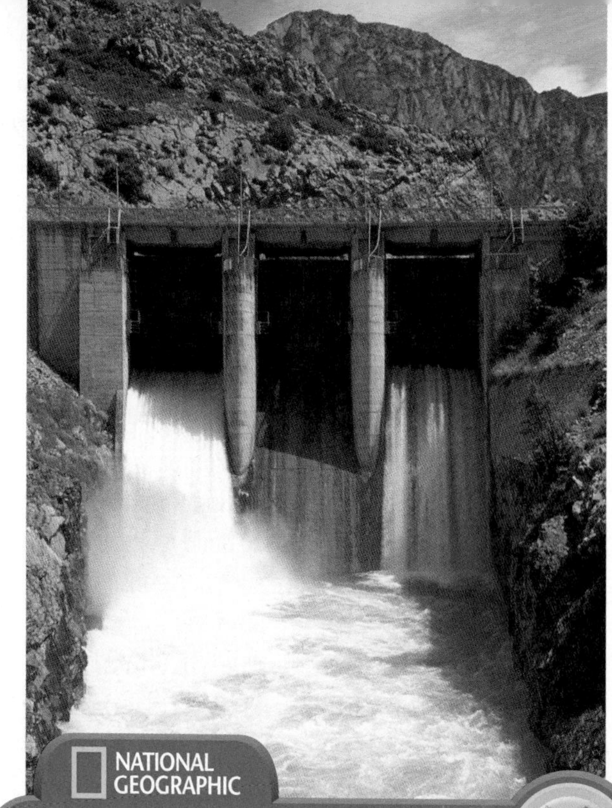

NATIONAL GEOGRAPHIC

Changing the Environment

Dams, like this one in Spain, can be built to control flooding, manage water flow, and supply electricity. ***Human-Environment Interaction*** **How and why do people affect the environment?**

Movement

Movement explains how and why people, ideas, and goods move from place to place. For example, people might leave a country that is involved in a war. Such movements can lead to great cultural change.

Regions

Regions refers to areas of the Earth's surface that have several common characteristics, such as land, natural resources, or population. For example, the Rocky Mountain region is a large area in the United States that is known for ranching and mining.

Reading Check **Explaining** Explain the difference between *place* and *location.*

A Geographer's Tools

Main Idea **Geographers use many different tools to help them study and analyze Earth's people and places.**

Geography and You Suppose a company wanted to build a new shopping center in your community. How would its managers know where to build it? Read to find out how geographers help make such decisions.

Geographers study the physical and human features of Earth. They rely on various tools to study people and places.

Types of Geography

When geographers study physical geography, they examine Earth's land areas, bodies of water, plant life, and other physical features. Physical geographers also study natural resources that are available in an area and the ways people use those resources. They help people make decisions about managing different types of resources such as water, forests, land, and even the wind.

Other geographers study human geography, focusing on people and their activities. Human geographers look at people's religions, languages, and ways of life. They may examine a specific location, or they may study entire countries or continents. They also compare different places to see how they are similar and different. Human geographers help plan cities and aid in international business.

Places in Time

Geographers use knowledge from other subject areas. History, for example, helps them understand how places appeared in the past. Geographers learn about places by studying the changes that have occurred over time.

Using Geography

Surveyors, like this woman in Canada, use specialized equipment to measure land areas.
Human-Environment Interaction **What tasks might geographers hired by the government carry out?**

History is divided into blocks of time known as periods. For example, a period of 10 years is called a **decade.** A period of 100 years is known as a **century.** A period of 1,000 years is a **millennium.**

In Western societies, it is common to group history into four long periods. The first of these periods is called Prehistory. Prehistory refers to the time before people developed writing, about 5,500 years ago. This time is followed by the period known as Ancient History, which lasted until about 1,500 years ago. The next thousand years is called the Middle Ages, or the medieval period. About 500 years ago, Modern History began and continues to the present.

Map Systems

Maps can provide geographers with different types of information about a place. Information for a map can be collected by using modern technology, or tools and methods that help people perform tasks. Satellites circling the Earth provide detailed digital images and photographs to create maps. Satellites can also measure changing temperatures and the amount of pollution in the air or land. This information can then be added to maps.

Another group of satellites makes up the **Global Positioning System (GPS).** This system uses radio signals to determine the exact location of places on Earth. Many hikers and truckers carry GPS equipment to avoid getting lost.

Geographic Information Systems (GIS) are computer hardware and software that gather, store, and analyze geographic information and then display it on a screen. It can display maps, but it also can show information that does not usually appear on maps, such as types of vegetation, types of soil, and even water quality.

Careers in Geography

Governments at all levels hire geographers for many kinds of tasks. Geographers help decide how land and resources might be used. For example, they analyze population trends, including why people live in certain areas and not in others.

In the business world, geographers often work as researchers and analysts. They can help companies decide where to locate new buildings. They also provide information about places and cultures where companies do business. Many geographers teach in high schools, colleges, and universities. As more schools recognize the importance of geography education, the demand for geography teachers is expected to grow.

✓ Reading Check **Explaining** How does modern technology make maps more precise?

Section 1 Review

Social Studies ONLINE
Study Central™ To review this section, go to glencoe.com.

Vocabulary

1. **Explain** the significance of:
   - **a.** geography
   - **b.** absolute location
   - **c.** relative location
   - **d.** environment
   - **e.** decade
   - **f.** century
   - **g.** millennium
   - **h.** Global Positioning System (GPS)
   - **i.** Geographic Information Systems (GIS)

Main Ideas

2. **Explaining** Use a web diagram like the one below to summarize information about the Five Themes of Geography.

3. **Contrasting** How is physical geography different from human geography?

Critical Thinking

4. **Drawing Conclusions** Describe how helpful you think GIS would be in deciding where to build a gas station.
5. **BIG Idea** What factors might influence where a city would develop?
6. **Challenge** Give three examples of how someone might use geography to plan for the future.

Writing About Geography

7. **Using Your FOLDABLES** Use your Foldable to write a paragraph that describes the uses of geography.

NATIONAL GEOGRAPHIC Geography Skills Handbook

How Do I Study Geography?

Geographers have created these broad categories and standards as tools to help you understand the relationships among people, places, and environments.

- **5 Themes of Geography**
- **6 Essential Elements**
- **18 Geography Standards**

5 Themes of Geography

1 Location
Location describes where something is. Absolute location describes a place's exact position on the Earth's surface. Relative location expresses where a place is in relation to another place.

2 Place
Place describes the physical and human characteristics that make a location unique.

3 Regions
Regions are areas that share common characteristics.

4 Movement
Movement explains how and why people and things move and are connected.

5 Human-Environment Interaction
Human-Environment Interaction describes the relationship between people and their environment.

6 Essential Elements

18 Geography Standards

I. The World in Spatial Terms

Geographers look to see where a place is located. Location acts as a starting point to answer "Where Is It?" The location of a place helps you orient yourself as to where you are.

1 How to use maps and other tools
2 How to use mental maps to organize information
3 How to analyze the spatial organization of people, places, and environments

II. Places and Regions

Place describes physical characteristics such as landforms, climate, and plant or animal life. It might also describe human characteristics, including language and way of life. Places can also be organized into regions. **Regions** are places united by one or more characteristics.

4 The physical and human characteristics of places
5 How people create regions to interpret Earth's complexity
6 How culture and experience influence people's perceptions of places and regions

III. Physical Systems

Geographers study how physical systems, such as hurricanes, volcanoes, and glaciers, shape the surface of the Earth. They also look at how plants and animals depend upon one another and their surroundings for their survival.

7 The physical processes that shape Earth's surface
8 The distribution of ecosystems on Earth's surface

IV. Human Systems

People shape the world in which they live. They settle in certain places but not in others. An ongoing theme in geography is the movement of people, ideas, and goods.

9 The characteristics, distribution, and migration of human populations
10 The complexity of Earth's cultural mosaics
11 The patterns and networks of economic interdependence
12 The patterns of human settlement
13 The forces of cooperation and conflict

V. Environment and Society

How does the relationship between people and their natural surroundings influence the way people live? Geographers study how people use the environment and how their actions affect the environment.

14 How human actions modify the physical environment
15 How physical systems affect human systems
16 The meaning, use, and distribution of resources

VI. The Uses of Geography

Knowledge of geography helps us understand the relationships among people, places, and environments over time. Applying geographic skills helps you understand the past and prepare for the future.

17 How to apply geography to interpret the past
18 How to apply geography to interpret the present and plan for the future

Understanding the BIG Ideas of Geography

The 15 Big Ideas will help you understand the information in *Exploring Our World: People, Places, and Cultures.* The Big Ideas are based on the Essential Elements and the Geography Standards. They help you organize important ideas, and they make it easier to understand patterns and relationships.

The World in Spatial Terms

- Geographers study how people and physical features are distributed on Earth's surface.

Places and Regions

- Places reflect the relationship between humans and the physical environment.
- Geographers organize the Earth into regions that share common characteristics.
- Culture influences people's perceptions about places and regions.

Physical Systems

- Physical processes shape Earth's surface.
- All living things are dependent upon one another and their surroundings for survival.

Human Systems

- The characteristics and movement of people impact physical and human systems.
- Culture groups shape human systems.
- Patterns of economic activities result in global interdependence.
- Geographic factors influence where people settle.
- Cooperation and conflict among people have an effect on the Earth's surface.

Environment and Society

- People's actions can change the physical environment.
- The physical environment affects how people live.
- Changes occur in the use and importance of natural resources.

The Uses of Geography

- Geography is used to interpret the past, understand the present, and plan for the future.

The World in Spatial Terms: Maps help you locate places on Earth's surface.

Physical Systems: Physical processes, such as hurricanes, shape the face of the Earth.

Human Systems: Technology impacts people and economies.

Environment and Society: Recycling is a choice people make to protect Earth's physical environment.

Using the BIG Ideas of Geography

You can find the Big Ideas throughout *Exploring Our World: People, Places, and Cultures.*

Look for the Big Ideas that will be presented in the chapter.

Think about what you expect to read in the section using the Big Idea.

Use the section's Big Idea to help you answer assessment questions.

Globes and Maps

What Is a Globe? ►

A **globe** is a round model of the Earth that shows its shape, lands, and directions as they truly relate to one another.

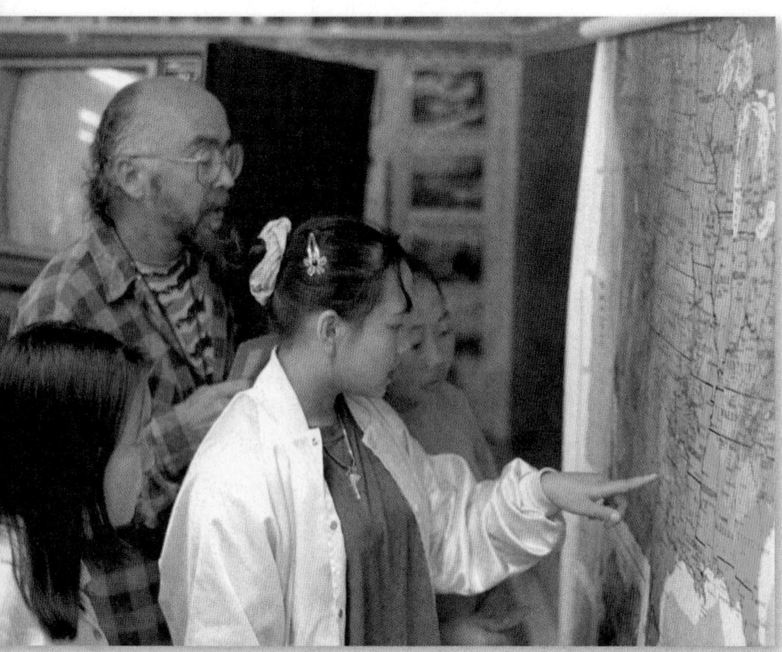

◄ What Is a Map?

A **map** is a flat drawing of all or part of the Earth's surface. Cartographers, or mapmakers, use mathematical formulas to transfer information from the round globe to a flat map.

Globes and Maps ►

Globes and maps serve different purposes, and each has advantages and disadvantages.

	Advantages	Disadvantages
Globes	• Represent true land shape, distances, and directions	• Cannot show detailed information • Difficult to carry
Maps	• Show small areas in great detail • Display different types of information, such as population densities or natural resources • Transport easily	• **Distort,** or change, the accuracy of shapes and distances

Map Projections

When the Earth's surface is flattened on a map, big gaps open up. Mapmakers stretch parts of the Earth to show either the correct shapes of places or their correct sizes. Mapmakers have developed different projections, or ways of showing the Earth on a flat piece of paper. Below are different map projections.

Goode's Interrupted Equal-Area Projection ▼

A map with this projection shows continents close to their true shapes and sizes. This projection is helpful to compare land area among continents.

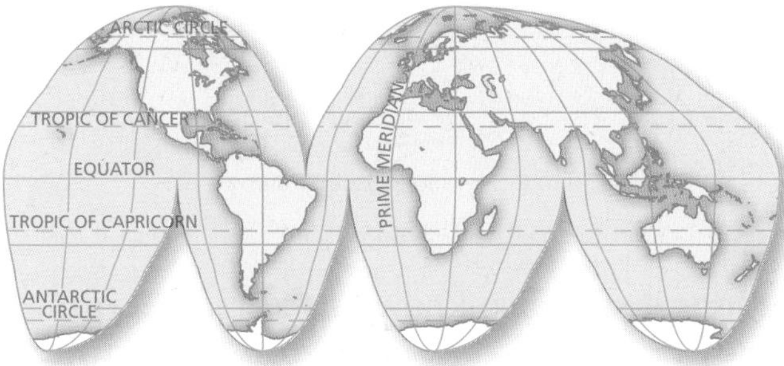

Mercator Projection ▼

The Mercator projection shows land shapes fairly accurately but not size or distance. Areas that are located far from the Equator are quite distorted. The Mercator projection shows true directions, however, making it useful for sea travel.

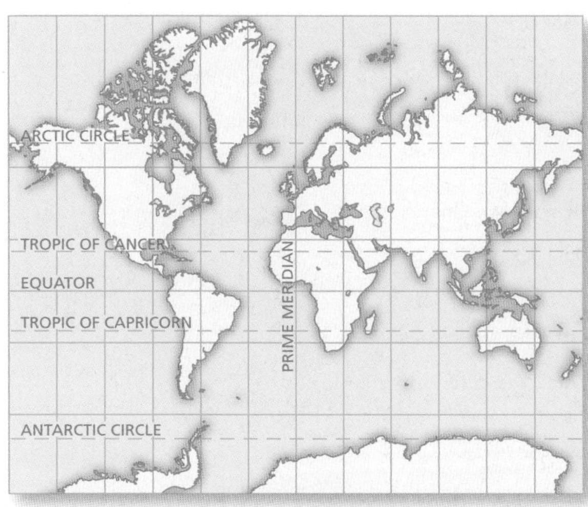

Robinson Projection ▼

The Robinson projection has minor distortions. Continents and oceans are close to their sizes and shapes, but the North and South Poles appear flattened.

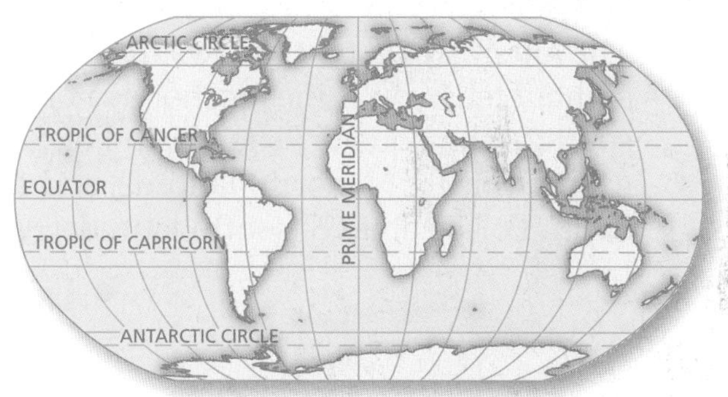

Winkel Tripel Projection ▼

This projection gives a good overall view of the continents' shapes and sizes. Land areas are not as distorted near the poles as they are in the Robinson projection.

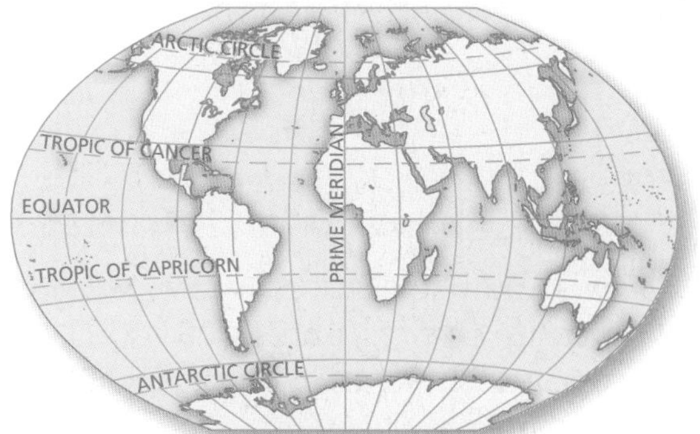

Skills Practice

1. **Comparing and Contrasting** Explain similarities and differences between globes and maps.
2. **Describing** Why do map projections distort some parts of the Earth?

Location

To locate places on Earth, geographers use a system of imaginary lines that crisscross the globe. These lines are called *latitude* and *longitude.*

Latitude ▶

- Lines of **latitude** are imaginary circles that run east to west around the globe. They are known as *parallels*. These parallels divide the globe into units called degrees.
- The **Equator** circles the middle of the Earth like a belt. It is located halfway between the North and South Poles. The Equator is 0° latitude.
- The letter *N* or *S* that follows the degree symbol tells you if the location is north or south of the Equator. The North Pole, for example, is 90°N (north) latitude, and the South Pole is at 90°S (south) latitude.

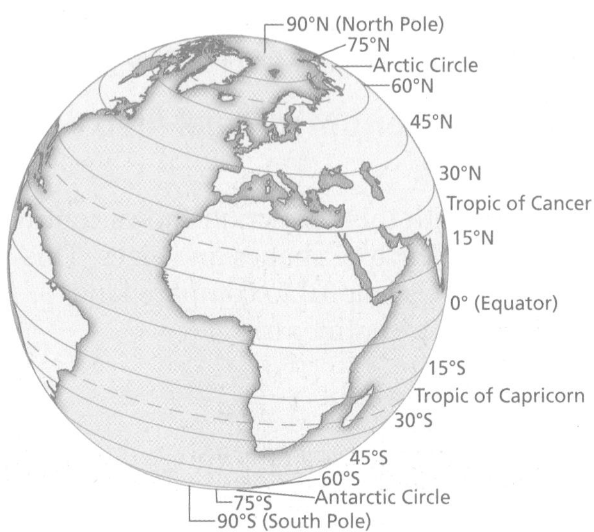

◀ Longitude

- Lines of **longitude,** also known as *meridians*, run from the North Pole to the South Pole. The **Prime Meridian** (also called the Meridian of Greenwich) is 0° longitude and runs through Greenwich, England.
- The letter *E* or *W* that follows the degree symbol tells you if the location is east or west of the Prime Meridian.
- On the opposite side of the Earth is the 180° meridian, also known as the International Date Line.

Absolute Location ▶

A place's exact location can be identified when you use both latitude and longitude. For example, Tokyo, Japan, is 36°N latitude and 140°E longitude.

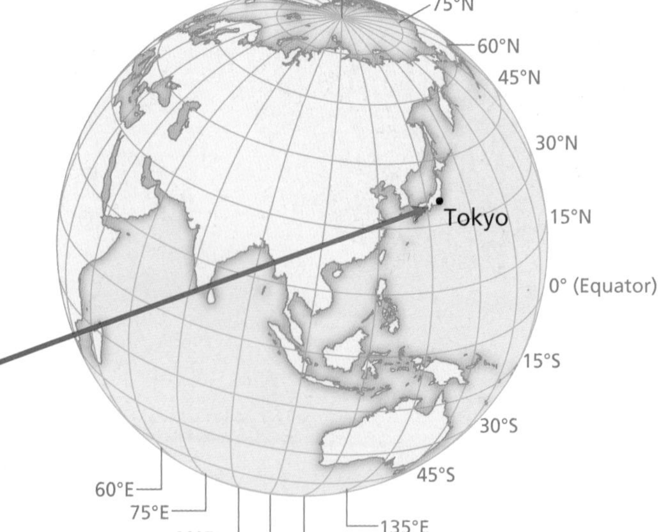

Hemispheres

The Equator divides the Earth into Northern and Southern Hemispheres. Everything north of the Equator is in the Northern Hemisphere. Everything south of the Equator is in the Southern Hemisphere.

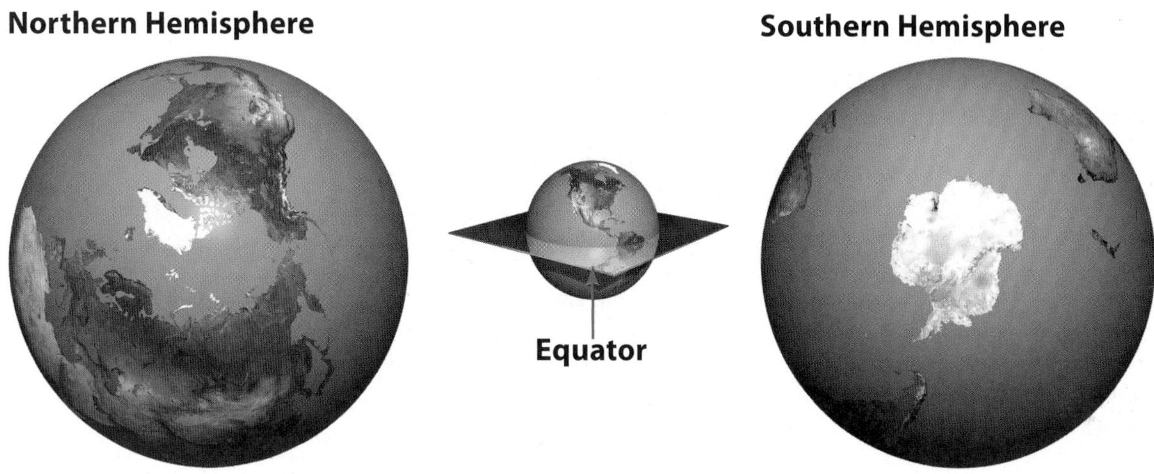

The Prime Meridian divides the Earth into Eastern and Western Hemispheres. Everything east of the Prime Meridian for 180 degrees is in the Eastern Hemisphere. Everything west of the Prime Meridian for 180 degrees is in the Western Hemisphere.

Eastern Hemisphere

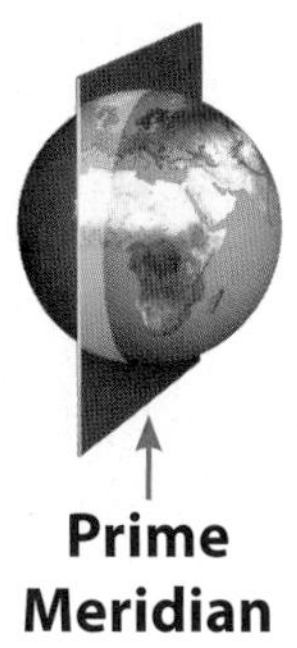

Prime Meridian

Western Hemisphere

Skills Practice

1. **Identifying** What country is located at 30°S and 120°E?
2. **Analyzing Visuals** In which hemispheres is Europe located?

Parts of a Map

Title
The title tells you what information the map is showing.

Key
The key explains the symbols, colors, and lines on the map. The key is also called a *legend*.

Scale Bar
A measuring line, often called a **scale bar,** helps you figure distance on the map. The map scale shows the relationship between map measurements and actual distances on the Earth.

Compass Rose
The compass rose is a symbol that tells you where the **cardinal directions**—north, south, east, and west—are positioned.

Cities
Cities are symbolized by a solid circle (●). This symbol is found in the key and on the map.

Capitals
Capitals are symbolized by a star (✪). This symbol is found in the key and on the map.

Boundary Lines
Boundary lines show the extent of an area's territory or political influence.

Using Scale

All maps are drawn to a certain **scale.** The scale of a map is the size of the map compared to the size of the actual land surface. Thus, the scale of a map varies with the size of the area shown.

Small-Scale Maps ▼

A small-scale map, like this political map of Mexico, shows a large land area but little detail.

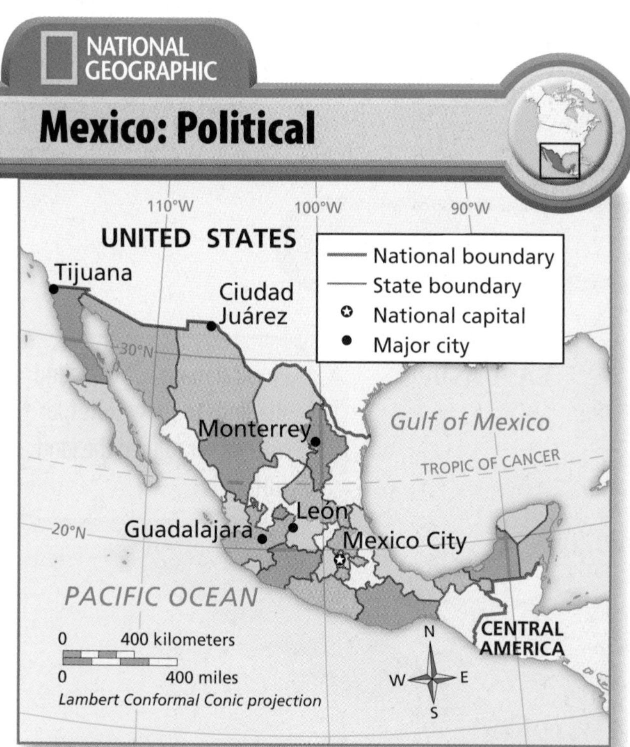

Large-Scale Maps ▼

A large-scale map, like this map of Mexico City, shows a small land area with a great amount of detail.

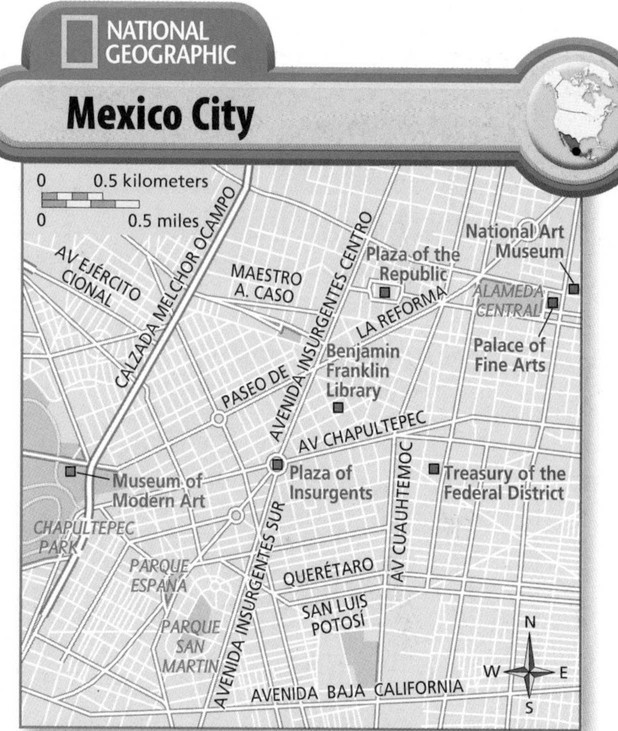

How Do I Use a Scale Bar?

Use the scale bar to find actual distances on a map. The scale bar tells you how many kilometers or miles are represented in that length. You can use a ruler, then, to calculate distances based on the scale bar's length.

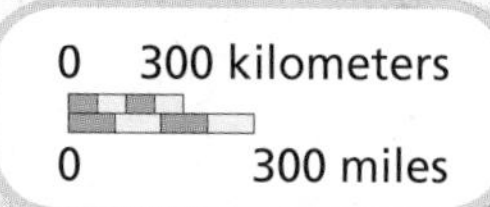

About ½ of an inch equals 300 miles. A little more than ½ of a centimeter is equal to 300 kilometers.

Skills Practice

1. **Defining** What is scale?
2. **Contrasting** What is the difference between a small-scale map and a large-scale map?
3. **Identifying** What are the four cardinal directions?
4. **Describing** Would you use a small-scale or a large-scale map to plan a car trip across the United States? Why?

Types of Maps

General Purpose Maps

Maps are amazingly useful tools. You can use them to show information and to make connections between seemingly unrelated topics. Geographers use many different types of maps. Maps that show a wide range of information about an area are called **general purpose maps.** Two of the most common general purpose maps are physical maps and political maps.

Physical Maps ▼

Physical maps call out landforms and water features. The map key explains what each color and symbol stands for.

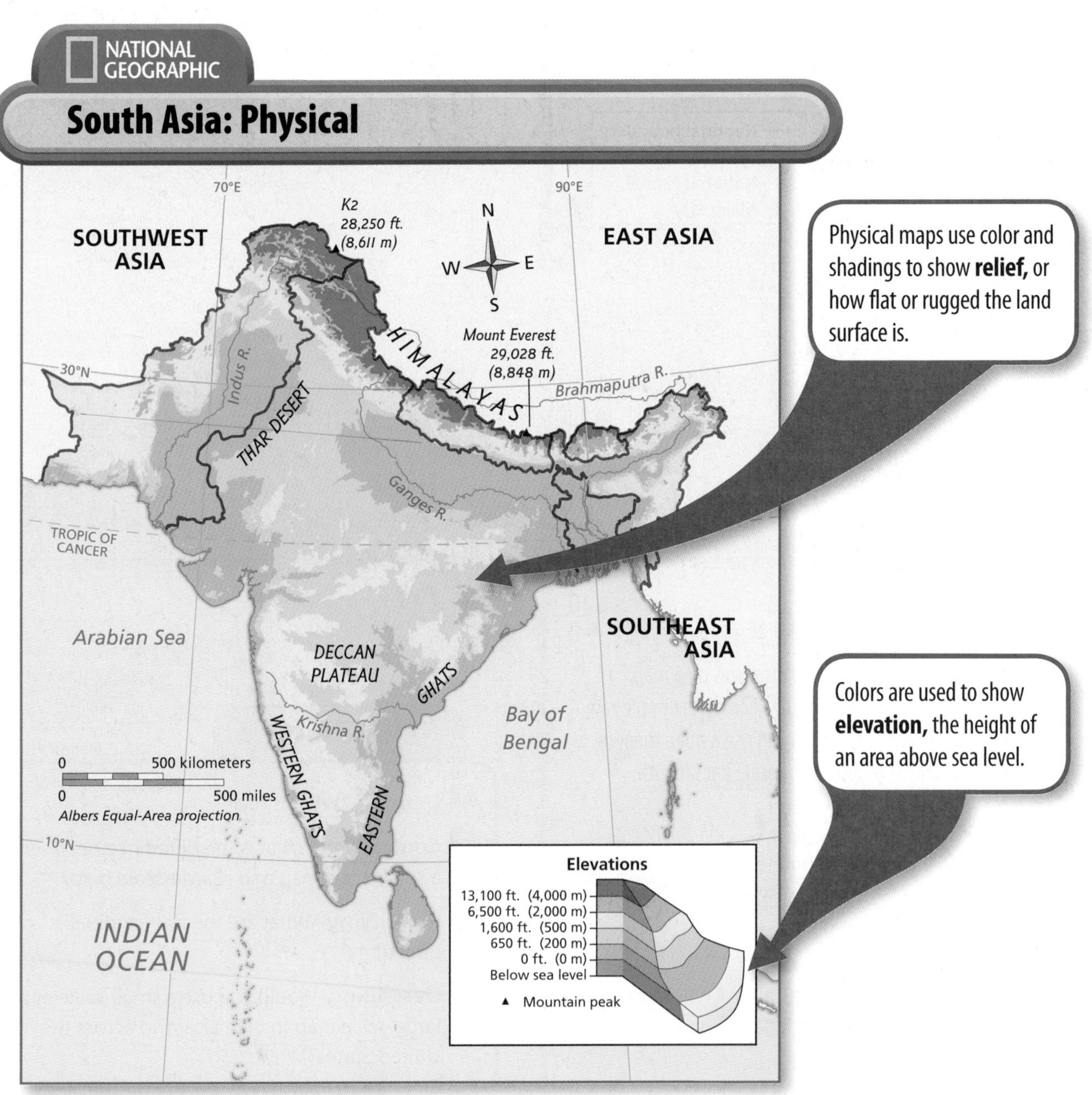

Political Maps ▼

Political maps show the names and political boundaries of countries, along with human-made features such as cities or transportation routes.

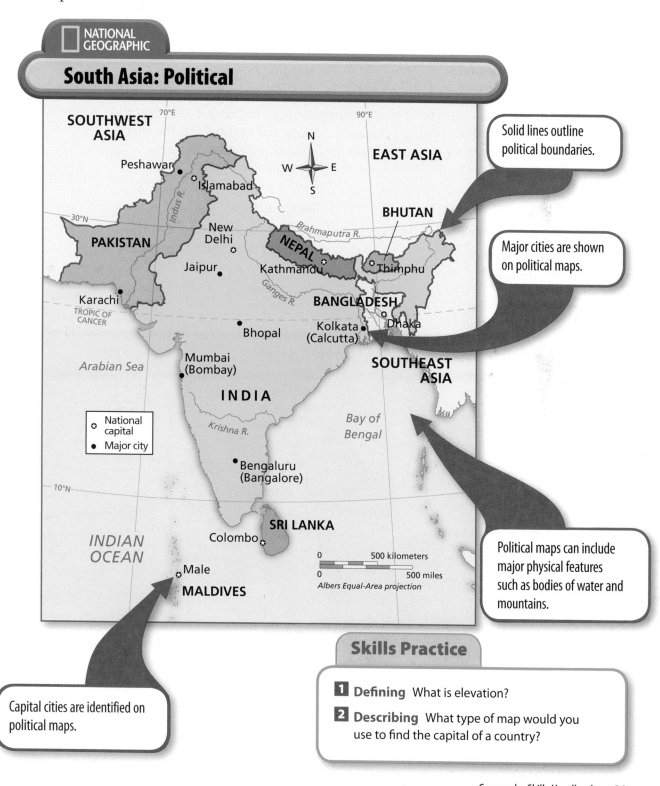

NATIONAL GEOGRAPHIC

South Asia: Political

Solid lines outline political boundaries.

Major cities are shown on political maps.

Political maps can include major physical features such as bodies of water and mountains.

Capital cities are identified on political maps.

Skills Practice

1 Defining What is elevation?

2 Describing What type of map would you use to find the capital of a country?

Types of Maps

Special Purpose Maps

Some maps are made to present specific types of information. These are called **thematic** or **special purpose maps.** These maps usually show specific topics in detail. Special purpose maps may include information about:

- climate
- vegetation
- natural resources
- population density
- historical expansion

Look at some of the types of special purpose maps on these pages. The map's title is especially important for a special purpose map because it tells you the type of information that is being presented. Colors and symbols in the map key are also important tools to use when you read these types of maps.

Historical Maps ▼

Historical maps show events that occurred in a region over time. On the map below, you can see where Europeans settled on the North American continent in the past.

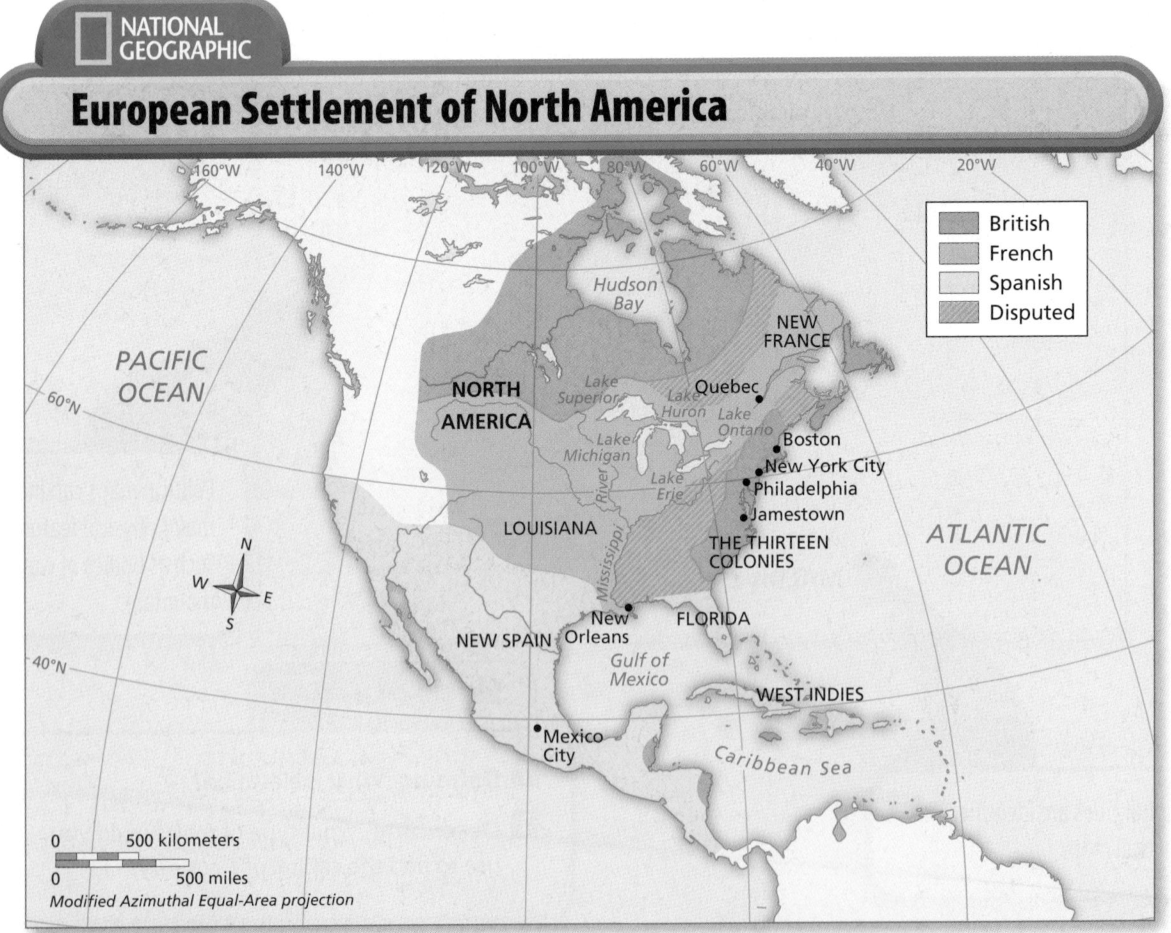

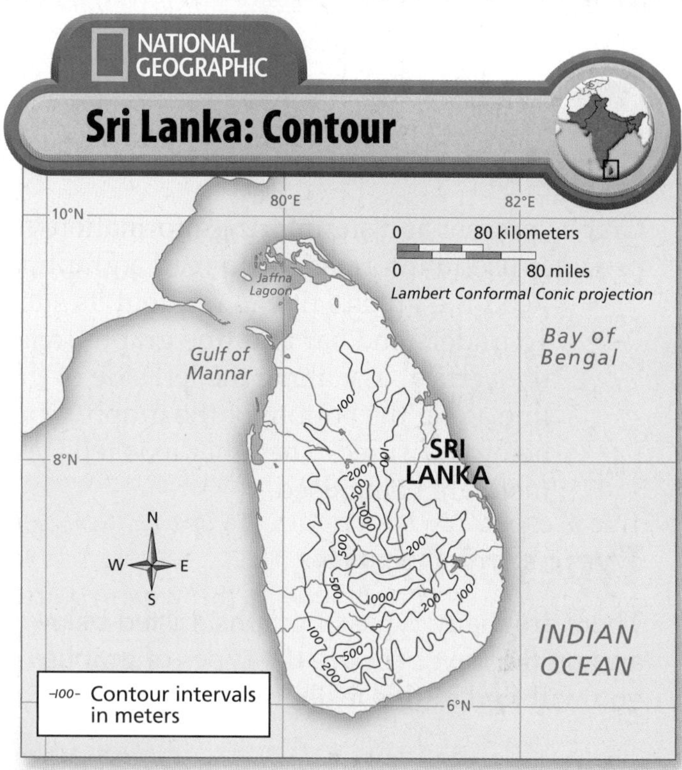

Contour Maps ▶

A contour map has **contour lines**—one line for each major level of elevation. All the land at the same elevation is connected by a line. These lines usually form circles or ovals—one inside the other. If contour lines are close together, the surface is steep. If the lines are spread apart, the land is flat or rises gradually.

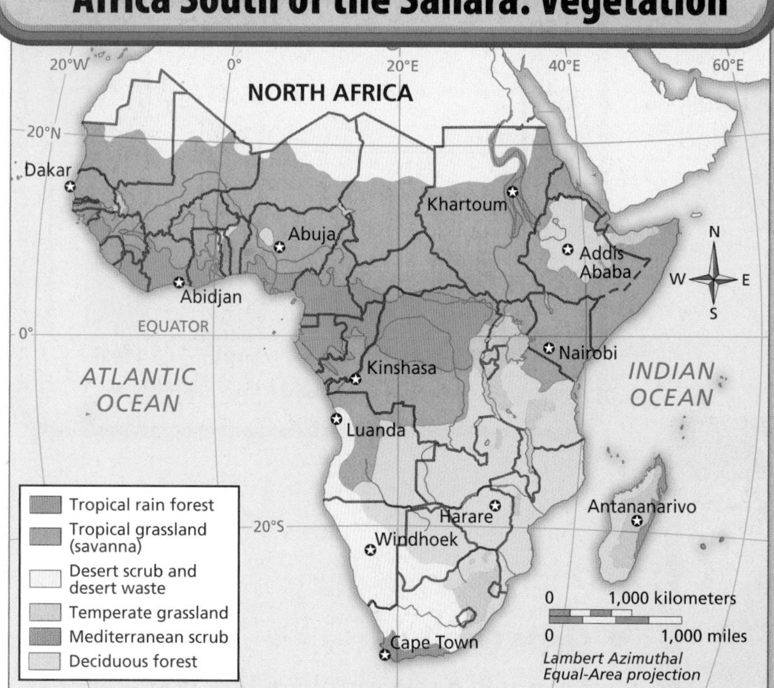

◀ Vegetation Maps

Vegetation maps are special purpose maps that show the different types of plants that are found in a region.

Skills Practice

1. **Identifying** What type of special purpose map might show battles during World War II?
2. **Contrasting** What is the difference between a general purpose map and a special purpose map?

Graphs, Charts, and Diagrams

Graphs

Graphs present and summarize information visually. Each part of a graph provides useful information. To read a graph, follow these steps:

- Read the graph's title to find out its subject.
- To understand bar and line graphs, read the labels along the **axes**—the vertical line along the left side of the graph and the horizontal line along the bottom of the graph. One axis will tell you what is being measured. The other axis tells you what units of measurement are being used.

Types of Graphs

There are many types of graphs. Listed below and on the next page are the types of graphs you will find in this textbook.

Bar Graphs ▶

Graphs that use bars or wide lines to compare data visually are called bar graphs.

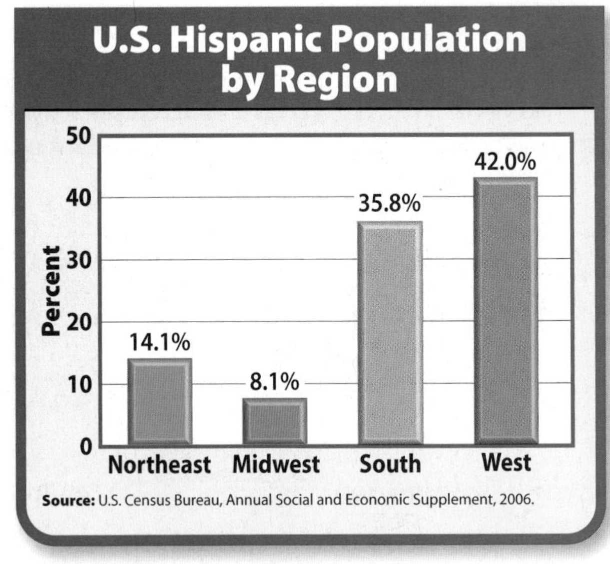

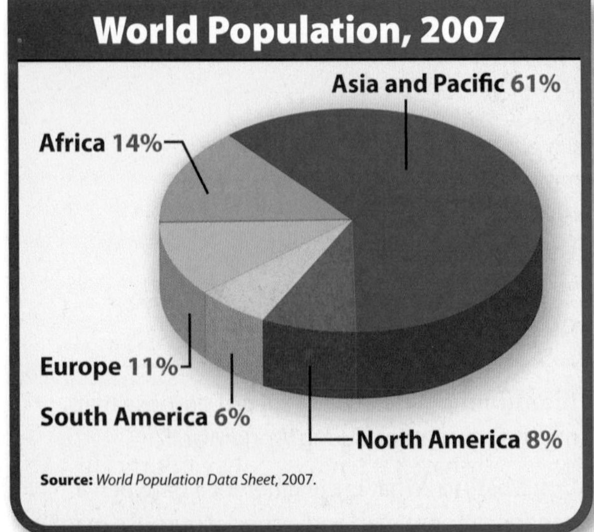

◀ Circle Graphs

You can use circle graphs when you want to show how the whole of something is divided into its parts. Because of their shape, circle graphs are often called *pie graphs.* Each slice represents a part or percentage of the whole pie. The complete circle represents a whole group—or 100 percent.

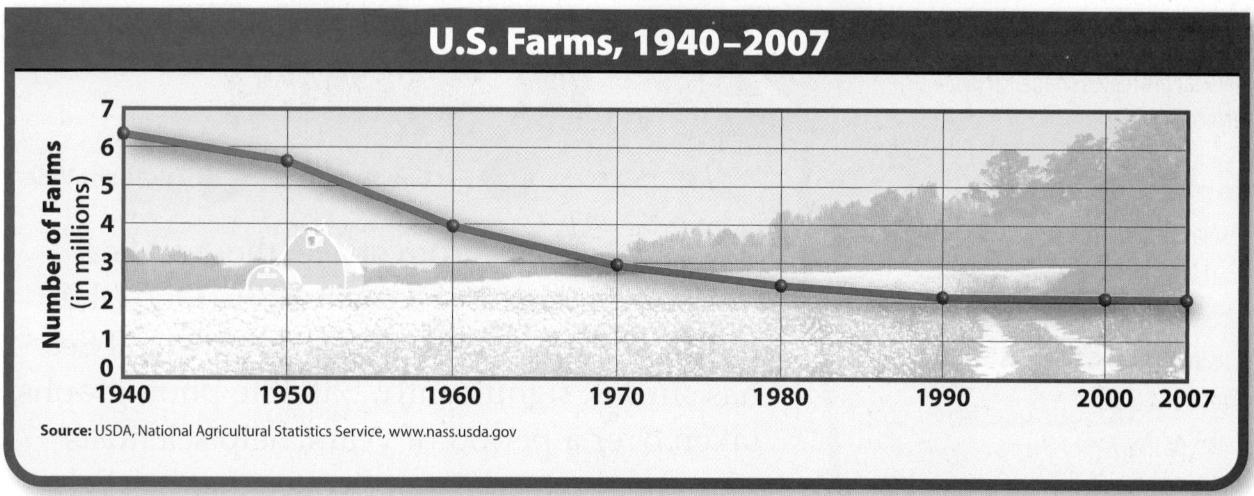

Source: USDA, National Agricultural Statistics Service, www.nass.usda.gov

Line Graphs ▲

Line graphs help show changes over a period of time. The amounts being measured are plotted on the grid above each year and then are connected by a line.

Charts

Charts present related facts and numbers in an organized way. They arrange data, especially numbers, in rows and columns for easy reference.

Island Populations

Island	Population
Aruba	71,891
Bermuda	65,773
British Virgin Islands	23,098
Jamaica	2,758,124

Source: *CIA World Factbook*, 2006.

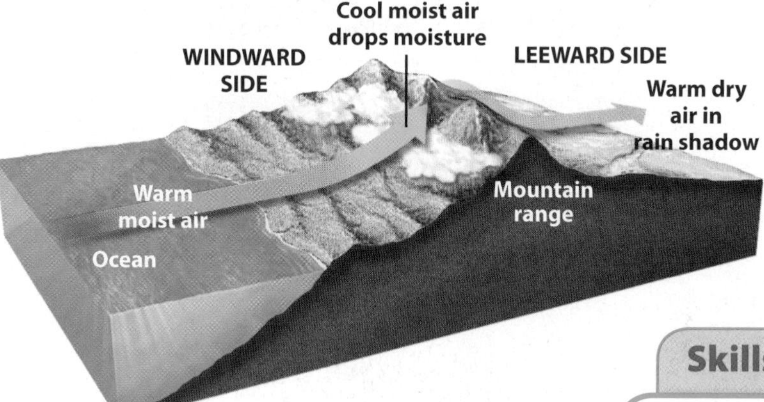

Diagrams

Diagrams are drawings that show steps in a process, point out the parts of an object, or explain how something works.

Skills Practice

1. **Identifying** What percentage does the whole circle in a circle graph represent?
2. **Analyzing Information** What type of graph would best show the number of Republicans and Democrats in the U.S. House of Representatives?

Guide to Reading

BIG Idea

Physical processes shape Earth's surface.

Content Vocabulary

- solar system *(p. 35)*
- orbit *(p. 35)*
- revolution *(p. 36)*
- leap year *(p. 36)*
- rotate *(p. 36)*
- axis *(p. 36)*
- atmosphere *(p. 36)*
- summer solstice *(p. 37)*
- winter solstice *(p. 38)*
- equinox *(p. 38)*
- Tropics *(p. 38)*

Academic Vocabulary

- significant *(p. 37)*
- reverse *(p. 38)*
- identical *(p. 38)*

Reading Strategy

Determining Cause and Effect
Use a diagram like the one below to show the effects of latitude on Earth's temperatures.

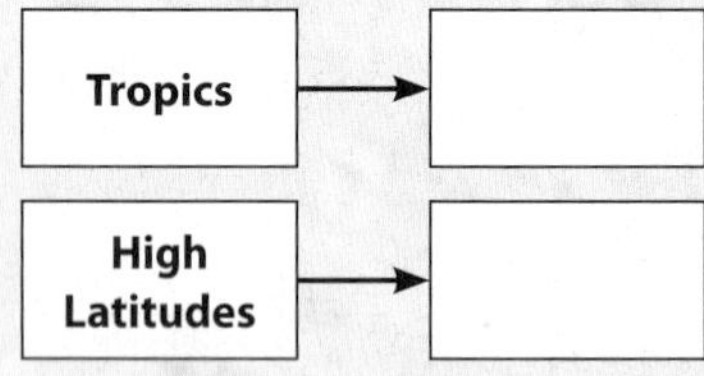

The Earth in Space

Picture This From space, the Aral Sea in Central Asia can be easily seen. Once the fourth-largest lake in the world, the Aral Sea has shrunk significantly. Satellite photographs, taken over a period of years, help scientists measure the total area of water that has been lost. These images, in addition to other information, help scientists understand what has caused the sea to change size. Scientists also continue to explore space and how the Earth's location in the solar system affects our planet. Read the next section to learn how the Earth's rotation, orbit, tilt, and latitude affect life on Earth.

▼ The Aral Sea from space

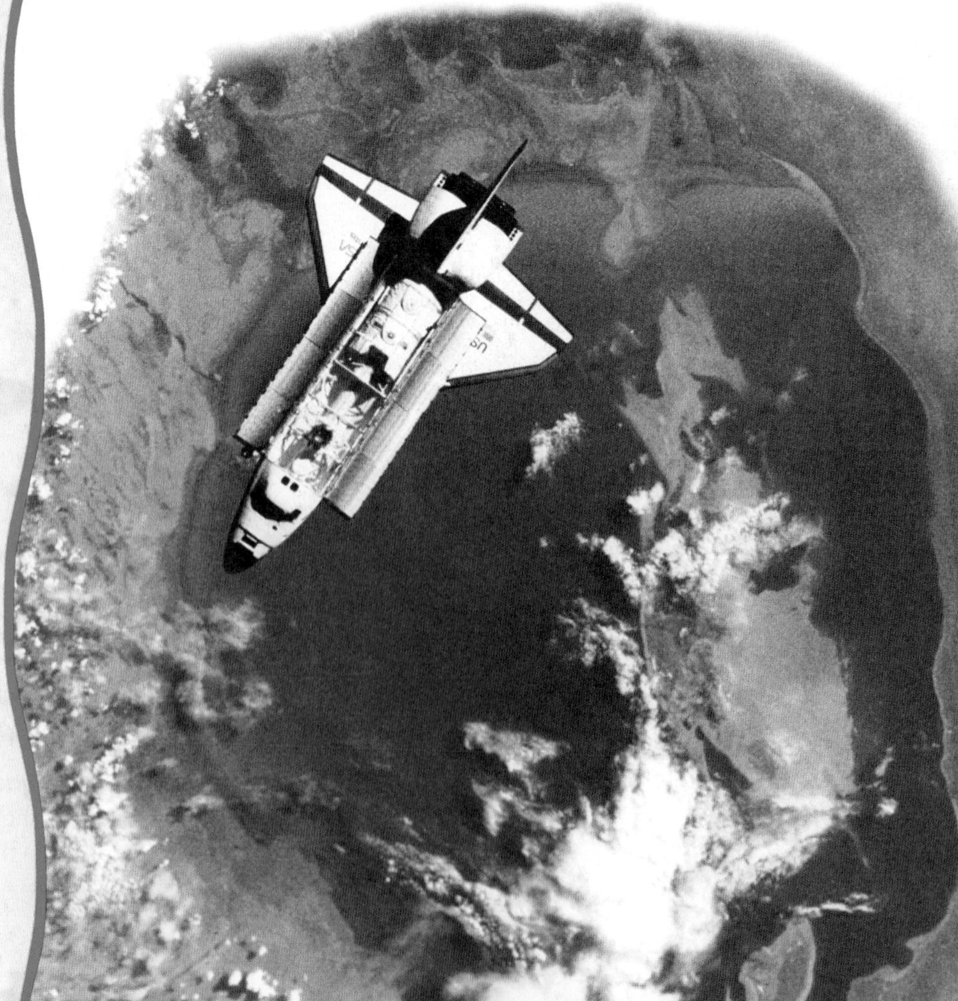

The Solar System

Main Idea **The Earth is one of eight planets in the solar system. It rotates on its axis every 24 hours and takes a year to orbit the sun.**

Geography and You Have you watched a sunrise or sunset and wondered why the sun seems to move across the sky each day? Read to find out about the Earth and its place in our solar system.

The sun provides the heat necessary for life on our planet. Earth, seven other major planets, and thousands of smaller bodies all revolve around the sun. Together with the sun, these bodies form our **solar system.**

Major Planets

The major planets differ from one another in size and makeup. Look at **Figure 1.** It shows that the inner planets—Mercury, Venus, Earth, and Mars—are relatively small and solid. The outer planets—Jupiter, Saturn, Uranus, and Neptune—are larger and composed mostly or entirely of gases. Pluto, once considered a major planet, is now called a minor planet.

Each planet follows its own path, or **orbit,** around the sun. The orbits vary from nearly circular to elliptical, or oval shaped.

Social Studies ONLINE
Student Web Activity Visit glencoe.com and complete the Chapter 1 Web Activity about the solar system.

NATIONAL GEOGRAPHIC

Figure 1 **The Solar System**

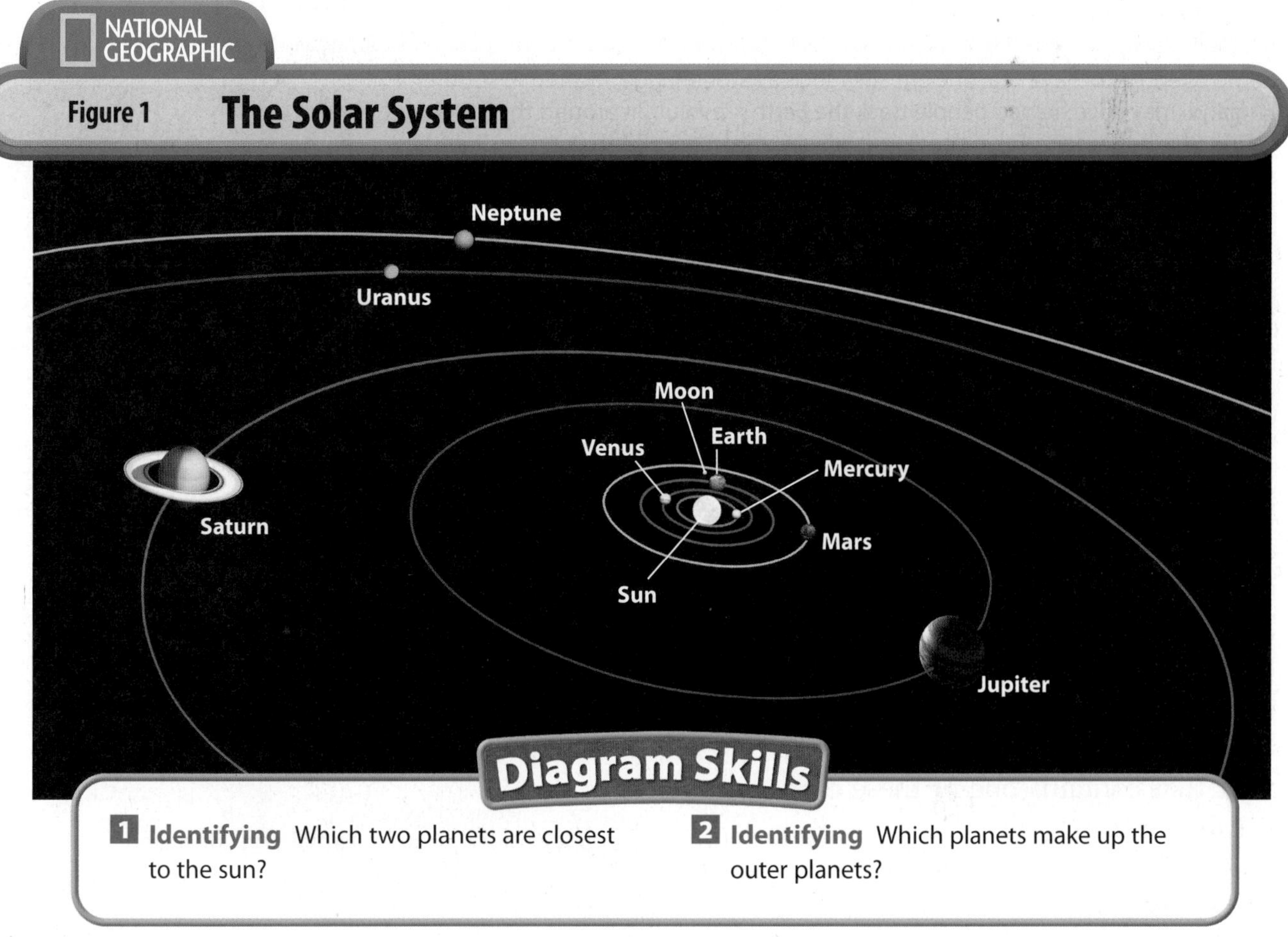

Diagram Skills

1. **Identifying** Which two planets are closest to the sun?
2. **Identifying** Which planets make up the outer planets?

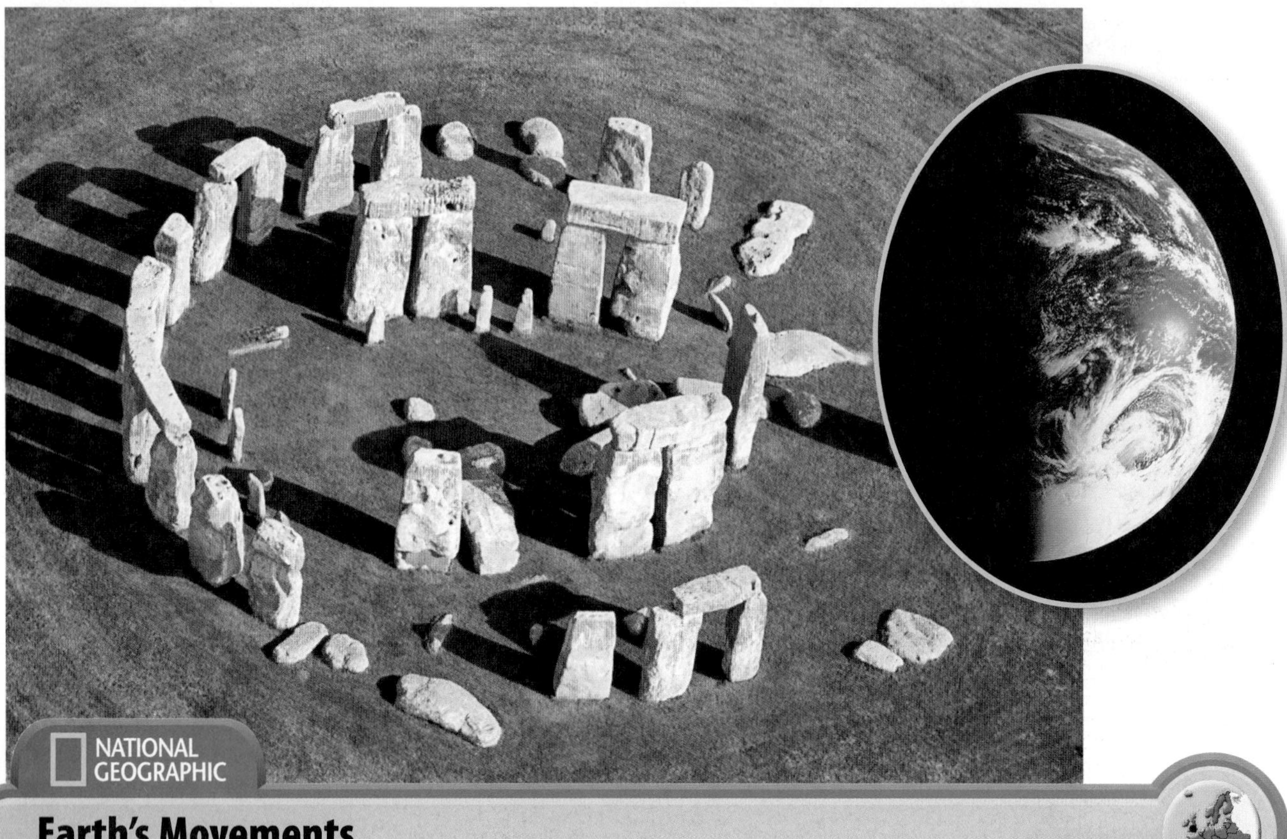

Earth's Movements

Some scientists believe that ancient sites such as Stonehenge, located in southern England, may have helped people track the Earth's revolution around the sun and the change of the seasons. From space, astronauts can see the Earth (inset) in light and shadow at the same time. ***Movement*** **Why do different parts of the Earth experience sunlight or darkness?**

The time necessary to complete an orbit differs, too. Mercury needs only 88 days to circle the sun, but faraway Neptune takes 165 years.

Earth's Movement

Earth takes almost 365¼ days to make one **revolution,** or a complete circuit, around the sun. This period is what we define as one year. Every four years, the extra fourths of a day are combined and added to the calendar as February 29th. A year that contains one of these extra days is called a **leap year.**

As Earth orbits the sun, it **rotates,** or spins, on its axis. The **axis** is an imaginary line that passes through the center of Earth from the North Pole to the South Pole. Earth rotates in an easterly direction, making one complete rotation every 24 hours. As Earth turns, different parts of the planet are in sunlight or in darkness. The part facing the sun experiences daytime, and the part facing away has night.

Why do we not feel Earth moving as it rotates? The reason is that the **atmosphere,** the layer of oxygen and gases that surrounds Earth, moves with it.

Explaining Describe Earth's two principal motions—revolution and rotation.

Sun and Seasons

Main Idea **The tilt of Earth and its revolution around the sun lead to changing seasons during the year.**

Geography and You Did you know that when it is winter in the United States, it is summer in Australia? Read to learn why seasons differ between the Northern and Southern Hemispheres.

Earth is tilted 23½ degrees on its axis. As a result, seasons change as Earth makes its year-long orbit around the sun. To see why this happens, look at **Figure 2.** Notice how sunlight falls directly on the northern or southern half of Earth at different times of the year. Direct rays from the sun bring more warmth than indirect, or slanted rays. When the people in a hemisphere receive direct rays, they enjoy the warmth of summer. When they receive only indirect rays, they experience the cold of winter.

Solstices and Equinoxes

Four days in the year are **significant,** or important, because of the position of the sun in relation to Earth. These days mark the beginnings of the four seasons. On or about June 21, the North Pole is tilted toward the sun. On noon of this day, the sun appears directly overhead at the Tropic of Cancer (23½°N latitude). In the Northern Hemisphere, this day is the **summer solstice**–the day with the most hours of sunlight. It is the beginning of summer—but only in the Northern Hemisphere. In the Southern Hemisphere, that same day is the day with the fewest hours of sunlight and marks the beginning of winter.

NATIONAL GEOGRAPHIC **Maps In MOtion** See StudentWorks™ Plus or glencoe.com.

Figure 2 **The Sun and Earth's Seasons**

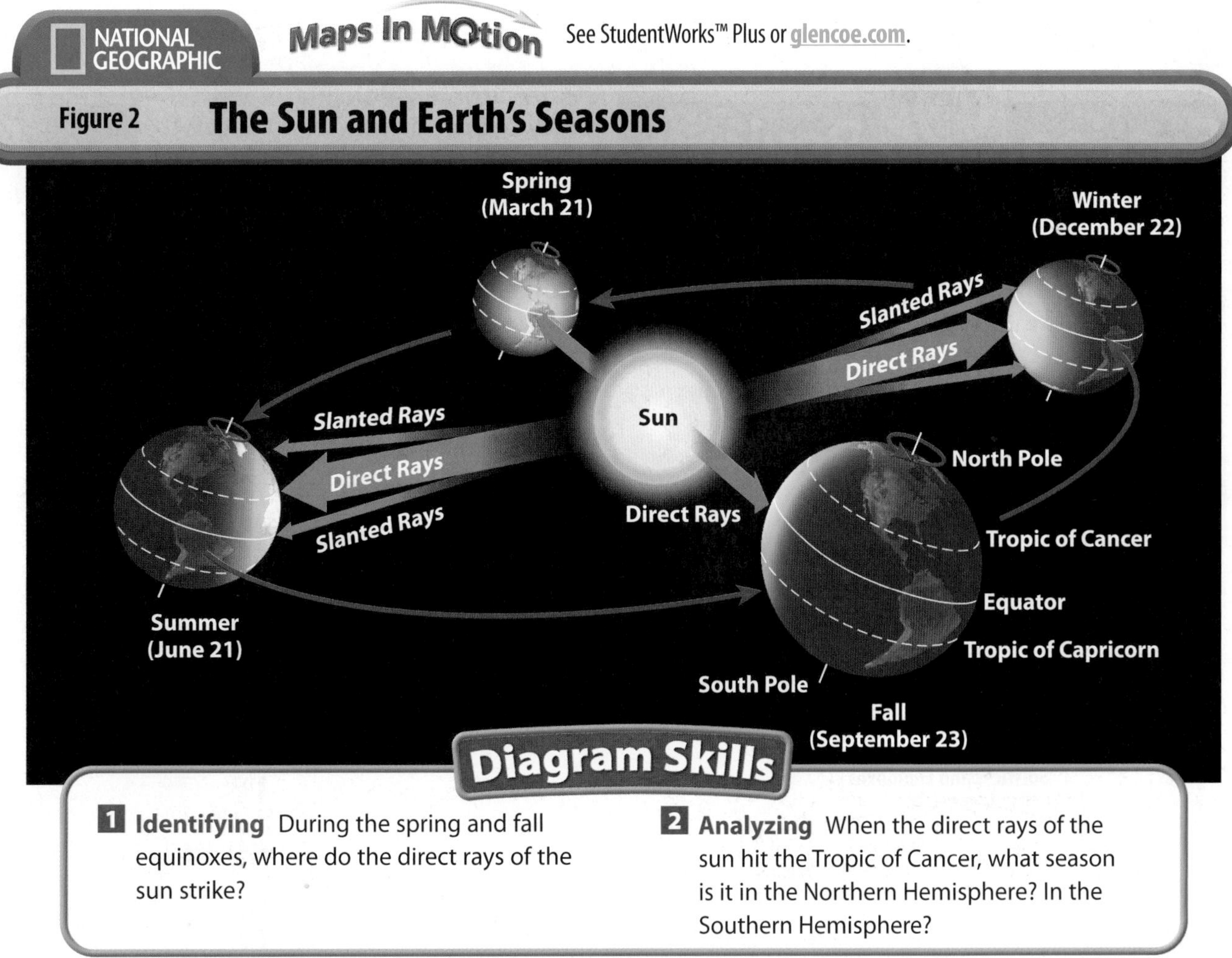

Diagram Skills

1. **Identifying** During the spring and fall equinoxes, where do the direct rays of the sun strike?
2. **Analyzing** When the direct rays of the sun hit the Tropic of Cancer, what season is it in the Northern Hemisphere? In the Southern Hemisphere?

Six months later—on or about December 22—the situation is **reversed,** or the opposite. The North Pole is tilted away from the sun. At noon, the sun's direct rays strike the Tropic of Capricorn. In the Northern Hemisphere, this day is the **winter solstice**—the day with the fewest hours of sunlight and the beginning of winter. This same day, however, marks the beginning of summer in the Southern Hemisphere.

Spring and autumn each begin on a day that falls midway between the two solstices. These two days are the **equinoxes,** when day and night are of **identical,** or equal, length in both hemispheres. On or about March 21, the spring equinox occurs. On or about September 23, the fall equinox occurs. On both days, the noon sun shines directly over the Equator.

Effects of Latitude

Earth's temperatures also are affected by the sun. Look again at **Figure 2.** The sun's rays directly hit places in the **Tropics,** the low-latitude areas near the Equator between the Tropic of Cancer and the Tropic of Capricorn. As a result, temperatures in the Tropics tend to be very warm.

At the high latitudes near the North and South Poles, the sun's rays hit indirectly. Temperatures in these regions are always cool or cold. In the midlatitudes—the areas between the Tropics of Cancer and Capricorn and the polar regions—temperatures, weather, and the seasons vary greatly. This is because air masses from both the high latitudes and the Tropics affect these areas.

Reading Check **Analyzing Information** Why are the Tropics the Earth's warmest regions?

Social Studies ONLINE
Study Central™ To review this section, go to glencoe.com.

Vocabulary

1. **Explain** the significance of:
   - **a.** solar system
   - **b.** orbit
   - **c.** revolution
   - **d.** leap year
   - **e.** rotate
   - **f.** axis
   - **g.** atmosphere
   - **h.** summer solstice
   - **i.** winter solstice
   - **j.** equinox
   - **k.** Tropics

Main Ideas

2. **Identifying** Name the inner and outer planets, and describe the differences between the two groups.
3. **Comparing** Use a diagram like the one below to compare the days that mark the beginnings of the seasons.

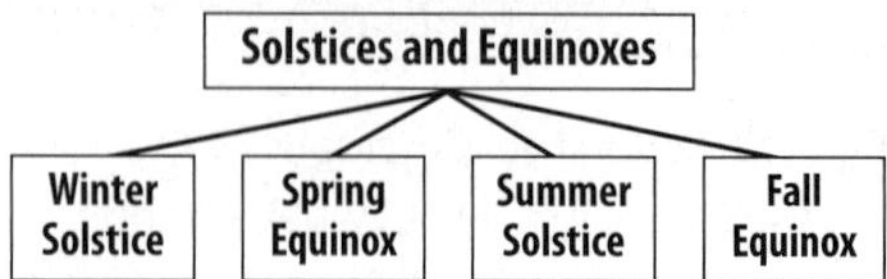

Critical Thinking

4. **Analyzing** Why do we not feel the Earth's movement as it rotates?
5. BIG Idea What causes different seasons on Earth?
6. **Challenge** How might latitude affect the population of a region?

Writing About Geography

7. **Expository Writing** Write a paragraph explaining why seasons in the Southern Hemisphere are the opposite of those in the Northern Hemisphere.

Visual Summary

Themes of Geography

- Geography is the study of the Earth and its people.
- In their study of people and places, geographers use five themes: location, place, human-environment interaction, movement, and regions.

Volcano, Costa Rica

Kinds of Geography

- Physical geography examines physical aspects of the Earth, such as land areas, bodies of water, and plant life.
- Human geography focuses on people and their activities, including religions, languages, and ways of life.

Hiker using GPS

Geographers at Work

- To study the Earth, geographers use maps, globes, photographs, the Global Positioning System (GPS), and Geographic Information Systems (GIS).
- People can use information from geographers to plan, make decisions, and manage resources.

Solar System

- The sun, eight planets, and many smaller bodies form our solar system.
- Earth takes almost 365¼ days to make one revolution around the sun.
- Earth spins on its axis, causing day and night.

Sun and Seasons

- The Earth's tilt and its revolution around the sun cause the changes in seasons.
- Four days in the year mark the beginning points of the four seasons.

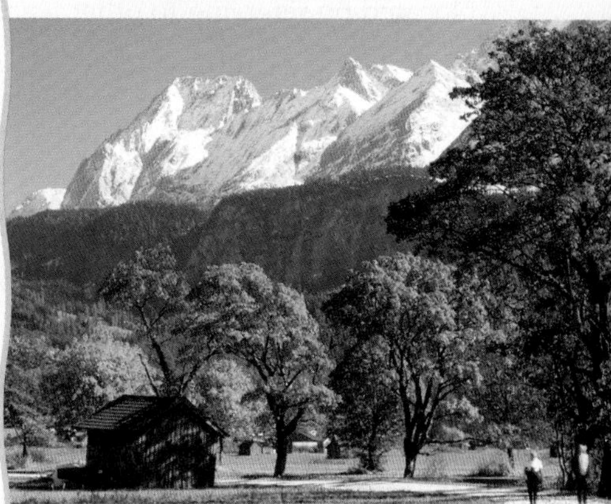

Autumn, Tyrol, Austria

The Earth from space

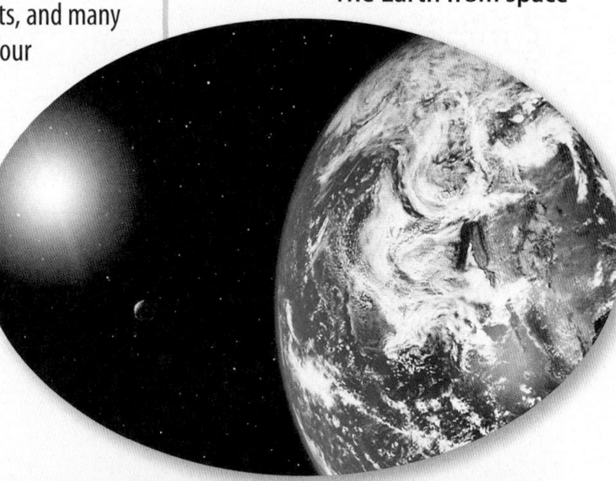

STUDY TO GO Study anywhere, anytime! Download quizzes and flash cards to your PDA from glencoe.com.

STANDARDIZED TEST PRACTICE

Read every exam question twice to make certain you know exactly what it is asking.

Reviewing Vocabulary

Directions: Choose the word(s) that best completes the sentence.

1. _______ describes where a geographic feature is located by referring to other features around it.

A Relative location

B Absolute location

C The Global Positioning System

D A Geographic Positioning System

2. According to historians, the period known as Modern History began about five _______.

A years ago

B decades ago

C centuries ago

D millennia ago

3. The path each planet follows around the sun is called its _______.

A axis

B orbit

C revolution

D solar system

4. In the Northern Hemisphere, the day of the year with the fewest hours of sunlight is the _______.

A fall equinox

B winter solstice

C spring equinox

D summer solstice

Reviewing Main Ideas

Directions: Choose the best answer for each question.

Section 1 *(pp. 14–17)*

5. What geographic theme involves characteristics that make a location unique?

A place

B regions

C location

D movement

6. _______ provide(s) detailed photographs for creating maps.

A A globe

B A satellite

C A Global Positioning System

D Geographic Information Systems

Section 2 *(pp. 34–38)*

7. The Earth circles the sun every _______.

A 24 hours

B 365¼ days

C 88 years

D 165 years

8. Temperatures are always cool or cold near the North and South Poles because

A the Poles face the sun in daytime.

B the rays of the sun hit the Poles directly.

C the Poles turn away from the sun at night.

D the Poles receive only slanted rays from the sun.

GO ON

Critical Thinking

Directions: Base your answers to questions 9, 10, and 11 on the map below and your knowledge of Chapter 1. Choose the best answer for each question.

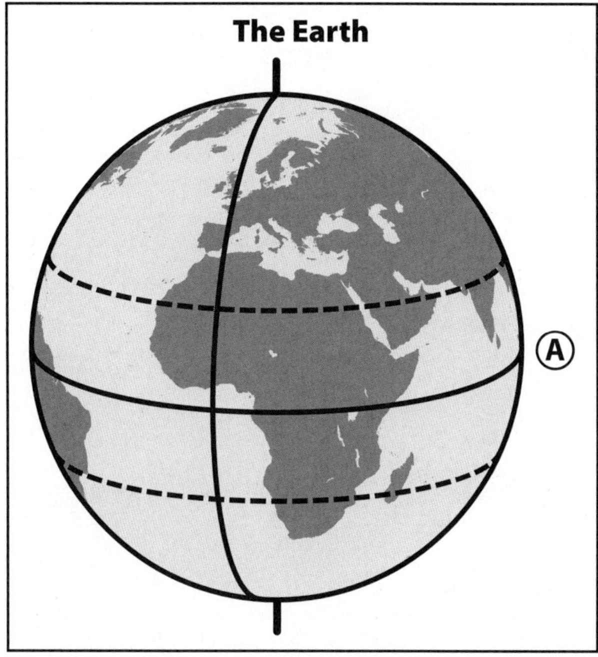

9. Label A is showing ________.

A Earth's axis

B the Equator

C the Prime Meridian

D the Tropic of Cancer

10. The Equator divides the Earth into

A the North Pole and South Pole.

B the Eastern and Western Hemispheres.

C the Tropics of Cancer and Capricorn.

D the Northern and Southern Hemispheres.

11. The sun's rays directly hit places in the ________.

A midlatitudes

B high latitudes

C Tropics

D spring equinox

Document-Based Questions

Directions: Analyze the following document and answer the short-answer questions that follow.

The following passage explains why scientists no longer consider Pluto a major planet.

> *Once known as the smallest, coldest, and most distant planet from the Sun, Pluto has a dual identity, not to mention being enshrouded in controversy since its discovery in 1930. On August 24, 2006, the International Astronomical Union (IAU) formally downgraded Pluto from an official planet to a dwarf planet. According to the new rules a planet meets three criteria: it must orbit the Sun, it must be big enough for gravity to squash it into a round ball, and it must have cleared other things out of the way in its orbital neighborhood. The latter measure knocks out Pluto and 2003UB313 (Eris), which orbit among the icy wrecks of the Kuiper Belt, and Ceres, which is in the asteroid belt.*
>
> —National Aeronautics and Space Administration, "Pluto"

12. In what two ways is Pluto like the major planets?

13. What other two bodies travel around the sun but are not considered planets?

Extended Response

14. Write a paragraph explaining how the concerns of physical geography and human geography often overlap.

STOP

For additional test practice, use Self-Check Quizzes—Chapter 1 at glencoe.com.

Need Extra Help?

If you missed question...	1	2	3	4	5	6	7	8	9	10	11	12	13	14
Go to page...	15	16	35	38	15	17	36	38	25	25	38	35	35	16

Earth's Physical Geography

Essential Question

Place Think about the characteristics of the area where you live. How does the land look? Is there a large body of water nearby? What is the climate like? Each place on the Earth is unique, with its own special characteristics. What kinds of geographic characteristics define the region where you live?

Canyon on the Colorado Plateau, Arizona

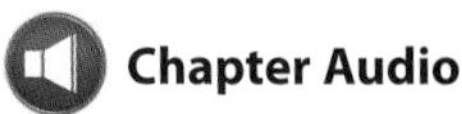

Section 1: Forces Shaping the Earth

BIG IDEA Physical processes shape the Earth's surface. Forces from within and the actions of wind, water, and ice have shaped Earth's surface.

Section 2: Landforms and Water Resources

BIG IDEA Geographic factors influence where people settle. Physical features determine where people live.

Section 3: Climate Regions

BIG IDEA Geographers organize the Earth into regions that share common characteristics. Geographers use climate to define world regions.

Section 4: Human-Environment Interaction

BIG IDEA All living things are dependent upon one another and their surroundings for survival. Human actions greatly affect the natural world.

Organizing Information Use this four-tab Foldable to help you record what you learn about the Earth's physical geography.

Step 1 Fold the top and bottom of a sheet of paper into the middle.

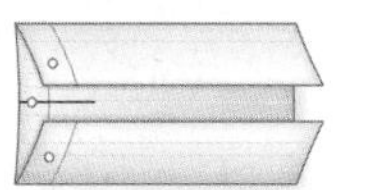

Step 2 Cut each flap at the midpoint to form 4 tabs.

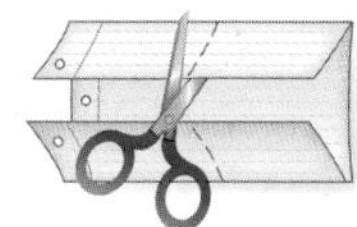

Step 3 Label the tabs as shown.

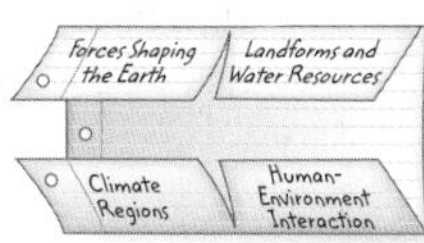

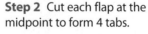

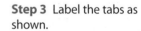

Reading and Writing As you read the chapter, take notes about each section under the appropriate head. Use your Foldable to help you write a summary for each section.

Visit glencoe.com and enter ***QuickPass***™ code EOW3109c2 for Chapter 2 resources.

Guide to Reading

BIG Idea

Physical processes shape the Earth's surface.

Content Vocabulary

- core *(p. 45)*
- mantle *(p. 45)*
- magma *(p. 45)*
- crust *(p. 45)*
- continent *(p. 45)*
- plate tectonics *(p. 46)*
- earthquake *(p. 47)*
- fault *(p. 47)*
- weathering *(p. 47)*
- erosion *(p. 48)*

Academic Vocabulary

- release *(p. 45)*
- constant *(p. 47)*
- accumulate *(p. 48)*

Reading Strategy

Determining Cause and Effect
As you read, use a diagram like the one below to list the forces shaping the Earth and the effects of each.

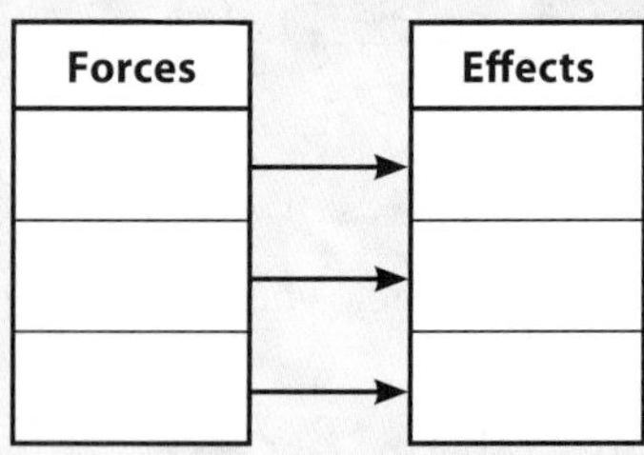

Forces Shaping the Earth

Picture This This spectacular gash is California's San Andreas Fault. The San Andreas Fault is about 800 miles long and extends 10 miles beneath the Earth's surface. It is the source of the deadly earthquakes that occurred in California in 1906 and 1989. Read this section to learn more about processes that have shaped the surface of the Earth.

▼ The San Andreas Fault, located 100 miles north of Los Angeles, California

Inside the Earth

Main Idea **The Earth is made up of several layers that have different characteristics.**

Geography and You What do you see when you cut a melon in half? Like a melon, the Earth has distinct sections or layers.

The ground feels solid when you walk on it and downright hard if you should happen to fall. Yet Earth is not a large rock, solid through the middle. Beneath our planet's solid shell lies a center that is partly liquid. As **Figure 1** shows, the Earth has different layers, much like a melon or a baseball.

At the center of the Earth is a dense solid **core** of hot iron mixed with other metals and rock. The inner core lies about 3,200 miles (5,150 km) below the surface. Scientists think it is made up of iron and nickel. They also believe the inner core is under tremendous pressure. The next layer, the outer core, is so hot that the metal has melted into a liquid. The temperature in the outer core can reach an incredible 8,500°F (about 4,700°C).

Surrounding the core is the **mantle,** a layer of hot, dense rock about 1,770 miles (2,850 km) thick. Like the core, the mantle has two parts. The section nearest the core is solid. The rock in the outer mantle, however, can be moved, shaped, and even melted. If you have seen photographs of an active volcano, then you have seen this melted rock called **magma.** It flows to the surface during a volcanic eruption. Once it reaches the surface, magma is called lava. This movement of the matter in the mantle **releases** much of the energy generated in the Earth's interior.

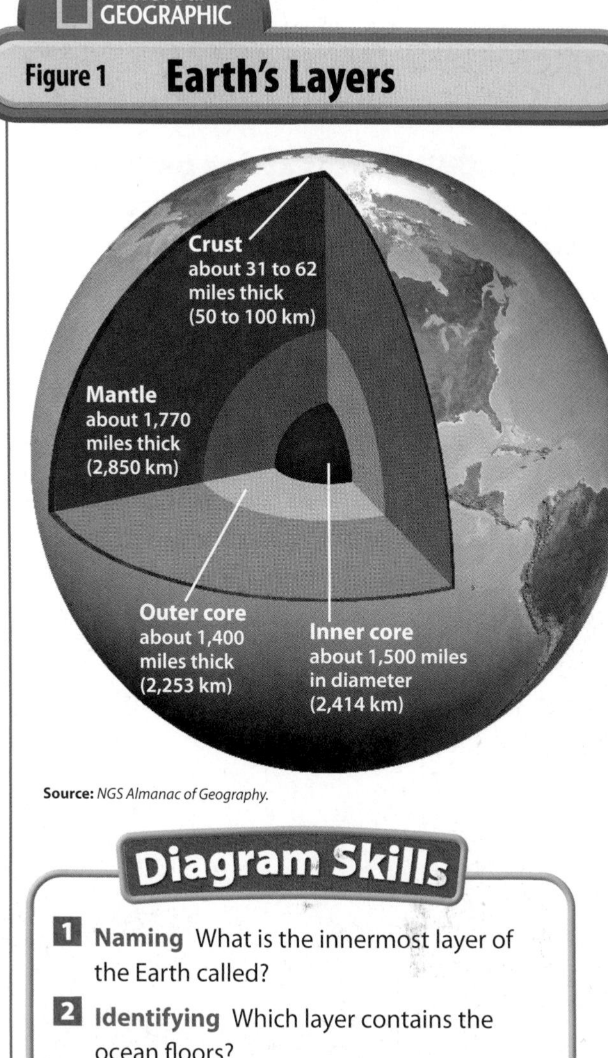

Earth's upper layer is the **crust,** a thin rocky shell that forms the surface. It reaches only 31 to 62 miles (50 to 100 km) deep. The crust includes ocean floors and seven large land areas known as **continents.** The continents are North America, South America, Europe, Asia, Africa, Australia, and Antarctica. The crust is just a few miles thick on the ocean floor, but is much thicker below the continents.

Explaining What is magma, and where does it originate?

Shaping the Earth's Surface

Main Idea **Forces acting both inside and outside the Earth work to change the appearance of the Earth's surface.**

Geography and You Have you been in an earthquake? Or, do you know anyone who has? Read on to discover what causes earthquakes.

The Earth's crust is not a fixed layer. It changes over time as new landforms are created and existing ones change forms. For hundreds of millions of years, the Earth's surface has been in constant motion, slowly transforming. Old mountains are worn down, while new mountains grow taller. Even the continents move.

Earthquake in Japan

NATIONAL GEOGRAPHIC

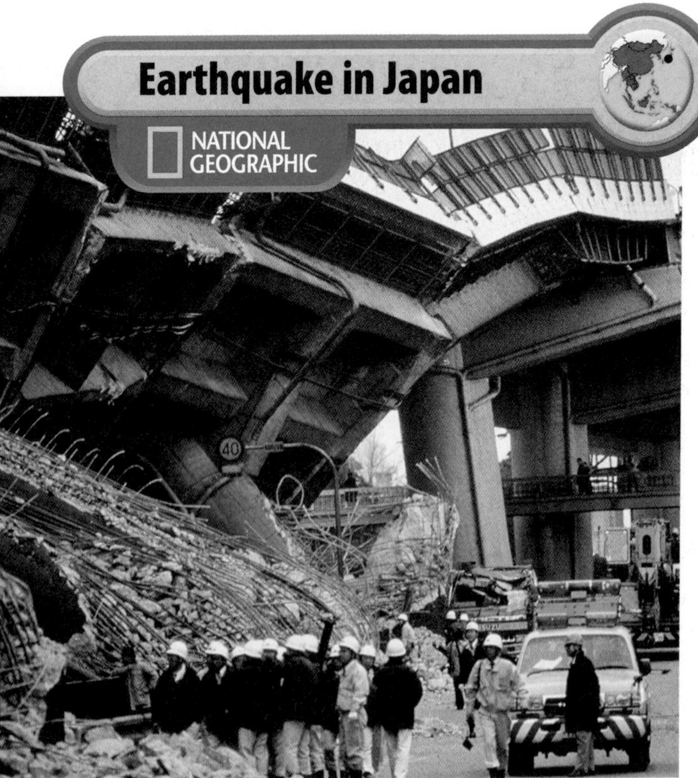

▲ City officials look over damage to an expressway that fell on one side during the 1995 earthquake in Kobe, Japan. ***Location*** **Where in the world are earthquakes common?**

Plate Movements

The theory of **plate tectonics** explains how the continents were formed and why they move. As **Figure 2** shows, each continent sits on one or more large bases called plates. As these plates move, the continents on top of them move. This movement is called continental drift.

The rate of movement varies from just under 1 inch (2.3 cm) to 7 inches (17 cm) per year. This movement is too slow for people to notice, but over millions of years, it can have dramatic effects.

Look at a map of the world. If you think of the eastern coast of South America as a giant puzzle piece, you will see that it seems to fit into the western coast of Africa. This is because these two continents were once joined together in a gigantic landmass that scientists call Pangaea. About 200 million years ago, however, the continents began to break and move apart because of tectonic activity.

When Plates Meet

The movements of Earth's plates have actually shaped the surface of the Earth. Sometimes the plates pull away from each other. Plates usually pull apart in ocean areas, but this kind of plate activity also occurs in land areas, such as Iceland and East Africa.

Plates can also collide. When two continental plates collide, they push against each other with tremendous force. This causes the land along the line where the plates meet to rise and form mountains. The Himalaya mountain ranges of Asia, the highest on Earth, were formed from such a collision.

Collisions of continental and oceanic plates produce a different result. The thinner ocean plate slides underneath the thicker continental plate. The downward

Figure 2 **Tectonic Plate Boundaries**

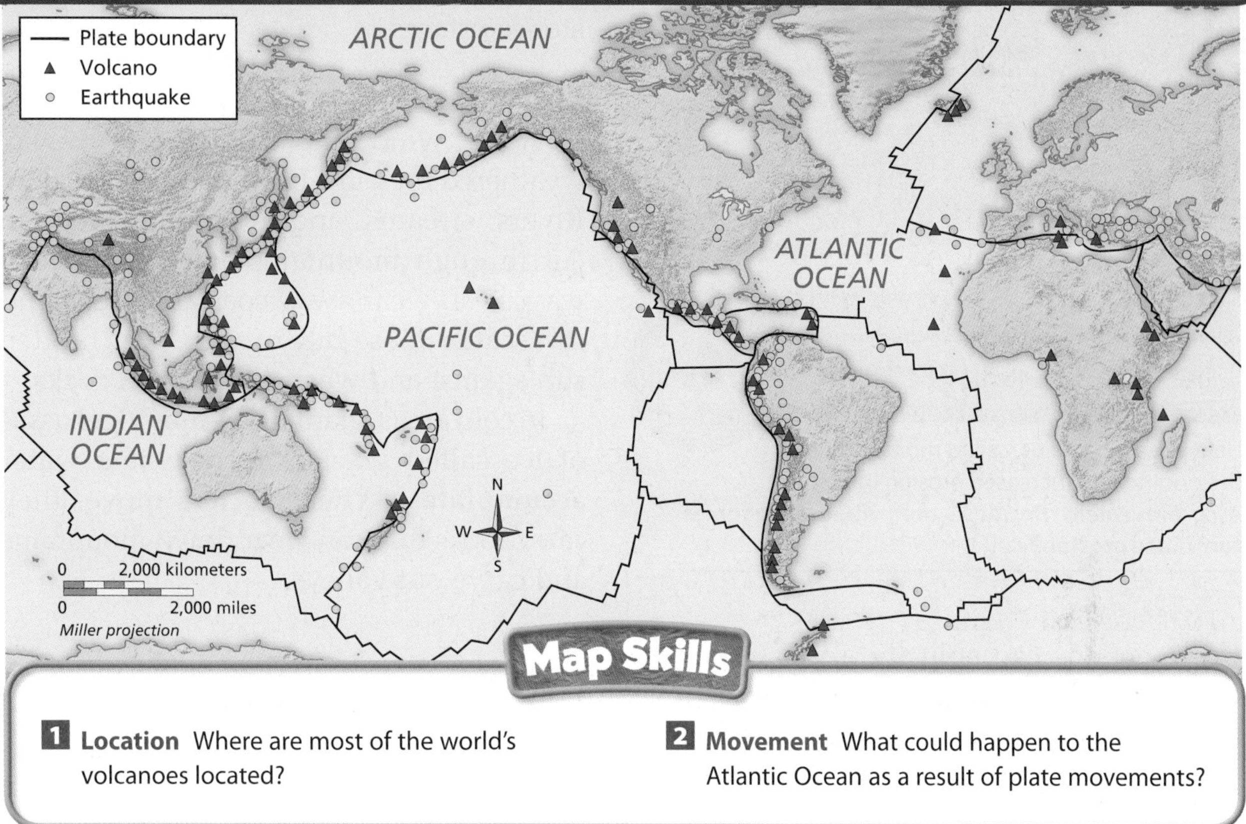

Map Skills

1 Location Where are most of the world's volcanoes located?

2 Movement What could happen to the Atlantic Ocean as a result of plate movements?

force of the lower plate causes magma to build up. Then the magma erupts and slowly hardens, forming volcanic mountains. This is how the Andes of South America were created.

Earthquakes are sudden and violent movements of the Earth's crust. They are common in areas where the collision of ocean and continental plates makes the Earth's crust unstable. For example, so many earthquakes and volcanoes occur around the edge of the Pacific Ocean that people call this region the Ring of Fire.

Sometimes two plates do not meet head-on but move alongside each other. This movement makes cracks in the Earth's crust called **faults.** Movements along faults do not take place **constantly,** but occur in sudden bursts that cause earthquakes. One of the most well-known faults in the United States is California's San Andreas Fault. A number of very destructive earthquakes have occurred in the region, and the threat of more still exists.

Weathering

The movement of tectonic plates causes volcanoes and earthquakes to change the Earth's landforms. Once created, however, these landforms will continue to change because of other forces that work on the Earth's surface.

One of these forces is called weathering. **Weathering** is when water and ice, chemicals, and even plants break rocks apart into smaller pieces. For example, water can run into cracks of rocks, freeze, and then expand.

Student Web Activity Visit glencoe.com and complete the Chapter 2 Web Activity about plate tectonics.

Erosion in Bangladesh

Heavy seasonal rains, called monsoons, lead to flooding and increased erosion in South Asia. ***Movement*** **Besides water, what other forces can cause erosion?**

These actions can split the rock. Chemicals, too, cause weathering when acids in air pollution mix with rain and fall back to Earth. The chemicals eat away rock and stone surfaces.

Erosion

Water, wind, and ice can move away weathered rock in a process called **erosion.** Rivers, streams, and even rainwater can cut through mountains and hills. Ocean waves can wear away coastal rocks. Wind can scatter loose bits of rock, which often rub against and wear down larger rocks.

In cold areas, giant, slow-moving masses of ice called glaciers form where water **accumulates.** When glaciers move, they carry rocks that can wear down mountains and carve out valleys.

Reading Check **Synthesizing** Why are earthquakes common where plates meet?

Section Review

Social Studies ONLINE
Study Central™ To review this section, go to glencoe.com.

Vocabulary

1. **Illustrate** the meaning of *core, mantle, magma, crust, continent, plate tectonics, earthquake, fault, weathering,* and *erosion* by drawing and labeling one or more diagrams.

Main Ideas

2. **Summarizing** Which layers of the Earth are solid? Which layers are liquid?
3. **Describing** Use a chart like the one below to list and describe the different results when plates meet.

Type of Plate Meeting	Results
1.	1.
2.	2.
3.	3.

Critical Thinking

4. **Drawing Conclusions** Where do you think an earthquake is more likely to occur—along North America's Pacific coast or along North America's Atlantic coast? Why?
5. **BIG Idea** How was the formation of the Himalaya and the Andes similar and different?
6. **Challenge** How do the shapes of South America and Africa support the theory of plate tectonics? Find another example of land areas that once might have been joined together but separated as plates moved apart.

Writing About Geography

7. **Using Your FOLDABLES** Use your Foldable to write a paragraph explaining how forces both beneath and on the surface help shape the surface of the Earth.

Guide to Reading

BIG Idea

Geographic factors influence where people settle.

Content Vocabulary

- continental shelf *(p. 50)*
- trench *(p. 50)*
- groundwater *(p. 52)*
- aquifer *(p. 52)*
- water cycle *(p. 53)*
- evaporation *(p. 53)*
- condensation *(p. 54)*
- precipitation *(p. 54)*
- collection *(p. 54)*

Academic Vocabulary

- occur *(p. 50)*
- define *(p. 50)*
- availability *(p. 52)*

Reading Strategy

Identifying Use a diagram like the one below to identify the various bodies of water that can be found on the Earth's surface.

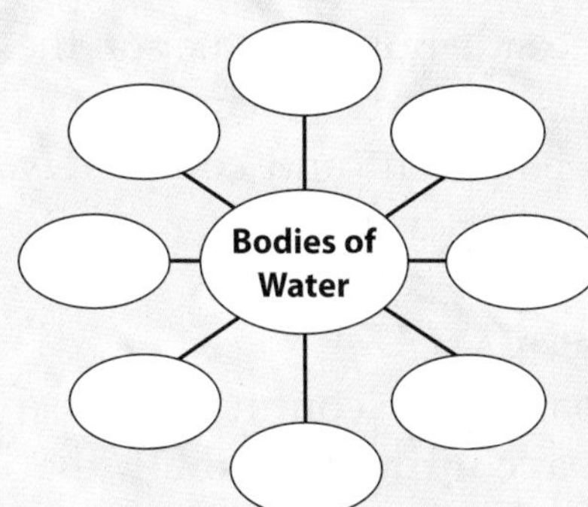

Landforms and Water Resources

Picture This This fisherman in Indonesia uses a hand dredge to catch fish. He lowers the dredge into the water and drags it along the bottom of the shallow, sandy ocean floor. There it scoops up fish, scallops, and oysters. Read this section to learn how landforms and water influence human activities.

▲ Dredging for seafood, Indonesia

Types of Landforms

Main Idea **Earth has a variety of landforms, and many of the landforms can be found both on the continents and the ocean floors.**

Geography and You Do you know that there are mountains underwater? If the area where you live was underwater, what would it look like?

The Earth has a great variety of landforms—from mountains that soar miles high to lowlands that barely peek above the sea. These landforms appear not only on continents but also under the oceans.

On Land

Mountains are huge towers of rock and are the highest landforms. Some mountains may be only a few thousand feet high. Others can soar higher than 20,000 feet (6,096 m). The world's highest mountain is Mount Everest in South Asia's Himalaya ranges. It rises more than 29,028 feet (8,848 m), nearly five and a half miles high!

Hills are lower and more rounded than mountains. Between mountains and hills lie valleys. A valley is a long stretch of land that is lower than the land on either side. Flatlands **occur** in one of two forms, depending on their height above sea level. Plains are flat lowlands, typically found along coasts and lowland river valleys. Plateaus are flatlands at higher elevations.

Geographers **define** some landforms by their relationship to other landforms or to bodies of water. Look back at the geographic dictionary in the Reference Atlas to see examples of the following landforms.

An isthmus is a narrow strip of land that connects two larger landmasses and has water on two sides. An example is Central America, which connects North and South America. A peninsula, such as Florida, is a piece of land that is connected to a larger landmass on one side but has water on the other three sides. A body of land that is smaller than a continent and completely surrounded by water is an island.

▲ The Karakoram Range in South Asia is home to more than 60 peaks above 23,000 feet (7,000 m). ***Location*** **Where is Mount Everest, the world's tallest peak, located?**

Under the Oceans

Off each coast of a continent lies a plateau called a **continental shelf** that stretches for several miles underwater. At the edge of the shelf, the land drops down sharply to the ocean floor.

On the ocean floor, tall mountains thousands of miles wide line the edges of ocean plates that are pulling apart. Tectonic activity also makes deep cuts in the ocean floor called **trenches.** The Mariana Trench in the western Pacific Ocean is the deepest. It plunges 36,198 feet (11,033 m) below sea level.

Roughing It

By Mark Twain

We jumped into the **stage,** the driver cracked his whip, and we bowled away. . . . It was a superb summer morning, and all the landscape was brilliant with sunshine. There was a freshness and breeziness, too, and an **exhilarating** sense of **emancipation** from all sorts of cares and responsibilities, that almost made us feel that the years we had spent in the close, hot city, toiling and slaving, had been wasted and thrown away. We were spinning along through Kansas, and in the course of an hour and a half we were fairly abroad on the great Plains. Just here the land was rolling—a grand sweep of regular elevations and depressions as far as the eye could reach—like the stately heave and swell of the **ocean's bosom** after a storm. And everywhere were cornfields, **accenting** with squares of deeper green this limitless expanse of grassy land. But presently this sea upon dry ground was to lose its "rolling" character and stretch away for seven hundred miles as level as a floor! . . .

There is not a tree of any kind in the deserts, for hundreds of miles—there is no vegetation at all . . . except the sage-brush and its cousin the "greasewood," which is so much like the sage-brush that the difference amounts to little. Camp-fires and hot suppers in the deserts would be impossible but for the friendly sage-brush.

From *Roughing It,* Mark Twain. New York: Harper & Brothers Publishers, 1899.

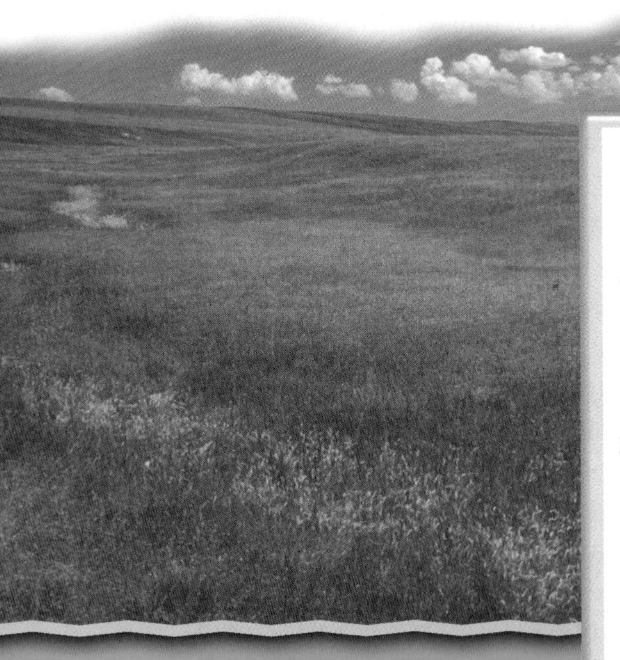

Analyzing Literature

1. **Making Inferences** What landform is Twain describing? What details make that clear?
2. **Read to Write** Think about the landforms in the area where you live. Write a letter describing what it would be like to travel over those landforms by foot or on a bicycle.

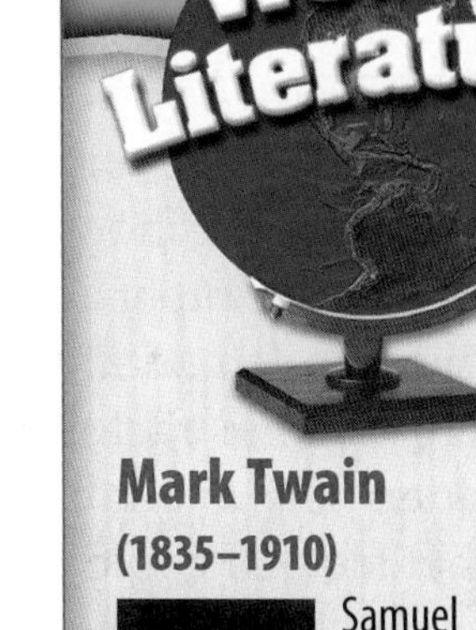

Mark Twain
(1835–1910)

Samuel Langhorne Clemens, who used the pen name "Mark Twain," was born in a Missouri river town along the banks of the Mississippi River. He held many jobs, including working as the pilot of a riverboat, before becoming a writer and humorist. He was one of the most popular American authors of the late 1800s.

Background Information

In *Roughing It,* Twain describes his experiences living and traveling in Nevada, California, and Hawaii in the 1860s. In this excerpt, he describes his trip from Missouri to Nevada. Twain traveled by stagecoach, a horse-drawn vehicle for carrying passengers.

Reader's Dictionary

stage: horse-drawn stagecoach

exhilarating: exciting

emancipation: freedom

ocean's bosom: ocean's surface

accenting: standing out

Humans and Landforms

Humans settle on all types of landforms. People choose a place to live based on a number of factors. Climate—the average temperature and rainfall of a region—is one factor that people must consider. The **availability** of resources is another factor. People settle where they can get freshwater and where they can grow food, catch fish, or raise animals.

Explaining What forces form ocean trenches?

Black River Delta, United States

NATIONAL GEOGRAPHIC

▲ The Black River forms a delta as it flows into the Mississippi River in Wisconsin. River deltas are often rich in wildlife, including birds and mammals. ***Place*** **How are deltas formed?**

The Water Planet

Main Idea Water covers much of the planet, but only some of this water is usable.

Geography and You Have you ever watched steam rise from a boiling pot of water? Read to learn how water changes from a solid, to a liquid, to a gas on Earth.

Earth is sometimes called the "water planet" because so much of it—about 70 percent of the surface—is covered with water. Water exists in many different forms. Streams, rivers, lakes, seas, and oceans contain water in liquid form. The atmosphere holds water vapor, or water in the form of gas. Glaciers and ice sheets are masses of water that have been frozen solid.

Salt Water

All of the oceans on Earth are part of a huge, continuous body of salt water. Almost 97 percent of the planet's water is salt water. Oceans have smaller arms or areas that are called seas, bays, or gulfs. These larger bodies of salt water can be linked to oceans by the more narrow bodies called straits or channels.

Freshwater

Only 3 percent of the water on Earth is freshwater. Much of this freshwater is frozen in ice that covers polar regions and parts of mountains. Some is **groundwater,** which filters through the soil into the ground. Groundwater often gathers in **aquifers** (A·kwuh·fuhrz). These are underground layers of rock through which water flows. People can pump the freshwater from aquifers. Only a tiny amount of all

Figure 3 **The Water Cycle**

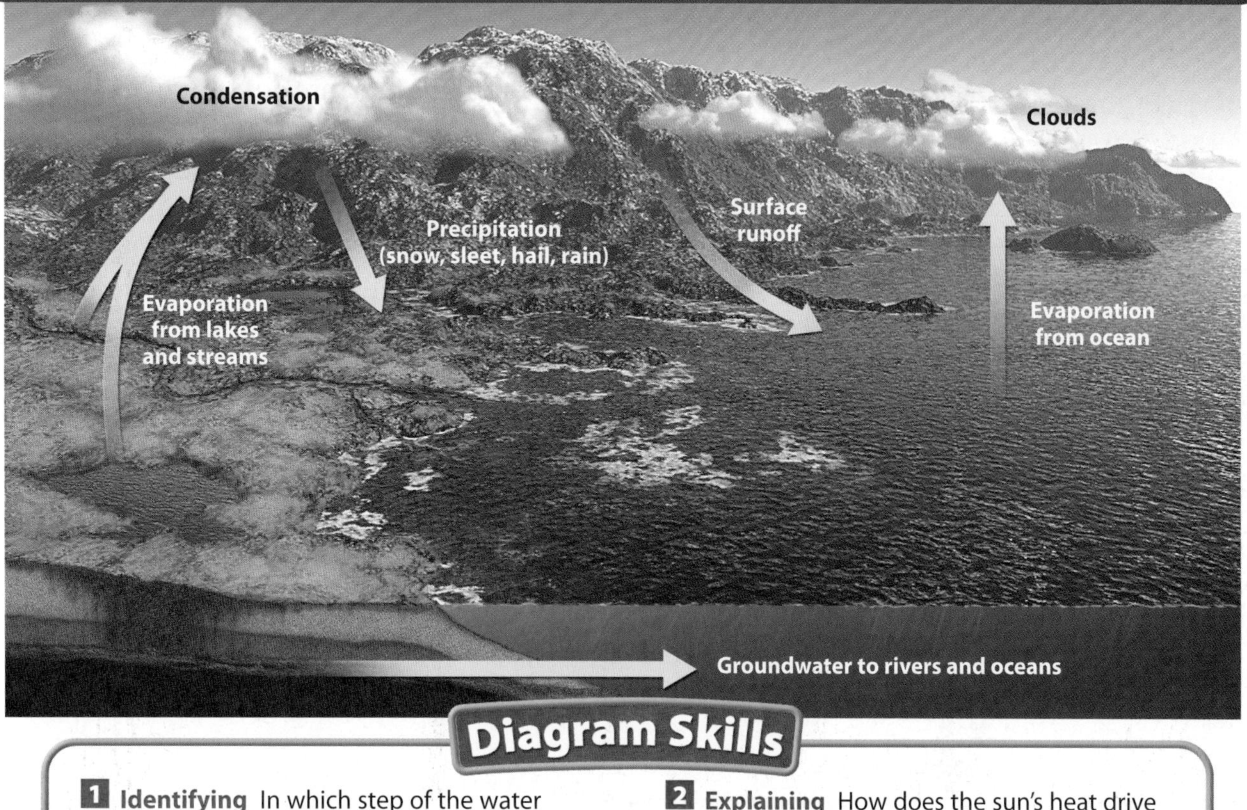

Diagram Skills

1. **Identifying** In which step of the water cycle does water vapor form clouds?
2. **Explaining** How does the sun's heat drive the water cycle?

the water in the world is found in lakes and rivers. This water is often not safe to drink until it has been purified.

Large inland bodies of water are called lakes. Most lakes are freshwater lakes. Long, flowing bodies of water are called rivers. They begin at a source and end at a mouth. The mouth is the place where a river empties into another body of water, such as an ocean or a lake.

The largest rivers often have many tributaries, which are separate streams or rivers that feed into them. Many rivers form deltas at their mouths. A delta is an area where a river breaks into many different streams flowing toward the sea. Rivers often carry rich soil to their deltas and deposit it, building up the land.

The Water Cycle

The total amount of water on Earth does not change. It does not stay in one place, either. Instead the water moves constantly. In a process called the **water cycle,** the water goes from the oceans, to the air, to the ground, and finally back to the oceans.

Look at **Figure 3** to see how the water cycle works. The sun's heat drives the water cycle because it evaporates the water on the Earth's surface. This **evaporation** changes water from liquid to a gas, called water vapor. Water vapor rises from the Earth's oceans and other bodies of water, and then circulates in the atmosphere. The air's temperature determines how much water the air holds. Warm air holds more water vapor than cool air.

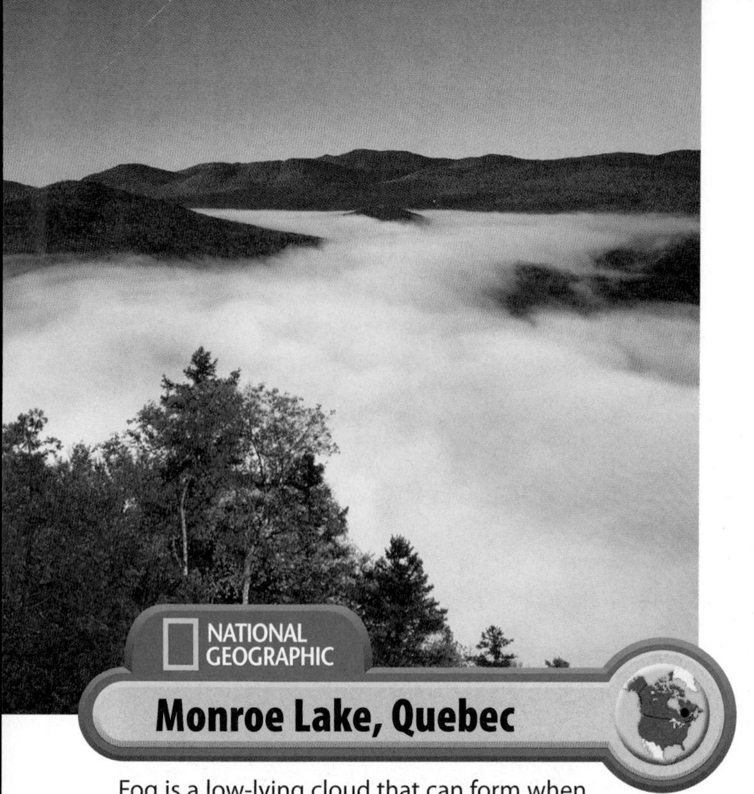

NATIONAL GEOGRAPHIC

Monroe Lake, Quebec

Fog is a low-lying cloud that can form when moist air blows over a cool surface. ***Place*** **How does air temperature affect water vapor in the air?**

When the air temperature drops low enough, **condensation** takes place. In this process, water changes from gas back to a liquid. Tiny droplets of water form in the air, although they are suspended in clouds.

When conditions in the atmosphere are right, these water droplets fall to the ground as some form of **precipitation.** This can be rain, snow, sleet, or hail. The form of precipitation depends on the temperature of the surrounding air.

Completing the cycle is the process called **collection.** The water collects on the ground and in rivers, lakes, and oceans. There it evaporates to begin the cycle again.

Reading Check **Making Inferences** Why is very little of the Earth's freshwater usable?

Section 2 Review

Social Studies ONLINE
Study Central™ To review this section, go to glencoe.com.

Vocabulary

1. **Explain** the meaning of the following terms by using each one in a sentence.
   - **a.** continental shelf
   - **b.** trench
   - **c.** groundwater
   - **d.** aquifer
   - **e.** water cycle
   - **f.** evaporation
   - **g.** condensation
   - **h.** precipitation
   - **i.** collection

Main Ideas

2. **Contrasting** How do an isthmus, a peninsula, and an island differ?
3. **Summarizing** Use a diagram like the one below to summarize the water cycle.

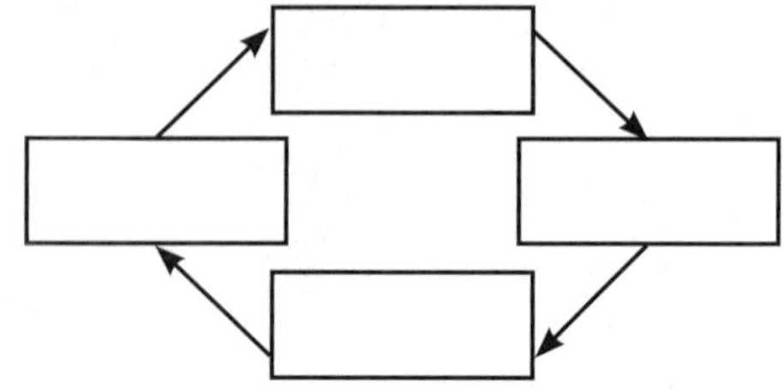

Critical Thinking

4. **Comparing and Contrasting** How are plains and plateaus similar and different?
5. **BIG Idea** Describe several factors that people consider when choosing a place to settle.
6. **Challenge** Which landforms do you think attracted people to settle in the area where you live? Which landforms, if any, may have kept people away?

Writing About Geography

7. **Expository Writing** Write a paragraph describing the major landforms found in the state where you live.

Guide to Reading

BIG Idea

Geographers organize the Earth into regions that share common characteristics.

Content Vocabulary

- weather *(p. 56)*
- climate *(p. 56)*
- prevailing wind *(p. 57)*
- current *(p. 57)*
- El Niño *(p. 58)*
- La Niña *(p. 58)*
- local wind *(p. 59)*
- rain shadow *(p. 59)*
- climate zone *(p. 59)*
- biome *(p. 60)*
- urban climate *(p. 61)*

Academic Vocabulary

- distribute *(p. 56)*
- alter *(p. 57)*

Reading Strategy

Identifying Central Issues Use a diagram like the one below to identify the effects of both El Niño and La Niña.

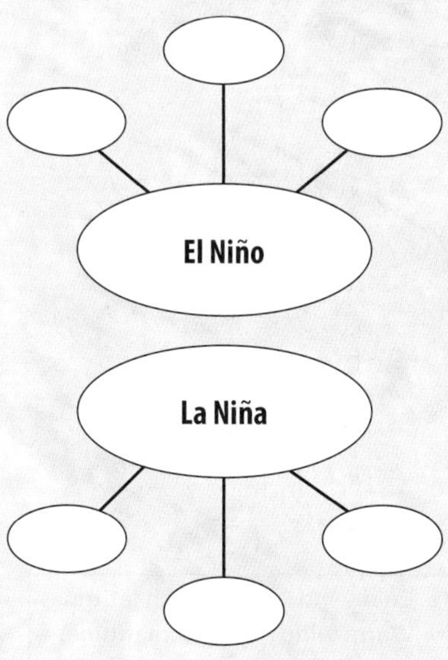

Climate Regions

 Section Audio  Spotlight Video

Picture This Residents rush to escape the swirling winds and pelting rain during the annual typhoon season in China. Typhoons are hurricanes that can topple buildings, snap power lines, and uproot trees. These violent thunderstorms draw their power from warm ocean waters and are common in the Tropics of southeast China. Read this section to learn about the variety of climates that are found on Earth.

Fleeing Typhoon Haitang, July 2005

Effects on Climate

Main Idea **Sun, wind, and water influence Earth's climate.**

Geography and You What is the weather today in your area? Is it typical of the particular season you are in, or is it unusual? Read to find out about the difference between weather and climate.

When you turn on the television to find out the day's high and low temperatures, you are checking the local weather. **Weather** refers to the changes in temperature, wind direction and speed, and air moisture that take place over a short period of time. When geographers look at the usual, predictable patterns of weather in an area over many years, they are studying **climate.**

The Sun

Earth's climate is linked directly to the sun. As you recall from Chapter 1, the Earth does not heat evenly. The Tropics receive more of the sun's heat energy and the Poles receive less. The movement of air and water over the Earth helps to **distribute** the sun's heat more evenly around the globe.

Maps In MOtion See StudentWorks™ Plus or glencoe.com.

Figure 4 **Prevailing Wind Patterns**

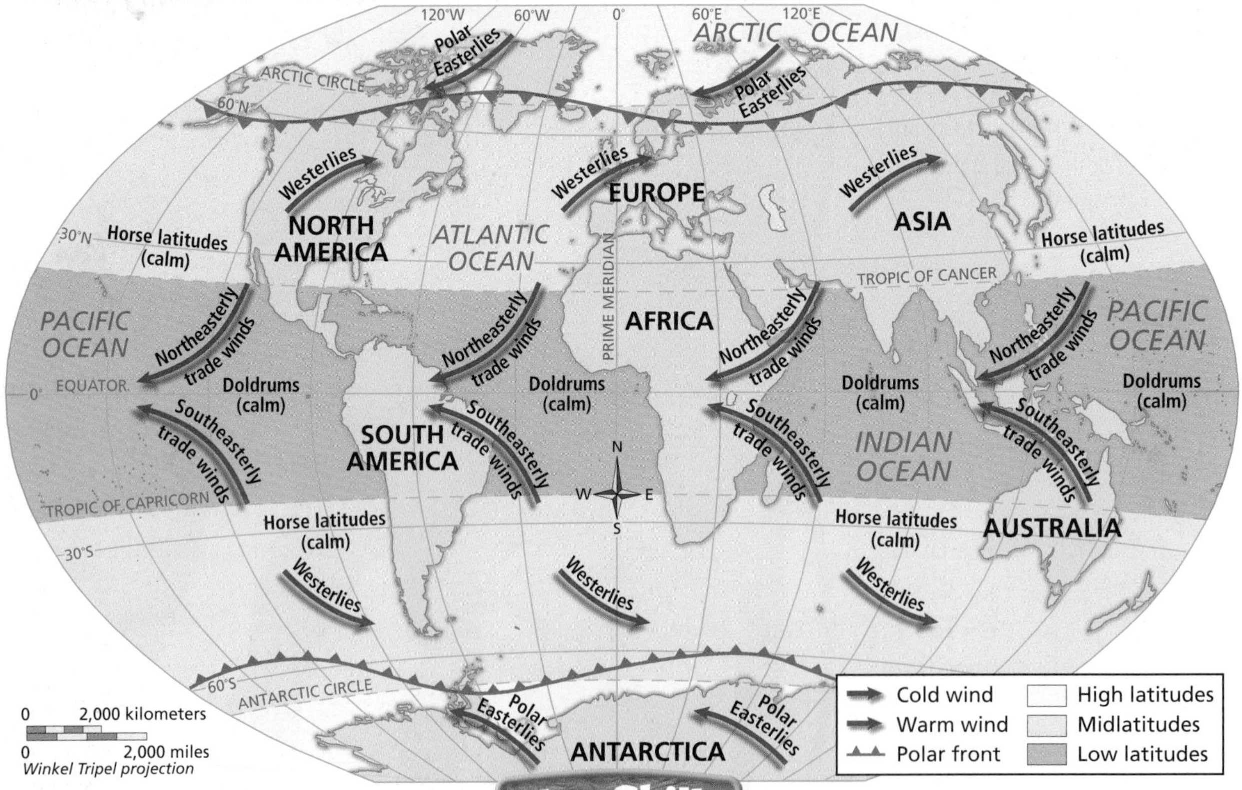

0 2,000 kilometers
0 2,000 miles
Winkel Tripel projection

Cold wind
Warm wind
Polar front
High latitudes
Midlatitudes
Low latitudes

Map Skills

1 Movement In which general direction does the wind blow over North America and Europe?

2 Regions Which two areas of the world experience calm winds?

Winds

Air in the Tropics, which is warmed by the sun, moves north and south toward the Poles of the Earth. Colder air from the Poles moves toward the Equator. These movements of air are winds. Major wind systems follow patterns that are similar over time. These patterns, shown in **Figure 4,** are called **prevailing winds.**

Because the planet rotates, winds curve across Earth's surface. The winds that blow from east to west between the Tropics and the Equator are called the trade winds. Long ago sailing ships used these winds to carry out trade. The westerlies, which blow over North America, move from west to east in the area between the Tropics and about 60° north latitude.

Storms

When moist, warm air rises suddenly and meets dry, cold air, major storms can develop. In the summer, these storms can include thunder and lightning, heavy rain, and, sometimes, tornadoes. Tornadoes are violent, funnel-shaped windstorms with wind speeds up to 450 miles (724 km) per hour. In the winter, storms can become blizzards that bring much snow.

Other types of destructive storms are hurricanes and typhoons. Hurricanes occur in the western Atlantic and eastern Pacific Oceans. Typhoons occur in the western Pacific Ocean. These storms arise in the warm ocean waters of the Tropics and can reach great size and power. Some are as much as 300 miles (483 km) across and create strong winds and heavy rains.

Ocean Currents

The steadily flowing streams of water in the world's seas are called **currents.** Like winds, they follow patterns, which are shown in **Figure 5,** on the next page.

NATIONAL GEOGRAPHIC

Effects of El Niño

South America can experience dramatic changes in weather due to El Niño and La Niña. Forest fires, like this one in Brazil, occur during periods of drought. ***Place*** **How do El Niño and La Niña differ?**

Currents that carry warm water to higher latitudes can affect the climates in those latitudes. For example, the North Atlantic Current carries warm water from the Tropics to western Europe. Winds blowing over the warm water bring warmth and moisture to western Europe, which enjoys an unexpectedly mild climate.

El Niño and La Niña

Every few years, changes in normal wind and water patterns in the Pacific Ocean cause unusual weather in some places. In one of these events, weakened winds allow warmer waters to reach South America's coast. This change **alters** weather there and beyond.

These conditions are called **El Niño,** Spanish for "the boy."

In an El Niño, very heavy rains fall on western South America, causing floods. Meanwhile, little rain falls on Australia, southern Asia, and Africa. Also, North America may see severe storms.

In some years the opposite occurs, producing conditions called **La Niña,** Spanish for "the girl." La Niña causes unusually cool waters and low rainfall in the eastern Pacific. In the western Pacific, rains are heavy and typhoons can occur.

Explaining How are winds formed?

Landforms and Climate

Main Idea **Landforms, especially mountains, can affect winds, temperature, and rainfall.**

Geography and You Have you ever felt a cooling sea breeze on a hot summer's day? Read on to learn how the sea can affect climate.

Sun, wind, and water affect climate, but the shape of the land has an effect on climate as well. The distance between landforms as well as their nearness to water influence climate.

NATIONAL GEOGRAPHIC

Figure 5 **World Ocean Currents**

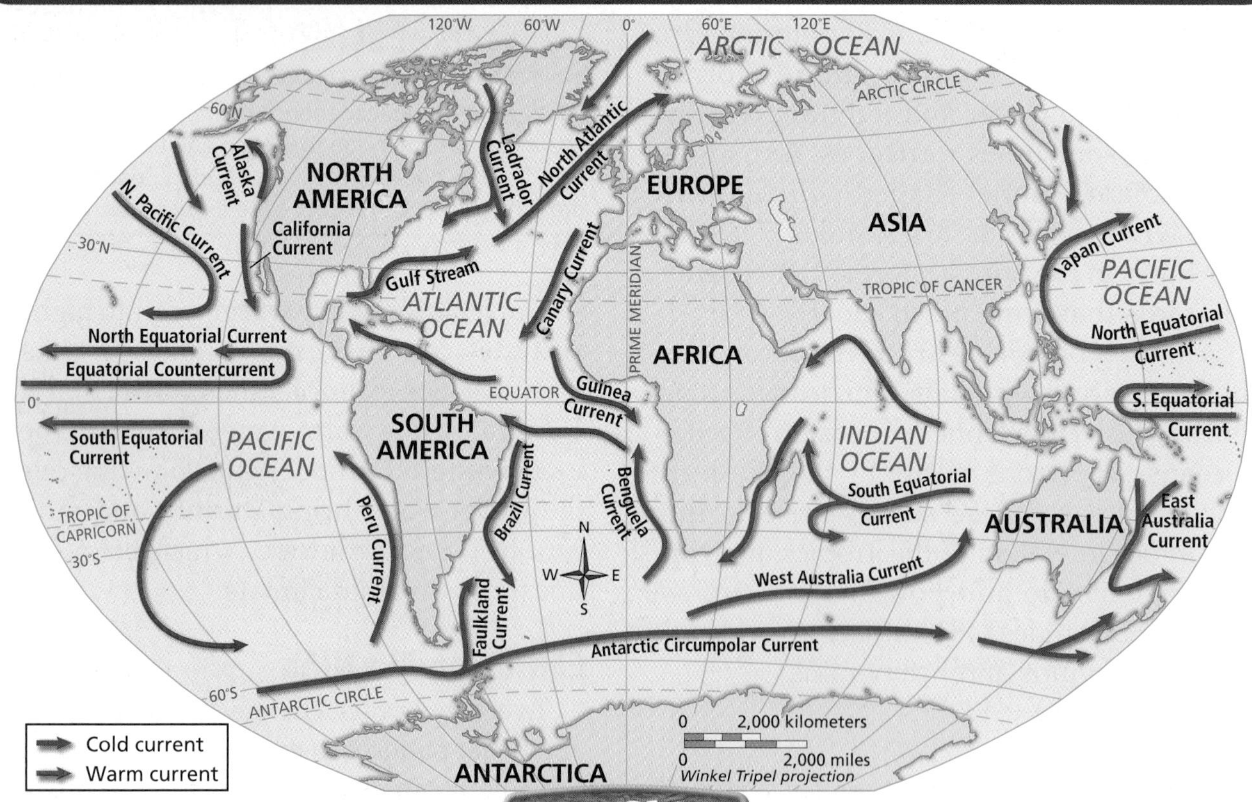

Map Skills

1. **Movement** What kind of climate is the North Atlantic Current likely to bring to Europe?
2. **Regions** Which area generally has warmer waters, western South America or eastern South America?

Landforms and Local Winds

Some landforms cause **local winds,** or wind patterns that are typical only in a small area. Some local winds occur because land warms and cools more quickly than water does. As a result, cool sea breezes keep coastal areas cool during the day. After the sun sets, the opposite occurs. The air over the land cools more quickly than the air over the water. At night, then, a cool breeze blows from the land out to sea.

Local winds also occur near tall mountains. When the air along a mountain slope is warmer than the air in the valley below, it rises and a cool valley breeze moves up the mountain.

Mountains, Temperature, and Rainfall

The slopes of a mountain facing the sun can heat more quickly than nearby land. Higher up in the mountains, however, the air is thin and cannot hold the heat very well. As a result, mountain peaks are cold. This explains why some mountains in the Tropics are covered with snow.

Mountains have an effect on rainfall called a **rain shadow** that blocks rain from reaching interior regions. As warm, moist ocean air moves up the mountain slopes, it cools and releases its moisture. As a result, the side of mountains facing the wind, called the windward side, receives large amounts of rainfall.

As the air passes over the mountain peaks to the other side, called the leeward side, it becomes cool and dry. As a result, the land on the leeward side of the mountains is often very dry. Deserts can develop on the leeward side of mountain ranges.

Determining Cause and Effect How do mountains cause the rain shadow effect?

Climate Zones

Main Idea The effects of wind, water, latitude, and landforms combine to create different climate zones.

Geography and You Suppose you visited two islands that were thousands of miles apart. Read to find out how similar their climates might be.

As you have read, the effects of wind, water, latitude, and landforms combine to shape the climate of an area. Scientists have found that many parts of the world, even though they are very distant from one another, have similar climates. Southern California, for instance, has a warm, dry climate similar to that around the Mediterranean Sea in Europe. These areas have the same **climate zone,** or similar patterns of temperature and precipitation. These regions would also have similar vegetation.

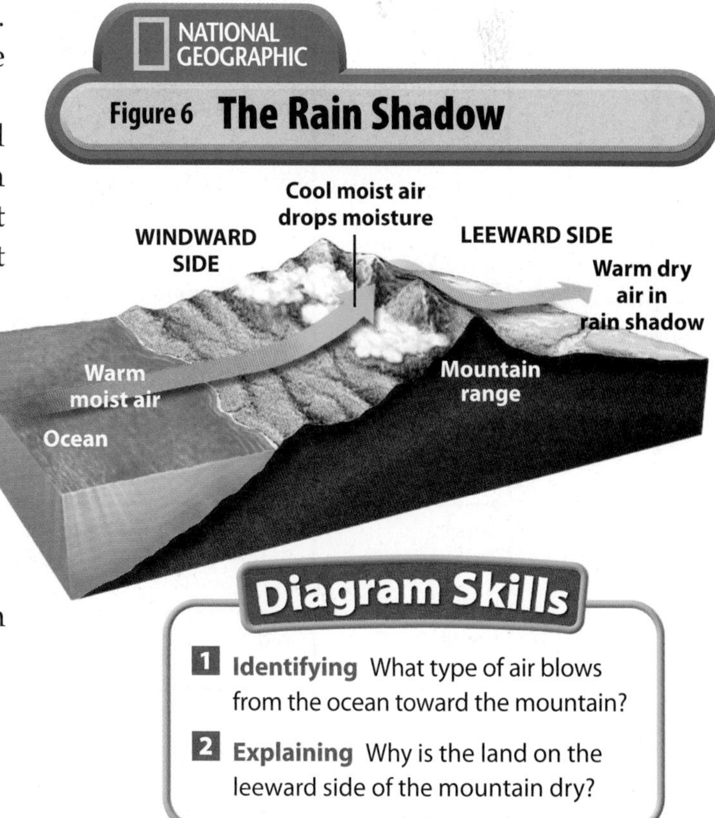

Figure 6 The Rain Shadow

Diagram Skills

1. **Identifying** What type of air blows from the ocean toward the mountain?
2. **Explaining** Why is the land on the leeward side of the mountain dry?

Climate zones include **biomes,** or areas such as rain forest, desert, grassland, and tundra in which particular kinds of plants and animals have adapted to particular climates.

Major Climates

Scientists have identified five major climate zones, which are described in the chart below. Four of these zones have several subcategories. For example, the dry climate zone is subdivided into steppe and desert subcategories. These generally dry climates differ slightly in rainfall and temperature. Locations in the highland zone show great variation. In these areas, altitude, the position of a place toward or away from the sun, and other factors can make large differences in climate even though two locations may be near each other.

World Climate Zones

Category	Subcategory	Characteristics	Vegetation	Example
Tropical	Tropical rain forest	Warm temperatures; heavy rainfall throughout year	Dense rain forests	Amazon basin (South America); Congo basin (Africa)
	Tropical savanna	Warm temperatures throughout year; dry winter	Grasslands dotted by scattered trees	Southern half of Brazil; eastern Africa
Dry	Steppe	Temperatures can be warm or mild; rainfall low and unreliable	Grasses, shrubs	Western Great Plains (United States); Sahel region south of the Sahara (Africa)
	Desert	Temperatures can be warm or mild; rainfall very low and very unreliable	Drought-resistant shrubs and bushes	Sonoran Desert (southwestern United States, Mexico); Sahara (Africa)
Midlatitude	Marine west coast	Cool summers, mild winters; ample rainfall	Deciduous or evergreen forests	Northwestern United States; northwestern Europe
	Mediterranean	Warm, dry summers; mild, wet winters	Shrubs, low trees, drought-resistant plants	Southern California; Mediterranean region (Europe)
	Humid subtropical	Hot, wet summers; mild, wet winters	Mixed forests	Southeastern United States; eastern China
	Humid continental	Hot, wet summers; cold, somewhat wet winters	Deciduous forests	Northeastern United States; eastern Europe; western Russia
High Latitude	Subarctic	Short, mild summers; long, cold winters; light precipitation	Coniferous forests	Most of Alaska, Canada; western Russia
	Tundra	Short, cool summers; long, cold winters; precipitation varies	Low-lying grasses, mosses, shrubs	Extreme north of North America; Europe
	Ice cap	Cold all year long	None to very little	Greenland; Antarctica
Highland		Varies depending on local conditions	Changes with altitude	Northern Rocky Mountains (United States); the Himalaya (Asia)

Urban Climates

Large cities show significant climate differences from surrounding areas in their climate zone. These **urban climates** are marked by higher temperatures and other differences. Paved streets and stone buildings soak up and then release more of the sun's heat energy than areas covered by plants. This absorption leads to higher temperatures—as much as 10° to 20°F (6° to 11°C) higher—than in the nearby countryside. These different heat patterns cause winds to blow into cities from several directions instead of the prevailing direction experienced in rural areas. Some scientists believe cities also have more precipitation than rural areas.

✓ Reading Check **Drawing Conclusions** How do large cities affect climate?

NATIONAL GEOGRAPHIC

Shanghai, China: City Heat

City temperatures can soar in the summer. Buildings and pavement absorb the sun's heat, raising temperatures within the city. ***Location*** **How does urban heat affect winds in the urban area?**

Section 3 Review

Social Studies ONLINE
Study Central™ To review this section, go to glencoe.com.

Vocabulary

1. **Explain** the meaning of the following terms by writing three paragraphs that include all of the terms: *weather, climate, prevailing wind, current, El Niño, La Niña, local wind, rain shadow, climate zone, biome,* and *urban climate.*

Main Ideas

2. **Explaining** How do wind and water affect the Earth's climates?
3. **Reviewing** Describe two types of local winds and why they form.
4. **Identifying** Use a diagram like the one below to identify the main characteristics of the climate zone in which you live.

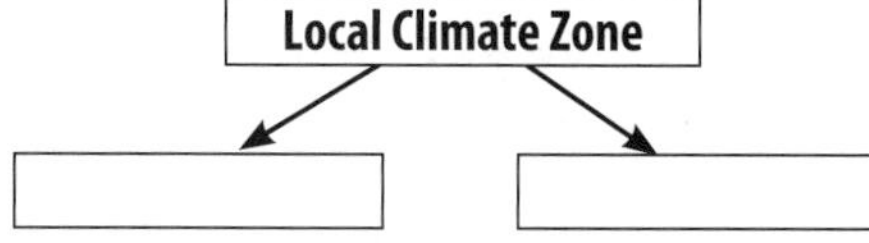

Critical Thinking

5. **BIG Idea** Choose two climate zones, and compare and contrast their characteristics.
6. **Challenge** How might El Niño affect weather conditions in the central United States?

Writing About Geography

7. **Expository Writing** Choose a place in the world you would like to visit because of its climate. Write a paragraph describing the climate of that area.

Invaders From Another Land

When plants and animals move from their natural environment to one in which they do not belong, they can cause great harm.

How They Arrive In the 1800s, a settler in Australia released about a dozen European rabbits onto his land. He brought the rabbits to Australia in order to hunt other animals. Over the years, the number of rabbits grew beyond control. They eventually damaged plant and animal life throughout Australia.

The rabbits brought to Australia are an example of an invasive species. These are plants and animals introduced to new areas where they increase rapidly and crowd out local plants and animal life.

▲ **European rabbit**

Invasive species are a threat in other parts of the world as well as in Australia. Zebra mussels, for example, came to the United States during the 1980s attached to the bottoms of ships. They fell off the ships and spread throughout lakes, rivers, and streams. These mussels have blocked city water treatment systems and destroyed some fish populations.

▲ **Zebra mussels clustered on wood**

Finding a Solution The costs of invasive species can be great. Within a short time of arriving in a new location, invasive species can cause billions of dollars in damage. Costs to the environment are high too. Invasive species can cause the extinction of local animals and plants not used to them.

Experts believe that invasive species are becoming more common. Increasing world trade means more contact among the world's peoples and environments. This contact means more chances for species to move from one place to another. Some governments are working both to keep invasive species from arriving and to restore environments that have been harmed.

NATIONAL GEOGRAPHIC

The Spread of Zebra Mussels

Think About It

1. Why are some plants and animals called *invasive species*?
2. Why are invasive species becoming more common?

Guide to Reading

BIG Idea

All living things are dependent upon one another and their surroundings for survival.

Content Vocabulary

- smog *(p. 64)*
- acid rain *(p. 64)*
- greenhouse effect *(p. 64)*
- crop rotation *(p. 65)*
- deforestation *(p. 65)*
- conservation *(p. 66)*
- irrigation *(p. 66)*
- pesticide *(p. 66)*
- ecosystem *(p. 66)*
- biodiversity *(p. 66)*

Academic Vocabulary

- layer *(p. 64)*
- technique *(p. 65)*

Reading Strategy

Solving Problems Use a chart like the one below to identify environmental problems and what people are doing to solve them.

Problem	Solution
1.	1.
2.	2.
3.	3.

Human-Environment Interaction

 Section Audio  Spotlight Video

Picture This Imagine guiding hundreds of logs through rough waters in a tugboat. In Deception Pass State Park in Washington, boats move newly cut logs along the waters of the park to reach the highway. The logs are loaded on trucks and taken to lumberyards. Read this section to learn about the effects of human activities on the Earth.

Logs moving through Deception Pass State Park in Washington

The Atmosphere

Main Idea **Human activity can have a negative impact on the air.**

Geography and You Have you ever seen a blanket of dirty air hanging over a large city? Read to find out how human actions affect the atmosphere.

Throughout the world, people burn oil, coal, or gas to make electricity, to power factories, or to move cars. These actions often cause air pollution.

Air Pollution

Air pollution has serious effects on people and the planet. Some polluting chemicals combine with ozone, a form of oxygen, to create **smog.** This is a thick haze of smoke and chemicals. Thick smog above cities can lead to serious breathing problems.

Chemicals in air pollution can also combine with precipitation to form **acid rain.** Acid rain kills fish, eats away at the surfaces of buildings, and destroys trees and entire forests. Because the chemicals that form acid rain come from the burning of coal and oil, solving this problem has proved difficult.

Some human-made chemicals, particularly chlorofluorocarbons (CFCs), destroy the ozone **layer.** Ozone forms a shield high in the atmosphere against damaging rays from the sun that can cause skin cancer. Nations today are working to limit the release of CFCs.

The Greenhouse Effect

Like the glass in a greenhouse, gases in the atmosphere trap the sun's warmth. Without this **greenhouse effect,** the Earth would be too cold for most living things. **Figure 7** shows the greenhouse effect.

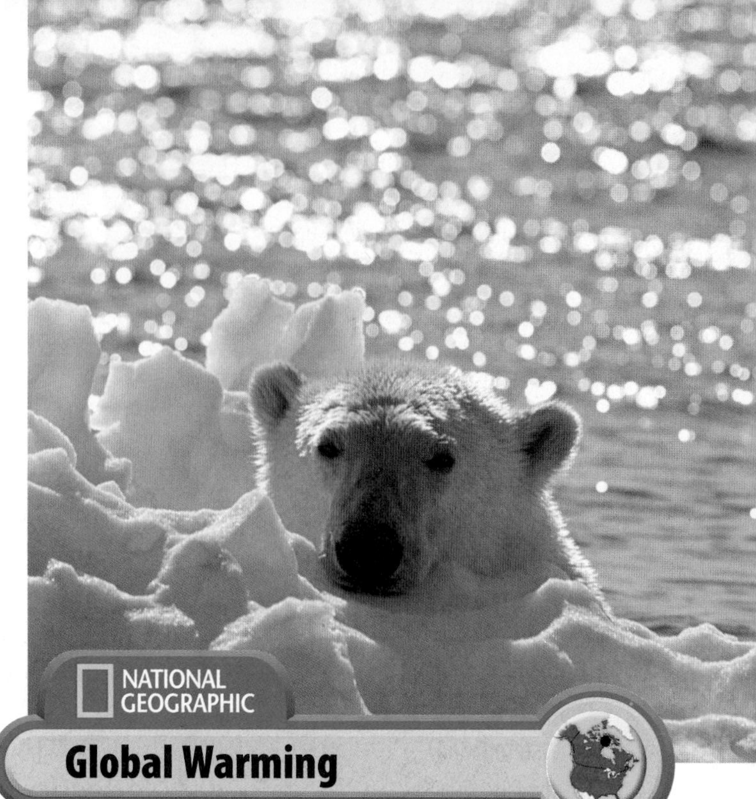

NATIONAL GEOGRAPHIC

Global Warming

Scientists are concerned that global warming might be harming wildlife, such as this polar bear. ***Human-Environment Interaction*** **What human activities might contribute to global warming?**

Some scientists, however, say that pollution is strengthening the greenhouse effect. They claim that the increased burning of coal, oil, and natural gas has released more gases into the atmosphere. These greenhouse gases have trapped more of the sun's heat near the Earth's surface, raising temperatures around the planet. Such warming could cause climate changes and melt polar ice. Ocean levels could rise and flood low-lying coastal areas.

The issue of global warming is debated. Critics argue that computer models showing global warming are unrealistic. Many nations, however, are addressing the problem. They are trying to use energy more efficiently, burn coal more cleanly, and adopt nonpolluting forms of energy such as wind and solar power.

Reading Check **Explaining** Why do some scientists debate the issue of global warming?

The Lithosphere

Main Idea **Some human activity damages our environment.**

Geography and You How might your community have looked 200 years ago? Read to discover how human actions have affected the land.

The lithosphere is another name for the Earth's crust. It includes all the land above and below the oceans. Human activities, such as farming, logging, and mining can have negative effects on the lithosphere.

Rich topsoil is a vital part of the lithosphere that, if not carefully managed, can be carried away by wind or water. Some farmers use contour plowing to limit the loss of topsoil. With this **technique,** farmers plow along the curves of the land rather than in straight lines, preventing the soil from washing away. **Crop rotation,** or changing what is planted from year to year, also protects topsoil. Planting grasses in fields without crops holds the soil in place.

Deforestation, or cutting down forests without replanting, is another way in which topsoil is lost. When the tree roots are no longer there to hold the soil, wind and water can carry the soil away. Many rain forests, such as the Amazon rain forest, are being cut down at high rates. This has raised concerns because the forests support the water cycle and help replace the oxygen in the atmosphere. Forests also are home to many kinds of plants and animals.

Reading Check **Identifying Central Issues** Why is deforestation a problem?

NATIONAL GEOGRAPHIC

Figure 7 **The Greenhouse Effect**

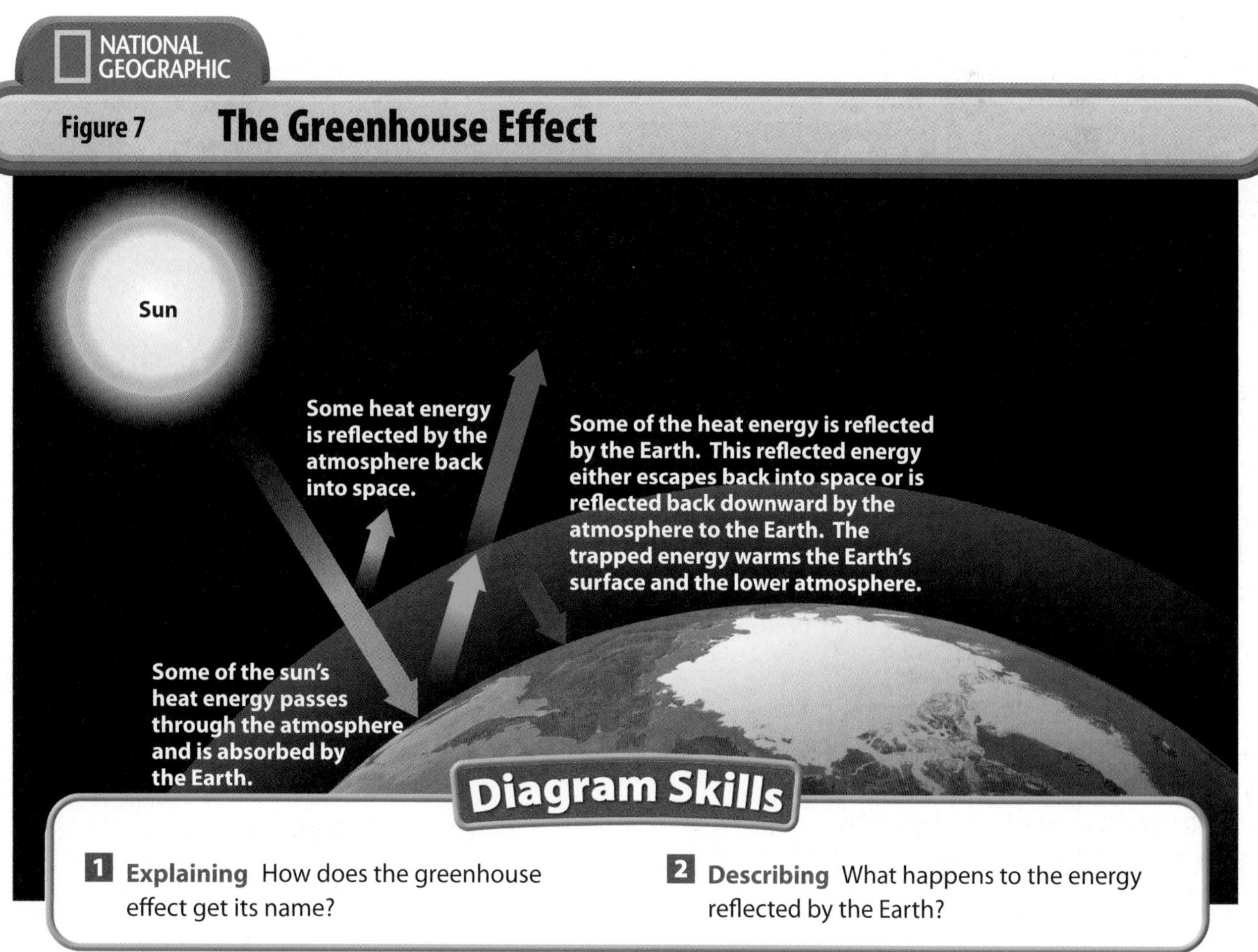

Sun
Some heat energy is reflected by the atmosphere back into space.
Some of the heat energy is reflected by the Earth. This reflected energy either escapes back into space or is reflected back downward by the atmosphere to the Earth. The trapped energy warms the Earth's surface and the lower atmosphere.
Some of the sun's heat energy passes through the atmosphere and is absorbed by the Earth.

Diagram Skills

1. **Explaining** How does the greenhouse effect get its name?
2. **Describing** What happens to the energy reflected by the Earth?

The Hydrosphere and Biosphere

Main Idea **Water pollution poses a threat to a vital and limited resource.**

Geography and You How much water do you use each day? How much of that water is wasted? Read to find out how people use water resources.

The hydrosphere refers to the Earth's surface water and groundwater. Water is vital to human life. Because the amount of freshwater is limited, people should practice **conservation,** the careful use of a resource, to avoid wasting water.

Throughout the world, farmers use **irrigation,** a process in which water is collected and distributed to crops. Irrigation is often wasteful, however, as much of the water evaporates or soaks into the ground before it reaches the crops. Pollution also threatens water supplies. Chemicals from industrial processes sometimes spill into waterways. **Pesticides,** or powerful chemicals that farmers use to kill crop-destroying insects, can also be harmful.

The biosphere is the collection of plants and animals of all types that live on Earth. The entire biosphere is divided into many **ecosystems.** An ecosystem is a place shared by plants and animals that depend on one another for survival.

Shrinking **biodiversity,** or the variety of plants and animals living on the planet, is also a concern. Changes to the environment can lead to decreasing populations of plants and animals in an ecosystem.

Reading Check **Explaining** Why is the conservation of water important?

Section 4 Review

Social Studies ONLINE
Study Central™ To review this section, go to glencoe.com.

Vocabulary

1. **Explain** the significance of
   - **a.** smog
   - **b.** acid rain
   - **c.** greenhouse effect
   - **d.** crop rotation
   - **e.** deforestation
   - **f.** conservation
   - **g.** irrigation
   - **h.** pesticide
   - **i.** ecosystem
   - **j.** biodiversity

Main Ideas

2. **Organizing** Use a diagram like the one below to identify problems related to air pollution.

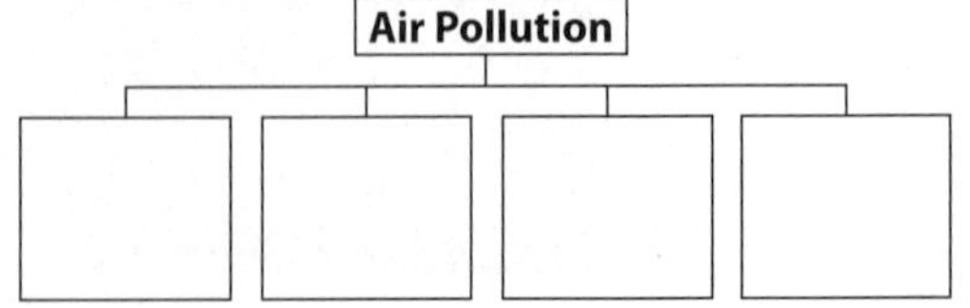

3. **Explaining** How do contour plowing and crop rotation preserve topsoil?

4. **Identifying** What is the biosphere?

Critical Thinking

5. **BIG Idea** What might happen to the animals of the rain forest if large areas of trees are cut down? Why?

6. **Challenge** Do you think countries should cooperate to solve problems like air and water pollution? Why?

Writing About Geography

7. **Persuasive Writing** Write a brief essay identifying the environmental issue you think is most important and what people can do about it.

Visual Summary

Inside the Earth

- Earth has four layers: the inner and outer cores, the mantle, and the crust.
- The continents are on large plates that move.
- Plates colliding or pulling apart reshape the land.

Shaping Landforms

- Water, chemicals, and plants break rock apart into smaller pieces.
- Water, wind, and ice can cause erosion.

Windstorm in West Africa

Types of Landforms

- Mountains, plateaus, valleys, and other landforms are found on land and under oceans.
- Climate and availability of resources affect where humans settle.

The Water Planet

- About 70 percent of the Earth's surface is water.
- In a process called the water cycle, water travels from the oceans to the air to the ground and back to the oceans.

Boaters on Inle Lake, South Asia

Climate

- Climate is the usual pattern of weather over a long period of time.
- Sun, winds, ocean currents, landforms, and latitude affect climate.
- Geographers divide the world into different climate zones.

AREA
BEYOND
THIS
SIGN
CLOSED
All public entry prohibited
By Foot, Bike, Or Vehicle
Violations Punishable

Hawk in protected area, United States

Humans and the Environment

- A delicate balance exists among the Earth's atmosphere, lithosphere, hydrosphere, and biosphere.
- Human actions, such as burning fuels and clearing rain forests, affect the environment.

Hills in Italy

STUDY TO GO Study anywhere, anytime! Download quizzes and flash cards to your PDA from glencoe.com.

STANDARDIZED TEST PRACTICE

As you read the first part of a multiple-choice question, try to anticipate the answer before you look at the choices. If your answer is one of the choices, it is probably correct.

Reviewing Vocabulary

Directions: Choose the word(s) that best completes the sentence.

1. The theory of ______ explains how continents were formed and why they move.

A magma formation
B erosion
C plate tectonics
D mantle disbursement

2. A plateau called a ______ lies off the coast of each continent and stretches for several miles underwater.

A continental aquifer
B continental shelf
C continental water cycle
D continental trench

3. Areas that have similar patterns of temperature and precipitation are known as ______.

A climate zones
B biomes
C El Niño
D currents

4. Chemicals in air pollution can combine with precipitation to form ______.

A chlorofluorocarbons
B the ozone layer
C the greenhouse effect
D acid rain

Reviewing Main Ideas

Directions: Choose the best answer for each question.

Section 1 *(pp. 44–48)*

5. Surrounding Earth's core is a layer of hot, dense rock called the ______.

A mantle
B crust
C magma
D core

Section 2 *(pp. 49–54)*

6. Almost 97 percent of the planet's water is ______.

A groundwater
B freshwater
C salt water
D frozen in glaciers and ice sheets

Section 3 *(pp. 55–61)*

7. The usual, predictable patterns of weather in an area over many years are called ______.

A climate
B current
C El Niño
D biome

Section 4 *(pp. 63–66)*

8. The careful use of resources to avoid wasting them is called ______.

A deforestation
B biodiversity
C irrigation
D conservation

GO ON

Critical Thinking

Directions: Base your answers to questions 9 and 10 on the graph below and your knowledge of Chapter 2. Choose the best answer for each question.

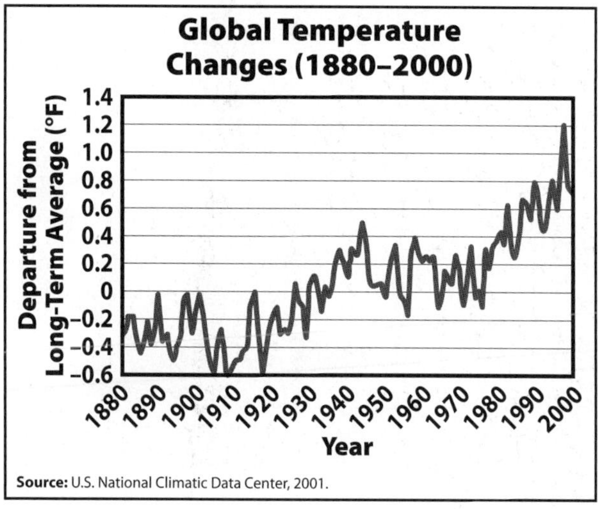

Source: U.S. National Climatic Data Center, 2001.

9. What is the overall trend of global temperature change in the twentieth century?

A There has been a stable or flat trend throughout the century.

B There has been an overall upward trend.

C There has been an overall downward trend.

D There was an upward trend early in the century followed by a downward trend.

10. During what twenty-year period of time did the sharpest rise in global temperatures take place?

A 1880–1900

B 1910–1930

C 1950–1970

D 1980–2000

Document-Based Questions

Directions: Analyze the document and answer the short-answer questions that follow.

> *Under the Kyoto Protocol, industrialized countries are to reduce their combined emissions of six major greenhouse gases during the five-year period 2008–2012 to below 1990 levels. The European Union, for example, is to cut its combined emissions by eight percent, while Japan should reduce emissions by six percent. For many countries, achieving the Kyoto targets will be a major challenge that will require new policies and new approaches. . . .*
>
> *Developing countries, including Brazil, China, India and Indonesia, are also Parties to the Protocol but do not have emission reduction targets. Many developing countries have already demonstrated success in addressing climate change.*
>
> —*UNEP, "Kyoto Protocol to Enter into Force 16 February 2005"*

11. According to this press release, what is the purpose of the Kyoto Protocol?

12. Compare how industrialized and developing countries would be affected by the Kyoto Protocol.

Extended Response

13. Which part of Earth's environment—the atmosphere, lithosphere, hydrosphere, or biosphere—do you feel is most threatened by human activity? In several paragraphs, define the part that you chose, explain why you think it is threatened, and describe what actions may help decrease the threat to that area.

STOP

For additional test practice, use Self-Check Quizzes—Chapter 2 at glencoe.com.

Need Extra Help?

If you missed question...	1	2	3	4	5	6	7	8	9	10	11	12	13
Go to page...	46	50	59	64	45	52	56	66	64	64	64	64	64

CHAPTER 3

Earth's Human and Cultural Geography

Essential Question

Movement The human population is growing rapidly, but the world in which people live is, in many ways, becoming a smaller place. In the past, many cultures were isolated from each other. Today, individuals and countries are linked in a global economy and by forms of communication that can instantly bring them together. What factors bring about changes in cultures?

Canal market, Thailand

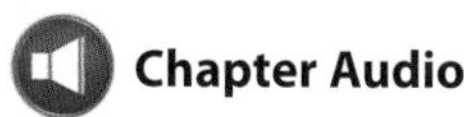

Chapter Audio

Section 1: World Population

BIG IDEA Geographers study how people and physical features are distributed on Earth's surface. Although the world's population is increasing, people still live on only a small part of the Earth's surface.

Section 2: Global Cultures

BIG IDEA Culture influences people's perceptions about places and regions. The world's population is made up of different cultures, each of which is based on common beliefs, customs, and traits.

Section 3: Resources, Technology, and World Trade

BIG IDEA Patterns of economic activities result in global interdependence. Because resources are unevenly distributed, the nations of the world must trade with each other. New technologies make the economies of nations more dependent on one another.

FOLDABLES™ Study Organizer

Categorizing Information Make this Foldable to organize information about Earth's population; cultures; and resources, technology, and trade.

Step 1 Place two sheets of paper about 1 inch apart.

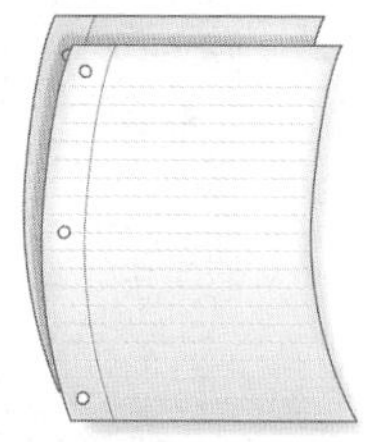

Step 2 Fold the paper to form four equal tabs.

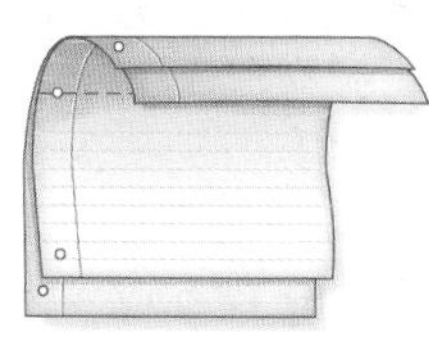

Step 3 Staple the sheets, and label each tab as shown.

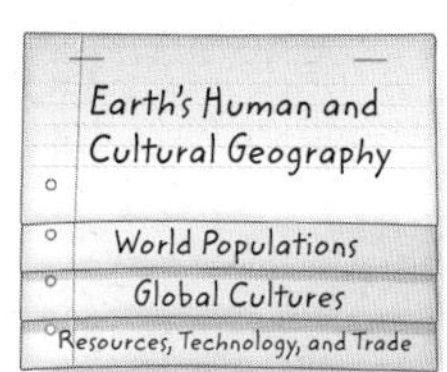

Reading and Writing
As you read the chapter, take notes under the appropriate tab. Write a main idea for each section using your Foldable.

Visit glencoe.com and enter ***QuickPass***™ code EOW3109c3 for Chapter 3 resources.

Guide to Reading

BIG Idea

Geographers study how people and physical features are distributed on Earth's surface.

Content Vocabulary

- death rate *(p. 73)*
- birthrate *(p. 73)*
- famine *(p. 73)*
- population density *(p. 74)*
- urbanization *(p. 75)*
- emigrate *(p. 75)*
- refugee *(p. 76)*

Academic Vocabulary

- technology *(p. 73)*
- internal *(p. 75)*

Reading Strategy

Determining Cause and Effect Use a diagram like the one below to show the causes and effects of global migration.

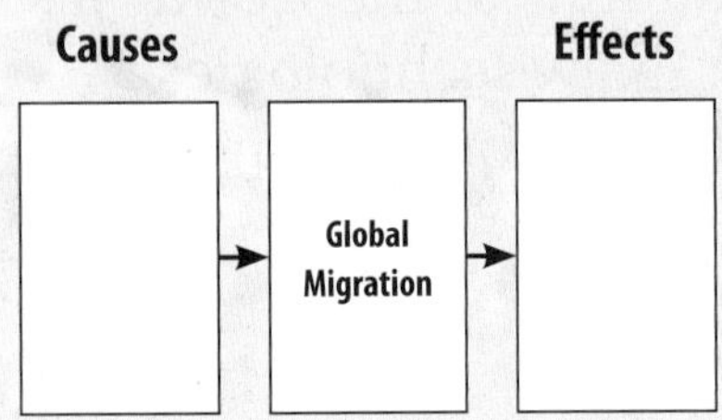

World Population

Spotlight Video

Picture This Forty years ago, for every car in China, there were 250 bicycles, earning the country the nickname "Bicycle Kingdom." Today, however, China, which is the world's most populous country, has a new love—the automobile. People are earning more money, and the number of people who own cars is increasing. Because of this, it is feared that China's cities will become more polluted and congested with traffic. Read this section to learn about the world's population and the effects it has on the Earth.

▼ Residents of Shanghai, China

Population Growth

Main Idea **The world's population has increased rapidly in the past two centuries, creating many new challenges.**

Geography and You Has the population in your community increased or decreased in recent years? Are new schools being built, for example? Read to find out why the world's population has grown so fast.

In the past 200 years, the world's population has increased rapidly. Around 1800, a billion people lived on Earth. Today the population is more than 6 billion.

Reasons for Population Growth

One reason the population has grown so fast in the last 200 years is that the death rate has gone down. The **death rate** is the number of deaths per year for every 1,000 people. Better health care and living conditions, as well as more plentiful food supplies, have decreased the death rate.

Another reason why the population has grown is high birthrates in Asia, Africa, and Latin America. The **birthrate** is the number of children born each year for every 1,000 people. High numbers of healthy births combined with lower death rates have increased the population growth, especially in these areas of the world.

Challenges of Population Growth

More food is needed for a growing population. Advances in **technology,** such as improved irrigation systems and the creation of hardier plants, will continue to increase food production. On the other hand, warfare and crop failures can lead to **famine,** or a severe lack of food. Some countries may also face shortages of water and housing. Additionally, growing populations require more services, like those provided by hospitals and schools.

Reading Check **Identifying** What has caused population growth in the last 200 years?

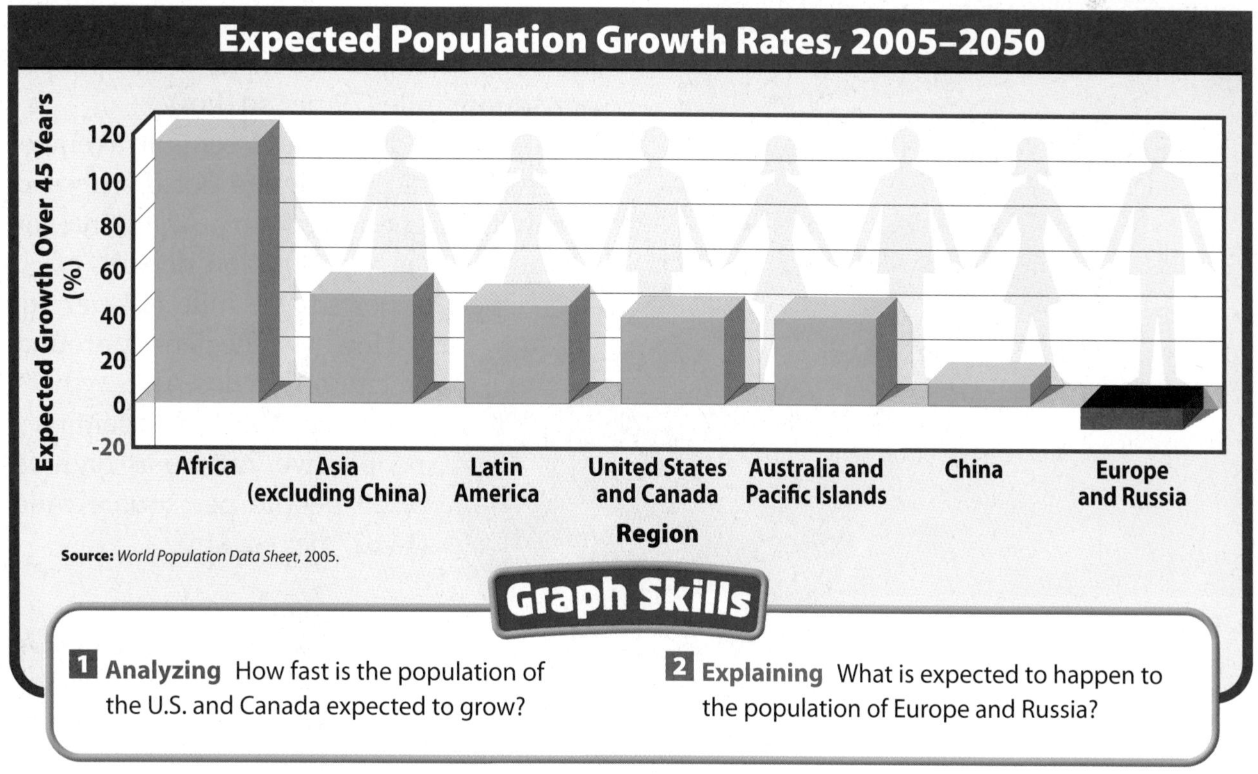

Graph Skills

1 **Analyzing** How fast is the population of the U.S. and Canada expected to grow?

2 **Explaining** What is expected to happen to the population of Europe and Russia?

Where People Live

Main Idea **The Earth's population is not evenly distributed.**

Geography and You Do you live in a city, a suburb, a small town, or a rural area? What are the advantages and disadvantages of your location? Read to find out where the world's people choose to live.

People live on a surprisingly small part of the Earth. Land covers only about 30 percent of the Earth's surface, and only half of this land is usable by humans. Deserts, high mountains, and ice-covered lands cannot support large numbers of people.

Population Distribution

On the usable land, population is not distributed, or spread, evenly. People naturally prefer to live in places that have fertile soil, mild climates, natural resources, and water resources, such as rivers and coastlines. Two-thirds of the world's people are clustered into five regions with these resources—East Asia, South Asia, Southeast Asia, Europe, and eastern North America. In most regions, more people live in cities than in rural areas because of the jobs and resources found there.

Population Density

Geographers have a way to figure out how crowded a country or region is. They measure **population density**—the average number of people living in a square mile or square kilometer. To arrive at this figure, the total population is divided by the total land area.

As you have just read, the world's population is not evenly distributed. Malaysia and Norway, for example, have about the same total land area, around 130,000 square miles (336,697 sq. km). Norway's population density is about 37 people per square mile (14 per sq. km). Malaysia, on the other hand, has a density of 205 people per square mile (79 per sq. km).

Population density represents an average. Remember that people are not distributed evenly throughout a country. Argentina, for example, has a population density of 36 people per square mile (14 per sq. km). However, the density around the city of Buenos Aires, where nearly one third of Argentina's people live, can be as high as 5,723 people per square mile (14,827 per sq. km).

Population Density

NATIONAL GEOGRAPHIC

▲ Population density is low on the grasslands of Mongolia. In contrast, Tokyo, Japan (inset), has a high population density. ***Regions*** **In what regions are most of the world's people clustered?**

Reading Check **Determining Cause and Effect** Why does much of the world's population live on a relatively small area of the Earth?

Population Movement

Main Idea **Large numbers of people migrate from one place to another.**

Geography and You Have you and your family ever moved? Read to learn some of the reasons why people all over the world move from one place to another.

Throughout history, millions of people have moved from one place to another. People continue to move today, sometimes as individuals, sometimes in large groups.

Types of Migration

Moving from place to place in the same country is known as **internal** migration. One kind of internal migration is the movement of people from farms and villages to cities. Such migrants are often in search of jobs. This type of movement results in **urbanization,** or the growth of cities. Urbanization has occurred rapidly in Asia, Africa, and Latin America.

Movement between countries is called international migration. Some people **emigrate,** or leave the country where they were born and move to another. They are emigrants in their homeland and immigrants in their new country. **Figure 1** shows the immigrant populations in regions of the world. Immigration has increased greatly in the past 200 years, partly due to better transportation.

NATIONAL GEOGRAPHIC

Figure 1 **World Immigrant Populations**

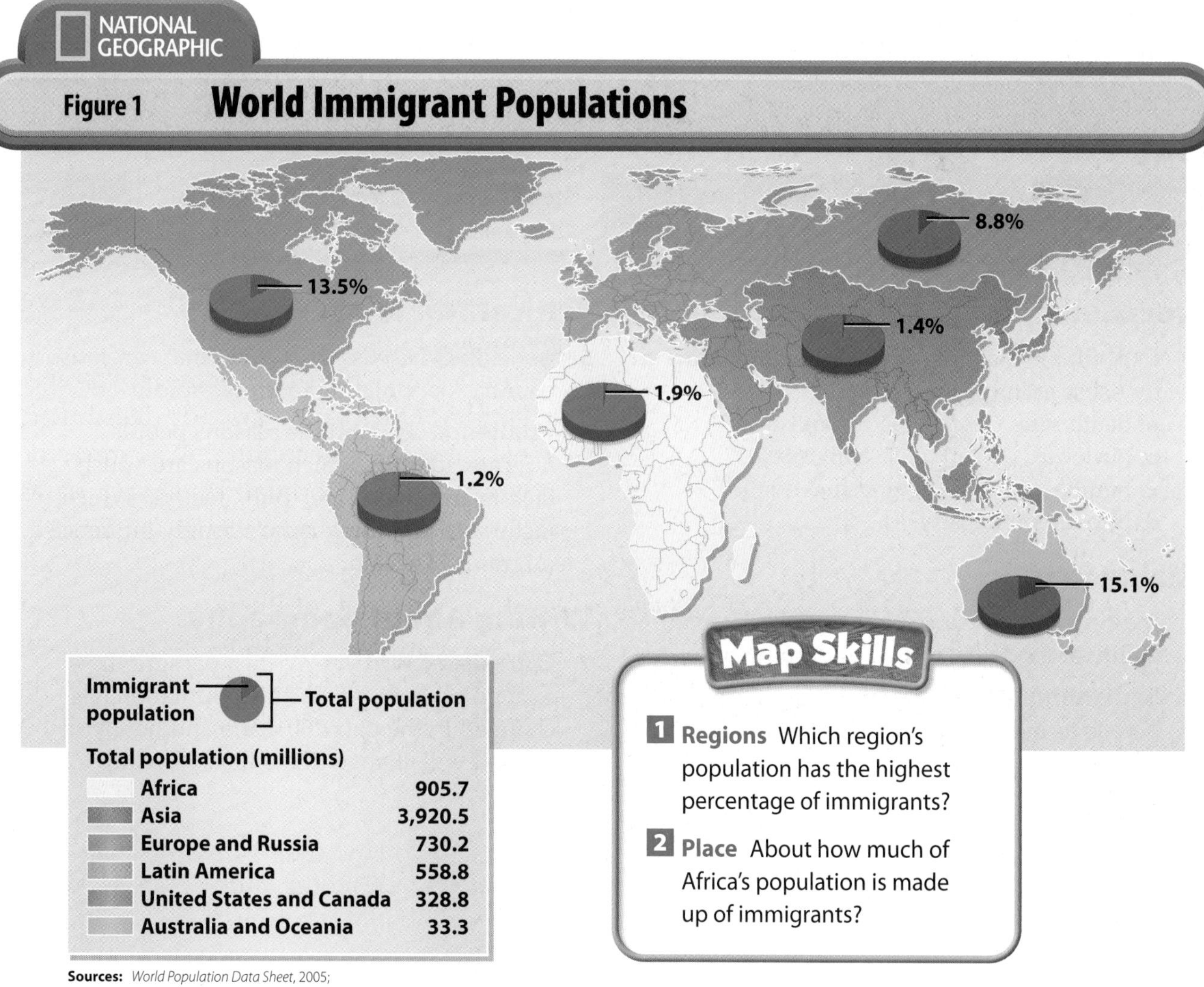

Total population (millions)

Region	Total population (millions)
Africa	905.7
Asia	3,920.5
Europe and Russia	730.2
Latin America	558.8
United States and Canada	328.8
Australia and Oceania	33.3

Sources: *World Population Data Sheet*, 2005; World Migrant Stock, United Nations.

Map Skills

1. **Regions** Which region's population has the highest percentage of immigrants?
2. **Place** About how much of Africa's population is made up of immigrants?

Reasons People Move

People migrate for a variety of reasons. Historians say that "push" factors convince people to leave their homes and "pull" factors attract them to another place. A shortage of farmland or few jobs in a region or country may "push" residents to emigrate. The lure of jobs has worked as a "pull" factor, attracting many immigrants to the United States.

People who are forced to flee to another country to escape wars, persecution, or disasters are called **refugees.** For example, 2.5 million refugees have fled mass killings in Sudan's Darfur area since 2003.

Impact of Migration

Mass migrations of people have major impacts—both on the region they leave and on the region where they settle. When emigrants leave a country, its population decreases or does not increase as quickly. This can ease overcrowding. However, if skilled or educated workers leave, emigration may hurt the country's economy. Emigration can also divide families.

Migration also affects the country to which people move. Immigrants bring with them new forms of music, art, foods, and language. Some native-born citizens, however, fear or resent immigrants and the changes that they bring. This has led to violence and unjust treatment toward newcomers in some instances.

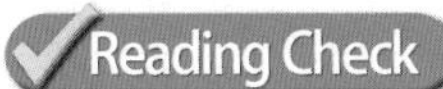

Making Generalizations

Why have so many rural citizens moved to cities in Asia, Africa, and Latin America?

Social Studies ONLINE
Study Central™ To review this section, go to glencoe.com.

Vocabulary

1. **Explain** the meaning of the following terms by using each one in a sentence.
   - **a.** death rate
   - **b.** birthrate
   - **c.** famine
   - **d.** population density
   - **e.** urbanization
   - **f.** emigrate
   - **g.** refugee

Main Ideas

2. **Making Connections** How might the availability of food affect population growth?
3. **Explaining** What geographic factors lead people to live in certain areas of the world?
4. **Summarizing** Use a diagram like the one below to summarize the positive and negative effects of emigration on a country.

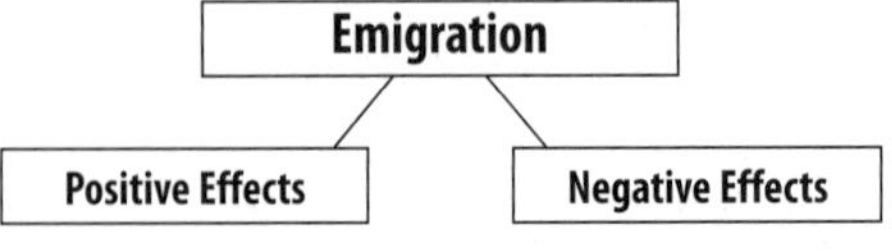

Critical Thinking

5. BIG Idea Discuss the factors that can cause a country's population to grow rapidly.
6. **Challenge** Explain the reasons people migrate. Identify which reasons are "push" factors and which are "pull" factors. Which factors do you think most strongly influence migrants? Explain.

Writing About Geography

7. **Expository Writing** Write a paragraph explaining how the Earth's population has changed in the past 200 years and how you think it will change in the next 50 years.

THE WORLD GOES GLOBAL

Technology and new methods of trade are affecting how the world interacts.

A local Inuit uses a laptop in the Canadian Arctic.

Around the world, technological advances are changing the way we live and work. Every day, new technologies make it possible for billions of e-mails and trillions of dollars to crisscross national borders. Communication between people and businesses and the movement of goods and money is done more quickly than ever before because of the Internet.

As technology continues to change, what might the world look like ten years from now? Inventions that create faster ways to communicate might make the world seem even smaller than it does today. And as globalization connects the world's economies as never before, people everywhere will learn about other nations and cultures.

TIME PERSPECTIVES

Workers at a call center in India answer questions from American customers.

A GLOBAL MARKETPLACE

Venugopla Rao Moram is a highly sought after worker. Recently, the computer software engineer who lives in Bangalore, India, was offered five jobs during a two-week period. All of the offers were from companies whose headquarters are located thousands of miles from India.

Luckily, Moram will not have to travel that far to get to work. Computer companies from around the world are opening offices in Bangalore in order to hire Indian workers. Many Indians speak English and are well educated. This makes them valuable to foreign companies that are establishing workplaces in countries where labor is inexpensive. This type of labor helps manufacturers keep their production costs low.

As a result of **globalization**, a trend that is linking the world's nations through trade, thousands of Indians are working for foreign companies. In Moram's case, a business in California hired him to create software that makes the characters in video games jump and run. The software Moram produces becomes part of a product that is assembled in other countries and sold all over the world. All types of products, from toys to clothes to TVs, are being made and traded this way. As a result, economies are becoming much more connected—or global.

The United States and the Global Economy

The United States trades with countries all over the world. It sells, or exports, some products, and buys, or imports, others. Here are the countries the U.S. did the most business with in 2005.

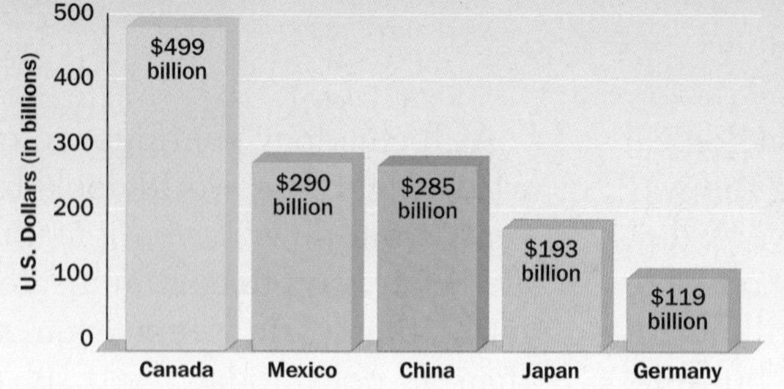

Source: U.S. Census Bureau, Foreign Trade Statistics Division.

INTERPRETING GRAPHS

Making Inferences Why might Canada and Mexico be the United States's top trading partners?

The Internet

The Internet has fueled globalization. The Internet is a giant electronic network that links computers all over the world. It was developed in the late 1960s when the U.S. military worked to connect its computers with those of college researchers so that they could share their ideas more easily. Over time, the Internet became available to everyone, and the way the world interacts changed forever.

An anti-globalization demonstrator protests in Japan.

People all over the world can trade stocks on the New York Stock Exchange.

A Thai woman uses a bank machine.

The Internet also changed how people and companies buy goods. Today, just like you can shop online for games or CDs, so can businesses. For example, a business in need of computer software can use the Internet to research the products of computer companies from all over the world. With the click of a mouse, the buyer can research and compare prices for software products on a computer company's Web site or at an online store. Then, in seconds, the buyer can purchase the product.

Before the days of the Internet, a company in need of software could not have learned about suppliers and products as easily. As a result, business tended to be conducted more locally and at a slower pace. Today, a buyer can shop and trade online in minutes without leaving his or her desk. Companies can conduct business in less time and from anywhere in the world.

Sharing Globalization's Gains

The impact of globalization has been amazing, but its benefits have not been shared equally. **Developed countries**, or countries in which a great deal of manufacturing is carried out, have more goods to trade than **developing countries** that are still trying to industrialize. Also many companies prefer to build factories in wealthier countries rather than in poor ones, where support systems like roads and airports are often unavailable. As a result, some of the poorer nations in Asia and Africa have had a hard time creating any new jobs.

What steps can be taken to spread the benefits of globalization? International businesses and wealthy developed nations can be part of the solution. By 2005, businesses and governments together had spent about $2.6 trillion to help poorer countries develop their economies. Investing in developing countries could help businesses trade more effectively and grow. There is still much work to be done. Finding ways to help every nation share the gains of globalization is one of biggest challenges the world faces.

Lumber is processed at a Canadian mill for shipment to the United States.

EXPLORING THE ISSUE

1. **Making Inferences** Why do you think companies are concerned about how much money it costs to make a product?

2. **Analyzing Information** How might investing in transportation systems help developing countries?

MUSIC GOES GLOBAL

It has been said that music is the universal language. This has never been more true than in the Internet age. Today, music lovers can listen to music from all over the world. Online music stores and portable music players make it easy to listen to what you want, when you want.

In the past, listeners had much less control over the music they heard. Record producers and companies recorded the music of homegrown musicians, and radio stations played their songs. Artists and songs from different regions of the world were rarely played.

In the Internet age, however, music lovers are being exposed to sounds from around the world. West African drumming or Latin American dance music, for example, is available to anyone online. Listeners can just search for a **genre**, or style of music, and download a song for a small fee.

With such easy access to global sounds, it is not uncommon for a portable player to include a list of songs and artists from several countries. As a result, musicians are working to please the public by blending "international" material and elements into their acts. The American pop singer Christina Aguilera sings in English, but she has also recorded a CD completely in Spanish. Hip-hop artist Wyclef Jean mixes **Creole**, the language of Haiti, into his songs. Madonna has worn traditional costumes from Japan and Scotland during her tours.

In the twenty-first century, musicians and music lovers are no longer tied to the sounds of one nation. In fact, cross-cultural appeal in the music industry is becoming a key to success.

Wyclef Jean uses Haitian elements in his music.

REUTERS/GARY HERSHORN

EXPLORING THE ISSUE

1. **Determining Cause and Effect** How does the Internet help people learn about the music styles of performers from other countries?

2. **Making Inferences** List three reasons why it may be easier to buy music online than in a store that sells CDs.

REVIEW AND ASSESS

UNDERSTANDING THE ISSUE

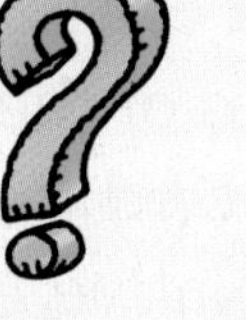

1. **Making Connections** How has globalization affected the way some products are produced?

2. **Writing to Inform** Write a short article about how the Internet has changed the way that businesses shop for and buy goods.

3. **Writing to Persuade** Do you think that American musicians who combine music from other countries and cultures can become stars in the United States? Defend your answer in a letter to the president of a record company.

INTERNET RESEARCH ACTIVITIES

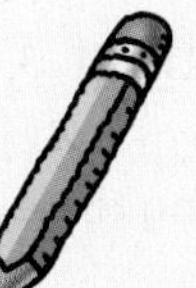

4. Go online to research the history of the Internet. Write an essay explaining why the Internet was created. Develop a time line that notes important developments.

5. With your teacher's help, use the Internet to research how many homes have access to the Internet in developed and developing nations. Compare the information and create a bar graph showing the top three countries in both categories.

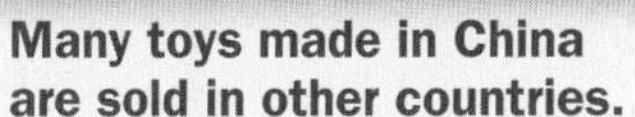

BALDEV/CORBIS

Many toys made in China are sold in other countries.

BEYOND THE CLASSROOM

6. **Organize the class into three teams.** One group should represent developed nations, and another should represent developing nations. Debate this resolution: "Globalization is good for everyone." The third group of students will decide which team has the most convincing arguments.

7. **Take an inventory of your home.** Look for products that were made in other countries. Count the items that were imported from different countries. Make a chart to show how many countries are represented in your home.

The Universal Language

The Internet is changing the way people listen to music. In the Internet age, music lovers around the world are shopping online. Here is a look at the number of people visiting music sites.

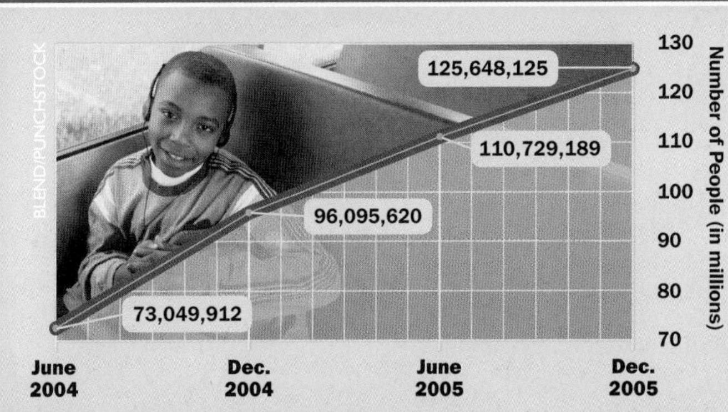

Source: Nielsen/Net Ratings.

Building Graph Reading Skills

1. **Comparing** How many more people visited music Web sites in December 2005 than in December 2004?

2. **Making Inferences** How might the increase of shopping online for music affect traditional music stores?

Global Cultures

 Section Audio  Spotlight Video

Guide to Reading

BIG Idea

Culture influences people's perceptions about places and regions.

Content Vocabulary

- culture *(p. 83)*
- ethnic group *(p. 84)*
- dialect *(p. 84)*
- democracy *(p. 85)*
- dictatorship *(p. 86)*
- monarchy *(p. 86)*
- civilization *(p. 86)*
- cultural diffusion *(p. 87)*
- culture region *(p. 88)*
- globalization *(p. 89)*

Academic Vocabulary

- widespread *(p. 86)*
- unique *(p. 89)*

Reading Strategy

Identifying Use a diagram like the one below to identify the elements of culture.

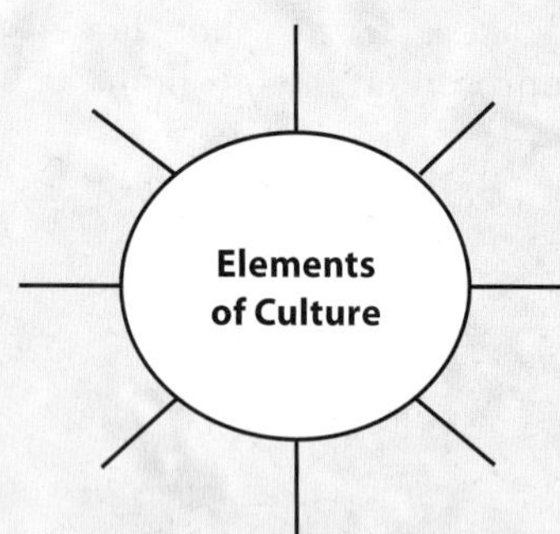

Picture This The eagles that soar through the skies of the American southwest have long been sacred to the native peoples of the area. Many Native Americans believe that eagles have special qualities such as wisdom and courage. Eagle feathers are treated with respect and are often given as rewards for great deeds. Native American groups, such as the Tewa of New Mexico, perform dances to honor this beautiful bird. To learn more about how traditions reflect a culture's beliefs, read Section 2.

▼ Honoring the eagle

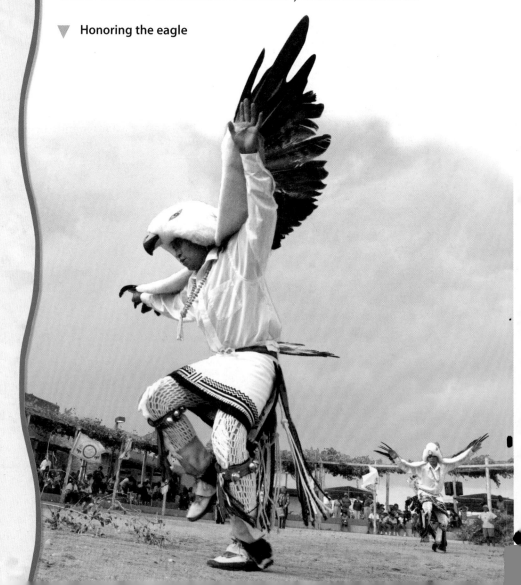

What Is Culture?

Main Idea **Culture refers to the many shared characteristics that define a group of people.**

Geography and You Think about the clothes you wear, the music you listen to, and the foods you eat. Read to learn about the many things that make up culture.

Culture is the way of life of a group of people who share similar beliefs and customs. A particular culture can be understood by looking at various elements: what languages the people speak, what religions they follow, and what smaller groups are part of their society. The study of culture also includes examining people's daily lives, the history they share, and the art forms they have created.

Geographers, anthropologists, and archaeologists all study culture. For example, geographers look at physical objects, such as food and housing. They also study elements such as religion, social groups, types of government, and economies. Anthropologists analyze cultures today to learn how different elements of culture are related. Archaeologists use the physical and historical objects of a culture, such as pottery and tools, to try to understand how people lived in the past. The work of all of these experts helps us better understand the world we live in.

NATIONAL GEOGRAPHIC

Figure 2 **World Language Families**

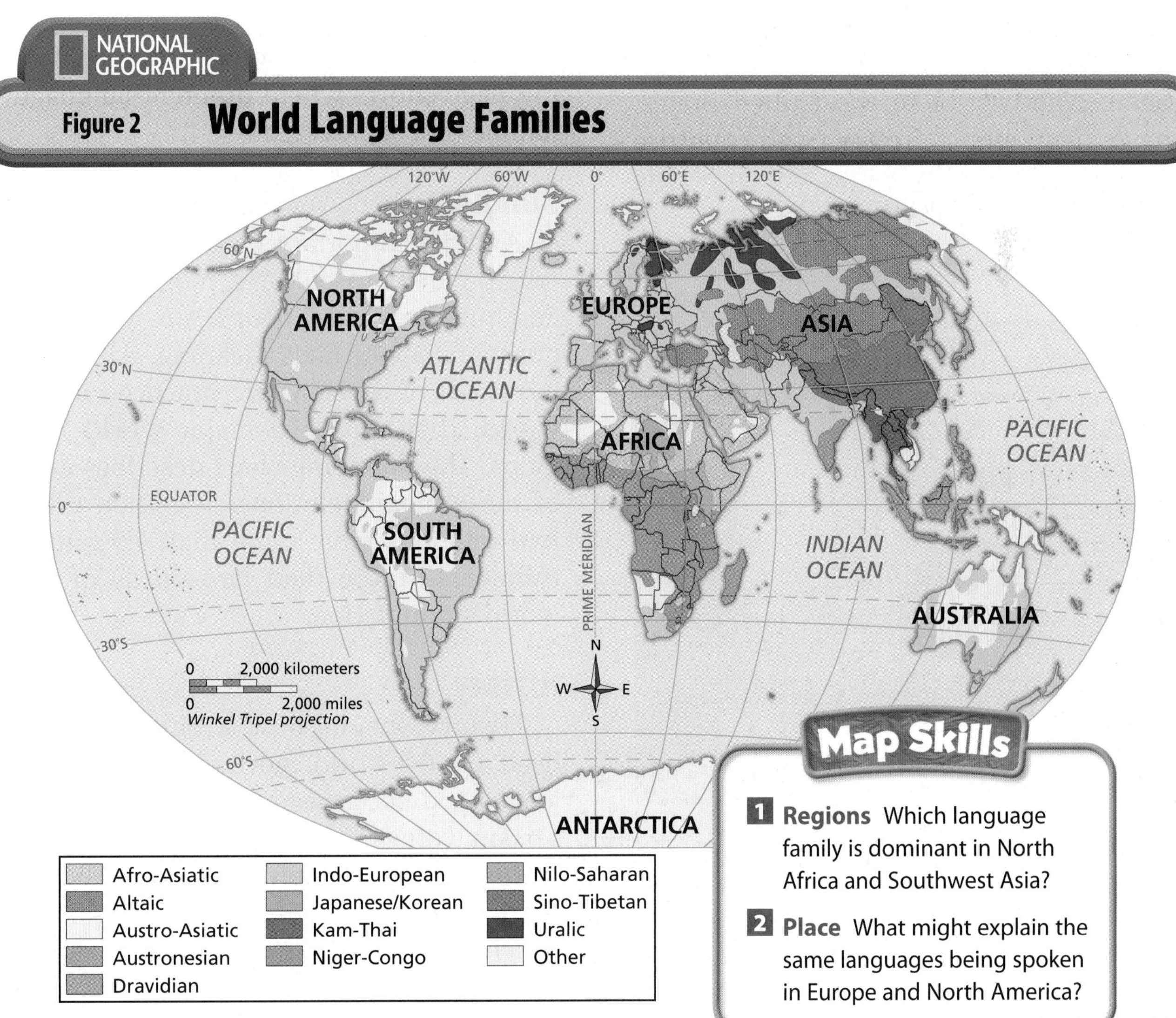

Map Skills

1. **Regions** Which language family is dominant in North Africa and Southwest Asia?
2. **Place** What might explain the same languages being spoken in Europe and North America?

Social Groups

One way scientists study culture is by looking at different groups of people in a society. Each of us belongs to many social groups. For example, are you old or young? Male or female? A student, a worker, or both? Most social groups have rules of behavior that group members learn. The process by which people adjust their behavior to meet these rules is called socialization. Within society, each person has a certain status. Status refers to a person's importance or rank. In all cultures, the family is the most important social group. Although family structures vary from culture to culture, most of us first learn how to behave from our families.

People also belong to an **ethnic group.** This is a group that shares a language, history, religion, and some physical traits. Some countries, like the United States, have many ethnic groups. Such countries have a national culture that all their people share, as well as ethnic cultures.

▲ Households in Japan can include several generations. ***Place*** **What elements of culture are found in this family gathering?**

In some cases, people come to believe that their own culture is superior to, or better than, other cultures. This attitude is called ethnocentrism. If carried to extremes, ethnocentrism may cause hatred and persecution of other groups.

Language

Sharing a language is one of the strongest unifying forces for a culture. A language, however, may have different variations called dialects. A **dialect** is a local form of a language that may have a distinct vocabulary and pronunciation. Despite different dialects, speakers of the same language can usually understand one another.

More than 2,000 languages are spoken around the world today. Most can be grouped with related languages into a specific language family. **Figure 2** on the preceding page shows where different language families are spoken today.

Religion

Another important cultural element is religion. In many cultures, religious beliefs and practices help people answer basic questions about life's meaning. Although hundreds of religions are practiced in the world, there are five major world religions. The following chart describes each of these major religions. Together, these five religions have more than 4.5 billion followers—more than two-thirds of the world's population.

History

History shapes how a culture views itself and the world. Stories about the challenges and successes of a culture support certain values and help people develop cultural pride and unity. Cultural holidays mark important events and enable people to celebrate their heritage.

Major World Religions

Religion	Major Leader	Followers	Beliefs
Buddhism	**Siddhartha Gautama, the Buddha**	**385.4 million**	Buddhism is based on the teachings of Siddhartha Gautama, known as the Buddha. The Buddha taught that the goal of life is to escape the cycle of birth and death by achieving a state of spiritual understanding called nirvana. Buddhists believe that they must follow an eight-step path to achieve nirvana.
Christianity	**Jesus Christ**	**2.2 billion**	Christianity is based on the belief in one God and the teachings and life of Jesus as described in the New Testament of the Bible. Christians believe that Jesus was the Son of God and was sent to Earth to save people from their sins.
Hinduism	**Unknown**	**875.1 million**	Hinduism is based on the belief in a supreme spiritual force known as Brahman as recorded in sacred texts, including the Upanishads. Hindus believe that to unite with Brahman, they must first pass through many lives, being reborn into new forms. To move closer to Brahman they must make improvements in each of their lives.
Islam	**Muhammad**	**1.4 billion**	Islam is based on the belief in one God, Allah, as revealed through the prophet Muhammad. The Muslim sacred text is the Quran. Muslims follow five major acts of worship known as the Five Pillars of Islam.
Judaism	**Abraham**	**15.2 million**	Judaism is based on the belief in one God and the spiritual and ethical principles handed down by God. These principles, including the Ten Commandments, are presented in Jewish sacred texts collected in the Hebrew Bible.

Source: *The World Factbook*, 2008.

Chart Skills

1 Identifying Which two religions include the belief that people are reborn into new forms?

2 Explaining What help do these religions give to their followers?

Daily Life

Food, clothing, and shelter are basic human needs. The type of food you eat and how you eat it reflect your culture. Do you use chopsticks, a fork, or bread to scoop up your food? The home you live in and the clothing that you wear reflect your culture and your physical surroundings. For example, the clothing people wear in the high, chilly Andes of South America differs greatly from the clothing people wear on the warm savannas of Africa.

Arts

Through music, painting, sculpture, dance, and literature, people express what they think is beautiful and meaningful. The arts can also tell stories about important figures and events in the culture.

Government

People need rules in order to live together without conflict. Governments fulfill this need. They can be either limited or unlimited. A limited government restricts the powers of its leaders. For example, in a **democracy**, power is held by the people.

Social Studies ONLINE

Student Web Activity Visit glencoe.com to learn more about forms of government around the world.

Most democracies today are called representative democracies because the people choose leaders to represent them and make decisions. In unlimited governments, leaders are all-powerful. In a **dictatorship,** for instance, the leader, or dictator, rules by force. Dictators often limit citizens' freedoms.

A **monarchy** is a government led by a king or queen who inherits power by being born into the ruling family. For much of history, monarchies had unlimited power. Today, most monarchies are constitutional monarchies in which elected legislatures hold most of the power.

Economy

People in every culture must earn a living. Geographers study economic activities to see how a culture uses its resources and trades with other places. An economy's success can be seen in people's quality of life—how well they eat and live and what kind of health care they receive.

Reading Check **Describing** Describe three elements that help unify a culture.

The Growth of Industry

NATIONAL GEOGRAPHIC

▲ Some of the earliest factories, like this one in Lowell, Massachusetts, used machines to make cloth. ***Movement*** **What recent technological advancements have led to cultural changes?**

Cultural Change

Main Idea **Cultures are constantly changing and influencing each other.**

Geography and You What influences from other cultures can you see in your community? Read on to see how cultures relate to each other and change.

Over time, all cultures experience change. Sometimes that change results from inventions and innovations, or technological improvements that bring about new ways of life. Sometimes change results from the influence of other cultures.

Inventions and Technology

Thousands of years ago, humans were hunters and gatherers who lived and traveled in small groups. After 8000 B.C., people learned to farm. Planting crops led to more reliable food supplies and larger populations. It also allowed people to settle in one place. Historians call this change the Agricultural Revolution. It had a huge impact on human culture because it led people to create **civilizations,** or highly developed cultures, in river valleys found in present-day Iraq, Egypt, India, and China. The people of these civilizations made a number of important advancements including building cities, forming governments, founding religions, and developing writing systems.

The world remained largely agricultural through the A.D. 1700s. Around that time, some countries began to industrialize, or use machines to make goods. The **widespread** use of machines made economies more productive. Industrial nations produced more food, goods, and wealth, which caused sweeping cultural changes.

The world has changed greatly in the past three decades. Computers have transformed businesses and households. Advances in communications allow people throughout the world to send and receive information almost instantly. Medical technology has dramatically increased human life expectancy. Each of these developments has sparked cultural changes.

Cultural Diffusion

The other major cause of cultural change is influence from other cultures. The process of spreading ideas, languages, or customs from one culture to another is called **cultural diffusion.** In the past, diffusion has taken place through trade, migration, and conquest. In recent years, new methods of communication have also led to cultural diffusion.

Historically, trade began with the exchange of goods, often over great distances. Soon trade also brought new ideas and practices to an area. Buddhist merchants brought their religion to China along trade routes, and Muslim traders shared their religious beliefs with people in West Africa. Trade continues to be a major means of cultural diffusion.

The movement of people from one place to another also leads to cultural diffusion. When Europeans arrived in North America, they brought horses, which were new to the continent. Native Americans living on the Great Plains quickly adopted the horse because it made hunting easier.

The conquest of one group by another is a third way culture can spread. Conquerors bring their culture to conquered areas. For example, the Romance languages, such as Italian, French, Spanish, and Portuguese, reflect the influence of the Roman Empire. These languages are based on Latin, the language of ancient Rome. In turn, conquered peoples can influence the culture of the conquerors. Christianity arose among the Jews, a people conquered by the Roman Empire. In time, Christianity became a major religion in the empire.

NATIONAL GEOGRAPHIC

Cultural Influences

Children around the world, such as these students in China, anticipate each new Harry Potter book and movie. The books have been translated into 47 languages. ***Movement*** **What is cultural diffusion?**

Today television, movies, and the Internet contribute to cultural diffusion. For example, movies made in the United States, Mexico, Brazil, and India are seen around the world, introducing people to different ways of life. The Internet allows people to have contact with and be influenced by people from other cultures.

Reading Check **Analyzing Information**
Describe one way that cultural diffusion takes place.

Regional and Global Cultures

Main Idea **As countries and regions share cultural traits, a global culture is emerging.**

Geography and You What do you have in common with a student who lives across town or across the country? Read to learn how similarities help to define cultural regions.

As you recall, geographers use the term *regions* for areas that share common physical characteristics. Likewise, geographers divide the world into several culture regions, as shown in **Figure 3.** A **culture region** is an area that includes different countries that share similar cultural traits.

Culture Regions

The countries in each culture region generally have similar social groups, governments, economic systems, religions, languages, ethnic groups, and histories. One example of a culture region is North Africa, Southwest Asia, and Central Asia. In that area, Islam is the dominant religion. Another culture region is Canada and the United States. These countries have similar languages, histories, and ethnic groups.

As you study the world, you will begin to recognize the characteristics shared by the

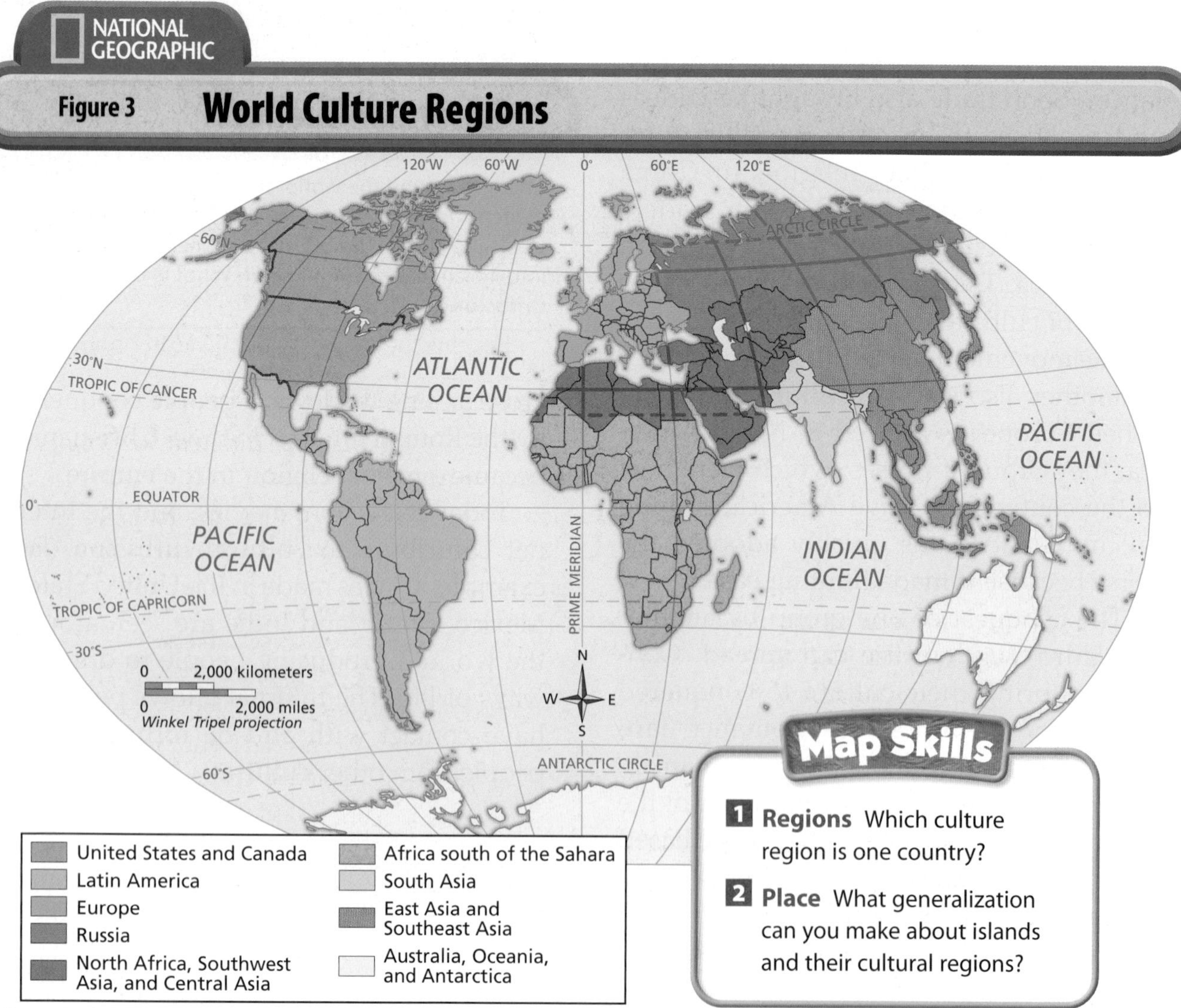

Figure 3 **World Culture Regions**

Map Skills

1. **Regions** Which culture region is one country?
2. **Place** What generalization can you make about islands and their cultural regions?

countries in each culture region. Although these countries are similar, they also have **unique** traits that set them apart.

Global Culture

Recent advances in communications and technology have helped break down barriers between culture regions. The result is **globalization,** or the development of a worldwide culture with an interdependent economy.

With globalization, individual economies rely greatly upon one another for resources and markets. Some people believe that as the global culture grows, local cultures will become less important. They point out that globalization might even erase the traditions and customs of smaller groups.

Reading Check **Defining** What are culture regions?

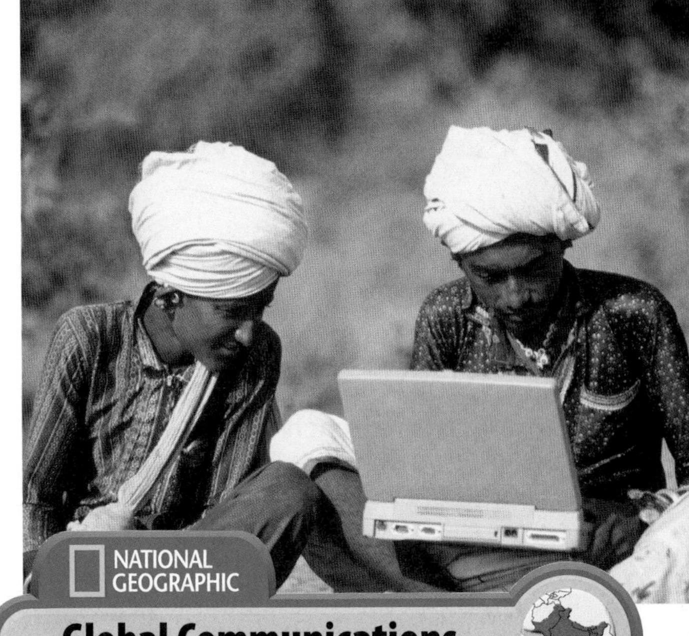

Global Communications

The Internet and other forms of communications have helped link people around the world, such as these boys in rural India. ***Movement*** **What might happen as the global culture grows?**

Section 2 Review

Social Studies ONLINE
Study Central™ To review this section, go to glencoe.com.

Vocabulary

1. **Explain** the meaning of *culture, ethnic group, dialect, democracy, dictatorship, monarchy, civilization, cultural diffusion, culture region,* and *globalization* by writing three to four paragraphs that use all of the terms.

Main Ideas

2. **Explaining** What is an ethnic group, and how do ethnic groups relate to a region's culture?
3. **Summarizing** Use a diagram like the one below to identify the advancements made by the world's earliest civilizations.

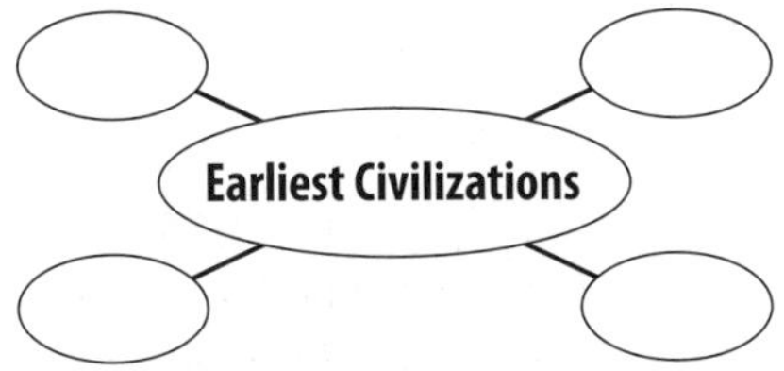

4. **Explaining** Why is globalization occurring?

Critical Thinking

5. **BIG Idea** Explain the different ways that cultural change can occur.
6. **Challenge** How do local and national differences affect culture on a regional or global level?

Writing About Geography

7. **Personal Writing** Write a journal entry describing examples of globalization that you have witnessed. Then add your predictions about how globalization might affect your community in the future.

Is Globalization Good for Everyone?

Globalization is sometimes defined as the linking together of the world's nations through trade. This trade among nations allows people from different cultures to interact with each other. As a result, cultures begin sharing traits with others. People disagree about the effects of globalization on economies and cultures. Some people think that globalization helps countries by providing them with jobs and new technologies. However, others believe that globalization destroys the cultural traditions and customs of smaller groups.

Globalization

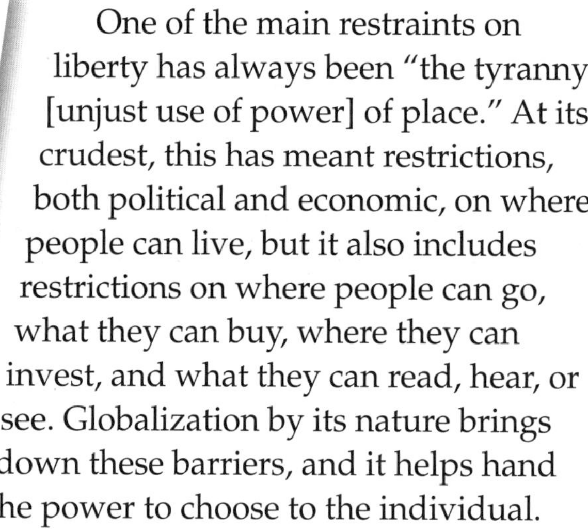

One of the main restraints on liberty has always been "the tyranny [unjust use of power] of place." At its crudest, this has meant restrictions, both political and economic, on where people can live, but it also includes restrictions on where people can go, what they can buy, where they can invest, and what they can read, hear, or see. Globalization by its nature brings down these barriers, and it helps hand the power to choose to the individual.

—John Micklethwait and Adrian Wooldridge
A Future Perfect: The Essentials of Globalization

Globalization

For millions of people globalization has not worked. Many have been actually made worse off, as they have seen their jobs destroyed and their lives become more insecure. They have felt increasingly powerless against forces beyond their control. They have seen their democracies undermined, their cultures eroded.

If globalization continues to be conducted in the way that [it] has been in the past, if we continue to fail to learn from our mistakes, globalization will not only not succeed in promoting development but will continue to create poverty and instability.

—Joseph Stiglitz
Globalization and Its Discontents

You Be the Geographer

1. **Identifying** Choose a sentence from each opinion that best summarizes the authors' views about globalization.
2. **Critical Thinking** What does Stiglitz mean when he writes "... globalization ... will continue to create poverty and instability"? Use the definition of *globalization* to explain your answer.
3. **Read to Write** Write one paragraph that identifies how globalization might benefit a nation. Then write a paragraph that describes how globalization might harm a nation.

Guide to Reading

BIG Idea

Patterns of economic activities result in global interdependence.

Content Vocabulary

- natural resource *(p. 93)*
- renewable resource *(p. 93)*
- nonrenewable resource *(p. 93)*
- economic system *(p. 94)*
- developed country *(p. 94)*
- developing country *(p. 94)*
- newly industrialized country *(p. 94)*
- export *(p. 95)*
- import *(p. 95)*
- tariff *(p. 95)*
- quota *(p. 95)*
- free trade *(p. 96)*
- interdependence *(p. 96)*

Academic Vocabulary

- finite *(p. 93)*
- finance *(p. 95)*

Reading Strategy

Categorizing Information Use a diagram like the one below to list three specific examples of each type of natural resource.

Renewable Resources	Nonrenewable Resources
1.	
2.	
3.	

Resources, Technology, and World Trade

 Section Audio  Spotlight Video

Picture This It might not have temperature controls, but this solar stove is one of the most important household appliances in Chinese homes. China is a world leader in the use of solar energy. As China's economy has grown, the demand for fuel has driven energy costs up, increasing the desire to use alternative energy sources. Read this section to learn more about other resources and how the world's people use them.

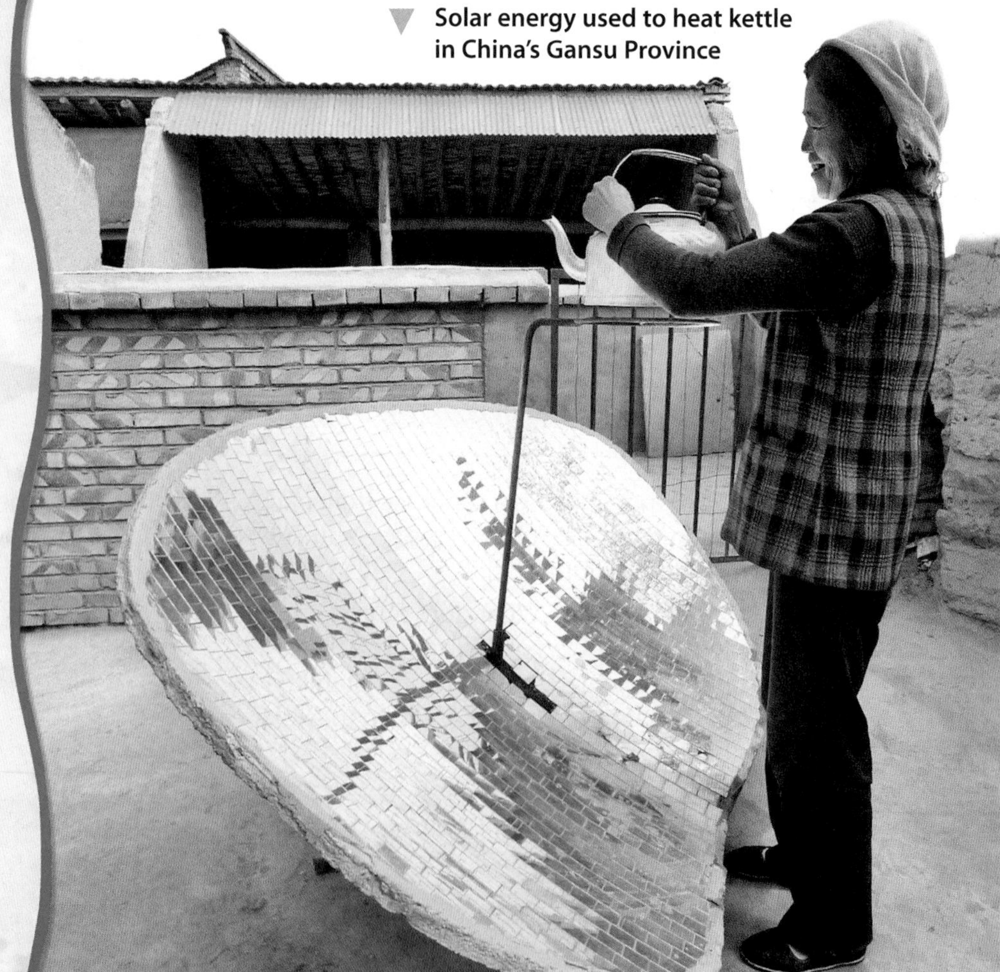

Solar energy used to heat kettle in China's Gansu Province

Natural Resources

Main Idea **Earth's resources are not evenly distributed, nor do they all exist in endless supply.**

Geography and You What natural resources can you name? Read to learn about two kinds of natural resources.

Natural resources are materials from the Earth that people use to meet their needs. Soil, trees, wind, and oil are examples of natural resources. Such resources can provide food, shelter, goods, and energy.

Renewable resources are natural resources that cannot be used up or that can be replaced. For example, the sun, the wind, and water cannot be used up, and forests can replace themselves. Some renewable resources, such as rivers, the wind, and the sun, can produce electricity and are important sources of energy.

Most natural resources are **finite,** or limited in supply. They are called **nonrenewable resources.** Once humans use up these resources, they are gone. Minerals like iron ore and gold are nonrenewable, as are oil, coal, and other fossil fuels. Fossil fuels heat homes, run cars, and generate electricity.

Reading Check **Identifying** Which energy resources are renewable? Nonrenewable?

NATIONAL GEOGRAPHIC Maps In MOtion See StudentWorks™ Plus or glencoe.com.

Figure 4 **World Energy Production and Consumption**

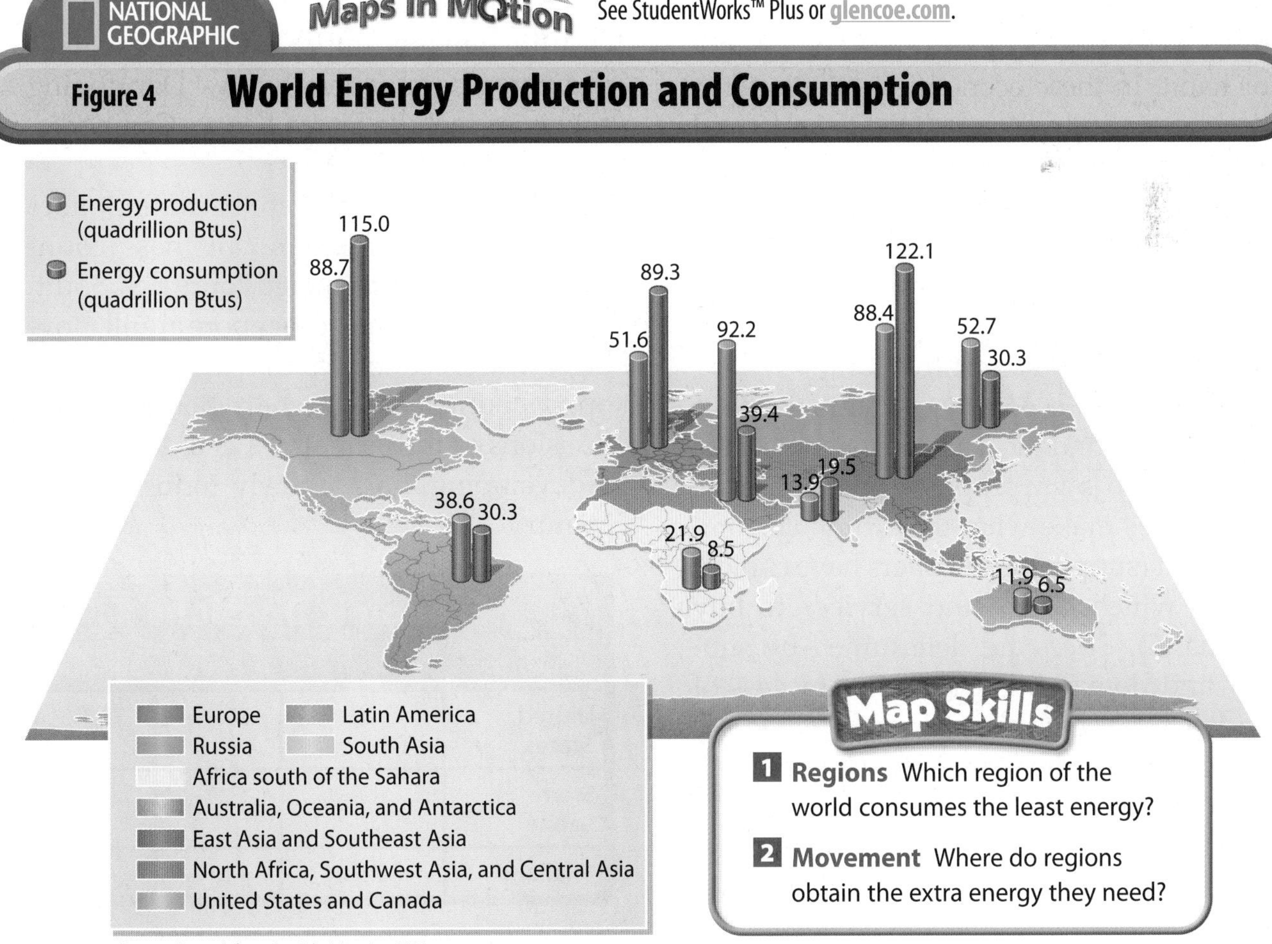

Source: Energy Information Administration, 2007.

Map Skills

1. **Regions** Which region of the world consumes the least energy?
2. **Movement** Where do regions obtain the extra energy they need?

Economies and Trade

Main Idea An economy is the way people use and manage resources.

Geography and You What kinds of goods and services do the people in your community produce? Read to find out about how economic decisions are made.

Economic Systems

To help make economic decisions, societies develop economic systems. An **economic system** is the method used to answer three key questions: what goods and services to produce, how to produce them, and who will receive them.

There are four kinds of economic systems. In a traditional economy, individuals decide what to produce and how to produce it. These choices are based on custom or habit. In these economies, people often do the same work as their parents and grandparents. Technology is often limited.

In a command economy, the government makes the key economic decisions about resources. It decides the costs of products and the wages workers earn, and individuals have little economic freedom.

In a market economy, individuals make their own economic decisions. People have the right to own property or businesses. Businesses make what they think customers want (supply). Consumers have choices about which goods and services to buy (demand). Prices are determined by supply and demand. People will buy less of an item as it gets more expensive. On the other hand, if the price is low, people will tend to buy more of an item.

Most nations have mixed economies, which is the fourth type of economic system. China, for example, has mostly a command economy, but the government allows some features of a market economy. The United States has mainly a market economy with some government involvement.

Developed and Developing Countries

Geographers look at economies in another way—how developed they are. A **developed country** has a mix of agriculture, a great deal of manufacturing, and service industries. Service industries, such as banking and health care, provide services rather than making products. Developed economies tend to rely on new technologies, and workers have relatively high incomes. Examples of developed countries include the United States, France, and Japan.

Countries with economies that are not as advanced are called **developing countries.** These countries have little industry. Agriculture remains important, and incomes per person are generally low. Developing countries include Sierra Leone, Cambodia, and Guatemala.

Still other countries are becoming more industrial. Geographers call these countries **newly industrialized countries.** South Korea, Thailand, and Singapore are all moving toward economies like those in developed countries. The chart below shows divisions in the economies of a developed, a developing, and a newly industrialized country.

Economic Divisions

Country	Agriculture	Industry	Services
United States	1%	20.4%	78.7%
Sierra Leone	49%	31%	21%
Thailand	9.9%	44.1%	46%

Source: *World Factbook*, 2006.

World Trade

Resources, like people, are not distributed evenly around the world. Because most countries have more than they need of some resources and not enough of others, trade is important.

Trade allows nations to **export,** or sell to other countries, the resources they have in abundance or the products made from those resources. They also **import,** or buy from other countries, the resources they do not have or the products they cannot make themselves.

Trade is important for both developed and developing nations. For example, the countries of Europe import what they need—food, energy resources, and minerals—to maintain their successful economies. The developing nations, in turn, rely on the sale of their products and resources to **finance,** or pay for, efforts to further industrialize and build their economies.

Barriers to Trade

Nations try to manage trade in order to boost their own economies. Some nations use **tariffs,** or taxes, to increase the price of imported goods. By making imported items more expensive, tariffs encourage consumers to buy less expensive items that are manufactured in their own country.

Quotas are another barrier to trade. A **quota** is a limit on how many items of a particular product can be imported from a certain nation.

TIME GLOBAL CITIZENS

NAME: BONO **HOME COUNTRY:** Ireland

ACHIEVEMENT: The lead singer of the mega-rock band U2 has proven himself to be one of the world's most effective voices for the poor. In 2005, he convinced leaders from the world's wealthiest countries, such as the United States and Japan, to approve a $50 billion aid package—including $25 billion for Africa. Thanks largely to Bono, the leaders pledged to make lifesaving drugs available to poor people with HIV and also agreed that the 18 poorest African nations did not have to pay back money they had borrowed from several nations and organizations. Now they can spend the money on health care and schools rather than on paying back loans.

QUOTE: "There is a goal out there worthy of our generation. . . . It is the defeat of humanity's oldest foe: disease."

Bono sings for children in Ghana, while U.S. Treasury Secretary Paul O'Neil looks on.

GEORGE PIMENTEL/WIREIMAGE.COM; (INSET) AP WIDE WORLD

CITIZENS IN ACTION How might Bono's actions today help people 10 years from now?

Free Trade

In recent years, many countries have agreed to get rid of trade barriers. The removal of trade limits so that goods flow freely among countries is called **free trade.** Often countries sign treaties agreeing to free trade. For example, in 1994 Canada, the United States, and Mexico joined together in the North American Free Trade Agreement (NAFTA). This pact removed most trade barriers between the three nations.

Interdependence and Technology

Growing trade among the world's countries has resulted in the globalization of the world's economies. As a result, the world's people and economies have become more interdependent. **Interdependence** means that countries rely on each other for ideas, goods, services, and markets, or places to sell their goods. When economies are linked together, a drought or a war in one region can cause price increases or shortages in another region far away.

Interdependence has come about in part because of new technologies. During the past 200 years, the invention of new technologies has occurred much faster than at any other time in history. Advances in transportation, such as trains and airplanes, and in communication, such as telephones and the Internet, have contributed greatly to globalization.

Reading Check **Explaining** Explain why trade barriers exist, and describe two types of trade barriers.

Section 3 Review

Social Studies ONLINE
Study Central™ To review this section, go to glencoe.com.

Vocabulary

1. **Explain** the significance of:
   - **a.** natural resource
   - **b.** renewable resource
   - **c.** nonrenewable resource
   - **d.** economic system
   - **e.** developed country
   - **f.** developing country
   - **g.** newly industrialized country
   - **h.** export
   - **i.** import
   - **j.** tarriff
   - **k.** quota
   - **l.** free trade
   - **m.** interdependence

Main Ideas

2. **Explaining** Why do people need natural resources?
3. **Comparing and Contrasting** Use a Venn diagram like the one below to compare and contrast developed and developing countries.

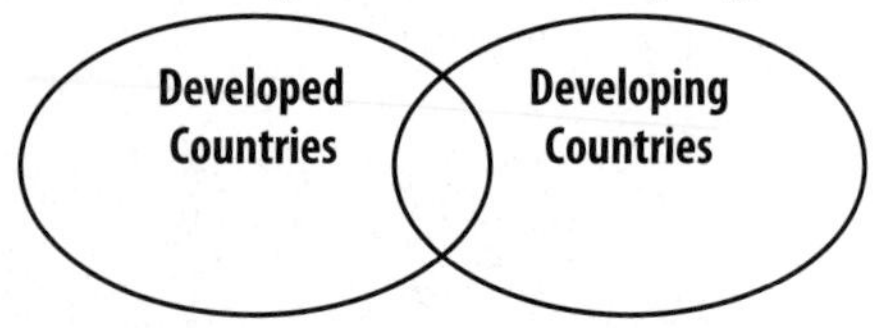

Critical Thinking

4. **Analyzing** Why has the world become more interdependent in recent years?
5. **BIG Idea** Explain how the distribution of natural resources relates to world trade.
6. **Challenge** In what ways might interdependence influence a place's cultural identity? Explain in two paragraphs.

Writing About Geography

7. **Using Your FOLDABLES** Use your Foldable to write a paragraph that predicts how population patterns might affect world resources in the future.

Visual Summary

World Population

- Low death rates and high birthrates have led to rapid population growth.
- Some areas of the world are more densely populated than others.
- Nearly half of the world's population lives in cities.

Commuters, New York City

Culture

- Culture is the way of life of a group of people who share similar beliefs and customs.
- Cultures change over time and influence one another.
- Modern technology has broken down barriers and helped create a global culture.

Oil worker, Iraq

Natural Resources

- Renewable resources either cannot be used up or can be replaced.
- Some resources—such as fossil fuels and minerals—are nonrenewable.

World Economies

- The four kinds of economic systems are traditional, command, market, and mixed.
- Developed countries use advanced technology and are highly productive.
- Developing countries have less advanced technology and are generally less productive.

World Trade

- In recent years, many countries have agreed to eliminate trade barriers.
- Growing trade among countries has made the world's people more interdependent.

Mexican president Vicente Fox (left), Canadian prime minister Jean Chrétien (center), and U.S. president George W. Bush (right) celebrate a trade agreement.

Grocery store in Yogakarta, Indonesia

Study anywhere, anytime! Download quizzes and flash cards to your PDA from glencoe.com.

STANDARDIZED TEST PRACTICE

Think of answers in your head before looking at the possible answers so that the choices on the test will not throw you off or trick you.

Reviewing Vocabulary

Directions: Choose the word(s) that best completes the sentence.

1. Geographers measure _______ to determine how crowded a country or region is.

A refugees
B population density
C death rates
D birthrates

2. _______ are people who are forced to flee to another country to escape wars, persecution, or natural disasters.

A Immigrants
B Free traders
C Refugees
D Importers

3. A(n) _______ group shares a language, history, religion, and some physical traits.

A democratic
B global
C social
D ethnic

4. Countries with a mix of agriculture and a great deal of manufacturing and service industries are called _______ countries.

A developed
B underdeveloped
C overdeveloped
D developing

Reviewing Main Ideas

Directions: Choose the best answer for each question.

Section 1 *(pp. 72–76)*

5. One reason for the rapid increase in world population over the last two centuries is _______.

A increased migration
B increased population density
C improved health care
D urbanization

6. An example of a "push factor" for migration is _______ in the homeland.

A a shortage of jobs
B an abundance of jobs
C low population density
D an abundance of farmland

Section 2 *(pp. 82–89)*

7. In recent years more and more countries and regions are sharing cultural traits resulting in a(n) _______ culture.

A isolated
B global
C refugee
D ethnic

Section 3 *(pp. 92–96)*

8. To answer the questions of what goods and services to produce, how to produce them, and who will receive them, societies develop _______.

A quota systems
B trading systems
C manufacturing systems
D economic systems

GO ON

Critical Thinking

Directions: Choose the best answer for each question.

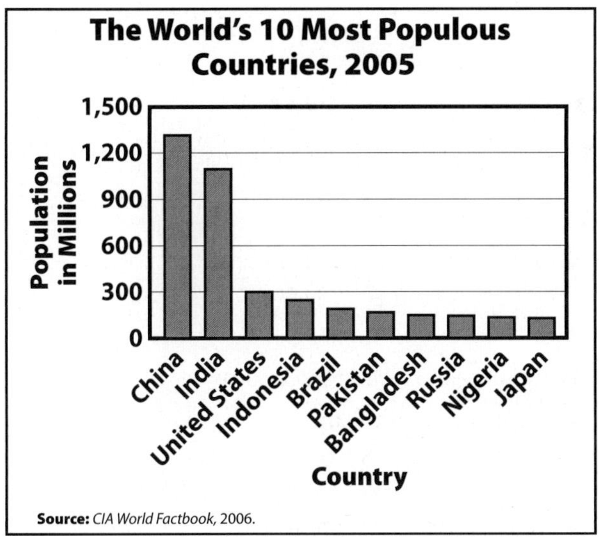

9. Based on the graph, what continent would likely be the most densely populated?

A North America

B South America

C Asia

D Africa

10. In which two countries would you expect to see the highest birthrates?

A United States and Russia

B India and China

C Pakistan and Bangladesh

D Brazil and Nigeria

Document-Based Questions

Directions: Analyze the document and answer the short-answer questions that follow.

> *[I]magine . . . that the world really is a 'global village.' . . . Say this village has 1,000 individuals, with all the characteristics of today's human race distributed in exactly the same proportions. . . .*
>
> *Some 150 of the inhabitants live in [a wealthy] area of the village, about 780 in poorer districts. Another 70 or so live in a neighborhood that is [changing]. The average income per person is $6,000 a year. . . . But just 200 people [own] 86 percent of all the wealth, while nearly half of the villagers are eking out an existence on less than $2 per day. . . .*
>
> *Life expectancy in the affluent district is nearly 78 years, in the poorer areas 64 years—and in the very poorest neighborhoods a mere 52 years. . . . Why do the poorest lag so far behind? Because in their neighborhoods there is a far higher incidence of infectious diseases and malnutrition, combined with an [serious] lack of access to safe water, sanitation, health care, adequate housing, education, and work.*
>
> —*Kofi Annan,* Millennium Report, *2000*

11. Describe the differences in income in the village.

12. According to the writer, where is life expectancy higher and why is this so?

Extended Response

13. Write a letter to a government leader in which you try to persuade him or her to invest taxpayer money into research on how to better use our energy resources. Explain why you think either renewable or nonrenewable resources deserve more funds for research.

STOP

Social Studies ONLINE

For additional test practice, use Self-Check Quizzes—Chapter 3 at glencoe.com.

Need Extra Help?

If you missed question. . .	1	2	3	4	5	6	7	8	9	10	11	12	13
Go to page. . .	74	76	84	94	73	76	89	94	73	73	94	94	93

TIME JOURNAL

It may be the middle of the night where you live, but in many parts of the world, people are well into their day. It's all because of the 24 time zones that divide up Earth. So while one part of the world sleeps, somewhere, kids are at school, workers are at their jobs, and some folks are having dinner. Take a look at what is happening on Earth at exactly the same moment during one day in April.

Monday, 7 a.m. **LOS ANGELES, CALIFORNIA** Some people are just waking up. Others are sitting down to breakfast. Early birds are headed to their jobs hoping to avoid traffic jams on the state's freeways. ▶

◀ **Monday, 10 a.m.** **WASHINGTON, D.C.** Workers are at their desks. And at the White House, the wheels of government have been turning since 7 a.m. or even earlier, where 12-hour workdays are routine.

◀ **Monday, 11 a.m.** **RIO DE JANEIRO, BRAZIL** Almost every day is a beach day in Rio. While beachgoers are enjoying sun and sand, traffic jams clog the city's streets, students are at their desks, and Rio's stores are filled with shoppers.

◀ **Monday, 2 p.m.** **DAKAR, SENEGAL** Outdoor markets are packed in this west African nation. School is winding down for the day, and fishers are returning home with their day's catch from the Atlantic Ocean.

Monday, 5 p.m. **CAIRO, EGYPT** This capital city is filled with the sounds of people being called to prayer, vendors selling their goods at outdoor bazaars, and the blare of car and bus horns on traffic-clogged streets. Tourists and residents alike can marvel at the Pyramids of Giza built almost 5,000 years ago. ▼

Monday, 4 p.m. **PARIS, FRANCE** School is out and some kids are playing soccer, a favorite pastime. Other students are studying for exams to get into special high schools. Some tourists are having their pictures taken in front of the Eiffel Tower while others are visiting the city's famous museums, perhaps catching a glimpse of the *Mona Lisa*. ▶

TOP TO BOTTOM: CREATAS/SUPERSTOCK; AP PHOTO; JIM ZUCKERMAN/CORBIS; KURT SCHOLZ/SUPERSTOCK; LISA ENGLEBRECHT/DANITADELIMONT.COM; GARY COOK/ALAMY; GLOBE: NASA

A DAY IN THE LIFE OF THE WORLD'S PEOPLE

Tuesday, 2 a.m. WELLINGTON, NEW ZEALAND What do Kiwis (a nickname for New Zealanders) do when they can't sleep? They might count sheep. That's because the nation's 45 million woolly animals outnumber the island-nation's human inhabitants 11 to 1. ▶

Monday, 10 p.m. BEIJING, CHINA The day is winding down for most of the 15 million residents of the nation's capital. China, with its more than one billion people, has one of the world's fastest growing economies. Night workers, including people who work with American companies, are starting their day, keeping to a U.S. time schedule. ▶

Monday, 6 p.m. MOSCOW, RUSSIA This huge country has 11 time zones. The nation, which has turned from communism to democracy, is undergoing a construction boom. Workers are going home for dinner. ▼

◀ **Monday, 8 p.m. DHAKA, BANGLADESH** Some residents of this city are sitting down to a dinner of fish or spicy curries. Meanwhile, fans of cricket, a popular sport in this country, are cheering for their favorite team.

TOP TO BOTTOM: ILLUSTRATION BY BOOKMAPMAN; JOSE AZEL/GETTY IMAGES; MACDUFF EVERTON/CORBIS; AP PHOTO; ITAR-TASS/VITALY BELOUSOV/NEWSCOM

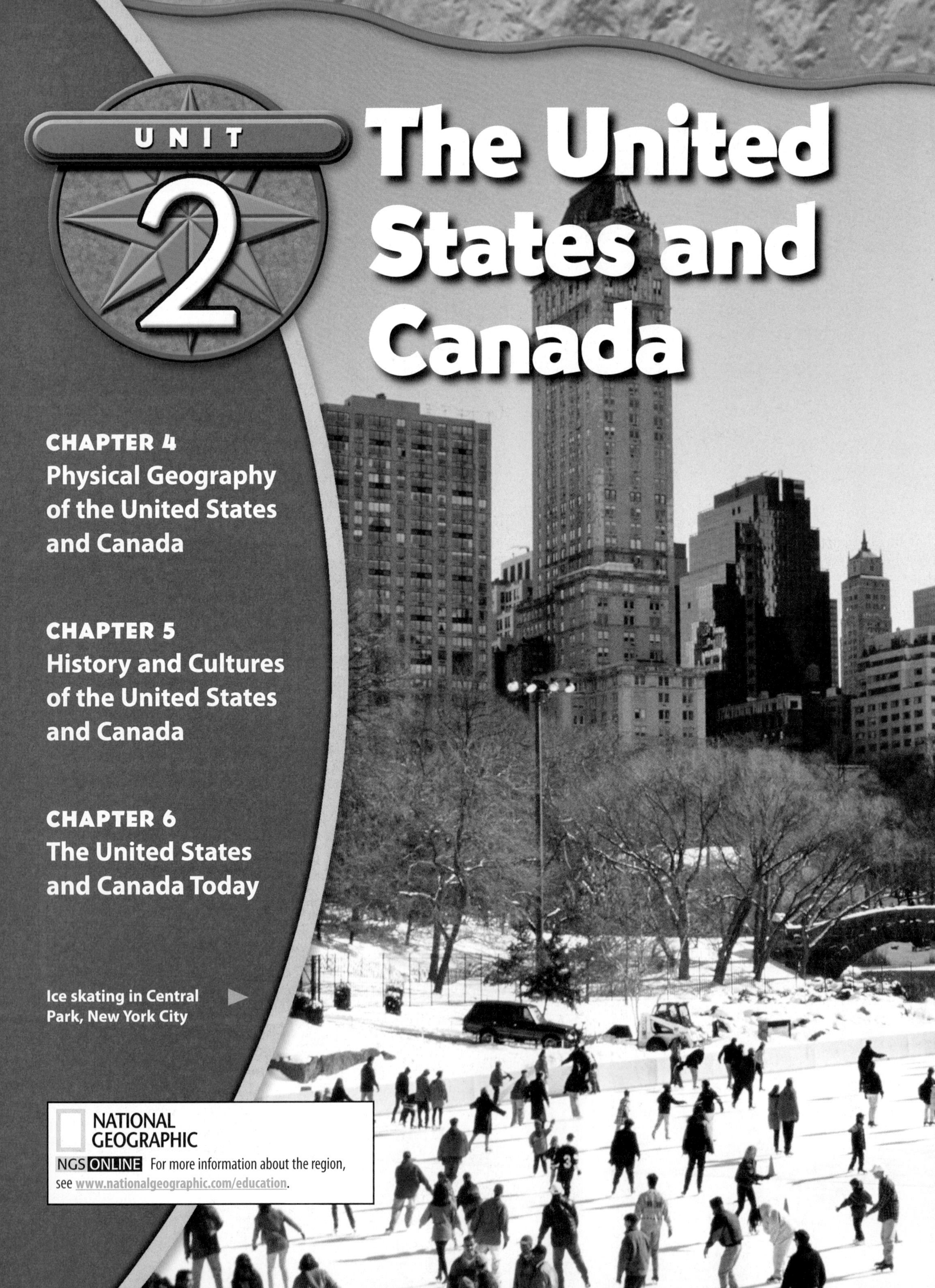

UNIT 2

The United States and Canada

CHAPTER 4
Physical Geography of the United States and Canada

CHAPTER 5
History and Cultures of the United States and Canada

CHAPTER 6
The United States and Canada Today

Ice skating in Central Park, New York City ▶

NATIONAL GEOGRAPHIC

NGS ONLINE For more information about the region, see www.nationalgeographic.com/education.

Regional Atlas

The United States and Canada

Where Is It?

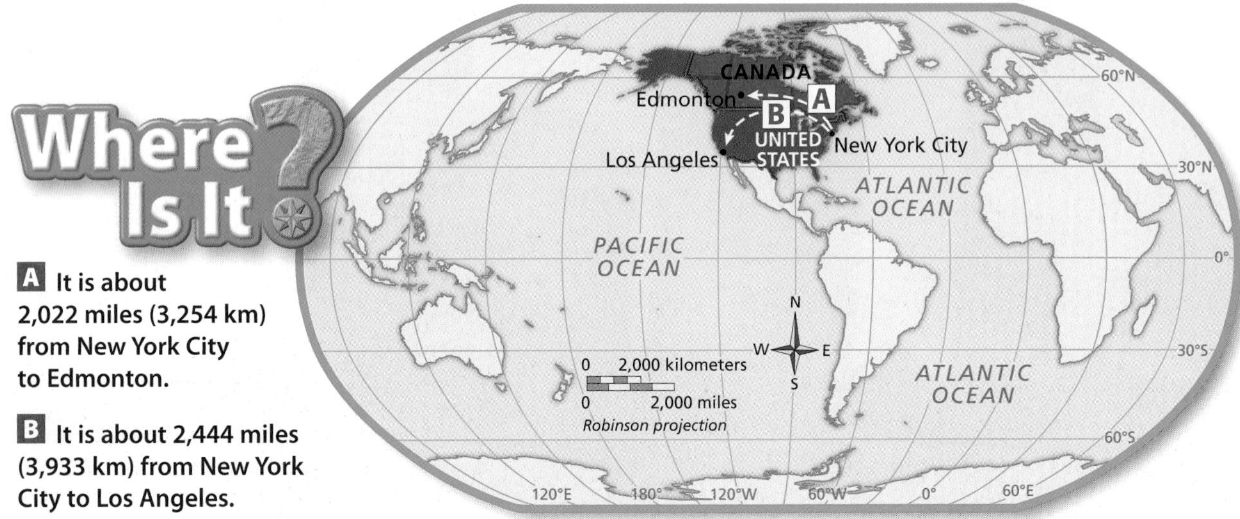

A It is about 2,022 miles (3,254 km) from New York City to Edmonton.

B It is about 2,444 miles (3,933 km) from New York City to Los Angeles.

The land area of the region of the United States and Canada is about 7.7 million square miles (19.9 million sq. km). In area, Canada is the second-largest country in the world, and the United States is the third largest. The United States has many more people than Canada, however, and is the third-most-populous country in the world, after China and India.

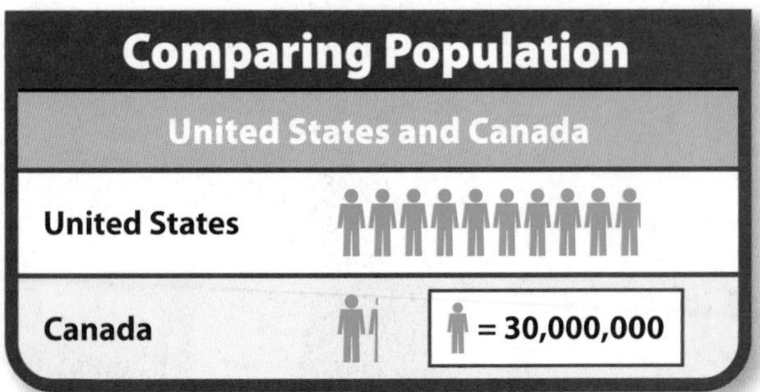

Comparing Population

United States and Canada	
United States	10 figures
Canada	1 figure

(1 figure = 30,000,000)

Source: *World Population Data Sheet*, 2005.

GEO Fast Facts

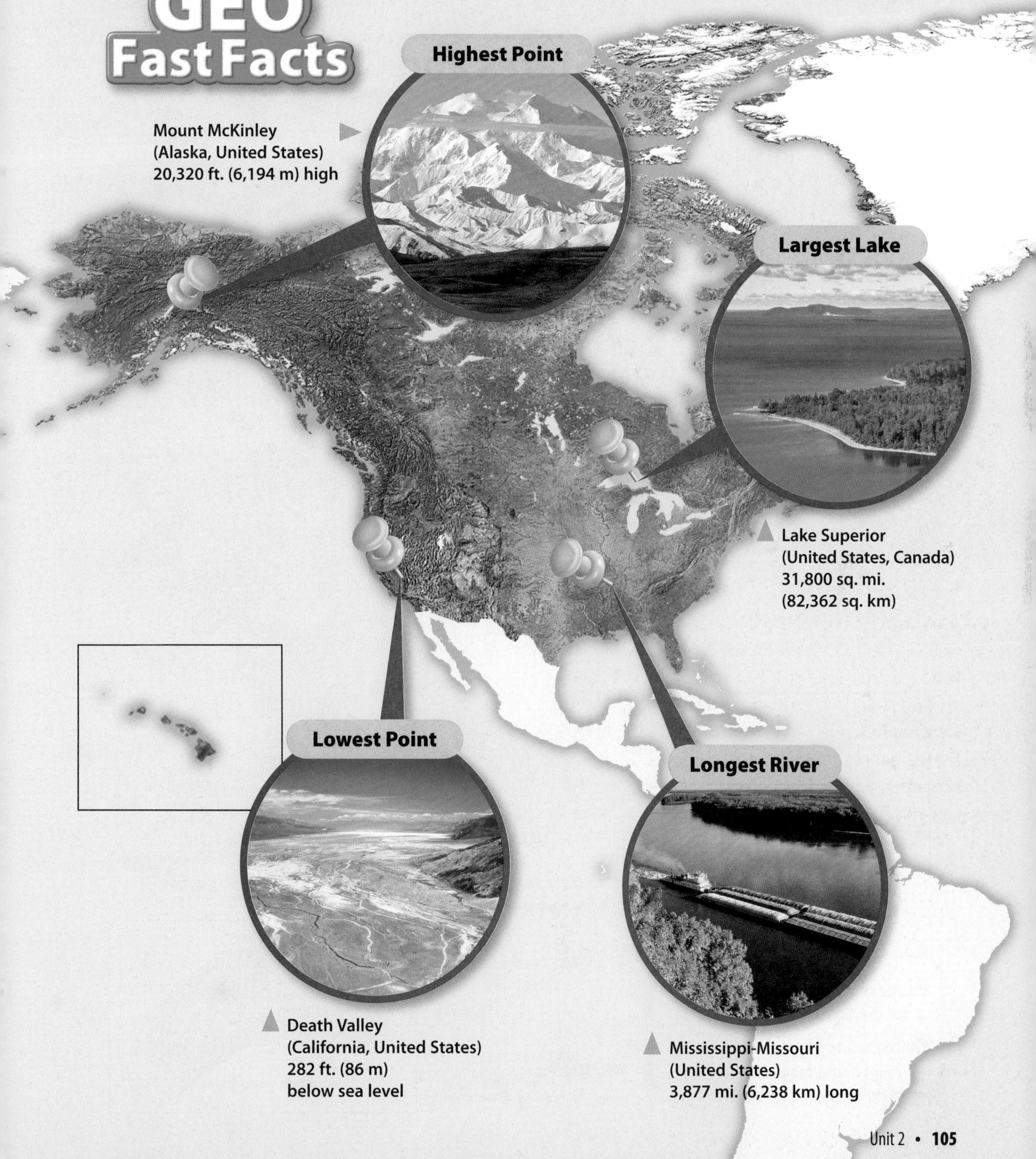

Highest Point

Mount McKinley
(Alaska, United States)
20,320 ft. (6,194 m) high

Largest Lake

Lake Superior
(United States, Canada)
31,800 sq. mi.
(82,362 sq. km)

Lowest Point

Death Valley
(California, United States)
282 ft. (86 m)
below sea level

Longest River

Mississippi-Missouri
(United States)
3,877 mi. (6,238 km) long

Unit 2 Regional Atlas

The United States and Canada

PHYSICAL

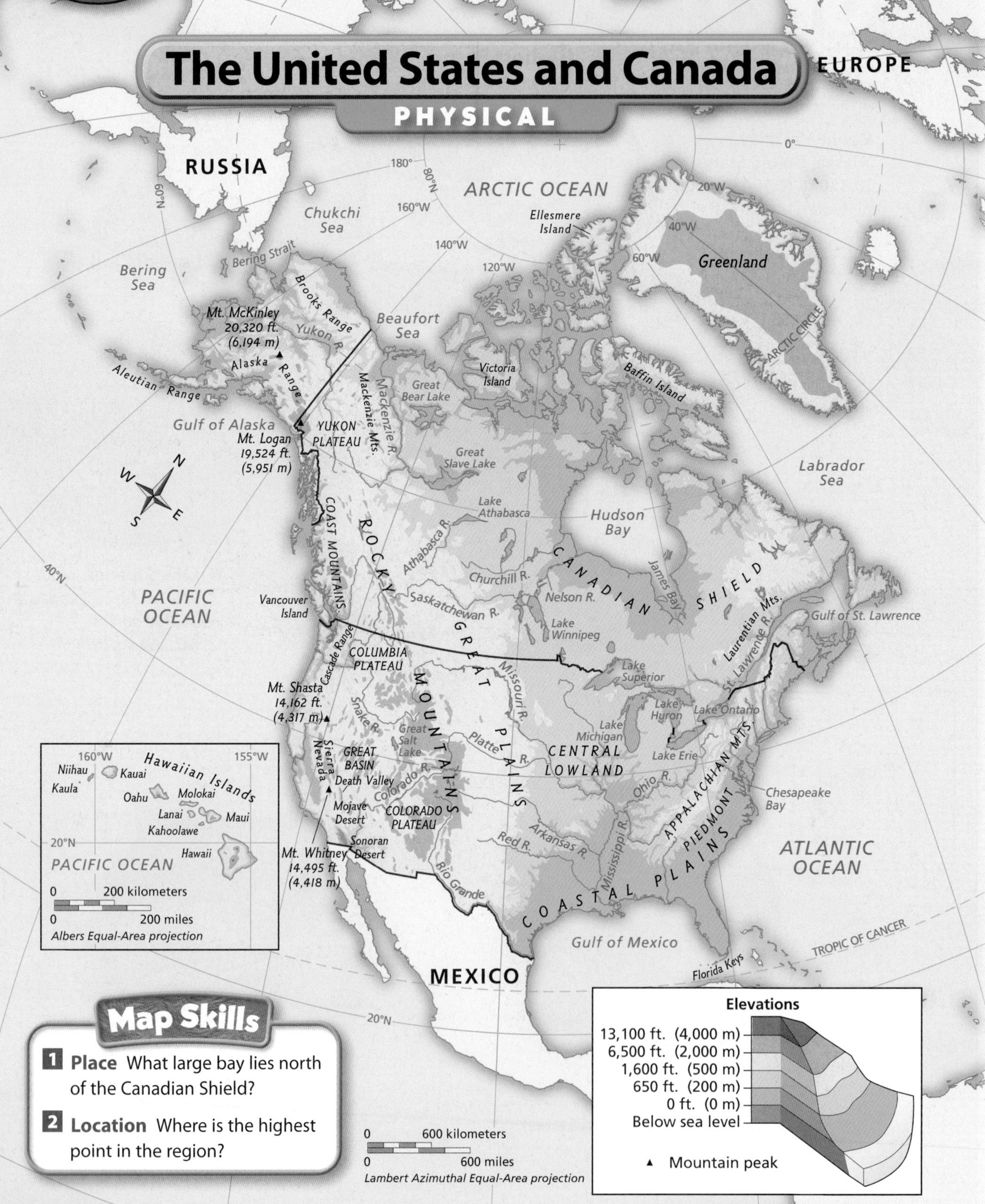

Map Skills

1. **Place** What large bay lies north of the Canadian Shield?
2. **Location** Where is the highest point in the region?

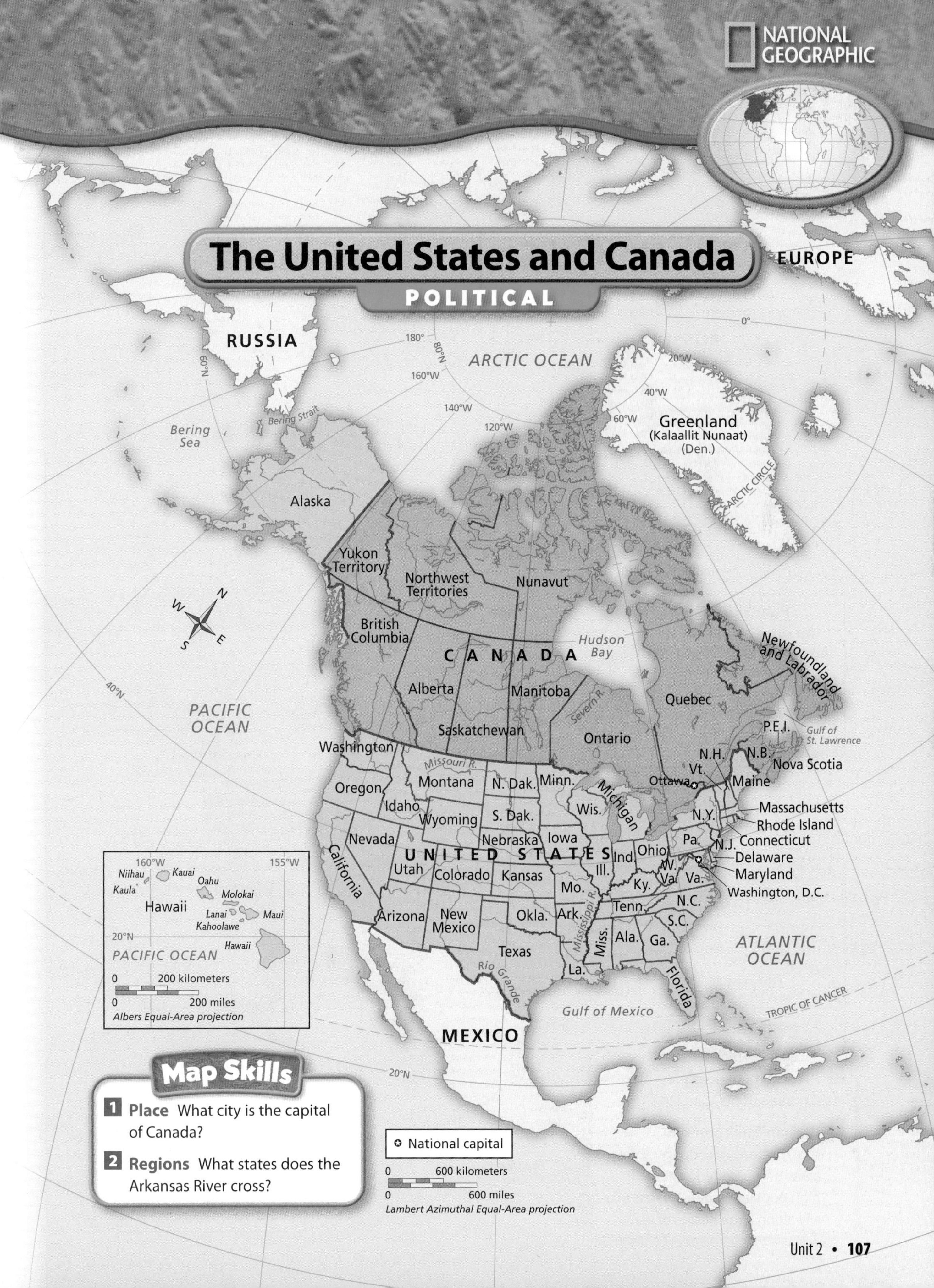
NATIONAL GEOGRAPHIC
The United States and Canada
POLITICAL
EUROPE
RUSSIA
ARCTIC OCEAN
Bering Sea
Bering Strait
Greenland
(Kalaallit Nunaat)
(Den.)
ARCTIC CIRCLE
Alaska
Yukon Territory
Northwest Territories
Nunavut
British Columbia
CANADA
Hudson Bay
Newfoundland and Labrador
Alberta
Manitoba
Saskatchewan
Severn R.
Quebec
Ontario
PACIFIC OCEAN
P.E.I.
Gulf of St. Lawrence
N.B.
Nova Scotia
N.H.
Vt.
Ottawa
Maine
Washington
Missouri R.
Montana
N. Dak.
Minn.
Oregon
Idaho
Wyoming
S. Dak.
Wis.
Michigan
N.Y.
Massachusetts
Rhode Island
Connecticut
Delaware
Maryland
Washington, D.C.
Nevada
Nebraska
Iowa
Pa.
N.J.
UNITED STATES
California
Utah
Colorado
Kansas
Ill.
Ind.
Ohio
W. Va.
Va.
Mo.
Ky.
Arizona
New Mexico
Okla.
Ark.
Tenn.
N.C.
S.C.
Mississippi R.
Miss.
Ala.
Ga.
Texas
La.
Rio Grande
Florida
ATLANTIC OCEAN
Gulf of Mexico
TROPIC OF CANCER
MEXICO
Hawaii
Niihau
Kauai
Kaula
Oahu
Molokai
Lanai
Maui
Kahoolawe
Hawaii
PACIFIC OCEAN
0 200 kilometers
0 200 miles
Albers Equal-Area projection
National capital
0 600 kilometers
0 600 miles
Lambert Azimuthal Equal-Area projection

Map Skills

1. **Place** What city is the capital of Canada?
2. **Regions** What states does the Arkansas River cross?

UNIT 2

Regional Atlas

The United States and Canada

POPULATION DENSITY

POPULATION

Per sq. mi.	Per sq. km
1,250 and over	500 and over
250–1,249	100–499
62–249	25–99
25–61	10–24
2.5–24	1–9
Less than 2.5	Less than 1

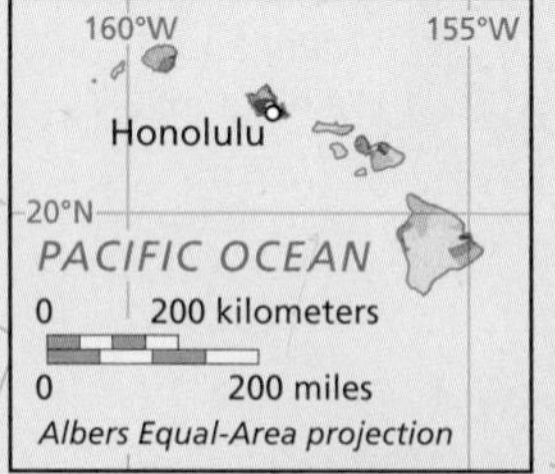

Cities
(Statistics reflect metropolitan areas.)

- ■ Over 10,000,000
- □ 5,000,000–10,000,000
- ⊙ 3,000,000–5,000,000
- • 2,000,000–3,000,000
- ○ Under 2,000,000

Map Skills

1 Place Where does most of Canada's population live?

2 Human-Environment Interaction Why do you think areas of the United States with high population densities are usually along large bodies of water?

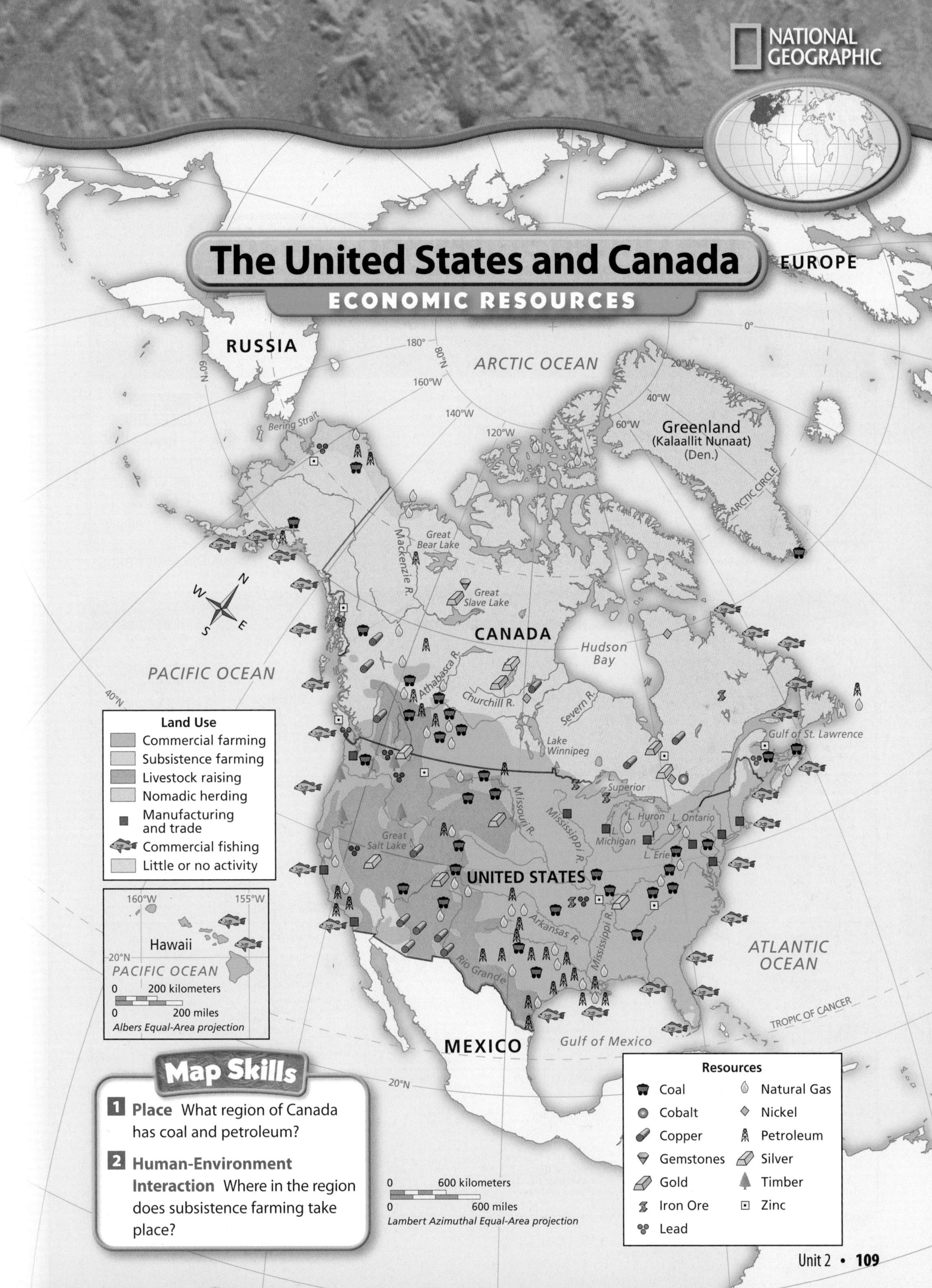

Map Skills

1. **Place** What region of Canada has coal and petroleum?
2. **Human-Environment Interaction** Where in the region does subsistence farming take place?

Regional Atlas

The United States and Canada

Country and Capital	Literacy Rate	Population and Density	Land Area	Life Expectancy (Years)	GDP* Per Capita (U.S. dollars)	Television Sets (per 1,000 people)	Flag and Language
UNITED STATES Washington, D.C.	97%	296,500,000 80 per sq. mi. 31 per sq. km	3,717,796 sq. mi. 9,629,047 sq. km	78	$40,100	844	English
CANADA Ottawa	97%	32,000,000 8 per sq. mi. 3 per sq. km	3,849,670 sq. mi. 9,970,599 sq. km	80	$31,500	709	English, French

Sources: *CIA World Factbook*, 2005; Population Reference Bureau, *World Population Data Sheet*, 2005.

*Gross Domestic Product

U.S. State Names: Meaning and Origin

State	Capital	Meaning and Origin
ALABAMA	Montgomery	"thicket clearers" (Choctaw)
ALASKA	Juneau	"the great land" (Aleut)
ARIZONA	Phoenix	"small spring" (O'odham/Pima)
ARKANSAS	Little Rock	"south wind" (Ohio Valley Native Americans' name for the Quapaws)
CALIFORNIA	Sacramento	named after Calafia, a place in a romantic Spanish story
COLORADO	Denver	"colored red" (Spanish)
CONNECTICUT	Hartford	"long river place" (Mohegan)
DELAWARE	Dover	named for Virginia's colonial governor, Thomas West, Baron De La Warr
FLORIDA	Tallahassee	"feast of flowers" (Spanish)
GEORGIA	Atlanta	named for England's King George II
HAWAII	Honolulu	unknown (Native Hawaiian)
IDAHO	Boise	unknown
ILLINOIS	Springfield	"tribe of superior men" (Algonquian)
INDIANA	Indianapolis	"land of Indians" (European American)
IOWA	Des Moines	name of a Native American group
KANSAS	Topeka	"people of the south wind" (Sioux)
KENTUCKY	Frankfort	"land of tomorrow," "cane and turkey lands," or "meadow lands" (Native American/ Iroquoian)
LOUISIANA	Baton Rouge	named for France's King Louis XIV
MAINE	Augusta	nautical term distinguishing the mainland from islands
MARYLAND	Annapolis	named in honor of the wife of England's King Charles I
MASSACHUSETTS	Boston	"at or about the great hill" (Native American)
MICHIGAN	Lansing	"great lake" (Ojibwa)
MINNESOTA	Saint Paul	"water that reflects the sky" (Dakota)
MISSISSIPPI	Jackson	"father of waters" (Chippewa)

Land areas and flags not drawn to scale

State	Capital	Meaning and Origin
MISSOURI	Jefferson City	"town of the large canoes" (Sioux)
MONTANA	Helena	"mountainous" (Spanish)
NEBRASKA	Lincoln	"flat water" (Oto)
NEVADA	Carson City	"snowcapped" (Spanish)
NEW HAMPSHIRE	Concord	named for Hampshire, a county in England
NEW JERSEY	Trenton	named for Isle of Jersey, a British territory
NEW MEXICO	Santa Fe	named for the state's former colonial ruler, Mexico
NEW YORK	Albany	named in honor of the English Duke of York
NORTH CAROLINA	Raleigh	named in honor of England's King Charles I
NORTH DAKOTA	Bismarck	"friend" (Sioux); the Dakota were a Sioux people
OHIO	Columbus	"great river" (Iroquoian)
OKLAHOMA	Oklahoma City	"red people" (Choctaw)
OREGON	Salem	unknown meaning (Native American)
PENNSYLVANIA	Harrisburg	"Penn's woodland," named for the father of Pennsylvania's founder, William Penn
RHODE ISLAND	Providence	named for the Greek island of Rhodes
SOUTH CAROLINA	Columbia	named for England's King Charles I
SOUTH DAKOTA	Pierre	"friend" (Sioux); the Dakota were a Sioux people
TENNESSEE	Nashville	named for tana-see, "the meeting place" (Yuchi)
TEXAS	Austin	"friends" (Caddo)
UTAH	Salt Lake City	"people of the mountains" (Ute)
VERMONT	Montpelier	"green mountain" (French)
VIRGINIA	Richmond	named for the unmarried Queen Elizabeth I of England, known as "the Virgin Queen"
WASHINGTON	Olympia	named in honor of George Washington
WEST VIRGINIA	Charleston	began as the western part of Virginia before becoming a state in 1863
WISCONSIN	Madison	"river of red stone" (Algonquian)
WYOMING	Cheyenne	"upon the great plain" (Delaware)

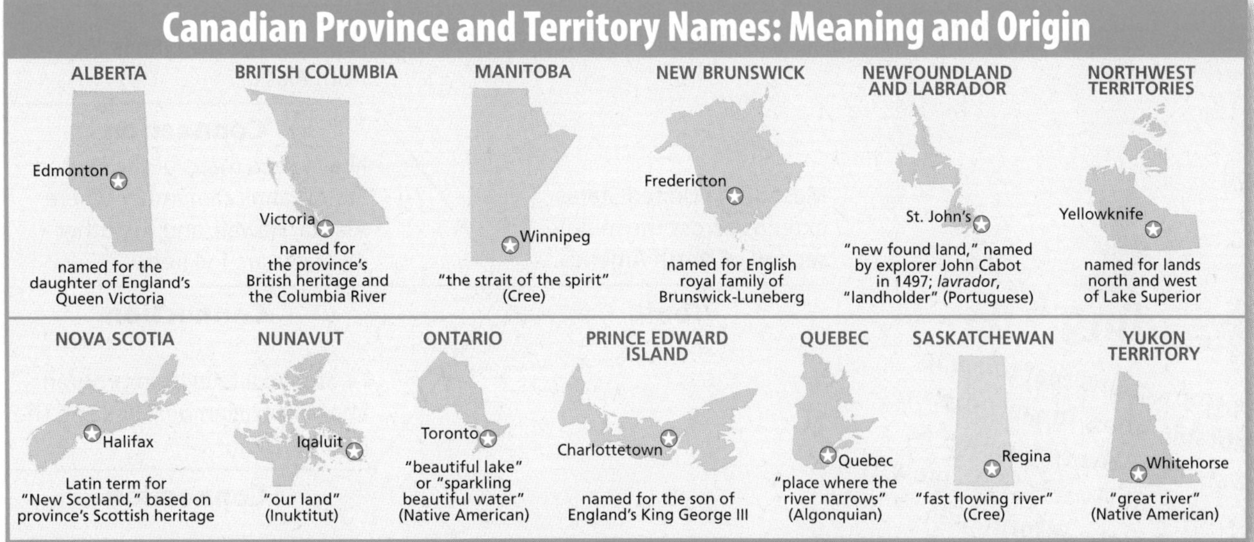

For more country facts, go to the **Nations of the World Databank** at glencoe.com.

Reading Social Studies

Making Connections

Learn It!

Making connections between what you read and what you already know is an important step in learning. Connections can be based on personal experiences (text-to-self), what you have read before (text-to-text), or events in other places (text-to-world).

As you read, ask connecting questions. Are you reminded of a personal experience? Have you read about the topic before?

- Read the paragraph below.
- Can you make one or more connections to the information?
- Look at the diagram for some sample connections.

> Most of the United States stretches across the middle part of North America. The 48 states in this part of the country are contiguous, or joined together inside a common boundary. Two states lie apart from the other 48. Alaska lies in the northwestern part of North America, adjacent to Canada. Hawaii is an island group in the Pacific Ocean, about 2,400 miles (3,862 km) southwest of California.
>
> *—from page 117*

Topic: Most of the United States extends across the middle section of North America.

Connection: I know someone who is from Alaska.

Connection: I have seen maps of the United States, and I remember where Alaska, Hawaii, and the other 48 states are located.

Connection: I watched a television program about Hawaiian beaches and the Pacific Ocean.

Reading Tip

Make connections that relate to memorable times in your life. The stronger the connection is, the more likely it is that you will remember the information.

Read to Write Activity

As you read Chapters 4, 5, and 6, choose five words or phrases from each chapter that make a connection to something you already know.

2 Practice It!

Read the following paragraph from this unit.
- Draw a graphic organizer like the one below.
- List the topic of the reading along with connections to the information.
- Share your connections with a partner.
- Compare your list with your partner's, and discuss their similarities and differences.

> Canadians are enthusiastic about hockey—a sport that began in Canada—as well as lacrosse, which began as a Native American game. Many Canadians also enjoy hunting and fishing.
>
> —*from page 150*

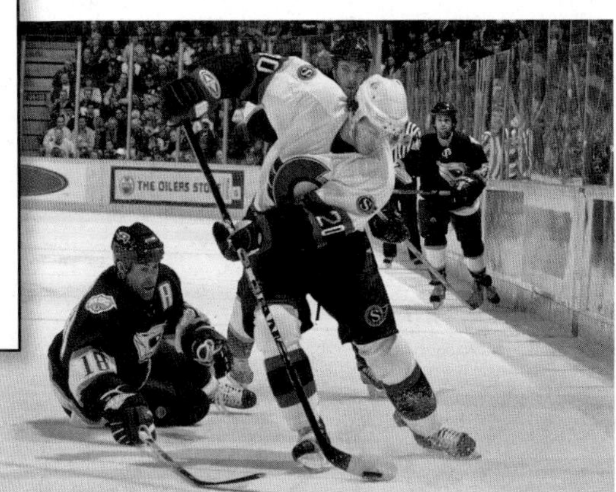

Ice hockey in Canada

Topic

Connection

Connection

Connection

3 Apply It!

As you read the chapters in the unit, try to identify one concept that makes the following connections.

Chapter	Connection
4	**Text-to-self:**
5	**Text-to-text:**
6	**Text-to-world:**

Physical Geography of the United States and Canada

Essential Question

Regions The United States and Canada cover most of the land area of North America, stretching from the Pacific Ocean to the Atlantic Ocean. These two huge countries share many of the same physical features, resources, and climates. How do landforms and climate help or hinder transportation in a vast region?

◀ Wheat harvest, Michigan

BIG Ideas

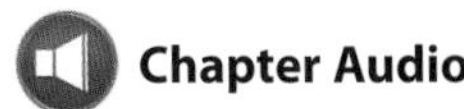

Section 1: Physical Features

BIG IDEA Geographers organize the Earth into regions that share common characteristics. The United States and Canada share a long border and many landforms. Their economies are closely linked by trade. Their governments have also worked together on major projects that have changed the land and benefited both countries.

Section 2: Climate Regions

BIG IDEA The physical environment affects how people live. A diversity of climates in the United States and Canada leads to different ways of life. Some parts of this region experience natural hazards that can threaten people's safety.

FOLDABLES™ Study Organizer

Organizing Information Make this Foldable to help you organize information about the physical features and climates of the United States and Canada.

Step 1 Fold a sheet of paper in half, leaving a ½-inch tab along one edge.

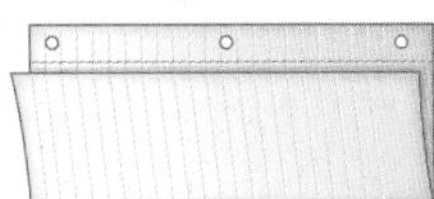

Step 2 Then fold into three sections.

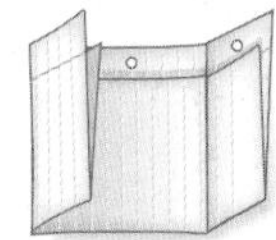

Step 3 Draw a Venn diagram like the one below and then cut along the folds to create three tabs.

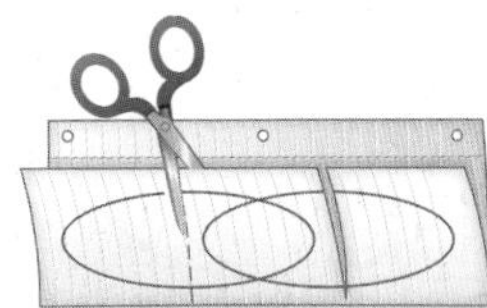

Step 4 Label your Foldable as shown.

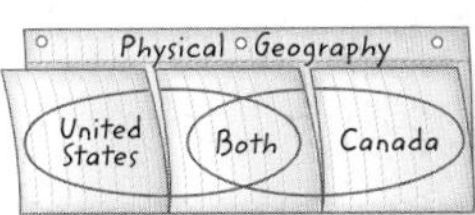

Reading and Writing Using the notes in your Foldable, write several short journal entries from an imaginary trip through Canada and the United States. In your entries, describe the landforms and climates you encounter.

Visit glencoe.com and enter ***QuickPass***™ code EOW3109c4 for Chapter 4 resources.

Guide to Reading

BIG Idea

Geographers organize the Earth into regions that share common characteristics.

Content Vocabulary

- contiguous *(p. 117)*
- megalopolis *(p. 117)*
- prairie *(p. 118)*
- cordillera *(p. 118)*
- canyon *(p. 119)*
- navigable *(p. 119)*
- glacier *(p. 119)*
- divide *(p. 120)*

Academic Vocabulary

- constrain *(p. 117)*
- route *(p. 119)*

Reading Strategy

Analyzing Information Use a Venn diagram like the one below to compare landforms in the eastern, western, and interior parts of the United States and Canada.

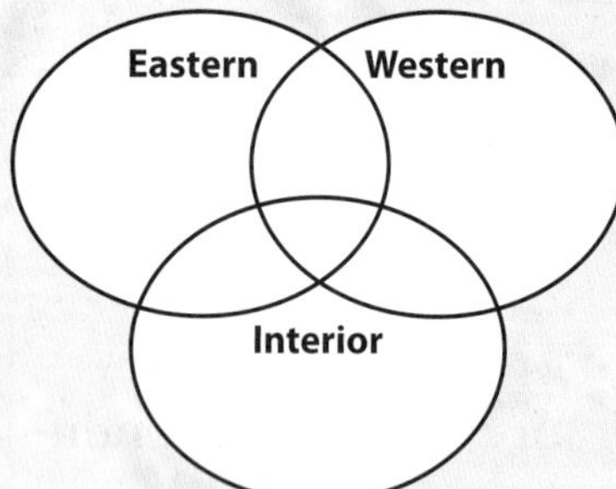

Physical Features

 Section Audio  Spotlight Video

Picture This Standing at the Grand Canyon's edge, you can see for miles. Its sheer size—277 miles (445 km) long, with walls rising up to 6,000 feet (1,829 m)—is almost mind-boggling. The Grand Canyon was formed by the Colorado River over a period of 6 million years. To learn more about the physical features of the United States and Canada, read Section 1.

▼ Grand Canyon

Major Landforms

Main Idea **The region rises in elevation from east to west.**

Geography and You Do you live in an area that is flat, hilly, or mountainous? Read to find out about the major landforms of the United States and Canada.

The United States and Canada form a region that covers most of North America. This region is bordered by the cold Arctic Ocean in the north, the Atlantic Ocean to the east, and the warm waters of the Gulf of Mexico in the southeast. The Pacific Ocean borders the western coast.

Canada occupies most of the northern part of North America. Canada's vast size makes it the second-largest country in the world, after Russia. The United States is the third-largest country. Most of the United States stretches across the middle part of North America. The 48 states in this part of the country are **contiguous,** or joined together inside a common boundary. Two states lie apart from the other 48. Alaska lies in the northwestern part of North America, adjacent to Canada. Hawaii is an island group in the Pacific Ocean, about 2,400 miles (3,862 km) southwest of California.

Eastern Lowlands and Highlands

The United States and Canada have a variety of landforms. A broad lowland runs along the Atlantic and the Gulf of Mexico coasts. In northeastern areas, the thin and rocky soil **constrains,** or limits, farming. A fertile, hilly area called the Piedmont, however, stretches inland from the coastal plain. Excellent harbors along the Atlantic coast have led to the growth of shipping ports.

The cities of Halifax, Boston, New York City, Philadelphia, and Washington, D.C., all lie along or near the Atlantic coast. In the United States, Atlantic coastal cities and their suburbs form an almost continuous line of settlement. Geographers call this connected area of urban communities a **megalopolis.** The Atlantic megalopolis has long been an important economic, cultural, and political center of the United States.

The coastal plain along the Gulf of Mexico is wider than the Atlantic plain. Soils in this region are better than those along the Atlantic coast. Large cities here include Houston and New Orleans.

West and north of the Atlantic coastal plain spread a number of highland areas.

NATIONAL GEOGRAPHIC

New York City

New York City is one of several huge cities that developed along the Atlantic coastal plain.
Location **Where is the area called the Piedmont located?**

Social Studies ONLINE

Student Web Activity Visit glencoe.com and complete the Chapter 4 Web Activity about the Piedmont.

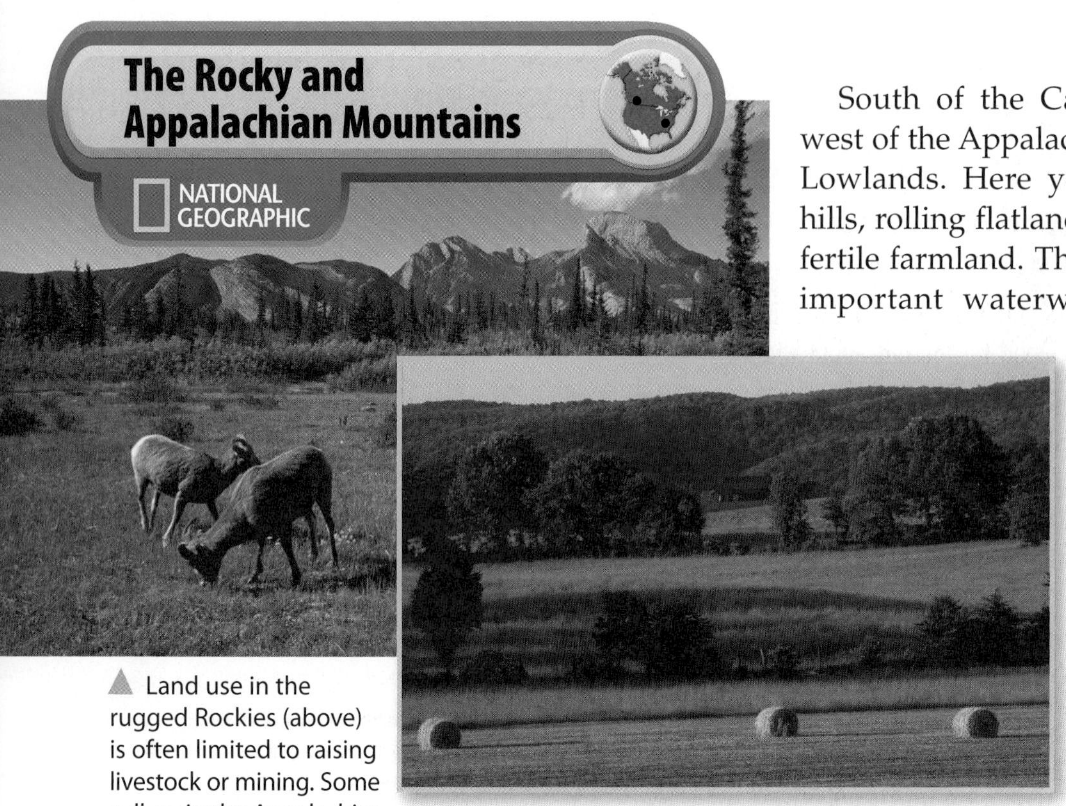

▲ Land use in the rugged Rockies (above) is often limited to raising livestock or mining. Some valleys in the Appalachian Mountains, like this one in Tennessee (right), have fertile soil and are good for farming. ***Place*** **Why do the Appalachian and Rocky Mountains appear physically different?**

These include the Appalachian Mountains, which run from eastern Canada to Alabama. The Appalachians are the oldest mountains in North America. Their rounded peaks show their age. Erosion has worn them down over time. The highest peak, Mount Mitchell in North Carolina, reaches 6,684 feet (2,037 m). Rich coal deposits in the Appalachians fueled industrial growth in the late 1800s and early 1900s.

Interior Lowlands

West of the eastern highlands are vast interior lowlands. In the north lies the Canadian Shield. This horseshoe-shaped area of rocky hills, lakes, and evergreen forests wraps around the Hudson Bay. With poor soil and a cold climate, the Canadian Shield is not farmable. It does, however, contain many mineral deposits, such as iron ore, copper, and nickel.

South of the Canadian Shield and west of the Appalachians lie the Central Lowlands. Here you will find grassy hills, rolling flatlands, thick forests, and fertile farmland. This area also contains important waterways, including the Great Lakes and the Mississippi River. Large cities, such as Chicago, Detroit, Cleveland, and Toronto, are located in the Central Lowlands.

The Great Plains stretch west of the Mississippi River, gradually rising in elevation from east to west. Much of this vast region is a **prairie,** or rolling inland grasslands with fertile soil. The Great Plains once provided food for millions of buffalo and the Native Americans who lived there. Today farmers grow grains, and ranchers raise cattle on the land. The Great Plains also have reserves of coal, oil, and natural gas.

Western Mountains and Plateaus

West of the Great Plains is a **cordillera,** which is a group of mountain ranges that run side by side. Millions of years ago, collisions between tectonic plates created these towering mountains. At the eastern edge of the cordillera, the Rocky Mountains begin in Alaska and run south to New Mexico. Although they are younger and higher than the Appalachians, the Rockies have not been a barrier to travel. The Rockies contain passes, or low areas in the mountains, that allow people to cross them.

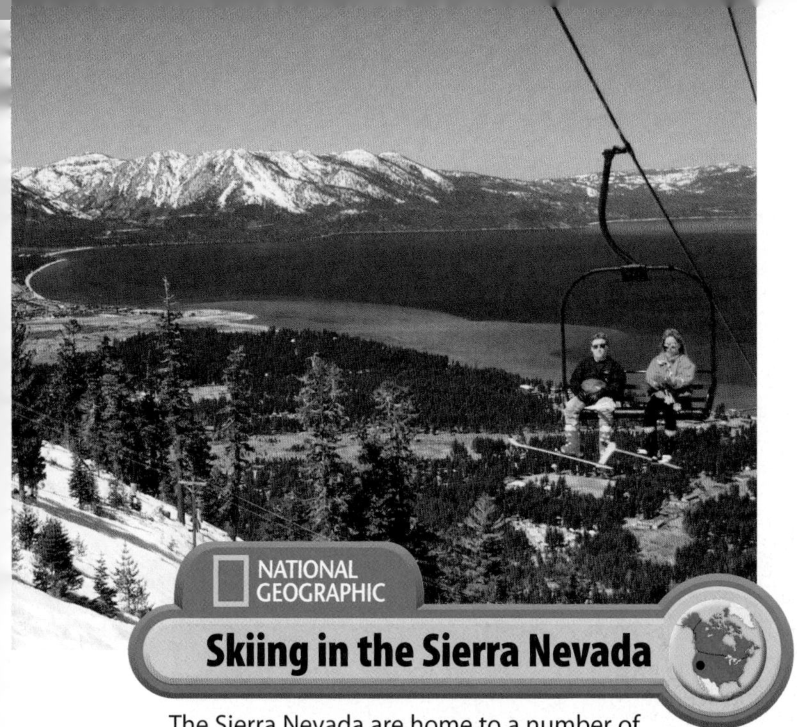

Skiing in the Sierra Nevada

The Sierra Nevada are home to a number of popular ski resorts, including those around Lake Tahoe on the California-Nevada border. ***Place*** **What other mountain chains are found near the Pacific coast of North America?**

Near the Pacific coast is a series of mountain chains that make up the western part of the cordillera. They are the Sierra Nevada, the Cascade Range, the Coast Ranges, and the Alaska Range. Mount McKinley in the Alaska Range rises to 20,320 feet (6,194 m) and is the highest point in North America.

Between these Pacific ranges and the Rocky Mountains is a stretch of dry basins and high plateaus. In the southern part of this area, rivers have worn through rock to create magnificent **canyons,** or deep valleys with steep sides. The most famous of these is the Grand Canyon of the Colorado River.

In the Pacific Ocean, eight large islands and 124 smaller islands make up the American state of Hawaii. The islands of Hawaii extend over a distance of about 1,500 miles (2,400 km). Volcanoes on the ocean floor erupted and formed these islands.

 **Making Generalizations** Describe the areas that make up the interior lowlands.

Bodies of Water

Main Idea The region's waterways provide transportation and electric power.

Geography and You Do you live near a river, lake, or ocean? What are the advantages and disadvantages of living by a body of water? Read to find out about the importance of waterways in the United States and Canada.

The United States and Canada have numerous freshwater lakes and rivers. Many of the region's rivers are **navigable,** or wide and deep enough to allow the passage of ships.

The Great Lakes

The Great Lakes—the world's largest group of freshwater lakes—lie in the central part of the region. Thousands of years ago, **glaciers,** or giant sheets of ice, formed Lake Superior, Lake Michigan, Lake Huron, Lake Erie, and Lake Ontario. The waters of these connected lakes flow into the St. Lawrence River, which empties into the Atlantic Ocean.

The St. Lawrence River is one of Canada's most important rivers. It flows for 750 miles (1,207 km) from Lake Ontario to the Gulf of St. Lawrence in the Atlantic Ocean. The Canadian cities of Quebec, Montreal, and Ottawa developed along the St. Lawrence River and its tributaries. They depend on the St. Lawrence as an important transportation link.

For many years, rapids, waterfalls, and other obstructions kept ships from navigating the entire **route,** or journey, from the Great Lakes to the Atlantic Ocean. Then, in the mid-1900s, the United States and Canada built the St. Lawrence Seaway. As shown in **Figure 1,** the Seaway links the Great Lakes and the Atlantic Ocean.

Figure 1 **St. Lawrence Seaway and Locks**

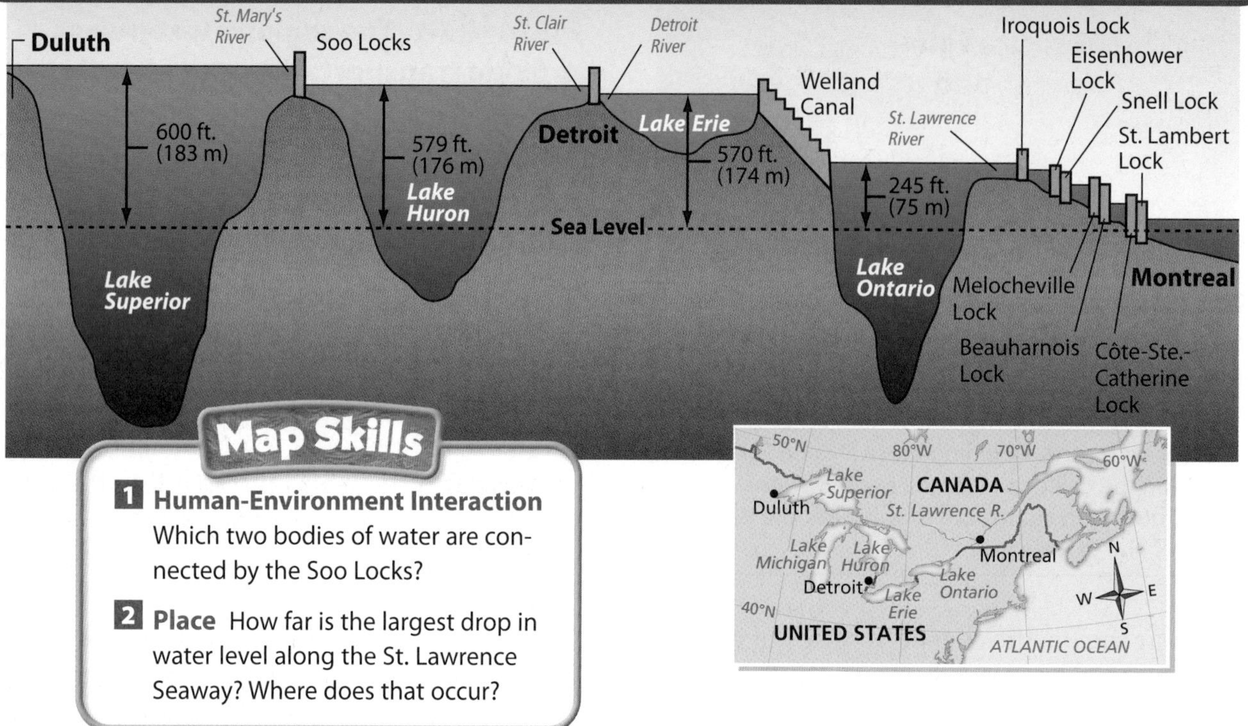

Map Skills

1 Human-Environment Interaction Which two bodies of water are connected by the Soo Locks?

2 Place How far is the largest drop in water level along the St. Lawrence Seaway? Where does that occur?

Today, ships carry raw materials and manufactured goods from Great Lakes cities, such as Chicago, Detroit, Cleveland, and Toronto, to the rest of the world.

The Mississippi River

The Mississippi River is North America's longest river. It flows 2,350 miles (3,782 km), beginning as a stream in Minnesota and enlarging to a width of 1.5 miles (2.4 km) before emptying into the Gulf of Mexico. Ships can travel on the Mississippi and some of its tributaries for great distances. Products from inland port cities, such as St. Louis and Memphis, are shipped down the river and on to foreign ports.

The Mississippi River system is the major waterway for the central part of the region. It drains about 1.2 million square miles (3.1 million sq. km) of land. This area includes all or part of 31 American states and much of central Canada.

The Continental Divide

Many rivers, such as the Colorado and Rio Grande, flow from the Rocky Mountains. A number of smaller rivers and streams connect with one of these rivers. The high ridge of the Rockies is called the Continental Divide. A **divide** is a high point that determines the direction that rivers flow. East of the Continental Divide, rivers flow toward the Arctic Ocean, the Atlantic Ocean, and the Mississippi River system into the Gulf of Mexico. To the west of the divide, rivers flow toward the Pacific Ocean.

Northeast of the Rockies, the Mackenzie River flows from the Great Slave Lake to the Arctic Ocean. It drains much of northern Canada's interior.

Reading Check **Identifying** What are some inland ports along the Mississippi River?

Natural Resources

Main Idea The region has many energy, mineral, and other natural resources.

Geography and You Think about if you like to eat canned, frozen, or fresh vegetables. Where were these foods grown and processed? Read to learn about the natural resources that provide products for the United States and Canada.

In addition to major river systems, the United States and Canada have a great variety of other natural resources. Energy sources and raw materials have made it possible for both countries to develop strong industrial economies.

Energy and Mineral Resources

The United States and Canada have major energy resources, such as oil and natural gas. Texas ranks first in oil and natural gas reserves in the United States. Alaska also has major oil reserves. The United States, however, uses nearly three times the amount of oil that it produces. So, even though the United States has a large reserve of oil, it must import more to meet the nation's needs.

Canada exports both oil and natural gas. Much of Canada's energy exports go to the United States. Most of Canada's oil and natural gas reserves lie in or near the province of Alberta. This province has the world's largest reserves of oil in the form of oil mixed with sands. Obtaining oil from these sands is more costly than working with liquid crude oil.

The United States and Canada also have significant amounts of coal. Coal is mined in the Appalachian Mountains, Wyoming, and British Columbia. The region has enough coal to supply energy for about 400 years, but using this energy source adds to air pollution.

In eastern areas, highlands drop to the lower Atlantic plain. Along this fall line, rivers break into waterfalls that provide hydroelectric power. Niagara Falls is a major source of hydroelectric power for both Canada and the United States. The falls lie on the Niagara River, which flows north from Lake Erie to Lake Ontario. The falls also form part of the border between Ontario, Canada, and the state of New York.

Mineral resources are also plentiful in the United States and Canada. Parts of eastern Canada and the northern United States have large iron ore deposits. The Rocky Mountains yield gold, silver, and copper. Deep within the Canadian Shield are iron ore, copper, nickel, gold, and uranium. Minerals from the shield helped create a manufacturing region in southern Ontario and Quebec.

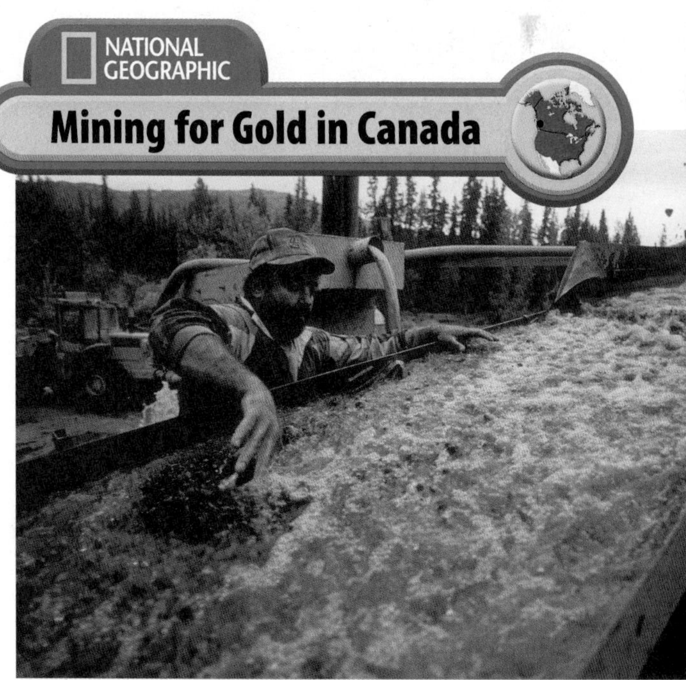

▲ Gold is still mined in British Columbia. Here, a stream of rushing water is used to wash away mud and rocks to help find gold nuggets. ***Place*** **What minerals are found in the Canadian Shield?**

Soil, Timber, and Fish

Rich soil in parts of the United States and Canada is excellent for farming. Crops vary throughout the region, depending on the local climate. Farmers grow corn in the Central Lowlands, which receive plentiful rainfall, and wheat on the drier Great Plains. The wet, mild climate of western Washington and Oregon supports dairy farming and the growing of fruits and vegetables. Irrigation is used in the drier eastern areas of these two states to grow grain. The warm, wet valleys of central California yield more than 200 different crops. In the south central part of British Columbia, fruits and vegetables are grown on irrigated land.

Timber is another important resource in the region. Forests once covered much of the United States and Canada. Today, however, forests cover less than 50 percent of Canada and about one-third of the United States. Still, lumber and wood products, such as paper, are major Canadian exports. The timber industry is also strong in the states of Oregon and Washington.

Coastal waters are important to the region's economies. Large fishing industries depend on the fish and shellfish in these waters. In recent years, however, the region's Atlantic fishing grounds have suffered from overfishing. The Grand Banks, located off Canada's southeast coast, was once one of the world's richest fishing grounds. Overfishing has harmed the area, and the Canadian government has banned fishing here for some species.

Reading Check **Explaining** What is unique about oil deposits in Alberta, Canada?

Section Review

Social Studies ONLINE
Study Central™ To review this section, go to glencoe.com.

Vocabulary

1. **Explain** the significance of:
   - **a.** contiguous
   - **b.** megalopolis
   - **c.** prairie
   - **d.** cordillera
   - **e.** canyon
   - **f.** navigable
   - **g.** glacier
   - **h.** divide

Main Ideas

2. **Describing** Describe the Canadian Shield and its resources.
3. **Summarizing** Use a diagram like the one below to summarize important facts about the Mississippi River.

4. **Comparing and Contrasting** Compare and contrast the agricultural conditions and crops grown in various parts of the region.

Critical Thinking

5. **BIG Idea** How did building the St. Lawrence Seaway change the land? How have the United States and Canada benefited from the St. Lawrence Seaway?
6. **Challenge** What conditions led to the formation of a megalopolis along the United States's Atlantic coast?

Writing About Geography

7. **Using Your FOLDABLES** Use your Foldable to make and write captions for a map of the region that describes the impact of landforms and waterways on people's lives.

NATIONAL GEOGRAPHIC **Geography & History**

Danger Zone

In August of 2005, a massive hurricane struck the southern United States. The damage it caused was overwhelming.

The Storm and the Damage Hurricane Katrina struck the Gulf coast of the United States on August 29, 2005. Katrina reached land as a category 4 hurricane, the second-strongest category of storm. The hurricane blasted the region with winds of more than 140 miles per hour (225 km/hr). It caused a storm surge—rising seas—of more than 30 feet (9 m) and brought as many as 16 inches (41 cm) of rainfall in a short time.

Storm conditions raged along the coasts of Louisiana, Mississippi, and Alabama. More than 1,800 people died, more than 500,000 were left homeless, and property damage exceeded $80 billion. Katrina was one of the worst natural disasters in American history.

Flooding from Hurricane Katrina, New Orleans

Katrina and New Orleans The city of New Orleans suffered extensive damage from Katrina. The strength of the storm and the geography of the city helped lead to disaster. New Orleans lies below sea level, and the city is almost surrounded by water. Lake Pontchartrain lies to the north. The Mississippi flows to the west and south of town. Many years ago, a complex system of high walls, called levees, was built along the lake and river to protect the city from flooding.

The power of Katrina overwhelmed the levees, some of which had weakened with age. Water rushed through breaks in the barriers. Floodwaters rose as high as 20 feet (6 m) in parts of the city. Fortunately, most of New Orleans's 450,000 residents were evacuated before the storm hit. Many months after Katrina, fewer than half of the city's people had returned to their homes.

Path of Hurricane Katrina

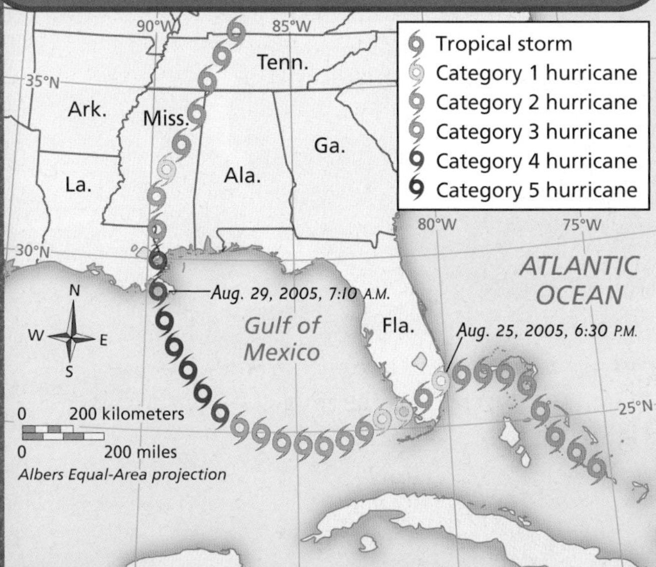

Think About It

1. **Regions** Where did Katrina strike? How much damage did it cause?
2. **Place** Why was the threat of flooding especially dangerous for New Orleans?

Guide to Reading

BIG Idea

The physical environment affects how people live.

Content Vocabulary

- drought *(p. 126)*
- tornado *(p. 127)*
- hurricane *(p. 127)*
- blizzard *(p. 128)*

Academic Vocabulary

- diverse *(p. 125)*
- adapt *(p. 125)*
- restore *(p. 126)*

Reading Strategy

Organizing Information Use a chart like the one below to organize key facts about at least three different climate zones in the region.

Climate Zones	Location	Description
1.		
2.		
3.		

Climate Regions

Spotlight Video

Picture This This sea of red is actually a sea of cranberries. The small, red fruit—also known as bounce berries, crane berries, and rubies of the pines—grows on ground-hugging vines in wetlands and bogs. To harvest the cranberries, farmers flood the bogs. Small air pockets in the cranberries cause them to rise to the surface, where they can be gathered by harvesting machines. Read this section to learn more about the climates of the United States and Canada and how they influence farming and other human activities.

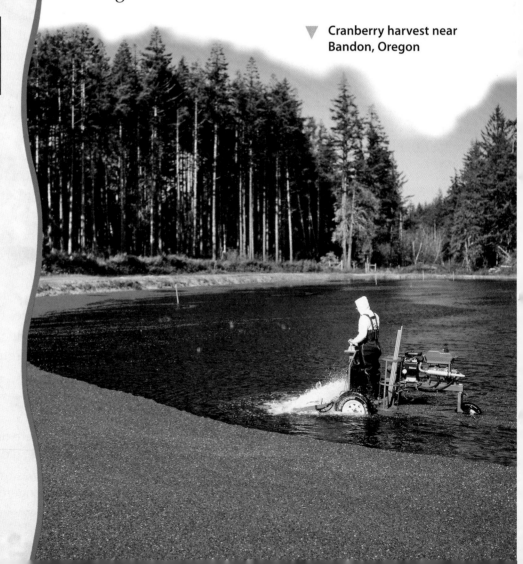

▼ Cranberry harvest near Bandon, Oregon

A Varied Region

Main Idea **Most people in the United States and Canada live in temperate climate regions.**

Geography and You What is the climate like in your area? Read to learn about the different climate regions of the United States and Canada.

The region of the United States and Canada stretches from cold Arctic wastelands in the far north to warm, sunny vacation areas near the Tropic of Cancer. This vast territory is **diverse** in both climate and vegetation. Most people in the United States and Canada, however, avoid the extremes of tropical and Arctic climates. They live in the middle latitudes where climates are more moderate. **Figure 2** shows all of the region's climates.

The Far North

Tundra and subarctic climates are found in the northern parts of Alaska and Canada. Winters are long and cold, while summers are brief and cool. As a result, few people live in this harsh environment.

Along the Arctic Ocean's coastline, the extremely cold tundra prevents the growth of trees and most plants. In the subarctic region farther south, dense forests of evergreen trees are specially **adapted,** or adjusted, to the climate. The waxy coating of evergreen needles keeps in moisture during the bitterly cold winters.

NATIONAL GEOGRAPHIC **Maps In MOtion** See StudentWorks™ Plus or glencoe.com.

Figure 2 **United States and Canada: Climate Zones**

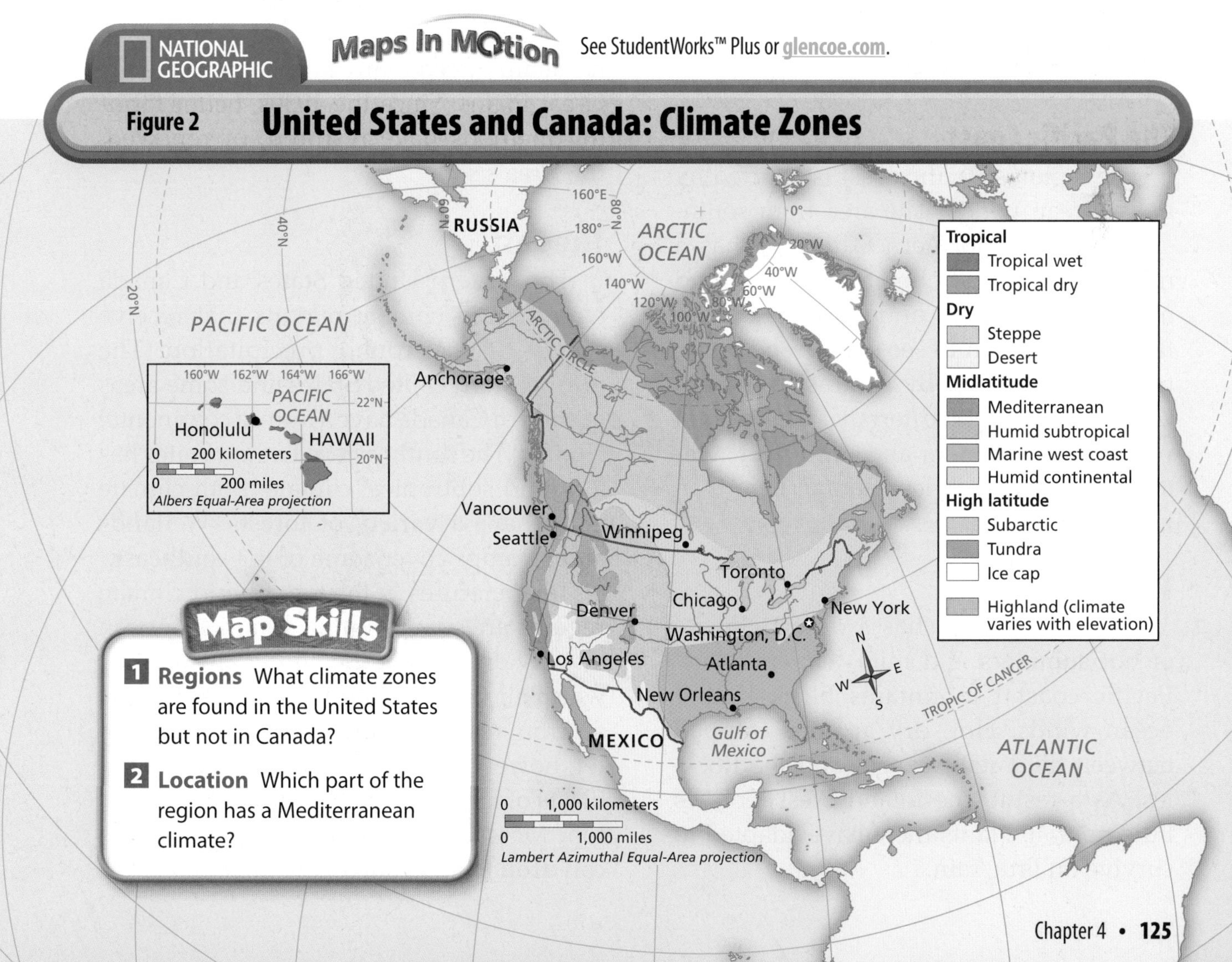

Map Skills

1 Regions What climate zones are found in the United States but not in Canada?

2 Location Which part of the region has a Mediterranean climate?

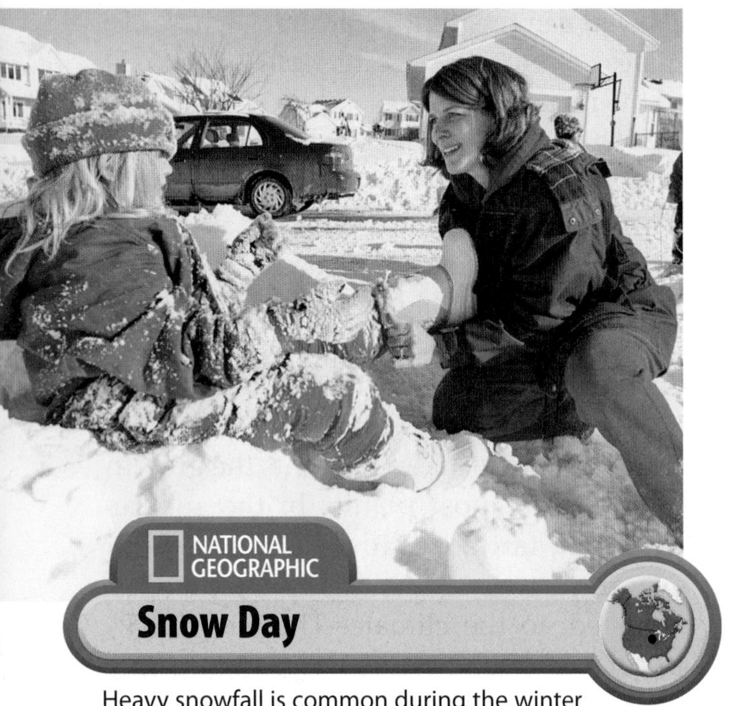

Snow Day

Heavy snowfall is common during the winter in Iowa, as well as in other states across the Great Plains. ***Human-Environment Interaction*** **How might a drought affect a farmer or rancher?**

The Pacific Coast

The region's Pacific coast is affected by moist ocean winds. The area from southern Alaska to northern California has a marine west coast climate of year-round mild temperatures and abundant rainfall. It is common to see evergreen forests, ferns, and mosses. By contrast, southern California has a Mediterranean climate of warm, dry summers and mild, wet winters. There is much less rainfall here than in northern areas.

The West

The inland West has a desert climate of hot summers and mild winters. Here, Pacific coastal mountains block humid ocean winds. Hot, dry air gets trapped between the Pacific ranges and the Rockies. As a result, the inland West receives little rainfall. Plants there have adapted to survive on little rain.

Areas on the eastern side of the Rockies have a partly dry steppe climate. **Droughts,** or long periods without rainfall, are a serious challenge, especially to farmers and ranchers who can lose crops and animals. In some areas, a growing population also strains water resources.

The Great Plains

The Great Plains area benefits from moisture-bearing winds from the Gulf of Mexico and from the Arctic. As a result, much of this area has a humid continental climate with cold, snowy winters and hot, humid summers. Enough precipitation falls to support prairie grasses and grains. Dry weather, however, sometimes affects the area. In the 1930s, winds eroded loose topsoil and turned the area into a wasteland called the Dust Bowl. Economic hardship forced many farmers to leave the Great Plains. Since the 1930s, better farming methods have **restored,** or renewed, this area's soil.

The East

The eastern United States and Canada have humid climate regions that receive plenty of year-round precipitation. The northeastern United States and some areas of eastern Canada have a humid continental climate. The southeastern United States has a humid subtropical climate. Both climate areas have a variety of forests. Wetlands and swamps cover some of the southeast.

Temperatures in the two humid climate regions are similar in the summer but can be very different in the winter. In summer, warm air from the south blocks cold Arctic air from reaching the eastern areas. In winter, however, the northeast receives strong blasts of icy Arctic air. For example, in Boston, Massachusetts, January temperatures can drop to an average low of 22°F (–6 °C).

Areas in the southeast still receive some warmth from the south. As a result, the average January temperature in Atlanta, Georgia, is 41°F (5°C).

Tropical Areas

Tropical climates are found in two areas of the United States. Southern Florida has a tropical dry climate. Temperatures are hot in summer and warm in winter. Rainfall occurs mainly during the summer. Hawaii, the other tropical area, has year-round temperatures that average above 70°F (21°C). The mild climate draws many visitors throughout the year. Rainfall, which varies throughout the state, supports tropical rain forests.

Reading Check **Explaining** What factors affect climate in the Great Plains?

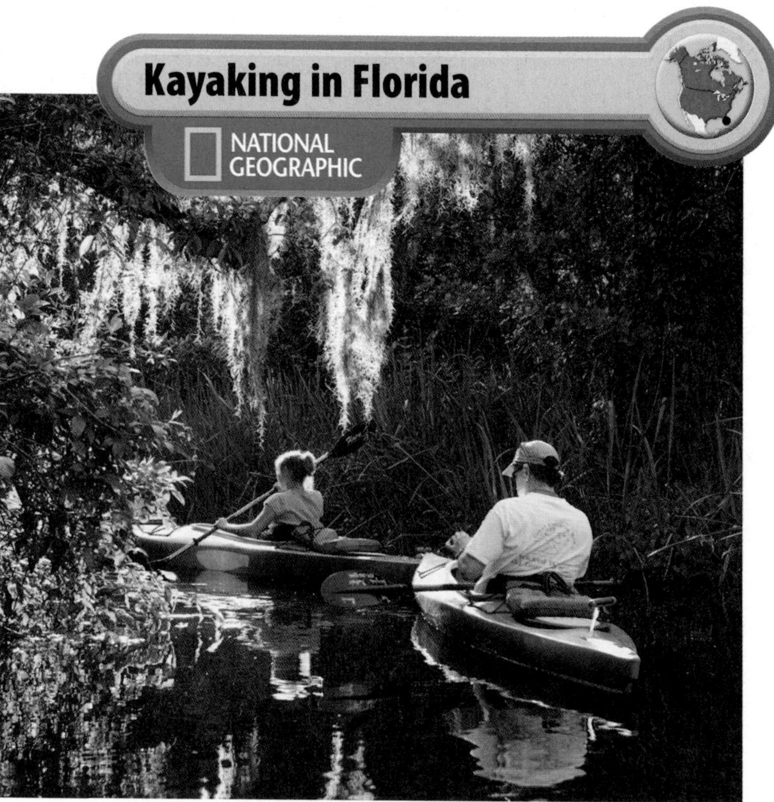

▲ In the large wetlands area known as the Everglades, south Florida's tropical climate produces lush vegetation. ***Location*** **Besides Florida, what other state has a tropical climate?**

Natural Hazards

Main Idea **Hurricanes, tornadoes, and earthquakes can threaten parts of the region.**

Geography and You Does the area where you live experience severe storms? Read to learn about the environmental challenges that affect the United States and Canada.

The landforms and climate of this region provide people with many benefits. They also pose challenges in the form of severe storms and other natural hazards.

Severe Weather

One hazard related to severe weather is a tornado. A **tornado** is a windstorm in the form of a funnel-shaped cloud that often touches the ground. The high winds of a tornado, which can reach more than 300 miles per hour (482 km per hour), can level houses, knock down trees, and hurl cars from one place to another. These storms can occur anywhere in the region and at any time of the year. The central United States sees more tornadoes each year than any other place in the world. As a result, this area has been nicknamed "Tornado Alley."

Another severe storm is a hurricane. **Hurricanes** are wind systems that form over the ocean in tropical areas and bring violent storms with heavy rains. As with tornadoes, high winds can do serious damage. In addition, hurricanes can create a storm surge, or high levels of seawater. The high waters can flood low-lying coastal areas. Hurricanes generally develop from June to September. They most often strike along the southeastern Atlantic coast and the Gulf of Mexico. However, northeastern states can also be affected by hurricanes.

One of the most damaging hurricanes in history, Hurricane Katrina, struck the coast of the Gulf of Mexico in August of 2005. It damaged a wide area from Mobile, Alabama, to New Orleans, Louisiana. More than 1,800 people died, and hundreds of thousands lost their homes. Most of New Orleans and many nearby towns were completely flooded. In Mississippi, entire towns were destroyed.

Winter weather can also be hazardous. **Blizzards** are severe winter storms that last several hours and combine high winds with heavy snow. The blowing snow limits how far people can see. "White-out" conditions, or snow that falls so heavily that a person cannot see very far, make driving dangerous. Also, the wind and snow can knock down electric power lines and trees and create icy road conditions. Blizzards can halt activity in a busy city for days as city workers attempt to clear the streets.

Earthquakes and Volcanoes

While earthquakes can occur anywhere in the region, most take place along the Pacific coast. This area lies along various fault lines, or areas of weakness in the Earth where two tectonic plates meet. A 1906 earthquake heavily damaged buildings in San Francisco. Many of the buildings that remained standing were destroyed by fires triggered by broken natural gas lines. Today, buildings in the region are often built using special techniques to protect them from damage.

The area where tectonic plates meet can also be the site of volcanoes. Volcanoes are found in the Pacific coast mountains, southern Alaska, and Hawaii. Most are now dormant, or unlikely to erupt soon. Several of Hawaii's volcanoes are still active.

Reading Check **Describing** Describe the types of damage that hurricanes can cause.

Section 2 Review

Social Studies ONLINE
Study Central™ To review this section, go to glencoe.com.

Vocabulary

1. **Describe** each of these weather conditions and where they are likely to occur: *drought, tornado, hurricane,* and *blizzard.*

Main Ideas

2. **Summarizing** Use a diagram like the one below to summarize information about one of the region's climate zones. Write the name of a climate zone in the large oval and details about it in the small ovals.

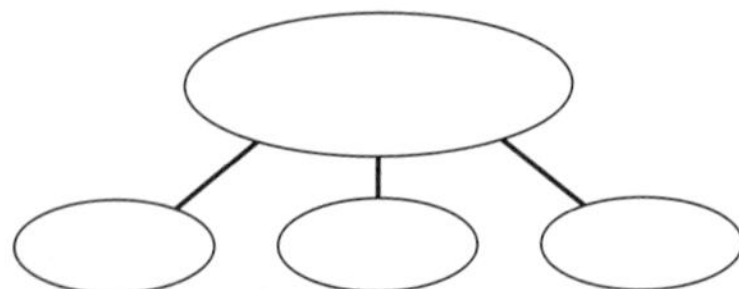

3. **Comparing and Contrasting** Compare and contrast tornadoes and hurricanes.

Critical Thinking

4. **Determining Cause and Effect** How do mountains in the region influence climate?
5. **BIG Idea** What were some of the effects of Hurricane Katrina?
6. **Challenge** Based on climate and the occurrence of natural hazards, which areas of the region do you think are most populated? Explain your answer.

Writing About Geography

7. **Personal Writing** Write a paragraph identifying which of the climates described in this section you think sounds most enjoyable to live in, and explain why.

Visual Summary

Major Landforms

- The East has low coastal plains and heavily eroded highlands.
- Lowland areas with minerals and rich soil make up the region's interior.
- The West has several parallel mountain ranges. Plateaus, basins, and valleys lie between the mountains.

Lowland marsh, Virginia

Major Bodies of Water

- The Great Lakes and the St. Lawrence Seaway support trade between the region's interior areas and other parts of the world.
- The Mississippi River is the most important waterway in the central part of the United States.

Riverboat, Mississippi River

Farming in Manitoba, Canada

Natural Resources

- The region's energy resources include oil, natural gas, and coal.
- Abundant mineral resources are found in the eastern highlands, the Canadian Shield, and the western mountains.
- Rich soils support farming in the Central Lowlands, the Great Plains, and western valleys.

Climate Regions

- Most Americans and Canadians live in moderate, middle-latitude climate areas.
- The inland West has dry and semidry climates because mountains block moist air.
- Pacific coastal areas generally have mild, wet climates.

Natural Hazards

- Tornadoes occur primarily in the central area of the region.
- Hurricanes can bring heavy winds and rain to the Atlantic and Gulf coasts.
- Earthquakes are a destructive threat along coastal fault lines in the West.
- Volcanoes are found in western coastal areas, Alaska, and Hawaii. Most are dormant.

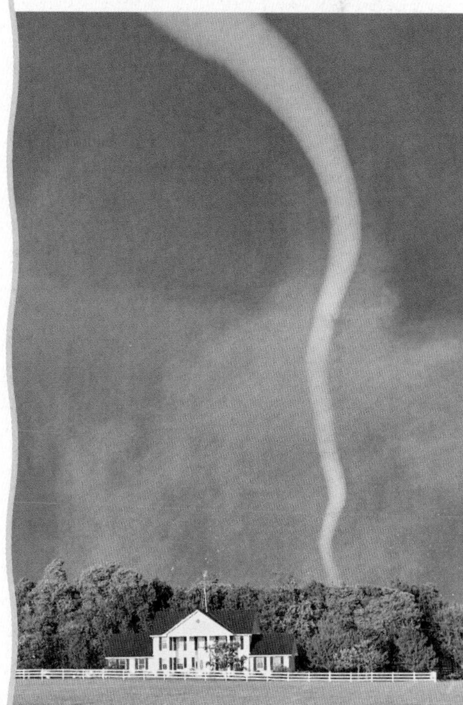

Tornado, Kansas

Study anywhere, anytime! Download quizzes and flash cards to your PDA from glencoe.com.

STANDARDIZED TEST PRACTICE

Do not wait until the last minute to study for an exam. Beginning about one week before the test, set aside some time each day for review.

Reviewing Vocabulary

Directions: Choose the word(s) that best completes the sentence.

1. Atlantic coastal cities and their suburbs form an almost continuous line of settlement called a ______.

A Piedmont
B megalopolis
C coastal plain
D coastal lowland

2. Much of the Great Plains is a ______, or rolling grassland.

A prairie
B canyon
C glacier
D cordillera

3. Ranchers in the dry steppe region east of the Rockies sometimes lose crops and animals due to ______.

A droughts
B blizzards
C hurricanes
D earthquakes

4. ______ most often strike along the southeastern Atlantic coast and the Gulf of Mexico.

A Droughts
B Tornadoes
C Hurricanes
D Earthquakes

Reviewing Main Ideas

Directions: Choose the best answer for each question.

Section 1 *(pp. 116–122)*

5. Which of the following areas of the United States has the highest elevation?

A Appalachian Mountains
B eastern Great Plains
C Rocky Mountains
D Atlantic coastal plain

6. ______ depend on the Mississippi River for shipping products on their way to foreign ports.

A Chicago and Detroit
B Quebec and Montreal
C Toronto and Cleveland
D St. Louis and Memphis

Section 2 *(pp. 124–128)*

7. The area from southern Alaska to northern California has a ______ climate with year-round mild temperatures and abundant rainfall.

A subtropical
B Mediterranean
C marine west coast
D humid continental

8. ______ has more tornadoes each year than any other place in the world.

A Eastern Canada
B Northern Canada
C The northeastern United States
D The central United States

GO ON

Critical Thinking

Directions: Choose the best answers to the following questions. Base your answers to questions 9 and 10 on the map below and your knowledge of Chapter 4.

Tornado Alley

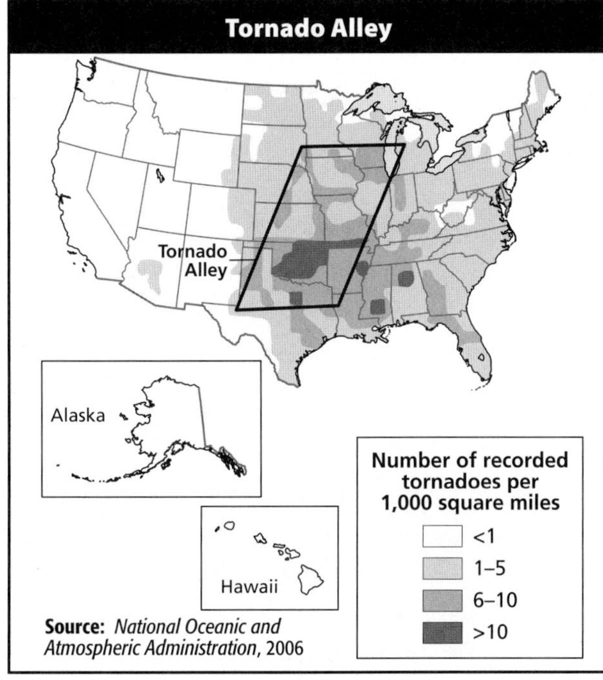

Source: *National Oceanic and Atmospheric Administration*, 2006

9. According to the map, which states have the most tornado activity?

A Texas and Oklahoma

B Texas and Louisiana

C Oklahoma and Kansas

D Missouri and Arkansas

10. According to the information in the map, which generalization is most accurate?

A The western United States never experiences tornadoes.

B Most tornadoes occur in the southeastern United States.

C "Tornado Alley" is an area of moderate tornado activity.

D The central United States has more tornadoes than the rest of the country.

Document-Based Questions

Directions: Analyze the following document and answer the short-answer questions that follow.

The news story below sums up findings in a 2005 study on recent hurricanes:

> *The number of Category 4 and 5 hurricanes worldwide has nearly doubled over the past 35 years, even though the total number of hurricanes has dropped since the 1990s. . . . The shift occurred as global sea surface temperatures have increased over the same period. . . .*
>
> *"What we found was rather astonishing," said [Peter Webster, professor at Georgia Tech's School of Earth and Atmospheric Sciences]. "In the 1970s, there was an average of about 10 Category 4 and 5 hurricanes per year globally. Since 1990, the number of Category 4 and 5 hurricanes has almost doubled, averaging 18 per year globally."*
>
> *Category 4 hurricanes have sustained winds from 131 to 155 miles per hour; Category 5 systems, such as Hurricane Katrina at its peak over the Gulf of Mexico, feature winds of 156 mph or more.*
>
> —*The National Center for Atmospheric Research*, 2005

11. How have hurricanes changed in the past 35 years?

12. What might be causing this hurricane intensity?

Extended Response

13. Suppose you are a foreign visitor to North America. Describe one region of the United States or Canada and write a letter to a friend, persuading him or her to visit.

For additional test practice, use Self-Check Quizzes—Chapter 4 at glencoe.com.

Need Extra Help?

If you missed question...	1	2	3	4	5	6	7	8	9	10	11	12	13
Go to page...	117	118	126	127	118	120	126	127	26	26	127	127	116–128

CHAPTER 5

History and Cultures of the United States and Canada

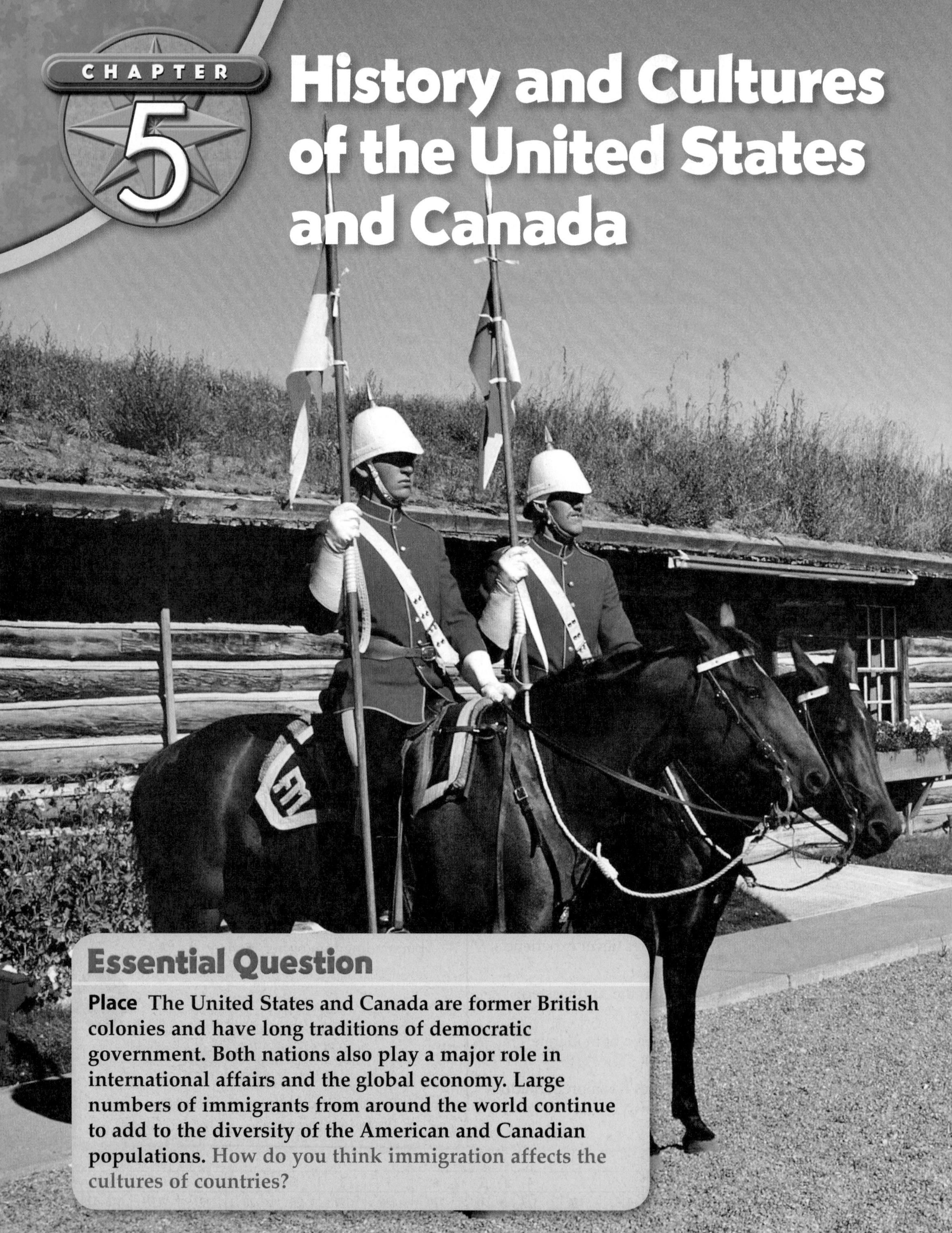

Essential Question

Place The United States and Canada are former British colonies and have long traditions of democratic government. Both nations also play a major role in international affairs and the global economy. Large numbers of immigrants from around the world continue to add to the diversity of the American and Canadian populations. How do you think immigration affects the cultures of countries?

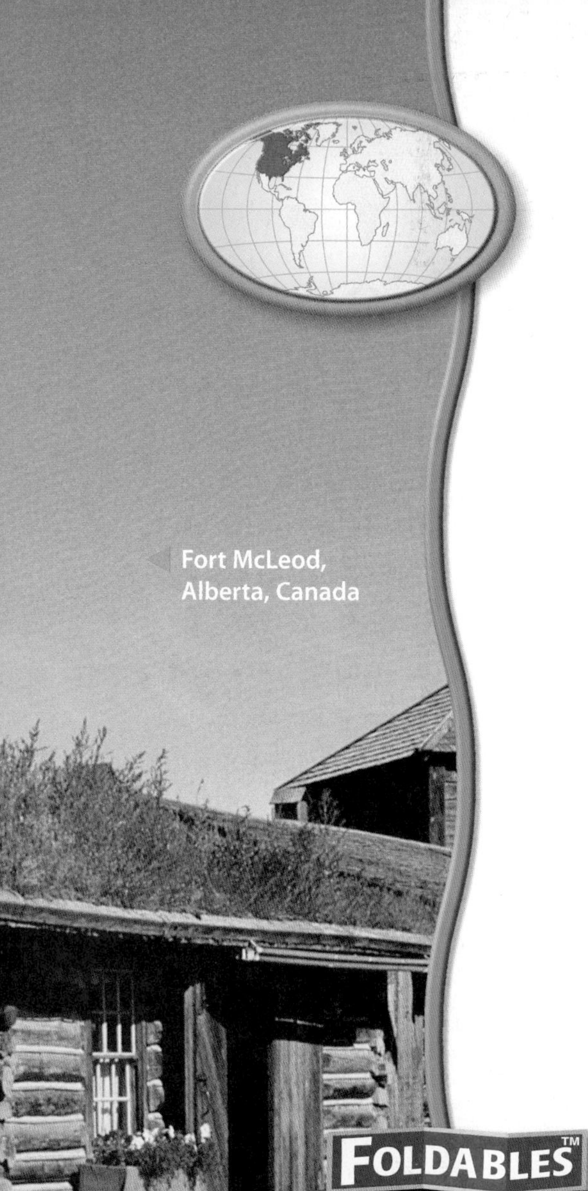

Fort McLeod, Alberta, Canada

Chapter Audio

Section 1: History and Governments

BIG IDEA The characteristics and movement of people impact physical and human systems. England established several colonies in North America during the 1600s and 1700s. These colonies later formed two large, independent democracies: the United States and Canada. Today they have become home to millions of people from around the world who moved to these lands to start new lives.

Section 2: Cultures and Lifestyles

BIG IDEA Culture influences people's perceptions about places and regions. The cultures of the United States and Canada reflect the influence of many different ethnic groups. These groups range from the Native Americans who first lived in the area to the most recent arrivals from all parts of the world.

FOLDABLES™ Study Organizer

Categorizing Information Make this four-tab Foldable to help you learn about the history, governments, and cultures of the United States and Canada.

Step 1 Fold the top and bottom of a piece of paper into the middle.

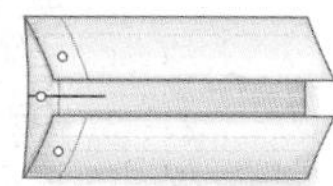

Step 2 Cut each flap at the midpoint to form 4 tabs.

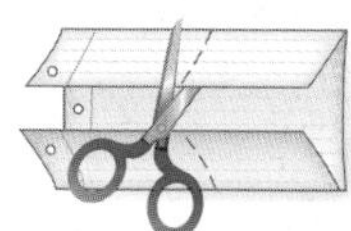

Step 3 Label the tabs as shown.

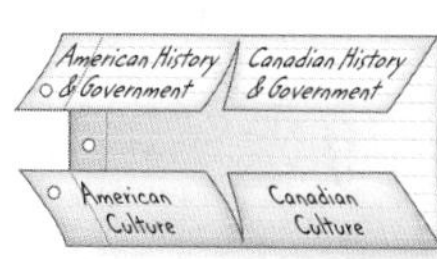

Reading and Writing As you read the chapter, take notes about each section under the appropriate head. Use your Foldable to help you write a summary for each section.

Social Studies ONLINE

Visit glencoe.com and enter ***QuickPass***™ code EOW3109c5 for Chapter 5 resources.

Guide to Reading

BIG Idea

The characteristics and movement of people impact physical and human systems.

Content Vocabulary

- colony *(p. 136)*
- annex *(p. 136)*
- terrorism *(p. 137)*
- dominion *(p. 139)*
- representative democracy *(p. 140)*
- federalism *(p. 140)*
- amendment *(p. 141)*
- parliamentary democracy *(p. 141)*

Academic Vocabulary

- economy *(p. 136)*
- regime *(p. 137)*
- principle *(p. 140)*
- core *(p. 141)*

Reading Strategy

Making a Time Line Use a diagram like the one below to list and compare key events and dates in the history of the United States and Canada.

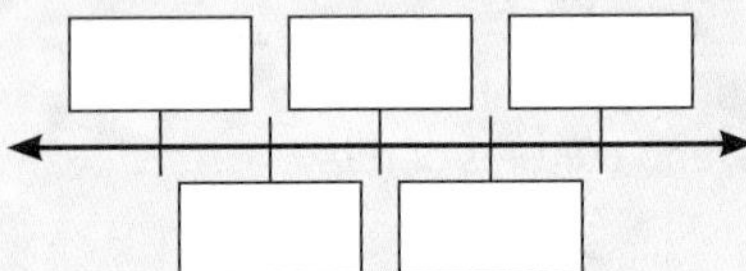

History and Governments

 Section Audio  Spotlight Video

Picture This This is not just any rock. It is the famous Plymouth Rock. Plymouth Rock, in the American state of Massachusetts, is believed to have been at the site where the Pilgrims first landed in 1620. The 10-ton rock is only about one-third of its original size. This is because, over the years, many souvenir hunters have broken off pieces to take home. To learn how people can affect the development of a nation, read Section 1.

▼ Plymouth Rock, Massachusetts

History of the United States

Main Idea **The United States emerged as a world power in the 1900s.**

Geography and You Do you like to make your own decisions without someone else telling you what to do? Read to find out why the Americans wanted to make their own decisions without British interference.

For thousands of years, people have been living in what is now the United States. They have come from many different parts of the world to settle a vast territory that stretches across much of North America. In the past few hundred years, Americans have made their country the most powerful nation in the world.

The First Americans

About 15,000 years ago, hunters in Asia followed herds of animals across a land bridge between eastern Siberia and Alaska. They are believed to be among the first people to settle the Americas. Eventually groups of people spread across the entire landmass. Their descendants today are called Native Americans. Over many centuries, Native American groups developed different ways of life using local resources.

The Colonial Era

Europeans were not aware of the Americas until 1492. That year, explorer Christopher Columbus sailed west from Europe in hopes of reaching Asia. Instead, he reached islands in the Caribbean Sea. His tales of these new lands excited many Europeans.

NATIONAL GEOGRAPHIC Maps In MOtion See StudentWorks™ Plus or glencoe.com.

Figure 1 **United States Expansion**

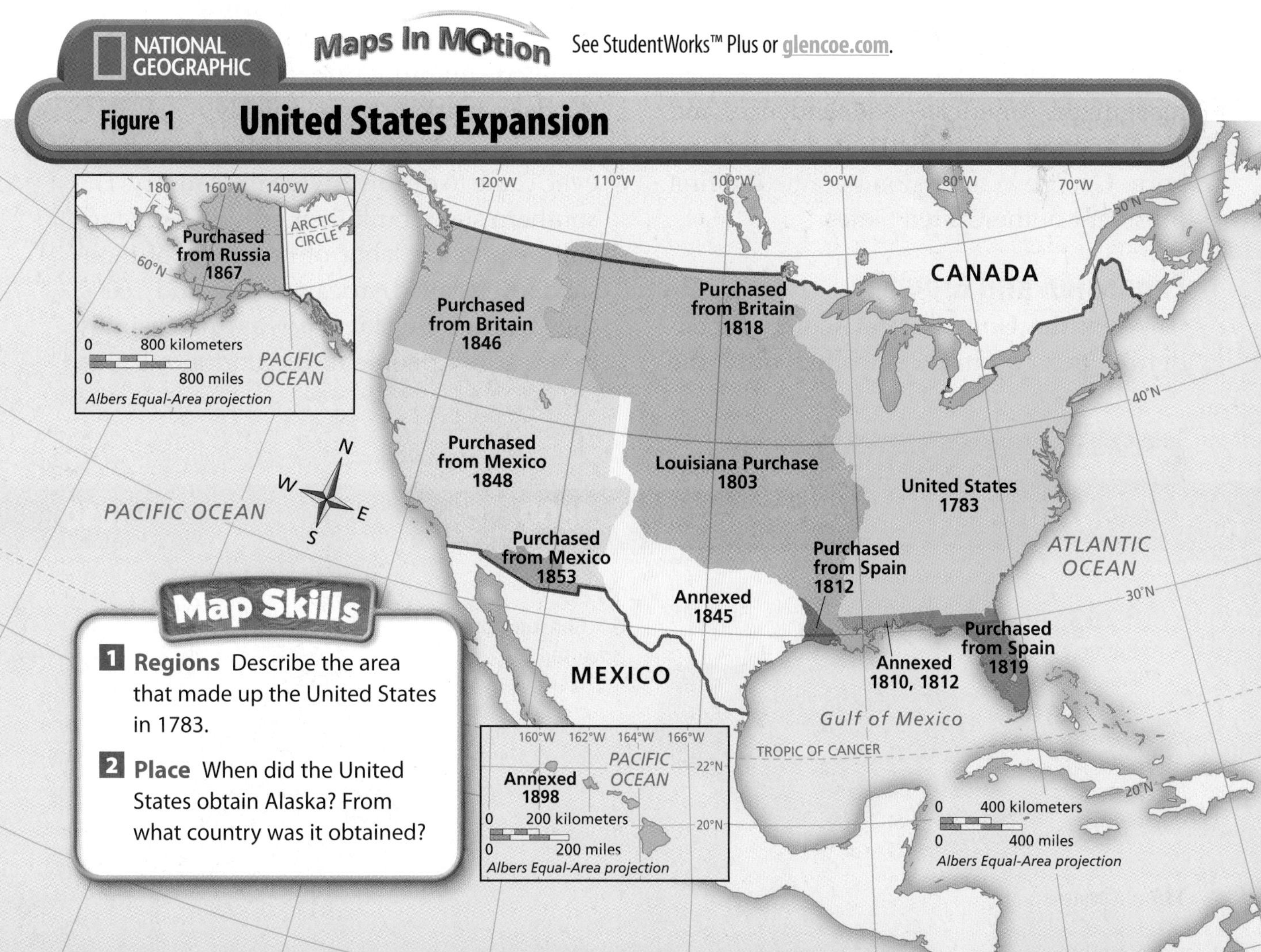

Map Skills

1. **Regions** Describe the area that made up the United States in 1783.
2. **Place** When did the United States obtain Alaska? From what country was it obtained?

Quickly, many other people began making their own trips. Spain soon set up **colonies,** which are overseas settlements tied to a parent country. The Spanish gained great wealth from gold and silver mines in Mexico and South America. They also had colonies in what is now the southern United States, but they focused primarily on the lands farther south.

France and Great Britain also established colonies in North America. The French controlled eastern Canada, the Great Lakes area, and the Mississippi River valley. The British settled along the Atlantic coast. In 1763 Great Britain defeated France in a war and won France's North American colonies.

Soon after, the people in Great Britain's 13 coastal colonies grew resentful of British taxes and trade policies. Discontent boiled over, and fighting broke out between the colonists and British forces. In 1776 the colonists declared their independence. In 1783, after several years of fighting, Britain recognized American independence, and a new nation called the United States was born. George Washington became the first president of the United States.

Expansion and Growth

When the United States won independence, the country controlled only the eastern coast from Maine to Georgia. During the 1800s, however, it expanded all the way to the Pacific Ocean, as seen in **Figure 1** on the previous page. Some of this growth came through treaties with other countries. Some came when the United States **annexed** lands, or declared ownership of a particular area. This expansion brought suffering, however, including loss of land, culture, and often life, to Native Americans who had lived in the region for centuries.

Throughout the 1800s, the United States grew in population as well. High birthrates, advances in public health, and the arrival of millions of European immigrants helped the population grow.

The American **economy,** or way of producing goods, also changed. New machines made planting and harvesting crops faster and easier. Manufacturers developed the factory system to produce many goods. Roads, canals, steamboats, and railroads allowed manufacturers to move their goods to markets more quickly.

Social and economic differences, however, came to deeply divide the country. The southern states built their economy on agriculture and the labor of hundreds of thousands of enslaved Africans. By the mid-1800s, the practice of slavery was increasingly criticized by people in the northern states.

History at a Glance
1750
1800
1850
1763
Britain gains France's North American colonies
1787
U.S. Constitution is written
1867
Dominion of Canada is formed

Native American belt, c. 1700s

Abraham Lincoln

Southerners worried that northerners would attempt to end slavery. In 1860 Abraham Lincoln, an opponent of slavery, was elected president. By early 1861, several southern states had withdrawn from the United States to set up a new country. This action led to a civil war. After four years of fighting, northern forces won. The country was reunited, and slavery ended. Racial tensions, however, continued.

The late 1800s saw the spread of industry in the United States. Waves of new immigrants provided workers for the growing economy. By 1900, the United States was one of the world's major industrial powers.

A World Leader

During the 1900s, the United States became a world leader. Its armies fought in World War I and World War II. During these conflicts, United States leaders urged the world's people to fight for freedom against oppressive **regimes,** or governments. American factories produced tanks and airplanes, while American soldiers helped win the wars.

After World War II, the United States and the Soviet Union became the world's two major powers. They competed for world leadership in a rivalry known as the Cold War. This rivalry was called the Cold War because the conflict never became "hot," with actual combat between the two powers. Nevertheless, tensions remained high as the countries competed politically and economically. The Cold War ended with the breakup of the Soviet Union in 1991.

During this period, African Americans, Latino Americans, Native Americans, and women became more active in seeking equal rights in the United States. Leaders such as Martin Luther King, Jr.; Rosa Parks; and César Chávez used peaceful methods that led to social changes for these groups.

Since 2000, the United States has faced challenges from the growth of terrorism both at home and throughout the world. **Terrorism** refers to the use of violence against civilians, by individuals or groups, to reach political goals. On September 11, 2001, terrorists seized four passenger planes. Two planes were crashed into the World Trade Center in New York City. A third aircraft damaged the Pentagon, the headquarters of the U.S. military, in Washington, D.C. A fourth plane crashed in rural Pennsylvania after what is believed to have been a passenger uprising against the terrorists. About 3,000 people died in the attacks.

Soon after, the United States sent troops to the southwest Asian country of Afghanistan.

1900
1950
2000
c. 1900
Immigration and industry transform the United States and Canada
1941
U.S. and Canada become allies in World War II
2001
Terrorists attack the United States
COME AND DO YOUR BIT
JOIN NOW
U.S. DEPARTMENT OF HOMELAND SECURITY

World War I poster

Seal, Department of Homeland Security

Afghanistan's rulers, the Taliban, had protected the Muslim terrorist group al-Qaeda, which carried out the September 11 attacks. The Taliban were overthrown, although many terrorists escaped.

In 2003 U.S. troops overthrew Iraq's dictator, Saddam Hussein, who was accused of hiding illegal weapons and helping terrorists. In the years that followed, the United States worked with Afghanistan and Iraq to set up democratic governments. Continued fighting within both countries, though, made these efforts difficult.

Reading Check **Explaining** What role did the United States play in global affairs after 1900?

Women in Uniform

NATIONAL GEOGRAPHIC

▲ Women make up about 20 percent of the U.S. military and are actively involved in operations around the world. ***Place*** **Why did the United States send troops into Afghanistan?**

History of Canada

Main Idea **Canada gradually won independence from British rule during the late 1800s and early 1900s.**

Geography and You What similarities do you share with your neighbors? In what ways are you different? Read to find out how Canada's history is similar to and different from that of the United States.

Canada and the United States share similar backgrounds. Nonetheless, the two countries traveled different paths to becoming nations.

Early Settlement

Like the United States, Canada was originally settled by Native American groups. The first Europeans to arrive in the area that is today Canada were Viking explorers from Scandinavia. They landed in about A.D. 1000. The Vikings briefly lived on the Newfoundland coast, but they did not stay in their settlements very long. Eventually they left the Americas.

In the 1500s and 1600s, both England and France claimed areas of Canada. French explorers, settlers, and missionaries founded several cities. The most important were Quebec and Montreal. For almost 230 years, France ruled the area around the St. Lawrence River and the Great Lakes. This region was called New France.

The French knew that the Spanish had gained riches in South America from gold and silver mines. They hoped to become wealthy in the same way in Canada. They did find riches, but not in gold and silver mines. Instead, the French traded with Native Americans for beaver furs, which were then sent back to Europe and sold.

During the 1600s and 1700s, the English and French fought each other for territory around the globe. In 1707 England and Scotland united to form Great Britain. This union laid the foundation for the British Empire. By the 1760s, the British had won control of most of France's Canadian colony.

Beginning in the late 1700s, British and American settlers began moving to Canada in greater numbers. They set up farms along the Atlantic coast and in what is now Ontario. French-speaking Canadians lived mostly in present-day Quebec. Tragically, European warfare and diseases had nearly destroyed many Native American cultures by this time.

An Independent Nation

For the next 75 years, Great Britain held various colonies in eastern Canada. These colonies constantly quarreled with one another over colonial government policies. Fears of a United States takeover, however, forced them together. In 1867 most of the colonies became one nation known as the Dominion of Canada. As a **dominion,** Canada had its own central government to run local affairs. Great Britain, though, still controlled Canada's relations with other countries.

Under Canada's central government, the colonies became provinces, much like states in the United States. At first, there were four provinces—Quebec, Ontario, Nova Scotia, and New Brunswick. Neighboring British-ruled areas—Manitoba, British Columbia, Saskatchewan (suh·SKA·chuh·wuhn), and Alberta in the west and Prince Edward Island and Newfoundland along the Atlantic coast—became provinces of Canada during the next 100 years. Today Canada is made up of 10 provinces and 3 additional territories—the Yukon Territory, the Northwest Territories, and Nunavut (NOO·nah·voot).

NATIONAL GEOGRAPHIC

Trade in Canada

As Europeans explored Canada, they made treaties and traded with the Native Americans who lived there. ***Place*** **What two European nations colonized Canada?**

At Canada's founding, the government promised to protect the French language and culture in Quebec. However, the English-speaking minority there was far wealthier and ran the economy. This led to tensions between the two groups. French speakers claimed that they were treated unfairly because of their heritage.

During the early 1900s, many immigrants arrived, and Canada's population and economy grew. Meanwhile, Canadians fought alongside the British and Americans in the two World Wars. Canada's support in these wars led to its full independence. In 1982 Canadians won the right to change their constitution without British approval. Today Canada still faces the possibility that Quebec will separate and become independent.

✓Reading Check **Summarizing** How did Canada's government change over time?

Governments of the United States and Canada

Main Idea **The United States and Canada are democracies, but their governments are organized differently.**

Geography and You Do you think democracy is a good government system? Why? Read to discover the similarities and differences between the democracies of the United States and Canada.

The United States and Canada are both **representative democracies,** in which voters choose leaders who make and enforce the laws. However, the two government systems have some important differences.

U.S. Democracy

The United States Constitution is the basic plan that explains how our national, or central, government is set up and how it works. The document was written in the late 1780s by leaders who wanted to create a government strong enough to guide the country. Those leaders also wanted a government with limited powers so that people's rights would be protected from government interference.

To achieve these goals, the writers of the Constitution applied the **principle,** or rule, of separation of powers. This means they divided the power of the national government among three branches: executive, legislative, and judicial. In addition, they gave each branch unique powers as a way to prevent the other branches from abusing their power. This idea is called checks and balances. With it, the Founders aimed to prevent one branch from becoming too powerful.

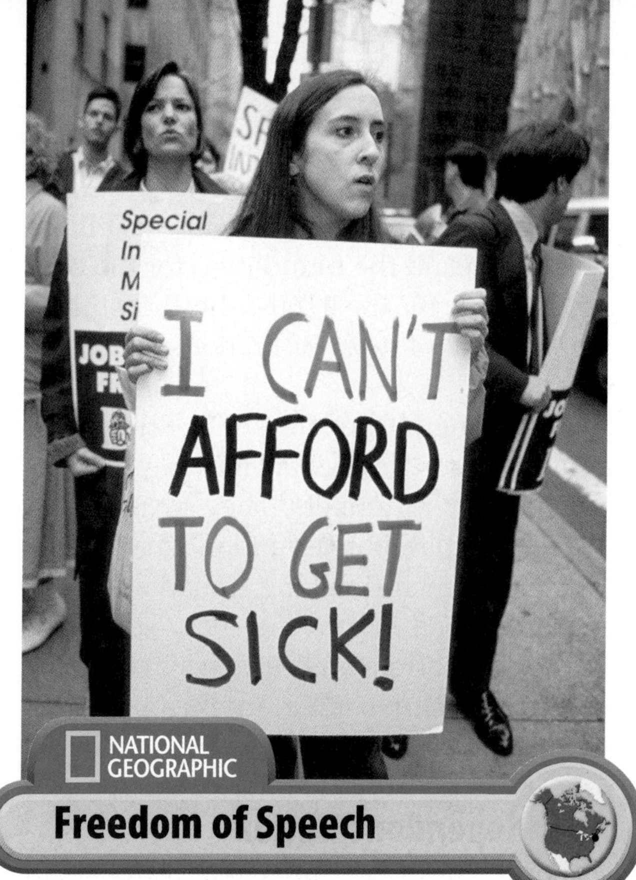

Freedom of Speech

Political freedoms are stated in the U.S. Constitution. The First Amendment allows citizens to disagree with the U.S. government. ***Place*** **What is the purpose of the Bill of Rights?**

The U.S. Constitution created a strong central government, but state governments were given certain responsibilities. This structure reflects the idea called federalism. In **federalism,** power is divided between the federal, or national, government and state governments. The national government makes treaties with other countries, coins money, and has the power to make laws about trade between states. State governments handle such issues as the health and education of their citizens.

In the U.S. federal system, people are citizens of both the nation and their state. As a result, citizens have the right to vote for both national and state leaders. The U.S. Constitution and state constitutions provide for that right. Citizens have the duty to make informed decisions when they vote. Citizens also have the responsibility to obey national and state laws.

In 1791 ten **amendments,** or additions, known as the Bill of Rights were added to the U.S. Constitution. Their purpose was to prevent the government from taking away people's freedoms. For example, the Bill of Rights ensures that the government cannot limit Americans' freedom of speech or religion.

Throughout its history, the Constitution has had other amendments added to it. Individual freedom is a **core,** or basic, value of the United States. Another is equality. The Constitution's Fourteenth Amendment requires the states to provide equal protection under the law to all persons within their boundaries.

Still, all Americans have not always enjoyed equal rights. Some groups suffered unfair treatment. For several decades after independence, only white males could vote. African American males gained the right to vote only when the Constitution was changed in 1870. All women were given the right to vote in 1920.

Canadian Democracy

Canada has a **parliamentary democracy** in which voters elect representatives to a lawmaking body called Parliament. These representatives then choose an official called the prime minister to head the government. The British monarch serves as king or queen of Canada.

Like the United States, Canada has a federal system. Power is divided between the central government and the provinces and territories. Canada's Charter of Rights and Freedoms is similar to the U.S. Bill of Rights. It protects the liberties of Canadian citizens.

Reading Check **Identifying** What powers do U.S. national and state governments have?

Section Review

Social Studies ONLINE
Study Central™ To review this section, go to glencoe.com.

Vocabulary

1. **Explain** the significance of:
   - **a.** colony
   - **b.** annex
   - **c.** terrorism
   - **d.** dominion
   - **e.** representative democracy
   - **f.** federalism
   - **g.** amendment
   - **h.** parliamentary democracy

Main Ideas

2. **Describing** What was the Cold War, and what brought it to an end?
3. **Categorizing** Use a chart like the one below to list five major events in American and Canadian history. Categorize each event as social, economic, or political, and explain why each is significant.

Event	Category	Significance

4. **Comparing and Contrasting** How are the governments of the United States and Canada similar? How are they different?

Critical Thinking

5. **BIG Idea** How did the French and British colonies in Canada differ from the Spanish colonies in Mexico and South America?
6. **Challenge** Why do you think it is important in a democracy for people to make informed decisions?

Writing About Geography

7. **Using Your FOLDABLES** Use your Foldable to write a paragraph explaining how the core American values of freedom and equality are evident in America's government and laws.

The Electoral College: Should It Be Changed?

When people vote for the president and vice president of the United States, they are really voting for electors who will vote for these people. These electors make up the Electoral College. The number of electors each state has is based on the total number of its U.S. senators and representatives. Electors cast votes for the candidate who wins the popular vote in their state. In every state except Maine and Nebraska, the winner of the popular vote will get all of that state's electoral votes.

For Change

[The Electoral College] gives more weight to votes cast in small states. . . . Second, . . . people who disagree with the majority in their state are not represented. Finally, the system allows the election of a President who does not have the support of a majority of voters. Without the Electoral College, candidates would campaign to get as many individual votes as possible in every state, instead of focusing on states that provide key electoral votes. Each vote would make a difference . . . which could lead to increased voting across the country.

—Kay Maxwell
President,
League of Women Voters

Change

[The Electoral College] has worked well in practice. . . . This system usually produces a decisive result, even when the popular vote is very close. Second, if America were to vote for President by direct popular vote, the campaigns might ignore smaller states and rural areas and concentrate only on the big metropolitan areas. Third, the Electoral College reinforces our two-party system. Because candidates are required to win elections in states in different parts of the country, the candidates' parties must have a broad base, not a regional one.

—John Fortier
American Enterprise Institute

You Be the Geographer

1. **Analyzing** How might presidential candidates change their campaign plans if the Electoral College were done away with?
2. **Critical Thinking** Do you think that candidates would change where they campaign if elections were based on the popular vote? Why or why not?
3. **Read to Write** Write a paragraph that explains whether you think the current process of electing a president and vice president is effective. You might want to give examples from what you know about American history and government.

Guide to Reading

BIG Idea

Culture influences people's perceptions about places and regions.

Content Vocabulary

- ban *(p. 145)*
- suburb *(p. 147)*
- indigenous *(p. 149)*
- bilingual *(p. 149)*

Academic Vocabulary

- evolve *(p. 147)*
- generate *(p. 147)*
- participate *(p. 147)*

Reading Strategy

Organizing Information Use a chart like the one below to organize key facts about the languages, cultures, and lifestyles of the United States and Canada.

	Key Facts
Language	
Art and Literature	
Daily Life	

Cultures and Lifestyles

 Section Audio  Spotlight Video

Picture This Cadillacs, instead of cattle, populate this dusty wheat field near Amarillo, Texas. In 1974 a Texas millionaire had this unusual display of cars created to honor the Cadillac. Layers of graffiti have been added to the Cadillacs over the years, and new guests are invited to add their own messages. Read Section 2 to learn more about the unique cultures in the United States and Canada.

Cadillac Ranch near Amarillo, Texas

Cultures and Lifestyles of the United States

Main Idea **The culture of the United States has been shaped by immigrants from around the world.**

Geography and You Have you ever decorated a Christmas tree or seen one at a store? This tradition actually comes from Germany. Read on to discover how immigration has transformed culture in the United States.

Becoming U.S. Citizens

Many immigrants eventually decide to become U.S. citizens. Once they meet the requirements and pass the citizenship test, they are officially sworn in as citizens and recite the Pledge of Allegiance. ***Movement*** **How have immigration trends changed since the late 1900s?**

About 300 million people live in the United States, making it the third-most-populous country. The population of the United States includes people of many different ethnic backgrounds.

Diverse Traditions

The United States has been called "a nation of immigrants." Throughout its history, the United States has attracted vast numbers of immigrants from around the globe. Yet the pattern of immigration has changed over time. During the late 1700s and early 1800s, the largest number of immigrants came from Great Britain, Ireland, western and central Africa, and the Caribbean. From the late 1800s to the 1920s, most immigrants came from southern, central, and eastern Europe. Many Chinese, Japanese, Mexican, and Canadian immigrants also arrived during this period.

The diverse backgrounds of so many late-1800s immigrants caused some Americans to become concerned about cultural change. As a result, they passed laws limiting immigration. In 1882 Congress passed a law that **banned,** or legally blocked, almost all immigration from China. In 1924 another law limited the numbers of immigrants from many countries. For many years, immigration to the United States slowed.

By the 1960s, opinions in the country had changed, in part due to growing support for civil rights. As a result, the country changed its immigration laws. A new law passed in 1965 based entry into the United States on a person's work skills and links to relatives already living in the United States. Changes in U.S. laws and in economic and political conditions worldwide led to increased immigration during the late 1900s. By 2000, nearly half of all the country's immigrants came from Latin America and Canada, and another third came from Asia. Less than 15 percent came from Europe.

Social Studies ONLINE

Student Web Activity Visit glencoe.com and complete the Chapter 5 Web Activity on the history of U.S. immigration.

Current immigration trends are changing the makeup of the American population. People of European descent still make up about two-thirds of the population, but the percentage of people from other areas is growing. Latinos, or Hispanics—who trace their heritages to the countries of Latin America and Spain—make up 15 percent of Americans. They are the fastest-growing ethnic group. African Americans, at 12 percent, are the next-largest ethnic group. Asian Americans make up 4 percent of the population, and Native Americans make up 1 percent.

The languages spoken in the United States reflect the diversity of its people. English is the primary language, but for one person in six, English is not their most familiar language. Spanish is the most widely spoken language after English. More than 1 million people speak one or more of the following: Chinese, French, Vietnamese, Tagalog, German, and Italian.

Diversity also extends to religion, which has long been an important factor in American life. Most Americans follow some form of Christianity. The largest number of Christians in the United States belongs to one of the many Protestant churches. These groups vary widely in their beliefs and practices. Roman Catholics make up the next-largest group of American Christians, followed by members of Eastern Orthodox churches. Judaism and Islam each have about 5 million followers in the United States. About 2 to 3 million Americans practice Buddhism, and another 2.5 million are followers of Hinduism.

Literature and the Arts

American artists, writers, and musicians have developed distinctly American styles. The earliest American artists used materials from their environments to create works of art. For centuries, Native Americans have carved wooden masks or created beautiful designs on pottery made from clay found in their areas. Later artists were attracted to the beauty of the landscape. For example, Winslow Homer painted the stormy waters of the North Atlantic. Georgia O'Keeffe painted the colorful deserts of the Southwest. In contrast, Thomas Eakins and John Sloan often painted the gritty, or rough, side of city life.

Two themes are common in American literature. One theme focuses on the rich diversity of the people in the United States. The poetry of Langston Hughes and the novels of Toni Morrison portray the triumphs and sorrows of African Americans. The novels of Amy Tan examine the lives of Chinese Americans. Oscar Hijuelos and Sandra Cisneros write about the country's Latinos.

▲ Originally from Wisconsin, O'Keeffe began her career as an art teacher. She moved to New Mexico in 1946, and lived and worked there until her death at the age of 98. ***Regions*** **What themes or subjects have attracted American artists?**

A second theme in American literature focuses on the landscape and history of particular regions. Mark Twain's books tell about life along the Mississippi River in the mid-1800s. Nathaniel Hawthorne wrote about the people of New England. Willa Cather and Laura Ingalls Wilder portrayed the struggles people faced in settling the Great Plains. William Faulkner wrote about life in the South.

Americans have created several new musical styles. Country music grew out of folk music from the rural South in the 1920s. The style gained many fans as it **evolved,** or developed, over the following decades. In the early 1900s, African Americans developed blues and jazz. Blues later inspired rock and roll in the 1950s. More recently, rap and hip-hop have gained popularity.

In the early 1900s, movies started attracting large audiences. Today, the movie industry **generates,** or makes, enormous profits and continues to entertain audiences around the world. In addition to movies, after 1950, television became an important part of American culture.

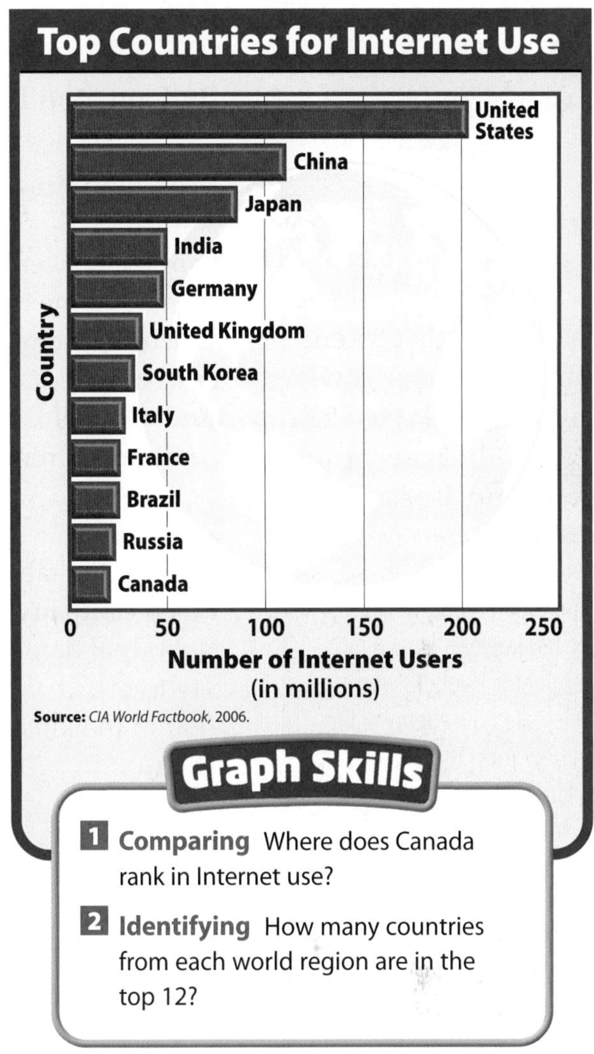

Source: *CIA World Factbook*, 2006.

Graph Skills

1. **Comparing** Where does Canada rank in Internet use?
2. **Identifying** How many countries from each world region are in the top 12?

Life in the United States

At one time, most Americans lived in rural areas. Today the United States is a land of urban dwellers. Many people, however, have moved from cities to **suburbs,** or smaller communities surrounding a larger city. They also have moved from one region to another. Since the 1970s, the fastest-growing regions have been the South and Southwest, often called the Sunbelt because of their sunny, mild climates.

Lifestyles vary across the United States. Americans live in different types of homes, from one-story houses in the suburbs to high-rise apartments in cities. About two-thirds of American families own their own homes. This is one of the highest home ownership rates in the world. Because of their economic well-being, Americans also lead the world in the ownership of cars and personal computers and in Internet use.

Many Americans watch movies and television, but they also exercise and play sports. Millions of young Americans **participate** in sports leagues devoted to games such as baseball and soccer. Important U.S. holidays include Thanksgiving, the Fourth of July, and other celebrations based on religious and ethnic traditions.

Reading Check **Summarizing** How has the United States controlled immigration?

Langston Hughes
(1902–1967)

Langston Hughes was one of the first African Americans to earn a living as a writer. He wrote plays, novels, essays, and children's stories, but he became best known for his poetry. Hughes's works celebrated the energy of life in New York City's Harlem, a largely African American neighborhood.

Background Information

Hughes was active in the Harlem Renaissance, an African American artistic movement of the 1920s. Inspired by jazz and blues, Hughes wove the rhythms of these forms of music into his poems. He urged African Americans to take pride in their culture and to stand up for their rights.

Reader's Dictionary

company: guests

ashamed: feel sorry or guilty for having done something wrong

I, Too

By Langston Hughes

I, too, sing America.

I am the darker brother.
They send me to eat in the kitchen
When company comes,
But I laugh,
And eat well,
And grow strong.

Tomorrow,
I'll be at the table
When **company** comes.
Nobody'll dare
Say to me,
"Eat in the kitchen,"
Then.

Besides,
They'll see how beautiful I am
And be **ashamed**—

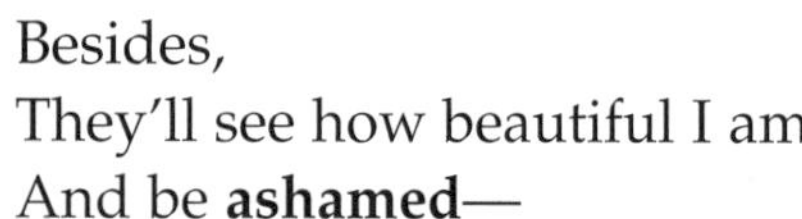

I, too, am America.

"I, Too" from *The Collected Poems of Langston Hughes* by Langston Hughes, edited by Arnold Rampersad with David Roessel, Associate Editor, copyright © 1994 by The Estate of Langston Hughes. Used by permission of Alfred A. Knopf, a division of Random House, Inc.

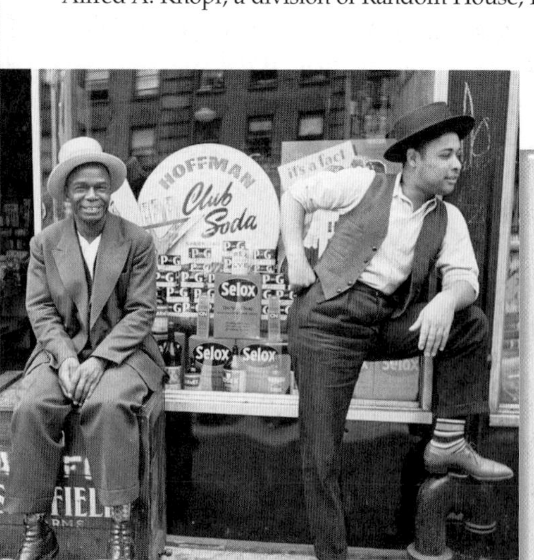

Analyzing Literature

1. **Making Inferences** How does Hughes view the role of African Americans in society at the time he wrote this poem?
2. **Read to Write** Write a poem or paragraph that responds to Hughes's feelings in "I, Too."

Cultures and Lifestyles of Canada

Main Idea **Canadians of many different backgrounds live in towns and cities close to the U.S. border.**

Geography and You Have you played hockey or seen a hockey game? Hockey is a Canadian sport. Read to find out how Canadians have developed unique cultures.

Like the United States, Canada is a nation formed by immigrants with many different cultures. Unlike the United States, Canada has had difficulty achieving a strong sense of national identity. Canada's vast distances and separate cultures cause some Canadians to feel more closely attached to their own region than to Canada as a country.

A Mix of Cultures

About one-fourth of Canadians are of French ancestry. Most of these people live in Quebec, where they make up 80 percent of that province's population. People of British ancestry form another fourth of Canada's population, and they live mainly in Ontario, the Atlantic Provinces, and British Columbia. People of other European backgrounds form about 15 percent of Canada's population.

Canada also is home to people of Asian, African, and Latin American backgrounds. Indigenous Canadians, similar to Native Americans, number more than a million. **Indigenous** refers to people who are descended from an area's first inhabitants. Many people in Canada call these indigenous groups the "First Nations."

Canada is a **bilingual** country, which means it has two official languages. In Canada, the languages are English and French.

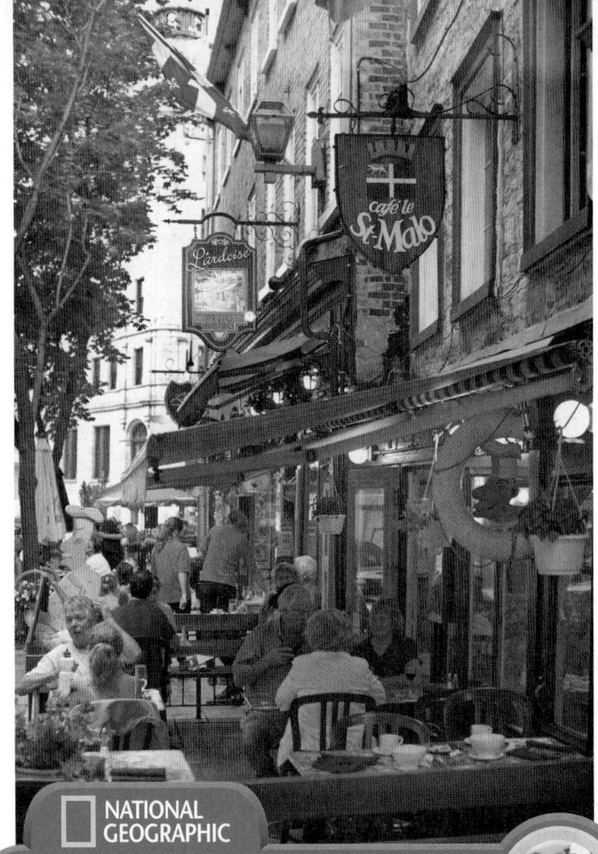

Quebec's French Culture

The French-speaking city of Quebec, founded in 1608, is one of Canada's oldest settlements.
Regions **What percentage of Canadians are of French ancestry?**

Despite promises of protection, many French speakers in Quebec do not believe that their language and culture can survive in largely English-speaking Canada. They would like Quebec to separate from Canada and become independent. So far, they have been defeated in two important votes on this issue. Canada's future as a united country, however, is still uncertain.

One cultural issue the country has been able to solve concerns the Inuit (IH·nu·wuht). The Inuit are a northern indigenous people who have wanted self-rule while remaining part of Canada. In 1999 the Canadian government created the territory of Nunavut for them. The name means "Our Land" in the Inuit language. There, the Inuit govern themselves, although they still rely on the national government for some services.

Art and Literature

The first Canadian artists were indigenous peoples who carved figures from stone and wood, made pottery, or were weavers. Today, Canadian art reflects both European and indigenous influences. The beauty of Canada's landscape has long been a favorite subject for many artists. Nature and history have been popular subjects for Canadian writers.

Centuries ago, indigenous peoples used song and dance as part of their religious rituals. European music, such as Irish and Scottish ballads, gained popularity after the 1700s. In recent decades, pop and rock have become as popular in Canada as in the United States.

Movies are also a major part of Canadian culture. The nation's film industry earns $5 billion each year. Theater is popular in Canada as well. Ontario's Stratford Festival is world famous for its productions of William Shakespeare's plays. The festival runs for a number of months each year and attracts over half a million fans.

Life in Canada

Like Americans, Canadians are a mobile people. Millions of them use cars to commute to work every day.

Certain foods are regional favorites. Seafood dishes are popular in the Atlantic Provinces, while French cuisine is preferred in Quebec. Ontario features Italian or Eastern European foods, reflecting the immigrant traditions there. British Columbia is known for salmon and Asian foods.

Canadians are enthusiastic about hockey—a sport that began in Canada—as well as lacrosse, which began as a Native American game. Many Canadians also enjoy hunting and fishing.

Canadians celebrate the founding of their country on July 1 and the fall Thanksgiving holiday in October. As in the United States, different ethnic groups celebrate their heritage at different times of the year.

Reading Check **Explaining** Why was the territory of Nunavut created?

Social Studies ONLINE
Study Central™ To review this section, go to glencoe.com.

Vocabulary

1. **Explain** the meaning of *ban, suburb, indigenous,* and *bilingual* by using each word in a sentence.

Main Ideas

2. **Identifying** Which styles of music did Americans invent?
3. **Summarizing** Use a diagram like the one below to show three ways immigration has shaped Canada.

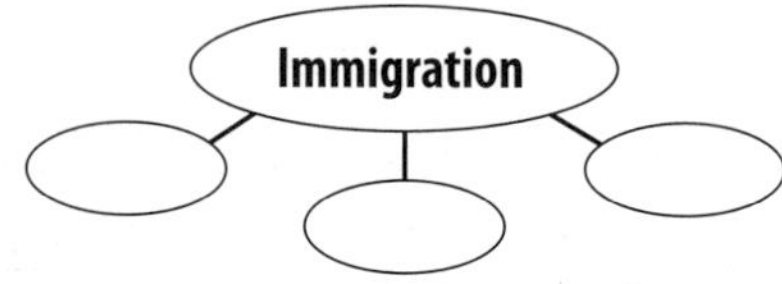

Critical Thinking

4. **Analyzing Information** How are immigration trends changing in the United States?
5. **BIG Idea** Explain why many people in Quebec have wanted to form a separate nation.
6. **Challenge** Why do you think the lifestyles of Americans and Canadians are similar?

Writing About Geography

7. **Expository Writing** Write a paragraph summarizing the styles of American art and literature.

Visual Summary

American History

- Native Americans, or indigenous peoples, are North America's earliest inhabitants.
- The 13 British colonies declared independence in 1776.
- During the 1800s, the United States grew in population and had a prosperous economy.
- The United States became a global power during the 1900s.

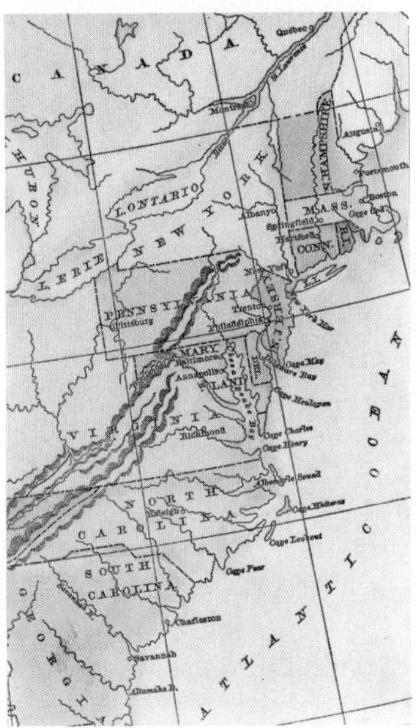

Map of original 13 colonies

Canadian History

- France and then Britain acquired control of the area that today is Canada.
- In 1867 the Dominion of Canada was founded.
- Canada grew through immigration and developed a modern economy.

American and Canadian Governments

- The United States and Canada are democracies based on federal systems.
- The U.S. government has three branches, each with special powers to check the power of the others.
- Canada has a parliamentary system. Legislative members choose a prime minister to head the government.

American Culture

- Immigration has created a diversity of groups, languages, and religions.
- American art and literature have focused on nature and freedom. American music includes country and jazz.
- American lifestyles reflect the economic well-being of the people.

Elvis Presley

Toronto, Canada

Canadian Culture

- Canadian culture reflects the diversity of the many peoples who settled the country.
- Canadian art and literature draw on nature and the history of the country.
- Foods and pastimes in Canada reflect regional life and the contributions of immigrants.

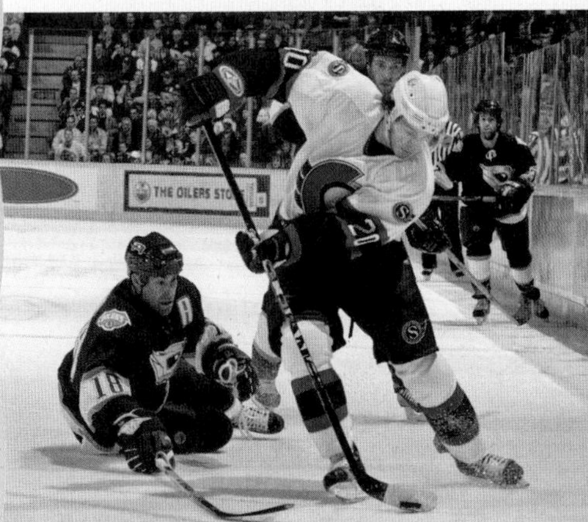

Hockey, Canada

STANDARDIZED TEST PRACTICE

Outline information from your textbook to study for an exam. Use bold type, questions, and summary paragraphs for headings and details in your outline.

Reviewing Vocabulary

Directions: Choose the word(s) that best completes the sentence.

1. Overseas settlements tied to a parent country are called ________.

A suburbs
B colonies
C dominions
D provinces

2. Both the United States and Canada have ________.

A monarchs
B colonial governments
C representative democracies
D parliamentary democracies

3. Smaller communities surrounding a larger city are known as ________.

A suburbs
B colonies
C dominions
D provinces

4. The term ________ is used to describe people who are descended from an area's first inhabitants.

A terrorist
B indigenous
C bilingual
D federalist

Reviewing Main Ideas

Directions: Choose the best answer for each question.

Section 1 *(pp. 134–141)*

5. How did the United States become a world leader during the 1900s?

A by developing the factory system
B by building an economy based on agriculture
C by enslaving hundreds of thousands of Africans
D by helping to win World War I and World War II

6. What accomplishment marked Canada's full independence?

A Canada's colonies became provinces.
B Canada established its own central government.
C Canadians fought in World War I and World War II.
D Canadians won the right to change their constitution.

Section 2 *(pp. 144–150)*

7. By 2000 where did most immigrants in the United States come from?

A Latin America, Canada, and Asia
B Ireland and Great Britain
C western and central Africa
D southern and eastern Europe

8. What percentage of Canadians has a European background?

A 15 percent
B 25 percent
C 50 percent
D 65 percent

GO ON

Critical Thinking

Directions: Base your answers to questions 9 and 10 on the circle graph below and your knowledge of Chapter 5. Choose the best answer for each question.

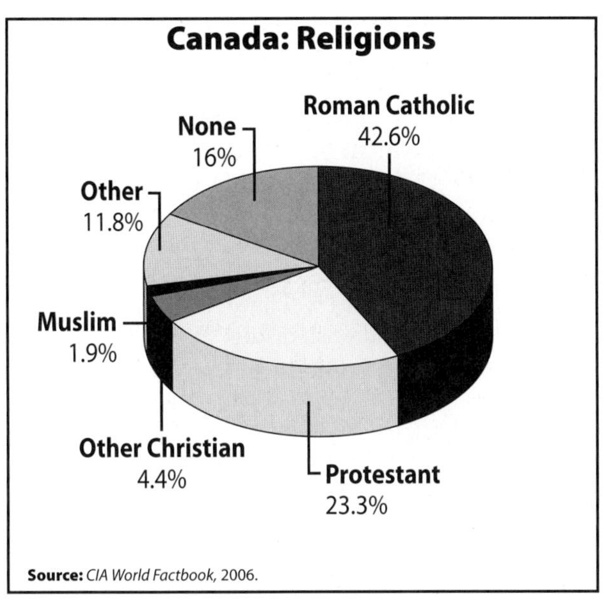

9. Which of the following generalizations is based on the circle graph?

A No Buddhists live in Canada.

B All Canadians are deeply religious.

C A majority of Canadians are Christian.

D Canadians are very tolerant of religious differences.

10. According to the graph, which of the following statements is correct?

A More Protestants than Catholics live in Canada.

B Muslims and other Christians (not including Catholics) outnumber Protestants.

C Islam has almost as many followers as Christianity.

D There are more Protestants than people who claim no religion.

Document-Based Questions

Directions: Analyze the following document and answer the short-answer questions that follow.

The following passage is from the *Canadian Charter of Rights and Freedoms* enacted in 1982.

> Guarantee of Rights and Freedoms
>
> *1. The* Canadian Charter of Rights and Freedoms *guarantees the rights and freedoms set out in it subject only to such reasonable limits prescribed by law as can be demonstrably justified in a free and democratic society.*
>
> Fundamental Freedoms
>
> *2. Everyone has the following fundamental freedoms:*
> *a) freedom of conscience and religion;*
> *b) freedom of thought, belief, opinion and expression, including freedom of the press and other media of communication;*
> *c) freedom of peaceful assembly; and*
> *d) freedom of association.*
>
> Democratic Rights
>
> *3. Every citizen of Canada has the right to vote in an election of members of the House of Commons or of a legislative assembly and to be qualified for membership therein.*

11. Do only Canadian citizens enjoy freedom of expression in Canada? Explain.

12. What qualification for membership in the Canadian House of Commons is mentioned?

Extended Response

13. Explain why Canada's future as a united country is uncertain.

For additional test practice, use Self-Check Quizzes—Chapter 5 at glencoe.com.

Need Extra Help?

If you missed question...	1	2	3	4	5	6	7	8	9	10	11	12	13
Go to page...	136	140	147	149	137	139	145	149	149	149	141	141	149

TIME JOURNAL

"Hello! My name is Taylor.

I'm 13 years old and live in Ashcroft, British Columbia. British Columbia is one of Canada's 10 provinces. My family runs a horse ranch and lodge there. Here's how I spend my day."

7:15 a.m. I wake up to the sound of my alarm clock. I shower and put on jeans, a T-shirt, and my tan cowboy boots. Then I head to the kitchen and say good morning to my parents and older brother, Daniel.

7:45 a.m. My mom has made pancakes and bacon for breakfast, and it smells great. I dig in. A hot breakfast is a nice treat. On most days, I just have cereal and toast.

8:25 a.m. Everyone goes their separate ways. My dad leaves for work in town. My brother goes off to high school, and my mom gets ready to work in the ranch office. I walk down our rural road and wait for the school bus.

9:00 a.m. My school day begins with math class. It's okay, but I am happy when the bell rings because I have English next. I love reading and writing and would like to become an author.

10:30 a.m. I stop at my locker, then go to French class. French is the official language in Quebec, Canada's biggest province. We speak English throughout the rest of Canada, but many kids study French as a second language.

11:40 a.m. It's time for lunch. Sometimes I walk to my friend's house for lunch (she lives near the school). Today, I stay at school and buy a grilled cheese sandwich and an apple.

12:20 p.m. In physical education class, we play basketball. In Canada, kids take "phys. ed." until tenth grade. We learn everything from softball to gymnastics.

1:20 p.m. I head to the school wood shop for woodworking class. I put on my safety goggles and use a power tool to carve a wooden sign. In a few weeks, woodworking will be over and I will start a new class—health.

3:00 p.m. I take the bus back home. There, I feed the horses and tackle some other ranch chores. In the summer, when the ranch is full of guests, I will be much busier! Then I will have to set up for meals and wash dishes.

4:45 p.m. I have some free time before dinner, so I grab my helmet, get my horse from the stable, and go riding.

6:15 p.m. Tonight, my grandfather and two uncles join us for dinner. They live in a separate house here on the ranch, so we see them all the time. We eat thick steaks and salad.

7:30 p.m. I help clear the table then do my homework. For one assignment, I have to use the family computer to log on to the Internet.

10:00 p.m. I read in my room for a while then get ready for bed. I'm tired!

ILLUSTRATIONS BY BOOKMAPMAN

SCHOOL TIME Taylor and her classmates share a laugh. Each semester Taylor studies four or five subjects. She gets three minutes between classes to get to her next lesson.

HORSING AROUND Ranching and tourism are big business in Taylor's village. Taylor's family owns a guest ranch. One of Taylor's chores is to feed the horses.

GREAT OUTDOORS Hiking is a popular pastime around Ashcroft. Here, Taylor and her dog take a break at a scenic spot overlooking the Thompson River.

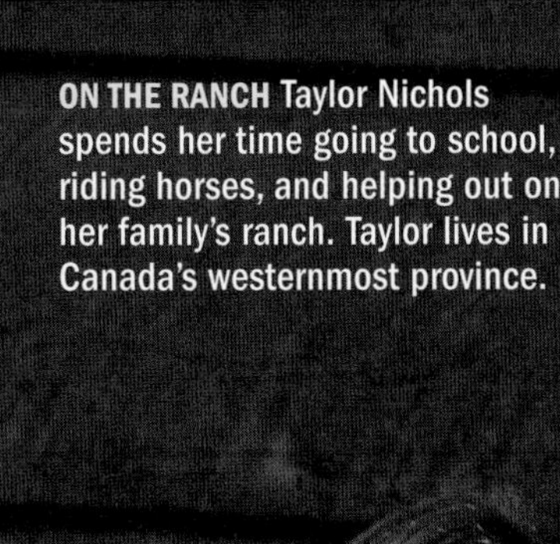

ON THE RANCH Taylor Nichols spends her time going to school, riding horses, and helping out on her family's ranch. Taylor lives in Canada's westernmost province.

What's Popular in Canada

Ice hockey Canadians are passionate about this national sport. At the professional level, there are intense team rivalries, like the one between the Toronto Maple Leafs and the Montreal Canadiens.

AP PHOTO/RYAN REMIORZ

Doughnuts Canada has more doughnut shops per person than any other country in the world! Apple fritters are a big favorite.

MELANIE ACEVEDO

Maple syrup Canada produces 85 percent of the world's maple syrup. Syrup makers tap the maple trees in early March. In one season, a single tree can make one liter of syrup.

ROY MORSCH/ AGE FOTOSTOCK

Say It in Canadian Slang

English and French are the national languages of Canada. In fact, all official signs, including road signs across the country, are printed in both languages. But Canada also has its share of unusual slang expressions. Try these examples.

How are you doing?
Whadda'yat? (This expression is from eastern Canada.)

A Canadian dollar
Loonie (The nickname comes from the picture of a loon that appears on the coin.)

Very good
Skookum (SKOO·kum)

PURESTOCK / ALAMY

AARON HUEY / POLARIS (4)

The United States and Canada Today

Essential Question

Regions The United States and Canada both have large land areas. Each has unique landforms and resources. Americans and Canadians have used their rich resources and technological skills to become leading economic powers. How might a region's economy influence the world economy?

San Francisco, California

Section 1: Living in the United States and Canada Today

BIG IDEA Places reflect the relationship between humans and the physical environment. Both the United States and Canada are often divided into economic regions. These regions are based on similar resources and climates. People in each region have developed distinctive ways of life based on the different physical characteristics of their area.

Section 2: Issues and Challenges

BIG IDEA Cooperation and conflict among people have an effect on the Earth's surface. The United States and Canada are peaceful neighbors, sharing the longest undefended border in the world. Landforms and weather patterns do not stop at the border, however, and environmental actions by one country can affect the other.

Organizing Information Make this Foldable to help you organize information about the economic regions and issues related to the United States and Canada.

Step 1 Fold an 11x17 piece of paper lengthwise to create 3 equal sections.

Step 2 Then fold it to form 3 columns.

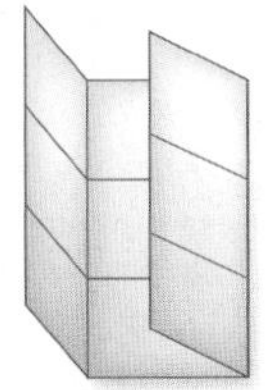

Step 3 Label your Foldable as shown.

Country	Regions/ Economies	Global/ Environmental Issues
United States		
Canada		

Reading and Writing As you read the chapter, take notes in the correct area of your Foldable. Use your notes to write the script of a short newscast describing the economies of the United States and Canada today.

Visit glencoe.com and enter ***QuickPass***™ code EOW3109c6 for Chapter 6 resources.

Guide to Reading

BIG Idea

Places reflect the relationship between humans and the physical environment.

Content Vocabulary

- free market *(p. 159)*
- profit *(p. 159)*
- stock *(p. 159)*
- biotechnology *(p. 159)*
- newsprint *(p. 162)*

Academic Vocabulary

- guarantee *(p. 159)*
- media *(p. 159)*
- reluctant *(p. 162)*

Reading Strategy

Categorizing Information Use a diagram like the one below to list the economic regions in the United States and provide several key facts about each.

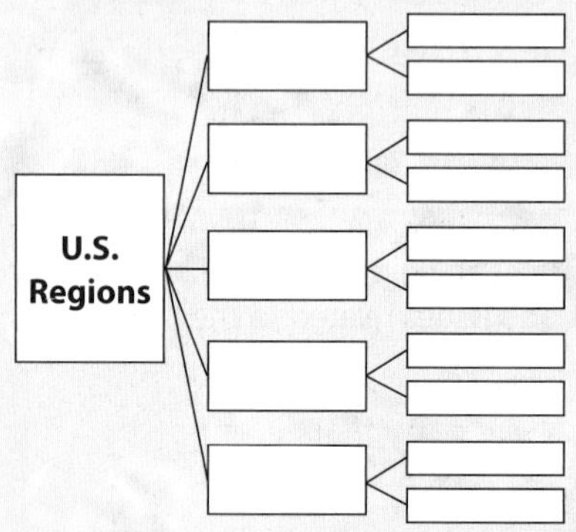

Living in the United States and Canada Today

Picture This "The doctor will see you now." Well, almost. This doctor of the future is actually a robot. It allows doctors who may be miles away to talk to patients, inspect wounds, and read charts through the use of a video camera and screen. A computer and a joystick allow the doctor to move through the hospital and visit patients. The robots have been developed for hospitals that are looking for ways to reduce expenses. Read Section 1 to find out more about other industries that shape the United States and Canada.

▼ A new type of doctor

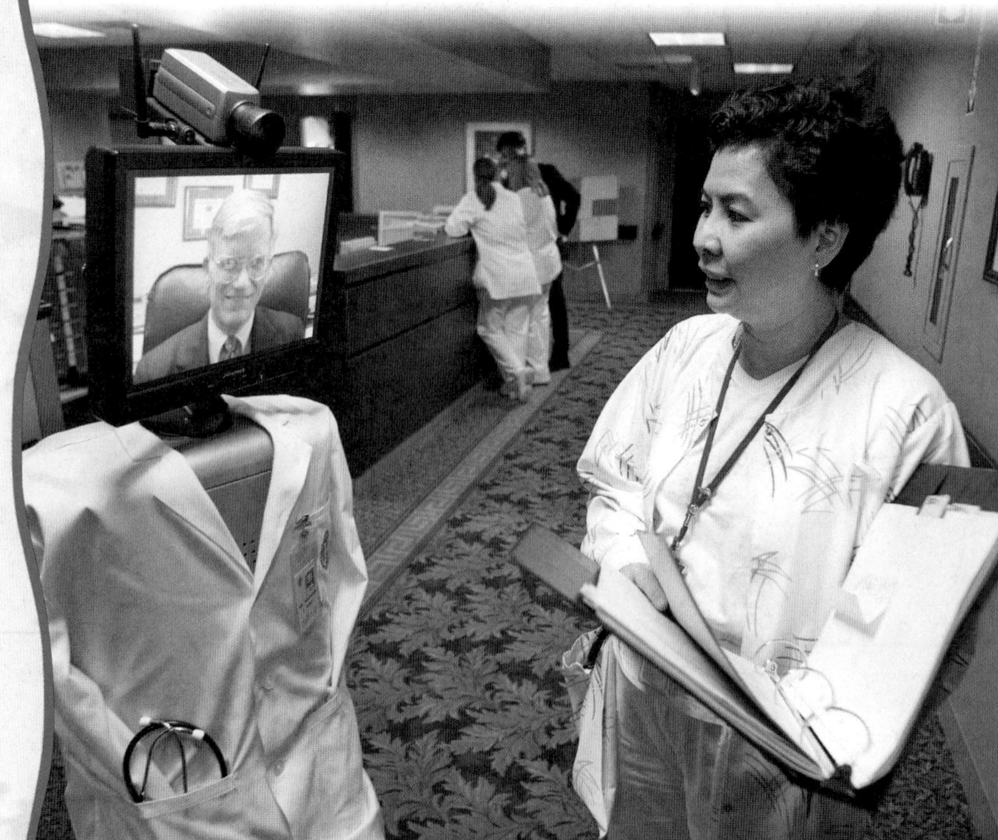

Economic Regions

Main Idea **The United States can be organized into economic regions.**

Geography and You Do you live in an area with many office buildings, factories, or farms? Read to discover why economic regions have developed.

Geographers group states together into five economic regions. These regions, and those of Canada, are shown in **Figure 1** on the next page. All of these regions are linked by the country's free market economy.

The Free Market Economy

In a **free market** economy, people are free to buy, sell, and produce whatever they want, with limited government involvement. They also can work wherever they want. A free market economy has two key groups: business owners and consumers. Business owners produce the products they think will make the most **profits,** or the most money after business expenses are paid. Consumers shop for the best products at the lowest prices.

People also take part in a free market economy by investing in businesses. People can buy **stock,** which represents part ownership in a company. When a company succeeds, it often pays some of its profits to the people who own its stock. Investing in stock involves risk, however. If the business fails, the stock becomes worthless. Consumers can also save their money in a bank, which is safer than buying stock. Because of government **guarantees,** or promises, savers have some of their money protected should a bank fail or go out of business. Although savings accounts are better protected, stocks provide a greater chance for high financial payoff.

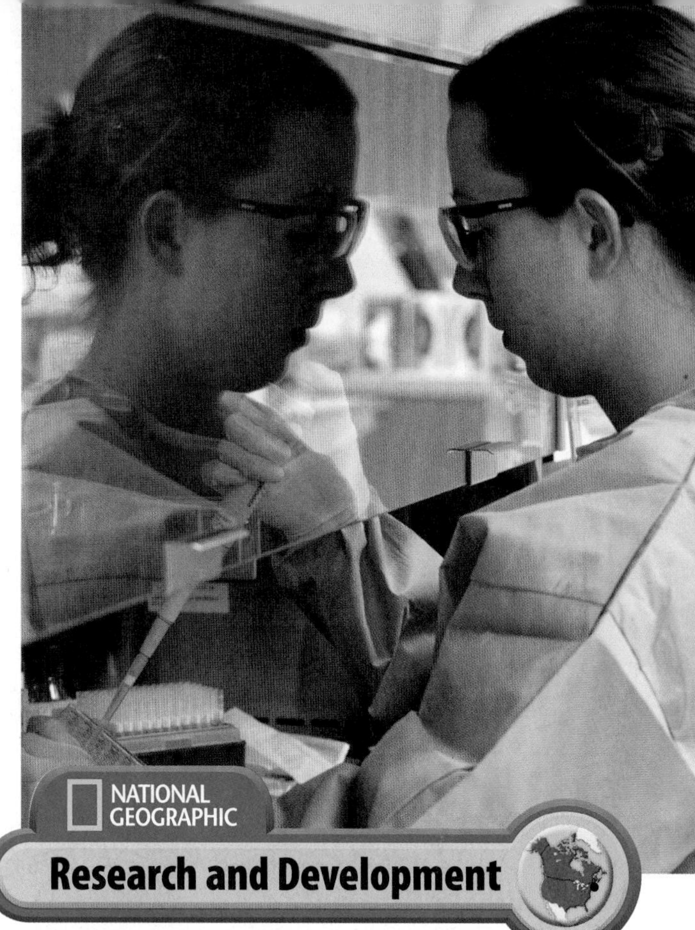

Research and Development

In Boston, biotechnology scientists study cells and how they respond to treatment. ***Regions*** **Why has the Northeast focused on business?**

A free market economy allows people to produce what they want. Resources are needed to produce a good or service, but some resources in a region are more available than others. As a result, economic regions have developed that specialize in producing products using their available resources.

The Northeast

The Northeast is made up of several large urban areas. Among them are New York City, the country's most populous city, and Boston, Massachusetts. With few mineral resources and poor soil for farming in many areas, the economic focus of the Northeast has been on business and trade. New York City has many financial and **media,** or communications, companies. Boston is an important center for **biotechnology** research. Biotechnology is the study of cells to find ways of improving health.

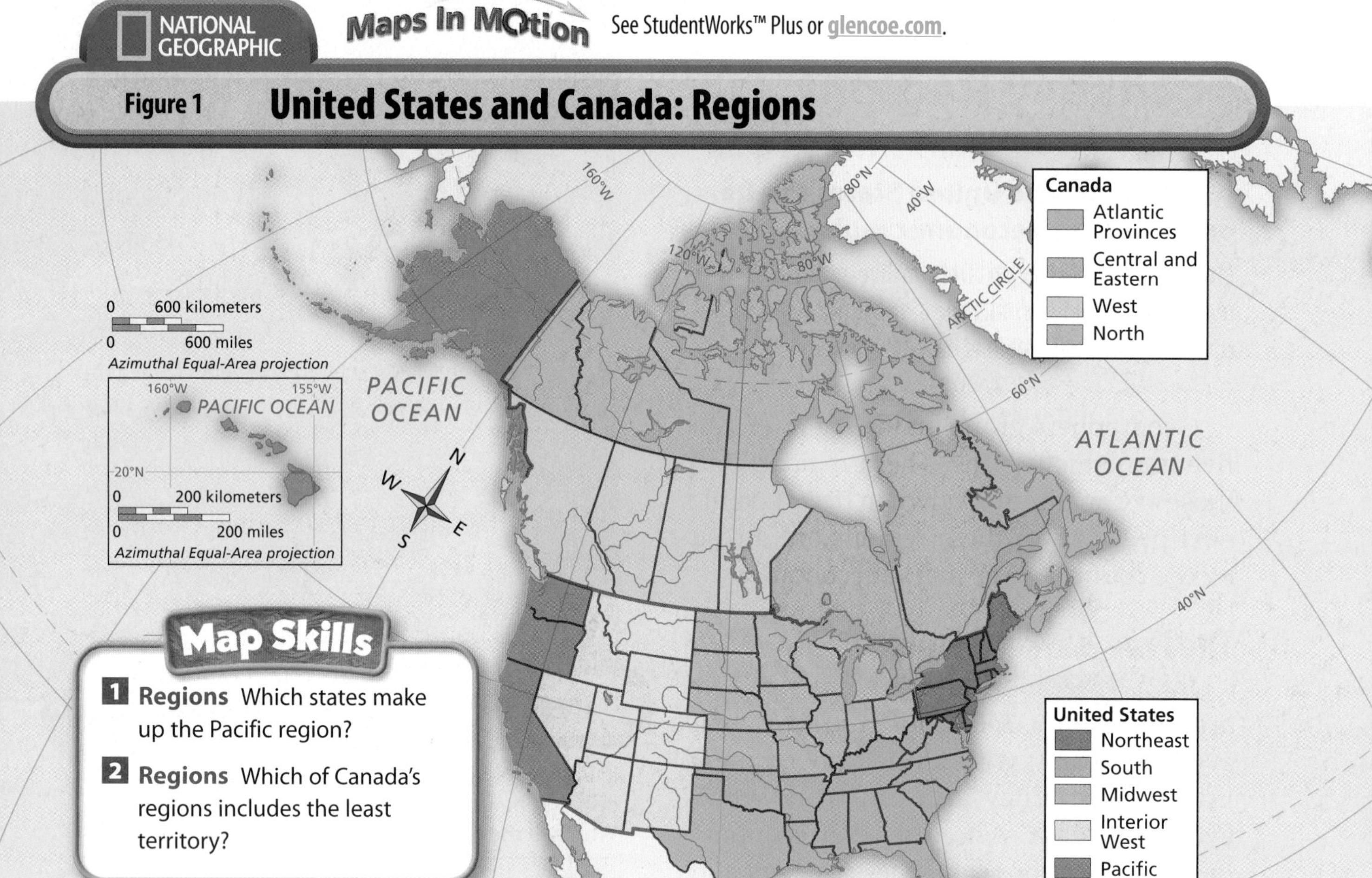

Figure 1 **United States and Canada: Regions**

The Midwest

The Midwest's rich soil enables farmers to grow crops such as corn, wheat, and soybeans. Mineral resources found here include iron ore, coal, lead, and zinc. Beginning in the 1800s, manufacturing developed in the Midwest. Towns like Cleveland and Detroit made steel and automobiles. Over time, however, the area's factories grew outdated. Many closed down, and thousands lost their jobs.

The South

With its rich soils, the South relied on agriculture. In recent decades the South has changed rapidly. The area now has expanding cities, growing industries, and diverse populations. Workers in cities such as Houston, Dallas, and Atlanta make textiles, electrical equipment, computers, and airplane parts. Texas, Louisiana, and Alabama produce oil and related products. In Florida, tourism and trade are major activities.

The Interior West

The Interior West has magnificent scenery and outdoor recreation that attracts many people. Although the region is dry, irrigation allows for some agriculture. For many decades, mining, ranching, and lumbering were the Interior West's main economic activities. In recent years, other parts of the economy have grown rapidly. The cities of Denver and Salt Lake City have growing information technology industries. Tourism and service industries are important to Albuquerque and Phoenix.

The Pacific

Fruits and vegetables are important crops in the fertile valleys of California, Oregon, and Washington. Sugarcane, pineapples, and coffee grow in the rich volcanic soil of Hawaii. Fish, timber, and mineral resources are important in the Pacific area as well. California has gold, lead, and copper, and Alaska has vast reserves of oil.

Many industries thrive in the Pacific area. Workers in California and Washington build airplanes and develop computer software. The city of Los Angeles is the world center of the movie industry. The area's booming economy draws many newcomers. California, the nation's most populous state, has wide ethnic diversity. Nearly half of its people are Latino or Asian American.

Reading Check **Summarizing** How has the south's economy changed?

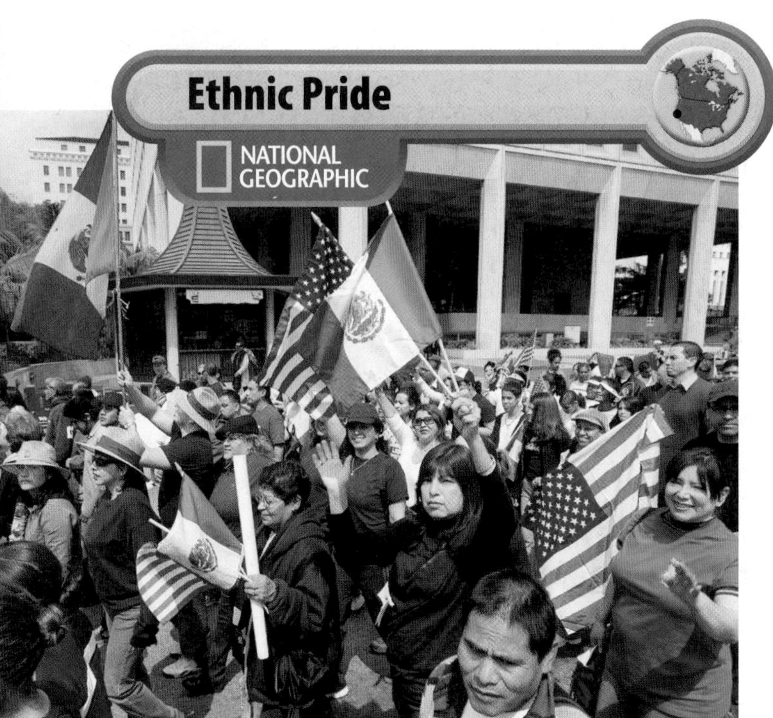

▲ This parade in Los Angeles, California, celebrates the participants' Latino heritage. ***Movement*** **Why are people drawn to the Pacific region?**

Regions of Canada

Main Idea **With a few exceptions, Canada's economic regions are similar to those in the United States.**

Geography and You Have you worked hard to build or make something? Did you feel a sense of accomplishment when you finished? Read to learn how Canadians have worked to build their economy despite geographic obstacles.

Like the United States, Canada consists of different economic regions. It also has a free market economy with limited government involvement. Canada's government, however, plays a more direct role in providing services. Unlike the United States, Canada's national and provincial governments provide health care for citizens. The government also regulates broadcasting, transport, and power companies.

Atlantic Provinces

Fishing was for many years a major industry in the Atlantic Provinces of Newfoundland and Labrador, Nova Scotia, Prince Edward Island, and New Brunswick. Overfishing, however, has weakened the industry. Today most people in the area hold jobs in manufacturing, mining, and tourism. The city of Halifax, in Nova Scotia, is a major shipping center in the region.

Central and Eastern Region

Canada's Central and Eastern Region includes the large provinces of Quebec and Ontario. The paper industry is important in Quebec, as is the creation of hydroelectric power. Montreal, on the St. Lawrence River, is a major port and leading financial and industrial center. Many in Quebec's largely French-speaking population would like the province to separate from Canada.

Because of the political and economic uncertainty this creates, many outside businesses have been **reluctant,** or hesitant, to invest in Quebec's economy.

Ontario has the largest population and greatest wealth of Canada's provinces. It is a major agricultural, manufacturing, forestry, and mining center. Ontario's capital, Toronto, is Canada's largest city and a major center of finance and business. Because of recent immigration, Toronto is now home to people from about 170 countries.

The West

Farming and ranching are major economic activities in the provinces of Manitoba, Saskatchewan, and Alberta. This area produces large amounts of wheat for export and contains some of the world's largest reserves of oil and natural gas.

The province of British Columbia has extensive forests. The forests help make Canada the world's largest producer of **newsprint,** the type of paper used for printing newspapers. Mining, fishing, and tourism also support British Columbia's economy. The city of Vancouver is Canada's main Pacific port.

The North

Canada's vast north covers about one-third of the country. This area includes the Yukon Territory, the Northwest Territories, and Nunavut. Many of the 25,000 people in this area are indigenous (ihn·DIH·juh·nuhs) peoples. The main resources in the North are minerals such as gold and diamonds.

Reading Check **Explaining** How has political uncertainty affected Quebec's economy?

Social Studies ONLINE
Study Central™ To review this section, go to glencoe.com.

Vocabulary

1. **Explain** the significance of:
   a. free market
   b. profit
   c. stock
   d. biotechnology
   e. newsprint

Main Ideas

2. **Analyzing** Describe the basic principles of a free market economy.
3. **Summarizing** Use a chart like the one below to list important facts about the economies of three regions of Canada, and identify a physical characteristic of the region that relates to that fact.

Economic Fact	Physical Characteristic
1.	1.
2.	2.
3.	3.

Critical Thinking

4. **BIG Idea** What economic activities in the Interior West of the United States are related to the area's natural resources?
5. **Comparing and Contrasting** Compare and contrast the Northeast region of the United States and the Atlantic Provinces of Canada.
6. **Challenge** Which region of the United States do you think has undergone the greatest economic change? Why?

Writing About Geography

7. **Descriptive Writing** Create a map that shows the economic regions of either the United States or Canada. Write a paragraph for each region describing the economy and people of that region.

TIME PERSPECTIVES

EXPLORING WORLD ISSUES

PROTECTING AMERICA FROM DISASTER

From terrorist attacks to hurricanes, the United States has faced a variety of deadly disasters. Will it be ready for the next crisis?

A victim of Hurricane Katrina is evacuated by helicopter.

Four years after the terrorist attacks of September 11, 2001, the United States was struck by three powerful hurricanes—Katrina, Rita, and Wilma. The massive storms brought a flood of despair as they tore through the U.S. Gulf Coast region. The hurricanes' strong winds and storm surges left death, destruction, and economic ruin in their paths.

As the government struggled to respond, thousands of survivors were left stranded. Many Americans wondered if the country had been too focused on preventing terrorist attacks and questioned whether officials had failed to plan for other types of disasters. People were also worried that the U.S. would be unprepared when the next catastrophe struck.

AP PHOTO

BETH DIXSON/ALAMY

Smoke poured from the World Trade Center's two towers after terrorists flew airplanes into them.

DISASTER STRIKES AMERICA

In recent years, the United States has been deeply affected by both human-made and natural disasters. On September 11, 2001, or "**9/11**," as the terrible events of that day are called, terrorists hijacked a jet airliner and crashed it into the north tower of the World Trade Center in New York City at 8:46 A.M. A second hijacked plane hit the south tower at 9:03 A.M. About an hour later, the south tower collapsed, crumbling to the ground.

At 9:43 A.M. outside Washington, D.C., a third hijacked plane crashed into the Pentagon, the headquarters of the U.S. military. About 20 minutes later, a fourth airliner went down in a field near Shanksville, Pennsylvania. Its target, which it never reached, might have been the White House or the U.S. Capitol. Back in New York City, the north tower of the World Trade Center collapsed at 10:28 A.M. In less than two hours, 19 terrorists had murdered thousands of people.

Crisis Along the Coast

The 9/11 attacks stunned the country. Nearly four years later, a series of monster hurricanes would do the same. On August 29, 2005, Hurricane Katrina slammed into the Gulf Coast. The storm packed 145-mile-per-hour winds that pushed powerful storm surges, or walls of water, inland from the Gulf of Mexico.

Katrina hit southern Florida first. Then, after strengthening while over the warm waters of the Gulf of Mexico, the hurricane struck Biloxi and Gulfport in Mississippi with deadly force. Homes, businesses, and entire towns were destroyed. The strong winds knocked down electrical lines. Power was lost across the Southeast, leaving millions without electricity.

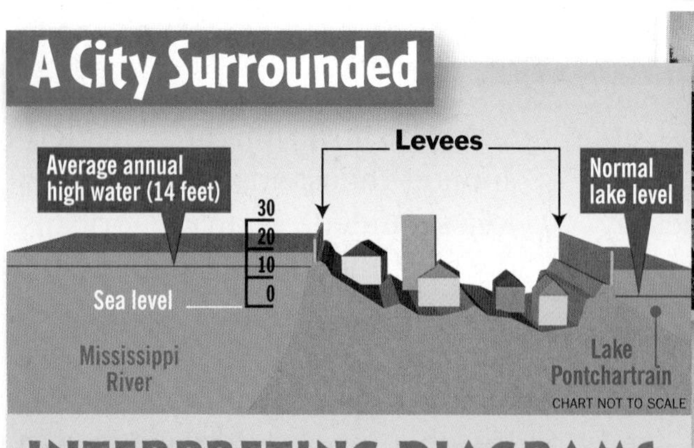

INTERPRETING DIAGRAMS

Making Inferences What does this diagram tell you about the need to maintain the New Orleans levee system?

AP PHOTO

Large portions of New Orleans are below sea level. Canals and 22 huge pumps are used to drain New Orleans when heavy rains cause flooding. Hurricane Katrina's powerful winds and waves, though, overwhelmed the city's flood prevention defenses.

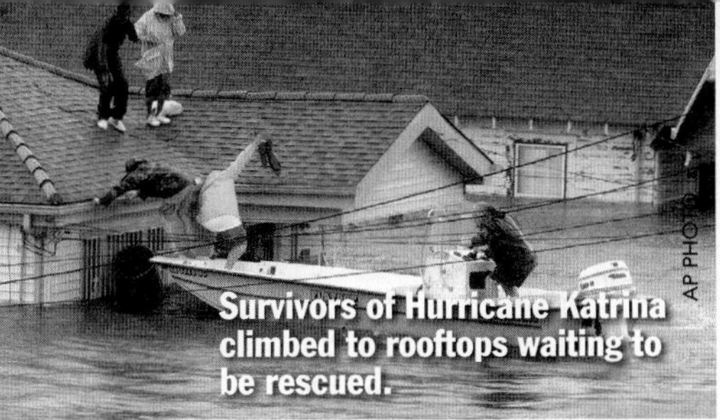
AP PHOTO
Survivors of Hurricane Katrina climbed to rooftops waiting to be rescued.

AP PHOTO
Young and old survivors of Hurricane Katrina hoped relief would come.

A roof was thrown onto the sidewalk by Hurricane Wilma's powerful winds.

Turmoil in New Orleans

Although New Orleans avoided a direct hit from Katrina, the storm punched four holes into the **levees**, or flood walls, that protect New Orleans from the sea. Soon, 80 percent of the city was under water.

Chaos followed. Officials called for all residents to leave the city. Across the city, thousands waited to be taken to safety. More than 20,000 people took shelter from the storm in New Orleans's Superdome. Conditions inside the stadium, however, had become unbearable as supplies of food and water ran out. People soon had to be evacuated.

The government was slow to respond with help. Evacuation plans were confusing, and coordination between local and national officials was poor. New Orleans would be a disaster area for months.

Soon after Katrina struck, another powerful storm, Hurricane Rita, ripped along the Gulf Coast with winds that reached 120 miles (193 km) per hour. More than 3 million people evacuated their homes. Then, in late October, Hurricane Wilma slammed into southwest Florida, leaving 6 million people without electricity.

A Stormy Economy

The devastation created by these deadly hurricanes affected Americans everywhere. Much of the country's oil and natural gas comes from the Gulf Coast. Oil platforms, refineries, and pipelines were damaged or shut down by Katrina. President George W. Bush authorized the release of oil from the nation's emergency reserve to help meet the country's energy demand.

Exports, or products sold outside the country, were affected too. Hundreds of barges that carried crops and products through the port of New Orleans to other countries sat backed up on the Mississippi River. Restoring the region would be costly. Experts estimated that the total price tag to rebuild the Gulf Coast would reach $200 billion.

Planning for the Future

Many Americans thought all levels of government—city, state, and federal—were unprepared to protect Gulf Coast residents. President Bush said, "Americans have every right to expect a more effective response in a time of emergency." Determining how to keep citizens safe will take time and hard work, but it is vital to the country's well-being.

EXPLORING THE ISSUE

1. **Understanding Cause and Effect** How did the 2005 hurricane season affect the broader U.S. economy?

2. **Interpreting Points of View** Do Americans have a right to expect the government to provide an "effective response" to emergencies? Why or why not?

AMERICA REACTS TO CATASTROPHES

The 2001 terrorist attacks and the 2005 hurricane season shocked many Americans. The tragic events also forced government leaders to take action. Finding new ways to protect the United States became a top priority.

Following 9/11, President George W. Bush signed a bill that created the **Department of Homeland Security**. The new department combined the duties and responsibilities of 22 different agencies. This included the Coast Guard and the **Federal Emergency Management Agency** (FEMA), the agency that responds to the nation's emergencies.

Lawmakers hoped the reorganization would improve communication between government agencies and help prevent future attacks. While some people thought the government changes were an effective response to terrorism, they were not useful for dealing with other types of disasters.

A Massive Cleanup

The aftermath of Hurricane Katrina made it clear that better coordination between local and national officials was required for disaster relief. The government was harshly criticized for its slow first response to Hurricane Katrina.

Congress approved $62 billion for relief work. These funds were used to find housing and provide health care for survivors. In the months after the hurricanes struck, local leaders and federal agencies began to reconstruct the Gulf Coast. By December 2005, the president pledged another $3.1 billion to repair New Orleans's levee system.

The task of rebuilding New Orleans and other areas of the Gulf Coast will take many years. As the government examines the results of these catastrophes, it hopes to make America safe from future disasters.

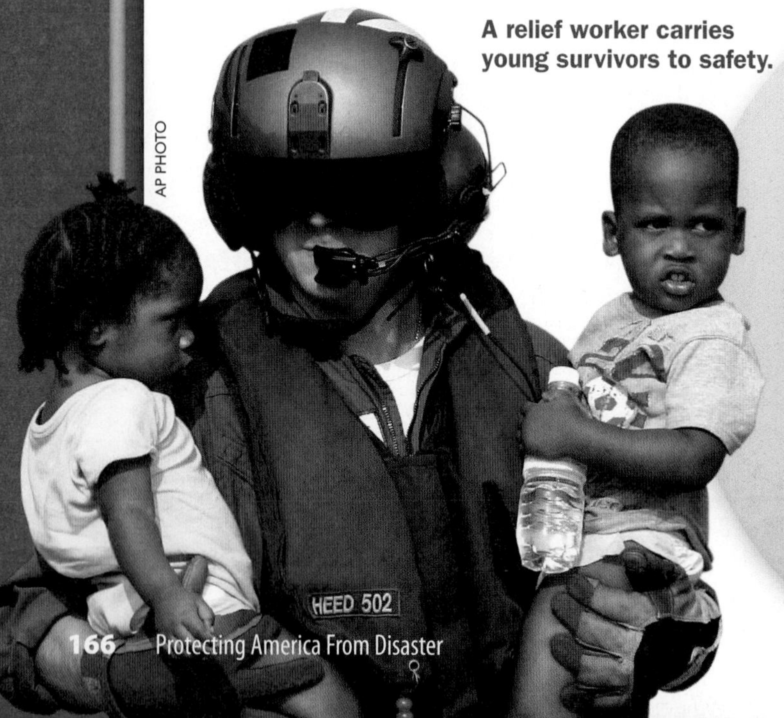

A relief worker carries young survivors to safety.

AP PHOTO

EXPLORING THE ISSUE

1. **Making Inferences** How might the Department of Homeland Security help officials fight terrorism? In what ways might the large department be ineffective in fighting natural disasters?

2. **Problem Solving** What steps do you think the government should take to fight terrorism and other types of catastrophes?

REVIEW AND ASSESS

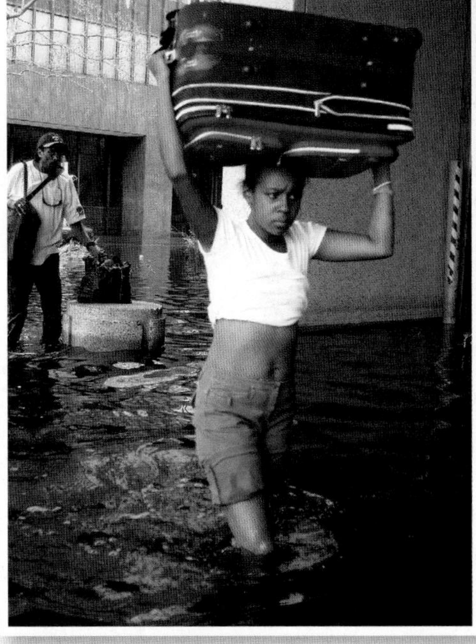

AP PHOTO

A survivor carries her belongings as she wades through water in New Orleans.

UNDERSTANDING THE ISSUE

1 **Making Connections** How did Hurricane Katrina affect the country's oil supply?

2 **Writing to Inform** In a brief essay, explain how the nation is preparing for natural and human-made disasters.

3 **Writing to Persuade** Write a letter to the editor of a newspaper expressing your views about the government's response to Hurricane Katrina and its readiness for future disasters.

INTERNET RESEARCH ACTIVITIES

4 With your teacher's help, go to www.ready.gov. Explore one of the three main items on the Web site's home page. Write a short essay that explains what you learned about preparing for a terrorist attack.

5 Navigate to www.fema.gov/kids. Click through some of the Web site's links to learn more about FEMA, how it is organized, and what its responsibilities are. Summarize your findings in a short written report.

BEYOND THE CLASSROOM

6 **Research the impact of terrorism on nations such as Sri Lanka and Ireland.** How have these nations tried to prevent terrorist attacks? Describe these ways, and discuss as a class whether they would be effective in preventing terrorist attacks in the United States.

7 **Work in Groups.** Create and display posters that explain how to prepare for any disaster, including hurricanes and terrorist attacks.

After Hurricane Katrina, the Costly Recovery Plan

Congress approved $62.3 billion in Gulf Coast aid for FEMA (though the final relief tab could well top $200 billion). Here is how FEMA distributed some of the funds.

- **$23 billion** for temporary housing, health care, clothing, household costs
- **$8 billion** for search-and-rescue missions, road and bridge safety, and adequate water
- **$3 billion** to the Defense Department for emergency repairs and evacuation
- **$5 billion** for FEMA to set up support and recovery work
- **$15 billion** for the Department of Homeland Security to monitor how relief aid was spent and distributed

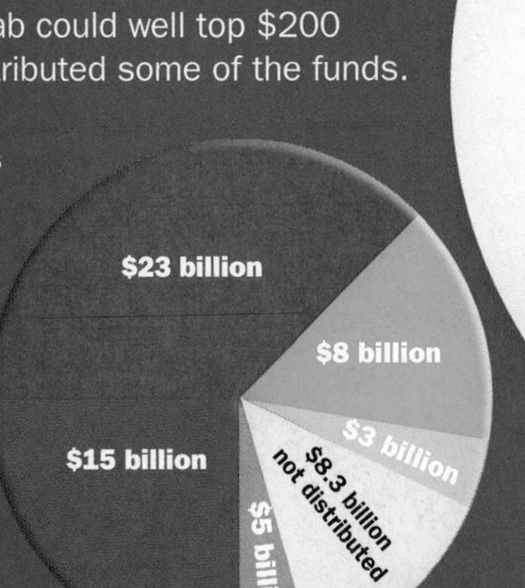

Source: TIME research, 2005.

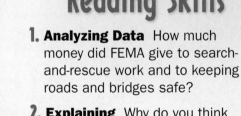

Building Graph Reading Skills

1. **Analyzing Data** How much money did FEMA give to search-and-rescue work and to keeping roads and bridges safe?
2. **Explaining** Why do you think most of the government relief funds were used to help Katrina survivors find housing, clothing, and health care?

Guide to Reading

BIG Idea

Cooperation and conflict among people have an effect on the Earth's surface.

Content Vocabulary

- trade deficit *(p. 169)*
- tariff *(p. 170)*
- trade surplus *(p. 170)*
- acid rain *(p. 171)*
- brownfield *(p. 172)*
- urban sprawl *(p. 172)*

Academic Vocabulary

- restrict *(p. 169)*
- community *(p. 172)*

Reading Strategy

Outlining Use a format like the one below to make an outline of the section. Write each main heading on a line with a Roman numeral, and list important facts below it.

I. First Main Heading
   A. Key Fact 1
   B. Key Fact 2
II. Second Main Heading
   A. Key Fact 1
   B. Key Fact 2

Issues and Challenges

 Section Audio  Spotlight Video

Picture This Long ago, ice canoeing was a way to deliver supplies across the icy Saint Lawrence River during winter. Ice canoes were small enough to dodge floating ice and light enough to carry over solid ice. Traveling between the river's islands was dangerous, difficult work, but it was necessary to carry vital items such as medicine and mail. Today, ice canoes are used for sport. Many Canadian teams compete in ice races at winter festivals. Learn more about Canada today in Section 2.

Ice canoe racing in Toronto, Canada

The Region and the World

Main Idea **The United States and Canada trade with countries throughout the world.**

Geography and You Think about all the trucks or trains you have seen carrying products to stores or to ports. Read to learn about our region's economic ties to the world.

Because of their large economies, the United States and Canada trade with many countries. In addition, the United States is one of the world's leading political and military powers.

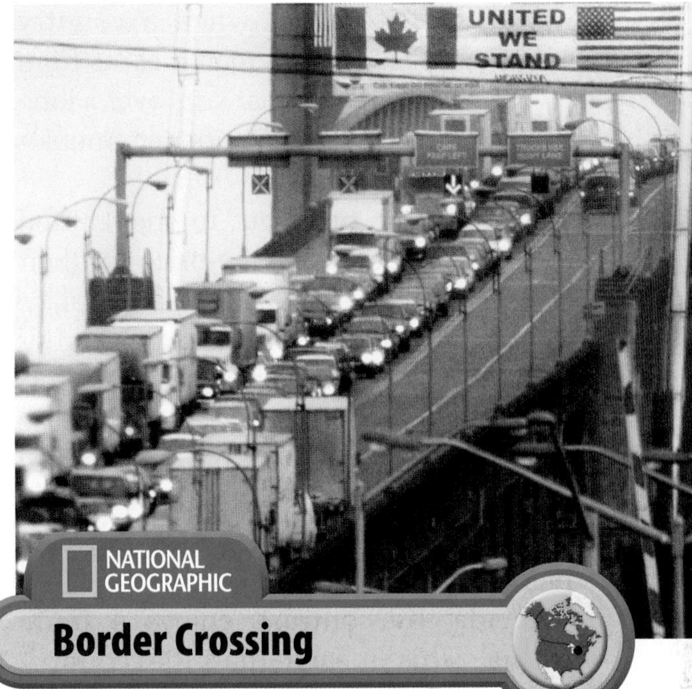

Border Crossing

The Ambassador Bridge is the busiest international border crossing in North America. More than 10,000 trucks cross the bridge daily between Detroit, Michigan, and Windsor, Canada. ***Regions*** **What did NAFTA establish?**

Economic Ties

The free market system has helped the United States and Canada build productive economies. In fact, the United States has the world's largest economy and is a leader in world trade. The United States exports chemicals, farm products, and manufactured goods, as well as raw materials such as metals and cotton fiber. Canada sends many of the same goods overseas, as well as large amounts of seafood and timber products. Both countries also import many items from all around the world.

In recent years, the United States and Canada have supported free trade. This means the removal of trade **restrictions,** or barriers, so that goods flow freely among countries. In 1994 the United States, Canada, and Mexico joined together in the North American Free Trade Agreement (NAFTA). They promised to take away most barriers to trade among the three countries. Canada now sends more than 80 percent of its exports to the United States and buys nearly 60 percent of its imports from it. Canada is the largest trading partner of the United States, and Mexico is the second largest.

The United States depends on trade to supply most of its energy resources. As you may recall, the country uses nearly three times the amount of oil it produces. The United States must import additional oil from major suppliers such as Canada, Mexico, Venezuela, Saudi Arabia, Nigeria, and Angola.

Americans buy many additional products from other countries. In fact, the United States spends hundreds of billions of dollars more on imports than it earns from exports. The result is a massive **trade deficit.**

Student Web Activity Visit glencoe.com and complete the Chapter 6 Web Activity about U.S. trade.

A trade deficit occurs when a country spends more on imports than it earns from exports. A trade deficit that lasts over a long period can cause serious economic troubles for a country.

To sell their products in the United States, some nations set the prices of their goods very low. Also, some countries place high **tariffs,** or taxes, on imports in order to protect their own industries from foreign competition. These tariffs then raise the price of U.S. products and thus reduce the sale of the products abroad. Such practices hurt American companies and cost American workers their jobs.

Canada, by contrast, enjoys a **trade surplus.** This means that the country earns more from exports than it spends for imports. Canada's smaller population makes its energy needs less costly. Also, Canada's export earnings have been growing.

Global Terrorism

Since the beginning of the 2000s, the United States and Canada have joined other nations to deal with increasing terrorism and violence around the world. As you may recall, the world changed dramatically on September 11, 2001, when terrorists attacked sites in New York City and Washington, D.C. To prevent further terrorist attacks, the United States and Canada have worked to increase security along their long border. They have also participated in international efforts to stop terrorism.

GLOBAL CITIZENS

NAMES: BILL AND MELINDA GATES

HOME COUNTRY: United States

ACHIEVEMENT: In 2008 the Bill and Melinda Gates Foundation gave more money away faster than anyone ever has. Bill Gates, a founder of Microsoft, and his wife, Melinda, created the foundation that is now the world's largest charitable organization. Many of the foundation's grants go toward global efforts—such as fighting disease—in more than 100 countries. Grants also are dedicated to improving people's lives in the U.S. The foundation is committed to giving away billions of dollars to support U.S. high schools and libraries and to provide college scholarships for African Americans. In 2008 Gates left his position at Microsoft to devote his full time to the Foundation.

QUOTE: “I felt I had a role to give some voice to the voiceless.”—Melinda Gates, referring to grants to developing countries

Melinda and Bill Gates visit a health research center in Mozambique.

CITIZENS IN ACTION Who are the “voiceless” that Melinda Gates refers to? Do they need champions such as the Gateses? Why or why not?

Guarding Against Terrorism

Since the 2001 terrorist attacks against New York City and Washington, D.C., new safeguards are in place to protect Americans. ***Place*** **How did the U.S. and Canada disagree over Iraq policy?**

Despite their generally close relations, the United States and Canada sometimes differ in their policies. In 2003 Canada opposed the U.S. decision to invade Iraq. Instead, it wanted the American government to continue seeking a peaceful solution through the United Nations (UN). The United Nations is the world organization that promotes cooperation among nations in settling disputes.

The United States and Canada both have strong roles in the United Nations. They provide much of the money that funds the organization. They also take part in UN agencies that provide aid to people in areas affected by war or natural disasters. The United States and especially Canada have sent soldiers to serve in UN forces that act as peacekeepers in troubled areas of the world.

Reading Check **Explaining** Why does the United States have a large trade deficit?

Environmental Issues

Main Idea **The United States and Canada face similar environmental issues.**

Geography and You Have you witnessed instances of air or water pollution in your area? Read to learn about challenges for the region's environment.

The United States and Canada face several environmental challenges. Because they are neighbors, some concerns are shared. Both countries have made progress in cleaning up their air and water. Still, pollution remains a problem.

Acid Rain and Climate Change

Americans and Canadians burn coal, oil, and natural gas to power their factories and run their cars. Burning these fossil fuels pollutes the air. The pollution also mixes with water vapor in the air to make **acid rain,** or rain containing high amounts of chemical pollutants. Acid rain damages trees and harms rivers, lakes, and the stone used in buildings. The United States and Canada have acted to reduce the amount of chemicals that are released into the air. They are particularly concerned about damage to some areas in the eastern parts of the region.

Many people in the United States and Canada also worry about global warming. Some scientists believe that global warming will lead to changing weather patterns. For example, a warmer climate could lead to drought conditions in some areas, or it could melt the polar ice caps. Melting would raise sea levels, which would be a problem for low-lying areas like Florida.

To address the issue of global warming, Canada has passed laws to reduce the amount of fossil fuels that are burned.

The United States has funded research to find new energy sources that are less harmful to the environment.

Pollution Issues

Changing climatic conditions and a rising demand for water have lowered water levels of the Great Lakes. Loss of water can cause a variety of problems. Lower lake levels decrease the amount of goods that can be shipped. Ships cannot carry extremely heavy loads into river channels or they might run aground. In addition, lower lake levels can harm fish populations. Tourism is also affected as water pulls back from the area's beaches. Government leaders of both countries have urged people to conserve water.

Brownfields are another challenge. **Brownfields** are places, such as old factories and gas stations, that have been abandoned and contain dangerous chemicals. These chemicals hinder any new development. Governments in the United States and Canada have given money to **communities,** or neighborhoods, for cleanup.

Urban sprawl, or the spread of human settlement into natural areas, also has created difficulties. Urban growth leads to the loss of farmland and wilderness areas. The building of homes and roadways can also produce traffic jams and increase air pollution. Growing populations put strains on water and other resources as well. People in some areas that are growing rapidly want to slow the rate of growth. They fear that having too many people will destroy the wide-open spaces and magnificent scenery that attracted them.

Reading Check **Explaining** How have the United States and Canada dealt with the issue of global warming?

Social Studies ONLINE
Study Central™ To review this section, go to glencoe.com.

Vocabulary

1. **Explain** the significance of:
   - **a.** trade deficit
   - **b.** tariff
   - **c.** trade surplus
   - **d.** acid rain
   - **e.** brownfield
   - **f.** urban sprawl

Main Ideas

2. **Describing** Describe the trade relationship between the United States and Canada.
3. **Determining Cause and Effect** Use a diagram like the one below to list the effects of falling water levels in the Great Lakes.

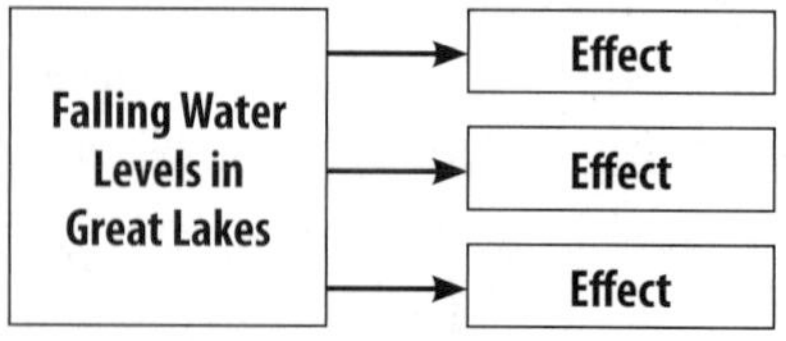

Critical Thinking

4. **Making Generalizations** Why is the U.S. economy important to the rest of the world?
5. **BIG Idea** Should the United States and Canada cooperate to try to reduce the problem of acid rain? Explain.
6. **Challenge** How would you describe overall relations between the United States and Canada? Why?

Writing About Geography

7. **Using Your FOLDABLES™** Use the notes in your Foldable to write a letter to a member of Congress explaining what you think should be done about the U.S. trade deficit.

Chapter 6 Visual Summary

Free Market Economies

- The United States and Canada have free market economies that allow people to own businesses and earn profits.
- Producers and consumers decide what to produce, how much to produce, and for whom to produce.
- Governments play a limited role in free market economies.

Small business owner, Los Angeles, California

U.S. Regions

- The five economic regions of the United States are the Northeast, the Midwest, the South, the Interior West, and the Pacific.
- The South and the Interior West are growing rapidly in population and economic strength.
- The Northeast focuses on business. The Midwest is rebuilding its industries.
- The Pacific area has diverse economies and populations.

Port at Seattle, Washington

Canadian wheelchair athlete

Canadian Regions

- Canada's main economic regions are the Atlantic Provinces, the Central and Eastern Region, the West, and the North.
- The Atlantic Provinces suffer from the decline of the fishing industry.
- Many people in French-speaking Quebec want their province to be independent.
- Ontario is Canada's most populous and economically prosperous province.
- The West includes the grain-producing areas and the Pacific coastal province of British Columbia.

Global Ties

- The United States and Canada have joined Mexico in promoting free trade among their countries.
- The United States is a major global trading power. Its trade deficits, however, could cause future economic problems.
- The United States and other countries are working to prevent terrorist attacks.

The Environment

- The United States and Canada are reducing the amount of chemicals released into the air to reduce acid rain.
- Declining water levels and rising demand for water are affecting the Great Lakes.
- The United States and Canada face the loss of farmland and wilderness areas as urban sprawl increases.

New home construction, Florida

STUDY TO GO Study anywhere, anytime! Download quizzes and flash cards to your PDA from **glencoe.com**.

CHAPTER 6 ASSESSMENT

STANDARDIZED TEST PRACTICE

During an exam, answer the questions you know first. Then go back to those that need more thought.

Reviewing Vocabulary

Directions: Choose the word(s) that best completes the sentence.

1. Business owners produce products they think will be most ______.

A beautiful

B affordable

C useful

D profitable

2. British Columbia's forests help make Canada the world's largest producer of ______.

A stocks

B newsprint

C acid rain

D biotechnology

3. A ______ occurs when a country spends more on its imports than it earns from its exports.

A trade deficit

B import tariff

C free market

D trade surplus

4. A country sometimes places ______ on imports to protect industries from foreign competition.

A trade deficits

B tariffs

C free markets

D trade surpluses

Reviewing Main Ideas

Directions: Choose the best answer for each question.

Section 1 *(pp. 158–162)*

5. Because it has abundant oil, the ______ economic region of the United States produces petroleum-based products.

A South

B Pacific

C Midwest

D Interior West

6. Most people in Canada's Atlantic provinces make a living through the ______ industries.

A fishing, farming, and forestry

B oil, natural gas, and exporting

C manufacturing, mining, and tourism

D finance, hydroelectricity, and ranching

Section 2 *(pp. 168–172)*

7. The largest trading partner of the United States is ______.

A Mexico

B Canada

C Nigeria

D Venezuela

8. To address the issue of ______, Canada passed laws to lower the amount of fossil fuels that are burned.

A low water levels

B urban sprawl

C global warming

D brownfields

GO ON

Critical Thinking

Directions: Base your answers to questions 9 and 10 on the graph below and your knowledge of Chapter 6.

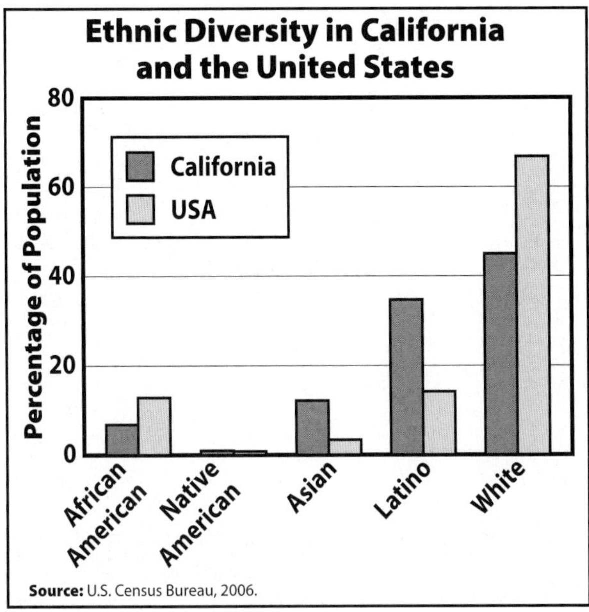

Source: U.S. Census Bureau, 2006.

9. What conclusion can you make from the data?

A More Latinos live in California than in the rest of the United States.

B California's population is about one-tenth of the nation's total population.

C California has a larger percentage of Japanese Americans than the country has as a whole.

D The U.S. population has a larger percentage of African Americans than does California's.

10. Which of these generalizations is accurate?

A Latinos are the fastest growing U.S. group.

B African Americans outnumber Asians 3 to 1 in California.

C Latinos and whites make up most of the nation's population.

D Whites outnumber the combined members of all other ethnic groups in the state.

Document-Based Questions

Directions: Analyze the following document and answer the short-answer questions that follow.

The following news story was broadcast in Canada on October 30, 2005.

> *Quebec Premier Jean Charest called federalism the "best choice for the future of Quebec" as provincial separatists vowed to continue to seek independence on Sunday—the 10th anniversary of the province's second referendum on sovereignty.*
>
> *On Oct. 30, 1995, more than 93 per cent of the province's 5.1 million registered voters cast a ballot. A razor-thin majority kept the province from heading towards independence, with 50.58 per cent voting No to 49.42 per cent voting Yes.*
>
> *Charest, in a closing speech to about 800 Liberals at the provincial party's general meeting on Sunday, said his most fervent [strong] desire was to avoid the holding of another sovereignty referendum [independence vote] so Quebecers don't once again "inflict such wounds upon themselves."*
>
> —Debate still rages on 10th anniversary of Quebec's sovereignty referendum, *CBC News.*

11. What statement indicates that Charest believes Quebec should remain part of Canada?

12. Did most Quebecois in 1995 have strong feelings about Quebec's independence? How do you know?

Extended Response

13. If you lived in Quebec, would you like to see it become independent or remain part of Canada? Explain your answer.

For additional test practice, use Self-Check Quizzes—Chapter 6 at glencoe.com.

Need Extra Help?

If you missed question...	1	2	3	4	5	6	7	8	9	10	11	12	13
Go to page...	159	162	169	170	160	161	169	171	161	161	161	161	161–162

UNIT 3 Latin America

CHAPTER 7
Physical Geography of Latin America

CHAPTER 8
History and Cultures of Latin America

CHAPTER 9
Latin America Today

Rio de Janeiro, Brazil ▶

NATIONAL GEOGRAPHIC

NGS ONLINE For more information about the region, see www.nationalgeographic.com/education.

Regional Atlas

Latin America

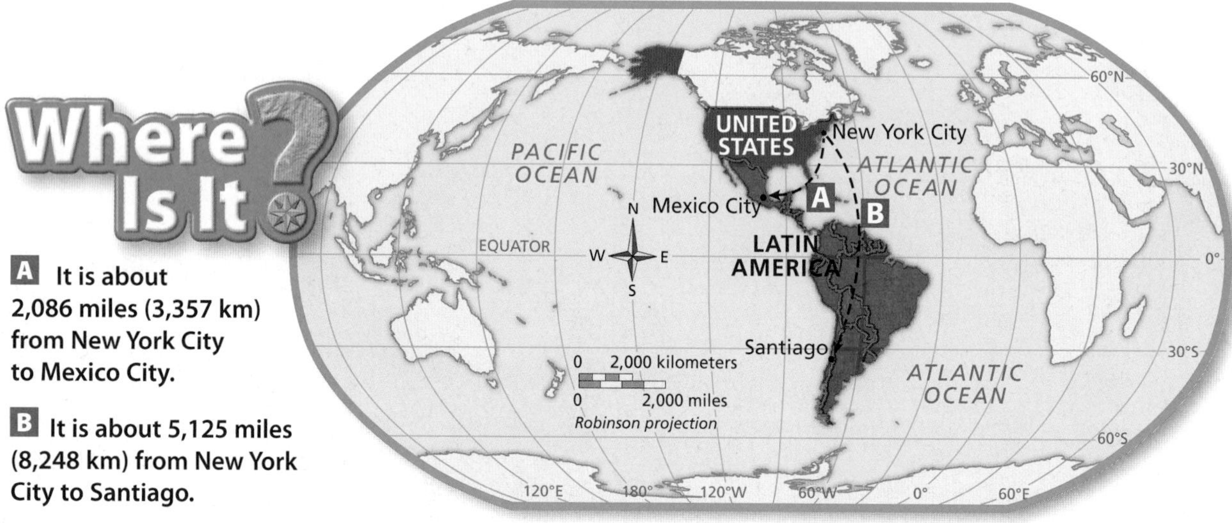

Where Is It?

A It is about 2,086 miles (3,357 km) from New York City to Mexico City.

B It is about 5,125 miles (8,248 km) from New York City to Santiago.

The region of Latin America is about two-and-a-half times the size of the continental United States. Its land area is about 7.9 million square miles (20.5 million sq. km). Latin America has about 558 million people, almost twice as many as the United States.

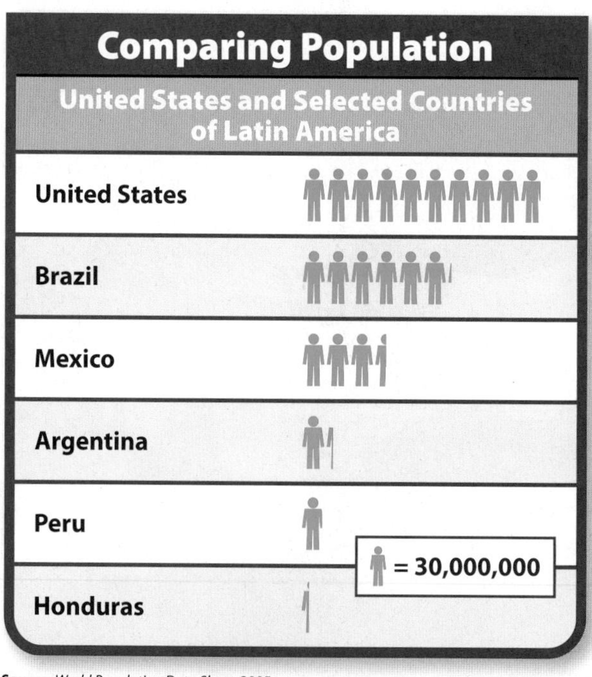

Comparing Population

United States and Selected Countries of Latin America

United States	10 figures
Brazil	6 figures (approx.)
Mexico	3½ figures (approx.)
Argentina	1½ figures (approx.)
Peru	1 figure
Honduras	less than ½ figure

= 30,000,000

Source: *World Population Data Sheet*, 2005.

GEO Fast Facts

Largest Lake

Lake Maracaibo (Venezuela) 5,217 sq. mi. (13,512 sq. km)

Highest Waterfall

Angel Falls (Venezuela) 3,212 ft. (979 m) high

Longest River

Amazon River (Peru, Brazil) 4,000 mi. (6,436 km) long

Highest Point

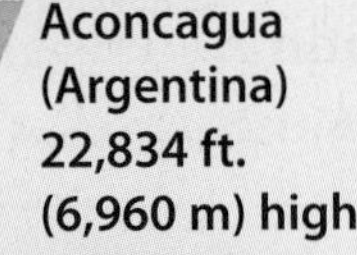

Aconcagua (Argentina) 22,834 ft. (6,960 m) high

UNIT 3

Regional Atlas

Latin America
PHYSICAL

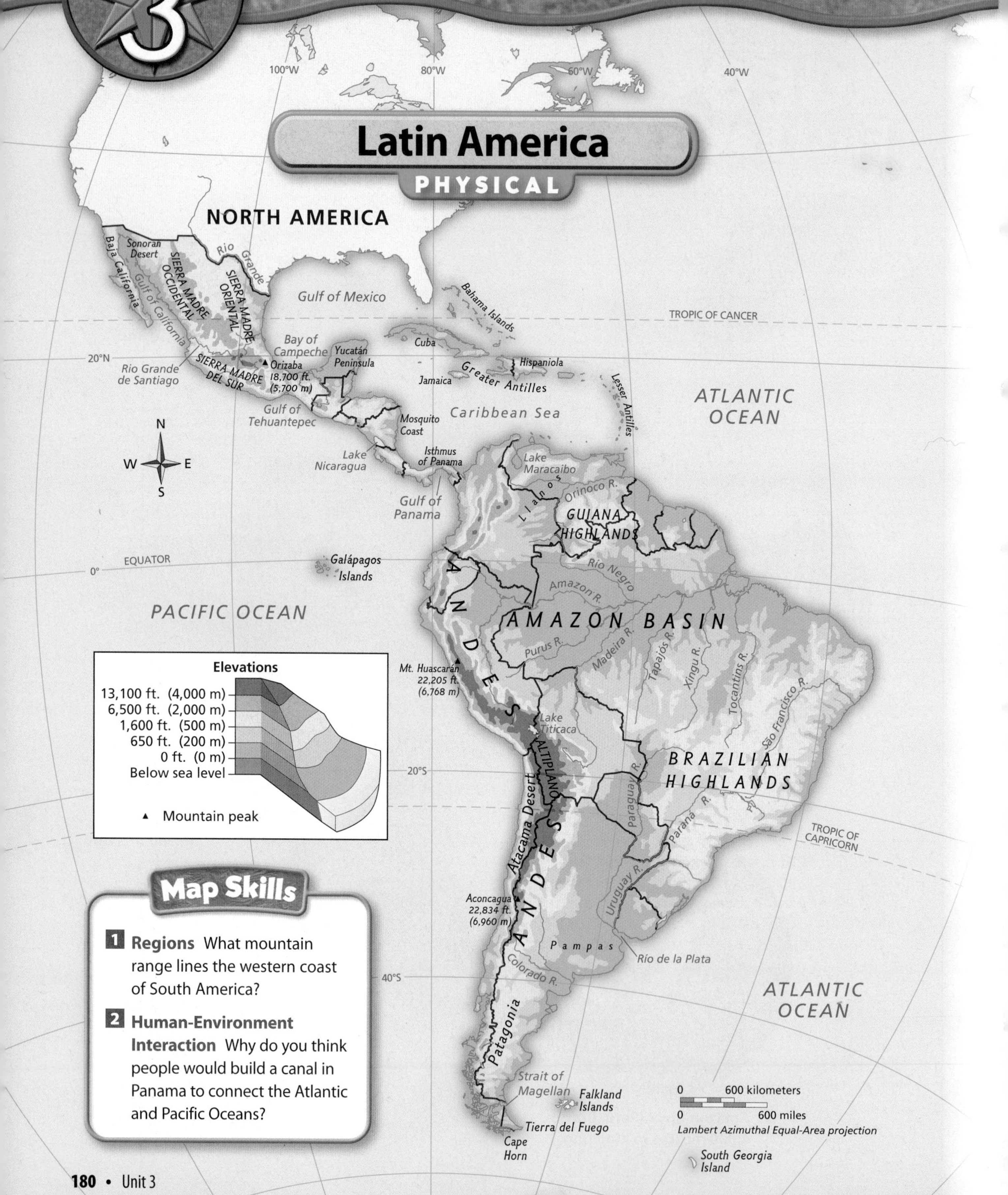

NORTH AMERICA
Sonoran Desert
Baja California
Gulf of California
SIERRA MADRE OCCIDENTAL
SIERRA MADRE ORIENTAL
Rio Grande
Gulf of Mexico
Bahama Islands
TROPIC OF CANCER
Bay of Campeche
Yucatán Peninsula
Cuba
Hispaniola
Jamaica
Greater Antilles
Lesser Antilles
Rio Grande de Santiago
SIERRA MADRE DEL SUR
Orizaba 18,700 ft. (5,700 m)
Gulf of Tehuantepec
Mosquito Coast
Caribbean Sea
ATLANTIC OCEAN
Lake Nicaragua
Isthmus of Panama
Gulf of Panama
Lake Maracaibo
Llanos
Orinoco R.
GUIANA HIGHLANDS
Galápagos Islands
EQUATOR
Rio Negro
Amazon R.
PACIFIC OCEAN
AMAZON BASIN
Purus R.
Madeira R.
Tapajós R.
Xingu R.
Tocantins R.
São Francisco R.
ANDES
Mt. Huascarán 22,205 ft. (6,768 m)
Lake Titicaca
ALTIPLANO
BRAZILIAN HIGHLANDS
Atacama Desert
Paraguay R.
Paraná R.
TROPIC OF CAPRICORN
Uruguay R.
Aconcagua 22,834 ft. (6,960 m)
Pampas
Río de la Plata
Colorado R.
ATLANTIC OCEAN
Patagonia
Strait of Magellan
Falkland Islands
Tierra del Fuego
Cape Horn
South Georgia Island
100°W
80°W
60°W
40°W
20°N
0°
20°S
40°S
N
S
E
W

Elevations

13,100 ft. (4,000 m)
6,500 ft. (2,000 m)
1,600 ft. (500 m)
650 ft. (200 m)
0 ft. (0 m)
Below sea level

▲ Mountain peak

0 600 kilometers
0 600 miles
Lambert Azimuthal Equal-Area projection

Map Skills

1. **Regions** What mountain range lines the western coast of South America?
2. **Human-Environment Interaction** Why do you think people would build a canal in Panama to connect the Atlantic and Pacific Oceans?

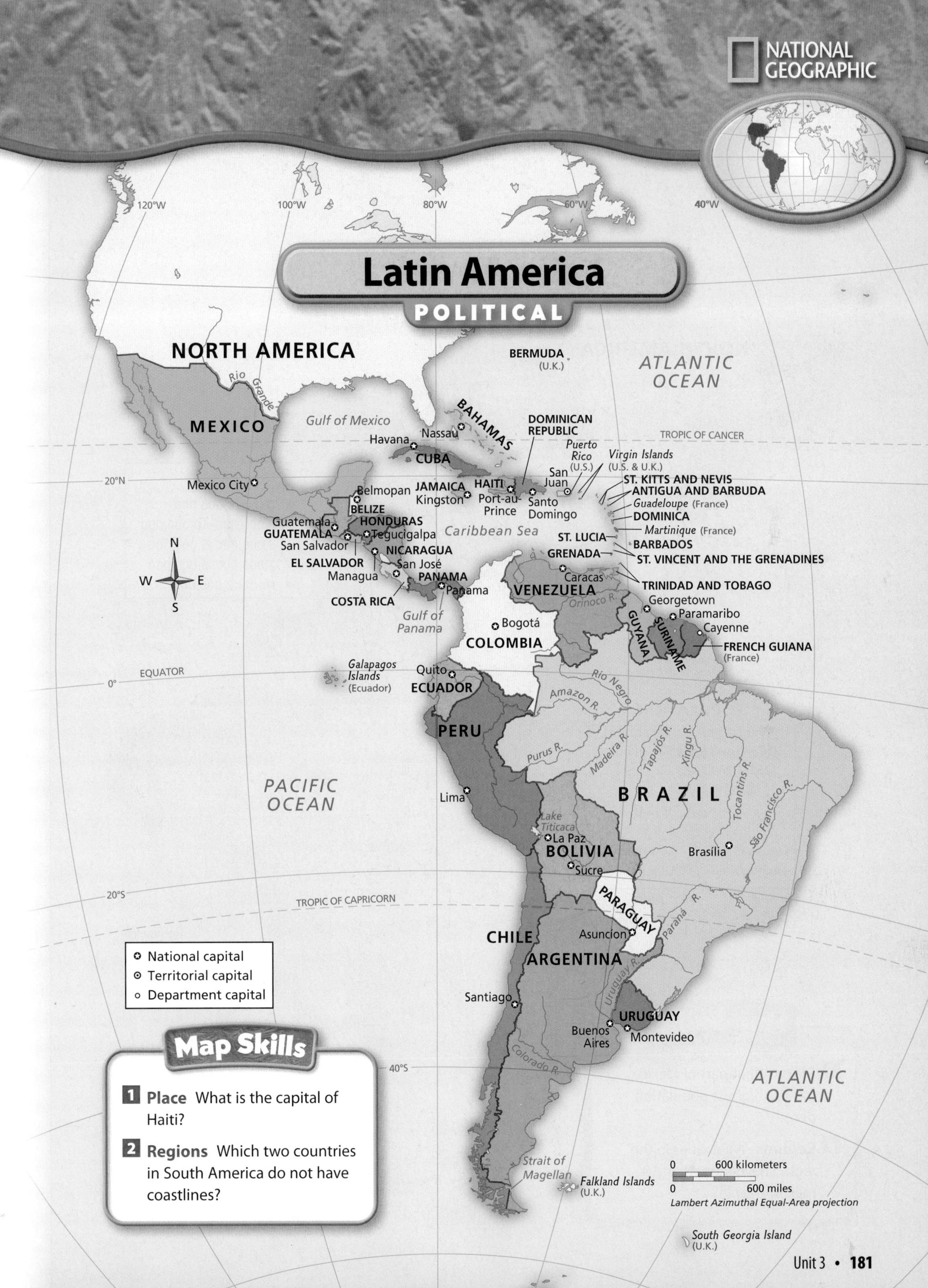

Map Skills

1. **Place** What is the capital of Haiti?
2. **Regions** Which two countries in South America do not have coastlines?

UNIT 3

Regional Atlas

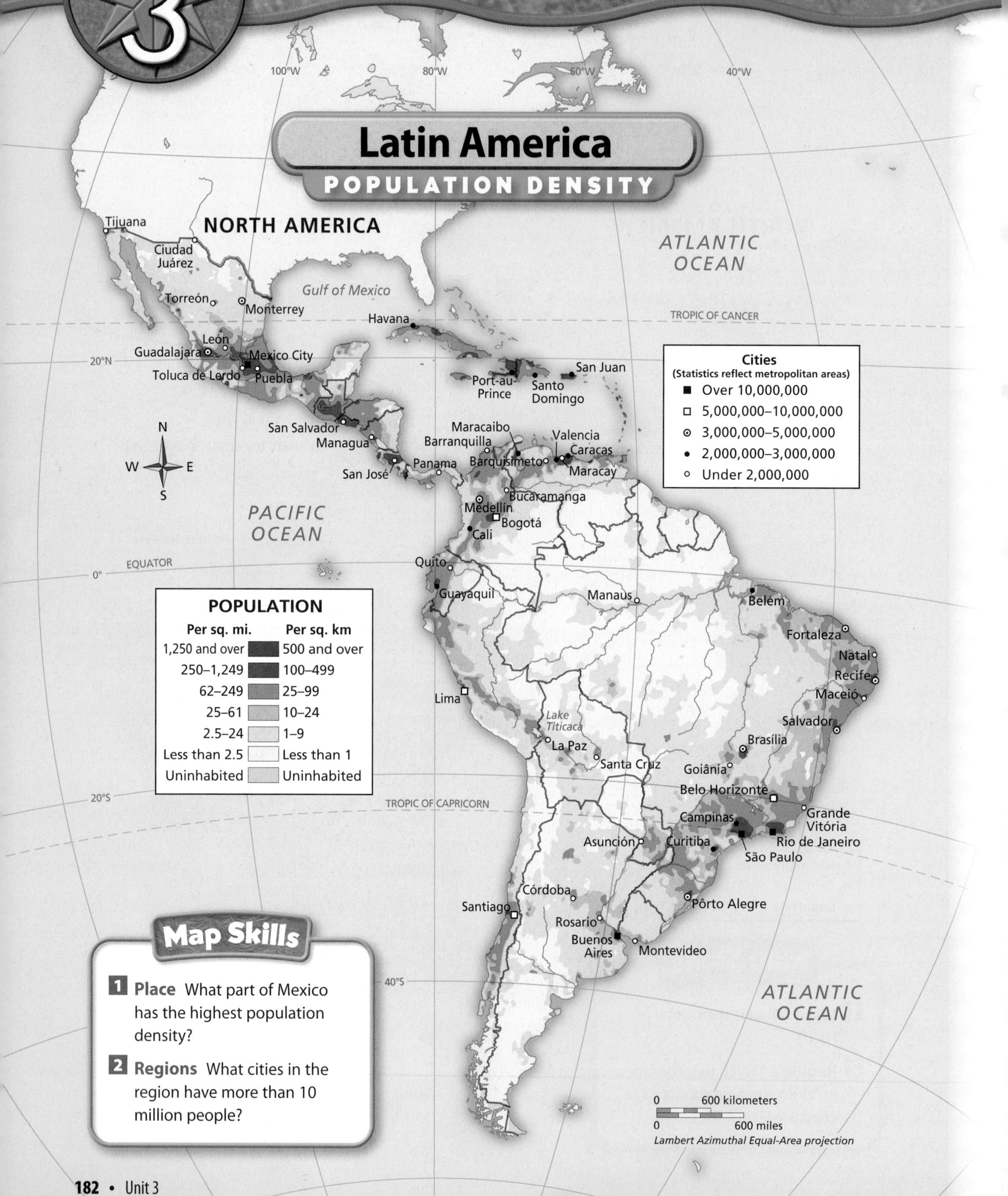

Latin America

POPULATION DENSITY

NORTH AMERICA

ATLANTIC OCEAN

PACIFIC OCEAN

Gulf of Mexico

100°W 80°W 60°W 40°W

20°N 0° 20°S 40°S

TROPIC OF CANCER

EQUATOR

TROPIC OF CAPRICORN

Tijuana, Ciudad Juárez, Torreón, Monterrey, León, Guadalajara, Mexico City, Toluca de Lerdo, Puebla, Havana, Port-au-Prince, Santo Domingo, San Juan, San Salvador, Managua, San José, Panama, Maracaibo, Barranquilla, Barquisimeto, Valencia, Caracas, Maracay, Bucaramanga, Medellín, Bogotá, Cali, Quito, Guayaquil, Manaus, Belém, Fortaleza, Natal, Recife, Maceió, Salvador, Lima, Lake Titicaca, La Paz, Santa Cruz, Brasília, Goiânia, Belo Horizonte, Campinas, Grande Vitória, Rio de Janeiro, São Paulo, Curitiba, Asunción, Córdoba, Pôrto Alegre, Santiago, Rosario, Buenos Aires, Montevideo

N S E W

Cities (Statistics reflect metropolitan areas)
■ Over 10,000,000
□ 5,000,000–10,000,000
⊙ 3,000,000–5,000,000
• 2,000,000–3,000,000
○ Under 2,000,000

POPULATION

Per sq. mi.	Per sq. km
1,250 and over	500 and over
250–1,249	100–499
62–249	25–99
25–61	10–24
2.5–24	1–9
Less than 2.5	Less than 1
Uninhabited	Uninhabited

0 600 kilometers

0 600 miles

Lambert Azimuthal Equal-Area projection

Map Skills

1. **Place** What part of Mexico has the highest population density?
2. **Regions** What cities in the region have more than 10 million people?

Latin America
ECONOMIC RESOURCES

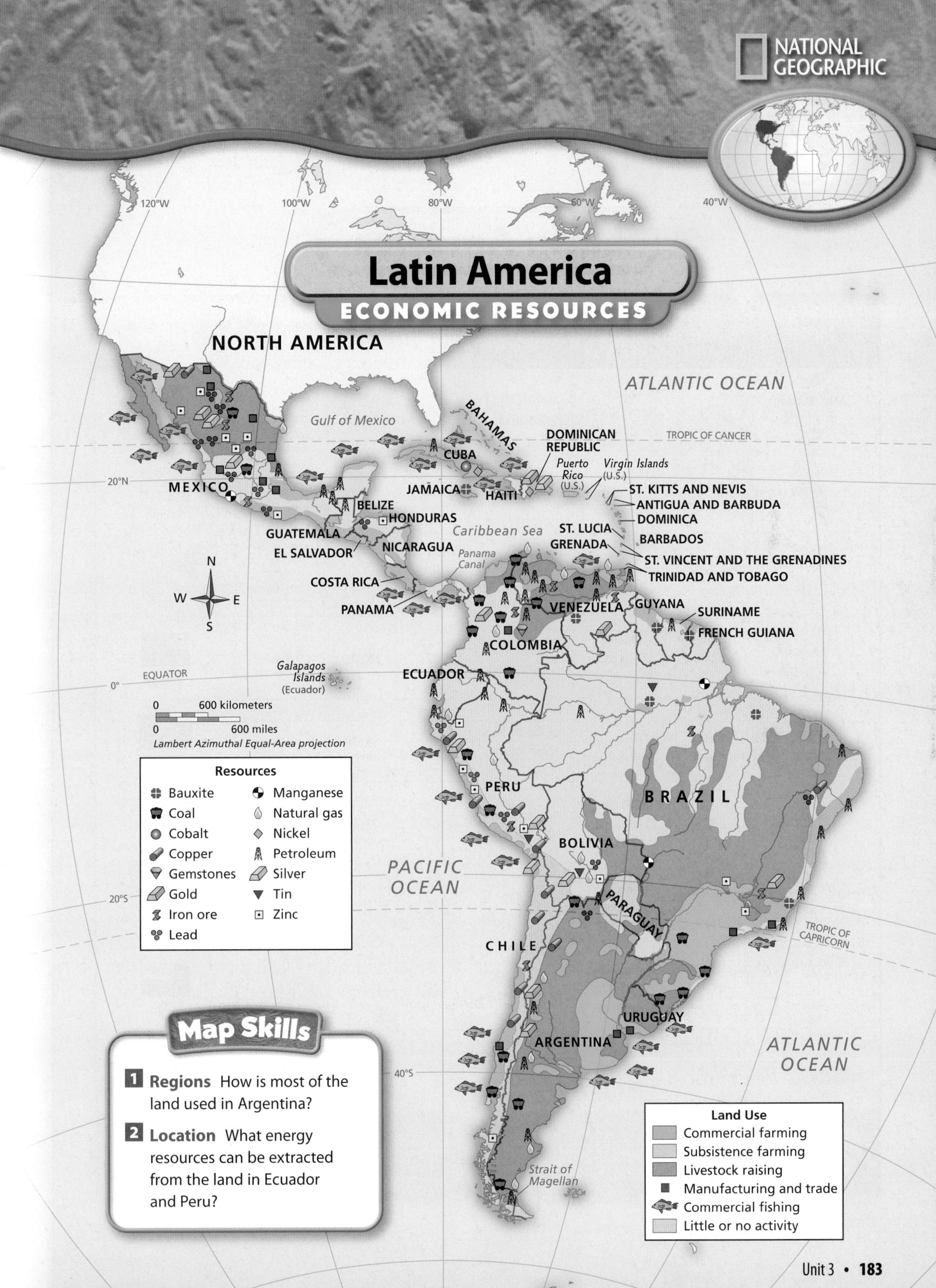

Map Skills

1. **Regions** How is most of the land used in Argentina?
2. **Location** What energy resources can be extracted from the land in Ecuador and Peru?

Regional Atlas

Latin America

Country and Capital	Literacy Rate	Population and Density	Land Area	Life Expectancy (Years)	GDP* Per Capita (U.S. dollars)	Television Sets (per 1,000 people)	Flag and Language
ANTIGUA AND BARBUDA St. John's	89%	100,000 588 per sq. mi. 227 per sq. km	170 sq. mi. 440 sq. km	71	$11,000	493	English
ARGENTINA Buenos Aires	97.1%	38,600,000 36 per sq. mi. 14 per sq. km	1,073,514 sq. mi. 2,780,388 sq. km	74	$12,400	293	Spanish
BAHAMAS Nassau	95.6%	300,000 60 per sq. mi. 22 per sq. km	5,359 sq. mi. 13,880 sq. km	70	$17,700	243	English
BARBADOS Bridgetown	97.4%	300,000 1,807 per sq. mi. 698 per sq. km	166 sq. mi. 430 sq. km	72	$16,400	290	English
BELIZE Belmopan	94.1%	300,000 34 per sq. mi. 13 per sq. km	8,865 sq. mi. 22,960 sq. km	70	$6,500	183	English
BOLIVIA La Paz Sucre	87.2%	8,900,000 21 per sq. mi. 8 per sq. km	424,162 sq. mi. 1,098,574 sq. km	64	$2,600	118	Spanish, Quechua, Aymara
BRAZIL Brasília	86.4%	184,200,000 56 per sq. mi. 22 per sq. km	3,300,154 sq. mi. 8,547,359 sq. km	71	$8,100	333	Portuguese
CHILE Santiago	96.2%	16,100,000 55 per sq. mi. 21 per sq. km	292,135 sq. mi. 756,626 sq. km	76	$10,700	240	Spanish
COLOMBIA Bogotá	92.5%	46,000,000 105 per sq. mi. 40 per sq. km	439,734 sq. mi. 1,138,906 sq. km	72	$6,600	279	Spanish
UNITED STATES Washington, D.C.	97%	296,500,000 80 per sq. mi. 31 per sq. km	3,717,796 sq. mi. 9,629,047 sq. km	78	$40,100	844	English

*Gross Domestic Product

Countries and flags not drawn to scale

Latin America

Country and Capital	Literacy Rate	Population and Density	Land Area	Life Expectancy (Years)	GDP* Per Capita (U.S. dollars)	Television Sets (per 1,000 people)	Flag and Language
COSTA RICA, San José	96%	4,300,000 218 per sq. mi. 84 per sq. km	19,730 sq. mi. 51,100 sq. km	79	$9,600	229	Spanish
CUBA, Havana	97%	11,300,000 264 per sq. mi. 102 per sq. km	42,803 sq. mi. 110,859 sq. km	77	$3,000	248	Spanish
DOMINICA, Roseau	94%	100,000 345 per sq. mi. 133 per sq. km	290 sq. mi. 751 sq. km	74	$5,500	232	English
DOMINICAN REPUBLIC, Santo Domingo	84%	8,900,000 471 per sq. mi. 168 per sq. km	18,815 sq. mi. 48,731 sq. km	68	$6,300	96	Spanish
ECUADOR, Quito	92.5%	13,000,000 119 per sq. mi. 46 per sq. km	109,483 sq. mi. 283,560 sq. km	74	$3,700	213	Spanish
EL SALVADOR, San Salvador	80.2%	6,900,000 849 per sq. mi. 328 per sq. km	8,124 sq. mi. 21,041 sq. km	70	$4,900	191	Spanish
FRENCH GUIANA, Cayenne	83%	200,000 6 per sq. mi. 2 per sq. km	34,749 sq. mi. 89,999 sq. km	75	$8,300	information not available	French
GRENADA, St. George's	98%	100,000 769 per sq. mi. 295 per sq. km	131 sq. mi. 339 sq. km	71	$5,000	376	English
GUATEMALA, Guatemala	70.6%	12,700,000 302 per sq. mi. 117 per sq. km	42,042 sq. mi. 108,888 sq. km	66	$4,200	61	Spanish
UNITED STATES, Washington, D.C.	97%	296,500,000 80 per sq. mi. 31 per sq. km	3,717,796 sq. mi. 9,629,047 sq. km	78	$40,100	844	English

Sources: *CIA World Factbook*, 2005; Population Reference Bureau, *World Population Data Sheet*, 2005.

For more country facts, go to the **Nations of the World Databank** at glencoe.com.

Regional Atlas

Latin America

Country and Capital	Literacy Rate	Population and Density	Land Area	Life Expectancy (Years)	GDP* Per Capita (U.S. dollars)	Television Sets (per 1,000 people)	Flag and Language
GUYANA, Georgetown	98.8%	800,000 10 per sq. mi. 4 per sq. km	83,000 sq. mi. 214,969 sq. km	63	$3,800	70	English
HAITI, Port-au-Prince	52.9%	8,300,000 775 per sq. mi. 299 per sq. km	10,714 sq. mi. 27,749 sq. km	52	$1,500	5	French, Creole
HONDURAS, Tegucigalpa	76.2%	7,200,000 166 per sq. mi. 64 per sq. km	43,278 sq. mi. 112,090 sq. km	71	$2,800	95	Spanish
JAMAICA, Kingston	87.9%	2,700,000 636 per sq. mi. 246 per sq. km	4,243 sq. mi. 10,989 sq. km	73	$4,100	191	English
MEXICO, Mexico City	92.2%	107,000,000 142 per sq. mi. 55 per sq. km	756,082 sq. mi. 1,958,243 sq. km	75	$9,600	272	Spanish
NICARAGUA, Managua	67.5%	5,800,000 116 per sq. mi. 45 per sq. km	50,193 sq. mi. 129,999 sq. km	69	$2,300	69	Spanish
PANAMA, Panama	92.6%	3,200,000 110 per sq. mi. 42 per sq. km	29,158 sq. mi. 75,519 sq. km	75	$6,900	192	Spanish
PARAGUAY, Asunción	94%	6,200,000 39 per sq. mi. 15 per sq. km	157,046 sq. mi. 406,747 sq. km	71	$4,800	205	Spanish, Guarani
PERU, Lima	87.7%	27,900,000 56 per sq. mi. 22 per sq. km	496,224 sq. mi. 1,285,214 sq. km	70	$5,600	147	Spanish, Quechua
UNITED STATES, Washington, D.C.	97%	296,500,000 80 per sq. mi. 31 per sq. km	3,717,796 sq. mi. 9,629,047 sq. km	78	$40,100	844	English

*Gross Domestic Product

Countries and flags not drawn to scale

Latin America

Country and Capital	Literacy Rate	Population and Density	Land Area	Life Expectancy (Years)	GDP* Per Capita (U.S. dollars)	Television Sets (per 1,000 people)	Flag and Language
PUERTO RICO, San Juan	94.1%	3,900,000 1,128 per sq. mi. 436 per sq. km	3,456 sq. mi. 8,951 sq. km	77	$17,700	information not available	Spanish
ST. KITTS-NEVIS, Basseterre	97%	50,000 360 per sq. mi. 139 per sq. km	139 sq. mi. 360 sq. km	70	$8,800	256	English
ST. LUCIA, Castries	67%	200,000 837 per sq. mi. 323 per sq. km	239 sq. mi. 619 sq. km	74	$5,400	368	English
ST. VINCENT AND THE GRENADINES, Kingstown	96%	100,000 737 per sq. mi. 256 per sq. km	151 sq. mi. 391 sq. km	72	$2,900	230	English
SURINAME, Paramaribo	93%	438,000 7 per sq. mi. 3 per sq. km	62,344 sq. mi. 161,470 sq. km	69	$4,300	241	Dutch
TRINIDAD AND TOBAGO, Port-of-Spain	98.6%	1,300,000 656 per sq. mi. 253 per sq. km	1,981 sq. mi. 5,131 sq. km	71	$10,500	337	English
URUGUAY, Montevideo	98%	3,400,000 50 per sq. mi. 19 per sq. km	68,498 sq. mi. 177,409 sq. km	75	$14,500	531	Spanish
VENEZUELA, Caracas	93.4%	26,700,000 76 per sq. mi. 29 per sq. km	352,143 sq. mi. 912,046 sq. km	73	$5,800	185	Spanish
VIRGIN ISLANDS (U.S.), Charlotte Amalie	information not available	108,708 799 per sq. mi. 309 per sq. km	136 sq. mi. 352 sq. km	79	$17,200	information not available	English
UNITED STATES, Washington, D.C.	97%	296,500,000 80 per sq. mi. 31 per sq. km	3,717,796 sq. mi. 9,629,047 sq. km	78	$40,100	844	English

Sources: *CIA World Factbook*, 2005; Population Reference Bureau, *World Population Data Sheet*, 2005.

For more country facts, go to the **Nations of the World Databank** at glencoe.com.

Reading Social Studies

Summarizing Information

Learn It!

Summarizing helps you focus on main ideas. By restating the important facts in a short summary, you can reduce the amount of information to remember. A summary can be a short paragraph that includes the main ideas.

Use these steps to help you summarize:

- Be brief—do not include many supporting details.
- Restate the text in a way that makes sense to you.

Read the text below. Then review the graphic organizer to see how you could summarize the information.

> A people called the Maya lived in Mexico's Yucatán Peninsula and surrounding areas between A.D. 300 and A.D. 900. The Maya built huge stone temples in the shape of pyramids with steps. They were skilled at astronomy and used their knowledge of the stars, moon, and planets to develop a calendar. They also had a number system based on 20. Using hieroglyphics, which is a form of writing that uses signs and symbols, the Maya recorded the history of their kings.
>
> —*from page 209*

Fact: The Maya used their knowledge of the stars, moon, and planets to develop a calendar.

Fact: The Maya had a number system based on 20.

Fact: The Maya recorded the history of their kings using hieroglyphics.

Summary:
The Maya developed a calendar, created a number system, and recorded their history using hieroglyphics.

Reading Tip

As you read and summarize in your own words, try not to change the author's original meanings or ideas.

Read to Write Activity

After you read each chapter in this unit, summarize its information. Then, use the chapter's Visual Summary to check to see if you identified the main ideas.

2 Practice It!

Read the following paragraph from this unit.

- Draw an organizer like the one shown below.
- Write the main facts from the paragraph in the top boxes.
- Write a summary of the paragraph in the bottom box.

> Family life is important in Latin America. Often several generations live together, and adults are expected to care for their aged parents. Adult brothers and sisters often live near each other, and their children—who are cousins—can form close relationships. Traditionally, the father is the family leader and the chief decision maker. In some parts of the Caribbean, however, the mother is the leader of the family.
>
> —*from page 223*

Latin American teenagers

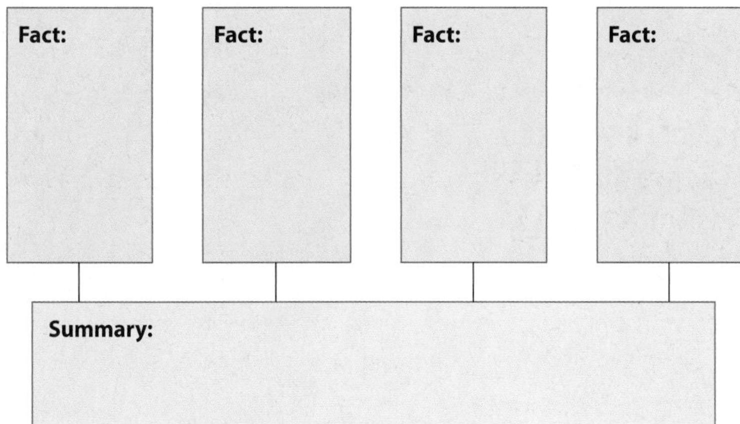

3 Apply It!

With a partner, choose a section to summarize. After each of you summarizes the section on your own, exchange your papers and check to see if the summaries are complete. Note whether any important ideas are missing. Return your summaries to each other and make the changes. Use your summaries to help you study for assessment.

CHAPTER 7

Physical Geography of Latin America

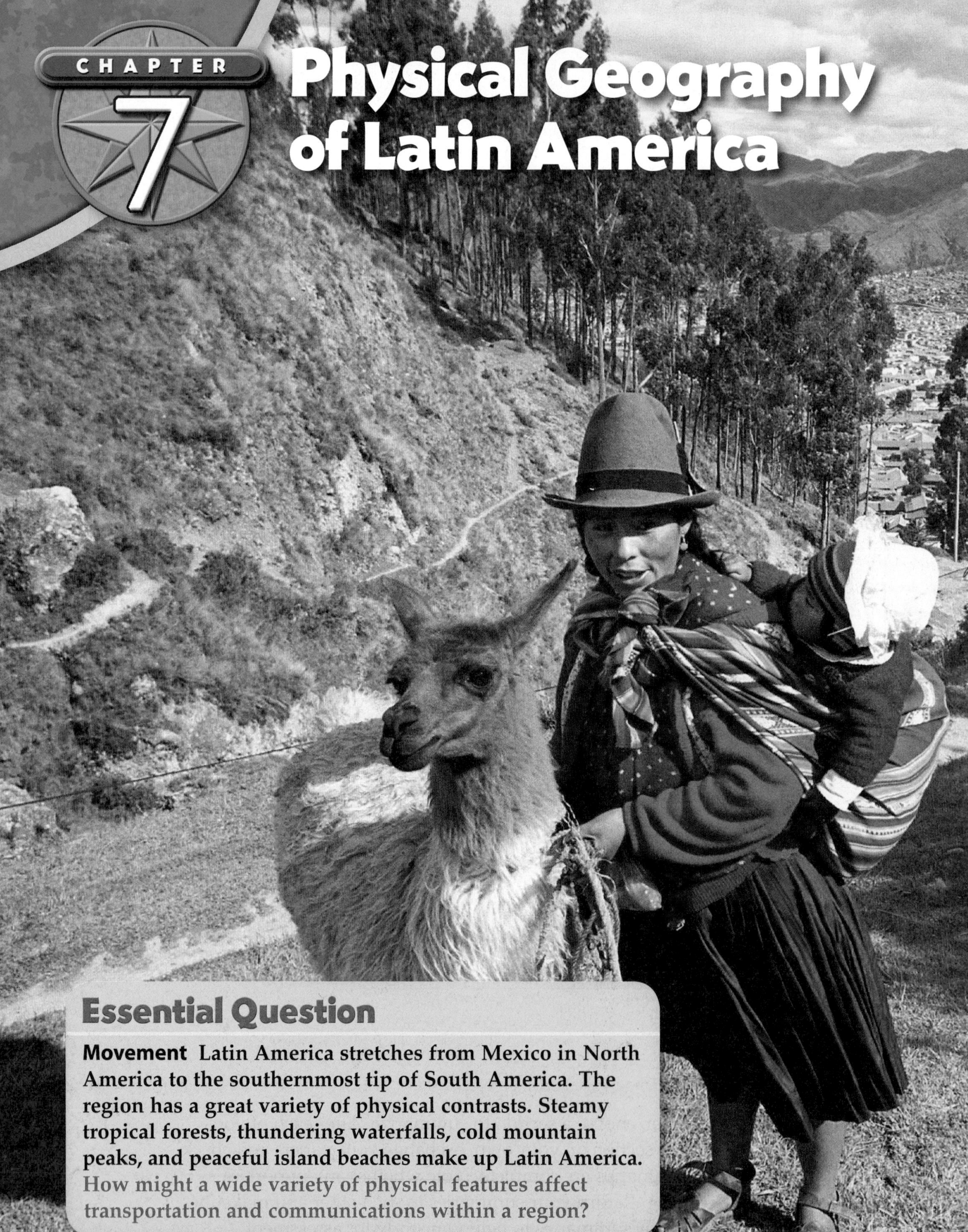

Essential Question

Movement Latin America stretches from Mexico in North America to the southernmost tip of South America. The region has a great variety of physical contrasts. Steamy tropical forests, thundering waterfalls, cold mountain peaks, and peaceful island beaches make up Latin America. How might a wide variety of physical features affect transportation and communications within a region?

Section 1: Physical Features

BIG IDEA Geographic factors influence where people settle. In Latin America, vast river systems provide transportation and support fishing. The region's rugged mountains and thick forests, however, have been obstacles to transportation and trade.

Section 2: Climate Regions

BIG IDEA The physical environment affects how people live. Latin America's vast expanse of rain forest is the largest in the world and contains valuable resources. In mountainous areas, climate and vegetation are affected more by altitude than by latitude.

FOLDABLES™ Study Organizer

Summarizing Information Make this Foldable to help you summarize information about the landforms, waterways, resources, and climates of Latin America.

Step 1 Fold the top of an 11 x 17 sheet of paper down about 2 inches.

Step 2 Then fold the paper to create 3 equal columns.

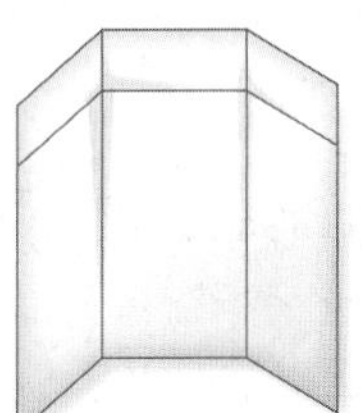

Step 3 Label each column of your Foldable as shown.

Landforms and Waterways	Resources	Climates

Reading and Writing Use the notes from your Foldable to write a travel pamphlet highlighting one of the subregions of Latin America. In your pamphlet, explain why the landforms, waterways, and climates of your chosen area are attractive to tourists.

◀ Quechuan Indian woman outside of Cuzco, Peru

Visit glencoe.com and enter ***QuickPass***™ code EOW3109c7 for Chapter 7 resources.

Guide to Reading

BIG Idea

Geographic factors influence where people settle.

Content Vocabulary

- subregion *(p. 193)*
- isthmus *(p. 193)*
- archipelago *(p. 193)*
- escarpment *(p. 194)*
- Llanos *(p. 194)*
- Pampas *(p. 194)*
- tributary *(p. 194)*
- estuary *(p. 194)*
- gasohol *(p. 195)*

Academic Vocabulary

- transport *(p. 193)*
- reside *(p. 194)*

Reading Strategy

Identifying Central Issues Use a diagram like the one below to identify and briefly describe six key landforms in the region.

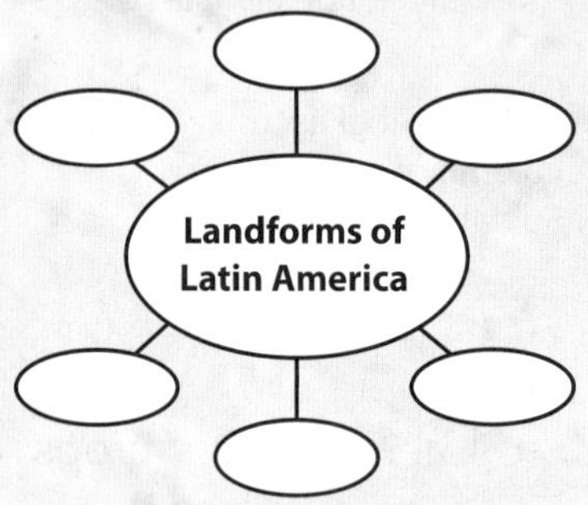

Physical Features

 Section Audio  Spotlight Video

Picture This How do you farm when there is no flat land? The Inca, an advanced civilization that existed hundreds of years ago in Peru, used a method called terracing. They carved layered fields, like wide steps, into the mountainsides. Today, descendants of the Inca still use this method to raise crops at high altitudes. To learn how the physical landscape has affected other human activities, read Section 1.

▼ A terraced hillside near Pisac, Peru

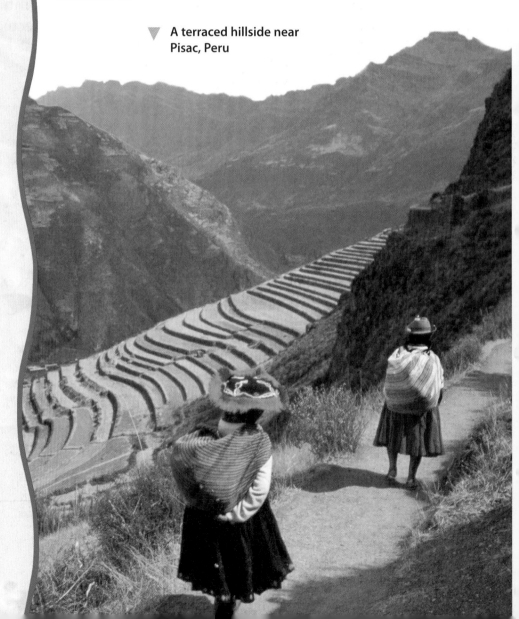

Landforms

Main Idea **Mountains are prominent features in many parts of Latin America.**

Geography and You If you traveled across your state, what geographic features would you see? Read on to learn about the landforms that Latin Americans would encounter if they crossed their region.

Geographers divide the region of Latin America into three **subregions,** or smaller areas. These subregions are Middle America, the Caribbean, and South America.

Middle America

Middle America is made up of Mexico and Central America. Central America is an **isthmus** (IHS·muhs), or a narrow piece of land that links two larger areas of land—North America and South America. Middle America lies where four tectonic plates meet. As a result, it has active volcanoes and frequent earthquakes. Deposits of ash and lava make the soil fertile.

Mexico has mountain ranges along its eastern and western coasts with a high plateau between. Farther south, mountains rise like a backbone through Central America. Lowlands along the coasts are often narrow. Thick forests, rugged mountains, and coastal marshes make it difficult to **transport** goods in Central America.

The Caribbean

The islands of the Caribbean Sea, also known as the West Indies, can be divided into three groups: the Greater Antilles, the Lesser Antilles, and the Bahamas. The Greater Antilles include the largest islands—Cuba, Hispaniola, Puerto Rico, and Jamaica. The Lesser Antilles is an **archipelago** (AHR·kuh·PEH·luh·GOH), or group of islands. It curves from the Virgin Islands to Trinidad. The third group is the Bahamas, another archipelago.

NATIONAL GEOGRAPHIC

Andean Village

In the Andes, most people live in valleys and work fields that have been cut into the hillsides.
Regions **Besides the Andes, what is South America's other major landform?**

Except for the largest islands, most Caribbean islands are small. Cuba alone has about half of the Caribbean's land area. Some islands are very low-lying. Others, formed by volcanoes, have rugged mountains. Some of the volcanoes are still active and can cause great damage. Farmers use the fertile volcanic soil here to grow crops such as sugarcane and tobacco.

South America

The Andes mountain ranges and the vast Amazon Basin are South America's major landforms. The Andes, the world's longest mountain system, are a cordillera (KAWR·duhl·YEHR·uh). They stretch along the Pacific coast of South America for about 5,500 miles (8,851 km). The high Andes ranges have many peaks that soar over 20,000 feet (6,096 m). Between the mountain chains lie plateaus and valleys.

That is where most people **reside,** or live, and the land can be farmed.

East of the Andes is the huge Amazon Basin. This low area contains the Amazon River and covers 2.7 million square miles (7.0 million sq. km). Highlands to the north and south border the basin. The Brazilian Highlands are so vast that they cross several climate zones. They end in an **escarpment,** a series of steep cliffs that drop down to the Atlantic coastal plain.

Other lowland plains are found north and south of the Amazon Basin. Tropical grasslands known as the **Llanos** (LAH·nohs) stretch through eastern Colombia and Venezuela. Another well-known plain, the **Pampas,** covers much of Argentina and Uruguay. Like North America's Great Plains, the Pampas is used for cattle herding and grain farming.

Reading Check **Identifying** What areas make up Middle America?

Herding cattle is a major economic activity on the Pampas of Argentina and Uruguay. ***Place*** **What are the Llanos?**

Waterways

Main Idea **Latin America's waterways provide important transportation routes.**

Geography and You What major rivers flow through your part of the country? How are they important to your area? Read on to find out about the amazing Amazon, one of the world's longest rivers.

Latin America has many natural rivers and lakes, most of which are in South America. The people of the region have used these waterways for transportation and water resources for ages.

Rivers

Latin America's longest river is the Amazon, which starts in the Andes and flows east about 4,000 miles (6,437 km) to the Atlantic Ocean. Heavy rains and many tributaries feed the Amazon. A **tributary** is a small river that flows into a larger river. Some ships can follow the Amazon as far west as Peru, more than 2,500 miles (4,023 km) inland. People also rely on the river for its fish.

Three rivers—the Paraná (PAH·rah·NAH), Paraguay (PAH·rah·GWY), and Uruguay (OO·roo·GWY)—form Latin America's second-largest river system. Together, they drain the rainy eastern half of South America. After winding through inland areas, the three rivers flow into a broad estuary. An **estuary** is an area where river currents and ocean tides meet. This estuary, the Río de la Plata, or "River of Silver," meets the Atlantic Ocean.

Student Web Activity Visit glencoe.com and complete the Chapter 7 Web Activity about the Amazon River.

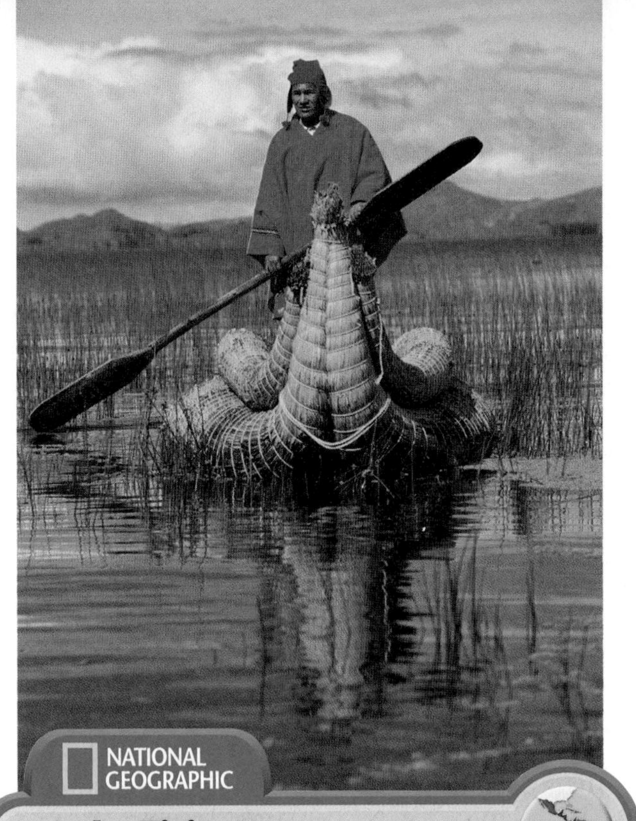

NATIONAL GEOGRAPHIC

Lake Titicaca

For hundreds of years, the native peoples around Lake Titicaca have traveled its waters using boats made from reeds. ***Place*** **What is unique about Lake Titicaca?**

The Orinoco is another important river. It flows through Venezuela to the Caribbean Sea. This river carries fertile soil into the Llanos region.

Other Waterways

Latin America has few large lakes. Lake Maracaibo, in Venezuela, is South America's largest lake. It contains some of Venezuela's oil fields. Lake Titicaca lies between Bolivia and Peru. About 12,500 feet (3,810 m) above sea level, it is the world's highest lake that can be used by large ships. The Panama Canal, a human-made waterway, stretches across the narrow Isthmus of Panama. Ships use the canal to shorten travel time between the Atlantic and Pacific Oceans.

Reading Check **Identifying** Why is Lake Maracaibo important to Venezuela?

A Wealth of Natural Resources

Main Idea **Latin America has vast natural resources, but political and economic troubles have kept some countries from fully using them.**

Geography and You Do you use aluminum foil to wrap dinner leftovers? Bauxite, a mineral used to make aluminum, is an important resource on the Caribbean island of Jamaica. Read to find out about the kinds of mineral wealth in Latin America.

Latin America has many natural resources. These include minerals, forests, farmland, and water. Not all of Latin America's countries, however, share equally in this wealth. Remote locations, lack of money for development, and the wide gap between rich and poor have kept many of the region's natural resources from being fully developed.

Brazil's Abundant Resources

Brazil, the largest country in Latin America, possesses a great wealth of natural resources. More than 55 percent of Brazil is covered in forests, including a large area of tropical rain forests. The rain forests provide timber and a range of products such as rubber, palm oil, and Brazil nuts.

Brazil also has great mineral wealth. It has large amounts of bauxite, gold, and tin. Its deposits of iron ore and manganese help support one of the world's largest iron and steel industries. Brazil's oil and natural gas reserves, however, are limited. They provide only some of the energy this huge country needs. To reduce its dependence on oil imports, Brazil uses alcohol produced from sugarcane and gasoline to produce a fuel for cars called **gasohol.**

Energy Resources

In addition to Brazil, other countries of Latin America have energy resources. Venezuela has the region's largest oil and natural gas reserves. Mexico has large amounts of oil and natural gas along the coast of the Gulf of Mexico. Both Mexico and Venezuela use the supplies for their own energy needs as well as for exports.

Bolivia and Ecuador also have valuable oil and natural gas deposits. In Bolivia, however, foreign companies have attempted to control the country's energy resources. Bolivia's government has struggled to prevent this. As a result, production has slowed, and Bolivia has not been able to fully benefit from exports.

Other Resources

Other mineral resources found in Latin America include silver mined in Mexico and Peru. Venezuela has rich iron ore deposits and is a major exporter of the mineral. Colombian mines produce the world's finest emeralds. Chile is the world's largest exporter of copper.

By contrast, the Caribbean islands have relatively few mineral resources, with a few important exceptions. Jamaica has large deposits of bauxite, a mineral used to make aluminum. In addition, Cuba mines nickel, and the Dominican Republic mines gold and silver.

Certain Central American countries, such as Nicaragua and Guatemala, have rich gold deposits. However, political conflicts and transportation difficulties make mining these deposits difficult.

Reading Check **Analyzing** Why are some Latin American countries unable to make full use of their natural resources?

Section Review

Social Studies ONLINE
Study Central™ To review this section, go to glencoe.com.

Vocabulary

1. **Explain** the significance of:
   - **a.** subregion
   - **b.** isthmus
   - **c.** archipelago
   - **d.** escarpment
   - **e.** Llanos
   - **f.** Pampas
   - **g.** tributary
   - **h.** estuary
   - **i.** gasohol

Main Ideas

2. **Describing** Describe the various mountains found throughout Middle America, the Caribbean, and South America.
3. **Explaining** Use a chart like the one below to note the significance of the listed waterways.

Waterway	Significance
Amazon River	
Paraguay, Paraná, Uruguay system	
Orinoco River	

4. **Identifying** Which Latin American country has the greatest resources? What are they?

Critical Thinking

5. **BIG Idea** What effects can volcanoes have on the peoples and economies of a region?
6. **Challenge** Based on Latin America's natural resources and physical geography, do you think the region will become more important economically in the future? Explain your answer.

Writing About Geography

7. **Using Your FOLDABLES** Use your Foldable to write a paragraph giving examples of how physical geography has affected the lives of people in the region.

NATIONAL GEOGRAPHIC **Geography & History**

The Columbian Exchange

What do corn, beans, and potatoes have in common? All of these foods were first grown in the Americas.

Separate Worlds For thousands of years, people living in the Eastern Hemisphere had no contact with those living in the Western Hemisphere. This changed in 1492 when European explorer Christopher Columbus arrived in the Americas. Columbus's voyages began what became known as "the Columbian Exchange"—a transfer of people, animals, plants, and even diseases between the two hemispheres.

For Better and for Worse The Europeans brought many new things to the Americas. They brought horses, which helped the Native Americans with labor, hunting, and transportation. European farm animals, such as sheep, pigs, and cattle, created new sources of income for people in the Americas. Europeans also brought crops—oats, wheat, rye, and barley. The sugarcane brought by Europeans grew well on plantations in the tropical Americas.

▲ **Tomato sauce on Italian pasta**

Some parts of the Exchange were disastrous, however. Europeans brought diseases that killed millions of Native Americans. Plantation owners put enslaved Africans to work in their fields.

From the Americas, explorers returned home with a wide variety of plants. Spanish sailors carried potatoes to Europe. Nutritious and easy to grow, the potato became one of Europe's most important foods. Corn from the Americas fed European cattle and pigs. Peanuts, tomatoes, hot peppers, and chocolate changed the diets of people in Europe, Asia, and Africa.

▼ **Mexican Indian making chocolate**

Think About It

1. **Place** What foods were unknown in Europe before 1492?
2. **Human-Environment Interaction** Why were some foods adopted from the Americas so important in other parts of the world?

Guide to Reading

BIG Idea

The physical environment affects how people live.

Content Vocabulary

- Tropics *(p. 199)*
- rain forest *(p. 199)*
- canopy *(p. 200)*
- altitude *(p. 201)*

Academic Vocabulary

- facilitate *(p. 199)*
- considerable *(p. 200)*

Reading Strategy

Comparing and Contrasting Use a Venn diagram like the one below to compare and contrast the tropical rain forest and tropical savanna climate zones.

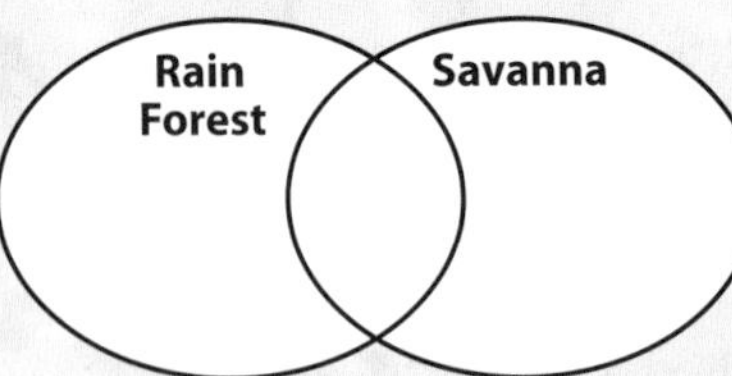

Climate Regions

Picture This These huge, 6-foot-wide water lilies are found deep in Brazilian rain forests near the mighty Amazon River. They are so strong that an average-sized adult could rest his or her full weight on them! The warm temperatures and heavy rains of the rain forest create an ideal growing environment for many exotic plants. To learn more about how climate affects the people, vegetation, and wildlife of Latin America, read Section 2.

▼ Rain forest water lilies

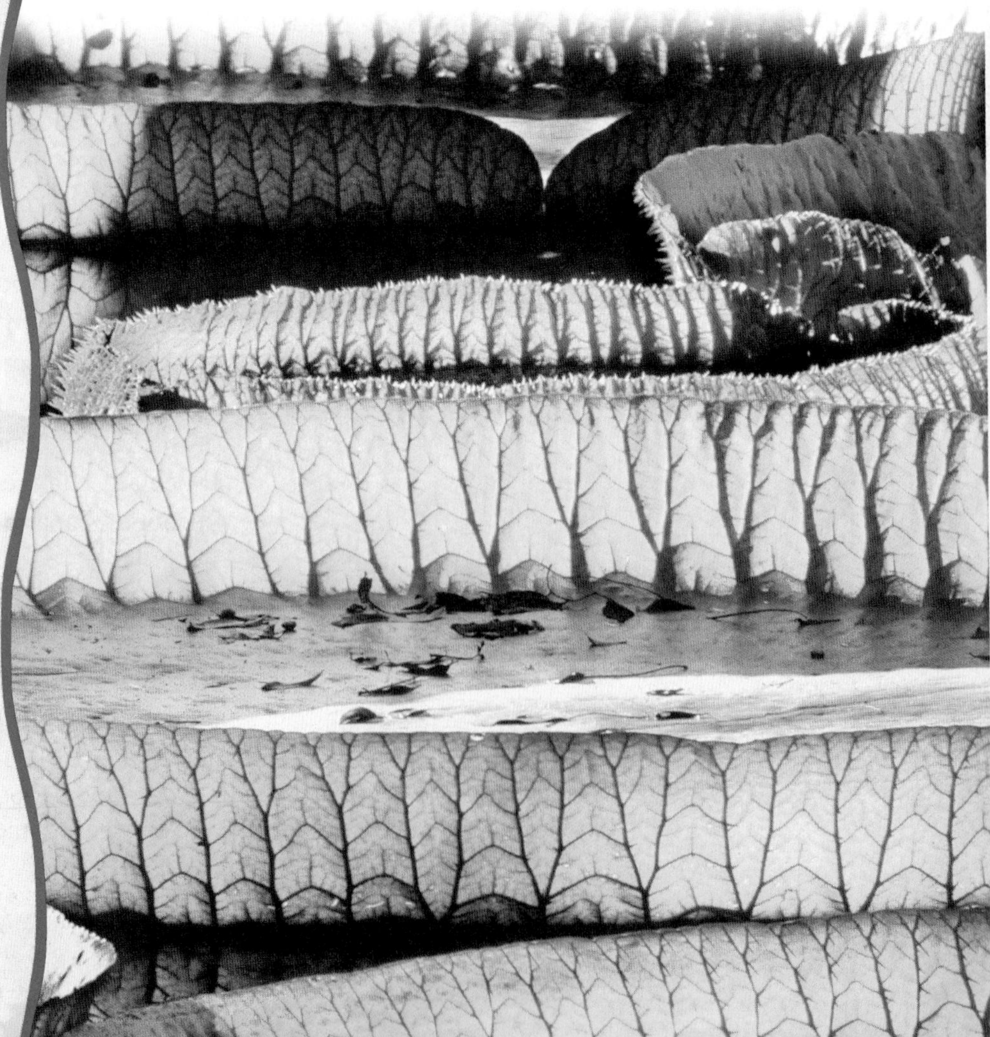

Hot to Mild Climates

Main Idea **Much of Latin America is located in the Tropics and has year-round high temperatures and heavy rainfall.**

Geography and You What might the view be like at the top of a rain forest tree 130 feet up? Read to find out why rain forests thrive in Latin America's tropical areas.

Most of Latin America lies within the **Tropics**—the area between the Tropic of Cancer and the Tropic of Capricorn. This area has generally warm temperatures because it receives the direct rays of the sun for much of the year. Yet even within the Tropics, mountain ranges and wind patterns create a variety of climates in the region. **Figure 1** shows Latin America's different climate zones.

Tropical Climates

A tropical wet climate is found in some Caribbean islands and much of Central America and South America. This climate is marked by year-round hot temperatures and heavy rainfall. Vast areas of rain forest cover much of this climate zone. A **rain forest** is a dense stand of trees and other plants that receive high amounts of precipitation. Warm temperatures and heavy rains **facilitate,** or make possible, the growth of rain forests.

South America's Amazon Basin is home to the world's largest rain forest.

NATIONAL GEOGRAPHIC **Maps In MOtion** See StudentWorks™ Plus or glencoe.com.

Figure 1 **Latin America: Climate Zones**

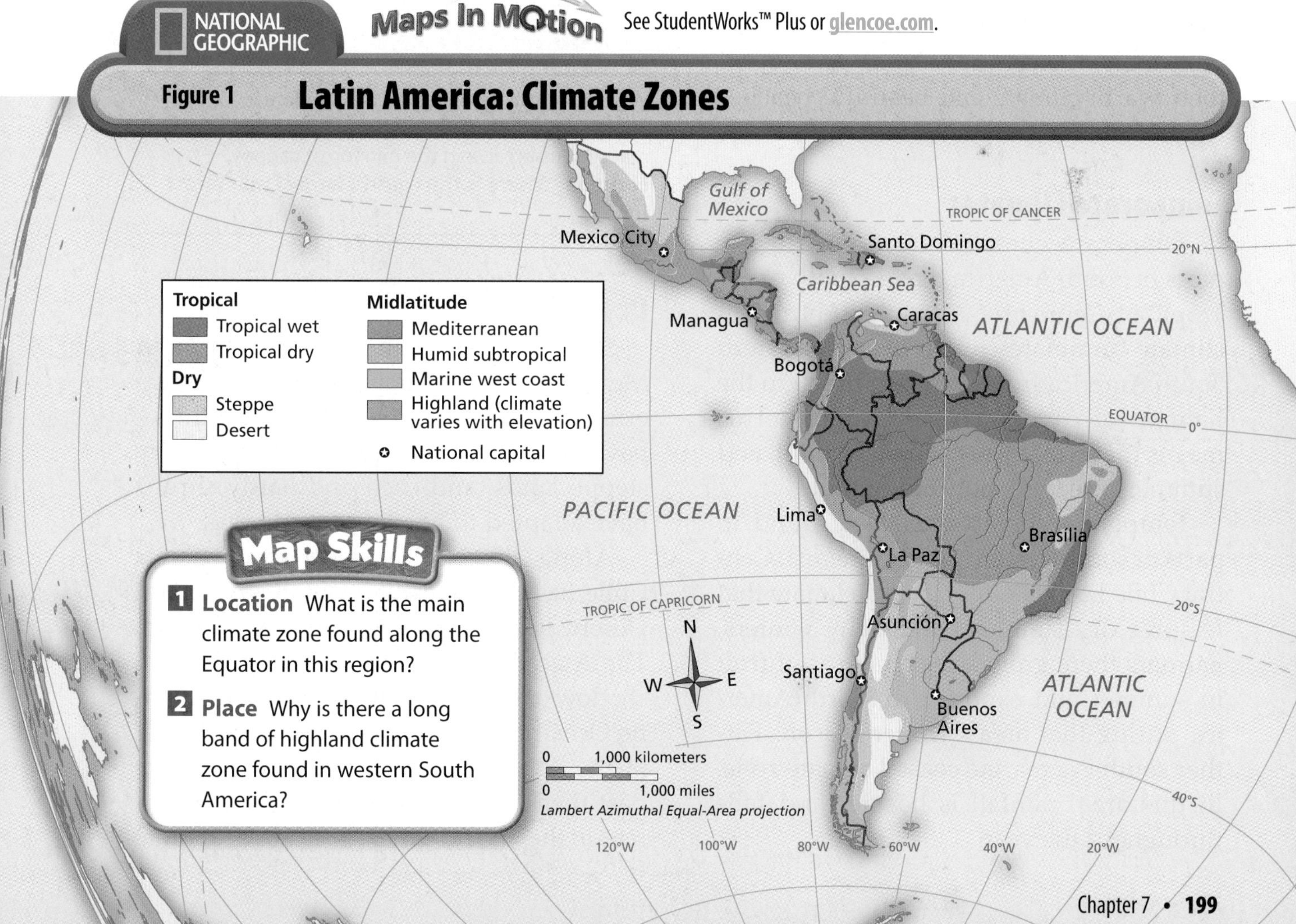

Map Skills

1. **Location** What is the main climate zone found along the Equator in this region?
2. **Place** Why is there a long band of highland climate zone found in western South America?

It shelters more species of plants and animals per square mile than anywhere else on Earth. Trees there grow so close together that their tops form a dense **canopy,** an umbrella-like covering of leaves. The canopy may soar to 130 feet (40 m) above the ground. It is so dense that sunlight seldom reaches the forest floor.

A tropical dry climate zone extends over parts of Middle America, most Caribbean islands, and north central South America. This savanna area has hot temperatures and abundant rainfall but also experiences a long dry season.

From June to November, powerful hurricanes often strike the Caribbean islands. The heavy winds and rain of these storms can cause **considerable,** or much, damage. Still, many Caribbean islands have used their warm climate and beautiful beaches to build a strong tourist industry.

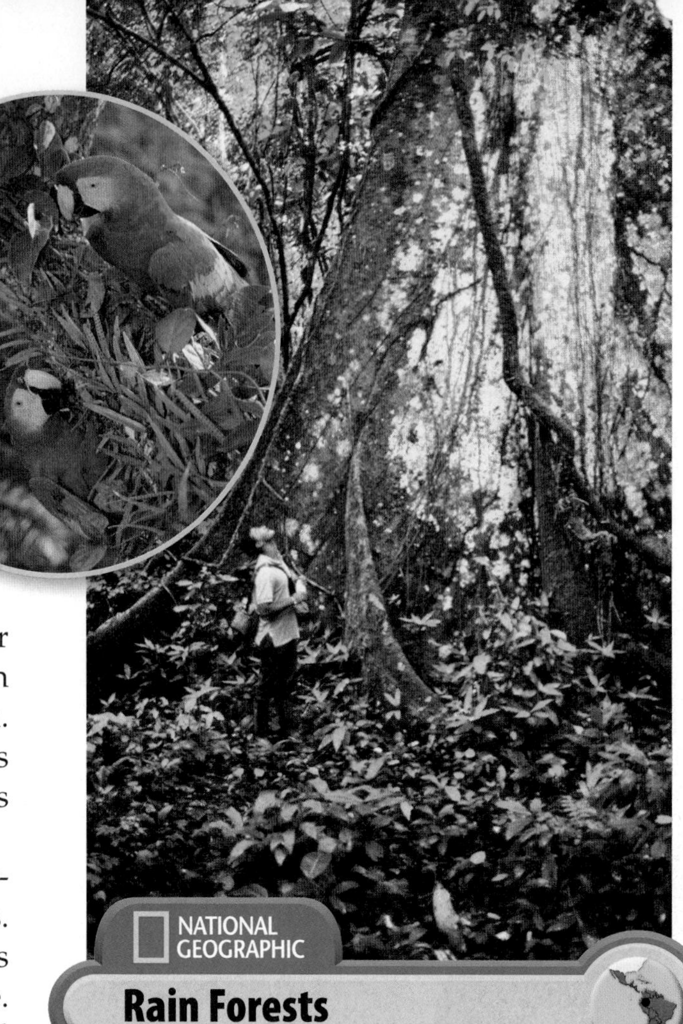

NATIONAL GEOGRAPHIC

Rain Forests

Vegetation can be dense on the rain forest floor. Many species of birds, including colorful macaws (inset), live in the rain forest canopy. ***Location*** **Where is the world's largest rain forest found?**

Temperate Climates

Temperate climates are found in the parts of South America that lie south of the Tropic of Capricorn. A humid subtropical climate dominates much of southeastern South America from southern Brazil to the Pampas of Argentina and Uruguay. This means that winters are short and mild, and summers are long, hot, and humid.

Temperate climates also are found in parts of southwestern South America. Central Chile has a Mediterranean climate that features dry summers and rainy winters. Farmers there grow large amounts of fruit in summer and export it to North America during that area's winter season. Farther south is a marine coastal climate zone. In this area, rainfall is heavier and falls throughout the year.

Dry Climates

Some parts of Latin America—northern Mexico, coastal Peru and Chile, northeastern Brazil, and southeastern Argentina—have dry climates. Grasses cover partly dry steppe lands, and cacti and hardy shrubs have adapted to harsher desert areas.

Along the Pacific coast of northern Chile lies the Atacama (AH·tah·KAH·mah) Desert. It is one of driest places on Earth. The Atacama Desert is located in the rain shadow of the Andes. Winds from the Atlantic Ocean bring rainfall to the regions east of the Andes, but they carry no moisture past them. In addition, the cold Peru Current in the Pacific Ocean does not evaporate as much moisture as a warm current does. As a result, only dry air hits the coast.

El Niño

As you may recall, weather in South America is strongly influenced by the El Niño effect. This is a set of changes in air pressure, temperature, and rainfall that begins in the Pacific Ocean.

When El Niño takes place, the Pacific waters off Peru's coast are unusually warm. As a result, winds blowing toward land carry heavy rains that lead to severe flooding along Peru's coast. El Niño can also bring a long dry season to northeastern Brazil, causing crop failures.

Reading Check **Summarizing** Why do the Tropics tend to have warm temperatures?

Elevation and Climate

Main Idea **In tropical Latin America, altitude causes great changes in climate and vegetation.**

Geography and You Have you ever traveled in the mountains and felt it getting cooler as you went higher? Read to find out how mountains affect climate in tropical areas of Latin America.

As you have read, mountains and highlands cover much of Latin America. **Altitude,** a place's height above sea level, affects climate in these rugged areas. The higher the altitude is, the cooler the temperatures are—even within the warm areas of the Tropics. The Andes, for example, have four altitude zones of climate.

NATIONAL GEOGRAPHIC

Figure 2 **Altitude Climate Zones**

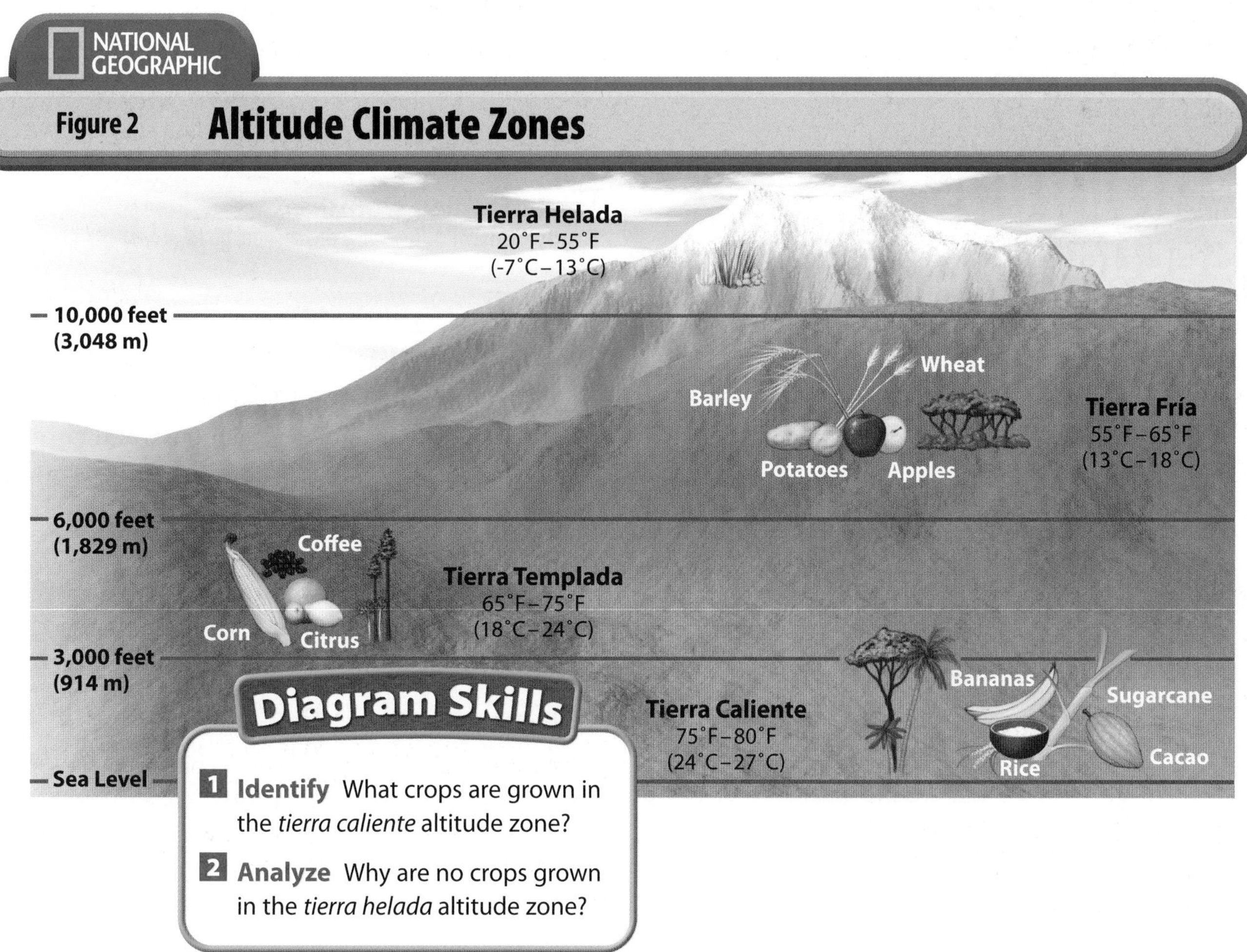

Diagram Skills

1. **Identify** What crops are grown in the *tierra caliente* altitude zone?
2. **Analyze** Why are no crops grown in the *tierra helada* altitude zone?

As **Figure 2** on the previous page shows, terms in the Spanish language are used to label the different zones.

The *tierra caliente,* or "hot land," refers to the hot and humid elevations near sea level. The average temperature range is between 75°F to 80°F (24°C to 27°C). There is little change from one month to another. In the *tierra caliente,* farmers grow a number of different tropical crops, including bananas, sugarcane, and rice.

Higher up the mountains—from 3,000 feet to 6,000 feet (914 m to 1,829 m), the air becomes cooler. Abundant rainfall encourages the growth of forests. This zone of moist, pleasant climates is called the *tierra templada,* or "temperate land." The mild temperatures—between 65°F and 75°F (18°C and 24°C)—make the *tierra templada* the most densely populated of the climate zones. Here, people grow corn and citrus fruits. Coffee, an important export crop in the region, is grown at this level.

The next zone is the *tierra fría,* or "cold land." It begins at 6,000 feet (1,829 m) and stretches up to 10,000 feet (3,048 m). Average yearly temperatures here can be as low as 55°F (13°C). The *tierra fría* has forested and grassy areas. Farming can take place in this zone in the warmer summers. The crops, however, are those that thrive in cooler, more difficult conditions. Potatoes, barley, and wheat are some of the major crops in this zone.

The *tierra helada,* or "frozen land," is the zone of highest elevation. It lies above 10,000 to 12,000 feet (3,048 m to 3,658 m). Conditions here can be harsh. The climate is cold, and the temperature can be as low as 20°F (–7°C). Vegetation throughout this zone is sparse. Relatively few people live at these heights.

Reading Check **Making Generalizations**
Why do people grow different crops at different altitudes?

Section 2 Review

Social Studies ONLINE
Study Central™ To review this section, go to glencoe.com.

Vocabulary

1. **Explain** the significance of *Tropics, rain forest, canopy,* and *altitude* by using each word in a sentence.

Main Ideas

2. **Identifying** Use a diagram like the one below to list the effects of El Niño.

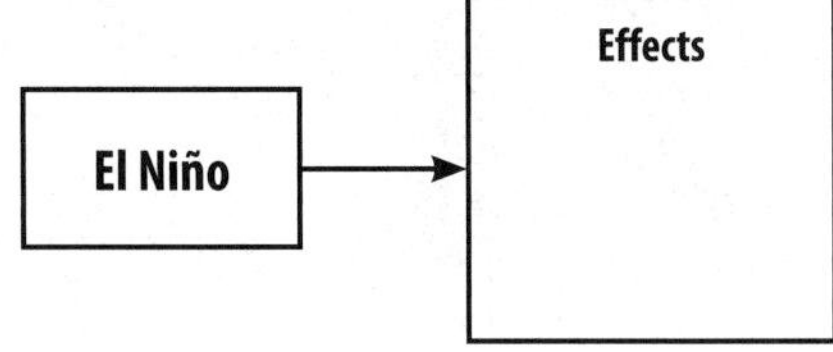

3. **Explaining** Why is the *tierra templada* the most populated altitude zone in the Latin American highlands?

Critical Thinking

4. **BIG Idea** How do some Caribbean countries benefit economically from their environment?
5. **Determining Cause and Effect** Why is the Pacific coast of northern Chile one of the driest places on Earth?
6. **Challenge** Why do parts of Latin America have mild temperatures even though they are located in the Tropics?

Writing About Geography

7. **Expository Writing** Create a chart that lists the climate zones of Latin America, explains where each zone is located, and describes the conditions and vegetation found in each zone.

Chapter 7 Visual Summary

Tortola, British Virgin Islands

Landforms

- Geographers divide Latin America into three subregions—Middle America, the Caribbean, and South America.
- Middle America, which joins North America and South America, has central mountains and narrow coastal plains.
- Caribbean islands can be low-lying or mountainous. Many have volcanoes.
- The towering Andes and the vast Amazon Basin are South America's major landforms.
- Highlands border the Amazon Basin. Lowland plains cross parts of Colombia, Venezuela, Uruguay, and Argentina.

Waterways

- Latin America's waterways provide food and transportation.
- The Panama Canal, a human-made waterway, links the Atlantic and Pacific Oceans.
- Large reserves of oil are found near Venezuela's Lake Maracaibo.

Fishing on the Amazon River

An emerald from Colombia

Resources of Latin America

- Venezuela, Mexico, and Bolivia export oil and natural gas.
- Mineral resources from Latin America include iron ore, copper, tin, silver, and emeralds.
- Political conflicts and transportation difficulties keep some countries from fully using their resources.

Climate Regions

- Latin America's tropical rain forest and savanna climates have warm temperatures.
- Rain forests, such as those in the Amazon Basin, have a great variety of plant and animal life.
- The El Niño effect brings heavy rain or drought to parts of South America.
- Climates tend to be drier and cooler at higher elevations, even within the Tropics.

Andes, Argentina

Study anywhere, anytime! Download quizzes and flash cards to your PDA from **glencoe.com**.

CHAPTER 7 ASSESSMENT

STANDARDIZED TEST PRACTICE

Look for words such as *usually, never, most,* and other qualifying words in exam questions. They indicate under what circumstances an answer is correct.

Reviewing Vocabulary

Directions: Choose the word(s) that best completes the sentence.

1. A group of islands is called a(n) ______.

A isthmus

B archipelago

C Pampas

D subregion

2. The Brazilian Highlands end with a(n) ______, or a steep cliff that drops down to the Atlantic coastal plain.

A Llanos

B estuary

C tributary

D escarpment

3. South American ______ have high temperatures and heavy rainfall year round.

A savannas

B rain forests

C steppes

D mountains

4. Cooler temperatures are found at higher ______.

A winds

B Tropics

C altitudes

D canopies

Reviewing Main Ideas

Directions: Choose the best answer for each question.

Section 1 *(pp. 192–196)*

5. The dominant landform along the Pacific coast of South America is ______.

A the Andes

B coastal marshes

C the Amazon Basin

D the Brazilian Highlands

6. ______, in South America, is the world's largest exporter of copper.

A Chile

B Brazil

C Bolivia

D Venezuela

Section 2 *(pp. 198–202)*

7. The Tropics have generally warm temperatures because

A hurricanes often strike the Tropics.

B the Tropics have a long dry season.

C they receive the direct rays of the sun.

D more plants grow there than anywhere else.

8. Vegetation is sparse in the ______ climate zone.

A *tierra fría*

B *tierra helada*

C *tierra caliente*

D *tierra templada*

GO ON

Critical Thinking

Directions: Base your answers to questions 9 and 10 on the chart below and your knowledge of Chapter 7.

Average Monthly Rainfall in Latin America		
	Manaus, Brazil	Lima, Peru
January	9.8 in. (24.9 cm)	0.1 in. (0.3 cm)
February	9.0 in. (23.1 cm)	0.0 in. (0.0 cm)
March	10.3 in. (26.2 cm)	0.0 in. (0.0 cm)
April	8.7 in. (22.1 cm)	0.0 in. (0.0 cm)
May	6.7 in. (17.0 cm)	0.2 in. (0.5 cm)
June	3.3 in. (8.4 cm)	0.2 in. (0.5 cm)
July	2.3 in. (5.8 cm)	0.3 in. (0.8 cm)
August	1.5 in. (3.8 cm)	0.3 in. (0.8 cm)
September	1.8 in. (4.6 cm)	0.3 in. (0.8 cm)
October	4.2 in. (10.7 cm)	0.1 in. (0.3 cm)
November	5.6 in. (14.2 cm)	0.1 in. (0.3 cm)
December	8.0 in. (20.3 cm)	0.0 in. (0.0 cm)
Source: BBC Weather Center, 2006.		

9. In which month does the average rainfall differ the least in Manaus and Lima?

A January

B May

C August

D December

10. Based on the chart, which of the following statements is accurate?

A Lima has a much dryer climate than Manaus.

B March is in the rainy season in both cities.

C During Lima's rainy season, it gets more rainfall than Manaus.

D Over the year, rainfall averages for both cities are about equal.

Document-Based Questions

Directions: Analyze the following document and answer the short-answer questions that follow.

The following passage is from an analysis by the Council on Foreign Relations of the struggle over control of Bolivia's natural gas.

> *[Bolivian President] Morales, a former coca farmer and union leader, won a resounding victory in the December 2005 elections. As the Movement to Socialism (MAS) candidate, he campaigned in favor of nationalizing, among other sectors of the economy, the gas and oil industries with the cooperation of foreign investors. Experts say that, given such promises, the nationalization was no surprise. But Peter DeShazo, director of the Center for Strategic and International Studies' Americas Program, says the move to occupy the gas fields with military forces lent a dramatic effect. "The confrontational nature of his move was certainly intended to get people's attention," he says, adding that Morales may be looking to garner [gain] votes in July elections for a[n] . . . assembly that will redraft Bolivia's constitution.*
>
> —Carin Zissis, "Bolivia's Nationalization of Oil and Gas"

11. What does it mean to "nationalize" the gas and oil industries?

12. What ideas expressed in the passage indicate that nationalization is popular among Bolivians?

Extended Response

13. Explain how El Niño affects the weather in South America.

STOP

For additional test practice, use Self-Check Quizzes—Chapter 7 at glencoe.com.

Need Extra Help?													
If you missed question...	1	2	3	4	5	6	7	8	9	10	11	12	13
Go to page...	193	194	199	201	193	196	199	202	199	199	196	196	201

History and Cultures of Latin America

Essential Question

Regions Common threads of language and religion unite Latin America. Once claimed as European colonies, Latin American countries today are primarily Roman Catholic, and most still use either Spanish or Portuguese as the official language. These two languages are based on Latin, which is how the region gets its name. In what ways can language and religion both unite and divide a region?

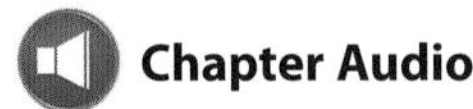

Chapter Audio

Section 1: History and Governments

BIG IDEA All living things are dependent upon one another and their surroundings for survival. Native American civilizations of Latin America developed ways of living that used the resources of their environment. People who lived in different areas depended on trade to obtain the goods they wanted. In colonial times, the people of Latin America exchanged goods with Europeans.

Section 2: Cultures and Lifestyles

BIG IDEA The characteristics and movement of people impact physical and human systems. The different groups who have settled Latin America include Native Americans, Europeans, Africans, and Asians. These groups have influenced the cultures and lifestyles of the region.

FOLDABLES™ Study Organizer

Organizing Information Make this Foldable to help you organize information about the history, peoples, cultures, and daily life of Latin America.

Step 1 Fold a sheet of paper in half lengthwise. Leave a ½-inch tab along the left edge.

Step 2 Cut the top layer only to make four equal tabs.

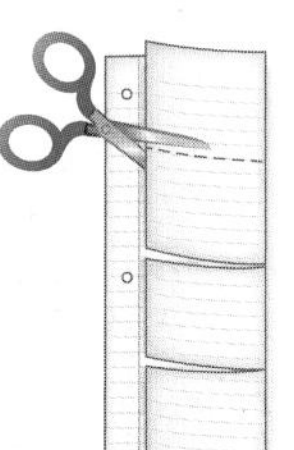

Step 3 Label the tabs as shown.

Reading and Writing Use the notes in your Foldable to write a short essay that describes the development of the countries and peoples of Latin America.

◀ Murals by Diego Rivera, Mexico City, Mexico

Social Studies ONLINE

Visit glencoe.com and enter ***QuickPass***™ code EOW3109c8 for Chapter 8 resources.

Guide to Reading

BIG Idea

All living things are dependent upon one another and their surroundings for survival.

Content Vocabulary

- maize *(p. 209)*
- jade *(p. 209)*
- obsidian *(p. 209)*
- hieroglyphics *(p. 209)*
- empire *(p. 210)*
- cash crop *(p. 211)*
- caudillo *(p. 213)*
- communist state *(p. 215)*

Academic Vocabulary

- complex *(p. 210)*
- transform *(p. 211)*
- stable *(p. 213)*
- revolution *(p. 215)*

Reading Strategy

Identifying Central Issues Use a chart like the one below to organize key facts about the Native American civilizations of the region.

	Key Facts
Olmec	
Maya	
Toltec	
Aztec	
Inca	

History and Governments

Section Audio

Spotlight Video

Picture This A Mayan village in Guatemala remembers its dead in a spectacular way. For their Day of the Dead celebration—when people remember relatives and friends who have died—villagers create enormous kites of tissue paper, bamboo, and wire. Finished kites can reach 40 feet across! Sailing above local cemeteries, the kites create a symbolic link between the living and the dead. Read this section to learn more about the historical traditions that have shaped Latin America.

▼ Ready to fly in Guatemala

Early History

Main Idea **Some Native Americans developed advanced civilizations in the region. Europeans later conquered much of the region and set up colonies.**

Geography and You What sorts of things do you like to read? History books, novels, comics? Read to find out what kinds of things the Maya wrote down.

The first people to arrive in Latin America were the ancestors of today's Native Americans. They came many thousands of years ago. Some settled and farmed. Eventually, some groups developed advanced civilizations. **Figure 1** on the next page shows these Native American civilizations.

Early Native American Civilizations

The Olmec of southern Mexico built Latin America's first civilization, which lasted from 1500 B.C. to 300 B.C. Each Olmec city focused on a certain activity, and they all depended on one another. Some cities were located near farming areas that grew **maize,** or corn, as well as squash and beans. Others controlled important mineral resources such as **jade,** a shiny green semiprecious stone; and **obsidian,** a hard, black, volcanic glass useful in making weapons. Some cities were religious centers with pyramid-shaped stone temples.

A people called the Maya lived in Mexico's Yucatán Peninsula and surrounding areas between A.D. 300 and A.D. 900. The Maya built huge stone temples in the shape of pyramids with steps. They were skilled at astronomy and used their knowledge of the stars, moon, and planets to develop a calendar. They also had a number system based on 20. Using **hieroglyphics** (HY·ruh·GLIH·fihks), which is a form of writing that uses signs and symbols, the

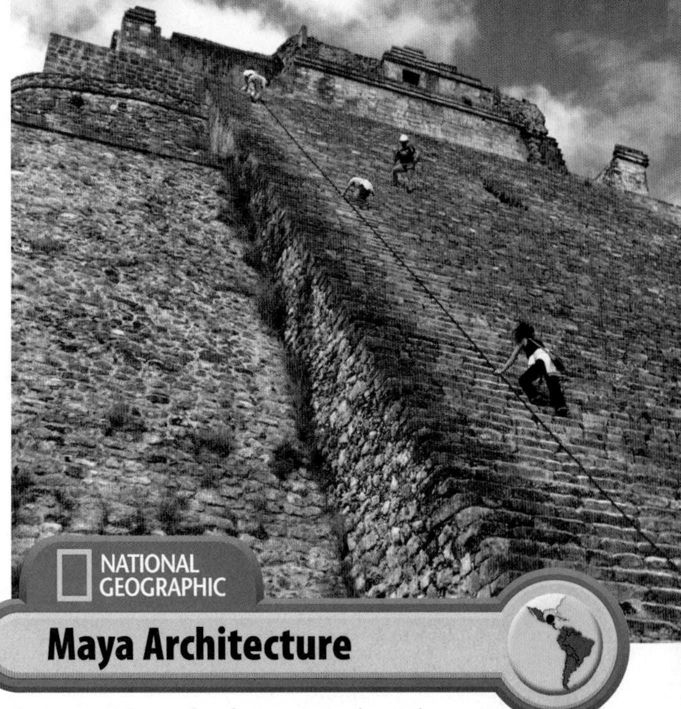

NATIONAL GEOGRAPHIC

Maya Architecture

A steep stairway leads to a temple at the top of the Pyramid of the Magician. The Maya built the pyramid about A.D. 800. ***Place*** **How did the Maya record their history?**

Maya recorded the history of their kings. About A.D. 900, the Maya civilization mysteriously collapsed. Despite intensive research, historians have not yet been able to determine what happened to the Maya.

Toltec, Aztec, and Inca

As the Maya civilization declined, a people called the Toltec seized what is now northern Mexico. These warriors built the city of Tula, northwest of present-day Mexico City. From Tula, they conquered lands all the way to the Yucatán (YOO·kah·TAHN) Peninsula.

Toltec rulers tightly controlled trade. They held a monopoly (muh·NAH·puh·lee), or sole right, to the trade in obsidian. As a result, the Toltec had the most powerful weapons in the surrounding areas. For many years this weaponry gave them the advantage they needed to maintain their rule.

Around A.D. 1200, the Aztec people from the north moved into central Mexico and captured Tula. They adopted Toltec culture, conquered neighboring peoples, and took control of the region's trade.

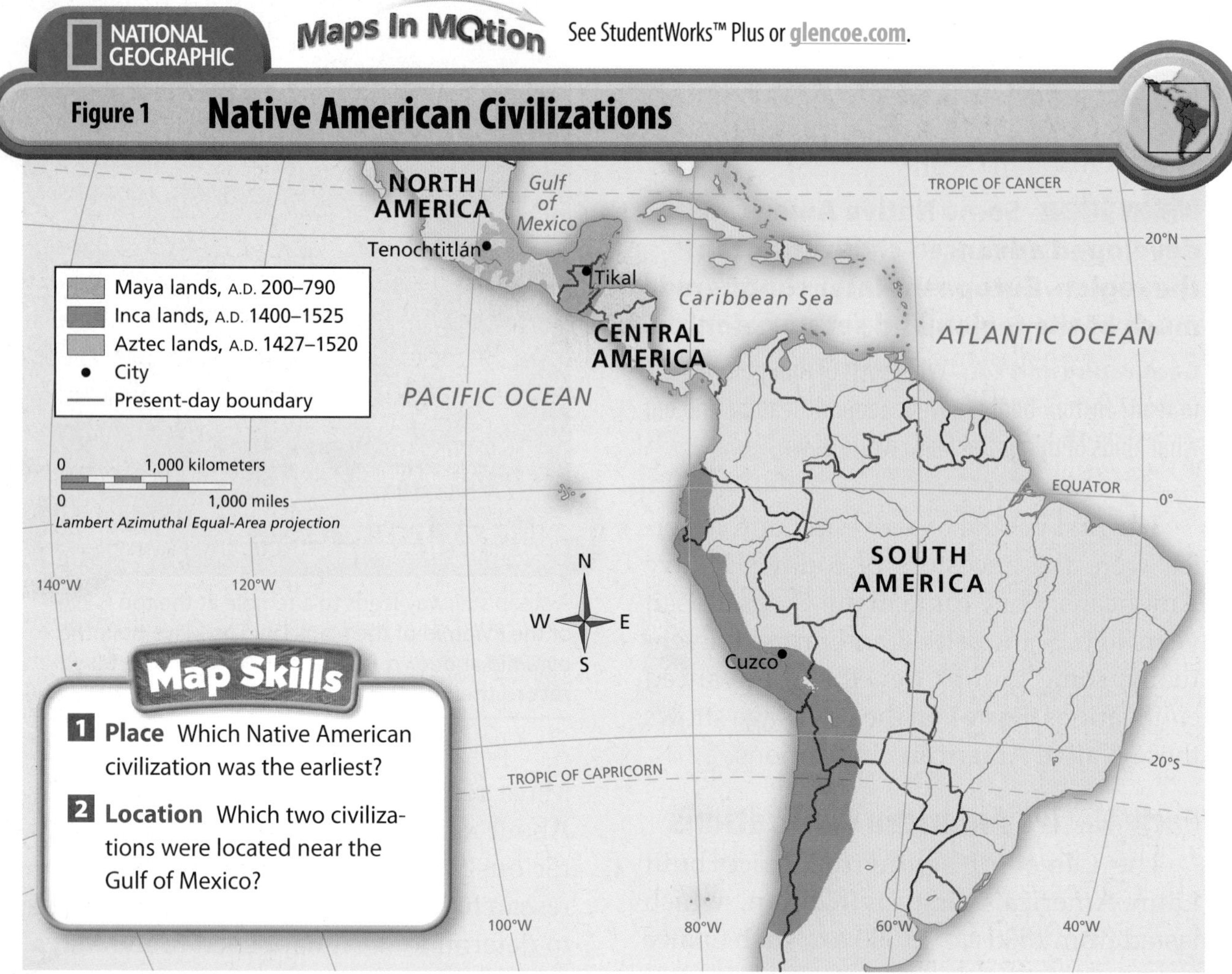

Tenochtitlán (tay·NAWCH·teet·LAHN), the Aztec capital, was a beautiful city built on an island in a lake. It held about 250,000 people, which was a large population at that time. Tenochtitlán had huge temples, including one that was more than 100 feet (30 m) tall. Roads and bridges joined the city to the mainland, allowing the Aztec to bring food and other goods to their busy markets. Aztec farmers grew their crops on "floating gardens," or rafts filled with mud. The rafts eventually sank to the lake bottom and piled up. Over time, many of these rafts formed fertile islands.

During the 1400s, the Inca had a powerful civilization in South America in what is now Peru. Their empire stretched more than 2,500 miles (4,023 km) along the Andes. An **empire** is a large territory with many different peoples under one ruler. The Inca ruler founded military posts and put in place a **complex,** or highly developed, system of record keeping. Work crews built irrigation systems, roads, and suspension bridges that linked regions of the empire to Cuzco, the capital. You can still see the remains of magnificent fortresses and buildings erected centuries ago by the skilled Inca builders.

European Conquests

In the late 1400s and early 1500s, Spanish explorers arrived in the Americas. They were greatly impressed by the magnificent cities and the great riches of the Native Americans.

In 1519 a Spanish army led by Hernán Cortés landed on Mexico's Gulf coast.

He and about 600 soldiers marched to Tenochtitlán, which they had heard was filled with gold. Some Native Americans who opposed the harsh rule of the Aztec joined forces with the Spanish. The Aztec's simple weapons were no match for the guns, cannons, and horses of the Spanish. The Spanish also had the help of unknown allies—germs that carried diseases, such as measles and smallpox. These diseases killed more Aztec than did the Spanish weapons. Within two years, Cortés's conquest of the Aztec was complete.

Another Spanish explorer named Francisco Pizarro desired the Inca's gold and silver. In 1532 Pizarro took a small group of Spanish soldiers to South America. The Spanish attacked the Inca with cannons and swords. Pizarro captured the Inca ruler and had him killed. The Inca had already been weakened by smallpox and other European diseases. After the death of their ruler, the Inca soldiers collapsed into disorder. Pizarro then quickly conquered the Inca Empire.

Colonial Latin America

The Aztec and Inca conquests provided Spain with enormous wealth and control over vast territories. Spain then built an empire that included much of South America, the Caribbean, Middle America, and parts of the present-day United States. Other European countries wanted the same power and influence that Spain had achieved. So, these countries seized different parts of the Americas. Portugal became the colonial ruler of what is today Brazil. France, Britain, and the Netherlands took control of some Caribbean areas and parts of North America.

European rule **transformed,** or greatly changed, the populations of these lands. Europeans settled the land, set up colonial governments, and spread Christianity among the Native Americans. They also used Native Americans as workers to grow **cash crops,** or farm products grown for export.

When hardship and disease greatly reduced the numbers of Native Americans, Europeans brought enslaved Africans to meet the labor shortage. A busy trade eventually resulted. Ships carried enslaved people from Africa and manufactured goods from Europe to the Americas. On the return trip, products including sugar, cotton, tobacco, gold, and silver went from the Americas to Europe. Despite European control, many Native American and African ways survived, leading to a blending of cultures.

Reading Check **Explaining** Why did the Spanish conquer Native American empires?

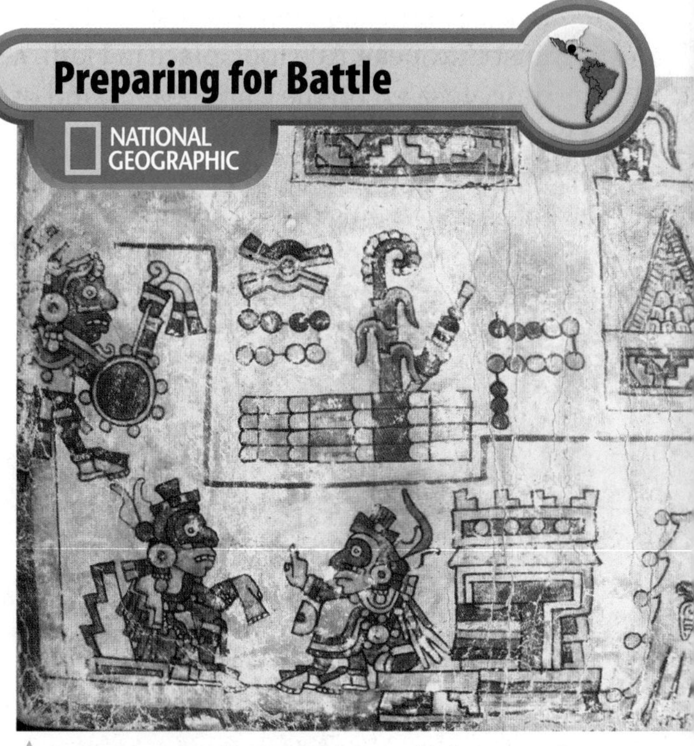

In this drawing, Native American chiefs opposed to the Aztec discuss whether to join with Cortés's forces. ***Regions*** **Why did some Native Americans support the Spanish?**

Forming New Nations

Main Idea **Most of Latin America gained independence in the 1800s, but hardships followed for many of the new nations.**

Geography and You Suppose you have just been elected class president. What challenges would you face? Read to find out what challenges faced new governments in Latin America.

In the late 1700s, revolutions in North America and France stirred the people of Latin America to action. Colonists tried to take charge of their own affairs. While European colonists called for self-rule, Native Americans and enslaved Africans wanted freedom from mistreatment and slavery.

Independence

Latin America's first successful revolt against European rule took place in Haiti, a territory located on the Caribbean island of Hispaniola. There, enslaved Africans under François-Dominique Toussaint-Louverture began a revolt that threw off French rule in 1804. Haiti, which was established as a republic, became the only nation ever created as a result of a successful revolt by enslaved people.

In Spanish and Portuguese Latin America, the fight for freedom increased in the next decade. In Mexico, two Catholic priests, Miguel Hidalgo and José María Morelos, urged poorer Mexicans to fight for freedom from Spanish rule. Both men were defeated and executed.

Despite many battles, Mexicans did not gain their independence until 1821. After a short period of rule under an emperor, Mexico became a republic in 1823. That same year, the countries of Central America won their freedom from Spain.

In northern South America, a wealthy military leader named Simón Bolívar (see·MOHN buh·LEE·VAHR) led the fight for independence. In 1819 Bolívar defeated the Spanish and won freedom for the present-day countries of Venezuela, Colombia, Ecuador, and Bolivia.

While Bolívar fought for self-rule in the north, a soldier named José de San Martín (hoh·SAY day SAN MAHR·TEEN) was fighting for freedom in the south. In 1817 San Martín led his army from Argentina across the Andes Mountains and into Chile. Although the crossing was difficult, San Martín was able to take Spanish forces by surprise, and

c. 1400s, Aztec mask

Atahuallpa, last Inca ruler

he began winning battles. A few years later, the armies of San Martín and Bolívar jointly defeated Spanish forces in Peru.

Political and Economic Challenges

By the end of 1824, all of Spain's colonies in Latin America had won their independence. The 1820s also saw Brazil break away from Portugal without bloodshed. Brazil was the only independent monarchy in Latin America before becoming a republic in 1889.

After winning independence, many of the new Latin American countries passed laws that ended slavery. Some people of African descent made economic and political gains. However, they generally did not have the advantages of Latin Americans of European background. On the other hand, African Latin Americans were better off than Native Americans, most of whom lived in poverty.

Although independent, many Latin American nations hoped their countries would become **stable,** or secure, democracies with prosperous economies. Because of a variety of problems, these goals proved hard to reach.

One major problem was frequent political conflict. Latin Americans quarreled over the role of religion in their society. Individual countries fought over boundary lines, and tensions developed between the rich and poor.

Meanwhile, strong leaders made it difficult for democracy and prosperity to develop. These leaders were known as caudillos (kow·THEE·yohs). **Caudillos** were usually high-ranking military officers or rich men supported by the upper class. They often ruled as dictators. Some built roads, schools, and new cities.

Many caudillos, however, favored the wealthy over the poor. Wealthy Latin Americans owned almost all of the land. The caudillos did nothing to help workers in the countryside. The workers remained landless and struggled to make a living.

Exporting Products

During the late 1800s, Latin America's economy depended on agriculture and mining. At this time, the United States and other industrial countries in Europe began to demand more of Latin America's food products and mineral resources. Businesspeople from these outside countries set up companies throughout Latin America. The companies exported Latin American products such as bananas, sugar, coffee, copper, and oil.

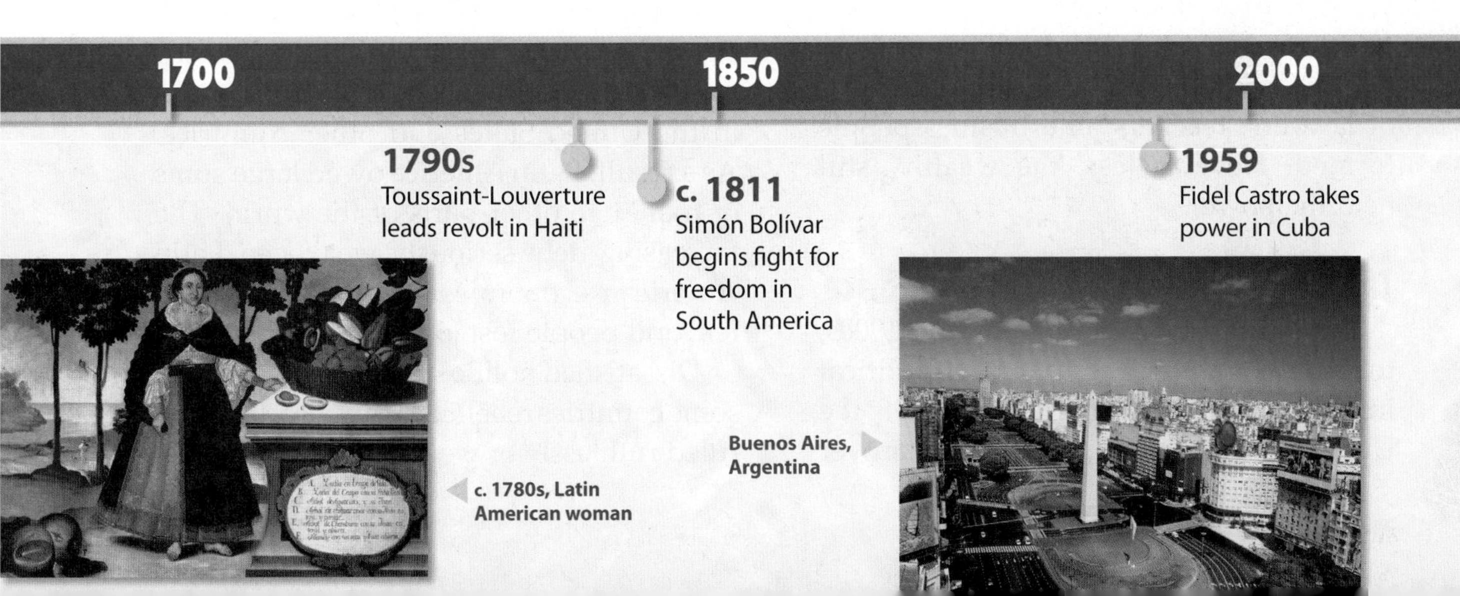

1700
1850
2000
1790s
Toussaint-Louverture leads revolt in Haiti
c. 1811
Simón Bolívar begins fight for freedom in South America
1959
Fidel Castro takes power in Cuba

c. 1780s, Latin American woman

Buenos Aires, Argentina

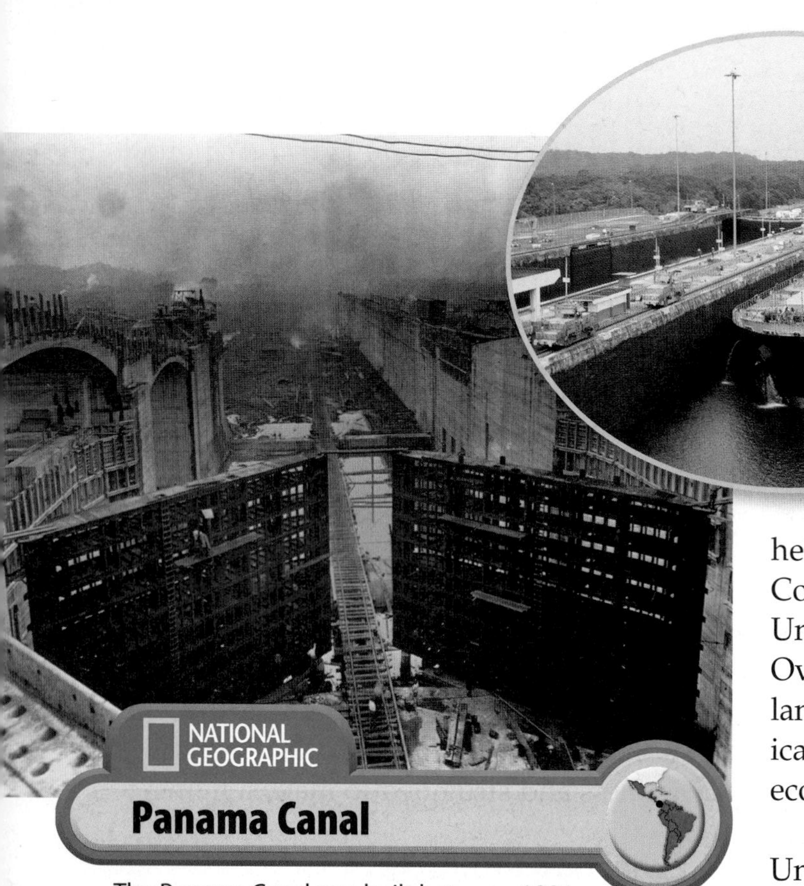

Panama Canal

The Panama Canal was built between 1904 and 1914. Today about 14,000 ships go through the canal each year (inset). ***Place*** **How did the Panama Canal affect the U.S. role in Latin America?**

As the number of exports rose, some Latin American countries chose to grow only one or two key products. Prices and profits increased as a result, but a decline in demand had serious effects. Prices dropped, and people lost income and jobs.

Despite the problems it caused, Latin America's dependence on exports also brought benefits. Foreign investors built ports, roads, and railroads. Cities grew in size and population, and a middle class of lawyers, teachers, and businesspeople formed. Nevertheless, the wealthy still held the power.

The United States and Latin America

During the late 1800s and early 1900s, the United States increased its political influence in Latin America. In 1898 the United States and Spain fought a war over Spanish-ruled Cuba. Spain was defeated, and Cuba became a republic under U.S. protection. The United States also gained control of the Caribbean island of Puerto Rico.

In 1903 the United States helped Panama win its freedom from Colombia. In return, Panama allowed the United States to build the Panama Canal. Over the next 25 years, American troops landed in Haiti, Nicaragua, and the Dominican Republic to protect U.S. political and economic interests.

Many Latin Americans distrusted the United States because of its great wealth and power. They thought the United States might try to control them as their former rulers had. To improve relations, the United States announced the Good Neighbor Policy toward Latin America in the 1930s. Under this policy, the United States promised not to send military forces to Latin America. It also pledged a greater respect for the rights of Latin American countries.

Modern Times

In the mid-1900s, agriculture was still important in Latin America, but many industries had developed there as well. To encourage economic growth, Latin American leaders borrowed heavily from banks in the United States and other countries. As a result, Latin America owed large sums of money to other parts of the world. The increasing debt seriously weakened Latin American economies. Prices rose, wages fell, and people lost jobs.

Dissatisfied political and social groups in some countries rebelled against leaders who ruled ruthlessly or were in power too long.

For example, in 1959 a young lawyer named Fidel Castro carried out a **revolution,** or a sudden, violent change of government, in Cuba. Instead of favoring democracy, Castro set up a **communist state,** in which the government controlled the economy and society.

At the same time, other countries were divided by civil wars among political, ethnic, or social groups. In El Salvador, fighters supported by Castro battled government troops armed by the United States. Thousands of people died before a settlement ended the fighting.

Difficult economic and political reforms made during the 1980s helped strengthen many Latin American countries. These changes were often very harsh, which turned many Latin Americans against dictators. During the 1990s, democratic movements succeeded in several countries.

Today's Latin American governments face many challenges. Population is growing rapidly, but resources are limited. Growing trade in illegal drugs has increased crime and corruption. Differences between rich and poor still create social tensions. In the early 2000s, angry voters in Venezuela, Bolivia, Peru, Mexico, and Chile elected new leaders. These leaders promised significant changes that would weaken the power of the wealthy and benefit the poor.

Analyzing Information Why were economies in Latin America hurt by focusing on one or two products?

Section 1 Review

Social Studies ONLINE
Study Central™ To review this section, go to glencoe.com.

Vocabulary

1. **Explain** the significance of:
   - **a.** maize
   - **b.** jade
   - **c.** obsidian
   - **d.** hieroglyphics
   - **e.** empire
   - **f.** cash crop
   - **g.** caudillo
   - **h.** communist state

Main Ideas

2. **Comparing and Contrasting** Use a Venn diagram like the one below to show similarities and differences between the Aztec and Inca civilizations.

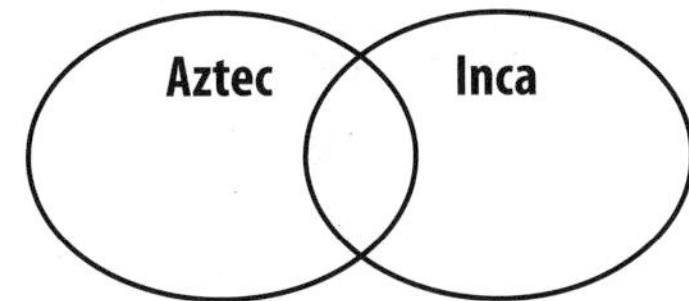

3. **Explaining** What was the social status of African Americans and Native Americans in the newly independent countries of Latin America?

Critical Thinking

4. **Drawing Conclusions** How did European colonial rule change the populations of the region?
5. **BIG Idea** How were the economies of Latin American colonies and European countries connected by trade?
6. **Challenge** Write a paragraph explaining whether you think U.S. involvement in Latin America has helped or hurt the region.

Writing About Geography

7. **Using Your FOLDABLES** Use your Foldable to write a paragraph that explains how political unrest in much of Latin America can be tied to social and economic challenges in the region.

The Darien Gap: Should a Highway Be Built?

The Pan-American Highway extends from Alaska to the tip of South America. The road stops short in Panama, at Darien National Park. In Colombia, the road starts again, where it continues for the length of South America. Roads have not been built through the Darien Gap because of its unique environment. Supporters of a road believe that it will help the region's economy and that the rain forest can still be preserved. Others, however, think that the forest could be lost forever.

For Construction

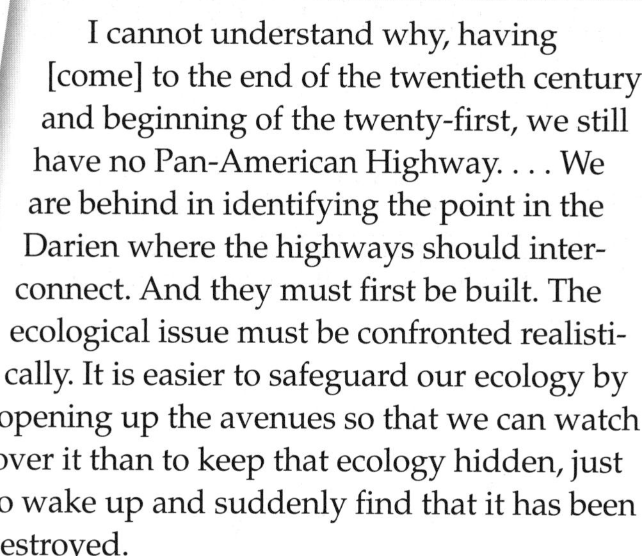

I cannot understand why, having [come] to the end of the twentieth century and beginning of the twenty-first, we still have no Pan-American Highway. . . . We are behind in identifying the point in the Darien where the highways should interconnect. And they must first be built. The ecological issue must be confronted realistically. It is easier to safeguard our ecology by opening up the avenues so that we can watch over it than to keep that ecology hidden, just to wake up and suddenly find that it has been destroyed.

—Alvaro Uribe Vélez
President of the
Republic of Colombia

Construction

I get very angry, seeing how the Panamanian economy . . . places value on felled trees and does not recognize the terrible damage to an area suffering constant deforestation. . . . One of the most comforting and encouraging sights that you can see today in the Darien is the presence of eco-tourists. . . . The worst enemy of a rain forest is the road. . . .

We should look at the Darien rain forest as a highly productive mine of eco-dollars. That is really the value of it. . . . If the Darien were to be lost, Panama would lose its soul, because nature is the base of everything.

—Hernan Arauz
Panamanian naturalist guide

You Be the Geographer

1. **Analyzing** What argument does Vélez make for building the highway?
2. **Critical Thinking** What does Arauz claim is an encouraging sight in the Darien Gap? Why do you think he feels that way?
3. **Read to Write** Write a paragraph describing how a road might benefit trade between North America and South America.

Cultures and Lifestyles

 Section Audio  Spotlight Video

Guide to Reading

BIG Idea

The characteristics and movement of people impact physical and human systems.

Content Vocabulary

- migration *(p. 219)*
- mestizo *(p. 221)*
- pidgin language *(p. 221)*
- carnival *(p. 224)*
- mural *(p. 224)*

Academic Vocabulary

- element *(p. 219)*
- comment *(p. 224)*
- style *(p. 224)*

Reading Strategy

Summarizing Use a diagram like the one below to summarize the cultures of Latin America by adding one or more facts to each of the outer boxes.

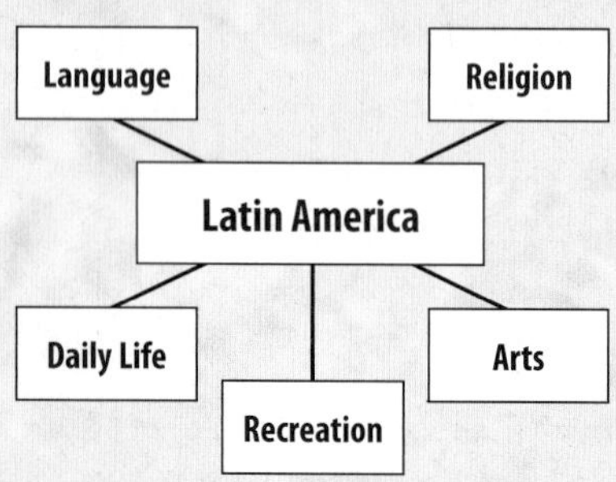

Picture This Teenage girls celebrate their African heritage during Trinidad's Children's Carnival Competition. Carnival is celebrated in the days before Lent begins. Lent is a time of prayer and fasting in the Roman Catholic Church. During Carnival, both young people and adults dress in costumes. Costumes include characters from nursery rhymes and movie superheroes. As you read this section, you will learn about the different cultures of the people of Latin America.

▼ Celebration in Port-of-Spain, Trinidad

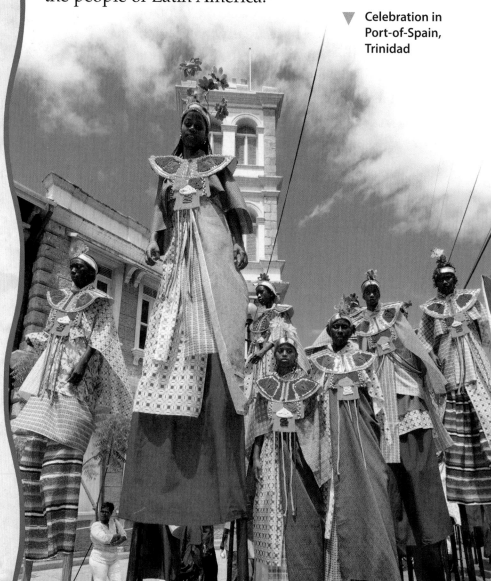

The People

Main Idea **Latin Americans come from a variety of cultures, but many share common characteristics.**

Geography and You Does anyone in your neighborhood speak a foreign language? Read to discover the mix of languages and cultures in Latin America.

Latin Americans come from many different backgrounds. Native Americans, Europeans, Africans, and others all have left their mark. Most Latin Americans today practice the Roman Catholic faith and speak either Spanish or Portuguese.

Population Patterns

Latin America has a high population growth rate. The region's highest birthrates are in Central America, except for Costa Rica, whose birthrate is relatively low. As a result, the Central American countries are growing most quickly in population. In fact, Guatemala and Honduras are expected to double in population by 2050.

Latin America's varied climates and landscapes affect where people live. Temperature extremes, rain forests, deserts, and mountains are common in many parts of Latin America. In these areas, harsh living conditions and poor soil limit where people live. As a result, most Latin Americans live in more favorable climates along the coasts of South America or in an area reaching from Mexico into Central America. These areas provide fertile land and easy access to transportation.

Migration

Migration, or the movement of people, has greatly shaped Latin America's population. In the past, Europeans, Africans, and Asians came to Latin America in large numbers, either willingly or by force. Today, people from places as far away as Korea and Syria come to Latin America looking for jobs or personal freedom.

NATIONAL GEOGRAPHIC

A Young Population

About 30 percent of people in Latin America are age 15 and younger. In the United States, 21 percent are 15 and younger. ***Regions*** **Which area of Latin America has the highest birthrate?**

In addition to people immigrating into the region, some leave Latin America for other parts of the world. Many Latin Americans move north to the United States to escape political unrest or to find a better way of life. Some go through the process of legally entering the United States, while others enter illegally. All of these new arrivals bring **elements,** or parts, of their culture with them. Most stay in close contact with family and friends in their home countries. Many also plan to return when economic conditions in their home countries improve.

Social Studies ONLINE

Student Web Activity Visit glencoe.com and complete the Chapter 8 Web Activity about Latin American populations.

Latin Americans also move within their country or the region. As in many parts of the world, Latin America's rural areas have increased greatly in population. In certain areas, this growth has resulted in a shortage of fertile land. Smaller farms cannot always support large families. People often leave to find jobs elsewhere, usually in cities. The result is urbanization, or the movement of people from the countryside to the cities.

Growth of Cities

In the past, most Latin Americans lived in the countryside and worked the land. Today most of them live in rapidly growing cities. Some of the largest urban areas in the world are in Latin America, including Mexico City, Mexico; São Paulo (sow POW·loo) and Rio de Janeiro (REE·oo dee zhah·NAY·roo) in Brazil; and Buenos Aires, Argentina.

The number of urban dwellers, however, varies throughout the region. In South America, about 80 percent of people live in cities—about the same as in the United States. In Central America and the Caribbean, only about 65 percent of people are urban dwellers.

Most Latin Americans leave villages for the cities to find better jobs, schools, housing, and health care. In many cases, people do not find what they seek. As city populations grow, jobs and housing become scarce. At the same time, rural dwellers often lack the education and skills to find good jobs. There have been too few schools and health care centers to serve the growing number of city dwellers. Unable to return to the countryside, many people have been forced by poverty to live in crowded neighborhoods with poor housing, lack of sanitation, and rising crime.

▲ Once the capital city of Brazil, Rio de Janeiro remains an important center for trade and industry. ***Place*** **What percentage of South Americans live in cities?**

Ethnic Groups and Languages

Latin America's people include Native Americans, Europeans, Africans, Asians, and mixtures of these groups. The blend of ethnic groups varies from area to area.

Most of Latin America's Native Americans live in Mexico, Central America, and the Andes countries of Ecuador, Peru, and Bolivia. Great Native American empires thrived in these places before Europeans arrived there. Today, Native Americans work to keep their languages and traditions alive while adopting features of other cultures.

Since the 1400s, millions of Europeans have settled in Latin America. Most settlers have been Spanish or Portuguese. Over the years, other Europeans—Italians, British, French, and Germans—have come as well. In the 1800s, many Spanish and Italian immigrants settled in Argentina, Uruguay, and Chile. As a result, these three nations today are mainly populated by people of European descent.

African Latin Americans form a high percentage of the populations in the Caribbean islands and northeastern Brazil. They are descended from enslaved Africans whom Europeans brought as laborers during colonial days. Over the years, Africans have added their rich cultural influences to the food, music, and arts of Latin America.

Large Asian populations live in the Caribbean islands and some countries of South America. Most Asians came during the 1800s to work as temporary laborers. They remained and formed ethnic communities. In Guyana about one-half of the population is of South Asian or Southeast Asian ancestry. Many people of Chinese descent make their homes in Peru, Mexico, and Cuba. About 1 million people of Japanese descent live in Brazil, making Brazil home to the largest number of Japanese in one place outside of Japan.

Over the centuries, there has been a blending of the different ethnic groups throughout Latin America. In countries such as Mexico, Honduras, El Salvador, and Colombia, **mestizos,** or people of mixed Native American and European descent, make up the largest part of the population. In Cuba, the Dominican Republic, and Brazil, people of mixed African and European descent form a large percentage of the population.

Because Spain once ruled most of Latin America, Spanish is the most widely spoken language in the region. In Brazil, which was once a colony of Portugal, most people speak Portuguese. Native American languages are still spoken in many countries. For example, Quechua (KEH·chuh·wuh), spoken centuries ago by the Inca, is an official language of Peru and Bolivia. In the Caribbean, where the British and French once ruled many islands, English and French are widely spoken. In some countries, people have developed a **pidgin language** by combining parts of different languages. An example is Creole, spoken in Haiti. Most Creole words are from French, but sentence structure, or organization, reflects African languages.

NATIONAL GEOGRAPHIC

Ethnic Diversity

São Paulo, Brazil, has a large Japanese community. Drummers in Barbados (inset) celebrate their African heritage. ***Place*** **What Latin American nations have populations that are mainly of European background?**

Reading Check **Analyzing** What challenges do Latin America's growing cities face?

Mexican Folktale

Folktales are stories that have no known author. They are the literature of the common people of a country or region. They express the views these people have about life, what is important to society, and how individuals are expected to behave.

Background Information

The folktales of Mexico reflect Mexican society. Like that society, they include a mix of Spanish and Native American cultures. Some tales point out the tensions between different social and ethnic groups. This folktale reflects basic Mexican values, including the respect that children owe to their parents.

Reader's Dictionary

crystal: expensive, high-quality glass

weevil: a kind of beetle

The Hard-Hearted Son

Mexican Folktale

There was an old couple who had a married son. They were very poor, and one day they went to visit their son to see if he would give them some corn and ask them to dinner. His corn bins were full, and his table was laid out with many good things. For dessert there was candy in a large dish made of **crystal.**

When he saw his parents coming, he told his wife, "There come those old people! Put the cover on the candy dish and hide the food, so we won't have to ask them to dinner."

The wife did so, and when his parents came in and saw it all, they asked their son for a few handfuls of corn. But he told them he didn't have anything, that he hadn't harvested his crop yet. "It's all right," his parents said. "God bless you and give you more." And they left.

When [the son and his wife] sat down to dinner, they found the food had spoiled. The man went to his corn bins and found it all eaten by **weevils.** He came back, and when he was going to eat the candy, a serpent came out and wound itself about his neck and strangled him.

It [wasn't] his parents' curse; rather, [it was] a punishment for his greed and hard-heartedness.

Analyzing Literature

1. **Making Inferences** What message does this tale give about how children should behave?
2. **Read to Write** Think about a kind of behavior or attitude that you think is important for people to have. Write a brief story like this one to illustrate why people should behave that way or have that attitude.

Daily Life

Main Idea **Many aspects of daily life in Latin America reflect the region's blend of cultures.**

Geography and You Do you enjoy eating tomatoes, potatoes, and chocolate? These foods were first eaten in Latin America. Read to find out about other features of Latin American daily life.

Religion and family play an important role in Latin American life. The region's history and politics are reflected in celebrations and art.

Religion

Religion has long played an important role in Latin American cultures. During colonial times, most Latin Americans became Christians, and Christianity still has the most followers. Roman Catholics form the largest Christian group. In recent years, however, Protestant missionaries have encouraged many people to convert, or to change their beliefs, to Protestant forms of Christianity.

Other faiths are also practiced in the region. For example, many traditional Native American and African religions thrive, often mixed with Christianity and other faiths. Islam, Hinduism, and Buddhism, brought by Asian immigrants, are practiced in the Caribbean region and coastal areas of South America. Judaism has followers in the largest Latin American cities.

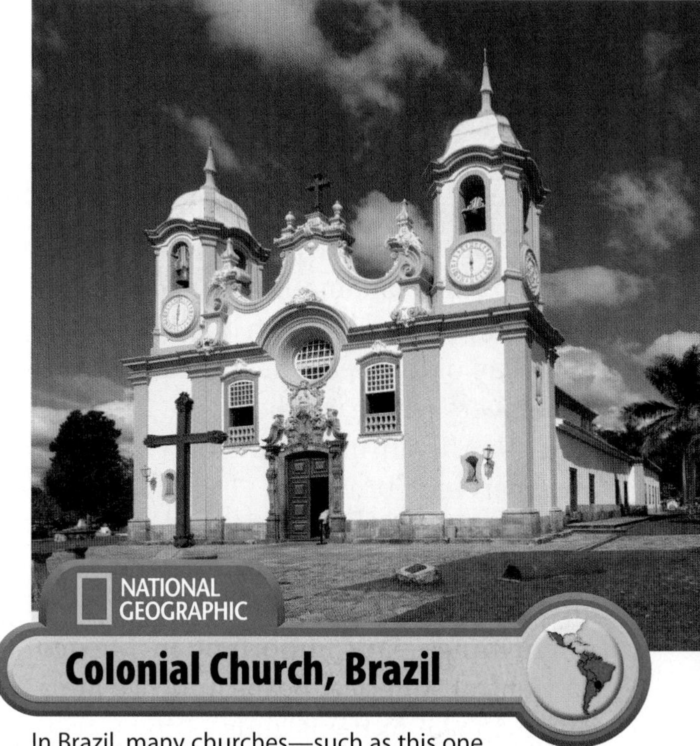

Colonial Church, Brazil

In Brazil, many churches—such as this one in the town of Tiradentes—were built during the 1700s under Portuguese rule. ***Regions*** **How is Christianity changing in Latin America?**

Family

Family life is important in Latin America. Often several generations live together, and adults are expected to care for their aged parents. Adult brothers and sisters often live near each other, and their children—who are cousins—can form close relationships. Traditionally, the father is the family leader and the chief decision maker. In some parts of the Caribbean, however, the mother is the leader of the family.

Recreation and Celebrations

Most Latin Americans are devoted sports fans. Soccer is popular throughout the region, and Brazil and Argentina have produced outstanding players and world championship teams. Cuba was the second country in the world—after the United States—to play baseball. This sport has taken hold throughout the Caribbean, Central America, and northern South America. Several countries have their own leagues, and many skilled players have gone to play in the United States. In Caribbean countries that were once ruled by the British, cricket is a favorite sport.

Religious and patriotic holidays are celebrated throughout Latin America. Each spring, many countries hold a large festival called **carnival** on the last day before the Christian holy period called Lent. The celebration is marked by singing, dancing, and parades. The Carnival held in Rio de Janeiro, Brazil, is famous for its color and excitement. On the Mexican holiday known as the Day of the Dead, people honor family members who have died.

Feasting is an important part of Latin American celebrations. The foods of Latin America blend the traditions of the region's many peoples. Corn and beans—crops grown by Native Americans since ancient times—are important in Mexico and Central America. Beans and rice are a standard meal in the islands of the Caribbean and in Brazil. Fresh fish from the sea is also popular in those areas. Beef is the national dish in Argentina, Uruguay, and Chile.

The Arts

Culture in Latin America shows the influence of its ethnic mix. Cuban music is famous for its use of African rhythms. During the 1930s, Mexican artists, such as Diego Rivera, painted **murals,** or large paintings on walls, that recall the artistic traditions of the ancient Maya and Aztec. In Latin America, many writers have used their work to **comment** on, or talk about, social and political conflicts.

Latin American artists have influenced those in other countries. The music of Cuba and Brazil has shaped American jazz. Latin American writers of the late 1900s invented a **style,** or form, of writing called magic realism that combined fantastic events with the ordinary. This style was adopted by European and Asian writers.

Reading Check **Identifying** What sports are popular in Latin America?

Social Studies ONLINE
Study Central™ To review this section, go to glencoe.com.

Vocabulary

1. **Explain** the significance of:
   a. migration
   b. mestizo
   c. pidgin language
   d. carnival
   e. mural

Main Ideas

2. **Describing** Describe patterns of migration in Latin America in the past and today.
3. **Summarizing** Use a diagram like the one below to summarize key facts about religion in Latin America.

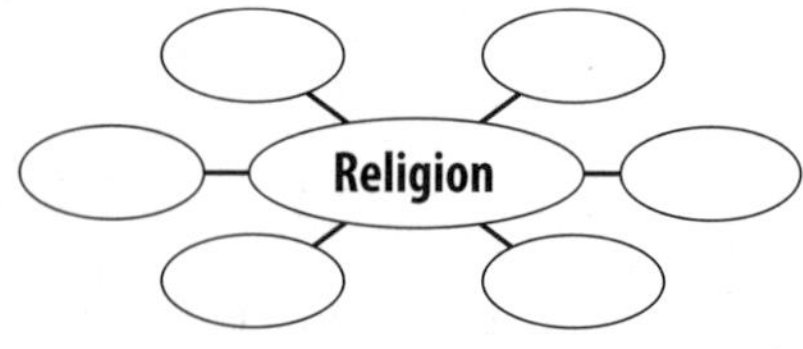

Critical Thinking

4. **BIG Idea** How do pidgin languages show the blending of different cultures in Latin America?
5. **Identifying Central Issues** In what parts of the region do most Latin Americans live? Why?
6. **Challenge** How has the region been influenced by other regions of the world?

Writing About Geography

7. **Expository Writing** Make a map of Latin America. Add labels that highlight facts about the population patterns, religions, and cultures of different countries and areas within the region. Then write a short paragraph describing the patterns you see.

CHAPTER 8 Visual Summary

Native American Civilizations

- The Olmec built the first civilization in Latin America.
- The Maya created a calendar and a complex number system.
- The Aztec set up a large empire in central Mexico.
- The Inca developed a network of roads to unite their territories.

Aztec stone calendar

Colonial Rule

- Spanish explorers conquered the Aztec and Inca Empires.
- Spain and Portugal ruled most of Latin America from the 1500s to the early 1800s.
- Colonial rule brought a mixing of different cultures.

Hernán Cortés

Forming New Nations

- Most Latin American countries achieved independence during the 1800s.
- Dictators, the military, or wealthy groups ruled Latin American countries, while most people remained poor and powerless.
- Many Latin American countries developed more democratic systems in the 1900s.

People

- About 80 percent of South Americans live in urban areas.
- Most people in Latin America are of European, Native American, or African background.
- Most Latin Americans speak Spanish or Portuguese, and most practice the Roman Catholic faith.

Culture

- Family life is important to most Latin Americans.
- Soccer and baseball are major sports in Latin America.
- Food, arts, and music reflect the diverse ethnic mixture of the region.
- Religious and patriotic holidays are important throughout Latin America.

Baseball players, Dominican Republic

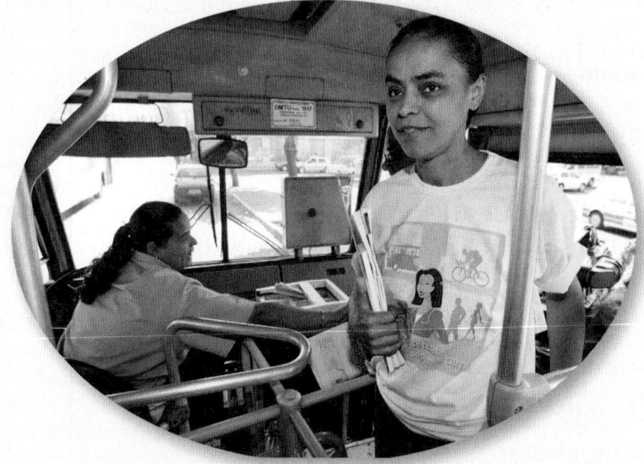

Bus rider in Brasília, Brazil

Study anywhere, anytime! Download quizzes and flash cards to your PDA from glencoe.com.

CHAPTER 8 ASSESSMENT

STANDARDIZED TEST PRACTICE

On answer sheets for standardized tests, neatly print information, such as your name, and carefully fill in ovals.

Reviewing Vocabulary

Directions: Choose the word(s) that best completes the sentence.

1. The Olmec made weapons with a volcanic glass called ______.

A jade
B obsidian
C maize
D copper

2. The Inca of Peru established a(n) ______ in the Andes.

A empire
B caudillo
C communist state
D Good Neighbor Policy

3. Creole is an example of a ______.

A mestizo
B cash crop
C hieroglyphic
D pidgin language

4. In Mexico, some artists painted ______, or large paintings on walls, recalling the artistic traditions of the Maya and Aztec.

A mestizos
B carnivals
C murals
D caudillos

Reviewing Main Ideas

Directions: Choose the best answer for each question.

Section 1 *(pp. 208–215)*

5. The Maya built their civilization in an area that is known today as ______.

A Brazil
B the Caribbean
C central Mexico
D the Yucatán Peninsula

6. What happened because Latin American leaders borrowed heavily from United States banks in the mid-1900s?

A Wages fell and people lost jobs.
B Prices in their countries dropped.
C Latin economies became stronger.
D American troops landed in Nicaragua.

Section 2 *(pp. 218–224)*

7. ______ is populated mainly by people of European descent.

A Bolivia
B Ecuador
C Guatemala
D Argentina

8. In ______, Quechua is an official language.

A Peru
B Brazil
C Honduras
D El Salvador

GO ON

Critical Thinking

Directions: Base your answers to questions 9 and 10 on the table below and your knowledge of Chapter 8. Choose the best answer for each question.

Internet and Cell Phone Users in Central America			
	Population	Internet Users	Cell Phone Users
Belize	287,730	35,000	93,100
Honduras	7,326,496	223,000	1,282,000
Guatemala	12,293,545	756,000	3,168,300
Costa Rica	4,075,261	1,000,000	1,101,000
Source: *CIA World Factbook.*			

9. Divide the population by the number of cell phone users to find out the number of people per cell phone in each country. Which country has the fewest cell phones in proportion to its population?

A Belize

B Honduras

C Guatemala

D Costa Rica

10. Divide the number of Internet users by the population. Which country has the most Internet users in proportion to its population?

A Belize

B Honduras

C Guatemala

D Costa Rica

Document-Based Questions

Directions: Analyze the following document and answer the short-answer questions that follow.

The following passage is by a Catholic priest who came to the Yucatán Peninsula in the 1560s.

> *Before the Spaniards subdued [overcame] the country the Indians lived together in well ordered communities; they kept the ground in excellent condition, free from noxious [harmful] vegetation and planted with fine trees. The habitation was as follows: in the center of the town were the temples, with beautiful plazas, and around the temples stood the houses of the chiefs and the priests, and next those of the leading men. Closest to these came the houses of those who were wealthiest and most [respected], and at the borders of the town were the houses of the common people. The wells, where they were few, were near the houses of the chiefs; their plantations were set out in the trees for making wine, and sown with cotton, pepper and maize. They lived in these communities for fear of their enemies, lest [for fear that] they be taken in captivity; but after the wars with the Spaniards they dispersed [scattered] through the forests.*
>
> —Friar Diego de Landa,
> *Yucatán Before and After the Conquest*

11. How did the friar feel about the Indian communities? Explain your answer.

12. How did conquest by the Spaniards affect communities in the Yucatán Peninsula?

Extended Response

13. Describe challenges and successes for Latin Americans in the 1990s and 2000s.

STOP

Social Studies ONLINE

For additional test practice, use Self-Check Quizzes—Chapter 8 at glencoe.com.

Need Extra Help?													
If you missed question...	1	2	3	4	5	6	7	8	9	10	11	12	13
Go to page...	209	210	221	224	209	214	221	221	219	219	210	210	213–224

"Hello! My name is Miguel.

I'm 14 years old and I live in San Cristóbal Ecatepec, a town near Mexico City, the capital of Mexico. I live in a small house with my mother, sister, and grandmother. Read about my day."

6:15 a.m. I wake up and get dressed then have breakfast with my family. This morning, my two young cousins come over to eat with us. We have quesadillas, which are corn tortillas with melted cheese. My grandmother also puts out a plate of bananas and papayas.

6:45 a.m. I comb my hair, brush my teeth, and put my books in my backpack. It's time to leave for school, even though it's still pretty dark outside. I walk to school with my sister, Areli (ah•ray•LEE).

7:00 a.m. The sun is starting to come up as we arrive at José María Morelos y Pavón Middle School. The school is named for a famous leader in Mexico's struggle for independence from Spain. (We have learned about him in history class!)

7:10 a.m. English is my first class of the day. Our teacher, Mr. Aranda, encourages us to speak mostly English during class.

7:45 a.m. It is time for my least favorite class—math. Today, though, I get a break. The local police visit our school to lead an assembly on crime prevention. They talk to us about staying safe and drug-free.

8:45 a.m. In physical education class, we play *fútbol*, or soccer. Then I move on to music, where I practice my skills on the recorder. I am learning to play a song from a musical.

10:30 a.m. During a short recess, I sit outside and talk with my friends Alejandra, Ismael, and José.

10:45 a.m. I go to history class, then to Spanish. I am working hard for a good grade in both classes. In Mexico, we are graded on a 10-point scale. A passing grade is a 6 or higher.

12:45 p.m. It is time for my elective class—family values. It's about respecting family and friends, and doing community volunteer work.

1:00 p.m. The school day is over. While I walk home, I chat with my dad on my cell phone. He and my mother are separated, so he does not live with us. I see him often though.

1:10 p.m. I change clothes and feed my dogs. Then I help my grandmother with lunch. I squeeze oranges for juice while she makes chicken and rice. My mom comes home for lunch from her job as a secretary.

2:30 p.m. I ride my bike to the hardware store that my family owns. My grandmother and uncles work there. In the back of the store, my Uncle Ricardo raises roosters. I help my uncle by feeding the birds and cleaning their cages.

4:30 p.m. I go back home and play soccer outside with my cousins. Then I start my homework.

6:30 p.m. For dinner, we have *pollo con mole*. It is chicken in a delicious black sauce made with chocolate and spices. Later, I watch some TV (I like to watch reality shows).

10:00 p.m. I am tired, so I go to bed.

ILLUSTRATIONS BY BOOKMAPMAN

WHAT'S THE WORD? Miguel works on a team project in Spanish class.

MAKING MUSIC Miguel plays the recorder in music class. He also knows how to play the flute.

AT HOME Miguel, his sister, cousins, and grandmother spend time together.

ALL SMILES In his spare time, Miguel Rodriguez (mee·GELL rod·REE·guez) helps care for his uncle's roosters.

What's Popular in Mexico

DANITA DELIMONT/ALAMY

Murals Mexico's early inhabitants, the ancient Maya, painted scenes from their daily lives on rock walls. Today, large murals or wall paintings are still a popular art form.

Turning 15 In Mexican tradition, a girl's fifteenth birthday calls for a special celebration. The event is called *quince años*, or "fifteen years."

Mariachi music Lively mariachi bands often play at Mexican festivals and weddings. The musicians play violins, guitars, and trumpets and often dress like traditional Mexican cowboys called *charros*.

KEN WELSH/AGE FOTOSTOCK

Say It in Spanish

There are still several native Indian languages spoken in Mexico, but the national language is Spanish. It was brought to Mexico in the sixteenth century by Spanish settlers. Try these everyday Spanish expressions.

Hello
Hola (OH · lah)

Good-bye
Adios (ah · dee · OHS)

My name is ______.
Me llamo (may YAH · moh) ______.

ADRIANA ZEHBRAUSKAS / POLARIS (4)

LUC NOVOVITCH/ALAMY

CHAPTER 9

Latin America Today

Essential Question

Human-Environment Interaction The climates and resources of Latin America have a great impact on the people who live there. Those countries with more resources tend to be wealthy, but many citizens never benefit from that wealth. At the same time, the use of resources has seriously affected the environment in certain areas. What human activities benefit the environment, and what activities harm it?

Chapter Audio

Section 1: Mexico

BIG IDEA Patterns of economic activities result in global interdependence. Many Mexicans now depend on factory jobs. Those who cannot find work at home migrate to the United States in search of work.

Section 2: Central America and the Caribbean

BIG IDEA The physical environment affects how people live. Many Caribbean islands have limited resources. Their warm climate and beautiful beaches, however, make tourism an important industry.

Section 3: South America

BIG IDEA People's actions can change the physical environment. The Amazon basin holds the world's largest rain forest. People are now using the rain forest's resources to boost economic growth. Their actions greatly affect the Amazon basin's fragile environment.

FOLDABLES™ Study Organizer

Organizing Information Make this Foldable to help you organize information about the countries of Latin America today.

Step 1 Place two 11x17 pieces of paper together.

Step 2 Fold the papers in half to form a booklet.

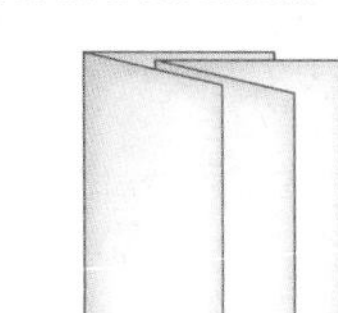

Step 3 Label your booklet as shown.

Reading and Writing As you read the chapter, take notes about each Latin American country. Use your notes to write five quiz questions for each section of the chapter.

Plaza de Mayo, Buenos Aires, Argentina

Visit glencoe.com and enter QuickPass™ code EOW3109c9 for Chapter 9 resources.

Guide to Reading

BIG Idea

Patterns of economic activities result in global interdependence.

Content Vocabulary

- plaza *(p. 233)*
- vaquero *(p. 234)*
- maquiladora *(p. 235)*
- subsistence farm *(p. 235)*
- plantation *(p. 235)*
- migrant worker *(p. 236)*

Academic Vocabulary

- reveal *(p. 234)*
- assemble *(p. 235)*

Reading Strategy

Summarizing Use a chart like the one below to organize key facts about Mexico's economic regions.

Region	Key Facts
North	
Central	
South	

Mexico

 Section Audio  Spotlight Video

Picture This They may not look like soccer balls, but these piles of plastic panels will be stitched together by workers in San Miguelito, Mexico, to create thousands of balls for the popular sport. The soccer balls are then sold to large companies that export them. Read this section to learn about Mexico's economy today and how it is connected to other regions of the world.

Soccer ball beginnings in San Miguelito, Mexico

Mexico's People, Government, and Culture

Main Idea **Mexico's culture reflects its Native American and Spanish past as well as modern influences.**

Geography and You Do you like tacos or enchiladas? These are tasty Mexican dishes. Read to learn about Mexico's people and culture.

Mexico is the United States's nearest southern neighbor. It is the third-largest country in area in Latin America, after Brazil and Argentina. Mexico also ranks second in population.

Mexico's People

Mexico's people reflect the blending of Spanish and Native American populations over the centuries. About two-thirds of Mexicans are mestizos. A quarter of Mexico's people are mostly or completely Native American.

In Mexico, rural traditions remain strong, but about 75 percent of Mexicans now live in cities. The largest city by far is Mexico City, the country's capital. With nearly 22 million people, Mexico City is one of the world's largest and most crowded urban areas.

Mexican cities show the influence of Spanish culture. Many of them are organized around large **plazas,** or public squares. City plazas serve as centers of public life. The main government buildings and the largest church are located alongside each city's plaza. Newer sections of the cities have glass office buildings and modern houses. In poorer sections, homes are built of boards, sheet metal, or even cardboard.

NATIONAL GEOGRAPHIC

Plaza in Mexico City

The plazas of Mexican cities are popular places to gather. The church on this plaza in Mexico City was built in 1536. ***Place*** **About what percentage of Mexico's people live in cities today?**

Mexico's Government

Mexico, like the United States, is a federal republic, where power is divided between national and state governments. A strong president leads the national government. He or she can serve only one six-year term but has more power than the legislative and judicial branches.

After a revolution in the early 1900s, one political party ruled Mexico for many decades. Then, in the 1990s, economic troubles and the people's lack of political power led to calls for change. In 2000 Mexican voters elected a president from a different political party for the first time in more than 70 years. In the next presidential election six years later, the vote count was too close to call. An election court finally ruled Felipe Calderón president of Mexico, despite bitter protests from supporters of the opposing candidate.

Mexican Culture

Mexican culture **reveals** both European and Native American influences. Folk arts, such as wood carving, are deeply rooted in Native American traditions. Favorite sports, such as soccer, were brought from Europe. Carved and painted religious statues display the mixing of the two cultures.

Mexican artists and writers have created many national treasures. In the early 1900s, Diego Rivera and his wife, Frida Kahlo, became well-known for their paintings. Carlos Fuentes and Octavio Paz have written works about the values of Mexico's people.

Mexicans enjoy celebrations called fiestas (fee·EHS·tuhs) that include parades, fireworks, music, and dancing. Food is an important part of Mexican fiestas. Tacos and enchiladas are now as popular in the United States as they are in Mexico.

Reading Check **Identifying** What are the sources of Mexico's culture?

Mexican Fiesta

NATIONAL GEOGRAPHIC

▲ Women wearing traditional clothes dance in a parade at a fiesta in Oaxaca, Mexico. ***Place*** **In what other ways do Mexicans celebrate at a fiesta?**

Mexico's Economy and Society

Main Idea **While Mexico's economy is improving, the country still faces significant challenges from poverty, overcrowded cities, and environmental issues.**

Geography and You Have you seen the brown haze of smog? Read to find out how economic growth in Mexico City has contributed to the increase of smog there.

With many resources and workers, Mexico has a growing economy. Mexico has tried to use its resources to improve the lives of its people. Although these efforts have brought some gains for Mexicans, they have also created some challenges for the future.

Economic Regions

Mexico's physical geography and climate together give the country three distinct economic regions. These regions are the North, Central Mexico, and the South.

Mexico's North has large stretches of land that are too dry and rocky to farm without irrigation. So farmers have built canals to carry water to their fields. As a result, they are able to grow cotton, grains, fruits, and vegetables for export. Areas in the North have grasslands that support cattle ranches. Mexican cowhands called **vaqueros** (vah·KEHR·ohs) developed tools and methods for raising cattle during Spanish colonial times. Their skills were later passed on to American cowhands. Vaqueros still carry on this work today.

In addition to farming and ranching, the North profits from rich deposits of copper, zinc, iron, lead, and silver. Manufacturing is located in cities near or along the Mexico–United States border. These cities include Monterrey, Tijuana (tee·HWAH·nah), and Ciudad Juárez (syoo·THAHTH HWAHR·EHS). In the North, many companies from the United States and elsewhere have built **maquiladoras** (muh·KEE·luh·DOHR·uhs). These are factories in which workers **assemble** parts made in other countries. The finished products are then exported to the United States and other countries.

Central Mexico holds more than half of Mexico's people. Although it is situated in the Tropics, this area has a high elevation that keeps it from being hot and humid. Temperatures are mild, and the climate is pleasant year-round. Fertile soil created by volcanic eruptions over the centuries allows for productive farming.

Large industrial cities, such as Mexico City and Guadalajara (GWAH·thuh·lah·HAH·rah), prosper in central Mexico. Workers in these cities make cars, clothing, household items, and electronic goods. The coastal area along the Gulf of Mexico is a center of Mexico's energy industry. This is because of the major oil and gas deposits that lie offshore.

Mexico's South is the poorest economic region. The mountains towering in the center of this region have poor soil. **Subsistence farms,** or small plots where farmers grow only enough food to feed their families, are common here. In contrast, coastal lowlands have good soil and abundant rain. Wealthy farmers grow sugarcane or bananas on **plantations,** large farms that raise a single crop for sale. Both coasts in the South have beautiful beaches and a warm climate. Tourists from all over the world flock to resort cities, such as Acapulco on the Pacific coast and Cancun on the Caribbean coast's Yucatán (YOO·kah·TAHN) Peninsula.

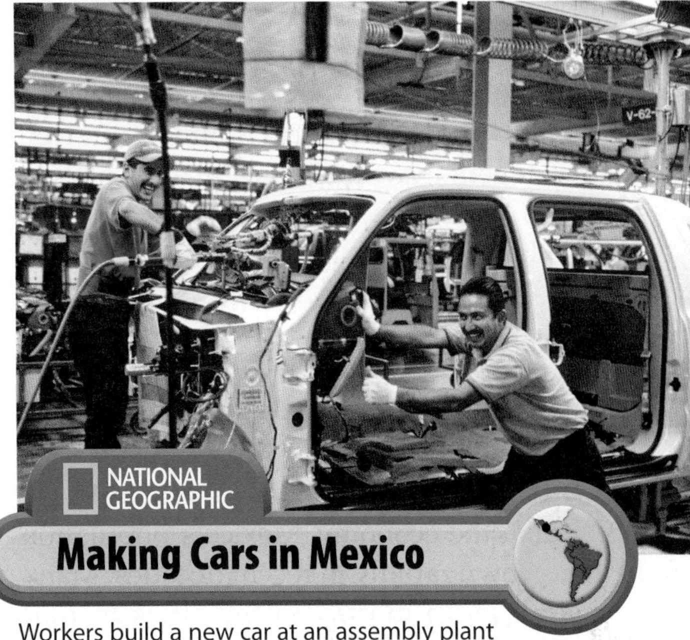

NATIONAL GEOGRAPHIC

Making Cars in Mexico

Workers build a new car at an assembly plant in central Mexico. American and Japanese companies have located hundreds of plants in Mexico. ***Regions*** **What agreement has helped Mexico's economy?**

Economic and Social Changes

For years Mexico's economy relied on agriculture. Today, Mexico still exports food products, but it relies less on farming and more on manufacturing. Much of the change has come about because of Mexico's closer ties with its northern neighbors: the United States and Canada. As you recall, Mexico, the United States, and Canada entered into the North American Free Trade Agreement (NAFTA) in 1994. Under NAFTA, the three countries decided to end barriers to trade among themselves.

Mexico's growing industries now provide a wide range of goods, such as steel, cars, and consumer goods. Many service industries, such as banking and tourism, also contribute to Mexico's economy.

Social Studies ONLINE

Student Web Activity Visit glencoe.com and complete the Chapter 9 Web Activity about economic changes in Mexico.

Economic advances have raised the standard of living, especially in the North. The speed of growth also has brought concerns about damage to the environment, as well as dangers to workers' health and safety.

As Mexico's economy has grown, pollution has increased. For example, the mountains that surround Mexico City trap car fumes and factory smoke. As a result, the city is often covered by unhealthy smog, a thick haze of fog and chemicals.

Population and Ethnic Challenges

Like the economy, Mexico's population has grown rapidly in recent decades. Many people have moved to the cities to find jobs. Because many jobs pay low wages, people have had to live crowded together in slums, or poor sections of the cities.

Many Mexicans who cannot find work become **migrant workers.** These are people who travel to find work when extra help is needed to plant or harvest crops. They legally and sometimes illegally cross Mexico's long border to work in the United States. Despite low pay, migrant workers can earn more in the United States than in Mexico. To reduce illegal immigration, the United States has tightened controls along the border. This has increased tensions with Mexico. Poorer Mexicans depend on money sent from relatives in the United States.

Many of Mexico's Native Americans are poor and live in rural areas. In the 1990s, Native Americans in southern Mexico rose up against the Mexican government. They demanded changes that would improve their lives. By the early 2000s, the struggle between Native Americans and the government had not been resolved.

Determining Cause and Effect Why have many Mexicans migrated to the United States?

Social Studies ONLINE
Study Central™ To review this section, go to glencoe.com.

Vocabulary

1. **Explain** the significance of:
   a. plaza
   b. vaquero
   c. maquiladora
   d. subsistence farm
   e. plantation
   f. migrant worker

Main Ideas

2. **Describing** Describe Mexico's form of government and recent events concerning the government.
3. **Identifying** Use a diagram like the one below to explain the challenges facing Mexico.

Critical Thinking

4. **Determining Cause and Effect** Why is irrigation needed to farm parts of the northern region?
5. **BIG Idea** Compare Mexico's three economic regions.
6. **Challenge** What problems might people in northern Mexico face if maquiladoras are closed, even if they do not work in the maquiladoras?

Writing About Geography

7. **Persuasive Writing** Choose one of the challenges facing Mexico. Write a newspaper editorial in which you suggest steps Mexico's government could take to improve the situation.

Guide to Reading

BIG Idea

The physical environment affects how people live.

Content Vocabulary

- literacy rate *(p. 239)*
- command economy *(p. 240)*
- remittance *(p. 240)*
- commonwealth *(p. 240)*

Academic Vocabulary

- shift *(p. 239)*
- fee *(p. 239)*

Reading Strategy

Comparing and Contrasting Use a Venn diagram like the one below to compare and contrast Guatemala and Costa Rica.

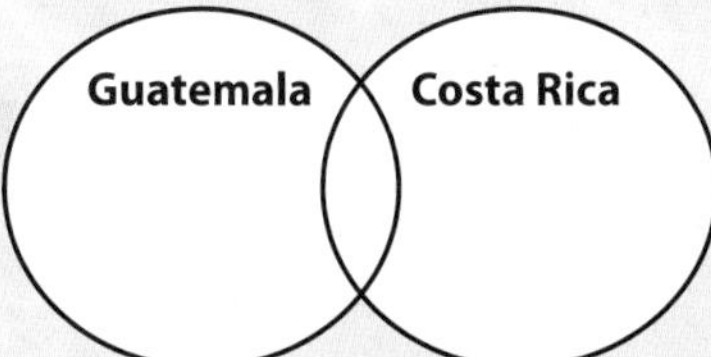

Central America and the Caribbean

Picture This What is it like to glide along a cable 230 feet (70 m) above a lagoon? Tourists can use this method to enjoy the spectacular views of Tiscapa Lagoon and the surrounding forest in Nicaragua. Opportunities like this show why ecotourism is fast replacing coffee, meat, and seafood as Nicaragua's primary source of income. Read this section to learn more about Central America and the Caribbean today.

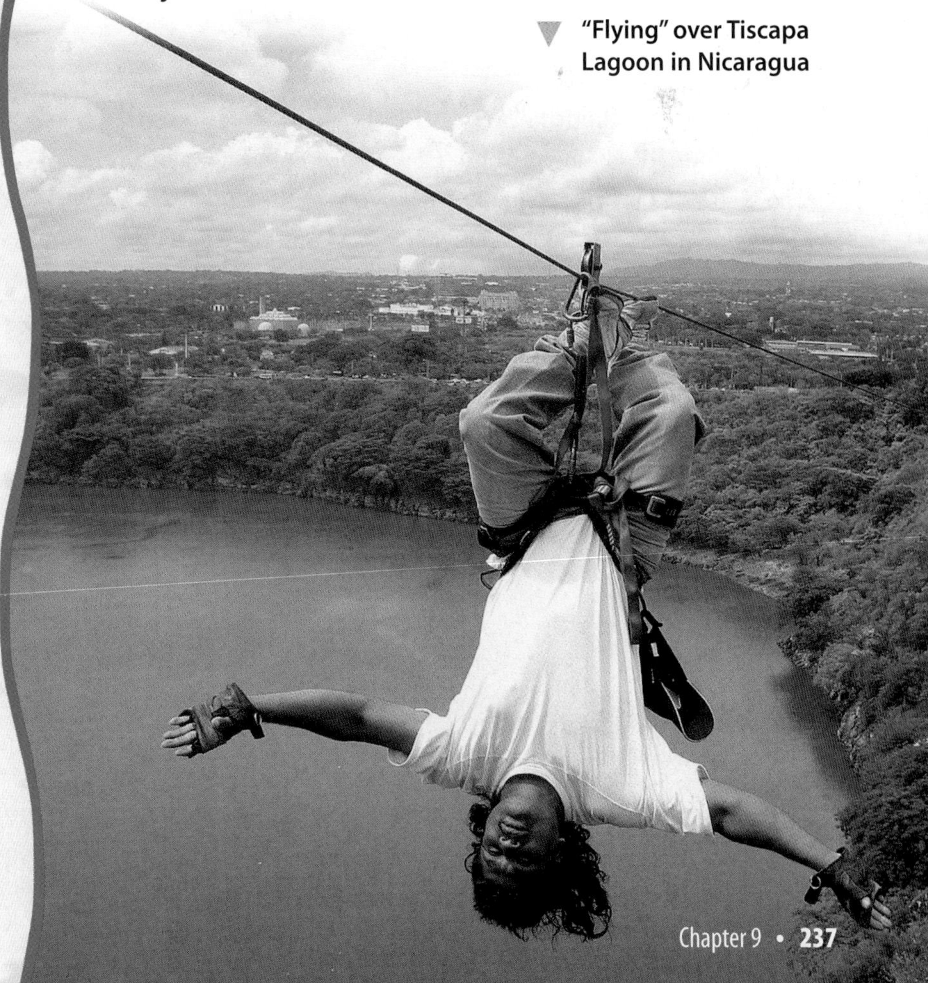

"Flying" over Tiscapa Lagoon in Nicaragua

Countries of Central America

Main Idea **Farming is the main way of life in Central America, where many people are poor.**

Geography and You Do you enjoy eating bananas at breakfast? They might have come from Central America. Read to find out about other ways in which Central Americans use their land and resources.

Central America is made up of seven countries: Belize, Guatemala, El Salvador, Honduras, Nicaragua, Costa Rica, and Panama. Most people in Central America depend on farming. For many decades, they have produced crops, such as bananas, sugarcane, and coffee, for export. In some Central American countries, conflict between ethnic or political groups has held back their economies.

Guatemala

Guatemala is a country of rugged mountains, thick forests, and blue lakes. About half of its people are descended from the ancient Maya. Many Guatemalans are also of mixed Maya and Spanish origin. Maya languages and Spanish are spoken.

Guatemala has fertile volcanic soil. Most of the land is owned by a small group of people who hold most of the wealth and power. During the late 1990s, rebel groups fought the government for control of the land.

TIME GLOBAL CITIZENS

NAME: MARIE CLAIRE PAIZ **HOME COUNTRY:** Guatemala

ACHIEVEMENT: Biologist Marie Claire Paiz directs a major preservation project for The Nature Conservancy in the Maya Forest, one of the world's largest rain forests. Here, for more than 1,000 years, ancient Maya flourished on both sides of the Usumacinta River, which divides Guatemala and Mexico. Maya ruins and writings that date back to 2300 B.C. attract tourists and scientists. Today, however, the forest is being destroyed by farmers clearing the land. And the possible construction of a hydroelectric dam threatens to flood the area. Paiz works to educate Latin Americans about the importance of the site's cultural heritage and what they can do to protect it.

QUOTE: "I hope that through conservation work, the wonders shared by Guatemala and Mexico can endure."

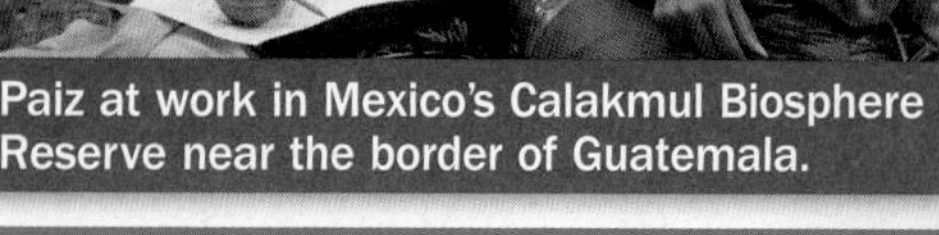

Paiz at work in Mexico's Calakmul Biosphere Reserve near the border of Guatemala.

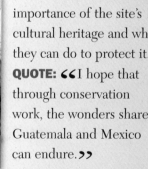

CITIZENS IN ACTION Paiz believes that it is important for human-made structures and nature to exist in harmony. How would that benefit both nature and humans?

When the conflict ended, more than 200,000 people had been killed or were missing.

Guatemala has shown recent signs of economic change. In the past, many farmers produced only bananas and coffee. Today they are **shifting** production to other crops that have higher values, such as fruits, flowers, and spices. In the early 2000s, Guatemala and its Central American neighbors agreed to free trade with the United States. This will remove trade barriers among these countries. Central Americans hope they can then sell more of their goods to the United States.

Costa Rica

Costa Rica has long stood out from its war-torn neighbors. A stable democracy is in place, and no wars have been fought within or outside the country since the 1800s. As a result, Costa Rica has no army—only a police force to keep law and order.

Costa Rica also has fewer poor people than other Central American countries. One reason is that Costa Rica has a higher literacy rate. **Literacy rate** is the percentage of people who can read and write. Workers with reading skills can be more productive and earn higher incomes.

Panama

Panama lies on the narrowest part of Central America. It is best known for the Panama Canal. The canal shortens distance and travel time between the Atlantic and Pacific Oceans. In 1999 the United States gave Panama control of the canal. Today, Panama profits from **fees,** or set charges, that ships pay to use the canal. Because of the commerce brought by the canal, Panama is an important banking center.

Determining Cause and Effect How does literacy rate affect income?

Countries of the Caribbean

Main Idea **Although most Caribbean island countries are poor, several are turning to tourism to help their economies grow.**

Geography and You Do many tourists visit the community where you live? Read to find out how several countries in the Caribbean are turning to tourism to boost their economies.

Many of the island countries of the Caribbean face political and economic challenges. For example, the people of Cuba and Haiti endure great poverty. Puerto Rico, which has connections to the United States, is more stable economically.

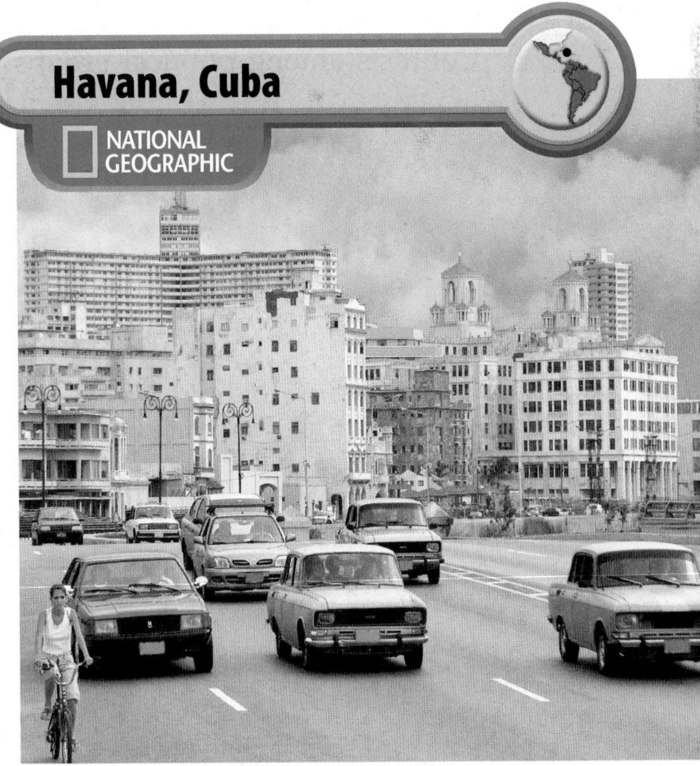

▲ Cars zoom down the Malecón, a famous avenue that runs along the coastline in Havana. ***Regions*** **What major economic challenge do Cuba and other Caribbean nations face?**

Cuba

Cuba lies about 90 miles (145 km) south of Florida. It has a **command economy,** in which the communist government decides how resources are used and what goods and services are produced. Many Cubans have not prospered under this system.

For years, Cuba relied on selling its chief crop, sugar. To end that dependence, Cubans are developing tourism and other industries. Meanwhile, Cuba's communist leaders tightly control society. People who criticize the government are often arrested and jailed. The United States has condemned Cuba for these actions. In 2008 Cuba's aging dictator Fidel Castro stepped down after nearly 50 years in power. With Castro's departure, Cuba's political future is uncertain.

Haiti

Located on the western half of the island of Hispaniola, Haiti has had a troubled history. Conflicts among political groups have made for an unstable government. In addition, most of Haiti's people are poor. A major source of income is **remittances,** or money sent back home by Haitians who work in other countries.

Puerto Rico

Since 1952, Puerto Rico has been a **commonwealth,** or a self-governing territory, of the United States. Puerto Ricans are American citizens. They can come and go as they wish between Puerto Rico and the United States mainland.

Puerto Rico has a high standard of living compared to most other Caribbean islands. It has industries that produce medicines, machinery, and clothes. Farmers there grow sugarcane and coffee. Puerto Rico makes more money from tourism than any other Caribbean island.

Drawing Conclusions Is Cuba's command economy effective?

Social Studies ONLINE
Study Central™ To review this section, go to glencoe.com.

Vocabulary

1. **Explain** the meaning of *literacy rate, command economy, remittance,* and *commonwealth* by using each term in a sentence.

Main Ideas

2. **Analyzing** How is Costa Rica different from its Central American neighbors?
3. **Comparing** Use a chart like the one below to examine the economies of Cuba, Haiti, and Puerto Rico.

Country	Economy
Cuba	
Haiti	
Puerto Rico	

Critical Thinking

4. **Contrasting** How does Cuba's government contrast with Costa Rica's?
5. **BIG Idea** Why are many people on the Caribbean islands poor?
6. **Challenge** Do you think farmers in Guatemala will earn more money by growing different crops? Why?

Writing About Geography

7. **Expository Writing** Write a paragraph explaining how specific countries in this region have achieved some economic success and how they have done it.

TIME PERSPECTIVES

EXPLORING WORLD ISSUES

PROTECTING NATURAL RESOURCES

People are learning how to profit from the land without harming it.

Activists camp out in an Ecuadorian forest to protest an oil pipeline being built through it.

For decades many natural environments and wildlife species have been damaged by human activity. Farmers and loggers in Brazil and Ecuador have cut down or set fire to countless trees in order to acquire the land for farming and other economic activities. Miners in Bolivia have also cleared land in search of minerals. As a result, thousands of miles of forests have disappeared and wildlife populations have suffered. Human activities have also hurt other environments. In Chile, fish farms that raise salmon in large tanks have harmed marine ecosystems.

In recent years, however, people have been working to protect but still profit from natural resources. Governments and citizens are working to limit the damage to forests and wildlife. Industries are developing alternative energy sources that are less harmful to natural environments. But is it too late?

AP PHOTO/DOLORES OCHOA

TIME PERSPECTIVES

EXPLORING WORLD ISSUES

Colorful macaws live in the Amazon rain forest.

ENVIRONMENT-FRIENDLY SOLUTIONS

South America has some of the world's largest and most beautiful forests. From the Amazon rain forest to the woodlands of the Andes mountain ranges, the region's green lush forests are home to many species of wildlife.

In recent decades, however, this fragile environment has changed. Since the 1960s, loggers, miners, and farmers have been clearing the trees from this region's forests. Some people cut down trees to produce wood and paper. Others burn the trees to clear land for mining, farming, and industry. The process of clearing an area of forest is called **deforestation**.

The Destruction of the Amazon Rain Forest

Since 1970, more than 232,000 square miles (600,000 square km) of the Amazon rain forest has been cleared. Here is a look at the amount of deforestation in recent years.

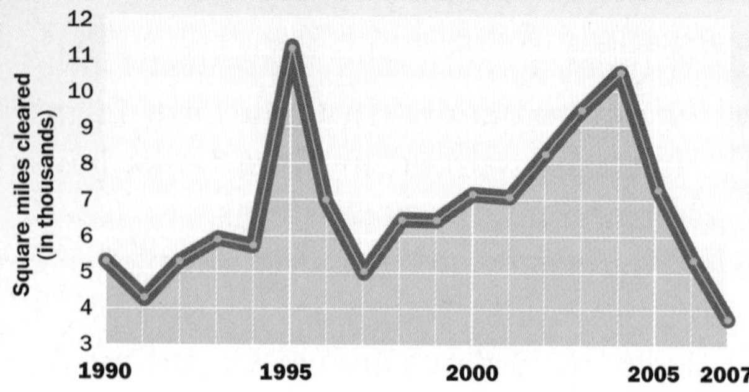

Source: National Institute of Space Research.

INTERPRETING GRAPHS

Analyzing Information During which four-year period did the largest amount of deforestation occur? About how many total square miles were cleared during this time?

Deforestation is a major challenge for South America and the world. Developing countries in this region need the land for industries that will help their economies grow. But deforestation destroys ecosystems and wildlife habitats. Deforestation also contributes to global warming. The burning of wooded areas sends large amounts of carbon dioxide into the atmosphere and speeds up the rate of global warming.

There is much work to be done to protect these lush forests. After years of neglect, the region's governments and citizens are beginning to realize how much is at stake. In recent years, people have been working to reverse decades of damage.

Amazon Alert!

The Amazon rain forest covers about 2.7 million square miles (7 million sq. km) of land in South America—mostly in Brazil. Parrots, jaguars, and piranhas are just some of the thousands of animals that make their home in the Amazon and the many rivers that run through it.

For decades, this tropical environment has been shrinking. In addition to farmers, cattle ranchers, and others clearing the land, the rain forest has also been cut down to make way for roads and highways that crisscross through the center of it. Since 1970, more than 232,000 square miles (600,000 sq. km) of the rain forest has been destroyed.

At a Brazilian ranch, cattle graze on cleared land.

This land was deforested by Brazilian farmers in order to grow soybeans.

The beauty and ecological diversity of the rain forest are at risk.

Stopping the Damage

Brazil's government has been working to preserve the Amazon rain forest. In order to slow the rate of deforestation, Brazil is studying ways to make land that has been cleared more productive. If deforested land can grow more crops or feed more cattle, it will lessen the need for more deforestation.

Legal limits on the amount of land that can be cleared have also been created. However, these laws have not always been enforced. In recent years, though, Brazilian officials are doing a better job at imposing and enforcing laws that protect the Amazon rain forest. Now companies and individuals who ignore the limits are punished with large fines.

Saving Wildlife Populations

Citizens throughout South America are also taking action to protect wildlife. In Chile, fish are often raised on fish farms in giant tanks, called cages. Breeding fish in captivity raises production. Chile is one of the world's biggest exporters of cage-bred salmon. In 2007 the country exported 1.3 million tons of fish (1.2 metric t).

But success has created its share of problems. Fish raised in crowded cages pollute the ocean floor and are prone to illness. Critics of the farms say that the fish are given large amounts of antibiotics and other chemicals to keep them from getting sick. When people eat the fish, the drugs may be passed on to them, which can be unhealthy.

A Chilean worker observes tanks full of farmed salmon.

REUTERS/CARLOS BARRIA

Juan Carlos Cárdenas is the director of Centro Ecocéanos, an organization that works to protect marine life. For years, the center has been working to improve the production methods of Chile's fishing industry. The center teaches local fisheries how to catch more fish using traditional methods. It also conducts research and educates the public about how fish farms affect ecosystems.

Cárdenas says there is still much work to do. He is encouraged that consumers are learning about the health risks associated with eating cage-bred fish. Cárdenas hopes that if people buy fewer farmed fish, the lower sales will force the fish industry to make changes in how it operates.

EXPLORING THE ISSUE

1. **Identifying Cause and Effect** How does deforestation speed up the process of global warming?
2. **Finding Solutions** Human activity can threaten the environment. What can you do in your community to help protect your natural environment?

A SWEET RIDE IN BRAZIL

Brazil's "Flex car" has a sweet tooth. *Flex* is short for "flexible," which describes the kinds of fuel the car uses. The Flex car looks and works like a regular vehicle, but it can run on gasoline or ethanol. Many Brazilians are filling up their gas tanks with ethanol—a fuel that is produced from sugarcane. The alternative fuel is pressed from sugarcane and then blended with gasoline. This "gasohol" mixture could eventually take the place of fossil fuels to keep cars running.

Ethanol-powered cars are not new in Brazil. The country developed them—and the fuel they operate on—in the 1980s, when the cost of buying oil from foreign nations began to soar. Over time, ethanol-powered cars zoomed onto the fast track. By 1988, more than 88 percent of cars sold each year in Brazil were running on a combination of ethanol and gasoline. Throughout Brazil there are now about 29,000 ethanol stations.

Today, Brazil is the world's largest producer of ethanol, and Flex cars are seen everywhere. In 2006 sales of Flex vehicles were higher than sales of cars that ran only on gasoline. Flex car technology is also spreading to other Brazilian industries. Small planes, such as crop dusters, are using ethanol because it is more widely available than conventional aviation fuel.

Added Mileage

Flex cars are also good for the environment and the economy. The ethanol they run on is cleaner than gasoline, so Flex cars create less air pollution. And ethanol is less expensive. Its price is almost half that of gasoline.

As gasoline prices continue to skyrocket, the nations of the world are expected to follow Brazil's example. In 2006 President George W. Bush called for the United States to develop more ethanol. "There is an enormous demand from abroad to know more," said the president of Brazil's carmakers' association. "This is an opportunity for Brazil." Perhaps it will be an opportunity for the rest of the world to have a sweet ride, too.

REUTERS/JAMIL BITTAR

A Flex car

EXPLORING THE ISSUE

1. **Explaining** Why did Brazil develop ethanol as an alternative fuel?
2. **Identifying Cause and Effect** How might Brazil's success with ethanol inspire other nations to develop and use alternative fuels?

This ethanol distillery produces fuel from sugarcane.

REVIEW AND ASSESS

UNDERSTANDING THE ISSUE

1 **Making Connections** How does deforestation affect wildlife?

2 **Writing to Inform** In a short article, explain some of the ways governments are working to preserve the natural environments in their countries.

3 **Writing to Persuade** In a letter to an editor of a newspaper, discuss your beliefs about driving vehicles that use alternative fuels.

INTERNET RESEARCH ACTIVITIES

4 Go to www.savethehighseas.org, the Web site of the Deep Sea Conservation Coalition. Click the "About Us" link and scroll down to the "Coalition Steering Group Members." Read about some of the organizations and how they work to protect marine ecosytems. Write a short essay describing one of these activities.

5 With your teacher's help, do an online search on alternative fuel sources, such as ethanol or solar power. Read about how the nations of the world are developing these energy sources. Write a brief article that explains your findings.

BEYOND THE CLASSROOM

6 **Work in groups** to create and display an ecological mural on paper that illustrates how people can protect natural environments in your community.

7 **At your school or local library,** research what other countries are doing to decrease their dependency on foreign oil imports. Do you think their strategies will succeed? Why or why not?

Major Producers of Ethanol

Ethanol can be made from sugarcane and corn. As oil prices soar, the nations of the world are expected to produce more ethanol. Here is a look at major producers in 2006.

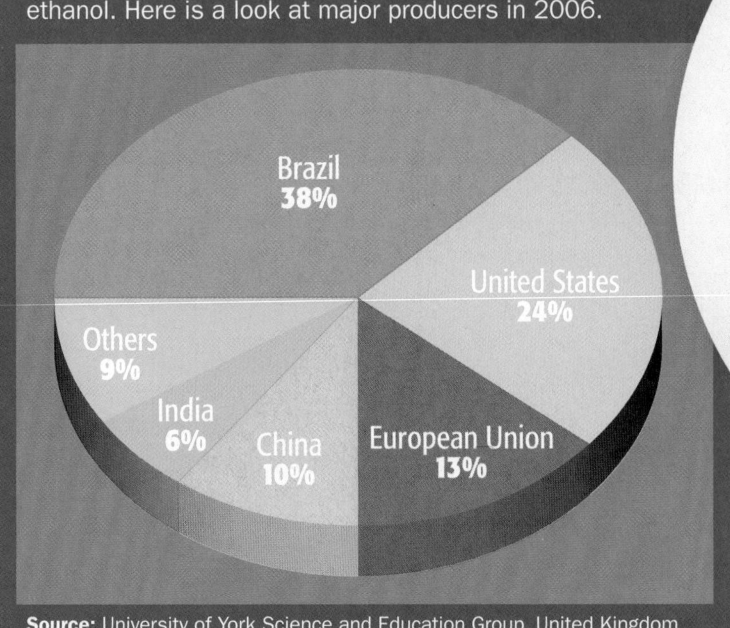

Source: University of York Science and Education Group, United Kingdom.

Building Graph Reading Skills

1. **Analyzing Data** What percentage of the world's ethanol is produced by the United States and the European Union?

2. **Identifying Cause and Effect** Brazil is the world's largest producer of ethanol. As the world searches for less expensive energy sources, how might Brazil's top ranking help its economy grow?

Guide to Reading

BIG Idea

People's actions can change the physical environment.

Content Vocabulary

- selva *(p. 247)*
- favela *(p. 247)*
- gaucho *(p. 249)*
- national debt *(p. 250)*
- default *(p. 250)*
- sodium nitrate *(p. 252)*

Academic Vocabulary

- maintain *(p. 248)*
- issue *(p. 248)*

Reading Strategy

Identifying Central Issues Use a diagram like the one below to describe Brazil's economy. Write the main idea on the line to the left and supporting details on the lines to the right. You can add as many additional lines as you have details.

South America

Section Audio

Spotlight Video

Picture This This giant dish is like an "eye" studying the universe. The Swedish ESO (European Southern Observatory) Submillimeter Telescope, or SEST, is not like some telescopes that use light from stars or planets to "see" them. SEST is able to study distant objects by gathering radio waves that radiate from them. The telescope is located in the southern Atacama Desert in Chile, where the clear sky conditions are ideal for this type of research. To learn more about South America today, read Section 3.

Learning about the universe in Chile

Brazil

Main Idea **Brazil is a leading economic power, but concerns have grown about its use of the Amazon rain forest.**

Geography and You Did you know that some of the best farmland in the United States was once forestland? The forests were cleared by farmers. Read to find out how Brazil's forests are being cut down for mining, logging, and farming.

Brazil is the fifth-largest country in the world and the largest in South America. The country is known for its Amazon rain forest, which Brazilians call the **selva.** This resource is threatened by Brazil's economic growth.

Brazil's People

With 187 million people, Brazil has the largest population of all Latin American countries. Brazil's culture is largely Portuguese because they were the first and largest European group to settle Brazil. Today Brazilians are of European, African, Native American, Asian, or mixed ancestry. Almost all of them speak a Brazilian form of Portuguese, which includes many words from Native American and African languages.

Most of Brazil's people live in cities along the Atlantic coast. São Paulo and Rio de Janeiro are among the largest cities in the world. In recent years, millions of Brazilians have moved from rural areas to coastal cities to find better jobs. Many of these migrants have settled in favelas. **Favelas** are overcrowded slum areas that surround many Brazilian cities. To reduce city crowding, the government now encourages people to move back to less-populated, inland areas. In 1960 Brazil moved its capital from Rio de Janeiro to the newly built city of Brasília 600 miles (966 km) inland. With more than 2 million people, Brasília is a modern and rapidly growing city.

NATIONAL GEOGRAPHIC

Harvesting Sugarcane

A truck is loaded with sugarcane at a farm in southeastern Brazil. ***Human-Environment Interaction*** **In what unique way does Brazil make use of its sugarcane?**

Brazil's Economy

Brazil is one of the world's leading producers of food crops. It grows more coffee, oranges, and cassava than any other country. Brazil's agricultural output has grown greatly in recent years. This is partly because Brazilian farmers have cleared more land in rain forest areas to grow crops. They also now use machinery to perform many tasks. Finally, farmers have planted crops that have been scientifically changed to produce more and to prevent disease.

In addition to productive farms, Brazil has valuable mineral resources, such as iron ore, bauxite, tin, manganese, gold, silver, and diamonds. Offshore deposits of oil, as well as hydroelectric power from rivers, supply the country with energy. Brazil also uses sugarcane to produce a substitute for gasoline.

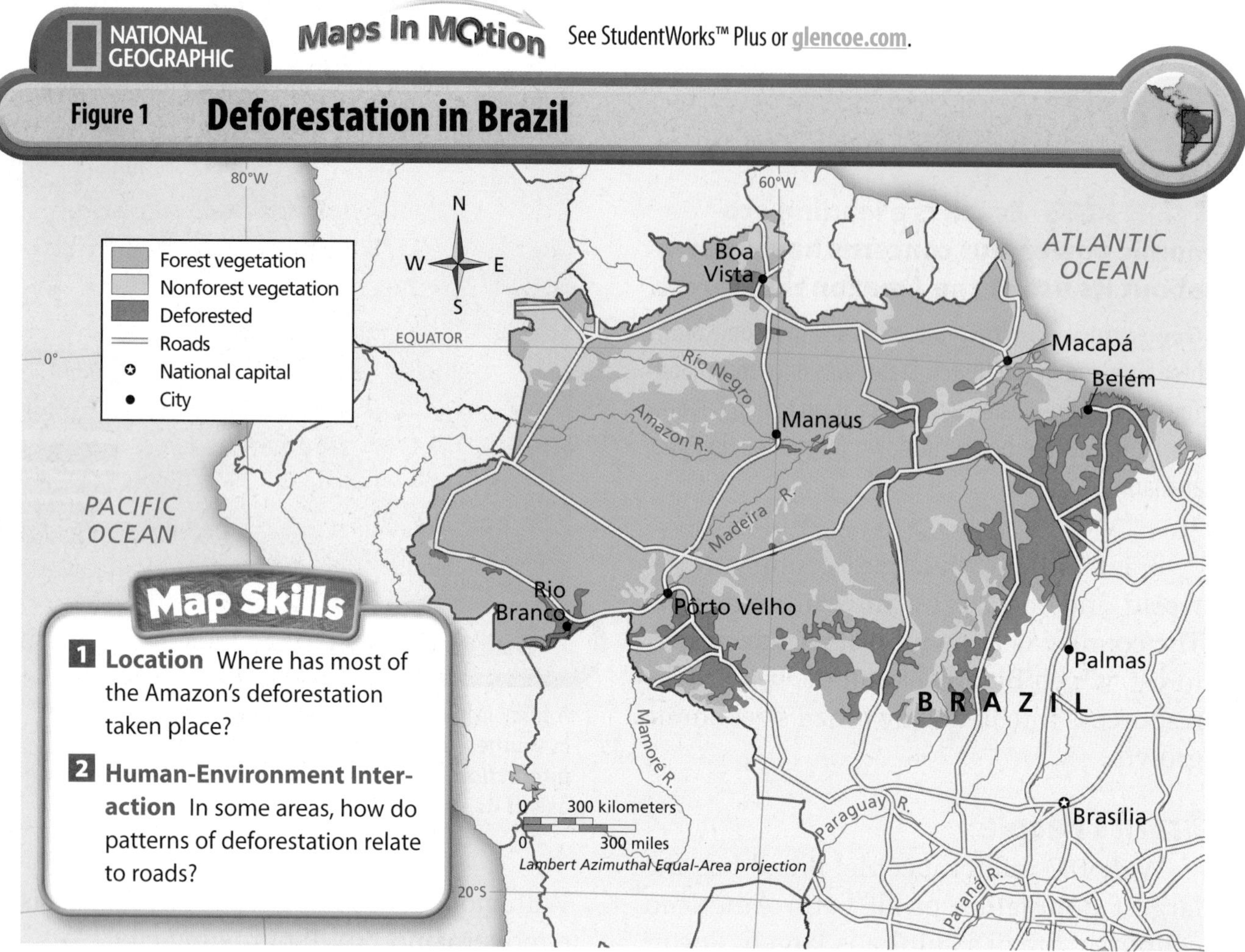

Figure 1 Deforestation in Brazil

Brazil has successful industries too. Most manufacturing takes place in São Paulo and other southeastern cities. Factory workers produce heavy industrial goods, such as machinery, airplanes, and cars. They also make food products, medicines, paper, and clothing.

The Rain Forest

Brazil's greatest natural resource is the Amazon rain forest. It is the world's largest rain forest area, yet it also has the highest rate of deforestation. Each year, the land deforested in the Amazon rain forest is equal in size to Ohio. **Figure 1** shows how much of the rain forest has been lost.

Why is the rain forest shrinking? To increase jobs and make products for export, Brazil's government has encouraged mining, logging, and farming in the rain forest. These activities lead to soil erosion and harm the rain forest's ecosystem and biodiversity.

As deforestation takes place, roads are built, bringing companies, farmers, and change. Native Americans who live in the rain forest find it difficult to follow their traditional cultures as this occurs.

In addition, tropical forests give off huge amounts of oxygen and play a role in **maintaining,** or keeping up, the Earth's climate patterns. They also provide shelter to many wildlife species that may not survive if deforestation continues. Thus, although the rain forest belongs to Brazil, the effects of deforestation are felt worldwide. Because deforestation is a global **issue,** or problem, other nations have convinced Brazil to protect at least part of the rain forest from economic development.

Brazil's Government

Brazil declared independence from Portugal in 1822. During most of the 1800s, emperors ruled the country. Today Brazil is a democratic federal republic, in which people elect a president and other leaders. Brazil has many political parties, not just two main ones, as does the United States.

The national government of Brazil is much stronger than its 26 state governments. Like the United States, Brazil's national government has three branches. The president heads the executive branch, which carries out the laws. The National Congress, which is similar to the U.S. Congress, makes the laws. A Supreme Federal Tribunal, or court, heads a judicial system that interprets the laws.

Reading Check **Analyzing Information** Why has Brazil's agricultural output greatly increased?

NATIONAL GEOGRAPHIC

Brazil's Government

▲ Brazil's President Luiz Inacio Lula da Silva (left) talks with Governor Rosinha Garotinho of the state of Rio de Janeiro. ***Place*** **How is Brazil's government like that of the United States?**

Argentina

Main Idea **Argentina has experienced harsh military rule but now has a democratic government.**

Geography and You How would you feel if the government seized a member of your family and you never saw him or her again? Read to find out how Argentina went through a period of violent rule in recent decades.

Argentina is South America's second-largest country after Brazil. It is about the size of the United States east of the Mississippi River. The Andes tower over western Argentina. South and east of the Andes lies a dry, windswept plateau called Patagonia. The center of Argentina has vast treeless plains known as the Pampas. More than two-thirds of Argentina's people live in this central area.

Argentina's People

About 85 percent of Argentina's people are of European ancestry, especially Spanish and Italian. European cultural traditions are stronger in Argentina than in most other Latin American countries.

The majority of people in Argentina are city dwellers. In fact, more than one-third of the country's population lives in the capital, Buenos Aires. This bustling city is a seat of government, a busy port, and a center of culture. Buenos Aires resembles a European city with its parks, beautiful buildings, wide streets, and cafés. It has been nicknamed "the Paris of the South."

Argentina's Economy

Argentina's economy depends heavily on farming and ranching. Huge ranches cover the Pampas. There, **gauchos** (GOW·chohs), or cowhands, raise livestock. Gauchos are Argentina's national symbol.

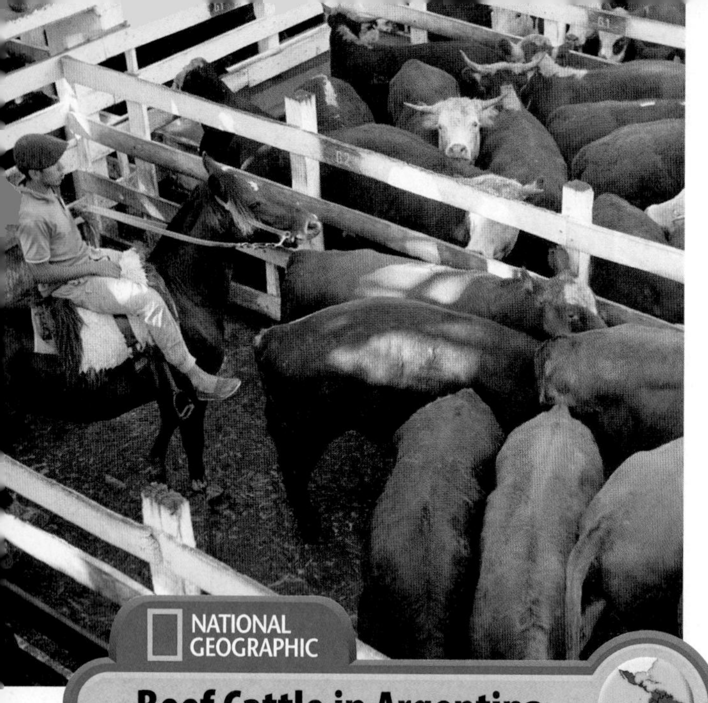

Beef Cattle in Argentina

Beef plays an important role in Argentina's foreign trade. Earnings from exports of animal products were about 1.9 billion dollars a year in the early 2000s. ***Location*** **Where are most of Argentina's livestock raised?**

They are admired for their independence and horse-riding skills. The livestock that the gauchos herd and tend are a vital part of the economy. Beef and beef products are Argentina's chief exports.

Argentina is one of the most industrialized countries in South America. Most factories are in or near Buenos Aires. They produce food products, cars, chemicals, and textiles. Zinc, iron ore, and copper are mined in the Andes. Oil fields also lie in the Andes as well as in Patagonia.

Despite these resources, Argentina's economy has struggled. To help its economy grow, Argentina borrowed money from foreign banks during the late 1900s. However, this led to a high **national debt,** or money owed by the government. A few years ago, Argentina had to default on its debts to the foreign banks. To **default** is to miss a debt payment to the company or person who lent the money. People in other countries stopped investing money in Argentina's businesses. This caused a severe economic slowdown in Argentina. Recently the economy has recovered, and the government has paid off part of the debt.

Argentina's Government

After independence in the early 1800s, Argentina was torn apart by civil war. By the mid-1850s, a strong national government had emerged, and Argentina prospered. During the early 1900s, though, the economy suffered, and the military took over. One of the military leaders, Juan Perón, became a dictator in the late 1940s. Perón tried to improve the economy and to help the workers. At the same time, he restricted freedom of speech and the press. These actions made people unhappy. In 1955 a revolt drove Perón from power and restored democracy.

Military officers again took control in the 1970s. They ruled harshly and secretly seized and killed thousands of people they believed opposed their policies. The families of these people did not know what had happened to them. It was a time of fear.

In 1982 Argentina suffered defeat in a war with the United Kingdom over control of the Falkland Islands. The Falklands, known to Argentinians as the Malvinas, lie in the Atlantic Ocean. After this loss, military leaders gave up power, and elected leaders gained control of the government.

Today, Argentina is a democratic federal republic. It has a central government and 23 state governments. A legislature with two houses makes the laws. A Supreme Court heads a system of judges. The nation is led by a powerful president elected every four years. In 2007, Cristina Fernandez was elected Argentina's first woman president.

Reading Check **Explaining** Why are food products among the leading manufactured items in Argentina?

Other Countries of South America

Main Idea **Economic growth for other countries of South America has been hindered by political and social troubles.**

Geography and You Can you recall hard times and good times in your life? Read on to learn which nations in South America are experiencing hard times and which are experiencing good times.

Many countries in South America face the same challenges as Brazil and Argentina. Some, such as Venezuela, Colombia, and Chile, have relatively strong economies. Others, however, face more difficult economic hardships.

Venezuela

Venezuela lies along the Caribbean Sea in northern South America. It is one of the world's leading producers of oil and natural gas. Although it relies mainly on oil production, Venezuela also benefits from mining bauxite, gold, diamonds, and emeralds. The country's factories make steel, chemicals, and food products. Farmers grow sugarcane and bananas or raise cattle. Most Venezuelans are poor, and some live in slums that sprawl over the hills around the capital, Caracas.

In 1998 Venezuelans elected a former military leader, Hugo Chávez, as president. Chávez promised to use oil money to better the lives of Venezuela's poor. His strong rule, however, split the country into opposing groups. Chávez also tried to spread his influence overseas. He became friendly with Cuba's leader, Fidel Castro, and frequently criticized the United States.

Colombia

Venezuela's neighbor, Colombia, has coasts on both the Caribbean Sea and the Pacific Ocean. The Andes rise in the western part of Colombia. Nearly 80 percent of Colombia's people live in the valleys and highland plateaus of the Andes. Bogotá (BOH·goh·TAH), the capital and largest city, lies on one of these plateaus.

Colombia has many natural resources, such as coal, oil, and copper. It is the world's leading supplier of emeralds. Colombian coffee, a major export, is famous for its rich flavor. Colombia also exports bananas, sugarcane, rice, and cotton.

Despite these economic strengths, Colombia has much political unrest. Wealth remains in the hands of a few, and many people are poor. Since the 1970s, rebel forces have fought the government and now control parts of the country.

Medical Technology

NATIONAL GEOGRAPHIC

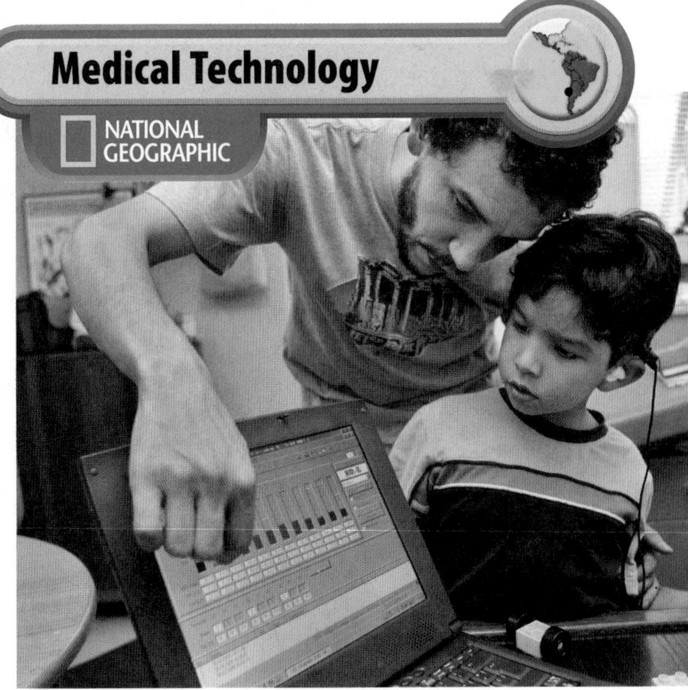

▲ A six-year-old boy hears for the first time as a result of new medical equipment provided by Chile's government. ***Regions*** **In addition to Chile, what other South American countries have relatively strong economies?**

Drug dealers are a major problem in Colombia. The dealers pay farmers to grow coca leaves, which are used to make the illegal drug cocaine. Much of the cocaine is smuggled into the United States and Europe. Drug dealers have used their profits to build private armies. The United States has lent Colombia support in an effort to break the power of the drug dealers.

Chile

Chile lies along the southern Pacific coast of South America. It has an unusual ribbonlike shape that is 2,652 miles long (4,268 km) and an average of 110 miles (177 km) wide. Chile's landscapes range from extremely dry desert in the north to ice formations in the south.

In recent years, Chile has had strong economic growth. Mining forms the backbone of Chile's economy. Chile is a major world producer of copper. It also mines and exports gold, silver, iron ore, and **sodium nitrate,** a mineral used in fertilizer and explosives.

Agriculture is also a major economic activity. Farmers produce wheat, corn, beans, sugarcane, and potatoes. The grapes and apples you eat in winter may come from Chile's summer harvest. Many people also raise cattle, sheep, and other livestock. Northern Chile's fishing industry is the largest in South America.

Like Argentina, Chile has emerged from a long period of military rule. During that time, the government treated its opponents harshly. Today, Chile is a democracy. In 2006 Michelle Bachelet was elected the country's first woman president.

Reading Check **Identifying** What resource is especially important to Venezuela?

Section 3 Review

Social Studies ONLINE
Study Central™ To review this section, go to glencoe.com.

Vocabulary

1. **Explain** the significance of:
   a. selva **b.** favela **c.** gaucho **d.** national debt **e.** default **f.** sodium nitrate

Main Ideas

2. **Explaining** In what ways has Brazil improved its economy?
3. **Sequencing** Use a diagram like the one below to show changes in Argentina's government following independence.

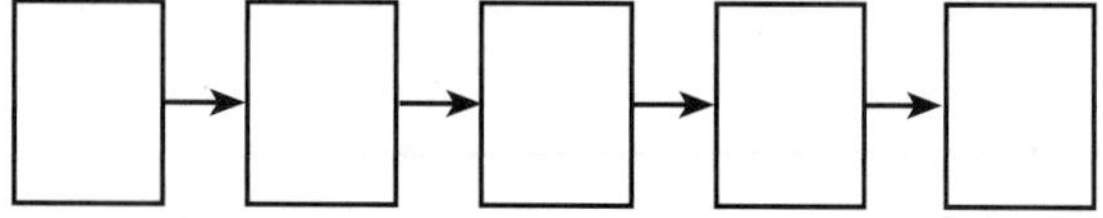

4. **Describing** Describe the problem of illegal drugs in Colombia.

Critical Thinking

5. **BIG Idea** How are Brazilians changing the rain forest, and why does that matter to people in other areas of the world?
6. **Challenge** Do you think Venezuela is likely to suffer from focusing on one major product? Why or why not?

Writing About Geography

7. **Using Your FOLDABLES™** Use your Foldable to write a paragraph comparing the roles that two governments of South America play in economic affairs. Be sure to analyze how effective you think their governments are.

Visual Summary

Mexico

- Mexico City is one of the world's largest cities.
- Mexico's culture reflects both European and Native American influences.
- Industry and farming dominate Mexico's North; agriculture leads in the South.
- Many Mexicans have migrated to cities and to the United States to find jobs.

Subsistence farming, Mexico

Central America and the Caribbean

- Civil wars have held back economic growth in parts of Central America.
- Costa Rica's citizens have a high literacy rate and enjoy a stable government.
- The Panama Canal enables ships to pass between the Atlantic and Pacific Oceans.
- Many Caribbean islands' economies rely on tourism.

Logs from Amazon forest, Brazil

Brazil

- Brazil is the biggest and most populous country in South America.
- Brazil's people, who speak Portuguese, are a mix of many different ethnic backgrounds.
- Brazil has many resources and a productive economy.
- Economic development threatens the Amazon rain forest.

Cruise ship docked in the British Virgin Islands

Argentina

- A large grassland called the Pampas covers much of Argentina.
- Argentina's economy depends on farming and ranching.
- More than a third of Argentina's people live in the capital, Buenos Aires.
- After years of military rule, Argentina is today a democracy.

Oil rig in Venezuela

Other Countries of South America

- Venezuela has relied on oil wealth to build a stronger economy.
- Colombia has been weakened by political unrest and illegal drug trade.
- Chile's economy depends on the export of copper and agricultural products.

Study anywhere, anytime! Download quizzes and flash cards to your PDA from glencoe.com.

CHAPTER 9 ASSESSMENT

STANDARDIZED TEST PRACTICE

TEST-TAKING TIP

Before answering essay questions, jot down a list of things you want to discuss.

Reviewing Vocabulary

Directions: Choose the word(s) that best completes the sentence.

1. Main government buildings in Mexican cities are located around central squares called ______.

A plazas

B murals

C selvas

D favelas

2. ______ are small plots where farmers grow only enough food for their families.

A Plantations

B Maquiladoras

C Commonwealths

D Subsistence farms

3. An important source of income in Haiti is ______.

A canal-use fees

B the literacy rate

C remittances

D the command economy

4. ______ take care of the livestock on ranches in Argentina.

A Gauchos

B Vaqueros

C Farmers

D Migrant workers

Reviewing Main Ideas

Directions: Choose the best answer for each question.

Section 1 *(pp. 232–236)*

5. Which expression of Mexican culture is rooted in Native American traditions?

A soccer

B bullfighting

C wood carving

D public squares

6. What helped change the Mexican economy in recent years?

A farmers' small plots

B foreign-built factories

C large sugarcane farms

D resort cities along the coast

Section 2 *(pp. 237–240)*

7. What country makes more money in tourism than any other country in the Caribbean?

A Cuba

B Haiti

C Puerto Rico

D El Salvador

Section 3 *(pp. 246–252)*

8. Why is deforestation in Brazil felt worldwide?

A Farmers grow crops in the rain forest.

B Native Americans live in the rain forest.

C The rain forest has large deposits of oil.

D The rain forest helps maintain climate patterns.

GO ON

Critical Thinking

Directions: Base your answers to questions 9 and 10 on the graph below and your knowledge of Chapter 9. Choose the best answer for each question.

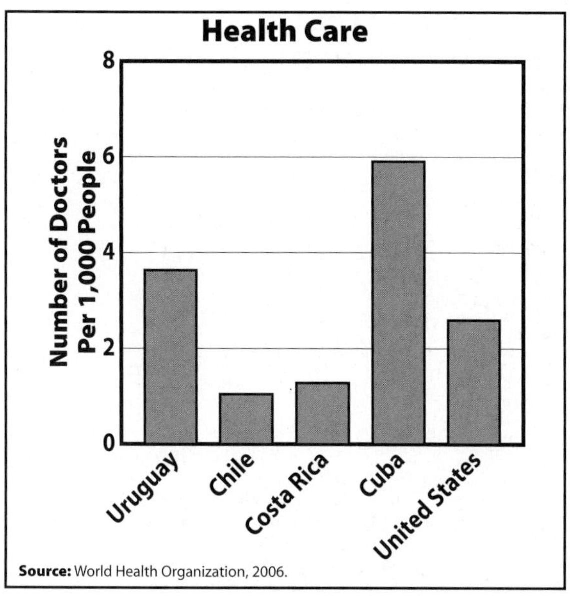

9. Which of the following countries has the fewest doctors per thousand population?

A Cuba

B Chile

C Uruguay

D Costa Rica

10. Which of the following generalizations does the graph support?

A Chile has more disease than the other countries.

B Chile is healthier than the other countries.

C Cuba has more doctors per 1,000 people than either the United States or Uruguay.

D The United States has more doctors per 1,000 people than all Latin American countries.

Document-Based Questions

Directions: Analyze the following document and answer the short-answer questions that follow.

The following passage discusses a movement that has helped strengthen Argentina's economy.

> *Since 1972 the Mil Hojas pasta factory [in Argentina] has churned out delicacies like ravioli and Italian desserts. But Mil Hojas' fortunes—along with those of the national economy—began to decline with the late 1990s as deep recession set in.*
>
> *The factory owners decided to abandon it amid a national epidemic of bankruptcies. Mil Hojas, like many other factories in Argentina, was to permanently close its doors.*
>
> *That was when its workers decided to act. They took back, or "recovered" Mil Hojas, transforming it into what today is a thriving cooperative, as Argentina emerges from one of the worst economic crises in its history.*
>
> *Today, thousands of workers are reactivating previously closed factories on their own terms and . . . breathing life into the national economy.*
>
> —Eduardo Stanley, "Argentina's Recovered Factories: A Story of Economic Success"

11. What happened to factories during economic hard times in Argentina?

12. Were the workers' actions consistent with a command economy or with free enterprise? Explain.

Extended Response

13. Compare and contrast the political systems of Brazil and the United States.

STOP

For additional test practice, use Self-Check Quizzes—Chapter 9 at glencoe.com.

Need Extra Help?

If you missed question. . .	1	2	3	4	5	6	7	8	9	10	11	12	13
Go to page. . .	233	235	240	249	234	235	240	248	239	239	249	249	249

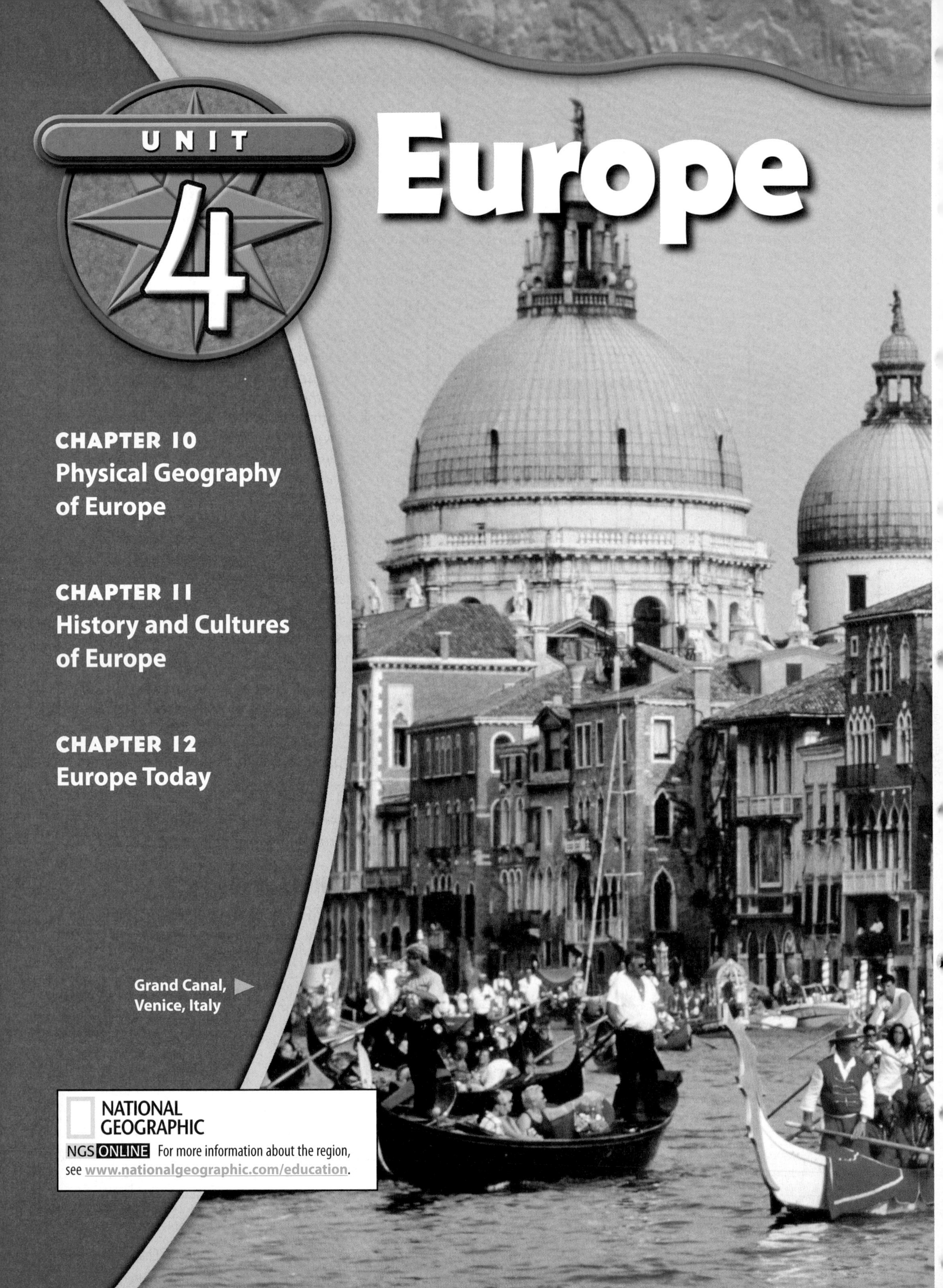

UNIT
4
Europe
CHAPTER 10
Physical Geography of Europe
CHAPTER 11
History and Cultures of Europe
CHAPTER 12
Europe Today
NATIONAL GEOGRAPHIC
NGS ONLINE For more information about the region, see www.nationalgeographic.com/education.

Grand Canal, Venice, Italy

VENEZIA

Regional Atlas

Europe

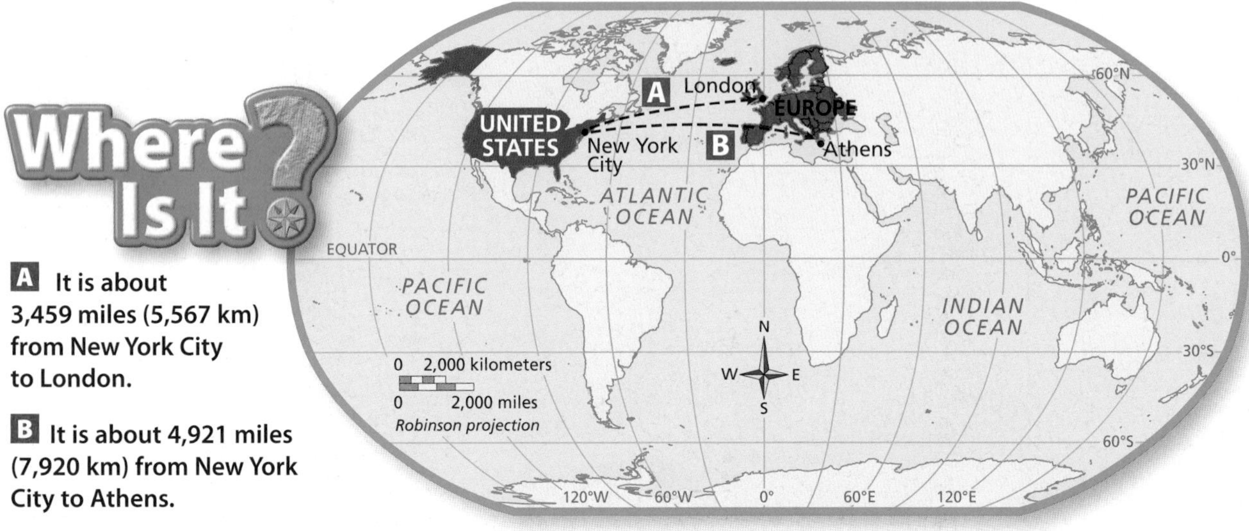

A It is about 3,459 miles (5,567 km) from New York City to London.

B It is about 4,921 miles (7,920 km) from New York City to Athens.

At about 2.3 million square miles (5.9 million sq. km), Europe is about three-fourths the size of the United States. More than 580 million people—almost twice the population of the United States—live in this area, which is one of the most densely populated regions on Earth.

Comparing Population

United States and Selected Countries of Europe	
United States	
Germany	
France	
United Kingdom	
Italy	
Ukraine	
Czech Republic	

(icon) = 30,000,000

Source: *World Population Data Sheet*, 2005.

GEO Fast Facts

Largest Island

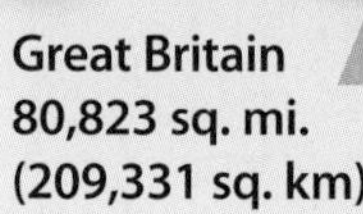

Great Britain
80,823 sq. mi.
(209,331 sq. km)

Largest Lake

Lake Vänern (Sweden)
2,156 sq. mi.
(5,584 sq. km)

Highest Point

Mont Blanc (France/Italy)
15,771 ft. (4,807 m) high

Longest River

Danube River
1,771 mi.
(2,850 km) long

UNIT 4

Regional Atlas

Europe

PHYSICAL

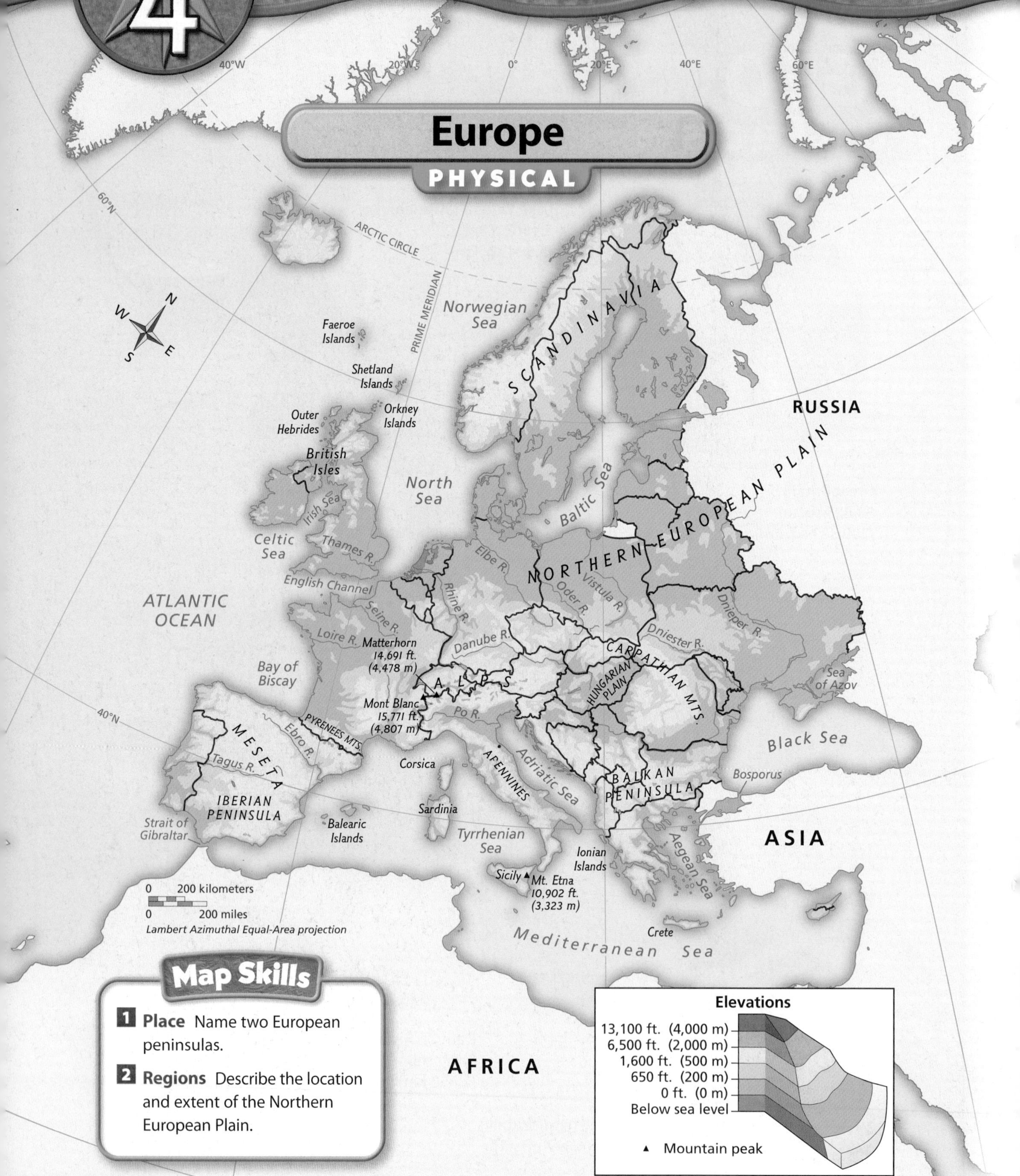

0 200 kilometers
0 200 miles
Lambert Azimuthal Equal-Area projection

Map Skills

1. **Place** Name two European peninsulas.
2. **Regions** Describe the location and extent of the Northern European Plain.

Elevations

13,100 ft. (4,000 m)
6,500 ft. (2,000 m)
1,600 ft. (500 m)
650 ft. (200 m)
0 ft. (0 m)
Below sea level

▲ Mountain peak

Matterhorn 14,691 ft. (4,478 m)
Mont Blanc 15,771 ft. (4,807 m)
Mt. Etna 10,902 ft. (3,323 m)

Map Skills

1. **Location** What countries are found on the British Isles?
2. **Movement** What two rivers could people use to ship goods from Germany to the North Sea?

UNIT 4

Regional Atlas

Europe

POPULATION DENSITY

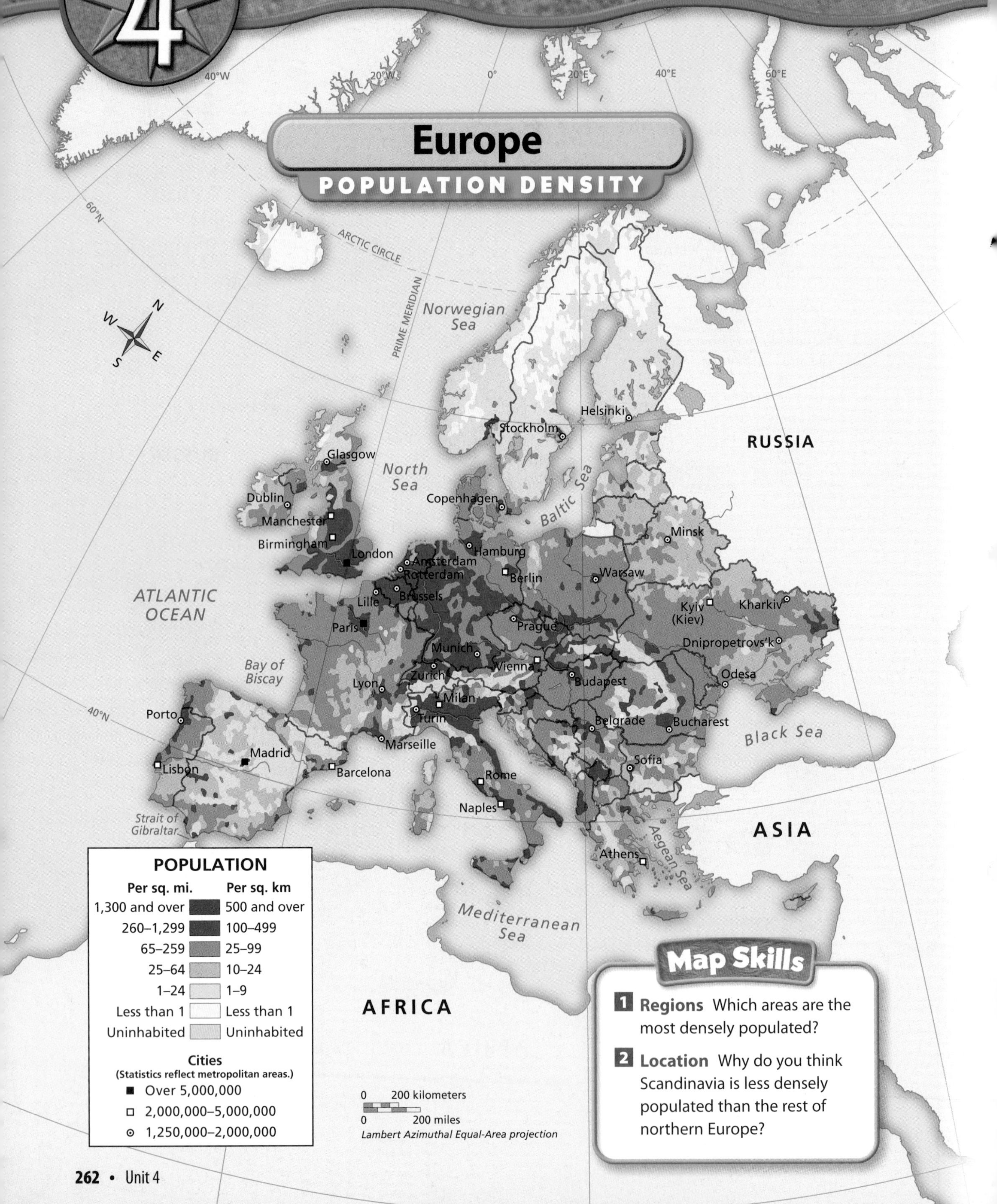

POPULATION

Per sq. mi.	Per sq. km
1,300 and over	500 and over
260–1,299	100–499
65–259	25–99
25–64	10–24
1–24	1–9
Less than 1	Less than 1
Uninhabited	Uninhabited

Cities
(Statistics reflect metropolitan areas.)

- ■ Over 5,000,000
- □ 2,000,000–5,000,000
- ⊙ 1,250,000–2,000,000

Map Skills

1. **Regions** Which areas are the most densely populated?
2. **Location** Why do you think Scandinavia is less densely populated than the rest of northern Europe?

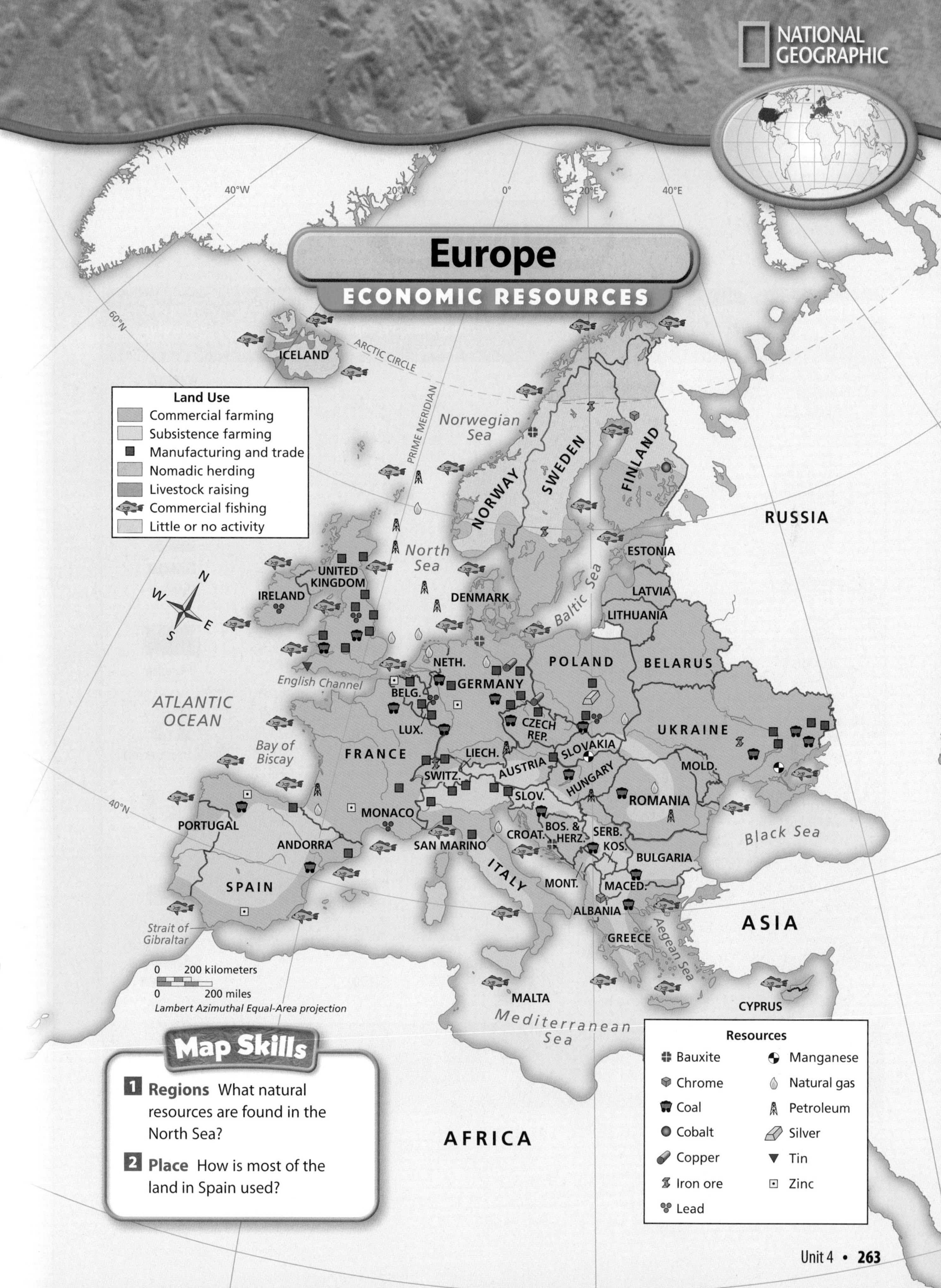

Map Skills

1 Regions What natural resources are found in the North Sea?

2 Place How is most of the land in Spain used?

Regional Atlas

Europe

Country and Capital	Literacy Rate	Population and Density	Land Area	Life Expectancy (Years)	GDP* Per Capita (U.S. dollars)	Television Sets (per 1,000 people)	Flag and Language
Tiranë, ALBANIA	86.5%	3,200,000 286 per sq. mi. 111 per sq. km	11,100 sq. mi. 28,749 sq. km	74	$4,900	146	Albanian
ANDORRA, Andorra la Vella	100%	100,000 426 per sq. mi. 222 per sq. km	174 sq. mi. 451 sq. km	—	$26,800	440	Catalan
Vienna, AUSTRIA	98%	8,200,000 252 per sq. mi. 98 per sq. km	32,378 sq. mi. 83,859 sq. km	79	$31,300	526	German
Minsk, BELARUS	99.6%	9,800,000 122 per sq. mi. 47 per sq. km	80,154 sq. mi. 207,598 sq. km	69	$6,800	331	Belarusian
Brussels, BELGIUM	98%	10,500,000 887 per sq. mi. 344 per sq. km	11,787 sq. mi. 30,528 sq. km	79	$30,600	532	Dutch, French, German
BOSNIA AND HERZEGOVINA, Sarajevo	—	3,800,000 195 per sq. mi. 74 per sq. km	19,741 sq. mi. 51,129 sq. km	74	$6,500	112	Bosnian
BULGARIA, Sofia	98.6%	7,700,000 181 per sq. mi. 69 per sq. km	42,822 sq. mi. 110,908 sq. km	72	$8,200	429	Bulgarian
Zagreb, CROATIA	98.5%	4,400,000 203 per sq. mi. 78 per sq. km	21,830 sq. mi. 56,539 sq. km	75	$11,200	286	Croatian
UNITED STATES, Washington, D.C.	97%	296,500,000 80 per sq. mi. 31 per sq. km	3,717,796 sq. mi. 9,629,047 sq. km	78	$40,100	844	English

*Gross Domestic Product

Countries and flags not drawn to scale

Europe

Country and Capital	Literacy Rate	Population and Density	Land Area	Life Expectancy (Years)	GDP* Per Capita (U.S. dollars)	Television Sets (per 1,000 people)	Flag and Language
CYPRUS Nicosia	97.6%	1,000,000 270 per sq. mi. 108 per sq. km	3,571 sq. mi. 9,249 sq. km	77	$20,300	154	Greek
CZECH REPUBLIC Prague	99.9%	10,200,000 335 per sq. mi. 129 per sq. km	30,448 sq. mi. 78,860 sq. km	75	$16,800	487	Czech
DENMARK Copenhagen	100%	5,400,000 326 per sq. mi. 125 per sq. km	16,637 sq. mi. 43,090 sq. km	77	$32,200	776	Danish
ESTONIA Tallinn	99.8%	1,300,000 77 per sq. mi. 29 per sq. km	17,413 sq. mi. 45,099 sq. km	72	$14,300	567	Estonian
FINLAND Helsinki	100%	5,200,000 40 per sq. mi. 15 per sq. km	130,560 sq. mi. 338,149 sq. km	79	$29,000	643	Finnish, Swedish
FRANCE Paris	99%	60,700,000 285 per sq. mi. 110 per sq. km	212,934 sq. mi. 551,497 sq. km	80	$28,700	620	French
GERMANY Berlin	99%	82,500,000 598 per sq. mi. 231 per sq. km	137,830 sq. mi. 356,978 sq. km	79	$28,700	581	German
GREECE Athens	97.5%	11,100,000 218 per sq. mi. 84 per sq. km	50,950 sq. mi. 131,960 sq. km	76	$21,300	480	Greek
UNITED STATES Washington, D.C.	97%	296,500,000 80 per sq. mi. 31 per sq. km	3,717,796 sq. mi. 9,629,047 sq. km	78	$40,100	844	English

Sources: *CIA World Factbook*, 2005; Population Reference Bureau, *World Population Data Sheet*, 2005.

For more country facts, go to the **Nations of the World Databank** at glencoe.com.

Regional Atlas

Europe

Country and Capital	Literacy Rate	Population and Density	Land Area	Life Expectancy (Years)	GDP* Per Capita (U.S. dollars)	Television Sets (per 1,000 people)	Flag and Language
HUNGARY, Budapest	99.4%	10,100,000 281 per sq. mi. 109 per sq. km	35,919 sq. mi. 93,030 sq. km	68	$14,900	447	Hungarian
ICELAND, Reykjavík	99.9%	300,000 8 per sq. mi. 3 per sq. km	39,768 sq. mi. 102,999 sq. km	81	$31,900	505	Icelandic
IRELAND, Dublin	98%	4,100,000 151 per sq. mi. 58 per sq. km	27,135 sq. mi. 70,279 sq. km	78	$31,900	406	English, Irish
ITALY, Rome	98.6%	58,700,000 505 per sq. mi. 195 per sq. km	116,320 sq. mi. 301,267 sq. km	77	$27,700	492	Italian
KOSOVO, Pristina	—	2,126,708 506 per sq. mi. 195 per sq. km	4,203 sq. mi. 10,887 sq. km	—	$1,800	—	Albanian, Serbian
LATVIA, Riga	99.8%	2,300,000 92 per sq. mi. 36 per sq. km	24,942 sq. mi. 64,599 sq. km	72	$11,500	757	Latvian
LIECHTENSTEIN, Vaduz	100%	40,000 645 per sq. mi. 248 per sq. km	62 sq. mi. 161 sq. km	80	$25,000	469	German
LITHUANIA, Vilnius	99.6%	3,400,000 135 per sq. mi. 52 per sq. km	25,174 sq. mi. 65,200 sq. km	72	$12,500	422	Lithuanian
LUXEMBOURG, Luxembourg	100%	500,000 501 per sq. mi. 193 per sq. km	999 sq. mi. 2,587 sq. km	78	$58,900	599	Luxembourgish, German, French
UNITED STATES, Washington, D.C.	97%	296,500,000 80 per sq. mi. 31 per sq. km	3,717,796 sq. mi. 9,629,047 sq. km	78	$40,100	844	English

*Gross Domestic Product
Kosovo's information is based on 2008 data.

Countries and flags not drawn to scale

Europe

Country and Capital	Literacy Rate	Population and Density	Land Area	Life Expectancy (Years)	GDP* Per Capita (U.S. dollars)	Television Sets (per 1,000 people)	Flag and Language
MACEDONIA, Skopje	—	2,000,000 201 per sq. mi. 78 per sq. km	9,927 sq. mi. 25,711 sq. km	71	$7,100	273	Macedonian
MALTA, Valletta	92.8%	400,000 3,278 per sq. mi. 1,246 per sq. km	124 sq. mi. 321 sq. km	79	$18,200	549	Maltese, English
MOLDOVA, Chişinău	99.1%	4,200,000 323 per sq. mi. 125 per sq. km	13,012 sq. mi. 33,701 sq. km	65	$1,900	297	Moldovan
MONACO, Monaco	99%	30,000 30,000 per sq. mi. 11,538 per sq. km	1 sq. mi. 2.6 sq. km	—	$27,000	758	French
MONTENEGRO, Podgorica	97%	650,000 122 per sq. mi. 47 per sq. km	5,333 sq. mi. 13,812 sq. km	73	$2,200	277	Montenegrin, Serbian, Albanian
NETHERLANDS, Amsterdam	99%	16,400,000 1,023 per sq. mi. 395 per sq. km	13,082 sq. mi. 33,883 sq. km	79	$29,500	540	Dutch
NORWAY, Oslo	100%	4,600,000 37 per sq. mi. 14 per sq. km	125,050 sq. mi. 323,878 sq. km	80	$40,000	653	Norwegian
POLAND, Warsaw	99.8%	38,200,000 306 per sq. mi. 118 per sq. km	124,807 sq. mi. 323,249 sq. km	71	$12,000	387	Polish
UNITED STATES, Washington, D.C.	97%	296,500,000 80 per sq. mi. 31 per sq. km	3,717,796 sq. mi. 9,629,047 sq. km	78	$40,100	844	English

Sources: *CIA World Factbook*, 2005; Population Reference Bureau, *World Population Data Sheet*, 2005.

For more country facts, go to the **Nations of the World Databank** at glencoe.com.

Regional Atlas

Europe

Country and Capital	Literacy Rate	Population and Density	Land Area	Life Expectancy (Years)	GDP* Per Capita (U.S. dollars)	Television Sets (per 1,000 people)	Flag and Language
PORTUGAL Lisbon	93%	10,600,000 298 per sq. mi. 115 per sq. km	35,502 sq. mi. 91,951 sq. km	78	$17,900	567	Portuguese
ROMANIA Bucharest	98.4%	21,600,000 235 per sq. mi. 91 per sq. km	92,042 sq. mi. 238,388 sq. km	68	$7,700	312	Romanian
SAN MARINO San Marino	96%	30,000 1,304 per sq. mi. 500 per sq. km	23 sq. mi. 60 sq. km	81	$34,600	875	Italian
SERBIA Belgrade	96%	8,032,338 269 per sq. mi. 104 per sq. km	29,913 sq. mi. 77,474 sq. km	75	$7,700	—	Serbian
SLOVAKIA Bratislava	—	5,400,000 285 per sq. mi. 110 per sq. km	18,923 sq. mi. 49,010 sq. km	70	$14,500	418	Slovak
SLOVENIA Ljubljana	99.7%	2,000,000 256 per sq. mi. 99 per sq. km	7,819 sq. mi. 20,251 sq. km	77	$19,600	362	Slovenian
SPAIN Madrid	97.9%	43,500,000 225 per sq. mi. 87 per sq. km	193,363 sq. mi. 500,808 sq. km	80	$23,300	555	Spanish
SWEDEN Stockholm	99%	9,000,000 52 per sq. mi. 20 per sq. km	173,730 sq. mi. 449,959 sq. km	81	$28,400	551	Swedish
UNITED STATES Washington, D.C.	97%	296,500,000 80 per sq. mi. 31 per sq. km	3,717,796 sq. mi. 9,629,047 sq. km	78	$40,100	844	English

*Gross Domestic Product
Serbia's information is based on 2008 data.

Countries and flags not drawn to scale

Europe

Country and Capital	Literacy Rate	Population and Density	Land Area	Life Expectancy (Years)	GDP* Per Capita (U.S. dollars)	Television Sets (per 1,000 people)	Flag and Language
Bern SWITZERLAND	96%	7,400,000 464 per sq. mi. 179 per sq. km	15,942 sq. mi. 41,290 sq. km	80	$33,800	457	German, French, Italian
Kyiv (Kiev) UKRAINE	93%	47,100,000 202 per sq. mi. 70 per sq. km	233,089 sq. mi. 603,698 sq. km	63	$6,300	433	Ukrainian
UNITED KINGDOM London	—	60,100,000 636 per sq. mi. 245 per sq. km	94,548 sq. mi. 244,878 sq. km	78	$29,600	661	English
VATICAN CITY	99.7%	1,000 1,000 per sq. mi. 385 per sq. km	1 sq. mi. 2.6 sq. km	—	—	—	Italian, Latin
UNITED STATES Washington, D.C.	97%	296,500,000 80 per sq. mi. 31 per sq. km	3,717,796 sq. mi. 9,629,047 sq. km	78	$40,100	844	English

Sources: *CIA World Factbook*, 2005; Population Reference Bureau, *World Population Data Sheet*, 2005.

For more country facts, go to the **Nations of the World Databank** at glencoe.com.

L'Arc de Triomphe, Paris, France

Reading Social Studies

Making Inferences

Learn It!

It is impossible for authors to write every detail about a topic in a textbook. Because of this, good readers must make inferences to help them understand what they are reading. To infer means to evaluate information and form a conclusion.

- Read the paragraph below.
- Think about what you already know about the topic.
- Then, look for clues that might explain what is happening in the passage even though it might not be stated.
- What inference can you make about Napoleon's skill as a military leader?

Clues

> A brilliant military leader named Napoleon Bonaparte gained power and made himself emperor. Napoleon was a small man with big ambitions. His armies conquered much of Europe, until several countries united to defeat him in 1815.
>
> *—from page 301*

Use the diagram below to help you make an inference about Napoleon's skill as a military leader.

What you already know:
What traits must brilliant military leaders have?

Clues in the text:
- Napoleon was able to gain power and lead armies to conquer much of Europe.
- The word *several* hints that it took many European countries to defeat him.

Inference:
Napoleon must have been a skillful military leader.

Reading Tip

Making inferences is an everyday part of life. For example, if you look at the sky and see dark clouds, you may infer that it is going to rain. As you read, use the facts in the text to make inferences by thinking beyond the words on the page.

Read to Write Activity

In Chapter 11, Section 1, read the paragraphs titled "Industry and Conflict." Then take notes about the Industrial Revolution. Write a statement in which you make an inference about what life was like for a teenager during the Industrial Revolution.

2 Practice It!

Read the following paragraph from this unit.

- Draw a diagram like the one shown below.
- Write what you know about the United Kingdom's constitutional monarchy, along with facts from the text.
- Make an inference about the power of the king or queen in the United Kingdom.

> The government of the United Kingdom is a constitutional monarchy. A king or queen serves as head of state and takes part in ceremonies, but elected officials actively run the government.
>
> *—from page 322*

What you already know:

Clues in the text:

Inference:

Queen Elizabeth II of the United Kingdom and members of the royal family

3 Apply It!

For each chapter in this unit choose a topic, and create a diagram like the one above. Write related information that you already know, along with facts from the text, in the diagrams. Make inferences using this information. Read your facts to a partner, and ask your partner to make inferences from them. Are your inferences the same?

CHAPTER 10

Physical Geography of Europe

Essential Question

Regions Europe's landforms include high, snowcapped mountains and broad, fertile plains that are good for farming. Europe might be most influenced, however, by its nearness to water. A number of oceans and seas border Europe's countries. Europe also has many important rivers. How do people use waterways?

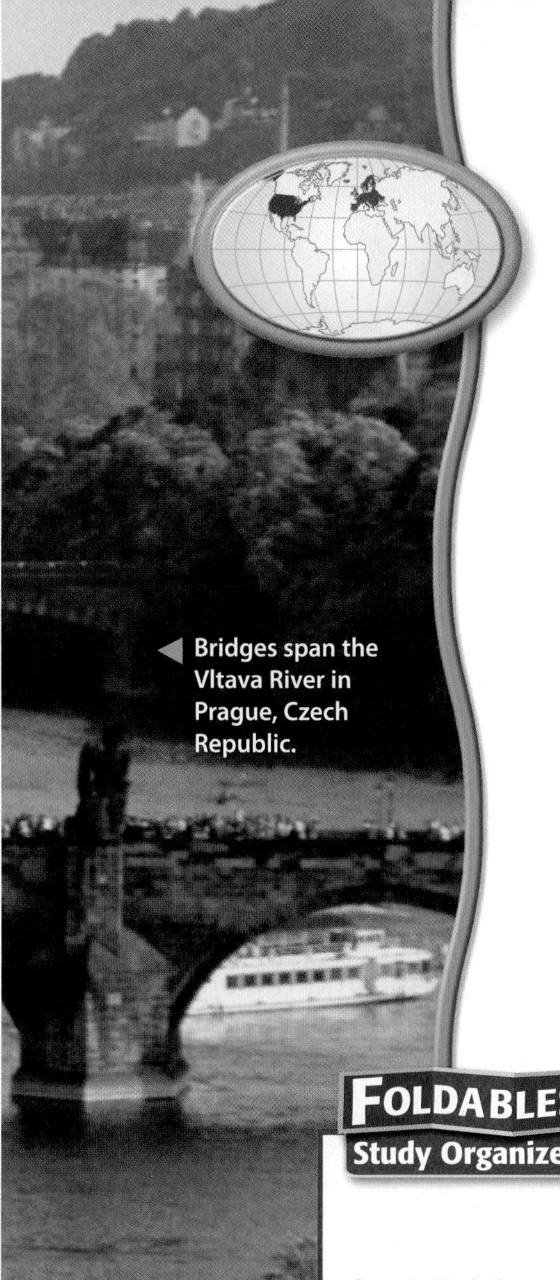

Bridges span the Vltava River in Prague, Czech Republic.

BIG Ideas

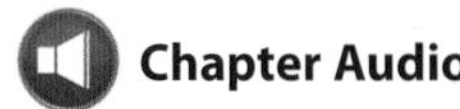
Chapter Audio

Section 1: Physical Features

BIG IDEA Geographic factors influence where people settle. Europe has a variety of landforms and plentiful natural resources that have attracted a large population. Most people live on Europe's plains, where industry and agriculture flourish. Such successes, however, have contributed to environmental problems in the region.

Section 2: Climate Regions

BIG IDEA The physical environment affects how people live. Although Europe is located relatively far north, much of the region has a mild climate that is ideal for farming and development. However, many Europeans are concerned that the climate is warming, which may have dangerous consequences.

FOLDABLES™ Study Organizer

Summarizing Information Make this Foldable to help you gather notes and summarize information about Europe's physical features, climate, and environmental issues.

Step 1 Mark the midpoint of the side edge of a sheet of paper.

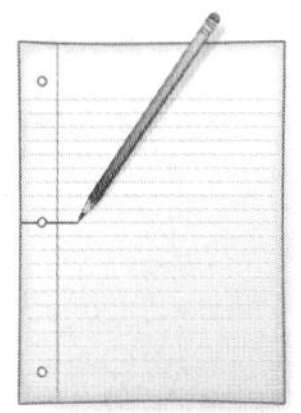

Step 2 Fold the top and bottom of the paper into the middle to make a shutter fold.

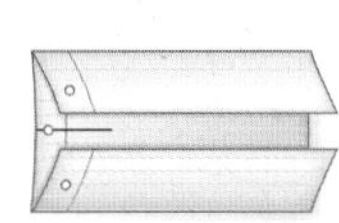

Step 3 Fold the paper in half from side to side.

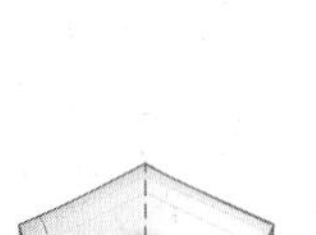

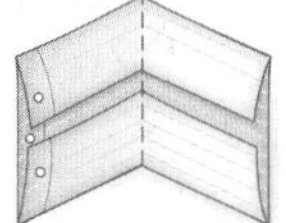

Step 4 Open and cut along the inside fold lines to form four tabs.

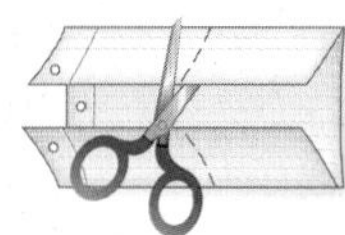

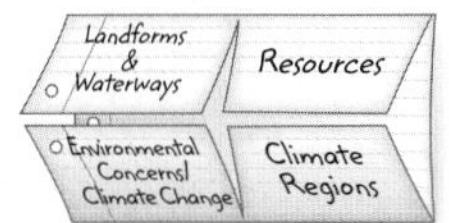

Step 5 Label the tabs as shown.

Reading and Writing As you read the chapter, fill in the Foldable. When you have finished the chapter, use your Foldable to write a list of the 10 most important facts about Europe's physical geography.

Social Studies ONLINE

Visit glencoe.com and enter ***QuickPass***™ code EOW3109c10 for Chapter 10 resources.

Guide to Reading

BIG Idea

Geographic factors influence where people settle.

Content Vocabulary

- landlocked *(p. 275)*
- pass *(p. 277)*
- navigable *(p. 277)*

Academic Vocabulary

- access *(p. 275)*
- affect *(p. 275)*
- impact *(p. 279)*

Reading Strategy

Organizing Information Use a diagram like the one below to organize key facts about each of Europe's major landforms (Islands and Peninsulas, Plains, Mountains and Highlands).

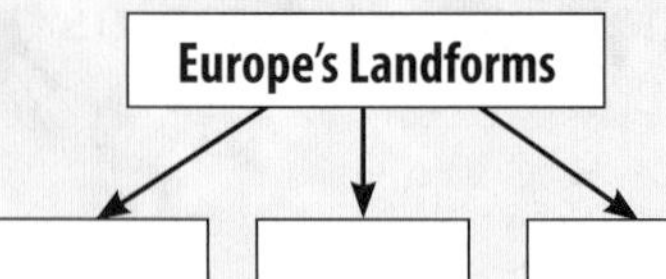

Physical Features

Spotlight Video

Picture This Snowdrifts? No, these snow-like mounds were formed about 1,500 years ago during a volcanic eruption on the island of Lipari, off the coast of Sicily, in Italy. The mounds are made of pumice, a stone formed from the cooling of lava, which rained down on the island during the eruption. Today the volcano is quiet, but Lipari hums with the sounds of open-pit pumice mines. Pumice is used to polish smooth surfaces. The stone is often used to give "stonewashed" jeans their worn look. In Section 1, you will learn about the different European landforms and the effect they have had on people living in the region.

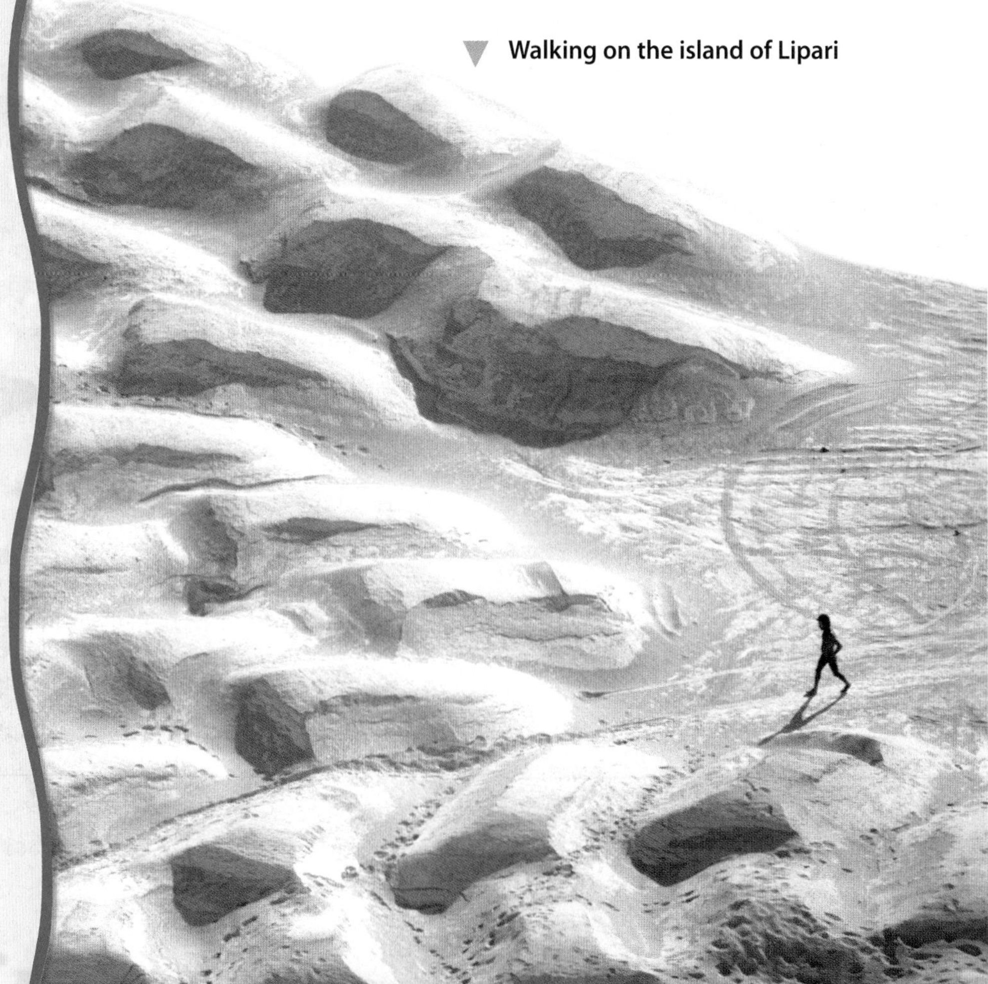

▼ Walking on the island of Lipari

Landforms and Waterways

Main Idea **Europe's landforms and waterways have greatly influenced where and how Europeans live.**

Geography and You What landforms can you find near your community? In Europe you would find impressive mountains, shimmering seas, and rolling farmland. Read to learn more about the variety of landscapes on this small continent.

When you look at a map of Europe, one of the first things you notice is that the continent is not a separate landmass. Instead, Europe and Asia share a common landmass called Eurasia. Europe extends to the west, from Asia to the Atlantic Ocean.

Europe's long coastline is framed by the Atlantic and by several seas. These include the Baltic, North, Mediterranean, and Black Seas. Most land in Europe lies within 300 miles (483 km) of a coast. Only a few countries are **landlocked,** meaning they do not border an ocean or a sea. Relatively long rivers, however, do give these inland countries **access** to coastal ports.

This nearness to water has shaped the lives and history of Europe's people. Europeans developed skills in sailing and fishing, which encouraged trade and helped Europe's economy grow. The closeness to the sea also allowed people to move easily between Europe and other continents. As a result, European culture has both influenced and been influenced by the cultures of Asia, Africa, and the Americas.

Peninsulas and Islands

Look at the physical map in this unit's Regional Atlas. You can see that Europe is a huge peninsula, with many smaller peninsulas branching out from it. Europe also includes many islands. Some of the major islands are Great Britain, Ireland, and Iceland in the Atlantic Ocean. Other large islands, such as Sicily, Crete, and Cyprus, are located in the Mediterranean Sea.

The large number of peninsulas and islands has **affected** Europe's history. Groups of people were separated by Europe's many seas, rivers, and mountains. As a result, many different cultures developed. Today, Europe is home to more than 40 independent countries. That is a remarkable number of neighbors squeezed together on a relatively small continent.

Plains

Europe's major landform is the Northern European Plain. This large lowland area stretches like a rumpled blanket across the northern half of mainland Europe.

The Northern European Plain

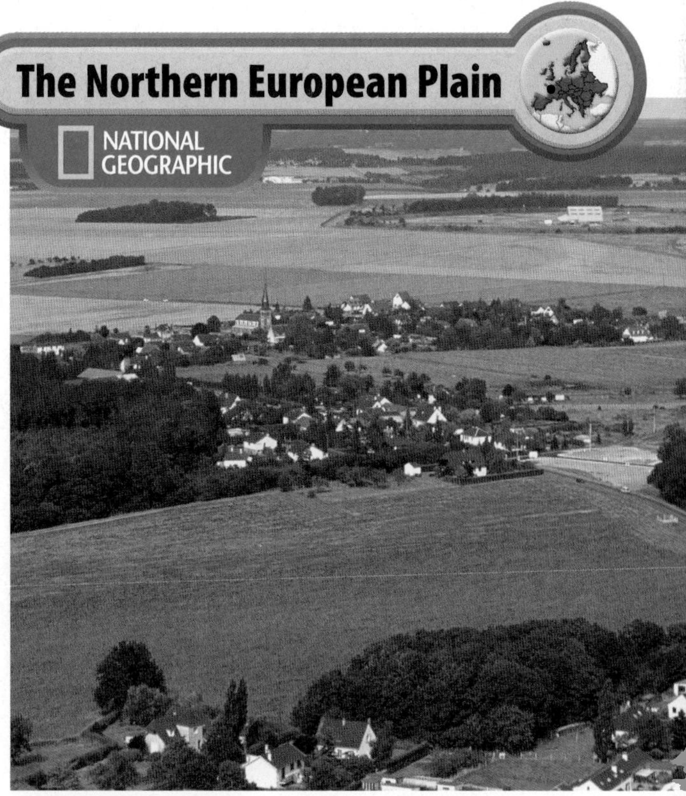

▲ Rich farmland in the region of Normandy in France is part of the Northern European Plain. ***Regions*** **Describe the Northern European Plain.**

It has rolling land with isolated hills. The Northern European Plain reaches from Belarus and Ukraine westward to France and also extends to the British Isles.

The plain's rich soil makes its farms highly productive. Farmers grow a great variety of grains, fruits, and vegetables. Some farmers raise dairy cattle to produce milk used in making cheese and other dairy products.

The Northern European Plain also has important energy and mineral resources. Deposits of coal, iron ore, and other minerals lie underground. These resources aided Europe's industrial growth.

Because the plain is so rich agriculturally and industrially, it is densely populated. Today, most of Europe's people live in this area. The landscape is dotted with villages, towns, and cities, including the busy capital cities of Warsaw, Berlin, Paris, and London.

Europe has other lowlands in addition to the Northern European Plain. For example, narrow plains rim the coasts of southern Europe. Two larger lowlands in the east—the Hungarian Plain, east of the Alps, and the Ukrainian Steppe, a broad, grassy plain north of the Black Sea—have rich soil that supports farming.

Mountains and Highlands

Highlands mark the northern border of the Northern European Plain. Steeper mountains lie south of the plain. Europe's highest mountain ranges form the Alpine Mountain System, which stretches from Spain to the Balkan Peninsula. The system takes its name from the Alps of south-central Europe. It also includes the Pyrenees, which lie between France and Spain, and the Carpathians, in east-central Europe.

The region's highest peak is Mont Blanc in the Alps of France. It rises to 15,771 feet (4,807 m). Most of Europe's mountains, however, are not very tall compared to those of Asia.

NATIONAL GEOGRAPHIC

The Alps

▲ Rugged beauty and good skiing conditions make the Alps a popular tourist destination. ***Place*** **What mountain ranges form the Alpine Mountain System?**

Like some of Europe's other landforms, mountains have helped isolate certain countries and peoples. Switzerland, for example, is located high in the Alps. While European wars have raged around it, the country has remained free from conflict and invasion for many centuries. Europe's mountains have never completely blocked movement though. **Passes,** or low areas between mountains, allow the movement of people and goods.

Less dramatic than Europe's mountains are three older highland areas that have eroded over time. Uplands in the northwest extend from Sweden through northern Great Britain to Iceland. Stripped of soil by glaciers, the land here is poor for farming, so many people raise sheep. A second highland area, the Central Uplands, contains much of Europe's coal. The area reaches from southern Poland to France. The third highland, the Meseta in Spain, is a plateau on which people grow grains and raise livestock.

Waterways

In addition to plains and mountains, Europe has an abundance of rivers, lakes, and other waterways. Europe's major rivers flow from inland highlands and mountains into the oceans and seas surrounding the region.

Many European rivers are **navigable,** or wide and deep enough for ships to use. People and goods can sail easily from inland areas to the open sea and, from there, around the world. The Danube and the Rhine, two of Europe's longest rivers, are important for transporting goods. Canals link these rivers, further improving Europe's water transportation network.

Rivers carry rich soil downstream, creating productive farmland along their banks and at their mouths. For this reason, river valleys have long been home to large numbers of people. Today, fast-flowing rivers are also used to generate electricity to support these large populations.

NATIONAL GEOGRAPHIC

Hydroelectric Power

This hydroelectric power plant is located on the Danube River between Austria and Germany. ***Human-Environment Interaction*** **How do Europeans use their rivers?**

Lakes cover only a small fraction of Europe. They are valuable for recreation, though, and for tourism. Most lakes are located on the Northern European Plain and in Scandinavia. The highland lakes in northern Great Britain and the Alps, however, are among the most beautiful and frequently visited. People flock to these lakes to boat, fish, swim, and appreciate nature.

Reading Check **Explaining** How have Europeans improved their water transportation network?

Social Studies ONLINE

Student Web Activity Visit glencoe.com and complete the Chapter 10 Web Activity about the Northern European Plain.

Europe's Resources

Main Idea **Europe has valuable resources that strengthen its economy.**

Geography and You Think of the products that you use every day. What are these products made of? As you read, think about how Europe's natural resources benefit people around the world.

Europe is a leader in the world economy. Part of this success comes from Europe's rich supply of natural resources.

Energy Resources

Coal has been a major energy source in Europe for many decades. By burning coal, Europeans fueled the development of modern industry in the 1800s. Today, almost half of the world's coal comes from Europe. Coal mining provides jobs for people from the United Kingdom in the west to Ukraine, Poland, and the Czech Republic in the east.

Petroleum and natural gas are other important energy resources found in Europe. The region's most productive oil fields lie beneath the North Sea, in areas controlled by the United Kingdom and Norway. To discourage dependence on oil, though, European governments tax gasoline heavily. Drivers in Europe pay some of the highest gasoline prices in the world.

Europe also relies on several "clean" energy sources that cause less pollution than burning coal or oil. In the highlands and mountains, swift-flowing rivers are used to create hydroelectric power. Europeans also make use of the wind's power. Germany, Spain, and Denmark are leaders in building wind farms, which use large turbines to create electricity from the wind.

Wind Farm in Spain

NATIONAL GEOGRAPHIC

▲ Europe produces more electricity from wind turbines than any other world region. ***Human-Environment Interaction*** **Why is wind power considered a "clean" energy source?**

Other Natural Resources

Besides energy resources, Europe has many other important natural resources. European mines produce about one-third of the world's iron ore, which is used in making steel. Ukraine has deposits of manganese, another important ingredient of steel. The United Kingdom exports a special clay used to make fine china dishes. Marble from Italy and granite from Norway and Sweden provide fine building materials. Of course, stone has always been used for building. Many European towns have narrow cobblestone streets and quaint stone houses that are centuries old.

Forests once covered a large part of Europe. Long ago, however, people cleared the land for farms and used much of the wood for building and for fuel. Today, only small pockets of forest remain. Sweden and Finland have the most forestland and produce the most lumber.

Air Pollution

Air pollution from industrial plants, such as this one in Wales, can destroy forests in Poland (inset). ***Human-Environment Interaction*** **What is acid rain, and why is it harmful?**

Fertile soil is another valuable resource, providing Europe with some of the best farmland on the planet. European farmers grow large amounts of grains, including nearly all of the world's rye, most of its oats, and nearly half of its wheat. Europe also produces more potatoes than any other region in the world.

The waters around Europe contain yet another resource—fish. From the Mediterranean Sea to the North Atlantic, Europeans catch salmon, cod, and other varieties of fish. Fishing is important to the economy, but many Europeans also eat a lot of fish and value its health benefits.

Reading Check **Describing** Where are Europe's most productive oil fields?

Environmental Issues

Main Idea Europe's plentiful resources have helped its economy, but environmental problems are a growing concern.

Geography and You Do you recycle at home and try to use energy wisely? As you read, see how Europeans are taking similar steps to protect their resources.

By taking advantage of its natural resources, Europe has become an economic powerhouse. The **impact** on the environment, however, has sometimes been harmful. For instance, in deforested areas of Southern Europe, tree roots no longer hold the soil in place. Valuable topsoil can be washed away.

Air Pollution and Acid Rain

Industrial growth in Europe has also hurt the environment—and created health risks. For example, car exhaust and smoke from burning oil and coal create air pollution. This pollution causes breathing problems, eye irritation, and lung disease.

Air pollution has another serious effect. When pollutant particles mix with precipitation, acid rain falls to Earth. Acid rain can make trees vulnerable to attack from insects and disease. Forests in eastern Europe are especially threatened. In that region, lignite coal is a major fuel source because it is cheap. It burns poorly, however, and pollutes heavily. The resulting acid rain has destroyed many forest areas in Hungary, Poland, the Czech Republic, and Slovakia.

Acid rain falls on Europe's waterways as well as its forests. As acids build up in lakes, rivers, and streams, fish and other wildlife are poisoned and die.

Acid rain is also a problem for Europe's historic buildings. The famous Tower of London, Germany's Cologne Cathedral, and ancient buildings and monuments dating from Greek and Roman times all show damage from acid deposits.

Water Pollution

Water pollution is another challenge for Europe. Sewage, garbage, and industrial waste have all been dumped into the region's seas, lakes, and rivers. As populations and tourism have increased, the problem has worsened.

Runoff from farms is also a problem for Europe's waters. Runoff is precipitation that flows over the ground, often picking up pesticides and fertilizers along the way. When chemicals from runoff enter a river, they encourage the growth of algae. Algae rob the river of so much oxygen that fish cannot survive. Runoff spilling into the Danube River, for example, has killed much of its marine life.

Finding Solutions

European leaders are trying to solve environmental problems in a number of ways. Many are working to prevent air pollution and acid rain by limiting the amount of chemicals that factories and cars can release into the air. Norway and Sweden are adding lime to their lakes. This substance stops the damage caused by acid rain and allows fish to multiply again.

Europeans are also making their lakes and rivers cleaner by treating waste and sewage. In addition, some countries encourage farmers to use less fertilizer to reduce damaging runoff.

Recycling is another strategy for protecting the environment. Europeans now recycle more paper, plastics, and glass than in the past. This saves energy that would otherwise be needed to produce these goods and cuts down on wastes.

Reading Check **Explaining** How does runoff contribute to water pollution?

Social Studies ONLINE
Study Central™ To review this section, go to glencoe.com.

Vocabulary

1. **Explain** the significance of:
   a. landlocked
   b. pass
   c. navigable

Main Ideas

2. **Organizing Information** Use a diagram like the one below to explain the importance of rivers in Europe.

3. **Analyzing** Why are Europe's clean energy sources important to the region?

4. **Explaining** Why and where is acid rain especially a problem?

Critical Thinking

5. **BIG Idea** Which resources helped industry and farming develop in Europe?

6. **Challenge** Describe how Europe's landforms and bodies of water influenced where people settled.

Writing About Geography

7. **Expository Writing** Write a paragraph explaining which physical feature you think has most helped Europe's economy to prosper.

Disaster at Chernobyl

In the modern world, we depend on technology to survive. What happens when technology goes wrong?

The Accident On April 26, 1986, the world saw its worst nuclear disaster. That day, a reactor at the Chernobyl nuclear power plant in Ukraine exploded. Dangerous radioactive material shot into the sky. The explosion also caused a fire that raged for 10 days, pouring out more radioactive dust and ash. During that time, radioactive material—called fallout—fell to the Earth over large parts of Ukraine and Belarus, as well as other parts of Europe.

The Impact Fewer than a hundred people died from the high levels of radiation that resulted from the explosion and fire. About 4,000 more, however, are expected to die from cancers caused by the accident. Fortunately, these numbers are far below what had originally been feared.

▲ **Abandoned amusement park, Pripyat', Russia**

Nevertheless, more than 20 years after the accident, its effects are ongoing. Around the area contaminated by fallout, officials created an exclusion zone of about 1,545 square miles (4,002 sq. km). People are prohibited from living within this zone. More than 350,000 people were forced to leave their homes following the accident. The disaster scarred the land as well. More than 1.8 million acres (728,435 hectares) of farmland and 1.7 million acres (687,966 hectares) of forest were abandoned because of contamination from fallout.

Meanwhile, a threat remains at Chernobyl. A protective concrete shell, built around the reactor to contain the contamination, could collapse. Also, rainwater leaks into the shell. When the water seeps back into the ground, it carries radioactive material with it, further poisoning the land.

NATIONAL GEOGRAPHIC

Chernobyl Area

Think About It

1. **Movement** How many people were forced to leave the contaminated area?
2. **Human-Environment Interaction** What effect did the accident have on the environment?

Guide to Reading

BIG Idea

The physical environment affects how people live.

Content Vocabulary

- deciduous *(p. 285)*
- coniferous *(p. 285)*
- mistral *(p. 287)*
- sirocco *(p. 287)*

Academic Vocabulary

- major *(p. 283)*
- feature *(p. 284)*

Reading Strategy

Categorizing Information Use a chart like the one below to describe Europe's climate zones.

Climate Zone	Characteristics
1.	1.
2.	2.
3.	3.
4.	4.

Climate Regions

Picture This These carefully balanced baskets will carry grapes that are handpicked in the Côte d'Or ("Golden Hill") region of Burgundy, France. The region has been producing wine since A.D. 900, and the grape harvest is vital to the local economy. Because of this, and because grapes are highly sensitive to the climate, big changes in temperature are always cause for concern. Read this section to find out about climate conditions in Europe and the concern over the warming trend.

Carrying grape baskets in Burgundy, France

Wind and Water

Main Idea **Wind patterns and water currents shape Europe's climate.**

Geography and You Doesn't a cool breeze feel great on a hot day? Read to learn how winds are helpful to Europe too.

Look at the physical map of the world in the Reference Atlas. Because Europe is farther north than the United States, you might expect Europe's climate to be colder than ours. In fact, much of Europe enjoys a milder climate. Why?

As **Figure 1** shows, the North Atlantic Current carries warm waters from the Gulf of Mexico toward Europe. Winds from the west pass over this water and carry more warmth to Europe. These prevailing winds, known as westerlies, are a **major** influence on warming the European climate.

Other wind patterns also affect the climate in parts of Europe. For example, warm winds from Africa contribute to the high temperatures in southern Europe. Blustery winter winds from Asia lower temperatures throughout much of eastern Europe.

The water surrounding Europe also affects the region's climate. Winds blowing off the water cool the hot land in the summer and warm the cold land in the winter. For this reason, coastal areas tend to have a more moderate climate than inland areas.

Reading Check **Explaining** Why does northwestern Europe have a mild climate?

NATIONAL GEOGRAPHIC **Maps In MOtion** See StudentWorks™ Plus or glencoe.com.

Figure 1 **Europe: Currents and Wind Patterns**

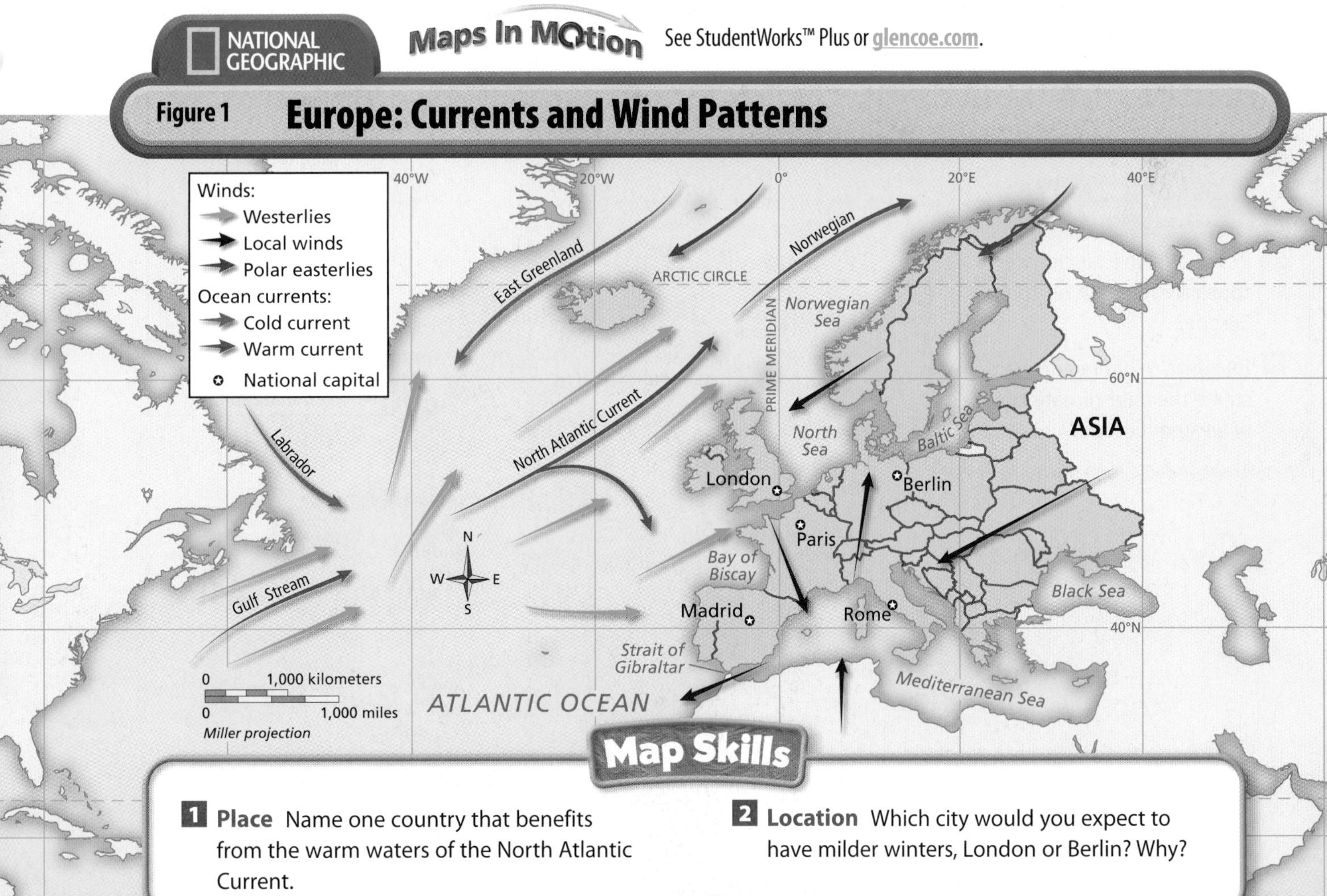

Map Skills

1. **Place** Name one country that benefits from the warm waters of the North Atlantic Current.
2. **Location** Which city would you expect to have milder winters, London or Berlin? Why?

Climate Zones

Main Idea **Europe has eight climate zones, each with different vegetation.**

Geography and You What is your ideal climate? Chances are, you can find it in Europe! Read to learn how Europe's climate varies from area to area.

Most of Europe falls into three main climate zones—marine west coast, humid continental, and Mediterranean. **Figure 2** also shows five other climate zones that appear in smaller areas—subarctic, tundra, highland, steppe, and humid subtropical.

Marine West Coast

Much of northwestern and central Europe has a marine west coast climate. This climate has two **features,** one of which is mild temperatures. The North Atlantic Current carries so much warmth that southern Iceland has mild temperatures, even though it is near the Arctic Circle. Because of the mild temperatures, this climate zone has surprisingly long growing seasons. In the United Kingdom, for example, farmers have a window of 250 or more days for planting and harvesting—nearly 60 more days than in eastern Canada—which is located at the same latitude.

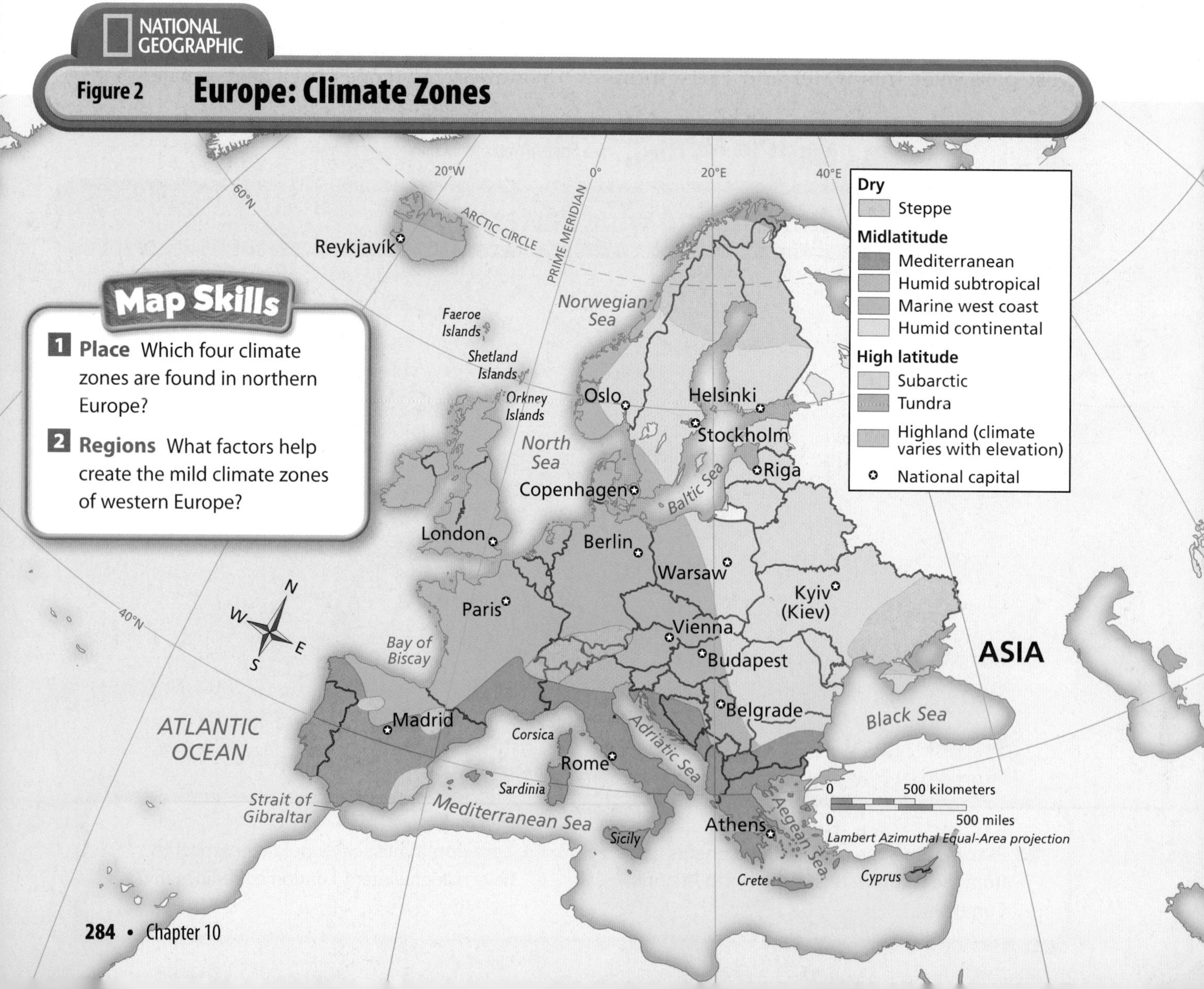

Although temperatures stay mild, differences do exist across the region. In the north, summers are shorter and cooler. Also, the farther away you get from water, the wider the range of temperatures will be. For example, in the coastal city of Brest, France, a December day might be only 20 degrees cooler than a July day. However, in Strasbourg, France, which is more than 400 miles (644 km) from the Atlantic Ocean, the temperature can differ by 40 degrees between summer and winter.

The second feature of the marine west coast climate, besides mild temperatures, is abundant precipitation. This typically falls in autumn and early winter. Although the zone as a whole gets plenty of rain, certain mountainous areas stay dry because of the rain shadow effect. For example, the coastal area of Norway, on the western edge of highlands, receives a yearly average of 90 inches (229 cm) of precipitation. The eastern slopes of those same highlands receive only one-third that amount.

Forests thrive in much of Europe's marine west coast climate zone. Some forests consist of **deciduous** trees, which lose their leaves in the fall. **Coniferous** trees, also called evergreens, grow in cooler areas of the marine west coast climate zone.

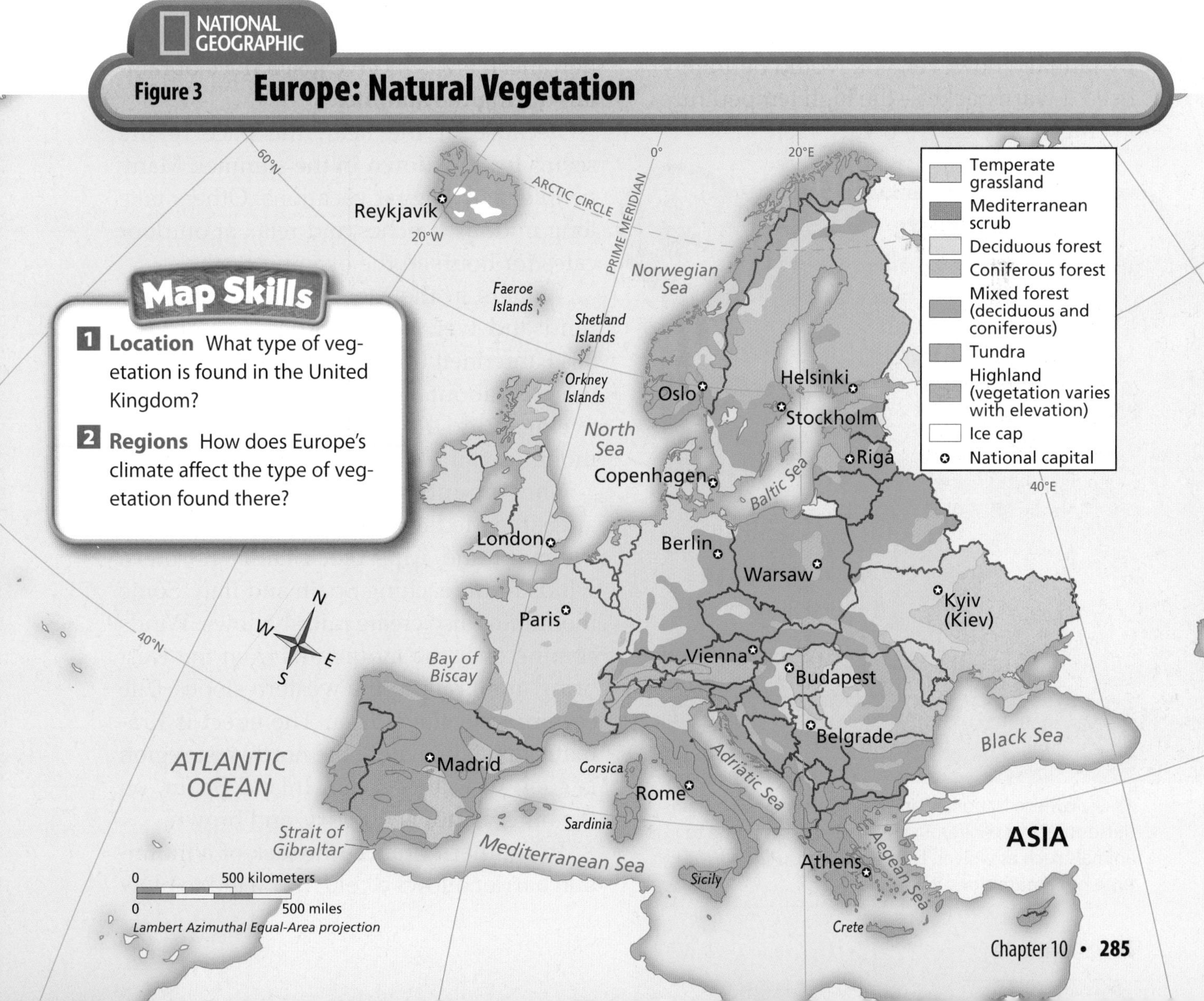

They dominate the landscape in southern Norway, Sweden, and parts of eastern Europe.

Although forests no longer blanket the continent, many people still earn their living from forest-related industries. They cut timber and produce a huge array of products, from lumber and paper to charcoal and turpentine.

Humid Continental

Eastern Europe and some areas of northern Europe have a humid continental climate. Cool, dry winds from the Arctic and Asia give this zone cooler summers and colder winters than the marine west coast zone. The city of Minsk in Belarus does not get much warmer than 70°F (21°C) in July. By January, however, you would definitely need a warm jacket—the high temperature averages only 22°F (–6°C)!

Poland's Forests

NATIONAL GEOGRAPHIC

▲ Bialowieza National Park in eastern Poland contains deciduous and coniferous trees. The park is home to animals such as wolves, lynx, and bison. ***Place*** **What type of climate does Poland have?**

Because of drier winds, the humid continental zone gets less rain and snow than the marine west coast zone. Nonetheless, some low-lying areas are wet and marshy. This is because the precipitation that does fall can be slow to evaporate in the cool climate. The humid continental zone supports mixed forests of deciduous and evergreen trees. However, only evergreens grow farther north and in higher elevations.

Mediterranean

Europe's third major climate zone, the Mediterranean zone, includes much of southern Europe. With average high temperatures in July ranging from 83°F to 98°F (28°C to 37°C), Mediterranean summers are hot. They are also very dry. Many Mediterranean areas receive just a trace of rainfall during the summer.

Because of the heat, the pace of life seems to slow down in the summer. Many people take August vacations. Others take long midday lunches and relax at outdoor cafes for hours in the evening.

Winters in the Mediterranean zone are mild and wet. With temperatures in the 50s Fahrenheit (low teens Celsius), nobody worries about snow. Rainfall, however, averages 3 to 4 inches (7.6 to 10.2 cm) per month, so an umbrella comes in handy.

The mountains of southern Europe affect the Mediterranean climate zone. The Pyrenees and Alps block chilly northern winds from reaching Spain and Italy. Some mountains also create rain shadows. Winds coming over the mountains from the west bring more rain to the western slopes. The eastern side stays drier. The effect is dramatic in Spain, where the northwest region is cool, wet, and green. Inland Spain, on the other hand, is hot, dry, and brown.

In southern France, the lack of a mountain barrier allows a cold, dry wind to blow

Greek Herder

Herding goats and sheep is a common economic activity in areas with a Mediterranean climate. ***Regions*** **What types of vegetation grow in Europe's Mediterranean climate zone?**

in from the north. This wind, the **mistral** (MIHS·truhl), occurs in winter and spring. It also creates waves, making southern France a popular site for windsurfing.

Countries in the Mediterranean climate zone are also affected by hot, dry winds from Africa to the south. In Italy, these winds are called **siroccos** (suh·RAH·kohs). They pick up moisture as they cross the Mediterranean, bringing uncomfortably humid conditions to southern Europe.

Because of the Mediterranean zone's low rainfall, plants that grow there must be drought resistant. Vegetation includes low-lying shrubs and grasses, as well as the olive trees and grapevines that the region is known for. Forests are rare, and stands of trees appear only on rainy mountainsides or along rivers.

Subarctic and Tundra

Farther north, Europe has two zones of extreme cold. The subarctic zone covers parts of Norway, Sweden, and Finland. Evergreens grow in this region at low altitudes. The tundra zone is found in the northern reaches of these countries and in Iceland. The tundra is an area of vast treeless plains near the North Pole. With cool summers that reach only about 40°F (4°C) and frigid winter temperatures that plunge as low as –25°F (–32°C), only low shrubs and mosses can grow in this region.

Because of Earth's tilt, the sun shines on the far north for up to 20 hours per day in late spring and early summer. As a result, the nights are extremely short. In the deep of winter, however, conditions are reversed. The days are short, and nights can last as long as 20 hours. Some people are so affected by the scarce light in winter that they lose energy and feel depressed.

Highland

The highland zone is found in the higher altitudes of the Alps and Carpathians where the climate is generally cool to cold. However, temperatures and precipitation vary greatly from place to place, depending on three factors—wind direction, orientation to the sun, and altitude. As an example, consider two peaks in the Swiss Alps—Säntis and Saint Gall. Säntis receives more than twice the precipitation of Saint Gall even though the two mountains are only 12 miles (19 km) apart. The difference is due to altitude—Säntis is about three times higher than Saint Gall.

Sturdy trees add color to the highland zone, but they grow only so far up the mountainsides. The point at which they stop growing is called the timberline. Above the timberline, where the sun barely warms the ground, only scrubby bushes and low-lying plants can survive.

Other Climate Zones

Europe's last two climate zones cover a relatively small part of the region. The steppe zone includes the southern part of Ukraine.

Steppes are dry, treeless grasslands, much like prairies but with shorter grass. Here the climate is not dry enough to be classified as desert, but not wet enough for forests to flourish.

A small sliver of land north of the Adriatic Sea falls into the humid subtropical zone. This zone has hot, wet summers and mild, wet winters.

Climate Change

Most scientists agree that the world's climate is growing warmer. Average temperatures have been inching upward for several decades. Measurements and photos show that glaciers are steadily eroding. In 2003 western Europe suffered its worst heat wave since the Middle Ages.

People debate whether this global warming is just part of nature's cycle or is instead related to human activities. Many scientists, though, believe that burning fossil fuels, such as coal and oil, contributes to the greenhouse effect. Gases build up in the atmosphere and trap large amounts of warm air near Earth's surface.

People also debate what this warming means for the planet. Many European leaders are worried. They fear that melting glaciers will produce higher ocean levels that will flood low-lying areas, such as the Netherlands and coastal cities like Venice, Italy. Such flooding would affect millions of people.

As a result, European officials are taking action. They are trying to slow global warming by encouraging changes in energy use. Most European governments have signed the Kyoto Treaty. This is an international agreement to limit the output of greenhouse gases, but its terms are not yet fully in effect.

Reading Check **Explaining** How do mountains affect southern Europe's Mediterranean climate zone?

Social Studies ONLINE
Study Central™ To review this section, go to glencoe.com.

Vocabulary

1. **Explain** the meaning of *deciduous, coniferous, mistral,* and *sirocco* by using each term in a sentence.

Main Ideas

2. **Identifying** What factors affect Europe's climates?
3. **Summarizing** Using a diagram like the one below, identify each of Europe's climate zones and the vegetation found in each zone.

Critical Thinking

4. **Comparing and Contrasting** How are Norway and Spain similar in climate? How are they different?
5. **BIG Idea** How does the North Atlantic Current affect farming in the marine west coast climate zone?
6. **Challenge** Describe how latitude and altitude affect climate and vegetation in Europe.

Writing About Geography

7. **Using Your FOLDABLES™** Use your Foldable to write a paragraph comparing and contrasting two of Europe's climate zones.

Visual Summary

Landforms

- The Northern European Plain is a rich farming region and has a high population density.
- Mountains separate much of northern and southern Europe.
- Uplands regions are found in northwest and central Europe and in Spain.

Berlin, Germany, on the Northern European Plain

Waterways

- Waterways have had a major impact on Europe's population and ways of life.
- Rivers provide transportation, good soil for farming, and hydroelectric power.

European Resources

- Europe's energy resources include coal, petroleum, natural gas, and hydroelectric and wind power.
- In some areas, good soil promotes farming and dairy farming.
- Fishing is important to coastal Europe.

Fishers in Spain

Environmental Issues

- The European environment has been damaged by deforestation, pollution, and acid rain.
- Europeans are working to protect and improve their environment through recycling and limiting forms of chemical pollution.

Tulip harvest in the Netherlands

Climate Regions

- Europe's nearness to water and its wind patterns greatly affect its climates.
- Europe has eight main climate zones: marine west coast, humid continental, Mediterranean, subarctic, tundra, highland, steppe, and humid subtropical.
- Europeans are concerned about the negative effects of global warming.

Reindeer on Norwegian tundra

Study anywhere, anytime! Download quizzes and flash cards to your PDA from glencoe.com.

CHAPTER 10 ASSESSMENT

STANDARDIZED TEST PRACTICE

Eliminate answers that you know for certain are incorrect. Then choose the most likely answer from those remaining.

Reviewing Vocabulary

Directions: Choose the word(s) that best completes the sentence.

1. Europe has many _______ rivers—rivers that are deep and wide enough for ships to travel.

A polluted

B passable

C navigable

D deciduous

2. In Europe, _______, or low areas between mountains, have allowed the movement of people and goods.

A ports

B siroccos

C runoffs

D passes

3. Hot, dry winds that pick up moisture as they cross the Mediterranean Sea and bring uncomfortable humidity to southern Europe are called _______.

A siroccos

B the mistral

C currents

D El Niño

4. Evergreens, or _______ trees, grow in cooler areas of Europe's marine west coast climate zone.

A palm

B coniferous

C deciduous

D banyan

Reviewing Main Ideas

Directions: Choose the best answer for each question.

Section 1 *(pp. 274–280)*

5. Europeans developed skills in fishing, sailing, and trading because

A much of Europe is landlocked.

B most Europeans live close to a coast or a navigable river.

C Europeans were more adventurous than other people.

D Europeans did not like farming or manufacturing.

6. One of Europe's most important resources has been _______, which helped modern industry grow there in the 1800s.

A clay

B granite

C soil

D coal

Section 2 *(pp. 282–288)*

7. The _______ has a major warming effect on Europe's climate.

A North Atlantic Current

B Asian landmass

C polar easterly

D region's position on the Equator

8. Much of northwest and central Europe has a(n) _______ climate.

A tundra

B highland

C marine west coast

D equatorial

GO ON

Critical Thinking

Directions: Choose the best answer for each question.

Base your answers to questions 9 and 10 on the bar graph below and your knowledge of Chapter 10.

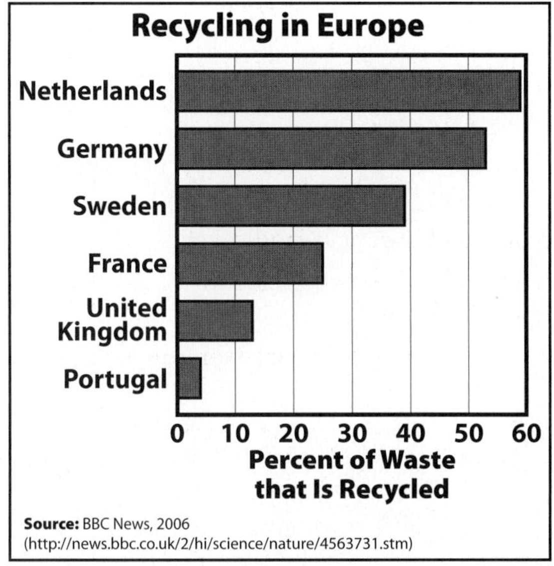

9. Which two European countries recycle more than 50 percent of their waste?

A Sweden and Portugal

B Netherlands and Germany

C Sweden and the United Kingdom

D France and Germany

10. Based on the graph, which of the following statements is accurate?

A Most European countries recycle more than half their waste.

B The six countries represented in the graph pollute more than any other European nations.

C The Netherlands relies more on recycling to handle waste than does France.

D The Netherlands produces more waste than the other five countries represented on the graph.

Document-Based Questions

Directions: Analyze the document and answer the short-answer questions that follow.

Eyewitness Account of Natalia Ivanovna Ivanova, Deputy Director, Vesnova Orphanage, Mogilev Oblast, Belarus:

> *It was terrible having to knock on the door or window in the middle of the night to tell the parents that their children should be evacuated the next morning. We said it was because of the radioactivity, which could have bad consequences for all of them. We arranged a place for everyone to gather to be put on buses. It was a dreadful sight.*
>
> —Natalia Ivanovna Ivanova, "Return to Chernobyl: 20 Years 20 Lives"

11. What did Natalia Ivanova have to do following the accident?

12. Describe Ivanova's reaction to the Chernobyl accident. How did she feel about the situation?

Extended Response

13. Write a short essay describing how Europe's successful use of its many resources has led to environmental problems in the region. Also describe what is being done to solve those problems.

STOP

For additional test practice, use Self-Check Quizzes—Chapter 10 at glencoe.com.

Need Extra Help?

If you missed question...	1	2	3	4	5	6	7	8	9	10	11	12	13
Go to page...	277	277	287	285	275	278	283	284	280	280	281	281	279

History and Cultures of Europe

Essential Question

Regions Europe is rich in history and culture. Like the United States, most countries in Europe are industrialized and have high standards of living. Unlike the United States, however, the people of Europe do not share a common language and government. What forces have helped unify Europeans at different times?

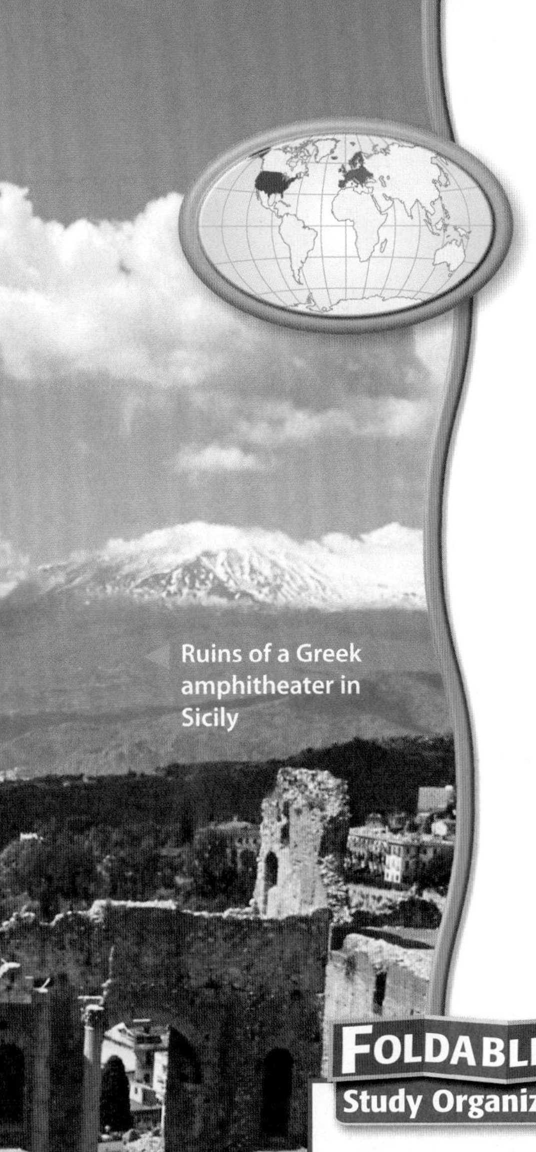
Ruins of a Greek amphitheater in Sicily

BIG Ideas

Section 1: History and Governments

BIG IDEA The characteristics and movement of people impact physical and human systems. Over the centuries, migrations and wars have brought different groups to power in Europe. As modern nations have taken the place of empires and kingdoms, ways of living and thinking have also changed.

Section 2: Cultures and Lifestyles

BIG IDEA Culture groups shape human systems. Europe is a region of many peoples with different ethnic backgrounds, languages, religions, and traditions. Despite their differences, Europeans lead similar lifestyles and share a rich cultural heritage.

FOLDABLES™ Study Organizer

Organizing Information Make this Foldable to help you organize information about Europe's history, population, and cultures.

Step 1 Place three sheets of paper on top of one another about 1 inch apart.

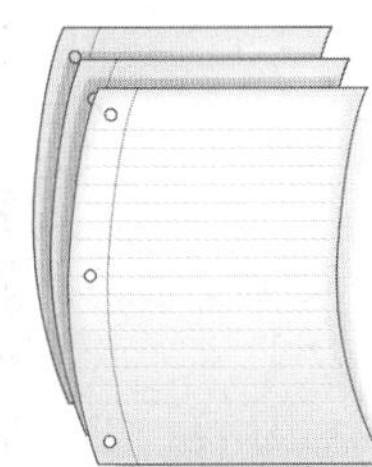

Step 2 Fold the papers to form six equal tabs.

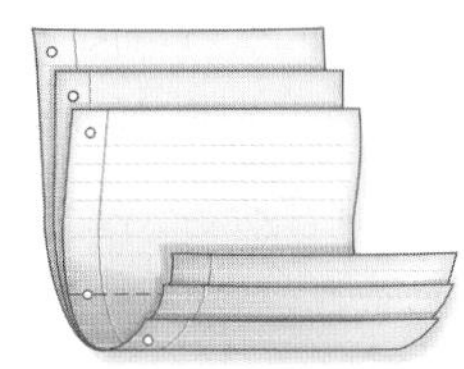

Step 3 Staple the sheets, and label each tab as shown.

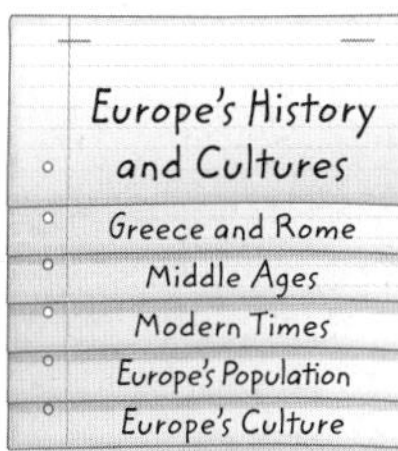

Reading and Writing As you read, use your Foldable to write down facts related to each period of European history, Europe's population, and European cultures. Then use the facts to write brief summaries for each of the tabs on the Foldable.

Visit glencoe.com and enter ***QuickPass***™ code EOW3109c11 for Chapter 11 resources.

Guide to Reading

BIG Idea

The characteristics and movement of people impact physical and human systems.

Content Vocabulary

- classical *(p. 295)*
- city-state *(p. 295)*
- democracy *(p. 295)*
- republic *(p. 296)*
- emperor *(p. 297)*
- pope *(p. 298)*
- feudalism *(p. 298)*
- nation-state *(p. 299)*
- revolution *(p. 301)*
- Holocaust *(p. 302)*
- communism *(p. 303)*

Academic Vocabulary

- dominant *(p. 296)*
- authority *(p. 300)*
- currency *(p. 303)*

Reading Strategy

Making a Time Line Use a time line like the one below to list at least five key events and dates in Europe's history.

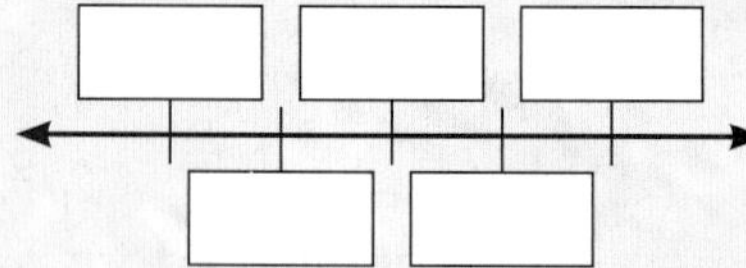

History and Governments

Picture This Who is that giant? Is it a warrior? A farmer? A king? One thing is certain—at almost 230 feet (70 m) high, the Long Man of Wilmington, in England, is one of the world's largest carved figures. Originally a chalk outline that became overgrown by grass, the Long Man was restored in 1969 with 770 concrete blocks. As scientists study the earth around the giant, they will be better able to judge when it was made—and maybe even why it was made! Read this section to learn more about the history of Europe.

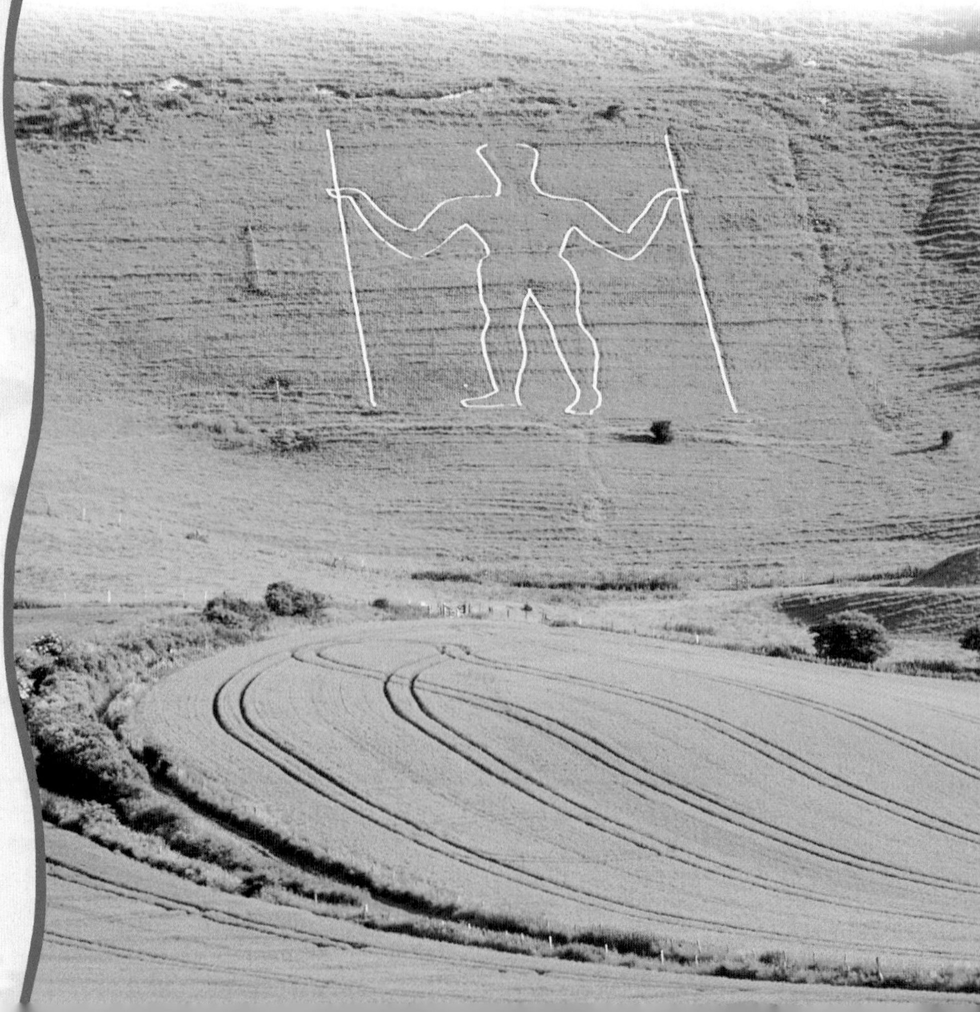

Ancient Long Man in hills of southern England

Ancient Europe

Main Idea **Ancient Greece and Rome laid the foundations of European civilization.**

Geography and You Do you get to vote on family decisions or elect leaders to your student government? Read to find out how voting rights arose with the ancient Greeks and Romans.

The story of European civilization begins with the ancient Greeks and Romans. More than 2,500 years ago, these peoples settled near the Mediterranean Sea. Eventually their cultures spread throughout Europe and beyond. Even today, the influence of the **classical** world—meaning ancient Greece and Rome—lingers.

Ancient Greece

Physical geography naturally shaped the development of ancient Greece. The people felt deep ties to the land, which is ruggedly beautiful. At the same time, Greece's many mountains, islands, and the surrounding seas isolated early communities and kept them fiercely independent.

The earliest Greek civilizations began among farming and fishing peoples who lived near the Aegean Sea. These civilizations became wealthy through trade. After warfare led to their decline, independent territories called **city-states** developed throughout Greece. Each city-state was made up of a city and its surrounding area. Although separated by geography, the Greek city-states shared the same language and culture.

One of the most prosperous and powerful city-states was Athens. The people of Athens introduced the world's first **democracy**, a political system in which all

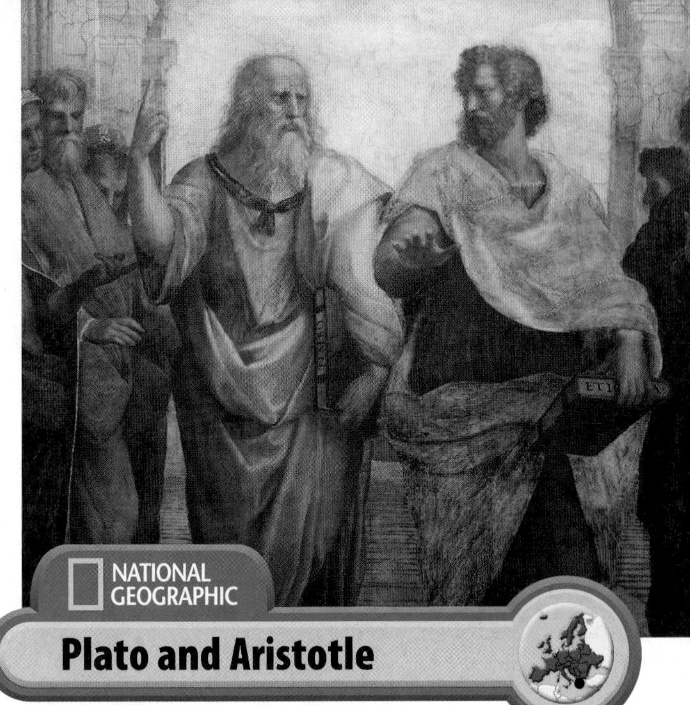

NATIONAL GEOGRAPHIC

Plato and Aristotle

Plato, on the left, set up a school in ancient Greece. His most famous student was Aristotle, another great Greek thinker. ***Place*** **In which Greek city-state did democracy develop?**

citizens share in running the government. Although women and enslaved persons could not vote because they were not citizens, Athenian democracy set an example for later civilizations. Learning and the arts also thrived in Athens. Among the city-state's great thinkers were Socrates, Plato, and Aristotle. Their ideas about the world and humankind had a major influence on Europe.

During the mid-300s B.C., warfare weakened the Greek city-states. Soon an invader from the north, Philip II of Macedonia, conquered Greece. Philip's son earned the name Alexander the Great by making even more conquests. As shown in **Figure 1** on the next page, his empire included Egypt and Persia and stretched eastward into India. Trade boomed, Greek culture mixed with Egyptian and Persian cultures, and scientific advances spread. Alexander died young, however, and his empire quickly broke into several smaller kingdoms. By about 130 B.C., the Romans had conquered most of the Greek kingdoms.

NATIONAL GEOGRAPHIC **Maps In MOtion** See StudentWorks™ Plus or glencoe.com.

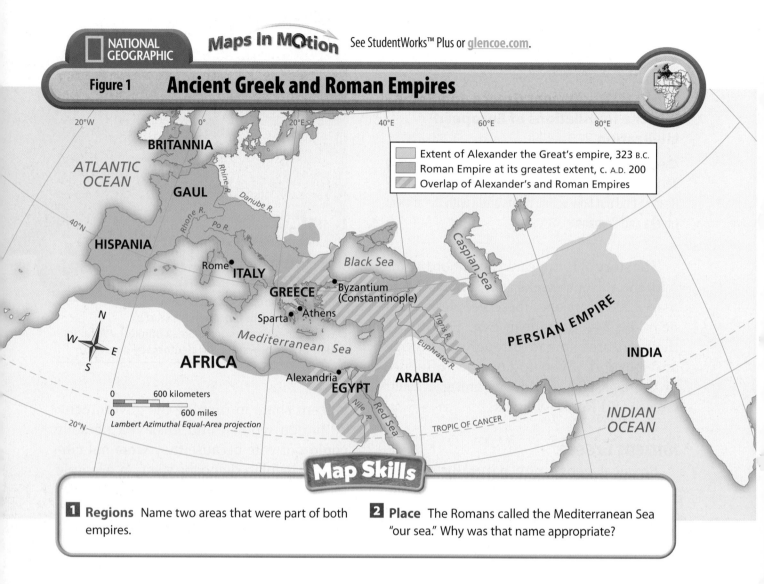

Figure 1 **Ancient Greek and Roman Empires**

Map Skills

1 Regions Name two areas that were part of both empires.

2 Place The Romans called the Mediterranean Sea "our sea." Why was that name appropriate?

The Roman Empire

While Greece ruled the eastern Mediterranean, Rome became a **dominant** power on the Italian Peninsula. Rome began as a monarchy but changed to a republic in 509 B.C. In a **republic,** people choose their leaders. Rome was led by two consuls who were elected by the citizens. The consuls reported to and were advised by the Senate, an assembly of rich landowners who served for life. One of the government's great achievements was the development of a code of laws. Written on bronze tablets known as the Twelve Tables, the laws stated that all free citizens had the right to be treated equally. Roman law led to standards of justice still used today. For example, a person was regarded as innocent until proven guilty. Also, judges were expected to examine evidence in a case.

About 200 B.C., Roman armies began seizing territory throughout the Mediterranean region. Instead of ruling only by force, though, the Romans allowed many of the people they conquered to become Roman citizens. By granting people citizenship, the Romans were able to build a strong state with loyal members.

Student Web Activity Visit glencoe.com and complete the Chapter 11 Web Activity on ancient Rome.

As the Roman Republic expanded, it evolved into the massive Roman Empire. The first **emperor,** or all-powerful ruler, was Augustus, who gained that position in 27 B.C. His rule brought order to Rome's vast lands. This period, called the *Pax Romana,* was a time of peace, artistic growth, and expanding trade that lasted about 200 years.

Christianity

During the *Pax Romana,* Christianity was developing in Palestine in the eastern part of the Roman Empire. There, a Jewish teacher, Jesus of Nazareth, preached a message of love and forgiveness. Jesus soon attracted followers as well as enemies. Fearing public unrest, the Roman authorities had Jesus executed. Yet within days, Jesus' followers, known as Christians, reported that he had risen from the dead. They took this as proof that Jesus was the son of God.

Eager to spread Jesus' teachings, two early Christian leaders, Peter and Paul, established the Christian Church in Rome. Roman officials at first persecuted, or mistreated, Christians. Despite this abuse, the new religion grew in popularity. In A.D. 392, Christianity became Rome's official religion.

Rome's Decline

By the late A.D. 300s, the Roman Empire was in decline. Rivals struggled to become emperor, and Germanic groups attacked from the north. About A.D. 395, the empire was divided into eastern and western parts. The eastern part remained strong and prosperous. Known as the Byzantine Empire, it lasted another thousand years. The western part was occupied by Germanic groups. In A.D. 476, Germanic leaders overthrew the last emperor in Rome and brought the Western Roman Empire to an end.

Despite its fall, Rome had great influence on Europe and the West. It helped spread classical culture and Christianity. Roman law shaped the legal systems in many countries. The Roman idea of a republic later influenced the founders of the United States. The Latin language of Rome became the basis for many modern European languages known as the Romance languages, such as Italian, French, and Spanish. Ancient Rome also influenced architectural styles in the Western world. For example, the U.S. Capitol and many other buildings have domes and arches inspired by Roman architecture.

Reading Check **Analyzing** Describe ancient Rome's influences on the modern world.

Roman Influences

NATIONAL GEOGRAPHIC

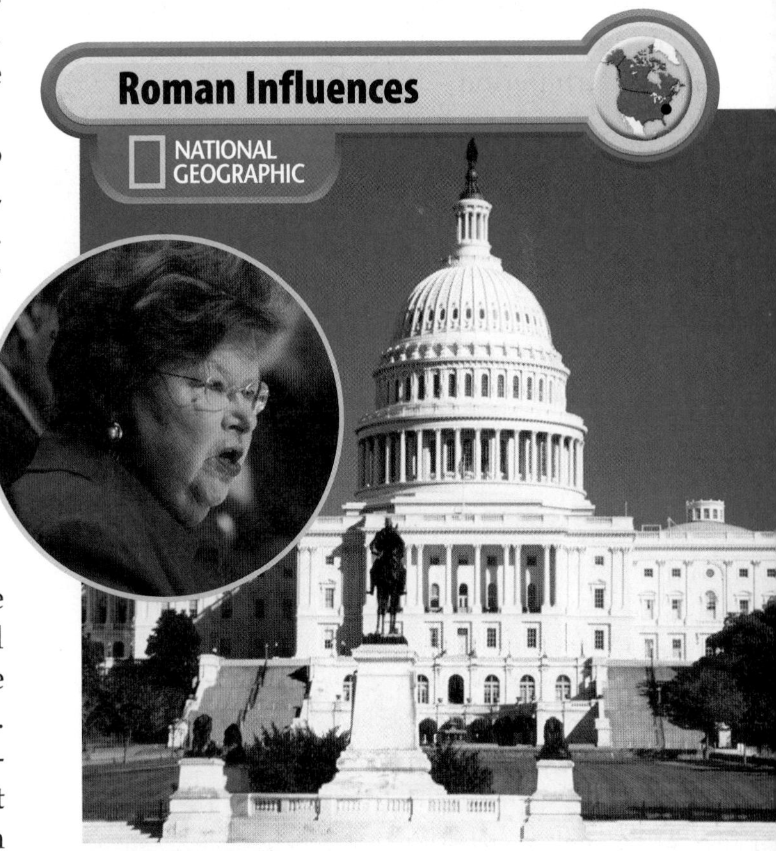

▲ The style of the U.S. Capitol was influenced by Roman architecture. Barbara Mikulski (inset) is a member of the U.S. Senate, which has powers similar to the ancient Roman Senate. ***Place Who made up the Roman Senate?***

Expansion of Europe

Main Idea **During the Middle Ages, European society, religion, and government underwent great changes.**

Geography and You Are there still parts of the world left to explore? Read to learn about changes in Europe, including how Europeans began to explore the far reaches of the world in the 1400s.

After Rome's fall, Europe entered the Middle Ages, a 1,000-year period between ancient and modern times. Christianity strongly influenced society during this period. In the 1300s, though, interest in education, art, and science exploded. Questions began to arise about earlier beliefs and practices. By the 1500s, Europe was experiencing changes that gave birth to the modern period.

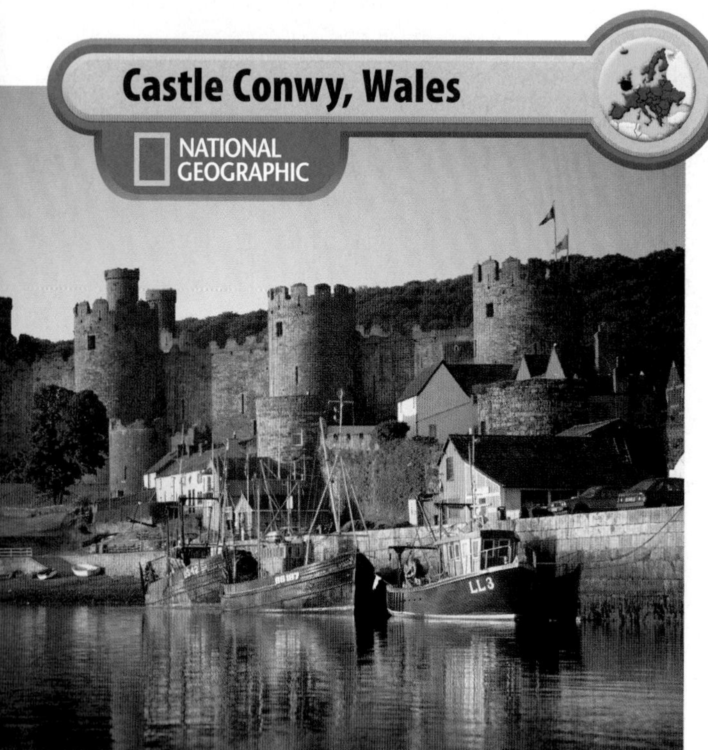

▲ During the Middle Ages, nobles were forced to build thick-walled castles for protection against invaders. ***Place*** **How can you tell that this castle was built for defense?**

A Christian Europe

During the Middle Ages, Christianity held a central place in people's lives. Two separate branches of the religion had formed, though. The Roman Catholic Church, based in Rome, was headed by a powerful **pope.** The Eastern Orthodox Church was centered in the Byzantine Empire. The Roman Catholic Church spread Roman culture and law to the Germanic groups living in western and central Europe. In eastern Europe, the Byzantine Empire passed on Eastern Orthodoxy and Greek and Roman culture to Slavic groups.

The Middle Ages

About A.D. 800, a Germanic king named Charlemagne united much of western Europe. After his death, this empire broke up. At that point, no strong governments existed to help western Europeans withstand invaders. To bring order, a new political and social system arose by the 1000s. Under this system, called **feudalism,** kings gave land to nobles. The nobles in turn provided military service, becoming knights, or warriors, for the king. As romantic as this may sound, life was hard for the masses. Most western Europeans were poor peasants. They farmed the lands of kings, nobles, and church leaders, who housed and protected them but who also limited their freedom.

The Crusades

In feudal times, the Christian faith united Europeans. Yet the religion of Islam, founded in the A.D. 600s by an Arab named Muhammad, was on the rise. Followers of Islam, called Muslims, spread through Southwest Asia to North Africa and parts of Europe. They also gained control of Palestine, alarming Christians who considered this the Holy Land. Beginning in the

1000s, nobles from western Europe gathered volunteers into large armies to win back the Holy Land. These religious wars, called the Crusades, were only partly successful. Muslims eventually recaptured much of the region.

The Crusades, however, had a major impact on Europeans. Goods began to flow more steadily between Europe and the Muslim lands. This trade benefited European kings, who taxed the goods that crossed their borders. Kings also took over land from nobles who left to fight in the Crusades. As a result, feudalism gradually withered, and Europe's kingdoms grew stronger and larger. Many of them later became modern Europe's **nation-states.** A nation-state is a country made up of people who share a common culture or history.

Muslim-Christian conflict arose again in Spain in the late 1400s. Since the A.D. 700s, Muslim groups had controlled parts of Spain. In what came to be known as the Reconquest, Spanish rulers forced out Muslims and united the country in 1492.

Meanwhile, in the 1300s, people all across Europe were battling a frightful disease. The bubonic plague, or Black Death, spread rapidly and killed perhaps a third of Europe's population. One consequence was a shortage of labor. Although the shortage hurt the economy, it helped workers earn higher wages and gain more freedom. In this way, the Black Death became another force that weakened feudalism.

The Renaissance

As parts of Europe recovered from the Black Death, interest in art and learning revived. Ways of thinking changed so much between about 1350 and 1550 that this period is called the Renaissance, from the French word for "rebirth."

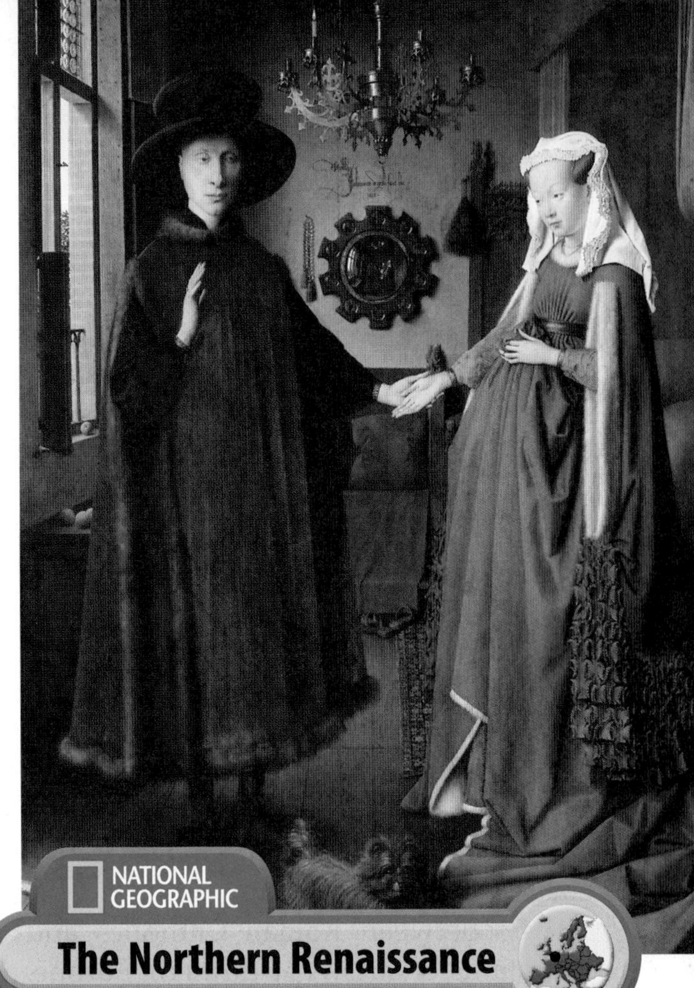

NATIONAL GEOGRAPHIC

The Northern Renaissance

Several great Renaissance artists were from northern Europe, including Jan van Eyck, who painted this portrait of a married couple. ***Location*** **Why did the Renaissance begin in Italian city states?**

The Renaissance thrived in Italian city-states, such as Florence, Rome, and Venice. Merchants in these city-states had gained great wealth through trade with Asia and the Mediterranean world. They then used this wealth to support scholars and artists. Poets, sculptors, and painters, such as Michelangelo and Leonardo da Vinci, created stunning masterpieces. People also took an interest in the cultures of ancient Greece and Rome. Another important element of the Renaissance was humanism, a way of thinking that gave importance to the individual and human society. Humanism held that reason, as well as faith, was a path to knowledge. Over time, Renaissance ideas and practices spread from Italy to other parts of Europe.

The Reformation

During the 1500s, the Renaissance idea of humanism led people to think about religion in a new way. Some people felt there were problems in the Roman Catholic Church that needed to be corrected.

In 1517 Martin Luther, a German religious leader, set out to reform, or correct, certain church practices. The pope in Rome, however, did not accept Luther's ideas, and Luther broke away from the Roman Catholic Church. Luther's ideas sparked a religious movement called the Reformation, which led to a new form of Christianity called Protestantism. By the mid-1500s, different Protestant groups dominated northern Europe, while the Roman Catholic Church remained strong in southern Europe.

Wars between Roman Catholics and Protestants soon swept through Europe. The Reformation thus shattered the religious unity of Europeans. It also strengthened the power of monarchs. As the **authority** of church leaders was challenged, kings and queens claimed more authority for themselves.

European Explorations

As Europe's kingdoms grew stronger, European seafarers began a series of ocean voyages that led to a great age of exploration and discovery. During the 1400s, Portugal wanted an easier way to get exotic spices from India and other parts of East Asia. Portuguese navigators developed new trade routes by sailing south around the continent of Africa to Asia. Sailing for Spain in 1492, the Italian-born explorer Christopher Columbus tried to find a different route to Asia. Instead of sailing south around the coast of Africa, Columbus attempted to sail west, across the Atlantic Ocean. Columbus's voyage took him to the Americas, continents unknown in Europe at the time.

In the Americas, Spain found gold and other resources and grew wealthy as a result of its overseas expeditions. Its success made other European countries, such as England, France, and the Netherlands, eager to send forth their own explorers. Conquests followed these voyages. Europeans began founding colonies, or overseas settlements, in the Americas, Asia, and Africa. Trade with the colonies brought Europe great wealth and power. Sadly, though, the Europeans often destroyed the local cultures in the lands they claimed.

Reading Check **Determining Cause and Effect** What changes did the Black Death bring to Europe?

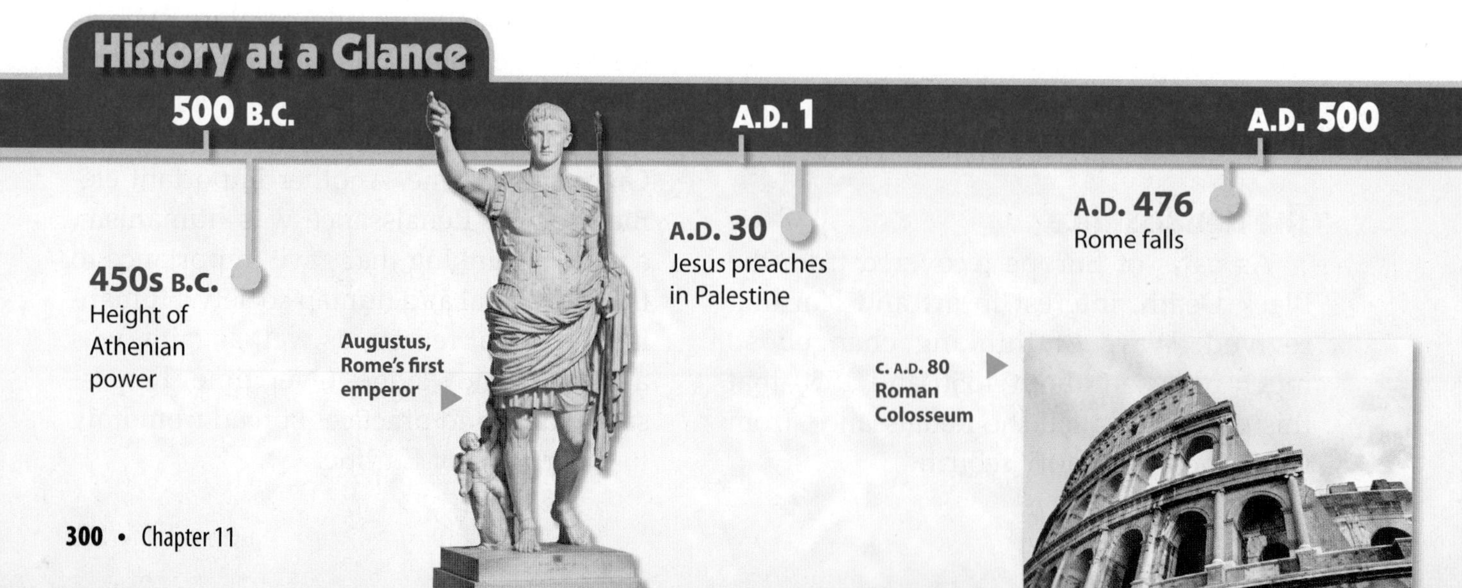

Modern Europe

Main Idea **From the 1600s to the 1800s and beyond, new ideas and discoveries helped Europe become a global power.**

Geography and You How would your life be different without computers, cell phones, or other modern technologies? Read on to discover how new technology changed Europe after 1600.

From the 1600s to the 2000s, Europe experienced rapid change. New machines helped economies grow. Powerful new weapons, however, made for deadly wars.

The Enlightenment

After the Renaissance, educated Europeans turned to science as a way to explain the world. Nicolaus Copernicus, a Polish mathematician, concluded that the sun, not the Earth, is the center of the universe. An Italian scientist named Galileo Galilei believed that new knowledge could come from carefully observing and measuring the natural world. These and other ideas sparked a **revolution**, or sweeping change, in the way people thought. During this Scientific Revolution, many Europeans relied on reason, rather than faith or tradition, to guide them. Reason, they believed, could bring both truth and error to light. As a result, the 1700s became known as the Age of Enlightenment.

Englishman John Locke was an important Enlightenment thinker. He said that all people have natural rights, including the rights to life, liberty, and property. He also said that when a government does not protect these rights, citizens can overthrow it. The American colonists later used Locke's ideas to support their war for independence from Britain in 1776.

Inspired by the American example, the people of France carried out their own political revolution in 1789. They overthrew their king, executed him three years later, and set up a republic. The French republic did not last long, however. A brilliant military leader named Napoleon Bonaparte gained power and made himself emperor. Napoleon was a small man with big ambitions. His armies conquered much of Europe, until several countries united to defeat him in 1815.

Political revolutions continued to erupt in Europe in the 1800s. By 1900, most countries had responded by limiting the powers of rulers and guaranteeing at least some political rights to citizens.

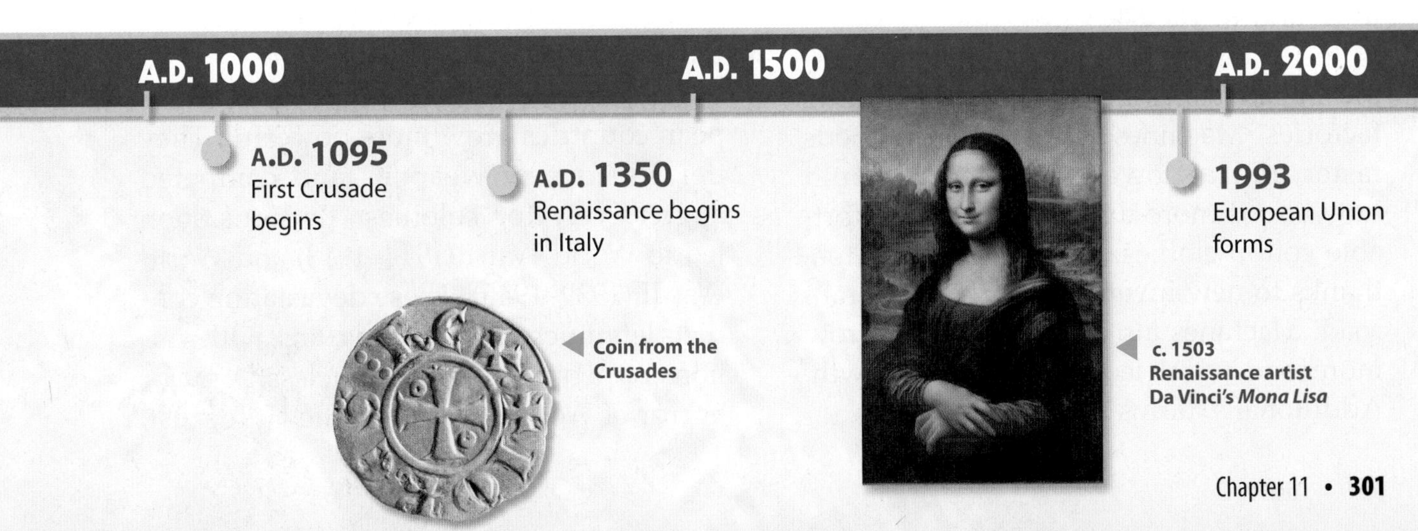

Coin from the Crusades

c. 1503 Renaissance artist Da Vinci's *Mona Lisa*

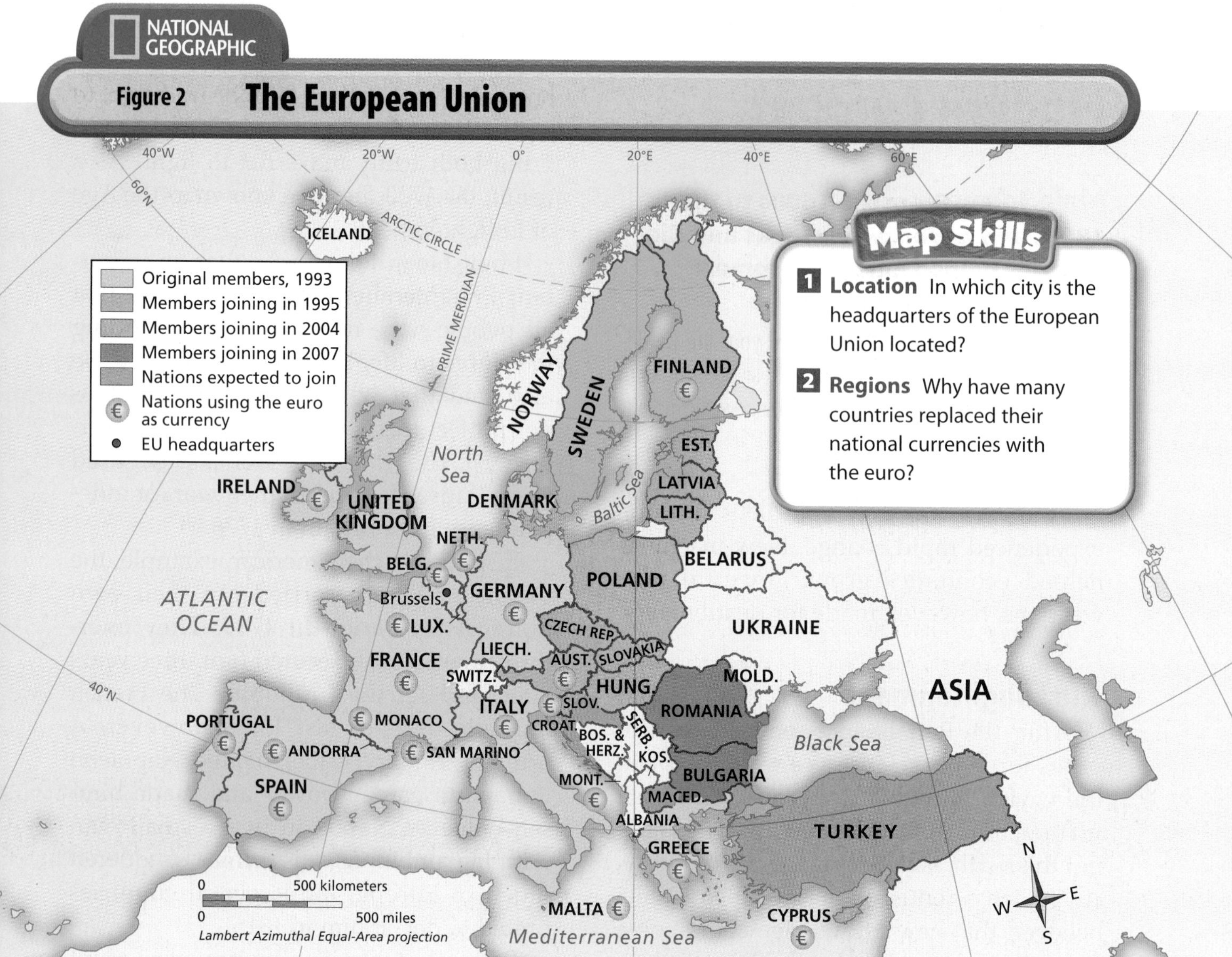

Industry and Conflict

Meanwhile, an economic revolution was also transforming Europe. The Industrial Revolution began in Britain. As it spread, it changed the way that people across Europe lived and worked.

Instead of making goods by hand, people began using machines and building factories. Machines could produce goods faster and at lower cost. People could now afford more things, such as comfortable cotton clothes. Travel improved, too, thanks to new inventions such as the railroad. Machines also helped farmers grow more food, which led to population growth. Additionally, farms required less labor.

Many people left farms to find work in cities. Cities became crowded, industries spewed out pollution, and diseases spread. Urban life remained grim for many Europeans until the mid-1800s. In the long run, though, the achievements of the Industrial Revolution benefited most people.

Industrial advances also helped European countries grow more powerful. They developed new weapons and competed aggressively for colonies. Tensions soon led to World War I (1914–1918) and World War II (1939–1945). These devastating conflicts left much of Europe in ruins, with millions of people dead or homeless. A major horror of World War II was the **Holocaust,** the mass killing of 6 million European Jews by Germany's Nazi rulers.

After World War II, the United States and the Soviet Union became rivals in the Cold War. This was not a war of bullets and bombs, but a struggle for world power. Much of Western Europe allied with the United States. Most lands in Eastern Europe allied with the Communist Soviet Union. **Communism** is a system in which the government controls the ways of producing goods.

A New Era for Europe

In 1989 people in Eastern Europe forced several Communist governments from power and set up new democracies. A year later, East and West Germany merged to become one democratic state. Then in 1991, the Soviet Union broke apart.

In 1993 several democracies in Western Europe formed the European Union (EU). The goal of the organization, which now also includes eastern European countries, is a united Europe. The EU allows goods, services, and workers to move freely among member countries. It has also created a common EU **currency** called the euro. Member countries, shown in **Figure 2,** can now trade more easily among themselves because there is no need to exchange, for example, French francs for German marks.

Reading Check **Describing** What was the Age of Enlightenment?

Section Review

Social Studies ONLINE
Study Central™ To review this section, go to glencoe.com.

Vocabulary

1. **Explain** the significance of:

a. classical	**e.** emperor	**i.** revolution
b. city-state	**f.** pope	**j.** Holocaust
c. democracy	**g.** feudalism	**k.** communism
d. republic	**h.** nation-state	

Main Ideas

2. **Describing** Describe the political system of ancient Athens.
3. **Explaining** How did the Crusades help lead to the creation of modern Europe's nation-states?
4. **Summarizing** Use a diagram like the one below to summarize the changes brought about by the Industrial Revolution.

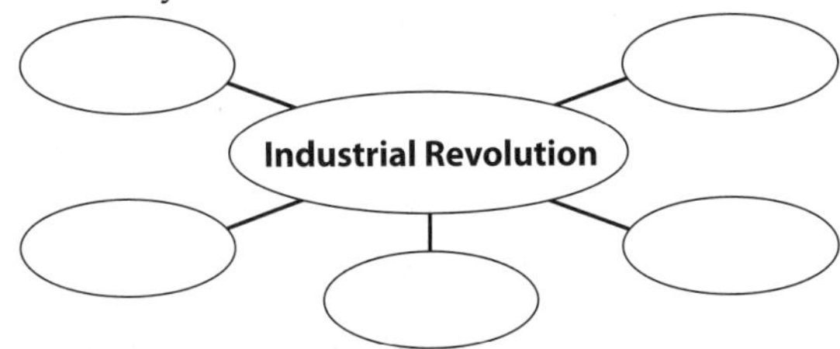

Critical Thinking

5. **Analyzing** How did Rome build a large, strong empire?
6. **Challenge** Which of Europe's revolutions do you think was most important for the creation of modern Europe? Explain your answer.
7. **BIG Idea** Provide an example of how political or social ideas, such as democracy or Christianity, spread in Europe.

Writing About Geography

8. **Using Your FOLDABLES™** Use your Foldable to write a summary describing how governments in Europe have changed over time.

Learning in School: Should All Students Speak the Same Language?

In Europe, migration between countries is common. As a result, many students do not speak the local language where they live. Some educators believe that all students should speak the nation's official language in school. For example, in Berlin, Germany, several schools allow only the German language to be spoken during class, on school property, and on school trips. However, others disagree, arguing that students should be allowed to learn in their own languages.

For: Speaking the Same Language

I believe that knowledge of the German language is the key to integration [becoming part of society] and to success both at school and in a future profession. . . . The pupils themselves are very satisfied with [this rule], because they know that speaking correct German increases their opportunities. . . .

When children start school and don't speak the language correctly, they . . . receive worse grades. That continues throughout their schooling and in the end they aren't able to get a vocational [job] training place. That's why we are in favour of introducing language tests starting from the age of four and thereby promoting language skills from kindergarten on.

—Armin Laschet
Minister for Generations, Family, Women, and Integration
North Rhine-Westphalia, Germany

Speaking the Same Language

Banning pupils from speaking their [traditional] languages in the schoolyard is not the answer, even if it were workable, which it is not. Other means of developing their linguistic [language] skills, such as pre-school instruction in German, are much more likely to be effective and should be fully supported.

. . . It is perfectly acceptable to ban other languages within the classroom. But outside the classroom pupils should be free to speak whichever language they like. Banning pupils' [traditional] languages sends a message that they are somehow "second class" citizens, which is likely to promote resentment rather than integration. . . .

Many children of immigrants choose to communicate in German in any case. . . .

There also appears to be a fallacy [mistaken belief] that speaking another language somehow [takes away] from pupils' ability with German. This is not the case. Humans have an almost unlimited ability to learn languages and in general there is no reason why the average person cannot master two or more languages.

—David Gordon Smith
Editor, Expatica: Germany

You Be the Geographer

1. **Identifying** What reasons do Laschet and Smith give to support their opinions?
2. **Critical Thinking** What might be some challenges for a student who speaks a different language than that of the other students in his or her class?
3. **Read to Write** Write a paragraph that explains your own opinion about students speaking only one language at school.

Guide to Reading

BIG Idea

Culture groups shape human systems.

Content Vocabulary

- ethnic group *(p. 307)*
- welfare state *(p. 307)*
- fertility rate *(p. 307)*
- urbanization *(p. 308)*
- secular *(p. 310)*

Academic Vocabulary

- bond *(p. 307)*
- attitude *(p. 310)*

Reading Strategy

Organizing Information Use a diagram like the one below to list key facts about Europe's population patterns.

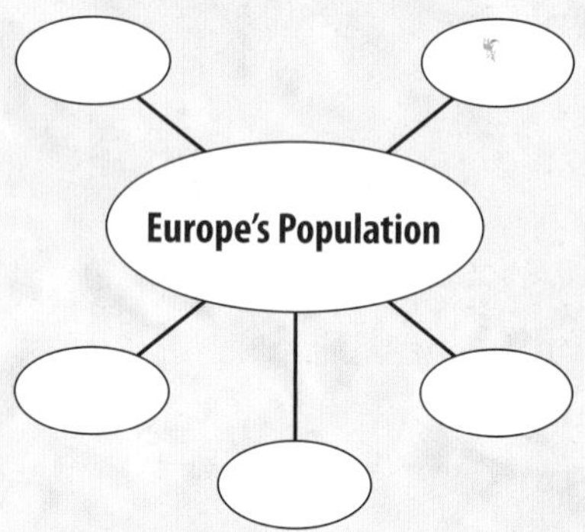

Cultures and Lifestyles

Picture This Bog snorkeling? For more than twenty years, competitors wearing snorkels and flippers have met in the small town of Powys, Wales, to swim in its slimy bog. The challenge is to swim the fastest without using any standard swimming strokes. The just-for-fun event has attracted swimmers from as far away as South Africa and Australia. Read Section 2 to learn more about the cultures and lifestyles of Europeans.

▼ **Decorated in blue paint, this swimmer hopes to win first place.**

Population Patterns

Main Idea Ethnic differences and population changes pose challenges for Europe.

Geography and You How do you treat a new person who joins your class? What kind of challenges does he or she face? Read to discover how Europe is responding to its new immigrant populations.

Europe's people are crowded into a relatively small space. The population is not distributed evenly, however, and it continues to undergo change.

A Rich Ethnic Mix

Today Europe is home to many ethnic groups. An **ethnic group** is a group of people with shared ancestry, language, and customs. Europe's ethnic mix has resulted from migrations, wars, and changing boundaries over the centuries.

Many Europeans identify strongly with their particular country or ethnic group. Having a common heritage or culture creates **bonds** among people. National and ethnic loyalties, however, have also led to conflict. By 2008, disputes among ethnic groups had split Yugoslavia into seven separate countries. In some of these new countries, ethnic hatred sparked the worst fighting in Europe since World War II.

Despite divisions, Europeans have a growing sense of unity. They realize that because their countries are linked by geography, cooperation can help bring peace and prosperity. In addition, the people share many values that go beyond ethnic or national loyalties. For example, Europeans value democracy and human rights. They also believe that a government must care for its citizens. Many European countries are **welfare states** in which the government is the main provider of support for the sick, the needy, and the retired.

NATIONAL GEOGRAPHIC

A Changing Population

These schoolgirls in London demonstrate Europe's growing immigrant population and ethnic diversity. ***Movement*** **From where have people immigrated to Europe?**

Population Changes

Because of recent immigration, Europe's population is still undergoing change. Since World War II, many people from Asia, Africa, and Latin America have settled in Europe. Tensions have risen as immigrants and local residents compete for jobs, housing, and social services. As a result, immigrants have not always felt welcome in many places in Europe.

European countries deal with immigrants in various ways. Some want immigrants to adapt quickly, so they require newcomers to learn the national language. Other countries pass laws to keep immigrants from settling within their borders. Still others try to improve educational and job opportunities for newcomers.

You might be surprised to learn that although the number of immigrants is increasing, the region's overall population is decreasing. Europe has a low **fertility rate,** which is the average number of children born to each woman.

As a result, Europe is expected to have 10 percent fewer people by 2050. This is worrisome, because there will be fewer workers to keep Europe's economy growing. Meanwhile, Europeans are living longer because of better health care. As older people increasingly account for a greater share of the population, young workers will face higher taxes to support them.

Reading Check **Making Generalizations** How do national and ethnic loyalties benefit and harm Europeans?

NATIONAL GEOGRAPHIC

European Fashion

With their generally high incomes, many Europeans can afford the latest fashions, like the one worn by this model at a fashion show in Paris. ***Regions*** **How might a shrinking population affect Europe's economy?**

Life in Europe

Main Idea **European lifestyles today reflect the region's urban society and level of wealth.**

Geography and You Does the idea of living in a city appeal to you? Read to discover how cities play an important part in the lives of Europeans.

Customs, languages, and religions have always differed among Europeans. In recent decades, however, differences in lifestyles among Europe's peoples have lessened as a result of industrial and economic growth, urban growth, and improved standards of living. Today, most Europeans are well-educated city dwellers with comfortable incomes.

Cities

Beginning in the late 1700s, the Industrial Revolution changed Europe from a rural, farming society to an urban, industrial society. Rural villagers moved in large numbers to urban areas. This concentration of people in towns and cities is known as **urbanization.** Many of Europe's cities grew quickly and became some of the world's largest.

Today, three of every four Europeans live in cities. Paris and London rank among the largest urban areas on the globe. The next biggest cities are Milan, Italy; Madrid, Spain; and Berlin, Germany.

Many European cities blend the old and the new. Ancient landmarks often stand near modern highways and skyscrapers. European cities are also crisscrossed by public transportation systems that bring people to jobs and urban attractions. In recent decades, however, more Europeans have bought cars and have chosen to live in suburbs outside the cities.

Transportation

Most of Europe's transportation systems are government owned. Standards differ from country to country, but overall, Europe boasts one of the world's most advanced transportation networks. The rail system links cities and towns across the continent. Trains travel underwater between England and France via a 31-mile (50-km) tunnel under the English Channel, known as the Chunnel. France developed the use of high-speed trains, which cause less damage to the environment than most other forms of transportation. High-speed rail lines now operate in many European countries.

Highways also allow high-speed, long-distance travel. Cars can zip along Germany's autobahns at more than 80 miles (129 km) per hour. Trucks use the roadways to carry the great majority of freight within Europe.

Canals and rivers are also used to transport goods. The Main-Danube Canal in Germany links hundreds of inland ports between the North Sea and the Black Sea. Europe's long coastline is dotted with other important ports, such as Rotterdam, in the Netherlands. This is one of the busiest ports in the world.

Airports connect European cities too. Planes fly both people and goods to their destinations all around Europe.

Education and Income

Europeans take schooling seriously. They tend to be well educated and have some of the highest literacy rates in the world. More than three-quarters of young people complete high school.

Because of their high levels of education, Europeans earn more money than people in many other parts of the world. There are differences, however, from place to place. Incomes are higher in northern and western Europe than in southern and eastern Europe. Many eastern European countries are still struggling to rebuild economies that were damaged by conflicts or slowed by years of Communist rule. Throughout Europe, service industries, such as banking, provide more jobs than any other economic activity.

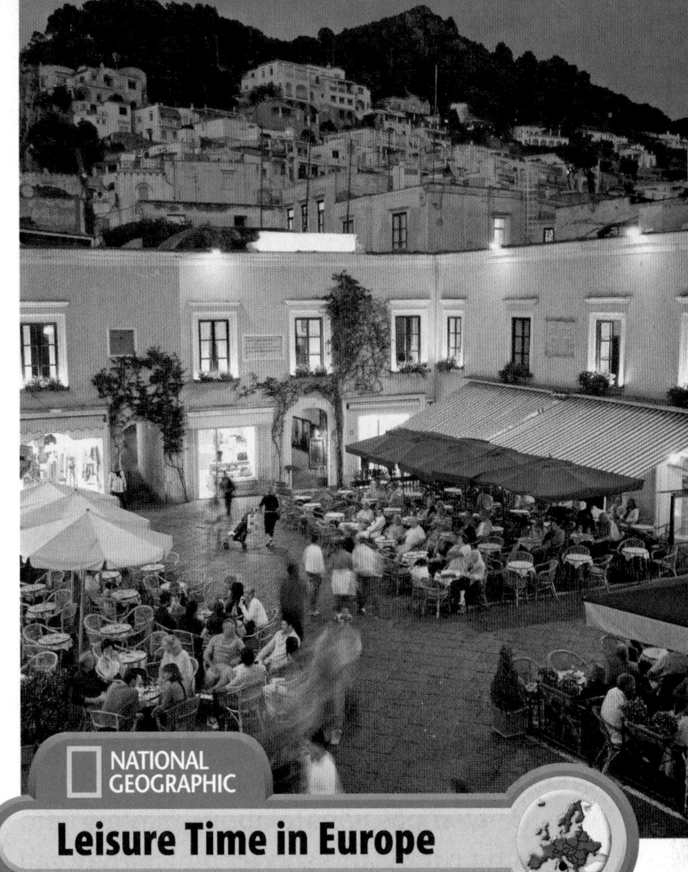

Leisure Time in Europe

Europeans enjoy dining with friends at outdoor cafés, such as this one on the island of Capri, Italy. ***Regions*** **What other leisure activities are popular with Europeans?**

Income can also vary greatly within a country. For example, unemployment and poverty are common in southern Italy. Mountains and a lack of natural resources in the area have slowed the development of industry. Workers are better off in northern and central Italy, where rich farmland and modern industries provide many jobs.

Leisure

Their relatively high incomes allow many Europeans to enjoy their leisure time. They have a generous amount of it too!

In a number of European countries, workers receive four weeks of paid vacation each year. Many Europeans use this vacation time to travel. France and Italy are popular vacation spots because of their lively cities, beautiful countryside, mild climate, and fine food.

Europeans also take full advantage of their natural surroundings. The region's mountains, seas, lakes, and rivers provide great opportunities for recreation. Winter sports such as ice skating and skiing had their beginnings in Scandinavia about 5,000 years ago. In summer, Europeans lace up their hiking boots, hop on their bikes, or take to the water. Many Europeans are also passionate about playing and watching rugby and soccer, which they call football.

Reading Check **Making Connections** What type of industry provides the most jobs in Europe?

Football in Europe

NATIONAL GEOGRAPHIC

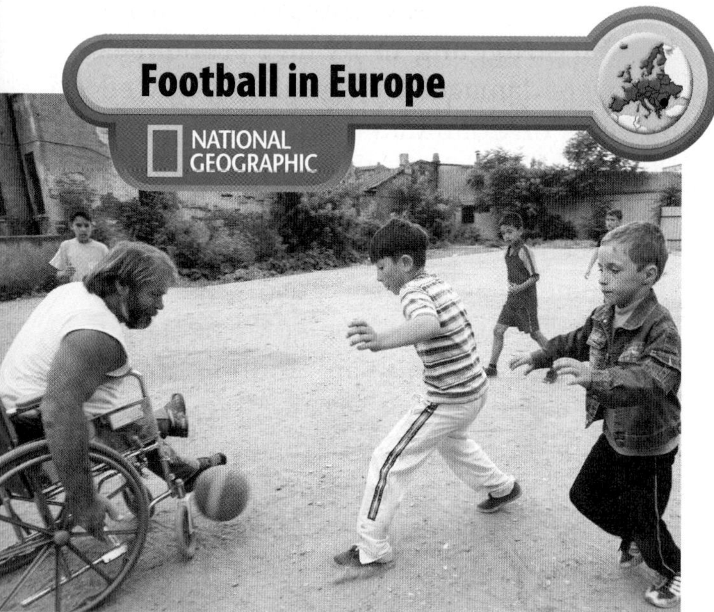

Many Europeans, like these Romanians, enjoy playing soccer. Rules for the game were first established in England in the 1800s. ***Regions*** **What winter sports developed in Europe?**

Religion and the Arts

Main Idea **Religion, especially Christianity, has had an important effect on European society and arts.**

Geography and You If you enjoy creative writing or making art or music, what ideas inspire you? Read on to find out how religion, nature, and other influences shaped the arts in Europe.

As in other parts of the world, religion has shaped European culture, including its arts. Today, though, European art reflects a variety of influences.

Religion

For centuries, Christianity was a major influence in European life. Since the 1700s, however, European **attitudes** have become more **secular,** or nonreligious. Today many Europeans do not belong to a particular religious group. Still, Christian moral teachings, such as respect for human life and compassion for others, remain core values throughout the region.

Most of Europe's Christians are Roman Catholic. As you can see in **Figure 3,** Roman Catholics are heavily concentrated in the southern part of western Europe and in some eastern European countries. Protestants are dominant in northern Europe. Eastern Orthodox churches are strongest in the southern part of eastern Europe.

Judaism and Islam have also influenced European culture. Judaism, like Christianity, reached Europe during Roman times. Despite eras of persecution, Jews have made major contributions to European life. Today, Jewish communities thrive in all major European cities. Thousands of Muslim immigrants are also settling in the region.

For the most part, Europeans of different faiths live together peacefully. In some cases, though, Europe's religious differences have led to violence. For years, hostility between Catholics and Protestants created conflict in Northern Ireland, a part of the United Kingdom. Since 1998, both sides have agreed to share political power, but the situation remains unstable. Religious and ethnic differences were also at the heart of troubles on the Balkan Peninsula. There, Roman Catholic, Eastern Orthodox, and Muslim groups fought over land and political power during the 1990s.

Arts

The arts have flourished in Europe for centuries. In ancient times, the Greeks and Romans constructed stately temples and public buildings with huge, graceful columns. During the Middle Ages, a new style known as Gothic architecture arose. Europeans built majestic churches called cathedrals, designing them with pointed arches and large, stained-glass windows.

Religion also inspired European art, literature, and music. From ancient times to the Middle Ages, artists and writers focused on holy or heroic subjects.

NATIONAL GEOGRAPHIC

Figure 3 **Europe's Religions**

Map Skills

1. **Place** Which religions are found in France?
2. **Regions** Why do you think most Christians in southern Europe are Catholic?

Composers wrote pieces to accompany religious services. In eastern Europe, Christian art included icons, or symbolic religious images painted on wood.

During the 1500s and 1600s, Renaissance artists continued to create religious art, but their art also portrayed life in the everyday world. Renaissance artists tried to make their works more realistic. When you study a painting by Leonardo da Vinci or a statue by Michelangelo, you will see lifelike figures. In the writings of England's William Shakespeare or Spain's Miguel de Cervantes, you will encounter believable characters with timeless problems.

Artistic creativity continued to soar in Europe. In the 1600s and 1700s, new types of music, such as opera and symphony, emerged. In the 1800s, musicians, writers, and artists developed a style known as Romanticism, which aimed to stir strong emotions. This style drew inspiration not from religion but from nature and historical events. Later in the 1800s, the Impressionist movement began. Impressionist painters used bold colors and brushstrokes to create "impressions" of the natural world.

In the 1900s, European artists moved away from portraying the world as it appeared to the human eye. They turned to abstract painting to express feelings and ideas. Architects began to create sleek, modern buildings using materials such as glass and concrete. New kinds of music, such as rock and roll, caught on. Today Europe's artists remain creative and influential.

Reading Check **Analyzing** How has religion influenced the arts in Europe?

Section 2 Review

Social Studies ONLINE
Study Central™ To review this section, go to glencoe.com.

Vocabulary

1. **Explain** how the terms *ethnic group, welfare state, fertility rate, urbanization,* and *secular* relate to Europe's population by writing a sentence containing each word.

Main Ideas

2. **Explaining** How do individual European countries deal with immigration?
3. **Describing** How does Europe's generally high level of education affect life there?
4. **Identifying** Use a chart like the one below to identify Europe's major religions, including the major forms of Christianity, and where each religion is generally located in Europe.

Major Religion	Where Found

Critical Thinking

5. **Challenge** Will immigration benefit Europe in the future? Explain your answer.
6. **Drawing Conclusions** What factors have slowed economic development in certain areas of Europe?
7. **BIG Idea** What factors help unify Europe's different ethnic groups today?

Writing About Geography

8. **Expository Writing** Write a paragraph comparing European and American cultures.

Visual Summary

Ancient Europe

- The Greek city-state of Athens introduced the world's first democracy.
- Rome influenced later civilizations through its legal system, its language, and its role in the spread of Christianity.
- Invasions by Germanic peoples led to the Roman Empire's decline.

Caesar Augustus

Europe's Expansion

- Christianity shaped Europe's society and culture during the Middle Ages.
- The Renaissance, which began in Italy, brought about a new interest in learning.
- European countries controlled various parts of the world as a result of overseas explorations.

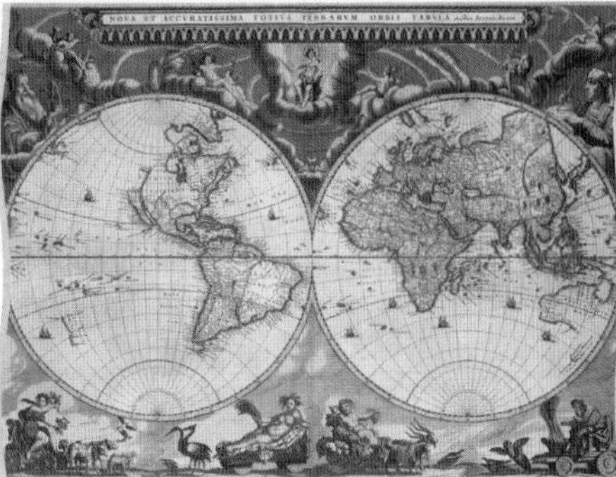

Map of the world, c. 1620

Modern Europe

- Through revolutions people challenged the power of kings and demanded certain rights for citizens.
- As industries grew, many people left rural areas to find work in city factories.
- Two costly world wars led European countries to seek peace and greater unity.

Eurostar train, London

Population Patterns

- Europe is densely populated in many areas.
- Europe's population is aging, and the total population is declining.
- Many people have immigrated to Europe from Asia, Africa, and Latin America.

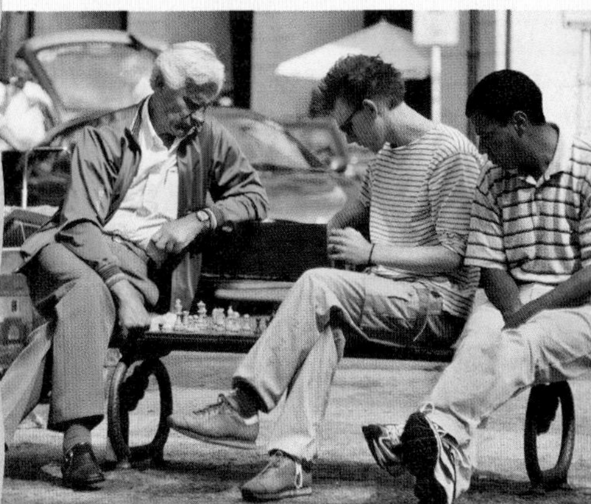

Playing chess in Prague, Czech Republic

Life and Culture

- Europeans tend to live in urban areas and have relatively high levels of education and income.
- With more leisure time, Europeans enjoy sports such as soccer.
- European society and culture have become more secular.

Study anywhere, anytime! Download quizzes and flash cards to your PDA from glencoe.com.

CHAPTER 11 ASSESSMENT

STANDARDIZED TEST PRACTICE

Eliminate answers that do not make sense. For instance, if an answer refers to a region of the world other than Europe, you know it cannot be correct.

Reviewing Vocabulary

Directions: Choose the word(s) that best completes the sentence.

1. In the system known as ______, nobles provided service to a king in exchange for land.

A communism
B feudalism
C a republic
D a city-state

2. The head of the Catholic Church is called the ______.

A emperor
B king
C pope
D archbishop

3. The ______ world refers to ancient Greece and Rome.

A classical
B Hellenistic
C democratic
D dominant

4. In a(n) ______ the government is the main provider of support for the sick, needy, and retired.

A democracy
B corporate donor
C urbanized center
D welfare state

5. Because Europe has a low ______, it is expected to have 10 percent fewer people by 2050.

A urbanization rate
B fertility rate
C ethnic group
D secular group

Reviewing Main Ideas

Directions: Choose the best answer for each question.

Section 1 *(pp. 294–303)*

6. The prosperous Greek city-state of Athens introduced the world's first ______.

A democracy
B artistic thinkers
C free society
D army

7. The ______ resulted in the rise of a new form of Christianity called Protestantism.

A Crusades
B Reformation
C Enlightenment
D Renaissance

Section 2 *(pp. 306–312)*

8. Europe's population shifts may pose challenges because

A few jobs exist for the growing population.
B few people immigrate into the region.
C Europeans are no longer living as long.
D there will be fewer workers to keep Europe's economy growing.

9. Which of the following most clearly explains how European lifestyles reflect the region's level of wealth?

A Europeans are able to enjoy their leisure time.
B Transportation standards vary from country to country.
C More than three-quarters of Europeans complete high school.
D many eastern Europeans are still struggling from years of Communist rule.

GO ON

Critical Thinking

Directions: Choose the best answer for each question.

10. Which of the following was a result of the Renaissance?

A A revival in art and literature began.

B Workers gained higher wages and more freedoms.

C A new form of Christianity developed.

D Many European countries began to send out explorers.

Base your answer to question 11 on the map below and your knowledge of Chapter 11.

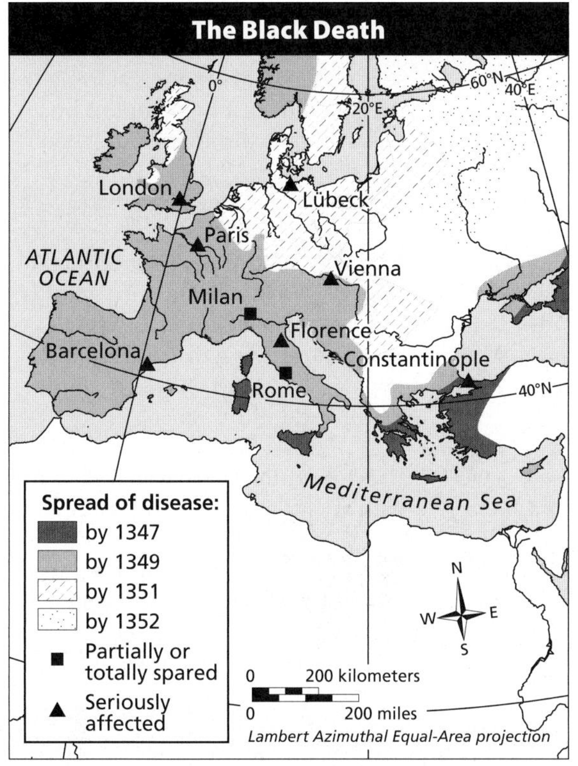

11. Which city was first affected by the Black Death?

A Constantinople

B Paris

C Lübeck

D Barcelona

Document-Based Questions

Directions: Analyze the document and answer the short-answer questions that follow.

In 1215 nobles in England forced King John to sign the Magna Carta. This document gave common people some freedoms and limited the power of the king. It was an important influence in the rise of democratic governments.

> *TO ALL FREE MEN OF OUR KINGDOM*
>
> *we have also granted, for us and our heirs for ever, all the liberties written out below . . .*
>
> *(20) For a trivial offence, a free man shall be fined only in proportion to the degree of his offence. . . .*
>
> *(30) No sheriff, royal official, or other person shall take horses or carts for transport from any free man, without his consent. . . .*
>
> *(40) To no one will we sell, to no one deny or delay right or justice.*
>
> —Magna Carta

12. What does article 20 of the Magna Carta mean? Rephrase it in your own words.

13. What complaints do you think the Magna Carta attempted to fix?

Extended Response

14. Take the role of a citizen of a European country that is not in the European Union but is debating whether to apply for membership. Write a short speech explaining why you think your country should or should not join the EU.

STOP

For additional test practice, use Self-Check Quizzes—Chapter 11 at glencoe.com.

Need Extra Help?

If you missed question...	1	2	3	4	5	6	7	8	9	10	11	12	13	14
Go to page...	298	298	295	307	307	295	300	308	309	299	299	299	299	303

TIME JOURNAL

"Hello! My name is Kade.

I am 13 years old and live in Paris, the capital of France. My family moved here from Guinea, a country in Africa. Like other immigrant families, we blend some of our own traditions with France's rich culture. Here's how I spend my day."

8:45 a.m. My mom wakes me up. I sleep late today because school does not start until 10 o'clock on Mondays! (On other days, it begins at 8 o'clock.) I shower and get dressed.

9:15 a.m. I eat breakfast with my parents and little brother and sister. We have warm chocolate milk, which we drink out of bowls. We also have flaky rolls called croissants. Croissants are delicious. I like them best when they are filled with chocolate.

9:40 a.m. I meet my friends and walk to school. We can see the Eiffel Tower from our building.

10:00 a.m. It's the start of a long day—and a long week. Like many French kids, I go to school six days a week. Wednesdays and Saturdays are half days, though. I look forward to them!

10:15 a.m. In history, my first class, we are learning about ancient Greece. Then I study geography. Our geography classroom is decorated with flags from all over the world.

12:00 p.m. In music class, I take an exam on the flute. I hope I did well!

1:00 p.m. It's time for *déjeuner* (day•zhuh•NAY), or lunch. Many students go home for lunch, but I buy my meal in the cafeteria. Today I choose a grapefruit, chicken nuggets, and pasta.

2:00 p.m. I go to the computer lab for technology class. After that, we have a short recess period. My friends and I play dodgeball. I enjoy sports. I think I would like to be a handball instructor one day.

3:00 p.m. In English class, we practice saying sentences that begin with the phrase, "Do you like…?" My English is already strong because we often speak it at home, but this lesson is fun.

4:00 p.m. School is over. I walk back home with my friends. Today, some of them come to my apartment. We watch music videos and listen to CDs. I like most kinds of music, including rock and reggae.

6:00 p.m. My dad will be home from work soon. He designs and sells clothing. My mom is starting to prepare dinner. I help her out by picking up my brother from school.

6:30 p.m. I do my homework.

7:30 p.m. Dinner is ready. We are having rice with fish and vegetables in a spicy sauce. It is a dish that is popular in Guinea. Now my family has brought it to France!

8:30 p.m. I play and watch cartoons with my little brother and sister.

9:30 p.m. I brush my teeth and go to bed. I listen to music until I get sleepy. I use earphones so I do not wake my sister.

ILLUSTRATIONS BY BOOKMAPMAN

BEFORE CLASS Kade meets up with her friends. French students go to school six days a week.

DODGEBALL At Kade's school, this sport is popular at recess. Soccer is still the number one sport in France, as it is in most of Europe.

MAP TIME Kade's teacher checks her work. Geography is a required subject in France, a nation that borders several countries.

BACKPACKED Kade Diallo (kahd dee•AH•low) passes the Eiffel Tower on her way to school. The teen moved to France with her parents and brother and sister in search of better economic opportunities.

RICHARD HARBUS/POLARIS (4)

PATRICK SHEANDELL O'CARROLL/GETTY IMAGES

What's Popular in France

STEVEN MARK NEEDHAM/PICTUREARTS/NEWSCOM

Cheese France produces hundreds of varieties. The average French person eats about 50 pounds of cheese each year.

Cycling Every July, France hosts a three-week, 2,000-mile bicycle race called the Tour de France. More people come to watch the race than any other sporting event in the world.

FRANCK FIFE/AFP/NEWSCOM

Fashion France is home to some of the world's top clothing designers. Styles that start in workrooms here end up in stores all over the world.

Say It in French

France's 60 million people are united by a common language, French. Like English and many other languages, the French language has roots in Latin. Try these simple French phrases.

Hello
Bonjour (bohn·ZHOOR)

Good-bye
Au revoir (oh reh·VWAH)

My name is ______.
Je m'appelle (zhuh mah·PELL) ______.

CHAPTER 12

Europe Today

Essential Question

Human-Environment Interaction Europe is one of the economic powerhouses of the world, home to many large companies that sell goods in the United States. Europe is also an important market for goods and services produced in North America, such as movies and computer programs. What factors help make a region an important world economic center?

Trafalgar Square, London

BIG Ideas

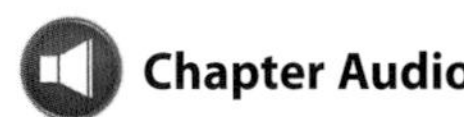
Chapter Audio

Section 1: Northern Europe

BIG IDEA Geographers organize the Earth into regions that share common characteristics. The countries of northern Europe have developed diverse economies and high standards of living.

Section 2: Europe's Heartland

BIG IDEA People's actions can change the physical environment. Today the countries of Europe's heartland are agricultural and manufacturing centers.

Section 3: Southern Europe

BIG IDEA Places reflect the relationship between humans and the physical environment. Seas and mountains have influenced where people live and how they work in southern Europe.

Section 4: Eastern Europe

BIG IDEA Geography is used to interpret the past, understand the present, and plan for the future. After changes in government, eastern Europe's economies are struggling to recover.

FOLDABLES™ Study Organizer

Summarizing Information Make this Foldable to help you collect information about Europe's people, politics, and economies.

Step 1 Fold the sides of a piece of paper into the middle to make a shutter fold.

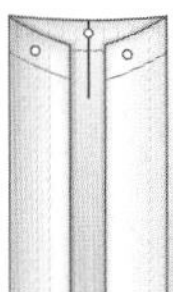

Step 2 Cut each flap at the midpoint to form four tabs.

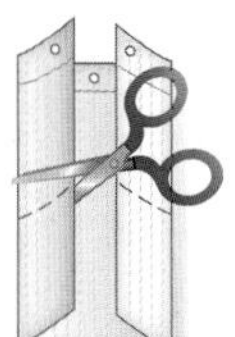

Step 3 Label the tabs as shown.

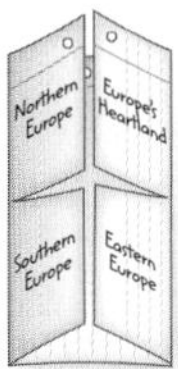

Reading and Writing Use the notes in your Foldable to write a short essay comparing the economies of Europe's four subregions.

Visit glencoe.com and enter ***QuickPass***™ code EOW3109c12 for Chapter 12 resources.

Northern Europe

Section Audio

Spotlight Video

Guide to Reading

BIG Idea

Geographers organize the Earth into regions that share common characteristics.

Content Vocabulary

- constitutional monarchy *(p. 322)*
- parliamentary democracy *(p. 323)*
- peat *(p. 325)*
- bog *(p. 325)*
- productivity *(p. 325)*
- geyser *(p. 327)*
- fjord *(p. 327)*
- geothermal energy *(p. 327)*

Academic Vocabulary

- differentiate *(p. 321)*
- document *(p. 322)*
- vary *(p. 326)*

Reading Strategy

Organizing Information Use a graphic organizer like the one below to organize key facts about the people and cultures of northern Europe.

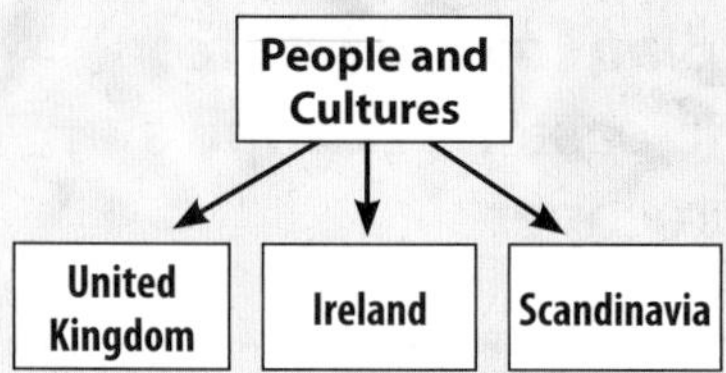

Picture This Iceland's huge chunks of moving ice are centuries old. Iceland, however, is not a bitter cold wasteland. It has a relatively mild climate even though it is near the Arctic Circle. The people of Iceland have adjusted to living in this climate and have made efficient use of the country's resources. Learn more about Iceland and other countries of northern Europe by reading Section 1.

A glacial wall in Iceland

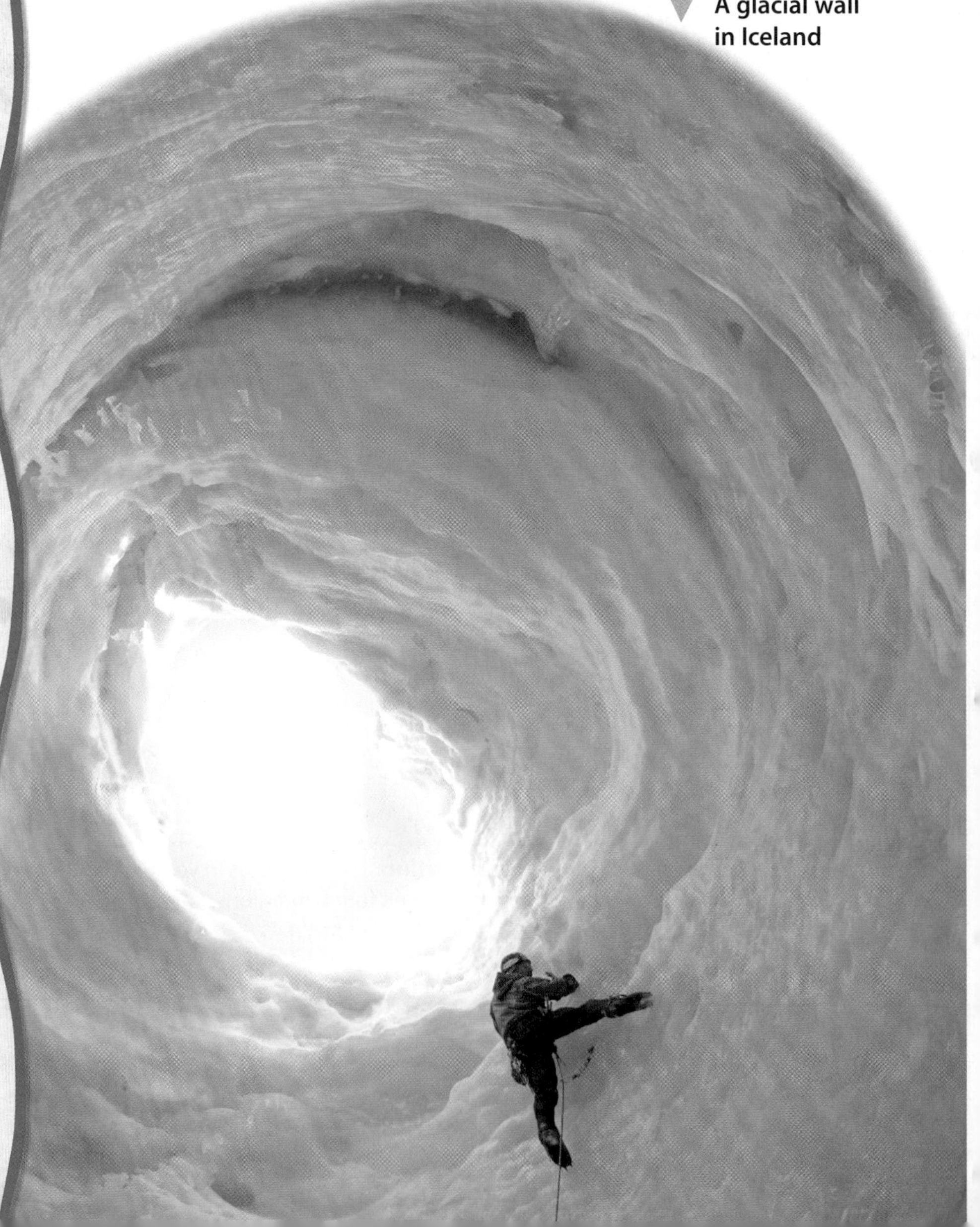

The United Kingdom

Main Idea **Once the center of a worldwide empire, the United Kingdom has had a great impact on the rest of the world.**

Geography and You Have you ever seen a picture of Big Ben, the large clock tower located in London? Big Ben is a symbol of the United Kingdom. Read to find out more about this country in the North Atlantic.

It is easy to be confused by the different names for the island nation off the northwest coast of mainland Europe. People sometimes call it Great Britain, the British Isles, or simply England. The true name, though, is the *United Kingdom of Great Britain and Northern Ireland,* or the *United Kingdom.*

The country includes four separate regions, which you can see in **Figure 1.** Three of them—England, Scotland, and Wales—make up the island of Great Britain. The fourth region, Northern Ireland, occupies a corner of the nearby island of Ireland. (The rest of that island is a completely independent country known as the Republic of Ireland.)

All the people of the United Kingdom can be described as British. Sometimes, though, people **differentiate** among them by referring to the English, the Scots, the Welsh, or the Irish.

The Land

Great Britain is separated from the rest of Europe by the English Channel. Historically, this body of water both connected and protected the British. They were close enough to the mainland to share in European culture. At the same time, they were far enough away to be largely safe from foreign invasions and free to develop their own government and economy.

Map Skills

1. **Regions** Which four regions make up the United Kingdom?
2. **Place** Which region probably has a larger percentage of people in rural areas, England or Scotland? Why do you think so?

Rolling fertile plains cover the southern and eastern areas of England. These plains support productive farms. Rough highlands and mountains are found to the north and west in Scotland and Wales. Poor soil and a cold climate make farming difficult in these areas, but many people herd sheep.

Government in the United Kingdom

The Palace of Westminster, with the clock tower known as Big Ben, lies in the heart of London. It is home to Parliament, the lawmaking body of the United Kingdom.
Place **How did the United Kingdom become a parliamentary democracy?**

In southeastern England, the Thames (TEHMZ) River helped make London a center for world trade. Today, shipping is much less important than it once was, and the Thames riverbanks in London are lined with apartment buildings rather than warehouses. London, however, remains a world center of finance and business.

The Economy

More than 250 years ago, British inventors and scientists sparked the Industrial Revolution. Today, the United Kingdom is still a major industrial and trading country. Manufactured goods and machinery are its leading exports. New computer and electronics industries, however, are gradually replacing these older industries. Service industries, such as banking and health care, are now a major part of the economy.

Coal once powered the British economy, but oil and natural gas are now the leading energy sources. These fossil fuels come from fields beneath the North Sea. These oil and gas fields meet most of the United Kingdom's energy needs. They also provide fuel exports that give the country a valuable source of income.

Government

The government of the United Kingdom is a **constitutional monarchy.** A king or queen serves as head of state and takes part in ceremonies, but elected officials actively run the government.

The British trace the roots of this form of government to the early 1200s. At that time, nobles forced King John of England to sign the Magna Carta, a **document** that took away some of the king's powers. For example, the king could no longer collect taxes unless a group of nobles agreed. Also, people accused of crimes had a right to fair trials by their peers, or equals.

Gradually, a lawmaking body called Parliament arose. In 1628 Parliament decided that King Charles I had misused his power. It forced him to sign the Petition of Right, which said that taxes could be enacted only if Parliament approved. In addition, the king could not imprison people unless they were convicted of a crime. As time passed, more limits were placed on the ruler's authority. The English Bill of Rights, passed in 1689, gave Parliament the power to tax and stated that monarchs could not suspend the laws or form their own armies. That document later helped shape the thinking of the men who wrote the U.S. Constitution.

Today, the United Kingdom is a **parliamentary democracy** as well as a constitutional monarchy. Voters elect members of Parliament, and the leader of the party with the most elected officials becomes prime minister, or head of the government. The prime minister can propose new laws, but only Parliament can put them into action. The prime minister must appear in Parliament regularly to explain and defend his or her decisions. Parliament also has the power to force the prime minister out of office and require new elections. This is a power the U.S. Congress does not have over the U.S. president.

Scotland, Wales, and Northern Ireland have regional legislatures that have control over matters such as health care and education. The Scottish Parliament even has the power to raise or lower taxes in Scotland.

The People

With more than 60 million people, the United Kingdom is the third-most-populous country in Europe. Nearly 9 of every 10 people live in cities. London is by far the largest city, with some 7.6 million residents.

The British people speak English, although Welsh and Scottish Gaelic (GAY·lihk) are spoken in some areas. Most people in the United Kingdom are Protestant Christians. Immigrants from South Asia, Africa, and the Caribbean area, however, practice religions such as Islam, Sikhism, and Hinduism.

In the 1700s and 1800s, when the United Kingdom had a powerful empire, British culture spread to many lands. As a result, the British sport of cricket is now played in Australia, South Asia, and the Caribbean. The English language is spoken in the United States, Canada, South Africa, and a number of other countries. Britain's rich literature of poems, plays, and novels is enjoyed worldwide, too.

Reading Check **Determining Cause and Effect** How has the location of the United Kingdom shaped its history?

Cricket: A British Sport

NATIONAL GEOGRAPHIC

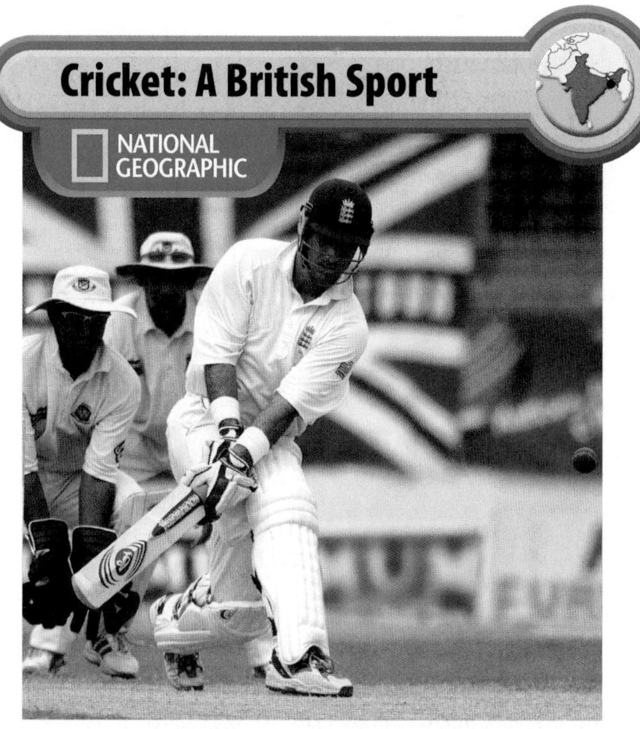

▲ A batsman, or player, hits the ball in a cricket match between England and the South Asian country of Bangladesh. ***Movement*** **How did British sports and culture spread to other lands?**

Charles Dickens
(1812–1870)

One of Britain's most famous novelists, Charles Dickens, had a difficult childhood because of family financial problems. As a result, Dickens developed a deep sympathy for the lower classes and for the young children who sometimes suffered from society's strict rules. These feelings are evident in many of his books.

Background Information

In *Hard Times,* Charles Dickens explores the problems raised by the Industrial Revolution. His book harshly criticizes the people who promoted this new way of working and the effects it had on the environment. In this passage, Dickens describes an industrial city, which he names Coketown.

Reader's Dictionary

interminable: unending

melancholy: sad

workful: useful

infirmary: hospital

dearest: for the highest price

HARD TIMES

By Charles Dickens

It was a town of red brick, or of brick that would have been red if the smoke and ashes had allowed it; but as matters stood it was a town of unnatural red and black. . . .

It was a town of machinery and tall chimneys, out of which **interminable** serpents of smoke trailed themselves for ever and ever, and never got uncoiled.

It had a black canal in it, and a river that ran purple with ill-smelling dye, and vast piles of buildings full of windows where there was a rattling and a trembling all day long, and where the piston of the steam-engine worked monotonously up and down, like the head of an elephant in a state of **melancholy** madness. It contained several large streets all very like one another, and many small streets still more like one another inhabited by people equally like one another, who all went in and out at the same hours, with the same sound upon the same pavements, to do the same work, and to whom every day was the same as yesterday and to-morrow, and every year the counterpart [duplicate] of the last and the next. . . .

You saw nothing in Coketown but what was severely **workful**. . . . All the public inscriptions in the town were painted alike, in severe characters of black and white. The jail might have been the **infirmary,** the infirmary might have been the jail, the townhall might have been either, or both, or anything else. . . . What you couldn't state in figures, or show to be purchasable in the cheapest market and salable in the **dearest,** was not [to be found there], and never should be. . . .

From: *Hard Times,* Charles Dickens, New York Books, Inc., n.d.

Analyzing Literature

1. **Making Inferences** How would you describe Coketown?
2. **Read to Write** Suppose you were a person who moved from a farm to work in a factory in Coketown. Write a letter to a family member that contrasts life on the farm with life in the city.

The Republic of Ireland

Main Idea **Ireland is growing economically, but a territorial dispute remains unsettled.**

Geography and You Why do you think Ireland is called the Emerald Isle? Read to find out about Ireland and its resources.

When people speak of Ireland, they usually mean the Republic of Ireland. This is the Catholic country that occupies the southern five-sixths of the island of Ireland. The country won its independence from the United Kingdom in 1922. The British, meanwhile, keep control of Northern Ireland, where most people are Protestants.

The Land

Ireland has the shape of a shallow bowl. The interior is a lowland plain with gently rolling hills. The coastal areas are rocky highlands and towering cliffs.

Ireland's regular rainfall produces lush, green fields. The landscape stays so green year-round that the country is nicknamed the Emerald Isle. Low-lying areas are rich in **peat,** or plants that have partly decayed in water. Peat is dug from **bogs,** or low swampy lands. It is then dried and can be burned for fuel.

The Economy

Irish farmers raise sheep and cattle and grow vegetables such as sugar beets and potatoes. Potatoes were Ireland's chief food in the 1800s. When disease destroyed the potato crop in the 1840s, more than one million people died. Another million left for the United States and other countries.

Today, manufacturing employs more of Ireland's people than farming does.

NATIONAL GEOGRAPHIC

Irish High-Tech Industry

Irish workers make computer parts in a laboratory "clean room." ***Place*** **How has Ireland's economy changed in recent years?**

The Irish work in industries that produce clothing, pharmaceuticals, and computer equipment. In recent years, the increased productivity of Irish workers has helped Ireland's economy. **Productivity** is a measure of how much work a person does in a specific amount of time. When workers produce more goods, companies earn higher profits and the workers earn higher incomes.

The People

The Irish trace their ancestry to the Celts, who settled the island hundreds of years ago. Irish Gaelic, a Celtic language, and English are Ireland's two languages. About 60 percent of the Irish live in cities or towns. Nearly one-third live in Dublin, the capital.

The Irish are very proud of their culture. Irish music and folk dancing are performed around the world. Of all the arts, however, the Irish have had their greatest influence on literature. Playwright George Bernard Shaw, poet William Butler Yeats, and novelist James Joyce are some of Ireland's best-known writers.

Conflict Over Northern Ireland

The Irish are also strong Catholics, and many of their Catholic neighbors in Northern Ireland would like to unite with them. However, most Protestants in Northern Ireland—who are the dominant group there—wish to remain part of the United Kingdom. This dispute over Northern Ireland has led to violence, especially from the 1960s to the 1990s. In 1998 leaders of the United Kingdom and the Republic of Ireland met with leaders of both sides in Northern Ireland. They signed an agreement to end the violence. In 2007 the heads of Northern Ireland's political parties agreed to share power in a new regional government.

✓ Reading Check **Identifying Cause and Effect** What happened as a result of the potato crop failure in the 1840s?

Irish Folk Dance

NATIONAL GEOGRAPHIC

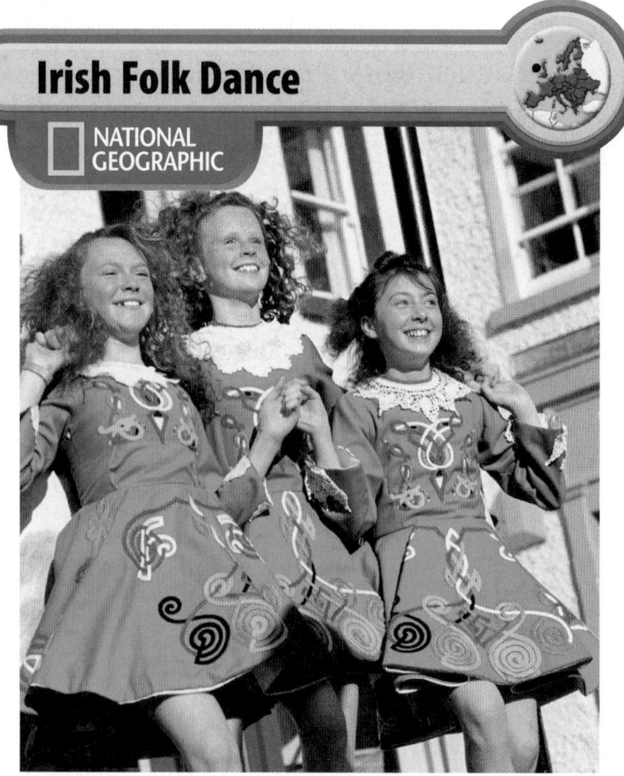

▲ The féis (FESH) is a celebration of Irish culture that includes dances such as the jig, reel, and hornpipe. ***Place*** **To whom do the Irish trace their ancestry?**

Scandinavia

Main Idea **The Scandinavian countries have similar cultures and high standards of living.**

Geography and You How would you like to live in a place where the sun never sets in midsummer? The Land of the Midnight Sun lies in the far north of Europe. Read on to see how the people there, known as Scandinavians, adapt to their environment.

Scandinavia, the northernmost part of Europe, is made up of five nations: Norway, Sweden, Finland, Denmark, and Iceland. These countries have related histories and, except for Finland, share similar cultures. They also have standards of living that are among the world's highest.

The Land

Although Scandinavia lies north, warm winds from the North Atlantic Current give its southern and western areas a relatively mild climate. Central Scandinavia has long, cold winters and short, warm summers. The northernmost part of Scandinavia near the Arctic Circle, however, has a very cold climate. Because this rugged area is so far north, there are summer days when the sun never sets. Many people have to darken their windows to sleep. In midwinter, though, these same people may battle depression, because there are days when the sun never rises.

Scandinavia's physical landscape is quite **varied** because of its large size. Many islands dot the jagged coastlines. Lowland plains stretch over Denmark and the southern part of Sweden and Finland. Mountains form a backbone along the border of Norway and Sweden. Forests and lakes cover much of Sweden and Finland. In the far north, above the Arctic Circle, the land is barren tundra that remains frozen for most of the year.

Two countries—Iceland and Norway—have special features. The island of Iceland sits in an area of the North Atlantic Ocean where two of Earth's tectonic plates meet. The two plates are pulling away from each other, allowing hot magma to rise to the surface. This creates hot springs and **geysers** (GY·zuhrs), which are springs that shoot hot water and steam into the air. Iceland also has about 200 volcanoes, though many are not active.

Norway, meanwhile, is known for its many beautiful **fjords** (fee·AWRDS), or narrow inlets of the sea. Steep cliffs or slopes surround the fjords, which were carved by glaciers long ago. Fjords provide inland waterways that supply fish for food and export.

Hot Lake in Iceland

People in Iceland swim in a human-made lake. The lake's warm water comes from the nearby plant, which produces energy from hot springs. ***Place*** **Why are hot springs numerous in Iceland?**

The Economies

The countries of Scandinavia are wealthy and prosperous. Their economies are based on a mix of agriculture, manufacturing, and service industries. Although farmland is limited, most Scandinavian countries produce most of the food they need. Fishing is an important industry, especially in Norway and Iceland.

For energy, Norway relies on its own oil and natural gas, taken from fields under the North Sea. Iceland taps the molten rock beneath its surface to make **geothermal energy.** This is electricity produced by natural underground sources of steam. Iceland also uses hydroelectric power. Finland takes advantage of its fast-running rivers to generate hydroelectric power as well. Sweden uses a combination of nuclear power and oil.

Some Scandinavian countries have abundant mineral and forest resources that support various industries. Sweden has reserves of iron ore that it uses to produce steel for a variety of products, including cars such as Saabs and Volvos. Shipbuilding is important in Finland and Denmark, as are wood and wood product industries in Finland and Sweden.

Denmark plays an important role in world trade. Copenhagen, its capital, sits at the entrance to the Baltic Sea. The largest ships cannot enter that sea because it is not deep enough for them. As a result, many ships stop in Copenhagen, where workers transfer cargoes to other vessels.

People and Culture

Most of the Scandinavian countries are less densely settled than other European countries. Large parts of Scandinavia are located in the cold north or are too mountainous to attract many people. Only Denmark, the smallest of the five countries, has a high population density. Denmark has a mild climate and relatively flat land that supports much agriculture.

The peoples of Norway, Sweden, Denmark, and Iceland share ethnic ties and speak related languages. They mostly descend from Germanic peoples who settled Scandinavia thousands of years ago. The ancestors of Finland's people, however, probably came from what is now Siberia in Russia. As a result, the Finnish language and culture differ from those of the other Scandinavian countries. Still, Finland shares close historic and religious links to the rest of Scandinavia. For example, most Finns—like most other Scandinavians—belong to the Protestant Lutheran Church.

During the Middle Ages, Scandinavian sailors and traders known as Vikings raided areas of western Europe and explored the North Atlantic Ocean, even reaching America. They also laid the foundation of the modern nations of Denmark, Norway, Sweden, and Iceland. For several hundred years, Sweden ruled its neighbor, Finland. Finland later was controlled by Russia for many years before gaining independence.

Today, Denmark, Norway, and Sweden are constitutional monarchies with governments similar to that of the United Kingdom. Finland and Iceland are republics with elected presidents. Iceland's parliament, the Althing, first met in A.D. 930, making it one of the oldest surviving legislatures in the world.

The Scandinavian countries take pride in providing extensive services to their citizens. As welfare states, they not only help the needy, but they also offer health care, child care, elder care, and retirement benefits to all. In return for these services, the people pay some of the highest taxes in the world.

Reading Check **Identifying** What energy resources are found in Scandinavia?

Section 1 Review

Social Studies ONLINE
Study Central™ To review this section, go to glencoe.com.

Vocabulary

1. **Explain** the meaning of:
   - **a.** constitutional monarchy
   - **b.** parliamentary democracy
   - **c.** peat
   - **d.** bog
   - **e.** productivity
   - **f.** geyser
   - **g.** fjord
   - **h.** geothermal energy

Main Ideas

2. **Summarizing** Use a graphic organizer like the one below to summarize important details about the United Kingdom's government, its history, and how it has influenced governments around the world.

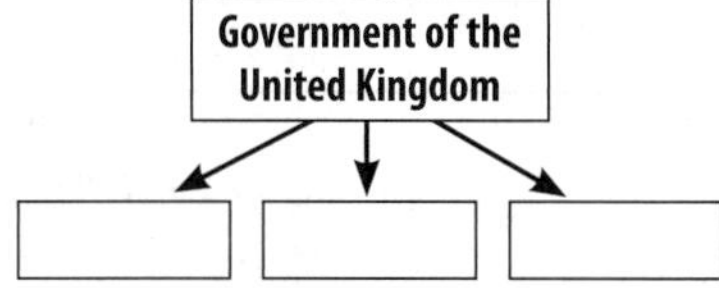

3. **Explaining** Why is the island of Ireland divided, and how has that led to conflict?
4. **Making Generalizations** What do the Scandinavian countries have in common?

Critical Thinking

5. **BIG Idea** Why are the United Kingdom, Ireland, and the countries of Scandinavia considered a subregion of Europe?
6. **Challenge** How are the constitutional monarchies of northern Europe similar to the government of the United States? How are they different?

Writing About Geography

7. **Expository Writing** Write a paragraph comparing the economies of the countries of northern Europe.

Guide to Reading

BIG Idea

People's actions can change the physical environment.

Content Vocabulary

- specialization *(p. 330)*
- high-technology industry *(p. 330)*
- bilingual *(p. 332)*
- polder *(p. 332)*
- multinational company *(p. 332)*
- reunification *(p. 336)*
- neutrality *(p. 336)*

Academic Vocabulary

- rely *(p. 330)*
- invest *(p. 335)*

Reading Strategy

Comparing and Contrasting
Use a Venn diagram like the one below to compare and contrast two countries in Europe's heartland.

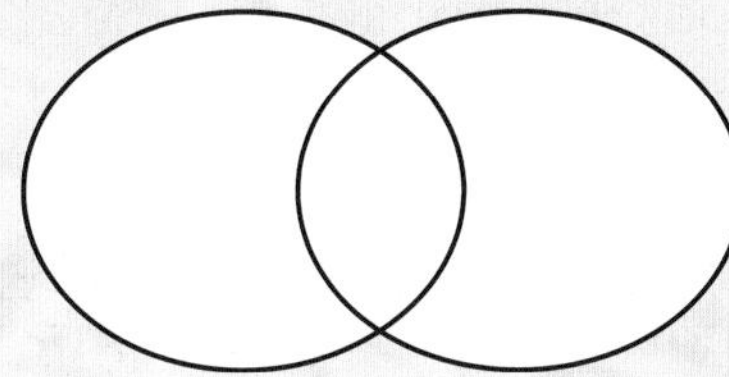

Europe's Heartland

 Section Audio  Spotlight Video

Picture This Mont Blanc, near the French-Italian border, is the highest point in Europe. Glacial hazards and frequent avalanches make hiking and skiing in this area dangerous. Torchlight parades are held to honor those who have lost their lives on the mountain. Mountains in Europe influence how and where people live. Read this section to learn how major landforms affect people living in the heartland of Europe.

▼ Skiers carrying torches descend Mont Blanc

France and the Benelux Countries

Main Idea France and the Benelux countries are important cultural, agricultural, and manufacturing centers of Europe.

Geography and You When you think of France, perhaps you picture the Eiffel Tower in Paris. There is, of course, much more to the country, as you will read.

France is in the heart of western Europe. Its three small neighbors to the northeast are known as the Benelux countries. The group name comes from the first syllables of the individual country names—*Be*lgium, the *Ne*therlands, and *Lux*embourg.

France's Land and Economy

France is the second-largest country in Europe. It is slightly smaller than Texas. The landscape in France varies widely. Most of northern France is part of the vast Northern European Plain. In the south, high mountain ranges separate the country from Spain, Italy, and Switzerland. Rivers, such as the Seine (SAYN) and the Loire (LWAHR), link France's different regions.

Most of France has a mild or warm climate and rich soil that is ideal for farming. France's agriculture is characterized by **specialization.** This means focusing efforts on certain activities to make the best use of resources. One area of specialization for the French is growing grapes and making wines. Farmers also use the milk of dairy cattle and sheep to produce about 250 kinds of world-famous cheese. France sells these cheeses and other food products to countries that cannot produce them on their own. In turn, France imports goods that it cannot easily make.

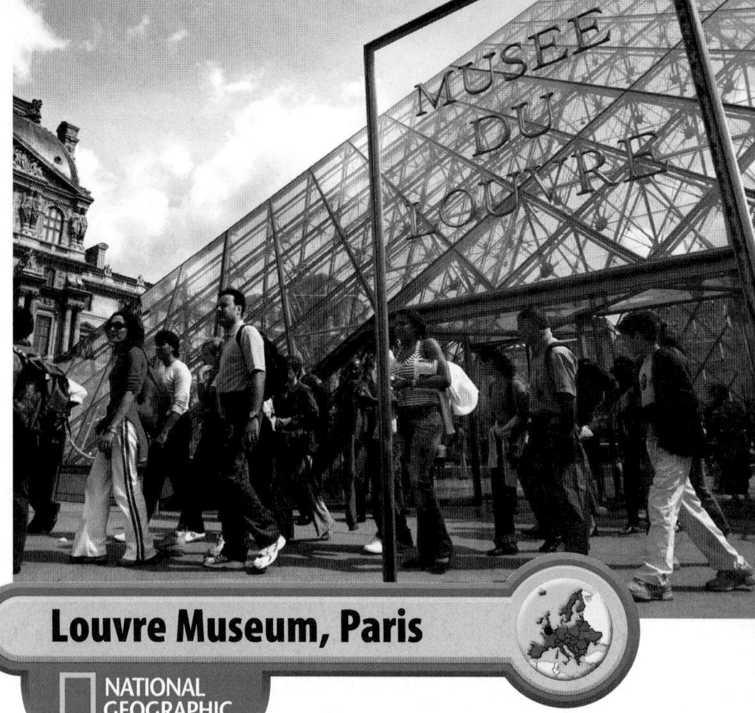

Louvre Museum, Paris

NATIONAL GEOGRAPHIC

The Louvre, in Paris, houses some of the world's most famous paintings and sculptures. ***Place*** **What other attractions in France draw tourists?**

France **relies** on industry as well as agriculture. Workers in traditional industries make cars and trucks, chemicals, textiles, and processed foods. France also has new **high-technology industries,** which include making computers and other products that require sophisticated engineering. Tourism is an important industry in France. It provides jobs to about 1 in 12 French workers. Millions of people come each year to visit Paris, France's vibrant capital. Other tourists vacation on sunny Mediterranean beaches, ski in the snowy Alps, and tour historic castles called châteaux (sha·TOHZ).

France's People and Culture

Most French trace their ancestry to the Celts, Romans, and Germanic peoples of early Europe. The majority speak French and consider themselves to be Roman Catholic. Islam is France's second religion, because so many people have migrated from Muslim countries in Africa.

Most of France's 60.7 million people live in urban areas. Almost 10 million make their homes in Paris, one of Europe's largest cities.

There, people can enjoy museums, universities, fine restaurants, and charming cafes. The Seine River and landmarks like the Eiffel Tower and Notre Dame Cathedral add to the city's beauty.

The French take great pride in their culture, which has greatly influenced the Western world. French cooking and French fashion are admired far and wide. France also boasts famous philosophers, writers, artists, composers, and film directors.

The French Revolution of the late 1700s also influenced the Western world. It brought about the decline of powerful monarchies and the rise of democracies. Today France is a democratic republic with both a president, elected by the people, and a prime minister, appointed by the president. The president has a great deal of power and can even dismiss the legislature, forcing new elections to be held.

The Benelux Countries

The small Benelux countries—Belgium, the Netherlands, and Luxembourg—have much in common. Their lands are low, flat, and densely populated. Most people live in cities, work in businesses or factories, and enjoy a high standard of living. All three nations are also parliamentary democracies with monarchs as heads of state.

Belgium has long been a trade and manufacturing center. With relatively few natural resources of its own, the country imports the raw materials to make and export vehicles, chemicals, and textiles.

GLOBAL CITIZENS

NAME: THIERRY HENRY **HOME COUNTRY:** France

ACHIEVEMENT: This soccer player helped power the French national team to years of success. Now Henry is using his hero status on the soccer field to fight racism in European society. Henry has been the target of racist slurs and has witnessed racial abuse by players and fans at European sporting events. So in January 2005, Henry launched the Stand Up Speak Up campaign to fight racism. In one year, Henry raised nearly $6 million to be distributed to groups in Europe dedicated to fighting racism. The funds also support teen athletic groups that emphasize sportsmanship and respect for others.

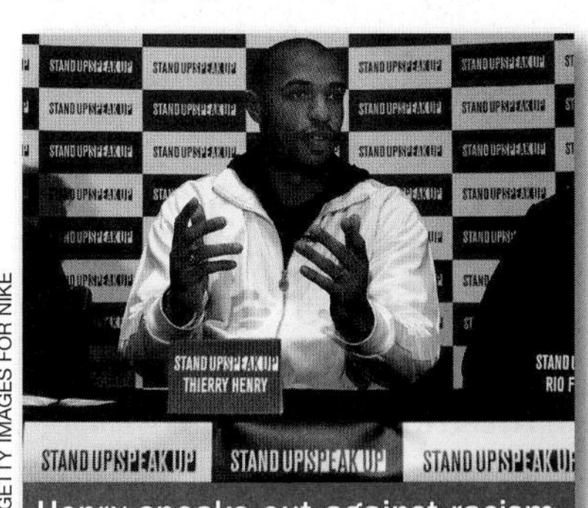

Henry speaks out against racism at sporting events.

EDDIE KEOGH/REUTERS PHOTO ARCHIVE/NEWSCOM; (INSET) GETTY IMAGES FOR NIKE

QUOTE: "I want to be able to watch football [soccer] on TV or attend a match and not hear a single racist insult. That's what I'd like to do for future generations of players."

Why might some people respect the views of athletes and celebrities more than those of other citizens? How should athletes handle this "power"?

Most Belgians live in crowded urban areas. Antwerp is a busy port and the center of the world diamond industry. Brussels is the capital and headquarters of the European Union (EU).

Belgium is made up of three regions—Flanders, Wallonia, and Brussels. In Flanders, to the north and west of Brussels, most people speak Dutch and are known as Flemings. In Wallonia, the areas south and east of Brussels, most people speak French and are known as Walloons. The population of the Brussels region comes from both language groups. As a result, the Brussels region is officially **bilingual,** using two languages. While each region practices self-rule, tensions sometimes arise between Flemings and Walloons.

NATIONAL GEOGRAPHIC

Antwerp's Diamond Trade

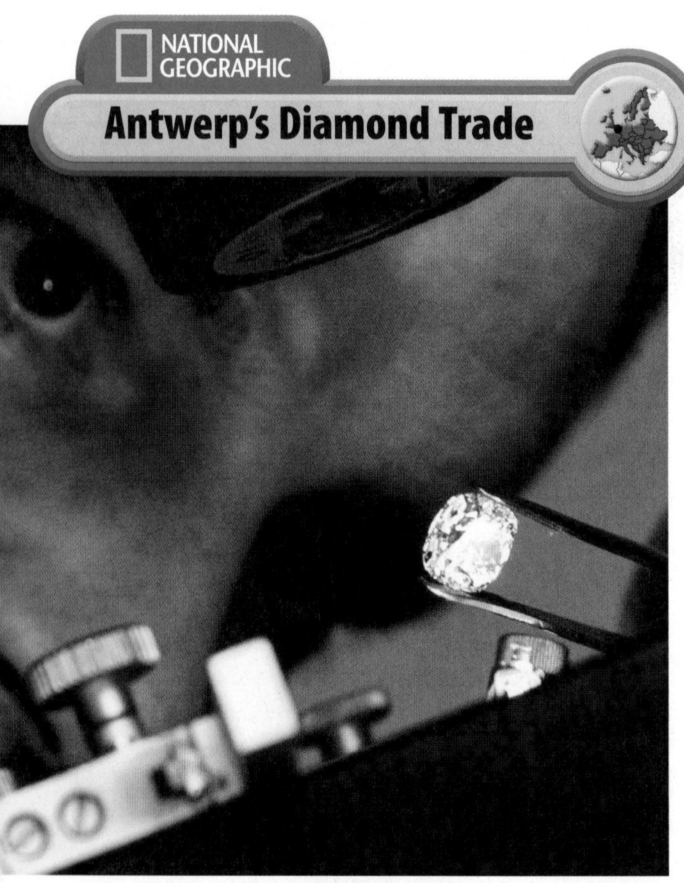

▲ Antwerp has been a center of the world's diamond trade for more than 500 years. Some $20 billion in diamond sales occur there annually. ***Movement*** **What goods does Belgium export?**

To the north of Belgium is the Netherlands, whose people are known as the Dutch. About 25 percent of the Netherlands lies below sea level. Without defenses against the sea, high tides would flood much of the country. The Dutch have built dikes, or banks of soil, to control and confine the sea as seen in **Figure 2.** They drain and pump the wetlands dry. Once run by windmills, pumps are now driven by steam or electricity. The drained lands, called **polders,** have rich farming soil.

About 90 percent of the Dutch live in cities and towns. Amsterdam is the capital and largest city. Living in a densely populated country, the Dutch make good use of their space. Houses are narrow but tall, and apartments are often built on canals and over highways. The Dutch work in service industries, manufacturing, and trade. The major exports of the Netherlands are cheese, vegetables, and flowers. Acres and acres of tulip fields bloom in the spring, and each year the Dutch export about two million tulip bulbs.

Southeast of Belgium lies Luxembourg, one of Europe's smallest countries. Centrally located in Europe, Luxembourg thrives as a center of trade and finance. Many **multinational companies,** or firms that do business in several countries, have their headquarters here. The people of Luxembourg have a mixed French and German background.

Challenges

France and the Benelux countries are challenged by population changes. First, their populations are aging. An aging population puts pressure on workers who must pay taxes to provide retirement and health care benefits for older people. Second, France and the Benelux countries have fairly large African and Asian minority

Maps In MOtion See StudentWorks™ Plus or glencoe.com.

Figure 2 **Areas of the Netherlands Reclaimed From the Sea**

Windmills in the Netherlands

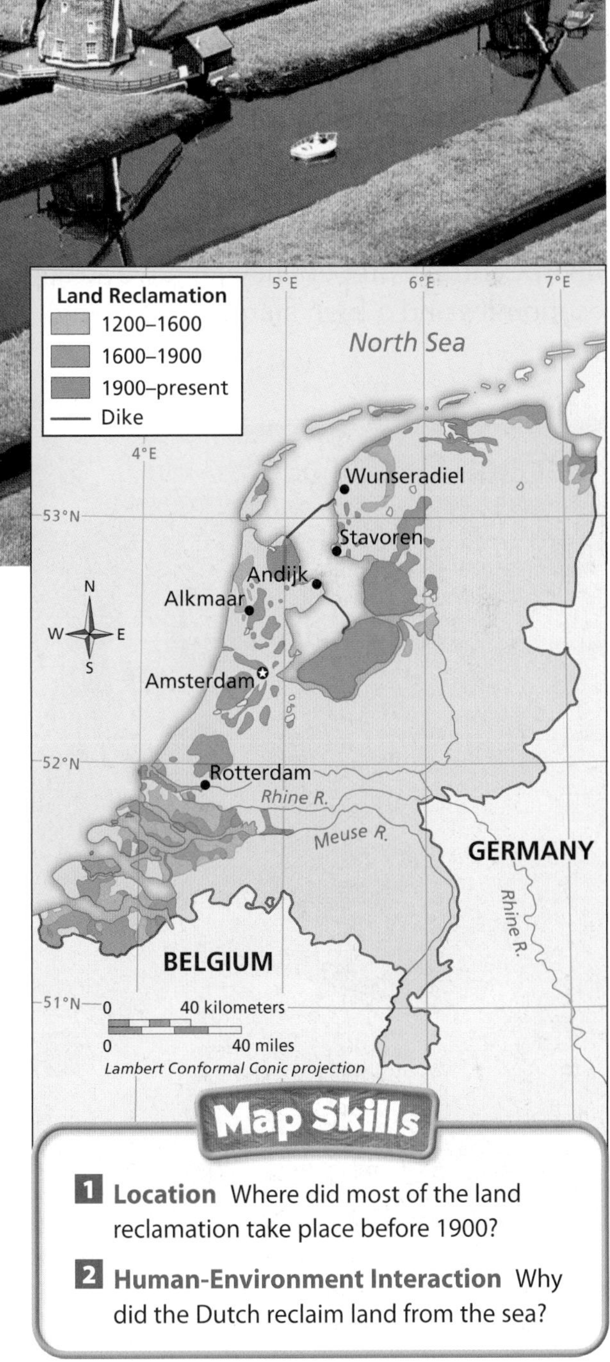

populations. Some people in the majority culture fear that many in these groups appear unwilling to accept European culture and customs. However, many members of minority groups live in crowded neighborhoods with poor schools, high unemployment, and little contact with the majority culture. In 2005 frustration among North Africans in France boiled over into almost three weeks of rioting. French government leaders have vowed to fight discrimination and improve conditions in ethnic communities.

Reading Check **Explaining** How does France's physical geography contribute to its agriculture?

Map Skills

1. **Location** Where did most of the land reclamation take place before 1900?
2. **Human-Environment Interaction** Why did the Dutch reclaim land from the sea?

Germany and the Alpine Countries

Main Idea **Germany, Switzerland, and Austria are known for their mountain scenery and prosperous economies.**

Geography and You Have you ever found yourself working alongside someone you used to compete against? Germans are in that position now that the two halves of their country are reunited. Read to learn more.

Germany and the Alpine countries—Switzerland, Austria, and Liechtenstein—lie in Central Europe. They all have strong economies and a high standard of living.

German Clock Maker

NATIONAL GEOGRAPHIC

▲ Germany's Black Forest region is famous for its finely crafted clocks, including cuckoo clocks. ***Place*** **Describe Germany's industry and agriculture today.**

Germany's Land

A large country encircled by nine other nations, Germany sits snugly in the heart of Europe. The flat Northern European Plain extends across northern Germany. Rocky highlands, some of which contain rich coal deposits, cover the central part of the country. The majestic Alps rise in the far south. The Alps are famous for their beauty, but many forests on the lower slopes of these mountains are threatened by acid rain.

Rivers have been vital to Germany's economic growth. They are used to transport raw materials to factories and to carry manufactured goods to market. The Danube River, one of Europe's most important waterways, begins in the Black Forest and winds eastward across southern Germany. Another river, the Elbe, flows from the central highlands to the North Sea. Hamburg, Germany's largest port city, is located on the Elbe River.

The most important of Germany's rivers—the Rhine—actually begins in the Swiss Alps. It then passes through Germany and the Netherlands before spilling into the North Sea. The Rhine is long and deep, allowing oceangoing ships to travel far inland.

History and Government

Germany's central location in Europe has long made it a crossroads for peoples, ideas, and armies. For centuries, Germany was a collection of states that were deeply involved in Europe's wars and religious struggles. In 1871 these states joined together to form the modern nation of Germany.

During the 1900s, Germany's efforts to dominate Europe helped spark two world wars. Allied countries—the United States, the Soviet Union, the United Kingdom, and France—defeated Germany in World War II.

In 1945 the Allies divided Germany into four zones of occupation. The Soviet zone later became Communist-ruled East Germany. The three other zones, controlled by the United States, the United Kingdom, and France, became democratic West Germany. After the collapse of communism, the two parts of Germany were reunited in 1990.

Today, Germany—like the United States—is a federal republic. This means that the national government and state governments share power. An elected president serves as Germany's head of state, but he or she performs only ceremonial duties. The country's chancellor, chosen by parliament, is the real head of government.

Germany's People

Germany has the largest population of the European countries—82.5 million. Nearly 90 percent of the people live in urban areas. The largest city, and the nation's capital, is Berlin. With many museums, concert halls, and theaters, Berlin is a cultural center as well as the seat of government. Germans are proud to have produced many brilliant thinkers and writers, as well as composers such as Johann Sebastian Bach and Ludwig van Beethoven.

About 90 percent of the country's people are native Germans, and German is the main language. Most of the rest of the population has immigrated from eastern Europe and Turkey. These immigrants came to Germany to find work or to escape political troubles in their homelands. The newcomers include many Muslims and Jews, but most Germans are Protestant or Catholic.

The Economy

Today, Germany is a global economic power and a leader in the European Union. This is due in part to Germany's highly productive agriculture. In the river valleys and plains areas, the fertile land and mild climate are well suited for farming. Germany produces enough food to feed its people and export its surplus.

NATIONAL GEOGRAPHIC

Germany's Auto Industry

New cars are placed in a huge storage tower near an automobile plant in Wolfsburg, Germany. ***Place*** **What role does Germany have in the global economy?**

It is industry, though, that is most responsible for Germany's strong economy. The country is a leading producer of steel, chemicals, cars, and electrical equipment. During the late 1900s, many Western industrialized countries experienced a decline in manufacturing. In Germany, however, the decrease was not dramatic. German firms had **invested** money to research and develop desirable, competitive products.

One of Germany's economic challenges has come as a result of **reunification,** when East and West Germany united under one government in 1990. At the time, workers in East Germany had less experience and less training in modern technology than workers in West Germany. Old and inefficient factories in the east could not compete with the more advanced industries in the west. Many businesses closed, and economic activities in the eastern part of Germany continue to lag behind those in the prosperous west.

The Alpine Countries

The Alpine countries take their name from the Alps of central Europe. These mountainous countries include Switzerland, Austria, and Liechtenstein. Liechtenstein is a tiny country of only 62 square miles (161 sq. km)—smaller than Washington, D.C. The whole population—about 40,000 people—would not even fill a major league baseball stadium.

Switzerland is also a small country, but it is much bigger than Liechtenstein and far more important internationally. The few travel routes that cut through the Alps lie in Switzerland. So for centuries, the Swiss have been "gatekeepers" between northern and southern Europe. That role helped Switzerland decide long ago to practice **neutrality,** or refusal to take sides in wars. As a result, for more than 700 years the Swiss have enjoyed a stable democratic government, even when fighting has raged around them. Today many international organizations, such as the International Red Cross, are based in the Swiss city of Geneva.

Switzerland's geography also affected the growth of individual communities. The rugged mountains isolated groups of people from one another. As a result, each town and city treasures its unique traditions and independence. Today the people of Switzerland represent many different ethnic groups and religions. The country also has not one but four national languages: German, French, and Italian—which are the native tongues of Switzerland's neighbors—and Romansch. Most Swiss speak German, and many speak more than one language.

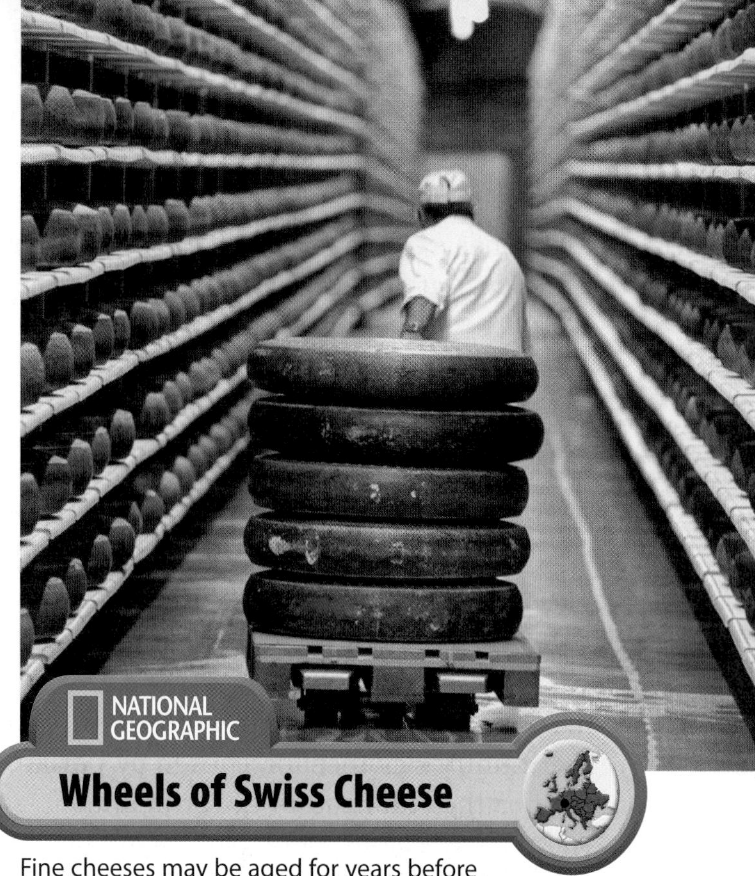

Wheels of Swiss Cheese

Fine cheeses may be aged for years before they are ready to eat. Switzerland exports more than 50,000 tons of cheese each year. ***Place*** **How did Switzerland's geography affect its communities?**

Although it has few natural resources, Switzerland is a thriving industrial nation. Dams on Switzerland's rivers produce great amounts of hydroelectric power for industries and homes. Using imported materials, Swiss workers make high-quality electronic equipment, chemicals, and other goods. The country is also known for its fine clocks and watches, excellent chocolate and cheeses, and its multipurpose Swiss army knives. A large part of the Swiss economy is dependent upon its banking and other financial services. Because Switzerland's neutrality is honored by other countries, people from around the world consider Swiss banks to be safe and secure.

East of Switzerland is landlocked Austria. The Alps cover the western three-quarters of Austria, so there is little good farmland. The beautiful mountain scenery does, however, attract many skiers and tourists. The mountains also provide valuable timber and iron ore and, as in Switzerland, fast-moving rivers generate hydroelectric power. With these resources, Austria's factories produce machinery, chemicals, metals, and vehicles. Austria also has strong banking and insurance industries.

The people of Austria mainly speak German and are Roman Catholic. Most Austrians live in cities and towns. Vienna, on the Danube River, is the capital and largest city, and about one-fifth of Austrians live there. Before World War I, Vienna was the heart of the vast Austro-Hungarian Empire that covered much of central and southeastern Europe. Vienna was also a center of culture and learning. Some of the world's greatest composers, including Wolfgang Amadeus Mozart, lived or performed in Vienna. The city's concert halls, historic palaces, and churches continue to draw music lovers and other visitors today.

Reading Check **Contrasting** How do the economies of the western and eastern parts of Germany differ?

Social Studies ONLINE
Study Central™ To review this section, go to glencoe.com.

Vocabulary

1. **Explain** the significance of the following terms:
   - **a.** specialization
   - **b.** high-technology industry
   - **c.** bilingual
   - **d.** polder
   - **e.** multinational company
   - **f.** reunification
   - **g.** neutrality

Main Ideas

2. **Explaining** How has French culture influenced the world?
3. **Analyzing** Draw a Venn diagram to analyze how agriculture is similar and different in France and Germany.

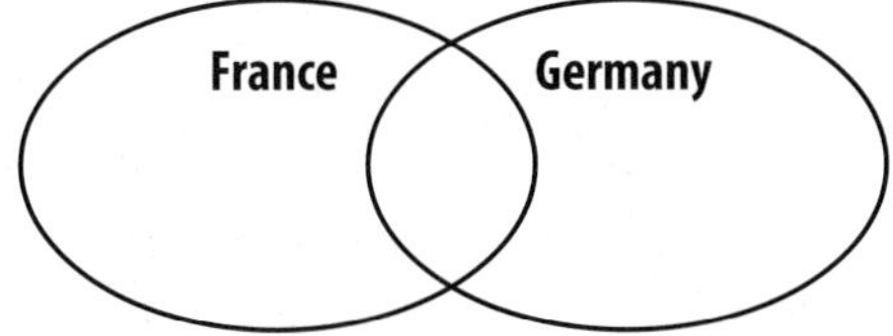

Critical Thinking

4. **Making Generalizations** How does specialization in French agriculture and food production lead to interdependence with other countries?
5. **BIG Idea** Give three examples of how people in this part of Europe have changed the environment. Explain if you think those changes are positive or negative.
6. **Challenge** Do you think the economic successes of the countries of Europe's heartland can continue in the future? Explain your answer fully.

Writing About Geography

7. **Persuasive Writing** Write a letter to a friend trying to persuade him or her to visit a specific country in Europe's heartland with you. Describe why that country interests you.

Guide to Reading

BIG Idea

Places reflect the relationship between humans and the physical environment.

Content Vocabulary

- dry farming *(p. 339)*
- autonomy *(p. 339)*
- subsidy *(p. 340)*

Academic Vocabulary

- similar *(p. 339)*
- militant *(p. 340)*

Reading Strategy

Making Generalizations Use a diagram like the one below to write three characteristics shared by the countries in this region.

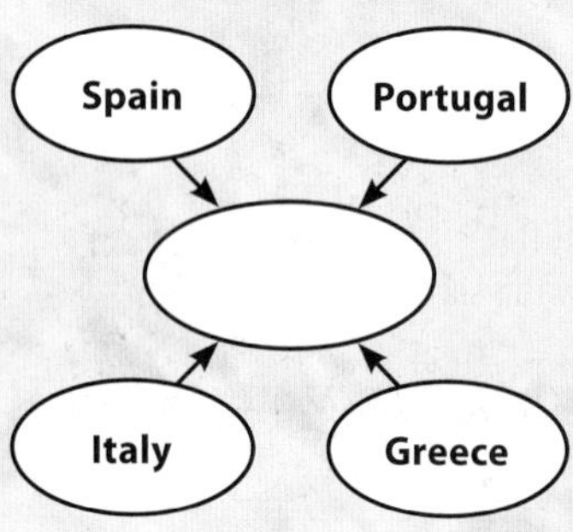

Southern Europe

Picture This Lunchtime lineup! Visit Antiparos, Greece, and you are likely to see octopuses draped over fishing lines to dry in preparation for a later meal. The boneless octopus has a parrot-like beak, a doughnut-shaped brain, eight arms, three hearts, and—it can change colors. Octopuses thrive in the clear, blue waters of the Mediterranean Sea. The sea and the lands surrounding it have supported numerous cultures. Read this section to learn about today's cultures of southern, or Mediterranean, Europe.

▼ Octopuses drying on line, Antiparos, Greece

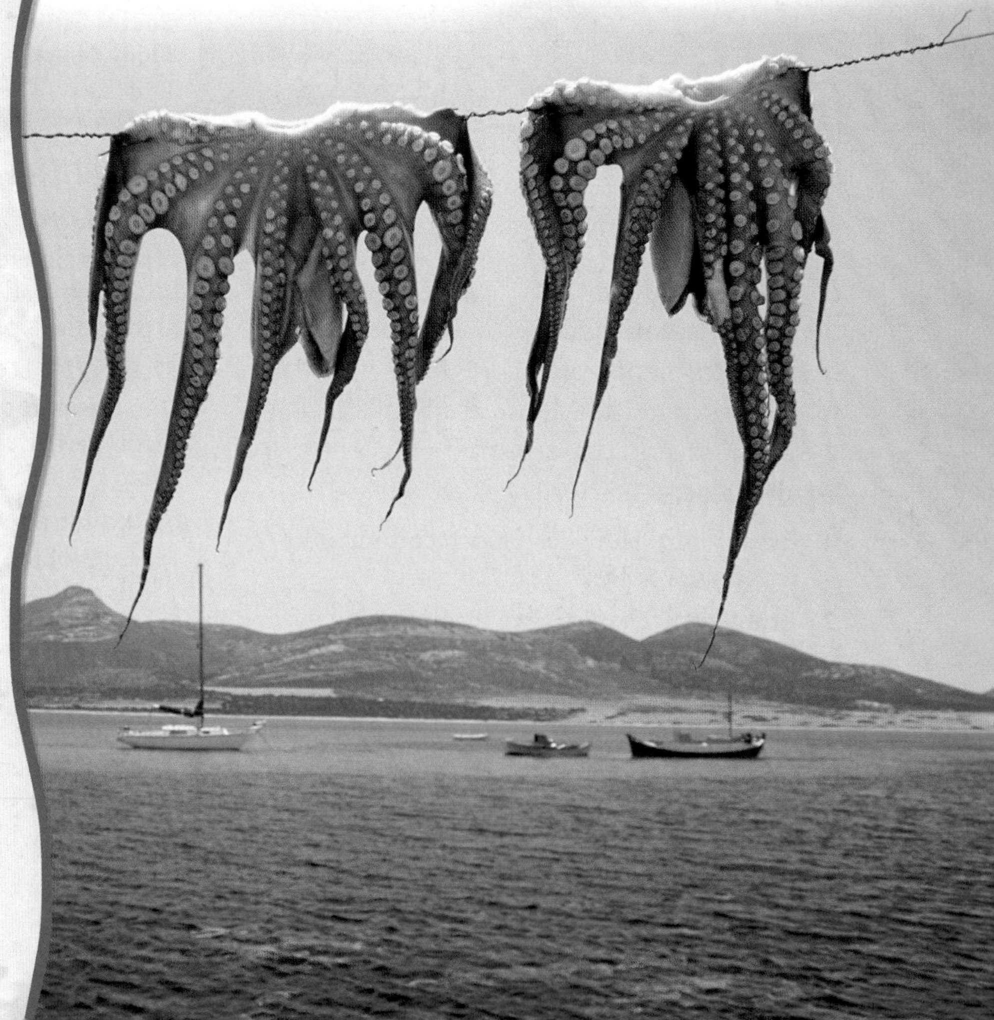

Spain and Portugal

Main Idea **Spain and Portugal are young democracies with growing economies.**

Geography and You Can you imagine chasing bulls down the main streets of your hometown? People in Pamplona, Spain, do this every year as part of a summer festival. Keep reading to discover more about colorful Spain and its neighbor, Portugal.

Spain and Portugal occupy the Iberian Peninsula in southwestern Europe. They share it with the tiny nation of Andorra, nestled in the Pyrenees Mountains.

Spain

Most of Spain is covered by the Meseta, a dry plateau surrounded by mountain ranges. The reddish-yellow soil there tends to be poor, and rain is scarce. However, crops such as wheat and vegetables are grown by **dry farming.** This technique does not depend on irrigation. Instead the land is left unplanted every few years so that it can store moisture.

Farming is easier in other parts of the country. Northwestern Spain, which borders the Atlantic Ocean, has mild temperatures and plenty of rain. Southern Spain, which borders the Mediterranean Sea, has wet winters and dry summers. In this area, farmers use irrigation to grow citrus fruits, olives, and grapes—Spain's leading agricultural products.

In the late 1900s, Spain's manufacturing and service industries grew rapidly. Today they dominate the economy. Spanish workers produce processed foods, clothing, footwear, steel, and cars. Spain also benefits greatly from tourism. The country's attractions include castles, cathedrals, and sunny Mediterranean beaches. Tourists also enjoy Spain's cultural traditions, such as bullfighting and flamenco dancing.

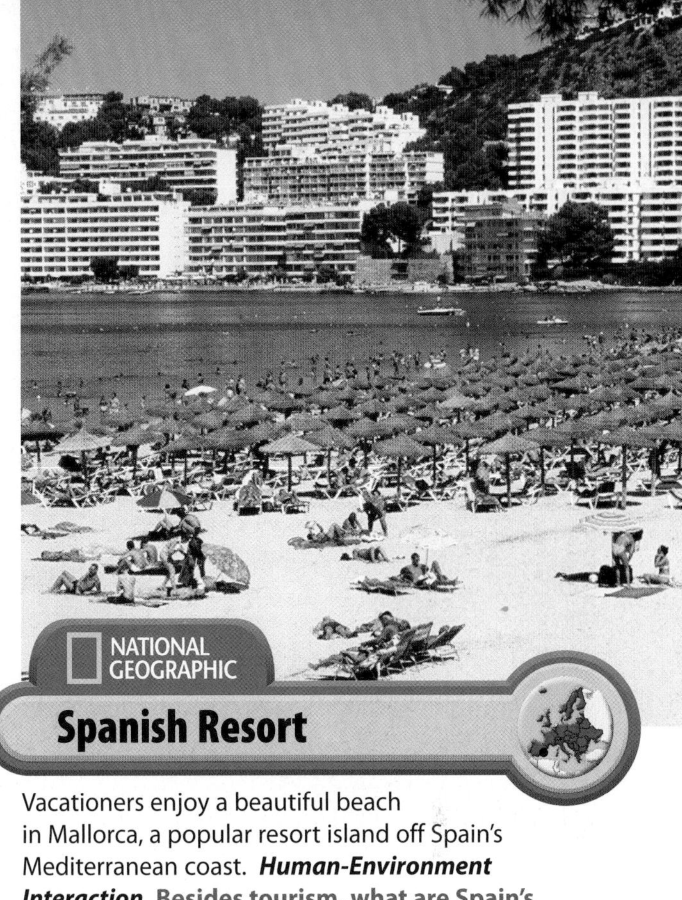

NATIONAL GEOGRAPHIC

Spanish Resort

Vacationers enjoy a beautiful beach in Mallorca, a popular resort island off Spain's Mediterranean coast. ***Human-Environment Interaction*** **Besides tourism, what are Spain's other major industries?**

Most of Spain's people speak Castilian Spanish, the country's official language. Some regions of Spain, however, are home to distinct groups with their own languages. The people of Catalonia, in the northeast, speak Catalan, which is **similar** to an old language of southern France. In the Pyrenees, the Basques speak Euskera, a language unrelated to any other in the world.

After years of rule by a dictator, Spain became a democracy in the late 1970s. In recent times, Spain's democratic government has given the different regions of Spain greater **autonomy,** or self-rule. In the Basque region, though, many people want to be completely separate from Spain. Some Basque separatists have used terrorism to try to achieve this goal.

NATIONAL GEOGRAPHIC

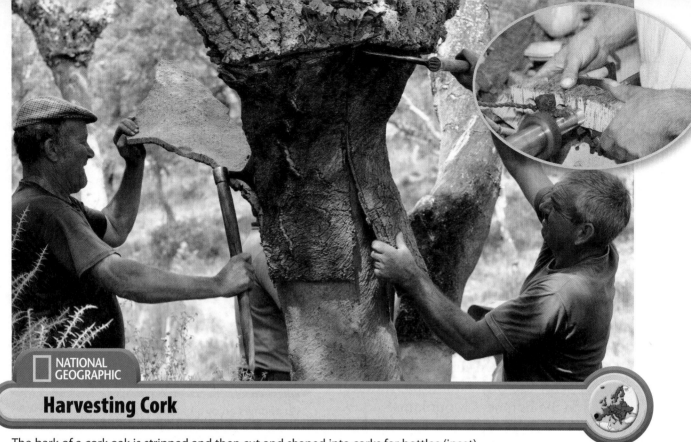

Harvesting Cork

The bark of a cork oak is stripped and then cut and shaped into corks for bottles (inset). The trees will grow new bark within 10 years. ***Place*** **How have subsidies from the European Union impacted agriculture in Portugal?**

Most of Spain's 43.5 million people live in urban areas. The main cities are Madrid, the capital, and Barcelona, the leading seaport and industrial center. The cities of Seville, Granada, and Córdoba, in the south, show the influence of the Muslims who ruled Spain for much of the Middle Ages.

Most people in Spain today are Roman Catholic. A large number of Muslims from North Africa have migrated to Spain in recent years. Tensions have sometimes developed between the Spanish population and Muslim immigrants. Spain was shaken in 2004 when terrorist attacks by suspected Muslim **militants** killed 191 people on Madrid trains.

Portugal

Spain's smaller neighbor to the west is Portugal. Most of Portugal's land is a low coastal plain split in half by the Tagus River. In both the north and the south, people grow a variety of crops. The most important are grapes used for wine making and oak trees that provide cork. Most Portuguese live in small villages on the coast, near the cities of Lisbon and Porto. Many people earn a living there by fishing in the Atlantic Ocean.

Closeness to the ocean helped Portugal become a sea power during the 1500s. The Portuguese built an empire that included Brazil and parts of Asia and Africa. Today Portugal has a democratic government, and its shaky economy is growing stronger with the help of subsidies from the European Union. **Subsidies** are special payments a government makes to support a group or industry. With this help, manufacturing and service industries have become more important than agriculture to Portugal's economy.

Identifying What languages are spoken in Spain?

Italy

Main Idea **Italy's north and south form two distinct economic regions.**

Geography and You Do you have a favorite Italian food? When people think of Italy, they often think of delicious pasta. Read to learn what else the country produces.

Italy juts out from Europe into the Mediterranean Sea. The mainland looks like a boot about to kick a triangular football. The "football" is Sicily (SIH·suh·lee), an island that is also part of the country.

In Italy's north, the Alps tower over the broad Lombardy plain. In central and southern areas, the Apennine Mountains form a backbone that stretches into Sicily. Volcanoes also dot the landscape. Throughout history, southern Italy has experienced volcanic eruptions and earthquakes.

The Economy

Since the mid-1900s, Italy has changed from a mainly agricultural country into a leading industrial economy. Most of this growth has taken place in northern Italy. Workers in northern manufacturing cities, such as Milan, Turin, and Genoa, produce cars, technical instruments, appliances, clothing, and high-quality goods. The north's fertile Po River valley is also the country's richest farming region. Farmers there raise livestock and grow grapes, olives, and other crops.

Southern Italy is poorer and less industrialized than northern Italy. Much of the terrain is mountainous, with limited mineral deposits, poor land for farming and grazing, and few navigable rivers. As a result, unemployment is high. Unemployment has led many southern Italians to seek a better life in northern Italy or other parts of Europe.

The People

About 90 percent of Italy's 58.7 million people live in urban areas. In the cities, modern life is mingled with the past. Rome, Italy's largest city, was once the center of the Roman Empire. Today, Rome is Italy's capital and home to the country's democratic republic form of government.

The people of Italy speak Italian, and nearly all are Roman Catholic. In fact, the Roman Catholic Church is based in Rome. The Church rules tiny Vatican City, where the pope and other Church leaders live and work. Although Vatican City is within Rome's boundaries, it is an independent country—the smallest in the world.

Reading Check **Explaining** How do the economies of northern and southern Italy differ?

An Evening in Rome, Italy

NATIONAL GEOGRAPHIC

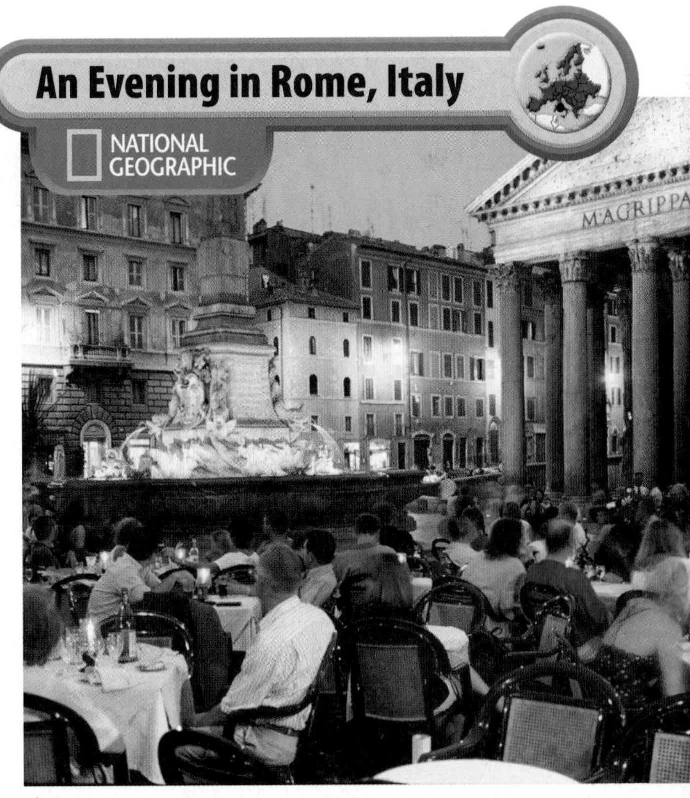

People dine at an outdoor restaurant near the Pantheon, a public building built by the ancient Romans. ***Place*** **What percentage of Italy's people live in urban areas?**

Greece

Main Idea **Mountains, seas, and islands have shaped Greece's people and economy.**

Geography and You Do you ever go boating or fishing? You will understand why these are popular activities for the Greeks when you read about Greece's geography.

East of Italy, Greece extends from the Balkan Peninsula into the Mediterranean Sea. The country includes not only a mainland, but about 2,000 islands. Like other Mediterranean areas, Greece is often shaken by earthquakes.

Because of mountains and poor, stony soil, agriculture plays a declining role in the Greek economy. In the highlands, people raise sheep and goats. On farms in plains and valleys, farmers grow wheat, olives, and other crops. In recent decades, Greece has developed new industries, such as textiles, footwear, and chemicals. Shipping is a major business. Greece has one of the world's largest shipping fleets, including oil tankers, cargo ships, and passenger vessels. Tourism is another key industry. Each year millions of tourists come to visit historic sites such as the Parthenon, an ancient temple in the city of Athens.

About 60 percent of Greeks are urban dwellers. Nearly a third live in or around Athens, the capital. The Greeks speak a form of Greek similar to that spoken in ancient times. Most of them are Greek Orthodox Christian. Today, Greece is a democratic republic and a member of the European Union.

Reading Check **Explaining** How has geography affected the way Greeks earn a living?

Section 3 Review

Social Studies ONLINE
Study Central™ To review this section, go to glencoe.com.

Vocabulary

1. **Define** *dry farming, autonomy,* and *subsidy,* and use each word in a sentence.

Main Ideas

2. **Explaining** How did Portugal benefit from joining the EU?
3. **Identifying** Create a graphic organizer like the one below to identify Italy's agricultural and industrial products.

4. **Comparing and Contrasting** How is Greece similar to and different from the other countries of southern Europe?

Critical Thinking

5. **BIG Idea** Write two generalizations describing the connection between the physical geography of southern Europe and the lives of the region's people.
6. **Challenge** Write a paragraph explaining how countries in southern Europe have worked to improve their economies.

Writing About Geography

7. **Descriptive Writing** Write the text for a travel brochure that encourages visitors to take a cruise that stops in the countries of southern Europe. Describe the landscapes, cities, and activities that visitors could see in those countries.

TIME PERSPECTIVES

EXPLORING WORLD ISSUES

WHOSE EUROPE IS IT?

As millions of immigrants relocate to Europe, the region's democracies struggle to redefine themselves.

Muslim immigrants gather on Westminster Bridge in London.

People in the United States, a nation formed by immigrants from around the world, understand the concept "out of many, one." Today, the countries of Europe are struggling to comprehend the idea too. For more than 60 years, millions of immigrants—many of them Muslims—have emigrated to some of Europe's oldest democracies.

It has not been easy to get so many different people to respect each other and live together in harmony. In recent years, cultural and religious clashes have developed as immigrants and Europeans struggle to understand each other. As that work continues, there is no doubt that the struggle will have an enormous impact on Europe's future.

In a restaurant in Paris, a Muslim immigrant and a woman born in France work together.

THE NEW MULTICULTURAL EUROPE

From Paris to Amsterdam and Brussels to Berlin, Europe is changing. Not long ago, most citizens of European countries shared certain characteristics. They were mostly all born in Europe, and the majority of them were white and Christian. These similarities helped create a **national identity** for countries like France, Great Britain, and Germany.

After World War II, many of Europe's immigrants were from countries that had been European colonies, like Algeria, India, and Pakistan. Friendly immigration policies following World War II welcomed the newcomers. Governments also established favorable labor policies in the 1960s that were designed to bring much-needed foreign workers to Europe.

But in recent years, the population of Europe has become more diverse. Millions of immigrants, many of them Muslims from North Africa, Turkey, and Southwest Asia, have left their homelands to start new lives in European nations.

Muslim Populations in European Countries

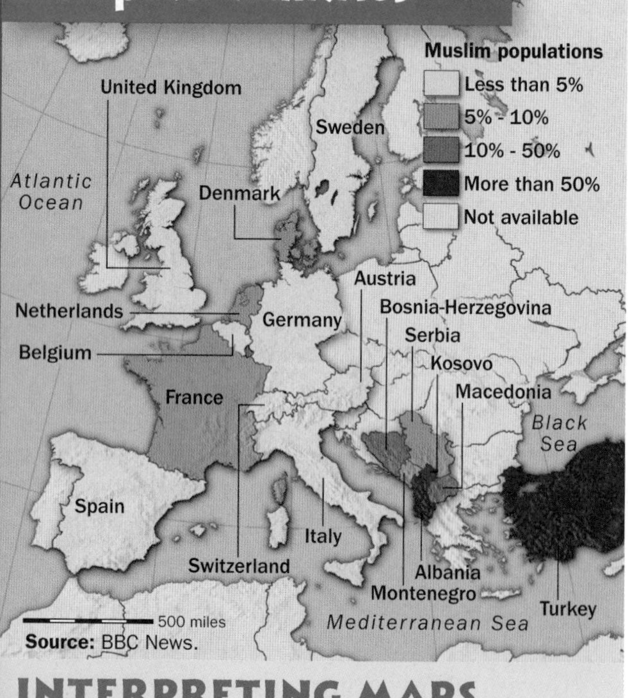

INTERPRETING MAPS

Categorizing Which countries have the largest Muslim populations?

Creating a New Identity

The number of immigrants living in Europe has greatly increased. In 2006, for example, there were about 7 million non-Germans living in Germany. Many of these immigrants are from Turkey. The large population of immigrants and their offspring have transformed and challenged traditional European beliefs.

Europe's immigrant **populations**, or groups of people, often view the world differently from people who were born in Europe. Many of the differences deal with culture, religious freedom, and the rights of women. At times, these different perceptions have caused conflict and bad feelings between Europe's older populations and its new ones. "We feel unwelcome," said a Muslim immigrant in Denmark. Some of the conflicts have been violent and have had a global impact.

Many Turks support their country's proposed admission to the European Union.

AP PHOTO

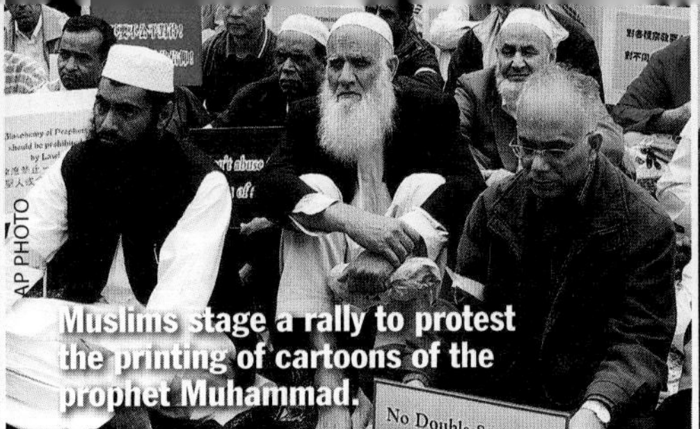

Muslims stage a rally to protest the printing of cartoons of the prophet Muhammad.

AP PHOTO

Danish Prime Minister Fogh Rasmussen, center, discusses the cartoons with a Pakistani diplomat.

AP PHOTO

When Cultures Collide

Early in 2006, Muslims around the world protested cartoons that were published in Europe. The cartoons showed the Muslim religion's prophet Muhammad in a negative way. They were first published in a newspaper in Denmark and later reprinted in various papers throughout Europe.

Muslims across Europe and the world were angry. They thought the cartoons were disrespectful of their religion, **Islam**. This is because Islam does not allow the publication of any images of Muhammad.

Religion and a Free Press

Anger over the cartoons led thousands of Muslims to protest worldwide. Angry protestors marched in several European cities, including London and Copenhagen. Demonstrations were also held throughout Southwest Asia. Many of the protests turned violent as demonstrators burned Danish flags and set fire to Denmark's embassy in Beirut, Lebanon. The riots killed at least 11 people in Afghanistan.

Denmark's prime minister, Anders Fogh Rasmussen, called the protests a global crisis and called for "calm and steadiness." But Fogh Rasmussen would not apologize for what the newspapers did. Like many European leaders and citizens, Fogh Rasmussen believed in the right of a free press. He defended the newspaper's right to print the cartoons of Muhammad even if their publication caused controversy and protest.

A Search for Common Ground

Learning to live with—and absorb—new ethnic groups is one of the greatest challenges facing Europe. Some experts believe that Turkey's proposed admission to the European Union, or the EU, may help bridge the gap between Muslims and traditional Europeans. Turkey would be the first Muslim country in the EU, a group of European nations that have joined together to solve common problems and create economic opportunities.

Experts believe that the European nations will have to learn to compromise and be tolerant of the cultural and religious differences of all of their citizens. Learning how to do that will be a challenge in a multiethnic Europe.

EXPLORING THE ISSUE

1. **Comparing** How were the opinions of Muslim immigrants and European-born citizens different concerning publishing the cartoons of Muhammad?

2. **Making Inferences** How might Turkey's admission into the EU change the way citizens born in Europe view Muslim immigrants?

FRANCE'S CLASH OVER SYMBOLS

In January 2004, thousands of Muslim women and men in France took to the streets of Paris to send a message to the French government. Marching hand in hand, many of the women wore head scarves. Some covered their hair with the French flag.

The protesters were angry over a proposed French law. If passed, the law would stop students from wearing head scarves and other noticeable religious symbols in public schools. French president Jacques Chirac (ZHAHK shee•RAHK) proposed the ban on the head scarves, called ***hijab*** (HEH•JAB), worn by Muslim women and girls. Chirac said the ban was created to make sure French children are not exposed to differences that will "drive people apart." Some Jewish and Christian religious symbols were also included in the proposed ban.

A Heated Debate

Despite the protests, the French government voted the bill into law in 2005. The law continued to be controversial. There are about 5 million Muslims in France, nearly 8 percent of the country's population. Critics of the law say wearing a head scarf in school is a personal choice and a basic right. "The government should not be in the business of telling a woman how to dress," said Salam Al-Marayati, of the Muslim Public Affairs Council.

French officials say the law is not directed against any religion. "The idea is to keep the influence of religion away from public schools," said one French diplomat. "A teacher does not have to know whether students are Muslim, Christian, Jewish, or whatever."

Soon after the ban went into effect, some French Muslims who arrived at school wearing scarves pushed them off their heads when they entered the school grounds. Will cultural differences continue to divide Europe's people? That question remains to be answered by European nations as their populations become more diverse.

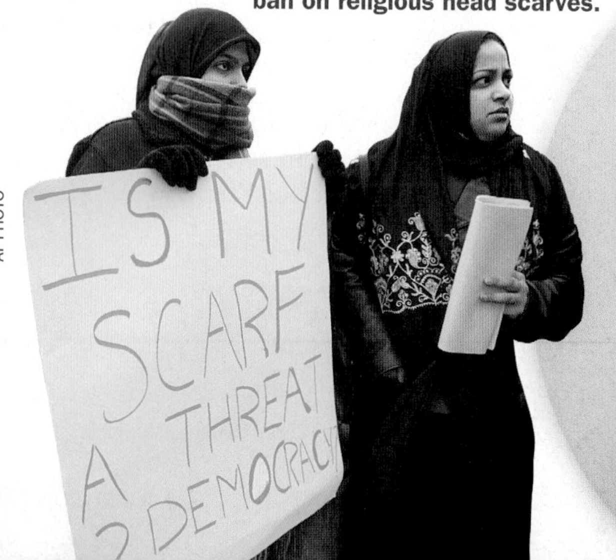

AP PHOTO

Muslim women protest France's ban on religious head scarves.

EXPLORING THE ISSUE

1. **Predicting** How might the ban on head scarves impact the French government's relations with Muslim countries around the world?

2. **Making Inferences** Why do you think France's law banning *hijab* also forbids the wearing of other religious symbols?

The EU flag

REVIEW AND ASSESS

UNDERSTANDING THE ISSUE

1 **Making Connections** What might be some reasons that people leave their homeland to live in another country?

2 **Writing to Inform** Write a short article describing the protests of cartoons about Muhammad. Include information about demonstrations around the world.

3 **Writing to Persuade** Write a letter to a friend in Denmark. Convince your friend that newspapers did or did not have the right to publish the cartoons of Muhammad.

INTERNET RESEARCH ACTIVITIES

4 Use Internet resources to find information about the European Union. Read about the EU's three main governing organizations. Choose one and write a brief description of it in your own words.

5 With your teacher's help, use Internet resources to find information about a former European colony such as India, Algeria, or Hong Kong. Read about how the colony was formed and the relationship it had with the colonizing country. Why do you think immigrants from former colonies might want to live in Europe? Write your answer in a 250-word essay.

BEYOND THE CLASSROOM

6 **Research the history of immigration in the United States during the early 1900s.** What were some of the challenges immigrants faced as they settled in the U.S.? Write your answer in an article appropriate for a school newspaper.

7 **Debate the issue.** Debate this resolution: "Wearing a religious head scarf in school is a personal choice and a basic right." A panel of student judges should decide which team has the most compelling arguments.

How the European Union Grew

For nearly 60 years, European nations have been forming an ever-closer economic and political union. Here's a look at the steps they have taken along the way.

1948 — The Council of Europe is established, creating a forum for Europe's leaders to discuss ways to work together.

1951 — France, Germany, Italy, the Netherlands, Belgium, and Luxembourg unite their coal and steel industries.

1957 — The European Economic Community is formed. The EEC is the first step toward a common economic market.

1967 — EEC merges with other European organizations to form the European Community (EC).

1973 — The United Kingdom, Denmark, and Ireland join the EC.

1981 — Greece becomes the EC's tenth member.

1991 — The Maastricht Treaty creates plans for a common currency and for cooperation in foreign affairs.

2002 — Most EU members agree to use a common currency, the euro. Britain, Sweden, and Denmark refuse.

2004 — The EU admits ten new member nations. EU leaders sign a new constitution for Europe. A year later, voters in France and the Netherlands reject it.

Building Time Line Reading Skills

1. **Analyzing Information** How many years does this time line cover? When did Greece join the EC?
2. **Making Inferences** Why do you think European nations cooperated economically before they worked together politically?

Guide to Reading

BIG Idea

Geography is used to interpret the past, understand the present, and plan for the future.

Content Vocabulary

- command economy *(p. 349)*
- market economy *(p. 350)*
- potash *(p. 350)*
- ethnic cleansing *(p. 356)*

Academic Vocabulary

- income *(p. 350)*
- medical *(p. 356)*

Reading Strategy

Summarizing Use a chart like the one below to summarize key facts about each group of countries.

Country Groups	Key Facts
Poland, Belarus, Baltic Republics	
Czech Republic, Slovakia, Hungary	
Countries of Southeastern Europe	

Eastern Europe

Section Audio

Spotlight Video

Picture This Show your colors! A young person in Kyiv (Kiev), Ukraine, shows his support for the new government of the Orange Revolution. The Orange Revolution took place during the 2004 presidential elections. Orange was the color of the victorious political party. The party chose orange to represent the change in Ukraine's government. Orange represents the changing color of the leaves in autumn—a process that is peaceful and unstoppable. Read this section to learn more about life in eastern Europe today.

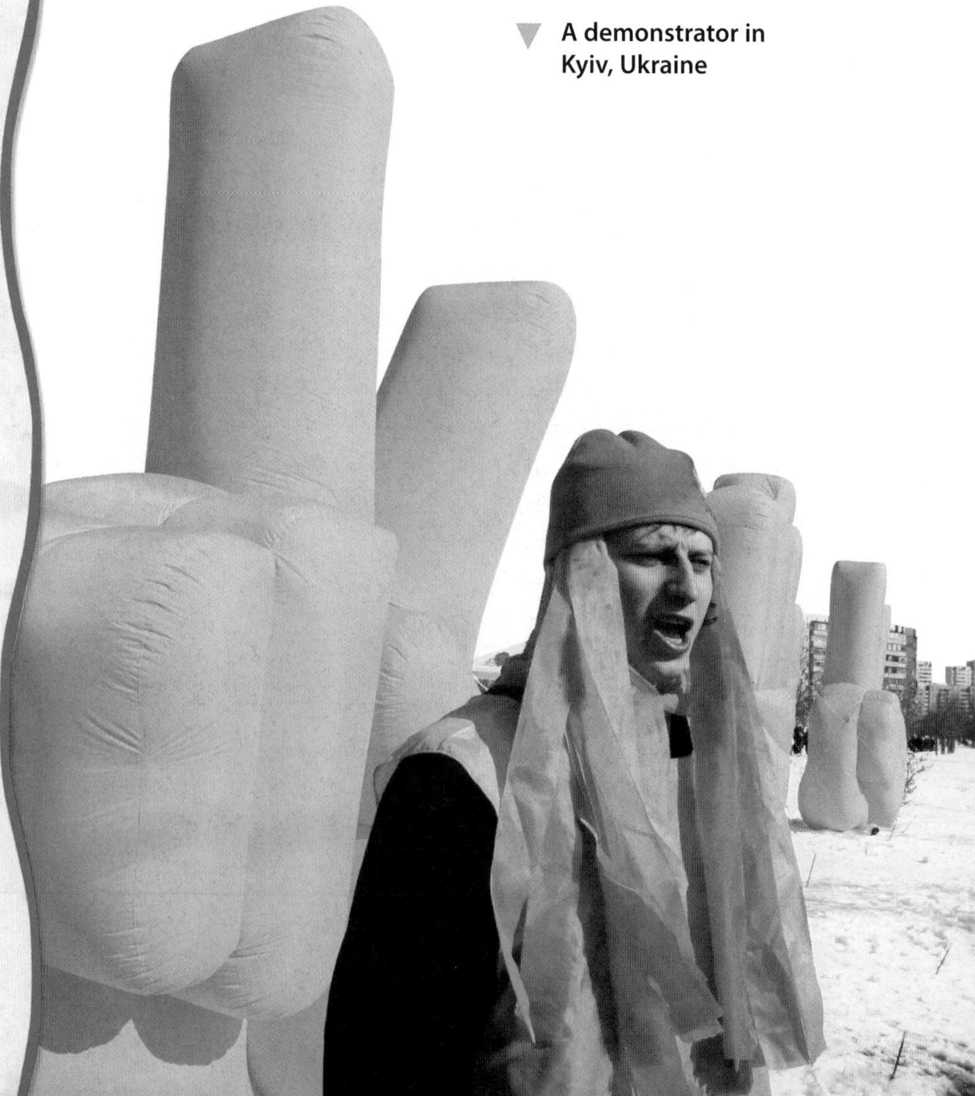

A demonstrator in Kyiv, Ukraine

Poland, Belarus, and the Baltic Republics

Main Idea **Poland and the Baltic Republics have become democratic, while Belarus is still influenced by its Communist past.**

Geography and You How do you feel when someone orders you to do something? The people of Poland, Belarus, and the Baltic states were all under foreign control at one time. Read what happened to these countries.

Poland, Belarus, and the three Baltic Republics—Lithuania, Latvia, and Estonia—are located in northeastern Europe on or near the Baltic Sea. Although they are neighboring countries, they have distinct histories and cultures.

Poland's Land and History

The sizable country of Poland lies east of Germany. The Carpathian Mountains and other highlands rise on its southern and western edges. Most of Poland, however, is a fertile lowland plain, and the majority of its people live there.

Throughout Poland's history, the largely flat landscape made the country an easy target for invading armies. By the 1800s, Poland had fallen victim to stronger neighbors—Germany, Russia, and Austria.

Poland established its independence again after World War I. But in 1939, German troops once more attacked the country, starting World War II. Poles suffered greatly during the war. Warsaw, the capital, was bombed to ashes.

Struggle for Freedom

After World War II, a Communist government came to power in Poland. Its leaders set up a **command economy,** in which the government decided what, how, and for whom goods would be produced. Poland's postwar government wanted heavy industry and military goods, so few products were made for consumers. This led to food shortages, causing Poland's people to become dissatisfied and demand huge changes.

The Poles wanted a better life, complete with political liberties and religious freedom. Deeply Roman Catholic, most Poles rejoiced when a Polish church leader was chosen to be the head of the Roman Catholic Church in 1978. Pope John Paul II not only stirred national pride in Poland, but he also encouraged the Poles to resist Communist rule.

In the 1980s, Polish workers and farmers formed Solidarity, a labor group that supported peaceful democratic change. Communist leaders finally allowed free elections in 1989 that brought about a democracy.

Auschwitz Memorial

Auschwitz was a World War II German prison camp located in Poland. At the camp's entrance was a sign in German that read "Work Sets You Free." Today the camp is a memorial to those who suffered there. ***Movement*** **Why was it easy for armies to invade Poland?**

Social Studies ONLINE

Student Web Activity Visit glencoe.com and complete the Chapter 12 Web Activity about Poland.

This event helped bring about the fall of Communist governments that had long ruled in eastern Europe.

Poland Today

Poland's democratic leaders quickly moved Poland toward a **market economy.** In this system, individuals and businesses make the decisions about how they will use resources and what goods and services to make.

Economic change caused great hardship at first, and many people lost their jobs. Within a few years, however, the economy began to improve. Agriculture remains important, with Poland among the world's top producers of rye and potatoes. Industries, however, are growing. As Poland's economy changes, more people are moving from rural areas to cities, such as Warsaw and Kraków.

▲ This shipyard in Gdansk, Poland, was the birthplace of the labor group Solidarity in the 1980s. Shipbuilding is still an important industry in Poland today. ***Place*** **What was Solidarity? What was its goal?**

Belarus

East of Poland is Belarus, which also is covered by a lowland plain. Belarus was once part of the Soviet Union, and it still has close ties to Russia. Its leaders favor strong government and a command economy.

Belarus has few resources other than **potash,** a mineral used in making fertilizer. Industries include processing fertilizer and manufacturing trucks, radios, televisions, and bicycles. Government-controlled farms produce vegetables, grain, and other crops.

Most people in Belarus are Eastern Orthodox Slavs. Two-thirds live in cities, such as Minsk, the capital.

The Baltic Republics

The small countries of Lithuania, Latvia, and Estonia lie on the shores of the Baltic Sea. Until 1991, the Baltic Republics were part of the Soviet Union. Today, all three countries have large Russian minority populations. Most people in Estonia and Latvia are Protestant, while Roman Catholics make up the majority in Lithuania.

All three Baltic Republics have seen strong economic growth since the mid-1990s. Their economies are based on dairy farming, beef production, fishing, and shipbuilding. Estonia has done especially well, and its people have the Baltic group's highest average **incomes.** Estonia's major export is telecommunications equipment.

Reading Check **Comparing and Contrasting** What do the Baltic Republics, Belarus, and Poland have in common? How are they different?

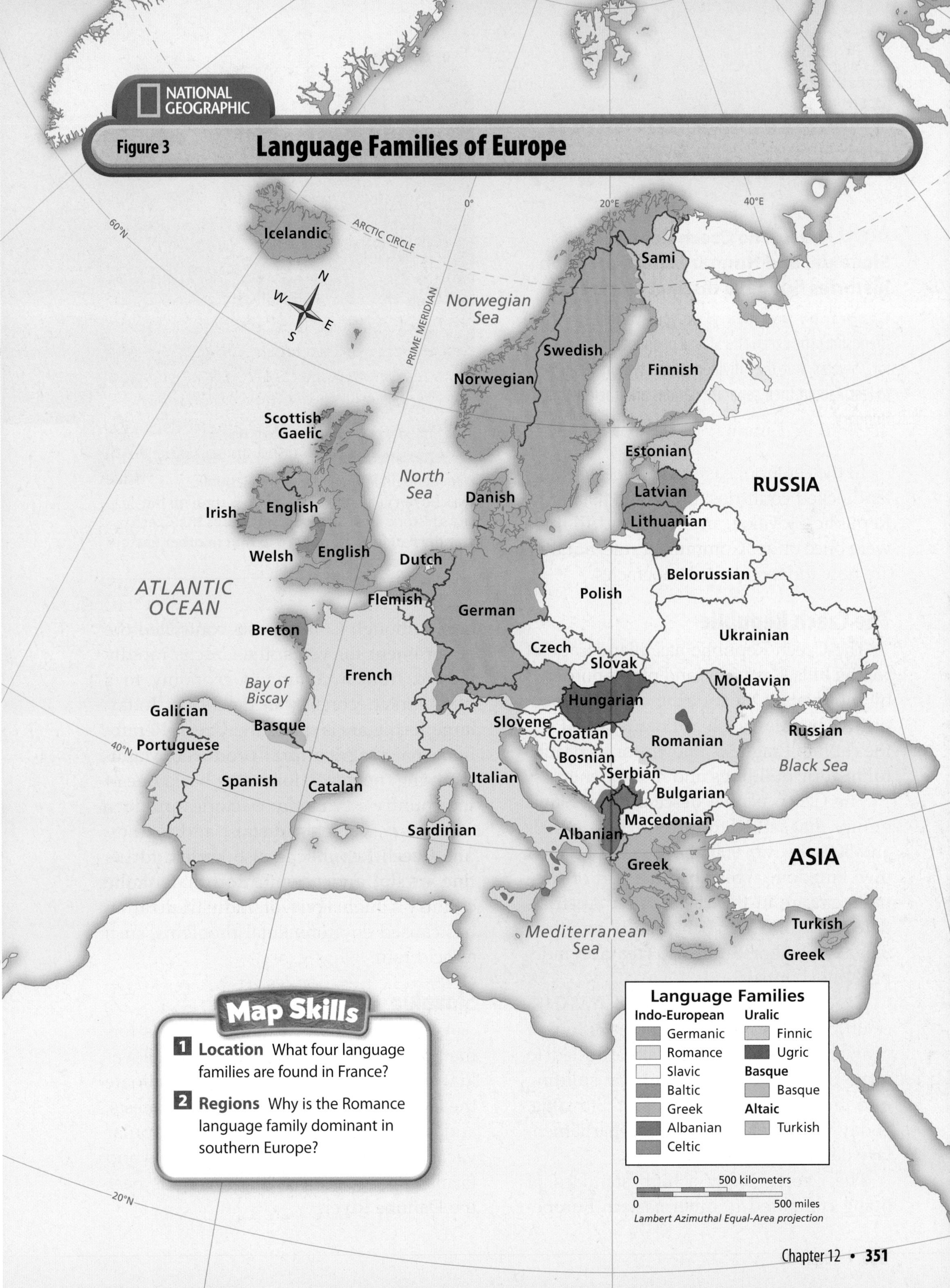

Map Skills

1. **Location** What four language families are found in France?
2. **Regions** Why is the Romance language family dominant in southern Europe?

The Czech Republic, Slovakia, and Hungary

Main Idea The Czech Republic, Slovakia, and Hungary share common histories but have distinct cultures.

Geography and You Have you ever ended a close friendship? The Czech and Slovak people shared a country for 75 years, but they divided it into two in 1993. Read to learn about their separate nations and their neighbor, Hungary.

In the center of eastern Europe are three landlocked countries: the Czech (CHEHK) Republic, Slovakia, and Hungary. All three were once under Communist rule, but they are now independent democracies.

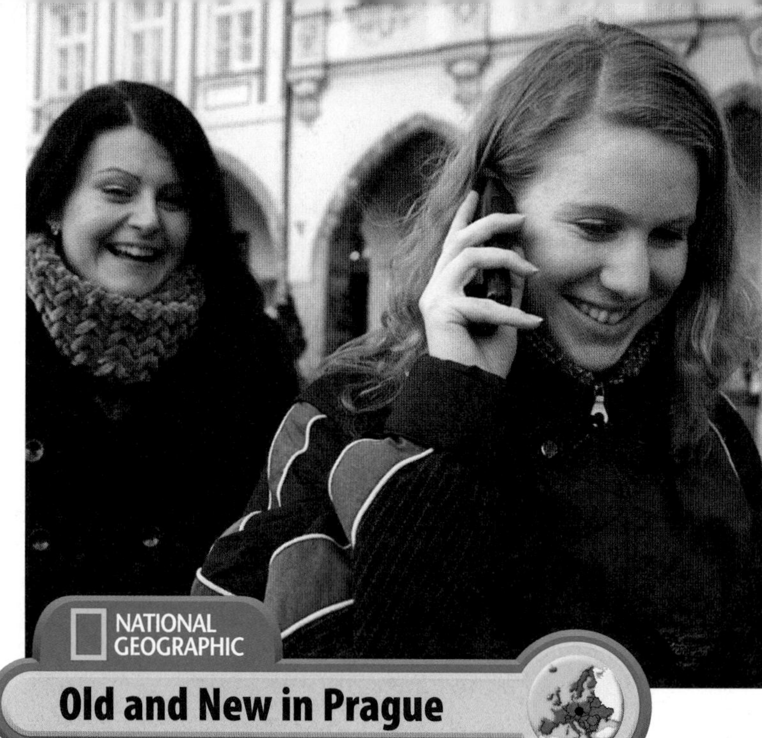

NATIONAL GEOGRAPHIC

Old and New in Prague

A teenager in Prague, the Czech Republic, talks on her cell phone as she strolls through the city's historic area. ***Regions*** **How does the Czech standard of living compare to that in other eastern European countries?**

The Czech Republic

The Czech Republic has a landscape of rolling hills, lowlands, and plains bordered by mountains. Most of the country's people live in cities, such as Prague (PRAHG), the capital. Prague is known for its beautiful historic buildings and monuments.

The Czech people descend from Slavic groups that settled the area in the A.D. 400s and 500s. By A.D. 900, the Czechs had their own kingdom, which became part of Austria's empire in the 1500s. After Austria's defeat in World War I, the Czechs and their Slovak neighbors united to create an independent country.

Czechoslovakia (CHEHK·uh·sloh·VAHK·ee·uh) lasted from 1918 until 1993. In that year, the Czechs and Slovaks decided to settle ongoing disagreements by splitting into the Czech Republic and Slovakia. Today the Czech Republic is a parliamentary democracy.

The Czechs enjoy a high standard of living compared to other eastern Europeans. Although Communists controlled the government for years, the Czechs rapidly moved from a command economy to a free market economy in the 1990s. Today, large fertile areas make the Czech Republic a major agricultural producer. Manufacturing, however, forms the backbone of the country's economy. Factories produce machinery, vehicles, metals, and textiles. The Czech Republic is also known for its fine crystal and beer. Unfortunately, the country's high level of industrialization has caused environmental problems, such as acid rain.

Slovakia

East of the Czech Republic lies its former partner, Slovakia. In northern Slovakia, the Carpathian Mountains dominate the landscape. Rugged peaks, thick forests, and blue lakes make this area a popular vacation spot. In the south, vineyards and farms spread across fertile lowlands near the Danube River.

Independent since 1993, Slovakia is a democracy today. The Slovaks, however, have moved more slowly to a free market economy than the Czechs have. Slovakia has fewer factories than the Czech Republic, and the country is much less industrialized.

The Slovaks also have a language and culture different from the Czechs. While most Czechs are nonpracticing Catholics or are not religious at all, most Slovaks are devout Catholics. Nearly 60 percent of Slovaks live in towns and cities. Bratislava, on the Danube River, is Slovakia's capital.

Hungary

Hungary is located on a large lowland area south of Slovakia and east of Austria. The capital city, Budapest, straddles Hungary's most important waterway, the Danube River.

The Hungarians are not related to the Slavic and Germanic peoples who live in most of eastern Europe, and the language spoken in Hungary is unique (see **Figure 3**). Their ancestors are the Magyars, who moved into the area from Central Asia about a thousand years ago. Like the Czechs and Slovaks, though, the Hungarians were once part of the Austro-Hungarian Empire. Later, after becoming an independent nation, Hungary too was led by Communists. Today it is a democracy headed by a president.

Hungary has few natural resources besides its fertile land, which is valuable for farming. By importing the necessary raw materials, though, the country began to industrialize after World War II. Today Hungary is an exporter of chemicals, food products, and other goods.

Reading Check **Explaining** Why did Czechoslovakia split into two countries?

Countries of Southeastern Europe

Main Idea **Because of limited natural resources, political upheaval, and ethnic conflict, many countries in southeastern Europe face challenges.**

Geography and You Do you adapt easily to change? The countries of southeastern Europe have been through major political and economic changes in recent times. Read to find out how they have responded.

A number of countries are clustered in southeastern Europe. Ukraine lies north of the Black Sea. Romania, Moldova, and Bulgaria are also on or close to the shores of the Black Sea. Their neighbors on the Balkan Peninsula include Albania, Slovenia, Croatia, Bosnia and Herzegovina, Serbia, Montenegro, Kosovo, and Macedonia.

▲ Workers harvest hay on a farm in western Ukraine, an area known for its productive agriculture.
Human-Environment Interaction **What food crops are harvested in Ukraine?**

With the exception of Albania, these are young countries with newly drawn borders. It is not surprising, then, that many of them are struggling for stability and economic success.

Ukraine

Slightly smaller than Texas, Ukraine is the largest country in all of Europe. It lies on a lowland plain with the Carpathian Mountains rising along its southwestern border. The most important waterway, the Dnieper (NEE·puhr) River, has been made navigable to allow the shipping of goods.

The Dnieper River divides Ukraine into two regions. To the west, the lowland steppes have rich, black soil that is ideal for farming. Farmers in this "breadbasket of Europe" grow grains, fruits, and vegetables and raise cattle and sheep. The people living here are of Ukrainian descent. To the east of the Dnieper lies a plains area that has coal and iron ore deposits. Heavily industrialized, this eastern area produces cars, ships, locomotives, and airplanes. Many of the people in eastern Ukraine are of Russian descent.

Ukraine was one of the original republics in the Soviet Union, but it became an independent nation after the Soviet Union dissolved in 1991. Since then, ethnic divisions have grown sharper. Ethnic Ukrainians in the west want to link the country to western Europe and join the European Union. Ethnic Russians in the east want closer ties to Russia.

NATIONAL GEOGRAPHIC

Figure 4 Percentage of Workforce in Agriculture

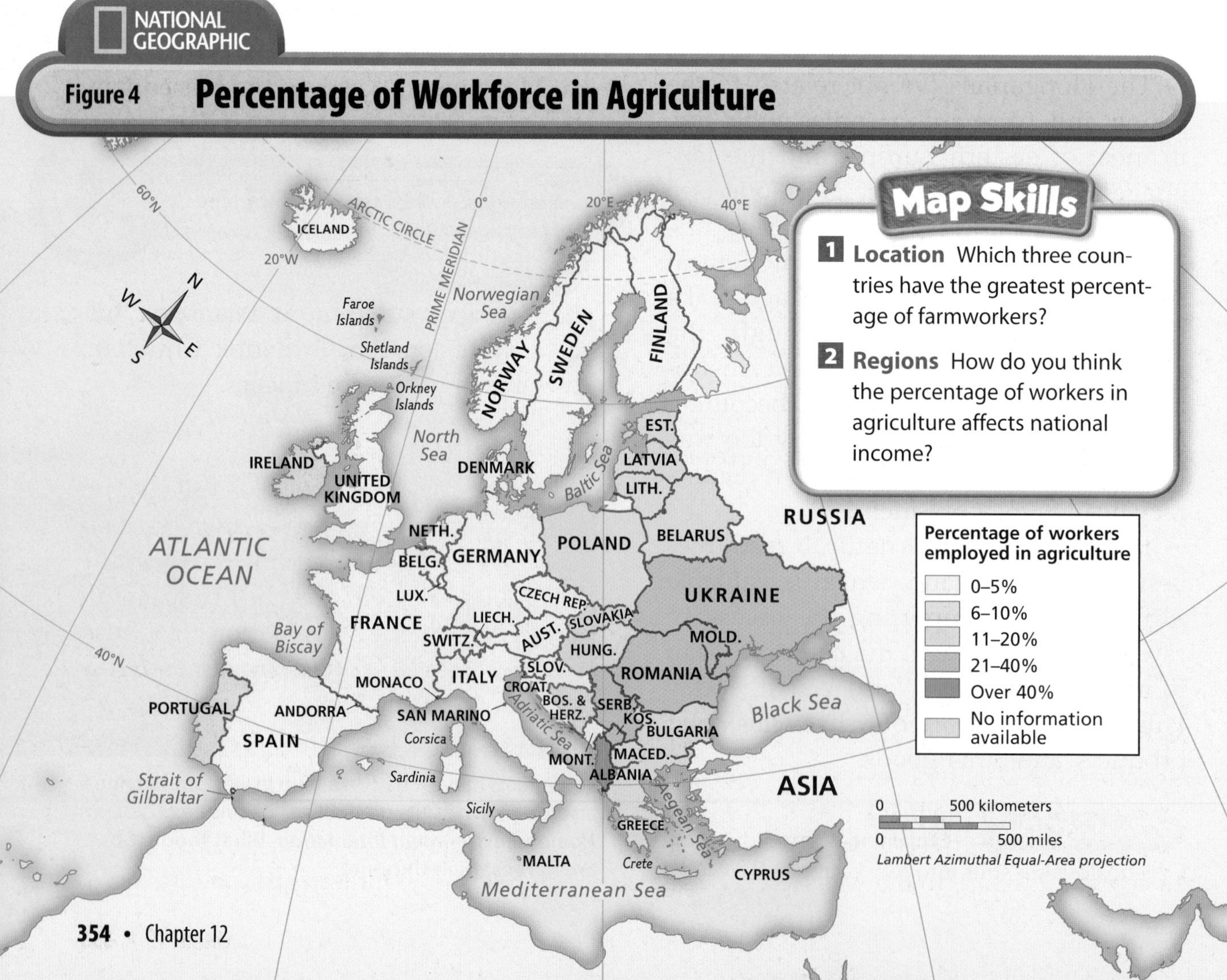

Romania and Moldova

Unlike other eastern European countries, which ended Communist rule peacefully, Romania drove out the Communists in a bloody revolt in 1989. Soon after, the country fell into a deep economic slump. Romania has a wealth of natural resources, however, and the country is now rebounding. Thanks to deposits of coal, petroleum, and natural gas, industry output is growing. Bucharest (BOO·kuh·REHST), the capital, is the major economic and commercial center in the country. Farming also contributes to the economy, and many Romanians are employed in agriculture, as shown in **Figure 4.** Farmers here grow grains, grapes, and other crops.

As the name *Romania* suggests, the Romans once ruled this region and influenced its culture. The Romanian language, for example, is based on the Latin spoken in ancient Rome. In religion, though, Romanians take after their Slavic neighbors. Many are Eastern Orthodox Christian.

Moldova is a small, landlocked country sandwiched between Ukraine on the east and Romania on the west. Moldova's people are mainly Romanian, but Ukrainians and Russians also make up part of the population. Moldova's farms are productive as a result of its fertile soil. Because there are few mineral resources and limited industry, however, Moldova ranks as Europe's poorest country.

Bulgaria

Bulgaria lies south of Romania. It is a mountainous country, but people farm in the fertile river valleys between the peaks. Manufacturing employs many people in Sofia, the capital, and other cities. Tourism is growing as visitors seek out Bulgaria's scenic resorts on the Black Sea.

Most of Bulgaria's 7.7 million people trace their ancestry to the Slavs, Turks, and other groups from Central Asia. Most Bulgarians are Eastern Orthodox Christian. A sizable minority are Muslim.

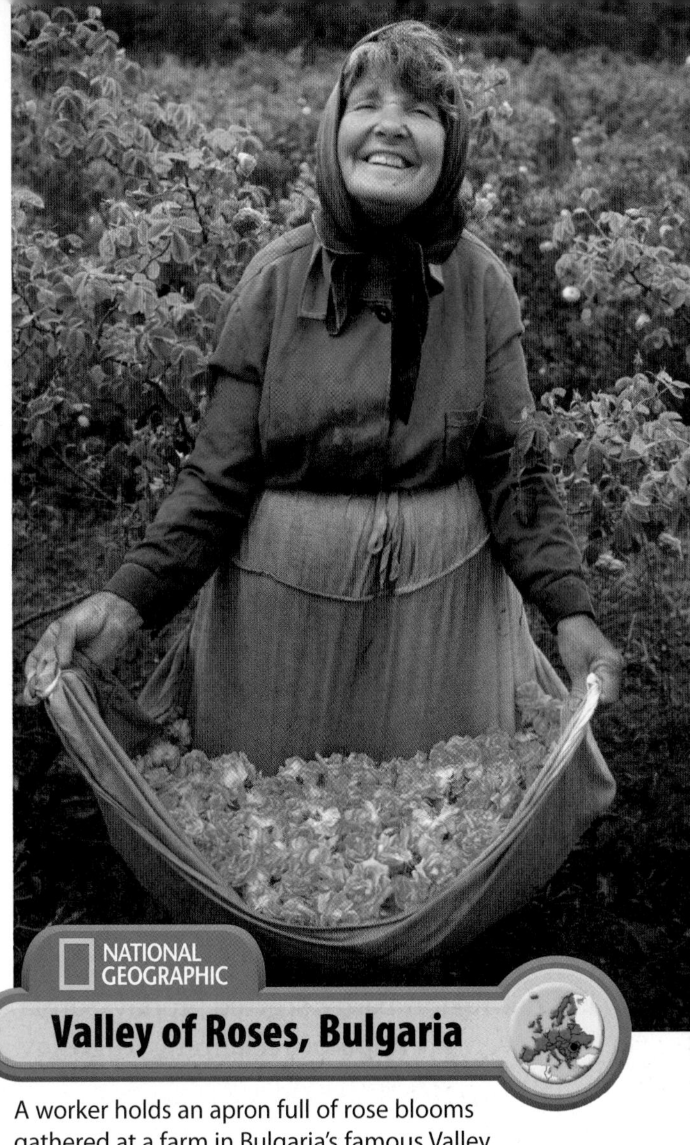

Valley of Roses, Bulgaria

A worker holds an apron full of rose blooms gathered at a farm in Bulgaria's famous Valley of Roses. ***Place*** **What are Bulgaria's major economic activities?**

Other Balkan States

To the south and west of Bulgaria, a number of other countries crowd the Balkan Peninsula. Albania, on the Adriatic Sea, is the only country in Europe with a majority Muslim population. It is also unique because farmers outnumber factory workers. With its economy still heavily agricultural, Albania is one of the poorest countries in Europe.

Other Balkan countries include Slovenia, Croatia, Bosnia and Herzegovina, Macedonia, Serbia, Montenegro, and Kosovo.

None of these nations were even on the map until the 1990s. Before then, they were all part of a Communist country called Yugoslavia.

When communism collapsed in Eastern Europe, the different ethnic groups of Yugoslavia struggled for power. In the early 1990s, four parts of the country—Slovenia, Croatia, Bosnia and Herzegovina, and Macedonia—declared their independence. Meanwhile, another strong part, Serbia, wanted to keep Yugoslavia together under Serbian rule. Serbia's leader used force to try to build power.

The heaviest fighting took place in Bosnia and Herzegovina. There, Serbs carried out **ethnic cleansing**—removing or killing an entire ethnic group—against the Bosnian population. Many people died or became refugees. This and other conflicts left the region badly scarred. Today Serbia has given up hope of reclaiming Yugoslav lands. Montenegro and Kosovo split away from Serbia, and where Yugoslavia once was, there are now seven separate nations.

These Balkan countries were relatively poor during Communist rule, and they continue to be among the poorest in Europe. The mountainous landscape of the Balkan Peninsula makes farming difficult, and there are few natural resources to support economic growth. Ethnic conflict has added to these problems.

Despite these challenges, some countries are moving forward. Slovenia has experienced steady economic growth since it gained independence. Slovenian industries produce machinery, appliances, vehicles, and **medical** supplies. Croatia's economy has also improved, although not as much as Slovenia's.

Reading Check **Explaining** Why have conflicts been fought in the Balkan Peninsula?

Section 4 Review

Social Studies ONLINE
Study Central™ To review this section, go to glencoe.com.

Vocabulary

1. **Explain** the differences between *command economy* and *market economy*. Define *potash* and *ethnic cleansing*.

Main Ideas

2. **Sequencing** List the events that led to democracy in Poland.
3. **Comparing** Create a Venn diagram like the one below to compare the Czech Republic and Slovakia.

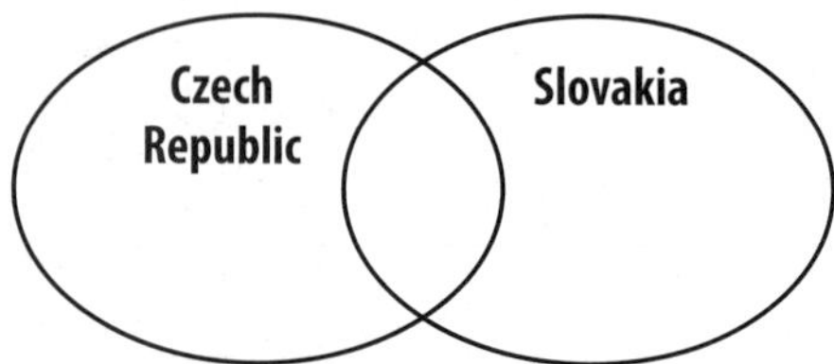

4. **Explaining** Why are many of the countries of southeastern Europe struggling for economic success?

Critical Thinking

5. **BIG Idea** How have eastern Europe's history and physical geography led to differences between that region and other parts of Europe today?
6. **Challenge** Based on what you have read, what do you think will happen politically in Ukraine over the next few years? Why?

Writing About Geography

7. **Using Your FOLDABLES** Use your Foldable to write a paragraph comparing and contrasting conditions in two countries in eastern Europe.

Visual Summary

Northern Europe

- The United Kingdom is a major industrial and trading country.
- The Republic of Ireland's economy is becoming more industrial.
- Fishing is an important industry in the Scandinavian countries.

Members of the British royal family

France and the Benelux Countries

- France is a world center of art, learning, and culture.
- The Benelux countries are low, flat, and densely populated.

Schoolgirls in France

Germany and the Alpine Countries

- Rivers have been vital to Germany's economic growth.
- The Alps dominate Switzerland, Austria, and Liechtenstein.

Southern Europe

- Spain's historic sites and sunny beaches attract many tourists.
- Most of Italy's industry lies in the northern part of the country.
- Greece consists of a mountainous mainland and more than 2,000 islands.

Eastern Europe

- Poland is a large country with northern plains and southern highlands.
- The Danube River flows through Budapest, Hungary's historic capital.
- Ukrainians disagree on whether to strengthen ties to western Europe or to Russia.
- Ethnic conflict has torn apart Balkan countries in recent years.

Industry in eastern Europe

Fisher in Malta

Europe at night

STUDY TO GO Study anywhere, anytime! Download quizzes and flash cards to your PDA from glencoe.com.

STANDARDIZED TEST PRACTICE

Consider carefully before changing your answer to a multiple-answer test question. Unless you misread the question, your first answer choice is often correct.

Reviewing Vocabulary

Directions: Choose the word(s) that best completes the sentence.

1. In a ______, a king or queen serves as a ceremonial head of state but elected officials actively run the government.

A constitutional dictatorship
B constitutional monarchy
C multinational organization
D parliamentary democracy

2. Focusing on certain economic activities in order to make the most advantageous use of a nation's resources is called ______.

A reunification
B neutrality
C specialization
D productivity

3. A country allowing a region within its borders a certain amount of self-rule is known as ______.

A autonomy
B neutrality
C specialization
D subsidy

4. In a ______, the government decides what, how, and for whom goods will be produced.

A market economy
B parliamentary democracy
C constitutional monarchy
D command economy

Reviewing Main Ideas

Directions: Choose the best answer for each question.

Section 1 *(pp. 320–328)*

5. ______ has been troubled by a history of violence between Catholics and Protestants.

A Finland
B Northern Ireland
C Sweden
D Iceland

Section 2 *(pp. 329–337)*

6. In the Netherlands, wetlands that have been drained, called ______, have rich farming soil.

A Benelux
B chateaus
C polders
D Walloons

Section 3 *(pp. 338–342)*

7. In ______, the southern part of the country is poorer and less industrialized than the northern part.

A Portugal
B Ireland
C Greece
D Italy

Section 4 *(pp. 348–356)*

8. In Eastern Europe, Poland and the Baltic Republics have moved toward democracy and market economies, but ______ maintains a command economy and strong central government.

A Belarus
B Lithuania
C Hungary
D Bosnia

GO ON

Critical Thinking

Directions: Choose the best answer for each question.

Base your answers to questions 9 and 10 on your knowledge of Chapter 12 and the map below.

Internet Use in Europe

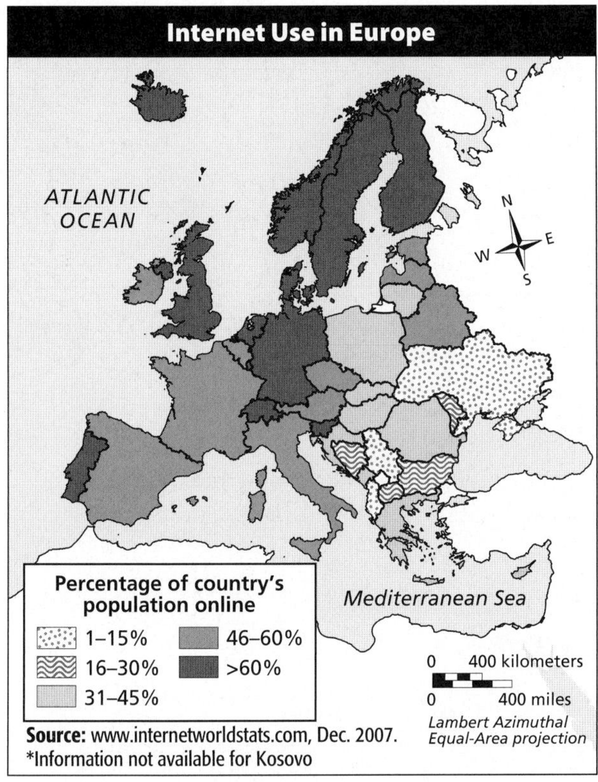

Source: www.internetworldstats.com, Dec. 2007.
*Information not available for Kosovo

9. Which of the following countries has the lowest percentage of residents online?

A Ukraine
B Czech Republic
C Bulgaria
D Poland

10. In general, which subregion of Europe has the highest percentage of people online?

A Southern Europe
B Northern Europe
C Eastern Europe
D the Balkan countries

Document-Based Questions

Directions: Analyze the document and answer the short-answer questions that follow.

> *The EU [European Union] views enlargement as a historic opportunity to help in the transformation of the countries involved, extending peace, stability, prosperity, democracy, human rights and the rule of law throughout Europe. The carefully managed process of enlargement is one of the EU's most powerful policy tools that has helped to transform the countries of Central and Eastern Europe into more modern, functioning democracies.*
>
> —from "European Union Enlargement," CRS Report for Congress, Congressional Research Service, 2006

11. According to this report, how does the European Union view its own enlargement? What effects does the EU's enlargement have on Europe?

12. How do you think the EU's carefully managed process of enlargement can be a powerful policy tool?

Extended Response

13. Imagine you have been invited to be on a panel that is going to pick "Europe's Most Promising Economy of the Future." Prepare a speech in which you nominate a European country for this award. Explain why you believe this country's economy will either grow or remain strong in the future. Be sure to use geographical, population, and economic factors to support your position.

STOP

Social Studies ONLINE

For additional test practice, use Self-Check Quizzes—Chapter 12 at glencoe.com.

Need Extra Help?

If you missed question...	1	2	3	4	5	6	7	8	9	10	11	12	13
Go to page...	322	330	339	349	326	332	341	350	26	26	355	355	321

UNIT 5

Russia

CHAPTER 13
Physical Geography of Russia

CHAPTER 14
History and Cultures of Russia

CHAPTER 15
Russia Today

St. Basil's Cathedral, Moscow, Russia ▶

NGS ONLINE For more information about the region, see www.nationalgeographic.com/education.

Победа деда-наша

Regional Atlas

Russia

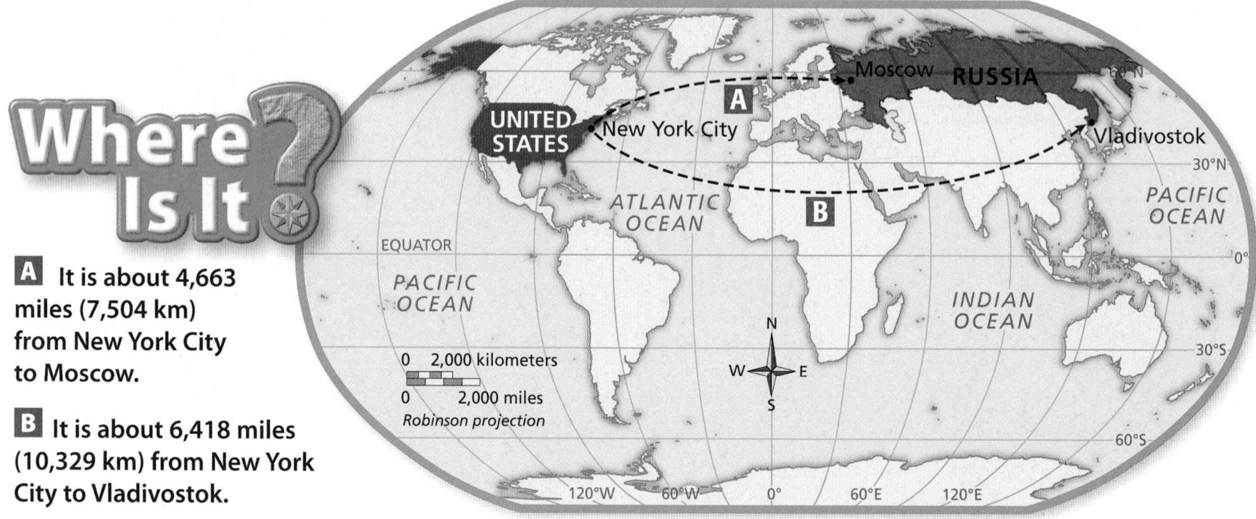

Where Is It?

A It is about 4,663 miles (7,504 km) from New York City to Moscow.

B It is about 6,418 miles (10,329 km) from New York City to Vladivostok.

How Big Is It?

The region of Russia is more than twice the size of the continental United States. Its land area is about 6,592,819 square miles (17,075,322 sq. km). Russia is the third-largest culture region and the largest single country in the world.

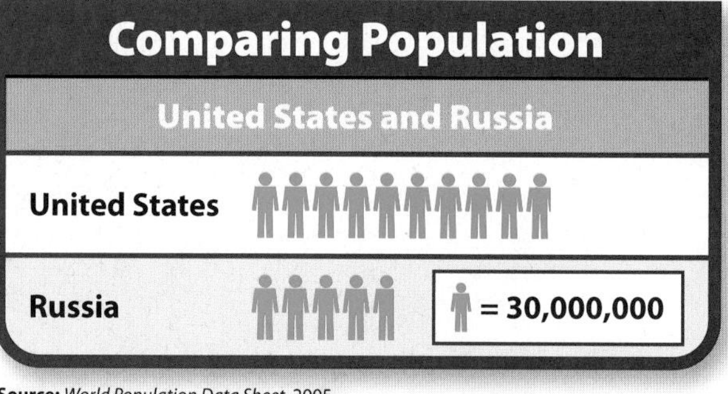

Comparing Population

United States and Russia	
United States	10 figures
Russia	5 figures

(1 figure = 30,000,000)

Source: *World Population Data Sheet*, 2005.

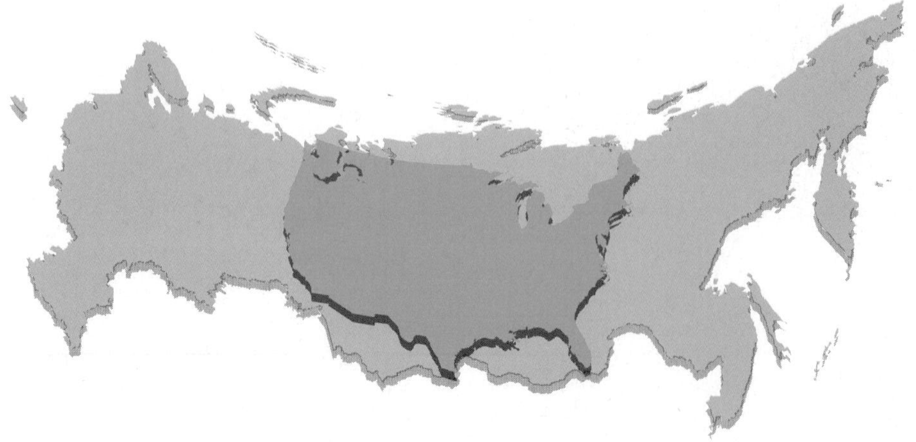

GEO Fast Facts

Longest River

Ob-Irtysh River
3,461 mi.
(5,569 km) long

Deepest Lake

Lake Baikal
5,715 ft. (1,742 m) deep

Lowest Point

Caspian Sea
92 ft. (28 m)
below sea level

Highest Point

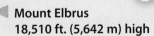

Mount Elbrus
18,510 ft. (5,642 m) high

UNIT 5

Regional Atlas

Russia

PHYSICAL

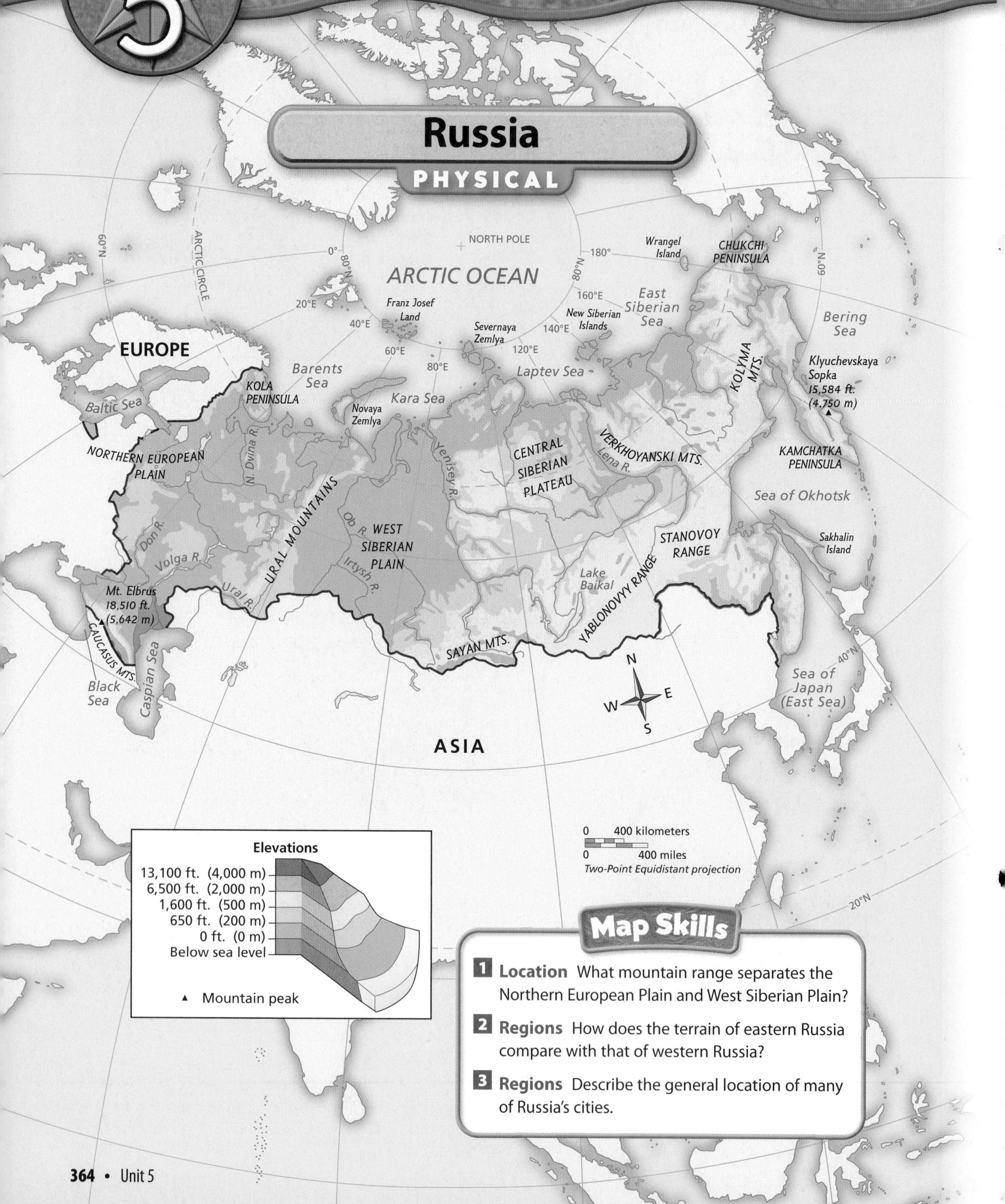

Map Skills

1. **Location** What mountain range separates the Northern European Plain and West Siberian Plain?
2. **Regions** How does the terrain of eastern Russia compare with that of western Russia?
3. **Regions** Describe the general location of many of Russia's cities.

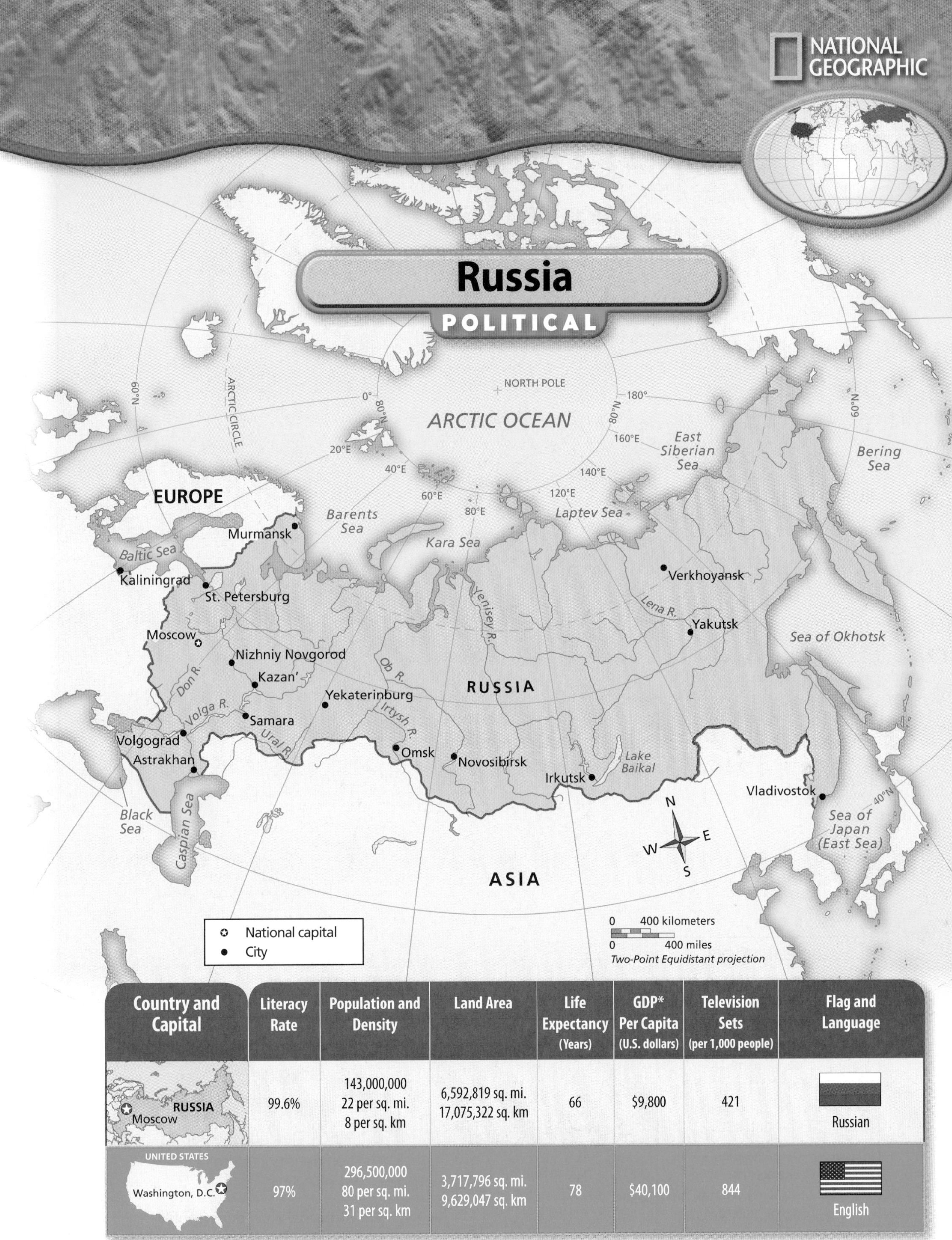

Country and Capital	Literacy Rate	Population and Density	Land Area	Life Expectancy (Years)	GDP* Per Capita (U.S. dollars)	Television Sets (per 1,000 people)	Flag and Language
RUSSIA Moscow	99.6%	143,000,000 22 per sq. mi. 8 per sq. km	6,592,819 sq. mi. 17,075,322 sq. km	66	$9,800	421	Russian
UNITED STATES Washington, D.C.	97%	296,500,000 80 per sq. mi. 31 per sq. km	3,717,796 sq. mi. 9,629,047 sq. km	78	$40,100	844	English

Sources: *CIA World Factbook*, 2005; Population Reference Bureau, *World Population Data Sheet*, 2005.

For more country facts, go to the **Nations of the World Databank** at glencoe.com.

Countries and flags not drawn to scale

*Gross Domestic Product

UNIT 5

Regional Atlas

Russia

POPULATION DENSITY

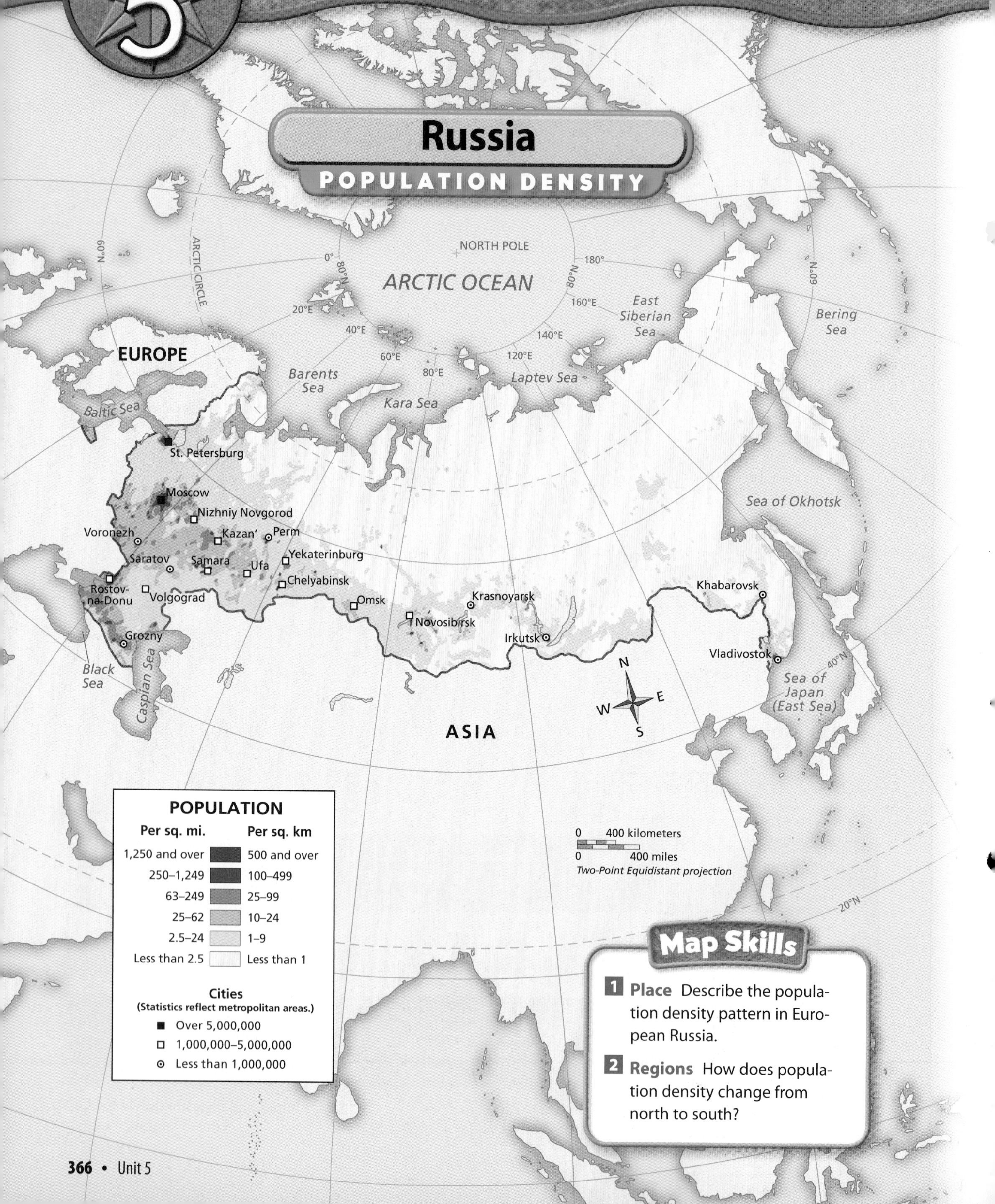

NORTH POLE
ARCTIC OCEAN
ARCTIC CIRCLE
60°N
80°N
0°
20°E
40°E
60°E
80°E
120°E
140°E
160°E
180°
East Siberian Sea
Bering Sea
Laptev Sea
Barents Sea
Kara Sea
EUROPE
Baltic Sea
St. Petersburg
Moscow
Nizhniy Novgorod
Voronezh
Kazan'
Perm
Saratov
Samara
Ufa
Yekaterinburg
Chelyabinsk
Rostov-na-Donu
Volgograd
Omsk
Krasnoyarsk
Novosibirsk
Irkutsk
Khabarovsk
Vladivostok
Grozny
Black Sea
Caspian Sea
Sea of Okhotsk
Sea of Japan (East Sea)
40°N
20°N
ASIA
N
E
S
W

POPULATION

Per sq. mi.	Per sq. km
1,250 and over	500 and over
250–1,249	100–499
63–249	25–99
25–62	10–24
2.5–24	1–9
Less than 2.5	Less than 1

Cities
(Statistics reflect metropolitan areas.)

- ■ Over 5,000,000
- □ 1,000,000–5,000,000
- ⊙ Less than 1,000,000

0 400 kilometers
0 400 miles
Two-Point Equidistant projection

Map Skills

1 Place Describe the population density pattern in European Russia.

2 Regions How does population density change from north to south?

Map Skills

1. **Location** In what part of Russia are gold and silver deposits found?
2. **Regions** Describe the general location of commercial farming areas in Russia.

Reading Social Studies

Comparing and Contrasting

When you *compare* people, things, or ideas, you look for the similarities among them. When you *contrast* people, things, or ideas, you identify their differences.

Textbook authors sometimes use this structure to help readers see the similarities and differences between topics.

- Read the following paragraph.
- Then determine how Russia and the United States are similar and how they are different.

> Russia's official name is the Russian Federation. This name reflects the fact that Russia comprises, or is made up of, many different regions and territories. Like the United States, Russia is a federal republic, with power divided between national and regional governments. In the United States, some powers belong to the states, and others belong only to the national government. Some powers are shared by both levels of government. In Russia, the division of powers is less clear because the new Russian government is still developing.
>
> —*from page 409*

In a Venn diagram, differences are listed in the outer parts of the circles. Similarities are described where the circles overlap.

The Governments of Russia and the United States

Russia	Similarities	United States
The division of powers between regional governments and the national government is not clear.	• Both countries are federal republics. • Both countries divide power between national and regional governments.	Some powers are given solely to states and solely to the national government, while some powers are shared by both governments.

Reading Tip

As you read, look for signal words that show comparisons, such as *similarly*, *at the same time*, and *likewise*. Contrast signal words include *however*, *rather*, and *on the other hand*.

2 Practice It!

Read the following sentences from this unit that describe Russian leaders Ivan IV and Joseph Stalin.

- Draw a Venn diagram like the one shown below.
- List the differences and similarities between the two leaders.

> In 1547 Ivan IV declared himself czar, or emperor. He ruled harshly, using secret police to carry out his will, and earned the name "Ivan the Terrible." By conquering neighboring territories, Ivan expanded his empire south to the Caspian Sea and east past the Ural Mountains.
>
> —*from page 390*
>
> Lenin's policies were later continued by Joseph Stalin, who ruled the Soviet Union after Lenin's death in 1924. A harsh dictator, Stalin prevented the Soviet people from practicing their religions and had religious property seized. His secret police killed or imprisoned anyone who disagreed with his policies.
>
> —*from page 392*

Read to Write Activity

Read and take notes about Russia's physical geography in Chapter 13. Then find similar information about another region in your book. Write an essay that describes the similarities and differences between the two regions.

▲ Ivan the Terrible

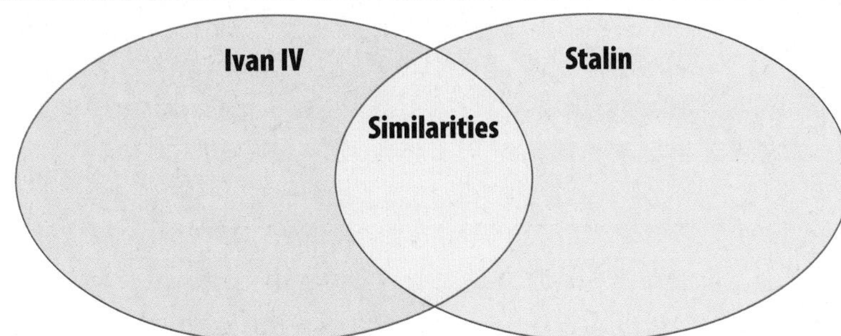

3 Apply It!

As you read Chapters 13, 14, and 15, look for people, things, or ideas that you can compare and contrast. As you read, put notes in a Venn diagram like the one above to help you compare and contrast.

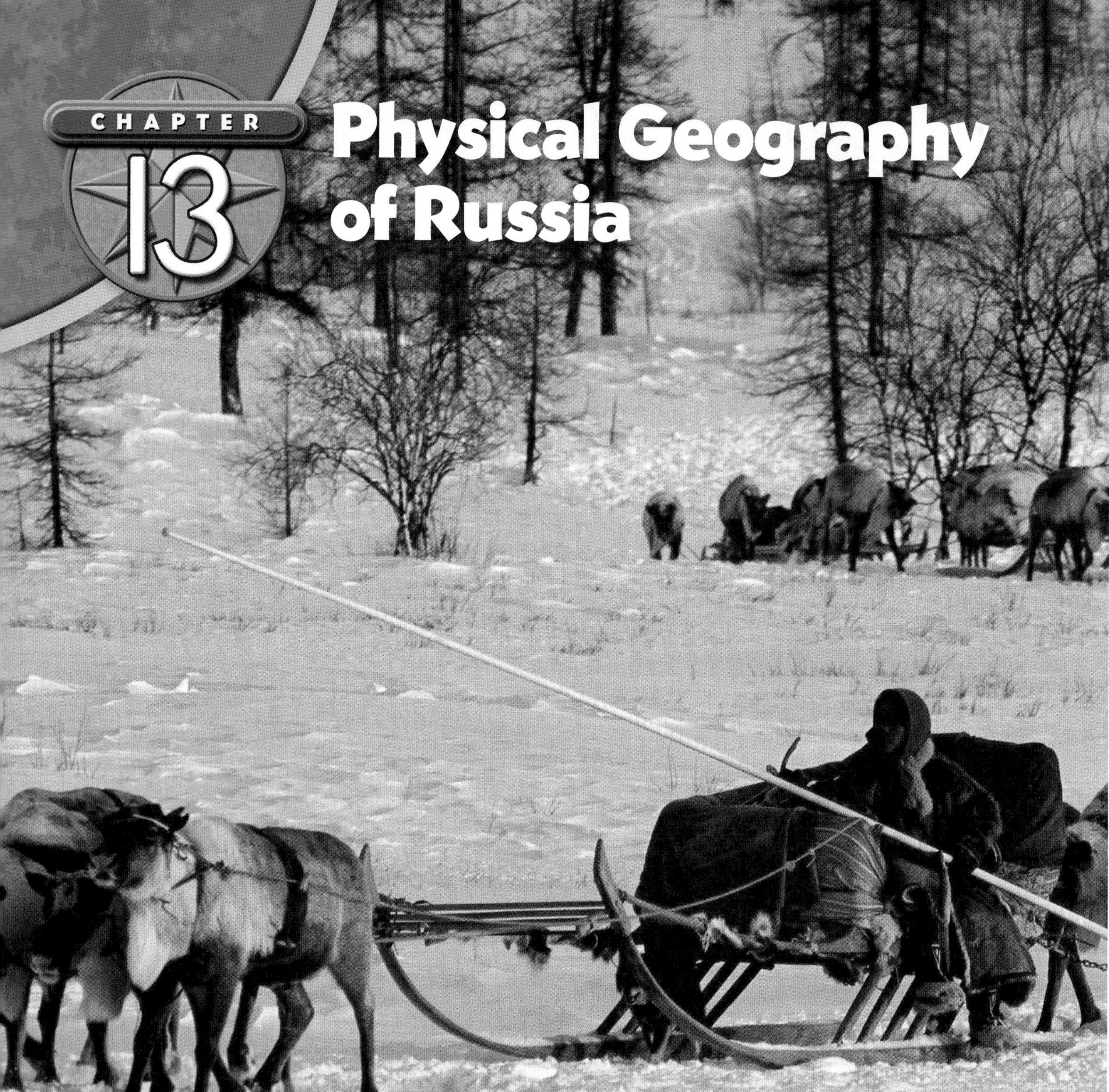

CHAPTER 13 Physical Geography of Russia

Essential Question

Human-Environment Interaction Russia's vast, cold landscapes include mountain ranges, plains, and evergreen forests. The country is also rich in natural resources, especially those used to create energy. To take advantage of these resources, however, Russia's people must overcome problems created by the country's landforms and climate. How do Russia's location and landforms affect its population and its use of resources?

Reindeer in Siberia

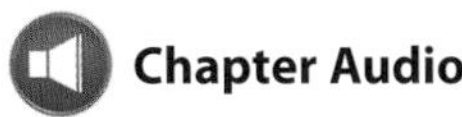

Section 1: Physical Features

BIG IDEA Changes occur in the use and importance of natural resources. Russians have used their soil, water, and timber for their own needs. As global demand for energy rises, Russia's rich supply of energy resources will be increasingly important.

Section 2: Climate and the Environment

BIG IDEA People's actions change the physical environment. Because much of Russia has a harsh climate, most Russians live where the climate is milder. The Russian people have adapted to their surroundings, but some of their actions have damaged the environment. Planning is necessary to take advantage of the country's great resources while preserving the environment.

Organizing Information Make this Foldable to help you organize facts about Russia's physical features, climate, and environment.

Step 1 Fold a piece of 11x17 paper in half.

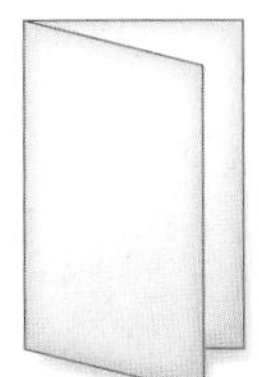

Step 2 Fold the bottom edge up two inches. Glue the outer edges of the tab to create two pockets.

Step 3 Label each pocket as shown. Use these pockets to hold notes taken on index cards or quarter sheets of paper.

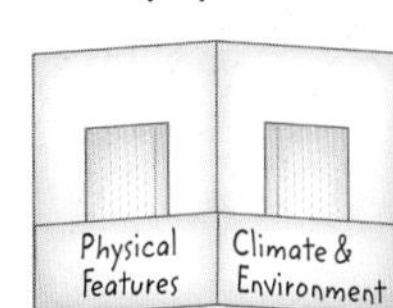

Reading and Writing As you read the chapter, write down facts on separate cards or sheets of paper and place them in the correct pocket. Use your cards to write quiz questions for each section topic.

Visit glencoe.com and enter *QuickPass*™ code EOW3109c13 for Chapter 13 resources.

Guide to Reading

BIG Idea

Changes occur in the use and importance of natural resources.

Content Vocabulary

- fossil fuel *(p. 375)*
- softwood *(p. 375)*
- infrastructure *(p. 375)*

Academic Vocabulary

- benefit *(p. 373)*
- inhibit *(p. 375)*

Reading Strategy

Identifying Use a diagram like the one below to list six of Russia's major landforms.

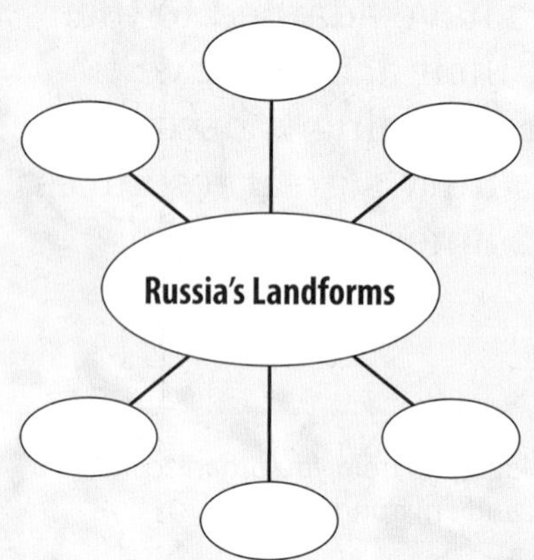

Physical Features

Picture This The bubbling goo in this volcanic pool is like a pot of boiling soup. Here in far eastern Russia, a huge volcano exploded tens of thousands of years ago. The area now contains cold, rushing rivers, hot springs, and pools filled with steaming, toxic mud. As volcanic gases push their way up through the thick ooze, they create bubbles. The bubbles are evidence that forces under Earth's crust are always in motion. Read this section to learn more about the different types of physical features found in Russia.

Volcanic crater on Kamchatka Peninsula in eastern Russia

Landforms of Russia

Main Idea **Russia is a huge country with a location and landforms that greatly affect how people live.**

Geography and You If you have ever traveled across the United States, you know that it takes a long time. It would take twice as long to travel across Russia. Read to learn about the landforms of Russia's vast terrain.

Russia is the world's largest country. From east to west, it measures some 6,200 miles (9,980 km). Nearly twice as large as the United States, Russia straddles both Europe and Asia. The Ural Mountains serve as a dividing line between the European and Asian parts of Russia.

A Vast Northern Land

Because of its enormous size, Russia has a long coastline. Russia does not **benefit** from its closeness to the sea, though, because of its northern location. Most of its coast lies along waters that are frozen for much of the year. As a result, Russia has few seaports that are free of ice year-round. Through the Black Sea in the southwest, though, Russian ships have a warm-water route to the Mediterranean Sea.

Russia contains a variety of landforms. Rugged mountains and plateaus lie in the south and east. In the north and west, vast lowland plains reach to the horizon.

European Russia

Most of European Russia lies on the Northern European Plain. This fertile area has Russia's mildest climate, and about 75 percent of Russians live here. Moscow, the capital, and St. Petersburg, a large port city near the Baltic Sea, are located in this region. Much of Russia's agriculture and industry are found on the Northern European Plain.

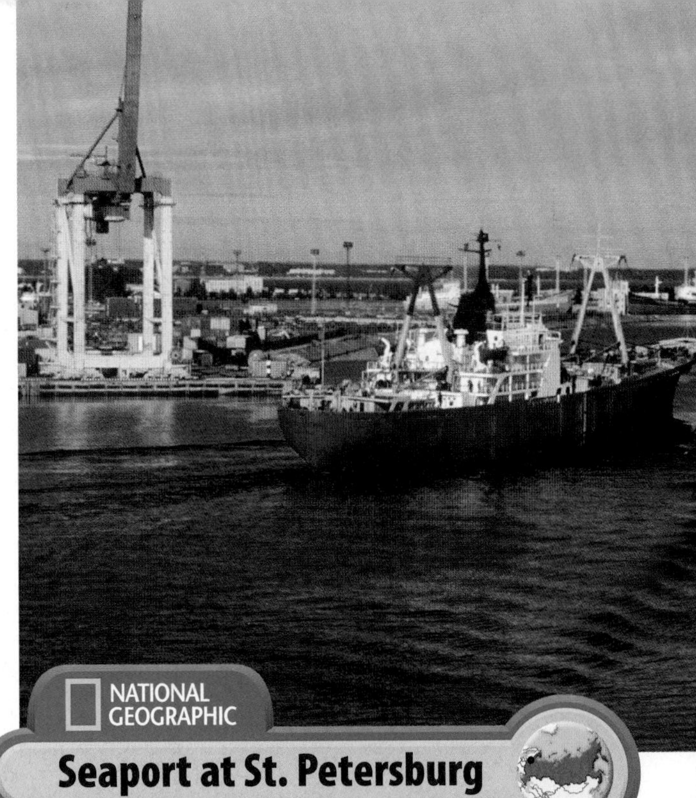

NATIONAL GEOGRAPHIC

Seaport at St. Petersburg

The port at St. Petersburg is one of the busiest in Russia, even though it is frozen three to four months every year. ***Location*** **Which sea provides a warm water outlet for Russian ships?**

Good farmland also lies farther south, along the Volga and other rivers. This area consists of a nearly treeless grassy plain. To the far south of European Russia are the rugged Caucasus Mountains. Located near a fault line, the Caucasus area is prone to destructive earthquakes. Other mountains—the Urals—divide European and Asian Russia. Worn down by erosion, the Urals are not very tall.

Asian Russia

East of the Urals is Asian Russia, which includes Siberia. Northern Siberia has one of the coldest climates in the world. It is a vast treeless plain that remains frozen much of the year. The few people who live here make their living fishing, hunting seals and walruses, or herding reindeer.

Social Studies ONLINE

Student Web Activity Visit glencoe.com and complete the Chapter 13 Web Activity about Siberia.

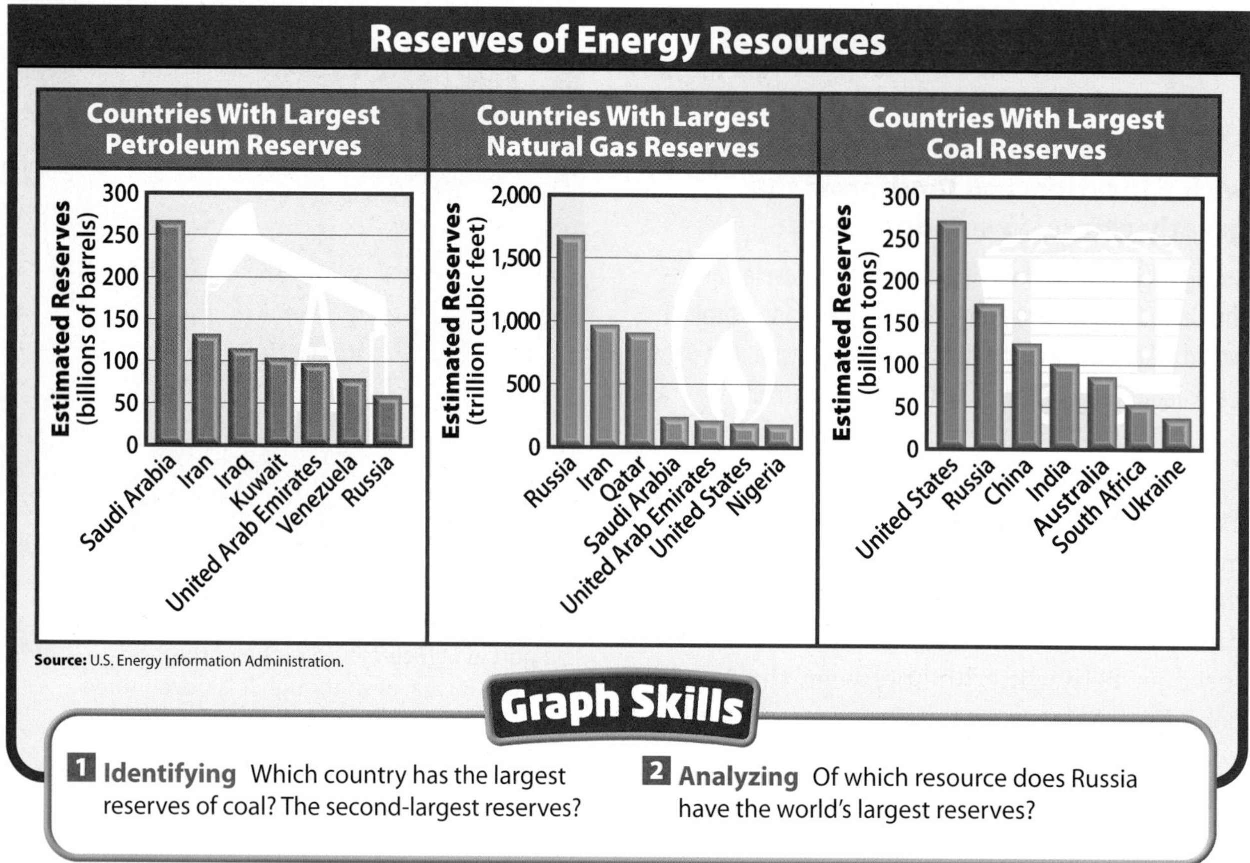

Graph Skills

1 Identifying Which country has the largest reserves of coal? The second-largest reserves?

2 Analyzing Of which resource does Russia have the world's largest reserves?

To the south of the northern plains area is a region of dense forests where people make their living by lumbering or hunting. Plains, plateaus, and mountain ranges cover the southern part of Siberia.

Mountains also rise on the far eastern Kamchatka (kam·CHAHT·kuh) Peninsula. These mountains are part of the Ring of Fire. This is a region along the rim of the Pacific Ocean where tectonic plates meet and cause Earth's crust to be unstable. As a result, Kamchatka has many volcanoes.

Inland Waters

Russia has many rivers. The Volga is European Russia's major river. Russians have long relied on the Volga for transportation. In Siberia, many rivers begin in mountains in the south and flow north across marshy lowlands, emptying into the frigid Arctic Ocean. The Lena (LEE·nuh), the Yenisey (YIH·nih·SAY), and the Ob' (AWB) are among the longest rivers in the world.

Russia includes or borders many inland bodies of water. Almost the size of California, the Caspian Sea in southwestern Russia is the largest inland body of water in the world. It is actually a saltwater lake that is an important resource for fishing. Major oil and gas deposits are found near or under the Caspian Sea.

In southern Siberia lies Lake Baikal—the world's deepest freshwater lake. The lake holds one-fifth of the world's supply of unfrozen freshwater. Baikal is home to many kinds of aquatic life, including Baikal seals, or nerpa, the only seals that live in freshwater.

Reading Check **Describing** Describe the landforms of the Kamchatka Peninsula.

Natural Resources

Main Idea **Although Russia has plentiful resources, many of them are in remote Siberia and are difficult to obtain.**

Geography and You Have you ever worked or played outside in the extreme cold of winter? Read to find out how a cold climate affects Russians' use of resources.

Russia is rich in natural resources. As the graphs on the previous page show, Russia is a leader in reserves of the **fossil fuels**—oil, natural gas, and coal. The country also has major deposits of iron ore, which the Russians have used to develop a large steel industry. Other important metals mined in Russia include copper and gold.

Russia's other great resource is timber. Trees cover much of Siberia, and Russia produces about a fifth of the world's **softwood.** This wood from evergreen trees is used in buildings and for making furniture.

Russia's large size and cold climate **inhibit,** or limit, humans' ability to use its many resources. Siberia is vast and remote, and its resources are difficult to use because of the area's lack of **infrastructure.** Infrastructure is the system of roads and railroads for transporting materials.

In addition to transportation difficulties, Siberia's cold climate brings other challenges. Besides trying to stay warm, workers must keep their equipment from freezing. Some advances in technology have made it easier to collect Siberia's resources. For example, a pipeline now carries natural gas from Siberia to Europe.

Reading Check **Explaining** Why are Russia's natural resources difficult to gather?

Section Review

Social Studies ONLINE
Study Central™ To review this section, go to glencoe.com.

Vocabulary

1. **Explain** the meaning of *fossil fuel, softwood,* and *infrastructure* by using each term in a sentence.

Main Ideas

2. **Comparing and Contrasting** Use a Venn diagram like the one below to compare and contrast physical features in European Russia and Siberia.

3. **Explaining** Why are Siberia's resources valuable?

Critical Thinking

4. **Analyzing Information** In what part of the country does most of Russia's population live? Why?
5. **BIG Idea** Name some natural resources found in Russia. How do these resources contribute to the Russian and world economies?
6. **Challenge** How might a country's far-north location affect its people and economy? Consider factors such as agriculture and transportation.

Writing About Geography

7. **Expository Writing** Write a paragraph describing what you believe is Russia's most prominent physical feature. Be sure to explain your choice.

Russia's Forests: Should They Be Used for Economic Development?

The boreal forests are located south of the Arctic Circle. They are found in northern Canada, Europe, and Russia. The forests contain most of the world's unfrozen freshwater. Many species of birds and animals live there or migrate through the forests. The forests are also home to various indigenous peoples. The Russian boreal forest, also known as the taiga, extends about 4,000 miles (6,436 km) across northern Russia. Some Russians want to develop the resources of the forest, such as lumber, oil, and natural gas. Many environmentalists, though, would like to preserve the forest and its biodiversity.

Development

These environmentalists . . . who criticize me have no idea what's really happening in the forest. . . . I was shocked by the extent of destruction I saw here 30 years ago. . . . But I've been surprised by the way nature has reacted and how well the trees have grown back. The law of the north woods is, the more you destroy them, the stronger they recover. . . . The world's environmentalists say don't touch the boreal forest—leave it alone, leave it natural. . . . But my philosophy with the forests is to use them. You want to watch TV, drive a car, live well, right? And how do you do that if you don't touch nature?

—Vladimir Sedykh
Chief Scientist
Sukachev Forest Institute, Novosibirsk

Development

The current economic model of forest use is oriented [directed] mainly towards the export of raw materials and cannot provide long-term economic growth for the region. This model . . . destroys the last old-growth forests in Russia, which have now become the most threatened natural ecosystems. If the current situation does not change, we will lose the most valuable old-growth areas in [the] next 5–10 years. Logging the last European old-growth forests will not solve any social or economical problem in Russia. In its best light it will only postpone the social and economic crash in the forest industry and logging settlements by a matter of several years.

—from *The Last of the Last: The Old-growth Forests of Boreal Europe*
Dmitry Aksenov, Mikhail Karpachevskiy, Sarah Lloyd, and Alexei Yaroshenko

You Be the Geographer

1. **Summarizing** In your own words, summarize the opinion of each writer about how the forest should be used.
2. **Critical Thinking** What argument does Sedykh make to appeal to people who want to live a modern lifestyle?
3. **Read to Write** Do you think developers and environmentalists can work together to use and preserve the boreal forest? Explain your opinion in a one-page essay.

Guide to Reading

BIG Idea

People's actions change the physical environment.

Content Vocabulary

- permafrost *(p. 380)*
- taiga *(p. 380)*
- smog *(p. 380)*
- pollutant *(p. 380)*

Academic Vocabulary

- period *(p. 379)*
- decline *(p. 382)*

Reading Strategy

Analyzing Information Use a diagram like the one below to list factors that lead to Russia's cold climate, especially those related to location and landforms.

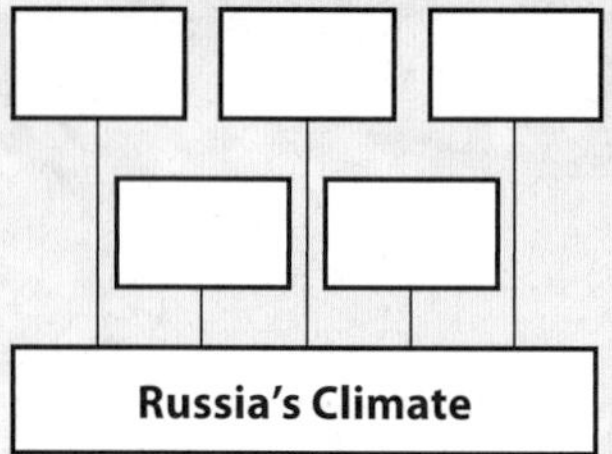

Climate and the Environment

Picture This For many Russians, ice fishing is a favorite pastime. In this photo, an ice fisher is shielded from the cold winds blowing along the Tom River in Siberia. The fisher must often reach into the icy water and remove slush from the hole to keep it from freezing over. Read this section to learn how Russia's many climate zones have influenced its people.

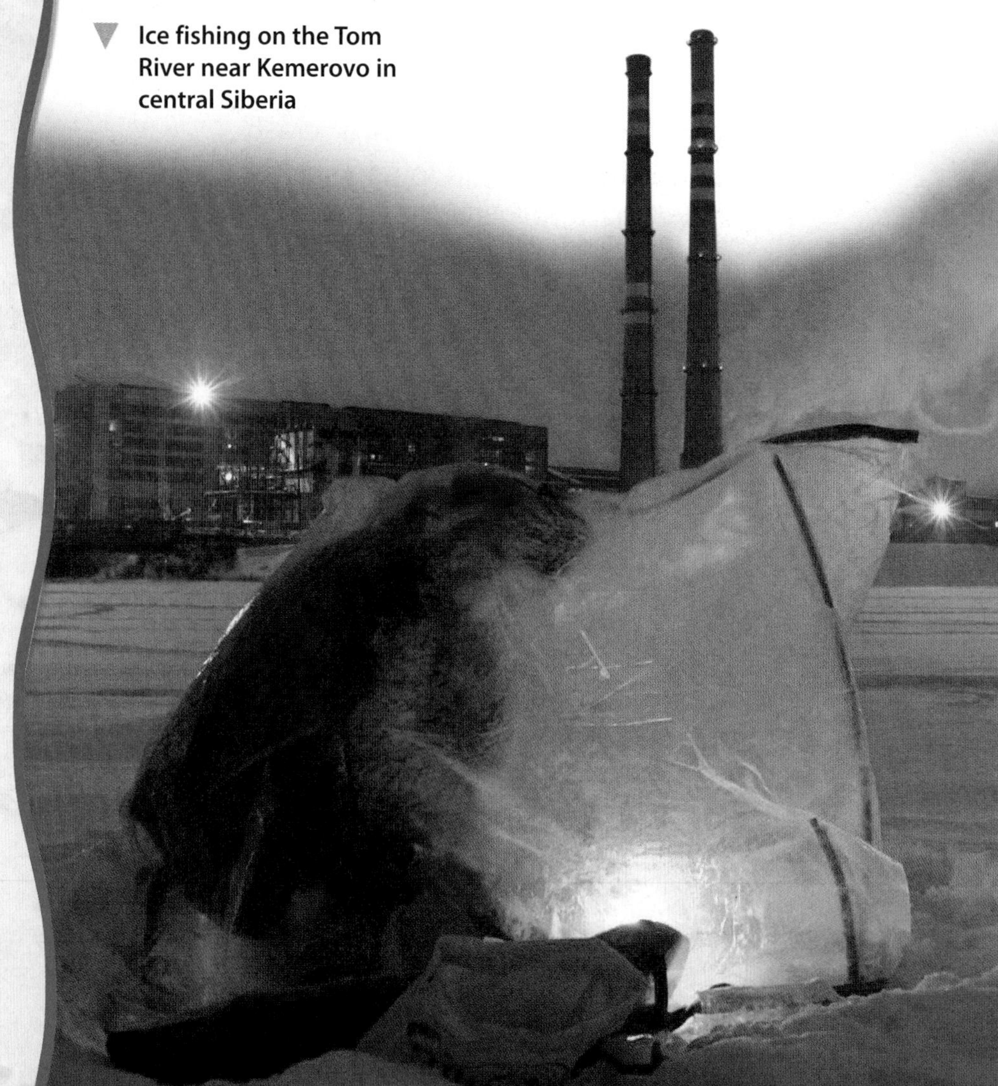

▼ Ice fishing on the Tom River near Kemerovo in central Siberia

A Cold Climate

Main Idea **Russia has a generally cool to cold climate because of its northern location.**

Geography and You Have you ever experienced a winter that was extremely cold and snowy and seemed to go on too long? Read to find out why much of Russia has long, cold winters and short summers.

As **Figure 1** shows, most of Russia lies in the high latitudes. As a result, Russia receives very little of the sun's heat even during summer. In addition, much of Russia lies inland, far from the moist, warm currents of the Atlantic and Pacific Oceans that help moderate temperature in other parts of the world. In Russia's far north, elevations are generally too low to prevent the southerly flow of icy Arctic air. In the country's south and east, tall mountains stop the warm air coming from the lower latitudes. Consequently, Russia has a generally cool to cold climate. Large areas of the country experience only winter- and summerlike conditions. Spring and autumn are simply brief **periods** of changing weather.

NATIONAL GEOGRAPHIC **Maps In MOtion** See StudentWorks™ Plus or glencoe.com.

Figure 1 **Russia: Climate Zones**

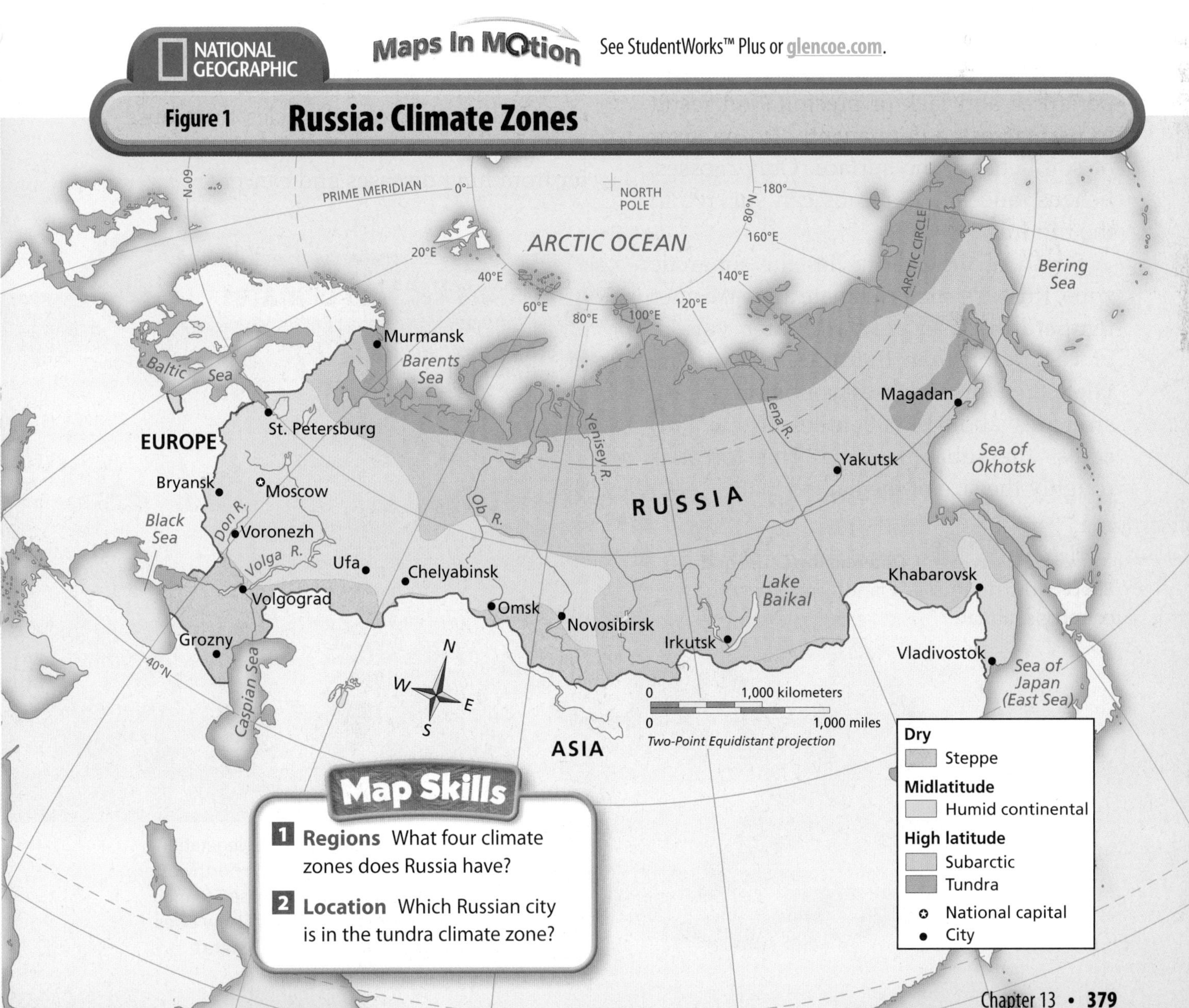

Map Skills

1. **Regions** What four climate zones does Russia have?
2. **Location** Which Russian city is in the tundra climate zone?

Most of western Russia has a humid continental climate. Summers are warm and rainy, and winters are cold and snowy. Moscow's average July temperature is just 66°F (19°C), while its average January temperature can plunge as low as 16°F (–9°C). The cold winters have played an important role in Russia's history. During World War II, bitter cold halted the German army's advance into Russia. Better-prepared Russian troops soon forced the Germans to retreat.

In contrast, the northern and eastern areas of Russia have short, cool summers and long, snowy winters. The northern tundra climate zone is so cold that moisture in the soil cannot evaporate. Cold temperatures and lack of precipitation result in **permafrost,** a permanently frozen layer of soil beneath the surface. Only mosses, lichens, and small shrubs can survive in the tundra.

South of the tundra lies the subarctic zone, Russia's largest climate area. Warmer temperatures support a greater variety of vegetation than the tundra does. The **taiga,** the world's largest coniferous forest, stretches about 4,000 miles (6,436 km) across the subarctic zone. This forest is roughly the size of the United States.

Reading Check **Determining Cause and Effect** How do Russia's landforms affect the country's climate?

Russia's Environment

Main Idea As Russia's economy expanded, the country's environment was poorly cared for.

Geography and You What might happen if the garbage in your neighborhood was not disposed of properly? Read to learn about Russia's efforts to clean up its environment.

For most of the 1900s, Russia's leaders stressed economic growth. They ignored the damage this growth caused to the environment. Today, **smog**—a thick haze of fog and chemicals—blankets many of Russia's cities. Factories pour **pollutants,** which are chemicals and smoke particles that cause pollution, into the air. Many Russians suffer from lung diseases and cancer.

Russia's Varying Climates

NATIONAL GEOGRAPHIC

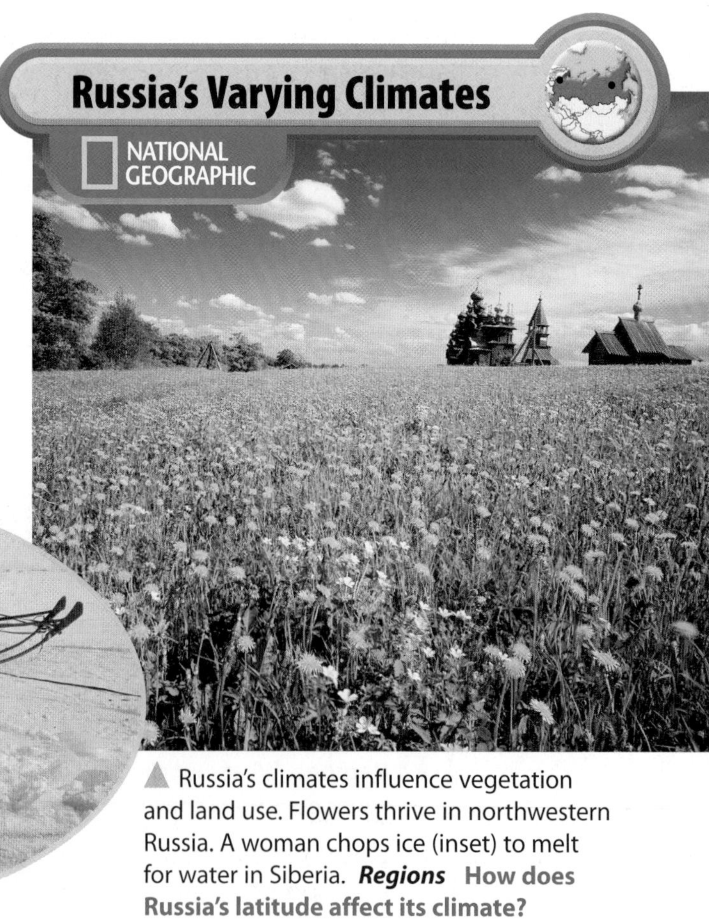

▲ Russia's climates influence vegetation and land use. Flowers thrive in northwestern Russia. A woman chops ice (inset) to melt for water in Siberia. ***Regions*** **How does Russia's latitude affect its climate?**

Farewell to Matyora

By Valentin Rasputin

And so the village had lived on in its lean and simple way, clinging to its spot on the bluff by the left bank [of the Angara River], greeting and seeing off the years, like the water that joined [the villagers] to other settlements and that had helped feed them since time **immemorial.** And just as the flowing water seemed to have no end or limit, the village seemed ageless: some went off to their Maker, others were born, old buildings collapsed, new ones were built. And the village lived on, through hard times and troubles, for three hundred and more years . . . until the rumor thundered down on them that the village would be no more. A dam was being built downriver on the Angara for a hydroelectric power station, and the waters would rise in the rivers and streams, flooding much land, including, first and foremost, Matyora. Even if you were to pile five islands like Matyora one on top of the other, they would still be flooded and you wouldn't be able to show the spot where people once lived. They would have to move. It wasn't easy to believe that it really would come to pass. . . . A year after the first rumors an evaluating **commission** came by [boat] and began assessing the **depreciation** of the buildings and determining how much money they were worth. There was no more doubt about the fate of Matyora, it was living out its last years. Somewhere on the right bank they were building a new settlement for the . . . state-owned farm, into which they were bringing all the neighboring **kolkhozes,** and some that were not so neighboring, and it was decided, so as not to have to deal with rubbish, to set fire to the old villages.

From: *Farewell to Matyora,* by Valentin Rasputin, translated by Antonina W. Bouis, Evanston, Ill.: Northwestern University Press, 1991.

Valentin Rasputin

(1937–)

Born in Siberia, Valentin Rasputin often celebrates the region in his writing. The short novel *Farewell to Matyora* expresses his fear that modern life can erase the traditions of the villagers there.

Background Information

In Soviet times, huge power plants and hydroelectric projects were built to provide energy for heavy industry. Decisions to build factories and power plants were made by the Communist government without consulting the local communities that would be affected by such actions. The government's attempts at modernization often had serious effects on Russia's people.

Reader's Dictionary

immemorial: before memory

commission: official government body

depreciation: lowered value

kolkhozes: government-owned farming villages

Analyzing Literature

1. **Identifying Central Issues** What attitude do you think the government has toward the village?
2. **Read to Write** Take the role of a local government official and write an editorial explaining why you either support or oppose the building of the dam.

Water Pollution

Water pollution is also a problem. Chemicals used in agriculture and industry often end up in rivers and lakes. Pollution entering Lake Baikal may be causing a **decline** in the populations of some animal species in the area. Another source of water pollution is poor sewer systems. Because of these problems, more than half of Russia's people do not have safe drinking water.

Cleaning Up

Steps have been taken to solve Russia's pollution problems. Other countries are providing Russia with aid to improve sewage systems and clean up heavily polluted sites. Cities are building more efficient power plants that use less energy and burn fuel more cleanly. Still, it will be a long time before Russia has a healthy environment.

Reading Check **Explaining** What government policies harmed Russia's environment?

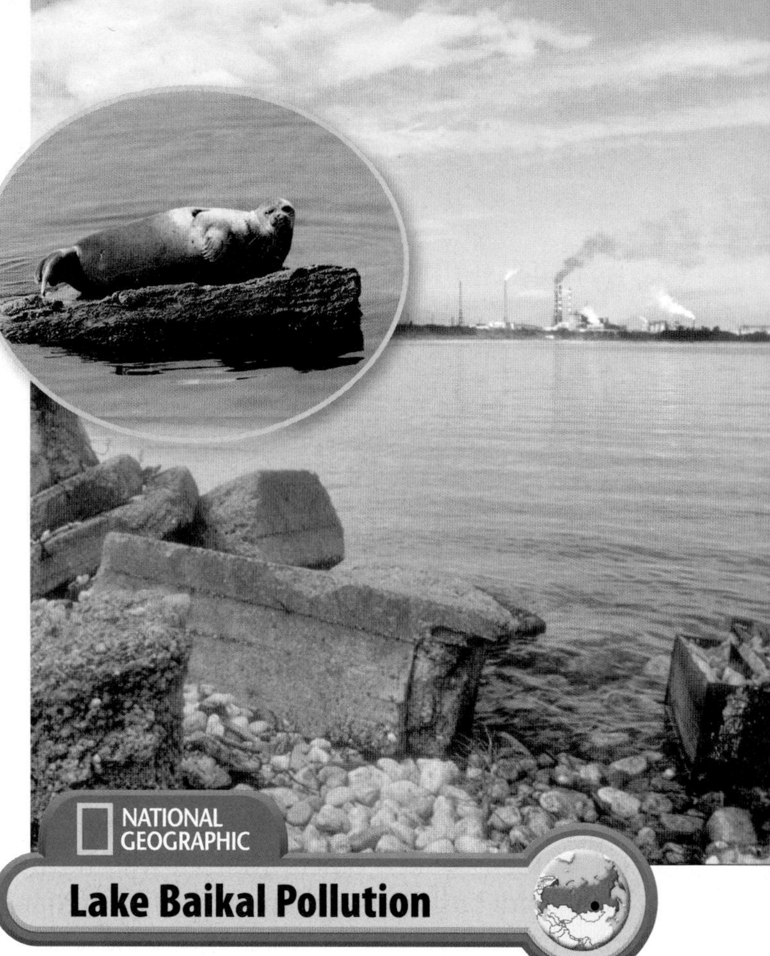

NATIONAL GEOGRAPHIC

Lake Baikal Pollution

Russian industries have polluted their environments, threatening species like the Baikal seal (inset), or nerpa. ***Human-Environment Interaction*** **What are sources of water pollution in Russia?**

Section 2 Review

Social Studies ONLINE
Study Central™ To review this section, go to glencoe.com.

Vocabulary

1. **Explain** the significance of:
   - **a.** permafrost
   - **b.** taiga
   - **c.** smog
   - **d.** pollutant

Main Ideas

2. **Summarizing** Use a chart like the one below to describe characteristics of three major climate zones in Russia.

Climate Zone	Characteristics
Humid Continental	
Tundra	
Subarctic	

3. **Explaining** What steps are being taken to solve Russia's pollution problems?

Critical Thinking

4. **Determining Cause and Effect** How did Russia's climate lead to the defeat of the German army's invasion during World War II?
5. **BIG Idea** How might Russia's vast size, climate, and pollution problems be related?
6. **Challenge** Why do you think the vast wooded areas of the Russian taiga still exist?

Writing About Geography

7. **Use Your FOLDABLES™** Use your Foldable to write a letter to a Russian government official explaining why you think improving Russia's environment should be a priority.

Visual Summary

A Vast Northern Land

- Straddling Europe and Asia, Russia is the world's largest country.
- Most of Russia's long coast lies along waters that are frozen for many months of the year.

Russian icebreaker in the Arctic Ocean

Russia's Landforms

- Northern and western parts of Russia are mostly plains. Eastern and southern areas of the country are covered with mountains and plateaus.
- Inland waterways are important for moving goods through Russia. Many long rivers flow north, however, into the cold Arctic Ocean and freeze in winter.
- Russia has many inland bodies of water, including the Caspian Sea and Lake Baikal.

Natural Resources

- Russia is rich in natural resources, including fossil fuels, metals, and timber.
- Russia's large size and generally cold climate make it difficult for Russians to use their resources.

Climate

- Most of western Russia has a humid continental climate of warm, rainy summers and cold, snowy winters.
- Northern and eastern parts of Russia have cold high latitude climates. The far north of Russia is so cold that moisture in the soil cannot evaporate.
- The country's cold winters helped the Russians defeat German forces during World War II.

Mount Elbrus

Russian coal miner

Environment

- Communist leaders paid little attention to the damage that economic growth was causing to Russia's environment.
- Other countries are providing Russia with aid to clean up heavily polluted areas.

STUDY TO GO Study anywhere, anytime! Download quizzes and flash cards to your PDA from glencoe.com.

STANDARDIZED TEST PRACTICE

When an answer contains multiple items, such as in question 10, make sure that all the items in the answer are correct.

Reviewing Vocabulary

Directions: Choose the word(s) that best complete the sentence.

1. Russia is among the world's leaders in reserves of ________, which include oil, natural gas, and coal.

A pollutants
B renewable resources
C fossil fuels
D ores

2. Siberia lacks a(n) ________, a system of roads and railroads for transporting materials.

A infrastructure
B trade barrier
C tariff system
D commercial sector

3. In Russia's northern tundra, cold temperatures and lack of precipitation result in a permanently frozen layer of soil beneath the surface called ________.

A taiga
B smog
C steppes
D permafrost

4. Russia has the world's largest coniferous forest called the ________.

A tundra
B taiga
C Baikal
D steppe

Reviewing Main Ideas

Directions: Choose the best answer for each question.

Section 1 *(pp. 372–375)*

5. The great majority of Russians live ________.

A in Siberian Russia
B in European Russia
C east of the Urals
D in Asian Russia

6. Many of Russia's plentiful resources are difficult to obtain because of their location ________.

A in European Russia
B west of the Urals
C in Siberia
D south of the Caucasus

Section 2 *(pp. 378–382)*

7. Russia has a generally ________ climate.

A cool to cold
B warm to hot
C dry to desert
D low latitude

8. Russia's largest climate zone is called the ________ zone.

A tundra
B taiga
C arctic
D subarctic

GO ON

Critical Thinking

Directions: Use the following results of a Russian public opinion survey on the environment to answer questions 9 and 10. Choose the best answer for each question.

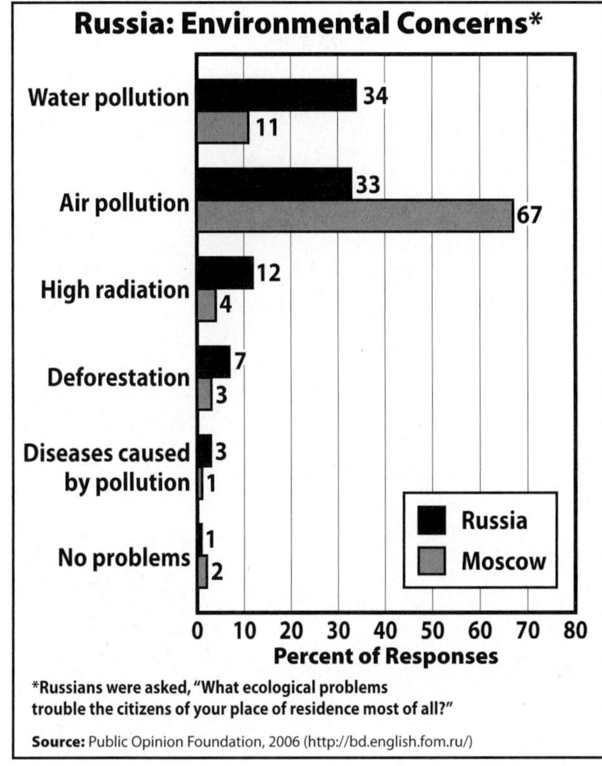

9. Residents of Moscow, Russia's largest city, are more concerned than others about ______.

A high radiation

B water pollution

C air pollution

D ecologically caused diseases

10. Russian citizens living outside of Moscow are almost equally concerned about both ______.

A high radiation and disease

B water and air pollution

C air pollution and high radiation

D water pollution and high radiation

Document-Based Questions

Directions: Analyze the document and answer the short-answer questions that follow.

Mark Sergeev, a Russian poet from the Siberian city of Irkutsk close to Lake Baikal, wrote the following in an essay about the lake.

> *[Native] Siberians have a mystical feeling for [Lake Baikal]. They believe that this is not simply 23 thousand cubic kilometers of water in some enormous stone bowl, but a wizard and healer who should neither be jested with nor enraged. This is why they never call Baikal a lake, only—the sea, or the Old Man, but more often than not they say—He!*
>
> —Mark Sergeev, Irkutsk poet

11. How do native Siberians feel about Lake Baikal? How do they show their feelings?

12. Why might Lake Baikal produce such feelings in the local Siberians?

Extended Response

13. Write an essay describing Asian Russia, including its landforms, waters, climates, and resources. Explain why Asian Russia may play an important role in the country's future economic development in your essay.

STOP

For additional test practice, use Self-Check Quizzes—Chapter 13 at glencoe.com.

Need Extra Help?

If you missed question...	1	2	3	4	5	6	7	8	9	10	11	12	13
Go to page...	375	375	380	380	373	375	379	380	380	380	374	374	373

History and Cultures of Russia

Essential Question

Movement Today, Russia is the world's largest country. Early in its history, however, it was a small territory on the edge of Europe. Strong rulers gradually expanded Russia's borders. While Russia's government has undergone a number of changes, the country's culture and traditions have remained strong. Why do countries often wish to expand their territory?

Section 1: History and Governments

BIG IDEA The characteristics and movement of people impact physical and human systems. Russia grew from a small trading center into a large empire. Although Russia's leaders and a few citizens enjoyed great power, most Russians remained poor and had few rights. In time, these conditions led to great changes in Russia's government.

Section 2: Cultures and Lifestyles

BIG IDEA Culture groups shape human systems. The Russian people have created a rich culture that reflects strong national feelings. Cultural traditions have shaped many areas of daily life, such as housing, recreation, and celebrations.

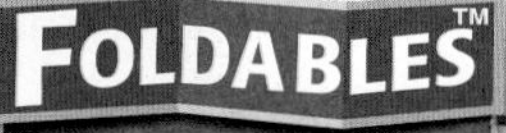

Categorizing Information Make this Foldable to help you collect and organize information about Russia's history and cultures.

Step 1 Place three sheets of paper on top of one another about 1 inch apart.

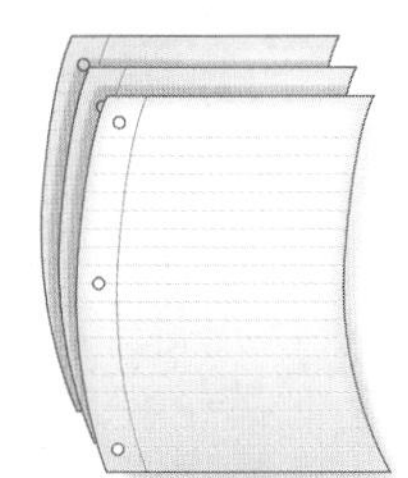

Step 2 Fold the papers to form six equal tabs.

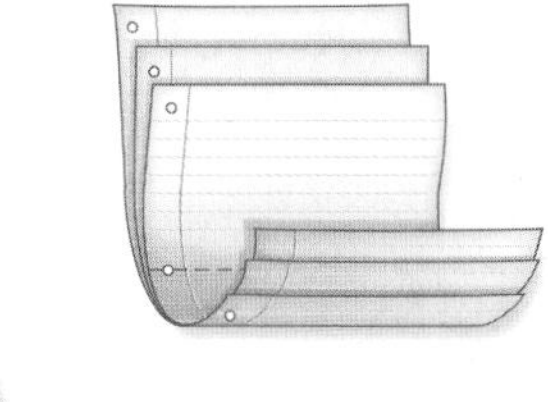

Step 3 Staple the sheets, and label each tab as shown.

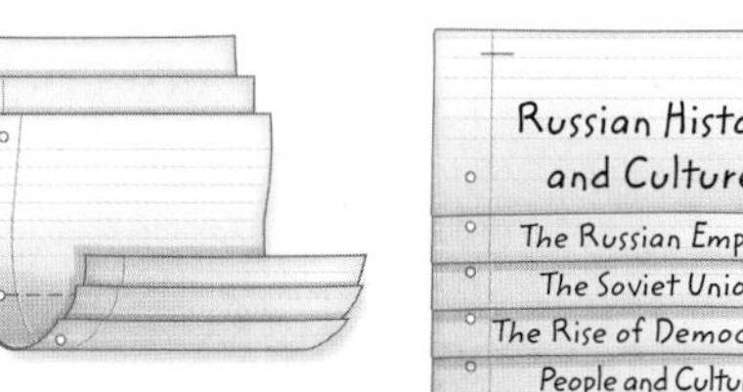

Reading and Writing Use the notes in your Foldable to write a short essay comparing population, culture, and everyday life in Russia and the United States.

Visit glencoe.com and enter *QuickPass*™ code EOW3109c14 for Chapter 14 resources.

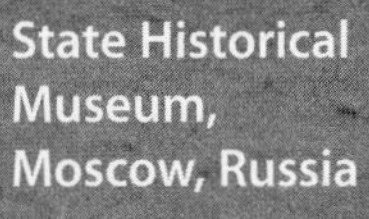

State Historical Museum, Moscow, Russia

Guide to Reading

BIG Idea

The characteristics and movement of people impact physical and human systems.

Content Vocabulary

- missionary *(p. 389)*
- czar *(p. 390)*
- serf *(p. 390)*
- communist state *(p. 391)*
- collectivization *(p. 392)*
- Cold War *(p. 393)*
- glasnost *(p. 393)*
- perestroika *(p. 393)*
- coup *(p. 394)*

Academic Vocabulary

- convert *(p. 389)*
- release *(p. 391)*
- eliminate *(p. 392)*

Reading Strategy

Outlining Use the model below to make an outline of the section. Use Roman numerals to number the main headings. Use capital letters to list important facts below each main heading.

I. **First Main Heading**
   A. **Key fact 1**
   B. **Key fact 2**
II. **Second Main Heading**
   A. **Key fact 1**
   B. **Key fact 2**

History and Governments

 **Section Audio**  **Spotlight Video**

Picture This A giant cat? No, it is actually a miniature city. The highest building in this model of the city of Königsberg stands only 3 feet (91 cm) high. *Königsberg* means "King's Mountain" and was once a German capital. In 1945, during the final days of World War II, German and Soviet armies fought a desperate four-day battle in the city. After the war, the Soviet Union took control of the city and renamed it Kaliningrad. Today, Kaliningrad is part of Russia, which became independent after the Soviet Union's fall in 1991. Read this section to find out more about Russia's history.

▼ Roaming around Königsberg

The Russian Empire

Main Idea **Strong leaders made Russia a vast empire, but widespread suffering eventually led to revolution.**

Geography and You What causes people to rise up and overthrow their government? Read on to learn about Russia's history up to the time of the Russian Revolution.

Russia today is a vast country, covering millions of square miles and spreading across two continents. This world power, however, began as a small trade center.

Early Russia

Modern Russians descend from Slavic peoples who settled along the rivers of what are today Ukraine and Russia. During the A.D. 800s, these early Slavs built a civilization around the city of Kiev (Kyiv), today the capital of Ukraine. This civilization, called Kievan Rus (KEE·eh·vuhn ROOS), prospered from river trade between Scandinavia and the Byzantine Empire.

In A.D. 988 the people of Kievan Rus **converted** to Eastern Orthodox Christianity. **Missionaries,** or people who move to another area to spread their religion, brought this form of Christianity from the Byzantine Empire to Kievan Rus. The missionaries also brought a written language.

In the 1200s, Mongol warriors from Central Asia conquered Kievan Rus. Under Mongol rule, Kiev lost much of its power. Many Slavs moved northward and built settlements. One new settlement was the small trading post of Moscow.

NATIONAL GEOGRAPHIC **Maps In MOtion** See StudentWorks™ Plus or glencoe.com.

Figure 1 **Expansion of Russia**

Map Skills

1. **Location** Which cities are located within the original Kievan territory?
2. **Movement** In which period did the Russian Empire expand beyond Europe and Asia? To what areas did it expand during this period?

It became the center of a new Slavic territory called Muscovy (muh·SKOH·vee). In 1480 Ivan III, a prince of Muscovy, rejected Mongol rule and declared independence. Because he was a strong ruler, Ivan became known as "Ivan the Great."

The Czars

Muscovy developed into the country known today as Russia. In 1547 Ivan IV declared himself **czar** (ZAHR), or emperor. He ruled harshly, using secret police to carry out his will, and earned the name "Ivan the Terrible." By conquering neighboring territories, Ivan expanded his empire south to the Caspian Sea and east past the Ural Mountains.

The expansion of Russian territory, shown in **Figure 1** on the preceding page, continued under later czars, such as Peter the Great and Catherine the Great. These czars wanted to obtain a warm-water port for trade. They also wanted to increase Russia's contact with Europe. In the early 1700s, Peter the Great built a new capital—St. Petersburg—close to Europe near the Baltic coast. By this time, the Russian Empire extended to the Pacific Ocean. Over the next hundred years, it came to include large parts of Central Asia. As a result, non-Russians, including many Muslims, became part of the Russian Empire.

Through the centuries, Russia remained largely rural and agricultural. The czars, large landowners, and wealthy merchants enjoyed comfortable lives. The majority of Russians, however, were poor peasants. Many were **serfs,** or farm laborers who could be bought and sold with the land. Serfs sometimes revolted, but the czars' armies put down their rebellions.

Revolution

Several times during Russia's history, the country's cold climate and huge size proved to be strong defenses against invasion. In 1812 a French army led by Napoleon Bonaparte invaded Russia. To conquer Russia, the French forces had to march hundreds of miles and capture Moscow. All along the French army's way, the Russian army burned villages and any supplies the French could use. The Russians then abandoned and burned Moscow, and the czar moved to St. Petersburg. Even though they had captured Moscow, French troops had few supplies and were forced to retreat during the brutal Russian winter. Thousands died from the harsh conditions and constant Russian attacks. Russia remained independent.

In the late 1800s, Russia entered a period of great change. In 1861 Czar Alexander II freed the country's 40 million serfs. Freedom did not **release** them from poverty, however. Alexander began to modernize Russia's economy, building industries and expanding railroads. Despite these changes, most Russians remained poor, and unrest spread among workers and peasants.

In 1914 Russia joined France and Britain to fight Germany and Austria in World War I. Poorly prepared, Russia suffered military defeats, losing millions of men between 1914 and 1916. Many Russians blamed Czar Nicholas II for the country's poor performance in the war and for food shortages. In early 1917 the people staged a revolution that forced the czar to step down from the throne. Later that year, Vladimir Lenin led a second revolt that overthrew the temporary government. He set about establishing a **communist state** in which the government controlled the economy and society. Fearing invasion, Lenin also moved Russia's capital from coastal St. Petersburg inland to Moscow.

Reading Check **Summarizing** Briefly describe the founding of Kievan Rus and its development into Russia.

The Rise and Fall of Communism

Main Idea The Communist system controlled many aspects of people's lives, but democratic ideas eventually took hold.

Geography and You What would it be like if the government told you what job you had to do and also greatly limited the choices of products you could buy in stores? Read to learn about the changes that the Communist government brought to Russia.

Vladimir Lenin and his followers created a new nation called the Union of Soviet Socialist Republics (U.S.S.R.), or the Soviet Union. This nation included 15 republics made up of different ethnic groups. Russia was the largest republic, and the Russian ethnic group dominated the Soviet Union's government.

Lenin followed the ideas of a German political thinker named Karl Marx. Marx believed that industrialization created an unjust system in which factory owners held great power, while the workers held very little. Lenin said that he wanted to make everyone in Soviet society more equal.

1600
1800
2000
1703
Building of St. Petersburg begins
1917
Revolution sweeps Russia
1991
Soviet Union falls
СССР
Russian peasants, c. 1900

Medal commemorating Trans-Siberian Railroad

So he ended private ownership, bringing all farms and factories under the control of the Soviet government. Lenin's policies were later continued by Joseph Stalin, who ruled the Soviet Union after Lenin's death in 1924. A harsh dictator, Stalin prevented the Soviet people from practicing their religions and had religious property seized. His secret police killed or imprisoned anyone who disagreed with his policies.

Agriculture and Industry

Soviet leaders set up a command economy in which the government ran all areas of economic life. They decided what crops farmers should grow and what goods factories should produce.

Leaders also introduced **collectivization**—a system in which small farms were combined into large, factorylike farms run by the government. Government leaders hoped these farms would be more efficient and reduce the need for farmworkers. Thousands of former peasants could then be put to work in factories to increase industrialization.

The Soviet economic plans had mixed success. The new farms were inefficient and did not produce enough food for the Soviet people. Industrial production was more successful. Huge factories produced steel, machinery, and military equipment. Strict government control, however, had drawbacks. The government **eliminated,** or did away with, competition, allowing only certain factories to make certain goods. This led to a lack of efficiency and poor-quality goods.

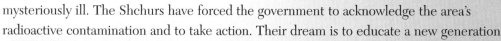

GLOBAL CITIZENS

NAMES: NIKOLAI AND TATYANA SHCHUR

HOME COUNTRY: RUSSIA

ACHIEVEMENT: As Russia becomes a democracy, this husband-and-wife team is trying to make sure that citizens' rights are protected. In 2002 the couple visited Karabolka, a village near the site of a nuclear plant accident 50 years ago. Many people moved after the accident—but some stayed. For decades, those who remained were not told why so many of them fell mysteriously ill. The Shchurs have forced the government to acknowledge the area's radioactive contamination and to take action. Their dream is to educate a new generation of journalists, teachers, and police with a respect for human rights, including the right to live in an area free of radiation.

The Shchurs stand by a Russian sign warning of high radiation levels.

YURI KOZYREV FOR TIME (2)

QUOTE: Tatyana says, “Nothing will really change until a generation grows up in Russia that is completely free from fear.”

CITIZENS IN ACTION Why do you think exposing past government cover-ups might be a difficult task?

Soviet Power

In 1941, during World War II, Germany invaded the Soviet Union. The Soviets joined Great Britain and the United States to defeat the Germans. About 20 to 30 million Russian soldiers and civilians died in the war. After the war, Stalin wanted to make sure the Soviet Union would not be invaded again. He kept Soviet troops in neighboring eastern European countries and established Communist governments in them.

The Soviet Union and the United States were allies during the war but became bitter rivals after it. From the late 1940s until about 1990, these superpowers, the two most powerful nations in the world, engaged in a struggle for world influence. Because the struggle never became "hot," with actual combat between the two opponents, the conflict became known as the **Cold War.**

Each superpower became the center of a group of nations. Members of each group pledged to come to one another's aid if a member country were attacked by a country from the other group. The United States was the chief member of the North Atlantic Treaty Organization, or NATO, which included most of western Europe's democracies. The Soviet Union led the Warsaw Pact, a group of Communist countries that included most of Eastern Europe.

The Soviet Union and the United States competed to produce military weapons and to explore outer space. With so many resources going to the military, the Soviet people had to endure shortages of basic goods, such as food and cars. By the 1980s, many Soviets were ready for change.

Student Web Activity Visit glencoe.com and complete the Chapter 14 Web Activity about the Cold War.

Woman in Space

In 1963 Valentina Tereshkova of the Soviet Union became the first woman to travel in space. ***Place*** **With whom did the Soviet Union compete in space exploration?**

Attempts at Reform

In 1985 Mikhail Gorbachev (mih·KAH·eel GAWR·buh·CHAWF) became the Soviet leader. He quickly began a number of reforms in the Soviet Union. Under the policy of **glasnost** (GLAZ·nohst), or "openness," Soviet citizens could say and write about what they thought without fear of being punished. Another policy, known as **perestroika** (PEHR·uh·STROY·kah), or "restructuring," aimed at boosting the Soviet economy. It gave factory managers more freedom to make economic decisions and encouraged the creation of small, privately owned businesses.

Instead of strengthening the country, Gorbachev's policies made the Soviet people doubt communism even more. Huge protests against Soviet control arose across Eastern Europe. By 1991 all of the region's Communist governments had fallen. Gorbachev hoped that giving some freedoms would win the people's support.

When Eastern Europeans rejected continued Communist rule, Gorbachev decided not to resort to force. He refused to send troops to Eastern Europe as other Soviet leaders had done in the past.

Collapse of the Soviet Union

As communism ended in Eastern Europe, the Soviet Union faced growing unrest among its ethnic groups. Gorbachev was criticized both by hard-liners who wished to maintain Communist rule as well as by reformers. The hard-liners wanted to stop changes. The reformers, on the other hand, felt that Gorbachev was not making changes fast enough. The reformers were led by a rising politician named Boris Yeltsin (BUHR·YEES YEHLT·suhn). He became the president of Russia, the largest of the Soviet republics.

In August 1991, hard-line Communists attempted a **coup** (KOO), an overthrow of the government by military force. Boris Yeltsin called on the people to resist. When many Russians stood firm, the hard-liners were forced to give up. The coup's failure was the beginning of the end of the Soviet Union. Within a few months, Russia and all of the other Soviet republics declared independence. By the end of 1991, the Soviet Union no longer existed as a nation.

In Russia, Yeltsin had some success in building democracy and a market economy. His successor, Vladimir Putin (vlah·DEE·meer POO·tuhn), however, increased government controls to deal with rising crime and violence. Challenges also came from some ethnic minorities. In the Chechnya (chehch·NYAH) region, a group trying to separate from Russia waged a bloody civil war.

Reading Check **Explaining** What were Gorbachev's reforms? Why were they introduced?

Section Review

Social Studies ONLINE
Study Central™ To review this section, go to glencoe.com.

Vocabulary

1. **Explain** the significance of:
   - **a.** missionary
   - **b.** czar
   - **c.** serf
   - **d.** communist state
   - **e.** collectivization
   - **f.** Cold War
   - **g.** glasnost
   - **h.** perestroika
   - **i.** coup

Main Ideas

2. **Identifying** What geographic goal drove the czars to expand their empire?
3. **Sequencing** Use a diagram like the one below to list events leading up to the collapse of the Soviet Union.

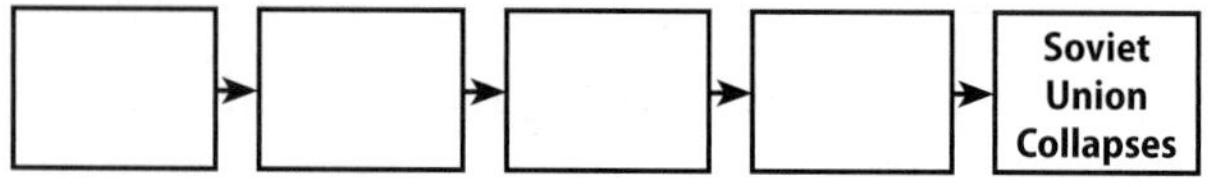

Critical Thinking

4. **Determining Cause and Effect** How did the Soviet Union's involvement in the Cold War affect the Soviet economy?
5. BIG Idea How did Russia become a society that included many different ethnic groups?
6. **Challenge** How did the economic systems of the Soviet Union and the United States differ?

Writing About Geography

7. **Using Your FOLDABLES** Use your Foldable to write a summary of how Russia's government changed after World War I.

NATIONAL GEOGRAPHIC **Geography & History**

The Longest Railroad

How long does it take to cross the world's largest country? Thanks to the world's longest railroad, you can make the trip in less than a week—barely.

Building the Railroad The Trans-Siberian Railroad starts in Moscow, in western Russia. Nearly 5,800 miles (9,334 km) later, it reaches Vladivostok, on the shores of the Pacific Ocean. Work to build the railroad began in 1891 and was completed in 1901.

The first route crossed Manchuria, a northern part of China. Russian leaders soon feared that Japan would seize Manchuria, however, which would threaten their rail line. As a result, they ordered work on a new route that went north of Manchuria and stayed entirely within Russian land. That route was finished in 1916.

The Trans-Siberian Railroad

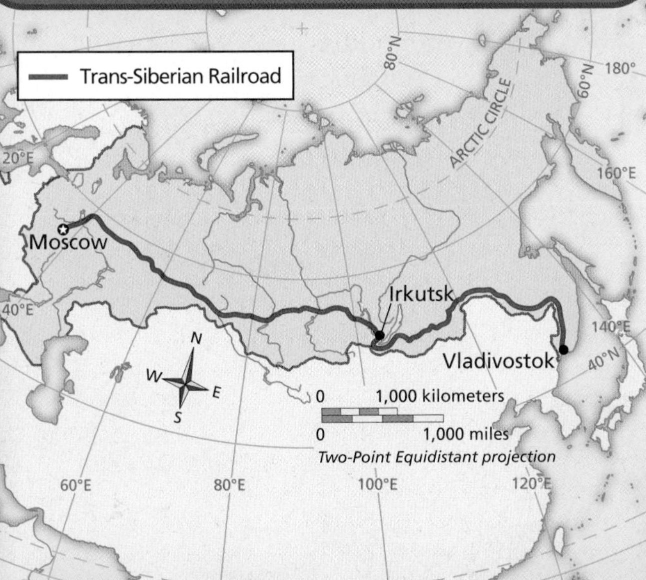

Construction of the railroad was difficult. Track had to be laid across swamps and permafrost and through dense forests and mountains. In eastern Russia, where the railroad crosses many rivers, dozens of bridges had to be built. To keep the railroad running, wells for water and mines to obtain coal and iron were needed.

▲ **A train on the Trans-Siberian Railroad**

The Impact of the Railroad The railroad had a huge impact on Russia. Even today, no highway connects western and eastern Russia. As a result, the railroad is the only land-based route linking these halves of the country. The railroad serves as a vital lifeline for bringing supplies to eastern settlements. It also moves Siberia's timber and mineral resources to factories in the west.

During World War II, the railroad helped save Russia. When Germany invaded western Russia in 1941, the government used the railroad to move equipment from that area into Siberia. There it set up new factories to make the vehicles, weapons, and ammunition needed to continue fighting the war.

Think About It

1. **Human-Environment Interaction** What obstacles did workers building the railroad have to overcome?
2. **Movement** Why was the railroad important during World War II?

Guide to Reading

BIG Idea

Culture groups shape human systems.

Content Vocabulary

- oral tradition *(p. 397)*
- nationalism *(p. 397)*
- autonomy *(p. 399)*

Academic Vocabulary

- promote *(p. 398)*
- primary *(p. 400)*
- exploit *(p. 400)*

Reading Strategy

Summarizing Use a diagram like the one below to list details about Russian music and dance.

SECTION 2

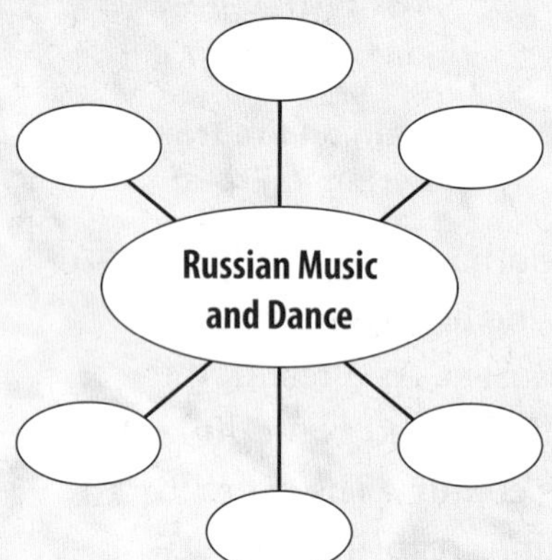

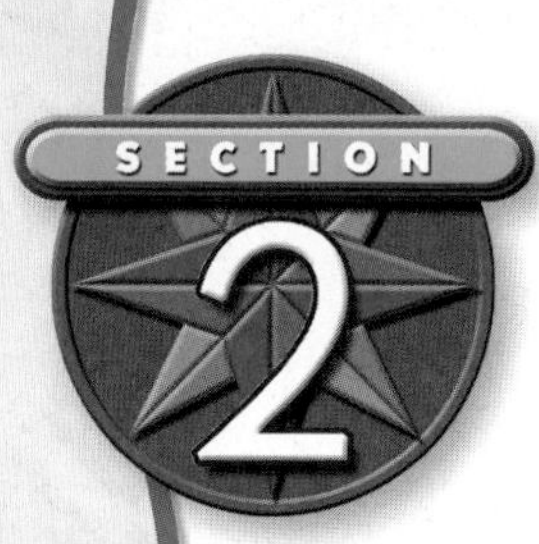

Cultures and Lifestyles

Section Audio   Spotlight Video

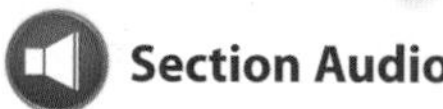

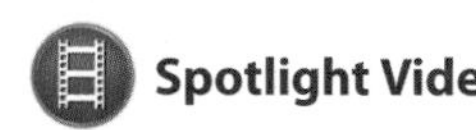

Picture This The model in this photograph is getting ready for Siberian Fashion Week in Krasnoyarsk, Russia. Once a fortress city that protected Siberia from invaders, the city is now home to a festival that highlights the world of beauty. Krasnoyarsk links the eastern and western regions of icy Siberia. In a similar way, Fashion Week combines the artistic traditions of the regions with the business of high fashion. Read Section 2 to learn more about the rich cultures of Russia.

▼ Siberian Fashion Week in Krasnoyarsk, Russia

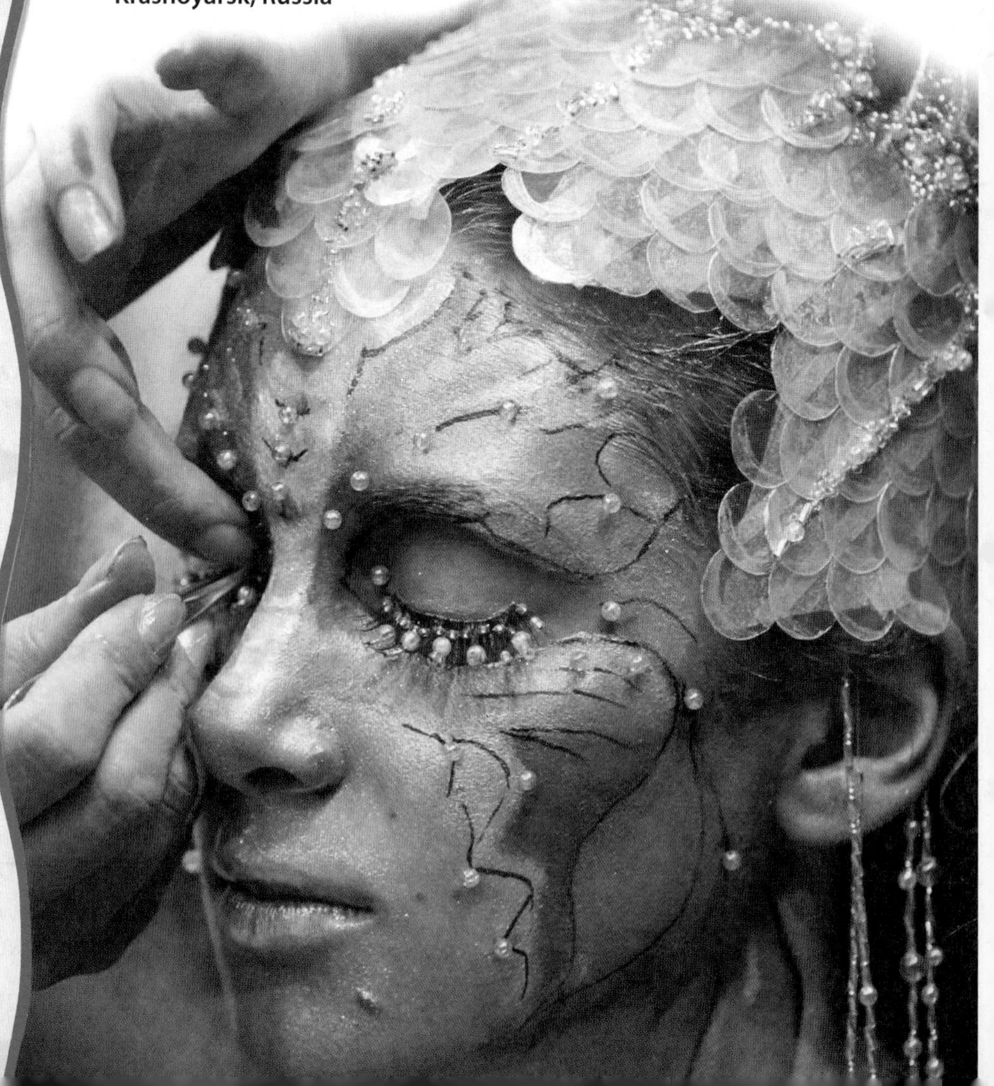

Russia's Cultures

Main Idea **Russia's many ethnic groups and a tradition of great achievements in the arts and sciences contribute to the country's cultures.**

Geography and You Have you ever watched the graceful motions of a ballet dancer? Read to find out how ballet and other arts are an important part of Russia's culture.

Russia has many different ethnic groups and religions. As the Russian Empire expanded, different peoples came under its control. Today dozens of ethnic groups live in Russia. Many of these groups speak their own language and have their own culture. Most people, however, speak Russian, the country's official language.

Russians, or Slavs who descended from the people of Muscovy, are the largest ethnic group. They live throughout Russia, although most Russians live west of the Urals. The next-largest groups include Tatars, who are Muslim descendants of the Mongols, and Ukrainians, who are descendants of Slavs that settled the area around Kiev (Kyiv). Smaller ethnic groups include the Yakut, who herd reindeer and also raise horses and cattle in eastern Siberia.

Under communism, Russia's people were not allowed to practice religion. The Soviet government officially promoted the position in its schools that there is no god or other supreme being. By the late 1980s, however, the Soviets began to relax their ban on religions. Today, people enjoy religious freedom, and about half of the population practices a faith. Eastern Orthodox Christianity is the country's major religion. Russia also has many Muslims, who live mainly in the Caucasus region. Lesser numbers of Roman Catholics, Protestants, and Jews also live in Russia.

NATIONAL GEOGRAPHIC

Kazan Cathedral

In 1936 Communist leaders destroyed the cathedral that stood on this site in Moscow. It was rebuilt in the 1980s. ***Regions*** **What is Russia's major religion today?**

The Arts

Russia has a rich tradition of literature, art, and music. Early Russians developed a strong **oral tradition,** or passing stories by word of mouth from generation to generation. Later, many writers and musicians drew on these stories or on folk music for their works. The Russian people's strong sense of **nationalism,** or feelings of loyalty toward their country, is reflected in many artistic works.

Russia has long been a center of music and dance. In the late 1800s, Peter Ilich Tchaikovsky (chy·KAWF·skee) wrote some of the world's favorite ballets, including *Swan Lake* and *The Nutcracker.* In the early 1900s, Igor Stravinsky wrote *The Firebird Suite* and other works. Today, the Bolshoi of Moscow and the Kirov of St. Petersburg are among the world's famous ballet companies.

The 1800s are often called the "golden age" of Russian literature. During this period, Leo Tolstoy, one of the greatest Russian writers, wrote *War and Peace.* This patriotic novel describes the Russians' defense against the French invasion in 1812.

During the Soviet era, writers did not enjoy freedom of expression. They were required to **promote** government policies in their works. Alexander Solzhenitsyn (SOHL·zhuh·NEET·suhn), who wrote about the harsh conditions of Communist society, spent time in Russian prison camps before he was forced to leave the country. Russian writers today are generally free to write about any idea or topic. Recently, however, the government placed new limits on freedom of speech.

The visual arts are also an important part of Russian culture. The Hermitage Museum in St. Petersburg has one of the most famous art collections in the world. Among its treasures are the jewel-encrusted Easter eggs crafted by jeweler Peter Carl Fabergé for the czars.

Scientific Advances

For decades, the Soviet Union emphasized education in the sciences. As a result, Russia has many scientists, mathematicians, and doctors. During the Cold War, Russian scientists helped the Soviet Union compete with the United States in space exploration. In 1961 Russian Yuri Gagarin was the first person to fly in space. Since 1998 the Russians have joined with Americans and people from other countries in building the International Space Station.

Reading Check **Explaining** How have religious practices in Russia changed since the fall of communism?

NATIONAL GEOGRAPHIC

Russian Treasures

▲ The State Russian Museum in St. Petersburg holds more than 400,000 exhibits of Russian art from the last 1,000 years. Ornate Fabergé eggs (inset) were handcrafted for the czar's family and others. ***Regions*** **For what other arts is Russia known?**

Life in Russia

Main Idea **Russian lifestyles are influenced by the region's cold climate and vast area, as well as the country's changing economic system.**

Geography and You Do you like to celebrate the arrival of spring after a long, cold winter? Read on to find out about Russia's spring festival and other aspects of Russian life.

While Russia continues to modernize after shedding Soviet control, it faces several challenges. Housing is still in short supply, and transportation and communication networks need updating.

Everyday Life

Most Russians live in cities located west of the Ural Mountains. City residents generally live in large apartment buildings rather than in single-family houses. Housing is scarce and often expensive. For this reason, grandparents, parents, and children frequently share the same apartment or house. Since many Russian mothers work outside of the home, grandparents often help take care of the household. Grandparents may cook, clean, shop, and care for young children in the family.

Middle-class and wealthy Russians may own country homes called dachas (DAH·chuhs). At their country homes, people often tend gardens, growing vegetables that they can either eat or sell in the cities. In rural areas, dachas and other homes are usually built of wood.

In the coldest areas of Russia, people take extra steps to keep their homes warm. For example, some homes in Siberia have three doors at each entrance. This prevents a cold blast of air from rushing in when the outside doors are opened.

Family Life in Russia

In Russia it is common for several generations of one family to share a home.
Location **Where do most Russians live?**

Sports and Holidays

It is not surprising that the country's most popular sports are associated with winter or are played indoors. During the Communist era, the Soviet Union placed great emphasis on training world-class athletes. Today, Russian hockey players, figure skaters, and gymnasts are strong competitors in international events.

Russians celebrate several holidays. The newest holiday, Independence Day, falls on June 12 and marks Russia's declaration of **autonomy,** or independence from the Soviet Union. New Year's Eve is one of the most festive holidays. Russian children decorate a fir tree and exchange presents with others in their families. In the spring, Russians celebrate *Maslenitsa.* This week-long holiday marks the end of winter and includes organized snowball fights, sleigh rides, and parties. Straw dolls that represent winter are burned to signal the beginning of spring.

Transportation and Communications

Russia is so large that people and goods must often be transported over great distances. Railroads were the **primary** means of transportation during the Soviet era and are still important today. The heavily populated area west of the Urals is covered by an extensive railroad network.

This railroad system is linked to the famed Trans-Siberian Railroad, which runs from Moscow in the west to Vladivostok in the east. Completed in the early 1900s, it is the longest rail line in the world. The railroad made it possible for Russians to **exploit,** or use, Siberia's natural resources.

Russia still lacks an effective highway system. No multilane highway system links major cities, and the roads that do exist are in poor condition. There are few gas stations or restaurants along the roads. The government is currently building a 6,600-mile (10,622-km) highway across the country. When completed, it will be the longest national highway in the world.

One reason that highway improvements are needed is because Russian car ownership is rising. In Soviet times, few families had cars. Now about half of Russian families own a car.

Russia's communications systems also need improvement. For years, telephones were less common in Russia than in most European countries. Since the early 1990s, major improvements have been made that will benefit Russian citizens and the country's economy. New phone lines allow for the rapid transfer of information, making it easier to use the Internet in Russia. Rural areas, however, still have poor phone service.

Reading Check **Making Connections** How does the *Maslenitsa* celebration reflect Russia's culture and environment?

Social Studies ONLINE
Study Central™ To review this section, go to glencoe.com.

Vocabulary

1. **Explain** the meaning of *oral tradition, nationalism,* and *autonomy* by using each term in a sentence.

Main Ideas

2. **Identifying** Name and describe four of Russia's ethnic groups.
3. **Organizing** Use a diagram like the one below to list details about Russia's transportation system.

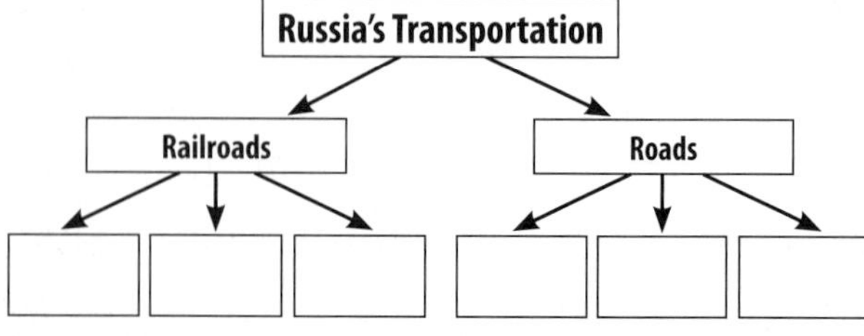

Critical Thinking

4. **Determining Cause and Effect** Why do many grandparents, parents, and children in Russia share the same apartment or house?
5. **BIG Idea** How are Russia's nationalist feelings represented in the country's arts?
6. **Challenge** Why do you think the Soviet Union competed with the United States in the area of space exploration?

Writing About Geography

7. **Expository Writing** Write a paragraph explaining how Russia's cold climate affects Russians' daily lives.

Visual Summary

The Russian Empire

- Russia had its beginnings in small trading centers built by Slavic peoples.
- Rulers known as czars governed the Russian Empire from 1547 to 1917.
- The czars expanded Russian territory to reach from Europe to the Pacific Ocean.
- Revolutions in 1917 overthrew the czar and brought the Communists to power.

Medieval Moscow

The Soviet Union

- Under the Communist Party, the government ran all areas of economic life.
- After World War II, the Soviet Union and the United States became bitter rivals.

Joseph Stalin

The Rise of Democracy

- In 1991 the Soviet Union broke up into Russia and other independent republics.
- Russia has struggled to build a democracy and a market economy.

Cold War parade with missiles

People and Culture

- Russia has many different ethnic groups.
- Russians practice a number of different religions, but most of the population is Eastern Orthodox Christian.
- Russian artists, musicians, and writers often used themes based on Russian history.

Dacha outside of Moscow

Life in Russia

- Most Russians live in apartments in large cities rather than in single-family homes.
- Railroads link heavily populated European Russia with sparsely settled Siberia.
- Russia is working to improve its highway and communications systems.

STUDY TO GO Study anywhere, anytime! Download quizzes and flash cards to your PDA from glencoe.com.

CHAPTER 14 ASSESSMENT

STANDARDIZED TEST PRACTICE

Read all the choices before choosing your answer. You may overlook the correct answer if you are hasty!

Reviewing Vocabulary

Directions: Choose the word(s) that best completes the sentence.

1. The emperors of Russia were called _______.

A serfs

B converts

C czars

D missionaries

2. Mikhail Gorbachev introduced _______, which was aimed at boosting the Soviet economy.

A perestroika

B collectivization

C Cold War

D glasnost

3. _______ were farm laborers who could be bought and sold with the land.

A Serfs

B Converts

C Missionaries

D Guilds

4. Early Russians developed a strong _______, which means passing stories by word of mouth from generation to generation.

A nationalism

B oral tradition

C autonomy

D glasnost

5. Russia's newest holiday, Independence Day, marks its declaration of _______ the Soviet Union.

A collectivization of

B dependence on

C alliance with

D autonomy from

Reviewing Main Ideas

Directions: Choose the best answer for each question.

Section 1 *(pp. 388–394)*

6. The Russian Empire eventually fell to a revolution due mostly to _______.

A threats from invaders

B widespread hunger and suffering

C outside agitators

D a decline in the military

7. When Communist leaders took control of Russia and renamed it the Soviet Union, they established a _______.

A free market economy

B democratic government

C government controlled economy

D new government controlled by a czar

Section 2 *(pp. 396–400)*

8. The Russian people's sense of nationalism _______.

A has little impact on Russia's arts

B has caused them to avoid foreign art

C is reflected in many artistic works

D is the only topic used in Russian literature

9. Russia's many good hockey players and gymnasts are partly a result of the country's _______.

A high mountains

B many factories and industries

C many professional sports teams

D cold climate

GO ON

Critical Thinking

Directions: Choose the best answer for each question.

10. Which of the following was a result of perestroika?

A The Soviet economy was weakening.

B The people were ready for change.

C Factory managers gained more freedoms.

D Citizens could say what they thought without punishment.

Base your answer to question 11 on the chart below.

Infant Mortality Rates per 1,000 Live Births		
Years	Soviet Union/Russia	United States
1950–1955	97	28
1955–1960	57	26
1960–1965	40	25
1965–1970	32	22
1970–1975	28	18
1975–1980	30	14
1980–1985	26	11
1985–1990	24	10
1990–1995*	21	9
1995–2000	17	8

*After 1991 the Soviet Union collapsed.

Source: http://world.britannica.com/analyst/chrono/table

11. Which of the following statements seems most likely?

A Between 1950 and 2000, medical care decreased in the United States.

B Between 1950 and 2000, infants in the Soviet Union had greater access to hospitals than in the United States.

C Between 1950 and 2000, infants in the Soviet Union had a better chance of surviving than in the United States.

D Between 1950 and 2000, medical care improved in the Soviet Union.

Document-Based Questions

Directions: Analyze the document and answer the short-answer questions that follow.

After it was first adopted in 1918, the constitution of the Soviet Union was rewritten several times. The following is part of the preamble from the 1977 version.

> *The supreme goal of the Soviet state is the building of a classless communist society in which there will be public, communist self-government. The main aims of the people's socialist state are: to lay the material and technical foundation of communism, to perfect socialist social relations and transform them into communist relations, to mould the citizen of communist society, to raise the people's living and cultural standards, to safeguard the country's security, and to further the consolidation of peace and development of international cooperation.*

12. How do you think the Soviet state will "lay the material and technical foundation of communism"?

13. Which parts of the preamble deal directly with Soviet citizens?

Extended Response

14. Citizens of the Soviet Union faced many problems during the period the Communist government was in control. Write an essay in which you explain what you think were the most difficult problems people had to face. Make sure to explain why these problems were worse than others.

For additional test practice, use Self-Check Quizzes—Chapter 14 at glencoe.com.

Need Extra Help?

If you missed question...	1	2	3	4	5	6	7	8	9	10	11	12	13	14
Go to page...	390	393	390	397	399	391	392	397	399	393	398	391	391	391

"Hello! My name is Irina.

I am 13 years old and I live in Moscow, the capital of Russia, with my parents and little brother, Misha. I go to Moscow School No. 429. This is how I spend a typical day."

7:30 a.m. I wake up and dress quickly in jeans and a sweater. Then I join my parents and Misha in the kitchen and have yogurt for breakfast. (On weekends, I like fried eggs, but today there is not enough time!)

8:00 a.m. My father drives me to school and Misha to kindergarten. In Russia, public schools do not provide buses. After dropping us off, my father will go to his job as a manager at a paper supply company.

8:10 a.m. I arrive at school. School No. 429 is a modern building, but Moscow is a very old city. Some of its buildings are more than 500 years old!

8:30 a.m. Classes begin. One of my first classes of the day is "work study," in which we learn different trades. Right now we are learning to sew clothes. It's interesting, but I don't think it's the right job for me. I would like to be a makeup artist one day.

9:30 a.m. It is time for geography class. I raise my hand to respond to a question, and I answer correctly. This makes me happy because I would like to get a grade of 4 or 5 in this class. A *4* means "very good." A 5 means "excellent."

11:30 a.m. I'm hungry! During a break between classes, I stop in at the canteen (cafeteria) for a small piece of pizza. Later, I will have a bigger lunch at home.

12:30 p.m. In computer class, we log on to the Internet. This is one of my favorite classes.

2:00 p.m. The school day is over, and I take the tram home. The tram is a street car that runs on electricity. Moscow's tramway is more than 100 years old.

3:00 p.m. My mother is home from her job at a print house and Misha is home from school. I help make a lunch of chicken and cold borscht. Borscht is a soup made of beets and other vegetables. It can be served cold or hot.

3:30 p.m. My mother, Misha, and I walk to a large park near our flat, or apartment. There I meet up with my friend Yulia. We roller skate and play with Misha.

5:30 p.m. Back at home, I help do the laundry and dust the flat. When I am done with my chores, I go to my bedroom and start my homework.

7:30 p.m. My father is home, and it is time for a light dinner. Tonight we have a salad made of eggs, cabbage, mayonnaise, and beets. Another dinner I enjoy is pelmeni. It is pieces of meat that are covered in dough and boiled in salty water.

8:00 p.m. I finish my homework and listen to music. I like Russian bands, but I also enjoy American pop music. I listen to it on the radio and on music videos.

10:30 p.m. I go to bed. Tomorrow is Friday, my last day of school for the week. I am looking forward to the weekend! We will spend it at our dacha, or country cottage. It is in the village of Istra, where my grandparents live. I enjoy spending time with them.

ILLUSTRATIONS BY BOOKMAPMAN

GLOBETROTTING Irina is tested on her geography skills. She got an "excellent"!

COME AND GET IT! Irina helps her mom make dinner. Russians often eat a big lunch and a light supper.

SOAP STORY Blowing bubbles in the park is fun for Irina and her friend Yulia.

GOING SKATING After school, Irina Timoshenko (uh • REE • nah teem • oh • SHENK • oh) loves to roller skate with her friends. Irina lives in Moscow, Russia's capital, which is filled with parks that are perfect for skating.

JEREMY NICHOLL / POLARIS (4)

What's Popular in Russia

Tea Time Many Russian families drink weak black tea all day long. It is brewed in a pot called a samovar.

BILL ARON / PHOTO EDIT

Steam baths This hot trend is actually an old Russian tradition. At a *banya,* or bath house, a visitor sits in a steam room, then takes an icy cold dip, and finally, receives a gentle massage.

Dachas Many Russian city dwellers—both rich and poor—own cottages, or dachas, in the country. They go on weekends and in summers to relax.

Ballet Russia is home to some of the world's most famous dancers. Moscow's Bolshoi Ballet brought the world *Swan Lake.*

ANATOLY RUKHADZE / ITAR-TASS PHOTOS / NEWSCOM

Say It in Russian

Most of Russia's 140 million people speak the official national language, Russian. Russian is also spoken in Belarus, Kazakhstan, and other nations that were once republics of the Soviet Union. Try saying these phrases in Russian.

Hello
Zdravstvuite (ZDRAHST · vet · yah)

Good-bye
Do svidanja (doh svee · DAH · nee · yah)

My name is ________.
Menya zavut (mee · NYAH zah · VOOT) ________.

CREATAS / PUNCHSTOCK

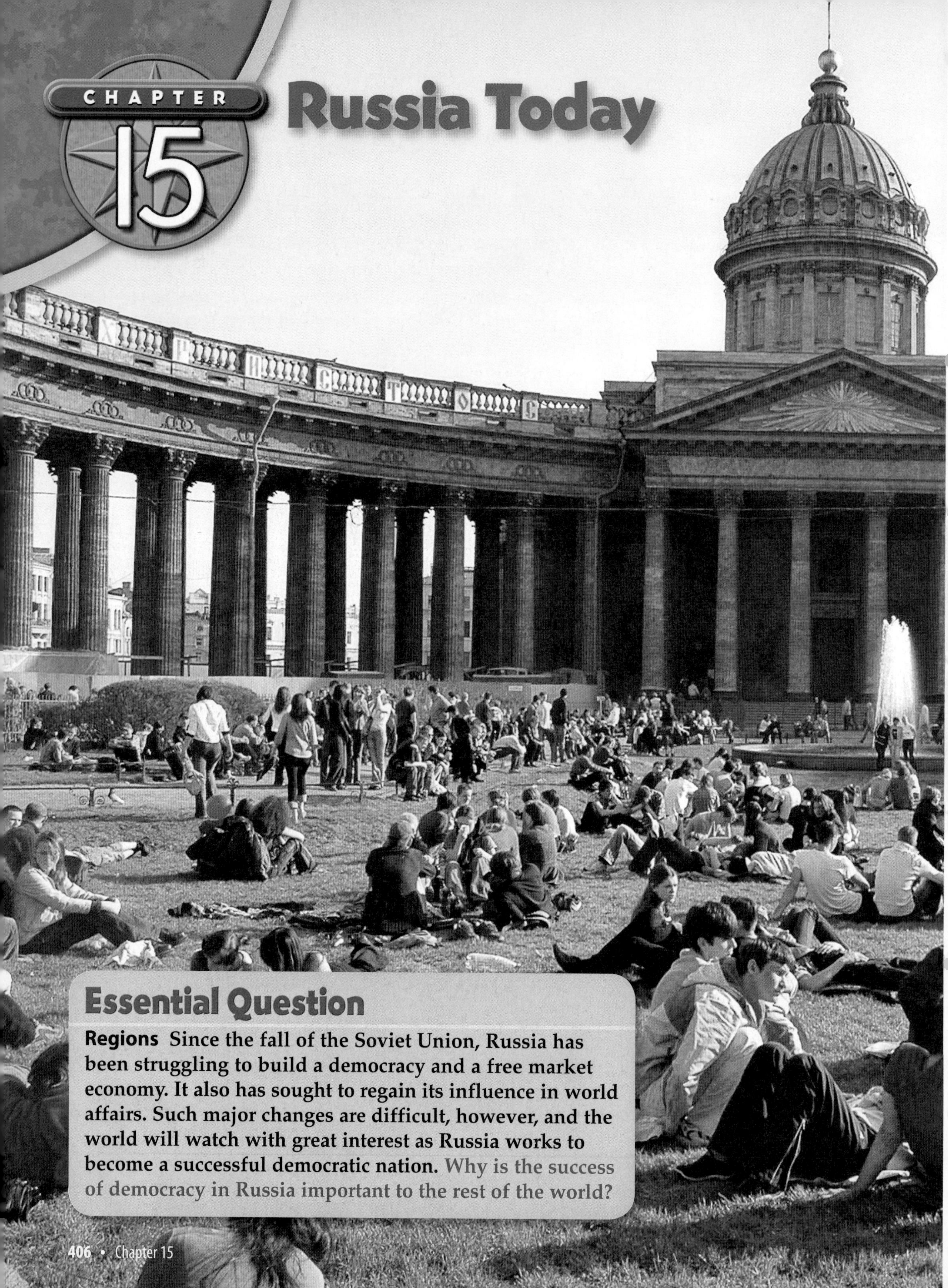

CHAPTER 15 Russia Today

Essential Question

Regions Since the fall of the Soviet Union, Russia has been struggling to build a democracy and a free market economy. It also has sought to regain its influence in world affairs. Such major changes are difficult, however, and the world will watch with great interest as Russia works to become a successful democratic nation. Why is the success of democracy in Russia important to the rest of the world?

BIG Ideas

Chapter Audio

Russians relaxing on May Day holiday, St. Petersburg

Section 1: A Changing Russia

BIG IDEA Geographers organize the Earth into regions that share common characteristics. New democratic institutions and a free market economy link the different parts of Russia. These positive changes, however, are threatened by the government's abuse of power, the spread of corruption in business, and a decline in population.

Section 2: Issues and Challenges

BIG IDEA Geography is used to interpret the past, understand the present, and plan for the future. The change to democracy and a market economy has been difficult for Russia because of its long history of all-powerful governments. The country also faces challenges from ethnic groups that want independence.

FOLDABLES™ Study Organizer

Organizing Information Make this Foldable to organize information about political and social changes in Russia and other issues in Russia today.

Step 1 Fold a sheet of paper in half from side to side so that the left edge lies about ½ inch from the right edge.

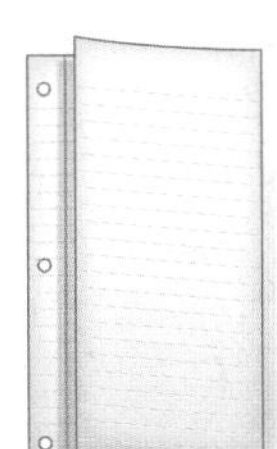

Step 2 Cut the top layer only to make five tabs.

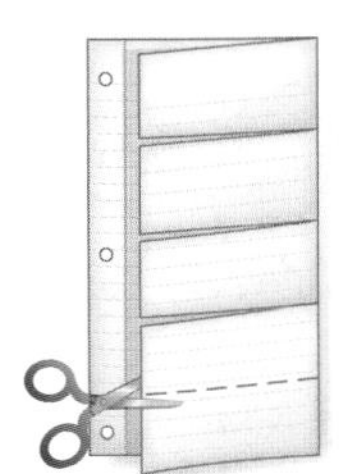

Step 3 Label the Foldable as shown.

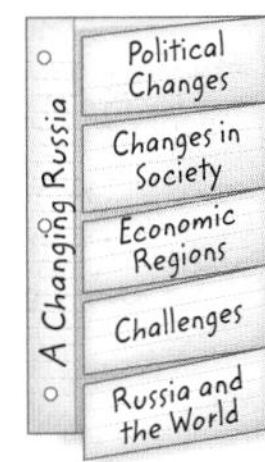

Reading and Writing Use the notes in your Foldable to create an outline showing the major issues facing Russia today.

Social Studies ONLINE

Visit glencoe.com and enter ***QuickPass***™ code EOW3109c15 for Chapter 15 resources.

Guide to Reading

BIG Idea

Geographers organize the Earth into regions that share common characteristics.

Content Vocabulary

- privatization *(p. 409)*
- middle class *(p. 410)*
- underemployment *(p. 410)*
- pensioner *(p. 410)*
- heavy industry *(p. 411)*
- light industry *(p. 411)*

Academic Vocabulary

- comprise *(p. 409)*
- invest *(p. 410)*
- volume *(p. 411)*

Reading Strategy

Identifying Central Issues Use a diagram like the one below to show three major effects of the fall of communism on Russia.

A Changing Russia

 Section Audio  Spotlight Video

Picture This The figures below are toys, art, and a history lesson all in one! *Matryoshka* dolls, which fit one inside the other, are popular toys and have been hand painted in Russia since the late 1800s. The traditional version of the *Matryoshka* shows a Russian woman wearing a babushka (scarf) and an apron. Here the colorful dolls represent Soviet and Russian leaders (from left to right) Mikhail Gorbachev, Boris Yeltsin, and Vladimir Putin. To learn more about recent changes in Russia, read Section 1.

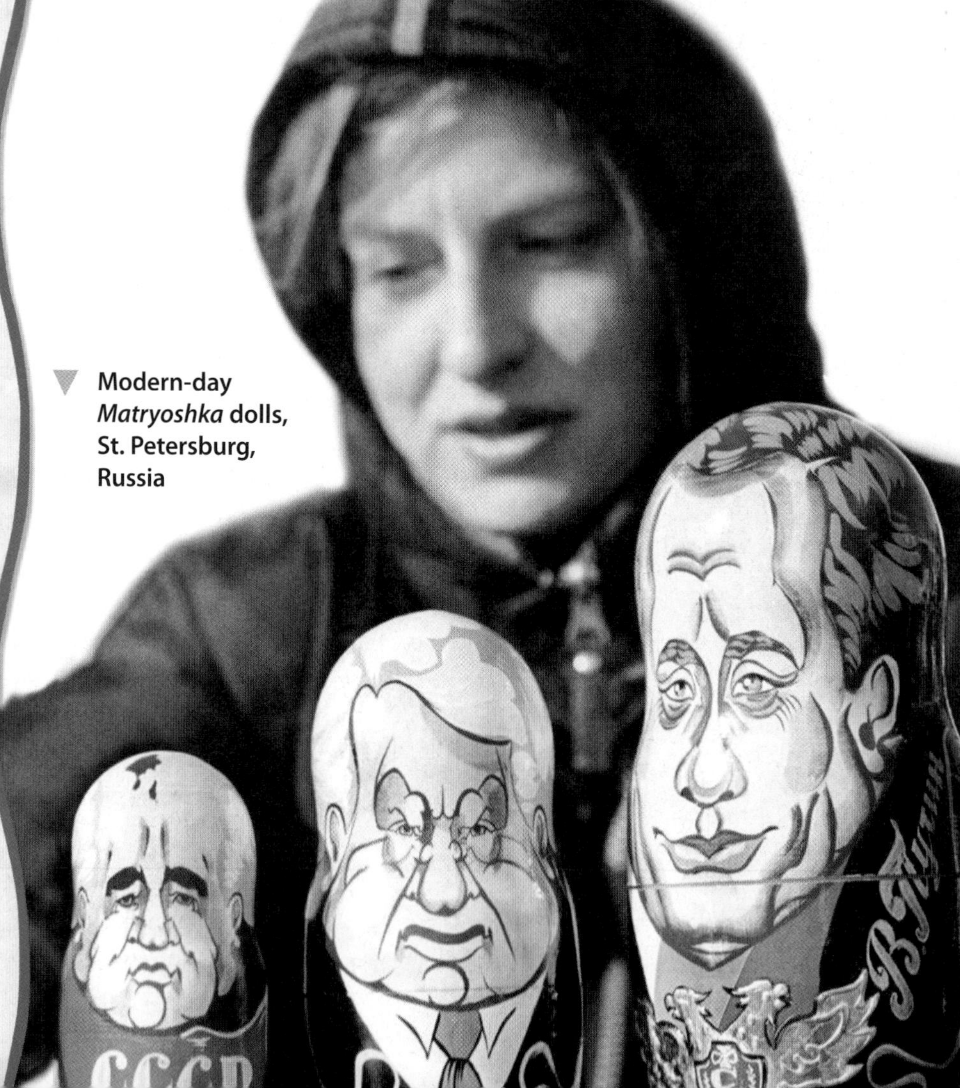

Modern-day *Matryoshka* dolls, St. Petersburg, Russia

Changing Politics and Society

Main Idea The fall of communism led to great changes in Russia's government, economy, and society.

Geography and You Can you imagine having to completely change your way of life? Read to learn how Russians faced that situation in the early 1990s.

Russia is still adjusting to the changes that occurred in the 1990s. When communism fell in 1991, Russia was forced to build a new government and economy. These ongoing changes continue to greatly affect the everyday lives of the Russian people.

A New Form of Government

The Communist Party ruled Russia when it was part of the Soviet Union. The Communists did not allow people to challenge their power, and everyday citizens had no voice in choosing their leaders. After the fall of communism, however, Russia became more democratic.

In a 1993 election, Russian voters approved a new constitution and elected members of a legislature to represent them. This new legislature included candidates from many different political parties. Boris Yeltsin, who was Russia's leader when it was a Soviet republic, was elected the first president of Russia.

Russia's official name is the Russian Federation. This name reflects the fact that Russia **comprises,** or is made up of, many different regions and territories. Like the United States, Russia is a federal republic, with power divided between national and regional governments. In the United States, some powers belong to the states, and others belong only to the national government. Some powers are shared by both levels of government. In Russia, the division of powers is less clear because the new Russian government is still developing.

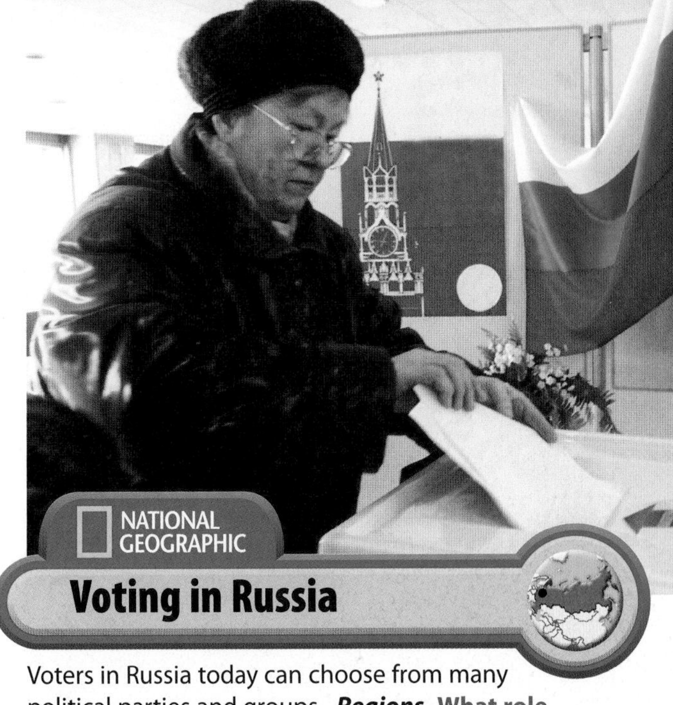

NATIONAL GEOGRAPHIC

Voting in Russia

Voters in Russia today can choose from many political parties and groups. ***Regions*** **What role did most citizens have in their government when Russia was part of the Soviet Union?**

A New Economic System

As part of the Soviet Union, Russia had a command economy. In a command economy, the central government makes all the economic decisions. Since the fall of communism, Russia has attempted to shift to a market economy.

To create a market economy, the government introduced **privatization** (PRY·vuh·tuh·ZAY·shuhn). This is the transfer of ownership of businesses from the government to individuals. In the new system, businesses have to compete with one another. As a result, Russian companies have begun to advertise to attract customers. The government has also dropped price controls, which were official prices set for different goods and services. In a market economy, prices result from competition among companies and from what Russian consumers need, want, and are willing to pay.

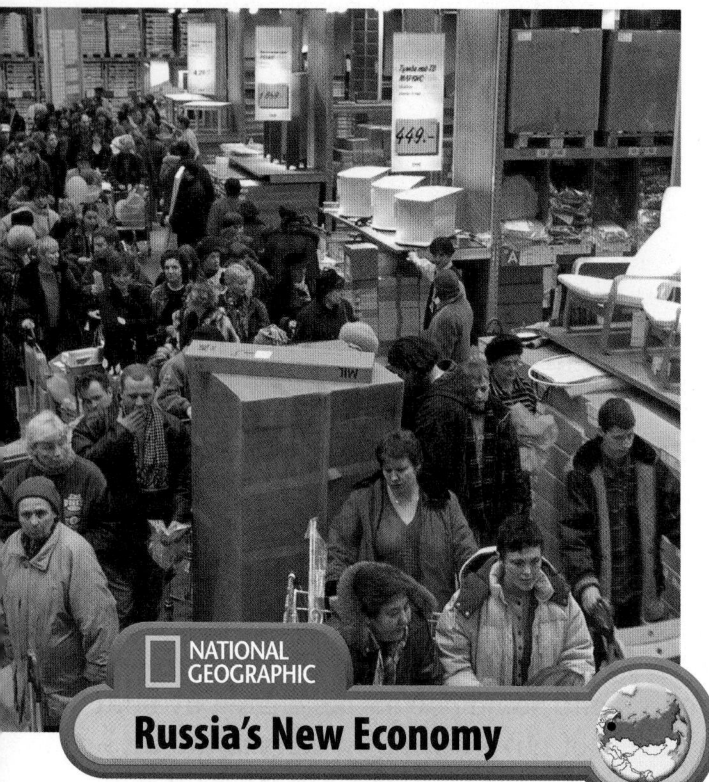

Russia's New Economy

Department stores and large discount stores are opening in Russia as the economy grows and people have more money to spend on consumer goods. ***Place*** **What is the middle class?**

Changes in Society

With the end of Communist rule, the government loosened its control on Russian society. Many different political parties were able to organize. Russians were allowed to criticize leaders and their policies. Additionally, the government no longer controlled the content of news reports or books.

Along with political freedom, Russians began to have more contact with other cultures. American and European books, television shows, and CDs became more readily available to Russians. Many people embraced the new ideas, music, and fashions that became available.

Russia's new economy led to the spread of consumerism—the desire to buy goods. Russians eagerly sought goods they had not been able to buy for years. Businesses prospered, and a Russian **middle class** emerged. This term refers to a social group that is neither very rich nor poor, but has enough money to buy cars, new clothing, electronics, and other luxury items.

The new economic system gave workers freedom to quit their jobs and seek employment elsewhere. Russians who were willing to take risks could open their own businesses. People also could **invest,** or put money into businesses run by others, in the hope of making even more money. This new economic freedom, however, did not guarantee success. Both new businesses and old businesses, which were no longer supported by the government, failed. Some tried to stay open by firing workers to cut costs. Other businesses simply could not compete and closed, putting more Russians out of work. Some skilled Russian workers still face **underemployment,** which means they are forced to take jobs that require lesser skills than they have. Many people must work second jobs to survive.

The unsettled economy is also difficult for **pensioners** (PEHN·shuh·nuhrs). Pensioners are people who receive regular payments from the government because they are too old or too sick to work. The amount of these payments is usually fixed, or remains the same. When prices rise but the amount of the payments does not increase, it becomes difficult for pensioners to buy goods. These problems caused many Russians to oppose privatization.

Population Changes

Russia's population also has experienced change. During Soviet times, many ethnic Russians moved to other parts of the Soviet Union. When these republics became independent, the ethnic Russians often were no longer welcome. About 3 million of them decided to return to Russia. People of other ethnic groups also left for Russia to escape hardships in the new republics.

Despite this arrival of immigrants, Russia's population declined from 150 million people in 1991 to 143 million in 2006. This decline is the result of a combination of low birthrates and rising death rates. The life expectancy of men has decreased to 60 years, compared with 74 years for women. These rates are well below those of other developed countries.

The sharp decline in life expectancy is a result of poor nutrition, alcoholism, and drug abuse. Also, pollution has led to more lung diseases. Meanwhile, government spending on health care has dropped.

Reading Check **Comparing** How is Russia's government similar to that of the United States?

Life Expectancies: Russia and Europe

Country	Average Life Expectancy	
	Men	Women
Sweden	78	83
Italy	77	83
France	76	84
Germany	76	82
Poland	71	79
Russia	60	74

Source: *CIA World Factbook*, 2006.

Chart Skills

1. **Identifying** Which country has the highest average life expectancy for women?
2. **Explaining** Why are life expectancies in Russia lower than those in most of Europe?

Russia's Economic Regions

Main Idea **Russia's four economic regions differ in the resources and products they supply.**

Geography and You How does your community contribute to your state's or the nation's economy? Read to learn how Russia's different regions contribute to the Russian economy.

The Moscow Region

Moscow is the political, economic, and transportation center of Russia. A large amount of manufacturing takes place in or near Moscow. Under Soviet rule, most of Russia's factories focused on **heavy industry,** or the production of goods such as machinery, mining equipment, and steel. After communism's fall, more factories shifted to **light industry,** or the production of consumer goods, such as clothing and household products. High technology and electronics industries have also developed in Moscow.

St. Petersburg and the Baltic Region

St. Petersburg and the Baltic region are located in northwestern Russia. St. Petersburg, once Russia's capital, is a major port and cultural center. Well-known for its palaces and churches, St. Petersburg attracts thousands of tourists from around the world each year.

Located near the Baltic Sea, St. Petersburg is an important trading center. A high **volume,** or amount, of goods passes through its port. The city is also a major industrial center. Factories here make machinery, ships, automobiles, and other items. St. Petersburg relies on other regions for food, fuel, and other resources.

Kaliningrad is another major Russian port along the Baltic Sea. It lies in a small piece of Russian land, about the size of the state of Connecticut, between Poland and Lithuania. This small area of Russia is isolated from the country's main area. Goods shipped to Kaliningrad must cross other countries to reach the nearest inland part of Russia. Kaliningrad is Russia's only port on the Baltic Sea that stays ice-free all year.

The Volga and Urals Region

The Volga and Urals region lies south and east of Moscow. It is a major center of manufacturing and farming. The Volga River is vital to these economic activities. This 2,300-mile (3,701-km) waterway carries nearly half of Russia's river traffic. The Volga River also supplies water for hydroelectric power and for irrigation. Farmers in the region grow large amounts of wheat, sugar beets, and other crops.

The area of the Ural Mountains is a major source for Russian resources. The mountains contain important minerals, including copper, gold, lead, nickel, and bauxite, as well as energy resources.

Siberia

Siberia's cold Arctic winds, rugged landscapes, and frozen ground make it difficult to take advantage of the region's many resources. The lands of Siberia hold valuable deposits of iron ore, uranium, gold, and coal. Timber from the sprawling taiga is also an important resource for Russia. Since resources in other parts of the world are being used up, Russia's economic future may depend on its ability to make use of Siberia's resources.

Categorizing Which of Russia's economic regions are important for manufacturing? For agriculture?

Social Studies ONLINE
Study Central™ To review this section, go to glencoe.com.

Vocabulary

1. **Explain** the significance of:
   - **a.** privatization
   - **b.** middle class
   - **c.** underemployment
   - **d.** pensioner
   - **e.** heavy industry
   - **f.** light industry

Main Ideas

2. **Describing** Describe some of the freedoms the Russian people gained after the fall of communism.
3. **Explaining** Create a chart like the one below to list at least two ways each economic region contributes to the Russian economy.

Region	Contributions

Critical Thinking

4. **Analyzing Information** Why have some Russians opposed the privatization of industries and businesses?
5. **BIG Idea** Compare and contrast the cities of Moscow and St. Petersburg in terms of location and economic activity.
6. **Challenge** What might happen to a country if it cannot obtain and make use of its own available resources?

Writing About Geography

7. **Persuasive Writing** Write an editorial to support or oppose the changes that the new democratic Russian government made after the fall of communism.

TIME PERSPECTIVES

EXPLORING WORLD ISSUES

RUSSIA'S CHALLENGING ROAD TO DEMOCRACY

Russians are enjoying new freedoms under democracy. But are their freedoms threatened?

Russians attend a political rally.

After the collapse of the Soviet Union in 1991, Russians began the difficult task of turning their country into a democracy with a free market economy. The new Russian republic allowed people more freedoms. Russian citizens were able to own their own businesses and to elect their leaders.

The new freedoms, however, have also brought challenges. Corruption has risen among officials and workers. Many people have died in a war between Russian forces and separatists in the region of Chechnya. As Russia's leaders deal with these problems, experts wonder if Russia's new freedoms are being lost.

AP PHOTO

TIME PERSPECTIVES

EXPLORING WORLD ISSUES

WOLFGANG KAEHLER/CORBIS

Increased housing construction is a sign of Russia's growing economy.

WHERE WILL REFORMS LEAD?

With territory that extends from Europe to Asia, Russia is the largest country in the world. Throughout its history, Russia has faced many great challenges. For centuries Russia was ruled by **czars**, or absolute rulers. In 1922, following a violent revolution, Russia became part of a group of republics called the Union of Soviet Socialist Republics (U.S.S.R.).

The U.S.S.R. had a Communist government that was very powerful and controlling. The U.S.S.R.'s Communist rulers often kept information about the country secret from the outside world. British prime minister Winston Churchill once called the U.S.S.R. "a riddle wrapped in a mystery."

The government controlled nearly every part of Soviet society. It owned all property and businesses and told citizens what they could do for a living and where they would live. Those suspected of disagreeing with the government were sent to labor camps in Siberia, a brutally cold area in eastern Russia. During the Soviet era, millions of people were imprisoned, executed, or tortured.

A New Chapter

Over time, the Soviet Union began to decline under its harsh system. Many citizens were assigned jobs they did not like. Many of those people were uninspired and did not work very hard. As a result, production suffered, and there were all types of shortages, including food and energy.

In the mid-1980s, the Soviet Union's leaders tried to reform the nation's Communist system. They introduced the policies of glasnost and perestroika. *Glasnost* is a Russian word that means "openness." *Perestroika* means "restructuring." Under glasnost, people were allowed to speak their opinions freely for the first time ever. Perestroika gave some of the government's decision-making power to private individuals and businesses.

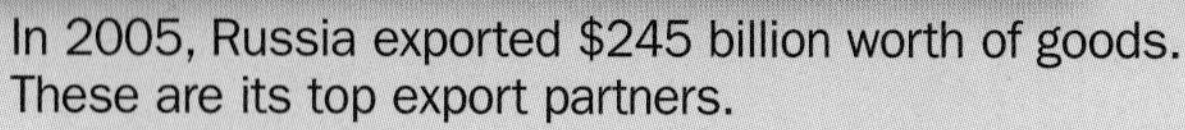

In 2005, Russia exported $245 billion worth of goods. These are its top export partners.

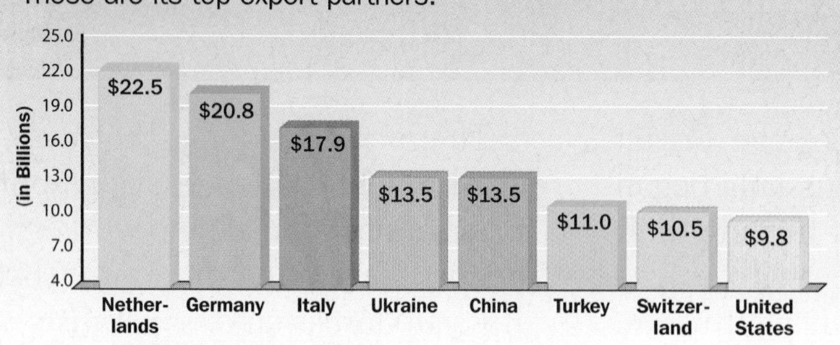

Source: *CIA World Factbook*, 2006.

INTERPRETING GRAPHS

Analyzing Information On what continent are Russia's three top export partners located?

U.S. president George W. Bush meets with Russian president Vladimir Putin.

Oil wells such as this one dot the Siberian landscape.

A Russian family shops at a modern mall in Moscow.

Not everyone, however, agreed with these reforms. Conservatives who supported the Communist Party and the military tried to take control of the Soviet government. Their attempt failed and eventually led to the Soviet Union's collapse in 1991. Since then, the Russian government and people have been working to change their country into a democracy.

Some of these changes have been successful. Russians now elect their leaders in free and open elections. Under the Soviet system, only members of the Communist Party could vote. Economic reforms have also been introduced to create a free market economy. Russians can now own their own factories, shops, and other businesses. Companies are buying materials to help them grow in the future. As a result of the reforms, the economy has grown.

A Long Road to Democracy

As successful as the economic reforms have been, many experts believe Russia has a long way to go before it has a stable democracy. The reforms have brought freedom, but they have also created new challenges. Criminal gangs and corruption thrive in Russia's open market. Many people must secretly pay officials to get drivers' licenses and permits to build houses and businesses.

Other challenges threaten Russia's security. In Chechnya, an area in southern Russia, rebels have been at war with the government since 1994. The rebels want Chechnya to be an independent nation. Some of them have carried out terrorist attacks against Russia. Hundreds of innocent people have died.

An Uncertain Future

Some experts think that these challenges threaten Russia's young democracy. They worry that the new freedoms are being lost as the government works to stop corruption and improve security. Critics of Russia's president Vladimir Putin and his successor Dmitry Medvedev have complained that, under their leadership, the government has abused **civil liberties**, or individual freedoms. They have also been accused of threatening democratic institutions like the free media.

Is a lasting democracy possible in Russia? The Russian people have a long history of dealing with difficult times and challenges. The future will tell whether history's lessons will be enough to establish democracy in the world's largest country.

EXPLORING THE ISSUE

1. **Making Inferences** How might widespread government corruption threaten democracy in Russia?

2. **Identifying Cause and Effect** How do you think glasnost and perestroika contributed to the collapse of the Soviet Union?

TERROR IN RUSSIA

In recent years, Russia has suffered a series of deadly terrorist attacks by Chechen rebels. In 2004 terrorists took control of a school and killed more than 300 people, including many children. Other bombings took place at bus and train stations. Altogether, more than 500 people were killed in 2004 as a result of terrorist attacks.

The attacks have shocked and angered the people of Russia. In September 2004, Russians filled the streets of St. Petersburg and Moscow to protest terrorism against their country. Many carried signs with antiterrorism slogans. Some read, "We won't give Russia to terrorists."

The demonstrators believed rebels from Chechnya were responsible for the attacks in 2004. In 2005, however, the violence continued as Chechen rebels set off another bomb in a village in the Caucasus region. The explosion killed 14 people.

Russians protested the terrorist attacks.

SMOLSKY SERGEI/ITAR-TASS/CORBIS

A Bitter History

The people of Chechnya want to form their own country. Most Chechens are Muslims, which sets them apart from Russians, who tend to be Christians. Russia first **conquered** Chechnya in 1858, but Chechens never accepted Russian rule. In 1991 Chechnya declared independence, but Russia would not allow it. Russians believe the territory belongs to them. In 1994 Russia went to war against the Chechen rebels. Hundreds of thousands of Chechens died in the fighting.

In 1996 Chechnya won the right to elect its own government, but it remained part of Russia. When Chechen attacks continued, Russian president Vladimir Putin sent troops back into the region in 1999.

Will there be peace? The Russian government refuses to deal with the rebels. Russian forces now control Chechen cities, which are being rebuilt. Despite Russian claims of victory, rebel attacks continue in some areas.

EXPLORING THE ISSUE

1. **Making Inferences** Why might the Chechen rebels feel that Chechnya should be free?
2. **Synthesizing** Why does Russia refuse to deal with the Chechen rebels?

A Russian worker produces steel at a metallurgy plant in the Ural Mountains.

AP PHOTO

REVIEW AND ASSESS

UNDERSTANDING THE ISSUE

1 **Making Connections** How might glasnost and perestroika have contributed to the creation of new ideas and ways of doing business?

2 **Writing to Inform** Suppose you are a Russian student. Write a letter to an American friend explaining some of the new freedoms that Russian citizens enjoy.

3 **Writing to Persuade** Russia has large amounts of natural resources, like oil and timber, and a well-educated population. Write a brief essay about how these strengths might help Russia build a strong democracy.

INTERNET RESEARCH ACTIVITIES

4 Like today's Russia, the United States had a weak banking system when it was first founded. With your teacher's help, browse the Internet to learn how the U.S. strengthened its early banking system. List ways Russia might learn from the U.S. experience.

5 Important industries in Russia, like steel and manufacturing, need to be modernized. Browse the Internet for information about how Russia plans to update one of these industries. Share your findings with the class.

BEYOND THE CLASSROOM

6 **Visit your school or local library** to learn more about the Soviet Union's labor camps in Siberia. Work in a group to find out what it was like to live in a camp. What hardships did those living in the camps have to endure? Present your findings to your classmates.

7 **Research other nations,** such as Great Britain and Spain, that have had to fight terrorism. What might Russia learn from those nations in dealing with Chechnya? Put your findings in a report.

Russia's Natural Resources

Russia is a vast country that holds many of the world's natural resources. The map shows where some of Russia's resources are located.

Building Map Reading Skills

1. **Analyzing Information** Based on the map's key, what type of work might prisoners in Siberian labor camps have done?
2. **Drawing Conclusions** Why might Russia want to protect its interests in its southwest?

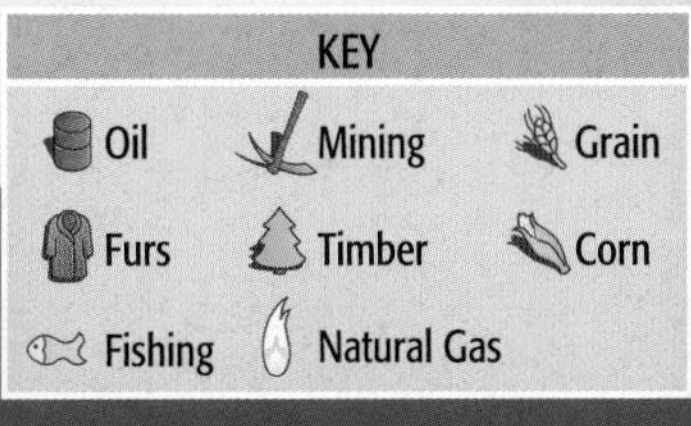

Guide to Reading

BIG Idea

Geography is used to interpret the past, understand the present, and plan for the future.

Content Vocabulary

- decree *(p. 419)*
- oligarch *(p. 420)*
- deposit insurance *(p. 421)*
- separatist movement *(p. 421)*

Academic Vocabulary

- prior *(p. 419)*
- unify *(p. 421)*
- conduct *(p. 421)*

Reading Strategy

Analyzing Information Use a diagram like the one below to identify the changes, both positive and negative, that have resulted from Russia's switch to a free market economy.

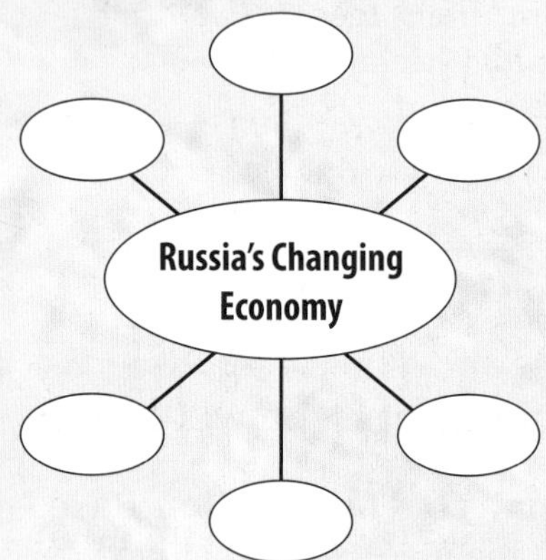

Issues and Challenges

 Section Audio  Spotlight Video

Picture This What do you think is the most popular possession in Russia? A car? A computer? No, it is most likely a cell phone. The popularity of cell phones has skyrocketed in Russia. Russia has a growing middle class with money to spend. Young business-savvy Russians are starting companies that provide trendy and modern products—like cell phones—to this middle class. As the number of cell phone businesses has increased, so has cell phone use. It is estimated that in 1996, only 10,000 people in Moscow owned cell phones, which cost about $2,000 each! Now 80 million Russians, or about 60 percent of the population, own cell phones, which cost about $100 each. The economy is just one part of Russia that is changing. Read on to learn more about modern Russia's challenges.

Most popular possession in Russia

Political and Economic Challenges

Main Idea **Russians face many challenges as they try to build a democracy and a market economy.**

Geography and You Do you think a country needs strong leaders to solve serious problems? Can leaders become too strong? Read to learn about growing challenges to Russia's democracy and free market economy.

Even with its many resources and industrial power, Russia is finding it difficult to make its new government and economy successful. **Prior** to 1991, Russians had little experience with democratic government. Now many political parties compete in free elections. Also, after communism's fall, Russians began to make their own economic decisions.

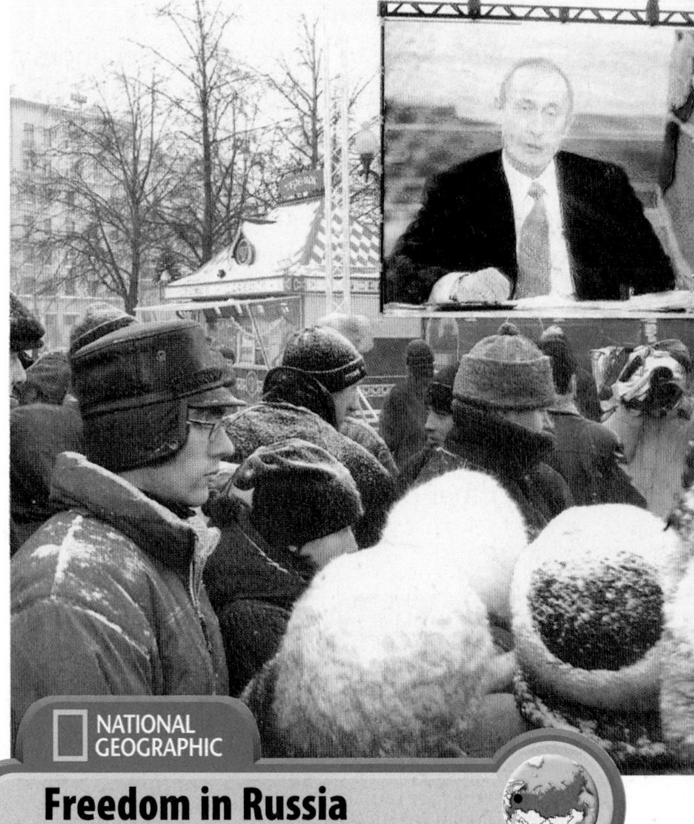

NATIONAL GEOGRAPHIC

Freedom in Russia

Russians gather in central Moscow to watch a televised news conference given by Russian President Vladimir Putin. ***Regions*** **How has Putin responded to criticisms of his rule?**

Roadblocks to Democracy

Becoming truly democratic has not been easy for Russia. Confusion over governmental powers is one problem. For example, the Russian president's power to issue **decrees**—rulings that have the force of law but do not need the approval of the legislature—might make that office too strong.

After taking office in 1999, Vladimir Putin strengthened presidential powers. He took harsh measures against his opponents. Newspapers remained free, but television news came under government control. In 2008, Putin's ally, Dmitry Medvedev, was chosen president in an election that critics claimed was neither free nor democratic.

Russia is a federal republic. Power is shared among national, regional, and local governments. To ensure that regional leaders would obey his wishes, Putin organized the country into seven large districts and appointed governors who would support his policies.

Throughout Russian government, many politicians disregard democratic ways. The courts and the legal system often favor rich, powerful citizens. In addition, many Russians still understand little about their government and, therefore, do not know how to make changes in the way it works.

Social Studies ONLINE

Student Web Activity Visit glencoe.com and complete the Chapter 15 Web Activity about Russia's government.

Shifting to a Market Economy

Russia's shift to a free market economy has brought many positive changes. New companies have been started, and some personal incomes have risen. Higher prices for Russia's oil and natural gas exports have brought the country more income.

Economic success, however, has brought an increase in crime and business corruption. A small group of people, often referred to as oligarchs, control various parts of the economy. An **oligarch** (AH·luh·gahrk) is a member of a small group of rulers that holds great power. In Russia, oligarchs are often corrupt business leaders. Putin has limited the power of some of these oligarchs, but to do so, he has had to strengthen government authority.

Another problem is that the benefits of economic change have not reached all of Russia's people. A few Russians have grown wealthy, but some have become even poorer. There are also strong regional differences in economic success. **Figure 1** shows the average income per capita—or per person—in Russia's seven districts.

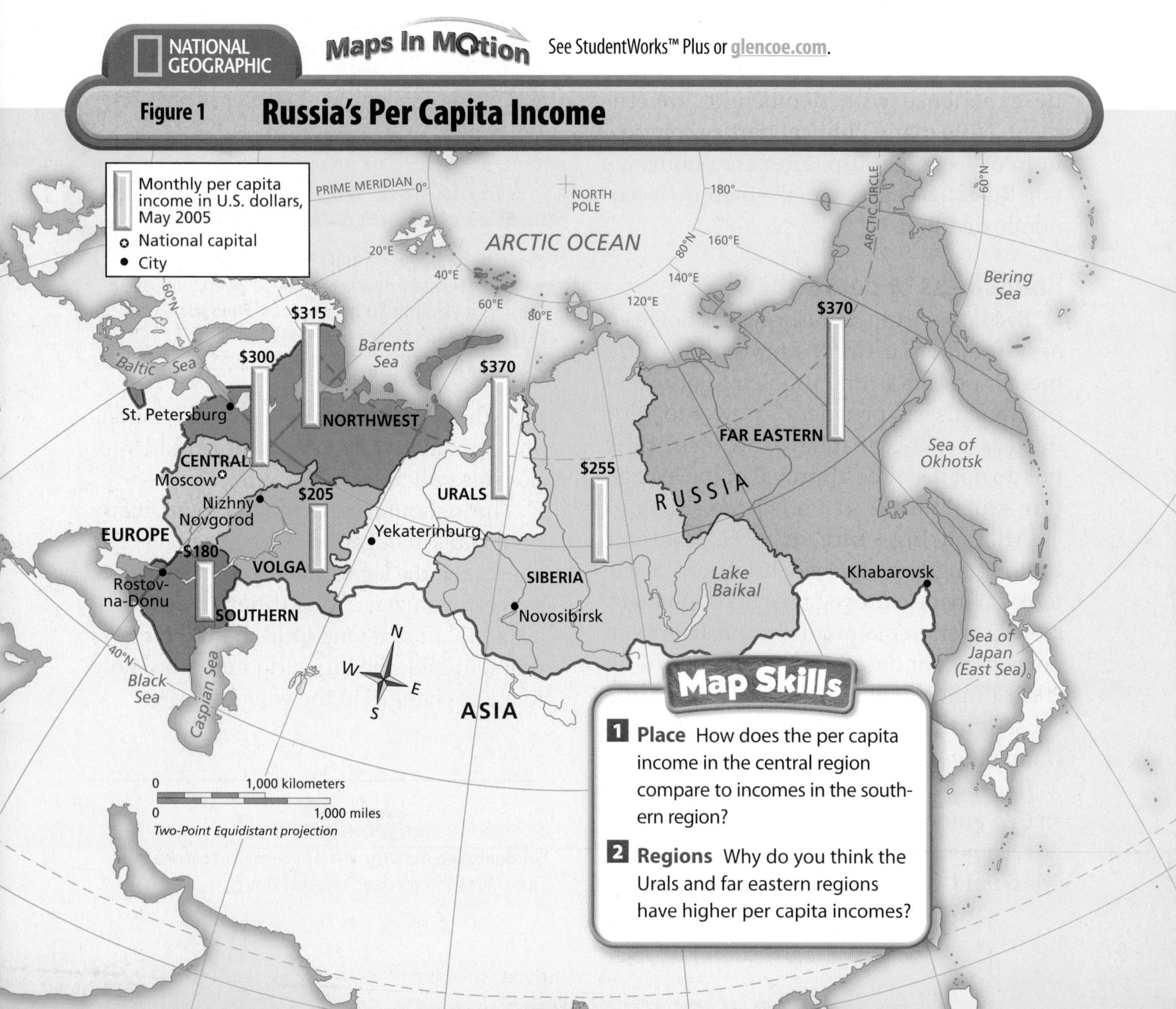

Map Skills

1. **Place** How does the per capita income in the central region compare to incomes in the southern region?
2. **Regions** Why do you think the Urals and far eastern regions have higher per capita incomes?

Monthly incomes in Moscow are much higher compared to other cities and areas of the country.

Russia's banking system also has not been able to fully contribute to economic growth. Banks play a vital role in an economy by collecting people's deposits and lending some of that money to other people. These people borrow the money to buy houses and cars or to start new businesses. All of these actions help create jobs within a region's economy. Greater savings means there is more money to loan, which in turn helps strengthen the economy. If people do not have enough money to deposit into savings, though, banks have fewer funds to lend and the economy suffers.

Many Russians, however, do not trust the country's banks. To encourage people to save their money, the government created a **deposit insurance** system, which will repay people who deposit their money in a bank if the bank goes out of business. Officials hope this system will make people feel safer and more willing to use the banking system.

Challenges to National Unity

While dealing with economic issues, Russians have had to face challenges to their country's political unity. Regional rivalries have increased in recent years. Such resentments have made it difficult to **unify,** or bring together, the country. An even larger challenge is the desire of many ethnic groups to form their own independent countries.

When the Soviet Union fell, several ethnic groups in Russia saw a chance for independence. They launched **separatist movements,** campaigns to break away from the national government and form independent countries. One of the most violent separatist movements began in Chechnya, a Muslim region near the Caucasus Mountains in southern Russia.

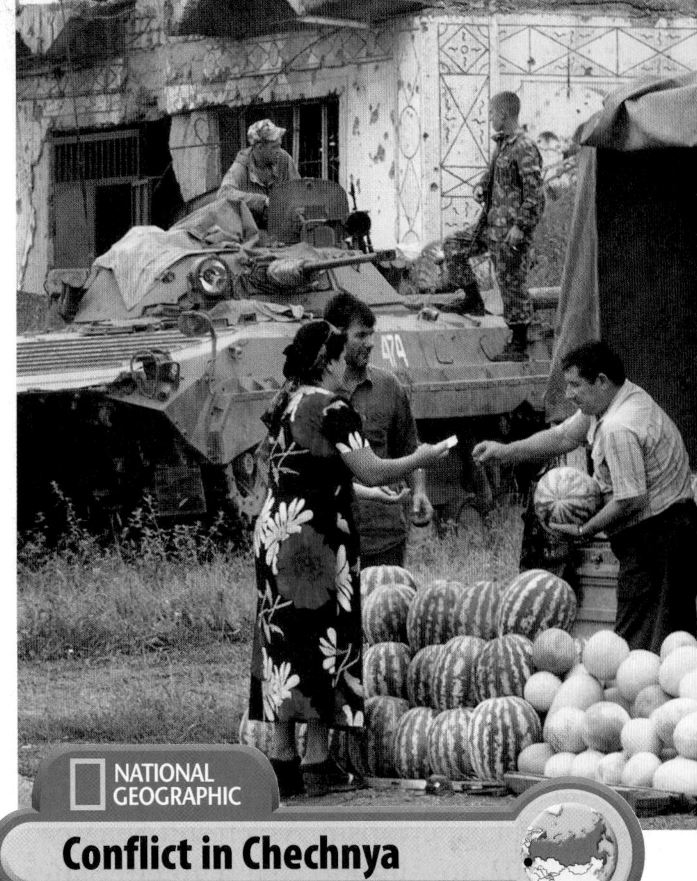

Conflict in Chechnya

Chechens buy watermelons as Russian soldiers guard a shopping area in Grozny, Chechnya's capital.
Place **Why did conflict break out in Chechnya?**

In the early 1990s, Russia's President Boris Yeltsin gave Chechnya more self-rule, but many Chechens (CHEH·chehnz) wanted complete independence from Russia. Yeltsin did not want to allow Chechen separatists to succeed, believing that other regions would also demand independence. In 1994 he sent a large Russian army into Chechnya to crush Chechen forces. Both sides suffered heavy losses.

The Chechen separatists continued to **conduct,** or carry out, terrorist attacks against the Russian government. Truces and agreements between the two sides have failed, and the situation remains unresolved.

Reading Check **Describing** Identify and describe a major challenge to Russian unity.

Russia and the World

Main Idea **Although Russia remains a world power, other nations have questioned some of its actions.**

Geography and You How do good relations among neighbors build a stronger community? Read to learn about Russia's relations with the rest of the world.

As a major world power, Russia plays an important role in world affairs. In recent years, it has worked to strengthen ties with other countries. Russia sees the war in Chechnya, for example, as a struggle against terrorism. As a result, it agreed in 2002 to support the United States and other NATO (North Atlantic Treaty Organization) countries in fighting global terrorist activities.

Still, the United States and other countries are concerned about the ongoing effects of Putin's rule and Russia's declining support for democracy. Meanwhile, Russia has uneasy relations with some of the countries that were once part of the Soviet Union. Some Russian leaders have said they would like to see Russian influence increase, which worries people in these former Soviet countries. In 2004 Putin supported a pro-Russian candidate in Ukraine's presidential election—a position many Ukrainians protested. While other neighboring countries are sometimes unhappy with Russia's actions, they also know they depend on Russia for certain resources, such as oil and natural gas.

Reading Check **Analyzing Information**
Why has Russia supported the global war against terrorism?

Section 2 Review

Social Studies ONLINE
Study Central™ To review this section, go to glencoe.com.

Vocabulary

1. **Explain** the meaning of the following terms by using each in a sentence.
   a. decree
   b. oligarch
   c. deposit insurance
   d. separatist movement

Main Ideas

2. **Identifying** Use a diagram like the one below to identify challenges to the growth of democracy in Russia.

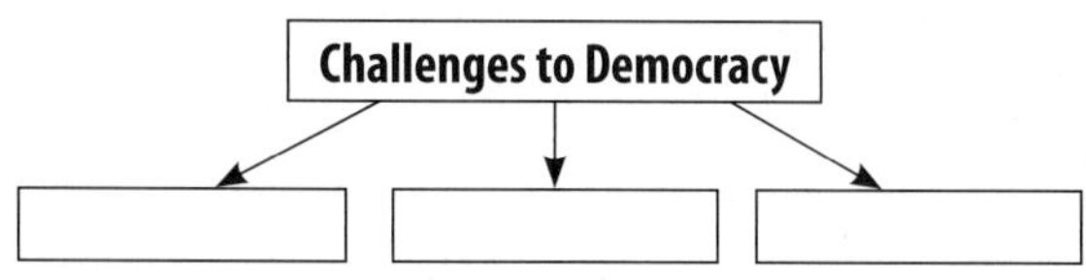

3. **Explaining** Why are some of Russia's neighbors concerned with recent Russian actions?

Critical Thinking

4. **Identifying Cause and Effect** How has the rapid change to a democratic government affected the Russian people's involvement in government? What drawback has this swift change created?
5. **BIG Idea** What factors create challenges to Russian unity?
6. **Challenge** Describe the influence of banking on a country's economy.

Writing About Geography

7. **Using Your FOLDABLES** Use your Foldable to write a paragraph predicting how successful Russia's change to a democracy and free market economy will be. Be sure to support your prediction with facts.

Visual Summary

Communism to Democracy

- After communism's fall, Russia became more democratic.
- Russia is a federal republic, with power divided among national, regional, and local governments.
- Russia has been moving from a command economy to a market economy.

Russia's duma, or parliament, in session

Changes in Society

- Russians now can vote freely and have increased contact with the cultures of other countries.
- The switch to a market economy has benefited some Russians while bringing hardships to others.
- Low birthrates and rising death rates have led to a decline in Russia's population.

Cargo ship on Volga River

Economic Regions

- Moscow, with its many industries, is the economic center of Russia.
- Ports in the St. Petersburg and Baltic region carry on trade between Russia and other countries.
- The Volga and Urals region is a center of manufacturing, mining, and farming.
- Siberia's resources are difficult to tap because of the area's remoteness and harsh climate.

Young Russians playing chess

Challenges

- The increasing power of Russia's president has placed limits on democracy.
- Crime and business corruption have grown in Russia.
- Some ethnic groups want to separate from Russia and form their own countries.

Russia and the World

- Russian leaders have worked to strengthen Russia's ties with the West.
- Russia has uneasy relations with some of the countries that were once part of the Soviet Union.

Vladimir Putin (right) meets with a South African leader

Study anywhere, anytime! Download quizzes and flash cards to your PDA from glencoe.com.

CHAPTER 15 ASSESSMENT

STANDARDIZED TEST PRACTICE

TEST-TAKING TIP

Do not pick an answer choice just because it sounds good. Sometimes a choice is meant to sound correct but is not. Read all of the answer choices very carefully before you select the best one.

Reviewing Vocabulary

Directions: Choose the word(s) that best completes the sentence.

1. In order to create a market economy, the Russian government legalized ______ .

A underemployment
B privatization
C heavy industry
D light industry

2. Due to changes in Russia's new economy, many people are ______ , which means they have to take jobs requiring lesser skills than they have.

A unemployed
B privatized
C underemployed
D invested

3. A(n) ______ is a member of a small ruling group that controls great power.

A decree
B pensioner
C separatist
D oligarch

4. When the Soviet Union fell, several ethnic groups in Russia launched ______ .

A separatist movements
B privatization efforts
C light industries
D oligarchies

5. A Russian president may issue ______ , or rulings that have the force of law and do not need the approval of the legislature.

A vetos
B decrees
C opinions
D considerations

Reviewing Main Ideas

Directions: Choose the best answer for each question.

Section 1 *(pp. 408–412)*

6. A new market economy took root in Russia after

A World War II.
B the fall of communism.
C separatist movements.
D a middle class emerged.

7. After the fall of communism, there was a manufacturing shift toward more

A heavy industry.
B oligarchy.
C light industry.
D separatist movements.

Section 2 *(pp. 418–422)*

8. A problem that has accompanied Russia's shift to a free market economy is

A starting up new companies.
B higher prices for gas and oil exports.
C the deposit insurance system.
D the rise of business oligarchs.

9. Other nations in the world have questioned some of Russia's actions, including the

A growing power of the Russian president.
B establishment of free elections.
C creation of a market economy.
D creation of a deposit insurance program.

GO ON

Critical Thinking

Directions: Base your answers to questions 10 and 11 on the chart below. Choose the best answer for each question.

GNP Per Capita in U.S. Dollars				
Year(s)	Russia	United States	France	Germany
1991	$3,470	$22,340	$20,460	$20,510
1992	2,820	23,830	22,300	23,360
1993	2,350	24,750	22,360	23,560
1994	2,650	25,860	23,470	25,580
1995	2,240	26,980	24,990	27,510
1996	2,410	28,020	26,270	28,870
1997	2,680	29,080	26,300	28,280
1998	2,260	29,240	24,210	26,570
1999	2,250	31,910	24,170	26,620
2000	1,660	34,100	24,090	25,120
2001	1,750	34,280	22,730	23,560

GNP per capita is the dollar value of a country's final output of goods and services in a year (its GNP), divided by its population. It reflects the average income of a country's citizens.

Source: http://world.britannica.com/analyst/chrono/table

10. What happened to Russia's GNP per capita during the period shown?

A It jumped rapidly between 1999 and 2000.

B It rose gradually.

C It declined drastically.

D It grew following 1997.

11. Which nation's GNP per capita increased the most between 1991 and 2001?

A United States

B Germany

C Russia

D France

Document-Based Questions

Directions: Analyze the document and answer the short-answer questions that follow.

> "Things Fall Apart–Russia After the Fall of Communism"
>
> *The break-up of the USSR and the failures of economic reform have hurt primarily the elderly and the children. In addition, the systems of social services, education, and health care also fell apart. The fabric of society changed with the disappearance of values and morals. Secrecy and [control] were quickly replaced by the power of money. . . . In the past there was a common expression–"Without papers you are a bug, but with papers, you are a man"–which meant that you were constantly required to ask permission from a countless army of [government officials] . . . . Now you cannot expect to be treated with respect unless your pocket is full of a wad of "greens." . . . It doesn't matter if you earned this money by [illegal activities or] selling drugs . . . . Your social status will be much higher than that of an engineer, a professor, or a doctor.*
>
> —by Nikolai Zlobin
> *World Affairs,* Winter, 1996

12. According to the writer, who has been most seriously affected by the "new" Russia after the fall of communism? Why?

13. How does the writer feel that Russian society has changed since the breakup of the USSR?

Extended Response

14. Do you believe Russia's transition to a market economy has been good for the country or caused more harm than good? Write a short essay in which you choose one side of the argument. Defend your position by using examples.

For additional test practice, use Self-Check Quizzes—Chapter 15 at glencoe.com.

Need Extra Help?														
If you missed question...	1	2	3	4	5	6	7	8	9	10	11	12	13	14
Go to page...	409	410	420	421	419	409	411	420	422	420	420	419	419	420

UNIT 6

North Africa, Southwest Asia, and Central Asia

CHAPTER 16
Physical Geography of North Africa, Southwest Asia, and Central Asia

CHAPTER 17
History and Cultures of North Africa, Southwest Asia, and Central Asia

CHAPTER 18
North Africa, Southwest Asia, and Central Asia Today

Bicycle race, Doha, Qatar ▶

NATIONAL GEOGRAPHIC

NGS ONLINE For more information about the region, see www.nationalgeographic.com/education.

Regional Atlas

North Africa, Southwest Asia, and Central Asia

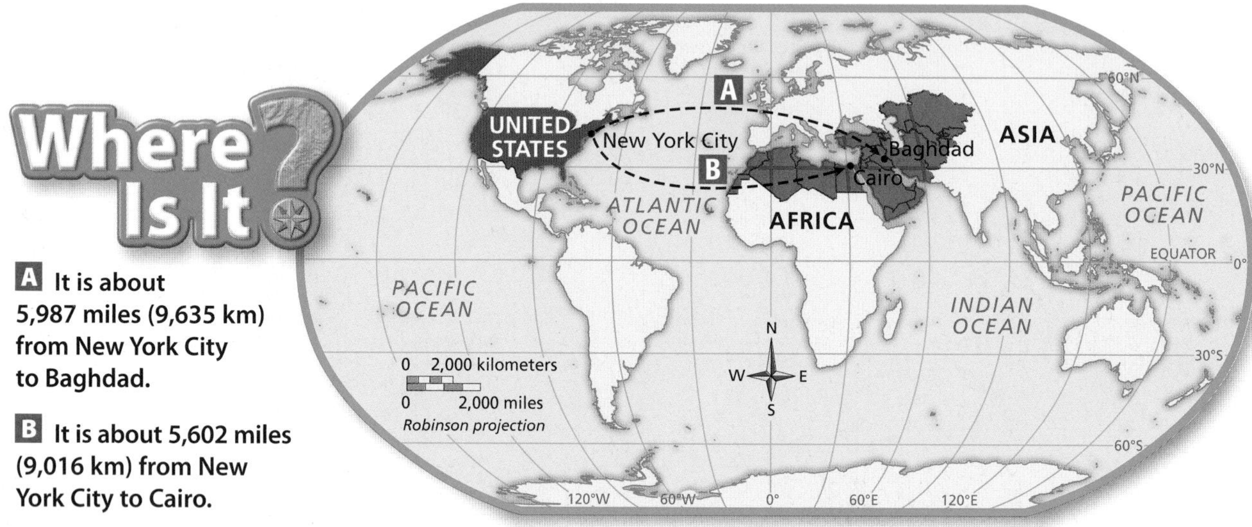

A It is about 5,987 miles (9,635 km) from New York City to Baghdad.

B It is about 5,602 miles (9,016 km) from New York City to Cairo.

Together, North Africa, Southwest Asia, and Central Asia occupy almost 6.6 million square miles (17 million sq. km), more than twice the area of the continental United States. At almost 530 million, the population of North Africa, Southwest Asia, and Central Asia is more than one and a half times that of the United States.

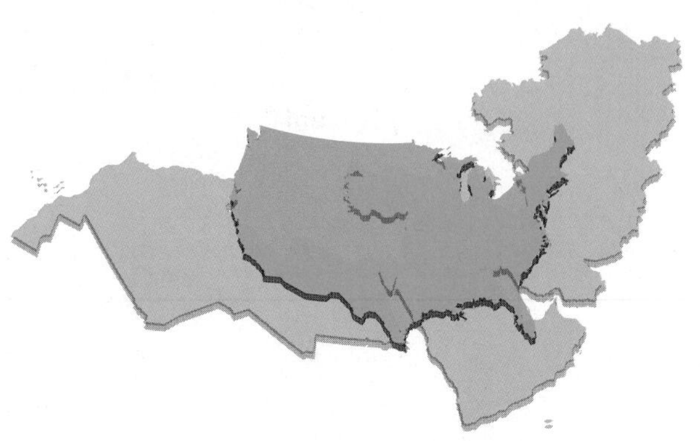

Comparing Population

United States and Selected Countries of North Africa, Southwest Asia, and Central Asia

Country
United States
Turkey
Egypt
Iran
Afghanistan
Israel

(person symbol) = 30,000,000

Source: *World Population Data Sheet*, 2005.

GEO Fast Facts

Largest Lake

Caspian Sea
143,550 sq. mi.
(371,795 sq. km)

Lowest Point

Dead Sea (Israel and Jordan)
1,312 ft. (400 m)
below sea level

Highest Point

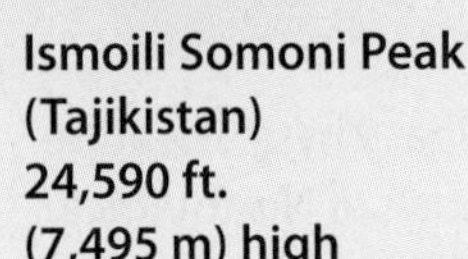

Ismoili Somoni Peak
(Tajikistan)
24,590 ft.
(7,495 m) high

Largest Desert

Sahara (northern Africa)
3,500,000 sq. mi.
(9,065,000 sq. km)

UNIT 6

Regional Atlas

North Africa, Southwest Asia, and Central Asia

PHYSICAL

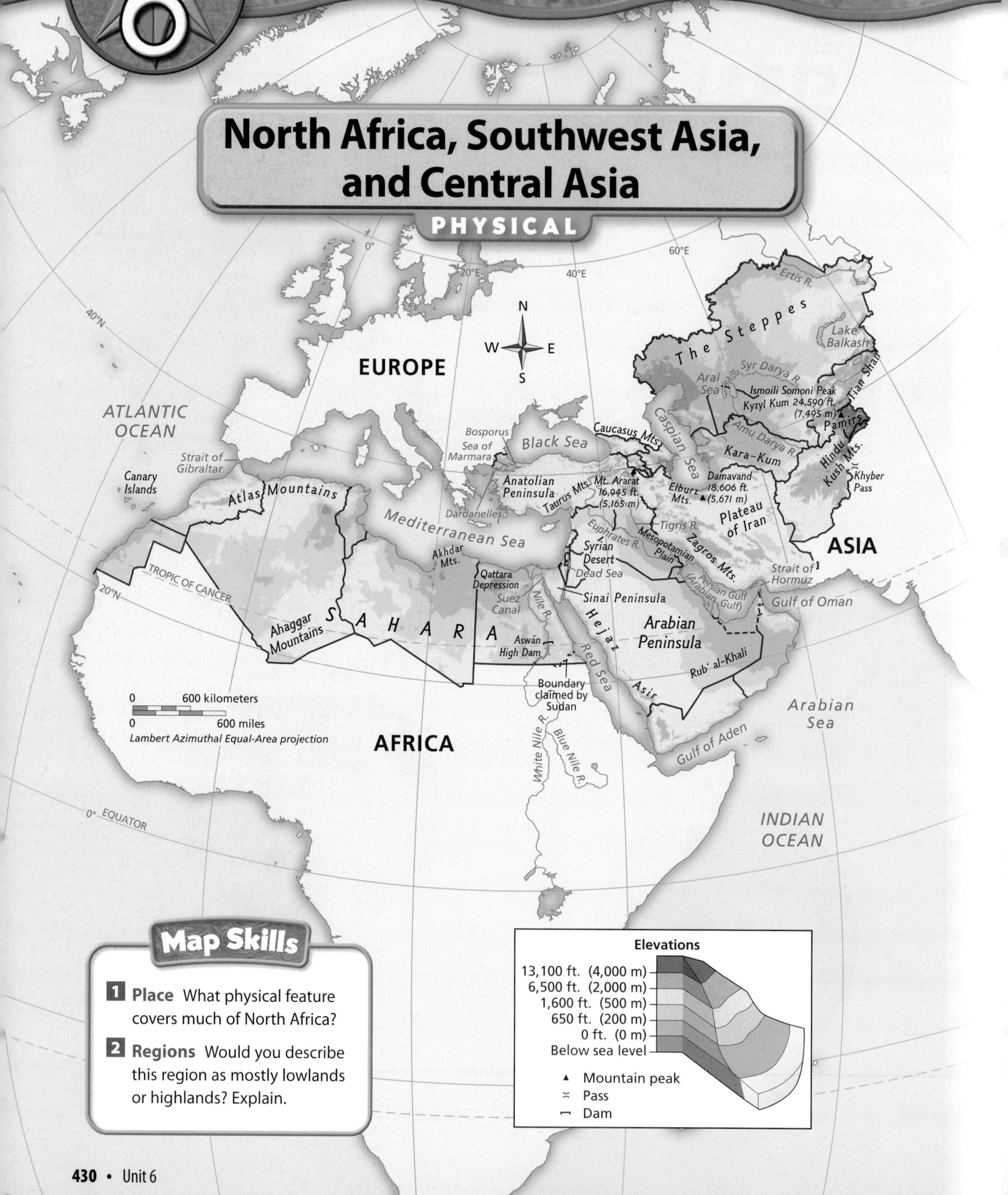

Map Skills

1. **Place** What physical feature covers much of North Africa?
2. **Regions** Would you describe this region as mostly lowlands or highlands? Explain.

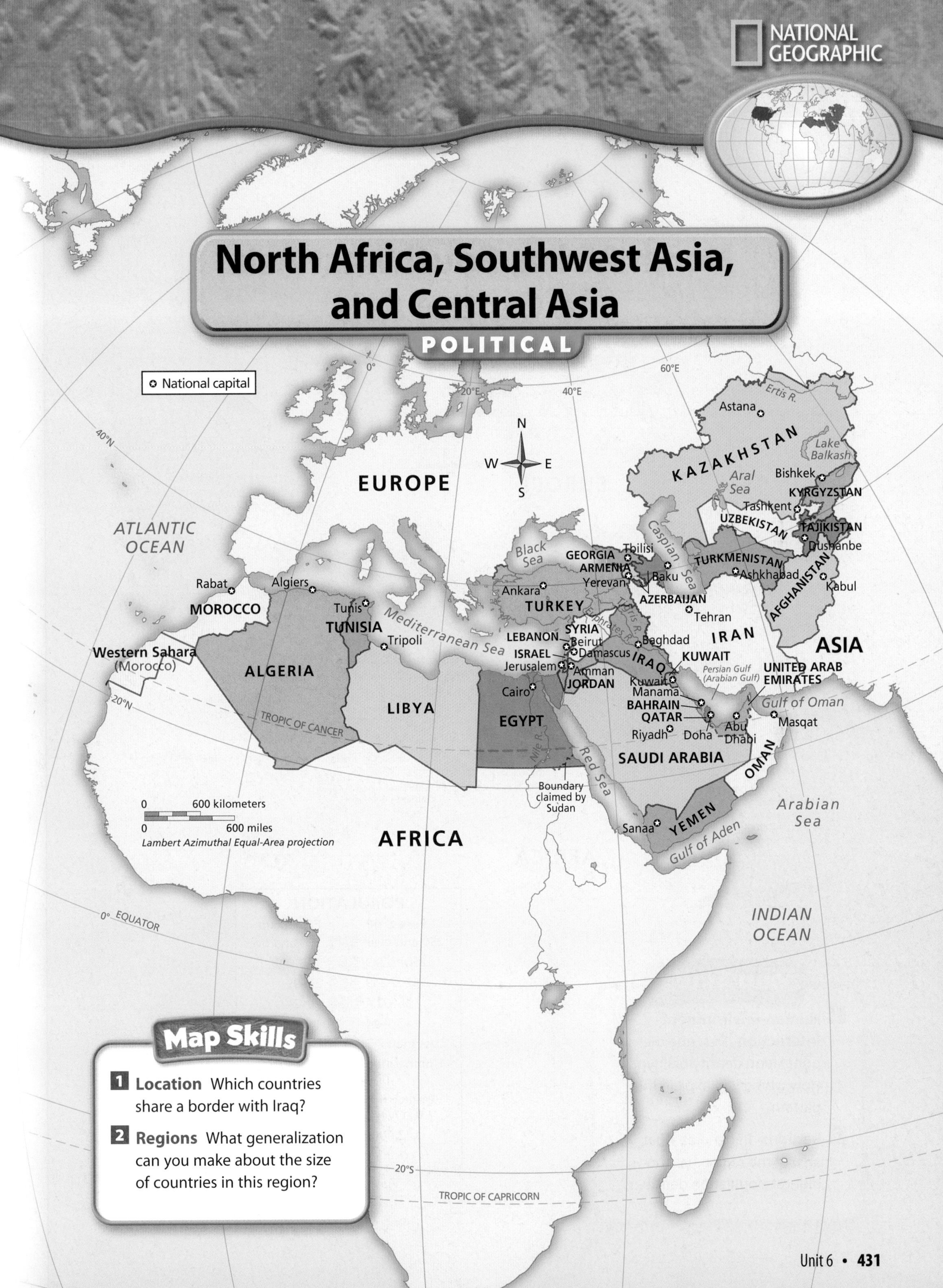

Map Skills

1. **Location** Which countries share a border with Iraq?
2. **Regions** What generalization can you make about the size of countries in this region?

UNIT 6

Regional Atlas

North Africa, Southwest Asia, and Central Asia

POPULATION DENSITY

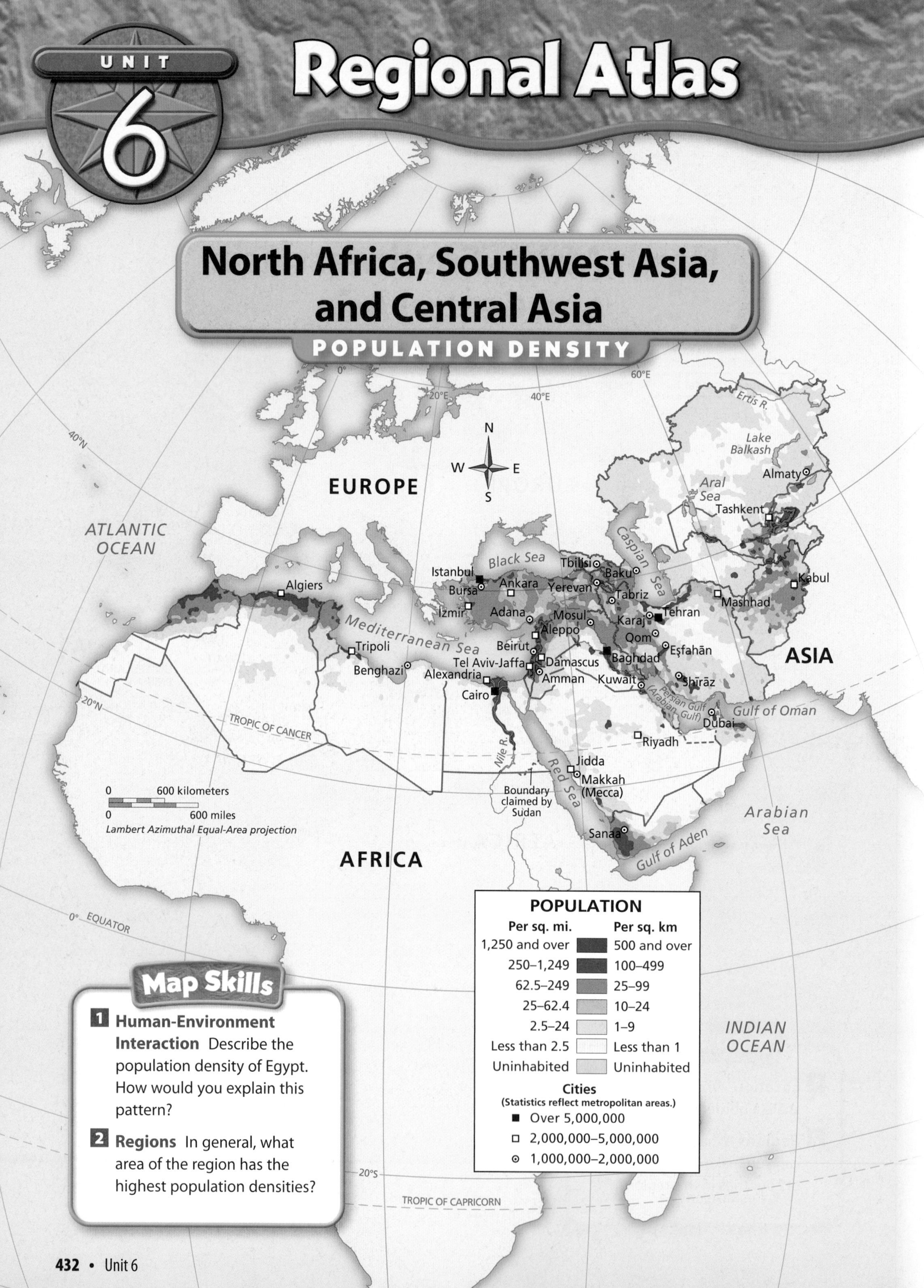

Map Skills

1. **Human-Environment Interaction** Describe the population density of Egypt. How would you explain this pattern?
2. **Regions** In general, what area of the region has the highest population densities?

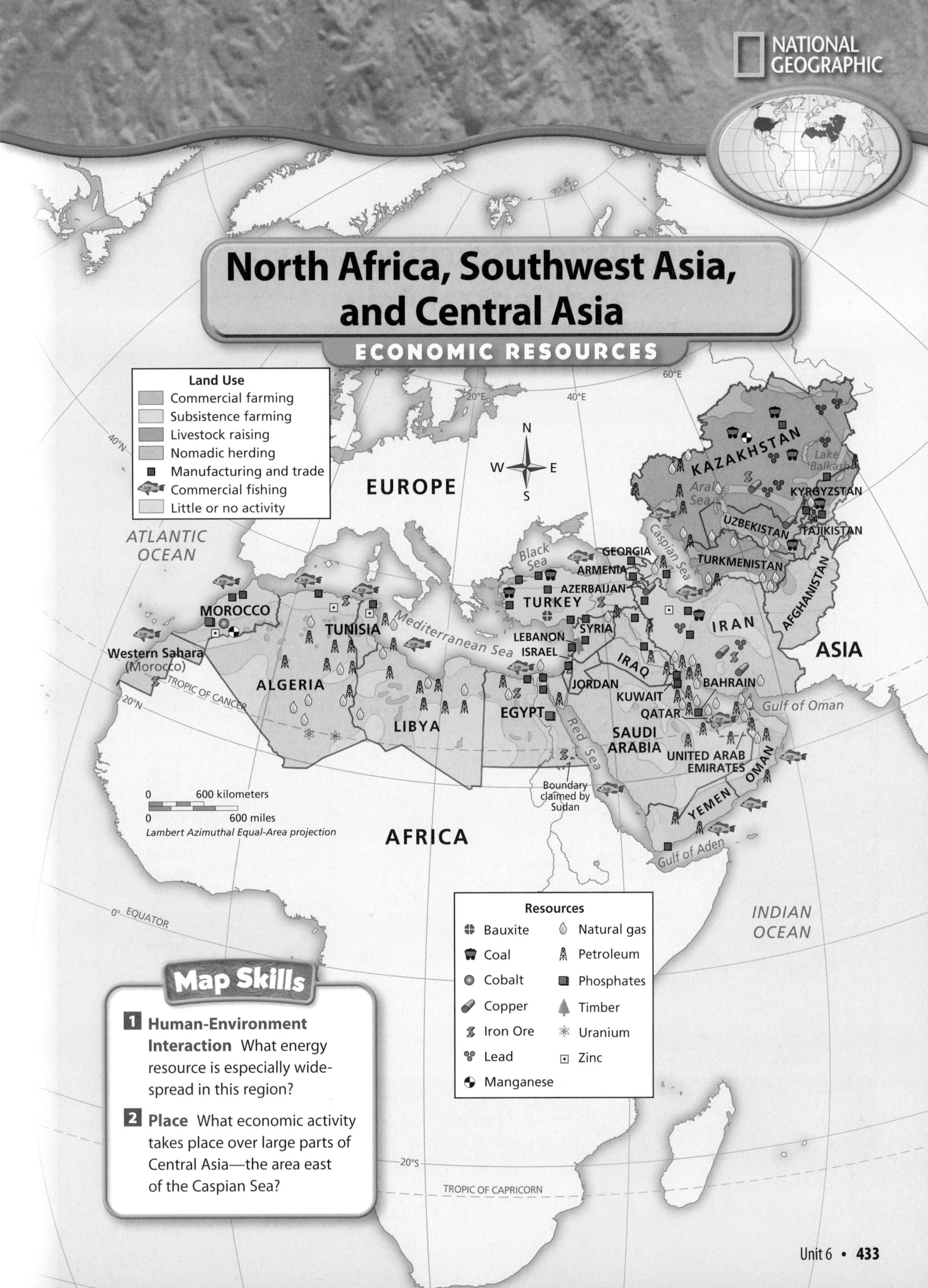

Map Skills

1. **Human-Environment Interaction** What energy resource is especially widespread in this region?
2. **Place** What economic activity takes place over large parts of Central Asia—the area east of the Caspian Sea?

Regional Atlas

North Africa, Southwest Asia, and Central Asia

Country and Capital	Literacy Rate	Population and Density	Land Area	Life Expectancy (Years)	GDP* Per Capita (U.S. dollars)	Television Sets (per 1,000 people)	Flag and Language
AFGHANISTAN Kabul	36%	29,900,000 119 per sq. mi. 46 per sq. km	251,772 sq. mi. 652,086 sq. km	42	$800	14	Dari, Pashto
Algiers ALGERIA	70%	32,800,000 36 per sq. mi. 14 per sq. km	919,591 sq. mi. 2,381,730 sq. km	73	$6,600	107	Arabic
ARMENIA Yerevan	98.6%	3,000,000 261 per sq. mi. 101 per sq. km	11,506 sq. mi. 29,800 sq. km	71	$4,600	241	Armenian
AZERBAIJAN Baku	97%	8,400,000 251 per sq. mi. 97 per sq. km	33,436 sq. mi. 86,599 sq. km	72	$3,800	257	Azerbaijan
Manama BAHRAIN	89.1%	700,000 2,632 per sq. mi. 1,016 per sq. km	266 sq. mi. 689 sq. km	74	$19,200	446	Arabic
Cairo EGYPT	57.5%	74,000,000 191 per sq. mi. 74 per sq. km	386,660 sq. mi. 1,001,445 sq. km	70	$4,200	170	Arabic
GEORGIA Tbilisi	99%	4,500,000 167 per sq. mi. 65 per sq. km	26,911 sq. mi. 69,699 sq. km	72	$3,100	516	Georgian
Tehran IRAN	79.4%	69,500,000 110 per sq. mi. 43 per sq. km	630,575 sq. mi. 1,633,182 sq. km	70	$7,700	154	Persian
UNITED STATES Washington, D.C.	97%	296,500,000 80 per sq. mi. 31 per sq. km	3,717,796 sq. mi. 9,629,047 sq. km	78	$40,100	844	English

*Gross Domestic Product

Countries and flags not drawn to scale

North Africa, Southwest Asia, and Central Asia

Country and Capital	Literacy Rate	Population and Density	Land Area	Life Expectancy (Years)	GDP* Per Capita (U.S. dollars)	Television Sets (per 1,000 people)	Flag and Language
IRAQ Baghdad	40.4%	28,800,000 170 per sq. mi. 66 per sq. km	169,236 sq. mi. 438,319 sq. km	59	$2,100	82	Arabic
ISRAEL† Jerusalem	95.4%	7,100,000 873 per sq. mi. 337 per sq. km	8,131 sq. mi. 21,059 sq. km	80	$20,800	328	Hebrew
JORDAN Amman	91.3%	5,800,000 184 per sq. mi. 71 per sq. km	31,444 sq. mi. 81,440 sq. km	72	$4,500	83	Arabic
KAZAKHSTAN Astana	98.4%	15,100,000 14 per sq. mi. 6 per sq. km	1,049,151 sq. mi. 2,717,289 sq. km	66	$7,800	240	Kazakh, Russian
KUWAIT Kuwait	83.5%	2,600,000 378 per sq. mi. 146 per sq. km	6,880 sq. mi. 17,819 sq. km	78	$21,300	480	Arabic
KYRGYZSTAN Bishkek	97%	5,200,000 68 per sq. mi. 26 per sq. km	76,641 sq. mi. 198,499 sq. km	68	$1,700	49	Kyrgyz, Russian
LEBANON Beirut	87.4%	3,800,000 947 per sq. mi. 365 per sq. km	4,015 sq. mi. 10,399 sq. km	74	$5,000	355	Arabic
LIBYA Tripoli	82.6%	5,800,000 9 per sq. mi. 3 per sq. km	679,359 sq. mi. 1,759,532 sq. km	76	$6,700	139	Arabic
UNITED STATES Washington, D.C.	97%	296,500,000 80 per sq. mi. 31 per sq. km	3,717,796 sq. mi. 9,629,047 sq. km	78	$40,100	844	English

† Israel has proclaimed Jerusalem as its capital, but many countries' embassies are in Tel Aviv. The Palestinian Authority has assumed all governmental duties in non-Israeli-occupied areas of the West Bank and Gaza Strip.

Sources: *CIA World Factbook*, 2005; Population Reference Bureau, *World Population Data Sheet*, 2005.

For more country facts, go to the **Nations of the World Databank** at glencoe.com.

Regional Atlas

North Africa, Southwest Asia, and Central Asia

Country and Capital	Literacy Rate	Population and Density	Land Area	Life Expectancy (Years)	GDP* Per Capita (U.S. dollars)	Television Sets (per 1,000 people)	Flag and Language
Rabat MOROCCO†	51.7%	30,700,000 178 per sq. mi. 69 per sq. km	172,413 sq. mi. 446,548 sq. km	70	$4,200	165	Arabic
Masqat OMAN	75.8%	2,400,000 29 per sq. mi. 11 per sq. km	82,031 sq. mi. 212,459 sq. km	74	$13,100	575	Arabic
QATAR Doha	82.5%	800,000 188 per sq. mi. 73 per sq. km	4,247 sq. mi. 11,000 sq. km	70	$23,200	866	Arabic
SAUDI ARABIA Riyadh	78.8%	24,600,000 30 per. sq. mi. 11 per sq. km	829,996 sq. mi. 2,149,680 sq. km	72	$12,000	263	Arabic
SYRIA Damascus	76.9%	18,400,000 257 per sq. mi. 99 per sq. km	71,498 sq. mi. 185,179 sq. km	72	$3,400	68	Arabic
Dushanbe TAJIKISTAN	99.4%	6,800,000 123 per sq. mi. 48 per sq. km	55,251 sq. mi. 143,099 sq. km	63	$1,100	328	Tajik
Tunis TUNISIA	74.2%	10,000,000 158 per sq. mi. 61 per sq. km	63,170 sq. mi. 163,610 sq. km	73	$7,100	190	Arabic, French
Ankara TURKEY	86.5%	72,900,000 244 per sq. mi. 94 per sq. km	299,158 sq. mi. 774,816 sq. km	69	$7,400	328	Turkish
UNITED STATES Washington, D.C.	97%	296,500,000 80 per sq. mi. 31 per sq. km	3,717,796 sq. mi. 9,629,047 sq. km	78	$40,100	844	English

*Gross Domestic Product

Countries and flags not drawn to scale

† Morocco claims the Western Sahara area, but other countries do not accept this claim.

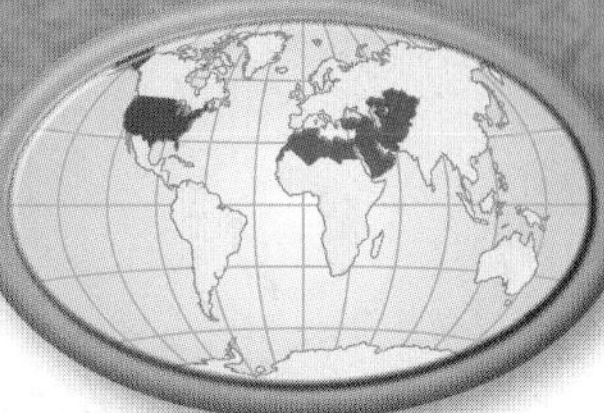

North Africa, Southwest Asia, and Central Asia

Country and Capital	Literacy Rate	Population and Density	Land Area	Life Expectancy (Years)	GDP* Per Capita (U.S. dollars)	Television Sets (per 1,000 people)	Flag and Language
TURKMENISTAN Ashkhabad	98%	5,200,000 28 per sq. mi. 11 per sq. km	188,456 sq. mi. 488,099 sq. km	63	$5,700	198	Turkmen
UNITED ARAB EMIRATES Abu Dhabi	77.9%	4,600,000 143 per sq. mi. 55 per sq. km	32,278 sq. mi. 83,600 sq. km	77	$25,200	309	Arabic
UZBEKISTAN Tashkent	99.3%	26,400,000 153 per sq. mi. 59 per sq. km	172,741 sq. mi. 447,397 sq. km	67	$1,800	280	Uzbek
YEMEN Sanaa	50.2%	20,700,000 102 per sq. mi. 39 per sq. km	203,849 sq. mi. 527,966 sq. km	61	$800	286	Arabic
UNITED STATES Washington, D.C.	97%	296,500,000 80 per sq. mi. 31 per sq. km	3,717,796 sq. mi. 9,629,047 sq. km	78	$40,100	844	English

Sources: *CIA World Factbook*, 2005; Population Reference Bureau, *World Population Data Sheet*, 2005.

For more country facts, go to the **Nations of the World Databank** at glencoe.com.

Reading Social Studies

Identifying Cause and Effect

Learn It!

To identify cause and effect, you need to know why an event occurred. A *cause* is the action or situation that produces an event. What happens as a result of a cause is an *effect*. For example, if you do not brush your teeth, you might get cavities. Not brushing your teeth is the cause of your cavities, which is the effect.

To identify cause-and-effect relationships, follow these steps:

- Identify two or more events or developments.
- Look for a logical relationship and decide whether one event caused the other.
- Identify the outcomes of events. Remember that some effects have more than one cause, and some causes lead to more than one effect. Also, an effect can become the cause of yet another effect.

Read the text below. The chart can then help you identify the cause-and-effect relationship.

> North Africa is made up of five countries. The easternmost of these is Egypt, a country covered by vast deserts. Because of its arid landscape, most Egyptians live within 20 miles (32 km) of the Nile River.
>
> —*from page 485*

Reading Tip

Some words or phrases, such as *because, led to, brought about, produced, as a result, so that, since,* and *therefore,* act as clues to help you identify cause-and-effect relationships.

Cause		Effect
Egypt is a country covered by vast deserts.	→	Because of its arid landscape, most Egyptians live within 20 miles (32 km) of the Nile River.

Read to Write Activity

Read Chapter 16 Section 1. Create a list of actions that have had negative effects on the environment. Try to link each effect with a cause.

2 Practice It!

Read the following paragraph from this unit. Use the chart to help you identify the cause-and-effect relationships. Remember, sometimes the effect of one event can be the cause of another.

> In 1947 the United Nations voted to divide Palestine into separate Jewish and Arab countries. Arabs within and outside of Palestine bitterly opposed the decision. In May 1948, the British left the area, and the Jews set up Israel in their part of Palestine.
>
> Since that time, Israel and its Arab neighbors have fought several major wars.
>
> —*from page 463*

An Israeli tank

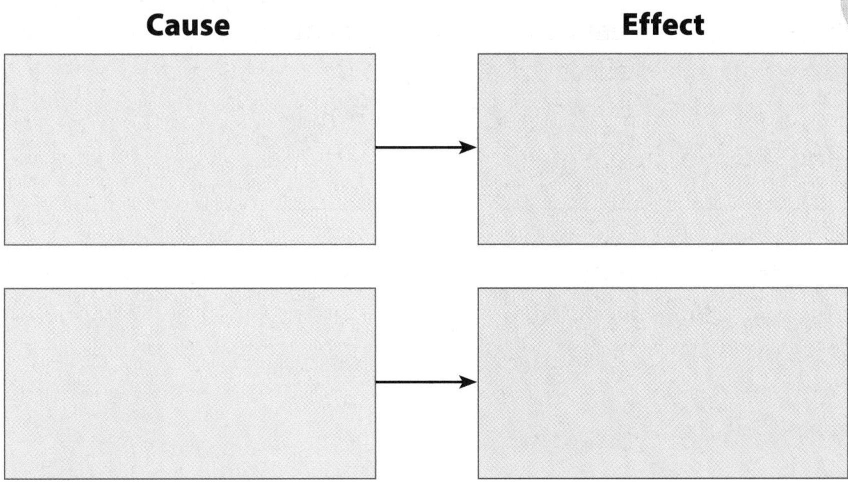

3 Apply It!

Write down five instances of cause-and-effect relationships from the text within this unit. Have a partner identify which actions are causes and which are effects.

Physical Geography of North Africa, Southwest Asia, and Central Asia

Essential Question

Human-Environment Interaction If you rode in a car or a bus to school today, that vehicle's fuel probably came from the region of North Africa, Southwest Asia, and Central Asia. This region is the world's leading producer of petroleum and natural gas—two of the main energy sources that power modern society. How have natural resources made this region a key player in world affairs?

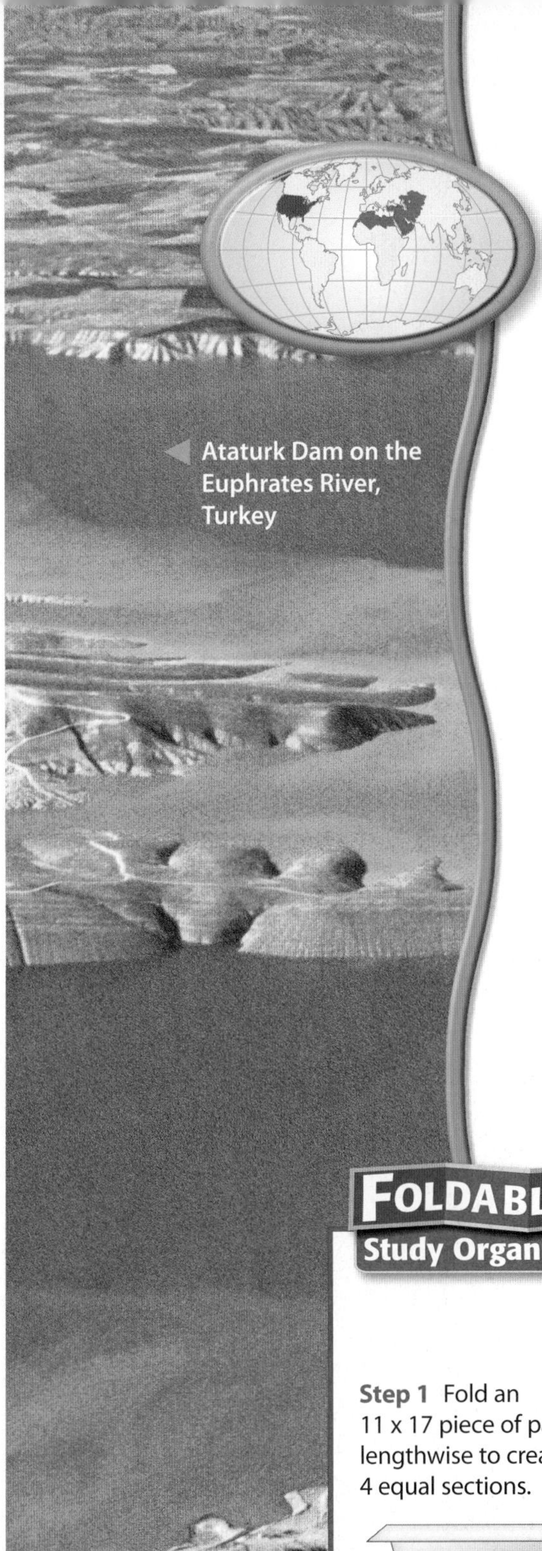

Ataturk Dam on the Euphrates River, Turkey

BIG Ideas

Chapter Audio

Section 1: Physical Features

BIG IDEA The physical environment affects how people live. For centuries, the people of North Africa, Southwest Asia, and Central Asia have adapted to survive in this dry region. During the past century, however, the increasing global need for petroleum and natural gas has brought new wealth and changing lifestyles to the area.

Section 2: Climate Regions

BIG IDEA Places reflect the relationship between humans and the physical environment. Many areas of North Africa, Southwest Asia, and Central Asia have harsh environments. As a result, people have settled in areas that can support large populations, such as river valleys. One of the most important challenges facing the region's people is how to manage water resources to meet their current needs while protecting supplies for the future.

FOLDABLES™ Study Organizer

Organizing Information Make this Foldable to help you organize information about the physical environment, including resources and climate, of North Africa, Southwest Asia, and Central Asia.

Step 1 Fold an 11 x 17 piece of paper lengthwise to create 4 equal sections.

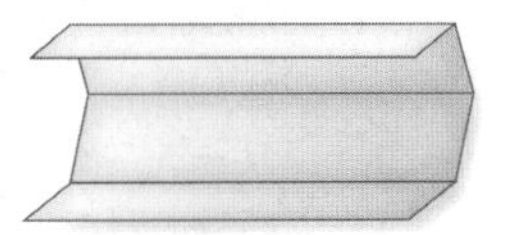

Step 2 Then fold it to form 5 columns.

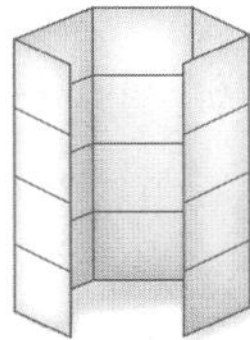

Step 3 Label your Foldable as shown.

Region	Landforms	Natural Resources	Environmental Issues/Water	Climates
North Africa				
Southwest Asia				
Central Asia				

Reading and Writing As you read the chapter, take notes in the correct place on the chart. Then use the notes on your Foldable to write a short essay describing how the region's physical landscapes and climates are related to growing environmental problems there.

Visit glencoe.com and enter QuickPass™ code EOW3109c16 for Chapter 16 resources.

Guide to Reading

BIG Idea

The physical environment affects how people live.

Content Vocabulary

- silt *(p. 444)*
- alluvial plain *(p. 444)*
- sedimentary rock *(p. 444)*
- phosphate *(p. 445)*
- poaching *(p. 445)*
- refinery *(p. 446)*

Academic Vocabulary

- intense *(p. 444)*
- expose *(p. 444)*

Reading Strategy

Analyzing Information Use a chart like the one below to list at least five bodies of water or landforms of the region and explain why each is important.

Physical Feature	Importance
1.	1.
2.	2.
3.	3.
4.	4.
5.	5.

Physical Features

Spotlight Video

Picture This The people of Turkey's Göreme Valley dwell in cliff-side apartments carved by nature. These mountain homes have storerooms on lower levels and family living quarters that include kitchens. Telephone-like devices allow family members to communicate from different floors. Wind and rain, as well as volcanoes and earthquakes, have shaped the rock and valleys of this land. Read on to find out about the landforms of North Africa, Southwest Asia, and Central Asia.

Sandstone dwellings in Turkey

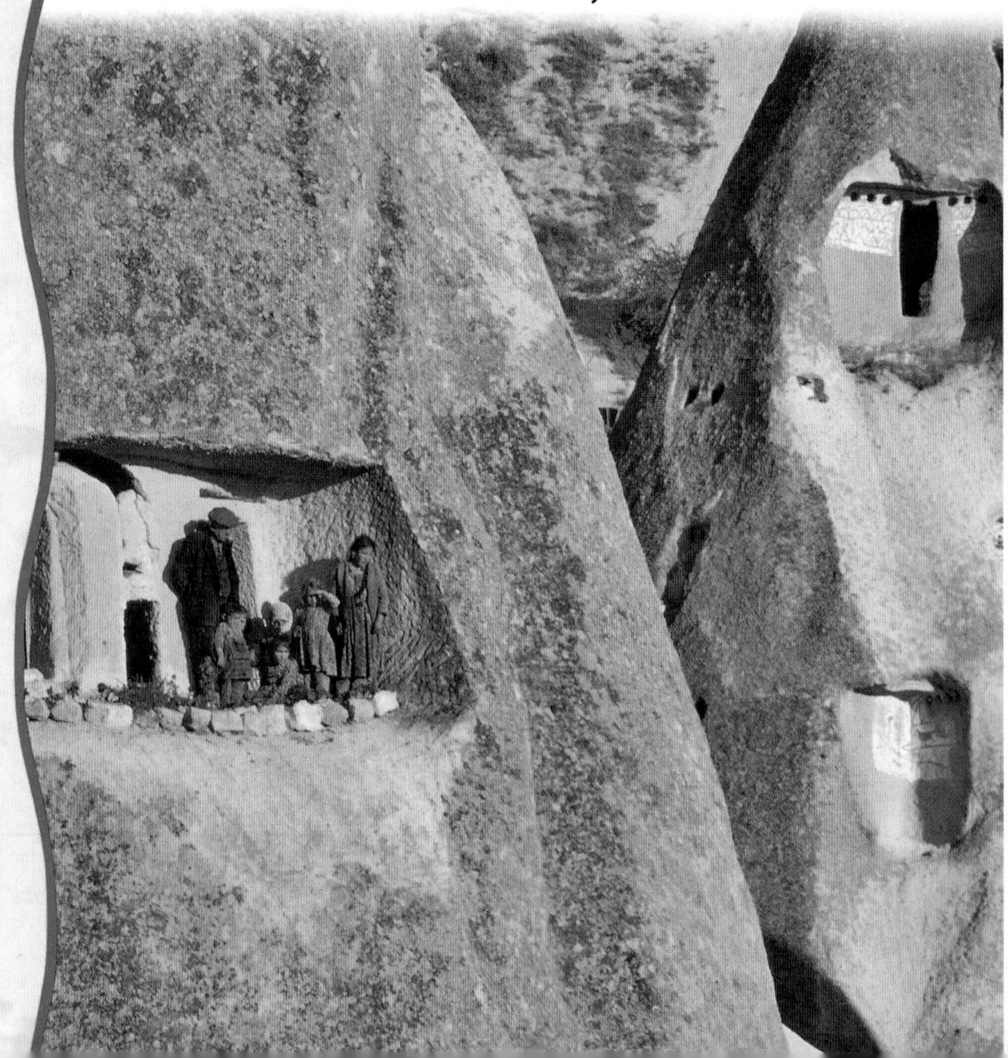

The Region's Landforms

Main Idea **This region includes a variety of landforms that affect how and where people live.**

Geography and You When you hear about floods, do you picture terrible damage and loss of life? Read to learn why people in ancient Egypt welcomed, rather than feared, river floods.

The region of North Africa, Southwest Asia, and Central Asia extends from the Atlantic coast of northwestern Africa to towering mountain ranges in the middle of Asia. Varied landforms and waterways lie in between.

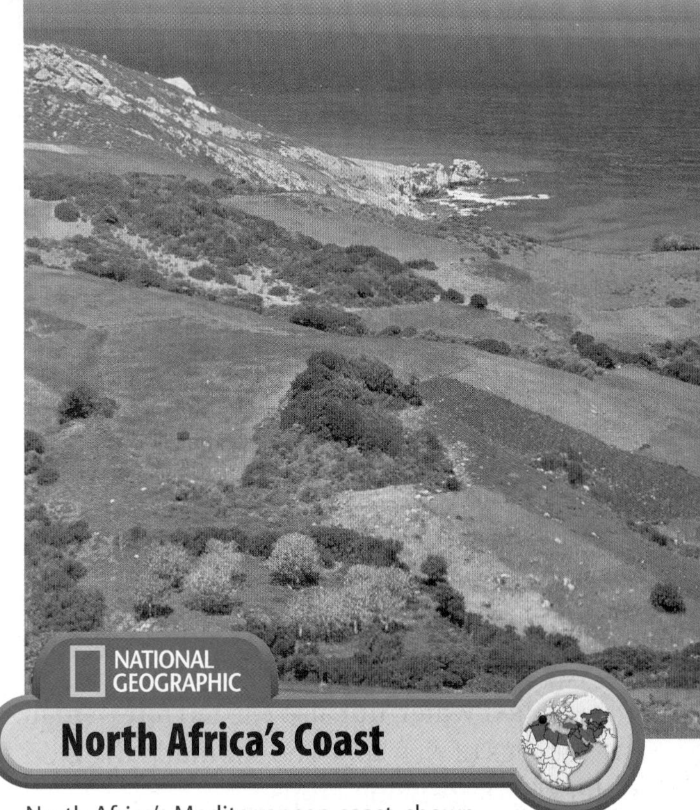

North Africa's Coast

North Africa's Mediterranean coast, shown here in Morocco, tends to be greener and support more agriculture than inland areas. ***Location*** **Why is the Strait of Gibraltar important?**

Seas and Waterways

Look at the physical map in the Regional Atlas to see the oceans and seas that surround North Africa, Southwest Asia, and Central Asia. These bodies of water have helped people trade more easily with the rest of Africa, Asia, and Europe. Four waterways control access to these seas. In the west, the Strait of Gibraltar—separating Africa and Europe—links the Mediterranean Sea with the Atlantic Ocean. In the north, the Dardanelles (DAHRD·uhn·EHLZ) Strait, the Sea of Marmara (MAHR·mah·RAH), and the Bosporus (BAHS·puh·ruhs) Strait together link the Mediterranean and Black Seas and separate Europe from Asia.

On North Africa's eastern edge, there is a human-made waterway called the Suez Canal. Ships use this canal to pass from the Mediterranean Sea to the Red Sea. North of the Arabian Peninsula, the Strait of Hormuz (HAWR·MUHZ) allows oil tankers to enter and leave the Persian Gulf.

Mountains, Plateaus, and Lowlands

All three areas within the region have somewhat similar landscapes. In North Africa, the Atlas and Ahaggar Mountains extend across much of the west. The rest of North Africa is covered by low plains and low-lying plateaus.

Like North Africa, Southwest Asia has a rugged landscape. Tectonic plate movements beneath the mountains still shake the ground and have caused deadly earthquakes. The Zagros Mountains stretch southeast through Turkey and Iran. The Hindu Kush range crosses Afghanistan. The Khyber (KY·buhr) Pass is a narrow gap, sometimes barely 50 feet (15 m) wide, between mountains in the Hindu Kush. It has been used as a trade route linking Southwest Asia to other parts of Asia.

Central Asia holds several lofty mountain ranges—the Pamirs and the Tian Shan. Lowlands here include a large area along the Caspian Sea and several desert areas.

Throughout North Africa, Southwest Asia, and Central Asia, coastal plains support agriculture.

Rivers

Because of the region's rugged and dry land, people have long settled in river valleys to take advantage of the rich soil. Early civilizations developed along Egypt's Nile River and Southwest Asia's Tigris and Euphrates Rivers.

The ancient Egyptians relied on the Nile's yearly flooding. The floods not only supplied water, but also carried **silt**—small particles of rich soil. The silt made the land fertile for growing crops.

Flooding of the Tigris and Euphrates Rivers, in a region called Mesopotamia, also led to a civilization based on farming. Mesopotamia was located on an **alluvial** (uh·LOO·vee·uhl) **plain,** an area of fertile soil left by river floods. Farmers built channels and ditches to bring water to their fields.

Reading Check **Identifying** Which two bodies of water does the Suez Canal connect?

Farming in the Nile Valley

NATIONAL GEOGRAPHIC

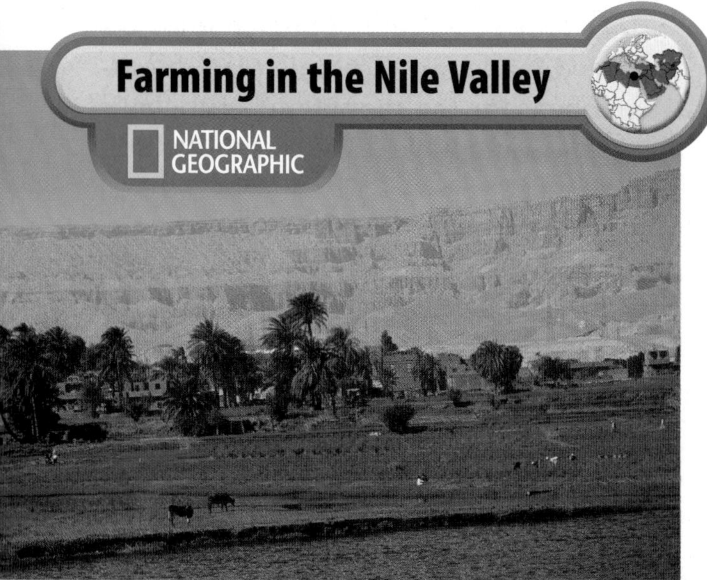

▲ Egyptian farmers today grow crops such as cotton and sugarcane along the banks of the Nile River.
Place **How did the Nile benefit early Egyptians?**

Natural Resources

Main Idea The land in this region is rich in energy resources.

Geography and You Have you ever depended on someone for something that you really needed? Read to find out why so much of the world is economically dependent on this region.

The region of North Africa, Southwest Asia, and Central Asia has many natural resources. Petroleum and natural gas, two of the region's most plentiful resources, are vital to economies throughout the world.

Oil and Gas

The largest reserves of petroleum and natural gas can be found in the area of the Persian Gulf. Oil is common here because the land is made up of **sedimentary rock.** This rock is created when layers of material are hardened by the **intense** weight of more materials piled above. Over millions of years, heat and pressure below the Earth's surface helped turn the remains of sea animals and plants into oil. The oil collected in the gaps between the rock.

Some of the region's countries have gained great wealth from selling oil. This wealth has been used to develop new industries and provide benefits to the region's people.

Oil wealth has brought challenges, too. Oil income has made some countries in the region very wealthy, while countries that are not rich in oil resources have remained poor. Also, because of television and the Internet, cultures in oil-rich countries have become more **exposed** to ideas from other parts of the world. As a result, conflicts sometimes develop in the region between people who support new ways and people who favor traditional customs and values.

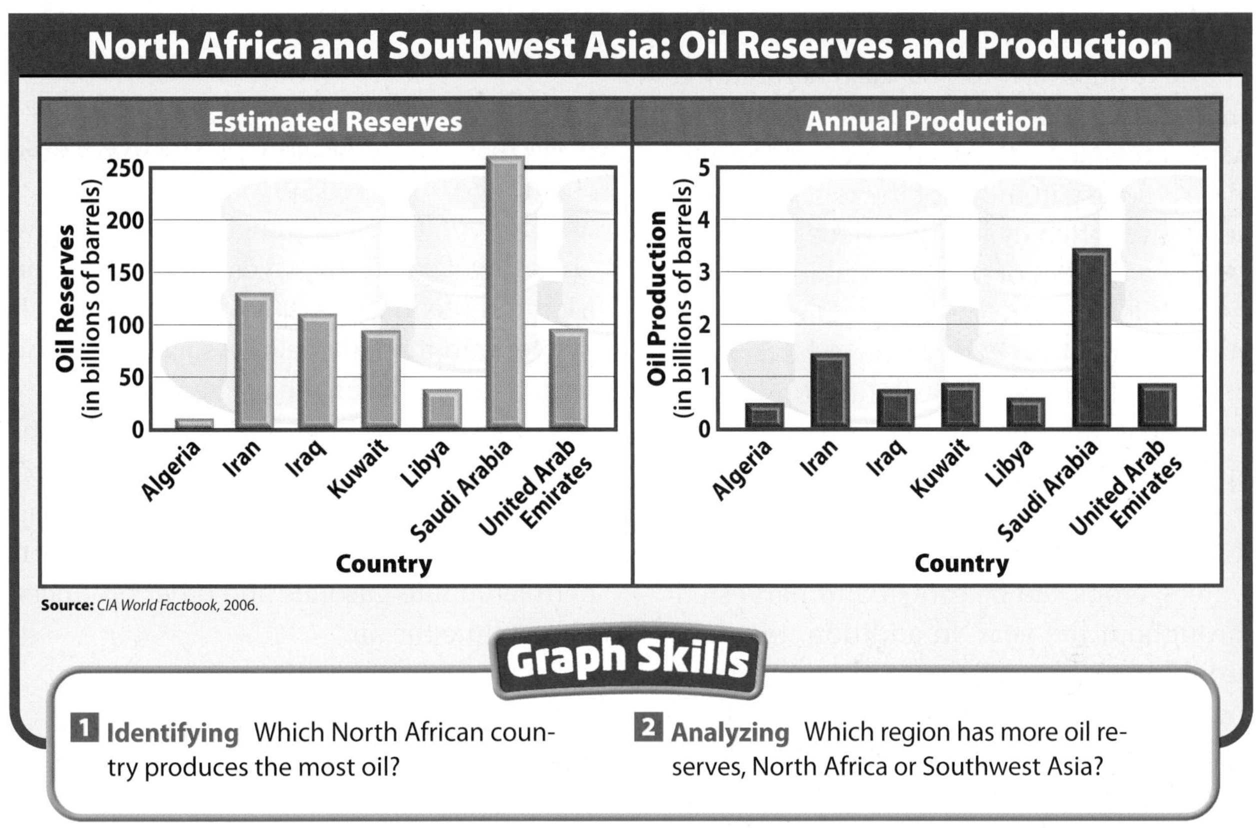

Graph Skills

1 Identifying Which North African country produces the most oil?

2 Analyzing Which region has more oil reserves, North Africa or Southwest Asia?

Other Resources

In addition to oil and natural gas, the region has other important resources. Both coal and iron ore are found in the region. Morocco and Kazakhstan have rich deposits of **phosphates,** which are mineral salts used to make fertilizer.

Few forests exist in this dry, rocky region. Only Lebanon has enough timber to support a lumber industry. Fish are an important resource in parts of the region, especially around Morocco.

Environmental Concerns—The Seas

Water is a scarce resource in North Africa, Southwest Asia, and Central Asia. While trying to improve their lives in the harsh environment, people have misused the area's waters.

The region's two inland seas, the Caspian (KAS·pee·uhn) and the Aral, have suffered greatly in recent decades. In the Caspian Sea, **poaching,** which is illegal fishing or hunting, has decreased the number of sturgeon, the fish whose eggs are used to make caviar. This drop has hurt the fishing industry in countries around the Caspian Sea.

The Aral Sea, which is shared by Kazakhstan and Uzbekistan, has also been badly damaged. During the 1960s, irrigation projects, mainly for cotton production, drained water from the two main rivers that feed the sea. Inadequate amounts of rainfall could not restore the sea's water level, and it began drying up. In addition, the sea's water became saltier and unfit for drinking. This change also harmed the sea's fish populations.

Recent efforts to save the Aral Sea have included building dams and dikes to help increase water levels, which, in turn, lowers the salt levels. Today the sea's water level is slowly beginning to rise and fish stocks are growing again.

Other Environmental Issues

The misuse of water has also harmed the land throughout North Africa, Southwest Asia, and Central Asia. Because the climate is so dry in much of the region, irrigation water often evaporates. This can leave behind a deposit of salt that makes the land less fertile. In some cases, the damage is so great that the land can no longer be farmed.

Dams built to control flooding have both benefited and harmed the environment. In 1968 Egypt's government opened the Aswān High Dam on the upper Nile River. By controlling the river's floodwaters, crops can be grown and harvested throughout the year. In addition, the dam provides electric power to Egypt's growing cities and industries.

The Aswān High Dam also has had negative effects on the environment. The dam has blocked the flow of silt down the river. Thus, the Nile cannot enrich the soil as it has in the past. Farmers have to use expensive fertilizers to make the land more productive. These chemicals run off the land back into the river and pollute it. In addition, the reduced flow of freshwater into the river has allowed the Mediterranean Sea's salty waters to reach deeper into the Nile delta. As a result, some farmlands in this area have been ruined.

Air pollution also is a growing problem. A large number of cars in the region are older, and they release more polluting fumes. **Refineries,** the facilities that turn petroleum into gasoline and other products, also pollute the air.

Reading Check **Explaining** What are some of the threats to the region's water supplies?

Social Studies ONLINE
Study Central™ To review this section, go to glencoe.com.

Vocabulary

1. **Explain** the meaning of the following terms by using each in a sentence.
   - **a.** silt
   - **b.** alluvial plain
   - **c.** sedimentary rock
   - **d.** phosphate
   - **e.** poaching
   - **f.** refinery

Main Ideas

2. **Explaining** How did the yearly flooding of several of the region's rivers aid early peoples?
3. **Identifying** Use a diagram like the one below to identify resources found in North Africa, Southwest Asia, and Central Asia.

Critical Thinking

4. **Comparing and Contrasting** How are landforms similar and different in the areas that make up North Africa, Southwest Asia, and Central Asia?
5. BIG Idea Describe how humans have affected the region's scarce water resources.
6. **Challenge** Based on the benefits and the costs, do you think the construction of the Aswān High Dam was necessary?

Writing About Geography

7. **Expository Writing** Considering the region's landforms, waterways, and resources, write a paragraph describing the economic challenges or advantages that a country in the region might have.

Guide to Reading

BIG Idea

Places reflect the relationship between humans and the physical environment.

Content Vocabulary

- wadi *(p. 448)*
- erg *(p. 448)*
- oasis *(p. 448)*
- steppe *(p. 448)*
- nomad *(p. 449)*
- dry farming *(p. 449)*
- aquifer *(p. 450)*
- rationing *(p. 450)*
- desalinization *(p. 450)*

Academic Vocabulary

- sparse *(p. 450)*
- adequate *(p. 450)*

Reading Strategy

Analyzing Information Use a web diagram like the one below to identify the climates in the region. Within each circle, list ways in which that particular climate affects people's lives.

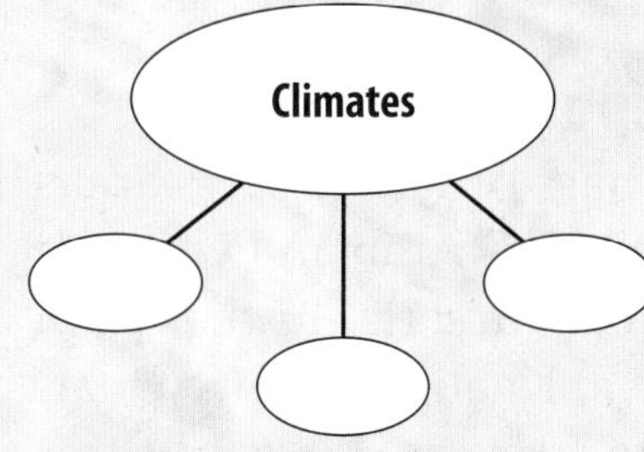

Climate Regions

 Section Audio  Spotlight Video

Picture This A shadoof looks like an abstract sculpture, but it serves a vital purpose: watering dry land areas in parts of North Africa. A bucket is attached by a rope at one end of the shadoof's beam and balanced by a weight at the other end. A person pulls on the rope, which lowers the bucket into a well, filling it with water. Releasing the rope raises the bucket, which can then be emptied into a ditch to water the soil. Read the section to learn more about how the region's dry climates affect those who live there.

▼ A shadoof in Algeria

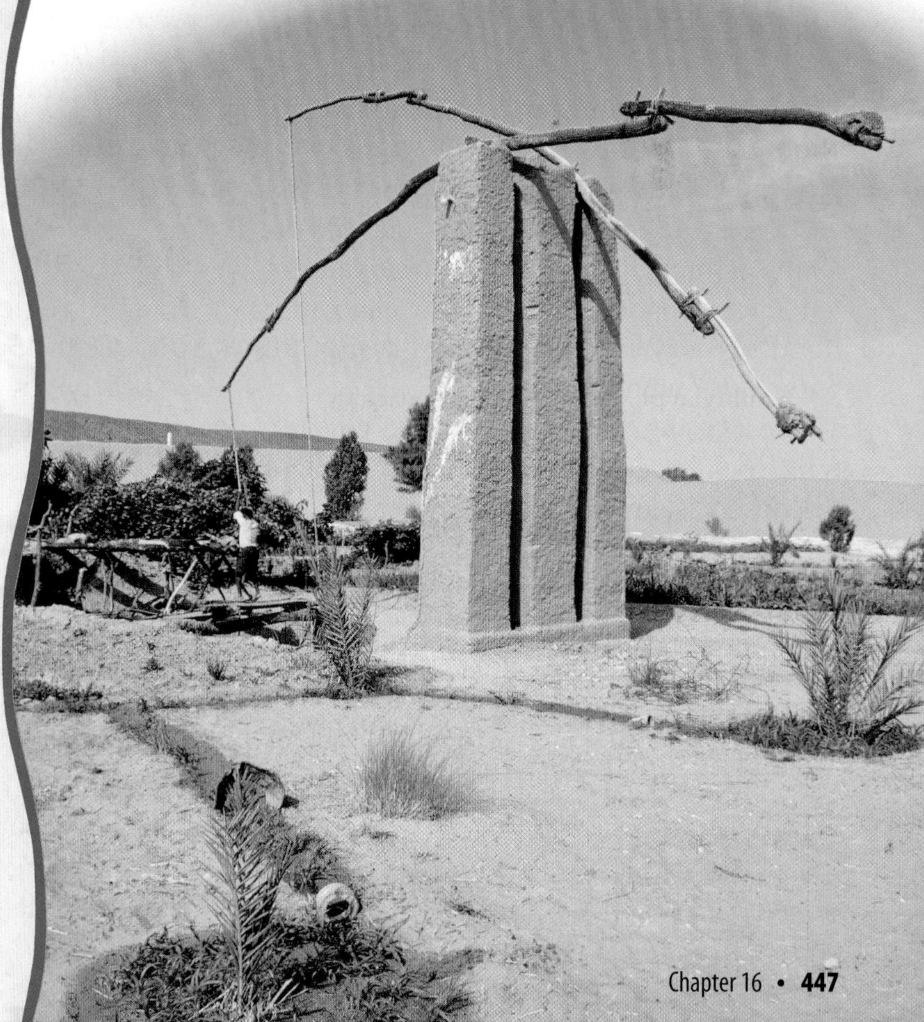

A Dry Region

Main Idea **Large areas of desert greatly affect life in the region.**

Geography and You What would it be like to live through a long period without rain? Read to find out how a generally dry climate affects vegetation and human activities.

Dry continental air masses warmed by the sun blow over much of North Africa, Southwest Asia, and Central Asia. As a result, much of the land is desert with a dry, hot climate.

The Sahara

The vast Sahara, the world's largest desert, covers much of North Africa. Summer temperatures can reach scorching highs. The highest temperature ever recorded, 136°F (58°C), was measured at Al-'Azīzīyah (AL·a·ZEE·zee·yuh), in Libya. Winter temperatures are cooler, averaging about 55°F (13°C).

Only about 3 inches (8 cm) of rain fall each year in the Sahara. Rain usually falls in the winter months, but occasional violent thunderstorms in summer can cause flooding. Dry riverbeds called **wadis** fill with water when it rains.

Most of the Sahara is dry land covered with rock or gravel. About 20 percent of the desert is covered by large areas of soft sands, which are called **ergs.** The Sahara also contains **oases** where the land is fertile as a result of water from a spring or well. Villages, towns, and cities have grown around many Saharan oases.

Student Web Activity Visit glencoe.com and complete the Chapter 16 Web Activity about the Sahara.

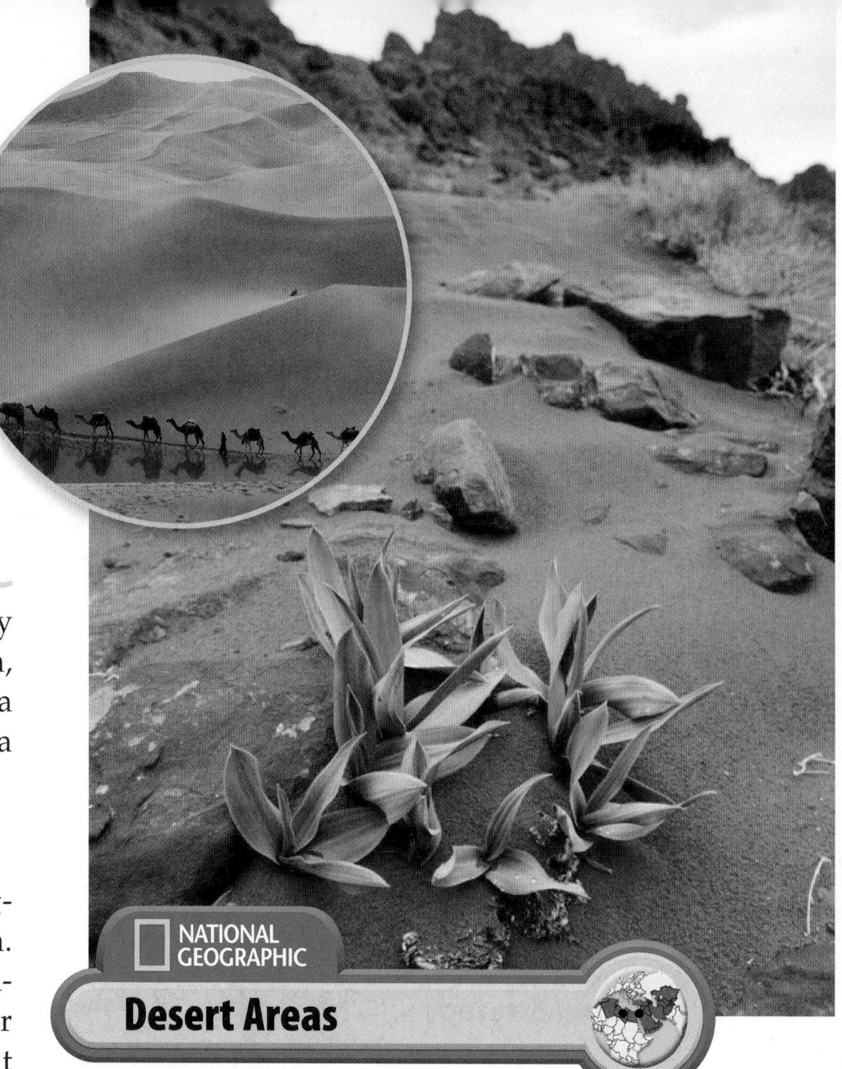

Desert Areas

A camel caravan crosses over sand dunes in the Sahara (inset). A wadi, or dry riverbed, cuts through a desert area in the Arabian Peninsula.
Place **What is the Rub' al Khali?**

Desert and Steppe Areas

The region has other desert areas besides the Sahara. The Arabian Peninsula is nearly covered by deserts. In the south lies the vast Rub' al Khali, or Empty Quarter. About the size of Texas, this sandy area averages only about 4 inches (10 cm) of rainfall per year.

In Central Asia, rain shadow areas created by high peaks along with dry continental winds have formed large deserts—the Kara-Kum and the Kyzyl Kum. Both deserts have hot summers but very cold winters. That is because these areas are in the middle latitudes.

Bordering the region's deserts are dry, treeless, but grassy plains called **steppes.**

Steppes are found in areas north of the Sahara, Turkey, and to the east in Central Asia. Steppe areas receive more rainfall—between 4 and 16 inches (10 and 41 cm) per year—than do deserts. Some people on the steppes live as **nomads,** moving across steppe areas to find food and water for their herds. Others practice **dry farming,** a method in which land is left unplanted every few years so that it can store moisture.

Other Climate Areas

Coastal areas in North Africa, the eastern Mediterranean, and Turkey have Mediterranean climates. These warm areas receive enough rain to support agriculture and therefore have more people than other parts of the region.

A small portion of Central Asia has a humid subtropical climate, with hot summers, mild winters, and plentiful rainfall. Highland climates are found in the region's mountainous areas. Because farming in the mountains is difficult, the few people who live in the highlands usually herd animals.

Reading Check **Contrasting** How do the region's desert and steppe areas differ?

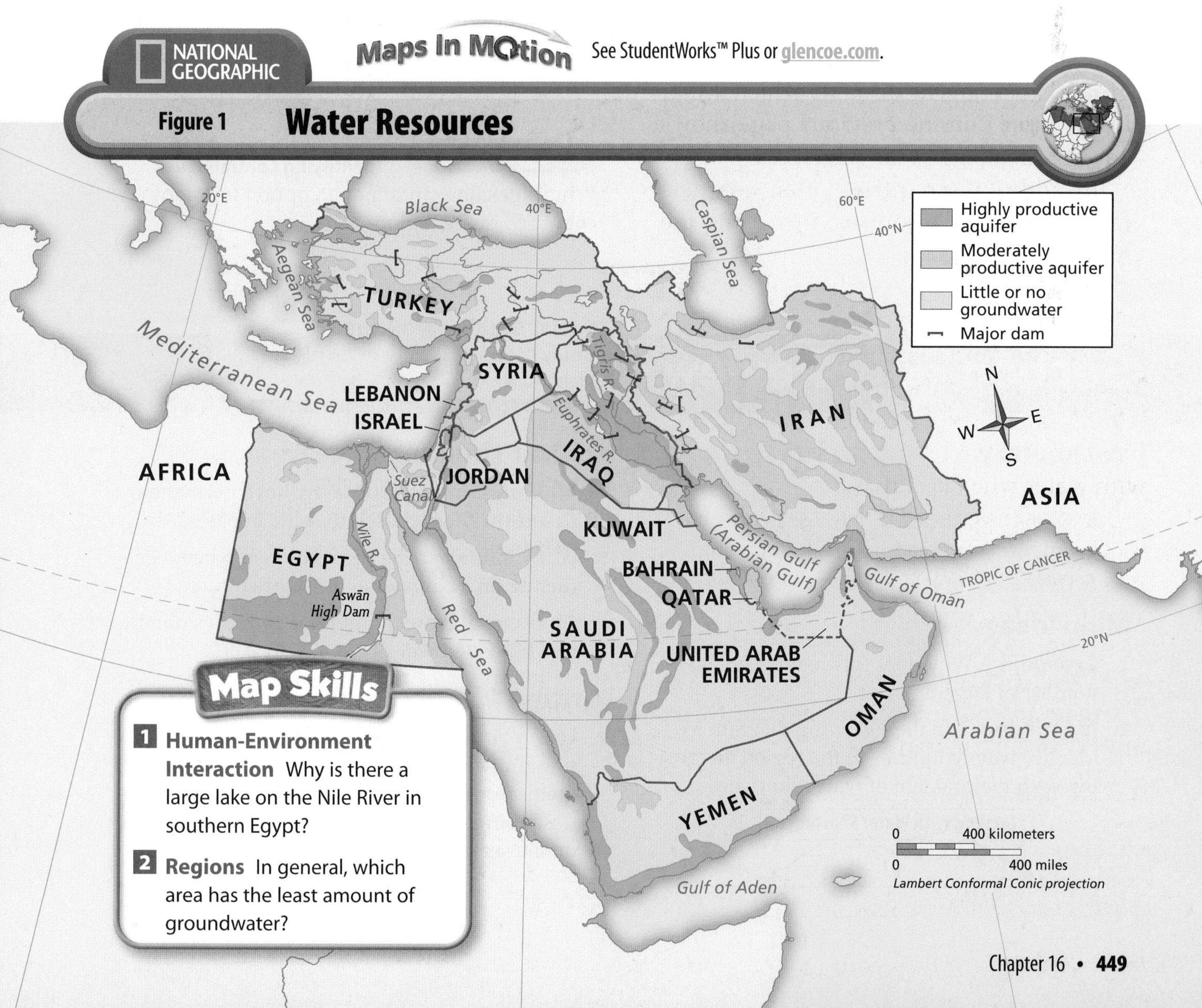

Map Skills

1. **Human-Environment Interaction** Why is there a large lake on the Nile River in southern Egypt?
2. **Regions** In general, which area has the least amount of groundwater?

The Need for Water

Main Idea The lack of water is a growing problem in this region.

Geography and You How much water do you use each day? How might your life be different if the amount of water you used daily was limited? Read to learn how people try to make the best use of limited water resources.

Rainfall is **sparse** over much of the region. Also, high temperatures cause surface water to evaporate rapidly. As a result, the growing population does not have **adequate** water to meet its needs. A large amount of water is used to irrigate dry farmland. Some countries, such as Libya, now draw water from **aquifers,** or underground rock layers through which water flows. **Figure 1** on the previous page shows aquifers in a part of the region.

In addition, countries often compete for scarce water resources, increasing the chance of conflict. For example, dams that Turkey is building on the Tigris and Euphrates Rivers redirect water that in the past would have flowed to Syria and Iraq. Some governments, such as those of Jordan and Syria, are dealing with water shortages by **rationing.** Rationing means that a resource is made available to people in only limited amounts.

Another approach to managing water use is **desalinization.** This process treats seawater to remove salts and minerals and make it drinkable. Oil-rich countries have built desalinization plants, making Southwest Asia the world's leader in creating usable water from seawater. Desalinization is costly, however, so poor countries are not able to use this process and will continue to face water shortages.

Identifying Central Issues Why is desalinization not an option for many countries in the region?

Section 2 Review

Social Studies ONLINE
Study Central™ To review this section, go to glencoe.com.

Vocabulary

1. **Explain** the significance of:
   - **a.** wadi
   - **b.** erg
   - **c.** oasis
   - **d.** steppe
   - **e.** nomad
   - **f.** dry farming
   - **g.** aquifer
   - **h.** rationing
   - **i.** desalinization

Main Ideas

2. **Explaining** Why does this region contain a number of large deserts?
3. **Identifying** Use the following diagram to identify ways countries in the region are dealing with the problem of water shortages.

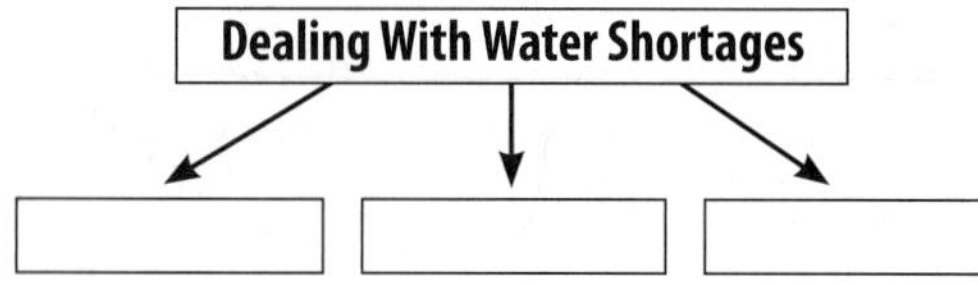

Critical Thinking

4. **Making Connections** Why have permanent desert dwellers chosen to settle in oases?
5. **BIG Idea** Describe the relationship between the region's climates and its agriculture.
6. **Challenge** How might poor countries obtain much-needed water?

Writing About Geography

7. **Using Your FOLDABLES** Use your Foldable to write several journal entries for a traveler going from North Africa to Central Asia, describing the climates and vegetation that he or she sees along the way.

Visual Summary

Landforms

- The seas bordering the region provide trade routes that connect Europe, Asia, and Africa.
- Physical features include mountains, low-lying plains and plateaus, and coastal plains.
- Early civilizations based on agriculture developed along several of the region's rivers.
- Tectonic plate movements can cause earthquakes in some parts of the region.

Earthquake survivor in Iran

Natural Resources

- The region holds much of the world's oil and natural gas reserves.
- Coal, iron ore, and phosphates are among the other valuable mineral resources found in the region.

Irrigation project, Saudi Arabia

Environment

- Human activities have damaged the Caspian and Aral Seas in recent decades.
- Dams and too much irrigation have made areas of land less useful for farming.
- Air pollution is a growing problem.

Fisher in Turkey

Climates

- Vast deserts with dry climates cover much of the region.
- The Sahara, the world's largest desert, covers almost all of North Africa.
- Most people live in steppe areas and coastal plains that receive adequate rainfall.
- The region's generally dry climate and a growing population have led to a water shortage.

Nomad family, Kazakhstan

Study anywhere, anytime! Download quizzes and flash cards to your PDA from **glencoe.com**.

CHAPTER 16 ASSESSMENT

STANDARDIZED TEST PRACTICE

On an extended-response question, read carefully. Think about what the question is asking. Are you being asked to explain something or to demonstrate what you know about a subject? Be sure you understand what you are being asked before you begin writing.

Reviewing Vocabulary

Directions: Choose the word(s) that best completes the sentence.

1. Mesopotamia was located on an area of fertile soil left by river floods called a(n) _______.

A aquifer
B oasis
C alluvial plain
D wadi

2. Oil is common in this region because the land is made up of _______, or layers of materials hardened by the intense weight of more materials piled above.

A wadi
B sedimentary rock
C silt
D steppes

3. _______ has decreased the number of sturgeon in the Caspian Sea.

A Poaching
B Rationing
C Selling oil
D Building canals

4. In this region of little rain, dry riverbeds called _______ fill with water when it rains.

A wadis
B ergs
C silt
D steppes

5. Bordering this region's deserts are dry, treeless, but grassy plains called _______.

A wadis
B oases
C ergs
D steppes

Reviewing Main Ideas

Directions: Choose the best answer for each question.

Section 1 *(pp. 442–446)*

6. The Aral Sea has been drying up over the past several years because

A the area has had an extended drought.
B irrigation projects took too much water from the rivers that feed the sea.
C people in cities use too much water.
D earthquakes damaged the sea floor allowing the water to leak out.

7. Two important waterways in the region include the Suez Canal and the Dardanelles. What is the main difference between the Suez and the Dardanelles?

A Ships can pass through the Dardanelles but not the Suez.
B The Dardanelles is human-made while the Suez is natural.
C The Suez is human-made while the Dardanelles is natural.
D Ships can pass through the Suez but not the Dardanelles.

Section 2 *(pp. 447–450)*

8. The world's largest desert, the Sahara, is located in _______.

A Central Asia
B Southwest Asia
C both Central and Southwest Asia
D North Africa

9. One of the most likely reasons for conflict in this region is competition over _______.

A water
B religion
C trade routes
D farming techniques

GO ON

Critical Thinking

Directions: Use the chart below to choose the best answer for questions 10 and 11.

Average Annual Precipitation		
City	Subregion	Precipitation In. (cm)
Algiers, Algeria	North Africa	30 (76)
Almaty, Kazakhstan	Central Asia	24 (60)
Benghazi, Libya	North Africa	11 (27)
Cairo, Egypt	North Africa	1 (3)
Damascus, Syria	Southwest Asia	3 (8)
Baghdad, Iraq	Southwest Asia	6 (14)
Riyadh, Saudi Arabia	Southwest Asia	3 (8)
Tangier, Morocco	North Africa	36 (90)
Tashkent, Uzbekistan	Central Asia	15 (37)
Tel Aviv, Israel	Southwest Asia	21 (54)

Source: National Oceanic and Atmospheric Administration.

10. If desert climates have 4 inches (10 cm) or less of precipitation a year, which city in North Africa has a desert climate?

A Benghazi
B Cairo
C Damascus
D Riyadh

11. If precipitation averages 4 to 16 inches (10 to 41 cm) in a steppe climate, which city has a steppe climate zone?

A Algiers
B Damascus
C Tangier
D Tashkent

Document-Based Questions

Directions: Analyze the document and answer the short-answer questions that follow.

> *The virtual disappearance of the Aral Sea, the great landlocked lake of Central Asia, has been condemned as one of the most catastrophic recent examples of human impact on the natural environment. . . .*
>
> *A new archaeological survey has shown, however, that . . . ancient shrines and settlements have been revealed by the lowered sea level. A medieval mausoleum at Kerderi in the northern part of the sea, which belongs to Kazakhstan, has been exposed recently, while its adjacent settlement is still under shallow water: in the 14th century the Aral Sea's level was even lower and the sea smaller than it is now.*
>
> *"The Aral Sea is an excellent location for tracing human reactions to past climate changes," Nikolaus Boroffka and his colleagues say in* Geoarchaeology. *"In this climatically sensitive area, which alternates between semi-arid and arid conditions, human influence can be traced back to the first millennium BC."*
>
> —Norman Hammond,
> "Aral Sea's Revealing Retreat," *Times* (London)

12. What recent discovery did archaeologists working in the Aral Sea make? What was their evidence?

13. How has the ecological tragedy of the Aral Sea been helpful to scientists and archaeologists?

Extended Response

14. If you were a world leader concerned about the environment, how would you try to persuade the Aral Sea countries to save the sea?

STOP

For additional test practice, use Self-Check Quizzes—Chapter 16 at glencoe.com.

Need Extra Help?

If you missed question. . .	1	2	3	4	5	6	7	8	9	10	11	12	13	14
Go to page. . .	444	444	445	448	448	445	443	448	450	450	450	445	445	445

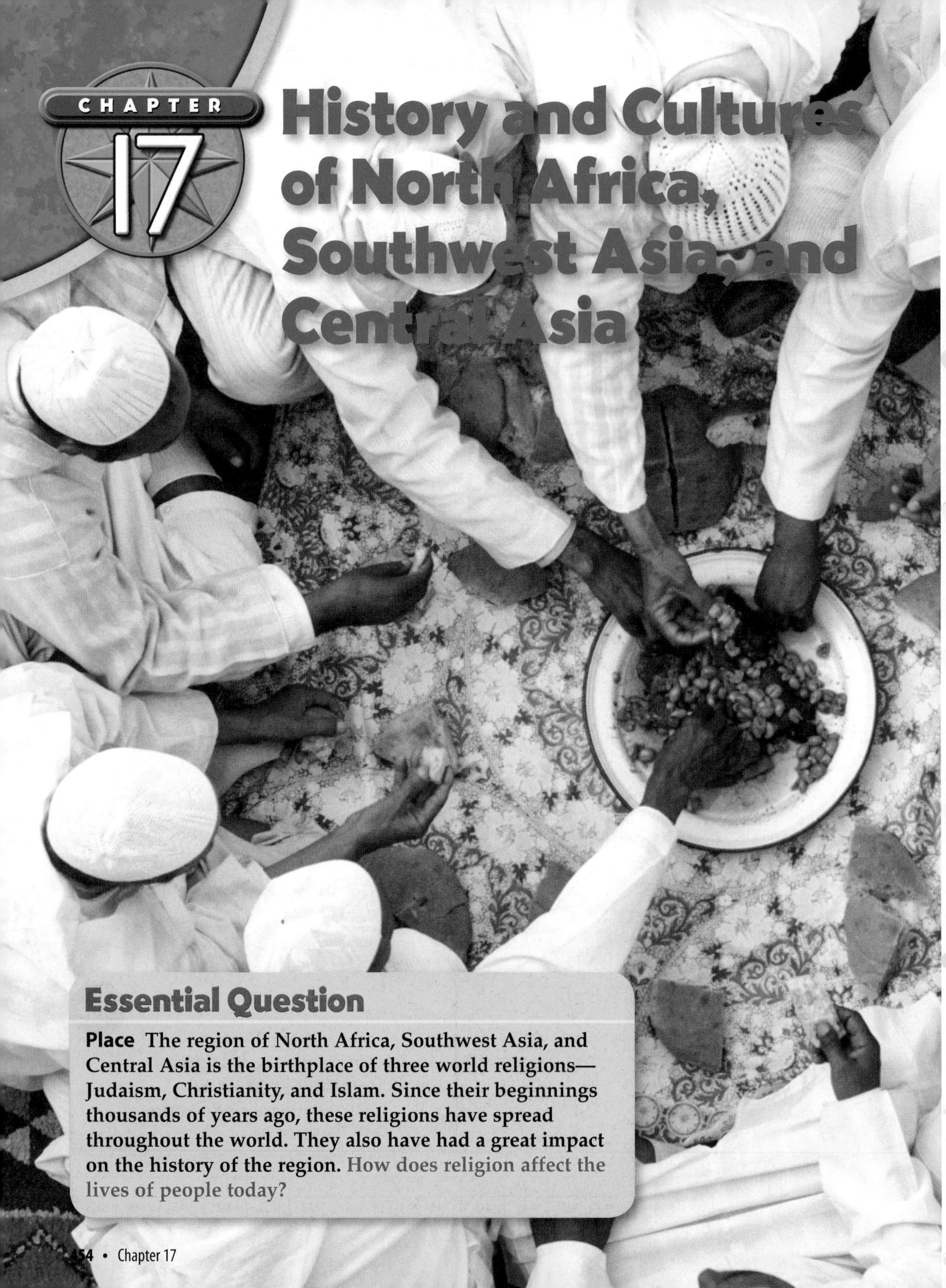

CHAPTER 17

History and Cultures of North Africa, Southwest Asia, and Central Asia

Essential Question

Place The region of North Africa, Southwest Asia, and Central Asia is the birthplace of three world religions—Judaism, Christianity, and Islam. Since their beginnings thousands of years ago, these religions have spread throughout the world. They also have had a great impact on the history of the region. How does religion affect the lives of people today?

Dinner in Marrakesh, Morocco

Section 1: History and Religion

BIG IDEA The characteristics and movement of people affect physical and human systems. The people of ancient Egypt and Mesopotamia built civilizations in fertile river valleys. They made many advances that spread to neighboring areas. Judaism, Christianity, and Islam have greatly influenced culture and politics over the centuries. In recent times, political unrest and conflict have troubled the region.

Section 2: Cultures and Lifestyles

BIG IDEA Culture groups shape human systems. Throughout North Africa, Southwest Asia, and Central Asia, religious beliefs and traditions have influenced the language, arts, and daily lives of people.

FOLDABLES™ Study Organizer

Summarizing Information Make this Foldable to help you summarize information about the history, people, culture, and daily life of North Africa, Southwest Asia, and Central Asia.

Step 1 Fold a paper in half lengthwise leaving a ½-inch tab on one edge.

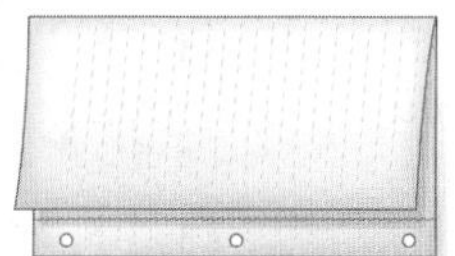

Step 2 Cut the top flap to make five equal tabs.

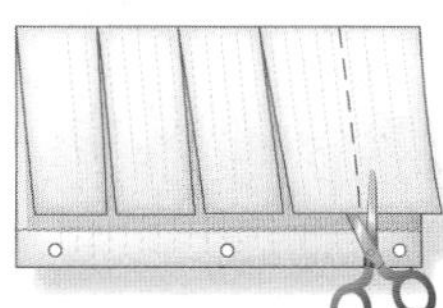

Step 3 Label the tabs as shown.

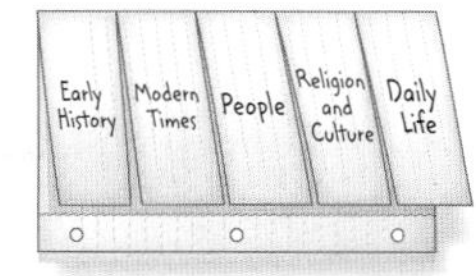

Reading and Writing Use the notes in your Foldable to write an essay describing similarities and differences in the history and cultures of the three subregions in this unit.

Visit glencoe.com and enter ***QuickPass***™ code EOW3109c17 for Chapter 17 resources.

History and Religion

 Section Audio  Spotlight Video

Guide to Reading

BIG Idea

The characteristics and movement of people affect physical and human systems.

Content Vocabulary

- irrigation *(p. 457)*
- city-state *(p. 457)*
- polytheism *(p. 457)*
- theocracy *(p. 457)*
- cuneiform *(p. 457)*
- pharaoh *(p. 458)*
- hieroglyphics *(p. 458)*
- monotheism *(p. 459)*
- covenant *(p. 459)*
- prophet *(p. 459)*
- caliph *(p. 460)*
- terrorism *(p. 464)*

Academic Vocabulary

- legal *(p. 458)*
- rely *(p. 458)*

Reading Strategy

Sequencing Use a flowchart like the one below to list key events that have occurred in Southwest Asia since World War I.

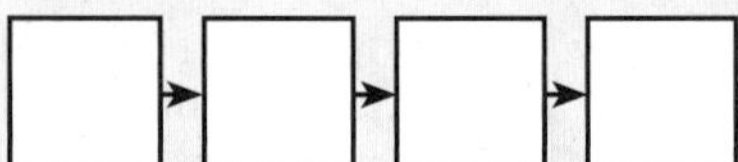

Picture This This giant rock tells a story. The words and pictures that are carved into the stone describe a great famine on the Egyptian island of Sehel. Stone markers, called stelae (STEE·lee), were often used in ancient Egypt to honor the dead or to remember special events. Egypt is one of the world's earliest civilizations and is known for its achievements in language, arts, and trade. To learn more about Egypt's history, read Section 1.

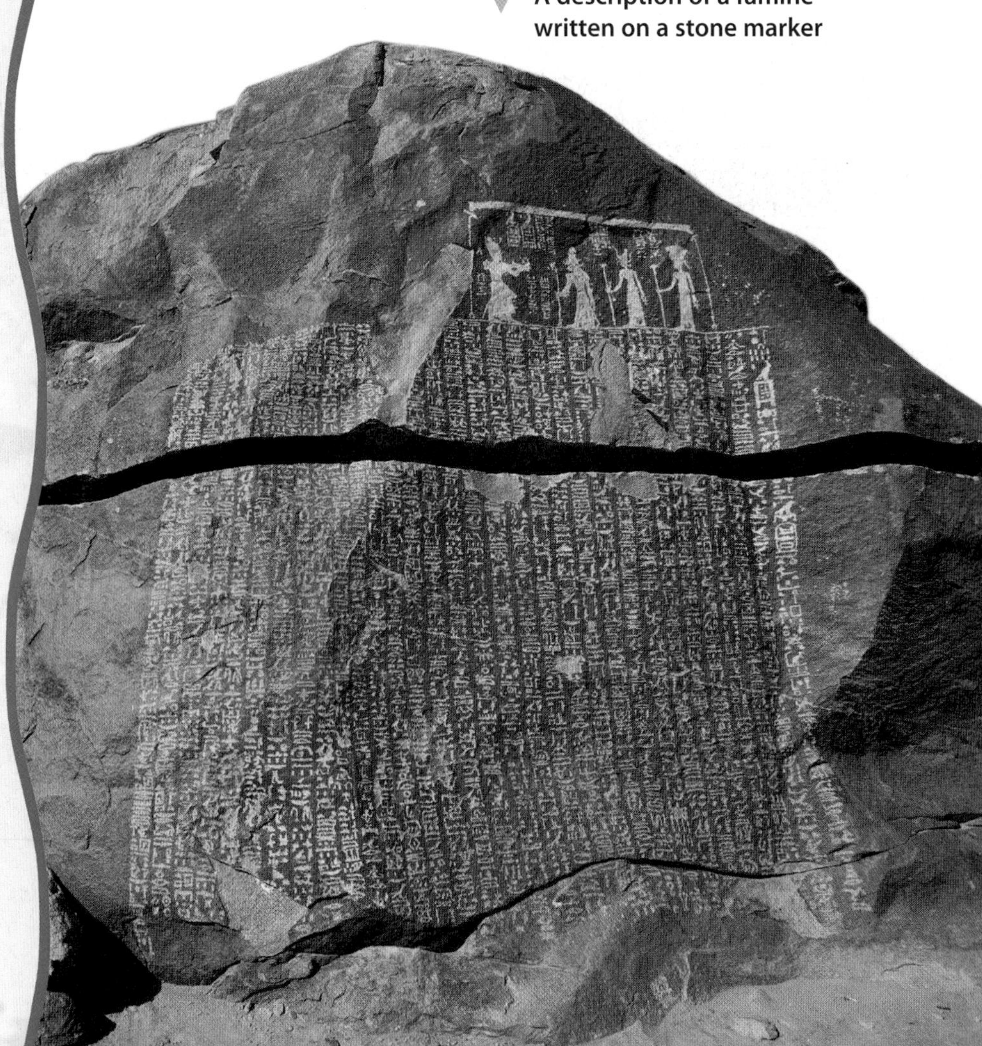

A description of a famine written on a stone marker

Early Civilizations

Main Idea **The early civilizations of Mesopotamia and Egypt had a great impact on later civilizations.**

Geography and You What do you view as the greatest human achievement? Sending people to the moon, perhaps, or inventing the computer? Read to learn about the accomplishments of two early civilizations.

Two of the world's oldest civilizations arose in Southwest Asia and North Africa about 5,000 years ago. Egypt, in North Africa, developed along the banks of the Nile River. The other civilization, Mesopotamia (MEH·suh·puh·TAY·mee·uh) in Southwest Asia, arose on a flat plain between the Tigris and Euphrates Rivers.

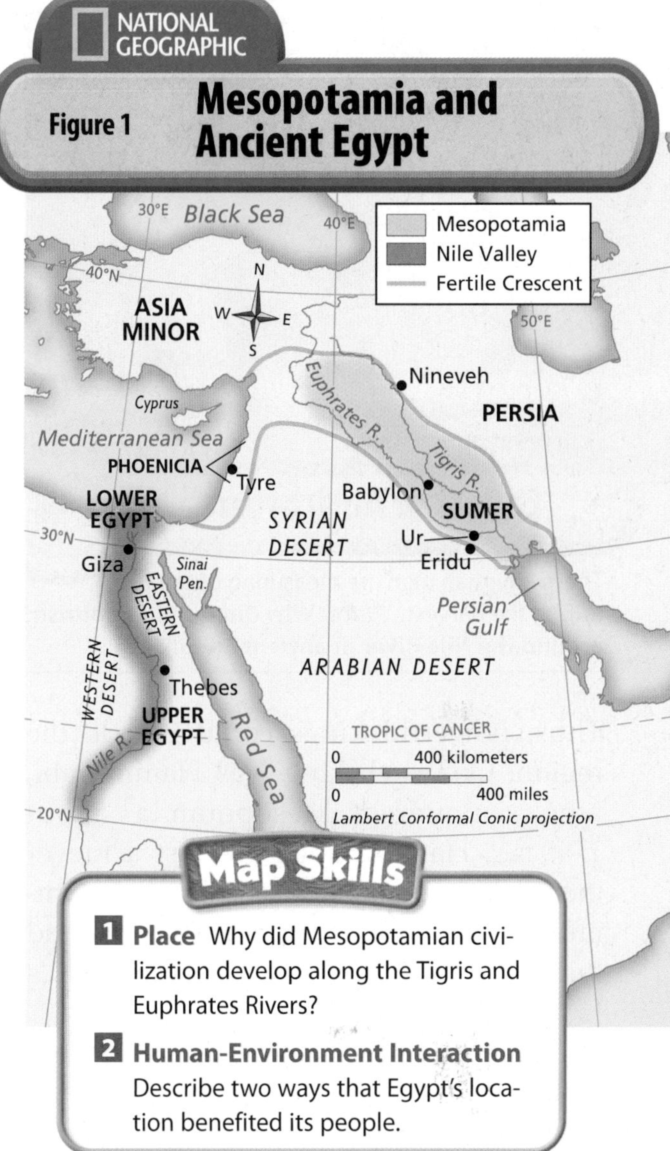

Figure 1 Mesopotamia and Ancient Egypt

Map Skills

1 Place Why did Mesopotamian civilization develop along the Tigris and Euphrates Rivers?

2 Human-Environment Interaction Describe two ways that Egypt's location benefited its people.

Mesopotamia

As **Figure 1** shows, Mesopotamia—located in present-day Iraq—lay in the Fertile Crescent, a strip of land that curves from the Mediterranean Sea to the Persian Gulf. Around 4000 B.C., people began settling along the Tigris and Euphrates Rivers. There they farmed the fertile soil left behind by yearly floods. To control flooding and carry water to their fields, they built walls, waterways, and ditches. This method of watering crops is called **irrigation.** Irrigation allowed the people to grow a larger, more stable supply of food. More food, in turn, supported a larger population. By 3000 B.C., cities had developed in southern Mesopotamia in a region known as Sumer (SOO·muhr).

Each Sumerian city and the land around it formed its own government and came to be called a **city-state.** At the center of each city was a large, steplike temple dedicated to the city's chief god. Mesopotamia's religion was based on **polytheism,** or the worship of many gods and goddesses. At first, each city-state was a **theocracy,** or a government controlled by religious leaders. Later, military leaders took power.

The Sumerians developed a number of remarkable inventions. They created one of the first calendars and were the first people known to use the wheel and the plow. They also developed **cuneiform,** which was an early form of writing. Cuneiform consists of wedge-shaped markings made with sharp reeds on clay tablets.

Mesopotamia faced certain challenges, however. Floods could damage crops and homes. In addition, the region had few natural barriers to prevent conquests.

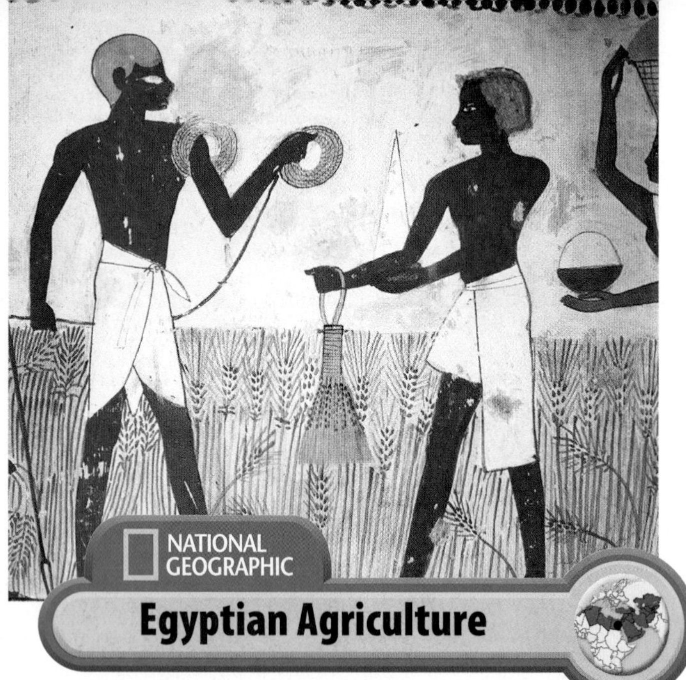

Egyptian Agriculture

These Egyptian men are measuring out grain during the harvest. ***Place*** **Why did farming flourish around the Nile River in ancient Egypt?**

Thus, powerful kings could invade the region. One such king was Hammurabi, who conquered Mesopotamia about 1790 B.C. Hammurabi's greatest achievement was a code of law. The Code of Hammurabi was meant to protect people and their property. It became one of the world's first written **legal,** or law, codes.

Ancient Egypt

Around 5000 B.C., farm villages began to develop along the Nile River in northeastern Africa. The ancient Egyptians **relied,** or depended, on the Nile's annual floods to bring water and enrich the soil. As in Mesopotamia, increased food supplies supported more people. For centuries, the Egyptians farmed and were protected from invaders by desert, the sea, and the Nile's waterfalls. Sometime around 3100 B.C., Egypt became a united kingdom under a single ruler.

Ancient Egyptians were keenly aware of social status, or the positions groups of people held in society. At the top were rulers, nobles, and priests. The middle group was made up of business- and craftspeople. At the bottom were most Egyptians—farmers and unskilled workers.

Religion was at the center of Egyptian society. Egyptians worshiped many gods and goddesses. Like Mesopotamia, Egypt was a theocracy. Egyptian rulers were called **pharaohs,** and the Egyptians believed that they were gods as well as rulers. The pharaohs owned the land and ordered thousands of people to build temples and tombs. The largest tombs, the pyramids, belonged to the pharaohs. Later, pharaohs were buried in tombs built into cliffs.

The Egyptians also developed a system of writing called **hieroglyphics,** which used pictures for words or sounds. The Egyptians carved and painted these symbols on the walls of the magnificent stone temples they built to honor their gods.

Connecting Ancient Lands

Trade was very important to the people of Mesopotamia. They traded with people from the eastern Mediterranean region to India. Through trade and conquest, the achievements and ideas of Mesopotamia spread to other lands. Some still shape life today. For example, the 60-second minute, 60-minute hour, 24-hour day, and 12-month year that we use today were developed by the Sumerians.

Trade helped spread Egyptian culture to other lands as well. The art of the eastern Mediterranean was influenced by Egyptian art. Egyptian religion, political ideas, and writing spread to other parts of Africa.

One of the greatest trading empires of the ancient world developed in the land of Phoenicia. Around 1000 B.C., the Phoenicians—who lived in what is now Lebanon—engaged in trade all across the Mediterranean Sea. They traveled and traded as far west as present-day Spain.

Reading Check **Explaining** How have Mesopotamian achievements influenced life today?

Three World Religions

Main Idea **Three major world religions began in Southwest Asia.**

Geography and You Is there a synagogue, a church, or a mosque in your community? These places of worship represent three major religions: Judaism, Christianity, and Islam. Read to find out about these three religions that began in Southwest Asia.

Judaism, Christianity, and Islam have become major world faiths. All three religions are examples of **monotheism,** or the belief in one God.

Judaism

The oldest of the three religions is Judaism. It was first practiced by a small group of people in Southwest Asia called the Israelites. The followers of Judaism today are known as Jews. We know about the early history of the Jewish people and their religion from their holy book—the Tanakh, or the Hebrew Bible.

According to Jewish belief, the Jews are descended from Abraham, a herder who lived in Mesopotamia about 1800 B.C. The Tanakh states that God made a **covenant,** or agreement, with Abraham. If Abraham moved to the land of Canaan, he and his descendants would be blessed. Abraham's descendants, later called the Israelites, believed they would continue to be blessed as long as they followed God's laws.

Jews believe that God revealed the most important of these laws to a **prophet,** or messenger of God, named Moses. According to the Hebrew Bible, Moses led the Israelites from Egypt. The Israelites had moved to Egypt to escape a long drought and were forced into slavery there. On their way from Egypt, at the top of Mount Sinai (SY·NY) in the desert, Moses received God's laws, including those known as the Ten Commandments. These rules differed from the laws of neighboring peoples because they were based on the worship of one god. The Israelites were not to worship other gods or human-made images. Also, all people—whether rich or poor—were to be treated fairly.

About 1000 B.C., the Israelites under King David created a kingdom in the area of present-day Israel. The kingdom's capital was the city of Jerusalem. By 922 B.C., the Israelite kingdom had split into two states—Israel and Judah. The people of Judah came to be called Jews. In later centuries, the Jews were conquered and many were forced to leave their homeland. Eventually, many Jewish people moved to countries in other parts of the world.

Prayer at the Western Wall

NATIONAL GEOGRAPHIC

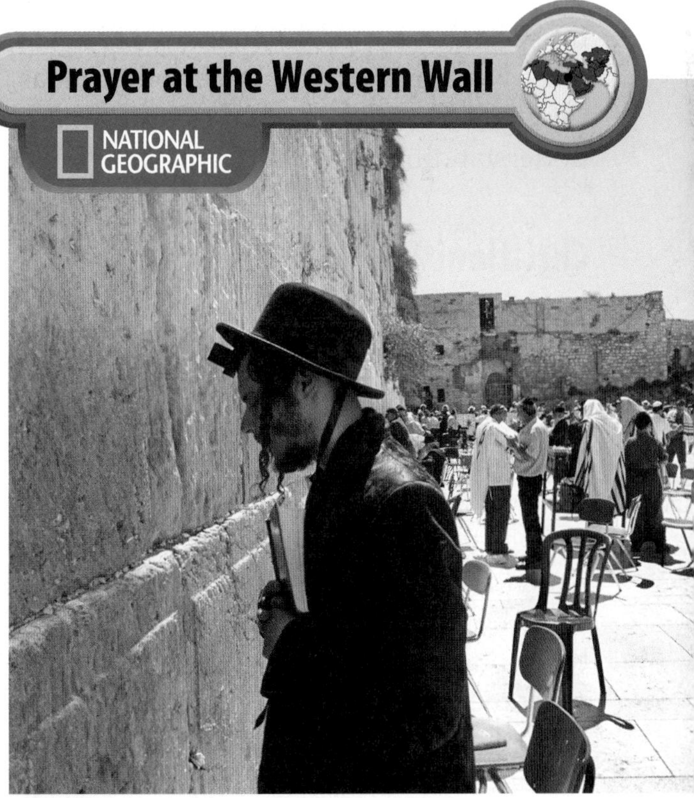

▲ The Western Wall in Jerusalem is believed to be a part of a great Jewish temple that was built in 512 B.C. ***Place*** **According to Jewish belief, what agreement did God make with Abraham?**

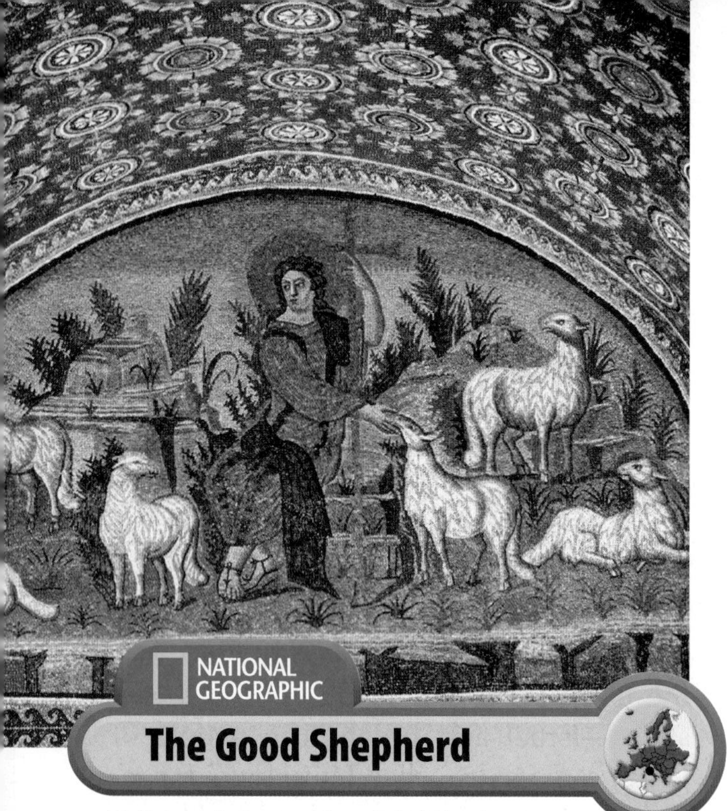

This mosaic from a church in Italy shows Jesus as the good shepherd, a popular theme in Christian tradition. ***Movement*** **Into what regions did Christianity spread?**

This scattering of the Jews was called the Diaspora (dy·AS·pruh). In many areas, the Jews were treated cruelly. In other areas, they were treated with tolerance and understanding.

Christianity

Judaism gave rise to another monotheistic religion—Christianity. About A.D. 30, a Jewish teacher named Jesus began preaching in what is today Israel and the West Bank. Jesus taught that God loved all people, even those who had sinned. He told people that if they placed their trust in God, their sins would be forgiven.

Some Jews greeted Jesus as a savior sent by God to help them. This acceptance worried other Jews, as well as the Romans who ruled their land. Jesus was convicted of treason under Roman law and was crucified, or executed on a cross. Soon afterward, Jesus' followers declared that he had risen from the dead and was the Son of God.

Jesus' followers spread his message throughout the Mediterranean world. Jews and non-Jews who accepted this message became known as Christians. They formed churches or communities for worship. Stories about Jesus and the writings of early Christians—known as the New Testament—became part of the Christian Bible.

In time, Christianity spread to Europe, where it became the dominant religion, and then around the world. Today, it is the world's largest religion, with about 2.1 billion followers.

Islam

The third major monotheistic religion to develop in Southwest Asia was Islam. Islam began in the A.D. 600s in the Arabian Peninsula with the teachings of Muhammad. Muslims, or followers of Islam, believe that Muhammad was the last and greatest prophet of Islam—following Abraham, Moses, and Jesus.

Born about A.D. 570, Muhammad was a merchant in Makkah (Mecca), a trading center in western Arabia. According to Islamic teachings, Muhammad heard messages from an angel telling him to preach about God. He told people that there is only one God, Allah, before whom all believers are equal. Muslims later wrote down Muhammad's messages. These writings became the Quran (kuh·RAN), or holy book of Islam.

After Muhammad died in A.D. 632, leaders known as **caliphs** ruled the Muslim community. Under their leadership, Arab Muslim armies conquered neighboring

Social Studies  **ONLINE**

Student Web Activity Visit glencoe.com and complete the Chapter 17 Web Activity about the origins of Judaism, Christianity, and Islam.

lands and created a vast empire. Over several centuries, Islam expanded into areas of Asia, North Africa, and parts of Europe, as shown in **Figure 2.** As time passed, many of the people in these areas accepted Islam and the Arabic language.

From the A.D. 700s to the A.D. 1400s, Muslims were the leading merchants in many parts of Asia and Africa. Muslim caravans passed overland from Southwest Asia to China, and Muslim ships sailed the Indian Ocean to India and Southeast Asia. Muslim traders enjoyed success for several reasons. Merchants all across the Islamic Empire used coins, which made trade easier. Also, Muslim merchants kept detailed records of their business deals and profits. In time, these practices developed into a new business—banking.

Trade helped leading Muslim cities grow. Located on important trade routes, the cities of Baghdad, Cairo, and Damascus became centers of government and learning. Muslim scholars made many contributions to mathematics, astronomy, chemistry, and medicine. They also preserved the writings of the ancient Greek philosophers. Later, Muslims passed on this knowledge. In fact, the system we use today for writing numbers and the concept of zero were brought by Muslims from India to Europe. Muslims also introduced the compass and several foods, including rice and sugar, to Europe.

Reading Check **Identifying Central Issues** At the time it developed, how did Judaism differ from other religions?

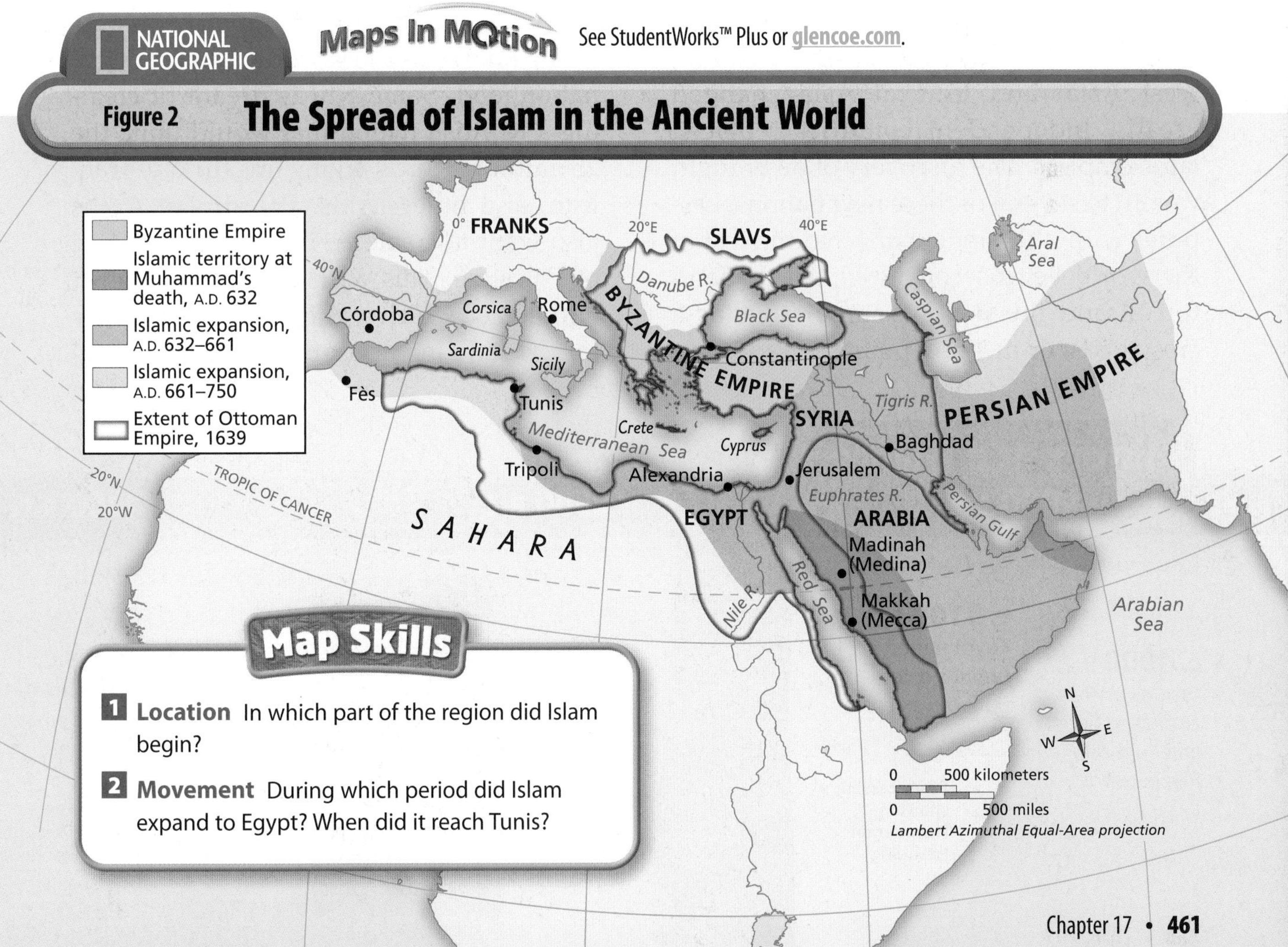

Figure 2 The Spread of Islam in the Ancient World

Map Skills

1. **Location** In which part of the region did Islam begin?
2. **Movement** During which period did Islam expand to Egypt? When did it reach Tunis?

The Region in the Modern World

Main Idea **In modern times, ethnic, cultural, and economic differences have led to conflict in the region.**

Geography and You Do you think people of very different backgrounds can live together peacefully? Read to see how this issue affects the region today.

By the end of the A.D. 900s, the Arab Empire had broken up into smaller kingdoms. During the next few centuries, waves of Mongol invaders swept into the Muslim world from Central Asia, ending the Arab Empire. In the late 1200s, another Muslim empire arose, led by a people known as the Ottoman Turks. The Ottoman Empire lasted until the early 1900s.

During the 1900s, North Africa, Southwest Asia, and Central Asia changed greatly. Independent countries replaced large empires. The discovery of oil brought wealth to several of these new nations. Disputes within the region affected the rest of the world because of the global demand for oil and the region's importance as a crossroads, or meeting point, of trade.

Independence

At the end of World War I, the Ottoman Empire broke apart as a result of its defeat in that conflict. By this time, European powers had gained control of large areas of North Africa, Southwest Asia, and Central Asia. The region's people, however, resented European rule and cultural influences. They turned to nationalism, or the belief that every ethnic group has a right to have its own independent nation. Through wars and political struggles, most countries in Southwest Asia and North Africa won political freedom by the 1970s. After the Soviet Union's fall in 1991, several Muslim nations in Central Asia also gained their independence.

Today, some groups in the region see themselves as stateless nations, or people with strong ethnic loyalties but no country of their own. For example, an ethnic group known as the Kurds has long sought nationhood. Some Kurds dream of creating a Kurdish nation that would unite the 25 million Kurds living in Turkey, Iraq, Iran, and other lands. Palestinian Arabs also want to create an independent nation of Palestine. This wish has led to conflict with Israel, which was formed as a country for the Jews in part of the same area.

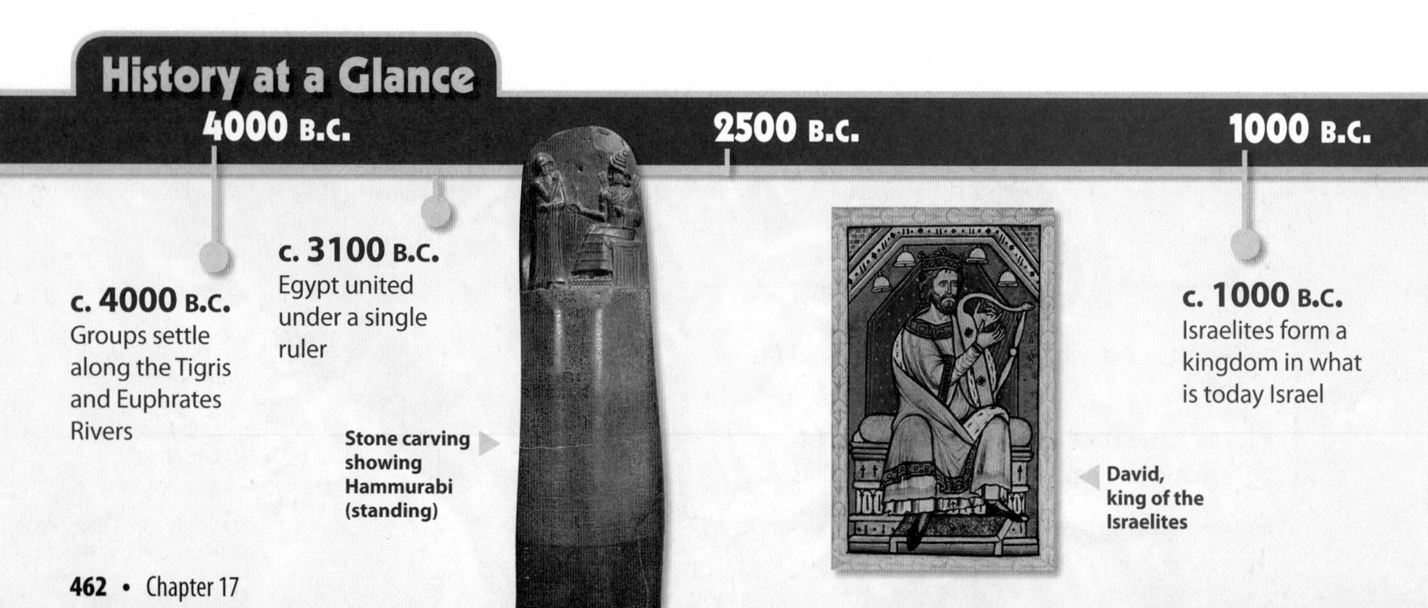

History at a Glance
4000 B.C.
2500 B.C.
1000 B.C.
c. 4000 B.C.
Groups settle along the Tigris and Euphrates Rivers
c. 3100 B.C.
Egypt united under a single ruler
c. 1000 B.C.
Israelites form a kingdom in what is today Israel

Stone carving showing Hammurabi (standing)

David, king of the Israelites

Arabs and Israelis

In recent years, the region has been torn by conflict between Arab countries and Israel, which was founded in 1948. As you read earlier, most Jews were driven from their homeland centuries ago and settled in other lands, where they often suffered hardships. In the late 1800s, some European Jews moved back to the area where Judaism began, which was called Palestine at the time. These settlers, known as Zionists, wanted to set up a safe homeland for the Jews.

After World War I, the British ruled Palestine. They supported a Jewish homeland there. Most of the people living in Palestine, however, were Arabs who also claimed the area as their homeland. To keep peace with the Arabs, the British tried to limit the number of Jews entering Palestine. Later, during World War II, the murder of 6 million Jews by the Nazis in the Holocaust increased support for the Zionists in Europe and the United States.

In 1947 the United Nations voted to divide Palestine into separate Jewish and Arab countries. Arabs within and outside of Palestine bitterly opposed the decision. In May 1948, the British left the area, and the Jews set up Israel in their part of Palestine.

Since that time, Israel and its Arab neighbors have fought several major wars. In one conflict, Israel won control of neighboring Arab areas, such as the West Bank and the Gaza Strip. Palestinian Arabs, many of whom were left homeless, demanded their own country.

Steps toward peace began when Israel signed peace treaties with Egypt in 1979 and Jordan in 1994. In a 1993 agreement, Palestinian Arab leaders said they would accept and work with Israel. Israel in return agreed to give the Palestinians control of the Gaza Strip and some areas of the West Bank. Both sides were to reach a permanent peace agreement within five years, but many issues remained unsettled. As frustration grew, peace efforts had halted by 2000. Violence and distrust continue between Israelis and Palestinians.

These Arab-Israeli conflicts have had an impact on the rest of the world. During a 1973 war, several oil-rich Arab nations raised oil prices and stopped the flow of oil to Western countries that supported Israel. This caused high gasoline prices and a gas shortage in the United States. Continuing American support for Israel has stirred anger toward the United States among many people in the region.

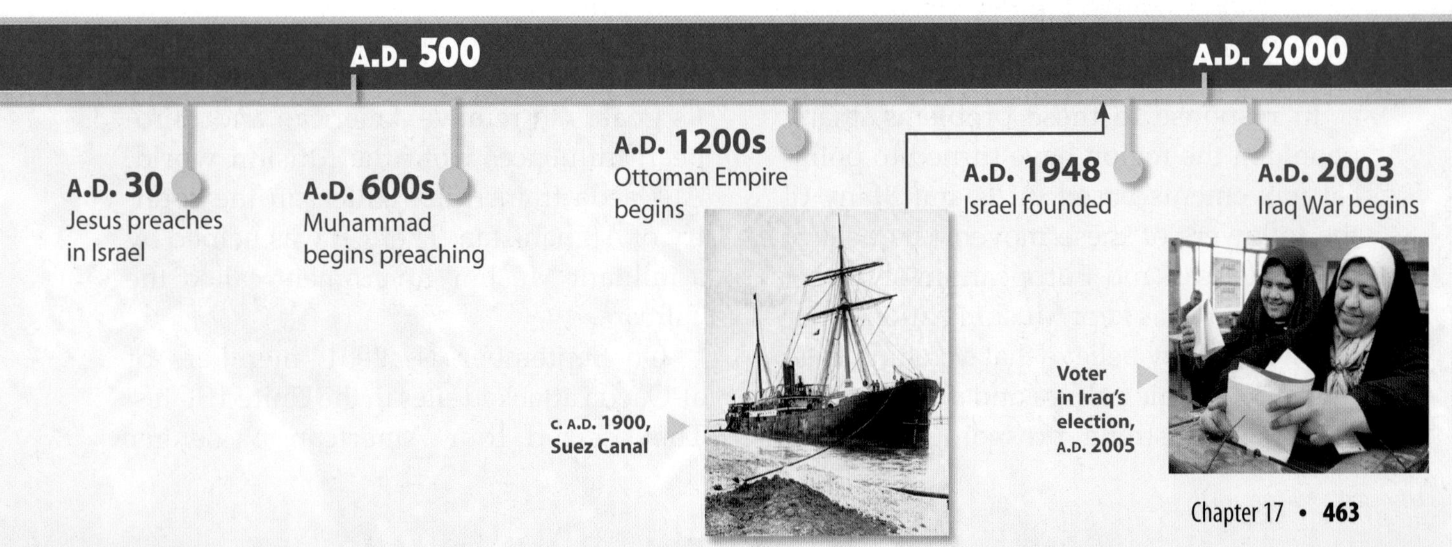

c. A.D. 1900, Suez Canal

Voter in Iraq's election, A.D. 2005

The Arab-Israeli Conflict

The conflicts between the Israelis and their Arab neighbors have caused concern around the world. In 1993, U.S. President Bill Clinton met with Israeli leader Yitzhak Rabin (left) and Palestinian leader Yasir Arafat to work toward peace in the region (inset). ***Regions*** **What other conflicts have occurred in the area?**

Revolution and Conflict

The struggle between Israel and the Palestinians is one of many challenges in the region. As in other parts of the world, a few people are rich, while many are poor. Some countries prosper because of oil, but others lack resources. Growing populations have especially strained water resources in the region.

In response to these problems, many people in the region have turned to political movements based on Islam. Many of the followers of these movements believe that American and European involvement in the region has kept Muslim nations poor and weak. They believe that Muslims must return to Islamic culture and values if they want to build strong, prosperous societies.

In 1979 an Islamic revolution in Iran overthrew that country's shah, or king. Muslim religious leaders, such as Ayatollah Ruhollah Khomeini (EYE·uh·TOH·luh ru·HAWL·la koh·MAY·nee), then made Iran an Islamic republic. They enforced the strict laws of a traditional Islamic society. Iran clashed with Iraq, which was led by a dictator named Saddam Hussein (hoo·SAYN). From 1980 to 1988, Iran and Iraq fought a war in which 1 million people died.

For Iraq, the long war with Iran proved costly. In 1990 Saddam Hussein sent his army to take over Kuwait (ku·WAYT), a small, neighboring country that is rich in oil. In response, the United States and other nations sent in military units to force the Iraqis out of Kuwait. Following a brief war, the Iraqis were defeated, although Saddam Hussein remained in power.

Terrorism

Since the 1990s, both Southwest Asia and other areas of the world have seen the dramatic growth of terrorism. **Terrorism** is the use of violence against civilians to achieve a political goal. A Muslim terrorist group called al-Qaeda (al·KY·duh) was formed by a Saudi Arabian named Osama bin Laden (oh·SAHM·uh bihn LAHD·uhn). Its goal is to remove American and European influences from the Muslim world. Al-Qaeda trained its fighters in the country of Afghanistan. There, it was helped by a militant Muslim government called the Taliban.

On September 11, 2001, members of al-Qaeda attacked sites in the United States. They seized four American passenger planes and flew two of them into the World Trade Center in New York City. A third plane was flown into the Pentagon—the headquarters of the U.S. Defense Department—outside Washington, D.C. The fourth plane crashed in a field in Pennsylvania. Almost 3,000 people were killed in the attacks. The United States responded to the attacks by declaring a war on terrorism.

In 2001 troops from the United States and other countries attacked Afghanistan. They defeated the Taliban and helped set up a democratic Afghan government. Within a few years, the United States went to war again with Iraq. President George W. Bush believed that Iraq's leader, Saddam Hussein, was hiding deadly chemical and biological weapons that could be given to terrorists. In 2003 a group of countries led by the United States invaded Iraq. The Iraqi army was quickly defeated. Saddam Hussein was later captured, put on trial, and executed.

For the United States and its partners, rebuilding Iraq was more difficult than overthrowing Hussein. Saddam Hussein's supporters, foreign terrorists, and Islamic militants all battled American forces. Meanwhile, two Muslim groups in the country, the Sunnis and the Shias, competed for power and began to fight each other. An ethnic group, the Kurds, demanded more self-rule. These conflicts and disputes made it difficult for Iraq to create a democracy and rebuild its society.

Determining Cause and Effect How have the different conflicts in Southwest Asia affected the rest of the world?

Section Review

Social Studies ONLINE
Study Central™ To review this section, go to glencoe.com.

Vocabulary

1. **Explain** the significance of:
   - **a.** irrigation
   - **b.** city-state
   - **c.** polytheism
   - **d.** theocracy
   - **e.** cuneiform
   - **f.** pharaoh
   - **g.** hieroglyphics
   - **h.** monotheism
   - **i.** covenant
   - **j.** prophet
   - **k.** caliph
   - **l.** terrorism

Main Ideas

2. **Describing** What physical conditions allowed civilizations to develop in Mesopotamia and Egypt?
3. **Summarizing** Use a diagram like the one below to summarize the features of one of the major religions that began in Southwest Asia.

   Major Religion

4. **Generalizing** What major change came to Iran in the late 1970s?

Critical Thinking

5. BIG Idea What were three reasons conflict increased in the region in the late 1900s?
6. **Challenge** Explain how the region helped shape the world in the ancient past and continues to do so today.

Writing About Geography

7. **Using Your FOLDABLES** Use your Foldable to write an essay describing the importance of Islam to the region.

Museum Artifacts: Who Should Have Them?

Many large museums own ancient artifacts that were taken out of the country in which they were found. For example, the Rosetta Stone, which helped archaeologists decode hieroglyphs, was taken from Egypt and is still on display in the British Museum. Today, countries are requesting the return of such artifacts. Their leaders believe that the artifacts are part of their national identity. Museums, however, are not always willing to give back these valuable items.

Returning Artifacts

The Rosetta Stone is one of the most important pieces in the British Museum, but it is more important for Egypt. It is an essential piece of our Egyptian national and historical identity and was disgracefully smuggled out of the country. . . .

If the British want to . . . restore their reputation, they should volunteer to return the Rosetta Stone because it is the icon of our Egyptian identity. . . .

Our previous attempts at returning the Rosetta Stone were ineffectual [did not work], but we hope that by organizing an international lobby, we can pressure with greater force the countries and museums in possession of such artifacts.

—Zahi Hawass
Secretary General of the
Supreme Council of Antiquities, Egypt

Returning Artifacts

While there are occasions when important cultural artifacts should be returned to their country of origin, this is far from the usual case.

Most of the major collections of Egyptian art . . . were built up through legal purchase and archaeological excavation. At the close of the nineteenth century, a French scholar, Auguste Mariette, created the Egyptian Antiquities Service to ensure that important pieces of Egypt's cultural patrimony [inheritance] remained safe and secure in the country. . . . This system, which lasted over a century, benefited everyone as the great, national collections in Egypt were built up at no cost to that nation, [and] objects were scientifically documented and studied.

. . . [N]ot every mummy, or every antiquity, in the world should go back to Egypt. For one thing, there would be no room for them all there. Most importantly, the artifacts from ancient civilizations are a reminder of the history we all share and the great achievements of humankind.

—Peter Lacovara
Michael C. Carlos Museum Curator
Emory University, Atlanta, Georgia

You Be the Geographer

1. **Explaining** What was the role of the Egyptian Antiquities Service?
2. **Critical Thinking** Why might a nation want to have its artifacts returned to it?
3. **Read to Write** Write a one-page editorial from the perspective of a French archaeologist who has recently uncovered a valuable artifact in Egypt. Do you think the artifact should remain in Egypt or be displayed in a museum in France?

Guide to Reading

BIG Idea

Culture groups shape human systems.

Content Vocabulary

- hajj *(p. 471)*
- saint *(p. 472)*
- dietary law *(p. 472)*
- epic *(p. 472)*
- mosque *(p. 473)*
- calligraphy *(p. 473)*
- bazaar *(p. 474)*

Academic Vocabulary

- distinctive *(p. 473)*
- revenue *(p. 474)*

Reading Strategy

Identifying Central Issues Use a diagram like the one below to summarize the culture of the region by adding one or more facts to each of the smaller boxes.

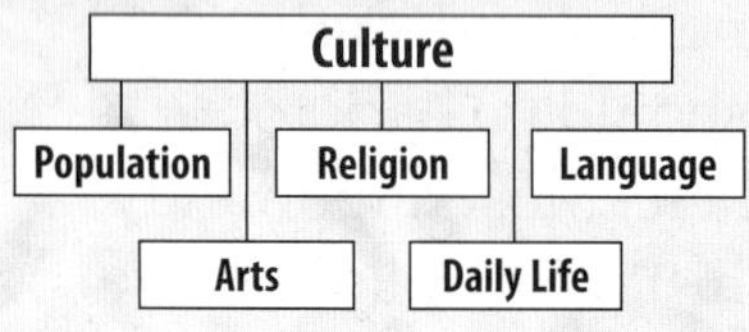

Cultures and Lifestyles

Picture This This solid pillar of sandstone was transformed into a tomb by stoneworkers more than 2,000 years ago. The tomb was part of the ancient city of Madain Salah in northwest Saudi Arabia. This once-bustling city was an important stop on the incense trade route to Syria and Egypt. Today, oil, not incense, drives the Saudi Arabian economy. Read this section to learn more about the culture and lifestyles of the people of North Africa, Southwest Asia, and Central Asia.

▼ A tomb in the Saudi Arabian desert

Population Changes

Main Idea **Rapid population growth has created challenges for the region.**

Geography and You What challenges in your life have forced you to make changes? Read to find out why rural people in North Africa, Southwest Asia, and Central Asia are moving to cities.

The region's harsh environment greatly affects where people settle. Because water is scarce, for centuries people have lived along seacoasts and rivers, near oases, or in rainy highland areas. The vast deserts covering much of the region remain largely empty of people, except where oil is plentiful. Nomads stay near oases where there is enough vegetation to feed their herds.

Overall, the region's population is growing rapidly. This is due to several factors. As a result of improved health care, fewer infants are dying than in the past. In addition, better medical care for adults means that people are living longer than before. At the same time, however, the birthrate has remained at fairly high levels.

Rural areas where farming is difficult cannot support the growing population. Because of this, many villagers are moving to cities in search of a better life. Large cities, such as Istanbul, Turkey; Cairo, Egypt; Tehran, Iran; and Baghdad, Iraq, are important economic, political, and cultural centers. They face many challenges, however, as a result of their fast growth. Overcrowding, lack of jobs, inadequate transportation, and poor housing are common. Many people live in poverty on the outskirts of the region's largest cities.

Reading Check **Explaining** Why are cities in the region growing, and what problems do they face?

NATIONAL GEOGRAPHIC

Kuwait

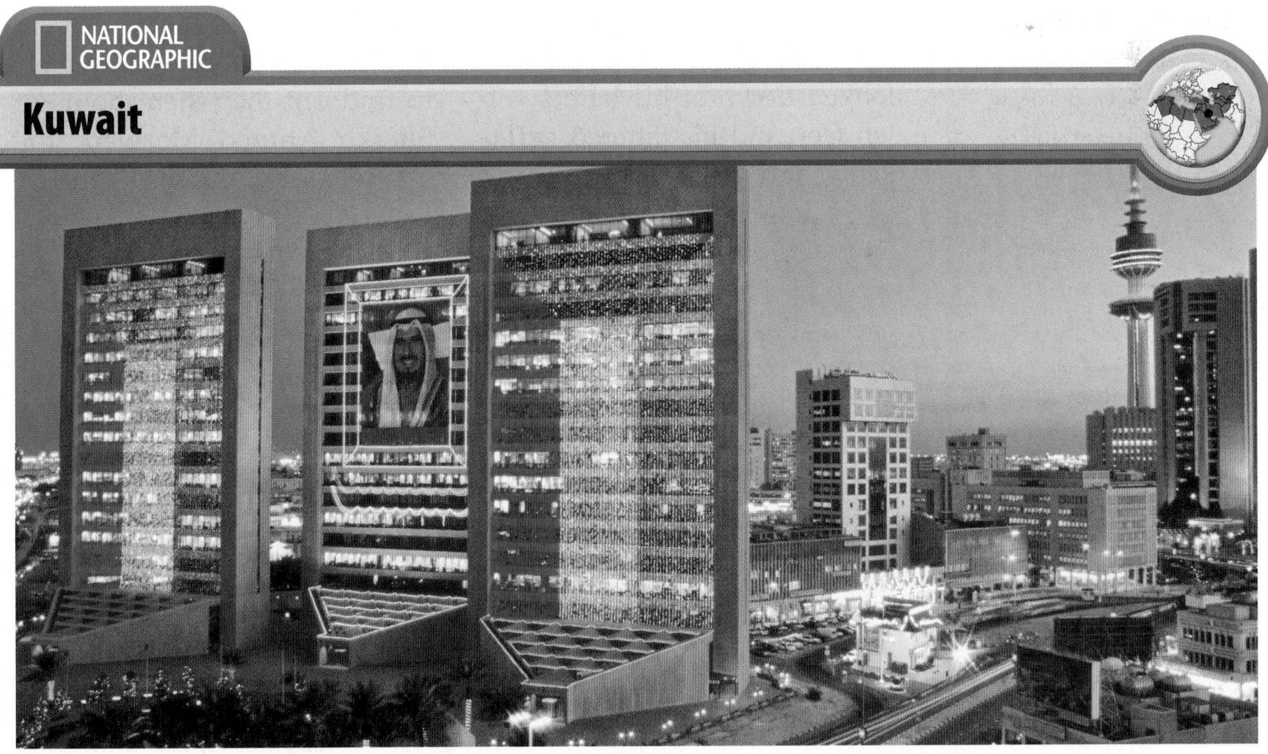

Kuwait's modern buildings and bright lights reflect the country's growing wealth from oil. ***Human-Environment Interaction*** **Why is the region's population growing?**

Mahmoud El-Saadani

(1927–)

Egyptian journalist and writer Mahmoud El-Saadani has written plays, novels, and stories. His works often show the poorer people of Egyptian society trying to adjust to the changes in their world.

Background Information

By the mid- to late 1900s, Egypt's population was growing rapidly. Rural areas were especially strained by the growth. As a result, many farmers left their homes for the cities, hoping to find work. They often had difficulty adapting to the sights, sounds, and pace of bustling Cairo.

Reader's Dictionary

suffocatingly: making it difficult to breathe

Mit el Hallagi: Haridi's home village

trams: streetcars

hawkers: people selling goods on the street

gallabeyya: traditional robe of Egyptian farmers

effendi: title of respect used in Muslim world

THE NIGHT TRAIN HOME

By Mahmoud El-Saadani

[Haridi] remembered the day he had come to Cairo—exactly a year ago. He had never been there before.

The time twelve o'clock, and the place Cairo. The crowd was **suffocatingly** dense. There were more cars there than all the buses that passed through **Mit el Hallagi** in a whole year. . . . The clanging of the **trams** rent [hurt] his ears. There were **hawkers** selling melons, watermelons and newspapers and there were so many people, more people than there were in the whole of Upper Egypt. But . . . [t]he people all had a tired look, and were pale. Everyone coughed. Yet they were all well-dressed, in clean tidy clothes. . . . Haridi had looked down at his bare swollen feet, and his tattered ***gallabeyya*** and . . . longed for a pair of clean shoes [and] a new *gallabeyya*. . . . He had lifted his sack on to his shoulder, and walked on, his fingers tight on a folded piece of paper. . . .

Before long he had stopped a gentleman who was also crossing the square, and showed him the folded note. The ***effendi*** had looked at it and then instructed him to go straight ahead, then turn right, then left, then . . . directions so complicated that Haridi was unable to understand a word of them. Then suddenly the *effendi* ran as quickly as he could to catch a huge bus which was lumbering past and in next to no time he was inside it, still holding the slip of paper; the bus was away in the distance, and Haridi had lost the address. Thus it took him four long days to find out the whereabouts of Sheikh Ahmed Marwan, the contractor for builders' gangs.

From "Night Train Home" by Mahmoud el Saadani, from *Arabic Writing Today: The Short Story*, edited by Mahmoud Manzalaoui. Reprinted by permission of the publisher, American Research Center in Egypt.

Analyzing Literature

1. **Analyzing Information** Describe how Haridi feels on his first day in Cairo.
2. **Read to Write** Write several paragraphs continuing Haridi's story and describing his experiences in the coming weeks.

Religion, Language, and Arts

Main Idea **Religion, especially Islam, remains extremely important throughout the region.**

Geography and You What are the main religions in your community? What practices do their believers follow? Read to find out how religion influences life in North Africa, Southwest Asia, and Central Asia.

As in the past, religion plays an important role in the region today. Islam is the major faith. It is divided into two major groups: Sunni (SU·nee) and Shia (SHEE·ah). Both groups follow the Quran and share many beliefs, but they disagree on how the Muslim faithful should be governed. Most Muslims in the region and throughout the world are Sunni. In Iran, Iraq, Azerbaijan, and parts of Lebanon and Syria, however, most Muslims are Shia.

Although Judaism and Christianity began in the region, their followers make up only a small percentage of the population. Most Jews in the area live in Israel. Christians are dominant in Armenia and Georgia. Large groups of Christians also live in Israel, Lebanon, Egypt, Syria, and Iran. Despite their long history, today Christian churches across the region are declining in numbers because of low birthrates and emigration to Europe and the United States.

Islam, Judaism, and Christianity have some beliefs in common. They all believe in one God who holds all power and who created the universe. They also believe that God determines right and wrong. People are expected to love God, obey God's will, and show kindness to others.

Religious Practices

Each of the three religions has its own unique practices. Muslims strive to fulfill the Five Pillars of Islam—acts of worship that are required from the faithful. The first duty is to make the statement of faith: "There is no God but Allah, and Muhammad is his messenger." Second, Muslims must pray five times a day while facing Makkah. The third duty is to give money to help people in need. The fourth duty is to fast, or not eat or drink, during daylight hours of the holy month of Ramadan. This is the month, according to Muslim beliefs, in which Muhammad received the first message from God. The last pillar of faith is to undertake a holy journey, or **hajj.** Once in each Muslim's life, he or she must, if able, journey to Makkah to pray.

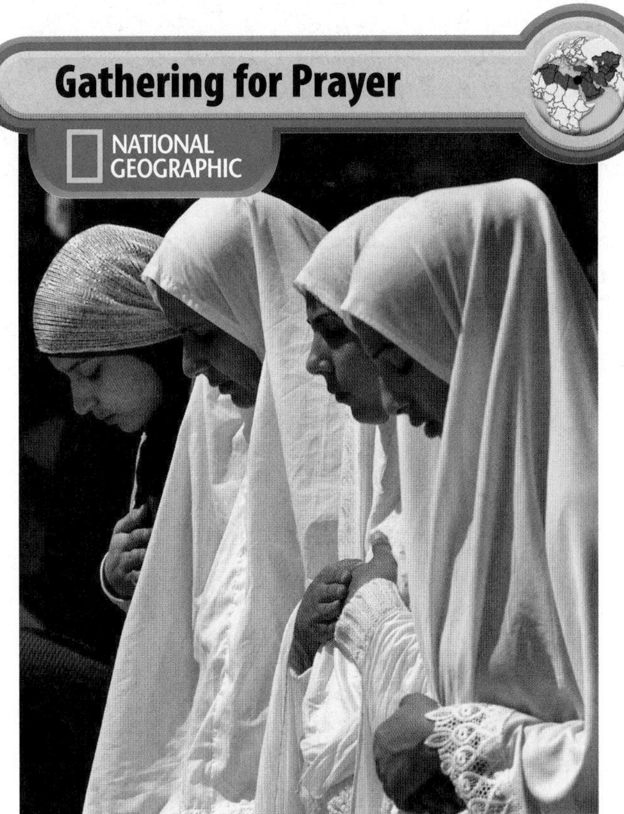

▲ At a mosque in Arab East Jerusalem, Muslim women gather to pray during the holy month of Ramadan. ***Regions*** **What are the Five Pillars of Islam?**

Dietary Laws

At a special school in Jerusalem, chefs are trained in preparing foods that meet Jewish dietary laws. ***Regions*** **What are some examples of dietary laws followed by Jews and Muslims?**

Christians in the region celebrate Easter as their major holy day. They also set aside special days to honor **saints,** or Christian holy people. Armenians and Georgians have their own Orthodox Christian churches with rituals that go back to the earliest days of Christianity. In Egypt, about 10 percent of people belong to the Coptic Orthodox Church. The Coptic language, which descends from the ancient Egyptian language, is still used in Coptic Orthodox services.

In Israel, where three-quarters of the population is Jewish, people follow the traditional practice of marking the Sabbath from sundown on Friday to sundown on Saturday. For Jews, the Sabbath is the weekly day of worship and rest. The holiest of Jewish holy days is Yom Kippur, the Day of Atonement. On this day, Jews fast from sunset to sunset and attend services, asking God's forgiveness for their sins.

Both Jews and Muslims have **dietary laws** that state which foods they can and cannot eat and how food should be prepared and handled. For instance, people of both religions are forbidden to eat pork. Muslims are also banned from drinking alcohol.

Languages and Literature

As in other parts of life, religion has had a strong influence on language in the region. As Islam spread through Africa and Asia, so did the Arabic language. Non-Arab Muslims learned Arabic in order to read the Quran. As more people became Muslim, Arabic became the major language in much of the region. Other major languages include Hebrew in Israel, Turkish in Turkey, and Farsi (FAHR·see) in Iran. Armenians and Georgians also have their own languages. The peoples of Central Asia generally speak Turkic languages that are related to Turkish.

A number of great works of literature have been written in the languages of the region. Many of these works are exciting **epics**—tales or poems about heroes and heroines. *The Thousand and One Nights,* a well-known collection of Arab, Indian, and Iranian stories, reflects life during the period of the Arab Empire. Today, writers such as Mahmoud El-Saadani of Egypt often describe the effects of change on traditional ways of life. Another Egyptian writer, Naguib Mahfouz, wrote of the conflict between traditional rural ways and modern city life. In 1988 Mahfouz became the first winner of the Nobel Prize in Literature whose primary language was Arabic.

The Arts

For many hundreds of years, the region's three religions—Judaism, Christianity, and Islam—provided inspiration for artists and architects. Today, the region's arts also reflect European and American secular, or nonreligious, influences.

Over the centuries, Muslims have developed a **distinctive,** or unique, style of architecture that includes large interiors, highly decorated surfaces, and brilliant colors. Islamic houses of worship, called **mosques** (MAHSKS), can be seen throughout the Muslim world.

According to early Islamic teachings, the human figure should not be shown in art. Muslim leaders believed that the practice would lead to idol worship. As a result, Muslim artists feature geometric patterns and floral designs in many of their works. They also use **calligraphy,** or the art of beautiful writing, for decoration. Passages from the Quran adorn the walls of many mosques.

In Georgia and Armenia, Christian architects built stone churches with domed roofs. Religious music is another art form that has long been prized by the Christians in these countries.

Another art form popular in the region is carpet making, especially in Iran, Turkey, Afghanistan, and the Central Asian countries. The region's handmade carpets are known for their complex designs and rich colors.

Reading Check **Identifying** What beliefs do Judaism, Christianity, and Islam share?

NATIONAL GEOGRAPHIC

Islamic Arts

In a traditional form of Islamic art, small pieces of glass or pottery are put together in elaborate designs to produce a mosaic. ***Regions*** **Why are geometric patterns and floral designs common in Muslim art?**

Daily Life

Main Idea **Living standards vary widely in the region, as do the effects of European and American culture.**

Geography and You How different do you think urban and rural families are in your community or state? Read on to find out more about lifestyles in North Africa, Southwest Asia, and Central Asia today.

North Africa, Southwest Asia, and Central Asia is a region of contrasts. Some people struggle to earn a living from nomadic herding and small-scale farming, while others live in great luxury. Another difference is the degree to which people have adopted modern culture or maintained traditional ways.

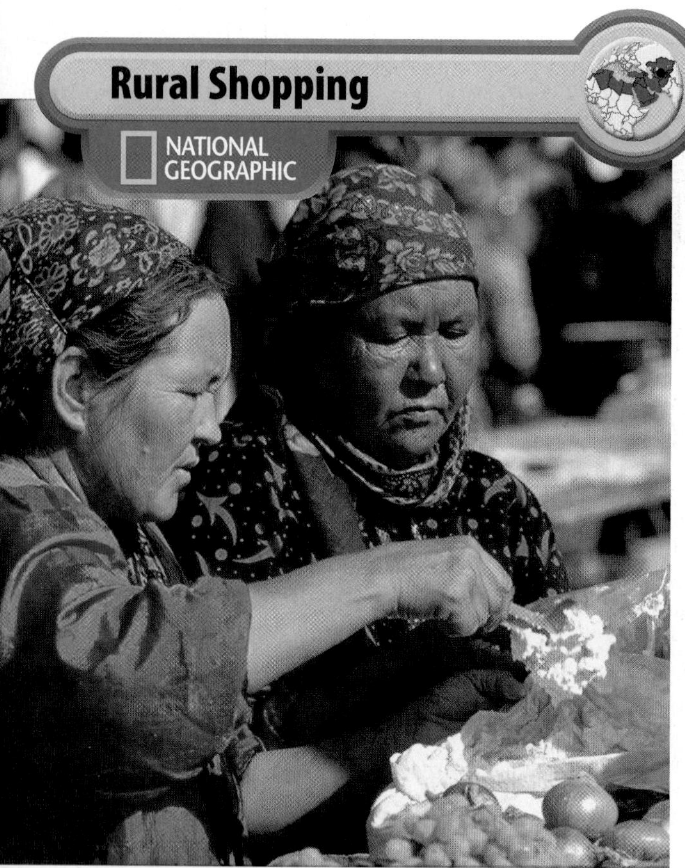

▲ Village bazaars remain important social gathering places in many parts of the region. ***Place*** **How does shopping differ from the countryside to the city?**

Rural and Urban Lifestyles

In the past, most people in the region lived in small villages and farmed the land. During the last century, many people moved to cities. Today, more than 50 percent of the region's people live in urban areas.

In the region's largest cities, many people live in high-rise apartments. In older city neighborhoods, however, people may live in stone or mud-brick buildings that are hundreds of years old. Some of these dwellings still lack running water and electricity.

People in the countryside often depend on their own farms or the village market for food. City dwellers can shop at supermarkets, but in many cities, the **bazaar** is still popular. This traditional marketplace is a busy area. It might be a single street of stalls or an entire district in a large city that extends along miles of walks and passages.

Living Standards

Standards of living vary widely across the region. Countries whose economies are based on manufacturing or oil production have relatively high standards of living. Israel, for example, has a strong economy. Its workers are highly skilled, and the country exports a number of high-technology products. In other wealthy countries, such as Saudi Arabia and Qatar (KAH·tuhr), citizens have prospered from oil production. Their governments have used oil **revenues,** or income, to build schools, hospitals, roads, and airports. Many people in these prosperous nations live in modern cities, work in manufacturing or service jobs, and receive free education and health care from their governments.

In the region's developing countries, however, there is little wealth to share. This is partly due to growing populations. High population growth in some North African countries, such as Egypt and Algeria, has greatly strained these countries' economies. In recent years, many North Africans have migrated to Europe to find work that is not available in their own countries. In some places, such as Afghanistan and Tajikistan, farming and herding are the leading economic activities. In these areas, daily life has changed little over hundreds of years. Many farmers still use simple tools to work their small plots of land. As a result, they are able to grow only enough food to feed their families.

Family Life and Education

People in this region place great value on family life. Many families gather at midday for their main meal. Leisure time is often spent visiting with family members. Generations may gather to watch movies or television. Playing board games, such as chess, is also a popular pastime in some areas. Holidays are important occasions for family celebrations. Most of these holidays are related to the Muslim, Jewish, or Christian religions.

In many countries of the region, men traditionally have the dominant role in the family. Wives are expected to obey their husbands, stay home, and raise children. Marriages arranged by parents are quite common. Also, women are expected to dress modestly. Many Muslim women, for example, wear a head scarf or veil in public, a practice that began in the area long before the rise of Islam.

GLOBAL CITIZENS

NAME: SHIRIN EBADI **HOME COUNTRY:** Iran

ACHIEVEMENT: This human rights champion was one of Iran's first female judges, but she was forced to leave her job after the Islamic revolution in 1979. The new conservative rulers declared that only men could be judges. Since then, Ebadi has argued that the law of Islam—the main religion in Iran—should exist in harmony with female equality, religious freedom, and freedom of speech. Her views have angered many people, and she has been the target of death threats. In 2003 she became the first Iranian and first Muslim woman to be awarded the Nobel Peace Prize for promoting peaceful solutions to serious problems in government and society. Today, she practices as a lawyer, teaches, and directs a program to protect children's rights.

QUOTE: "When you are hopeless, you are at a dead end."

Ebadi visits a home for street children that she founded in Tehran, Iran.

ANA E. FUENTES/ZUMA PRESS/NEWSCOM; (INSET) MICHELINE PELLETIER/CORBIS

CITIZENS IN ACTION What qualities do people need in order to stand up for their beliefs—however unpopular?

Most young people attend school. Primary education is free, and the number of children who attend school is increasing. Many students now complete both primary and secondary school, and a small percentage attends university.

Changes in Women's Roles

In the past 50 years, changes have occurred in the status of urban women in the region. Women in rural areas have always done farmwork alongside their husbands, but most urban women stayed at home to manage households. Today, many women in the cities have jobs in business, education, and government.

Some countries are more traditional than others in their attitudes toward women. In Saudi Arabia, for example, women are not allowed to vote, to drive, or to travel unless accompanied by a male relative. Women are allowed to attend universities but must go to classes separate from men. Although women are allowed to work, they can do so only in professions such as teaching and medicine in which they can avoid close contact with men.

Other countries have accepted different views. For example, in Turkey, women can vote and hold public office. In fact, Tansu Çiller (TAHN·suh CHEE·luhr) served as Turkey's first woman prime minister in the mid-1990s. Israel also had a woman prime minister, Golda Meir, who was in office from 1969 to 1974. In several countries, women have served in other government posts.

Reading Check **Describing** Describe how the status of women differs around the region.

Social Studies ONLINE
Study Central™ To review this section, go to glencoe.com.

Vocabulary

1. **Explain** the meaning of the following terms by using each one in a sentence.
   - **a.** hajj
   - **b.** saint
   - **c.** dietary law
   - **d.** epic
   - **e.** mosque
   - **f.** calligraphy
   - **g.** bazaar

Main Ideas

2. **Discussing** How does the region's environment affect settlement patterns?
3. **Identifying** Use a diagram like the one below to identify aspects of literature and the arts in the region.

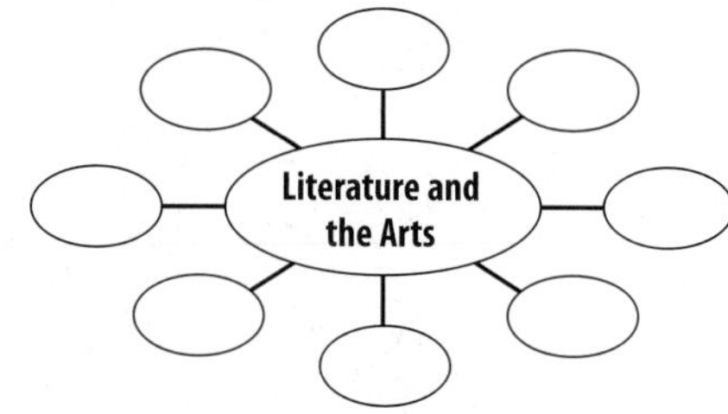

4. **Analyzing Information** Why do some countries in the region have generally high standards of living while others do not?

Critical Thinking

5. **BIG Idea** How has Islam shaped elements of the region's culture?
6. **Challenge** How might countries in this region deal with the challenges of population growth?

Writing About Geography

7. **Expository Writing** Write an essay examining recent economic changes in the Muslim world and the economic challenges that still exist.

Chapter 17 Visual Summary

Early History

- Two of the world's earliest civilizations developed in Mesopotamia and Egypt.
- Three of the world's major religions—Judaism, Christianity, and Islam—began in Southwest Asia.

Mesopotamian tablet with cuneiform

- The Arab Empire made contributions in mathematics, astronomy, medicine, and the arts.

Christian church, Armenia

Modern Times

- By the end of World War I, European powers had gained control of much of the region.
- Independent states arose during the 1900s.
- Oil reserves brought wealth to countries such as Saudi Arabia, Iraq, and Iran.
- Political unrest and conflict have troubled the region since World War II.

People

- Because water is scarce, most people live along seacoasts and rivers or in oases or highland areas.
- A high birthrate has led to rapid population growth.

Religion and Culture

- Islam and the Arabic language are dominant in most of the region.
- Muslim arts use decoration and calligraphy instead of showing the human form.
- Israel is mainly Jewish, while Armenia and Georgia are largely Christian.

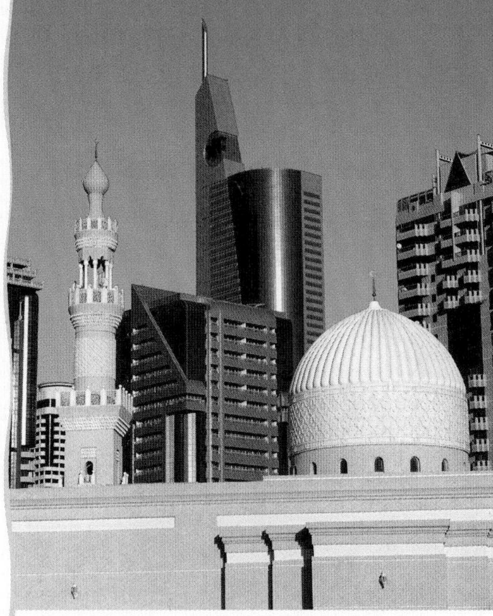

Dubai, United Arab Emirates

Daily Life

- Some governments have used oil money to build schools, hospitals, roads, and airports.
- Family life is important, and the status of women is changing.

Family meal, Saudi Arabia

Study anywhere, anytime! Download quizzes and flash cards to your PDA from glencoe.com.

CHAPTER 17 ASSESSMENT

STANDARDIZED TEST PRACTICE

For a short-answer or extended-response item, take time to review what you have written. Does it completely answer the question? Do you need to add any more information?

Reviewing Vocabulary

Directions: Choose the word(s) that best completes the sentence.

1. The worship of many gods or goddesses is known as ______.

A monotheism

B polytheism

C cuneiform

D theocracy

2. Egyptians believed their rulers, or ______, were also gods.

A prophets

B caliphs

C pharaohs

D covenants

3. The ______ ruled the Muslim community after Muhammad's death.

A prophets

B kings

C caliphs

D emperors

4. One of the Five Pillars of the Muslim faith is a holy journey called a ______.

A bazaar

B mosque

C calligraphy

D hajj

5. Islamic houses of worship are called ______.

A mosques

B covenants

C bazaars

D caliphs

Reviewing Main Ideas

Directions: Choose the best answer for each question.

Section 1 *(pp. 456–465)*

6. The most serious conflict in this region in recent years has been between Muslim Arab nations and ______.

A Lebanon

B Israel

C Syria

D Egypt

7. The most recent of the world's major religions to develop in this part of the world was ______.

A Islam

B Christianity

C Judaism

D monotheism

Section 2 *(pp. 468–476)*

8. The region has faced challenges recently because of ______.

A flooding

B a declining birthrate

C immigration

D rapid population growth

9. The majority of people living in this part of the world are ______.

A Christian

B Jewish

C Kurdish

D Muslim

GO ON

Critical Thinking

Base your answers to questions 10 and 11 on the chart below and your knowledge of Chapter 17.

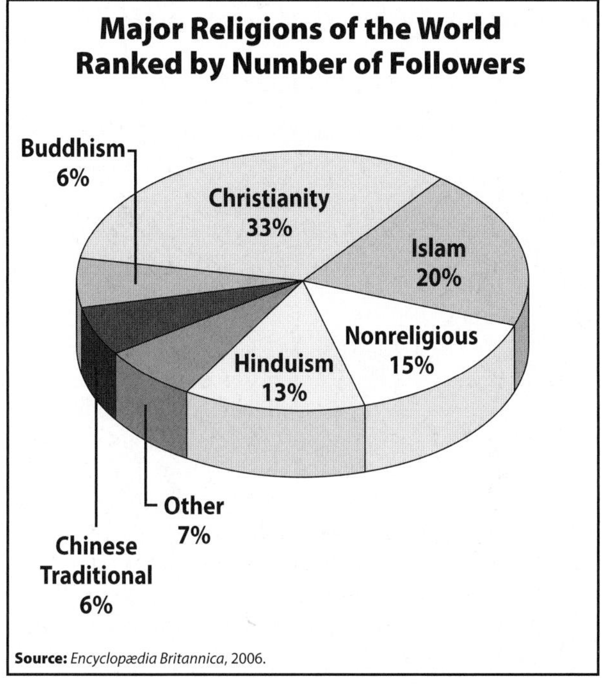

10. What can you learn from this graph?

A the largest religion in the United States

B the rate of growth for each religion

C the religion with the most followers in the world

D how much the percent of Hindus has changed since 1950

11. According to the graph, which of the following statements is accurate?

A Islam is the world's largest religion.

B Hinduism has more followers than Islam.

C Numbers of Hindus and Buddhists in the world are about equal.

D Islam and Hinduism together have about as many followers as Christianity.

Document-Based Questions

Directions: Analyze the document and answer the short-answer questions that follow.

The Code of Hammurabi, written under the direction of the king of Mesopotamia in about 1790 B.C., is one of the earliest written law codes.

> *#22. If any one is committing a robbery and is caught, then he shall be put to death.*
>
> *#53. If any one be too lazy to keep his dam in proper condition, and does not so keep it; if then the dam break and all the fields be flooded, then shall he in whose dam the break occurred be sold for money, and the money shall replace the corn which he has caused to be ruined.*
>
> *#267. If the herdsman overlook something, and an accident happen in the stable, then the herdsman is at fault for the accident which he has caused in the stable, and he must compensate the owner for the cattle or sheep.*
>
> —*Code of Hammurabi, The Avalon Project at Yale Law School*

12. Do you think Mesopotamian laws are similar to those of the United States? Explain.

13. Why do you think such strong punishments were used on people whose dams broke?

Extended Response

14. Take the role of a political analyst. Write an article in which you argue the best way for modern governments to peacefully solve the troubles between Israel and its neighbors.

For additional test practice, use Self-Check Quizzes—Chapter 17 at glencoe.com.

Need Extra Help?

If you missed question...	1	2	3	4	5	6	7	8	9	10	11	12	13	14
Go to page...	457	458	460	471	473	463	460	469	471	459	459	458	458	463

"Hello! My name is Jamshid.

I'm 15 years old and live in Kabul, Afghanistan. In 2005, I lost my leg in a car bombing near my home. It changed my life. But in many ways I am still a typical Afghani teen. Read about my day."

7:00 a.m. I wake up, wash my face, and brush my teeth. Then I greet my big family. I live with my parents, three brothers, and six sisters. My aunt and grandmother live with us as well.

7:30 a.m. For breakfast, we have green tea and nan. Nan is a traditional Afghan flat bread that my mother bakes at home. It is delicious!

8:00 a.m. I massage the area where my leg was amputated. The doctors say that will reduce swelling. I do not know who was responsible for the bomb that hurt me. Sadly, violence is not unusual here. (When I was younger, our government was run by a group called the Taliban. The war in 2001 removed the Taliban from power, but violence continues.)

8:15 a.m. Some of my brothers and sisters are getting ready for school. I have not attended school since the bombing, but I plan to go back next year. I think I might like to become a doctor, so education is important.

8:30 a.m. I put on the artificial leg that was given to me by the International Red Cross. Then I walk down the road to a small fruit and vegetable market. I chat with the owner for a while. Then I buy some tomatoes and oranges for my mother.

10:30 a.m. On some days at this time, I go to English and computer classes at an institute in Kabul. It is not a regular school, but it is the next best thing. I do not have a class today, so I sit at home and read. I especially like to read poetry collections and books written in English.

12:15 p.m. I eat lunch with my family. Today we have my favorite dish, kabuli palau. It is white rice topped with grilled goat meat. We also have nan, fried potatoes, salad, yogurt, and bamiya. Bamiya is a blend of cooked green beans and tomatoes. I'm stuffed!

2:00 p.m. I get ready to pray. At the mosque, they always pray at exactly one o'clock, but at home, we do it whenever we want to. Like all Muslims, I wash before praying. Then I face west and recite my prayers.

3:00 p.m. I sit and talk with my brother Mohammad. Later I am feeling tired, so I take a short nap.

5:15 p.m. My brothers and I go outside to play volleyball. I used to love soccer, but I can't play it anymore because of my leg. Now I play volleyball and cricket instead.

7:00 p.m. It is time for dinner. We eat leftovers from our huge lunch. We finish the meal with some of the oranges I bought this morning.

8:00 p.m. I relax with my family and watch TV. My favorite program is a sports show called *World of Youth.* We are lucky to have electricity tonight. The country's power system was destroyed during the war, but it is being rebuilt. In the meantime, our service is unreliable. Some parts of Kabul have no electricity at all.

9:00 p.m. I go to bed.

ILLUSTRATIONS BY BOOKMAPMAN

LUNCH Jamshid has lunch with his cousin, Atah, and older brother, Mohammad. Afghanis often use bread to scoop up food.

SPORTS Volleyball has become one of Jamshid's favorite sports. The game is very popular in Afghanistan—but not as popular as soccer.

PRAYER TIME Jamshid kneels on his prayer rug, facing west toward Makkah (Mecca)—Islam's holy city.

STANDING TALL After Jamshid (jam•sheed) lost his leg in a bombing, the teen had to learn to walk using an artificial leg. Sometimes he still needs to use crutches. Jamshid stands near his home in Kabul, Afghanistan's capital and largest city.

DAVID ROCHKIND / POLARIS (4)

What's Popular in Afghanistan

School Kids in Afghanistan view school as a privilege. Under Taliban rule, girls were forbidden to attend school, and boys' schools were often unsafe. Today, everyone wants to go.

Kite fights Here, kite flying is a national pastime. You can have a kite fight, or *jang*. In a *jang*, two kites are flown next to each other. You try to use the wire of your kite to cut the wire of your opponent's kite.

AFP PHOTO/JIMIN LAI/NEWSCOM

Head scarves Under Taliban rule, Afghani women had to wear *burkas,* which are garments that cover them from head to toe. Today, many women have traded *burkas* for scarves that cover only their head and shoulders.

REUTERS/BAZUKI MUHAMMAD/NEWSCOM

Say It in Dari

Dari (also called Farsi) and Pashto are Afghanistan's two main languages. Many Afghans can speak both. In the capital city of Kabul, most people speak Dari in their everyday lives. Here are some phrases in Dari.

Hello
salaam (sa·LAHM)

Good-bye
khoda hafez (koh·DAH HAH·fehz)

My name is ________.
Nom e ma ________ ast.
(nah·MEH·ma ________ ast)

EMMANUEL DUNAND/AFP/GETTY IMAGES/NEWSCOM

North Africa, Southwest Asia, and Central Asia Today

Essential Question

Regions Many countries in North Africa, Southwest Asia, and Central Asia are rich in oil and natural gas resources. They supply much of the energy for economies around the world. Political unrest and wars, however, have troubled this area for decades. What effects can conflict have on a region?

BIG Ideas

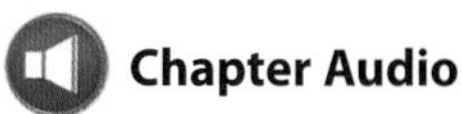

Section 1: North Africa

BIG IDEA Changes occur in the use and importance of natural resources. While some North African countries are enjoying income from oil resources, others still have struggling economies.

Section 2: Southwest Asia

BIG IDEA Cooperation and conflict among people have an impact on the Earth's surface. Religious and ethnic conflicts in Southwest Asia affect other parts of the world because of the area's oil and gas resources.

Section 3: Central Asia

BIG IDEA Places reflect the relationship between humans and the physical environment. Valuable natural resources are helping the people of Central Asia overcome the limitations of the area's harsh environment.

FOLDABLES™ Study Organizer

Organizing Information Make this Foldable to organize information about current topics in each of the three subregions discussed in this chapter.

Step 1 Fold the bottom edge of a piece of paper up 2 inches to create a flap.

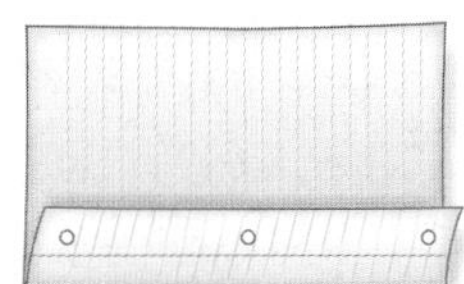

Step 2 Fold into thirds.

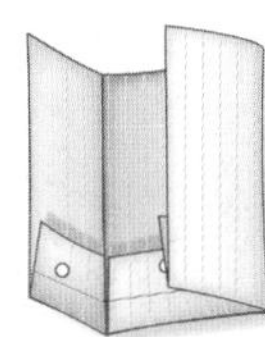

Step 3 Glue to form pockets and label as shown.

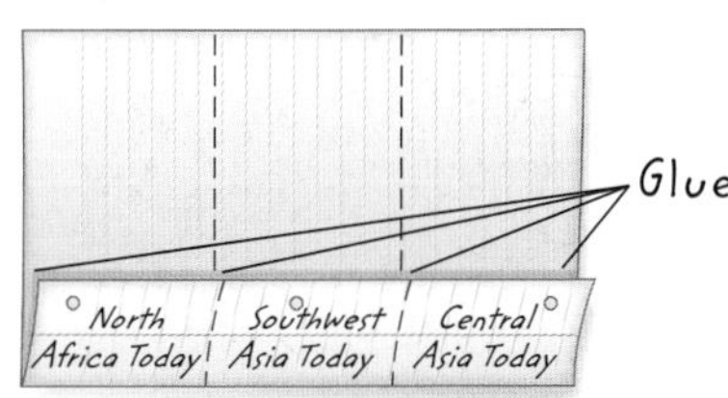

Reading and Writing As you read the chapter, take notes on index cards or slips of paper and place each note in the appropriate pocket of the Foldable. Use your notes to write two quiz questions for each section of the chapter.

◀ Dubai, United Arab Emirates

Social Studies ONLINE

Visit glencoe.com and enter **QuickPass**™ code EOW3109c18 for Chapter 18 resources.

Guide to Reading

BIG Idea

Changes occur in the use and importance of natural resources.

Content Vocabulary

- fellahin *(p. 485)*
- phosphate *(p. 485)*
- dictatorship *(p. 487)*
- trade sanction *(p. 487)*
- casbah *(p. 487)*
- civil war *(p. 488)*
- constitutional monarchy *(p. 488)*

Academic Vocabulary

- infrastructure *(p. 486)*
- policy *(p. 487)*

Reading Strategy

Making Generalizations Use a diagram like the one below to list the major products of agriculture and industry in the region and the country producing them. Add as many ovals as you need.

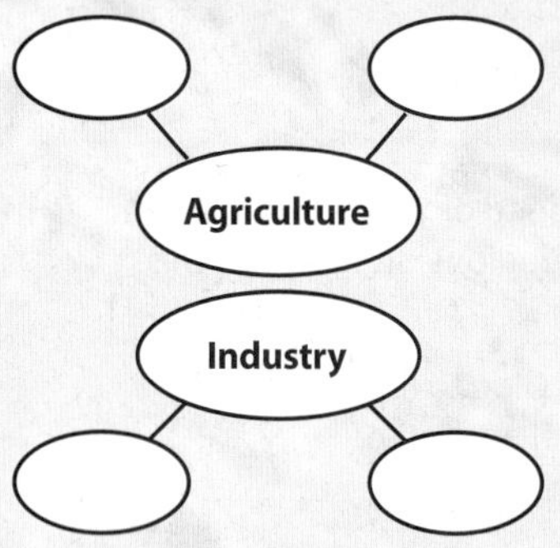

North Africa

 Section Audio  Spotlight Video

Picture This The flavors of these spices are as rich as their colors. Moroccan cooking is known for its use of spices. A home-cooked meal may include the familiar tastes of cinnamon and black pepper, as well as cumin, turmeric, and anise seed. To learn more about countries in North Africa today, read Section 1.

▼ Spices for sale in Morocco

Egypt

Main Idea **Egypt is an important and powerful country in the region, but it faces serious challenges.**

Geography and You How much farmland is in your area? Egypt has relatively little farmland, but many people there still make a living by farming. Read to find out more about modern Egypt's economy.

North Africa is made up of five countries. The easternmost of these is Egypt, a country covered by vast deserts. Because of its arid landscape, most Egyptians live within 20 miles (32 km) of the Nile River. Along the Nile's banks, you can see mud-brick villages, ancient ruins, and modern cities.

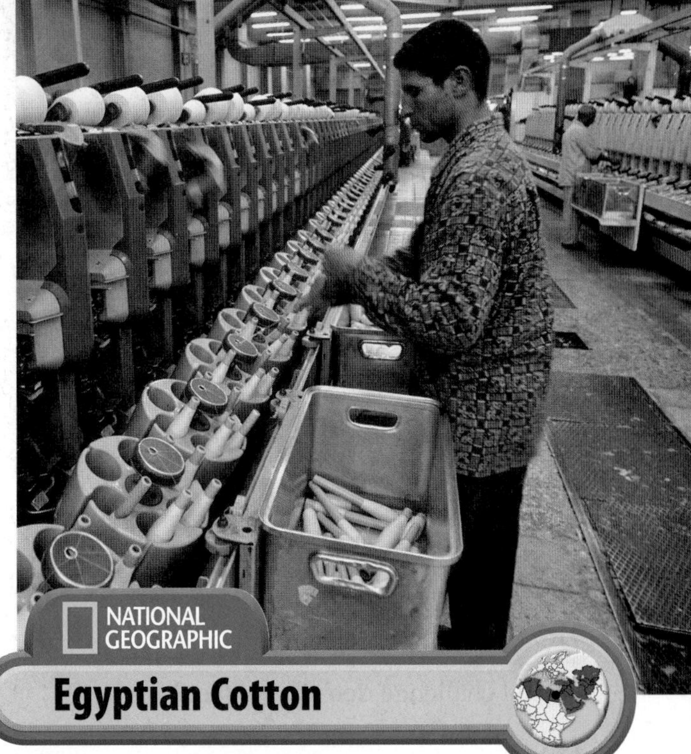

Egyptian Cotton

Cotton and cotton products make up almost 25 percent of Egypt's export trade. ***Human-Environment Interaction*** **Why is the Aswān High Dam important for Egyptian industry?**

Economy

Egypt has a developing economy. Only about 2 percent of Egypt's land is used for farming, but about a third of Egypt's people work in agriculture. The best farmland lies in the fertile Nile Valley.

Some Egyptian **fellahin** (FEHL·uh·HEEN), or peasant farmers, use the simple tools of their ancestors. Others rely on modern machinery. All, however, depend on the Aswān High Dam to control the Nile's floodwaters. Without the dam, the Nile would flood their lands yearly. Instead, controlled releases of the water stored behind the dam occur several times during the year. This allows farmers to harvest more than one crop per year without harmful flooding.

In addition to controlling floods, the Aswān High Dam provides hydroelectric power for Egypt's growing industries. Nevertheless, Egypt's main energy resource is oil, which is found in and around the Red Sea. Petroleum products and **phosphates,** which are minerals used in fertilizers, are Egypt's major exports. Egyptian workers also make food products, textiles, and other consumer goods.

Egypt's industries are drawing more people to the cities than in the past. Nearly 11 million people live in Cairo, Egypt's capital. Another 3.5 million live in Alexandria, a bustling port city on the Mediterranean Sea. Neither city can provide enough houses, schools, and hospitals for all of its people. Poverty, heavy traffic, and pollution now plague Cairo and Alexandria.

Past and Present

From about 300 B.C. to A.D. 300, Egypt was dominated politically by Greece and Rome. In A.D. 641, Arabs from Southwest Asia took control of Egypt. Most of Egypt's people began to speak the Arabic language and became Muslims.

In the 1800s, Europeans and Egyptians together built the Suez Canal, one of the world's most important waterways.

The Suez Canal eventually came under British control. In 1952 army officers overthrew Egypt's British-supported king, and Egypt became fully independent. Today Egypt is a republic with one political party controlling the government. In recent years, Muslim political groups have pushed for a voice in government affairs. Some have used violence to try to achieve their goals.

Reading Check **Drawing Conclusions** What political challenge does Egypt face?

Tourism in Egypt

NATIONAL GEOGRAPHIC

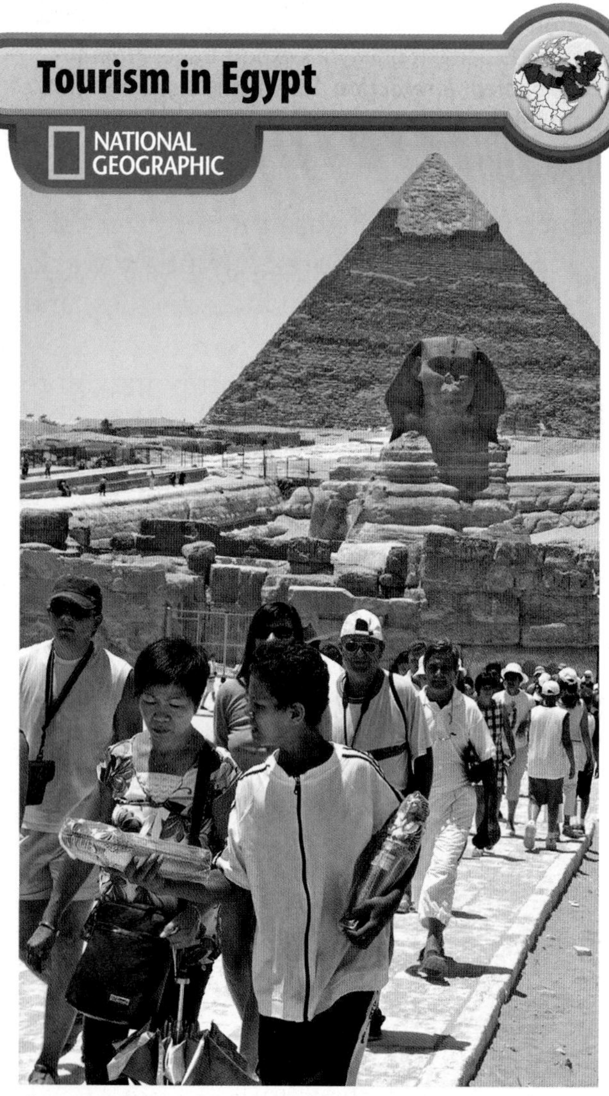

▲ The Great Pyramid and Great Sphinx, built nearly 4,500 years ago, are popular tourist sites in Egypt today. ***Place*** **What are Egypt's major exports?**

Libya and the Maghreb

Main Idea **Oil-rich Libya is improving ties with the outside world, while Tunisia, Algeria, and Morocco face political unrest and economic uncertainty.**

Geography and You Have you watched the *Star Wars* films? Many scenes were filmed in Tunisia. Read to discover how Tunisia and its neighbors are trying to blend traditional and modern ways of life.

In addition to Egypt, North Africa also includes Libya, Tunisia, Algeria, and Morocco. These countries have economies based on oil and other resources found in the Sahara. Tunisia, Algeria, and Morocco also form a smaller region known as the Maghreb (MUH·gruhb). In Arabic, *Maghreb* means "the land farthest west." These countries were given this name because they are the westernmost part of the Arabic-speaking Muslim world.

Libya

Much of Libya is desert, but aquifers lie beneath the sands. New pipelines carry underground water from the desert to Libya's Mediterranean coast. There, a growing population lives in the modern cities of Tripoli and Benghazi (behn·GAH·zee).

In recent decades, oil has brought Libya great wealth. This new wealth helps to build schools and hospitals and to improve the country's **infrastructure**—or roads, ports, and water and electric systems.

Almost all of Libya's people have a mixed Arab and Berber ethnic background. Berbers are a group that settled North Africa before the arrival of the Muslim Arabs in the A.D. 600s. Since the Arabs arrived, Libya has been a Muslim country, and most of its people speak Arabic. Through the centuries, Libya has often been ruled by foreign powers. In 1951 Libya finally established independence. Not long after, a military officer named Muammar al-Qaddafi (kuh·DAH·fee) took power. He set up a **dictatorship,** or a government under the rule of one all-powerful leader.

From the 1970s to the 1990s, Qaddafi supported terrorism and sought to acquire nuclear weapons. In response, the United States and the United Nations placed trade sanctions on Libya. A **trade sanction** is an effort to punish a country by using trade barriers, for example, refusing to buy the country's products. These sanctions successfully forced Qaddafi to change his **policy,** or plan of action. Libya's relations with other countries then improved.

Moroccan Market

Markets, like this one in Morocco where a vendor sells olives, are an important part of the economy and social life of North Africa. ***Region*** **What crops are grown in North Africa's Mediterranean climate?**

Tunisia

Tunisia is North Africa's smallest country. Almost all of Tunisia's people are of mixed Arab and Berber ancestry, speak Arabic, and practice Islam. Tunis is the country's capital and largest urban area.

Most of Tunisia's people live in coastal areas, where a mild Mediterranean climate provides some rainfall. Along the fertile eastern coast, farmers grow wheat, olives, fruits, and vegetables. Tunisia's developing industries produce food items, textiles, and petroleum products. Tourism is also a growing industry.

Tunisia's coastal location has attracted people, ideas, and trade for centuries. Like Libya, Tunisia was part of several Muslim empires. Later it became a colony of France, until gaining independence in 1956 and becoming a republic. Since that time, Tunisia has been governed by powerful presidents, some of whom limited political opposition. Many of these leaders, however, have tried to provide education and health care to poorer Tunisians. Because of these efforts, Tunisia has one of the lowest rates of poverty in Africa. Tunisian women also enjoy more rights than women in other Arab nations.

Algeria

Algeria is North Africa's largest country, with an area about one-and-a-half times the size of Alaska. The country's fertile Mediterranean coast produces wheat, barley, olives, oats, and grapes. Algiers, the country's capital and largest city, is a major Mediterranean port. Although a modern city, Algiers is known for its **casbahs,** or older sections with narrow streets and bazaars.

Farther inland, Algeria is made up of rugged mountain ranges as well as the vast Sahara. The country's economy depends on oil and natural gas pumped from the Sahara. These deposits help Algeria's economy, but there are not enough jobs for the growing population. Many Algerians have moved to Europe to find work.

Algeria's people are of Arab and Berber heritage and practice Islam. From 1830 to 1962, Algeria was a French colony.

In 1954 Algerian Arabs rose up against the French. A bloody civil war erupted. A **civil war** is a conflict between different groups within a country. When the fighting ended in 1962, Algeria won independence. Since then, Algeria has developed into a republic with a strong president and a legislature.

In the early 1990s, disagreements between Muslim political factions led to another civil war. Many lives were lost, but peace was restored in 1999. Since then, Algeria's government has been trying to bring order to the country.

Morocco

Morocco lies in Africa's northwest corner. Farmers on Morocco's fertile coastal plains raise livestock, grains, vegetables, and fruits. Morocco is a leading producer of phosphates. Recently tourism has become an important part of its economy as well. Morocco is known for its historic cities, such as Marrakesh (muh·RAH·kihsh) and Casablanca (KA·suh·BLANG·kuh).

Like other Maghreb countries, Morocco has an Arab and Berber heritage, and was a Muslim kingdom for many years. In the early 1900s, Europeans gained control, but in 1956 Morocco became independent once again. Today the country is a **constitutional monarchy,** in which a king or queen is head of state but elected officials run the government. The Moroccan king, however, still holds great influence.

Morocco controls Western Sahara, a mineral-rich desert territory to the south. Since the Moroccan government seized Western Sahara in 1975, groups of Western Saharans have fought for independence. The United Nations has tried to promote an election to decide Western Sahara's future, but nothing has been settled.

Reading Check **Describing** What conflicts have occurred in the countries of North Africa?

Section 1 Review

Social Studies ONLINE
Study Central™ To review this section, go to glencoe.com.

Vocabulary

1. **Explain** the significance of:
   - **a.** fellahin
   - **b.** phosphate
   - **c.** dictatorship
   - **d.** trade sanction
   - **e.** casbah
   - **f.** civil war
   - **g.** constitutional monarchy

Main Ideas

2. **Summarizing** What problems are Egypt's major cities facing?
3. **Comparing** Use a diagram like the one below to compare governments in North Africa.

Country	Form of Government	Citizens' Role

Critical Thinking

4. **BIG Idea** How has the discovery of oil affected the countries in the region?
5. **Drawing Conclusions** Why do you think Libya's leader recently tried to build better relations with other nations?
6. **Challenge** How do the strengths and weaknesses of Egypt's economy compare to those of other countries in the region?

Writing About Geography

7. **Using Your FOLDABLES** Use your Foldable to write a paragraph about the challenges facing North Africa.

NATIONAL GEOGRAPHIC **Geography & History**

A Divided People

Why might an ethnic group that is split into separate countries want a homeland?

The Kurds and Their History The Kurdish people have lived in Southwest Asia for a few thousand years. Their traditional homeland—called Kurdistan—is in eastern Turkey, northeastern Iraq, and western Iran. The Kurds adapted to this mountainous land by living as nomadic herders. Though they were known as fierce fighters, they never formed a powerful kingdom. For centuries, they lived under the rule of different Muslim empires.

Following World War I, the European nations that conquered the region agreed to create an independent Kurdish nation. Within a few years, though, the Europeans abandoned that idea. Instead, Kurdistan was split among three separate countries—Turkey, Iraq, and Iran—and the Kurds became a divided people.

Kurds Today The dream of an independent Kurdistan has remained strong among Kurds. Within Turkey, Iraq, and Iran, Kurds have been mistreated. Harsh laws have banned their language and traditional costumes. Governments have tried to force them to leave.

▲ Kurd family, Iraq

Kurds in some areas responded to these measures with armed attacks. A Kurdish uprising in Iran was crushed in the 1980s. Meanwhile, in Turkey, the government's response led to the deaths of thousands of Kurds. Once the violence lessened, Turkey's government eased some of the laws the Kurds objected to.

In Iraq, dictator Saddam Hussein's army attacked the Kurds several times, killing thousands. Since the early 1990s, though, the Kurds have largely been able to govern themselves. Following Hussein's fall from power in 2003, some Kurds again pushed for independence. The Kurds hoped to control oil-rich areas that would provide money to build a new nation. The fate of Iraq's Kurds, however, still remains unsettled.

Think About It

1. **Regions** How did the Kurds come to live in different countries?
2. **Movement** How has the situation of the Kurds changed in recent years?

Guide to Reading

BIG Idea

Cooperation and conflict among people have an impact on the Earth's surface.

Content Vocabulary

- secular *(p. 491)*
- bedouin *(p. 492)*
- kibbutz *(p. 493)*
- moshav *(p. 493)*
- clan *(p. 495)*
- alluvial plain *(p. 496)*
- embargo *(p. 497)*

Academic Vocabulary

- regime *(p. 492)*
- sole *(p. 494)*

Reading Strategy

Summarizing Use a diagram like the one below to list key facts about the economies of Southwest Asia.

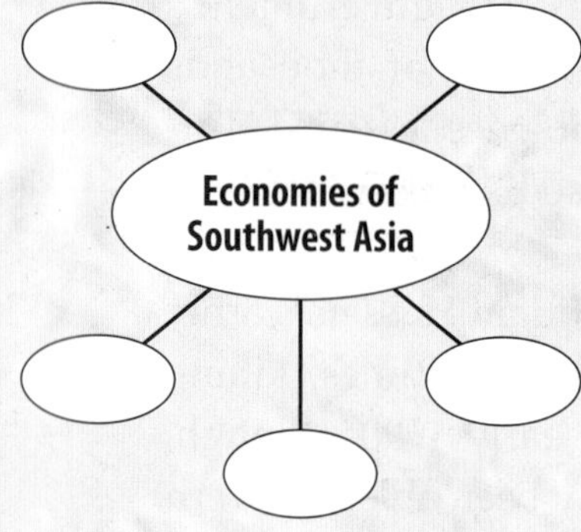

Southwest Asia

 Section Audio  Spotlight Video

Picture This Dubai is one of the seven states that form the United Arab Emirates. Dubai's business leaders are building the world in the Persian Gulf—the World Islands, that is. More than 300 artificial islands will be created to look, from above, like a map of the world, as seen in this computer-generated model. The islands will contain private homes, resorts, and businesses. Dubai is trying to expand its economy by relying more on tourism and less on oil production. To learn more about the diverse countries in Southwest Asia today, read Section 2.

The World Islands, Dubai, United Arab Emirates

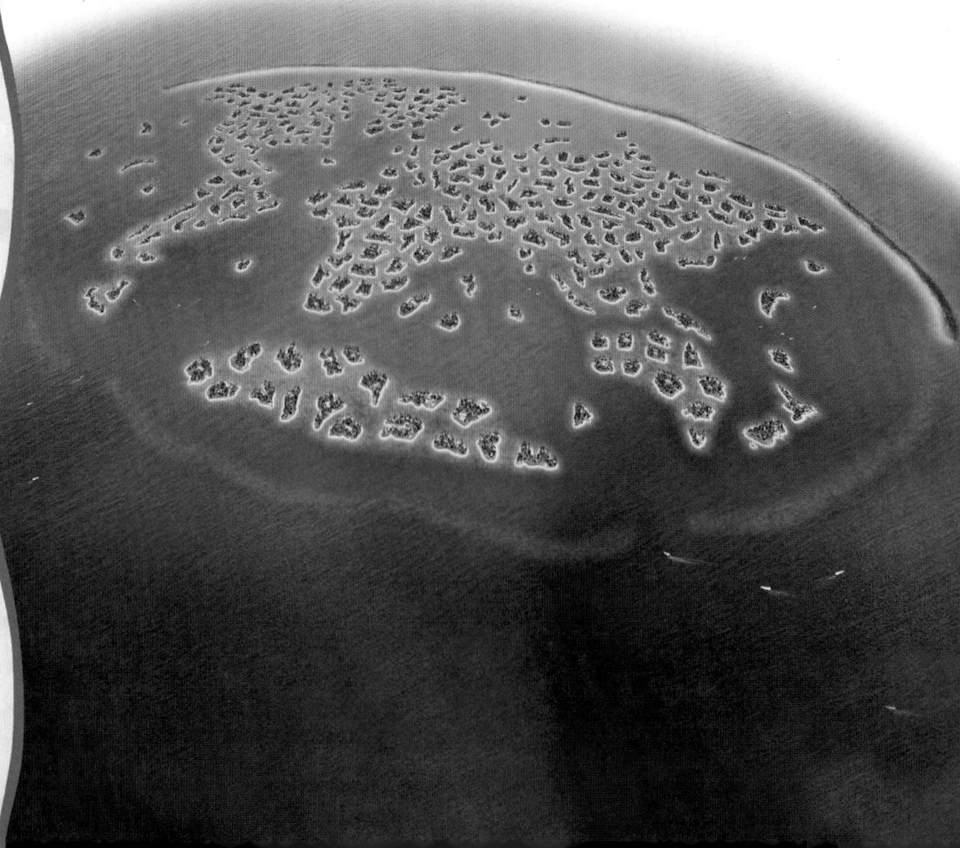

The Eastern Mediterranean

Main Idea **The countries in the eastern Mediterranean have faced many conflicts and are struggling to achieve peace.**

Geography and You Does anyone in your family have a special connection to your neighborhood or community? Read to find out how people in the eastern Mediterranean feel a strong connection with the area where they live.

Southwest Asia lies where Europe and Asia meet. Oil and natural gas are vital resources in much of this area, but the Southwest Asian countries along the eastern Mediterranean do not have much oil. They do, however, have areas with mild climates and fertile lands.

Turkey

Turkey is located at the entrance to the Black Sea and bridges the continents of Asia and Europe. Much of the country has a mild Mediterranean climate that allows farmers to grow food for local use, as well as cotton and tobacco for export. Turkey's economy is also supported by the production of textiles, steel, and cars.

Turkey has the largest population in Southwest Asia. Almost 70 percent of Turkey's people live in cities or towns. Most Turks are Muslims, and Turkish is the official language. The largest city and major economic center is Istanbul, a city filled with beautiful palaces and mosques. The capital and second-largest city is Ankara.

Turkey once was the heart of the Ottoman Empire. After the empire's fall, Turkey became a republic in 1923. Turkey's first president, Kemal Atatürk, did away with many Muslim traditions and made the country more European. Since the 1990s, Muslim political groups have gained support, although many Turks prefer a **secular**, or nonreligious, society. One minority group, the Kurds, wants to form its own country. Although Turkey has not allowed them to separate, Turkey's government has promised to respect the rights of Kurds and other non-Turkish groups.

Hagia Sophia

Hagia Sophia in Istanbul, Turkey, was built as a Christian church in the A.D. 500s, but became a mosque in 1453. It is a museum today. ***Place*** **How is the role of Islam changing in Turkish society?**

Syria

South of Turkey lie the mountains, deserts, and fertile coastal plains of Syria. Most of Syria's people live in rural areas, where they grow cotton, wheat, and fruit. Dams on the Euphrates River provide water for irrigation as well as electric power for cities and industries. Damascus, the capital, is one of the world's oldest continuously inhabited cities. It was founded as a trading center more than 4,000 years ago.

Like most of its neighbors, Syria has been controlled by many different foreign powers. In 1946 Syria became an independent country. Since the 1960s, one political party has controlled Syria. This **regime** (ray·ZHEEM), or government, does not allow many political freedoms.

Lebanon

Tiny Lebanon is about half the size of New Jersey. On fertile land along the Mediterranean Sea, farmers raise citrus fruits, vegetables, grains, olives, and grapes. Most Lebanese live in or near Beirut (bay·ROOT), the capital and major port. They work in banking, insurance, and tourism.

After being ruled by the Ottoman Turks and the French, Lebanon became independent in 1943. Most Lebanese speak Arabic, but their culture blends Arab, Turkish, and French influences. For many years, fighting between Lebanon's Christians and Muslims caused much destruction. Religious-based conflict is still a problem. In 2006 a Muslim group in Lebanon called the Hezbollah (hehz·BOW·lah) clashed with Israel, which increased tensions in the region.

Jordan

A land of contrasts, Jordan stretches from the fertile Jordan River valley in the west to dry, rugged desert in the east. Because Jordan lacks water resources, its farmers use irrigation to grow wheat, fruits, and vegetables in the Jordan River valley. People in cities, such as Amman, the capital, work in service and manufacturing industries. Jordan's desert is home to tent-dwelling **bedouin** (BEH·duh·wuhn), or nomads who traditionally raise livestock.

Jordan's people are mostly Arab Muslims. Their territory once was part of several empires. In 1946 Jordan gained independence as a constitutional monarchy. Jordan's rulers, such as King Hussein (hoo·SAYN) I, have worked to blend the country's traditions with modern ways.

Israel

Israel lies west of Jordan, on the Mediterranean Sea. As you may recall, in 1947 the United Nations gave the Jews control of land where Jewish communities had existed since ancient times. In 1948 Israel was proclaimed an independent Jewish republic by David Ben-Gurion (BEHN·gur·YAWN), Israel's first prime minister. At the time, however, Palestinian Arabs outnumbered the Jews living there. The Palestinian Arabs believed that Israel was founded on land that belonged to them. As a result, conflict between Israelis and Palestinian Arabs has taken place since 1948 and has claimed thousands of lives.

Today, about 76 percent of Israel's people are Jews. Many were encouraged to move to Israel because of the Law of Return.

Rebuilding in Lebanon

NATIONAL GEOGRAPHIC

▲ In Lebanon, residents have had to rebuild after a number of religious and political conflicts caused great destruction. ***Place*** **What cultural influences are found in Lebanon?**

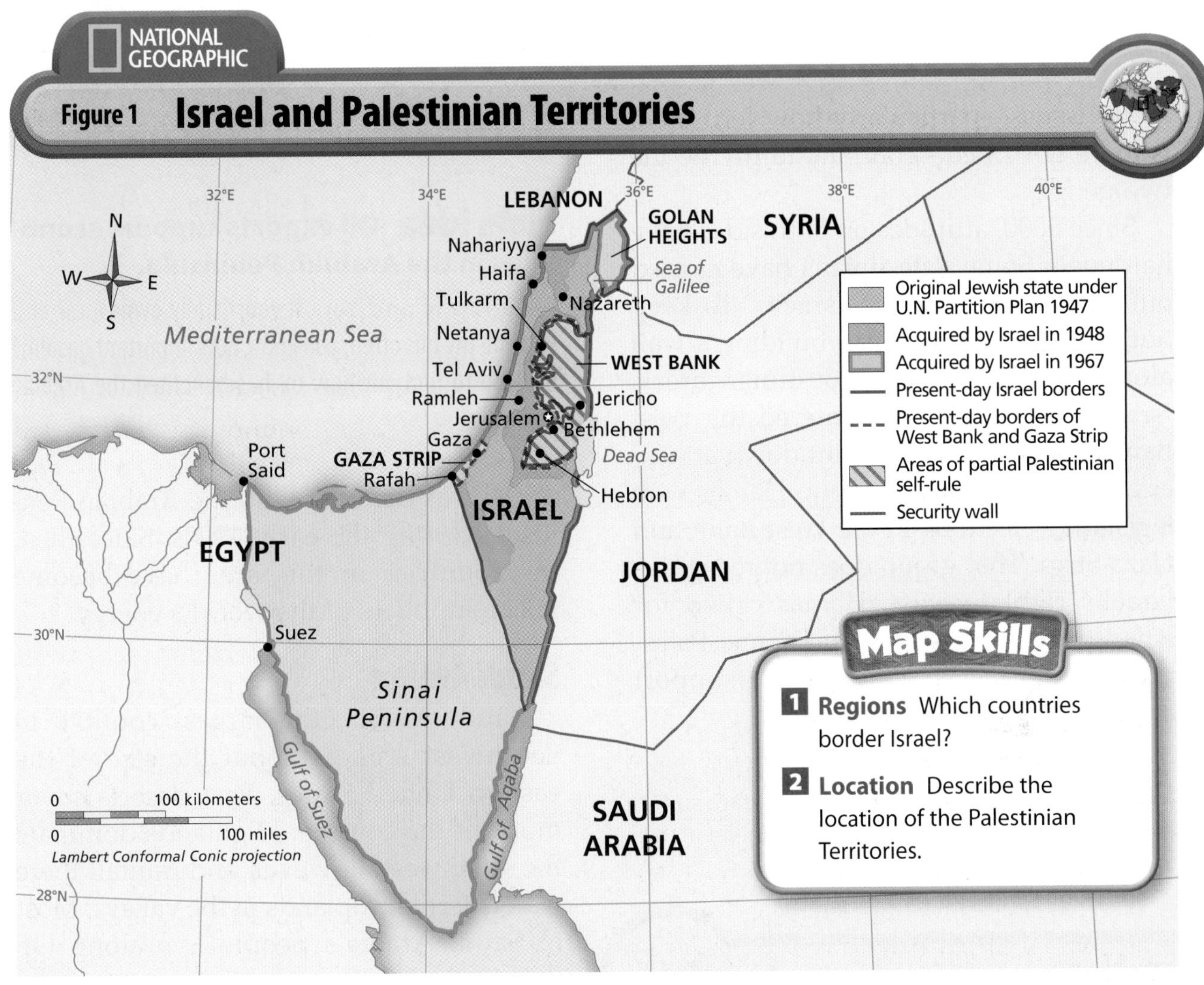

This law allows Jews anywhere in the world to immigrate to Israel and to become citizens. As a result, Jewish people have moved to Israel from many countries. Arabs, including both Muslims and Christians, make up 20 percent of Israel's population. More than 90 percent of Israel's people live in cities, such as Jerusalem, Tel Aviv, and Haifa (HY·fuh).

Israel has a developed industrial economy. Israelis produce high-technology equipment, clothing, chemicals, and machinery. The country also has productive agriculture. Advanced irrigation systems help grow citrus fruits, vegetables, and cotton. Some Israeli farmers live and work together in settlements. In a **kibbutz** (kih·BUTS), farmers share all of the work and property. In a **moshav** (moh·SHAHV), members share in the work, but each can also own some private property.

Palestinian Territories

A serious dispute exists between Israel and Palestinian Arabs. Nearly 8 million Palestinians are scattered throughout Southwest Asia today. **Figure 1** shows the Gaza Strip and the West Bank, two areas with many Palestinian residents that form the Palestinian Territories.

In 1993 Israel agreed to give the Palestinians limited self-rule in the Gaza Strip and parts of the West Bank. In return, moderate Palestinian leaders for the first time recognized Israel's right to exist. However, some Israeli Jews still lived in settlements in the West Bank, and tensions between Israelis and Palestinians remain.

Many issues—particularly how Jerusalem is to be governed—continue to divide the two peoples.

Since 2000, attitudes on both sides have hardened. Some Palestinians have carried out bombing attacks on Israelis. To keep out attackers, Israel began building a wall along its border with Palestinian areas. Israeli forces also have entered the West Bank and Gaza Strip to hunt down attackers. In 2006 the Islamic group Hamas won legislative elections in the West Bank and Gaza Strip. That group does not recognize Israel's right to exist. Hamas called for Israel to be replaced by an Islamic Palestinian state. It also continued to support armed attacks on Israeli territory.

Reading Check **Identifying** What are some crops grown in the eastern Mediterranean area?

▲ In the Palestinian Territories, daily life can be difficult. These women are waiting for medical aid at a clinic. ***Regions*** **What is the major source of conflict between Israel and the Palestinians?**

The Arabian Peninsula

Main Idea **Oil exports support economies in the Arabian Peninsula.**

Geography and You If your family owns a car or if you ride the bus often, you know how important gasoline is. Read to find out how oil has benefited the Arabian Peninsula.

Oil was discovered in the Arabian Peninsula during the early 1900s. Since then, the countries in this area have become major suppliers of the world's energy.

Saudi Arabia

Saudi Arabia, the largest country in Southwest Asia, is about the size of the eastern United States. Vast deserts cover much of the country. Highlands dominate the southwest, however, and rainfall there waters fertile croplands in the valleys. Most of Saudi Arabia's people live along the Red Sea and Persian Gulf coasts—the country's oil region—or around desert oases. The capital and largest city, Riyadh (ree·YAHD), sits amid a large oasis in central Saudi Arabia.

Saudi Arabia is one of the world's major oil producers. It has used money from oil to boost its standard of living. Schools, hospitals, roads, and airports have been built. Riyadh and other cities are filled with skyscrapers; freeways jammed with cars; and stores that carry cell phones, stereos, and other high-tech devices. Saudi Arabia's government has also been building new industries and improving agriculture so that the economy does not rely **solely** on oil, which will eventually run out.

Saudi Arabia has existed as a country since 1932. In that year, the Saud family established a monarchy that united the country's many **clans,** or groups of families related by blood or marriage. Under Saud rule, Islam has maintained a strong influence on life throughout the country. Government, business, school, and home life are organized to allow the five daily prayers required by Islam and to celebrate important holy days. The government also helps prepare the holy cities of Makkah (Mecca) and Madinah (Medina) for the several million Muslims who visit each year. Saudi customs concerning the roles of women in public life are stricter than in most Muslim countries.

The Persian Gulf Countries

Kuwait (ku·WAIT), Bahrain (bah·RAYN), Qatar (KAH·tuhr), and the United Arab Emirates are located along the Persian Gulf. These Persian Gulf countries have used profits from oil exports to build prosperous economies and provide free education, health care, and other services.

Like Saudi Arabia, these countries are also planning for a time when the oil runs out. Qatar has developed its natural gas industry, and Bahrain is now a banking center. Dubai (doo·BY), in the United Arab Emirates, is a large port, financial center, and tourist resort.

For decades, monarchs held all of the power in the Persian Gulf countries. Recently, some of these countries have moved toward democracy. Legislatures elected by voters now hold some of the power in Bahrain, Qatar, and Kuwait. Women in these nations also have voting rights.

Oman and Yemen

Oman lies at the southeastern corner of the Arabian Peninsula. Although the country is largely desert, it has important oil resources. Oman has used its oil wealth to build ports for oil tankers and has developed its tourist industry.

Oman's location has made the country important to world oil markets. The northern part of Oman lies along the strategic Strait of Hormuz. Oil tankers must go through this narrow waterway to pass from the Persian Gulf into the Arabian Sea.

Yemen, in the southwestern corner of the Arabian Peninsula, has little oil. Most of its people are farmers or sheep and cattle herders. They live in the high fertile interior where Sanaa, the capital, is located. Farther south lies Aden (AH·duhn), a major port for ships traveling between the Arabian Sea and the Red Sea.

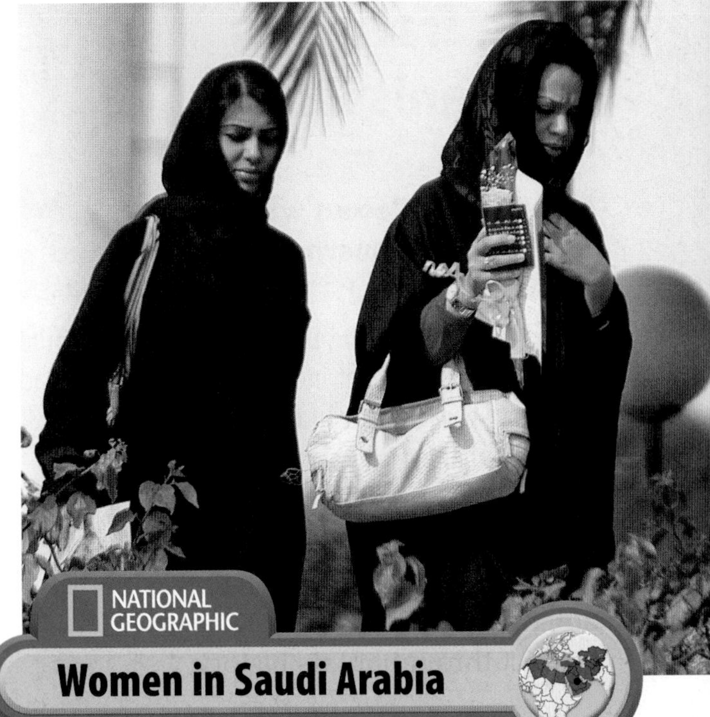

NATIONAL GEOGRAPHIC

Women in Saudi Arabia

Traditionally, Saudi Arabian women must wear an *abaya* (robe) and *niqab* (veil) when in public.
Regions **How does Islam influence Saudi culture?**

Reading Check **Comparing and Contrasting**
How does the role of women differ between Saudi Arabia and the Persian Gulf countries?

Iraq, Iran, and Afghanistan

Main Idea **Recent wars have changed the governments of Iraq and Afghanistan.**

Geography and You Do you know anyone in the military who has fought in Iraq or Afghanistan? Read to find out how these countries have been the focus of America's attention in recent years.

Iraq, Iran, and Afghanistan lie in an area where some of the world's oldest civilizations developed. This area has experienced conflict throughout its history.

Iraq

Some of the world's first-known cities arose between the Tigris and Euphrates Rivers. These rivers are the major geographic features of Iraq. Between the two rivers is an **alluvial plain,** or an area built up by rich soil left by river floods. Farmers here grow wheat, barley, rice, vegetables, dates, and cotton. Iraq's factories process foods and make textiles, chemicals, and building materials. Oil, however, is Iraq's major export.

About 70 percent of Iraqis live in urban areas. Baghdad, the capital, is the largest city. From the A.D. 700s to the A.D. 1200s, Baghdad was the center of an Arab Empire that made many advances in the arts and sciences. Muslim Arabs make up the largest group in Iraq's population. The majority are Shia, though a sizable portion are Sunnis. The third-largest group consists of the Kurds, who are Sunni Muslims and have long wanted to form their own country. **Figure 2** shows where Iraq's religious and ethnic groups live.

Modern Iraq gained its independence as a kingdom in 1932. In 1958 the last king was overthrown in a revolt. During the rest of the 1900s, Iraq was governed by dictators, including Saddam Hussein who ruled from 1979 to 2003. After Iraq invaded neighboring Kuwait, the United States successfully fought the Persian Gulf War in 1991. Following Iraq's defeat, the United Nations put an embargo on the country.

Maps In MOtion See StudentWorks™ Plus or glencoe.com.

Figure 2 **Iraq's Religious and Ethnic Groups**

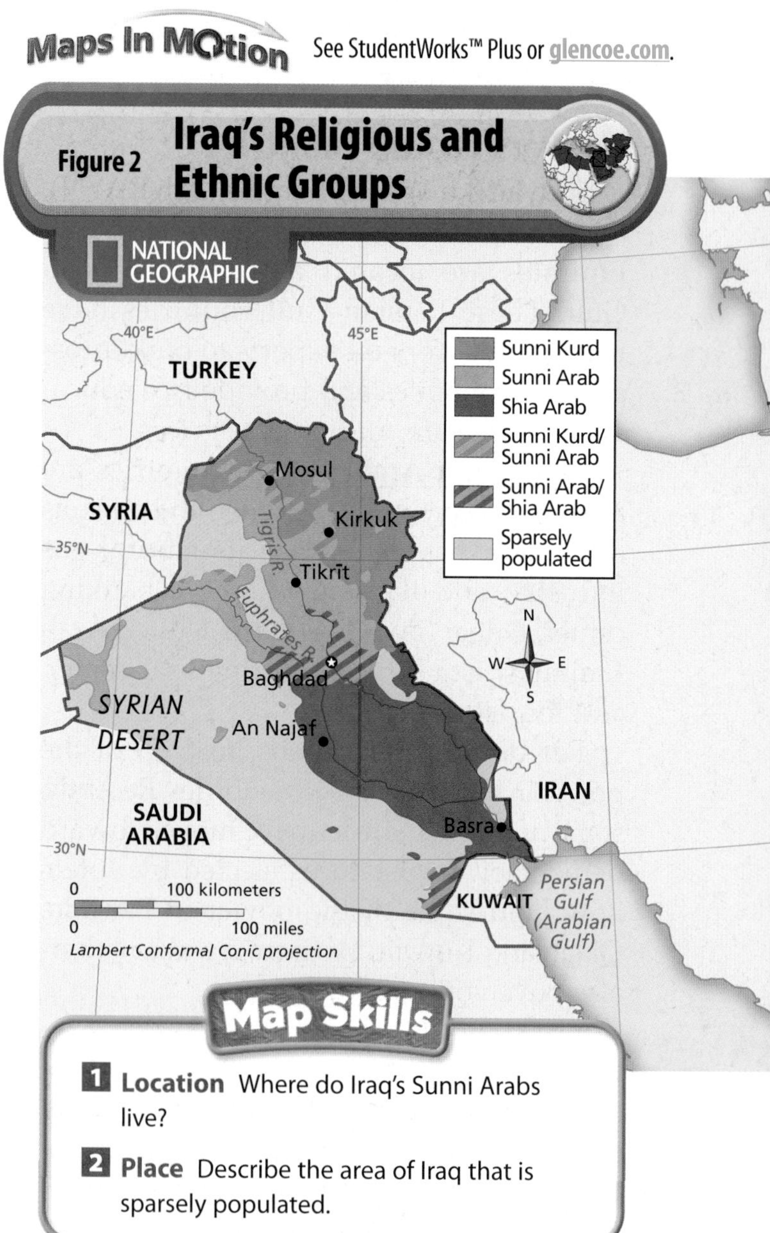

Map Skills

1. **Location** Where do Iraq's Sunni Arabs live?
2. **Place** Describe the area of Iraq that is sparsely populated.

Iran's Islamic Republic

Iranians ride past posters of Islamic religious leaders in Tehran, the capital. An elected president serves under Iran's religious leadership. ***Place*** **How was the Islamic republic established?**

An **embargo** is an order that restricts trade with another country. This embargo severely damaged Iraq's economy. Fearing that Iraq owned biological weapons, American and British forces invaded the country in 2003 and overthrew Saddam Hussein.

After Hussein's removal, rebels fought American forces in areas north and west of Baghdad. At the same time, intense disputes also grew among Iraq's Shias, Sunnis, and Kurds. The Shia majority had long been ruled by the Sunnis, who often treated them harshly. The Shia saw Hussein's downfall as a chance to gain power. Meanwhile, the Sunnis feared that a Shia-controlled government would attack them. Iraq's third major group—the Kurds—also pressed to have either their own country or self-rule within Iraq. These disputes sparked outbreaks of violence after Saddam's overthrow.

In the midst of the fighting, American and Iraqi officials tried to build a democracy in Iraq. In June 2004, American forces transferred power to a temporary Iraqi government. Elections held in December 2005 brought a high turnout of both Shia and Sunni voters. Although many Iraqis hope for stability, the future of their country remains uncertain.

Social Studies ONLINE

Student Web Activity Visit glencoe.com and complete the Chapter 18 Web Activity about U.S. involvement in Iraq.

Iran

Like many countries in Southwest Asia, Iran is a Muslim nation. In fact, nearly 90 percent of Iran's population is Shia Muslim. Unlike other populations in the region, however, most Iranians are not Arab. Three-fourths of Iran's people are Persians or Azeri. About 67 percent of Iranians live in cities, such as the capital, Tehran.

About 2,500 years ago, Iran was the center of the powerful Persian Empire that was ruled by monarchs. Through the centuries, control of the country has passed to various conquering peoples. In 1979 religious leaders overthrew the government, which was a monarchy. Iran is now an Islamic republic, a government run by Muslim religious leaders. The government has introduced laws based on its understanding of Islamic teachings. Many Western customs are seen as a threat to Islam and are forbidden in Iran.

Iran is an oil-rich nation. The first oil wells in Southwest Asia were drilled in Iran in 1908. Like other oil countries in the region, Iran is trying to build other industries to be less dependent on oil income. Major Iranian industries produce textiles, metal goods, and building materials. Farmers in Iran grow wheat, rice, sugar beets, nuts, and cotton.

Since 1995, the United States and other Western countries have accused Iran's leadership of attempting to develop nuclear weapons. Iran, however, claims it wants nuclear energy only to produce electrical power. So far, Iran has resisted foreign pressures to abandon its nuclear activities.

Afghanistan

To the east of Iran lies the landlocked, mountainous country of Afghanistan (af·GA·nuh·STAN). Afghanistan is covered by the rugged peaks of the Hindu Kush range. Its famous Khyber Pass has been a major trade route through the mountains for centuries. Afghanistan's capital, Kabul (KAH·buhl), lies in a valley.

The people of Afghanistan are divided into many different ethnic groups. The two largest groups are the Pashtuns and the Tajiks. About 70 percent of the people herd livestock or grow crops, such as wheat, fruits, and nuts. Wool and handwoven carpets are important export products.

In 2001 terrorists with support from groups in Afghanistan attacked the United States. The United States responded by invading and overthrowing Afghanistan's terrorist-supporting Taliban government. With American help, Afghanistan began to build a democracy. By 2006, the country had held elections for a president and parliament. The new government could not control the entire country, however. Powerful local leaders as well as Taliban fighters maintained control over small areas. Their attacks prevented the government from gaining control over the whole country.

Reading Check **Identifying** Who controls Iran's government?

Section 2 Review

Social Studies ONLINE
Study Central™ To review this section, go to glencoe.com.

Vocabulary

1. **Explain** the meaning of the following terms by writing a separate sentence for each one.
   - **a.** secular
   - **b.** bedouin
   - **c.** kibbutz
   - **d.** moshav
   - **e.** clan
   - **f.** alluvial plain
   - **g.** embargo

Main Ideas

2. **Sequencing** Use a flowchart like the one below to create a time line of selected events in Israel and the Palestinian territories since 1948.

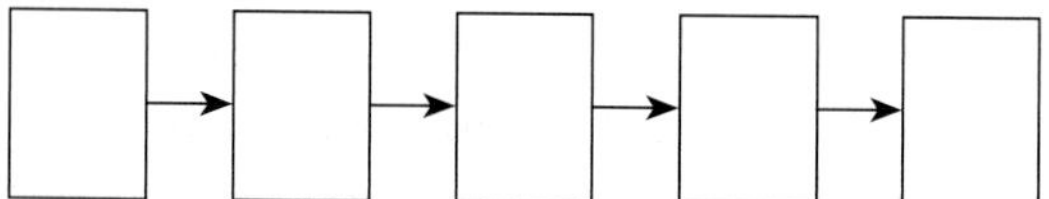

3. **Describing** How has oil production affected Saudi Arabia's standard of living?

4. **Identifying** What sort of government does Iran have, and how does it affect life there?

Critical Thinking

5. **BIG Idea** How has oil affected the economies of the region?

6. **Challenge** Considering the region's history, why is religious conflict a problem in Southwest Asia?

Writing About Geography

7. **Expository Writing** Choose two countries in Southwest Asia with different forms of government. Write several paragraphs comparing the governments of those countries. Be sure to discuss any religious influences on the governments as well as issues the governments must face.

TIME PERSPECTIVES

EXPLORING WORLD ISSUES

SOUTHWEST ASIA'S UNCERTAIN FUTURE

A war-torn region is working toward peace. But will it last?

People stand on the rubble of destroyed houses in Fallujah, Iraq.

Southwest Asia is one of the world's most important regions. Oil produced in countries like Saudi Arabia, Iraq, and Iran provides fuel for many of the world's economies, including the United States. Because of this, events in the region often have a major impact on global affairs.

In recent years, some of the nations in the region have been working to build peace. After decades of dictatorship and war, Iraq has begun to create a new government. In 2005, Israel withdrew from some of the land that Palestinians claimed was theirs. This was an important step toward establishing a permanent home for the Palestinians.

As promising as these important advances were, the troubled region still faces major challenges. Will a lasting peace be achieved?

Iraqi women cast their votes during the parliamentary election.

REBUILDING IRAQ

In April 2003, the United States led the military invasion of Iraq that toppled Saddam Hussein, the country's long-time dictator. Since then, Iraqis have been trying to rebuild their nation and turn their country into a democracy.

Rebuilding Iraq would take time and hard work. First, the nation's energy and transportation systems would need to be fixed. Many of Iraq's power plants, water stations, and roads had been in terrible shape for years. The fighting that drove Hussein from power caused even more damage. In many parts of the country, electricity and clean water were unavailable for long periods of time. Doing simple chores like shopping or cleaning became major challenges.

Iraq's Muslims

(As a percentage of Iraq's population)

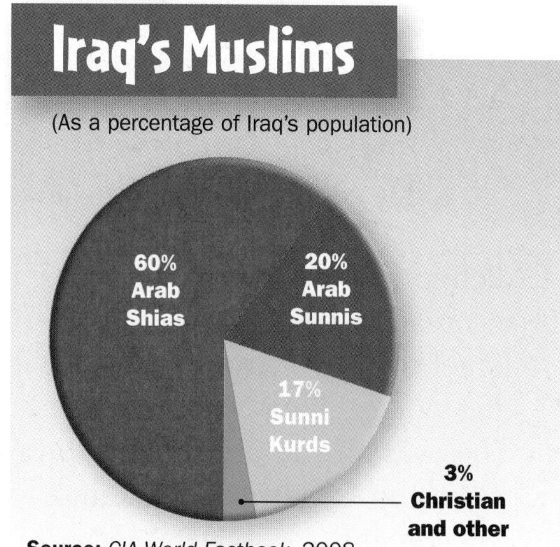

Source: *CIA World Factbook*, 2008.

INTERPRETING CHARTS

Analyzing Data Why might Shias feel they have more of a right to run Iraq's government than other groups?

Laying a Foundation

Creating a new democracy was even more challenging. Iraqis had to write a **constitution**, or body of laws, that described how each part of the government would work. They had to hold free elections and hire people to make the government function.

It has not been easy changing from a dictatorship to a democracy. Iraq's population is deeply divided along religious and ethnic lines. Iraqi Muslims are divided into different groups. **Shia** (SHEE-yah) Muslims belong to one of the two main branches of Islam. **Sunnis** (SOO-nees) are members of the second branch. **Kurds** are an ethnic minority who live mostly in northern Iraq. The three groups bitterly disagree about many issues—including the direction their country should take.

Iraq is also plagued by violence. **Insurgents**, or rebels who oppose the new democratic government, have fought a violent battle with other Iraqis and U.S. troops. Their terrorist acts have killed thousands of innocent people.

Shias attend prayers at a mosque.

AP PHOTO
U.S. troops patrol the streets of Fallujah, the scene of deadly insurgent attacks.

AP PHOTO
Iraqis distribute copies of the draft constitution to the public.

AP PHOTO
Iraqis work in an oil field near Basra.

By 2008, it was still too soon to know if Iraq's citizens would be successful at rebuilding their country. The Iraqis had, however, taken two key steps toward their future.

A Historic Election

In December 2005, Iraqis elected the country's first **parliament**, or lawmaking body. Shias, Sunnis, Kurds, and other groups voted for 275 members of parliament who would finalize a permanent constitution.

About 11 million of the nation's 15 million registered voters, including many women, took part in the election. It was one of the highest voter turnouts ever in an Arab country. "It's the beginning of our new life," said Buthana Mehdi, a schoolteacher.

Saddam on Trial

In 2005 Saddam Hussein and seven other former leaders stood trial before an Iraqi **tribunal**, or court. The eight were charged with violent crimes against their country. In late 2006, Hussein was found guilty and executed. Hussein, a Sunni, ruled the country from 1979 until he was forced from power in 2003 by the U.S.-led invasion. Many Iraqis say that Hussein tortured, jailed, and killed those who disagreed with him.

Iraqis hoped Hussein's trial would end a painful chapter in their country's history. "Saddam is gone and we are moving ahead," said Iraq's prime minister Ibrahim al-Jaafari.

What Will Happen Next?

In 2008, Iraq still faced many difficult challenges. Leaders in Iraq and the United States expected attacks by insurgents to continue across the nation. And Iraq's government still had to figure out how to extract and sell important resources, such as oil.

The United States had to make some tough decisions, too. Many Americans were wondering when U.S. troops would come home. President George W. Bush presented his plan for getting out of Iraq. "Our goal is to train enough Iraqi forces so they can carry the fight," said Bush. "This will take time and patience." Clearly, building a democratic and peaceful Iraq will not be an easy job.

EXPLORING THE ISSUE

1. **Making Connections** What do you think is the most important task of creating a new system of government?

2. **Summarizing** What steps have Iraqis taken since the start of the war to stabilize their country?

A STRUGGLE FOR PEACE

The ongoing conflict between Israelis and Arabs has a long—and often violent—history. When the British gave up control of Palestine after World War II, the United Nations divided the area into Arab and Jewish states. The Arabs rejected this plan and war broke out.

The two sides have battled for decades. In 1967, after one of the several wars between Israel and its Arab neighbors, Israel took control of the Gaza Strip, a sliver of land on the Mediterranean Sea. Israel also took parts of the West Bank. A violent cycle of attacks and revenge attacks followed, and thousands of people died.

Today's **Palestinians** are the descendants of Arab people who fled their homes during the Arab-Israeli wars. They believe that Israel took their land. The Israelis, on the other hand, argue that the territory is rightfully theirs.

A Fragile Plan

In 2005, Israel and the Palestinian Authority agreed to a **cease-fire**. By the year's end, Israel had withdrawn Jewish settlers from Gaza. Mahmoud Abbas, the Palestinian president, called the withdrawal a "historic step" toward peace.

A year later, however, Palestinians elected a parliament controlled by the Hamas Party. Hamas has called for Israel's destruction. Worried by the Hamas victory, world leaders backed Abbas and moderate Palestinians. In 2007, rising tensions between Hamas and moderate Palestinians led to fighting. Hamas seized the Gaza Strip, leaving Abbas in charge of the West Bank.

As the Palestinians split into two groups, the United States persuaded Israel and moderate Palestinians to hold talks. Their goal was to achieve peace and set up a fully independent Palestinian state. Meanwhile, Hamas staged rocket attacks on Israel from Gaza, and Israel continued to build Jewish settlements in the West Bank. Events once again threatened to halt peace efforts in the region.

AP PHOTO

Israeli police escort a family out of their home in a Jewish settlement in Gaza.

EXPLORING THE ISSUE

1. **Making Predictions** In your opinion, will the peace effort succeed?

2. **Problem Solving** What are some of the problems that keep the Israeli-Palestinian conflict alive? How might the region's leaders solve them?

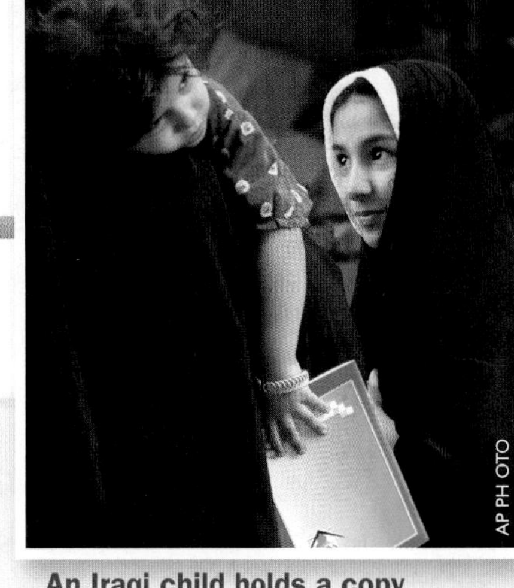

An Iraqi child holds a copy of the country's draft constitution.

REVIEW AND ASSESS

UNDERSTANDING THE ISSUE

1 **Making Connections** What types of issues are resolved in a constitution? Do you think a constitution will bring an end to the violence in Iraq? Why or why not?

2 **Writing to Inform** In a short essay, explain some of the important steps Iraqis have taken to rebuild their country.

3 **Writing to Persuade** Why was it important that Iraq's former leader, Saddam Hussein, stand trial? Put your answer in a brief letter to the editor of your local newspaper. Support your opinion with facts.

INTERNET RESEARCH ACTIVITIES

4 Go to the home page of the Iraq Foundation, www.iraqfoundation.org. Scroll down to the "Projects" link and read about the foundation's work to promote democracy and human rights in Iraq. Write two paragraphs explaining one project and share it with your classmates.

5 Navigate to the Web site of Columbia University's Middle East and Jewish Studies, www.columbia.edu/cu/lweb/indiv/mideast/cuvlm. Scroll down to the "Religion in the Middle East" link. Read about one religion and its history in the region. Write about what you learned.

BEYOND THE CLASSROOM

6 **At your school or local library,** find out more about Iraq's major religious and ethnic groups: Shias, Sunnis, and Kurds. What are some of the issues about which they disagree?

7 **Research the United Nations'** involvement in the creation of Israel. Find out how the Jewish and Arab people responded to the division of Palestine. Present your findings to the class.

Tension Between Neighbors

Religion plays a major role in the often-troubled relations between Iran and Iraq. This map shows the number of people who belong to each religious group in the two countries.

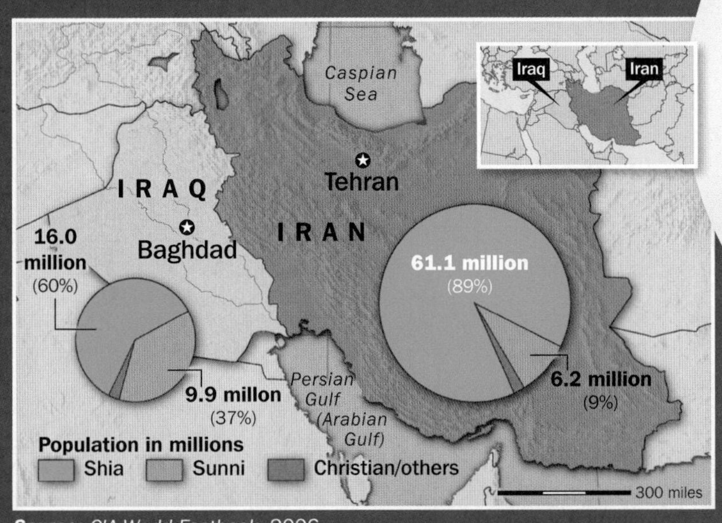

Source: *CIA World Factbook*, 2006.

Building Graph Reading Skills

1. **Analyzing Information** What religious group has the most followers in Iran?
2. **Comparing** How are the populations of Iran and Iraq alike?

Guide to Reading

BIG Idea

Places reflect the relationship between humans and the physical environment.

Content Vocabulary

- cash crop *(p. 505)*
- fault *(p. 507)*
- genocide *(p. 508)*
- enclave *(p. 508)*

Academic Vocabulary

- emphasis *(p. 505)*
- output *(p. 506)*

Reading Strategy

Making Generalizations Use a diagram like the one below to write three important facts about this region in the smaller boxes. Then, in the larger box, write a generalization that you can draw from those facts.

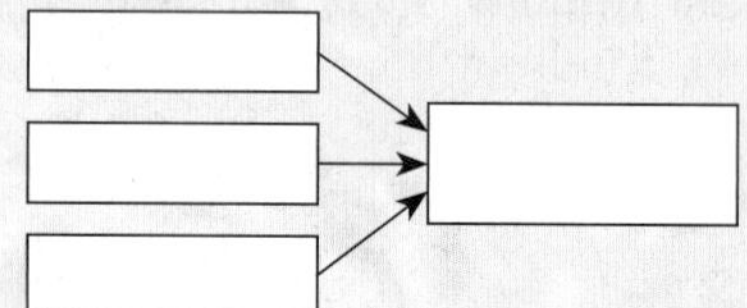

Central Asia

 Section Audio  Spotlight Video

Picture This 3 . . . 2 . . . 1 . . . The *Soyuz-U* rocket lifts off from the Baikonur Cosmodrome in Kazakhstan. The aerospace center in Kazakhstan was formerly a secret launch site for the Soviet Union's space program. Today, Russia leases the base from Kazakhstan and launches satellites, along with cargo ships that take supplies to the astronauts living in the International Space Station. Read this section to learn more about Central Asia today.

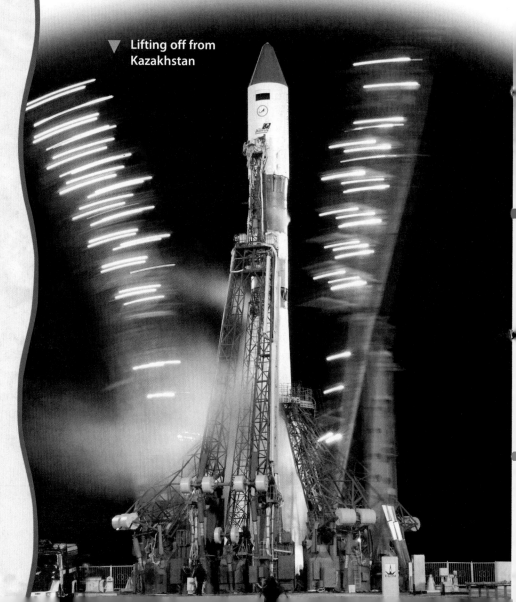

▼ Lifting off from Kazakhstan

The Central Asian Republics

Main Idea **The Central Asian Republics are working to improve their economies after years of Communist rule.**

Geography and You Have you ever heard of the "stan" republics? They are five Central Asian countries that were once part of the Soviet Union. Read to find out more about their place in the world today.

The republics of Central Asia include Kazakhstan (kuh·ZAKH·STAHN), Uzbekistan, Turkmenistan, Kyrgyzstan (KIHR·gih·STAN), and Tajikistan. These countries spread across a huge area of land east of the Caspian Sea. All are located in harsh environments but have a wealth of mineral resources. All are Islamic countries that once were part of the Soviet Union.

Kazakhstan

Kazakhstan is the largest of the Central Asian Republics. In fact, this vast country is nearly four times the size of Texas. Dry, treeless plains cover much of its landscape. Farming is limited, but raising livestock has become an important industry. The country is rich in minerals, such as copper and petroleum.

About half of Kazakhstan's people are ethnic Kazakhs. They were once horse-riding nomads. However, the Soviets forced the Kazakhs to live settled lives. The Soviet government also set up factories, and Russian workers poured into the country. As a result, Russians now form the second-largest ethnic group in the country.

After the Soviet collapse in 1991, Kazakhstan became independent but did not adopt democracy. Today Kazakhstan's leaders keep a tight grip on citizens and deny rights to their political opponents. Nevertheless, the leaders have set up a free market economy, and many government-run industries have been sold to individual buyers. Furthermore, foreign businesspeople have begun investing in industries in the country, which should boost its economy.

Kazakhstan's Economy

Under a free market economy, entrepreneurs in Kazakhstan are able to start and manage their own businesses. ***Place*** **Describe Kazakhstan's economic activities.**

Uzbekistan

South of Kazakhstan lies Uzbekistan, a country slightly larger than California. Most of the country's people are Uzbeks. Many live in fertile valleys and oases. Tashkent, the capital, is Central Asia's largest city and industrial center. About 2,000 years ago, the oases of Tashkent, Bukhara, and Samarqand were part of the busy trade route called the Silk Road that linked China and Europe.

Today Uzbekistan relies on agriculture. Cotton is the country's major **cash crop,** or farm product grown for sale as an export. The heavy **emphasis** on cotton production has had a negative impact on the environment. Cotton is a thirsty crop.

Nomadic Life

A nomadic Kyrgyz woman in the mountainous Tian Shan area of Kyrgyzstan cooks milk in order to make cheese. ***Place*** **What ethnic groups make up Kyrgyzstan's population?**

Large farms using irrigation have nearly drained away the rivers flowing into the Aral Sea, reducing its size and crippling its fishing industry. Today Uzbek leaders are trying to vary the economy by drawing on newly discovered deposits of oil, gas, and gold.

Turkmenistan

Turkmenistan is larger than Uzbekistan, but it has far fewer people. Most of the country is part of a huge desert called the Kara-Kum (KAHR·uh·KOOM). Turkmenistan has a largely ethnic Turkmen population. Most people live in oases, where they grow cotton and raise livestock.

Although the land is harsh, Turkmenistan is important to the world's energy markets. It contains abundant amounts of petroleum and natural gas. The government hopes to increase oil and natural gas **output,** or the amount produced, to boost the economy.

Turkmenistan's capital and major city is Ashkhabad. A powerful president runs the country from this city. Turkmenistan's government keeps strict control over education, religion, and printed materials.

Kyrgyzstan

Kyrgyzstan is largely mountainous. Farmers raise cotton, vegetables, and fruit in valleys and plains. Kyrgyzstan has valuable mercury and gold deposits but little industry. The government has sought foreign investment in an attempt to help small businesses grow.

More than half of Kyrgyzstan's people belong to the Kyrgyz ethnic group. Important ethnic groups in the country include Russians, Uzbeks, and Ukrainians.

Kyrgyzstan has faced much political unrest. In 2005 the government was overthrown during a revolt. Kyrgyzstan's new leaders have resisted democratic reforms.

Tajikistan

Mountainous Tajikistan lies south of Kyrgyzstan. The country has fertile mountain valleys in which farmers grow cotton, grapes, grain, and vegetables. Factory workers in urban areas produce aluminum, vegetable oils, and textiles. The largest city and industrial center is Dushanbe (DYOO·SHAHM·buh), the capital.

Most of Tajikistan's people are ethnic Tajiks or ethnic Uzbeks. In the 1990s, a bitter civil war between the government and certain Muslim political groups killed many people and damaged the economy. Since the fighting ended in 1997, recovery has been slow, and political tensions remain high.

Reading Check **Analyzing Information**
What might be the advantages and disadvantages of starting a business in a Central Asian country?

The Caucasus Republics

Main Idea **The Caucasus countries are new nations with diverse ethnic groups that often find themselves in conflict with each other.**

Geography and You How might ethnic diversity prevent a country from developing? Read to find out about the challenges facing the Caucasus Republics.

The tall Caucasus (KAW·kuh·suhs) Mountains extend across Armenia, Georgia, and Azerbaijan. Thus, these countries are known as the Caucasus Republics. Like the Central Asian Republics, the Caucasus Republics were once part of the Soviet Union.

Despite their mountainous landscape, the Caucasus Republics generally have mild climates. The climate supports commercial farming in river valleys. Farmers in these areas grow tea, citrus fruits, wine grapes, and vegetables.

▲ While part of the Soviet Union, Georgians had few political freedoms. Today, leaders are chosen through free and democratic elections. ***Place*** **When did Georgia become independent?**

Georgia

The northernmost Caucasus Republic is Georgia, a country that borders the Black Sea. Its mountains contain many mineral resources, such as copper, coal, manganese, and oil. Swift rivers provide hydroelectric power for Georgia's industries. Heavy industries here make railway locomotives and construction vehicles. Georgia's steel factories developed when the country was part of the Soviet Union. Other industries produce cotton, wool, and silk fabrics, as well as clothing made from those materials.

Tbilisi (tuh·bih·LEE·see), Georgia's capital, lies near the Caucasus Mountains. Because the city is located in an area where tectonic plates collide, it has warm mineral springs heated by high temperatures inside the Earth. Bathing in these hot springs has long been thought to provide positive effects for a person's health. The area's resorts, along with the warm beaches found on the Black Sea, have made the country popular with those seeking to improve their well-being.

Most of Georgia's people are ethnic Georgians who are proud of their unique language, culture, and Christian heritage. Georgia became independent after the fall of the Soviet Union in 1991. Since then, conflict has taken place between Georgians and other ethnic groups in the country who want independence.

Armenia

South of Georgia and east of Turkey lies landlocked Armenia. Armenia sits on top of many **faults,** or cracks in the Earth's crust caused by colliding tectonic plates.

As a result, the country suffers frequent, serious earthquakes. For example, in 1988 a major earthquake in Armenia caused thousands of deaths and widespread destruction.

Armenia's people are mostly ethnic Armenians who share a unique language and ancient culture. Yerevan, the capital, is one of the world's oldest cities. In A.D. 301 Armenia made Christianity its official religion—the first country in the world to do so. Throughout much of their long history, Armenians have been ruled by other peoples, including the Ottoman Turks and the Russians. During World War I, the Ottoman Turks killed hundreds of thousands of Armenians in a terrible **genocide**, or the deliberate killing of an ethnic group. Many Armenians who survived fled to Southwest Asia, Europe, and the United States. The Armenians who remained joined their country to Soviet Russia.

In 1991 Armenia became an independent republic. Shortly afterward, Armenia sent its army to protect ethnic Armenians living in a small enclave surrounded and ruled by neighboring Azerbaijan. An **enclave** is a small territory surrounded by a larger territory. Fighting over this land hurt the economies of both countries. Today the dispute remains unsettled.

Azerbaijan

Azerbaijan lies on the eastern edge of the Caucasus region. Most of its people are Azeris and practice Shia Islam. The largest city is the capital, Baku (bah·KOO), a port on the Caspian Sea.

Azerbaijan has a developing economy. Farmers use irrigation to grow grains, cotton, and wine grapes. Oil and natural gas deposits under the Caspian Sea promise a bright future for Azerbaijan. The country has made agreements with foreign companies to develop and transport these resources.

Reading Check **Drawing Conclusions** Why is Azerbaijan's economic future promising?

Section 3 Review

Social Studies ONLINE
Study Central™ To review this section, go to glencoe.com.

Vocabulary

1. **Explain** the significance of:
   a. cash crop
   b. fault
   c. genocide
   d. enclave

Main Ideas

2. **Analyzing** What do the five Central Asian Republics have in common?
3. **Comparing** Use a chart like the one below to compare one republic in Central Asia to another in the Caucasus.

	Country 1	Country 2
Economy		
People		

Critical Thinking

4. **Identifying Central Issues** What are the key issues facing the countries in Central Asia and the Caucasus Republics?
5. BIG Idea How does Central Asia's physical geography affect the region's economies?
6. **Challenge** What role does religion play in the Caucasus Republics?

Writing About Geography

7. **Persuasive Writing** Write a paragraph explaining which of the countries in Central Asia you think will have the strongest economy in the future.

Visual Summary

North Africa

- Most of Egypt's people live in the Nile River valley.
- Many Egyptians are farmers, although industries have grown in recent years.
- The landscape of Libya and the Maghreb is mostly desert and mountains.
- Most people in North Africa are Muslims and speak Arabic.

Suez Canal, Egypt

Eastern Mediterranean

- Turkey lies in both Europe and Asia.
- Damascus, Syria's capital, is one of the world's oldest cities.
- Farmers grow fruits and vegetables on fertile coastal land.
- Israel was founded in 1948 as an independent Jewish republic.
- Religious and political conflicts continue in the area.

Member of the Afghani legislature

The Arabian Peninsula

- Saudi Arabia is the world's leading oil producer.
- The holy cities of Makkah and Madinah make Saudi Arabia an important Islamic center.
- Some Persian Gulf states have recently adopted democratic reforms.

Iran, Iraq, and Afghanistan

- In 2003 U.S.-led forces overthrew Iraq's dictator. Despite turmoil, Iraq is trying to build a democracy.
- Most Iranians are Shia Muslims. Since 1979, Iran has been an Islamic republic.
- Mountainous Afghanistan has many different ethnic groups.

Central Asia and the Caucasus

- The Central Asian Republics and Azerbaijan are mostly Muslim. Armenia and Georgia are mostly Christian.
- Central Asia and the Caucasus Republics have been using natural resources to rebuild their economies since the Soviet collapse.
- The Caucasus Republics have faced ethnic conflicts in recent years.

Muslim pilgrims, Makkah, Saudi Arabia

Study anywhere, anytime! Download quizzes and flash cards to your PDA from glencoe.com.

STANDARDIZED TEST PRACTICE

Some questions based on visuals require you to apply outside knowledge of a subject. As you study the visual, think about what else you have learned about the subject.

Reviewing Vocabulary

Directions: Choose the word(s) that best completes the sentence.

1. Egyptian ________, or peasant farmers, rely on the Aswān High Dam to control the Nile's waters.

A casbahs
B fellahin
C Kurds
D mosques

2. Many Turks prefer a nonreligious, or ________, government.

A Kurdish
B regime
C dictatorial
D secular

3. During World War I, Ottoman Turks killed hundreds of thousands of Armenians in a terrible ________.

A enclave
B regime
C genocide
D embargo

4. Israeli settlements in which farmers live together and share the land and work is called a ________.

A kibbutz
B bedouin
C regime
D clan

Reviewing Main Ideas

Directions: Choose the best answer for each question.

Section 1 *(pp. 484–488)*

5. Egyptian farmers are able to produce more than one crop per year because of ________.

A heavy rainfall
B advanced agricultural technology
C flooding controls on the Nile River
D vast stretches of fertile land

Section 2 *(pp. 490–498)*

6. Conflict between Palestinian Arabs living in the West Bank and Gaza Strip and ________ has been an ongoing problem.

A Saudis
B Israelis
C Kurds
D Jordanians

7. ________, the largest country in Southwest Asia, is a major oil producer.

A Israel
B Kurdistan
C Palestine
D Saudi Arabia

Section 3 *(pp. 504–508)*

8. While the Central Asian Republics are located in harsh environments, they all ________.

A possess a wealth of mineral resources
B are highly industrialized
C are predominantly Christian
D have open democratic governments

GO ON

Critical Thinking

Directions: Base your answers to questions 9 and 10 on the circle graph below and your knowledge of Chapter 18. Choose the best answer for each question.

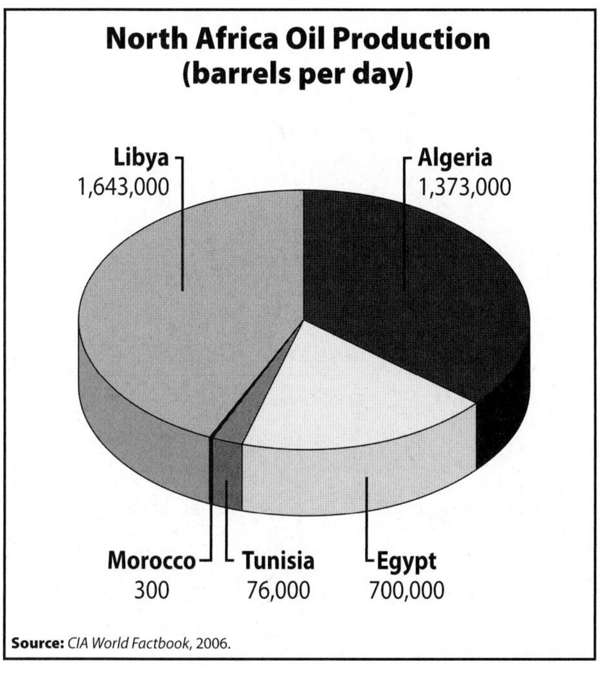

9. According to the circle graph, which country in North Africa produces the most oil?

   A Egypt

   B Morocco

   C Libya

   D Tunisia

10. How many barrels of oil per day are produced in Tunisia?

    A 700,000

    B 76,000

    C 1,640,000

    D 1,370,000

Document-Based Questions

Directions: Analyze the document and answer the short-answer questions that follow.

While Egypt's economy has—in general—been improving, certain portions have suffered setbacks.

> ***Cairo's Changing Face***
> *Mounir Ibrahim*
> *June 3, 2005*
>
> *Tourism was perhaps the most promising career in Egypt; getting a job with a large international hotel chain or tour group proved [well-paying] for the average Egyptian. . . . [T]ourism is still a huge industry in Egypt, but unfortunately it is stagnant and possibly even shrinking. Tourists are increasingly reluctant to travel to Egypt, due to the war in Iraq, the ongoing instability in Palestine, and the post 9/11 effect that still resides today. Consequently, tourism is no longer the promising profession it once was for Egyptians.*
>
> *—Mounir Ibrahim, "Cairo's Changing Face"*

11. According to the writer, why is tourism important to Egypt's workers?

12. What is happening to tourism in Egypt? Why?

Extended Response

13. Write an essay comparing and contrasting the economies of the three subregions discussed in the chapter. Be sure to describe the products that come from each subregion, how the political situation in the area is affecting the economy, and the possible economic futures for each subregion.

STOP

For additional test practice, use Self-Check Quizzes—Chapter 18 at glencoe.com.

Need Extra Help?

If you missed question...	1	2	3	4	5	6	7	8	9	10	11	12	13
Go to page...	485	491	508	493	485	494	494	505	485	485	485	485	484–508

Africa South of the Sahara

CHAPTER 19
Physical Geography of Africa South of the Sahara

CHAPTER 20
History and Cultures of Africa South of the Sahara

CHAPTER 21
Africa South of the Sahara Today

Waterfront in Cape Town, South Africa

NATIONAL GEOGRAPHIC

NGS ONLINE For more information about the region, see www.nationalgeographic.com/education.

Regional Atlas

Africa South of the Sahara

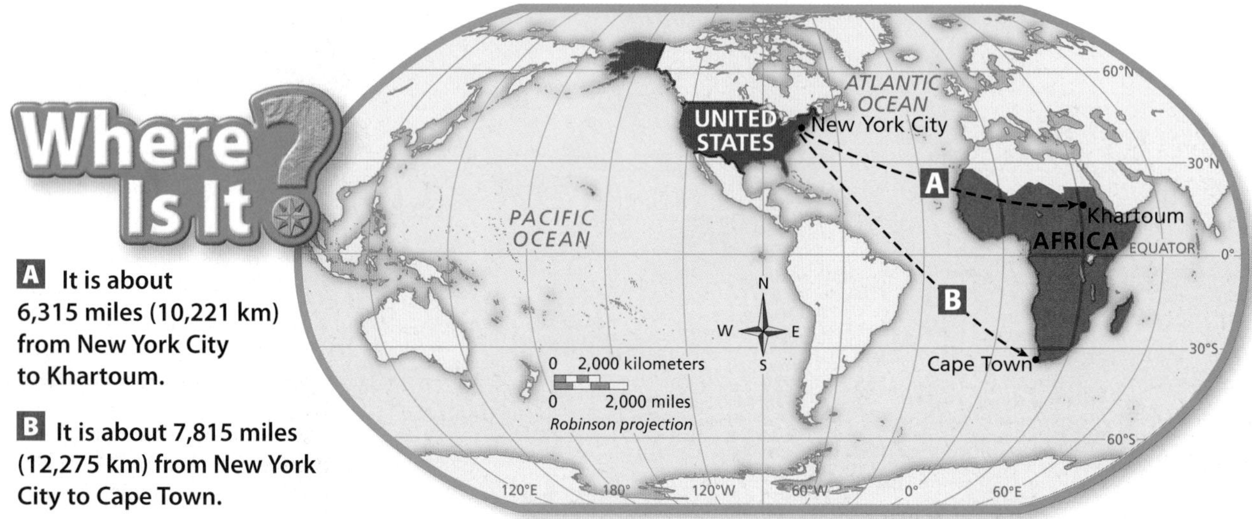

Where Is It?

A It is about 6,315 miles (10,221 km) from New York City to Khartoum.

B It is about 7,815 miles (12,275 km) from New York City to Cape Town.

At about 10.3 million square miles (26.8 million sq. km), Africa south of the Sahara accounts for about one-fifth of all the land in the world. The region is about three times larger than the United States.

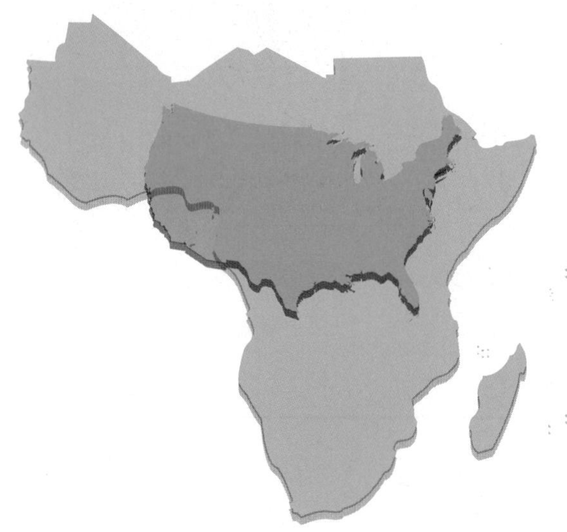

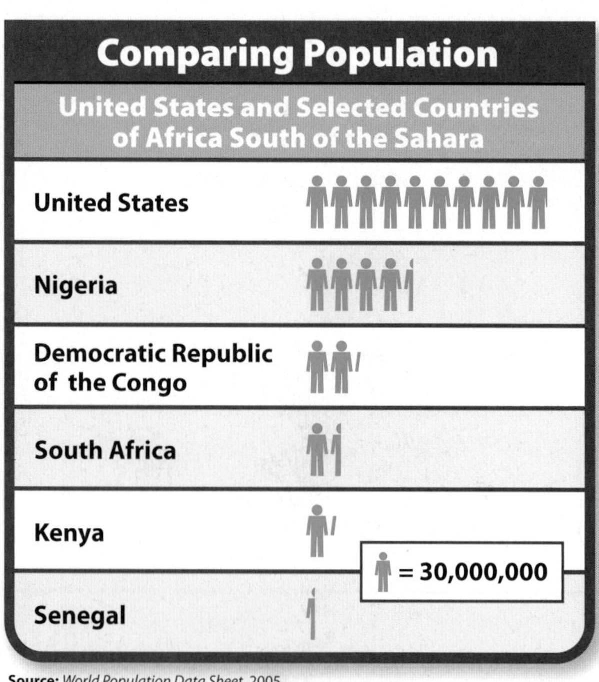

Source: *World Population Data Sheet*, 2005.

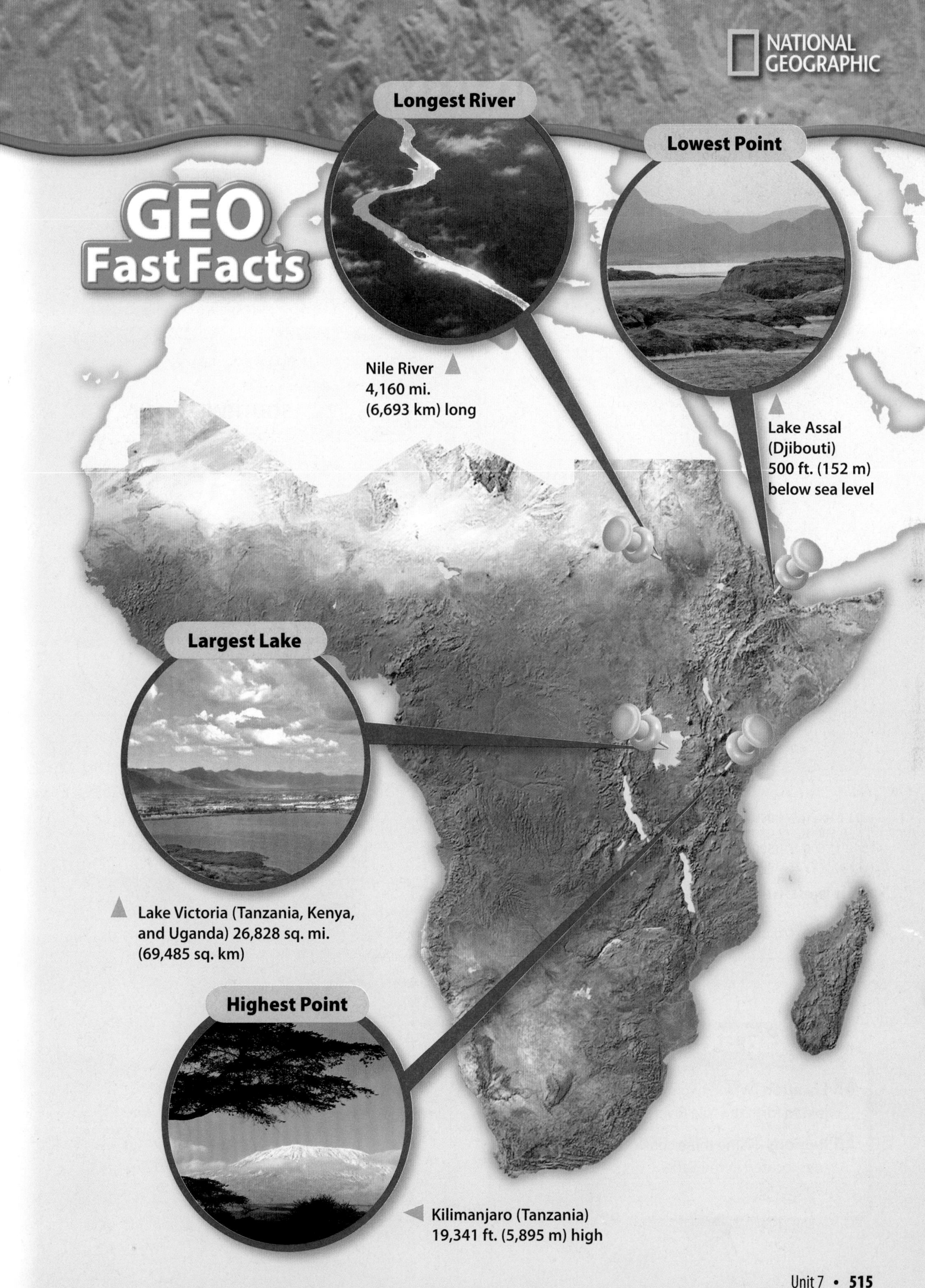
NATIONAL GEOGRAPHIC
GEO Fast Facts
Longest River
Nile River
4,160 mi.
(6,693 km) long
Lowest Point
Lake Assal
(Djibouti)
500 ft. (152 m)
below sea level
Largest Lake
Lake Victoria (Tanzania, Kenya, and Uganda) 26,828 sq. mi. (69,485 sq. km)
Highest Point
Kilimanjaro (Tanzania)
19,341 ft. (5,895 m) high

UNIT 7 Regional Atlas

Africa South of the Sahara

PHYSICAL

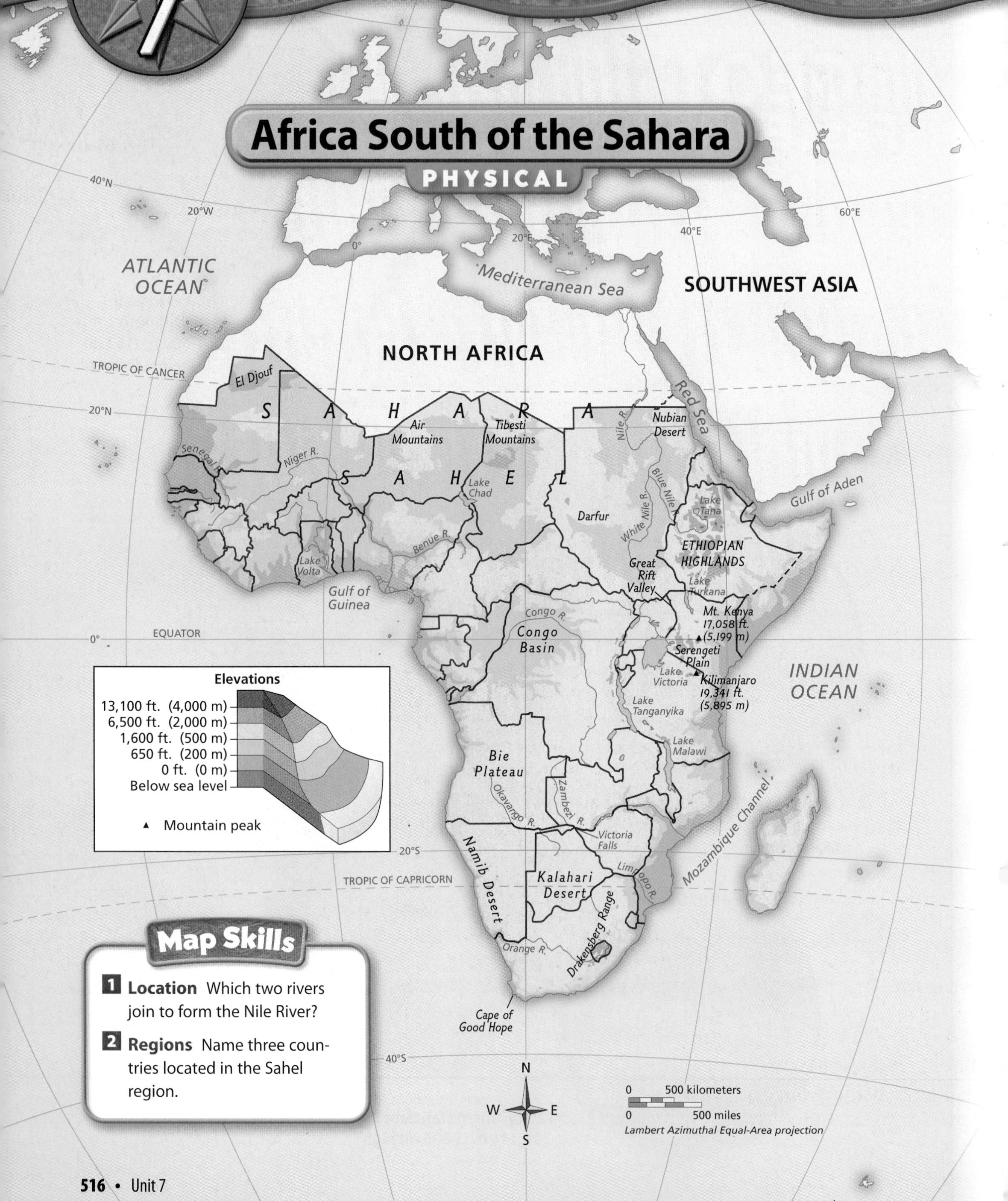

Map Skills

1 Location Which two rivers join to form the Nile River?

2 Regions Name three countries located in the Sahel region.

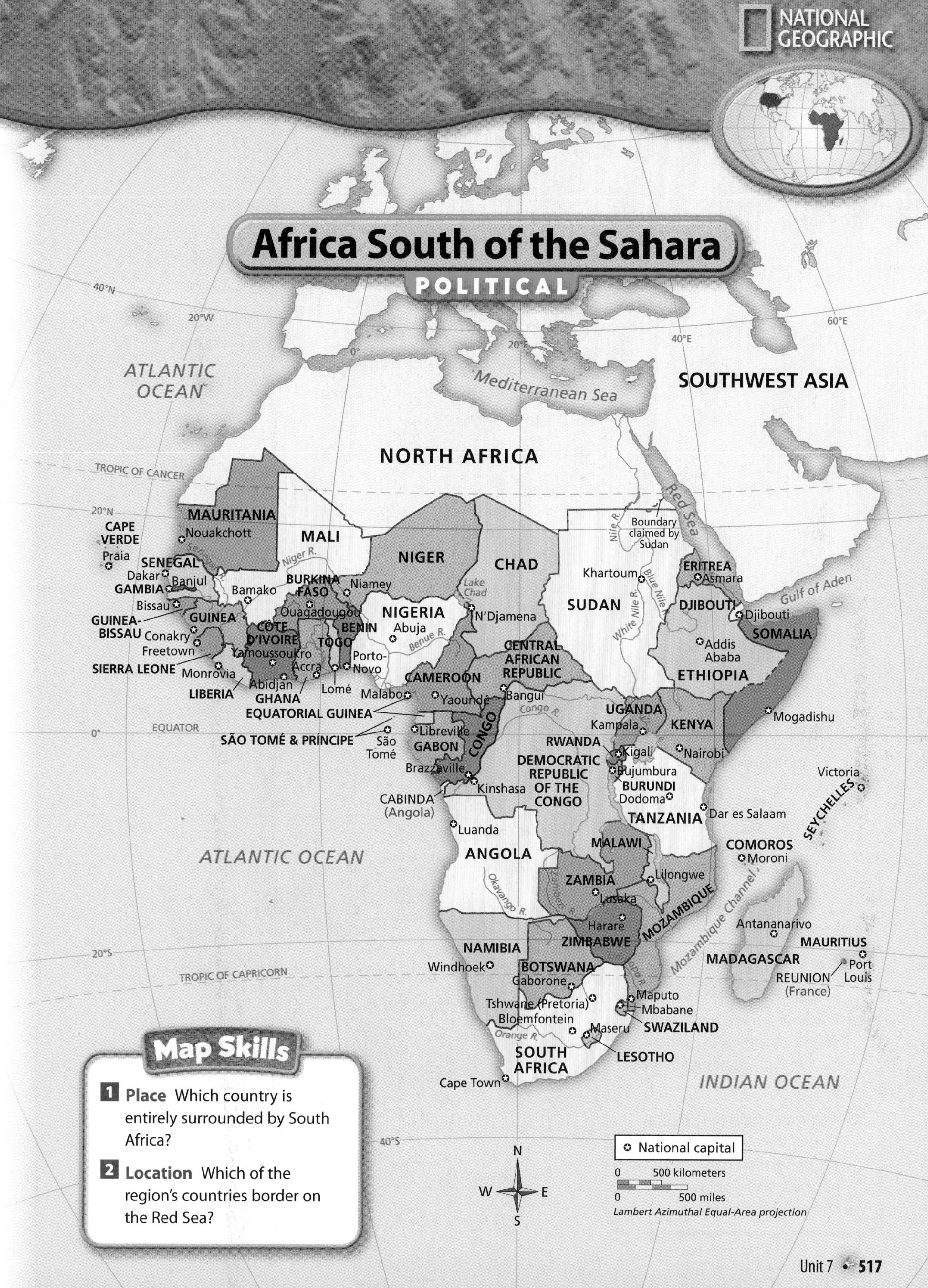

Map Skills

1 Place Which country is entirely surrounded by South Africa?

2 Location Which of the region's countries border on the Red Sea?

UNIT 7

Regional Atlas

Africa South of the Sahara

POPULATION DENSITY

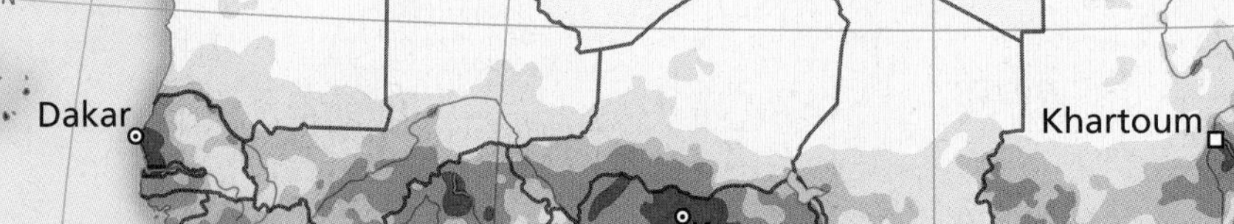

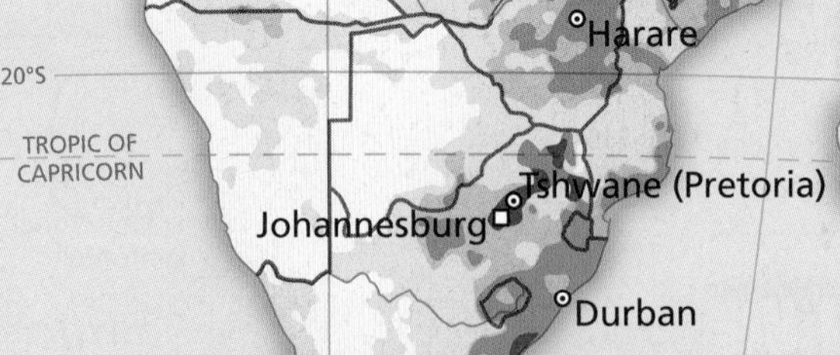

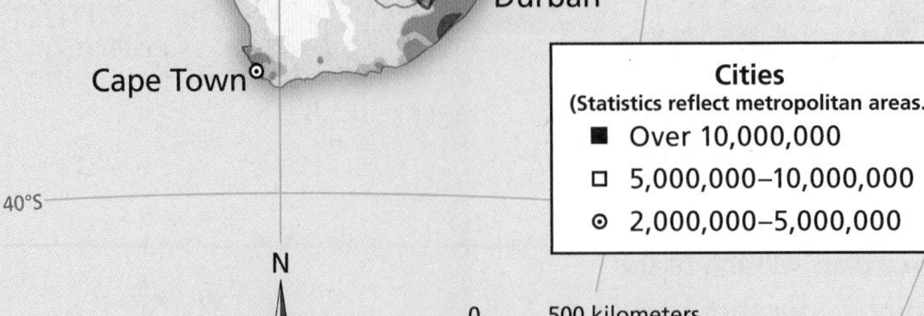

POPULATION

Per sq. mi.	Per sq. km
1,250 and over	500 and over
250–1,249	100–499
63–249	25–99
25–62	10–24
2.5–24	1–9
Less than 2.5	Less than 1

Cities
(Statistics reflect metropolitan areas.)

- ■ Over 10,000,000
- □ 5,000,000–10,000,000
- ⊙ 2,000,000–5,000,000

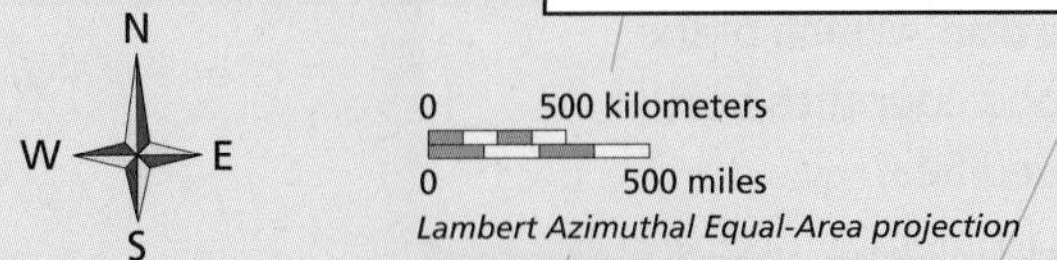

Map Skills

1 Place In general, what part of Africa south of the Sahara has the highest population density?

2 Regions Using the physical map of the region, explain the population patterns in the northern and southwestern areas of the region.

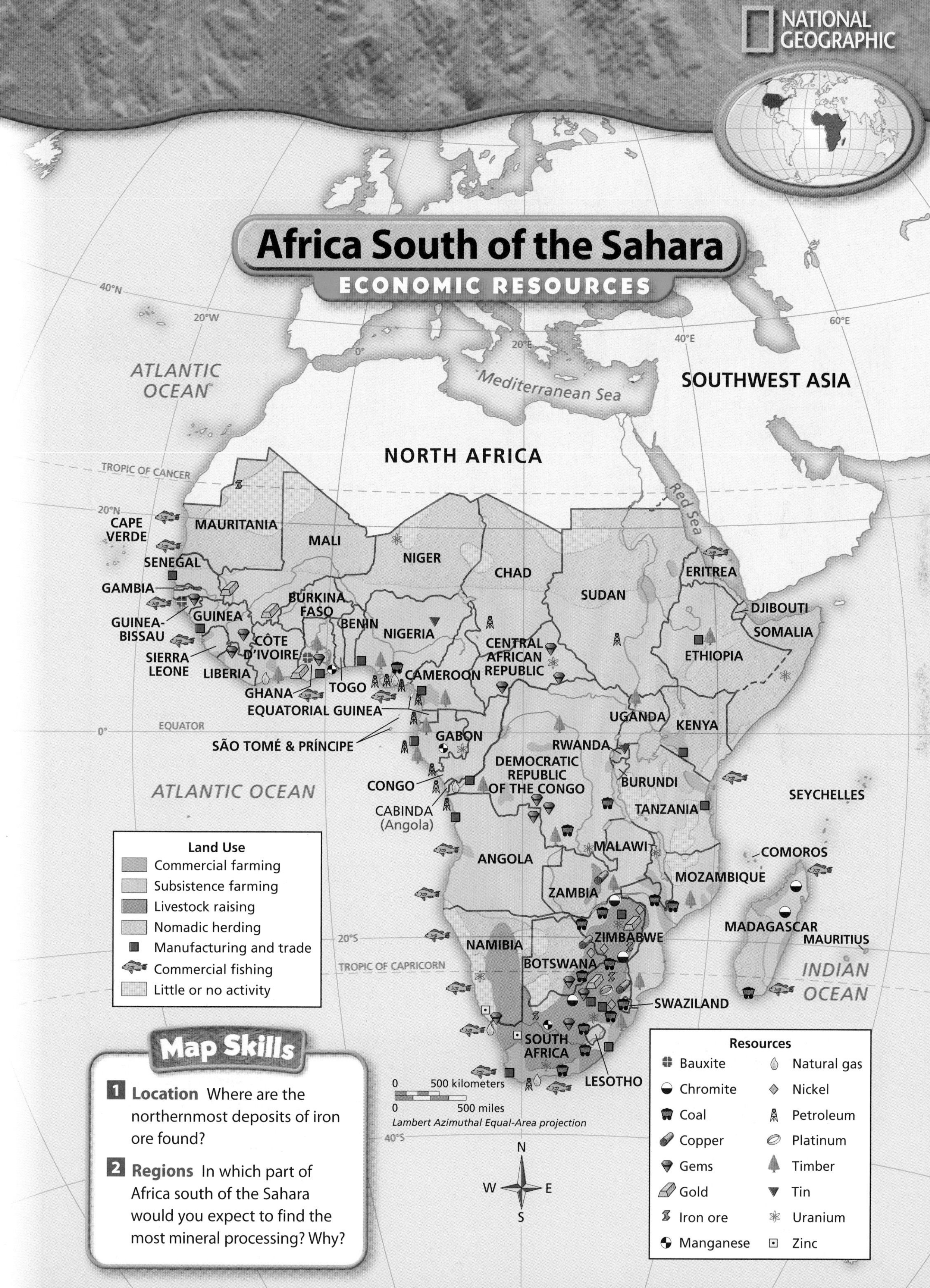

Map Skills

1. **Location** Where are the northernmost deposits of iron ore found?
2. **Regions** In which part of Africa south of the Sahara would you expect to find the most mineral processing? Why?

Regional Atlas

Africa South of the Sahara

Country and Capital	Literacy Rate	Population and Density	Land Area	Life Expectancy (Years)	GDP* Per Capita (U.S. dollars)	Television Sets (per 1,000 people)	Flag and Language
ANGOLA Luanda	42%	15,400,000 32 per sq. mi. 12 per sq. km	481,351 sq. mi. 1,246,693 sq. km	40	$2,100	15	Portuguese
BENIN Porto-Novo	40.9%	8,400,000 193 per sq. mi. 75 per sq. km	43,483 sq. mi. 112,620 sq. km	54	$1,200	44	French
BOTSWANA Gaborone	79.8%	1,600,000 7 per sq. mi. 3 per sq. km	224,606 sq. mi. 581,727 sq. km	35	$9,200	21	English, Setswana
BURKINA FASO Ouagadougou	26.6%	13,900,000 131 per sq. mi. 51 per sq. km	105,792 sq. mi. 274,000 sq. km	44	$1,200	11	French
BURUNDI Bujumbura	51.6%	7,800,000 726 per sq. mi. 280 per sq. km	10,745 sq. mi. 27,829 sq. km	49	$600	15	Kirundi, French
CAMEROON Yaoundé	79%	16,400,000 89 per sq. mi. 34 per sq. km	183,568 sq. mi. 475,439 sq. km	48	$1,900	34	English, French
CAPE VERDE Praia	76.6%	500,000 321 per sq. mi. 124 per sq. km	1,556 sq. mi. 4,030 sq. km	69	$1,400	5	Portuguese
CENTRAL AFRICAN REPUBLIC Bangui	51%	4,200,000 17 per sq. mi. 8 per sq. km	240,533 sq. mi. 622,978 sq. km	44	$1,100	6	French
UNITED STATES Washington, D.C.	97%	296,500,000 80 per sq. mi. 31 per sq. km	3,717,796 sq. mi. 9,629,047 sq. km	78	$40,100	844	English

*Gross Domestic Product

Countries and flags not drawn to scale

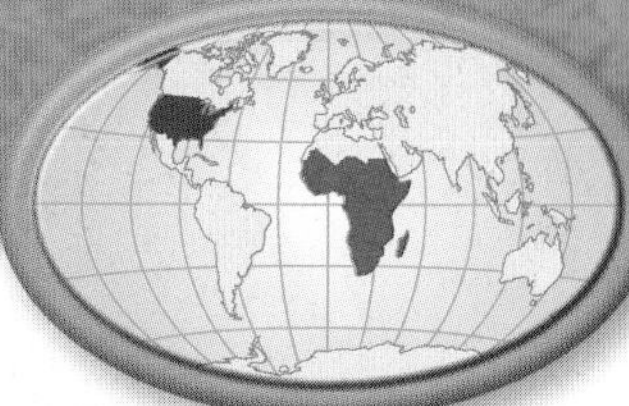

Africa South of the Sahara

Country and Capital	Literacy Rate	Population and Density	Land Area	Life Expectancy (Years)	GDP* Per Capita (U.S. dollars)	Television Sets (per 1,000 people)	Flag and Language
CHAD N'Djamena	47.5%	9,700,000 20 per sq. mi. 8 per sq. km	495,753 sq. mi. 1,283,994 sq. km	47	$1,600	1	French, Arabic
COMOROS Moroni	56.5%	700,000 813 per sq. mi. 314 per sq. km	861 sq. mi. 2,230 sq. km	60	$700	4	Arabic, French
DEMOCRATIC REPUBLIC OF THE CONGO Kinshasa	65.5%	60,800,000 67 per sq. mi. 26 per sq. km	905,351 sq. mi. 2,344,848 sq. km	50	$700	2	French
REPUBLIC OF THE CONGO Brazzaville	83.8%	4,000,000 30 per sq. mi. 12 per sq. km	132,046 sq. mi. 341,998 sq. km	52	$800	13	French
CÔTE D'IVOIRE Yamoussoukro	50.9%	18,200,000 146 per sq. mi. 56 per sq. km	124,502 sq. mi. 322,459 sq. km	47	$1,500	65	French
DJIBOUTI Djibouti	67.9%	800,000 89 per sq. mi. 34 per sq. km	8,958 sq. mi. 23,201 sq. km	52	$1,300	48	French, Arabic
EQUATORIAL GUINEA Malabo	85.7%	500,000 46 per sq. mi. 18 per sq. km	10,830 sq. mi. 28,050 sq. km	45	$2,700	116	Spanish, French
ERITREA Asmara	58.6%	4,700,000 104 per sq. mi. 40 per sq. km	45,405 sq. mi. 117,598 sq. km	58	$900	16	Afar
UNITED STATES Washington, D.C.	97%	296,500,000 80 per sq. mi. 31 per sq. km	3,717,796 sq. mi. 9,629,047 sq. km	78	$40,100	844	English

Sources: *CIA World Factbook*, 2005; Population Reference Bureau, *World Population Data Sheet*, 2005.

For more country facts, go to the **Nations of the World Databank** at glencoe.com.

Regional Atlas

Africa South of the Sahara

Country and Capital	Literacy Rate	Population and Density	Land Area	Life Expectancy (Years)	GDP* Per Capita (U.S. dollars)	Television Sets (per 1,000 people)	Flag and Language
ETHIOPIA Addis Ababa	42.7%	77,400,000 182 per sq. mi. 70 per sq. km	426,371 sq. mi. 1,104,296 sq. km	48	$800	5	Amharic
Libreville GABON	63.2%	1,400,000 14 per sq. mi. 5 per sq. km	103,347 sq. mi. 267,667 sq. km	56	$5,900	251	French
Banjul GAMBIA	40.1%	1,600,000 367 per sq. mi. 142 per sq. km	4,363 sq. mi. 11,300 sq. km	53	$1,800	3	English
GHANA Accra	74.8%	22,000,000 239 per sq. mi. 92 per sq. km	92,100 sq. mi. 238,538 sq. km	58	$2,300	115	English
GUINEA Conakry	35.9%	1,600,000 17 per sq. mi. 7 per sq. km	94,927 sq. mi. 245,860 sq. km	49	$2,100	47	French
GUINEA-BISSAU Bissau	42.4%	1,600,000 115 per sq. mi. 44 per sq. km	13,946 sq. mi. 36,120 sq. km	44	$700	information not available	Portuguese
KENYA Nairobi	85.1%	33,800,000 151 per sq. mi. 58 per sq. km	224,081 sq. mi. 580,367 sq. km	47	$1,100	22	English, Kiswahili
Maseru LESOTHO	84.8%	1,800,000 154 per sq. mi. 59 per sq. km	11,718 sq. mi. 30,349 sq. km	35	$3,200	16	Sesotho, English
UNITED STATES Washington, D.C.	97%	296,500,000 80 per sq. mi. 31 per sq. km	3,717,796 sq. mi. 9,629,047 sq. km	78	$40,100	844	English

*Gross Domestic Product

Countries and flags not drawn to scale

Africa South of the Sahara

Country and Capital	Literacy Rate	Population and Density	Land Area	Life Expectancy (Years)	GDP* Per Capita (U.S. dollars)	Television Sets (per 1,000 people)	Flag and Language
LIBERIA Monrovia	57.5%	3,300,000 77 per sq. mi. 30 per sq. km	43,000 sq. mi. 111,369 sq. km	42	$900	26	English
MADAGASCAR Antananarivo	68.9%	17,300,000 76 per sq. mi. 29 per sq. km	226,656 sq. mi. 587,036 sq. km	55	$800	23	French, Malagasy
MALAWI Lilongwe	62.7%	12,300,000 269 per sq. mi. 104 per sq. km	45,745 sq. mi. 118,479 sq. km	45	$600	3	Chichewa
MALI Bamako	46.4%	13,500,000 28 per sq. mi. 11 per sq. km	478,838 sq. mi. 1,240,185 sq. km	48	$900	13	French
MAURITANIA Nouakchott	41.7%	3,100,000 8 per sq. mi. 3 per sq. km	395,954 sq. mi. 1,025,516 sq. km	52	$1,800	95	Arabic
MAURITIUS Port Louis	85.6%	1,200,000 1,523 per sq. mi. 588 per sq. km	788 sq. mi. 2,041 sq. km	72	$12,800	248	Creole, French
MOZAMBIQUE Maputo	47.8%	19,400,000 63 per sq. mi. 24 per sq. km	309,494 sq. mi. 801,586 sq. km	42	$1,200	5	Portuguese
NAMIBIA Windhoek	84%	2,000,000 6 per sq. mi. 2 per sq. km	318,259 sq. mi. 824,287 sq. km	46	$7,300	38	English
UNITED STATES Washington, D.C.	97%	296,500,000 80 per sq. mi. 31 per sq. km	3,717,796 sq. mi. 9,629,047 sq. km	78	$40,100	844	English

Sources: *CIA World Factbook*, 2005; Population Reference Bureau, *World Population Data Sheet*, 2005.

For more country facts, go to the **Nations of the World Databank** at glencoe.com.

Regional Atlas

Africa South of the Sahara

Country and Capital	Literacy Rate	Population and Density	Land Area	Life Expectancy (Years)	GDP* Per Capita (U.S. dollars)	Television Sets (per 1,000 people)	Flag and Language
NIGER Niamey	17.6%	14,000,000 29 per sq. mi. 11 per sq. km	489,189 sq. mi. 1,266,994 sq. km	43	$900	15	French
NIGERIA Abuja	68%	131,500,000 369 per sq. mi. 142 per sq. km	356,668 sq. mi. 923,766 sq. km	44	$1,000	69	English
Kigali RWANDA	70.4%	8,700,000 855 per sq. mi. 330 per sq. km	10,170 sq. mi. 26,340 sq. km	44	$1,300	0.09	Kinyarwanda, English, French
SÃO TOMÉ AND PRÍNCIPE São Tomé	79.3%	200,000 539 per sq. mi. 208 per sq. km	371 sq. mi. 961 sq. km	63	$1,200	229	Portuguese
Dakar SENEGAL	40.2%	11,700,000 154 per sq. mi. 59 per sq. km	75,954 sq. mi. 196,720 sq. km	56	$1,700	41	French
Victoria SEYCHELLES	58%	100,000 575 per sq. mi. 222 per sq. km	174 sq. mi. 451 sq. km	71	$7,800	214	Creole, English
SIERRA LEONE Freetown	31.4%	5,500,000 199 per sq. mi. 77 per sq. km	27,699 sq. mi. 71,740 sq. km	40	$600	13	English
SOMALIA Mogadishu	37.8%	8,600,000 35 per sq. mi. 13 per sq. km	246,201 sq. mi. 637,658 sq. km	47	$600	14	Somali
UNITED STATES Washington, D.C.	97%	296,500,000 80 per sq. mi. 31 per sq. km	3,717,796 sq. mi. 9,629,047 sq. km	78	$40,100	844	English

*Gross Domestic Product

Countries and flags not drawn to scale

Africa South of the Sahara

Country and Capital	Literacy Rate	Population and Density	Land Area	Life Expectancy (Years)	GDP* Per Capita (U.S. dollars)	Television Sets (per 1,000 people)	Flag and Language
SOUTH AFRICA Tshwane (Pretoria), Bloemfontein, Cape Town	86.4%	46,900,000 99 per sq. mi. 38 per sq. km	471,444 sq. mi. 1,221,034 sq. km	52	$11,100	138	Afrikaans, English, Zulu
SUDAN Khartoum	61.1%	40,200,000 42 per sq. mi. 16 per sq. km	967,494 sq. mi. 2,505,798 sq. km	57	$1,900	173	Arabic
SWAZILAND Mbabane	81.6%	1,138,000 170 per sq. mi. 66 per sq. km	6,642 sq. mi. 17,203 sq. km	33	$5,100	112	English, siSwati
TANZANIA Dodoma, Dar es Salaam	78.2%	36,500,000 100 per sq. mi. 39 per sq. km	364,900 sq. mi. 945,087 sq. km	44	$700	21	Kiswahili, English
TOGO Lomé	60.9%	6,100,000 278 per sq. mi. 107 per sq. km	21,927 sq. mi. 56,791 sq. km	54	$1,600	22	French
UGANDA Kampala	69.9%	26,900,000 289 per sq. mi. 112 per sq. km	93,066 sq. mi. 241,040 sq. km	48	$1,500	28	English
ZAMBIA Lusaka	80.6%	11,200,000 39 per sq. mi. 15 per sq. km	290,583 sq. mi. 752,606 sq. km	37	$900	145	English
ZIMBABWE Harare	90.7%	13,000,000 86 per sq. mi. 33 per sq. km	150,873 sq. mi. 390,759 sq. km	41	$1,900	35	English
UNITED STATES Washington, D.C.	97%	296,500,000 80 per sq. mi. 31 per sq. km	3,717,796 sq. mi. 9,629,047 sq. km	78	$40,100	844	English

Sources: *CIA World Factbook*, 2005; Population Reference Bureau, *World Population Data Sheet*, 2005.

For more country facts, go to the **Nations of the World Databank** at glencoe.com.

UNIT
7

Reading Social Studies

Paraphrasing

Reading Skill

1

Learn It!

Did a friend ever tell you a story that you later told to another person? If so, you were probably paraphrasing. Paraphrasing is simply restating something in your own words. When you paraphrase a story, you describe the main events or characters using words that are different from those used in the story originally.

You can also paraphrase what you read. Once you identify the main idea and supporting details in a paragraph, summarize them in your own words. Remember that there is no one way to paraphrase as long as you include the important information from the text.

Follow these steps to paraphrase the text below.

- Identify the main idea and describe it in your own words.
- Describe how the main idea is supported in your own words.

> Some parts of Africa south of the Sahara have long droughts, or periods of time when there is no rain at all. Droughts can cause crop failures and widespread starvation.
>
> *—from page 539*

A web diagram like the one below will help you learn to paraphrase.

Reading Tip

Try to answer the questions *Who? What? Where? When? Why?* and *How?* to help you decide on important facts to include as you paraphrase.

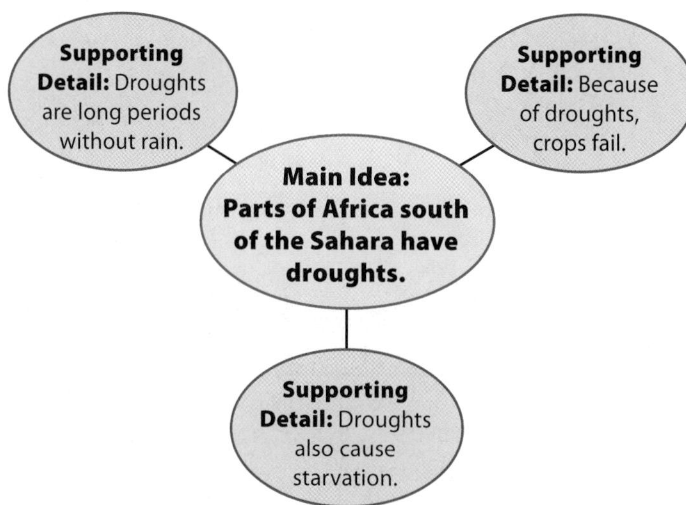
Supporting Detail: Droughts are long periods without rain.
Supporting Detail: Because of droughts, crops fail.
Main Idea: Parts of Africa south of the Sahara have droughts.
Supporting Detail: Droughts also cause starvation.

Read to Write Activity

Read the paragraphs under the heading "Health Care" in Chapter 20, Section 2, pages 559 and 560. In your own words, restate what you learned about AIDS. Remember to use facts from the text.

2 Practice It!

Read the following paragraph from this unit.

- Draw a graphic organizer like the one below.
- Paraphrase the main idea of the text by writing it in the center oval.
- Write four important facts from the paragraph in your own words in the surrounding ovals.

> The Democratic Republic of the Congo is a major source of copper, tin, and industrial diamonds. The country has not been able to take full advantage of its rich resources, however. One difficulty is transportation. Many of the minerals are found in the country's interior. Lack of roads and thick rain forests make it hard to reach those areas. Political unrest has also limited the mining of resources. For many years, a civil war hurt efforts to develop the country's economy.
>
> —*from page 577*

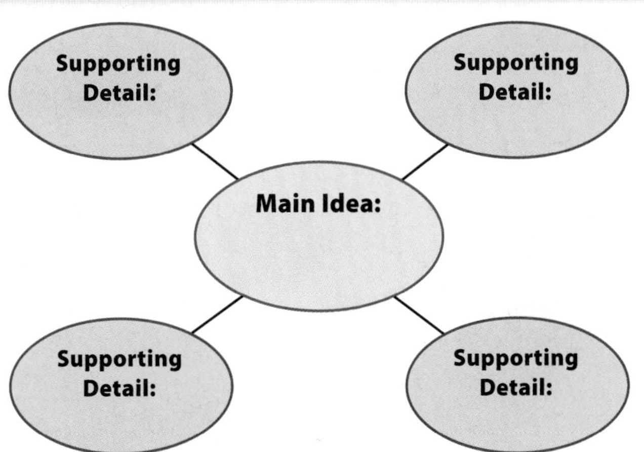

▲ **Diamonds mined in the Democratic Republic of the Congo**

3 Apply It!

Try to paraphrase several subsections from each chapter in the unit. Create web diagrams like the ones on these pages to help you identify the main ideas and supporting details in your own words.

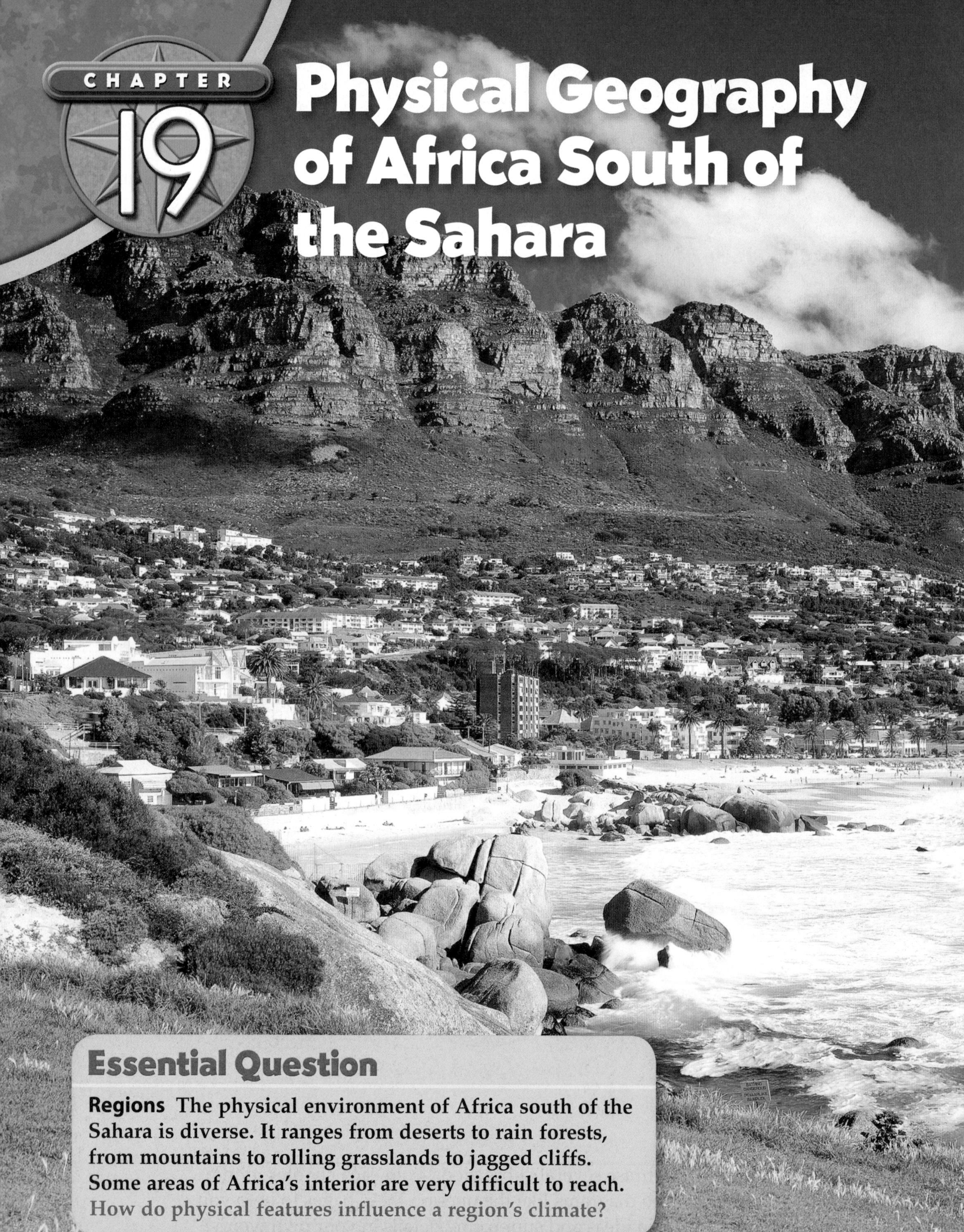

CHAPTER 19 Physical Geography of Africa South of the Sahara

Essential Question

Regions The physical environment of Africa south of the Sahara is diverse. It ranges from deserts to rain forests, from mountains to rolling grasslands to jagged cliffs. Some areas of Africa's interior are very difficult to reach. How do physical features influence a region's climate?

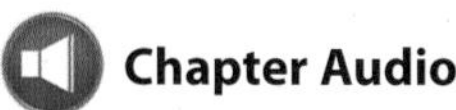

Section 1: Physical Features

BIG IDEA Physical processes shape Earth's surface. Over thousands of years, the movement of the Earth's tectonic plates has shaped the landforms of Africa south of the Sahara. The region's landscape includes large plateaus, rocky cliffs, and great, steep valleys.

Section 2: Climate Regions

BIG IDEA Geographers organize the Earth into regions that share common characteristics. Africa south of the Sahara has four main climate regions, each of which covers a large area. Similar climate zones appear north and south of the Equator in the region. Climates range from damp rain forests to vast grasslands to hot deserts.

FOLDABLES™ Study Organizer

Organizing Information Make this Foldable to determine what you already know, identify what you want to know, and organize information you learn about the physical geography of Africa south of the Sahara.

Step 1 Fold a sheet of paper into thirds from top to bottom.

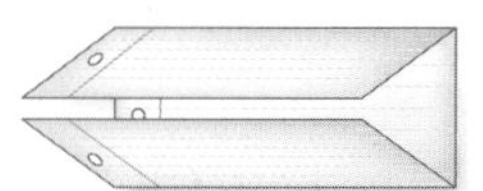

Step 2 Turn the paper horizontally, unfold, and label the three columns as shown.

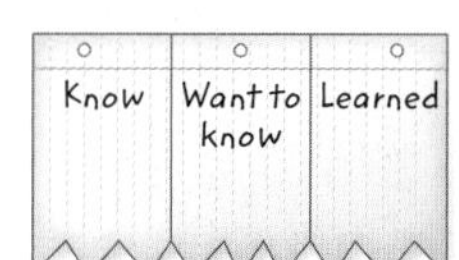

Reading and Writing Before you read the chapter, fill in the "Know" and "Want to Know" columns. Fill in the "Learned" column as you read the chapter. Then write a short summary explaining what you learned.

Visit glencoe.com and enter QuickPass™ code EOW3109c19 for Chapter 19 resources.

Guide to Reading

BIG Idea

Physical processes shape Earth's surface.

Content Vocabulary

- escarpment *(p. 531)*
- rift valley *(p. 532)*
- gorge *(p. 534)*
- industrial diamond *(p. 535)*

Academic Vocabulary

- series *(p. 531)*
- principal *(p. 534)*

Reading Strategy

Analyzing Information Complete a web diagram like the one below by listing the physical forces that have shaped the landforms of Africa.

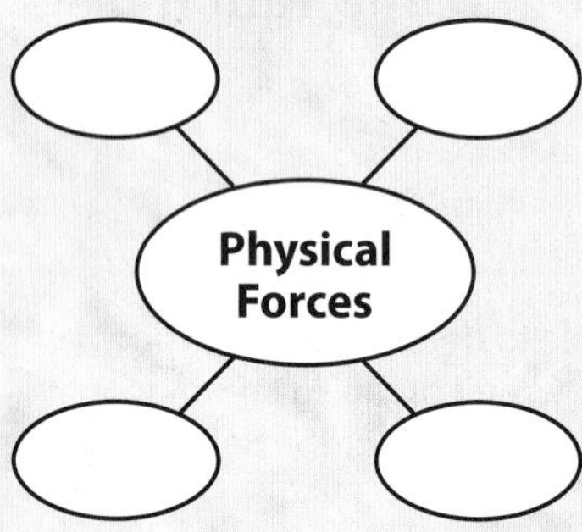

Physical Features

 Section Audio  Spotlight Video

Picture This A large camel can drink 25 gallons (95 L) of water in 10 minutes! It then stores the water in its bloodstream. Camels sweat very little, so the water they drink can last for weeks. Unlike a camel, a person in a desert can lose about two gallons of water a day by sweating! Humans must replace water frequently. Water is more plentiful in some regions of Africa south of the Sahara than in others. Read on to learn more about the physical features of this region.

▼ Water trough in Mali

Landforms of Africa South of the Sahara

Main Idea **Africa south of the Sahara consists mainly of vast plateaus with few mountains and lowlands.**

Geography and You Do you know what the landscape might look like if tectonic plates beneath the Earth's surface pulled apart? Read to learn about the amazing landscape of Africa's Great Rift Valley.

Africa south of the Sahara is more than two and a half times larger than the United States. As **Figure 1** shows, this enormous region is made up of four subregions, or smaller regions: West Africa, Central Africa, East Africa, and Southern Africa. Africa south of the Sahara extends from the Sahara in the north to Africa's southern tip at the Cape of Good Hope.

Africa south of the Sahara includes many islands off the African mainland. Madagascar is the largest of these and the fourth-largest island in the world. It lies in the Indian Ocean near Africa's southeastern coast. Scientists believe that it broke away from the African continent millions of years ago. Most of the other, smaller islands were formed by volcanoes.

Plateaus and Lowlands

Almost all of Africa south of the Sahara lies on a **series** of plateaus. The plateaus are formed from the solid rock that lies under most of the African continent. They rise like steps across the continent from west to east, as well as from the coasts into the interior. Many of these landforms rise from 1,000 to 2,000 feet (305 to 610 m) in western Africa to 7,000 feet (2,134 m) or more in the east. The plateaus give Africa south of the Sahara the highest overall elevation of any world region—more than 1,000 feet (305 m) above sea level.

In eastern and southern Africa, the edges of plateaus are often marked by escarpments. **Escarpments** are steep, jagged cliffs. Rivers that flow across plateaus drop suddenly at escarpments to become rushing rapids or tumbling waterfalls. Escarpments create barriers to trade by blocking ships from sailing between the interior and the sea.

Africa south of the Sahara also has some lowland areas. These include narrow plains that border the region's Atlantic and Indian Ocean coastlines. Among Africa's plateaus are low, sunken areas called basins. Basins formed when tectonic activity lifted up the land surrounding them.

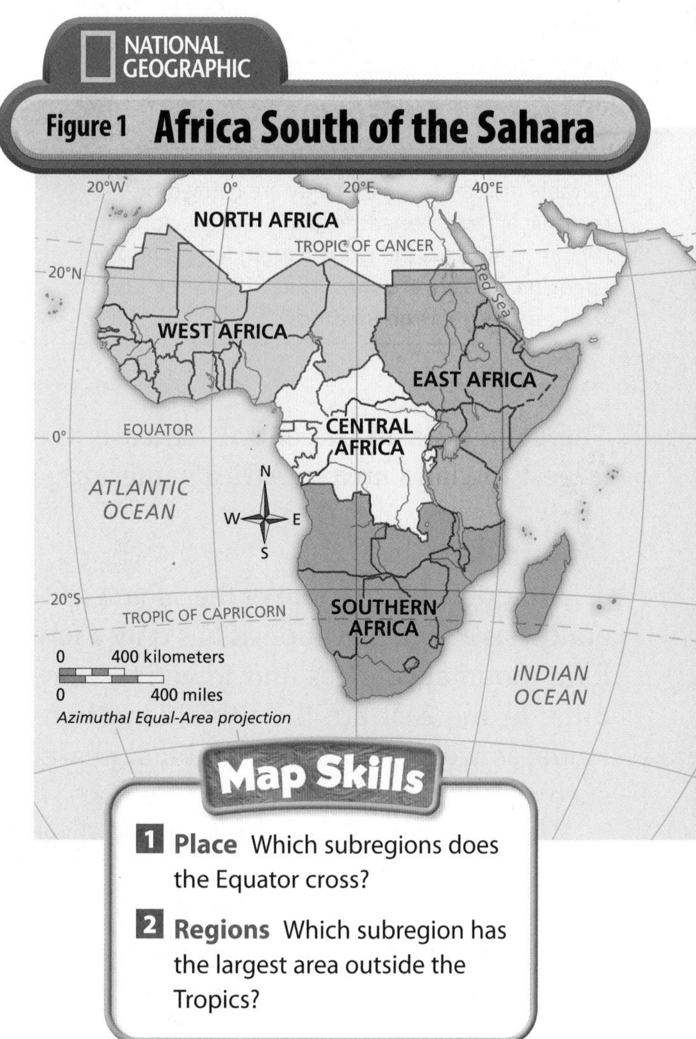

Diverse Landforms

Geysers in the Great Rift Valley are caused by tectonic activity beneath Earth's surface. Volcanic lava flows formed the Drakensberg Range (inset).
Location **Where are the Great Rift Valley and the Drakensberg Range located?**

The Congo Basin in central Africa is the largest lowland area in Africa's interior.

Mountains

Although Africa south of the Sahara generally has a high elevation, it has only a few long mountain ranges and towering summits. In the east are the Ethiopian Highlands, as well as volcanic mountain peaks, such as Kilimanjaro and Mount Kenya.

Kilimanjaro in Tanzania is the highest peak in the region, rising to a height of 19,341 feet (5,895 m). The name *Kilimanjaro* comes from a phrase in the Swahili language that means "shining mountain." From the steamy grasslands below, Kilimanjaro's snowcapped peak appears to be shining in the sun. Even though the mountain sits almost on the Equator, snow covers the summit year-round.

The Drakensberg (DRAH·kuhnz·BUHRG) Range is in southern Africa. It reaches about 11,400 feet (3,475 m) high and is about 700 miles (1,127 km) long. People call the eastern side of the Drakensberg the "barrier of pointed spears." This is because the mountains rise suddenly and look like giant spears sticking out of the ground.

The Great Rift Valley

Few features break the flatness of Africa's large plateau areas. In eastern Africa, however, an amazing natural wonder—the Great Rift Valley—cuts through the landscape. The Great Rift Valley stretches about 4,000 miles (6,437 km) from Southwest Asia to southern Africa.

A **rift valley** is a large break in the Earth's surface formed by shifting tectonic plates. Millions of years ago, plate movements created deep cuts in the Earth's crust where the Great Rift Valley now lies. Volcanic eruptions and earthquakes helped create the valley's striking landscape. The valley's floor lies below sea level in many places, while the valley's walls generally rise 2,000 to 3,000 feet (610 to 914 m) above sea level. In some places, the walls rise 9,000 feet (2,743 m) from the valley floor. Jagged mountains and deep lakes add to the region's beauty. Some areas of the Great Rift Valley have rich volcanic soil that supports farming.

Reading Check **Explaining** How was the Congo Basin formed?

Waterways of the Region

Main Idea Waterways provide transportation, freshwater, and electricity for Africans living south of the Sahara.

Geography and You Do people travel on rivers or fish in lakes where you live? Read to find out about the ways Africans south of the Sahara use their rivers and lakes.

Africa south of the Sahara has numerous waterways. People in the region rely on its lakes and rivers for freshwater and transportation.

Lakes

Most of the region's large lakes lie in or near East Africa's Great Rift Valley. One of these lakes, Lake Tanganyika (TAN·guhn·YEE·kuh), is 420 miles (676 km) in length, making it the longest freshwater lake in the world. Lake Victoria lies in a low basin and is Africa's largest lake. It is the world's second-largest freshwater lake, after Lake Superior in North America. Lakes in the Great Rift Valley provide freshwater and fish to people who live near them.

Some Great Rift Valley lakes also serve as the sources of rivers. Lake Victoria, for example, is the source of the White Nile, and Lake Tana is the source of the Blue Nile. The White Nile and the Blue Nile meet farther north in Sudan to form the Nile River, the world's longest river.

Another important body of water is Lake Chad, which lies in West Africa. Lake Chad changes dramatically in size from about 10,000 square miles (25,900 sq. km) in the rainy season to about 3,800 square miles (9,842 sq. km) in the dry season.

Rivers

Africa south of the Sahara has four large river systems—the Nile, the Congo, the Niger, and the Zambezi (zam·BEE·zee). All of these rivers begin in the interior plateaus and make their way to the sea. In some places these rivers and their many branches are useful for freshwater and transportation, but geographical barriers limit their use in other areas.

The same tectonic activity that produced the region's rugged landscape also affected the region's rivers. As you learned earlier, escarpments create waterfalls and rapids that make transportation on some rivers difficult. Along a section of the Congo River in Central Africa, more than 30 waterfalls make travel difficult. A railroad line was built to bypass the rapids.

Water Taxi in Senegal

NATIONAL GEOGRAPHIC

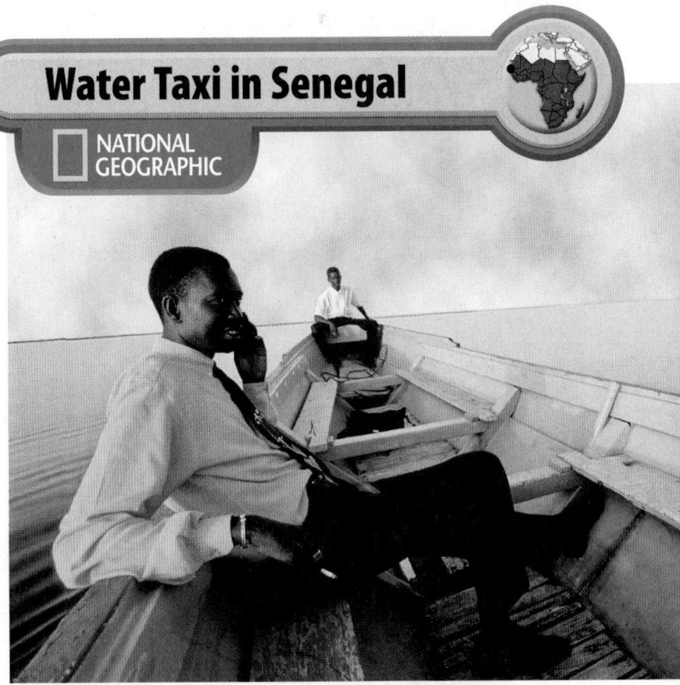

In western Senegal, people often hire water taxis for local travel along the area's rivers and coast. ***Regions*** **What four large rivers are located in Africa south of the Sahara?**

Social Studies ONLINE

Student Web Activity Visit glencoe.com and complete the Chapter 19 Web Activity about the Niger River.

The Zambezi River in southern Africa plunges over a cliff, creating Victoria Falls, a series of waterfalls that drop as much as 420 feet (128 m). The thick mist from the falls can be seen from miles away.

Rivers that begin in Africa's highlands shape the land. Many, like the Congo, flow through plateaus and carve deep **gorges,** or steep-sided valleys formed when rivers cut through the land. Other rivers are interrupted by inland lakes and marshes that can hinder travel. For example, West Africa's Niger River fans out into swamps in southern Mali before re-forming in a channel that continues to the sea.

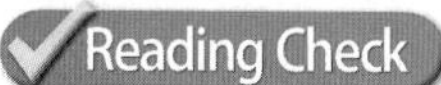

Reading Check **Determining Cause and Effect** How has tectonic activity affected waterways in Africa south of the Sahara?

Mineral Resources

Main Idea **Africa south of the Sahara holds both a great variety and large quantities of mineral resources.**

Geography and You Have you ever seen diamond rings displayed in the window of a jewelry store? Read to discover where many of the world's diamonds are found.

Africa south of the Sahara is rich in energy resources. Plentiful petroleum deposits are found along the Atlantic coast from Nigeria to Angola. Landlocked Chad and Sudan also have large petroleum deposits. Oil has replaced agricultural products as the **principal** export in many of these countries.

Kongou Falls, Gabon

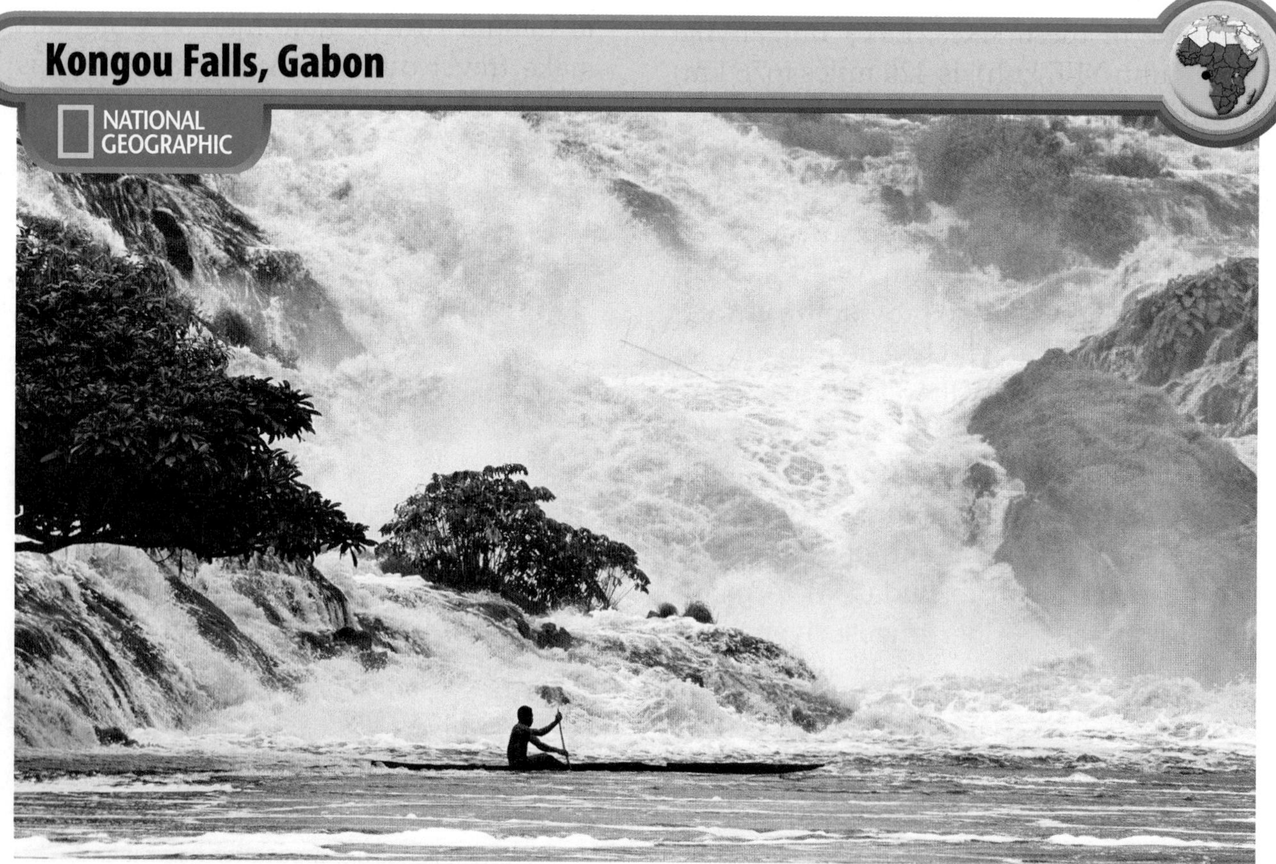

▲ Kongou Falls in Gabon is about 2 miles (3 km) wide. More than 275,000 gallons (1,040,988 L) of roaring water rush over the falls each second. ***Place*** **How do some African rivers shape the land?**

Other important resources include natural gas, which is also found in Central African countries along the Atlantic coast. Nigeria, the Democratic Republic of the Congo, and the Republic of South Africa have coal deposits. In addition, the region provides an important resource by means of its fast-flowing rivers—hydroelectric power. One of the most important sources of hydroelectric power is the Akosombo Dam in Ghana. It holds back two branches of the Volta River to form Lake Volta, one of the largest human-made lakes in the world. The dam supplies hydroelectric power to several West African countries.

Metals are among the region's most important mineral resources. Large reserves of iron ore exist throughout Africa south of the Sahara. Zimbabwe in Southern Africa holds vast amounts of iron ore. Chromium, which manufacturers mix with iron to make steel, is mined in Zimbabwe, South Africa, and several countries of both West and East Africa. Deposits of uranium—used to produce nuclear power—and copper are also found in the region.

Africa south of the Sahara has large deposits of precious materials. South Africa is believed to have half of the world's gold. A gold deposit more than 300 miles (483 km) long is located in the Transvaal, a grassy plateau. South Africa is also rich in platinum, chromium, and manganese.

Many gemstones are mined in Africa south of the Sahara, including diamonds, rubies, emeralds, and sapphires. South Africa is a major diamond producer. Not all of the diamonds are used for jewelry, however. Because diamonds are such a hard substance, **industrial diamonds** are used to make drills, saws, and grinding tools.

Reading Check **Identifying Central Issues**
Name three resources found in large quantities in this region.

Social Studies ONLINE
Study Central™ To review this section, go to glencoe.com.

Vocabulary

1. **Explain** the significance of:
   a. escarpment
   b. rift valley
   c. gorge
   d. industrial diamond

Main Ideas

2. **Explaining** How do escarpments affect river travel in the region?
3. **Illustrating** Use a diagram like the one below to describe key features of waterways in Africa south of the Sahara.

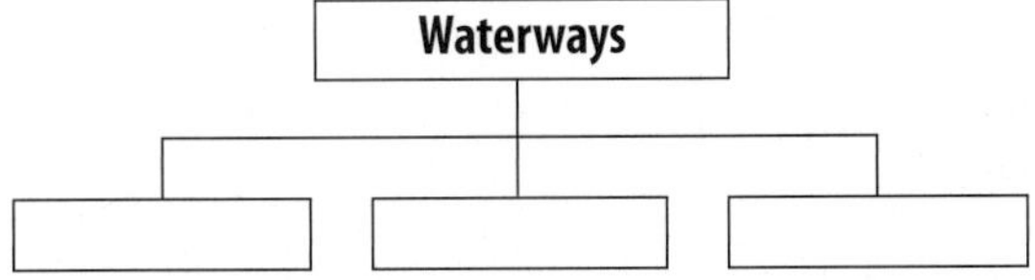

4. **Identifying** What resources make Africa important to the world economy?

Critical Thinking

5. **BIG Idea** How was the Great Rift Valley formed?
6. **Challenge** What effect do you think physical geography has on a region's ability to profit from its many energy and mineral resources?

Writing About Geography

7. **Using Your FOLDABLES** Use your Foldable to create a summary chart listing key facts about the landforms, waterways, or resources in West, Central, East, and Southern Africa.

Water Resources: Who Should Control Them?

More than 1 billion people in the world do not have access to clean water. In countries south of the Sahara, local and national governments are working to establish reliable water and sanitation systems. Some governments create and maintain their own systems, which are owned by the public and run by government employees. In other cases, private companies are contracted to provide the same services for a fee to the user. This system is known as privatization.

For Privatization

During the 1990s, it also became apparent that private participation could bring better oversight and management. The most detailed studies…concluded that well designed private schemes have brought clear benefits—but not perfection. For example, in water, the most difficult sector, in cities as diverse as…Abidjan and Conakry service coverage has increased significantly…. Extended coverage tends to bring the biggest benefits to households with lower incomes, as they previously had to pay much more for the service by small informal vendors.

—Michael Klein, World Bank,
Vice President for Private Sector
Development and Infrastructure

Privatization

Water is about life. The saying that "water is life" cannot be more appropriate. Privatizing water is putting the lives of citizens in the hands of a corporate entity that is accountable only to its shareholders. Secondly, water is a human right and this means that any philosophy, scheme, or contract that has the potential to exclude sections of the population from accessing water is not acceptable both in principle and in law. Privatization has that potential because the privateers are not charities: they are in for the profit. Price therefore becomes an important barrier to access by poor people. Water is the collective heritage of humanity and nature….Water must remain a public good for the public interest.

—Rudolf Amenga-Etego
Global Policy Forum

You Be the Geographer

1. **Analyze** According to the World Bank, what advantages does privatization provide in managing water resources?
2. **Critical Thinking** How does Amenga-Etego view people's relationship with water? Do you agree with him? Why or why not?
3. **Read to Write** Write a paragraph describing your feelings about whether private companies have the right to make a profit by providing water to citizens.

Guide to Reading

BIG Idea

Geographers organize the Earth into regions that share common characteristics.

Content Vocabulary

- drought *(p. 539)*
- rain forest *(p. 540)*
- canopy *(p. 540)*
- deforestation *(p. 540)*
- ecotourism *(p. 540)*
- savanna *(p. 540)*
- desertification *(p. 541)*
- succulent *(p. 541)*

Academic Vocabulary

- annual *(p. 539)*
- enormous *(p. 540)*

Reading Strategy

Making Generalizations Use a diagram like the one below to make a generalization about the climate of Africa south of the Sahara. In the smaller boxes, write three facts about the region's climate. Then, in the larger box, write a generalization you can draw from those facts.

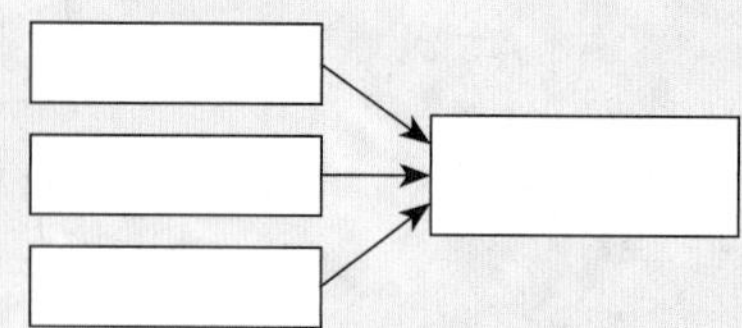

Climate Regions

Spotlight Video

Picture This Fishponds? Puddles? No! The holes in the ground are actually evaporation ponds in Niger. The water in the ponds evaporates, leaving behind salt. The plentiful sunshine and the mainly dry climate of Niger speed up the evaporation process. Then people gather the salt and take it to market to sell. Read this section to learn about other climates in Africa and how they affect the people who live there.

▼ Landscape in Niger

Factors Affecting Climate

Main Idea **Most of Africa south of the Sahara has warm or hot climates. Rainfall, however, varies greatly throughout the region.**

Geography and You Have you ever lived through a long period without much rain? Read to learn about how lack of rain affects the lives of people in parts of Africa south of the Sahara.

Africa south of the Sahara lies mainly in the Tropics. As a result, most of the region receives the direct rays of the sun year-round, producing generally high temperatures. At higher elevations in this latitude, the climate is very different, however. For example, places with high elevation, such as mountains, often are cooler than lowland plains at the same latitude.

Figure 2 shows that Africa south of the Sahara has wet, dry, and temperate climate zones. Rainfall varies greatly throughout the region. The rain forests of Central and West Africa receive more than 80 inches (203 cm) of rain **annually.** By contrast, the Namib Desert in Southern Africa often gets less than 10 inches (25 cm) of rain per year. Some parts of Africa south of the Sahara have long **droughts,** or periods of time when there is no rain at all. Droughts can cause crop failures and widespread starvation.

Reading Check **Identifying Central Issues** How does rainfall vary in Africa south of the Sahara?

NATIONAL GEOGRAPHIC **Maps In MOtion** See StudentWorks™ Plus or glencoe.com.

Figure 2 **Africa: Climate Zones**

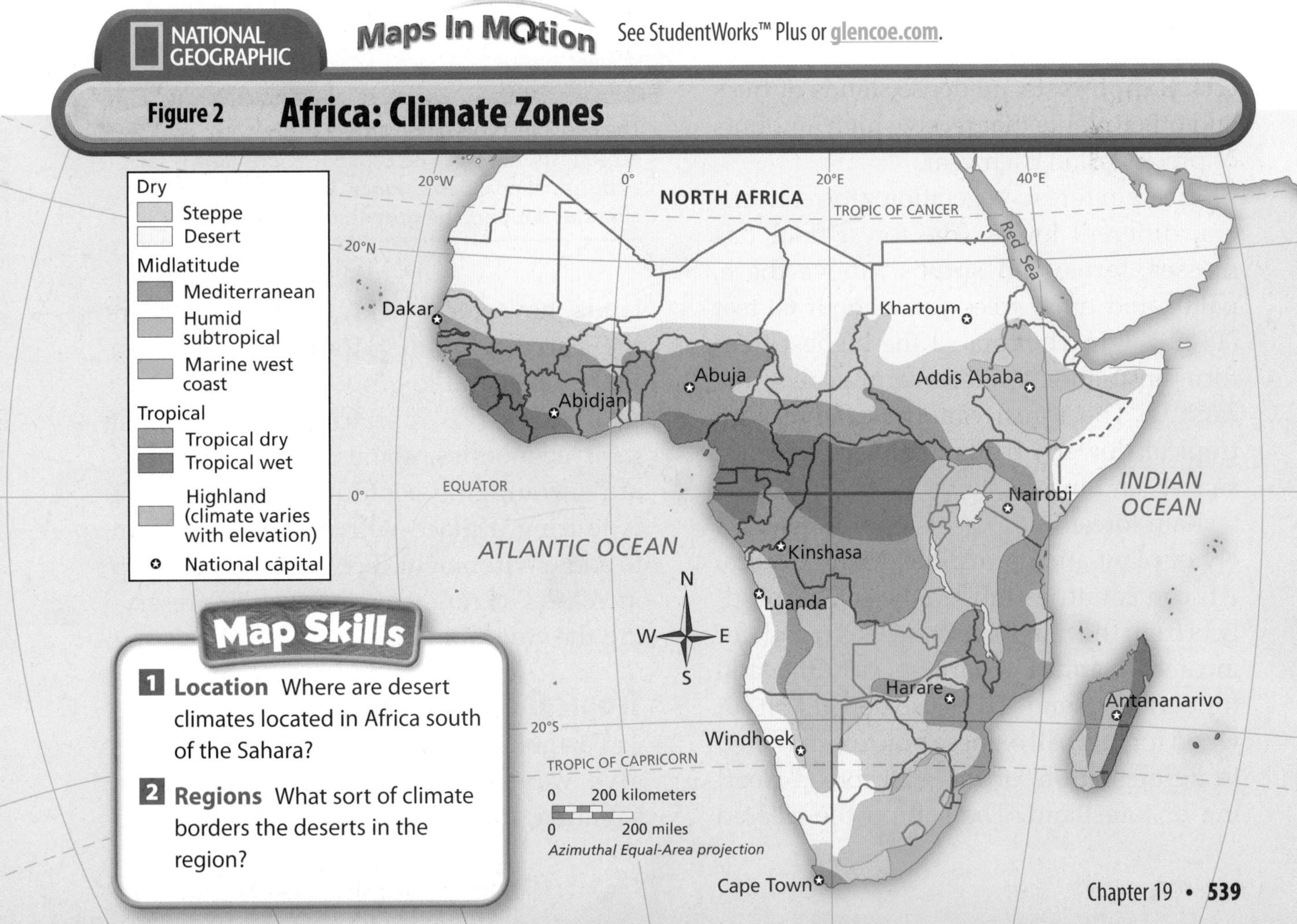

Map Skills

1. **Location** Where are desert climates located in Africa south of the Sahara?
2. **Regions** What sort of climate borders the deserts in the region?

Tropical and Dry Climates

Main Idea **Most of Africa south of the Sahara is covered by tropical or dry climate zones.**

Geography and You What kind of trees and plants grow best in your area? Read to find out how climate affects vegetation in Africa south of the Sahara.

Suppose you are standing at the Equator in Africa. Traveling either north or south, you would pass through the same pattern of climate zones: from tropical wet to tropical dry, then to steppe, and then desert.

Tropical Wet Climate

A tropical wet climate is found along the Equator in Central Africa and West Africa. Hot temperatures and plentiful rainfall in this zone support the growth of rain forests. **Rain forests** are dense stands of trees and other plants that receive high amounts of precipitation each year.

In a rain forest, vegetation grows at several different levels. The forest floor has mosses, ferns, and shrubs. Above these, palms and other trees grow about 60 feet (18 m) high. The tops of the highest trees form an umbrella-like covering called the **canopy.** The forest canopy is alive with tropical flowers, fruits, monkeys, parrots, snakes, and insects.

Rain forests support an **enormous** variety of plant and animal life. Many tropical African countries rely on the sale of products from the rain forests, such as wood, for income. In addition, farmers clear the land for new farmland. They also depend on cut wood for fuel. All of these practices have led to **deforestation,** or the widespread clearing of forestland. The soil on the cleared lands, however, quickly becomes less fertile. Farmers are then forced to clear even more forestland to grow their crops.

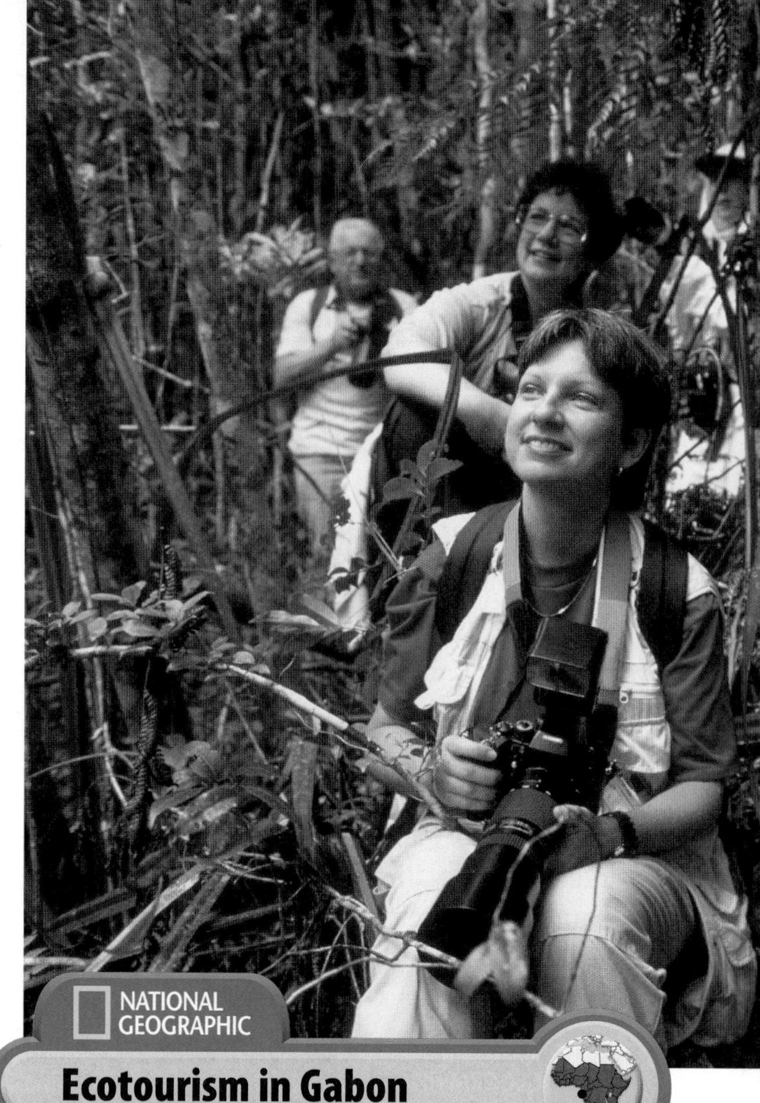

NATIONAL GEOGRAPHIC

Ecotourism in Gabon

The rain forests of Gabon attract many ecotourists. About 11 percent of Gabon's land is set aside as national parks. ***Place*** **Why do tropical wet climates support the growth of rain forests?**

To preserve rain forests, and boost their economies, some African countries are encouraging ecotourism. **Ecotourism** is touring a place without causing harm to the environment. Ecotourists thus help increase a region's revenue while preserving the environment.

Tropical Dry Climate

Farther from the Equator, rain forests give way to great stretches of tropical **savanna,** or grasslands with scattered woods. In this climate zone, temperatures remain hot all year, but rainfall amounts are much lower than in rain forest areas. Rains are heavy in the summer but light in the winter.

Savanna grasslands are home to some of Africa's most recognizable animals, including elephants, lions, rhinoceroses, and giraffes. Because hunting and human settlement threaten savanna plants and animals, several countries have set aside land as national parks to protect them.

Steppe

Continuing farther from the Equator, rainfall becomes more scarce, and savannas merge into drier steppes. In these areas, only about 8 to 15 inches (20 to 38 cm) of rain falls over the course of a few months each year. Vegetation includes different varieties of trees, thick shrubs, and grasses.

Steppe areas are threatened by **desertification,** the process that turns fertile land into land that is too dry to support life. Climate changes that bring long periods of extreme dryness and water shortages can lead to desertification. Clearing areas of trees and other vegetation or herding large amounts of livestock can also damage and dry out the land.

Deserts

In very dry areas of Africa, deserts dominate the landscape. The largest are the Sahara in the north and the Kalahari and the Namib in the south.

The Sahara has high temperatures and little rain. Instead of sandy dunes, it contains barren rock or stony plains covered by rocky gravel. Very little vegetation can live outside the oases and the highlands.

By contrast, the Kalahari in Southern Africa is covered by vast stretches of sand. It has high temperatures and little rainfall. When rains do fall, they are immediately absorbed by the sand, leaving the surface dry. Certain areas of the Kalahari have trees with long roots that reach the moisture in the deep sand.

The Namib, along the southwestern coast, is made up of rocks and dunes. This desert is arid, but temperatures tend to be cooler than in other African deserts because of breezes from the ocean. Fog that forms along the coast reaches the desert and provides moisture to many varieties of succulents. **Succulents** are plants such as cacti with thick, fleshy leaves that can conserve moisture.

Reading Check **Contrasting** How do the tropical dry and steppe climates of Africa differ?

Sand Dunes of the Namib

NATIONAL GEOGRAPHIC

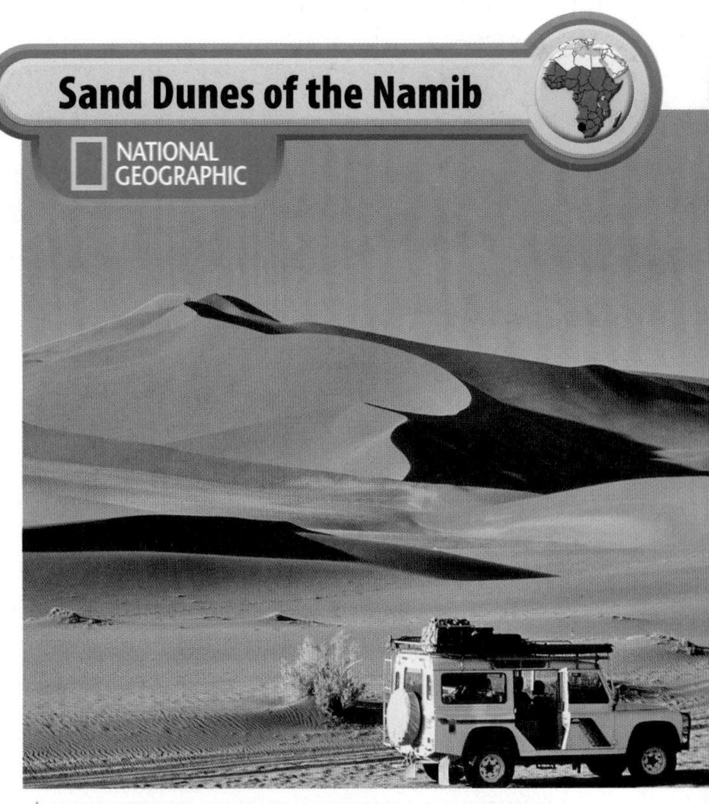

▲ The tallest dunes in the Namib are more than 1,280 feet (390 m) high and can be seen from space. ***Location*** **How does the location of the Namib Desert affect its climate?**

Moderate Climate Regions

Main Idea **Small areas of Africa south of the Sahara have moderate climate regions.**

Geography and You Is it extremely hot or cold where you live, or is the climate more moderate? Read to learn about the areas of Africa south of the Sahara that have moderate climates.

As **Figure 2** shows, moderate climates are found in coastal Southern Africa and the highlands of East Africa. These areas have comfortable temperatures and enough rainfall for farming.

Southeastern Africa has a humid subtropical climate of hot, wet summers and mild, wet winters. The farther south you go in Africa south of the Sahara, the farther you are from the Equator. As a result, temperatures become cooler.

Southwestern Africa has a Mediterranean climate. Here, winters are mild and wet, but the summers are warm and dry. Because the area is south of the Equator, seasons occur opposite of those in the United States. Autumn occurs in April, a period when some rain may fall. However, most rain falls during the area's winter months, which are June through August.

Highland climates are found in areas of higher elevation in East Africa. Temperatures in the highlands are cooler than in surrounding areas because of the higher altitude. Snow often falls at high elevations, and vegetation is abundant at lower elevations.

Reading Check **Determining Cause and Effect** Why does Southern Africa have cooler climates than other parts of the region?

Section 2 Review

Social Studies ONLINE
Study Central™ To review this section, go to glencoe.com.

Vocabulary

1. **Explain** the significance of:
   - **a.** drought
   - **b.** rain forest
   - **c.** canopy
   - **d.** deforestation
   - **e.** ecotourism
   - **f.** savanna
   - **g.** desertification
   - **h.** succulent

Main Ideas

2. **Explaining** Why does most of Africa south of the Sahara have generally high temperatures?
3. **Comparing and Contrasting** Use a chart like the one below to compare vegetation.

Area	Vegetation
Rain forest	
Savanna	
Steppe	
Desert	

4. **Describing** Describe features of southwestern Africa's climate.

Critical Thinking

5. **Comparing and Contrasting** How does the Namib Desert differ from the Sahara and the Kalahari? Why?
6. **BIG Idea** What characteristic of the steppe climate would make it difficult for people to farm there?
7. **Challenge** Do you think the creation of national parks in Africa south of the Sahara will protect its natural environment? Explain.

Writing About Geography

8. **Creative Writing** Write a weather report for a typical summer day and a typical winter day at a location within the tropical dry climate region.

Visual Summary

Landforms

- Most of Africa south of the Sahara lies on a series of plateaus.
- Africa's landforms include plateaus and volcanic peaks.
- Narrow plains hug Africa's coastlines. In some places, the plains spread deep into inland areas.

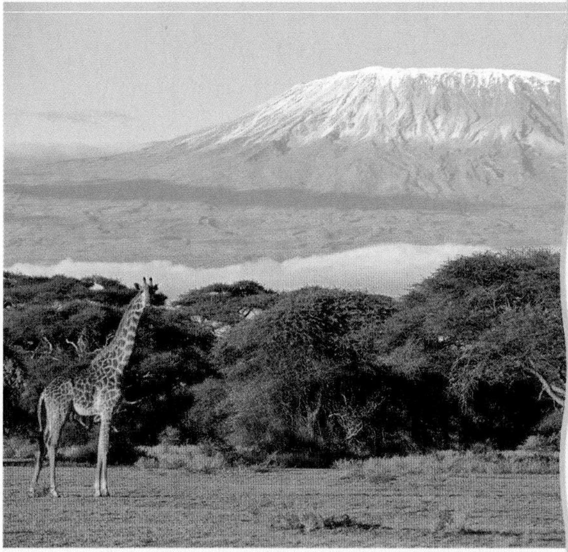

Kilimanjaro

Waterways

- Most lakes lie in the Great Rift Valley. Lakes are a source of freshwater and fish.
- The major rivers of Africa south of the Sahara are the Nile, the Congo, the Niger, and the Zambezi.

Impala calves, Kenya

A Tropical Region

- Tropical rain forests have hot temperatures and plentiful rains throughout the year.
- The amount of rainfall varies in the savannas. A variety of animals lives on these grasslands.
- Several countries have created national parks to protect forests and grasslands.

Niger River

Deserts and Steppes

- Deserts dominate the landscape in large areas of Africa south of the Sahara.
- The main deserts include the Sahara, the Kalahari, and the Namib.
- Partly dry grasslands near deserts are threatened by desertification.

Gathering water in Ethiopia

Moderate Climates

- Parts of southeastern and southwestern Africa lie outside the Tropics. They have moderate climates.
- Highland climates are found in mountainous areas of East Africa.
- Temperatures in the highlands are cooler than in surrounding areas because of higher altitude.

Study anywhere, anytime! Download quizzes and flash cards to your PDA from glencoe.com.

CHAPTER 19 ASSESSMENT

STANDARDIZED TEST PRACTICE

TEST-TAKING TIP

Skim through a test before you start to answer the questions. That way you can decide how to pace yourself.

Reviewing Vocabulary

Directions: Choose the word(s) that best completes the sentence.

1. In Africa south of the Sahara, the edges of plateaus are often marked by _______, or sharp, jagged cliffs.

A rifts

B gorges

C escarpments

D savannas

2. Steep-sided valleys that form when rivers cut through the land are called _______.

A savannas

B rift valleys

C gorges

D steppes

3. Parts of Africa south of the Sahara experience _______, or long periods of time when it does not rain at all.

A monsoons

B canopies

C savannas

D droughts

4. In rain forests, the tops of the highest trees form an umbrella-like covering called the _______.

A canopy

B succulent

C drought

D savanna

Reviewing Main Ideas

Directions: Choose the best answer for each question.

Section 1 *(pp. 530–535)*

5. Almost all of Africa south of the Sahara lies on a series of _______.

A plateaus

B escarpments

C mountain ranges

D rift valleys

6. In some nations south of the Sahara, _______ has replaced agricultural products as the principal export.

A steel

B diamonds

C gold

D oil

Section 2 *(pp. 538–542)*

7. The region's tropical wet climate supports the growth of _______.

A savannas

B rain forests

C steppes

D succulents

8. The moderate climate found in southwestern Africa is a _______ climate.

A tropical wet

B desert

C Mediterranean

D low latitude

GO ON

Critical Thinking

Directions: Base your answers to questions 9 and 10 on the map below and your knowledge of Chapter 19.

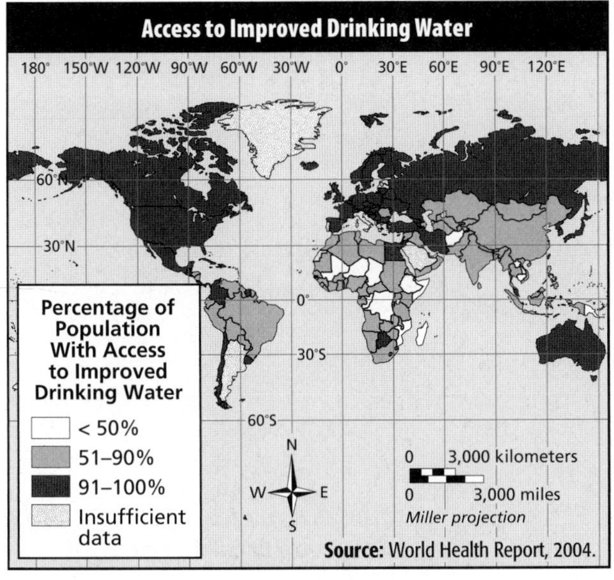

9. Which continents have the highest percentage of people using improved drinking water?

A South America, Asia, and Africa

B North America, Europe, and Australia

C North America, South America, and Africa

D Asia, Australia, and South America

10. Which continent has the greatest number of areas in which less than 50 percent of the population uses improved drinking water?

A Asia

B South America

C Africa

D Australia

Document-Based Questions

Directions: Analyze the document and answer the short-answer questions that follow.

Writer Paul Theroux set out on a journey that took him the length of the African continent, from Cairo, Egypt, to Cape Town, South Africa.

> *In places the [Shire] river twisted into bewildering marshland, dividing into many separate streams, softening, losing its [riverlike] look and becoming slow water in a mass of spongy weeds. The Shire ceased to be a river at the Ndinde Marsh, which was so dense with high grass and reeds we could not see ahead of us, so choked with hyacinths that our progress was slowed to hard paddling. In this marsh we could navigate only by occasionally going upstream, fighting the current . . . but after an hour in the marsh we emerged, with a view of Mozambique.*
>
> —Paul Theroux, *Dark Star Safari: Overland from Cairo to Cape Town*

11. How does the river on which the writer is traveling change?

12. Do you think the river described here is useful to local residents? Explain your answer.

Extended Response

13. Choose a landform, waterway, or combination of physical features in Africa south of the Sahara that you find especially fascinating. Write a short essay that describes the feature or features, explains why it is unique or interesting, and encourages others to visit the feature.

For additional test practice, use Self-Check Quizzes—Chapter 19 at glencoe.com.

Need Extra Help?

If you missed question...	1	2	3	4	5	6	7	8	9	10	11	12	13
Go to page...	531	534	539	540	531	534	540	542	26	26	533	533	530–535

History and Cultures of Africa South of the Sahara

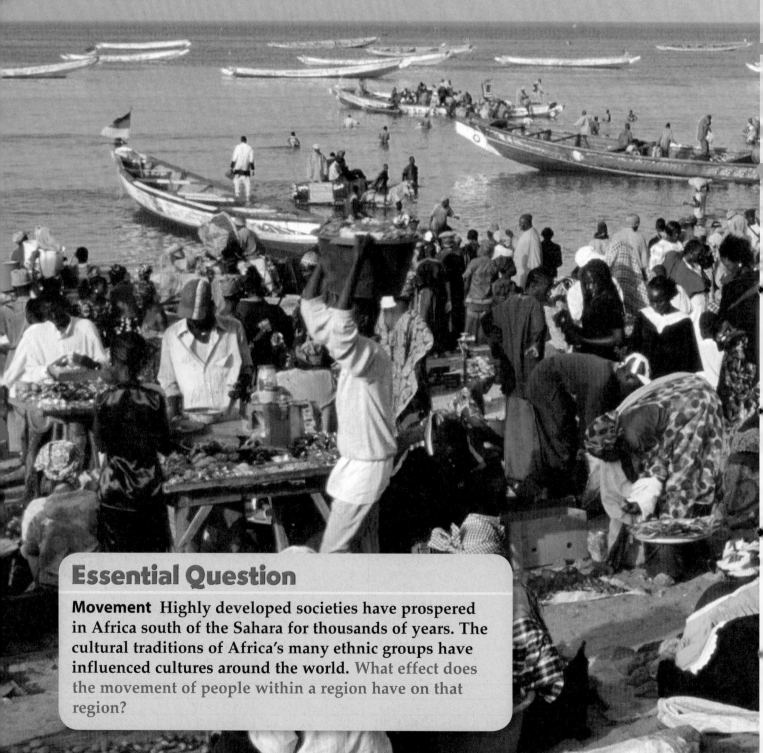

Essential Question

Movement Highly developed societies have prospered in Africa south of the Sahara for thousands of years. The cultural traditions of Africa's many ethnic groups have influenced cultures around the world. What effect does the movement of people within a region have on that region?

Petite Côte, Senegal

BIG Ideas

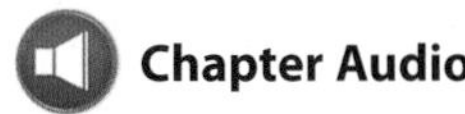
Chapter Audio

Section 1: History and Governments

BIG IDEA The characteristics and movement of people impact physical and human systems. In ancient Africa, the migrations of the Bantu people spread a common language and technology. Later, powerful kingdoms emerged in Africa. Beginning in the 1400s, Africans faced European colonial rule before they eventually achieved independence as a number of nation-states.

Section 2: Cultures and Lifestyles

BIG IDEA Culture groups shape human systems. Different ethnic groups and cultures have shaped Africa south of the Sahara. The region has also been influenced by non-African cultures. At the same time, African cultural influences have spread to other parts of the world.

FOLDABLES™ Study Organizer

Categorizing Information Make this Foldable to help you organize information about the history and cultures of Africa south of the Sahara.

Step 1 Fold a piece of 11 x 17 paper in half.

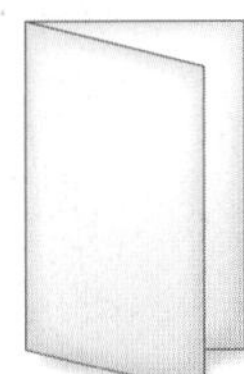

Step 2 Fold the bottom edge up two inches. Glue the outer edges of the flap to create pockets.

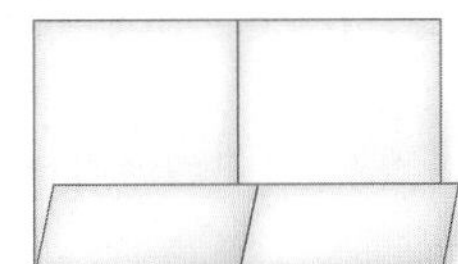

Step 3 Label each pocket as shown. Use these pockets to hold notes taken on index cards or quarter sheets of paper.

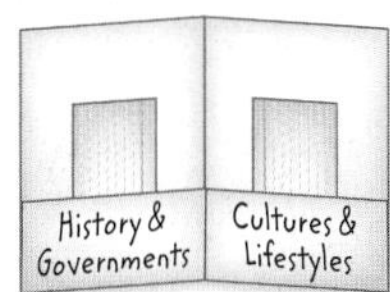

Reading and Writing As you read the chapter, take notes on index cards or slips of paper and place each note in the appropriate pocket of the Foldable. When you have finished the chapter, use your notes to write either a historical or cultural summary of the region.

Social Studies ONLINE

Visit glencoe.com and enter **QuickPass**™ code EOW3109c20 for Chapter 20 resources.

Guide to Reading

BIG Idea

The characteristics and movement of people impact physical and human systems.

Content Vocabulary

- hunter-gatherer *(p. 549)*
- plantation *(p. 552)*
- nationalism *(p. 552)*
- discrimination *(p. 553)*
- refugee *(p. 553)*
- apartheid *(p. 555)*

Academic Vocabulary

- isolate *(p. 549)*
- administrator *(p. 550)*

Reading Strategy

Identifying Central Issues Use a diagram like the one below to list features of the early kingdoms and empires that developed in Africa south of the Sahara.

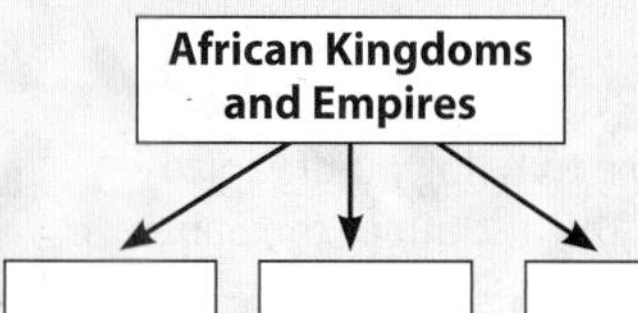

History and Governments

 Section Audio  Spotlight Video

Picture This The past speaks to the future through art. These statues represent the people who have received the Nobel Peace Prize for their efforts to promote human rights in South Africa. The statues are, from left to right, Albert Lutuli, Archbishop Desmond Tutu, F.W. de Klerk, and Nelson Mandela. To learn more about the history and governments of countries south of the Sahara, read Section 1.

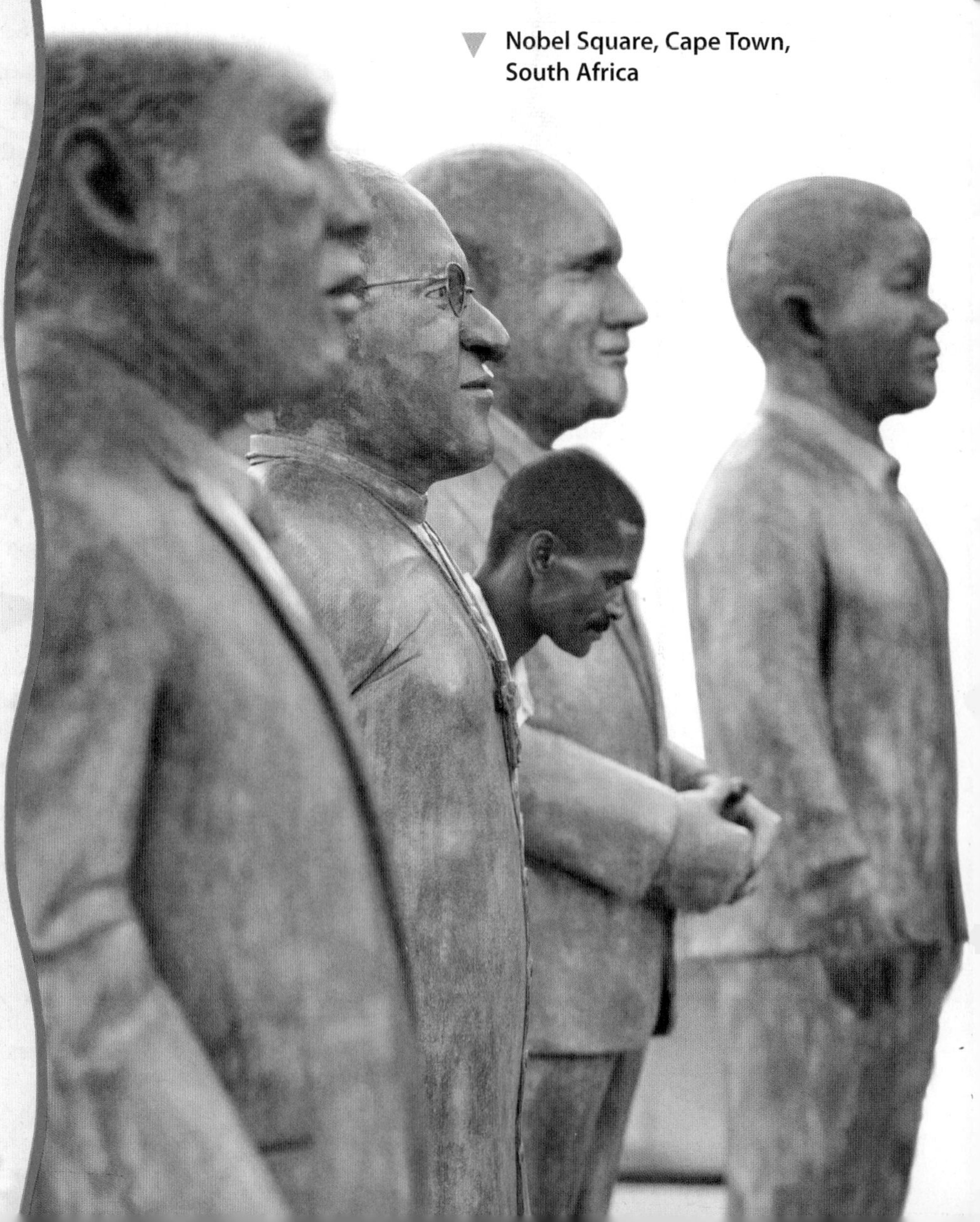

▼ Nobel Square, Cape Town, South Africa

Early African History

Main Idea **African peoples built successful societies in the region beginning in ancient times.**

Geography and You If you moved to a place far away, what would you bring to your new home? Read to learn how a large migration in ancient Africa spread a common culture throughout the continent.

People have been living in Africa for thousands of years. Great kingdoms and empires developed across the continent.

Early History

Early Africans lived as **hunter-gatherers,** or people who moved from place to place to hunt and gather food. Over time, people began to herd livestock and to farm. As northern Africa's climate became drier and hotter, many people began migrating southward to more fertile areas.

Around 3000 B.C., a migration that lasted several thousand years began. The migrants, a people known as the Bantu, shared a common language, culture, and technology. The Bantu migrated from modern-day Nigeria to the west and south, spreading their farming and ironworking skills, along with their language. Today, millions of Africans south of the Sahara speak Bantu languages.

East and Southern Africa

Some of Africa's earliest kingdoms developed in East and southern Africa. Around 800 B.C., Kush developed along the Nile River in present-day Sudan. The people of Kush grew wealthy from trade and ironworking. Kush gold, ivory, and iron products were traded as far as Egypt and Southwest Asia. As Kush's wealth increased, its rulers built temples and monuments like those of Egypt. During the A.D. 300s, however, Kush was defeated by the neighboring kingdom of Axum.

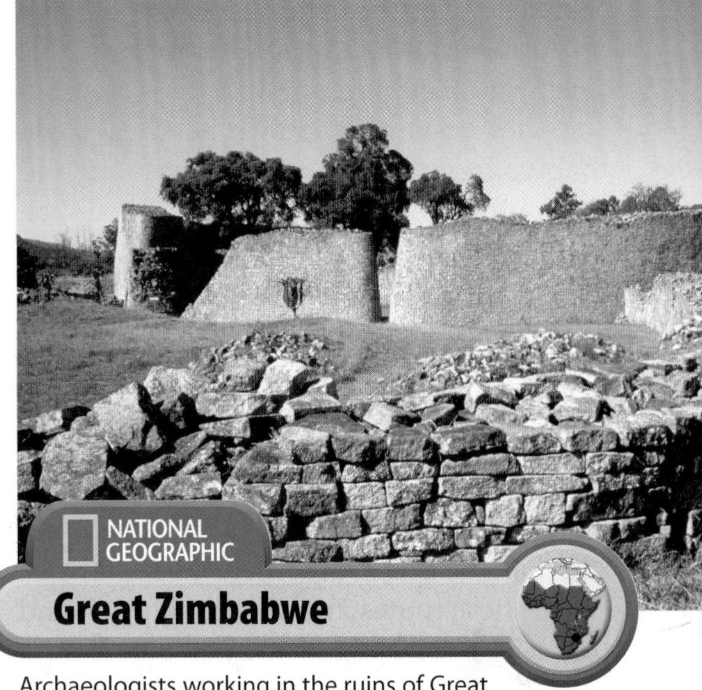

Great Zimbabwe

Archaeologists working in the ruins of Great Zimbabwe have unearthed beads and pieces of pottery from Arabia and China. ***Movement*** **How long ago did the Bantu migration begin?**

Axum, in what is now Ethiopia, prospered from trade. Goods from Africa, the Mediterranean area, and East Asia flowed through Axum. In the A.D. 300s, King Ezana accepted Christianity and made it the official religion of Axum. Arab Muslims in the 600s, however, gained control of much of the surrounding region. Axum became **isolated,** although it remained a center of African Christianity.

Farther south, people in several East African coastal cities—Kilwa, Mombasa, and Mogadishu (MOH·guh·DEE·shoo)—traded with Arabia, India, and China.

Student Web Activity Visit glencoe.com and complete the Chapter 20 Web Activity about the Swahili culture.

Between A.D. 1000 and A.D. 1500, a blend of Arab and African ways led to the rise of the Swahili culture.

Another empire known as Great Zimbabwe (zihm·BAH·bway) arose inland in southeastern Africa. During the 1400s, Zimbabwe supplied gold, silver, and ivory to the East African coast.

West Africa's Trading Empires

As **Figure 1** shows, three trading empires emerged in West Africa from the A.D. 800s to the A.D. 1500s. Ghana (GAH·nuh), the earliest empire, controlled trade between the Sahara and West Africa's rain forests. Over these routes, traders brought salt and cloth and exchanged them for gold and ivory. By taxing this trade, Ghana became very wealthy. In the late 1000s, North African invaders disrupted Ghana's trade and the kingdom collapsed.

The empire of Mali replaced Ghana. Like Ghana, Mali grew wealthy from farming and from control of the gold and salt trade. Its most famous ruler, Mansa Musa, was a skilled **administrator.** During his rule, Timbuktu (TIHM·buhk·TOO) became a center of trade, education, and Islamic culture.

In the 1400s, the kingdom of Songhai (SAWNG·HY) took over Mali. Muhammad Toure was the most famous Songhai leader. Like Mansa Musa, he was a follower of Islam. In about 1600, invaders from North Africa defeated the Songhai, ending their empire.

Reading Check **Analyzing Information** Why was Timbuktu important to early Africa?

NATIONAL GEOGRAPHIC **Maps In MOtion** See StudentWorks™ Plus or glencoe.com.

Figure 1 **The Trading Empires of West Africa**

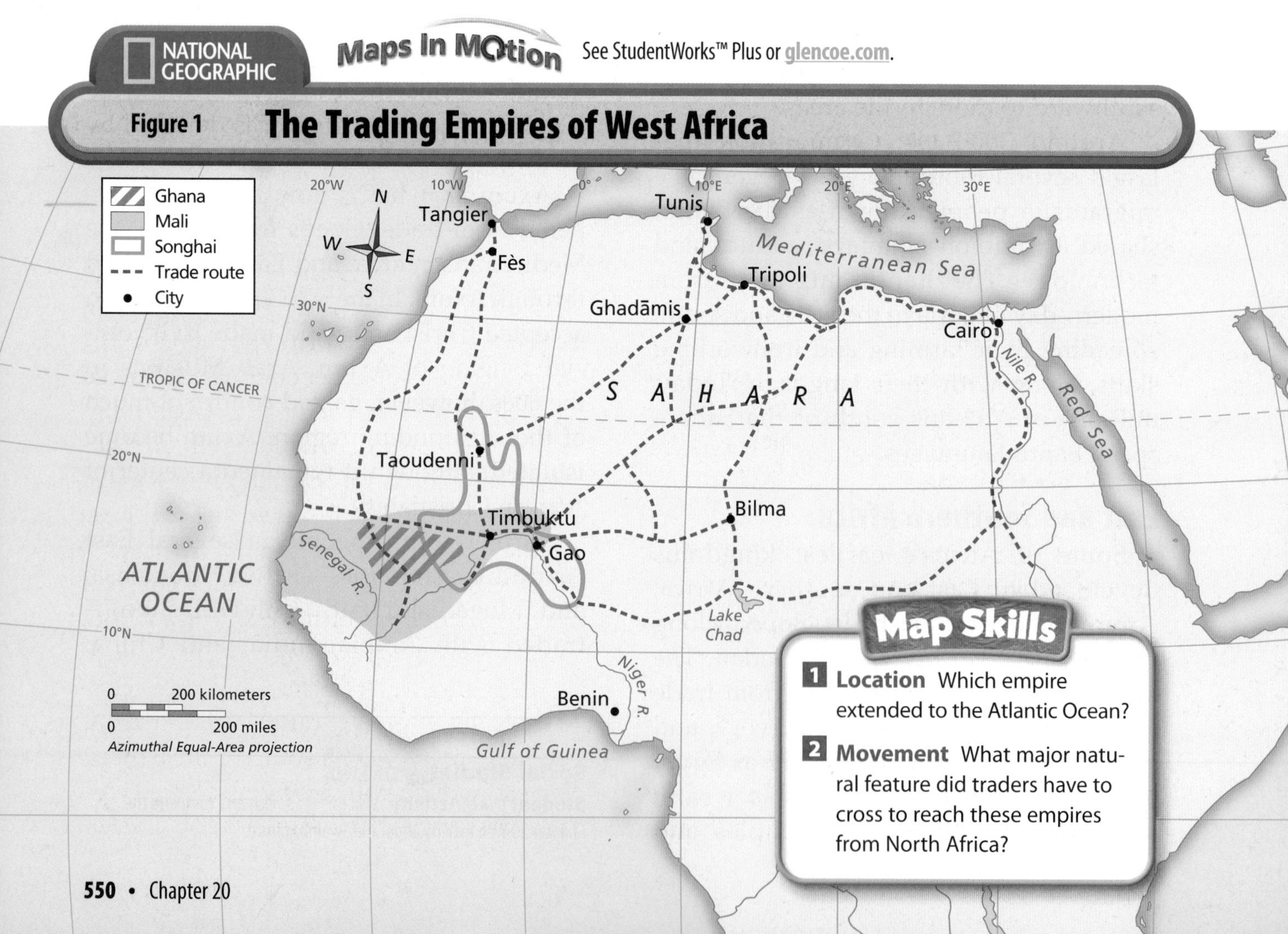

Map Skills

1. **Location** Which empire extended to the Atlantic Ocean?
2. **Movement** What major natural feature did traders have to cross to reach these empires from North Africa?

European Contact

Main Idea **After 1500, increased contact with Europeans led to great changes in Africa south of the Sahara.**

Geography and You Has a new business, store, or factory moved into your community? What effect did it have? Read to find out about the effects of European commerce on Africa's peoples.

In the 1400s and 1500s, Europeans began trading with African societies. Merchants from Portugal set up trading posts along Africa's western coast, and traders from other countries soon followed. The arrival of Europeans in Africa south of the Sahara dramatically changed the region.

The Slave Trade

One of the many changes was the growth of the slave trade. Europeans did not introduce slavery or the slave trade to the African continent. For centuries, African rulers had enslaved and traded prisoners. Arab traders had brought enslaved Africans to the Islamic world since the A.D. 800s.

The slave trade, however, greatly increased when Europeans began shipping Africans to the Americas. There, the Africans were forced to grow sugar and other cash crops. In Africa, African traders armed with European guns seized captives and delivered them to European trading posts on the coast. European traders then shipped the captured people across the Atlantic Ocean to be sold into slavery. Between about 1500 and the late 1800s, nearly 12 million Africans were sent to the Americas.

For Africans, the slave trade brought great suffering. Millions of people were torn from their homes and families. Entire villages disappeared, local economies collapsed, and kingdoms weakened.

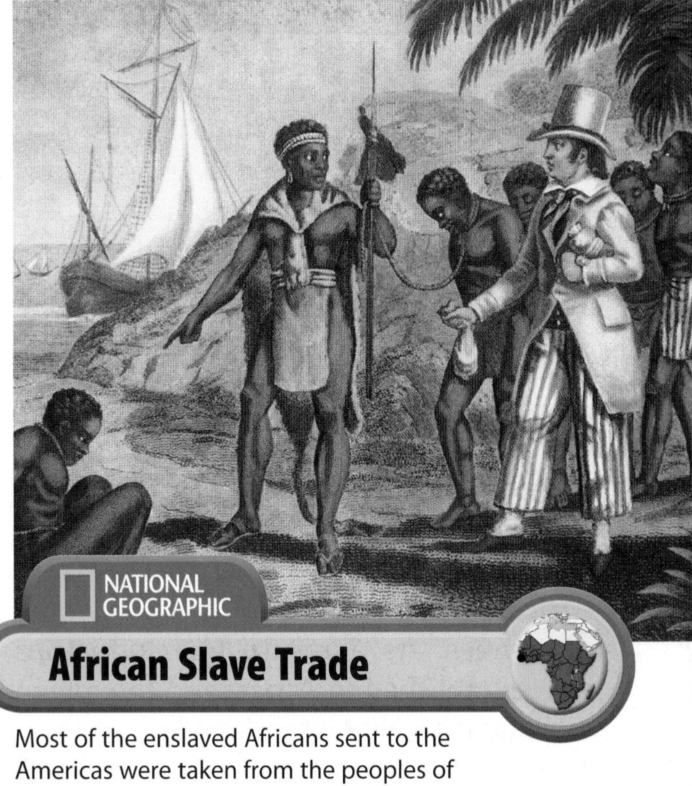

African Slave Trade

Most of the enslaved Africans sent to the Americas were taken from the peoples of West Africa. ***Regions*** **Who first enslaved Africans?**

European Rule

By the 1800s, many Europeans had decided slavery and the slave trade should be stopped. When Britain declared the slave trade illegal, other countries followed its lead. This did not stop European interest in the region, particularly in Africa's vast inland areas. Business leaders wanted Africa's gold, timber, hides, and palm oil for their growing industries. Military leaders wanted to protect the coastal areas their countries already controlled. Missionaries wanted to convert Africans to Christianity.

Beginning in the 1880s, European countries, such as Britain, France, and Germany, set out to claim Africa south of the Sahara. For economic profit and political advantage, they carved the region into colonies.

In doing so, they ripped apart once-unified regions and threw together ethnic groups that had little in common. European armies with advanced weapons defeated Africans who resisted the takeover. By 1914, almost the entire region was under European control. The only territories free of European rule were Ethiopia and Liberia.

European rule greatly changed Africa south of the Sahara. Europeans built railroads and roads and introduced new feeds and fertilizers to improve farming. They provided some Africans with European-style education and medical care. A chief goal of Europeans, however, was to export Africa's raw materials to help their own economies. As a result, few industries were established in Africa, and the region remained largely undeveloped.

In addition, Africans faced many hardships. In their own lands, Africans had fewer rights and economic opportunities than the Europeans who lived there. Many Africans were forced to work in harsh conditions in mines or on large farms called **plantations.** Any opposition to European rule was severely punished.

Reading Check **Drawing Conclusions** Why did Europeans establish colonies throughout much of Africa?

Independence

Main Idea **In the late 1900s, African countries won independence, but the new nations faced many challenges.**

Geography and You Think about how you feel when you are not treated fairly. How do you react? Read on to learn how Africans sought better treatment and independence from European rulers.

In the early 1900s, Africans helped European powers fight in both World Wars. Many hoped they would be rewarded with independence. Instead, European countries further increased the size of their empires. They held on to their African colonies, causing many Africans to demand freedom from European rule. During the last half of the 1900s, their efforts were successful, and many African colonies became independent countries.

Struggle for Freedom

As the 1900s began, feelings of nationalism arose among European-educated Africans. **Nationalism** is a people's desire to rule themselves and have their own independent country. Many Africans grew frustrated with their European rulers. Europeans wanted democracy for themselves

History at a Glance
500 B.C.
A.D. 1
A.D. 500
c. 800 B.C.
Kush develops along the Nile River
c. 700s B.C.
Pyramids, Kush
c. A.D. 330
Axum accepts Christianity
c. A.D. 300s
Axum crown

but refused the same freedoms to colonial people overseas. Eventually, leaders came forth who convinced greater numbers of Africans to demand freedom.

After World War I, more Africans became politically active. They staged protests against **discrimination,** or unfair and unequal treatment of a group. European governments responded with force and arrests, but they also made some reforms. Africans, however, were not satisfied with limited steps and demanded complete independence.

During World War II, the European powers weakened because of the economic and military strains of the war. After the war, African protests against European rule grew. The weakened European countries did not have the resources to stop the independence movements.

In the early 1950s, Kwame Nkrumah (KWAHM·eh ehn·KROO·muh) led a nationalist movement in Britain's colony of the Gold Coast in West Africa. In 1957 the Gold Coast, now renamed Ghana, became independent. Soon after, more independence movements swept through Africa south of the Sahara. As **Figure 2** shows, by the end of the 1960s, most African territories had thrown off European rule.

After Independence

Freedom brought many political challenges to Africa south of the Sahara. Previously you learned that European countries in the late 1800s split Africa into colonies. In doing so, they divided once-united regions and combined ethnic and religious groups that did not get along.

After independence many countries kept the old colonial borders. This led to violence, and many of the new African countries suffered from civil wars. Ethnic and religious conflicts divided people in such new countries as Nigeria, Sudan, Rwanda, and the Democratic Republic of the Congo.

During these civil wars, many people died or became **refugees,** people who flee to another country to escape mistreatment or disaster. Some of this unrest spilled over from one country to another. In some cases, United Nations (UN) peacekeeping troops were called in to restore and maintain peace.

South Africa and Apartheid

In some countries, white-run governments denied basic rights to much larger non-European populations. In South Africa, white South Africans strengthened their rule through a system known as apartheid.

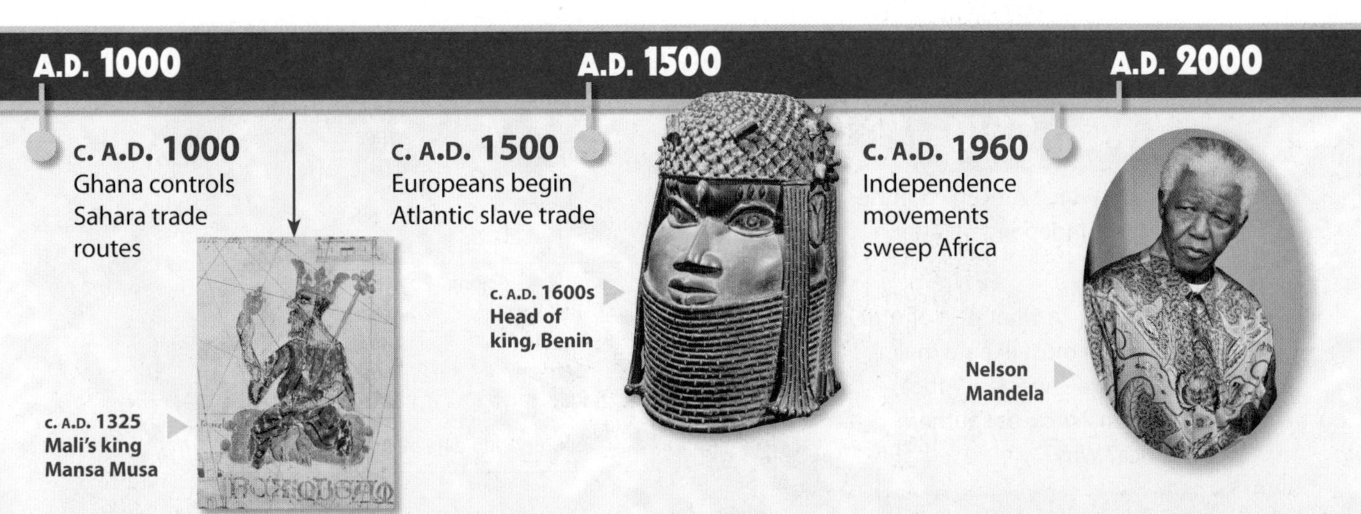
A.D. 1000
A.D. 1500
A.D. 2000
c. A.D. 1000
Ghana controls Sahara trade routes
c. A.D. 1500
Europeans begin Atlantic slave trade
c. A.D. 1960
Independence movements sweep Africa

c. A.D. 1325 Mali's king Mansa Musa

c. A.D. 1600s Head of king, Benin

Nelson Mandela

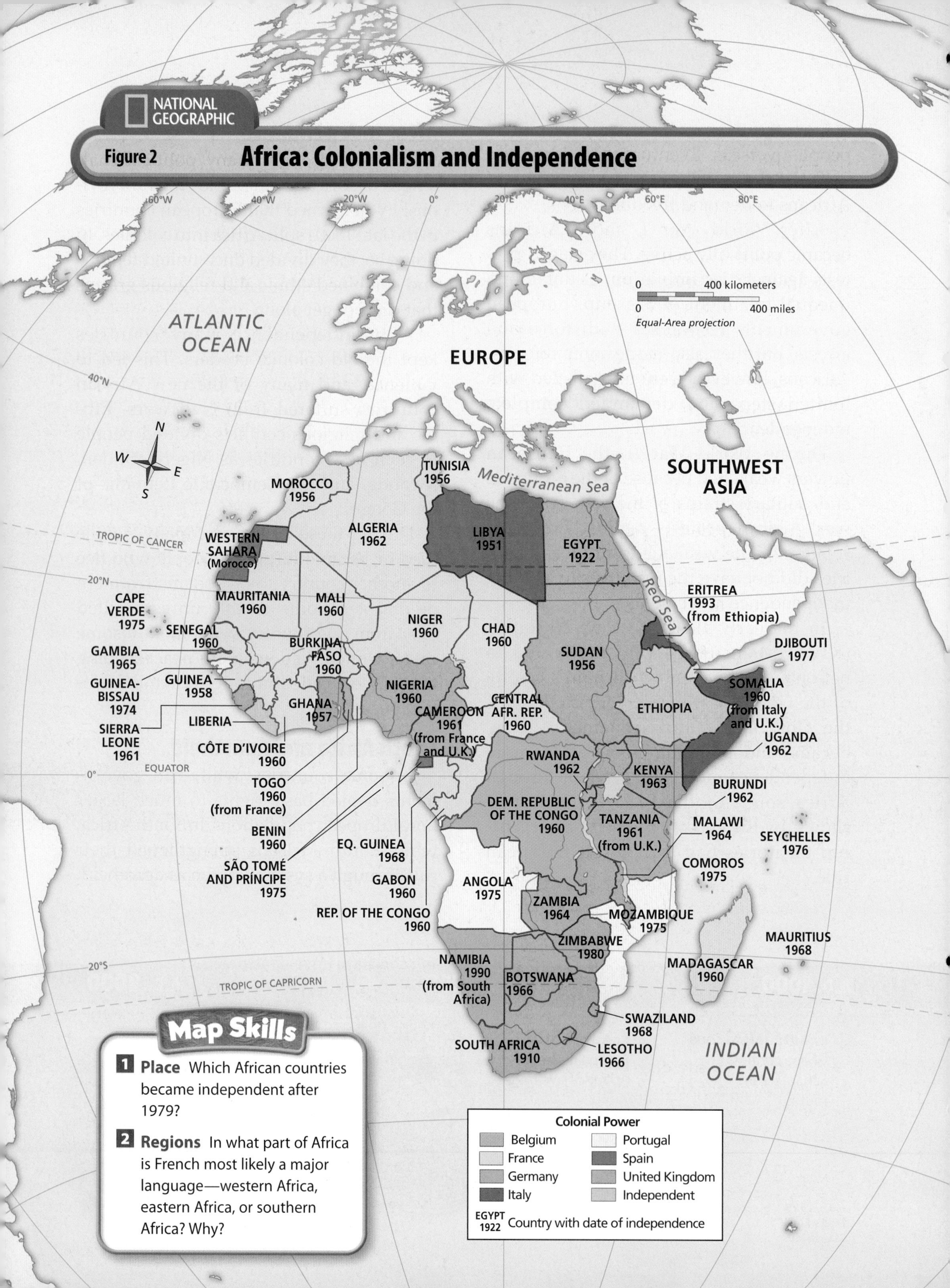

Map Skills

1. **Place** Which African countries became independent after 1979?
2. **Regions** In what part of Africa is French most likely a major language—western Africa, eastern Africa, or southern Africa? Why?

The policy of **apartheid** (uh·PAHR·TAYT), or "apartness," was carried out through laws that separated ethnic groups and limited the rights of black South Africans. For example, blacks had to live in separate areas, called "homelands," that had few resources. Also, people of non-European background were not allowed to vote.

Black South Africans protested the laws, and the white government responded by arresting protestors. Many black leaders, such as Nelson Mandela, were jailed. The United Nations condemned apartheid, and many countries cut off trade with South Africa.

Because of this pressure, the white-run government ended apartheid in the early 1990s. It also released Nelson Mandela from prison. All South Africans, regardless of race, were declared equal under the law.

In 1994 South Africa held its first democratic election in which people of different races were allowed to vote. South Africans elected Nelson Mandela as their nation's first black president. He worked to unite and rebuild the country.

To help heal South Africa's divisions, Nelson Mandela's government set up a Truth and Reconciliation Commission in 1995. Desmond Tutu, a church leader and winner of the 1984 Nobel Peace Prize, headed the commission. Over several years, the commission examined human rights crimes of the apartheid years. In 1998 the commission's report condemned human rights abuses by both white and black South Africans. It also gave amnesty, or forgiveness, to some people who had carried out crimes. The commission helped the country move forward and recover from its apartheid past.

Reading Check **Explaining** How did World War II lead to changes in European rule in Africa?

Social Studies ONLINE
Study Central™ To review this section, go to glencoe.com.

Vocabulary

1. **Explain** the meaning of:
   - **a.** hunter-gatherer
   - **b.** plantation
   - **c.** nationalism
   - **d.** discrimination
   - **e.** refugee
   - **f.** apartheid

Main Ideas

2. **Explaining** What were the effects of the Bantu migrations?
3. **Identifying** Use a diagram like the one below to identify the effects of European rule on Africa south of the Sahara.

4. **Summarizing** What challenges did many African nations face after independence?

Critical Thinking

5. **BIG Idea** What were the effects of the European slave trade on Africa?
6. **Challenge** How do you think the white-run government maintained control in South Africa for so long?

Writing About Geography

7. **Using Your FOLDABLES** Use your Foldable to write a paragraph about how the empires of West Africa rose and fell between A.D. 800 and the A.D. 1500s.

NATIONAL GEOGRAPHIC **Geography & History**

Africa's Salt Trade

Passing the salt at dinner may not be a big deal, but in parts of ancient Africa, salt built empires.

Good as Gold Salt is essential for life. Every person's body contains about 8 ounces (227 g) of salt—enough to fill several saltshakers. Salt helps muscles work, and it aids in digesting food. Salt also preserves food and adds flavor to meals. In hot climates, people need extra salt to replace the salt lost when they sweat. In tropical Africa, salt has always been precious.

Salt is plentiful in the Sahara and scarce in the forests south of the Sahara (in present-day countries such as Ghana and Côte d'Ivoire). These conditions gave rise to Africa's salt trade. Beginning in the A.D. 300s, Berbers drove camels carrying European glassware and weapons from Mediterranean ports into the Sahara. At the desert's great salt deposits they traded European wares for salt.

Rise and Decline Camels arrived in Africa from Asia in A.D. 300. In time, caravans of thousands of camels loaded with tons of salt arrived at southern markets. Local kings along the trade routes put taxes, which were payable in gold, on all goods crossing their realms. The ancient empires of Mali, Ghana, and Songhai rose to great power from wealth brought by the salt trade.

Trade routes also provided avenues for spreading ideas and inventions. By the A.D. 800s, Arab traders brought to Africa a system of weights and measures, a written language, and the concept of money. They also brought a new religion—Islam. Although salt no longer dominates trade, it is used as a seasoning throughout the world.

Member of a present-day caravan in Niger

NATIONAL GEOGRAPHIC

Salt Trade Routes

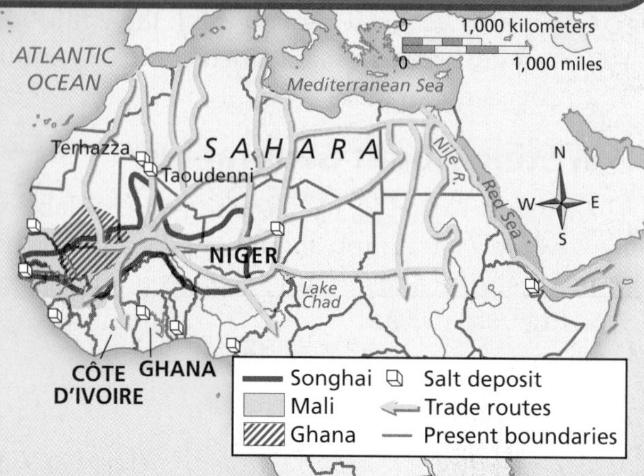

Think About It

1. **Movement** What goods were exchanged in the salt trade?
2. **Regions** How did the salt trade affect regions south of the Sahara?

Guide to Reading

BIG Idea

Culture groups shape human systems.

Content Vocabulary

- malnutrition *(p. 559)*
- sanitation *(p. 559)*
- life expectancy *(p. 560)*
- social status *(p. 562)*
- rite of passage *(p. 562)*
- griot *(p. 562)*
- compound *(p. 563)*
- extended family *(p. 564)*
- nuclear family *(p. 564)*
- clan *(p. 564)*
- lineage *(p. 564)*

Academic Vocabulary

- trend *(p. 559)*
- tradition *(p. 562)*
- require *(p. 564)*

Reading Strategy

Identifying Central Issues Use a diagram like the one below to list information about African society and culture.

Cultures and Lifestyles

 Section Audio  Spotlight Video

Picture This The vibrant art of South Africa's Ndebele people decorates their houses, clothing, and jewelry. Once powerful landowners, the Ndebele lost their land to white farmers in the late 1800s. The Ndebele painted their houses with colors and symbols to show their continued pride as a people. This tradition continues today. To learn more about the cultures of Africa south of the Sahara, read Section 2.

▼ Ndebele art in South Africa

The People of Africa South of the Sahara

Main Idea **Africa south of the Sahara has a rapidly growing population.**

Geography and You Has the population in your community increased or decreased in the last 10 years? Read to learn how cities in this region are growing larger.

Today Africa south of the Sahara has about 750 million people. Population throughout the region has grown rapidly in recent decades. In fact, the rate of population growth in Africa south of the Sahara is among the highest in the world.

A Growing Population

There are several reasons why the population in Africa south of the Sahara has grown. Better sanitation and medical care have lowered the death rates for infants and children. At the same time, the region has a high birthrate. Families in the region average five to seven children. Traditionally, African societies have seen large families as signs of wealth and prestige.

Rapid population growth has brought major challenges to Africa south of the Sahara. Overcrowding has led to poor living conditions in towns and cities. Governments find it difficult to provide shelter, water, and electricity for new city dwellers. In addition, much of Africa south of the Sahara has been unable to feed its people. Although about 70 percent of Africans work in agriculture, they cannot grow as much as they once could. Large areas of farmland in the region have been ruined by overuse, soil erosion, and serious droughts.

In the past, some African governments supported exporting goods to boost national incomes. These efforts did not

NATIONAL GEOGRAPHIC

Rapid Urbanization

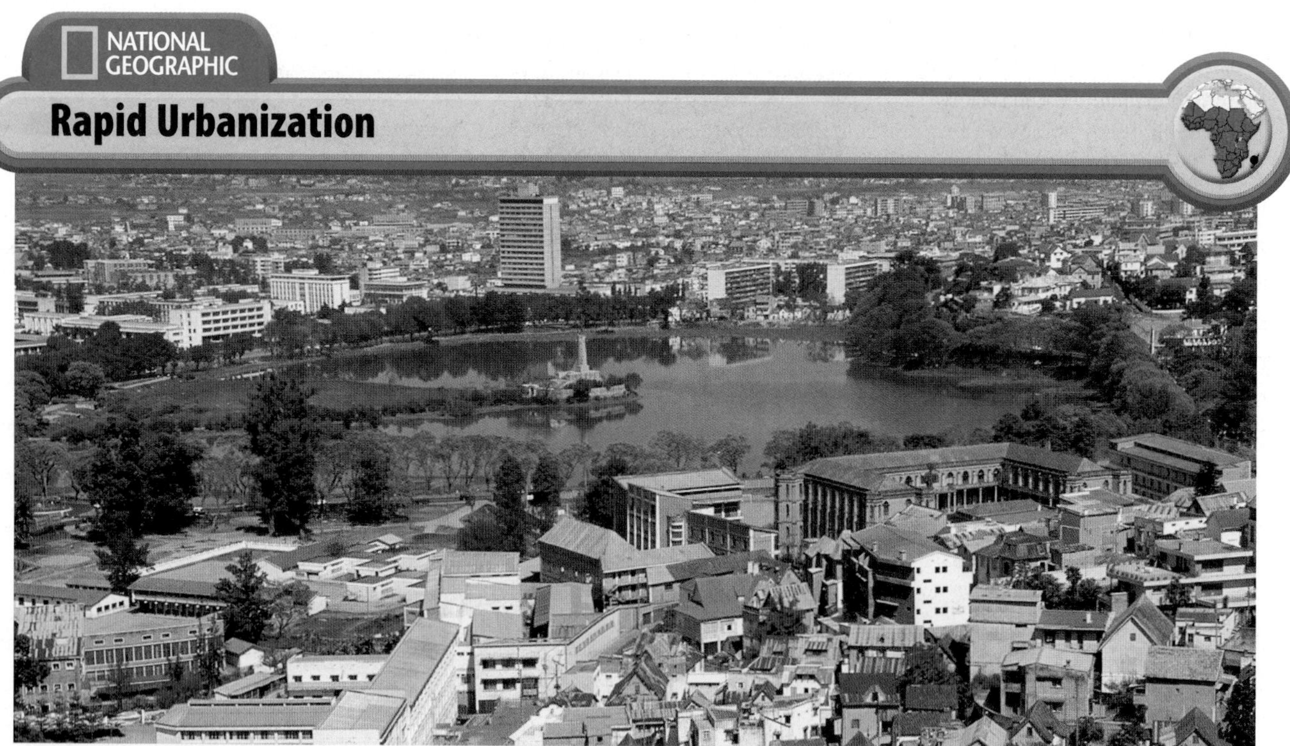

▲ Antananarivo is the capital of the island nation of Madagascar. It is home to about 1.7 million people. ***Human-Environment Interaction*** **What challenges do cities in Africa south of the Sahara face as a result of rapid population growth?**

leave enough food for local populations. As a result, many African countries must buy food from other countries. Often this food is expensive and not plentiful.

Throughout the region, governments have begun addressing these problems. They are now teaching people better ways to farm. To slow population growth, some governments have tried to promote the idea of having smaller families.

Where Africans Live

The region's growing population is not evenly distributed. This uneven distribution is a result of the climate and land features. Desert or steppe covers large portions of the region. In these areas, the land is generally too dry to support farming or raising livestock. Most of the region's people are crowded in coastal areas of West Africa, the lakes region of East Africa, and along the eastern coast of southern Africa. These areas have plentiful rainfall, milder temperatures, and fertile soil that attract large populations. Most of the region's farms, industries, and businesses are located in these areas.

Throughout Africa south of the Sahara, most people live in rural villages. Only about a third of the region's population lives in cities and towns. Cities in Africa south of the Sahara, however, are growing rapidly. A **trend,** or general tendency, in the region's population is toward urbanization, or the movement of people from rural areas to cities. The region has the world's fastest rate of urbanization.

Africans leave their rural villages for cities in order to find better jobs, health care, and education. Urban growth has its challenges however. Many African cities have towering skyscrapers and shopping areas, but city residents must endure traffic jams, overcrowded neighborhoods, pollution, and poor sanitation.

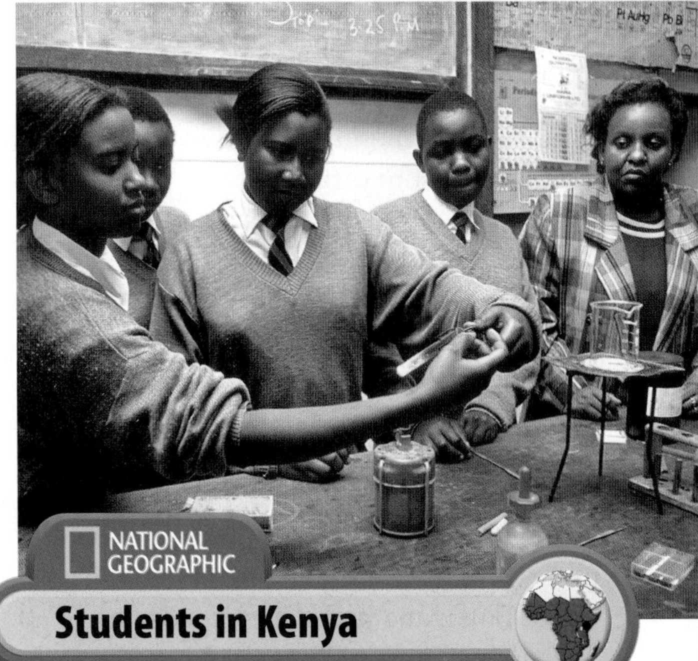

Students in Kenya

Students are more likely to have access to education in the cities of Africa than in poor, rural areas. ***Regions*** **Where does most of the population of Africa south of the Sahara live?**

Health Care

Population growth in Africa south of the Sahara has been helped by many advances in health care. More health clinics, hospitals, and medical centers have opened. Also, drugs and medical supplies are more readily available. Still, the region's death rate remains high compared to other world regions. People in many parts of Africa south of the Sahara suffer from **malnutrition,** or poor health due to not eating the right foods or enough food.

Many rural Africans lack clean water to drink, as well as adequate **sanitation,** or removal of waste products. Terrible famines have killed many people, especially in East Africa and areas bordering the Sahara. Diseases such as malaria are widespread. Insects such as the mosquito and tsetse (SEHT·see) fly transfer viruses to people and animals.

Because of these problems, health care is a major issue in Africa south of the Sahara.

Governments and private groups are trying to deal with health concerns, but providing good health care is very expensive. In addition, health emergencies have made solutions even more difficult.

Despite the region's rapid rise in population, one factor may drastically limit population growth in the future. Millions of Africans have been infected with the virus that causes AIDS. In some southern African countries, such as Botswana and South Africa, the disease has resulted in the decline of life expectancy. **Life expectancy** is the average number of years a group of people can expect to live. In the early 1990s the people of Botswana had a life expectancy of 60 years. This figure fell into the 30s in the early 2000s. The loss of large numbers of young adults to AIDS, as well as the children they would have had, means that the country's population will soon be decreasing.

AIDS has had a destructive impact on Africa south of the Sahara in other ways. Individuals and governments find it difficult to pay the costly medical bills that are associated with treating AIDS patients. Also, the loss of skilled workers to AIDS weakens economies. The death of parents causes enormous hardships for countless families. The United Nations predicts that the number of African children orphaned by AIDS could reach 20 million by 2010.

Reading Check **Determining Cause and Effect** Why has the population grown so rapidly in the region?

TIME GLOBAL CITIZENS

NAME: ZACKIE ACHMAT

HOME COUNTRY: Republic of South Africa

ACHIEVEMENT: The AIDS activist and founder of Treatment Action Campaign won his fight with the South African government to sell cheaper versions of very expensive life-saving medicines. As a result, many of the country's five million HIV/AIDS sufferers can now get those drugs at work and clinics at affordable prices. Achmat's courage and perseverance earned him a Nobel Peace Prize nomination in 2003.

QUOTE: “I won't stop my work until everyone who needs HIV medicines can get them easily and cheaply.”

Achmat meets regularly with public health workers to discuss public health issues.

ANNA ZIEMINSKI/AFP/GETTY IMAGES; (INSET) ANNA ZIEMINSKI/AFP/GETTY IMAGES/NEWSCOM

CITIZENS IN ACTION What character traits and qualities do you think Achmat has that drive his work to help people with HIV/AIDS?

Culture in Africa South of the Sahara

Main Idea **Africa south of the Sahara is home to many different ethnic and language groups.**

Geography and You Have you ever listened to jazz? Jazz developed from African music. Read to find out about culture and lifestyles in Africa south of the Sahara.

Africa south of the Sahara has a diverse population and distinct cultures. Africans belong to many different ethnic groups, speak thousands of different languages, and practice a variety of religions. African cultures also influence other cultures worldwide. In the past, enslaved Africans carried their culture and music to other parts of the world. As a result, the heritage of Africa is seen and heard today in the United States and other countries. For example, modern forms of music, such as jazz, rock and roll, and rap, have their roots in African music.

Ethnic and Language Groups

Many Africans identify themselves as members of a particular ethnic group. They sometimes feel a stronger loyalty to their ethnic group than to a national government. In Africa, a person's particular ethnic group is most commonly defined by the language he or she speaks.

Between 2,000 and 3,000 different languages are spoken in Africa south of the Sahara. Most of these languages are used by only a small number of people. About a dozen African languages, however, have more than one million speakers. The most widely spoken language is Swahili, which is spoken by about 50 million people in East Africa. In West Africa, the Hausa and Yoruba languages each have more than 20 million speakers.

Djenné, Mali

The area outside of this mosque in the town of Djenné often serves as a local marketplace. ***Movement*** **Which religions were brought to Africa?**

Because of the region's history of European rule, European languages are also spoken in the region. English and French are official languages in countries that were once colonies of Britain and France. For example, English is 1 of 11 official languages in South Africa, and French is the official language of Guinea. In some countries influenced by Islam and Arabic culture, such as Chad, Arabic is an official language.

Religion

A variety of religions are practiced in Africa south of the Sahara. Most people belong to the Christian or Muslim faiths. Many Christian Africans live in coastal areas where Africans had greater contact with Europeans during the colonial period.

Most Muslims in the region live in West Africa, where Islamic empires prospered in the 1400s and 1500s.

Hundreds of traditional African religions also are practiced. Despite differences among these religions, most of them include belief in a supreme being, other gods, and the spirits of dead ancestors, all of whom influence everyday life. These lesser gods and ancestral spirits often are believed to be connected to particular places, such as a river or a spring. They also may be linked to specific human activities, such as farming or fishing. Many Africans see positive events, such as a good harvest, as rewards from the spirits. Negative events, such as illness, are viewed as punishments resulting from the spirits' anger.

Throughout Africa, many people of different religions live together peacefully. Conflict, however, sometimes occurs. Recently, Nigeria and Sudan have experienced fighting among Christians, Muslims, and followers of traditional religions.

The Arts

African art has many forms and uses. Art from the region often has some religious meaning or use. Woodcarvers make masks and statues for religious ceremonies. Some works tell stories and serve practical purposes. Artists working in wood, ivory, or bronze show the faces of important and everyday people. Weavers design brightly colored textiles, such as West Africa's *kente* cloth, for people to wear.

Music and dance are vital elements in weddings, funerals, and religious ceremonies. For Africans, dance is an expression of a community's life. The roles that people have in these dances often reflect their **social status,** or position in the community. For instance, many Africans have special ceremonies called **rites of passage** that mark particular stages of life, such as when young boys or girls reach adulthood. The young men and women often perform particular dances as part of these rites.

In addition to music and dance, Africans also have a storytelling **tradition.** Stories are told aloud and passed down from generation to generation. In West Africa, **griots** (GREE·OHS), or storytellers, preserve a group's history by telling these stories. In modern times, written literature has become popular. African writers, including Chinua Achebe, Wole Soyinka, and Nadine Gordimer, have gained international fame.

Reading Check **Analyzing Information** In what ways are the arts important to Africans?

Spinning Tales

NATIONAL GEOGRAPHIC

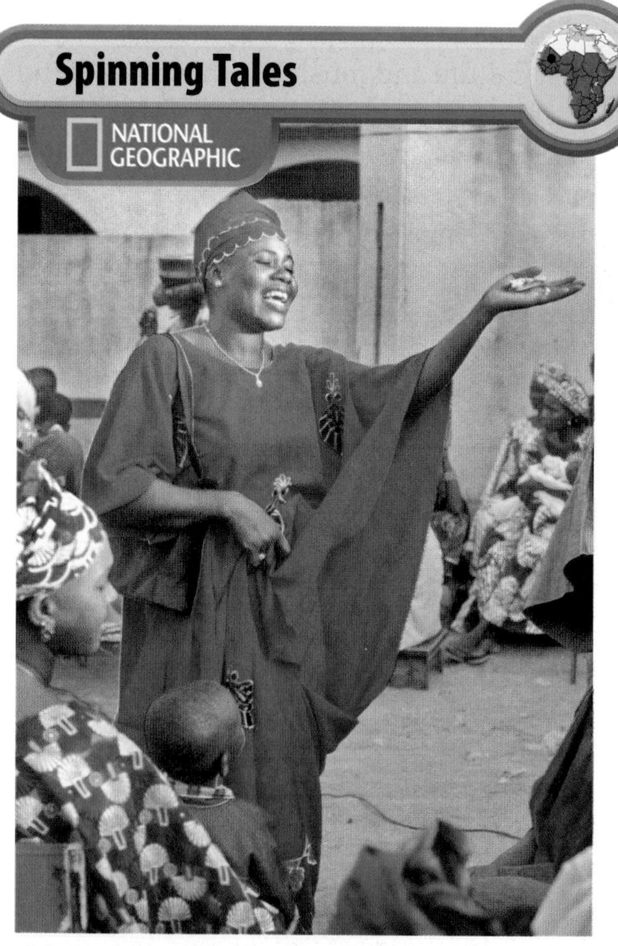

Wedding guests in Mali hear stories, poetry, and songs from the local griot. ***Place*** **Besides entertainment, what other role does a griot typically serve?**

Daily Life in Africa South of the Sahara

Main Idea **Traditional and modern ways exist together in Africa south of the Sahara.**

Geography and You How do you think daily life in your community has changed since your parents or grandparents were children? Read to learn about the changes that have occurred in the region in recent years.

Because of high population growth and greater outside influences, everyday life in the region is undergoing rapid change. Despite this, many traditional features of African life remain.

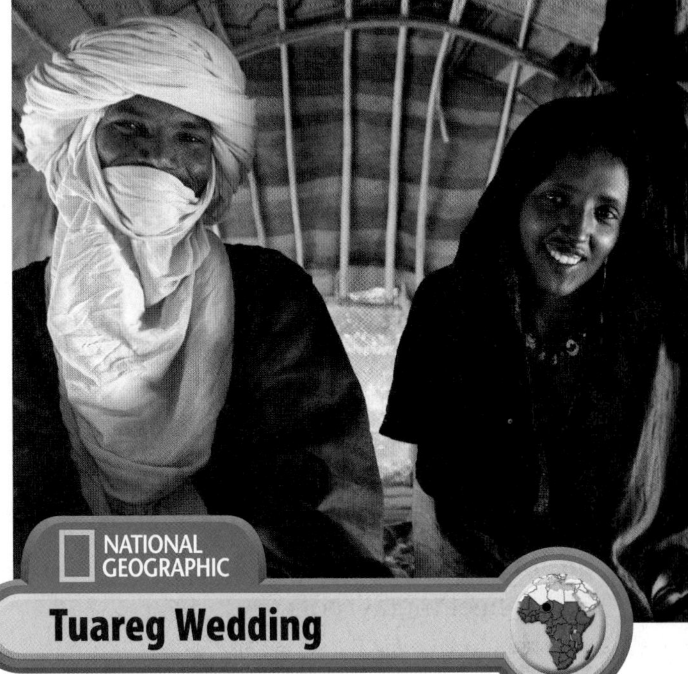

NATIONAL GEOGRAPHIC

Tuareg Wedding

Wedding celebrations preserve cultural traditions and provide a welcome break from daily routines. In the Tuareg culture of West Africa, a wedding can last as long as a week. ***Place*** **Describe a rural African village.**

Rural Life

About 70 percent of all Africans live in rural areas. The rural people depend on farming or livestock herding for their livelihood. African farmers usually are able to grow only enough food to feed their families. After family needs are met, some farmers sell extra crops or animals at a local market for cash. They also might trade for other goods they need or want.

Other farmers, however, work on large, company-run farms that grow cash crops to send overseas. These crops include coffee, cacao, cotton, tea, peanuts, and bananas. Many African economies rely on a single export crop to earn money. When prices for that crop drop, incomes fall and jobs are lost.

Rural families live in villages near the land they farm. Most villages are made up of a cluster of houses and maybe a few shops, a medical clinic, or a schoolhouse. In the past, rural families often lived in a **compound**, or a group of houses surrounded by walls. Such walls are less common today. The homes of rural Africans are often made from dried mud with straw or palm leaves for roofs.

Both men and women work long hours at farming to make a living. In many parts of the region, young men leave their farm villages and work at least a few years as laborers in the cities. The women who are left behind in the villages have had to take on new roles, such as running the farms.

City Life

City residents make up about 30 percent of Africa's total population south of the Sahara. Cities are centers of modern industry and commerce. In the past, Africa's colonial rulers left the region largely undeveloped. Since gaining independence, however, African leaders have wanted to depend less on agriculture and more on manufacturing and service industries.

They hope such changes will improve the standard of living in their countries. These leaders have encouraged new factories and business offices to be built in urban areas.

Because of better jobs, most city dwellers in Africa south of the Sahara have a higher standard of living than rural people. City lifestyles, however, vary widely throughout the region. Some people are wealthy and live in luxury apartments or large, modern houses. Most people, however, live in single-story homes. Many residents are crowded into communities built on the edge of cities. In these areas, homes are built of wood or concrete blocks and have sheet metal roofs.

Families

In Africa south of the Sahara, a person's family ties are extremely important. In rural areas, most people live in **extended families,** or households made up of several generations. An extended family usually includes grandparents, parents, and children. In the cities, however, nuclear families are becoming more common. A **nuclear family** includes a husband, a wife, and their children.

African families traditionally have been organized into clans. A **clan** is a large group of people who are united by a common ancestor in the far past. Many Africans also belong to a particular **lineage** (LIH·nee·ihj), or a larger family group with close blood ties. All members of a lineage might have the same grandmother or grandfather, for instance. Some ethnic groups **require** people to marry outside their own lineage.

Reading Check **Identifying Central Issues** How are family ties in cities different from those in rural areas?

Social Studies ONLINE
Study Central™ To review this section, go to glencoe.com.

Vocabulary

1. **Explain** the significance of:
   - a. malnutrition
   - b. sanitation
   - c. life expectancy
   - d. social status
   - e. rite of passage
   - f. griot
   - g. compound
   - h. extended family
   - i. nuclear family
   - j. clan
   - k. lineage

Main Ideas

2. **Discussing** What are some causes and effects of the trend toward urbanization in the region?
3. **Describing** What are some central beliefs of traditional African religions?
4. **Summarizing** Use a chart like the one below to summarize key facts about lifestyles of the region.

	Facts
Rural Life	
City Life	
Families	

Critical Thinking

5. **BIG Idea** Why are music and dance important in African cultures?
6. **Challenge** How do you think having a variety of ethnic groups and languages affects the unity of a region like Africa south of the Sahara?

Writing About Geography

7. **Expository Writing** Write a paragraph explaining the impact of AIDS on Africa south of the Sahara.

Visual Summary

Early African History

- Bantu groups spread their culture and language throughout most of the region.
- Some of Africa's earliest kingdoms arose in East and southern Africa.
- After A.D. 800, a series of trading empires ruled large areas of West Africa.

Islamic text from Timbuktu

European Contacts

- European traders transported millions of enslaved Africans to the Americas.
- Beginning in the late 1800s, European powers set up colonies in much of Africa.
- Europeans wanted to use Africa's resources and people for their own economies.

Voting in South Africa

Independence

- Most African colonies gained independence during the mid-1900s.
- Conflict among ethnic groups tore apart several newly free countries.
- Black South Africans won their freedom after years of hardship and struggle.

People and Culture

- In recent decades, the population has grown rapidly in Africa south of the Sahara.
- Africa's numerous ethnic groups speak many different languages.
- Stories, sculpture, music, and dance celebrate major events in African life.

Daily Life

- Standards of living are higher in cities than in rural villages.
- Africans value their ties to family, ethnic group, and clan.

Extended family, Nigeria

Dancer in Djibouti

Study anywhere, anytime! Download quizzes and flash cards to your PDA from glencoe.com.

STANDARDIZED TEST PRACTICE

In a table, each column has a particular set of information. Use the table headings to determine which column contains the data that is relevant to the question.

Reviewing Vocabulary

Directions: Choose the word(s) that best completes the sentence.

1. After World War I, many Africans staged protests against _______, or unfair and unequal treatment of a group.

A nationalism

B discrimination

C plantations

D slavery

2. In South Africa, the policy of _______ was carried out through laws that separated ethnic groups and limited the rights of black South Africans.

A nationalism

B independence

C urbanization

D apartheid

3. Many people in Africa south of the Sahara suffer from _______, or poor health from not eating the right foods or enough food.

A lineage

B urbanization

C malnutrition

D sanitation

4. In West Africa, storytellers called _______ preserve a group's history.

A griots

B clans

C lineages

D rites

Reviewing Main Ideas

Directions: Choose the best answer for each question.

Section 1 *(pp. 548–555)*

5. Which of the early African kingdoms was a center of Christianity?

A Kush

B Mali

C Axum

D Ghana

6. In the early 1950s, Kwame Nkrumah helped lead the colony then called _______ to independence.

A South Africa

B the Gold Coast

C Songhai

D Zimbabwe

Section 2 *(pp. 557–564)*

7. In Africa south of the Sahara, most people live in _______.

A the dry interior regions

B large cities

C modern suburbs

D rural villages

8. Many Africans feel a stronger loyalty to their _______ than to a national government.

A ethnic group

B plantation

C school

D political party

GO ON

Critical Thinking

Directions: Base your answers to questions 9, 10, and 11 on the table below and your knowledge of Chapter 20.

Expected Population Change in Africa South of the Sahara		
Area	**Population (millions)**	**Expected Change to 2050 (percent)**
Central Africa	112.0	175
East Africa	281.0	142
Southern Africa	54.0	0
West Africa	304.0	127
Entire Region	751.2	130
Source: *World Population Data Sheet*, 2005.		

9. Which area currently has the largest share of the region's people?

A Central Africa

B East Africa

C southern Africa

D West Africa

10. In which area is the population not expected to grow very much in the future?

A Central Africa

B East Africa

C southern Africa

D West Africa

11. How much greater is the expected rate of change for Central Africa than the expected rate for the entire region?

A 0%

B 45%

C 12%

D 3%

Document-Based Questions

Directions: Analyze the document and answer the short-answer questions that follow.

The description of Mansa Musa's visit to Cairo in 1324 was written by Al-Umari, who arrived in Cairo several years after Mansa Musa's visit.

> *I asked the emir Abu [about Mansa Musa's visit] . . . and he told me . . . . "When I went out to meet him . . . he did me extreme honour and treated me with the greatest courtesy. He addressed me, however, only through an interpreter despite his perfect ability to speak in the Arabic tongue. Then he forwarded to the royal treasury many loads of unworked native gold and other valuables. I tried to persuade him to go up to the Citadel to meet the sultan, but he refused persistently saying: "I came for the Pilgrimage and nothing else." . . . I realized that the audience was repugnant [disagreeable] to him because he would be obliged [forced] to kiss the ground and the sultan's hand."*
>
> —Al-Umari, as cited in *Corpus of Early Arabic Sources for West African History*

12. How did Mansa Musa treat the person telling the story?

13. How did Mansa Musa demonstrate his power and wealth?

Extended Response

14. Write a newspaper editorial that encourages independence for African nations in the mid-1900s, or that argues against the system of apartheid in South Africa in the early 1990s. Include historical information in your argument along with your reasons for calling for the change.

For additional test practice, use Self-Check Quizzes—Chapter 20 at glencoe.com.

Need Extra Help?

If you missed question...	1	2	3	4	5	6	7	8	9	10	11	12	13	14
Go to page...	553	555	559	562	549	553	559	561	558	558	558	550	550	552–555

"Hello! My name is Irewole.

I'm 13 years old and I go to Bishop James Yisa Secondary School. I live in Suleja, a town near the capital, Abuja, with my mum, my sister, and my aunt. Read about my day."

5:30 a.m. The alarm clock wakes me up. I freshen up and join Mum and my sister, Olu (OH•loo), in the living room.

6:00 a.m. I wash dishes, sweep around the outside of my house, and fetch water from a well for the toilet. Then I take a bath.

7:30 a.m. I wait for the school bus at a stop near our house. Olu takes a later bus. My bus is usually very crowded. I squeeze my way in, and 20 minutes later we arrive at school.

8:00 a.m. School starts with an assembly of all the students. The principal, Mr. Abraham, warns the senior students against bullying others and advises us to be obedient and respectful.

8:15 a.m. Our first class of the day is government. Today we are learning about the legislative process of drafting and approving bills. It is very interesting. Next, it is time for a lesson from our business teacher.

9:35 a.m. In biology class, we are studying the human digestive system, including enzymes and nutrition.

10:15 a.m. It's time for a break! For my snack I have one boiled egg and some popcorn. When we are finished eating, we have religion class.

12:05 p.m. It is a free period, but we are not allowed to leave the classroom. My friend Solomon joins me so we can practice math equations.

12:50 p.m. We go to the computer room. This afternoon we are learning a word processing program. Today, I have a computer all to myself. Most of the students in my school—and most families in Nigeria—don't have computers, so we are lucky.

4:00 p.m. School is out. The school bus drops my sister and me at home. I wash our school uniforms and socks. Then I'm off to the local market to buy ground-nut (peanut) oil for Mum. We use it for cooking.

6:00 p.m. On weekends I play football, which Americans call soccer. But today I'm just having a quick game with some friends.

7:00 p.m. It's dinnertime. We are having *amala* and palm-nut soup. *Amala* is made from ground dried cassava pellets. It is the favorite dish of the Yoruba people.

7:30 p.m. We watch a popular television show called *Super Story*. I still have to finish my math homework, so I pick up my math books and start to work.

10:00 p.m. I am very tired. I'm off to bed.

ILLUSTRATIONS BY BOOKMAPMAN

MARKET DAY Families shop at their local market.

SCHOOL DAY Irewole (center) studies hard at school. His favorite subject is biology.

FUN AND GAMES Soccer, known as football in Nigeria, is one of Irewole's favorite sports.

LEFT: AP PHOTO; MIDDLE: ABAYOMI ADESHIDA; RIGHT: CHRIS HONDROS—GETTY IMAGES

What's Popular in Nigeria

Drumming Music is an important part of Nigerian culture. Handmade drums are often used to accompany singing and dancing.

PATRICK OLEAR/PHOTO EDIT

Soccer The country's national team, the Super Eagles, won the gold medal at the 1996 Summer Olympics. The women's team, the Super Falcons, is also one of the top teams in Africa.

Afro beat This kind of music, which combines jazz, funk, and other styles, is popular throughout Africa. Juju, another type of music, was made famous by the Nigerian musician King Sunny Ade.

Fufu Made of yams, cassava, and plantains, *fufu* is a treat loved by kids.

ABAYOMI ADESHIDA

Say It in Pidgin

English is Nigeria's official language, but many people speak a form of pidgin, which mixes English and native languages. Pidgin is spoken in other countries too. Here are some phrases in Nigerian pidgin.

What time is it?
Na wetin be time?
(nah wet·in be time)

I am leaving.
I dey go. (eye day go)

I will see you later.
A go dey see yu.
(ah go day see you)

MORNING BELL Irewole Adelegan (ee•ree•WOH•lay ah•DELL•uh•gahn) arrives at school in time for the morning bell. All students must wear uniforms. (Irewole's tie is in his backpack!)

ABAYOMI ADESHIDA

JANE SWEENEY / LONELY PLANET IMAGES

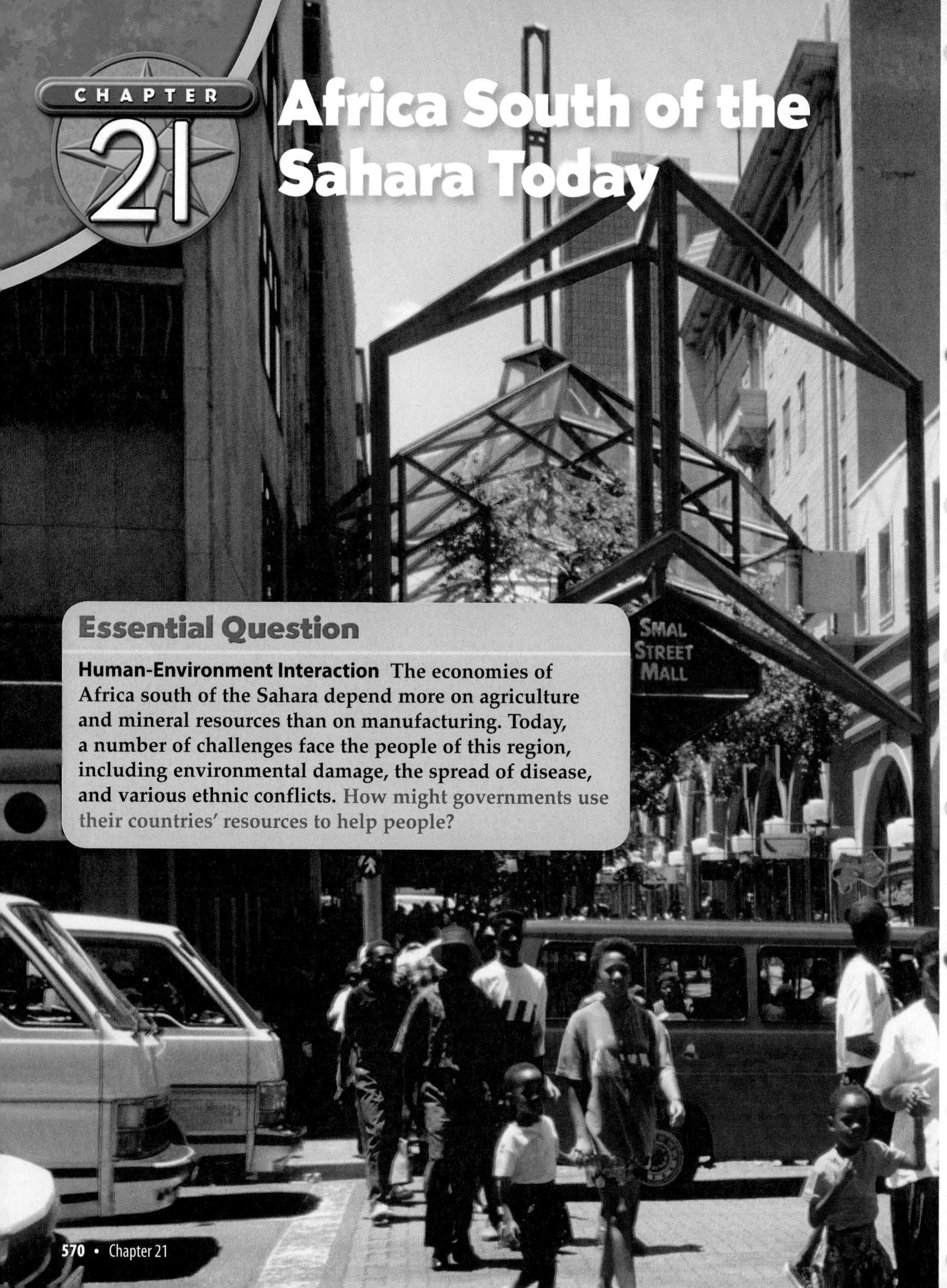

CHAPTER 21

Africa South of the Sahara Today

Essential Question

Human-Environment Interaction The economies of Africa south of the Sahara depend more on agriculture and mineral resources than on manufacturing. Today, a number of challenges face the people of this region, including environmental damage, the spread of disease, and various ethnic conflicts. How might governments use their countries' resources to help people?

Johannesburg, South Africa

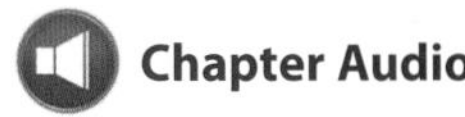

Section 1: West Africa

BIG IDEA Geographers study how people and physical features are distributed on Earth's surface. While some West African countries lie on the coast, others are located in the dry interior.

Section 2: Central and East Africa

BIG IDEA Cooperation and conflict among people have an effect on the Earth's surface. In various parts of Central and East Africa, ethnic conflicts have hurt both people and the environment.

Section 3: Southern Africa

BIG IDEA Patterns of economic activities result in global interdependence. The export of valuable minerals, such as diamonds and gold, is important to the economies of southern Africa.

FOLDABLES™ Study Organizer

Summarizing Information Make this Foldable to help you summarize information about the subregions of Africa south of the Sahara.

Step 1 Fold two sheets of paper in half lengthwise.

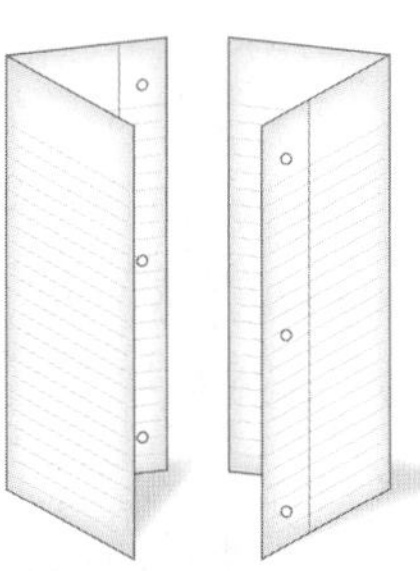

Step 2 Tape the long edges of the two sheets together to form a tall standing cube. Label the four sides "West Africa," "Central Africa," "East Africa," and "Southern Africa."

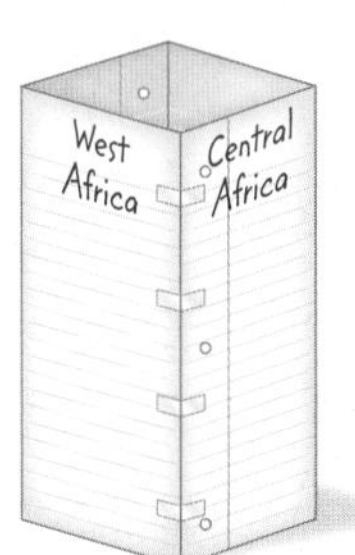

Reading and Writing As you read the chapter, take notes on the correct side of the Foldable. Then write a summary of Africa south of the Sahara today and include a prediction about what you think will happen in the region in the future.

Visit glencoe.com and enter ***QuickPass***™ code EOW3109c21 for Chapter 21 resources.

West Africa

 Section Audio  Spotlight Video

Guide to Reading

BIG Idea

Geographers study how people and physical features are distributed on Earth's surface.

Content Vocabulary

- subsistence farm *(p. 573)*
- cacao *(p. 573)*
- landlocked *(p. 574)*
- overgraze *(p. 575)*

Academic Vocabulary

- benefit *(p. 573)*
- stable *(p. 574)*

Reading Strategy

Identifying Central Issues Use a web diagram like the one below to organize key facts about Nigeria.

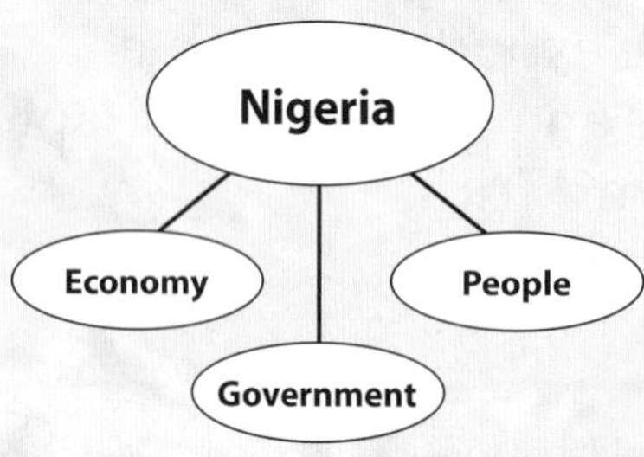

Picture This A woman carries food in a local market in Lagos, Nigeria. Lagos faces many challenges, including enormous population growth. Because of this growth, the city's leaders must find ways to supply freshwater, housing, and sanitation services for its residents. Even though the city's population is growing, its people still shop at the local market for their daily food. To learn more about West Africa today, read Section 1.

▼ Lagos, Nigeria

Nigeria

Main Idea **Nigeria is a large, oil-rich country that has more people than any other nation in Africa south of the Sahara.**

Geography and You Think about what it would be like to move every few years for a new job or to find new farmland. Read to discover what farmers in Nigeria must do to earn a living.

Nigeria is about the size of California, making it one of the largest nations in Africa south of the Sahara. Nigeria also has the largest population of any country in the region. Ethnic conflict and political uncertainty, however, have kept Nigeria from **benefiting** from its rich natural resources.

The Economy

Nigeria is one of the world's major oil-producing countries. Nearly all of the country's income comes from oil exports. Because Nigeria's economy relies largely on this one product, the country is affected by changes in world oil prices. Recently, oil prices have risen, and the economy has improved. The government has used oil profits to build roads, schools, and factories.

Despite oil revenues, Nigeria's people work mainly as farmers. Most have **subsistence farms,** or small plots where farmers grow only enough food to feed their families. Others work on larger farms that produce cash crops, such as rubber, peanuts, palm oil, and cacao. The **cacao** is a tropical tree whose seeds are used to make chocolate and cocoa. By focusing on cash crops like this, however, Nigeria has not grown enough food crops. As a result, food has to be imported to feed the country's growing population.

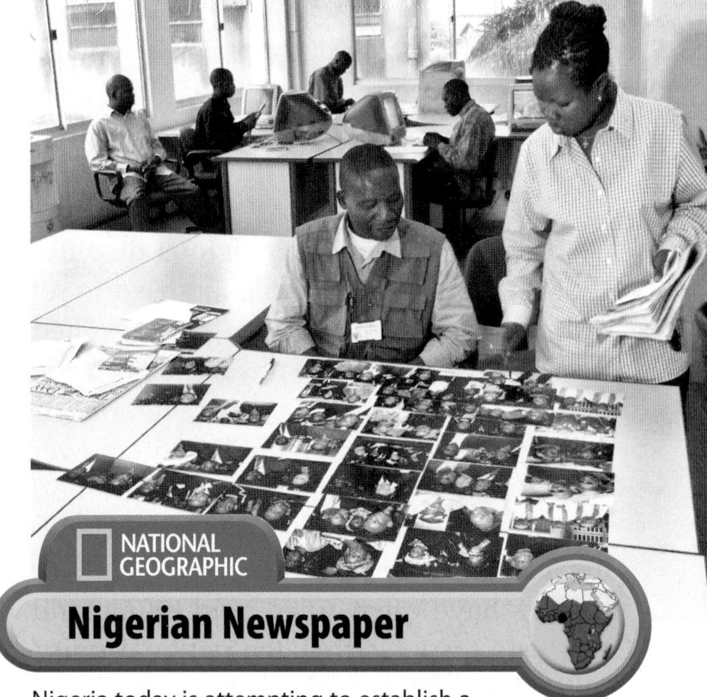

NATIONAL GEOGRAPHIC

Nigerian Newspaper

Nigeria today is attempting to establish a stable democracy in order to protect certain basic rights, such as freedom of the press. ***Human-Environment Interaction*** **What resource is responsible for much of Nigeria's income?**

The People

Nigeria has more than 250 ethnic groups. The four largest are the Hausa (HOW·suh), Fulani (FOO·LAH·nee), Yoruba (YAWR·uh·buh), and Ibo (EE·boh). Nigerians speak many different African languages, but they use English in business and government affairs. About one-half of Nigeria's people are Muslim, and another 40 percent are Christian. The remaining 10 percent of the population practices traditional African religions.

About 60 percent of Nigerians still live in rural areas. The cities, however, are rapidly growing because many people have left farms in search of better jobs.

Social Studies ONLINE

Student Web Activity Visit glencoe.com and complete the Chapter 21 Web Activity on Nigeria.

The largest city is the port of Lagos. Abuja (ah·BOO·jah), Nigeria's capital, is a planned city that was built during the 1980s.

The Government

Nigeria won freedom from the British in 1960. Ethnic and religious disputes soon broke out. In 1967 the Ibo tried to set up their own country. A civil war resulted that led to 2 million deaths. The Ibo were defeated and remained part of Nigeria.

Nigeria is a federal republic, with power shared by a national government and states. It has faced the challenge of building a **stable,** or secure, democracy. Military leaders have often ruled, and ethnic and religious disputes threaten national unity. In the early 2000s, violence between Christians and Muslims raised fears of another civil war in Nigeria.

Reading Check **Explaining** How has a focus on cash crops affected Nigeria's food supply?

The Sahel and Coastal West Africa

Main Idea **West Africa consists of inland grasslands and coastal rain forests, areas with different populations and resources.**

Geography and You Do you live in a coastal area or in an inland area? What do you like about the area in which you live? Read to find out how life in the inland Sahel contrasts with life in coastal West Africa.

West African countries fall into one of two groups. Some lie inland in the partly dry grasslands called the Sahel. The others lie along the Atlantic Ocean or on islands.

The Sahel

Five countries—Mauritania, Mali, Burkina Faso, Niger, and Chad—make up the Sahel countries. Except for Mauritania, the Sahel countries are **landlocked,** which

See StudentWorks™ Plus or glencoe.com.

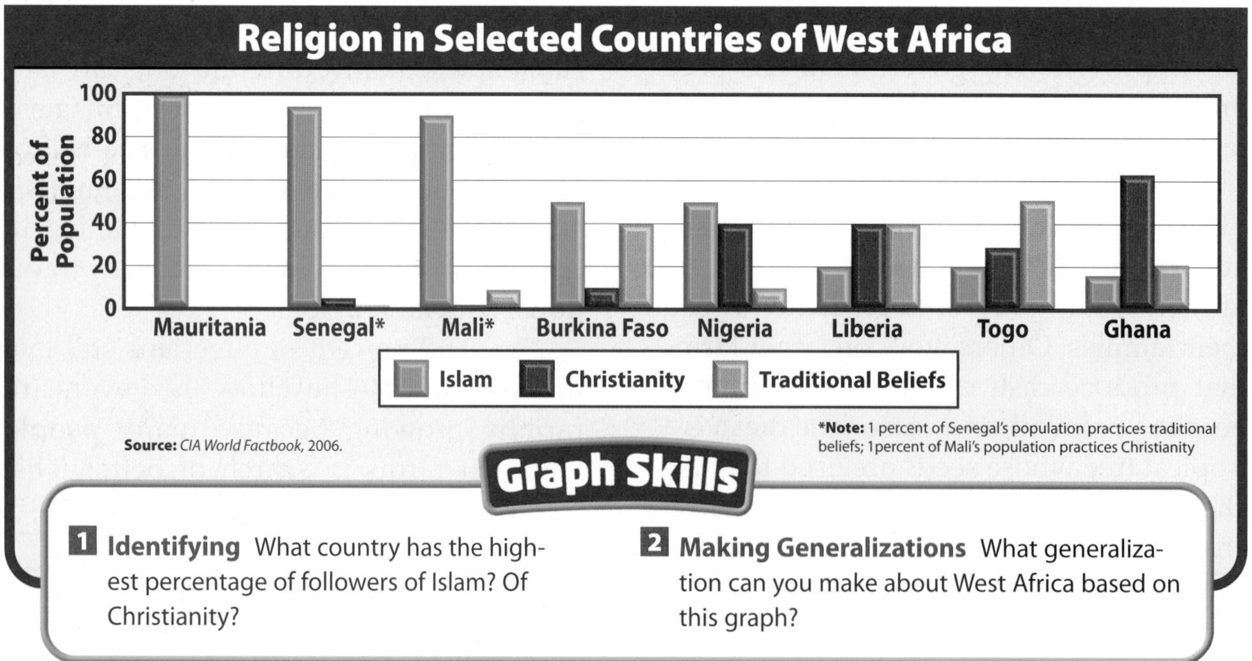

Source: *CIA World Factbook*, 2006.

***Note:** 1 percent of Senegal's population practices traditional beliefs; 1 percent of Mali's population practices Christianity

Graph Skills

1. **Identifying** What country has the highest percentage of followers of Islam? Of Christianity?
2. **Making Generalizations** What generalization can you make about West Africa based on this graph?

means they do not have a sea or an ocean border. The lack of a good transportation system and ports limits the ability of the Sahel countries to develop their valuable deposits of uranium, gold, and oil.

Only grasses and small trees grow in the Sahel, which receives little rainfall. Herding livestock is a major activity, but large numbers of animals have overgrazed the land in many places. When animals **overgraze** land, they strip areas so bare that winds can blow away the soil. Overgrazing and extreme dryness have contributed to desertification in the Sahel.

Because of the Sahel's difficult living conditions, populations there are small compared to the rest of Africa. In the past, many people were nomads. Today, most people live and work in small towns or rural villages. Most Sahel peoples are Muslim and speak Arabic, yet many African languages and French are also spoken.

Coastal West Africa

Coastal West Africa includes the Cape Verde Islands and the mainland countries that stretch from Senegal to Benin. In these lands, rain forests have been cleared for palm, coffee, cacao, and rubber plantations. This has led to deforestation along the densely settled coasts. Rural people in search of work have settled in port cities, such as Dakar (Senegal) and Accra (Ghana).

Coastal peoples belong to many ethnic and language groups. Traditional African religions, Christianity, and Islam all have followers in the area. Recent civil wars have cost many lives and destroyed economies in Liberia, Sierra Leone, and Côte d'Ivoire. Other countries, such as Ghana, Senegal, and Benin, have stable democracies and generally prosperous economies.

Reading Check **Naming** What five countries make up the Sahel?

Section 1 Review

Social Studies ONLINE
Study Central™ To review this section, go to glencoe.com.

Vocabulary

1. **Explain** the significance of *subsistence farm, cacao, landlocked,* and *overgraze* by using each term in a sentence.

Main Ideas

2. **Discussing** Why is Nigeria considered a country of great diversity?
3. **Describing** Use a fish-bone diagram like the one below to list key facts about the environment of the Sahel.

Sahel

Critical Thinking

4. **Comparing and Contrasting** How is Nigeria's government similar to that of the United States? How effective has it been with establishing national rule?
5. BIG Idea Explain population patterns in West Africa.
6. **Challenge** Do you think the Sahel countries could prosper from their mineral resources if those countries had better transportation systems? Why or why not?

Writing About Geography

7. **Using Your FOLDABLES** Use your Foldable to write a paragraph explaining how Nigeria is similar to both the Sahel and the other coastal countries of West Africa.

Guide to Reading

BIG Idea

Cooperation and conflict among people have an effect on the Earth's surface.

Content Vocabulary

- sisal *(p. 578)*
- habitat *(p. 578)*
- cassava *(p. 579)*
- genocide *(p. 580)*

Academic Vocabulary

- source *(p. 577)*
- shift *(p. 577)*
- restore *(p. 582)*

Reading Strategy

Comparing and Contrasting Use a chart like the one below to compare a Central African country with an East African country.

Country	Economy	People

Central and East Africa

Picture This Underwater farming? Off the coast of Zanzibar Island, Tanzania, people harvest seaweed. It is then sold and used as an ingredient in cosmetics, fertilizers, shampoos, and even cheeses. During the monsoon season, the seaweed is harvested more often to prevent it from being ripped out by high winds and rough currents. To learn more about Central and East Africa today, read Section 2.

▼ Farming off Zanzibar Island in the Indian Ocean

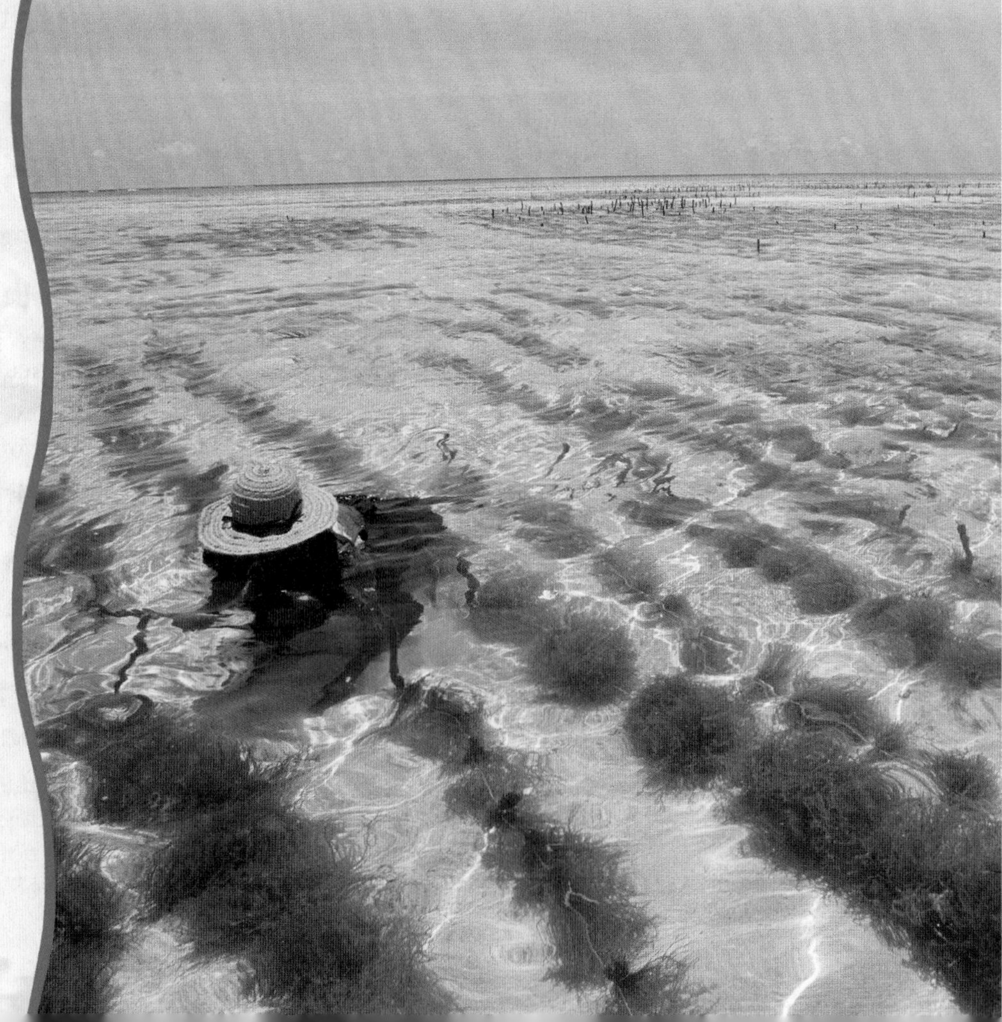

Central Africa

Main Idea **Although rich in natural resources, Central Africa remains largely undeveloped because of a difficult environment and political conflicts.**

Geography and You Do you know how groceries get to your local grocery store? Most likely they arrive by truck. Read to find out the importance of rivers in transporting goods in Central Africa.

Central Africa includes seven countries. They are the Democratic Republic of the Congo, Cameroon, the Central African Republic, Congo, Gabon, Equatorial Guinea, and São Tomé and Príncipe (sow too·MAY and PRIHN·sih·pee).

Democratic Republic of the Congo

The Democratic Republic of the Congo is a large country with many different landscapes, including rugged mountains and broad savannas. One of the world's largest rain forests spreads across its center. This rain forest, however, is being destroyed rapidly as it is cleared for timber and farmland.

The Democratic Republic of the Congo is a major **source** of copper, tin, and industrial diamonds. The country has not been able to take full advantage of its rich resources, however. One difficulty is transportation. Many of the minerals are found in the country's interior. Lack of roads and thick rain forests make it hard to reach those areas. Political unrest has also limited the mining of resources. For many years, a civil war hurt efforts to develop the country's economy.

The Democratic Republic of the Congo has more than 200 different ethnic groups. Many people speak African languages, but French is the country's official language. Although most Congolese people live in rural areas, the capital, Kinshasa, has more than 6 million people.

The residents of Gabon's capital, Libreville, enjoy a stable government and a growing economy. ***Location*** **How might the location of Gabon help its economy grow?**

Other Countries

Other Central African countries have moved toward economic growth. Gabon bases its economy on oil, manganese, uranium, and timber. Cameroon, to the north, produces cacao and coffee for export. Both Congo and the Central African Republic remain in poverty because of weak governments.

Equatorial Guinea, like Gabon, benefits from oil resources. The island country of São Tomé and Príncipe relies on cacao and coconut exports. Recently, it has **shifted** to oil production.

Reading Check **Identifying** What issues have slowed the development of resources in parts of Central Africa?

Southern East Africa

Main Idea **The highlands in the southern part of East Africa attract people and support thriving farms. The region has experienced much conflict, however.**

Geography and You Does your community or state draw tourists, perhaps with a local park or some other attraction? Read to find out about the role tourism plays in southern East Africa.

Southern East Africa includes the coastal countries of Kenya and Tanzania and the inland nations of Uganda, Rwanda, and Burundi. Landscapes here include white beaches on the Indian Ocean, savannas filled with wildlife, and the rugged beauty of the Great Rift Valley.

Tanzania

Tanzania, on the southern edge of East Africa, is the largest of the southern East African countries. Tanzania's population includes many different ethnic groups. Each group has its own language, but most people speak Swahili. Friendly relations among ethnic groups and a stable government have prevented conflict in Tanzania since independence.

Most Tanzanians work in farming or herding. Important export crops are coffee and **sisal,** a plant fiber used to make rope and twine. The island of Zanzibar, off Tanzania's coast, is a major producer of cloves, a spice used to flavor baked ham.

Tourism is a fast-growing industry in Tanzania. National parks here help to protect the habitats of the country's wildlife. A **habitat** is the type of environment in which a particular animal species lives. Serengeti (sehr·uhn·GEH·tee) National Park is home to lions, zebras, wildebeests, and other wildlife. The park attracts many ecotourists, or people who travel to another country to view its natural wonders.

NATIONAL GEOGRAPHIC

Harvesting Coffee

Coffee is an important crop in Tanzania. Coffee cherries must be hand picked and then dried to become coffee beans. ***Place*** **What other crops are grown in Kenya?**

Kenya

North of Tanzania is Kenya, a country about the size of Nevada. Most of Kenya's people live in the highlands in the center of the country. Nairobi (ny·ROH·bee), the country's capital and the largest city in East Africa, is located there. Mombasa (mohm·BAH·sah), on the Indian Ocean coast, is a large and busy port.

Kenya has a developing free market economy that has enjoyed prosperity. Nairobi serves as a major business center for all of East Africa. Foreign companies have set up regional headquarters there.

Most Kenyans, however, are farmers who raise corn, bananas, sweet potatoes, and cassava. **Cassava** (kuh·SAH·vuh) is a plant whose roots are ground to make porridge. Some larger farms raise coffee and tea for export. Tourism is also important to Kenya's economy. The country has many national parks to help protect its wildlife.

Like Tanzania, Kenya has many ethnic groups. The Kikuyu (kee·KOO·yoo) people are Kenya's main group, making up one-fourth of the population. Although most Kenyans live in rural areas, many have moved to the cities in search of a better life.

After independence from Britain in 1963, Kenya prospered despite rule by one political party. Recently, Kenyans have made advances toward democracy. In 2007, however, a disputed presidential election led to riots that claimed about 1,500 lives. The violence finally ended when the two rival candidates agreed to share power in a new government.

Highland Countries

West of Tanzania and Kenya lie Uganda, Rwanda, and Burundi. All three landlocked countries are located in the highlands of East Africa. Rich soil and plentiful rainfall make the land in these countries good for farming. Subsistence farms produce bananas, cassava, potatoes, corn, and grains. Some plantations grow coffee, cotton, and tea for export. Farming in this highland area is productive and supports large populations. Rwanda and Burundi have the highest population densities in Africa south of the Sahara.

Since independence, the three highland countries have faced unrest and tragedy.

NATIONAL GEOGRAPHIC

Cultural and Economic Diversity in Kenya

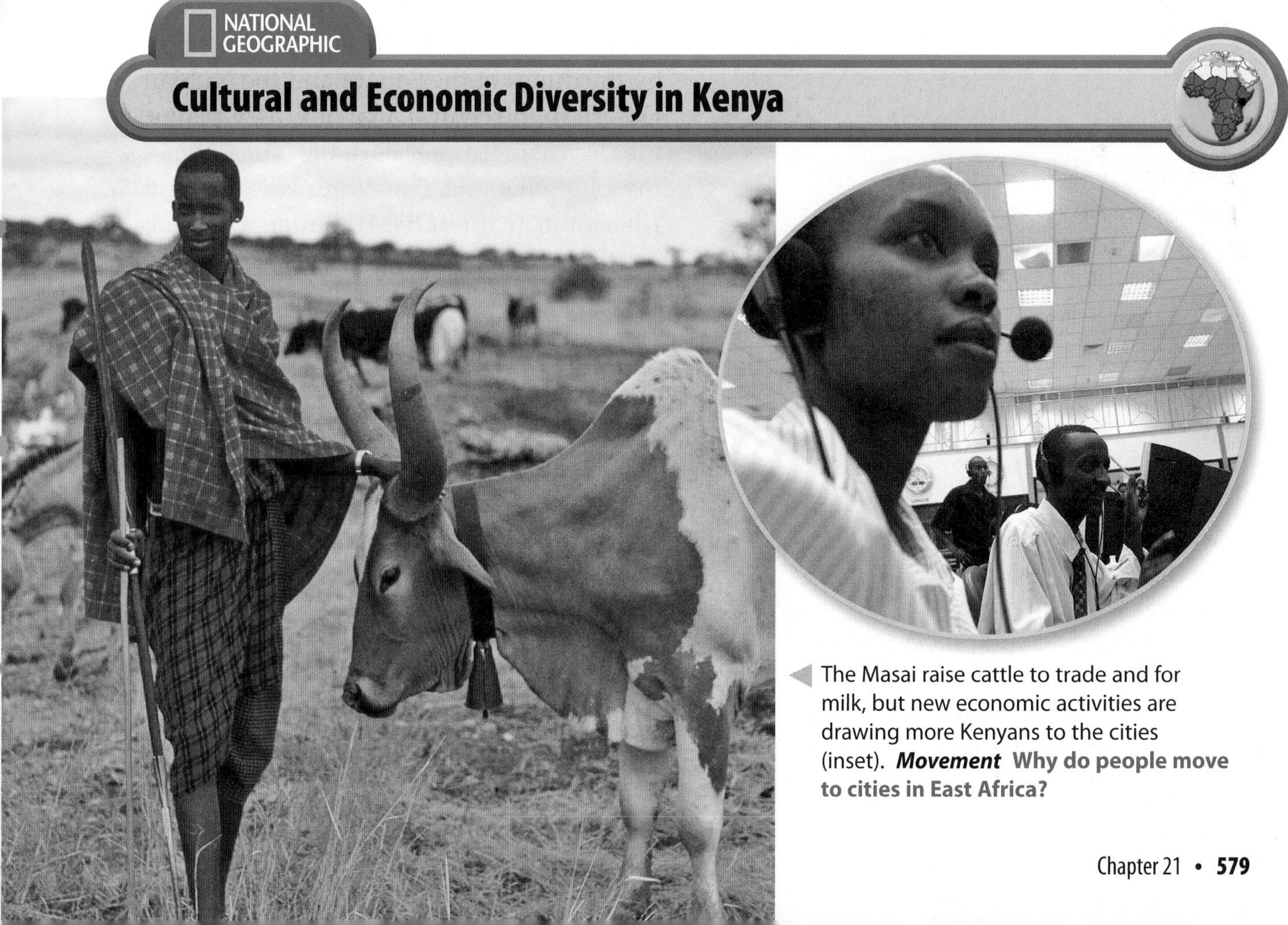

The Masai raise cattle to trade and for milk, but new economic activities are drawing more Kenyans to the cities (inset). ***Movement*** **Why do people move to cities in East Africa?**

Uganda was ruled for much of the 1970s by a cruel dictator, Idi Amin (EE·dee ah·MEEN). In recent years, Uganda has become more democratic and prosperous.

Rwanda and Burundi are made up of two ethnic groups—the Hutu and the Tutsi. About 80 percent of the people in both countries are Hutu, but the Tutsi ran the governments and economies for many years. A power struggle between the two groups erupted into civil war and genocide in the 1990s. **Genocide** is the deliberate murder of a group of people because of their race or culture. A Hutu-led government in Rwanda killed hundreds of thousands of Tutsi. Two million more Tutsi became refugees. When the fighting ends, Rwanda and Burundi will face many years of rebuilding.

Reading Check **Identifying** What crops are grown for export in this part of Africa?

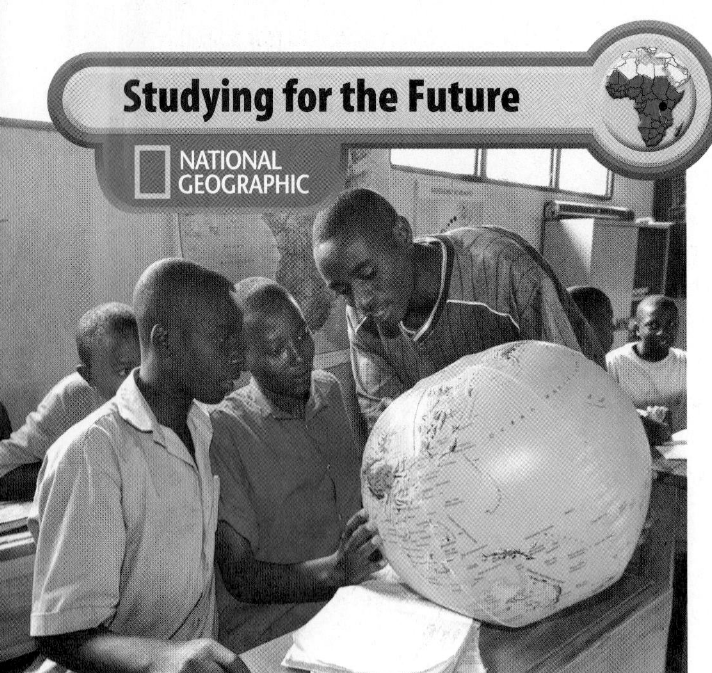

▲ Leaders in Rwanda are investing in the future by improving schools and education. ***Regions*** **Which two ethnic groups engaged in civil war in Rwanda in the 1990s?**

The Horn of Africa

Main Idea **The countries of the Horn of Africa have all been scarred by conflict in recent years.**

Geography and You Earlier, you read about the harsh steppe and desert climates that cover parts of East Africa. Read to learn how conflicts have made life especially hard in this part of Africa.

The northern part of East Africa is a region called the Horn of Africa. This region got its name because it is shaped like a horn that juts out into the Indian Ocean. The countries in the Horn of Africa are Sudan, Ethiopia, Eritrea (EHR·uh·TREE·uh), Djibouti (jih·BOO·tee), and Somalia.

Sudan

Sudan is the largest country in Africa—about one-third the size of the continental United States. Northern Sudan is covered by the dunes of the Sahara and the Nubian Desert. In the central area of grassy plains, the two main tributaries of the Nile River—the Blue Nile and the White Nile—join at Khartoum (kahr·TOOM), Sudan's capital. Southern Sudan receives plenty of rain and has fertile soil and swamplands.

Most of Sudan's people live along the Nile River or its tributaries. They use Nile waters to irrigate their fields. Farmers grow sugarcane, grains, dates, and cotton—the country's leading export. Sudan also has large reserves of oil in the south.

As you may recall, Sudan was the ancient center of the powerful Kush civilization. During the A.D. 500s, missionaries brought Christianity to the region. About 900 years later, Muslim Arabs entered northern Sudan and converted its people to Islam. From the late 1800s to the 1950s, the British and Egyptians together ruled the country.

Sudan became an independent nation in 1956. Since then, Sudan has been ruled mostly by military leaders.

Today, Sudan is divided by deep ethnic differences. The north is populated largely by Arab Muslims. The south is populated by black Africans who are Christians or followers of local religions. From 1983 to 2004, the north and the south fought a bitter civil war. About 1.5 million Sudanese died, and many more were forced to leave their homes. Another 200,000 people are believed to have been killed during a conflict in Sudan's western Darfur region that began in 2001.

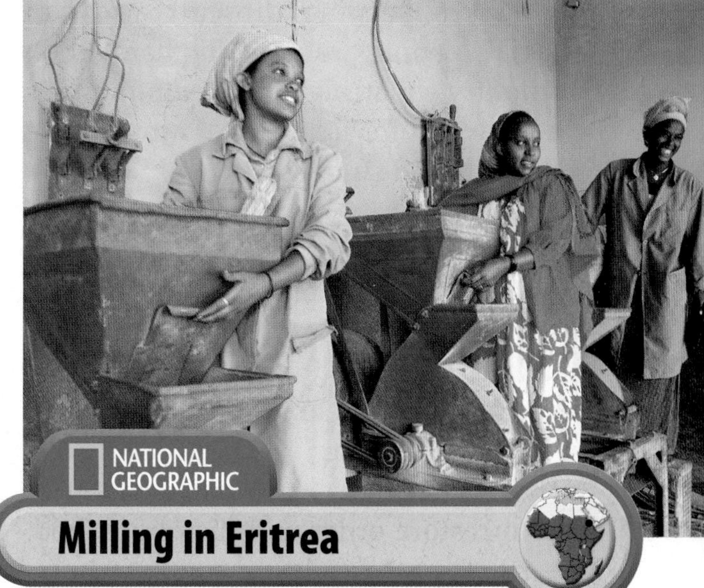

Milling in Eritrea

These Eritrean women are grinding roasted chickpeas. The powder will be used to make *shiro*, a staple food in Eritrea. ***Human-Environment Interaction*** **What climate challenge do farmers in Ethiopia and Eritrea face?**

Ethiopia and Eritrea

Landlocked Ethiopia is almost twice the size of Texas. Ethiopia's landscape varies from hot lowlands to highlands and rugged mountains. Mild temperatures and good soil make the highlands Ethiopia's best farming region. Farmers raise grains, sugarcane, potatoes, and coffee. Coffee is a major export crop. The southern highlands are believed to be the world's original source of coffee.

Rainfall amounts vary in many parts of Ethiopia. Low rainfall can lead to drought. When this occurs, Ethiopia's people suffer. In the 1980s, a drought caused a famine that attracted worldwide attention. Once-rich fields turned into seas of dust. Despite food aid, more than 1 million Ethiopians died from starvation and disease.

About 85 percent of Ethiopians live in rural areas. The capital, Addis Ababa (AHD·dihs AH·bah·BAH), however, is one of the largest cities in East Africa. Ethiopians practice Christianity, Islam, or traditional African religions. Almost 80 languages are spoken in Ethiopia. Amharic, similar to Hebrew and Arabic, is Ethiopia's official language.

Ethiopia is the oldest independent nation in Africa. The ancient Christian kingdom of Axum once ruled the area. For centuries, Ethiopia withstood Muslim and European attempts to control it. The last monarch was overthrown in 1974. The country then suffered under a military dictator. Now it is trying to build a democracy.

This goal has been hindered by warfare with neighboring Eritrea, a small Muslim country that broke away from Ethiopia in 1993. Eritrea sits on the shores of the Red Sea. Most of Eritrea's people farm, but farming is uncertain work because the climate is dry.

Somalia

East of Ethiopia lies Somalia, which is situated on the tip of the Horn of Africa. Somalia is shaped like the number seven, with a long coastline but few natural harbors. Much of the country's climate and landscape are hot and dry.

This makes farming difficult. Most of Somalia's people are nomadic herders on the country's plateaus. In the south, rivers provide water for irrigation. Farmers here grow fruits and sugarcane.

Nearly all of the people of Somalia are Muslims, but they are deeply divided. They belong to different clans. In the late 1980s, disputes among these clans led to civil war. When a drought struck a few years later, hundreds of thousands of people starved to death.

The United States and other countries tried to **restore** order and distribute food. Nevertheless, the fighting continued and often kept the aid from reaching the people who needed it. Today, armed groups control various parts of Somalia. The country does not have a truly functioning government.

Djibouti

Tiny Djibouti is the most stable country in the Horn of Africa. Djibouti is strategically located at a narrow water passage that links the Red Sea and the Gulf of Aden. This passage is the meeting point of a number of trade routes. These routes link the Indian Ocean with the Mediterranean Sea, as well as Africa with Southwest Asia.

In addition, Djibouti has an excellent harbor at its capital, the city of Djibouti. To take advantage of the city's favorable location, Djibouti has built a modern port there. Shipping and commerce have become the heart of Djibouti's economy.

Most of Djibouti's people are Muslims. In the past, they lived a nomadic life of herding. Djibouti is one of the hottest, driest places on the Earth. Because of the dry climate, farming and herding are difficult. In recent years, many people have moved to the city of Djibouti to find jobs.

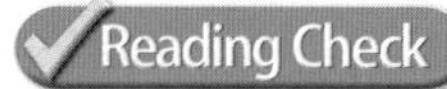

Comparing and Contrasting
How does Djibouti differ from other countries in the Horn of Africa?

Section 2 Review

Social Studies ONLINE
Study Central™ To review this section, go to glencoe.com.

Vocabulary

1. **Explain** the significance of:
   - **a.** sisal
   - **b.** habitat
   - **c.** cassava
   - **d.** genocide

Main Ideas

2. **Identifying** Identify resources found in specific countries of Central Africa.
3. **Comparing and Contrasting** Use a Venn diagram like the one below to compare and contrast the economies of Tanzania and Kenya.

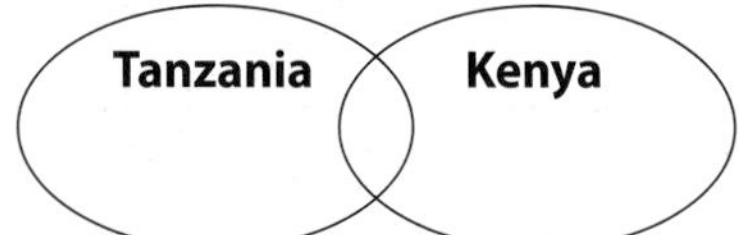

4. **Describing** Describe the recent history of Ethiopia and Eritrea.

Critical Thinking

5. **BIG Idea** Describe the reasons for and the effects of a conflict that has occurred in the region.
6. **Challenge** Why have some countries in this area adapted to independence better than others?

Writing About Geography

7. **Expository Writing** Write a paragraph describing how the area's physical geography has affected economic development.

TIME PERSPECTIVES

EXPLORING WORLD ISSUES

THE TRAGEDY OF REFUGEES IN AFRICA

Millions of Africans are fleeing their homes. Will they have anyplace to go back to?

Refugees line up outside an aid center in northern Darfur.

Africa has a long history of refugee issues. In recent years, armed conflict and natural disasters have forced millions to flee their homes. In Sudan alone, recent violence between African farmers and government-supported armed militia has killed at least 200,000 people and forced some 2.5 million more to seek safety outside their country.

Many have sought shelter in crowded refugee camps. Human-rights groups say the world needs to move rapidly to help refugees in Africa so that more people do not die violently or from starvation and disease.

MARCUS PRIOR/WFP/EPA/CORBIS

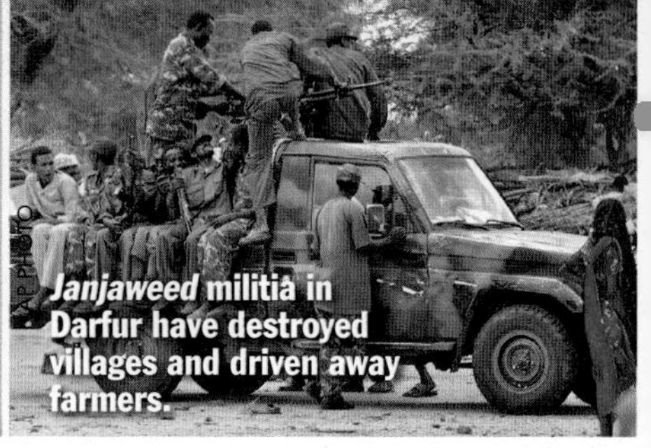
Janjaweed militia in Darfur have destroyed villages and driven away farmers.

HOW CAN AFRICA BE HELPED?

Halima was working in her family's field in Darfur, a region of Sudan, in Africa, when she heard "the voice of guns." The guns belonged to Arab bandits, called *janjaweed* (jahn•*jah*•weed), who were attacking African farmers. The *janjaweed* wanted the farmers' land and livestock for themselves.

"The attackers wanted to kill us," Halima said. She quickly picked up her young daughter and fled. "We had to hide and walk at night. We had nothing to eat." It took the mother and child weeks to reach a refugee camp in Chad, Sudan's western neighbor.

There are many stories like Halima's about Darfur. Aid workers say the violence in Sudan that began in 2003 has killed tens of thousands and forced about 2.5 million **refugees** into refugee camps. Refugees are people who are forced to find shelter outside their country. Life is very difficult in the camps, which are run by groups like the United Nations. The violence has made it difficult for aid to reach the camps. Nearly half of the refugees cannot get enough food, clean water, or medicines.

Violence in Darfur

In the Darfur region of Sudan, Arab militia have destroyed villages and displaced millions. Here is a close-up look at one trouble spot.

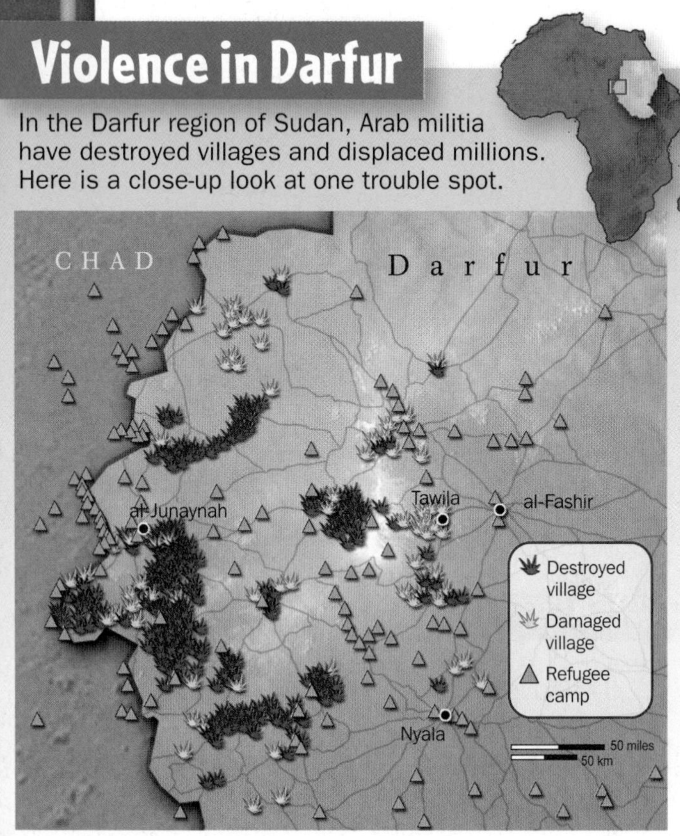

INTERPRETING MAPS

Analyzing Information Near what city in Darfur have most of the villages been destroyed?

All Too Ordinary

Halima's experience is not unique in the region. In recent decades, violence has caused millions of other Africans to become refugees. Four million people in Sudan were forced from their homes in 1983 when a violent civil war broke out. In 1994 **genocide**, or the deliberate attempt to kill or hurt a race or a group of people, became a problem in Rwanda. Militia groups killed up to 1 million people of the Tutsi and Hutu ethnic groups. Nearly 100,000 children were orphaned. This genocide in Rwanda forced more than 2 million frightened people to seek **asylum**, or safety, in other countries.

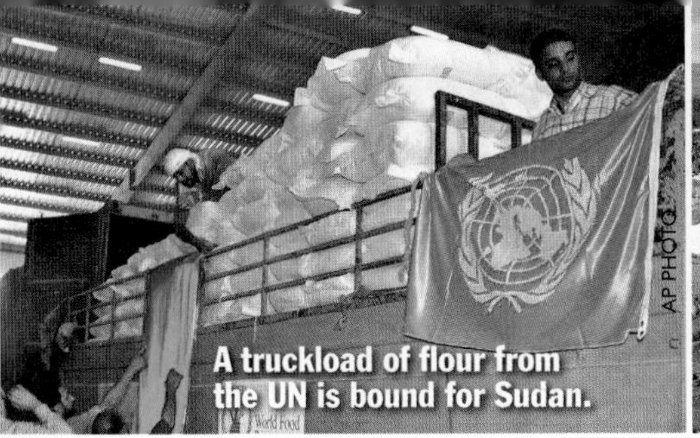
A truckload of flour from the UN is bound for Sudan.

AP PHOTO

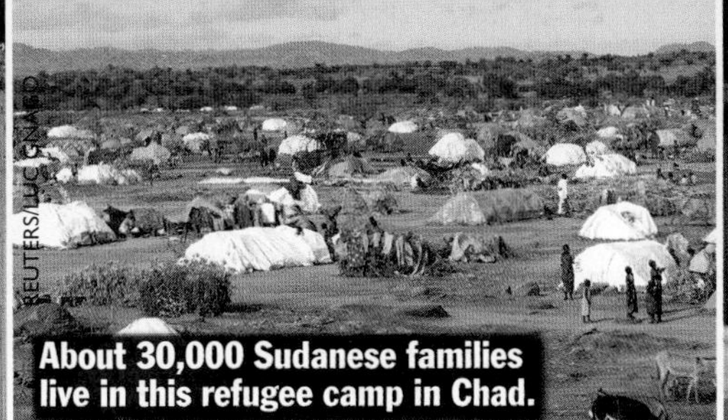
About 30,000 Sudanese families live in this refugee camp in Chad.

REUTERS/LUC GNAGO

Orphaned refugees wait for milk at a feeding center.

AP PHOTO

Africa's Troubled Past

Many experts trace Africa's refugee problem back to the late 1800s. During this time, European nations divided the continent into colonies. European colonizers, however, did not organize their colonies in a way that recognized the differences among Africans. For example, there are many ethnic groups in Africa that speak different languages. When boundaries were organized, similar **ethnic groups** were separated. Other ethnic groups, who were traditional enemies, such as Rwanda's Hutu and Tutsi, were forced to live together.

When Africa's colonies became independent nations nearly a century later, many of these ethnic groups had little interest in working together. This lack of agreement made it difficult for stable governments to be established. In many nations, armed groups overthrew these unstable governments. Such struggles for control turned millions of Africans into refugees. In addition to these issues, religious differences between groups also led to violence and instability.

Ethnic, political, and religious conflict is not the only reason for Africa's refugee problem. In 1984 a drought caused a severe famine in Ethiopia. Nearly 1 million people died and hundreds of thousands were forced to seek help far from their homes. People who flee natural disasters, such as floods and famines, are considered **environmental refugees**.

The Cost to Africa

The presence of so many homeless people impacts *all* of Africa. Away from their villages, refugees can no longer grow crops, resulting in food shortages. Deadly diseases like hepatitis and AIDS are easily spread in refugee camps. The large number of refugees also drains the resources of the poor **host countries**. The cost of helping so many refugees makes it more difficult for host countries to provide services, such as education and health care, to their own citizens.

What can be done to help Africa's refugees? Human rights groups say world leaders must establish economic and military punishments against governments that support violence. The African Union, which represents the continent's nations, agrees. The Union's leaders "welcomed the support of the international community" to resolve the problems in Sudan. The leaders believe that if nothing is done, many more refugees will die from violence, starvation, and disease.

EXPLORING THE ISSUE

1. **Explaining** Why was it difficult for stable governments to be established when the former colonies in Africa became independent nations?
2. **Determining Cause and Effect** How do Africa's refugee issues affect all Africans?

BAND AID FOR AFRICA

What does rock music have to do with helping Africa? Rock stars have been involved in the fight against poverty and disease in Africa for years. In 1985 rocker Bob Geldof organized an event called Live Aid. On July 13, 1985, live rock concerts were held in several cities, including Philadelphia and London. The Live Aid concerts were broadcast on TV and helped to raise money for starving Africans. More than 1.5 billion people watched Live Aid, which raised more than $100 million.

In 2005 Geldof organized a similar event, called Live 8, to call attention to poverty in Africa. During Live 8, concerts were held on four continents. Stars including Madonna, Bono from U2, and Destiny's Child attracted music fans from all over the world. The series of concerts drew more than a billion television viewers.

Sweet Songs for Africa

Rock stars are not the only musicians working to help Africa. A New York City band of young musicians and singers, called Creation, has partnered with relief-aid groups that build schools in developing countries. Creation donated the money it made on one album toward a school in Mali, a country in West Africa. "We're not doing this just to become famous and make money," said one of Creation's singers. "We're doing it to help other people."

The Agape Children's Choir is another group singing for Africa. In 2005 the South African choir performed in cities around the United States, raising money to rebuild an orphanage for children whose parents had died from AIDS. In 2004 the orphanage burned down, and some 90 children had to live in a trailer with no heat, water, or electricity. "The orphanage was like a home. We were a family," said an Agape singer. "When we sing, it's like we're together with the others back home."

Musicians of all types are proving that they can play music and make a difference at the same time.

AP PHOTO

During her Live 8 performance, Madonna shared the stage with a famine survivor from Ethiopia.

EXPLORING THE ISSUE

1. **Drawing Conclusions** How are rock stars and pop singers able to raise awareness of Africa's refugee and famine problems?

2. **Problem Solving** How might young volunteers in your hometown help Africa's refugees in their struggle against poverty and disease?

PATRICK ROBERT/SYGMA/CORBIS

Rwandan refugees receive aid from the United Nations World Food Program.

REVIEW AND ASSESS

UNDERSTANDING THE ISSUE

1 **Making Connections** How are refugees and environmental refugees different from each other? Describe some of the ways refugees can impact their host countries.

2 **Writing to Inform** Write a 250-word article about the causes of Africa's refugee problem. Be sure to use information from this report.

3 **Writing to Persuade** Write a letter to your favorite pop star explaining why he or she should help Africa's refugees.

INTERNET RESEARCH ACTIVITIES

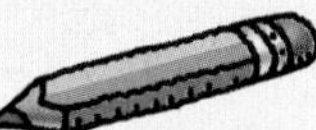

4 Navigate to the Web site of the United Nations Refugee Agency, www.unrefugees.org. Read about some of the work the agency does to help refugees around the world. Write a brief report that provides details on the agency's relief work.

5 With your teacher's help, use Internet resources to find information about Live Aid. Read about how the event was organized and how pop artists from around the world worked together to make it a success. Be prepared to report on the success of Live Aid in class.

BEYOND THE CLASSROOM

6 **Research a refugee problem outside of Africa.** Write a report that explains how it is similar to and different from refugee problems in Africa.

7 **Research an African country,** like Chad or Tanzania, which has been a host nation for refugees. Concentrate your research on how the country helped refugees. Did helping refugees drain resources from the country's own citizens? Report your findings to the class.

The Home Countries of Refugees

By 2006 there were more than 12 million refugees around the world. This graph illustrates the locations from which most of them fled.

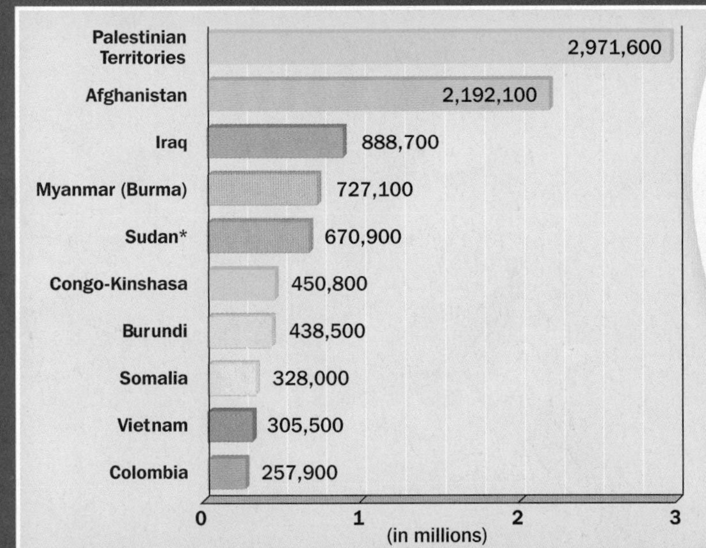

* Does not include displaced persons in Darfur

Source: U.S. Committee for Refugees and Immigrants.

Building Graph Reading Skills

1. **Transferring Data** Make a list that categorizes the nations in the graph under the appropriate region: Latin America, Africa, Central Asia, Southeast Asia, and Southwest Asia.

2. **Analyzing Information** Which of the above regions have the greatest and smallest number of refugees?

Guide to Reading

BIG Idea

Patterns of economic activities result in global interdependence.

Content Vocabulary

- constitution *(p. 589)*
- suffrage *(p. 589)*
- migrant worker *(p. 591)*
- enclave *(p. 591)*

Academic Vocabulary

- structure *(p. 589)*
- brief *(p. 591)*
- widespread *(p. 591)*

Reading Strategy

Identifying Central Issues Use a diagram like the one below to summarize key facts about southern Africa.

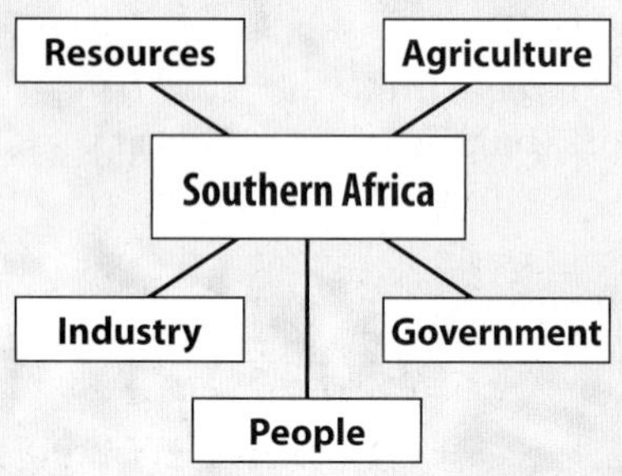

Southern Africa

Picture This Open wide! These jaws of a great white shark are evidence of its presence in the waters off the coast of Durban, South Africa. Tourists are drawn to Durban's sandy beaches and clear water. The water, however, is the natural habitat of different species of sharks. The Natal Sharks Board is an organization that maintains nets that line the coast. The nets prevent the sharks from swimming too close to beachgoers. The Board also studies shark behaviors and why they attack humans. To learn more about southern Africa, read Section 3.

The jaws of a great white shark

Republic of South Africa

Main Idea **The Republic of South Africa has great mineral wealth, and has experienced major political and social changes in recent decades.**

Geography and You What guarantees that you have the same rights as the person sitting next to you? Read to find how South Africans struggled to get the same rights.

The country officially named the Republic of South Africa is often simply called South Africa. It has a large land area, many resources, and a democratic form of government.

The Economy

South Africa has the most highly developed economy in Africa thanks to exports of certain minerals, such as gold, diamonds, and platinum. Industry, farming, and ranching are also developing successfully here.

Not all South Africans benefit from this prosperity. In rural areas, many people live in poverty and depend on subsistence farming. In the cities, industries have not grown fast enough to provide enough jobs.

People

South Africa has a diverse population. Black ethnic groups, such as the Zulu and Xhosa (KOH·suh), make up about 75 percent of the population. People of European descent make up about 10 percent of South Africa's population. They include Afrikaners, who are descendants of Dutch, German, and French settlers, and people of British origin.

South Africa also has other ethnic groups. The eastern part of the country has many citizens of South Asian descent. Other South Africans are of mixed European, Asian, and ethnic African background.

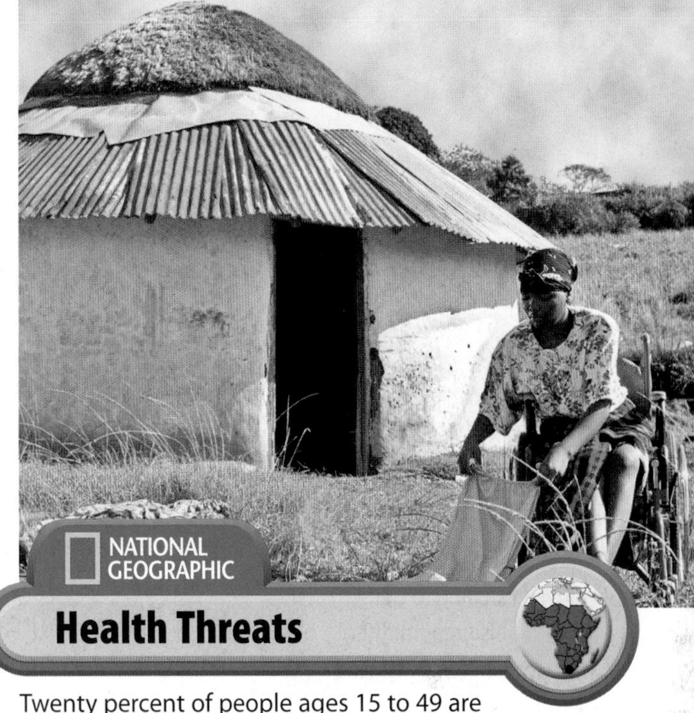

NATIONAL GEOGRAPHIC

Health Threats

Twenty percent of people ages 15 to 49 are infected with HIV, the virus that can lead to AIDS.
Place **What other problems does South Africa face?**

History and Government

In the early 1900s, British and Afrikaner settlers formed South Africa as a white-ruled country. They set up apartheid to control non-European groups. Black South Africans founded the African National Congress (ANC), hoping to gain power. Protests led to apartheid's end in the early 1990s. South Africans then wrote a constitution based on majority rule. A **constitution** is a document describing a government's **structure** and powers as well as the rights of citizens. South Africa's constitution stated that people of all races and both genders would have equality. It gave **suffrage,** or the right to vote, to all citizens who are 18 or older.

Reading Check **Analyzing Information** How does South Africa protect citizens' rights?

WHO WILL STOP THE DARK?

By Charles Mungoshi

Finally, [after reaching the river] they dug for worms in the wet clay on the river banks. . . .

"Worms are much easier to find," the old man said. "They stay longer on the hook. But a **maggot** takes a fish faster. . . . Locusts and **hoppers** are good too, but in bigger rivers, like Munyati where the fish are so big they would take another fish for a meal. Here the fish are smaller and cleverer. They don't like hoppers."

The old man looked into the coffee tin into which they were putting the worms and said, "Should be enough for me [for] one day. There is always some other place we can get some more when these are finished. No need to use more than we should."

"But if they should get finished, *Sekuru* [Grandfather]? Look, the tin isn't full yet," Zakeo looked **intently** at his grandfather. He wanted to fit in all the fishing that he would ever do. . . . The old man looked at him. He understood. But he knew the greed of thirteen-year-olds and the **retribution** of the land and the soil when well-known laws were not obeyed. . . .

"Why do you spit on the bait before you throw the line into the pool, *Sekuru*?"

The old man grinned. "For luck, boy, there is nothing you do that fate has no hand in. Having a good hook, a good line, a good rod, good bait or a good pool is no guarantee that you will have good fishing. So little is knowledge, boy. The rest is just mere luck."

From: "Who Will Stop the Dark?" *The Setting Sun and the Rolling World,* Charles Mungoshi. Boston: Beacon Press, 1989.

Charles Mungoshi
(1947–)

Charles Mungoshi was born in Zimbabwe (then Southern Rhodesia). As a young boy, Mungoshi enjoyed listening to the stories of his grandmother, who was a gifted storyteller. Mungoshi has received many awards for his poems and short stories about the Shona of Zimbabwe.

Background Information

The Shona have lived in the area that is now Zimbabwe for more than 1,000 years. These people are farmers who have struggled with invasions, disease, and drought. The Shona believe that everything on Earth contains a spirit. As Zimbabwe modernizes, however, the Shona way of life is in danger of being lost.

Reader's Dictionary

maggot: worm

hoppers: grasshoppers

intently: closely

retribution: revenge

Analyzing Literature

1. **Making Inferences** What does the grandfather say that might reflect the beliefs of the Shona?
2. **Read to Write** Write a paragraph that describes how Zakeo might respond to his grandfather's belief in luck.

Other Southern African Countries

Main Idea **Other southern African countries are rich in resources and are home to many different ethnic groups.**

Geography and You Do you know anyone who travels a long distance for a job, perhaps to another state? Read to find out how some people in southern Africa get to work.

In addition to South Africa, the subregion of southern Africa includes a number of countries. Some are located in inland areas north of South Africa. Others have either Atlantic or Indian Ocean coastlines or are island nations.

Inland Southern Africa

The six countries of inland southern Africa include Lesotho (luh·SOH·toh), Swaziland, Botswana (bawt·SWAH·nah), Zimbabwe, Zambia, and Malawi (mah·LAH·wee). Each is landlocked, has a mild climate, and is dominated by high plateaus. Most of the citizens practice subsistence farming in rural villages. Thousands move to the cities or to South Africa as **migrant workers.** They spend most of the year working in mines and factories, visiting their families only a few times each year for **brief,** or short, periods.

Within South Africa lie Lesotho and Swaziland. These two countries are **enclaves,** small territories located inside a larger country. Both are poor countries that depend on South Africa for goods and markets.

Directly north of South Africa lies Botswana, a country with swamplands and part of the vast Kalahari Desert. Its economy relies on the mining and exporting of diamonds and other minerals. The dry climate limits farming, so food must be imported from South Africa. Nevertheless, Botswana is one of Africa's strongest democracies.

Northeast of Botswana is Zimbabwe, which takes its name from the ancient African city and trading center—Great Zimbabwe. It is rich in gold, copper, iron ore, and asbestos. Some large plantations grow coffee, cotton, and tobacco. For years, Europeans owned Zimbabwe's richest farmland. In recent years, the government has tried to turn over this land to Africans, but this has led to disorder and violence. The economy has been hurt, and there are **widespread** shortages throughout the country. People in Zimbabwe have protested against the strong-handed rule of their president, Robert Mugabe.

Zimbabwe's northern neighbor, Zambia, relies on copper for most of its income.

Purifying Water

NATIONAL GEOGRAPHIC

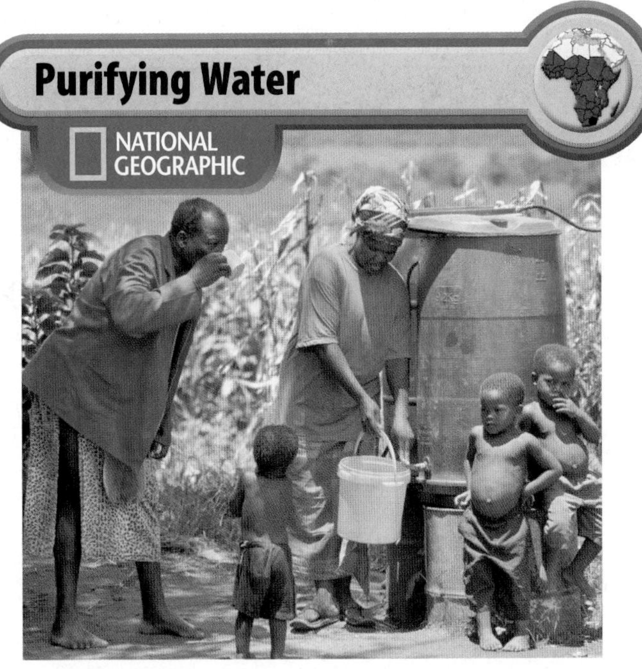

▲ Obtaining clean water is a problem in many countries in southern Africa. This barrel collects rainwater and then filters it through a layer of sand. ***Place*** **How do most people in inland southern Africa make their living?**

As a result, when world copper prices go down, Zambia's income goes down too. City dwellers work in mining and service industries. Villagers grow corn, rice, and other crops to support their families.

East of Zambia, Malawi boasts wetlands, lakes, mountains, and forests. Wildlife in Malawi's national parks attracts visitors from around the world. Malawi's people grow crops like tobacco, tea, and sugar for export. After years of harsh government, Malawi became democratic in the mid-1990s.

Coastal and Island Countries

Angola and Namibia have Atlantic Ocean coastlines. These countries are made up of hilly grasslands and rocky deserts. Both are rich in minerals. Angola is one of Africa's major oil producers, and Namibia mines diamonds, copper, gold, and zinc. Despite this mineral wealth, most people in Angola and Namibia live in poverty and practice herding and subsistence farming.

Mozambique borders the Indian Ocean. After achieving independence, the country's development was slowed by civil war and famine. Recently, Mozambique has begun to attract foreign investors.

Madagascar, Comoros, Mauritius (maw·RIH·shuhs), and Seychelles (say·SHEHL) are island nations in the Indian Ocean. They are populated by a mix of peoples from Asia as well as from Africa. Madagascar has a sizable population, but the others are relatively small, especially tiny Seychelles. All of these countries depend on agriculture, although Mauritius has a growing banking industry, and Seychelles has a strong tourist industry.

Reading Check **Explaining** What role do minerals play in southern African economies?

Social Studies ONLINE
Study Central™ To review this section, go to glencoe.com.

Vocabulary

1. **Explain** the significance of:
   - **a.** constitution
   - **b.** suffrage
   - **c.** migrant worker
   - **d.** enclave

Main Ideas

2. **Generalizing** Use a diagram like the one below to write three important facts about political events in South Africa during the past 100 years. Then, in the larger box, write a generalization that you can draw from those facts.

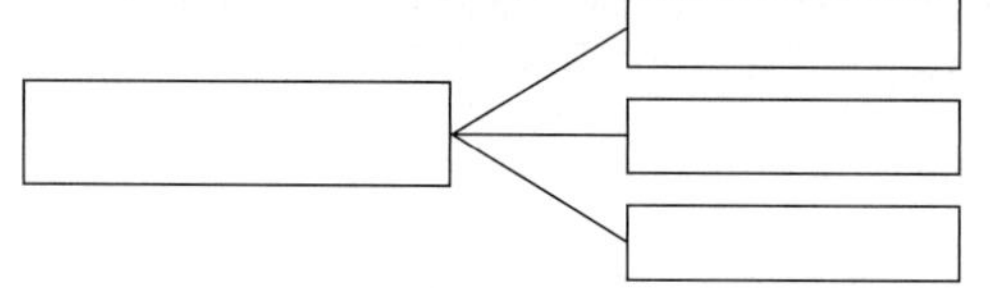

3. **Describing** Describe the economies of southern Africa's coastal countries.

Critical Thinking

4. **Comparing and Contrasting** How are the populations of island countries different from those of mainland countries?

5. **BIG Idea** How are neighboring countries connected economically to the Republic of South Africa?

6. **Challenge** How has the history of the countries of southern Africa both helped and hindered the area's economic development?

Writing About Geography

7. **Expository Writing** Write a paragraph describing the forms of agriculture and agricultural products of southern Africa.

Visual Summary

West Africa

- Nigeria is a major oil producer and is Africa's most populous country.
- Ethnic and religious differences threaten Nigeria's political stability.
- The Sahel countries have dry climates.
- Plentiful rainfall supports agriculture in coastal West Africa.

Harvesting rice, Gambia

Central Africa

- The Democratic Republic of the Congo has more than 200 distinct ethnic groups.
- Gabon's economy thrives on oil, timber, and minerals.
- Some Central African countries have suffered from years of political instability.

Tourist safari, Kenya

East Africa

- The economies of Tanzania and Kenya rely on tourism and farming.
- Ethnic conflict led to millions of deaths in Rwanda and Burundi during the 1990s.
- Droughts and warfare have often occurred in countries located on the Horn of Africa.

Oil drill, Equatorial Guinea

Southern Africa

- Many countries of southern Africa have large deposits of metal ores and gems.
- In the 1990s, apartheid ended in South Africa and a new democratic constitution was put in place.
- Most people in southern Africa live in rural villages and practice subsistence farming.

Voting in South Africa

Study anywhere, anytime! Download quizzes and flash cards to your PDA from **glencoe.com**.

CHAPTER 21 ASSESSMENT

STANDARDIZED TEST PRACTICE

When answering questions based on tables, study the table's title to help you understand the subject matter. Then carefully review each column heading to identify the information contained in each column.

Reviewing Vocabulary

Directions: Choose the word(s) that best completes the sentence.

1. Most Nigerian farmers have ______, small plots of land on which they grow only enough food to feed their families.

A plantations
B subsistence farms
C landlocked farms
D sisal farms

2. A Tanzanian export crop is ______, a plant fiber used to make rope and twine.

A cacao
B Tutsi
C cassava
D sisal

3. In the 1990s, the country of South Africa approved a new ______, or document that describes the structure and powers of its government.

A monarchy
B declaration
C democracy
D constitution

4. Small territories, like Lesotho and Swaziland, that are located inside a larger country are called ______.

A enclaves
B colonies
C habitats
D headquarters

Reviewing Main Ideas

Directions: Choose the best answer for each question.

Section 1 *(pp. 572–575)*

5. Despite oil revenues, most Nigerians work as ______.

A service workers
B farmers
C factory workers
D government officials

6. The landlocked countries of West Africa, along with Mauritania, are called the ______.

A cacao
B Ibo
C Côte d'Ivoire
D Sahel

Section 2 *(pp. 576–582)*

7. In Rwanda, conflict between the ______ led to hundreds of thousands of deaths.

A Hutu and Tutsi
B Uganda and Burundi
C Kikuyu and Kenyans
D Muslims and Christians

Section 3 *(pp. 588–592)*

8. The country of ______ has the most developed economy in Africa.

A Lesotho
B South Africa
C Swaziland
D Botswana

GO ON

Critical Thinking

Directions: Base your answers to questions 9, 10, and 11 on the table below.

Largest National Parks in Africa South of the Sahara

Park Name (Country)	Year Established	Area (sq. mi.)	Area (sq. km)
Salonga (Democratic Republic of the Congo)	1970	14,116	36,560
Gemsbok (Botswana)	1971	9,266	23,999
Southern (Sudan)	1939	8,880	22,999
Boma (Sudan)	1981	8,803	22,780
Kafue (Zambia)	1972	8,649	22,401

Source: Encyclopædia Britannica.

9. Which of these national parks is the largest?

A Boma

B Gemsbok

C Kafue

D Salonga

10. Which African country shown in the table was the first to establish a national park?

A Democratic Republic of the Congo

B Sudan

C Botswana

D Zambia

11. Based on the information in the table, which statement is most accurate?

A Most national parks are found in western Africa.

B Zambia is least concerned with conserving parkland.

C The area of Boma and Southern National Parks combined is larger than the area of the Salonga National Park.

D Several new national parks have been created in the past 20 years.

Document-Based Questions

Directions: Analyze the document and answer the short-answer questions that follow.

United Nations: Consolidated Appeal for West Africa 2006

Through the first half of 2006, some countries of West Africa have faced threats of political instability and nutritional and food insecurity, while others have benefited from progress towards stability and development. Political instability in Côte d'Ivoire and Guinea-Bissau has led to population movements and a continued need for assistance. . . . In the Sahel region, even with improved harvests predicted through 2006, the vulnerability of the poorest households to food and nutritional insecurity is still high. . . . [F]ollowing the successful elections in Liberia in November 2005, . . . all [refugee] camps have been officially closed . . . after the return of some 314,000 Internally Displaced Persons (IDPs) and over 69,000 refugees from neighbouring countries to their places of origin.

—United Nations

12. According to the United Nations report about West Africa, what has been the result of political instability in Côte d'Ivoire and Guinea-Bissau?

13. What country appears to be making progress toward stability? How do you know?

Extended Response

14. Choose a subregion of Africa south of the Sahara and write a summary of that subregion today. Include information about economies, governments, and any recent conflicts in the area.

For additional test practice, use Self-Check Quizzes—Chapter 21 at glencoe.com.

Need Extra Help?

If you missed question...	1	2	3	4	5	6	7	8	9	10	11	12	13	14
Go to page...	573	578	589	591	573	574	580	589	578	578	578	575	575	572–592

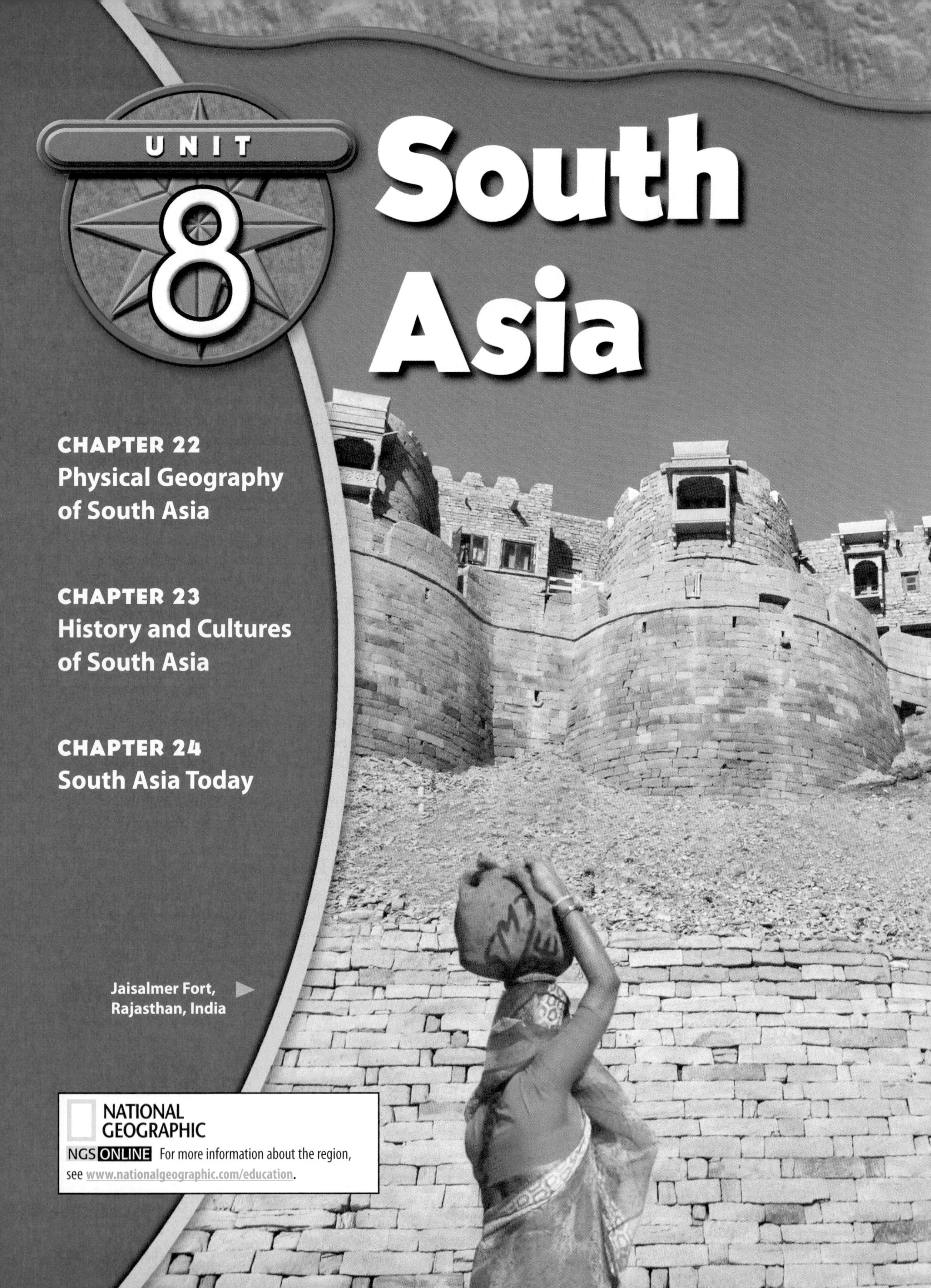

UNIT 8 South Asia

Jaisalmer Fort, Rajasthan, India ▶

NATIONAL GEOGRAPHIC

NGS ONLINE For more information about the region, see www.nationalgeographic.com/education.

Regional Atlas

South Asia

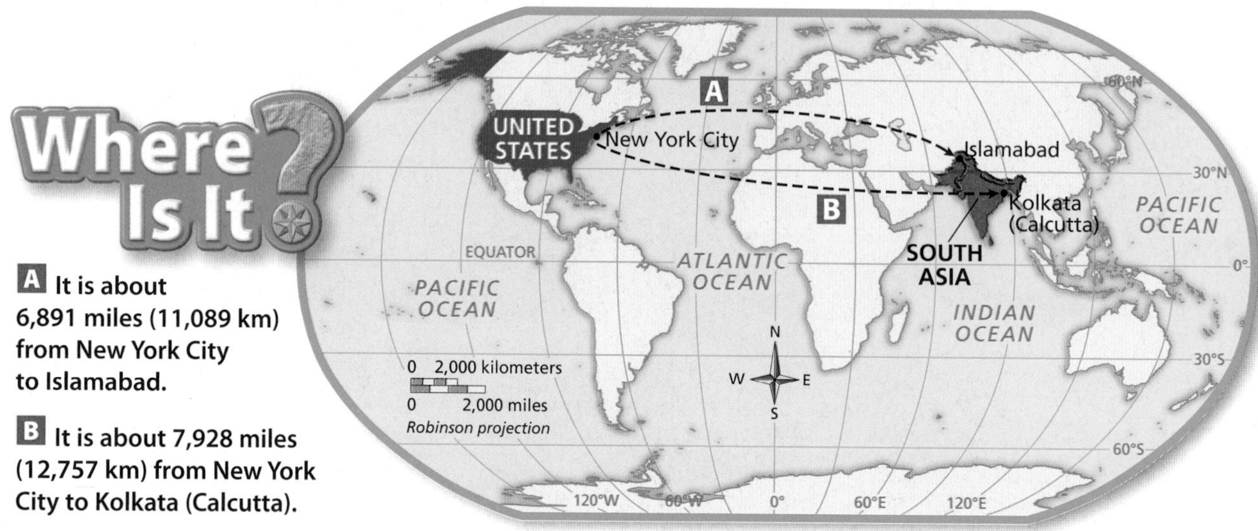

Where Is It?

A It is about 6,891 miles (11,089 km) from New York City to Islamabad.

B It is about 7,928 miles (12,757 km) from New York City to Kolkata (Calcutta).

How Big Is It?

The region of South Asia is more than half the size of the continental United States. Its land area is about 1.7 million square miles (4.5 million sq. km). Though smaller than the United States, South Asia has nearly five times the number of people as the United States and more than one-fifth of the people in the world.

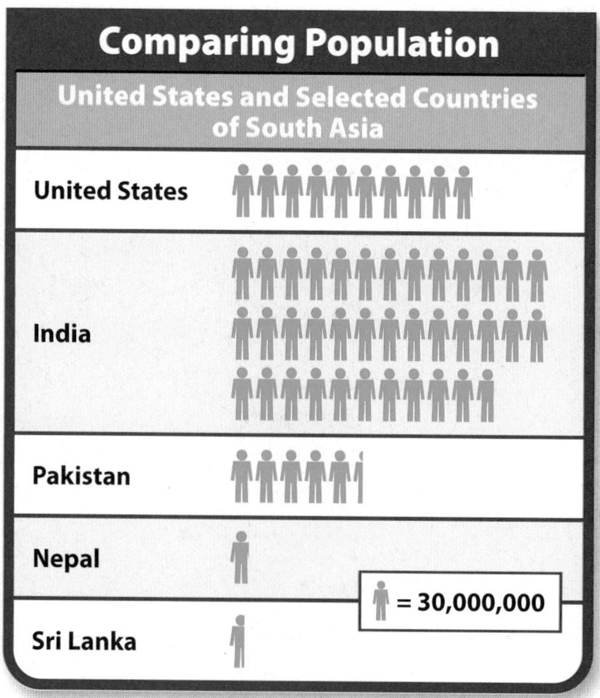

Source: *World Population Data Sheet*, 2005.

GEO Fast Facts

Longest Rivers

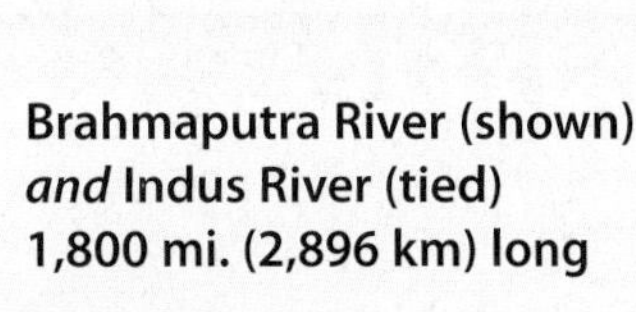

Brahmaputra River (shown) *and* Indus River (tied) 1,800 mi. (2,896 km) long

Highest Point

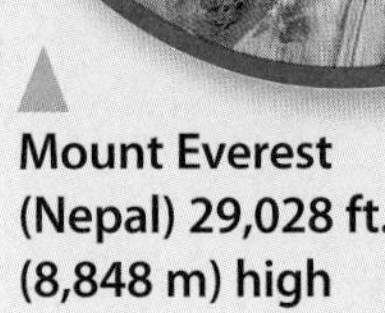

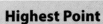

Mount Everest (Nepal) 29,028 ft. (8,848 m) high

Largest Island

Sri Lanka 25,332 sq. mi. (65,610 sq. km)

Lowest Point

Coast of Indian Ocean (Bangladesh) 0 ft. (0 m) high

UNIT 8

Regional Atlas

South Asia

PHYSICAL

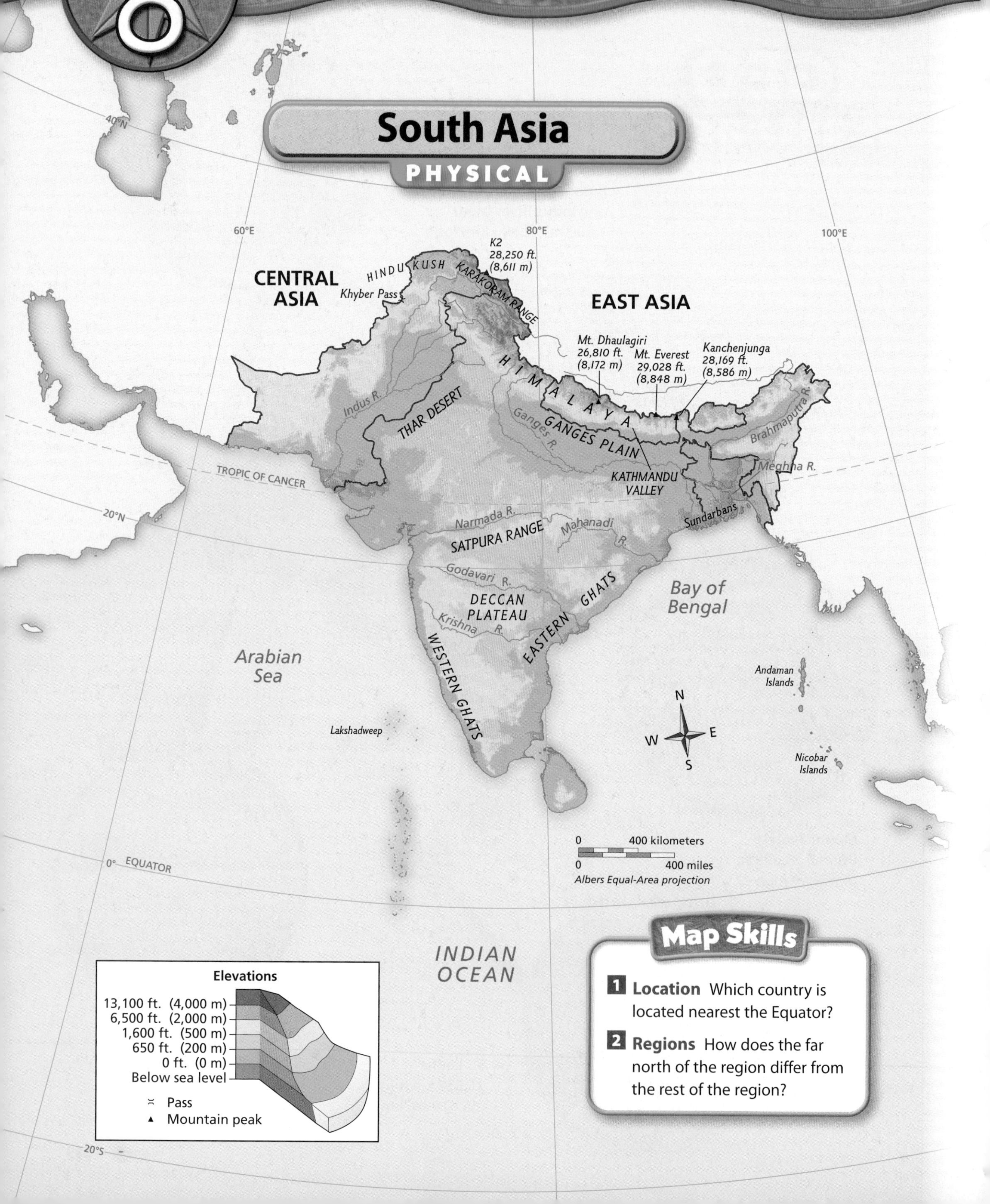

40°N
60°E
80°E
100°E
20°N
0° EQUATOR
20°S
TROPIC OF CANCER

CENTRAL ASIA
EAST ASIA
HINDU KUSH
Khyber Pass
K2 28,250 ft. (8,611 m)
KARAKORAM RANGE
HIMALAYA
Mt. Dhaulagiri 26,810 ft. (8,172 m)
Mt. Everest 29,028 ft. (8,848 m)
Kanchenjunga 28,169 ft. (8,586 m)
Indus R.
THAR DESERT
Ganges R.
GANGES PLAIN
KATHMANDU VALLEY
Brahmaputra R.
Meghna R.
Sundarbans
Narmada R.
SATPURA RANGE
Mahanadi R.
Godavari R.
DECCAN PLATEAU
Krishna R.
EASTERN GHATS
WESTERN GHATS
Bay of Bengal
Arabian Sea
Lakshadweep
Andaman Islands
Nicobar Islands
INDIAN OCEAN

N
W
E
S

0 400 kilometers
0 400 miles
Albers Equal-Area projection

Elevations

13,100 ft. (4,000 m)
6,500 ft. (2,000 m)
1,600 ft. (500 m)
650 ft. (200 m)
0 ft. (0 m)
Below sea level

⌅ Pass
▲ Mountain peak

Map Skills

1 Location Which country is located nearest the Equator?

2 Regions How does the far north of the region differ from the rest of the region?

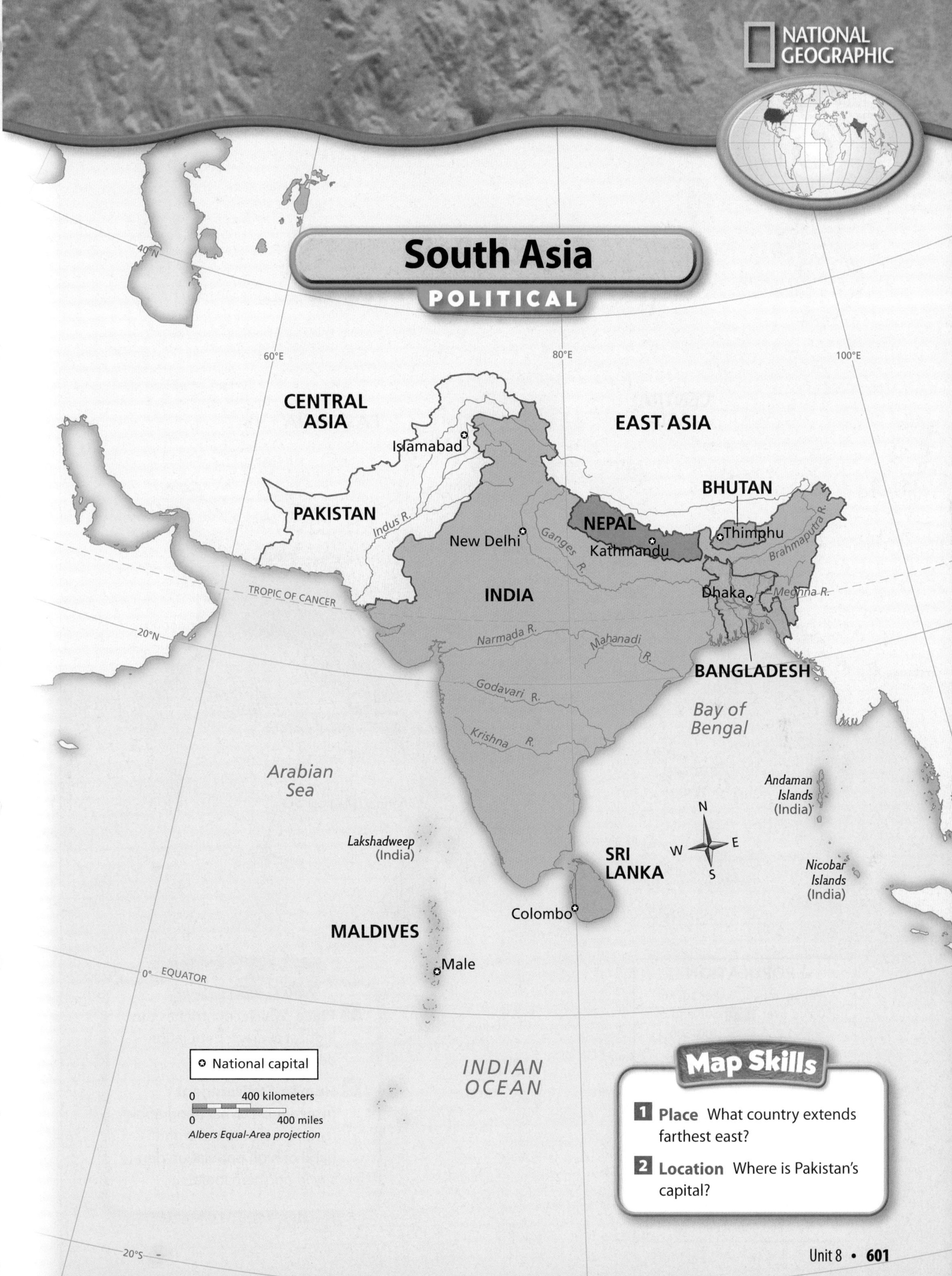

Map Skills

1 **Place** What country extends farthest east?

2 **Location** Where is Pakistan's capital?

UNIT 8

Regional Atlas

South Asia

POPULATION DENSITY

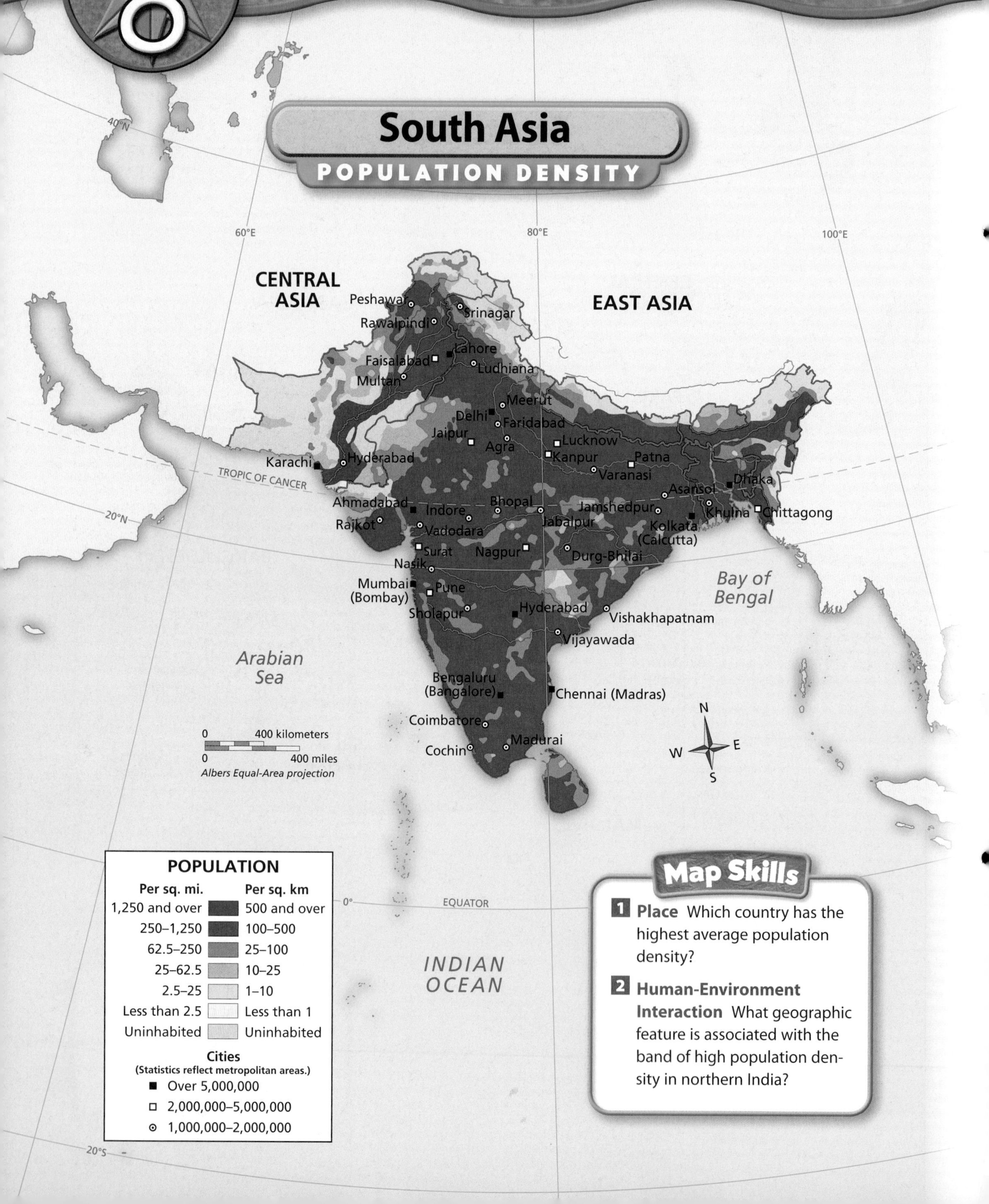

40°N 60°E 80°E 100°E 20°N 0° 20°S

TROPIC OF CANCER

EQUATOR

CENTRAL ASIA

EAST ASIA

Arabian Sea

Bay of Bengal

INDIAN OCEAN

Peshawar
Srinagar
Rawalpindi
Lahore
Faisalabad
Ludhiana
Multan
Meerut
Delhi
Faridabad
Jaipur
Agra
Lucknow
Karachi
Hyderabad
Kanpur
Patna
Varanasi
Dhaka
Asansol
Ahmadabad
Indore
Bhopal
Jamshedpur
Khulna
Chittagong
Rajkot
Vadodara
Jabalpur
Kolkata (Calcutta)
Surat
Nagpur
Durg-Bhilai
Nasik
Mumbai (Bombay)
Pune
Sholapur
Hyderabad
Vishakhapatnam
Vijayawada
Bengaluru (Bangalore)
Chennai (Madras)
Coimbatore
Madurai
Cochin

0 400 kilometers
0 400 miles
Albers Equal-Area projection

N S E W

POPULATION

Per sq. mi.	Per sq. km
1,250 and over	500 and over
250–1,250	100–500
62.5–250	25–100
25–62.5	10–25
2.5–25	1–10
Less than 2.5	Less than 1
Uninhabited	Uninhabited

Cities
(Statistics reflect metropolitan areas.)

- ■ Over 5,000,000
- □ 2,000,000–5,000,000
- ⊙ 1,000,000–2,000,000

Map Skills

1 Place Which country has the highest average population density?

2 Human-Environment Interaction What geographic feature is associated with the band of high population density in northern India?

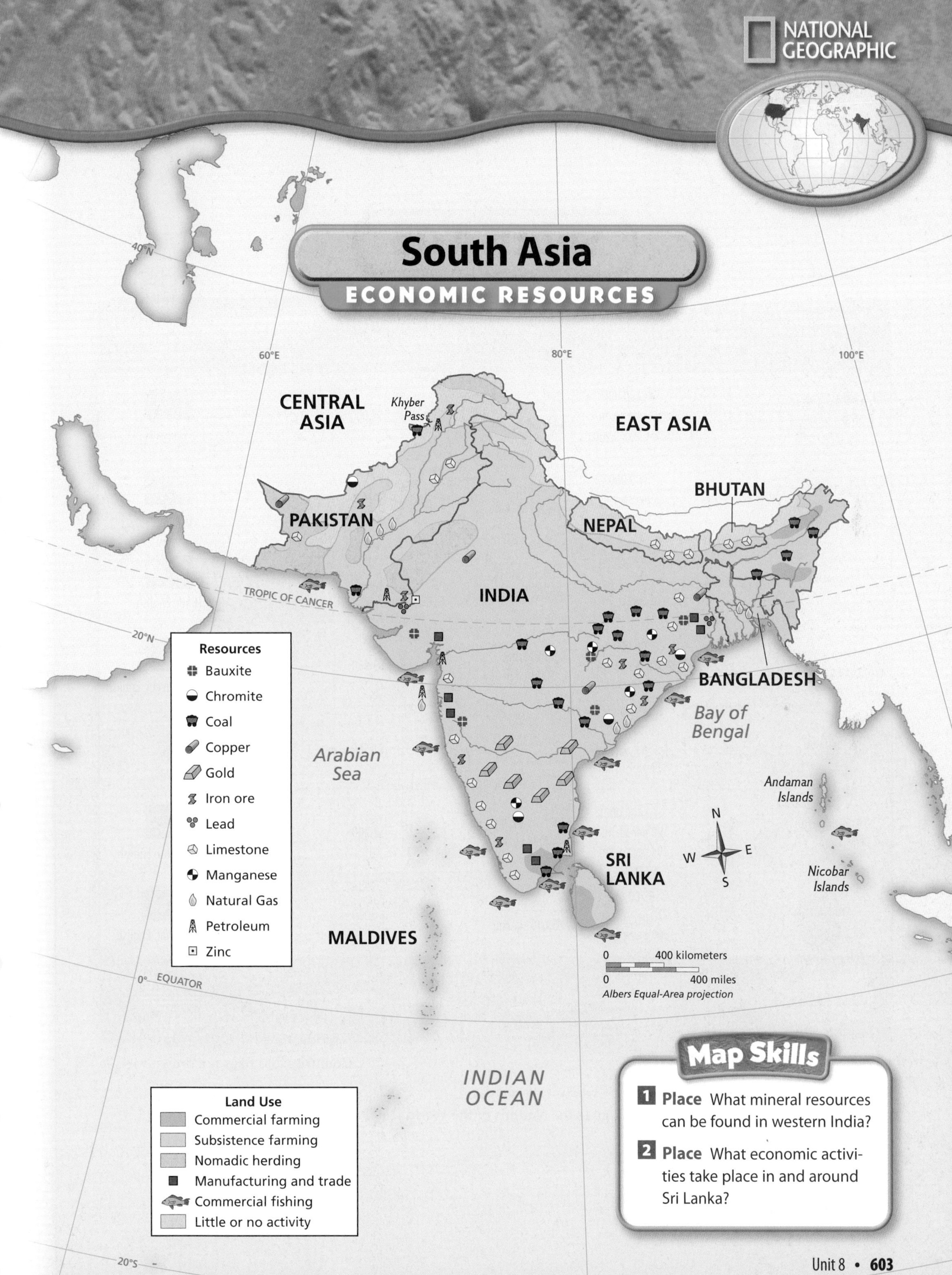

Map Skills

1. **Place** What mineral resources can be found in western India?
2. **Place** What economic activities take place in and around Sri Lanka?

Regional Atlas

South Asia

Country and Capital	Literacy Rate	Population and Density	Land Area	Life Expectancy (Years)	GDP* Per Capita (U.S. dollars)	Television Sets (per 1,000 people)	Flag and Language
BANGLADESH Dhaka	43.1%	144,200,000 2,594 per sq. mi. 1,001 per sq. km	55,598 sq. mi. 143,998 sq. km	61	$2,000	7	Bengali
Thimphu BHUTAN	42.2%	1,000,000 55 per sq. mi. 21 per sq. km	18,147 sq. mi. 47,001 sq. km	63	$1,400	6	Dzongkha
New Delhi INDIA	59.5%	1,103,600,000 869 per sq. mi. 336 per sq. km	1,269,340 sq. mi. 3,287,575 sq. km	62	$3,100	75	Hindi, English
Male MALDIVES	97.2%	300,000 2,586 per sq. mi. 1,000 per sq. km	116 sq. mi. 300 sq. km	72	$3,900	38	Maldivian Dhivehi, English
NEPAL Kathmandu	45.2%	25,400,000 447 per sq. mi. 173 per sq. km	56,826 sq. mi. 147,179 sq. km	62	$1,500	516	Nepali
Islamabad PAKISTAN	45.7%	162,400,000 528 per sq. mi. 204 per sq. km	307,375 sq. mi. 796,098 sq. km	62	$2,200	105	Punjabi, Urdu, English
SRI LANKA Colombo	92.3%	19,700,000 778 per sq. mi. 300 per sq. km	25,332 sq. mi. 65,610 sq. km	73	$4,000	102	Sinhala, Tamil, English
UNITED STATES Washington, D.C.	97%	296,500,000 80 per sq. mi. 31 per sq. km	3,717,796 sq. mi. 9,629,047 sq. km	78	$40,100	844	English

*Gross Domestic Product

Countries and flags not drawn to scale

Sources: *CIA World Factbook,* 2005; Population Reference Bureau, *World Population Data Sheet,* 2005.

For more country facts, go to the **Nations of the World Databank** at glencoe.com.

Braga, Nepal

Reading Social Studies

Distinguishing Fact From Opinion

Learn It!

A *fact* is something that can be proved by evidence such as records, documents, or historical sources. An *opinion* is based on a person's values or beliefs. Distinguishing fact from opinion can help you make reasonable judgments about what others say and write.

Follow these steps to identify facts and opinions.

- Read or listen to the information carefully. Which statements can be proved from a reliable source? These are facts.
- Identify opinions by looking for statements of feelings or beliefs. Do statements include words like *should* or *always?*

Read the following statements. The chart below can help you distinguish fact from opinion and explain why.

1. Call center jobs include answering customer questions or entering data online.
2. Many call centre employees answer telephones but some also do highly skilled back office jobs on-line.
3. Indeed, so glamoured are many of them [Indians] by the prospect of working for a multinational [worldwide corporation] . . . that they feel that they are already half-way to America.

—*from pages 658–659*

Reading Tip

Sometimes people use facts to support their opinions. Remember to check the sources for these facts to be sure they are reliable.

Facts	Opinions
1. This fact could be proven by checking employment advertisements or call center job descriptions.	**3.** The author describes many Indians as having the same feelings about work, which is not a proven fact.
2. This fact could be proven by checking employment advertisements or call center job descriptions.	

Read to Write Activity

Identify a problem that challenges South Asia today. In an editorial, discuss this challenge and how you think it could be resolved. Cite facts to support your opinion. Then, exchange your editorial with a partner. Above each sentence that is a fact, write "F." Above each sentence that is an opinion, write "O." Discuss the editorials as a class.

2 Practice It!

Read the following paragraph from this unit.

- Draw a chart like the one shown below.
- Write facts from the paragraph in the column on the left.
- Rewrite the paragraph so that it reflects your opinion about arranged marriages in South Asia.

> Marriage in South Asian countries is commonly viewed as the joining of two families. As a result, parents often arrange marriages for their children by choosing partners they consider suitable. After a woman marries, she becomes part of her husband's family. In India and Pakistan, several generations often live together in the same house.
>
> *—from page 643*

Facts	Opinions

Bride and groom in Pakistan

3 Apply It!

As you read the chapters in this unit, identify topics that you have an opinion about. Share your opinions with the class, using facts from your reading. Identify where you might be able to find additional information to support your opinion.

Physical Geography of South Asia

Essential Question

Place South Asia has a varied landscape that includes the highest mountains in the world as well as lowlands that rise just a few feet above sea level. The region also has a variety of climate zones. How do seasonal weather patterns affect a region?

◀ Sherpa agricultural workers, Nepal

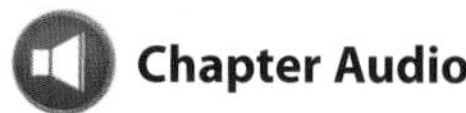

Chapter Audio

Section 1: Physical Features

BIG IDEA Geographic factors influence where people settle. Some parts of South Asia have mountains and deserts and are not heavily settled. Other areas of the region have fertile farmlands that support large populations.

Section 2: Climate Regions

BIG IDEA The physical environment affects people. The climate in much of South Asia is marked by contrasts—heavy rainfall during part of the year, and extreme dryness in other periods. If there is too little or too much rainfall, millions of lives are threatened.

FOLDABLES™ Study Organizer

Organizing Information Make this Foldable to help you organize information about South Asia's landforms and climates.

Step 1 Fold a piece of paper in half lengthwise.

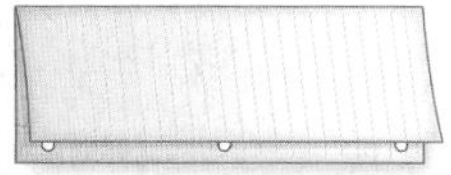

Step 2 Then fold the paper to form 5 equal sections.

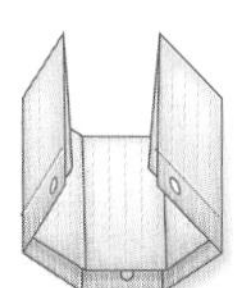

Step 3 Cut along the folds on the top flap to create tabs.

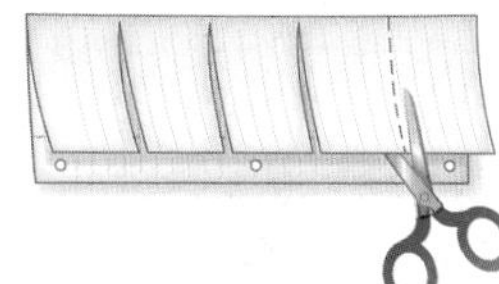

Step 4 Label the tabs as shown.

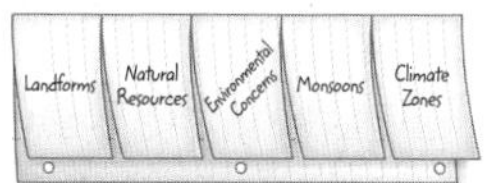

Reading and Writing As you read the chapter, write notes under the correct tab on the Foldable. Use your notes to write a short essay describing South Asia's various landforms, climates, and seasonal climate patterns.

Social Studies ONLINE

Visit glencoe.com and enter ***QuickPass***™ code EOW3109c22 for Chapter 22 resources.

Guide to Reading

BIG Idea

Geographic factors influence where people settle.

Content Vocabulary

- subcontinent *(p. 611)*
- delta *(p. 612)*
- atoll *(p. 613)*
- lagoon *(p. 613)*

Academic Vocabulary

- eventual *(p. 613)*
- concentration *(p. 614)*

Reading Strategy

Organizing Information Use a diagram like the one below to list key facts about the physical environment of South Asia.

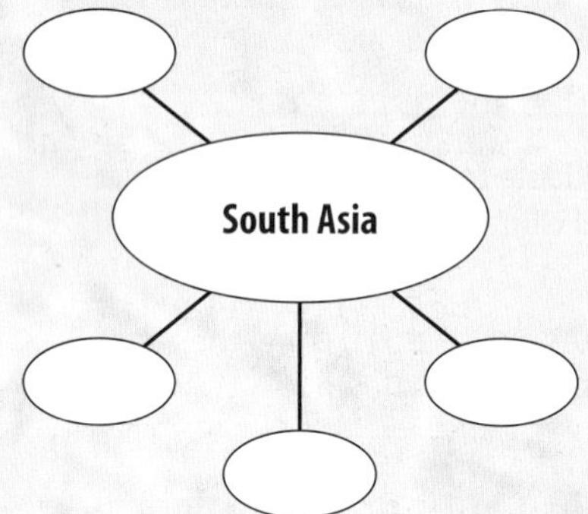

Physical Features

 Section Audio  Spotlight Video

Picture This Perched on thin poles driven into the seabed, fishermen in South Asia use baitless hooks without barbs to snare mackerel and herring. On a good day, a fisherman can catch up to 1,000 fish. Each village claims its own section of reef for fishing, and local law prohibits fishing from boats or using nets to catch fish. The stilt fishermen's poles are passed down from father to son. Read this section to find out how the geography of this region has shaped people's lives and the area's economy.

Stilt fishing, Sri Lanka

Landforms and Resources

Main Idea **The geography of South Asia varies from towering mountains to lowland river plains.**

Geography and You How would you like to feel truly "on top of the world"? You could if you climbed Mount Everest, the highest peak on Earth. Read to learn about this mountain in South Asia and the region's other physical features.

South Asia is made up of seven countries. India is the largest among them, covering three-fourths of the region. South Asia also includes Pakistan, Bangladesh, Nepal, Bhutan, Sri Lanka (SREE LAHNG·kuh), and Maldives (MAWL·DEEVZ). Most of these countries are located on the Indian subcontinent. A **subcontinent** is a large landmass that is a part of a continent.

Northern Mountains

Three huge walls of mountains form South Asia's northern boundary and separate the subcontinent from the rest of Asia. These mountain systems are the Hindu Kush, the Karakoram (KAH·rah·KOHR·ahm), and the Himalaya (HIH·muh·LAY·uh). The Himalaya range is the highest mountain system in the world. Among the snow-capped peaks of Nepal is Mount Everest, which, at 29,028 feet (8,848 m) is the tallest mountain in the world.

The Himalaya attract adventurous climbers and hikers, but their rugged terrain and harsh climate once kept travelers away. The mountains protected Nepal and Bhutan from outside influence until the 1900s. However, people from the north entered other parts of South Asia through narrow mountain passes in the Hindu Kush. The most famous of these is the Khyber Pass between Afghanistan and Pakistan. For centuries, trading caravans and conquering armies marched through the Khyber Pass and on to India.

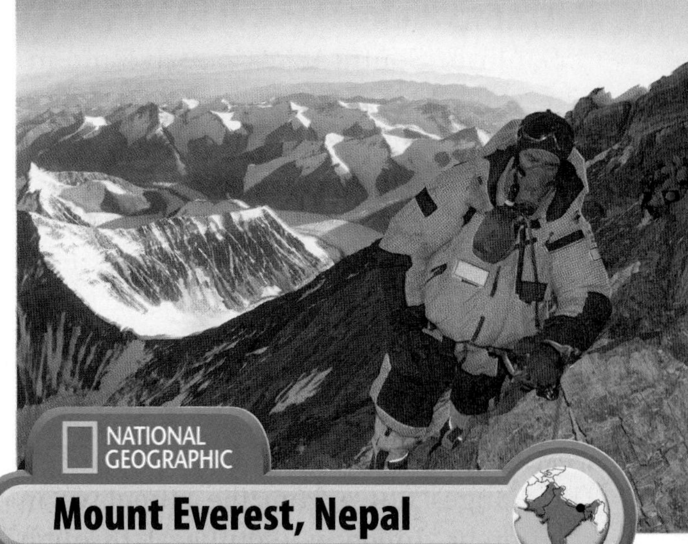

Mount Everest, Nepal

Most people use portable oxygen tanks when they climb Mount Everest. They need extra oxygen to maintain their ability to breathe comfortably. ***Place*** **Which three mountain systems make up South Asia's northern edge?**

Scientists believe that South Asia's northern mountain ranges were formed by tectonic plate movements. About 60 million years ago, the South Asian subcontinent was part of the same landmass as Africa. Then the subcontinent broke away, drifted across the Indian Ocean, and collided with the southern edge of Asia. The force of this collision thrust up the Hindu Kush, the Karakoram, and the Himalaya.

Plate movements are still going on. As a result, South Asia's northern mountains grow a tiny bit taller every year. Plate movements also cause destructive earthquakes throughout the region.

Social Studies ONLINE

Student Web Activity Visit glencoe.com and complete the Chapter 22 Web Activity about the Khyber Pass.

Northern Plains

South of South Asia's massive mountains are wide, fertile plains. These areas are watered by the region's three great rivers—the Indus, the Ganges (GAN·JEEZ), and the Brahmaputra (BRAHM·uh·POO·truh). The people of the region have long depended on these rivers for farming, transportation, and trade.

The Indus River begins north of the Himalaya in Tibet, China, and flows southwest through Pakistan to the Arabian Sea. The Ganges flows from the Himalaya in a different direction—southeast through India's Ganges Plain. This vast lowland area boasts some of the country's richest soil and is home to about 40 percent of India's population. In eastern India, the Ganges River turns south through Bangladesh. There it combines with the Brahmaputra River to form the world's largest delta. A **delta** is a soil deposit at the mouth of a river.

Southern Landforms

The landscape in the south is quite different from that in the north. At the base of the subcontinent are two chains of eroded coastal mountains—the Eastern Ghats and the Western Ghats. Between them lies a highland area known as the Deccan Plateau. The Western Ghats block seasonal rains from reaching this plateau, leaving it extremely dry. The Karnataka Plateau south of the Deccan Plateau receives these rains instead, so the hills there are lush

TIME GLOBAL CITIZENS

NAME: 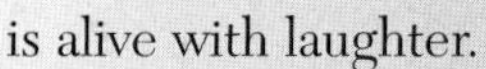

HOME COUNTRY: Maldives

ACHIEVEMENT: Zaeema Ismail, 14, lives on an island the size of a soccer field in the middle of the Indian Ocean. In 2004 a tsunami devastated her island and killed her grandmother. Ismail's mother was so grief stricken that she could not speak or eat, and her brother Mohammed, 2, had nightmares about the event. To find help, Ismail traveled to a nearby island to attend a UNICEF trauma workshop. She learned that her family's behavior was normal in tragedy. She encouraged her family to do chores together to keep them busy and distracted. The plan worked. Today, Ismail's mother eats normally, her brother sleeps soundly, and their tin hut is alive with laughter.

Ismail walks through her village with her mom and her brother and sisters.

PRASHANT PANJIAR (2)

PRAISE FROM OTHERS: Mohamed Naeem, a UNICEF officer who met Ismail at the trauma workshop, says, “Zaeema was a simple girl who did some simple things and achieved something extraordinary. She held her family together.”

Ismail found inner strength to help her family. Have there been situations in which you have found inner strength to help others?

and green. You can smell spices growing on plantations in this area. You can also see wild elephants moving through the plateau's dense rain forests.

Islands of South Asia

South Asia includes two island nations: Sri Lanka and Maldives. Sri Lanka, the larger of the two nations, lies off the southeast coast of India. Shaped like a teardrop, the country has a small pocket of highlands in the interior. This area is made up of ridges, valleys, and steep cliffs that offer spectacular scenery. Coastal lowlands encircle these highlands and cover more than 80 percent of the island.

Maldives, which lies off India's western coast, is one of the smallest countries in the world. Maldives includes about 1,200 islands, though people live on only about 200 of them. Many of the islands are **atolls,** circular-shaped islands made of coral. Coral is a rocklike material formed from the skeletons of tiny sea creatures. As coral deposits build up, many of them **eventually** become covered by soil and sand to make islands. Atolls have a shallow body of water in the center called a **lagoon.** The outer ring of the island protects the lagoon from the sea.

Natural Resources

South Asia is not a land of plenty. Even good farmland is scarce outside of India, Bangladesh, and Pakistan. Although most South Asians grow crops or tend livestock, plots of land are small, and many farmers barely earn a living.

India is luckier than its neighbors. As South Asia's largest country, it not only has productive land, but it also has most of the region's mineral resources. These include iron ore, manganese, and chromite, which are all used in making steel. Pakistan, too, has some valuable minerals, especially limestone, which is an ingredient for making cement.

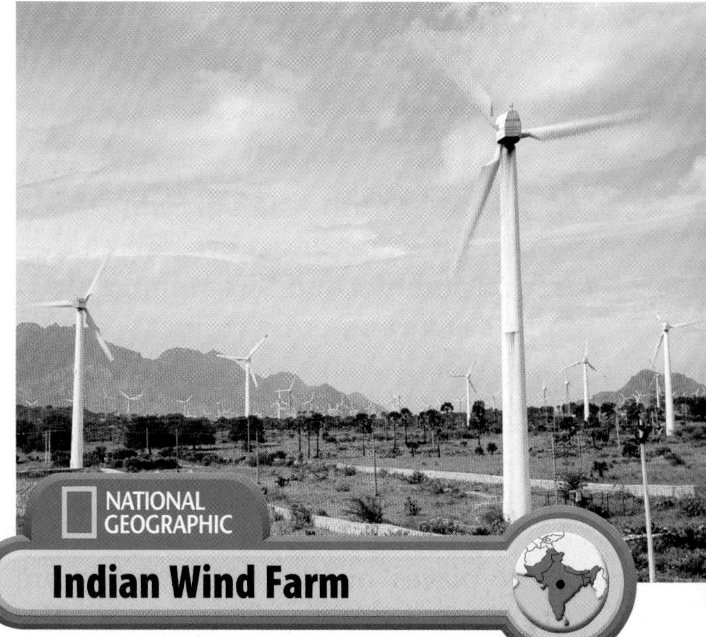

Indian Wind Farm

India is a world leader in generating power from wind energy. Leaders plan to use the technology to bring electricity to 25,000 rural villages. ***Human-Environment Interaction*** **What other energy resources are found in South Asia?**

To meet their energy needs, the countries of South Asia rely heavily on imported oil. Pakistan and Bangladesh also have reserves of natural gas, while India is rich in coal. Bangladesh has coal deposits too, but they lie so deep in the ground that mining them is difficult.

Another source of energy for South Asia is water. The mountainous landscape creates swift-flowing rivers that can be used to generate electricity. Bangladesh already has one hydroelectric plant, and India has several. Nepal and Bhutan, too, are pursuing hydroelectric projects. These plants provide power and also control flooding, which is a serious problem for South Asians.

Reading Check **Explaining** Why is the Ganges Plain important in India?

Environmental Concerns

Main Idea **South Asia's growing population is creating more demand for food and fuel and threatening the region's environment.**

Geography and You Have you ever been on a street or in a stadium crowded with people? What kind of an experience was that? Read to find out how the masses of people in South Asia affect the environment.

Few places on the planet are more densely settled than South Asia. The region is home to more than 20 percent of the world's people, but they live on only 3 percent of the world's land. To add to the pressure, South Asia's population is increasing.

This growth seriously affects the environment. For one thing, greater numbers of people mean greater demand for animal products. Farmers then raise more livestock. This leads to overgrazing, which causes grasslands to dry up. It is not just land that is at risk, though. South Asia's growing population also threatens the water, the forests, and the air.

Water

Because South Asia has such a huge **concentration** of people, supplies of freshwater are low. The climate, which brings long dry seasons to much of the region, contributes to water shortages. In addition, farmers, the largest consumers of water, often use wasteful irrigation methods. Much water is also wasted in cities because of old, leaky distribution pipes.

To meet the demand for water, South Asian countries are tapping underground aquifers. In urban areas, however, as fresh water is being pumped out, saltwater enters the aquifers. The higher salt content makes the water less useful. This problem is particularly troublesome in the cities of Dhaka in Bangladesh and Karachi in Pakistan.

Water pollution is increasing, too. The Ganges River is among the most polluted waterways in the world. The water it brings to urban areas is dirtied by sewage, runoff from factories, and waste products. Rural water supplies are often no cleaner. Even rural Nepal has seriously polluted rivers. Many farmers apply fertilizers to fields to increase crop yields. Runoff from fertilizers then makes the drinking water unsafe.

Deforestation

Only a small part of South Asia is forested. Most of the land was cleared centuries ago. However, many of the forests that

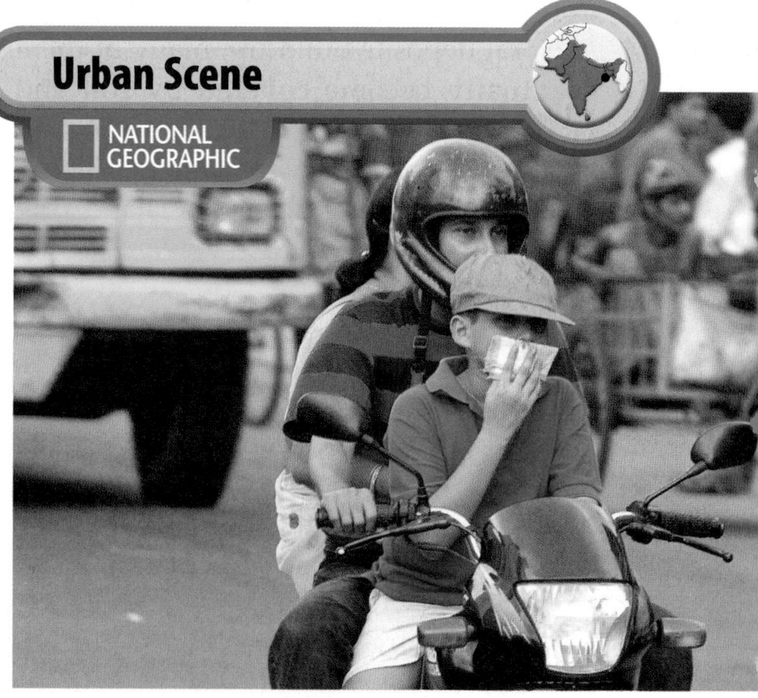

▲ The city of Kolkata (Calcutta) suffers from some of the worst air pollution in India. ***Human-Environment Interaction*** **What other environmental problems threaten South Asia?**

remain are now being cut down to provide building materials as well as wood for fuel. Rural people throughout South Asia rely on wood for heating their homes and for cooking. For example, almost 70 percent of the energy used in Nepal comes from burning wood.

When trees are cut down, new seeds are rarely planted. People need the land for crops instead. However, the clearing of trees has led to erosion and flooding. Nepal and India have now introduced programs at the local level to limit forest loss. Villages are given control of managing nearby woodlands. As encouragement to restore cut areas, they also are allowed to receive all the income from the sale of wood products.

Air Pollution

Air pollution is another challenge that affects parts of South Asia. The number of cars in the region's cities has risen rapidly in recent decades. More automobiles mean the release of more exhaust fumes that make the air in urban areas dangerous to breathe.

Air pollution is affecting rural areas as well. Many villagers cook and heat their homes by burning wood, kerosene, charcoal, or animal dung. These substances release smoke and chemicals that are harmful in closed spaces. As a result, many people develop breathing problems, and some die of lung diseases.

Air pollution from South Asia (and from Southeast Asia as well) is so severe that a brown cloud of chemicals, ash, and dust has formed over the Indian Ocean. The cloud decreases the sunlight reaching the Earth's surface there by 10 percent. Scientists worry that this clouding may be changing the region's climate and disrupting rain patterns. That, in turn, may cut crop yields and threaten people's livelihoods.

Reading Check **Analyzing** Why are South Asia's freshwater supplies low?

Social Studies ONLINE
Study Central™ To review this section, go to glencoe.com.

Vocabulary

1. **Describe** the physical geography of South Asia in a paragraph in which you use each of the following terms: *subcontinent, delta, atoll,* and *lagoon.*

Main Ideas

2. **Illustrating** Use a diagram like the one below to explain how the northern mountains of South Asia were formed.

3. **Explaining** Why is air pollution also affecting rural areas of South Asia?

Critical Thinking

4. **Identifying Central Issues** What effect does South Asia's growing population have on the environment?
5. **BIG Idea** Compare South Asia's Deccan Plateau and Karnataka Plateau.
6. **Challenge** Do you believe South Asian countries are dealing effectively with deforestation? Explain.

Writing About Geography

7. **Using Your FOLDABLES** Use your Foldable to create a map of South Asia that describes the region's physical geography for tourists.

NATIONAL GEOGRAPHIC **Geography & History**

Sacred Waters

What happens when a place people see as holy is being spoiled by pollution?

The Sacred River Millions of India's Hindus hold the Ganges River as the most sacred, or holy, of all waters. Called "Mother Ganges," the river is believed to have the power to wash away sins. Thousands of people bathe in the river each morning. Hindus also place the remains of deceased family members in the Ganges. The remains are either ashes after the body has been burned or the body itself. It is believed that the waters of the holy Ganges will ease the person's path into the next life.

The Polluted Ganges Unfortunately, the Ganges has become one of the most polluted rivers in the world. Besides human remains, the remains of dead cattle—animals that Hindus hold as sacred—are placed in the river. Waste from factories and fertilizer runoff from farms also pollute the Ganges.

▼ **Washing clothes in the Ganges River**

▲ **The Ganges River at Rishikesh, India**

The biggest source of pollution, though, is the waste, garbage, and trash from the millions of people who live along the Ganges. The germs in the Ganges pose a serious infection risk to people using the water for drinking and cooking.

Cleaning Up the River In the 1980s, India built new sewage treatment plants to clean up the river. These did not work well, partly due to India's wet monsoon season. In addition, government officials have found it difficult to enforce laws against industrial waste pollution. This is because of the small industrial workshops in the Ganges area. Today groups are trying to make cleanup efforts more citizen-based, encouraging Indians to protect the waters of "Mother Ganges."

Think About It

1. **Place** Why is the Ganges River important to India's Hindus?
2. **Human-Environment Interaction** Why is the Ganges so polluted?

Guide to Reading

BIG Idea

The physical environment affects people.

Content Vocabulary

- monsoon *(p. 618)*
- cyclone *(p. 619)*

Academic Vocabulary

- distinct *(p. 618)*
- vary *(p. 619)*
- contrast *(p. 619)*
- survive *(p. 620)*

Reading Strategy

Outlining Use an outline like the one below to summarize the monsoon cycle.

I. First Main Heading
  A. Key Fact 1
  B. Key Fact 2
II. Second Main Heading
  A. Key Fact 1
  B. Key Fact 2

Climate Regions

Section Audio

Spotlight Video

Picture This This long-haired, short-legged, oxlike mammal of the Himalaya is a yak. The Sherpas of Nepal call the male of the species "yak" and the females "nak." The yak is a valued animal in this part of the world. In a region where climate limits plant growth, the yaks can eat the low-quality scrub found in the area. The yak produces high-fat milk and is a source of lean meat. Its wool is used to make clothing and tents. Yaks are also a reliable source of transportation in this rocky, mountainous region. They are as stable on their feet as mountain goats. Read this section to learn more about the climates in South Asia and the effects they have on the animals and people who live there.

▼ Traveling through a mountain pass in the Himalaya

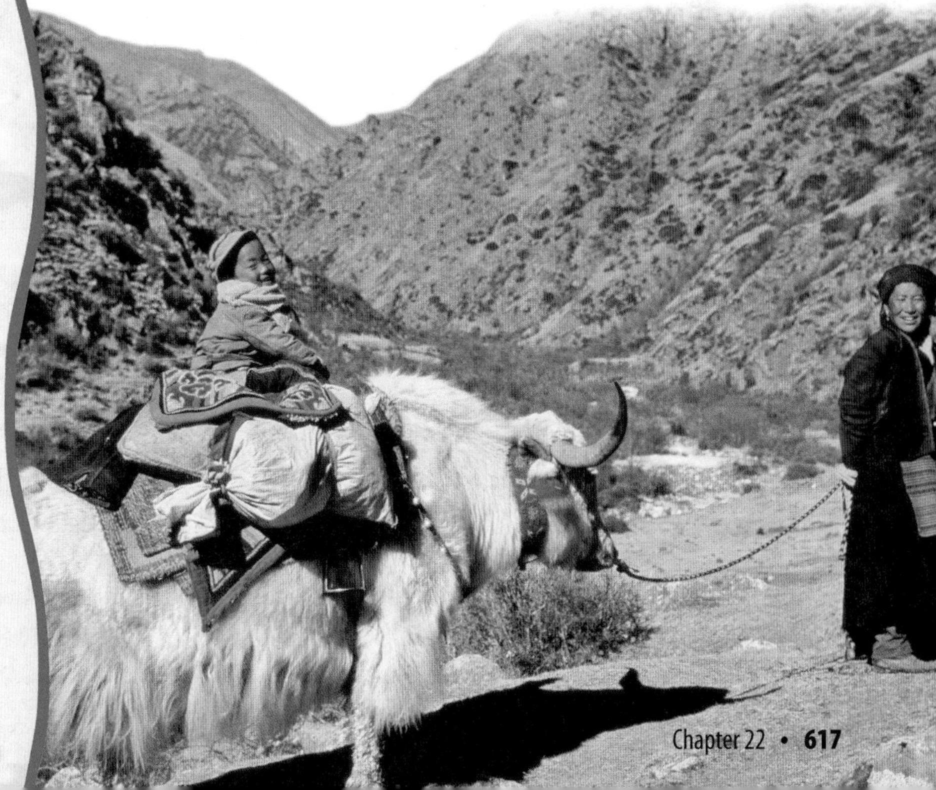

Monsoons

Main Idea Seasonal dry and wet winds are the major factor shaping South Asia's climate.

Geography and You How does the environment where you live change from season to season? The pattern in your area is probably quite different from that in South Asia, as you will read in this section.

Much of South Asia experiences three **distinct,** or unique, seasons—hot, wet, and cool. These three seasons depend on seasonal winds called **monsoons.** **Figure 1** shows the yearly pattern of the monsoons.

During the cool season, from October to late February, dry monsoon winds blow from the north and northeast. The hot season follows from late February to June. During this period, warm temperatures heat the air, which rises and causes a change in wind direction. Moist ocean air then moves in from the south and southeast, bringing monsoon rains. The wet season lasts from June or July through September.

The monsoon rains are heaviest in eastern South Asia. When the rains sweep over the Ganges-Brahmaputra delta, the Himalaya block them from moving north. Instead, the rains move west to the Ganges Plain, bringing water needed for farming.

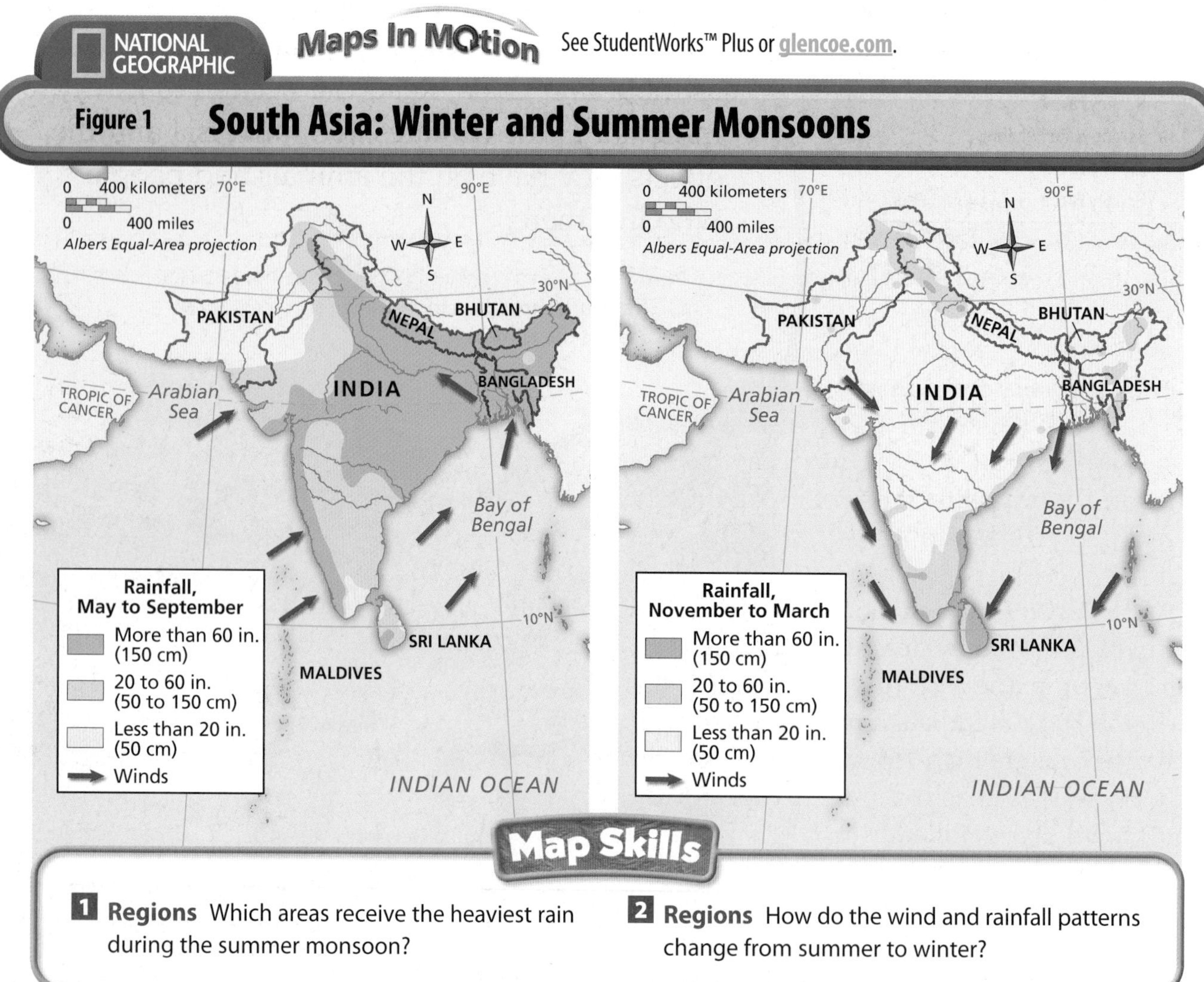

Figure 1 South Asia: Winter and Summer Monsoons

Map Skills

1. **Regions** Which areas receive the heaviest rain during the summer monsoon?
2. **Regions** How do the wind and rainfall patterns change from summer to winter?

Natural Disasters

The high temperatures of the hot season and the rains of the wet season have good and bad effects on South Asians. As long as water is plentiful, high temperatures allow farmers to grow crops, especially the rice that is a huge part of the people's diet. The extreme heat, however, causes water to evaporate quickly and dries out the soil.

The monsoon winds likewise have mixed effects. The rains they shower on Bangladesh and the Ganges Plain help crops there grow well. However, areas outside the monsoon's path—such as the Deccan Plateau and western Pakistan—may receive little or no yearly rainfall. If there is no rain, or not enough, some areas become scorched, or burnt, by drought.

Too much rain can also bring trouble. In the low-lying delta of Bangladesh, monsoons often cause devastating floods that drown the flat land. Water also runs down from deforested slopes upriver in northern India. Together, these violent flows of water kill thousands of people as well as livestock. They also ruin crops, destroy homes, and wipe out roads.

Another kind of weather disaster often strikes South Asia. A **cyclone** is an intense tropical storm with high winds and heavy rains. Cyclones are similar to hurricanes in the Atlantic Ocean and typhoons in the north Pacific Ocean. In South Asia, cyclones can be followed by deadly tidal waves that surge from the Bay of Bengal. In 1999 a cyclone struck India's northeast coast with winds of more than 160 miles (257 km) per hour. Waves reached over 20 feet (6 m) high. The storm killed nearly 10,000 people and left about 15 million people homeless.

Reading Check **Summarizing Information**
When do the wet and dry monsoons occur?

Climate Zones

Main Idea **South Asia's climate zones are affected by location, landforms, and monsoon winds.**

Geography and You Do you think it ever warms up at the top of the world's highest mountain? Read to find out about the climate on Mount Everest and in the rest of South Asia.

In many parts of South Asia, the climate is tropical and the plant life abundant. In some areas, however, climates **vary.** They range from cold in the Himalaya to hot in the deserts around the Indus River.

Tropical Areas

Much of south central India has a tropical dry climate. The region's grasslands and deciduous forests grow green in the short wet season and turn brown in the long dry season. Bangladesh and southern Sri Lanka, by **contrast,** have a tropical wet climate with warm temperatures year-round.

Monsoon Season, India

NATIONAL GEOGRAPHIC

▲ Heavy monsoon rains can cause flooding and landslides and leave thousands of people homeless. ***Regions*** **How does the monsoon climate help farmers in the region?**

South Asia's tropical regions receive the heaviest rainfalls from the wet monsoons. Most of Bangladesh gets 100 inches (254 cm) of rain per year. The city of Cherrapunji in northeastern India receives an annual rainfall averaging up to 450 inches (1,143 cm), making it one of the wettest spots on Earth.

Dry and Temperate Climates

The wet monsoons, of course, do not reach all of South Asia. As a result, some areas have dry climates. Along the lower Indus River, the land is dry and windswept. Farmers must use irrigation to grow wheat and other crops.

To the east of the Indus River lie the sand dunes and gravel plains of the Thar Desert. Surrounding this desert, except on the coast, is a steppe. Few trees grow on this partly dry grassland. Another steppe area crosses the Deccan Plateau, which sits between the Eastern and Western Ghats. The Western Ghats block rainfall in the area, making the central Deccan dry.

The climate becomes humid and subtropical as you travel north to the Ganges Plain. This area has high temperatures, with muggy summers but fairly dry winters.

Highlands

Highland climates are found along South Asia's northern edge, where towering mountains rise. Above 16,000 feet (4,877 m), temperatures are always below freezing. As a result, snow never disappears, and little vegetation can **survive.** Farther down the mountain slopes, the climate turns more temperate. In Nepal's Kathmandu Valley, January temperatures average a mild 50°F (10°C). The average July temperature is a pleasant 78°F (26°C).

Reading Check **Identifying** What areas of South Asia receive the most rainfall?

Social Studies ONLINE
Study Central™ To review this section, go to glencoe.com.

Vocabulary

1. **Explain** the roles of *monsoons* and *cyclones* in South Asia's climate.

Main Ideas

2. **Explaining** How do the high temperatures of the hot season both benefit and harm the people of South Asia?
3. **Categorizing** Use a main idea chart like the one below to identify four major climate zones of South Asia and some characteristics of each.

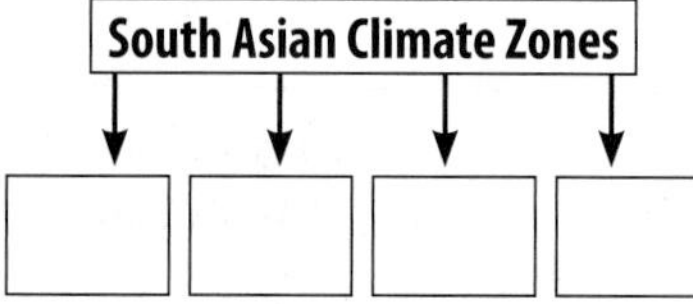

Critical Thinking

4. **Determining Cause and Effect** Why is the central area of the Deccan Plateau dry?
5. **BIG Idea** How do monsoons affect the lives of South Asians?
6. **Challenge** Is drought more likely to occur in Pakistan or Bangladesh? Why?

Writing About Geography

7. **Expository Writing** Write a paragraph identifying the natural disasters that can affect South Asia, the areas where they strike, and their characteristics.

Chapter 22 Visual Summary

Khyber Pass, Pakistan

Mountains and Plains

- Three of the world's largest mountain chains stretch across northern South Asia.
- The Indus, Ganges, and Brahmaputra Rivers bring water to South Asia's heavily populated plains.
- Highlands and lowlands dominate southern India.

Islands

- Sri Lanka has a highland interior and surrounding coastal lowlands.
- Maldives includes islands that are coral atolls.

Maldives atoll

Natural Resources

- India has most of South Asia's natural resources.
- South Asian countries need to import energy resources, such as oil and natural gas.
- Hydroelectric power is a promising energy source for South Asia.

Bengal tiger, Indian rain forest

Environment

- South Asia's large population has put pressure on limited water resources.
- South Asian countries are trying to protect their few remaining forests.
- Exhaust from more vehicles and burning wood for fuel have increased air pollution.

Tea plantation, Sri Lanka

Climate Patterns

- Monsoons, or seasonal winds, dominate South Asia's climate.
- Farmers depend on the monsoons to grow crops.
- Cyclones, or powerful storms, can cause destruction to coastal lowlands.

Climate Zones

- Much of South Asia is tropical, although the region also has temperate, desert, and highland climates.
- South Asia's tropical areas receive heavy rainfall.

Study anywhere, anytime! Download quizzes and flash cards to your PDA from glencoe.com.

CHAPTER 22 ASSESSMENT

STANDARDIZED TEST PRACTICE

TEST-TAKING TIP

After you have finished, review your test to make sure that you have answered all questions, followed directions carefully for each set of questions, and avoided simple mistakes.

Reviewing Vocabulary

Directions: Choose the word(s) that best completes the sentence.

1. Soil and sediment deposited at the mouth of a river forms a ______.

A lagoon
B delta
C peninsula
D silt

2. Circular-shaped islands made of coral are called ______.

A deltas
B lagoons
C atolls
D peninsulas

3. Much of South Asia experiences three distinct, or unique, seasons (hot, wet, and cool) that are caused by seasonal winds called ______.

A monsoons
B cyclones
C lagoons
D deltas

4. South Asia occasionally suffers from ______, or damaging storms with high winds and heavy rains.

A monsoons
B deltas
C lagoons
D cyclones

Reviewing Main Ideas

Directions: Choose the best answer for each question.

Section 1 *(pp. 610–615)*

5. The Himalaya are growing slightly taller each year because ______ are still occurring.

A earthquakes
B sedimentations
C tectonic plate movements
D climate changes

6. Greater numbers of people in South Asia mean greater demand for animal products, which leads farmers to raise more livestock. This can sometimes lead to overgrazing, which results in ______.

A higher average cholesterol
B more overweight people
C more large, corporate-owned farms
D dried up grasslands

Section 2 *(pp. 617–620)*

7. The heaviest monsoon rains in the region fall ______.

A over the ocean
B in eastern South Asia
C north of the Himalaya
D during the cool season

8. Highland climates are found in South Asia's ______, where towering mountains rise.

A southern region
B northern region
C central region
D eastern region

GO ON

Critical Thinking

Directions: Base your answers to questions 9 and 10 on the graph below. Choose the best answer for each question.

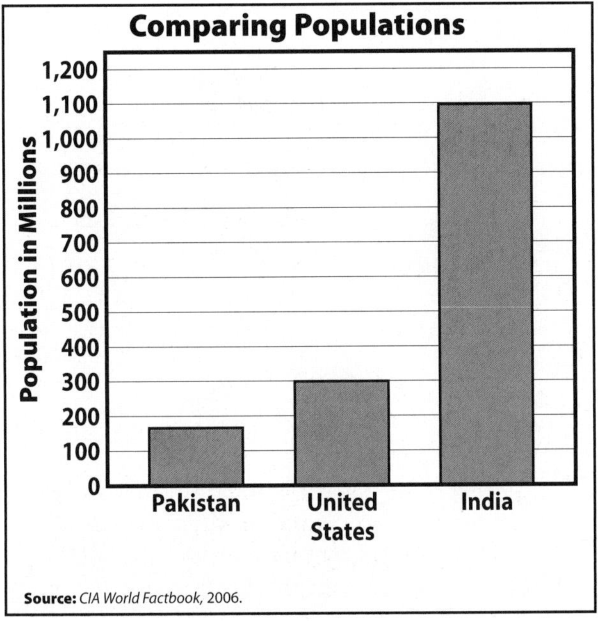

9. About how many more people live in India than in the United States?

A nearly 2 times as many
B nearly 4 times as many
C nearly 8 times as many
D nearly 10 times as many

10. Which of the following statements is true based on the bar graph information?

A The United States has a larger population than Pakistan.
B The United States and Pakistan have similar population sizes.
C India's population is about half the amount of the United States.
D India has the largest population in the world.

Document-Based Questions

Directions: Analyze the document and answer the short-answer questions that follow.

In 1999 the television program NOVA sent a team of experts to Mount Everest to solve a mystery. Liesl Clark, filmmaker and correspondent, posted online dispatches from base camp.

> *In the upcoming weeks, . . . we will attempt . . . to piece together . . . Mallory and Irvine's last day on Mount Everest. By determining, for example, the flow rate of Mallory's oxygen bottle, we can figure out when he may have run out of his last oxygen and discarded his second empty bottle. . . . By analyzing the photographs taken of Mallory's remains [we] may be able to reconstruct Mallory's final moments and the exact cause of his death. . . . Did the altimeter give accurate readings and is there a way to determine its highest rendering? Is there a small particle of rope left on the blade of the pocket knife to indicate that Mallory cut himself free from Irvine? After 75 years, . . . it is possible that a clue . . . may reveal what our heroes could never tell us—whether they were the first to reach the highest point on Earth.*
>
> — Liesl Clark, "Unanswered Questions," Nova Online Adventures

11. Based on the document, what are the writer and her team investigating?

12. What sort of clues will the team be looking at to help them solve the mystery?

Extended Response

13. Write a letter to a United Nations official discussing and offering possible solutions to the problem of malnutrition in India.

For additional test practice, use Self-Check Quizzes—Chapter 22 at glencoe.com.

Need Extra Help?

If you missed question...	1	2	3	4	5	6	7	8	9	10	11	12	13
Go to page...	612	613	618	619	611	614	618	620	614	614	611	611	614

History and Cultures of South Asia

Essential Question

Regions South Asia is the birthplace of several world religions, such as Hinduism, Buddhism, and Sikhism. Islam has a large following in the region, and there are also followers of Jainism, Christianity, and other faiths. How do religious beliefs and practices influence people's lives?

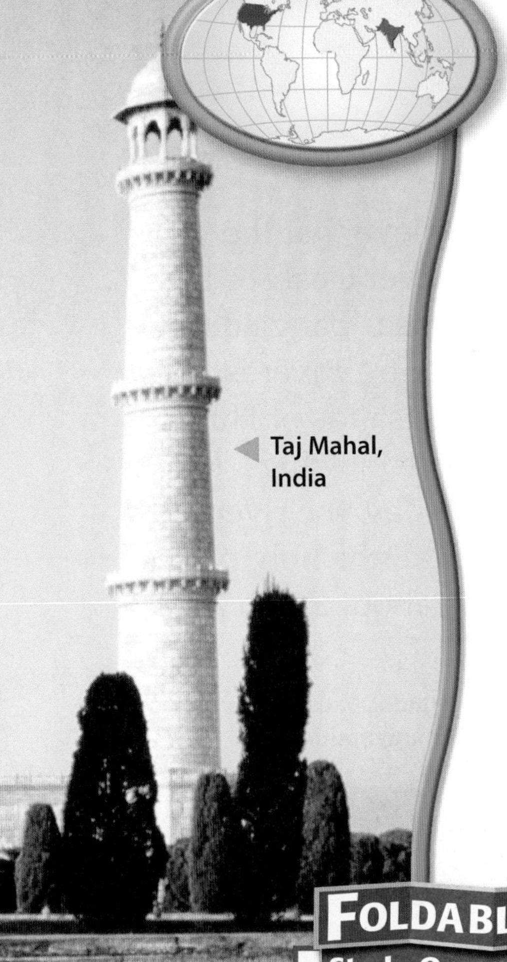

Taj Mahal, India

Section 1: History and Governments

BIG IDEA Geography is used to interpret the past, understand the present, and plan for the future. Civilizations and empires rose and fell in South Asia, but the religions that developed centuries ago are still influential. South Asia's countries were also shaped by a long period of British rule from the late 1700s until the mid-1900s.

Section 2: Cultures and Lifestyles

BIG IDEA The characteristics and movements of people impact physical and human systems. The people of South Asia belong to different ethnic groups, speak a variety of languages, and practice a number of religions. The rapid growth of this varied population is straining the region's resources and contributing to widespread poverty.

FOLDABLES™ Study Organizer

Organizing Information Make this Foldable to help you organize information about the history, peoples, cultures, and daily life of South Asia.

Step 1 Place three sheets of paper on top of one another about 1 inch apart.

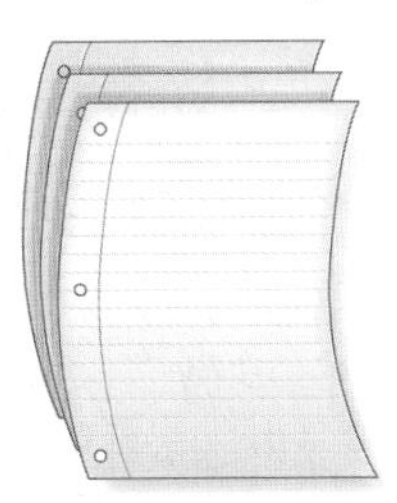

Step 2 Fold the papers to form six equal tabs.

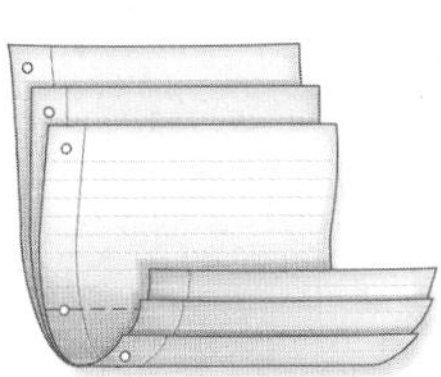

Step 3 Staple the sheets, and label each tab as shown.

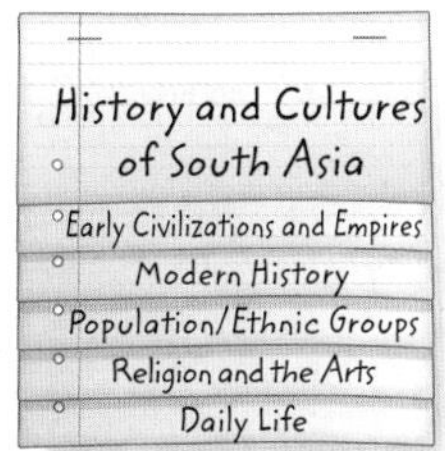

Reading and Writing Use the notes in your Foldable to write a short essay that describes the development of the countries and peoples of South Asia.

Visit glencoe.com and enter ***QuickPass***™ code EOW3109c23 for Chapter 23 resources.

Guide to Reading

BIG Idea

Geography is used to interpret the past, understand the present, and plan for the future.

Content Vocabulary

- varna *(p. 627)*
- caste *(p. 628)*
- reincarnation *(p. 628)*
- dharma *(p. 628)*
- karma *(p. 628)*
- nirvana *(p. 629)*
- civil disobedience *(p. 631)*
- boycott *(p. 631)*

Academic Vocabulary

- status *(p. 627)*
- consequence *(p. 628)*
- capable *(p. 629)*

Reading Strategy

Making a Time Line Use a diagram like the one below to list the key events and dates in the history of South Asia.

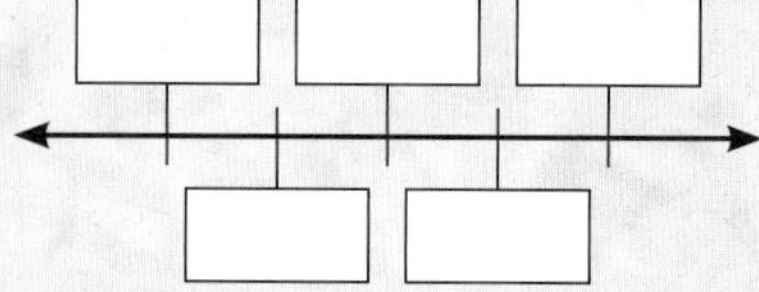

History and Governments

 Section Audio  Spotlight Video

Picture This Hindus believe that the water of the Brahmaputra River cleanses the body and the soul. Located near Bangladesh's capital, Dhaka, the Brahmaputra River is where, on a specific day, thousands of Hindu believers take baths to receive blessings. People believe that on this holy day, the river contains all the blessings of all the holy places in the world. To learn more about South Asia, read Section 1.

▼ Hindus bathe in the Brahmaputra River

Early History

Main Idea **Thousands of years ago, people in South Asia developed a complex social structure, two of the world's major religions, and powerful empires.**

Geography and You What kind of work do you want to do when you get older? In ancient India, people prepared for certain jobs from birth. Read to find out how the social order and religious beliefs governed South Asians' lives.

A great civilization with advanced building skills developed in South Asia. As in other parts of the world, this early civilization grew in a river valley.

Indus River Valley

By 2500 B.C., people in the Indus River valley had built what may have been South Asia's first cities: Harappa (huh·RA·puh) and Mohenjo Daro (moh·HEHN·joh DAHR·oh), which are shown in **Figure 1** on the next page. These cities, with brick buildings, were well planned. They had carefully laid-out streets, ceremonial gateways, and buildings to store grain. The cities also had plumbing, sewers, and other technology that would not be matched again for centuries.

As the population grew, farming, small industries, and trade brought wealth to the Indus Valley. The people made copper and bronze tools, clay pottery, and cotton cloth. They also developed a writing system.

After centuries of prosperity, the Indus Valley civilization declined between 1700 B.C. and 1500 B.C. Historians believe that earthquakes and floods may have damaged the cities. Also, the Indus River may have changed its course. These events were evidently severe enough to cause the fall of the Indus Valley civilization.

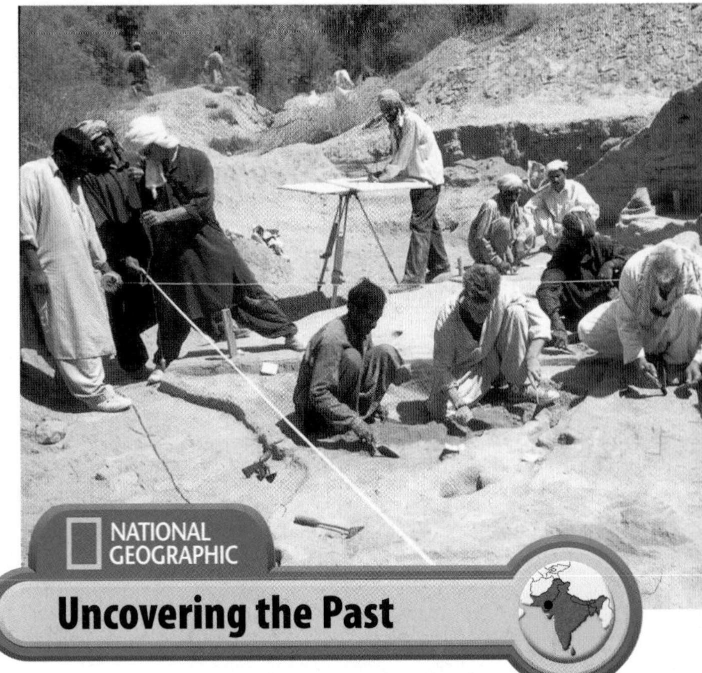

Archaeologists carefully search for artifacts as they explore ruins of the ancient civilization of Harappa. ***Place*** **How were Mohenjo Daro and Harappa unique?**

Aryans

About 1500 B.C., nomadic herders known as Aryans were settling in parts of northern South Asia. The Aryans developed a spoken language called Sanskrit (SAN·SKRIHT). They passed on hymns and religious teachings by word of mouth. When Sanskrit later became a written language, these traditions were recorded in sacred, or holy, texts called the Vedas.

The Vedas show that the Aryans were organized into four ***varnas,*** or broad social groups. Priests had the highest **status.** Warriors came next, followed by farmers. At the bottom were unskilled laborers and servants. At first, people of different groups could marry each other and change jobs.

Student Web Activity Visit glencoe.com and complete the Chapter 23 Web Activity about the Indus Valley civilization.

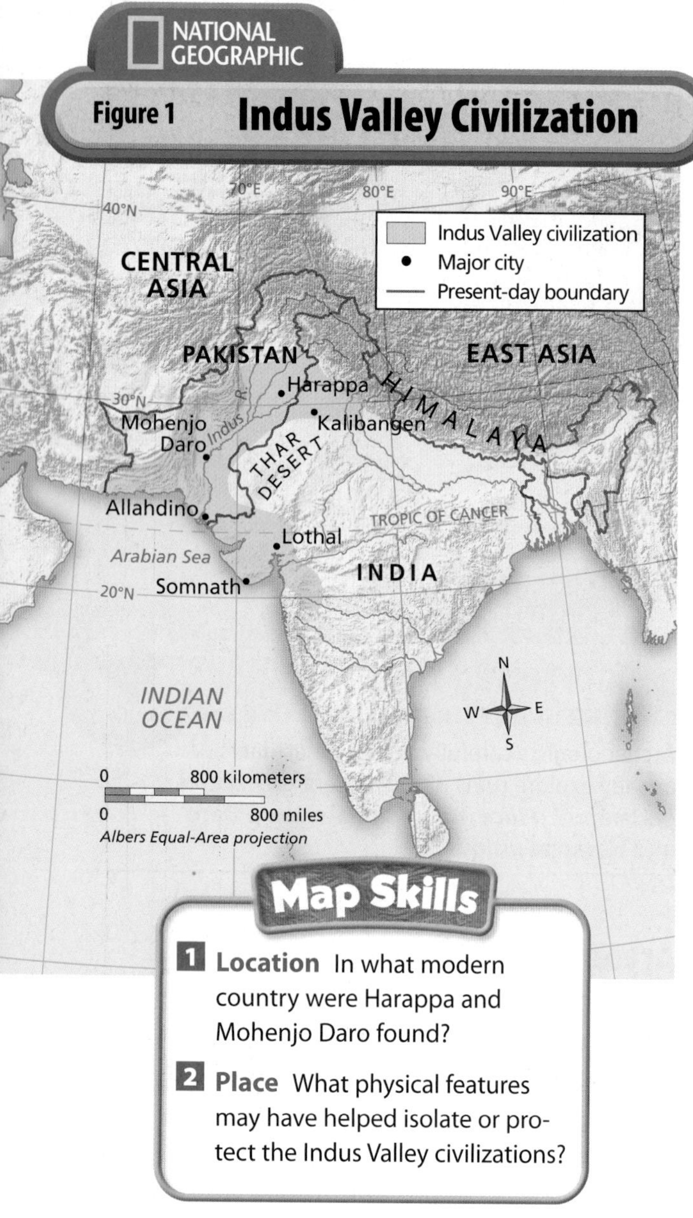

Over time, a caste system arose. A **caste** is a social group that someone is born into and cannot change. In South Asia, no one uses the word *caste,* which is the word Europeans later used to describe the region's social groups. South Asians call these groups *jati.* Thousands of *jati* still exist, especially in India, but as people adopt modern ways, *jati* are becoming less important.

Two Religions

The religions of Hinduism and Buddhism both developed in South Asia. They have had a lasting influence in the region. Hinduism is one of the world's oldest religions and the third largest. It developed gradually as the beliefs of the ancient Aryans mixed with the beliefs of other peoples in the region. This blending might explain why Hindus worship thousands of deities. They tend to think of all deities, however, as different parts of one eternal spirit. This eternal spirit is called Brahman (BRAH·muhn).

Hindus believe that every living being has a soul that wants to be reunited with Brahman. To achieve this reunion, a soul must repeatedly undergo **reincarnation** (REE·ihn·kahr·NAY·shuhn)—being born into a new body after dying. Thus Hindus believe that a soul passes through many lives, becoming purer each time, before reaching Brahman.

To ensure that their next lives are better, Hindus believe they must perform their duty, or **dharma** (DUHR·muh). Each caste has its own dharma. For example, a farmer has different duties than a priest, and a woman has different duties than a man. The **consequences,** or effects, of how a person lives are known as **karma** (KAHR·muh). Hindus believe that if they do their duty, they will have good karma. This will move them closer to Brahman in the next life.

In the 500s B.C., Buddhism arose in South Asia. It was founded by a young prince named Siddhartha Gautama (sih·DAHR·tuh GOW·tuh·muh). Born in a small kingdom near the Himalaya, Gautama gave up wealth and family in search of truth. After many years, he found what he was seeking. He became known as the Buddha, or "Enlightened One."

The Buddha taught that people suffer because they are too attached to material things, which are not lasting. He believed that people can be released from these attachments by following the Eightfold Path.

The eight steps include thinking clearly, working hard, and showing deep concern for all living things. By following the eight steps, people can escape suffering and reach **nirvana** (nihr·VAH·nuh), a state of endless peace and joy.

Buddhism won many followers among people who were poor or had no social standing. The religion eventually spread throughout South Asia and beyond to Southeast Asia and East Asia. In India, however, Buddhist ideas were absorbed into Hinduism, which remained the major religion.

South Asian Empires

In addition to new religions, powerful empires also arose in early South Asia. In the 300s B.C., a family called the Maurya (MAUR·yuh) founded the Mauryan Empire. The most famous Mauryan ruler, Aśoka (uh·SOH·kuh), brought much of the subcontinent under his control. About 260 B.C., Aśoka dedicated his life to peace and became a Buddhist. Aśoka sent Buddhist missionaries throughout Asia, but he also allowed his people to practice other religions. Trade and culture thrived under his **capable** rule. After Aśoka died, however, invasions led to the empire's fall.

About A.D. 320, a ruler named Chandragupta I (CHUHN·druh·GUP·tuh) set up the Gupta Empire in northern India. Under the empire's Hindu rulers, trade increased and ideas were exchanged with other parts of the world. As a result, science, mathematics, medicine, and the arts thrived. South Asian mathematicians developed the numerals 1 to 9 that we still use today. These symbols were later adopted by Muslim Arab traders, who brought them to Europe.

During the early 1500s, Muslim warriors, known as the Moguls (MOH·guhlz), who came from the mountains north of India, formed an empire in South Asia. Akbar (AK·buhr), the greatest Mogul ruler, added new lands to the empire, lowered taxes, and supported the arts. He brought peace to his empire by treating all of his people fairly. The majority of Hindus were allowed to worship freely and to serve in the government.

Later Mogul rulers were less capable. Heavy taxes led many people to rebel. Foreign invaders further weakened the empire. By the early 1700s, the Mogul rule was close to collapse.

Reading Check **Comparing** How were the empires of Aśoka and Akbar similar?

The Emperor Akbar

NATIONAL GEOGRAPHIC

Akbar became Mogul emperor when he was only 13 years old. In this court scene, Akbar passes the crown to his grandson. ***Regions*** **What happened to Akbar's empire after his reign?**

Modern South Asia

Main Idea **After a period of British rule, South Asians set up independent countries during the 1900s.**

Geography and You Think about how you might feel if someone made all your choices and decisions for you. Under British rule, South Asians had no control over their own lands. Read to learn how South Asians eventually won their independence.

During the 1600s, English traders from the East India Company arrived in India. They built a string of trading posts along the coasts, with forts to protect them. In 1707 the English and the Scots joined together to form the United Kingdom. Both peoples—known as the British—created the British Empire. Through trade and military might, the British became the dominant power in South Asia. By the mid-1800s, they had colonized most of the subcontinent.

British Rule

For many years, the task of governing South Asia was left to the British East India Company. As the company introduced European ideas and practices, resentment grew. Many local people felt that the British were trying to change their culture. In 1857 Indian soldiers in the company's army rebelled against their British officers. The revolt spread across northern India. Britain sent more troops and put down the rebellion. Soon afterward, the British government took direct control of India.

Over the years, the British brought many positive changes to the region. They set up a well-run government and founded schools. They built railroads, bridges, and ports. They also introduced the telegraph and a postal service throughout India.

At the same time, British rule caused great hardships for South Asians. Cheap British textiles flooded local markets and destroyed the local textile industry. Taxes fell heavily on the poor. Despite improvements in agriculture, a series of severe famines occurred during the late 1800s. With only minor advances in health care, death rates remained high and life expectancy was low. These developments made Indians even more opposed to British rule.

New Nations

By the early 1900s, independence movements had spread across South Asia. The most popular Indian leader was Mohandas Gandhi (MOH·huhn·DAHS GAHN·dee).

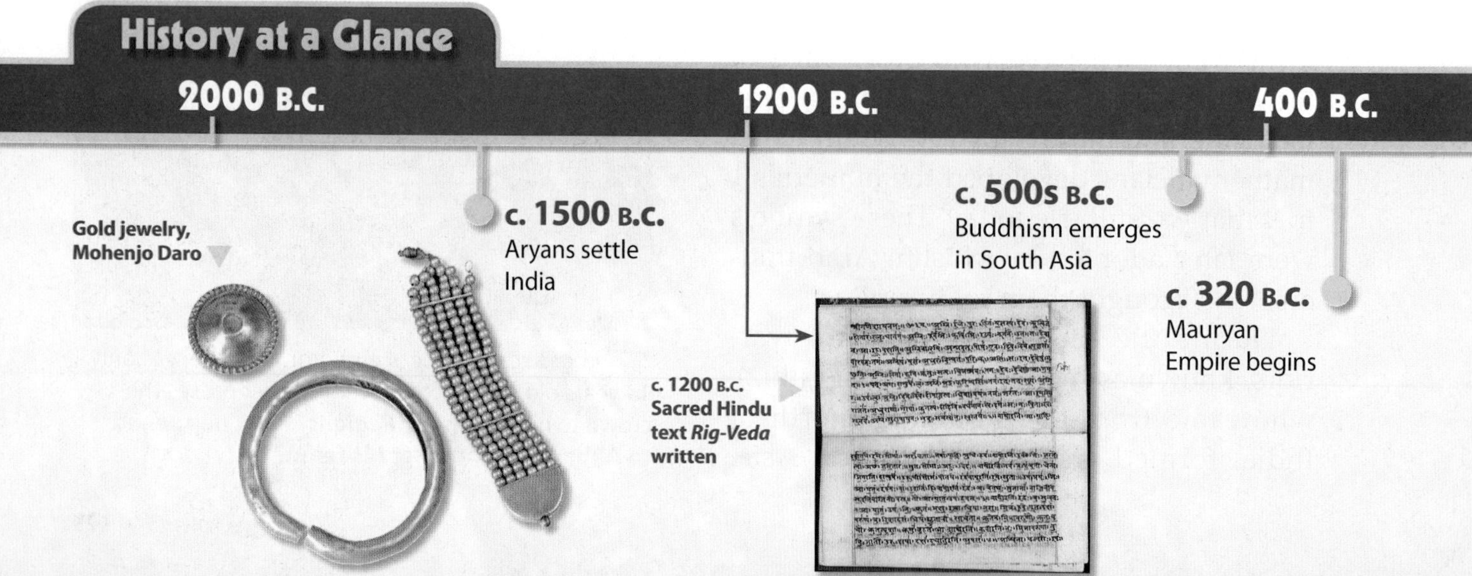

Gandhi opposed violence in all forms. Instead, he protested British rule using nonviolent **civil disobedience**—the refusal to obey unjust laws using peaceful protests. Gandhi and his followers held strikes and **boycotted,** or refused to buy, British goods. Their goal was to bring independence to the subcontinent. Gandhi's movement won widespread support among Hindus. Muslims, however, feared that the much-larger Hindu population might mistreat them in an independent India.

After World War II, Britain realized that it could not keep control of South Asia. Giving the people independence was difficult, though, because of the bitter divisions between Hindus and Muslims. In 1947 the British government divided India into two independent countries. Areas that were mostly Hindu became the country of India. Areas that were mostly Muslim became the country of Pakistan (PA·kih·STAN). Pakistan was made up of two areas geographically separated by India. West Pakistan was northwest of India, and East Pakistan was to the northeast.

Following this division, many Hindus in Pakistan fled to India, while many Muslims in India fled to Pakistan. Fighting erupted and as many as 500,000 people were killed.

Tensions soon surfaced between the two parts of Pakistan too. In 1971 East Pakistan declared its independence. After a brief civil war, it became the new country of Bangladesh. Pakistan now includes only the lands northwest of India.

Meanwhile, other political changes were occurring in South Asia. In 1948 Britain gave independence to the island of Ceylon. This country later took back its ancient name of Sri Lanka. Maldives, a group of islands in the Indian Ocean, won independence from Britain in 1965. Nepal and Bhutan, two countries in the Himalaya area, had always been free of European rule.

Conflict in South Asia

Tensions between India and Pakistan continue today. Religious differences play a part in this conflict. Another dispute involves land, with both countries claiming ownership of the region of Kashmir (KASH·MIHR) in the Himalaya and Karakoram mountains. India and Pakistan have fought several wars over this matter. Terrorists from Kashmir also have carried out attacks in India. Because both nations have nuclear weapons, people worry about the outbreak of a nuclear war.

However, India and Pakistan have held peace talks. They have developed better relations, but they have not been able to settle their dispute over Kashmir.

Political conflicts also trouble other parts of South Asia. In Sri Lanka, a civil war began in 1983 between the government and ethnic Tamil groups who want a separate Tamil nation. In 2002 both sides agreed to peace talks, but the violence has continued.

In Nepal, communist rebels have fought the government since 1996. In 2006 democratic groups forced Nepal's king to give up many political powers. When political leaders agreed to end the monarchy in 2008, the communist rebels joined the government. Meanwhile, ethnic groups in southern Nepal want local self-rule.

Explaining Why was the country of Pakistan created?

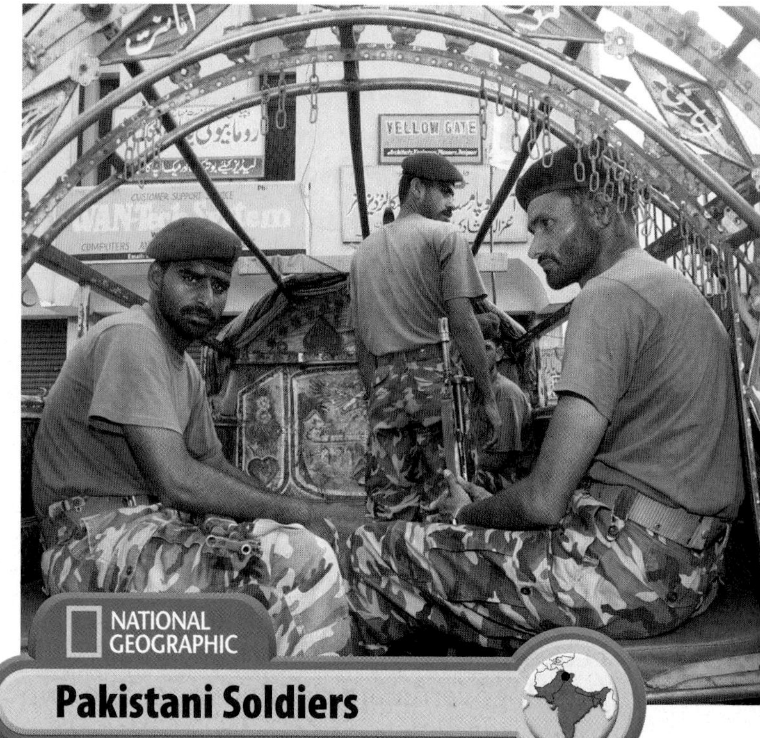

Pakistani Soldiers

A group of soldiers guards a city neighborhood in the part of Kashmir controlled by Pakistan.
Place **Why is the dispute between Pakistan and India over Kashmir a major concern in South Asia?**

Social Studies ONLINE
Study Central™ To review this section, go to glencoe.com.

Vocabulary

1. **Explain** the significance of:
   a. *varna*
   b. caste
   c. reincarnation
   d. dharma
   e. karma
   f. nirvana
   g. civil disobedience
   h. boycott

Main Ideas

2. **Explaining** How did Hinduism develop?
3. **Determining Cause and Effect** Use a diagram like the one below to list key facts about British rule and its effects in South Asia.

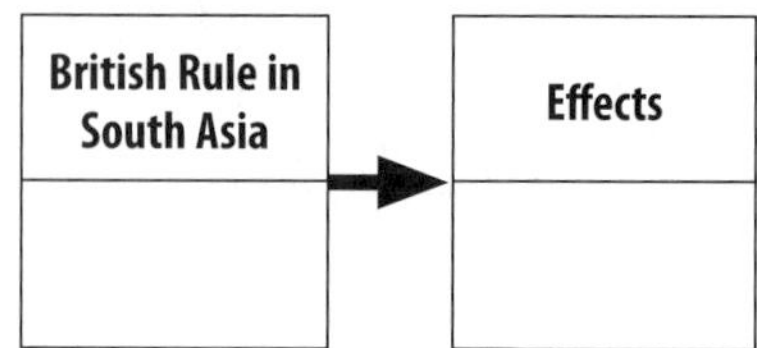

Critical Thinking

4. **BIG Idea** What lasting impact did the Mogul Empire have on the region?
5. **Analyzing Information** How effective was Gandhi's leadership?
6. **Challenge** Do you believe civil disobedience is an effective method for bringing about social change? Refer to Gandhi's methods in India to help explain your answer.

Writing About Geography

7. **Using Your FOLDABLES™** Use your Foldable to make a historical map that shows the various empires and governments that have ruled South Asia.

TIME PERSPECTIVES

EXPLORING WORLD ISSUES

INDIA ENTERS THE MODERN AGE

Economic change and a growing middle class are turning India into a modern nation. How will India's new prosperity affect the U.S.?

An Indian woman listens to music on an MP3 player.

With a population of more than one billion people, India is the world's largest democracy. The country is enormous, not only in size and population, but also in the number of cultures, languages, and religions it embraces. India has 15 official languages and hundreds of dialects.

In recent years, many international companies have hired Indians for jobs that used to be done by workers in the United States. As a result, the size of India's middle class has soared, and its society is experiencing great changes that will also impact U.S. workers and companies for many decades to come.

MEDIOIMAGES/GETTY IMAGES

TIME PERSPECTIVES

Middle-class Indians are shopping like never before.

INDIA'S GROWING MIDDLE CLASS

For Uma Satheesh, finding a good job in India's computer industry had not been easy. A college-educated software programmer, she lived in Bangalore, a city in central India. The only computer work she could find paid poorly and was not interesting. If Satheesh wanted a better job, she knew that she would have to leave India. "Until three years ago, the preference was to go overseas," she said.

In 2003, Satheesh found a job that allowed her to remain in India. Satheesh now works for Wipro, a company that uses outsourced workers. Companies throughout the world are turning to outsourcing to save money. **Outsourcing** is the practice of sending work outside a company's home country in order for the work to be done more cheaply.

By 2015, U.S. companies are expected to outsource 3.3 million jobs to other countries. American businesses of all types, including banks, insurance firms, and computer companies, are attracted to the low-cost labor that is available in foreign countries. For example, the average salary of a software programmer in India is $10,000, compared to $66,100 in the United States. Jobs are also outsourced to India because the country has a well-educated and large English-speaking population. These traits are important, especially for working in call centers. Indians who work in call centers answer questions from a company's customers. When you call a company's phone number for help, you just might be speaking to someone in India.

Wages in India and the U.S.

Hourly wages in India are much lower than in the U.S. for similar work.

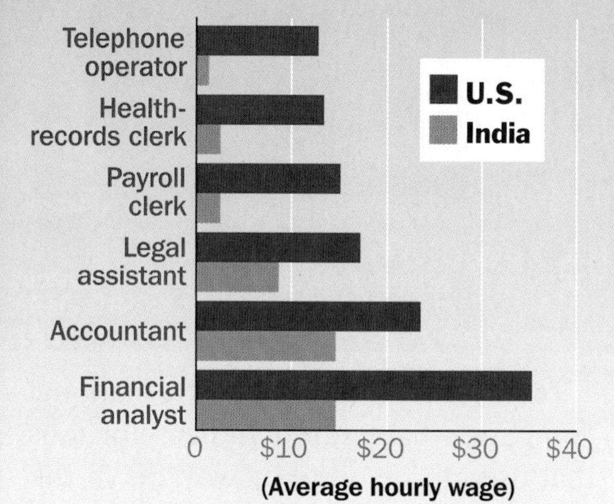

Source: Fisher Center for Real Estate and Urban Economics, University of California, Berkeley.

INTERPRETING GRAPHS

Analyzing Data How much more does an American financial analyst earn, per hour, than an Indian analyst?

A Changing Society

Satheesh and the thirty-eight workers she supervises are part of India's growing **middle class**. Outsourcing has created more jobs in India, and the size of the country's middle class has skyrocketed in recent years. By 2006, nearly 300 million Indians were considered middle class. That number is about equal to the total population of the United States.

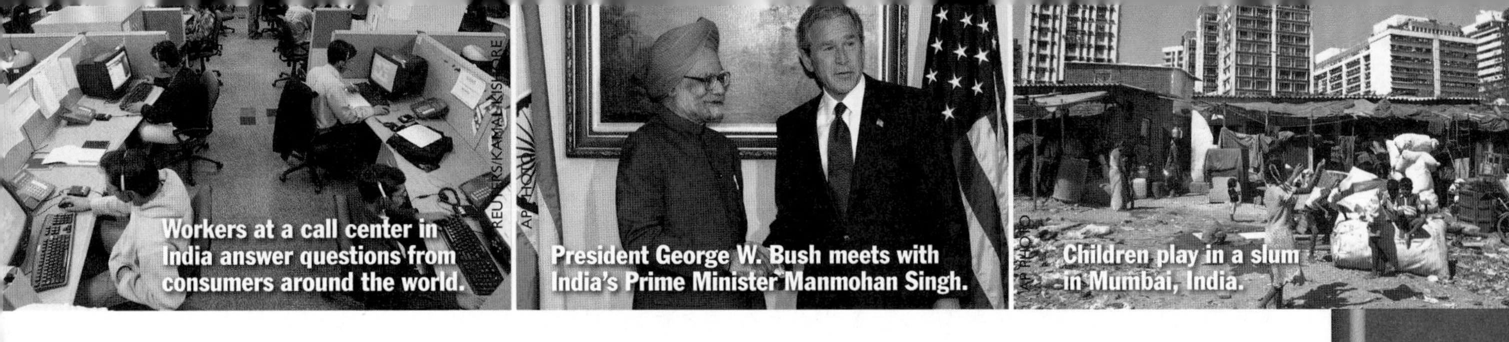
Workers at a call center in India answer questions from consumers around the world.

President George W. Bush meets with India's Prime Minister Manmohan Singh.

Children play in a slum in Mumbai, India.

This growing middle class has had a huge impact on India's economy and society. In the past, young Indian workers were encouraged to save their money. Inspired by new job opportunities, Indians are becoming world-class **consumers**, or buyers. Many Indians are spending their newfound money on products such as cell phones, motorbikes, televisions, and refrigerators. They are also buying cars and houses that were once unaffordable.

"The attitude is to enjoy life and spend money," said an American businessperson working in India. As more Indians spend more money, the country's economy grows. In 2005, it grew 7.6 percent, nearly double that of the U.S. economy.

Indian society, however, still faces major challenges. Although the country's economy has grown, there is still a huge gap between the rich and the poor. As the middle class grows, this gap between the two groups widens.

India has many resources, but they are not evenly distributed. As a result, the country's upper and middle classes have many benefits, but the poor do not always receive proper health care, food, or education. Many children are forced to drop out of school to earn money for their families.

A Loss or an Opportunity?

In the United States, the loss of jobs to India and other countries affects American workers. Many workers who lose their jobs to outsourcing have a hard time finding new ones. When they do, the jobs are often temporary and do not include much needed **benefits**, such as health care and retirement savings plans. Many American workers are forced to relocate and accept large pay cuts. "It's quite an upheaval," said a U.S. worker who lost his job to outsourcing.

Many American companies, on the other hand, see India's booming middle class as an opportunity to do more business and make money. Motorola, a U.S. company that makes communication equipment like cell phones, sells inexpensive phones to young Indians. Sales of its mobile phones tripled in six months. "We're looking at India as a high growing market," said a Motorola executive in India. With middle-class incomes growing, businesses are reacting quickly in order to meet this group's wants and needs.

EXPLORING THE ISSUE

1. **Analyzing Information** How is India's rising middle class contributing to the country's growing economy?

2. **Summarizing** How does outsourcing affect American companies and workers?

NEW MENUS FOR INDIA

As India's middle class expands, many American fast-food chains see an opportunity for growth. It is not, however, "business as usual." In India, fast-food chains such as McDonald's, Pizza Hut, and Domino's Pizza are changing the way they work to fit local cultures and traditions.

When McDonald's began doing business in India in 1996, executives knew that they could not serve the traditional American menu of Big Macs and Quarter Pounders. Cows are highly valued by followers of India's Hindu religion, and followers will not eat beef. Instead, McDonald's developed the McAloo Tikki, a vegetarian fried-potato patty with cheese. The Indian style veggie burgers are a huge attraction. McDonald's restaurants serve an average of 3,000 customers each day.

Ordering lunch at a Pizza Hut in New Delhi or Mumbai is also a uniquely Indian experience. Instead of sausage pizza, Pizza Hut serves vegetarian pies that are covered with cottage cheese and a spicy orange sauce. Some Pizza Huts do not serve meat at all.

KAPOOR BALDEV/SYGMA/CORBIS

A McDonald's worker serves up an Indian style meal.

Social Change Equals Sales

In addition to alternative menus, Domino's Pizza is taking advantage of a new social **trend** in India to bring in business. Traditionally, Indian families sit down to a home-cooked dinner. In recent years, however, high-paying jobs have taken young people to big cities and out of their parents' homes at an earlier age. As a result, more Indian women are working and not staying home to cook family meals. Instead, they are ordering more take-out food. To take advantage of that new trend, Domino's promises to deliver a hot pizza in 30 minutes—just like it does in the United States. The head of Domino's in India calls pizza a "universal food." He hopes the need for quick delivery is, too!

EXPLORING THE ISSUE

1. **Explaining** Many American fast-food companies in India are developing products and services that respect local cultures and traditions. How might that help their businesses?
2. **Summarizing** What social trends in India are helping Domino's improve its business?

Middle-class students use their cell phones.

REVIEW AND ASSESS

UNDERSTANDING THE ISSUE

1 **Making Connections** How is outsourcing affecting India's middle class? Why might it be important for companies that do business in different countries to respect the local cultures?

2 **Writing to Inform** Write a short article explaining how India's middle class is changing the country's economy.

3 **Writing to Persuade** Write a letter to the head of an American company that might outsource jobs to India. Describe whether you think this is a smart business decision. Use facts to support your opinion.

INTERNET RESEARCH ACTIVITIES

4 With your teacher's help, use Internet resources to find information about outsourcing. Read opinions about the benefits and drawbacks of outsourcing. Choose a side and present your opinion to the class.

5 Go to www.mcdonaldsindia.com and read about the company's menu in India. List items that McDonald's serves in the United States and items that it serves in India. Write a short essay explaining why U.S. fast-food chains offer different food choices overseas.

BEYOND THE CLASSROOM

6 **Research another nation,** such as China, that is experiencing rapid economic growth. How is that country's experience different from India's? How is it similar? Write your findings in a report.

7 **Visit your school or local library** to learn more about India's poor. Working in groups, find out what it is like to be poor in India. What is India's government doing to help its poor citizens? Discuss your findings with your classmates.

India's Many Religions

India's large population practices many different religions. The circle graph below shows religions in India and the percentage of the population belonging to each one.

Hindu **80.5%**
Muslim **13.4%**
Parsi, Jewish, other **0.7%**
Jain **0.4%**
Buddhist **0.8%**
Sikh **1.9%**
Christian **2.3%**

Source: Census of India, 2001.

Building Graph Reading Skills

1. **Analyzing Data** What percentage of India's population are Christians and Buddhists?

2. **Making Inferences** Based on the circle graph, which religion do you think influences Indian society and culture the most? Why?

Guide to Reading

BIG Idea

The characteristics and movements of people impact physical and human systems.

Content Vocabulary

- *dzong (p. 642)*
- sitar *(p. 643)*
- sari *(p. 644)*

Academic Vocabulary

- ongoing *(p. 639)*
- primary *(p. 640)*
- contemporary *(p. 643)*

Reading Strategy

Finding Examples Use a chart like the one below to identify examples of each element of culture in South Asia.

Element of Culture	Example
Religion	
Arts	
Daily Life	

Cultures and Lifestyles

 Section Audio  Spotlight Video

Picture This In Jaipur, India, elephants are decorated with bright colored paint, fancy cloth, and jewelry for the Elephant Festival. The festival is held on the day before the celebration of Holi. This is the Indian festival of colors when people welcome the coming of spring. Spectators watch elephant races and polo matches and even an elephant tug-of-war, in which elephants compete against men. Read the next section to learn more about the people and cultures of South Asia.

▼ Preparing for the Elephant Festival

The People of South Asia

Main Idea **South Asia's population has grown rapidly in the past 100 years.**

Geography and You How large is your community? What might life be like if many more people lived there? Read to find out how rapid population growth is affecting South Asia.

South Asia is home to nearly 1.5 billion people. The region includes three of the world's seven most populous nations—India, Pakistan, and Bangladesh. **Ongoing,** or continuing, population growth presents major challenges for South Asia.

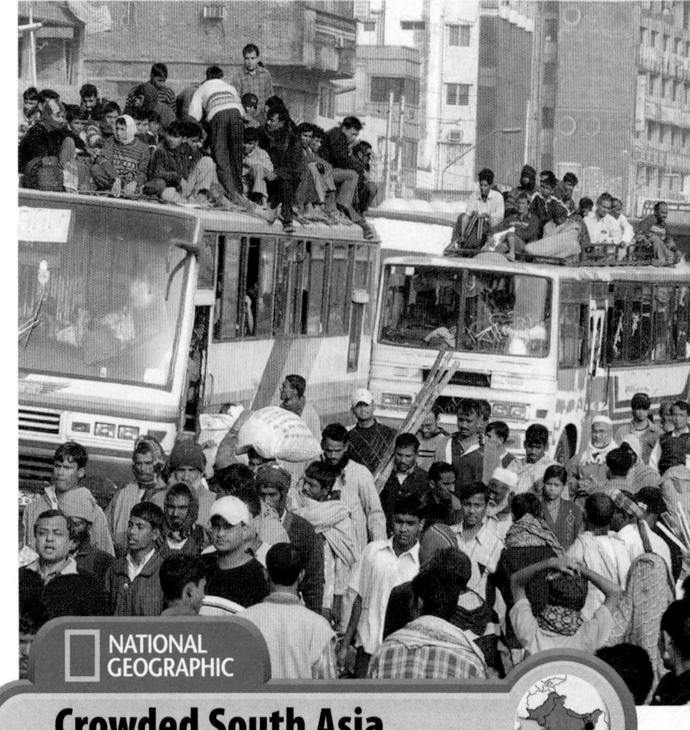

Crowded South Asia

Streets in South Asia's cities, such as Dhaka, Bangladesh, can be very crowded, especially on Hindu and Muslim holy days. ***Place*** **What is the population density of Bangladesh?**

Population Squeeze

The population of South Asia grew dramatically during the last century. One reason for this growth was improved medical and health care, which lowered death rates. Another factor was continued high birthrates. In the 1990s alone, India's population rose by 175 million people. Although growth rates have slowed in recent years, the number of people in South Asia is still climbing steadily.

Of course, while the population swells, South Asia's land area stays the same size. As a result, population densities in the region are very high. India averages 869 people per square mile (336 per sq. km). In comparison, the United States averages 80 people per square mile (31 per sq. km). Crowding is even worse in Bangladesh, South Asia's most densely populated nation. Bangladesh has a whopping 2,594 people per square mile (1,001 per sq. km).

The population of South Asia is not evenly distributed. River valleys are densely settled, but desert areas have few inhabitants. More than two-thirds of South Asians live in rural areas, but the region also has large and growing cities. The biggest, with more than 19 million people, is the Indian city of Mumbai (formerly Bombay). Four other urban areas have more than 12 million people. They are Kolkata (formerly Calcutta) and Delhi, which are both in India; Karachi in Pakistan; and Dhaka in Bangladesh. These megacities make New York City, with its 8 million people, seem small by comparison.

Urban and Rural Life

The growing cities of South Asia buzz with human activity. Sidewalks and shops are packed with people buying and selling items. People, animals, carts, bicycles, and cars move through crowded city streets.

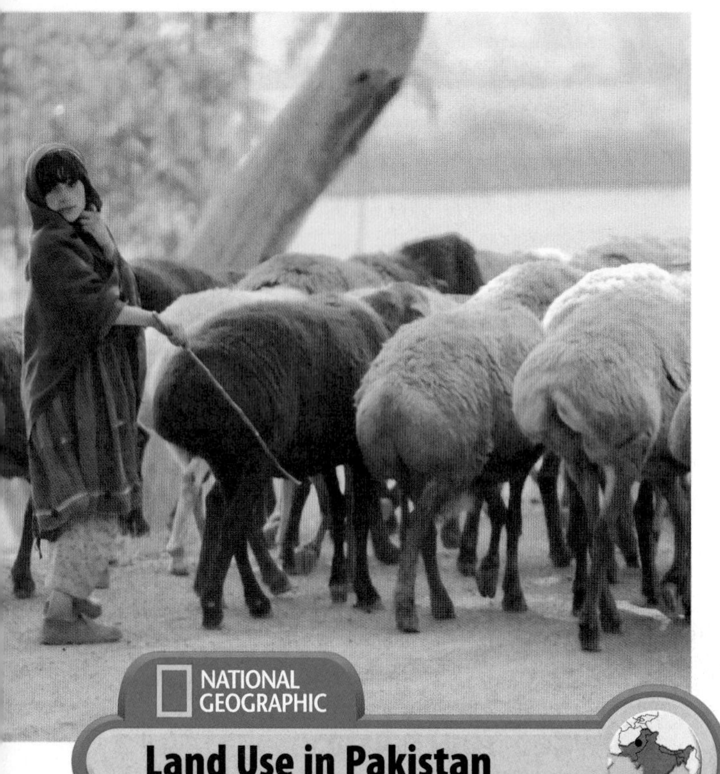

NATIONAL GEOGRAPHIC

Land Use in Pakistan

Parts of Pakistan are mountainous and more suitable for raising animals than for large-scale farming. ***Human-Environment Interaction*** **How is Pakistan using technology to support agriculture?**

Towering skyscrapers and modern apartments are signs of urban wealth and the growing middle class. At the same time, poverty is widespread in South Asia. Large numbers of people live in inadequate housing or are homeless. Children, many homeless or orphaned, are forced to beg in the streets for money to buy food. Unemployment, pollution, disease, crime, and lack of clean water are common problems in the region's urban slums.

People living in South Asia's rural areas also face challenges. Farmland is limited in mountainous Nepal and Bhutan and on the sandy islands of the Maldives. Elsewhere in the region, overcrowding has reduced the size of the land plots that farmers can work. In addition, inefficient farming methods lead to low crop yields. As a result, millions of people barely grow enough food to feed their own families. Rural villages in South Asia may also lack safe drinking water and electricity.

Technological advances have helped raise living standards in some parts of rural South Asia in recent years. In Pakistan, for example, an expanded irrigation system has boosted farm incomes. India's government has been working to provide villagers with electricity, drinking water, better schools, and paved roads. Still, many villagers move to cities to find jobs and a better standard of living.

Ethnic Groups and Languages

South Asia's population is rich with many different ethnic groups—people who share a common ancestry. The Sherpa, whose ancestors came from Tibet, are one of the best-known ethnic groups in South Asia. They live mainly in the Himalaya area of Nepal. The Sherpa are a farming people, but they are also known as strong mountain climbers. Because of their skills and remarkable endurance in high altitudes, many Sherpa are hired to guide climbing expeditions or to carry the gear and supplies on such trips.

South Asia's people speak 19 major languages and hundreds of local dialects. In India alone, the government officially recognizes 15 languages. About half of India's people, especially in northern and central areas, speak Hindi as their **primary,** or major, language. Urdu is Pakistan's official language, and Bengali is the official language of Bangladesh. English is also widely spoken in the parts of South Asia that were once under British rule.

Reading Check **Explaining** What problems do the people of South Asia's cities face?

Religion, the Arts, and Daily Life

Main Idea **Religious and cultural traditions in South Asia are thousands of years old.**

Geography and You Have you ever tried yoga? This popular form of exercise and meditation began with Hindus in India. Keep reading to learn about other religious and cultural traditions in South Asia.

As you have read, religion has been a major influence on the history of South Asia. Two major religions—Hinduism and Buddhism—had their beginnings here. Today, Hinduism, Islam, Buddhism, and other religions still affect daily life and the arts.

Major Religions of South Asia

Hinduism is the most widely practiced religion in South Asia. Most people in India and Nepal are Hindu. Hinduism is also practiced, to a lesser extent, in Bhutan, Sri Lanka, Pakistan, and Bangladesh.

Among India's large population, Hinduism is a powerful unifying force. It influences the daily life of about 800 million Indians. India's Hindus share the same sacred texts and perform common rituals, such as bathing in the Ganges River for spiritual healing.

Islam is the second-largest faith in the region. Pakistan, Bangladesh, and Maldives were each founded as Islamic countries. Most of their citizens are Muslims. India's 140 million Muslims form the country's second-largest religious group.

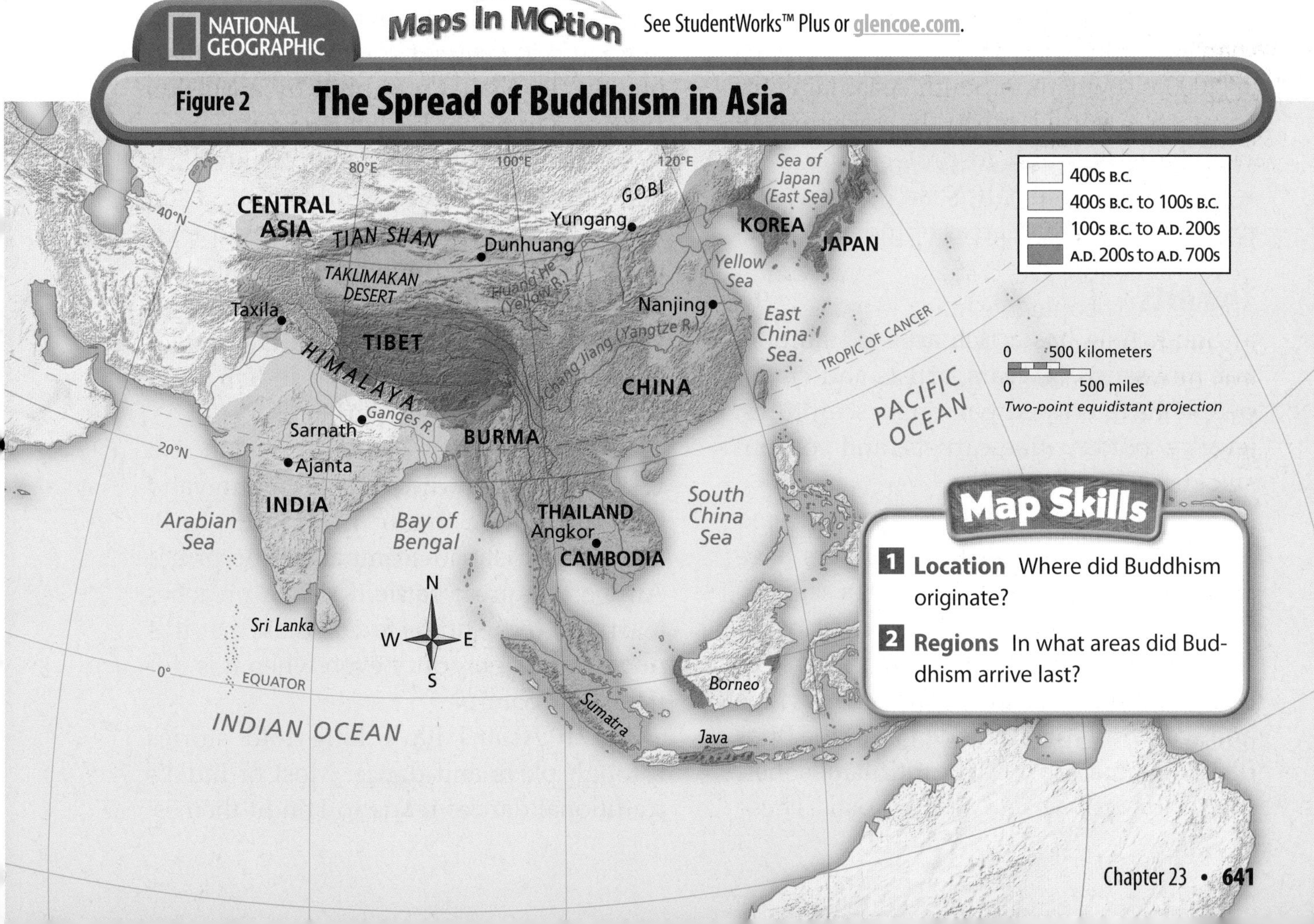

Figure 2 **The Spread of Buddhism in Asia**

Map Skills

1. **Location** Where did Buddhism originate?
2. **Regions** In what areas did Buddhism arrive last?

Muslims and the Hindu majority have sometimes clashed, but India has a secular, or nonreligious, government and Muslims' rights are protected under the law.

Buddhism, which spread from India beginning in the 400s B.C. as shown in **Figure 2** on the previous page, is no longer a major religion in that country. It remains strong, though, in Sri Lanka, Nepal, and Bhutan. In Bhutan, ***dzongs,*** or Buddhist centers of prayer and study, have been important in shaping the country's arts and culture.

The people of South Asia practice a number of other religions. Sikhism (SEE·KIH·zuhm) was founded in the early 1500s. It teaches belief in one God and stresses doing good deeds as the way to escape the cycle of reincarnation and join with God. Most of South Asia's Sikhs live in northwestern India. Many of them want an independent Sikh state there.

Another religion, Jainism, has about 4 million followers in India and perhaps 100,000 elsewhere in South Asia. Jains try to reach spiritual purity by rejecting all violence. They aim to protect every living creature. Small Christian communities exist in some urban areas of India.

The Arts

South Asia has a rich artistic tradition. For thousands of years, artists and craftspeople have created paintings, sculptures, jewelry, pottery, fine carpets, and colorful silk and cotton textiles.

Since early times, the arts have reflected a strong religious influence. Painters have been inspired by sacred writings. Hindu, Buddhist, and Sikh architects built beautiful temples across the region. Many of these holy places hold elaborate carvings and sculptures of Hindu deities or the Buddha. Muslims, too, built beautiful mosques, forts, and palaces in South Asia. These buildings include the famous Taj Mahal in Agra, India. Made of gleaming white marble, the Taj Mahal was built by a Muslim ruler as a tomb for his beloved wife.

Indian Pottery

This woman in Tamil Nadu, India, is carefully finishing her work on a hand-made ceramic water pot. ***Regions*** **How has religion influenced architecture in South Asia?**

Like much of the art and architecture, South Asian literature also has its roots in religion. One of India's major sacred texts is the *Mahabharata* (muh·hah·BAH·ruh·tuh). Written about 100 B.C., the *Mahabharata* is very long—about 88,000 verses. It describes a great war for control of an Indian kingdom about 1,000 years earlier.

The best known section in this ancient text is the Bhagavad Gita (BAH·guh·vahd GEE·tuh), or "Song of the Lord." In it, the deity Krishna accompanies the prince Arjuna to a great battle. Krishna preaches a sermon to Arjuna. He tells him that it is noble to do one's duty even when it is difficult and painful.

South Asians have long told stories through plays and dance. Most of India's traditional dances relate to Hindu themes.

The dancers communicate with movements and facial expressions that their audiences understand well.

Music is another important art form in South Asia. Classical Indian music usually features the **sitar** (sih·TAHR), a long-necked instrument with 7 strings on the outside and 10 inside the neck. The sitar helps give Indian music a distinctive sound. **Contemporary,** or present-day, South Asian music reflects the growing influence of Western styles. Rock music, for example, has recently gained popularity in Pakistan.

South Asians also greatly enjoy films, and moviemaking is a booming business in the region. India releases hundreds of new movies each year. Many of them have gained international popularity. The city of Mumbai, nicknamed "Bollywood," is the center of the Indian film industry. Traditional Bollywood movies are known for their grand spectacles with wild plots and lots of singing and dancing.

Daily Life

The life of South Asians centers on the family. Marriage in South Asian countries is commonly viewed as the joining of two families. As a result, parents often arrange marriages for their children by choosing partners they consider suitable. After a woman marries, she becomes part of her husband's family. In India and Pakistan, several generations often live together in the same house.

NATIONAL GEOGRAPHIC

South Asian Wedding Traditions

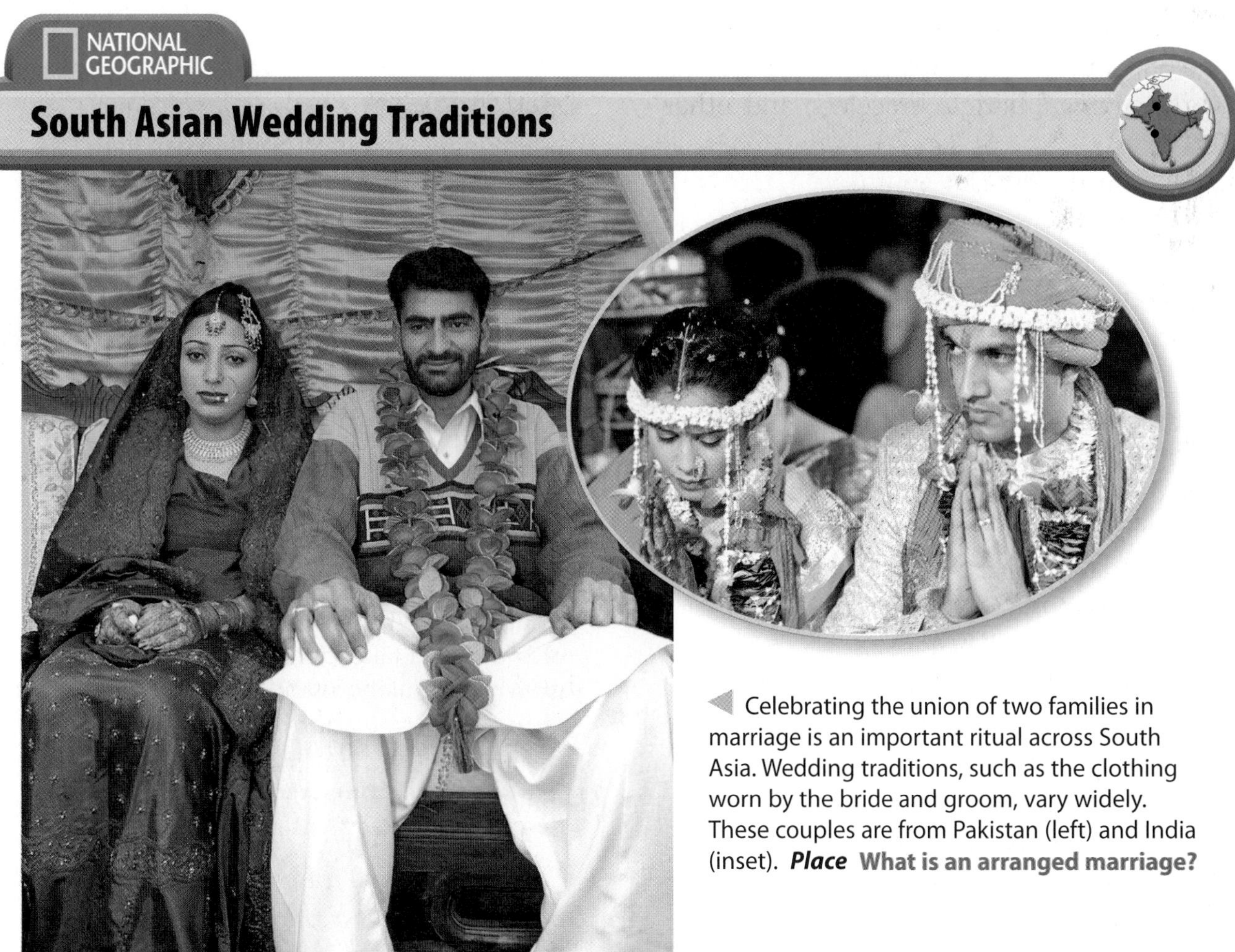

Celebrating the union of two families in marriage is an important ritual across South Asia. Wedding traditions, such as the clothing worn by the bride and groom, vary widely. These couples are from Pakistan (left) and India (inset). ***Place*** **What is an arranged marriage?**

Housing varies throughout South Asia. In thousands of villages across the region, people live in simple homes with mud walls and thatched roofs. Residents sometimes paint the outside walls with brightly colored patterns. In Bangladesh, people build their houses on stilts to protect against frequent floods. Homes in South Asian cities can be modern or traditional. People in the middle and upper classes generally live in comfortable houses or high-rise apartments. The millions of urban poor, however, crowd into cheap apartments or flimsy shacks, or even sleep on the streets.

Western-style clothing is popular in South Asian cities, but many people still dress in traditional garments. Indian women, for example, wear colorful saris. A **sari** (SAHR·ee) is a long, rectangular piece of cloth that is draped gracefully around the body. Women complement their outfits with earrings, bangle bracelets, and other jewelry.

Religious beliefs have a significant impact on diet in South Asia. Hindus view cows as sacred and do not eat beef. Muslims do not eat pork, which they consider unclean. Jains, among others, avoid eating any meat. As a result, vegetarian cooking is popular. Whether dishes are made with vegetables or meat, they are typically prepared in spicy sauces and served with rice or flat breads. Tea is a favorite beverage, as are flavored yogurt drinks.

Sports in the region reflect the area's more recent history. Two of the most popular sports, field hockey and cricket, were introduced by the British during the colonial era. India and Pakistan excel at both sports. India has won eight Olympic gold medals in field hockey, and Pakistan has won three. Teams from both countries have won several international cricket titles.

Reading Check **Describing** How do Western influences affect South Asian life?

Section 2 Review

Social Studies ONLINE
Study Central™ To review this section, go to glencoe.com.

Vocabulary

1. **Describe** how dzong, *sitar,* and *sari* relate to South Asian culture.

Main Ideas

2. **Determining Cause and Effect** Use a diagram like the one below to show the causes and effects of population growth in South Asia.

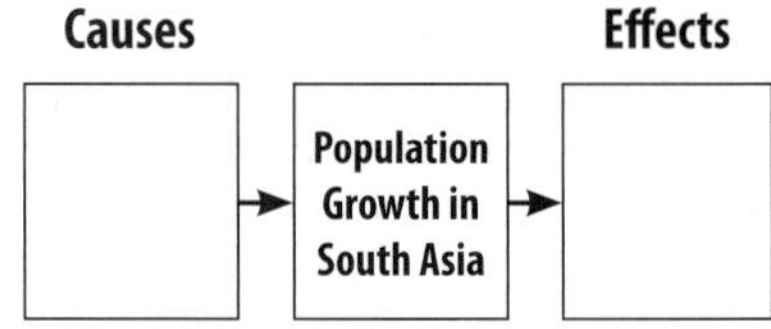

3. **Identifying** What are three examples of religious art in South Asia?

Critical Thinking

4. **Comparing** Compare urban and rural ways of life in South Asia.
5. **BIG Idea** How has religion influenced peoples' diets in South Asia?
6. **Challenge** What steps might the countries of South Asia take to solve the problems of growing populations and food shortages?

Writing About Geography

7. **Expository Writing** Write a paragraph describing the religious geography of South Asia.

Visual Summary

British colonialism in India

History

- Powerful Buddhist and Hindu empires ruled early South Asia.
- Britain controlled much of South Asia from the 1700s to the mid-1900s.
- British India was divided into the independent countries of India and Pakistan in 1947. Bangladesh, once East Pakistan, won its independence in 1971.
- Other countries of South Asia are Bhutan, Nepal, Sri Lanka, and Maldives.

Population

- Population growth presents major challenges to South Asia.
- Most South Asians live in rural areas, but many cities have large populations.
- Cities in South Asia reflect both wealth and poverty.

Ethnic and Language Groups

- South Asia is a region of many ethnic and cultural groups and languages.
- India has 15 official languages, but Hindi is the most widely used.

Muslims in Bangladesh

Religion and the Arts

- Hinduism, Islam, and Buddhism are major South Asian religions.
- Temples and mosques are examples of religious influences on South Asia's architecture.

Mumbai, India

Daily Life

- Family life is important in South Asia.
- Housing in the region varies from slum dwellings to modern high-rise apartments.
- Religious rules have an effect on diet in South Asia.
- South Asians enjoy sports such as field hockey and cricket.

Soccer players, India

STUDY TO GO Study anywhere, anytime! Download quizzes and flash cards to your PDA from **glencoe.com**.

CHAPTER 23 ASSESSMENT

STANDARDIZED TEST PRACTICE

Start thinking about studying as soon as you know that a test is scheduled. Think about what could be on the test and what you should study. Ask the teacher what the test will cover so there will not be any surprises.

Reviewing Vocabulary

Directions: Choose the word(s) that best completes the sentence.

1. A ______ is a social group that someone is born into and cannot change.

A *varna*
B caste
C karma
D boycott

2. Refusal to buy certain kinds of goods is known as ______.

A a karma
B a boycott
C a caste
D reincarnation

3. Buddhist centers of prayer and study in Bhutan are called ______.

A dharmas
B saris
C sitars
D *dzongs*

4. A ______ is a long-necked instrument with seven strings used in Indian classical music.

A sitar
B sari
C *dzong*
D *varna*

Reviewing Main Ideas

Directions: Choose the best answer for each question.

Section 1 *(pp. 626–632)*

5. The two major world religions that developed in South Asia are ______.

A Hinduism and Buddhism
B Hinduism and Islam
C Buddhism and Christianity
D Buddhism and Islam

6. Mohandas Gandhi led a movement to free India from ______ rule.

A American
B British
C French
D Chinese

Section 2 *(pp. 638–644)*

7. The nation of ______ in South Asia is one of the world's seven most populous countries.

A Nepal
B China
C Pakistan
D Sri Lanka

8. The most widely practiced religion in South Asia is ______.

A Sikhism
B Islam
C Buddhism
D Hinduism

GO ON

Critical Thinking

Directions: Use the chart below and your knowledge of Chapter 23 to help you choose the best answer for each question.

Religions of South Asia	
India	Hindu 80.5%, Muslim 13.4%, Christian 2.3%, Sikh 1.9%, other 1.8%, unspecified 0.1% (2001 census)
Pakistan	Muslim 97% (Sunni 77%, Shia 20%), Christian, Hindu, and other 3%
Bangladesh	Muslim 83%, Hindu 16%, other 1% (1998)
Maldives	Sunni Muslim
Nepal	Hindu 80.6%, Buddhist 10.7%, Muslim 4.2%, other 4.5% (2001 census) *note:* only official Hindu state in the world
Bhutan	Buddhist 75%, Hindu 25%
Sri Lanka	Buddhist 69.1%, Muslim 7.6%, Hindu 7.1%, Christian 6.2%, unspecified 10% (2001 census provisional data)
Source: CIA World Factbook (https://www.cia.gov/cia/publications/factbook/)	

9. In which two South Asian countries is Hindu the dominant religion?

A Sri Lanka and Nepal

B India and Nepal

C Bangladesh and Bhutan

D Pakistan and India

10. Which South Asian country is the least religiously diverse?

A Nepal

B Bangladesh

C Maldives

D Bhutan

Document-Based Questions

Directions: Analyze the document and answer the short-answer questions that follow.

> *"My teaching is not a philosophy. It is the result of direct experience.*
>
> *My teaching is a means of practice, not something to hold onto or worship.*
>
> *My teaching is like a raft used to cross the river. Only a fool would carry the raft around after he had already reached the other shore of liberation."*
>
> —The Buddha, *The Kalama Sutra*

11. Based on the reading above, how do you think Buddhism might be different from some other major religions?

12. What do you think the Buddha means when he says "My teaching is like a raft used to cross the river"?

Extended Response

13. Select one of the countries of South Asia. Imagine that you are on a trip to that country. Write a letter home to your parents or to a friend using what you have learned in this chapter to persuade them to visit this country also.

STOP

For additional test practice, use Self-Check Quizzes—Chapter 23 at glencoe.com.

Need Extra Help?

If you missed question...	1	2	3	4	5	6	7	8	9	10	11	12	13
Go to page...	628	631	642	643	628	631	639	641	641	641	628	628	624–644

"Hello! My name is Mandovi.

I'm 15 years old and live in Mumbai, one of India's busiest cities. I have just finished 11th grade at Saint Xavier's College. (Here, the last two years of high school are called college.) Now I am enjoying my summer holiday! Here's a look at my day."

7:00 a.m. I wake up early because I want to see my mom before she goes to work. I hug her good-bye and say good morning to my dad and my grandmother, who lives with us. Then I make my bed.

7:30 a.m. I take my dog, Caesar, for a walk. I love the sights in my neighborhood, like the beautiful Hindu temple behind our building. Afterward, I ride the elevator back up to our 18th-floor apartment.

8:30 a.m. Time for breakfast! I usually eat *dosa*—a crispy pancake made of rice and lentil flour. I eat it with my hands. Many Indians enjoy fresh fruit, yogurt, and eggs for breakfast.

9:15 a.m. My dad, who works in an office, has today off, so we spend some time together. I play the piano while Dad jams on the guitar. (He used to play in a rock band when he was in college!)

10:30 a.m. I take a taxi to the local animal shelter. I volunteer there in the summer, helping to care for stray animals. Before I go in, I pop into a small shop to buy crackers for the dogs.

1:00 p.m. I take a taxi back home and eat lunch. Today it is rice, vegetables, *dal*, and *papat*. *Dal* is a dish made of lentils. *Papat* is like a cracker.

1:30 p.m. I curl up on the deep windowsill in my bedroom and draw. Drawing is one of my favorite hobbies.

2:30 p.m. Since there are lots of traffic jams in Mumbai with its almost 13 million people, I take a train to a different part of the city. I'm meeting friends to go shopping. From a street vendor, I buy a fresh mango. It is a tangy golden fruit grown in India.

3:15 p.m. While we are shopping, my friends and I see a cow walking down the street. Cows are allowed to wander freely because they are highly valued by Hindus. Practicing Hindus do not eat beef.

4:00 p.m. I head to my piano lesson. I have been playing the piano since I was a little girl. I also play the harmonica and guitar.

5:30 p.m. When I return home, my dad is on the computer. There is an e-mail from my older sister, Vidula. She attends a university in the United States. In one year, I will go to a university too. My plan is to attend a school in India and study law.

7:00 p.m. Everyone gathers for dinner. I help to prepare the salad. We also have spicy fish curry and *paranthas*. *Paranthas* are pieces of wheat bread that are cooked on a griddle.

9:00 p.m. I chat with friends on the phone and watch an Indian movie on TV. It's nice to have time to relax. During the school year, I get a lot of homework in subjects such as world history, computers, and biology.

11:00 p.m. It has been a great day, and I'm exhausted! I head to bed.

ILLUSTRATIONS BY BOOKMAPMAN

HIGH NOTES Mandovi and her father, Madhu, play both Indian tunes and rock music.

REFRESHING PAWS Mandovi is a volunteer at an animal shelter for stray and homeless dogs.

SHOP TALK Shopping is always fun for Mandovi and her friends.

TRAIN RIDE Mandovi Menon (mahn • DOH • vee meh • NOHN) takes a commuter train from one part of crowded Mumbai to another. Mumbai changed its name from Bombay, which is what the city was called when it was ruled by the British.

DRAY/THE IMAGE WORKS

VAN HOLLENSHEAD/POLARIS (4)

What's Popular in India

SUSAN LIEBOLD

Tandoor In this Indian style of cooking, a clay pot is partially buried in the ground, then filled with hot coals. Lamb and other meats are soaked in yogurt and spices and cooked on skewers over the coals.

Bollywood India has the largest movie industry in the world. It produces more than 900 films per year! The nickname "Bollywood" is a blend of Hollywood and Bombay, another name for Mumbai.

Cricket One of the nation's top sports, cricket was brought to India by British colonists 250 years ago. This baseball-like sport is played with a ball and bat. Matches can last for days!

AP PHOTO

Say It in Hindi

People in India speak 22 different languages! For example, Indians on the eastern tip of the country speak Bengali, and those in the far north speak Urdu. To unify the country, India's government has two official national languages—Hindi and English. Here are some commonly used phrases in Hindi.

Hello/good-bye
Namaste (nuh · MUS · stay)

How are you?
Aap kaise hein? (ahp KAY · se hain)

My name is ______.
Mera nam ______ hein.
(MAY · ra nahm ______ hain)

CHAPTER 24

South Asia Today

Essential Question

Regions Home to nearly one-fourth of the world's people, South Asia plays an important role in world affairs. India is the world's most populous democracy and is becoming one of the world's largest economies. Pakistan, a major Muslim nation, is an ally of the United States in the war on terrorism, and its economy is growing as well. How do a country's resources affect its role in world affairs?

Outside Chennai, India

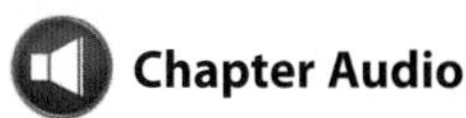

Section 1: India

BIG IDEA Patterns of economic activity result in global interdependence. India is a vast country with a large and varied population. In recent years, India has become a key player in the global economy.

Section 2: Muslim Nations

BIG IDEA All living things are dependent on one another and their surroundings for survival. Millions of people in Pakistan and Bangladesh make their living by farming. Natural disasters, such as flooding and drought, however, often threaten their livelihoods.

Section 3: Mountain Kingdoms, Island Republics

BIG IDEA Cooperation and conflict among people have an effect on the Earth's surface. Ethnic and religious conflicts continue to be a challenge in South Asia's mountain kingdoms and island republics.

FOLDABLES™ Study Organizer

Categorizing Information Make the Foldable below to help you organize information about the countries of South Asia today.

Step 1 Fold an 11 x 17 piece of paper lengthwise to create four equal sections.

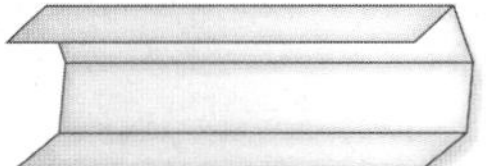

Step 2 Then fold it to form six columns.

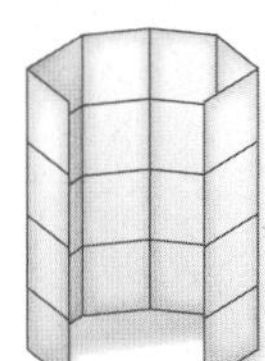

Step 3 Label your Foldable as shown.

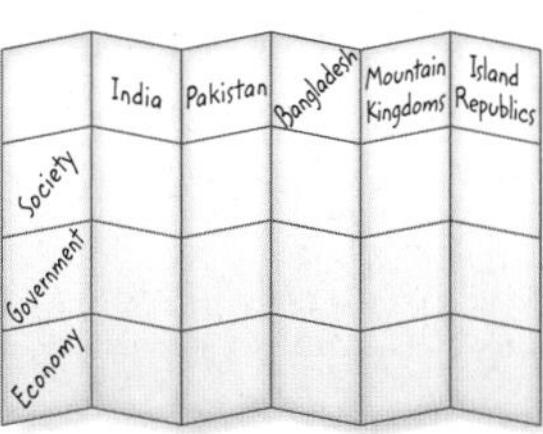

Reading and Writing On your Foldable, take notes for each category under each of the areas listed. Use your notes to write a generalization about the region in the future.

Visit glencoe.com and enter ***QuickPass***™ code EOW3109c24 for Chapter 24 resources.

Guide to Reading

BIG Idea

Patterns of economic activity result in global interdependence.

Content Vocabulary

- green revolution *(p. 656)*
- jute *(p. 656)*
- cottage industry *(p. 656)*
- outsourcing *(p. 657)*

Academic Vocabulary

- overlap *(p. 653)*
- fundamental *(p. 654)*
- professional *(p. 657)*

Reading Strategy

Organizing Information Use a diagram like the one below to list key facts about India's economy.

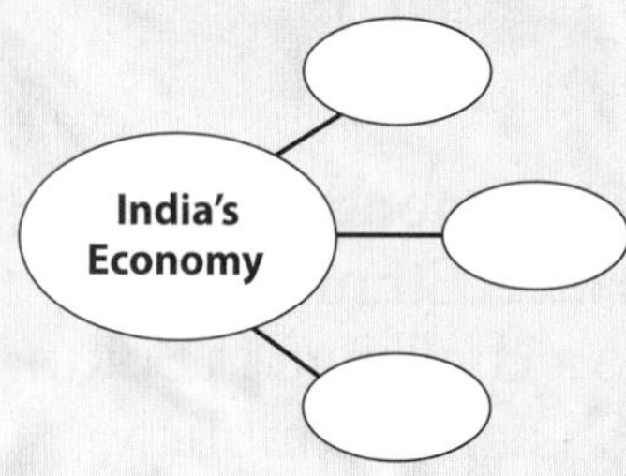

India

 Section Audio  Spotlight Video

Picture This What do you think it might be like to carry hundreds of bricks for up to 12 hours per day? In India, brickworkers need plenty of energy to get through a workday, since they are paid based upon the amount of work they do. The brick-making industry, however, provides men and women with steady work and allows families to live and work together at the plant site. India's economy has grown dramatically in the past 40 years. To learn more about India's economy and how it is connected to—and dependent upon—other nations, read this section.

▼ Indian workers carrying bricks

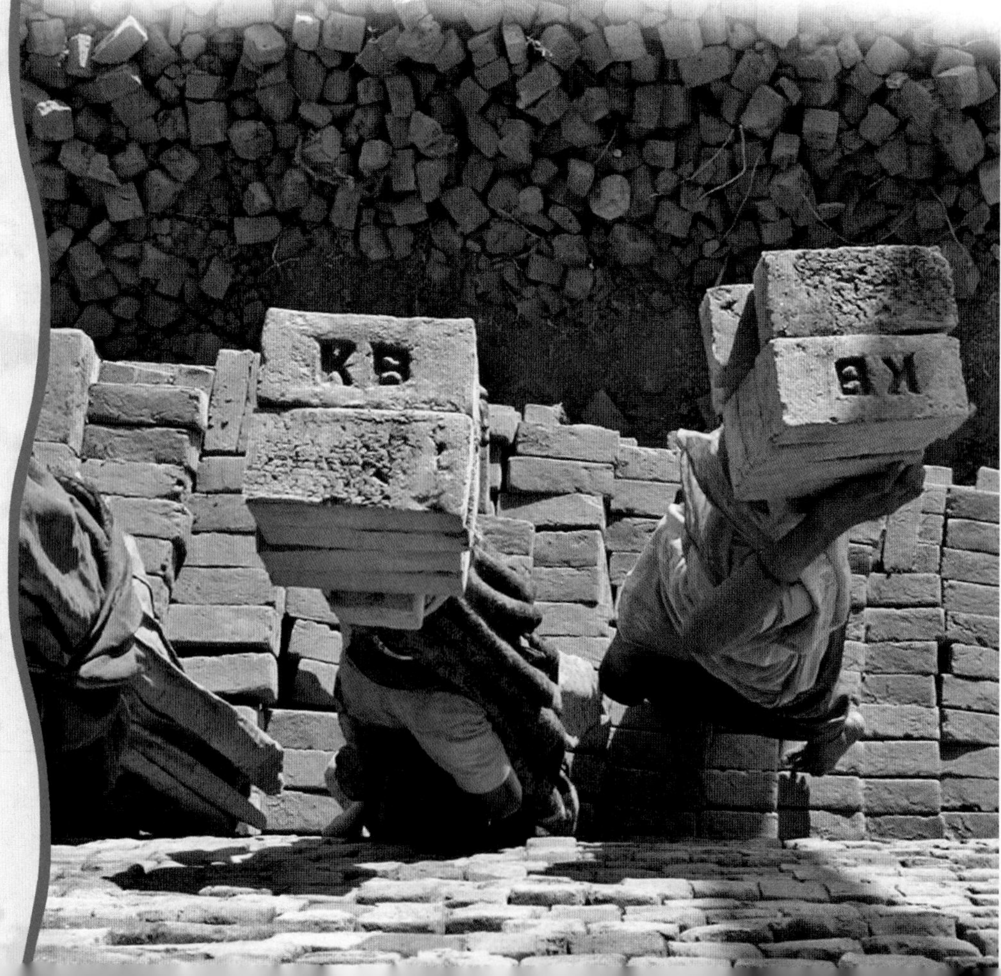

India's Government

Main Idea India has a democratic government in the form of a federal republic.

Geography and You Can you recite the first words of the U.S. Constitution? India's constitution begins exactly the same way: "We the people. . . ." Keep reading to learn more about the form, structure, and values of India's federal government.

With more than a billion people, India is the world's largest democracy. Its citizens rarely speak with one voice, because they come from many different ethnic, cultural, and religious backgrounds. Nevertheless, India's government and political system are remarkably stable.

A Federal System

India, like the United States, is a federal republic. In other words, power is shared between a national government and various state governments. The national government, located in the capital city of New Delhi, has certain clearly defined responsibilities. These include defending the country and dealing with other countries. The states have their own duties, such as carrying out energy policies and providing police protection.

The powers of the national and state governments sometimes **overlap,** or cover some of the same areas. When a state law conflicts with a national law, the national law must be followed.

India's federal system includes 28 states that vary widely in area and population.

NATIONAL GEOGRAPHIC **Maps In MOtion** See StudentWorks™ Plus or glencoe.com.

Figure 1 Languages of India

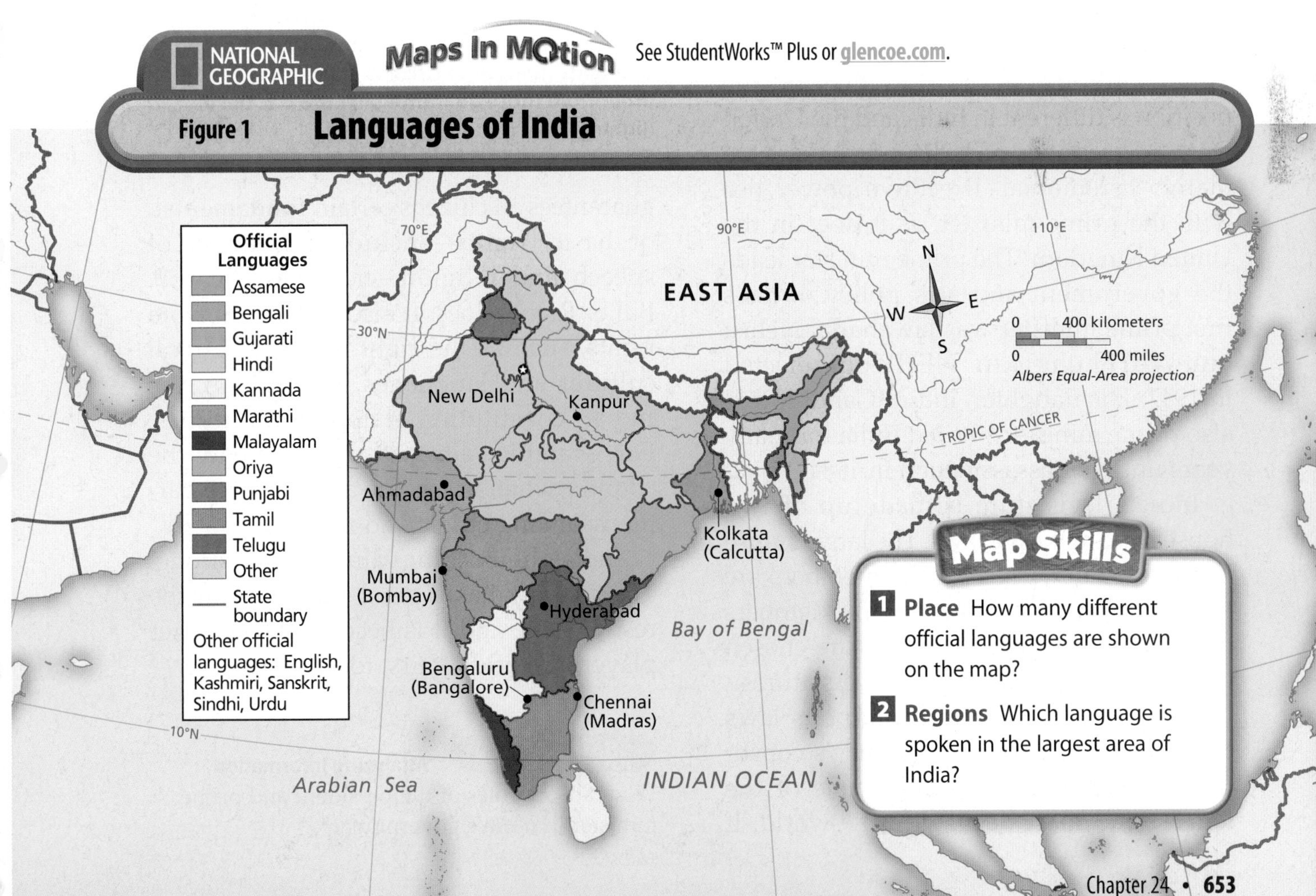

Map Skills

1. **Place** How many different official languages are shown on the map?
2. **Regions** Which language is spoken in the largest area of India?

Several states are dominated by a particular ethnic or religious group. Having their own states allows groups to focus on their unique needs and interests. India's many languages, as shown in **Figure 1** on the previous page, also vary among the states. In addition to the states, India has seven union territories. These are small political areas directly under the control of the national government. The union territories include some of India's cities and offshore islands.

Structure of the Government

India's national government has much in common with our own. There are three branches of government—executive, legislative, and judicial—that operate under the principle of separation of powers. This means that each branch of government has specific rights and responsibilities that the other branches cannot interfere with.

India's head of state is a president, but the position is different in India and the United States. The duties of India's president are mainly ceremonial. Executive power lies with the prime minister, as it does in the United Kingdom. The prime minister leads the government and sets policy. India's first prime minister was Jawaharlal Nehru (juh·WAH·huhr·LAHL NEHR·oo), elected in 1947. His daughter, Indira Gandhi, was also prime minister. She led India for many years until her assassination in 1984.

India's legislature is made up of two houses that make the laws. The larger house is the People's Assembly. Its members are elected directly by Indian voters. Members of the smaller Council of States are chosen by the prime minister or state legislatures.

India's Supreme Court interprets laws to see if they uphold the country's constitution. India's constitution is one of the longest and most detailed in the world. It guarantees all citizens certain **fundamental,** or basic, rights—including freedom of speech and religion—much as the U.S. Bill of Rights does. It also lists many more rights, such as the right to preserve local cultures and languages. In addition, the Indian constitution states certain duties. For example, citizens must defend the country when necessary and promote harmony among ethnic and religious groups.

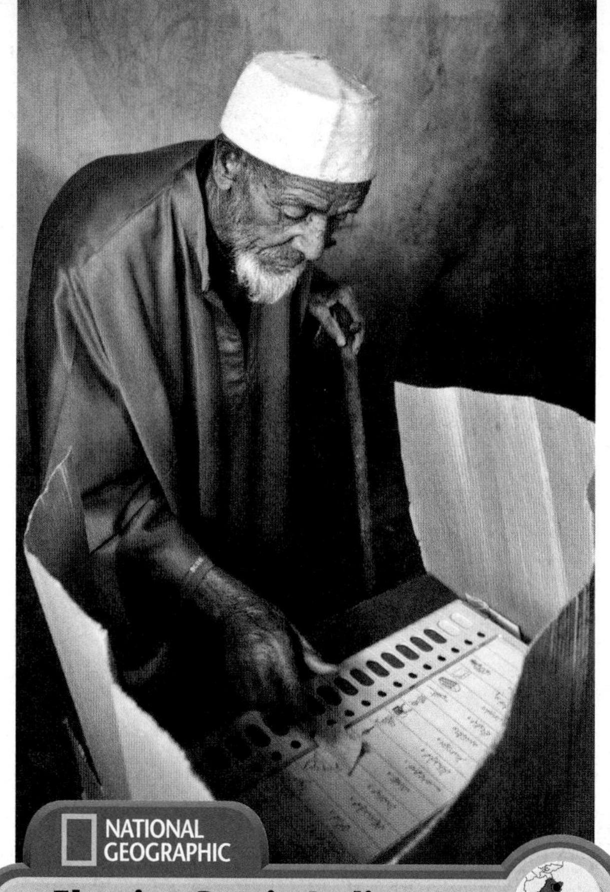

NATIONAL GEOGRAPHIC

Election Day in India

Election day in India is a national holiday. More than 670 million voters go to the polls to select members of the People's Assembly. ***Place*** **How do the rights granted to citizens in India's constitution differ from those guaranteed in the U.S. Bill of Rights?**

India's democratic values remain strong. The country has a great influence on the rest of Asia, and it is becoming a stronger player in world affairs, too.

Reading Check **Analyzing Information** What are the roles of the president and prime minister in India's government?

A STRANGE AND SUBLIME ADDRESS

By Amit Chaudhuri

There are several ways of spending a Sunday evening. You could drive to **Outram Ghat,** and then stroll with your family by the **River Hooghly**. . . . You could stay at home and listen to **plays** on the radio once the **football** commentary was over: comedies, melodramas, **whodunits.**

Sometimes **Chhaya** would come in and say excitedly:

"They're showing a **seenema** in the field!"

"Seenema! What seenema?" Mamima would ask.

"Street-Singer," she would reply, or the name of some other such film made forty years ago. . . .

The boys would run up to the terrace and lean out to look at the field that lay beyond the professor's house. This surprising piece of empty land, which builders and contractors had somehow overlooked, was usually a meeting place for fireflies. . . . [Many people] had now gathered in the field to watch the seenema; a great piece of white cloth had been hung between two poles at one end. After some time, giant black-and-white figures came alive on the piece of cloth, and a white funnel of light ran from the projector to the screen; the audience sat dwarfed by the indistinct majesty of the figures moving before them. Voices, loud and elemental as thunder, boomed from a scratchy soundtrack.

Analyzing Literature

1. **Making Inferences** Do you think the movie is a special event for the neighborhood? Explain.
2. **Read to Write** Write a paragraph describing what you might do on a warm summer evening.

Amit Chaudhuri
(1962–)

Amit Chaudhuri was born in Calcutta (now Kolkata) and educated in India and Great Britain. Chaudhuri has written several novels and short stories exploring life in modern India.

Background Information

In *A Strange and Sublime Address,* Chaudhuri describes the experiences of a boy named Sandeep from Bombay (now Mumbai). Sandeep is visiting Calcutta and finds life in that eastern city different from what he is used to.

Reader's Dictionary

Outram Ghat: a waterfront area in Calcutta

River Hooghly: branch of the Ganges River on which Calcutta is located

plays: radio programs

football: soccer

whodunits: mysteries

Chhaya: a serving woman

seenema: cinema; a movie

India's Economy

Main Idea **India has shifted from a largely government-run economy toward a free market economy.**

Geography and You Have you ever made a phone call asking for help with a computer software problem? If so, the person who helped you might have been living in India. Read to learn about India's growing role in the world economy.

After India became independent, the government worked to improve the economy. At first the government brought much of the country's industry under its control. It also increased the amount of land that could be farmed. During the 1970s, the economy slowed. In hopes of boosting growth, India began moving toward a free market economy. The government reduced its controls, and businesses were shifted to private ownership. Foreign investment was also encouraged in order to create jobs. Today India has one of the world's most rapidly growing economies. Even so, with such a large population, not enough jobs exist and many residents remain poor.

Agriculture and Related Industries

Farming is an important economic activity in India. Nearly 75 percent of Indian workers are farmers, and more than half of India's land is used for farming.

Today India produces most of the food it needs. It has benefited greatly from the **green revolution,** a set of changes that modernized agriculture and greatly increased food production in the 1970s. New strains of wheat, rice, and corn were developed that produce more grains. The government also built dams to store water for irrigation during the dry season.

NATIONAL GEOGRAPHIC

Bollywood

India's movie industry is big business. "Bollywood" produces hundreds of movies each year and generates about $1.5 billion annually. ***Location*** **How has a free market economy benefited India?**

Indian farmers raise a variety of crops, including rice, wheat, cotton, tea, sugarcane, and jute. **Jute** is a plant fiber used for making rope, burlap bags, and carpet backing.

India produces more than just agricultural products. The country has rich deposits of coal and ranks as one of the world's top coal producers. India also mines iron ore, manganese, bauxite, and diamonds.

Fishing is becoming another important industry in coastal areas and river valleys. In recent years, India's government has promoted deep-sea fishing. It has built processing plants and invested in oceangoing ships, and fish exports are increasing.

Manufacturing

There are two types of manufacturing industries in India: cottage industries and factory-based industries. **Cottage industries** involve people working in their homes and using their own equipment to make goods. They craft pottery, spin and weave cloth, or create metal or wooden items. These items can then be sold to individuals or to companies for resale or export.

Most of India's industrial goods, however, come from factories. Textile factories produce quantities of cotton, jute, and synthetic, or human-made, fabrics. This industry employs the most manufacturing workers. Food-processing plants also provide many jobs, although mainly around harvest time. Other factory workers are employed in heavy industry and make steel, locomotives, trucks, and chemicals. Factories in India also produce a variety of electronic products such as televisions.

Services

India's service industries are growing faster than any other part of the economy. Computer software services, in particular, are booming, especially in southern Indian cities such as Hyderabad and Bengaluru (Bangalore). Many of India's software developers and tech support people work for American companies. In a practice known as **outsourcing,** many American businesses hire overseas workers to do certain jobs. Outsourcing work to India is popular because wages there are low and because the country has large numbers of workers who are educated, skilled, and fluent in English.

India also has a large number of doctors, scientists, and engineers with skills to apply. These **professionals,** too, are increasingly doing outsourced work. They perform research, writing, and other tasks for American companies.

Reading Check **Making Connections** How is India's economy linked to the U.S. economy?

Social Studies ONLINE
Study Central™ To review this section, go to glencoe.com.

Vocabulary

1. **Explain** the significance of:
   - **a.** green revolution
   - **b.** jute
   - **c.** cottage industry
   - **d.** outsourcing

Main Ideas

2. **Organizing Information** Use a diagram like the one below to show the organization and powers of the branches of India's government.

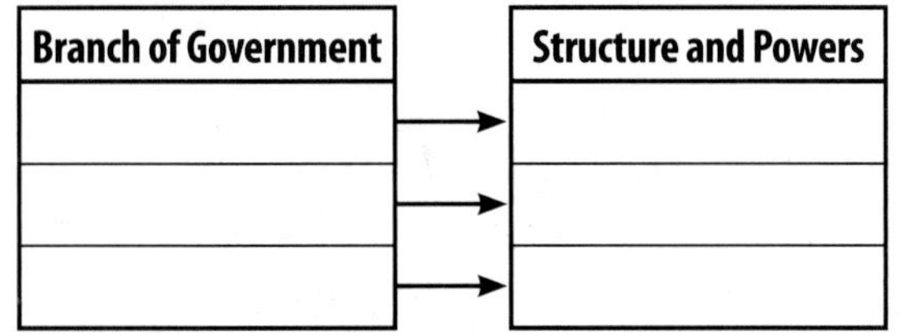

3. **Explaining** Why do many Indians remain poor, even with a growing economy?

Critical Thinking

4. **Analyzing Information** In what way does India's government reflect the principle of separation of powers?
5. **BIG Idea** What are three examples of economic links between India and other countries?
6. **Challenge** Why do you think the ability to speak more than one language is important in India's growing economy?

Writing About Geography

7. **Using Your FOLDABLES** Use your Foldable to analyze how India's economy makes use of the country's resources.

Call Centers: Are They Good for India's Workforce?

Because India has a young, well-educated workforce and low salaries, many foreign companies have established call centers there. Call center jobs include answering customer questions or entering data online. Many young Indians believe that call centers provide experience working for international businesses. Others think that these jobs take advantage of Indians and do not provide the skills that are important for working in a global economy.

For Call Centers

. . . I find [Harish] Trivedi's* depiction truly bizarre. What he sees as exploitation [unfair use] by multinationals [worldwide corporations], the young boys and girls see as an exciting chance to work with the world's top brands and acquire new skills to make a career in the global economy. It is true that many work the night shift but so do 21.2 percent of all American workers. Yes, it isn't much fun to persuade someone [in America] to pay his credit card bill, but it does build valuable negotiating skills. Many call centre** employees answer telephones but some also do highly skilled back office jobs on-line—for example, medical students prepare medical dictionaries . . . [and] accountants prepare payrolls. . . . Is it better to have an idle son at home or a productive one at work?

—Gurcharan Das
The Times of India

*Das is referring to Harish Trivedi's opinion on the next page.
***Centre* is the British spelling of *center.*

Against Call Centers

Indeed, so glamoured are many of them [Indians] by the prospect of working for a multinational [world-wide corporation] and so beguiled [excited] by what they imagine to be the American life-style swirling around their work-place, that they feel that they are already half-way to America. Except that, for most of them, the enchantment wears off sooner rather than later. Many find that they have no social life left to speak of, as they are at work when their friends and family are at home. Some develop long-term sleep disorders, and some take so much verbal abuse, day after day, from irate [angry] American customers that they actually need psychological help, which some call-centers have themselves [learned] to provide.

The burn-out is high, the turnover is rapid, and the [mental] scars . . . run deep. . . .

—Harish Trivedi
Little India Magazine

You Be the Geographer

1. **Summarizing** In your own words, summarize the opinions of Das and Trivedi.
2. **Critical Thinking** Why is India appealing to foreign companies?
3. **Read to Write** Do you think working in call centers is good for young Indians? Write a paragraph that explains your opinion.

Guide to Reading

BIG Idea

All living things are dependent on one another and their surroundings for survival.

Content Vocabulary

- nationalize *(p. 661)*
- ship breaking *(p. 664)*

Academic Vocabulary

- temporary *(p. 661)*
- cooperate *(p. 662)*
- resolve *(p. 664)*

Reading Strategy

Comparing and Contrasting Use a Venn diagram like the one below to compare and contrast the economies of Pakistan and Bangladesh.

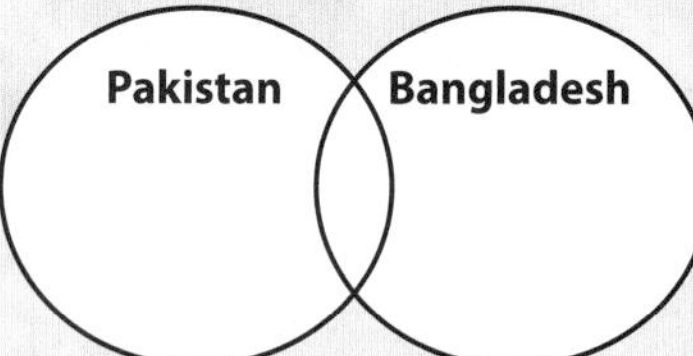

Muslim Nations

Spotlight Video

Picture This Rows of vermicelli noodles are hung out to dry in Bangladesh. The noodles are being prepared to celebrate the end of the Islamic holy month of Ramadan. During Ramadan, Muslims fast all day—they do not eat, drink, or even chew gum from dawn to sunset. Muslims are expected to use this time to reflect on their spiritual lives. When Ramadan ends, Muslims celebrate their blessings with family and friends at *Eid-al-Fitr,* or the "Festival of Breaking the Fast." Read this section to learn more about the Muslim nations in South Asia.

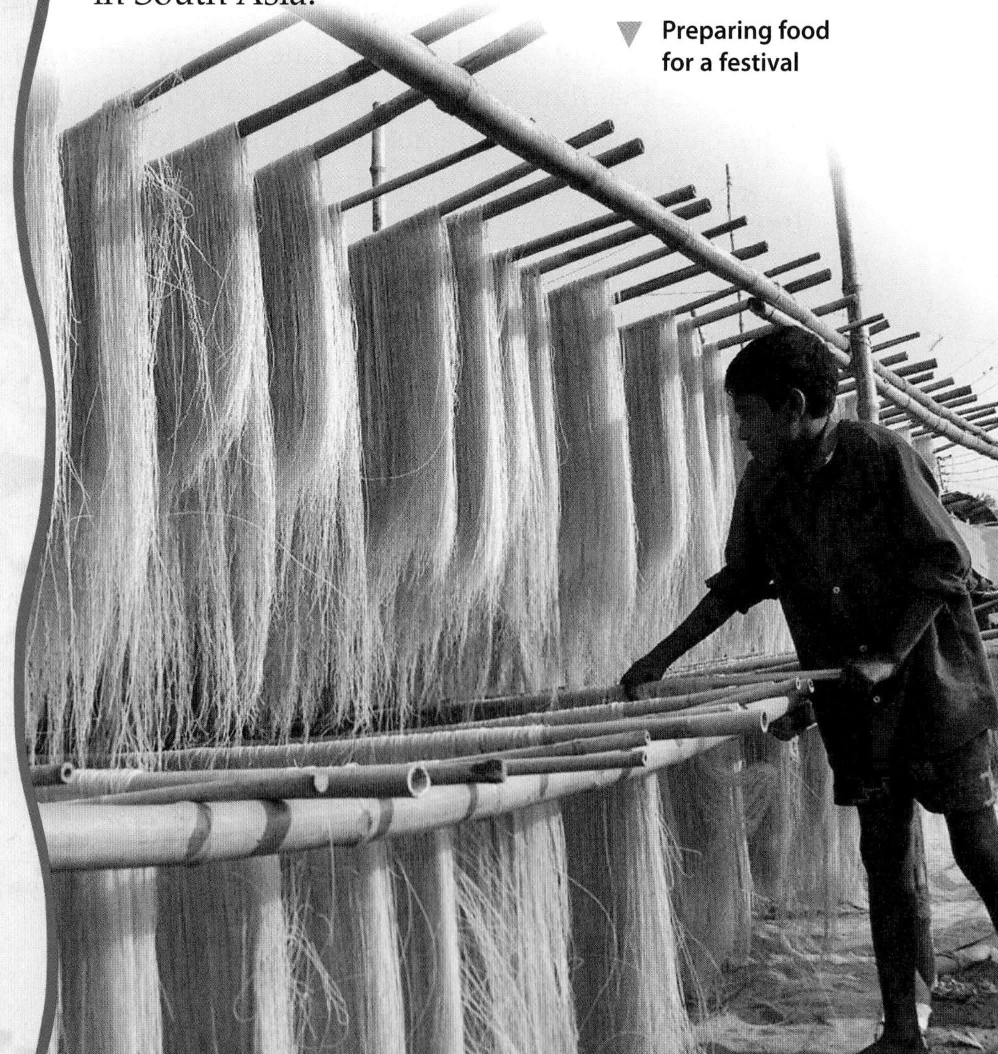

Preparing food for a festival

Pakistan

Main Idea **Pakistan is a Muslim country that is playing an increasingly important role in world affairs.**

Geography and You How well do you get along with your neighbors? Read to find out about relations between Pakistan and the countries it borders.

Pakistan is a long, wide country wedged between Afghanistan, Iran, and India. Tall mountains rise in the far north, and the Indus River valley is located to the south. This area provides the fertile land Pakistan needs to support its growing population.

The People

With more than 160 million people, Pakistan is one of the world's most populous nations. Its population continues to grow rapidly too. Although Pakistan's death rate has declined, its birthrate is still very high.

Almost all the people of Pakistan are Muslim. Their religion gives them a common bond, but it does not always bridge their cultural differences. Pakistanis come from many ethnic groups, and each one has its own language, territory, and identity.

The Economy

For many years, Pakistan's government had a strong role in the economy. In the 1970s, Pakistan's industries were **nationalized,** or put under government control. Since the 1990s, however, many government-owned industries have been sold to private owners. The government maintains control over certain parts of the economy, such as banks, hospitals, and transportation.

About half of Pakistan's people are farmers. A large irrigation system helps them grow crops such as sugarcane, wheat, rice, and cotton. Cotton cloth and clothing are among the country's major exports. Manufacturing and service industries are another important part of the economy. Many people also work in cottage industries making metalware, pottery, and carpets.

Even though Pakistan's economy has grown and incomes have risen, there are not enough jobs for everyone. Most Pakistanis are still poor. To escape poverty, millions of people leave Pakistan to become **temporary** workers in other countries. The money they send home helps support their families and also boosts the local economy in Pakistan.

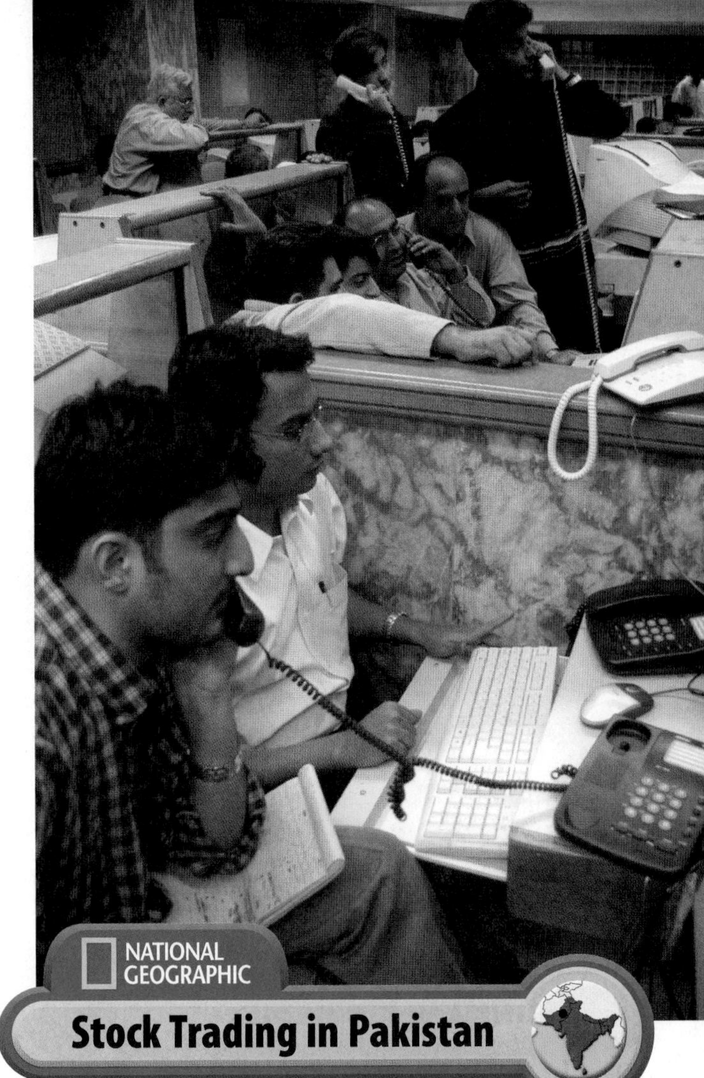

NATIONAL GEOGRAPHIC

Stock Trading in Pakistan

Pakistan's stock market is vital to its growing economy as leaders encourage local and foreign investment in business. ***Place*** **What are some of Pakistan's major exports?**

Conflict in Pakistan

Pakistani soldiers display weapons taken from Islamic militants. A crowd protests U.S. air strikes against militants based in Pakistan (inset). ***Place*** **What position does Pakistan's leader take in the war on terrorism?**

Government and Foreign Relations

Like India, Pakistan is a federal republic. Democracy, however, is limited in Pakistan. Since independence, the military has often forced elected leaders out of office and seized, or taken, power. This happened most recently in 1999, when General Pervez Musharraf (puhr·VAYS moo·SHAHR·uhf) took over the government. Three years later, Pakistan's people overwhelmingly voted to keep him as president.

One of Musharraf's most important decisions was to join with the United States in the struggle against terrorism. In 2001 Musharraf helped the United States overthrow Afghanistan's Taliban government, which had supported terrorists. He also sent troops to fight militant Muslim forces along Pakistan's border with Afghanistan.

Within Pakistan, however, many people opposed Musharraf's firm hold on power. In 2007 Musharraf's main political foe, Benazir Bhutto, was assassinated. Bhutto had served as Pakistan's first female leader during the 1990s. Many Pakistanis blamed Musharraf for her death. In early 2008 political parties opposed to Musharraf won parliamentary elections. The election results were a clear challenge to Musharraf's leadership.

Pakistan has also had trouble with its eastern neighbor, India. Both countries claim the territory of Kashmir, and they have fought two wars for control of the area. Each country occupies a part of Kashmir and keeps troops there. In 1998 tensions rose when both countries successfully tested nuclear weapons. The possibility that Pakistan and India could start a nuclear war worried many world leaders.

Since then, Pakistan and India have moved toward greater **cooperation.** In 2003 they agreed to a cease-fire in Kashmir. Two years later, they worked together to rebuild after a powerful earthquake struck northern Pakistan and Kashmir. The two countries also have agreed to closer trade ties.

Reading Check **Summarizing Information** Describe the relationship between Pakistan and India.

Bangladesh

Main Idea The problems facing Bangladesh include overpopulation, severe poverty, and deadly floods.

Geography and You Think about what might happen if a natural disaster, such as a flood, struck your community. Read to find out how natural disasters and other obstacles hinder Bangladesh's development.

Bangladesh, established in 1971, is the "youngster" in South Asia. It is struggling for success as an independent nation, but with a large population and few resources, it has not been easy.

Bangladesh sits surrounded by India on three sides, with the Bay of Bengal to the south. In area, Bangladesh is slightly larger than Wisconsin, but it holds 144 million people—about half the population of the entire United States. As a result, Bangladesh is one of the most densely populated countries in the world.

The People

Bangladesh's people are largely Muslim. They are also overwhelmingly poor. About 75 percent of the people live in rural villages. In recent years, however, many people have moved to crowded urban areas to find work in factories and workshops. Many go to Dhaka (DA·kuh), Bangladesh's capital and major port.

In urban and rural areas alike, people face serious threats from natural disasters. The country is made up of lush, low plains crossed by the Brahmaputra and Ganges Rivers. Heavy monsoon rains cause the riverbanks to overflow almost yearly. Powerful cyclones can also cause flooding. Because of the country's high population density, floodwaters can kill thousands of people at a time. Floods also drown crops and cause food shortages. As a result, malnutrition affects many people in Bangladesh.

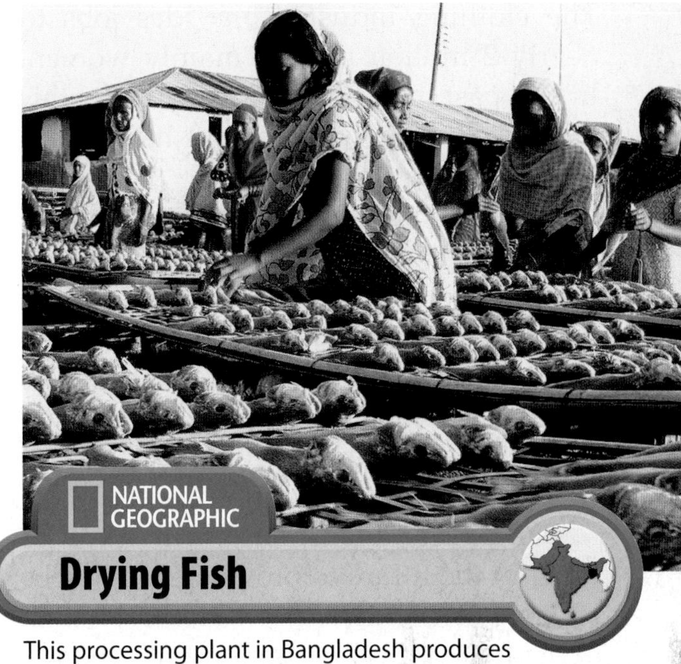

Drying Fish

This processing plant in Bangladesh produces more than 3,000 tons of dried fish every year.
Location **What are the strengths and weaknesses of agriculture in Bangladesh?**

The Economy

Most people in Bangladesh earn their living by farming. The warm climate, fertile soil, and plentiful water make it possible to plant and harvest three times a year. Rice is the country's most important crop. Other crops include sugarcane, jute, wheat, and tea. Despite favorable growing conditions, Bangladesh cannot produce enough food for its growing population. Farmers have few modern tools, and they use outdated farming methods.

Although Bangladesh has to import some foodstuffs, it has a thriving clothing industry that accounts for a large share of exports.

Student Web Activity Visit glencoe.com and complete the Chapter 24 Web Activity about Bangladesh.

The clothing industry provides jobs to nearly 2 million people, mainly women. In the past, children also worked in this industry, but the government has moved to end child labor.

Ship breaking is another profitable industry for Bangladesh. **Ship breaking** involves bringing ashore and tearing apart large, oceangoing ships that are no longer in service. The scrap metal is then sold for reuse in steelmaking or construction projects. Ship breaking is dangerous work, but it offers needed income to thousands of people.

Bangladesh's economy will have to grow in the future to provide enough jobs for its growing population. It is likely that agriculture will soon support only about a third of Bangladesh's workers. One option for improving the economy may be increasing natural gas production. Large reserves of natural gas were recently discovered in Bangladesh. The government has not yet decided whether to use these reserves to meet the country's own energy needs or to earn money by selling natural gas abroad.

Relations With Other Countries

Bangladesh tries to stay on good terms with its neighbors. Tensions have sometimes arisen with India, though, over use of the Ganges River, which flows through both countries. Thus far, the countries have **resolved** their disputes peacefully.

Bangladesh also takes pride in helping other world nations. About 10,000 of its soldiers serve as United Nations peacekeepers in twelve countries. In 2006, economist Muhammad Yunus and his Grameen Bank won the Nobel Peace Prize for providing loans to the world's poor.

Reading Check **Determining Cause and Effect** Why can Bangladesh produce three rice crops per year?

Section 2 Review

Social Studies ONLINE
Study Central™ To review this section, go to glencoe.com.

Vocabulary

1. **Explain** the meaning of *nationalize* and *ship breaking* by using each term in a sentence.

Main Ideas

2. **Explaining** How did President Musharraf of Pakistan help the United States in the struggle against terrorism?
3. **Summarizing Information** Use a diagram like the one below to summarize key facts about the people of Bangladesh.

Critical Thinking

4. **Making Inferences** How democratic is Pakistan's government? Explain your answer.
5. **BIG Idea** Why do most people in Bangladesh depend on the land to live?
6. **Challenge** Describe changes Pakistan could make to support its growing population.

Writing About Geography

7. **Expository Writing** Write a paragraph evaluating Pakistan's economic progress.

Guide to Reading

BIG Idea

Cooperation and conflict among people have an effect on the Earth's surface.

Content Vocabulary

- consumer goods *(p. 666)*
- tsunami *(p. 668)*

Academic Vocabulary

- link *(p. 666)*
- discriminate *(p. 667)*

Reading Strategy

Evaluating Information Use a chart like the one below to list an important issue facing each country, and explain its significance to that country.

Issue	Significance
1.	1.
2.	2.
3.	3.

Mountain Kingdoms, Island Republics

Section Audio

Spotlight Video

Picture This According to Buddhist teaching, sand mandalas are thought to bring positive energy to people who view them. Mandalas can be made of sand, ground marble, or powdered flowers, herbs, or grains. After the mandalas are created, the sand is poured into a nearby stream or river. It is believed that the water will transmit the positive power of the mandala to others. Continue reading to learn more about traditions in South Asia.

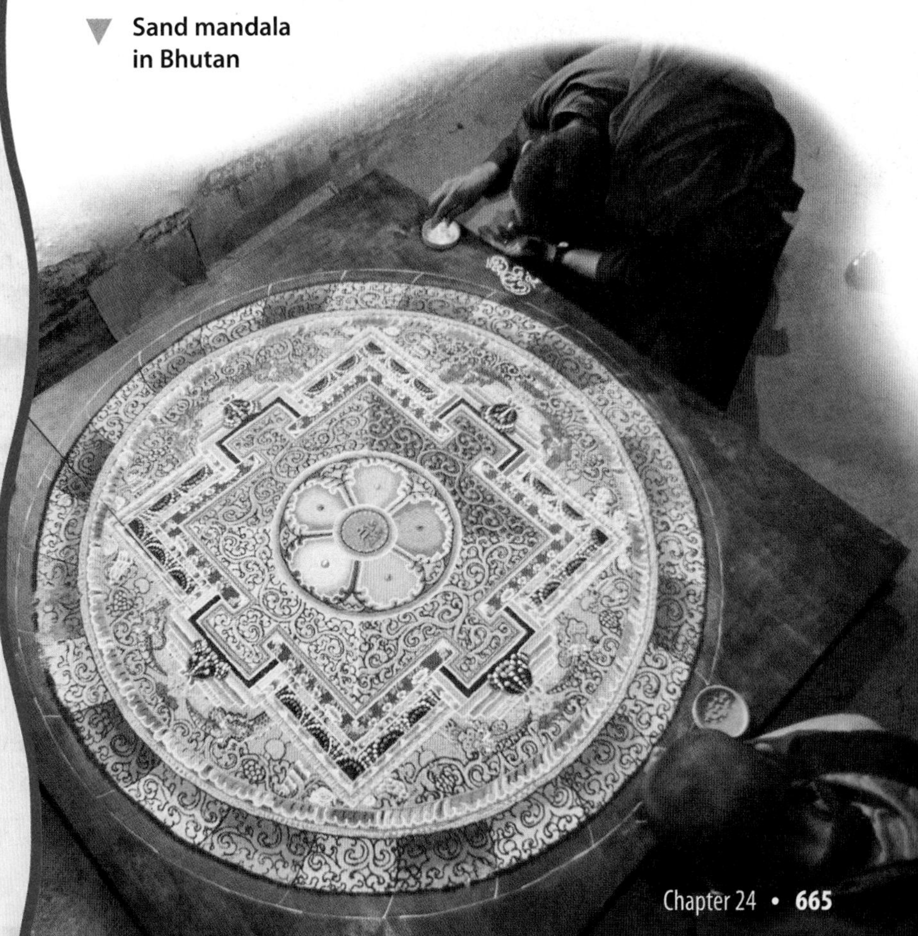

Sand mandala in Bhutan

Nepal and Bhutan

Main Idea **Limited resources and political unrest have held back development in Nepal and Bhutan.**

Geography and You Imagine not having books, paper, or pens when you go to school. How successful do you think you would be? Read to learn how a lack of resources affects Nepal and Bhutan.

Nepal and Bhutan are small, mountainous kingdoms to the north of India. Both are still largely rural and struggling to build stronger economies.

Nepal

Nepal forms a steep stairway to the Himalaya. In the north are 8 of the world's 10 highest mountains, including Mount Everest. Hills, valleys, and a fertile river plain are also part of the landscape.

More than 85 percent of Nepal's people live in rural villages. Kathmandu (KAT·man·DOO), the capital, is the only major city. Many ethnic groups make up the population. Hinduism is Nepal's official religion, but Buddhism is practiced as well.

Nepal's economy depends almost entirely on farming. Farmers grow rice and other crops on small patches of land. Unfortunately, the need for new fields leads to the clearing of forests, which causes erosion. Valleys often flood, fields are destroyed, and rivers fill with mud. Not only does the environment suffer, but so do the people who earn a living from the land.

Tourism and trade, however, help the economy. For centuries, Nepal had no **links** to other countries because the mountains formed a strong barrier. Today, there are roads and air service to India and Pakistan. Nepal exports clothing and carpets, and it imports gasoline, machinery, and **consumer goods**—products that people buy for personal use.

In recent years, Nepal has been torn by political conflict. After years of struggle, pro-democracy groups and communist rebels finally ended the monarchy in 2008. The instability in government, however, has made it difficult for Nepal to strengthen its economy. Even with substantial foreign aid, the country remains desperately poor.

▲ Business owners in Kathmandu benefit from the tourist industry, as more than 300,000 people flock to Nepal each year to explore the landscape. ***Place*** **How do you think political unrest in Nepal might impact its economy?**

Bhutan

East of Nepal is tiny Bhutan, about half the size of Indiana. As in Nepal, the Himalaya are Bhutan's major landform. Thick forests cover the foothills. To the south—along Bhutan's border with India—lie plains and river valleys.

Birthday Celebration

These marchers are taking part in an annual parade to celebrate the king of Bhutan's birthday. ***Place*** **What ethnic conflict is occurring in Bhutan?**

Once isolated by mountains, Bhutan is still difficult to travel to, and the country is struggling economically. Most of Bhutan's people live in remote rural villages and are subsistence farmers. However, roads now link Bhutan to the outside world. With India's help, Bhutan has built hydroelectric plants to create electricity from rushing mountain waters. Tourism is a growing industry, but the government limits the number of tourists to protect Bhutan's cultural traditions.

Most of Bhutan's people belong to the Bhutia ethnic group and are faithful Buddhists. Tensions are high between the Bhutia and the smaller Nepali group, who are mostly Hindu. The Nepali complain of **discrimination** during years of rule by powerful Buddhist kings. Recently, though, Bhutan has moved toward democracy.

Reading Check **Describing** Describe the economies of Nepal and Bhutan.

Island Republics

Main Idea **Sri Lanka and Maldives have growing economies sustained partly by tourism, but ethnic and political conflict is a problem in Sri Lanka.**

Geography and You What kind of place would you like to visit for a vacation? Read to find out why tourists enjoy Sri Lanka and Maldives.

South Asia includes two island republics: Sri Lanka and Maldives. Both lie south of India in the Indian Ocean.

Sri Lanka

Sri Lanka lies off the southeastern coast of India. Much of the country is rolling lowlands, with white sandy beaches that attract tourists. Highlands cover the center, and tourists come here, too, to hike on nature trails that are rich with wildlife.

For many years, Sri Lankans have farmed. In lowland areas, they grow food crops, especially the rice that people eat daily. At higher elevations are large plantations of rubber trees, coconut palms, and the Ceylon tea that is a famous export.

Sri Lanka's economy is becoming more industrialized. Factories produce textiles, fertilizers, cement, leather goods, and wood products for export. Sri Lanka also exports sapphires, rubies, and other gemstones. Colombo, the largest city, is a busy port on the country's western coast.

Sri Lanka's people are made up of two main groups. The Sinhalese (SIHNG·guh·LEEZ), who form about 74 percent of the population, live in the south and west and are mostly Buddhist. The Tamils (TA·muhlz), who make up about 17 percent of the population, live in other parts of the country and are mainly Hindu.

Since 1983, the Tamils and the Sinhalese have been fighting a violent civil war. The minority Tamils claim they have not been treated justly by the majority Sinhalese. They want to set up a separate Tamil nation in northern Sri Lanka. Thousands have died in the fighting.

Adding to its troubles, Sri Lanka suffered its worst natural disaster in December 2004. A **tsunami,** or huge ocean wave, was released by a powerful earthquake near Indonesia on the eastern edge of the Indian Ocean. The tsunami struck Sri Lanka two hours later, killing more than 30,000 people and leaving 850,000 homeless. Tourist areas were damaged and much of the country's fishing fleet was destroyed. The thousands of people who survived the disaster were in need of food, water, and medical care. Governments and international aid organizations responded to the tragedy with one of the largest relief efforts in modern history.

Maldives

About 1,200 coral islands make up the Maldive Islands, which lie southwest of India. None of the islands is more than 6 feet (1.8 m) above sea level. Some scientists believe that global warming will eventually cause ocean levels to rise and completely cover the Maldives.

About 360,000 people, mostly Muslims, live in Maldives. Some 80,000 of them reside in the capital city of Male (MAY·lay). Farmers in Maldives can grow only a few crops in the sandy soil, so most food must be imported. In recent years, Maldives's palm-lined, sandy beaches and coral formations have attracted many tourists. As a result, tourism is now the largest industry. Fishing and boatbuilding are other important economic activities.

Reading Check **Explaining** How did the tsunami of December 2004 affect Sri Lanka?

Social Studies ONLINE
Study Central™ To review this section, go to glencoe.com.

Vocabulary

1. **Explain** the meaning of *consumer goods* and *tsunami* by using each term in a sentence.

Main Ideas

2. **Summarizing** What economic and social changes have occurred in Bhutan recently?
3. **Describing** Use the following diagram to describe the two main ethnic groups in Sri Lanka.

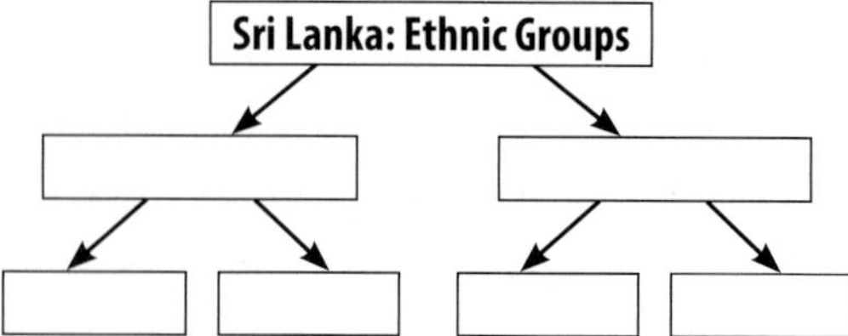

Critical Thinking

4. **Making Connections** How has India helped Bhutan's economy?
5. **BIG Idea** What effect might the civil war in Sri Lanka have on the country's political future? Why?
6. **Challenge** Compare the strengths and weaknesses of the economy of each country studied in this section.

Writing About Geography

7. **Expository Writing** Describe the similarities and differences in the ethnic situations in Bhutan and Sri Lanka.

Chapter 24 Visual Summary

India

- India is the world's most populous democracy.
- India is a federal republic with a central government, states, and territories.
- India's economy is based on farming as well as cottage and factory industries.
- The computer software industry is a growing part of India's economy.

Pakistan

- Pakistan has many ethnic groups but is overwhelmingly Muslim.
- Pakistan has fertile land and energy resources, but its economy is still developing.
- Democracy is limited, and military leaders have often ruled the country.
- Pakistan is a key player in the war against terrorism.

Mother and daughter praying, Pakistan

Bangladesh

- Bangladesh is one of the world's most densely populated countries.
- Most Bangladeshis live in rural areas and farm the land.
- Because of low elevation, Bangladesh often faces flooding from monsoon rains and cyclones.

Women picking tea, Bangladesh

Mountain Kingdoms

- The Himalaya dominate the landscapes of Nepal and Bhutan.
- Most people in Nepal farm the land.
- Buddhism has shaped Bhutanese culture.
- Political conflict has divided Nepal in recent years.

Precious gems, Sri Lanka

Island Republics

- Sri Lanka produces cash crops but also has developed new industries.
- Sri Lanka's Sinhalese and Tamil ethnic groups are engaged in a civil war.
- Tourism is a major industry in Maldives.

Political protest, Nepal

Study anywhere, anytime! Download quizzes and flash cards to your PDA from glencoe.com.

CHAPTER 24 ASSESSMENT

STANDARDIZED TEST PRACTICE

If you are stuck on a question, skip it temporarily. Return to the question once you have answered those you are more sure about. Do not forget to come back to the question before you turn in your test. If you are still not sure, take a guess.

Reviewing Vocabulary

Directions: Choose the word(s) that best completes the sentence.

1. India has benefited greatly from changes known as _______ that have modernized agricultural practices.

A the green revolution
B regulated production
C cottage farming
D outsourced farming

2. In a practice known as _______, many American businesses hire workers in India to provide tech support and other services.

A cottage industries
B insourcing
C outsourcing
D protectionism

3. Products that people buy for personal use are called _______.

A boycotted goods
B consumer goods
C exported goods
D foreign goods

4. In 2004 a huge ocean wave called a _______ killed more than 30,000 people in Sri Lanka.

A earthquake
B cyclone
C rip tide
D tsunami

Reviewing Main Ideas

Directions: Choose the best answer for each question.

Section 1 *(pp. 652–657)*

5. India's government has three branches that operate under the principle known as _______.

A communism
B separation of powers
C monopoly of powers
D combination of powers

6. In recent years, India's economy has shifted toward a _______ economy.

A government-run
B command
C free market
D socialistic

Section 2 *(pp. 660–664)*

7. India and Pakistan have fought two wars over a territory called _______, which both countries claim.

A Kashmir
B Musharraf
C Indus
D Taliban

Section 3 *(pp. 665–668)*

8. Eight of the world's ten highest mountains, including Mount Everest, are located in _______.

A Bhutan
B Pakistan
C Bangladesh
D Nepal

GO ON

Critical Thinking

Directions: Use the population pyramid below to help you choose the best answer for each question.

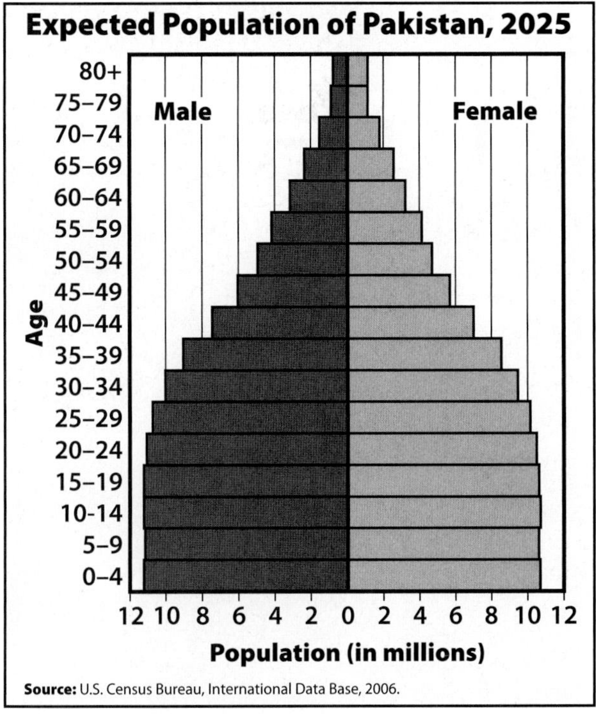

9. What will be true of Pakistan's population?

A Males and females in every age group will be equal.

B The population of males ages 65 to 69 will be more than 2 million.

C Expected populations of older age groups will outnumber those of younger age groups.

D Females under the age of 4 will number fewer than 8 million.

10. What will be true of age groups in Pakistan?

A Age groups will increase in size as they get older.

B The largest age group in 2025 will be 30–34.

C The 20–24 age group will outnumber the 40–44 age group.

D Males older than 75 will outnumber females of the same age.

Document-Based Questions

Directions: Analyze the document and answer the short-answer questions that follow.

> *As global warming seems to be more of an obvious reality, Maldivian scientists and government officials alike are concerned about the effects of rising sea levels. Since the Maldives islands are on average 5 feet (1.5 meters) above sea level, even a sea level rise of half a meter would cause severe problems. . . . Not only would flooding be a problem, but the seas may rise so quickly that they could erode the coral islands. If the reefs supporting an island fail to keep up with the rising waters, the island itself will inevitably disintegrate. To date the only recourse the Maldivians have . . . are concrete retainer walls. While such walls have effectively kept the sea at bay in a few key areas regularly struck by high waves, constructing them around dozens of inhabited islands would be . . . impossible . . . for the relatively poor country. And no amount of retainer wall would completely stave off [prevent] the erosion of an island.*
>
> —John Weier, "Amazing Atolls of the Maldives"

11. Why are rising sea levels such a concern to the people of the Maldives?

12. How are the Maldivians handling the problem of rising waters? Why is this ineffective?

Extended Response

13. If you lived in the Maldives, you would be concerned about global warming. Write a letter to a newspaper editor warning about the danger to your country.

For additional test practice, use Self-Check Quizzes—Chapter 24 at glencoe.com.

Need Extra Help?

If you missed question...	1	2	3	4	5	6	7	8	9	10	11	12	13
Go to page...	656	657	666	668	654	656	662	666	661	661	668	668	668

UNIT 9

East Asia and Southeast Asia

CHAPTER 25
Physical Geography of East Asia and Southeast Asia

CHAPTER 26
History and Cultures of East Asia and Southeast Asia

CHAPTER 27
East Asia and Southeast Asia Today

Lantern Festival, Singapore ▶

NATIONAL GEOGRAPHIC

NGS ONLINE For more information about the region, see www.nationalgeographic.com/education.

Regional Atlas

East Asia and Southeast Asia

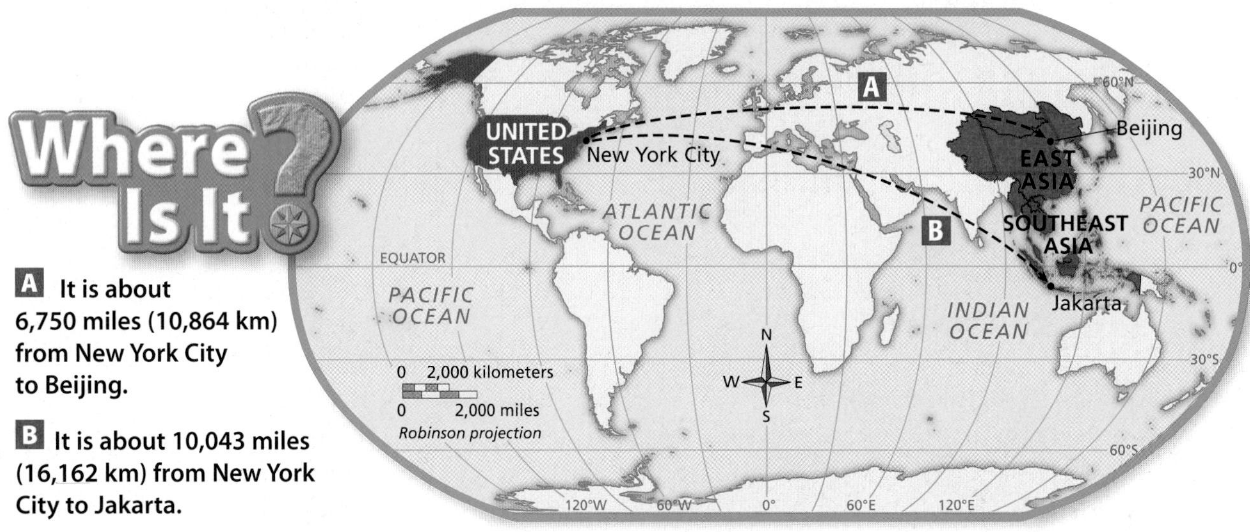

A It is about 6,750 miles (10,864 km) from New York City to Beijing.

B It is about 10,043 miles (16,162 km) from New York City to Jakarta.

The region of East Asia and Southeast Asia is more than two times the size of the continental United States. Its land area is nearly 6.3 million square miles (16.3 million sq. km). With a population of more than two billion people, almost one-third of the people of the world live in the region.

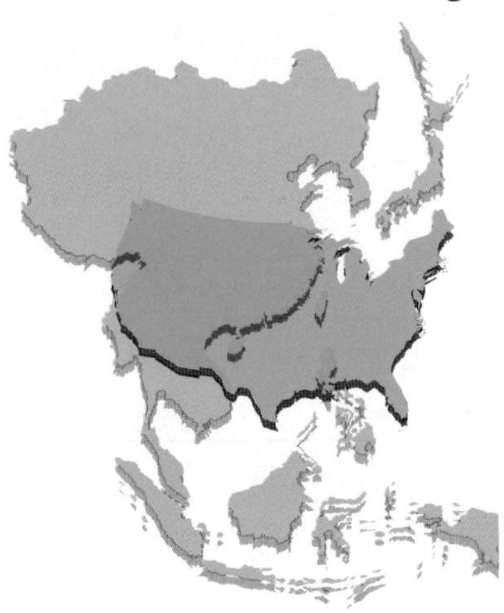

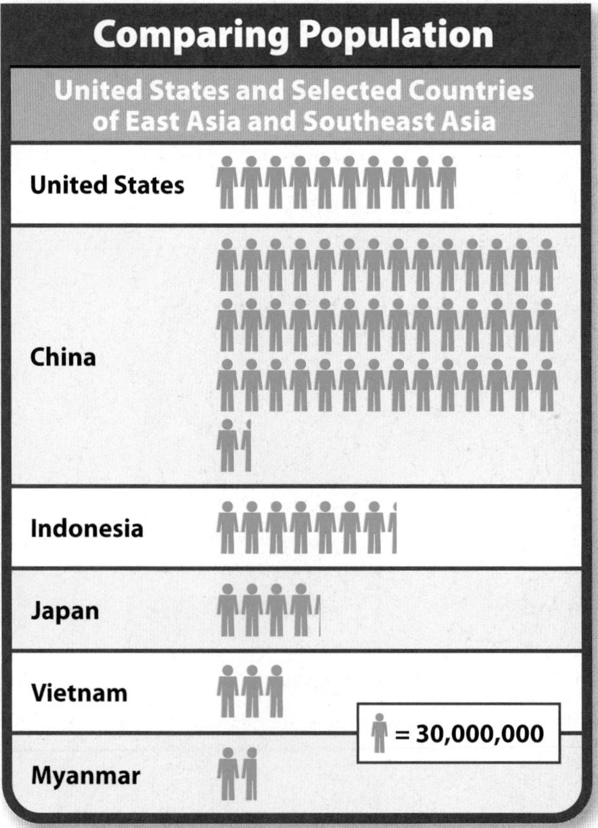

Source: *World Population Data Sheet,* 2005.

GEO Fast Facts

Largest Desert

Gobi (Mongolia and China) 500,000 sq. mi. (1,295,000 sq. km)

Longest River

Chang Jiang (Yangtze River) (China) 3,434 mi. (5,525 km) long

Largest Lake

Tonle Sap (Cambodia) 9,500 sq. mi. (24,605 sq. km)

UNIT 9

Regional Atlas

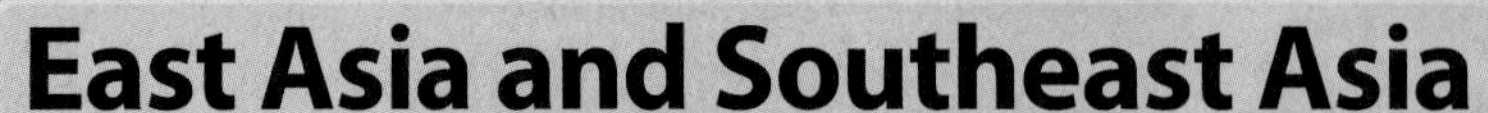

East Asia and Southeast Asia

PHYSICAL

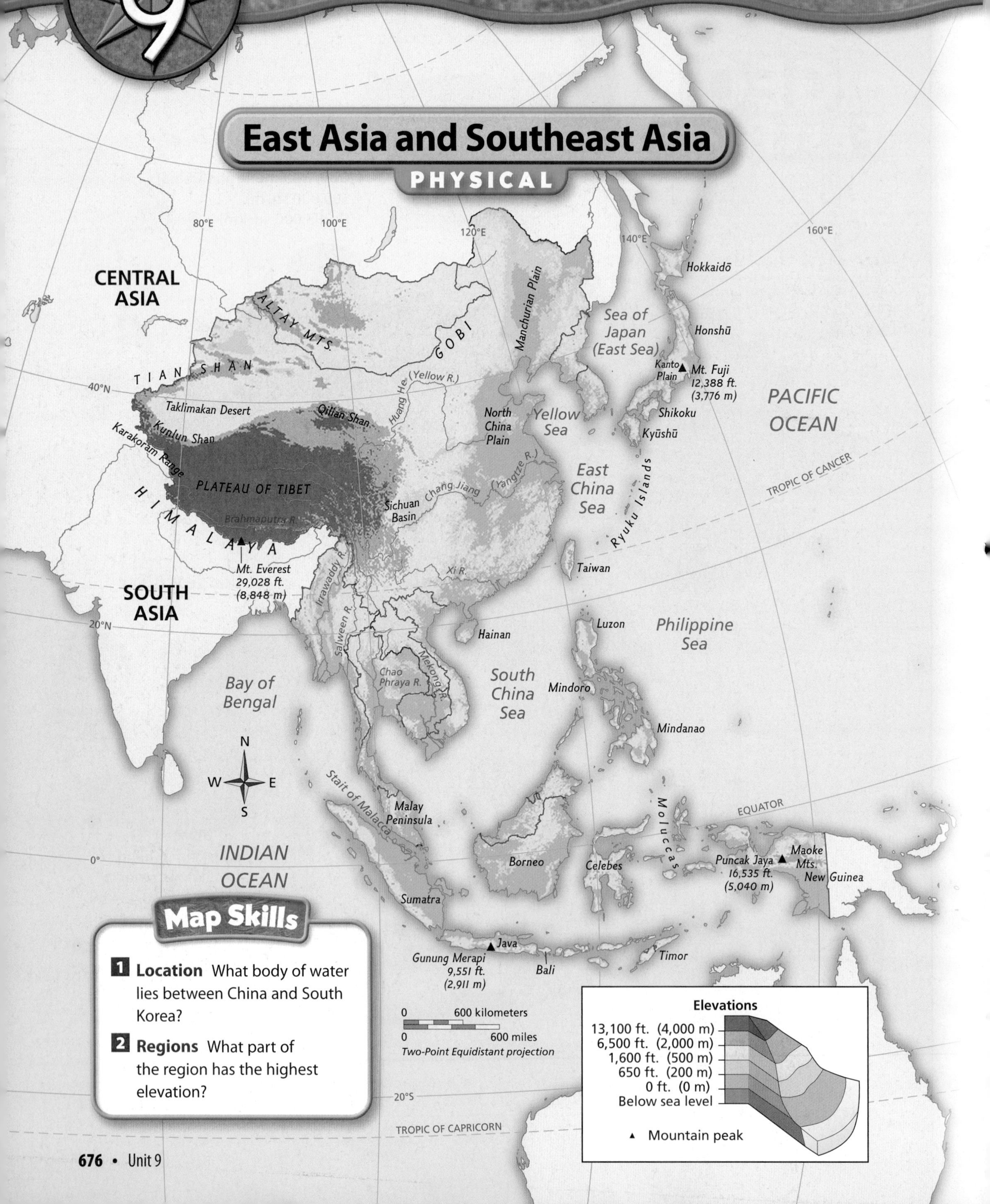

80°E
100°E
120°E
140°E
160°E
40°N
20°N
0°
20°S
CENTRAL ASIA
SOUTH ASIA
ALTAY MTS.
GOBI
Manchurian Plain
TIAN SHAN
Taklimakan Desert
Qilian Shan
Huang He (Yellow R.)
Kunlun Shan
Karakoram Range
PLATEAU OF TIBET
HIMALAYA
Brahmaputra R.
Mt. Everest 29,028 ft. (8,848 m)
Sichuan Basin
Chang Jiang (Yangtze R.)
North China Plain
Xi R.
Irrawaddy R.
Salween R.
Mekong R.
Chao Phraya R.
Hainan
Sea of Japan (East Sea)
Hokkaidō
Honshū
Kanto Plain
Mt. Fuji 12,388 ft. (3,776 m)
Shikoku
Kyūshū
Yellow Sea
East China Sea
Ryuku Islands
Taiwan
PACIFIC OCEAN
TROPIC OF CANCER
Luzon
Philippine Sea
Mindoro
Mindanao
South China Sea
Bay of Bengal
Stait of Malacca
Malay Peninsula
Borneo
Celebes
Moluccas
Puncak Jaya 16,535 ft. (5,040 m)
Maoke Mts.
New Guinea
EQUATOR
INDIAN OCEAN
Sumatra
Java
Gunung Merapi 9,551 ft. (2,911 m)
Bali
Timor
TROPIC OF CAPRICORN
N
S
E
W
0 600 kilometers
0 600 miles
Two-Point Equidistant projection
Elevations
13,100 ft. (4,000 m)
6,500 ft. (2,000 m)
1,600 ft. (500 m)
650 ft. (200 m)
0 ft. (0 m)
Below sea level
▲ Mountain peak

Map Skills

1 Location What body of water lies between China and South Korea?

2 Regions What part of the region has the highest elevation?

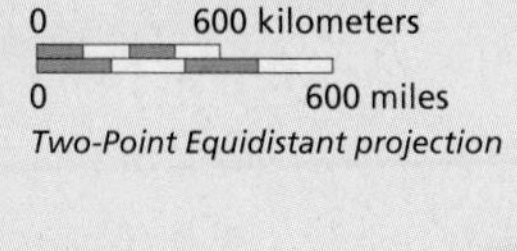

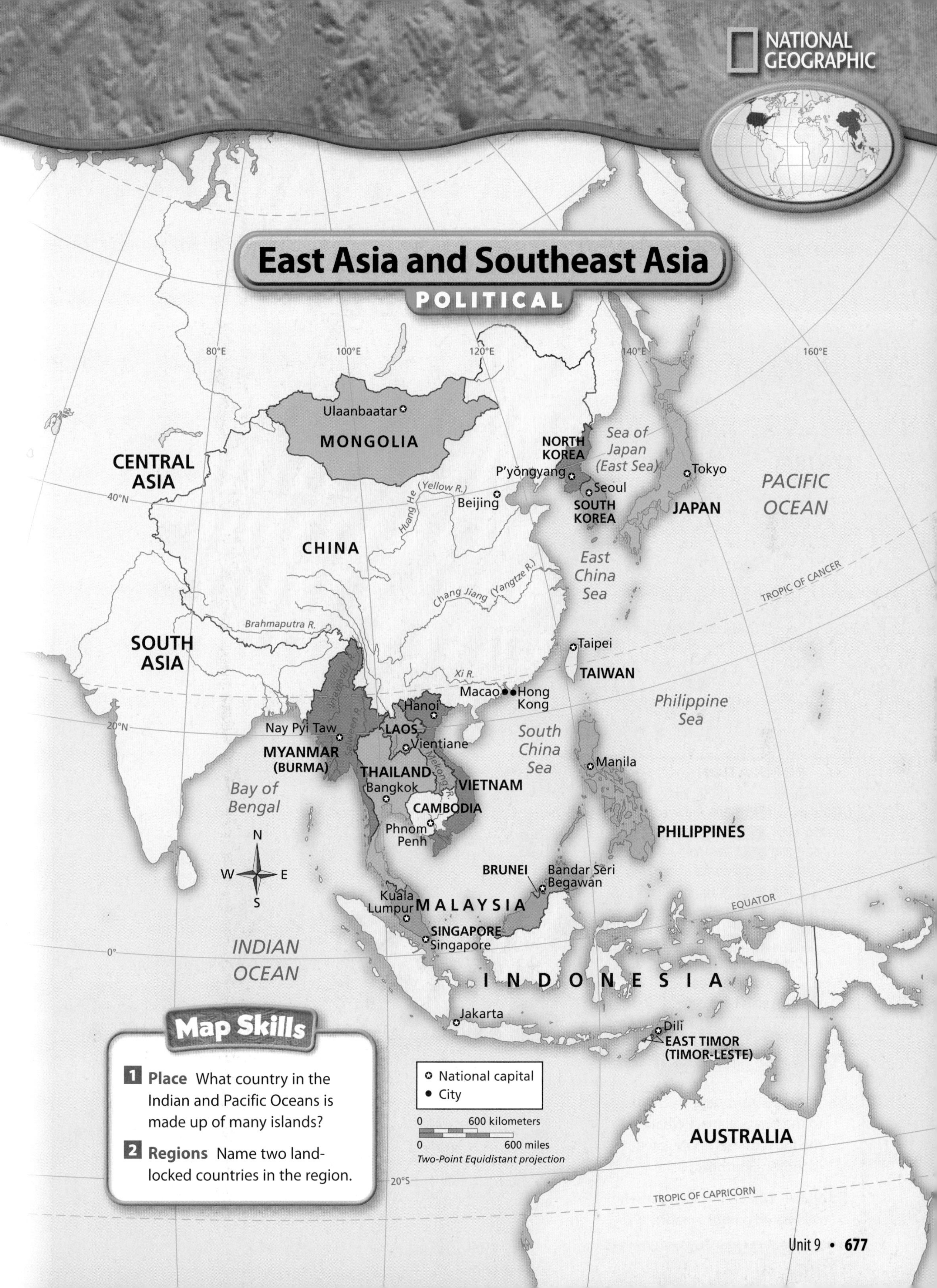

NATIONAL GEOGRAPHIC
East Asia and Southeast Asia
POLITICAL
80°E
100°E
120°E
140°E
160°E
40°N
20°N
0°
20°S
Ulaanbaatar
MONGOLIA
CENTRAL ASIA
NORTH KOREA
Sea of Japan (East Sea)
P'yŏngyang
Tokyo
Seoul
SOUTH KOREA
JAPAN
PACIFIC OCEAN
(Yellow R.)
Huang He
Beijing
CHINA
East China Sea
TROPIC OF CANCER
Chang Jiang (Yangtze R.)
Brahmaputra R.
SOUTH ASIA
Taipei
TAIWAN
Xi R.
Macao
Hong Kong
Philippine Sea
Irrawaddy R.
Salween R.
Hanoi
Nay Pyi Taw
LAOS
Vientiane
South China Sea
MYANMAR (BURMA)
THAILAND
Mekong R.
Manila
Bangkok
VIETNAM
Bay of Bengal
CAMBODIA
Phnom Penh
PHILIPPINES
N
W
E
S
BRUNEI
Bandar Seri Begawan
Kuala Lumpur
MALAYSIA
EQUATOR
SINGAPORE
Singapore
INDIAN OCEAN
INDONESIA
Jakarta
Dili
EAST TIMOR (TIMOR-LESTE)
National capital
City
0 600 kilometers
0 600 miles
Two-Point Equidistant projection
AUSTRALIA
TROPIC OF CAPRICORN
Map Skills
1 Place What country in the Indian and Pacific Oceans is made up of many islands?
2 Regions Name two land-locked countries in the region.

UNIT 9

Regional Atlas

East Asia and Southeast Asia

POPULATION DENSITY

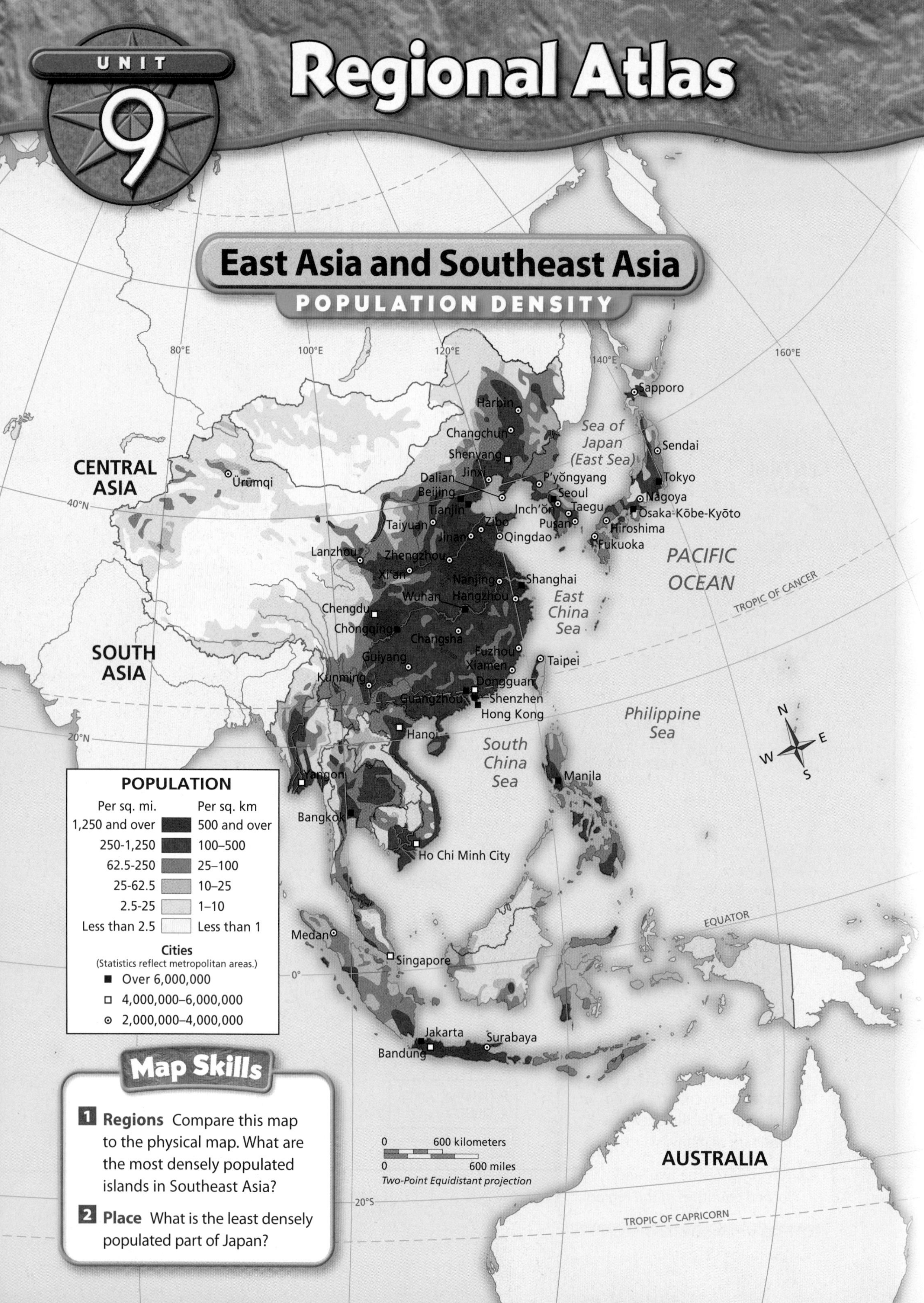

Map Skills

1. **Regions** Compare this map to the physical map. What are the most densely populated islands in Southeast Asia?
2. **Place** What is the least densely populated part of Japan?

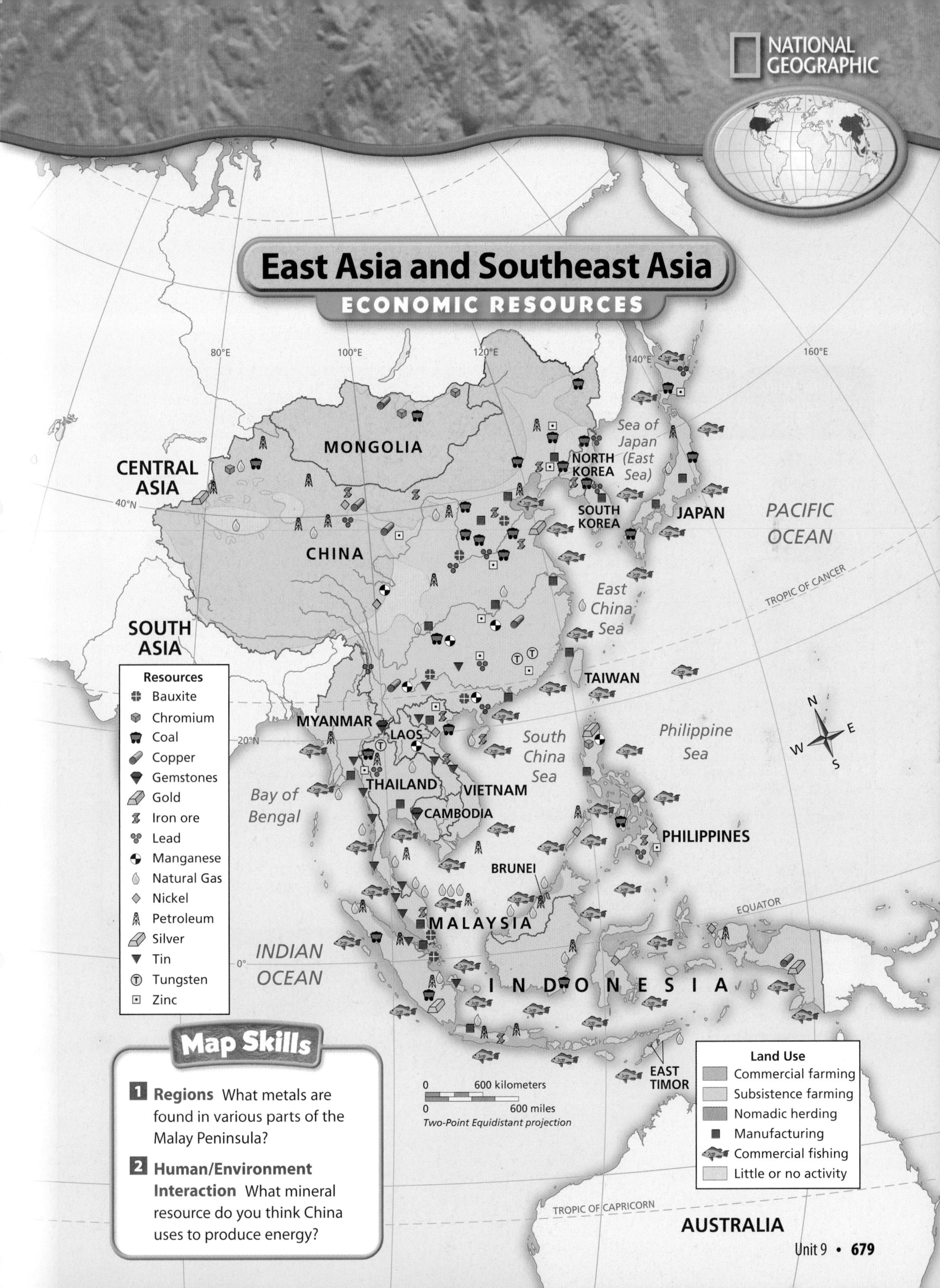

Map Skills

1. **Regions** What metals are found in various parts of the Malay Peninsula?
2. **Human/Environment Interaction** What mineral resource do you think China uses to produce energy?

Regional Atlas

East Asia and Southeast Asia

Country and Capital	Literacy Rate	Population and Density	Land Area	Life Expectancy (Years)	GDP* Per Capita (U.S. dollars)	Television Sets (per 1,000 people)	Flag and Language
BRUNEI, Bandar Seri Begawan	93.9%	400,000 180 per sq. mi. 72 per sq. km	2,228 sq. mi. 5,570 sq. km	74	$23,600	637	Malay
CAMBODIA, Phnom Penh	69.4%	13,300,000 190 per sq. mi. 73 per sq. km	69,900 sq. mi. 181,040 sq. km	56	$2,000	9	Khmer
CHINA, Beijing	90.9%	1,303,700,000 353 per sq. mi. 136 per sq. km	3,696,100 sq. mi. 9,572,855 sq. km	72	$5,600	291	Mandarin Chinese
EAST TIMOR (TIMOR-LESTE), Dili	58.6%	900,000 157 per sq. mi. 61 per sq. km	5,741 sq. mi. 14,869 sq. km	55	$400	information not available	Tetum, Portuguese
INDONESIA, Jakarta	87.9%	221,900,000 302 per sq. mi. 117 per sq. km	735,355 sq. mi. 1,904,561 sq. km	68	$3,500	143	Bahasa Indonesia
JAPAN, Tokyo	99%	127,700,000 875 per sq. mi. 338 per sq. km	145,869 sq. mi. 377,799 sq. km	82	$29,400	719	Japanese
UNITED STATES, Washington, D.C.	97%	296,500,000 80 per sq. mi. 31 per sq. km	3,717,796 sq. mi. 9,629,047 sq. km	78	$40,100	844	English

*Gross Domestic Product

Countries and flags not drawn to scale

East Asia and Southeast Asia

Country and Capital	Literacy Rate	Population and Density	Land Area	Life Expectancy (Years)	GDP* Per Capita (U.S. dollars)	Television Sets (per 1,000 people)	Flag and Language
LAOS Vientiane	66.4%	5,900,000 65 per sq. mi. 25 per sq. km	91,429 sq. mi. 236,800 sq. km	54	$1,900	10	Lao
MALAYSIA Kuala Lumpur	88.7%	26,100,000 205 per sq. mi. 79 per sq. km	127,317 sq. mi. 329,750 sq. km	73	$9,700	174	Bahasa Melayu
MONGOLIA Ulaanbaatar	97.8%	2,600,000 4 per sq. mi. 2 per sq. km	604,826 sq. mi. 1,566,492 sq. km	64	$1,900	58	Khalkha Mongol
MYANMAR Nay Pyi Taw	85.3%	50,500,000 193 per sq. mi. 75 per sq. km	261,228 sq. mi. 676,577 sq. km	60	$1,700	7	Burmese
NORTH KOREA P'yŏngyang	99%	22,900,000 492 per sq. mi. 190 per sq. km	46,541 sq. mi. 120,541 sq. km	71	$1,700	55	Korean
PHILIPPINES Manila	92.6%	84,800,000 732 per sq. mi. 283 per sq. km	115,830 sq. mi. 299,998 sq. km	70	$5,000	110	Filipino, English
UNITED STATES Washington, D.C.	97%	296,500,000 80 per sq. mi. 31 per sq. km	3,717,796 sq. mi. 9,629,047 sq. km	78	$40,100	844	English

Sources: *CIA World Factbook*, 2005; Population Reference Bureau, *World Population Data Sheet*, 2005.

For more country facts, go to the **Nations of the World Databank** at glencoe.com.

Regional Atlas

East Asia and Southeast Asia

Country and Capital	Literacy Rate	Population and Density	Land Area	Life Expectancy (Years)	GDP* Per Capita (U.S. dollars)	Television Sets (per 1,000 people)	Flag and Language
SINGAPORE Singapore	92.5%	4,300,000 17,992 per sq. mi. 6,947 per sq. km	239 sq. mi. 619 sq. km	79	$27,800	341	Mandarin
Seoul SOUTH KOREA	97.9%	48,300,000 1,260 per sq. mi. 487 per sq. km	38,324 sq. mi. 99,259 sq. km	77	$19,200	364	Korean
Taipei TAIWAN†	96.1	22,700,000 1,625 per sq. mi. 627 per sq. km	13,969 sq. mi. 36,180 sq. km	76	$25,300	327	Mandarin Chinese
THAILAND Bangkok	92.6%	65,000,000 328 per sq. mi. 127 per sq. km	198,116 sq. mi. 513,118 sq. km	71	$8,100	274	Thai
Hanoi VIETNAM	90.3%	83,300,000 650 per sq. mi. 251 per sq. km	128,066 sq. mi. 331,689 sq. km	72	$2,700	184	Vietnamese
UNITED STATES Washington, D.C.	97%	296,500,000 80 per sq. mi. 31 per sq. km	3,717,796 sq. mi. 9,629,047 sq. km	78	$40,100	844	English

*Gross Domestic Product

† The People's Republic of China claims Taiwan as its 23rd province.

Countries and flags not drawn to scale

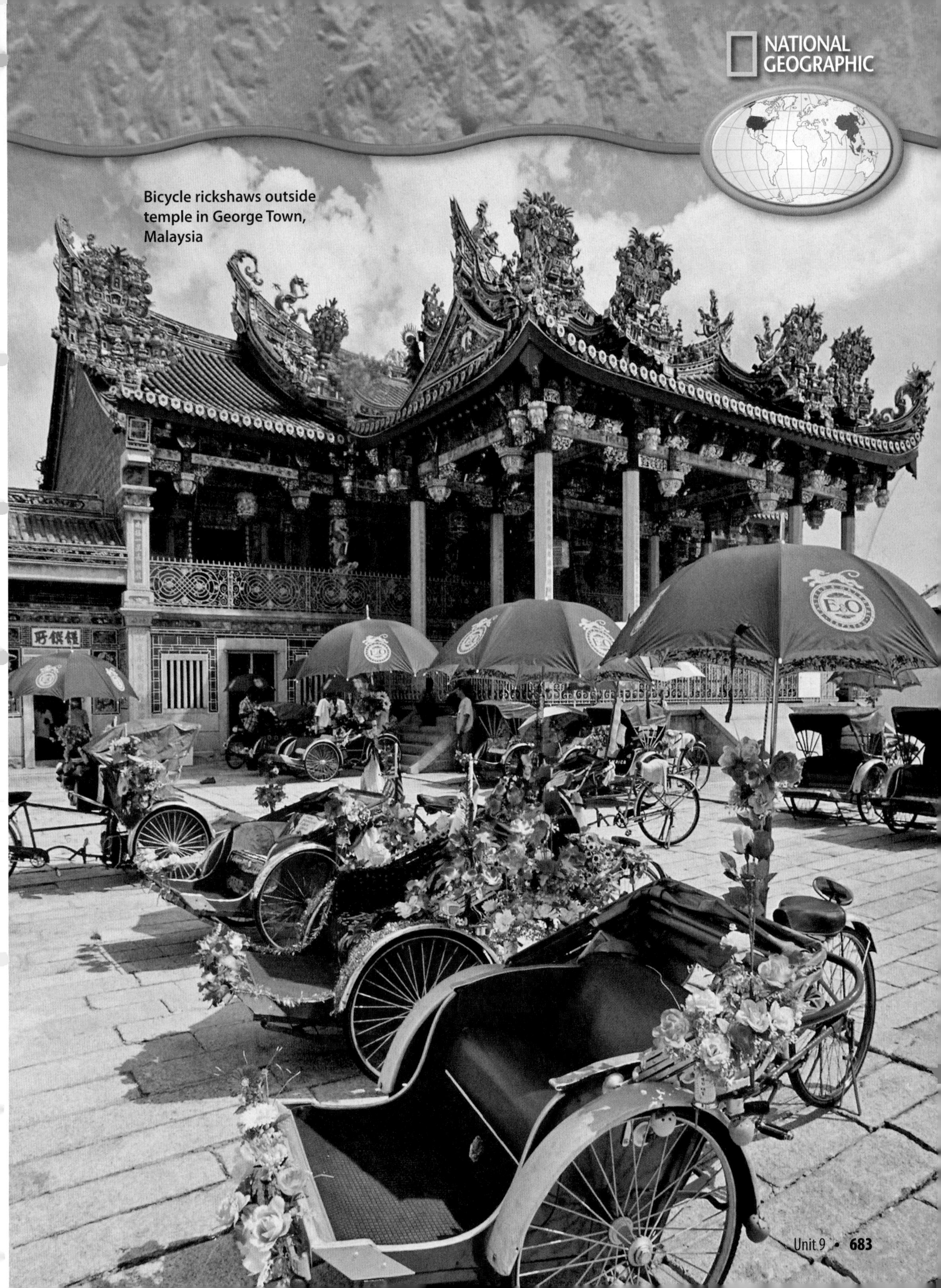

Bicycle rickshaws outside temple in George Town, Malaysia

Reading Social Studies

Identifying Problems and Solutions

Learn It!

Have you ever had a problem that was hard to solve? What did you do? Did you ask a family member or friend to help you find a solution, or a way to solve your problem? Textbooks often describe problems that people face and how they try to solve them. As you read this unit, look for problems for which people had to find solutions.

- Read the paragraph below.
- Ask yourself these questions:
  —What is the problem?
  —How did the Chinese solve this problem?

> In 1998 China's government was forced to act to protect the environment. That year, the flooding of the Chang Jiang killed thousands of people. The cutting of trees and the draining of wetlands for farming had kept land upriver from soaking up the floodwaters. In response, the government began to turn farmland back into forests and wetlands.
>
> —*from page 732*

Problem
The flooding of the Chang Jiang killed thousands. The cutting of trees and draining of wetlands prevented floodwaters from being absorbed.

→

Solution
The Chinese government began to turn farmland back into forests and wetlands.

Reading Tip

As you read the chapters, identify a problem that people have faced and think about what would be the best way to solve it. Continue to read to find out if the writer mentions your solution.

The Land

Main Idea Tectonic plate movements have created mountains and caused powerful earthquakes in parts of the region.

Geography and You Imagine living in a city where the ground could shake at any time—and often does. Read to find out why parts of East Asia and Southeast Asia are likely to have earthquakes and volcanic eruptions.

The vast region of East Asia and Southeast Asia extends from the mountains of inland China eastward to the Pacific shores of Japan. The region also sweeps north to south from the highlands of northeastern China to the tropical islands of Indonesia.

Landforms of East Asia

East Asia occupies much of the Asian continent south of Russia. China and Mongolia extend over most of East Asia's landmass. The other East Asian countries—North Korea, South Korea, Japan, and Taiwan—lie on peninsulas or islands.

Mountain ranges, such as the Himalaya and the Kunlun Shan, slice through western East Asia. Between these ranges is the Plateau of Tibet. About 15,000 feet (4,572 m) in height, the Plateau is so high that it is called the Roof of the World. Plate movements cause earthquakes in these mountain areas. In 2008 a major earthquake in southwestern China killed nearly 70,000 people.

East of the mountains lie East Asia's major lowlands—the North China and Manchurian Plains. Narrow lowlands also line the coasts of Korea and Japan. Most East Asians live in these fertile areas.

In the Pacific Ocean off East Asia's coast lies an arc, or curve, of mountainous islands. These islands include Japan, which forms an **archipelago** (AHR·kuh·PEH·luh·GOH), or chain of islands, and Taiwan off the coast of southeastern China. Formed millions of years ago by undersea volcanic activity, East Asia's islands are part of the Ring of Fire. This is an area bordering the Pacific Ocean where plate movements cause many earthquakes and volcanic eruptions.

The city of Hong Kong, with a population of more than 6 million, is located along the coast of southern China. ***Human-Environment Interaction*** **Why does much of East Asia's population live in fertile lowlands?**

Landforms of Southeast Asia

South of China lies Southeast Asia. This area has a mainland of peninsulas as well as thousands of islands. Several countries—Myanmar (Burma), Thailand, Laos, Cambodia, and Vietnam—lie entirely on Southeast Asia's mainland. Countries that are partly or entirely made up of islands are Indonesia, East Timor, Malaysia, Singapore, Brunei, and the Philippines. Both Indonesia and the Philippines are archipelagoes.

Mainland Southeast Asia is crossed by **cordilleras,** or mountain ranges that run side by side. Fertile river plains and deltas separate the ranges. These lowlands are home to most mainland Southeast Asians.

South and east of the mainland are thousands of mountainous islands. Part of the Ring of Fire, these islands hold many active volcanoes. **Despite** dangers, these volcanoes provide rich soil for farming.

Earthquakes also challenge the islands of Southeast Asia. A great tsunami in 2004 was caused by an earthquake on the Indian Ocean floor. The tsunami washed over the coastal lowlands of more than a dozen countries. More than 300,000 people died, and thousands of homes and businesses were destroyed.

Reading Check **Explaining** What is the Ring of Fire?

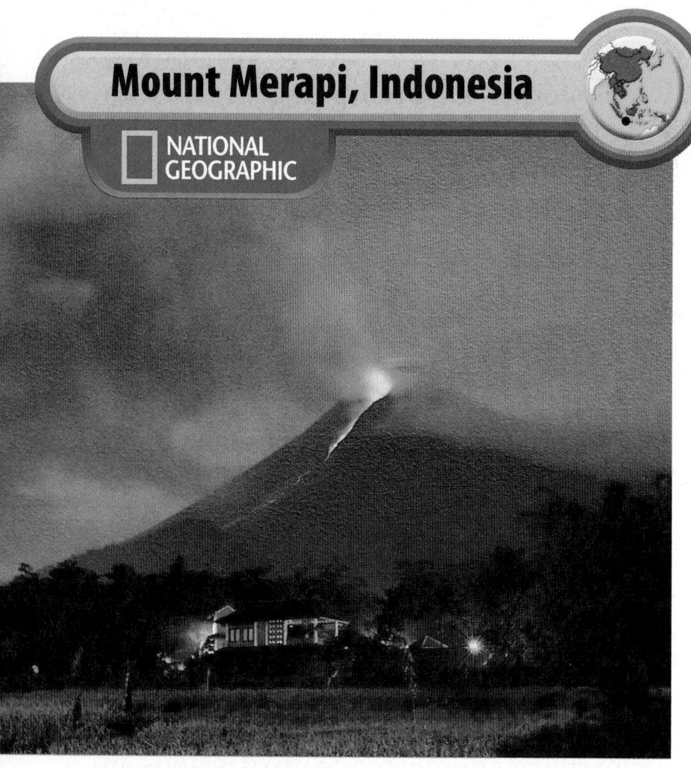

▲ Mount Merapi in Indonesia rumbles often, threatening to erupt. Villagers living nearby frequently evacuate as a precaution. ***Regions*** **How do volcanoes benefit people?**

Seas and Rivers

Main Idea Seas and rivers play an important role in agriculture and trade in the region.

Geography and You What bodies of water are located near your community? How do people use that water? Read to find out why waterways are important to the people of the region.

The long, winding coastlines of East Asia and Southeast Asia are washed by two major oceans—the Indian and the Pacific—as well as by many seas. Oceans and seas have influenced the region's history. Japan, for example, developed a unique culture partly because its surrounding seas kept it isolated from the Asian mainland.

Oceans and seas have also been important for the region's economies. They have served as routes for trade. The South China Sea, stretching south from Taiwan to Southeast Asia, carries much of the world's shipping traffic. Singapore has become one of the world's busiest ports because it controls the Strait of Malacca, which connects the South China Sea and the Indian Ocean.

The region's island and coastal areas also depend on the sea for food. Japan, South Korea, Taiwan, and China have the world's biggest deep-sea fishing industries.

Rivers of East Asia

The most important rivers in East Asia flow through China. They begin in the Plateau of Tibet and flow eastward to the Pacific Ocean. The Huang He (Yellow River) is northern China's major river system. This river is called "yellow" because it carries tons of fine, yellow-brown soil called **loess** (LEHS) that blows in from deserts in western China. When deposited, the rich soil—along with the river's water—makes

Three Gorges Dam: 2000 and 2006

The photo on the right shows the massive areas of land that were flooded as the Three Gorges Dam neared completion in 2006. More than 1 million people will have to move from areas near the dam as some 1,200 towns and villages will be covered by water.
Human-Environment Interaction **Why is the Chang Jiang important to China?**

the North China Plain a major wheat-growing area. Throughout China's history, the Huang He has regularly flooded the land, destroying homes and drowning many people. As a result, the Chinese called the Huang He "China's sorrow."

China also has another great river called the Chang Jiang (Yangtze River). The Chang Jiang is Asia's longest river. It flows for about 3,400 miles (5,741 km) through spectacular **gorges,** or canyons, and broad plains. It then empties into the ocean at the port city of Shanghai. The Chang Jiang provides water for a large farming area where more than half of China's rice and other grains grow. It is also an important trade route. Ocean-going ships can travel far upriver, and barges carry goods even farther.

The Chinese are **constructing,** or building, a huge dam on the Chang Jiang. When completed in 2011, the Three Gorges Dam—607 feet (185 m) high and 1.4 miles (2.3 km) wide—will be the world's largest dam. The Three Gorges Dam is expected to prevent floods and to supply a large amount of hydroelectric power. When finished, however, it will create a lake that will force over a million people from their homes and cover vast areas with water.

Rivers of Southeast Asia

Southeast Asia's major rivers are located on the mainland. They begin in northern highlands and in southern China. Several of them flow southward toward the Gulf of Thailand, which is an arm of the South China Sea.

Southeast Asia's major rivers include the Irrawaddy (IHR·ah·WAH·dee) and the Salween in Myanmar, and the Chao Phraya (chow PRY·uh) in Thailand. Another river, the Mekong (MAY·KAWNG), begins its 2,600-mile (4,184-km) journey in China and forms the border between Thailand and Laos. It twists and turns through Cambodia and southern Vietnam before emptying into the South China Sea. Warm temperatures and heavy rains make the Mekong region a fertile rice-growing area. This region has a large population.

Identifying Central Issues Why are the Chinese building the Three Gorges Dam?

A Wealth of Natural Resources

Main Idea **The region's valuable resources support its growing economies.**

Geography and You What do you think is the most important natural resource today? Why? Read to find out about the resources of East Asia and Southeast Asia.

East Asia and Southeast Asia have rich natural resources. Mineral resources have helped develop the economies in the region. Some countries have important oil reserves. China is the region's economic giant and has the most resources.

Energy Resources

There is an abundant supply of energy resources in East Asia and Southeast Asia. China has large oil deposits in the South China Sea as well as in the western part of the country. Other countries with rich petroleum resources are Indonesia, Brunei (bru·NY), and Malaysia. Vietnam also has rich oil reserves offshore in the South China Sea. Major coal producers are China, Indonesia, North Korea, South Korea, and Japan. The Philippines and Vietnam also mine coal.

Several countries in the region use hydroelectric power to meet their energy needs. China produces electricity from the Three Gorges Dam on the Chang Jiang.

TIME GLOBAL CITIZENS

NAME: SANGDUEN "LEK" CHAILERT

HOME COUNTRY: Thailand

ACHIEVEMENT: Though her nickname Lek means "small one," Sangduen Chailert runs two healing centers for Asia's largest mammals: the Elephant Nature Park and Elephant Haven. At the nature park, Chailert and her staff of mahouts (elephant handlers), animal doctors, and volunteers care for elephants that have suffered wounds or abuse by their owners. After the elephants heal, Chailert transfers them to Elephant Haven, a 2,000-acre preserve in which they live out their days. Today, 25 elephants live in Haven's forest. Despite their respected place in Thai culture, the country's elephants badly need such a champion. Logging has destroyed much of their habitat, and their numbers have dwindled from 100,000 a century ago to between 2,500 and 5,000 today.

QUOTE: "I can't turn my back on the elephants. I can look in their eyes and see fear. Somebody has to stand up for them."

PALANI MOHAN FOR TIME

CITIZENS IN ACTION Why do some people choose to defend the rights of animals? Do you think that such causes are important? Why or why not?

Dams on Japan's swift, short rivers provide hydroelectric power for that country.

Minerals

The region has a wealth of minerals. Indonesia, Malaysia, and China are leading producers of tin. China has one of the largest iron ore deposits in the world. North Korea, Vietnam, Malaysia, Indonesia, and the Philippines also mine iron ore. Chromium, manganese, nickel, and tungsten, which are used to make high-quality steel, are found in China and the Philippines. **Tungsten** is also used to make lightbulbs and rockets.

Gems are plentiful in East Asia and Southeast Asia. Sapphires and rubies are mined in Myanmar, Thailand, Cambodia, and Vietnam. Pearls are harvested in Japan and the Philippines.

Other Resources

Some trees in the region's forests are valued for their wood. **Teak** is used to make buildings and ships because it is strong and durable. Myanmar, Indonesia, and Thailand have much teak. Mahogany from the Philippines is used in paneling and furniture.

NATIONAL GEOGRAPHIC

Bamboo Scaffolding

Bamboo, a grass with a woody stem, is strong enough to be used as scaffolding for construction work. ***Human-Environment Interaction*** **What other valuable trees grow in the region?**

Reading Check **Identifying** Which metals in the region help make high-quality steel?

Section Review

Social Studies ONLINE
Study Central™ To review this section, go to glencoe.com.

Vocabulary

1. **Explain** the meaning of:
   a. archipelago **b.** cordillera **c.** loess **d.** gorge **e.** tungsten **f.** teak

Main Ideas

2. **Determining Cause and Effect** Use a diagram like the one below to show the effects of tectonic activity within the region.

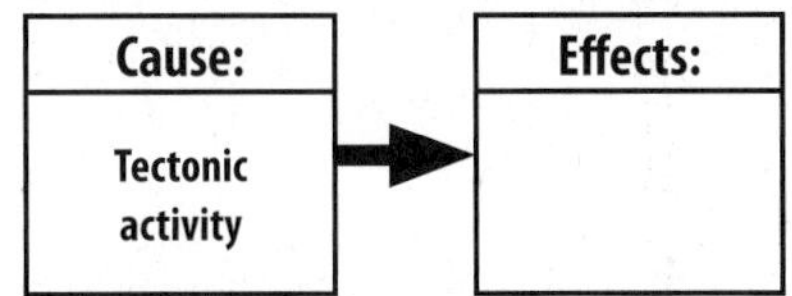

3. **Explaining** Name three major rivers in the region. Why are they important?
4. **Identifying** What energy resources are found in East Asia and Southeast Asia?

Critical Thinking

5. BIG Idea How does the landscape of western China differ from that of much of eastern China?
6. **Challenge** How have natural resources helped some countries develop their economies?

Writing About Geography

7. **Expository Writing** Write a paragraph comparing the physical geography of East Asia with the physical geography of Southeast Asia.

Guide to Reading

BIG Idea

Geographic factors influence where people settle.

Content Vocabulary

- *dzud (p. 696)*
- landslide *(p. 698)*

Academic Vocabulary

- series *(p. 697)*
- site *(p. 697)*

Reading Strategy

Making Connections Use a diagram like the one below to identify the air masses that affect climate in the region. Also, briefly explain the effects of each air mass.

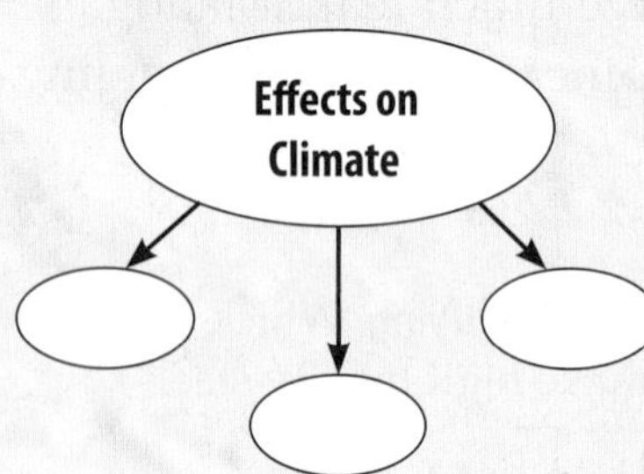

Climate Regions

Spotlight Video

Picture This The rafflesia is the largest—and maybe the smelliest—flower in the world. The rafflesia is a type of lily that grows in the rain forests of Malaysia and Indonesia. The flower measures up to 3 feet (1 m) wide and gives off the smell of rotting meat to attract flies that pollinate it. Rafflesia and other flowers in this region are used to make lifesaving medicines. Read Section 2 to learn more about East Asia and Southeast Asia's climate and vegetation.

▼ Measuring the world's largest flower in Borneo

Effects on Climate

Main Idea Wind patterns influence the climates in East Asia and Southeast Asia.

Geography and You Do you live in a windy area? Read to find out how different wind patterns greatly affect climates in East Asia and Southeast Asia.

NATIONAL GEOGRAPHIC

Differing Climates

The cold climate of Mongolia requires warm clothing. In tropical Myanmar (inset), monsoons can bring warm rains. ***Place*** **What helps cause Mongolia's cold winter temperatures?**

The climates of East Asia and Southeast Asia are shaped by three different air masses. One brings cold, dry air from the Arctic region. The second brings cool, dry air from the west eastward across Asia. The third carries warm, moist air from the Pacific Ocean.

These winds, along with the region's landforms, shape the climates of East Asia and Southeast Asia. In winter, cold Arctic winds sweep across Siberia and lower temperatures in Mongolia and northern China. The average January temperature around Ulaanbaatar (OO·LAHN·BAH·TAWR), Mongolia's capital, is about –15°F (–26°C).

The cold Siberian winds do not reach the vast parts of the region that lie below northern China. In these areas, monsoon winds like those in South Asia are common. In the summer, warm, moist Pacific air blows from the southeast into East Asia, bringing rain. In some inland parts of East Asia, the summer monsoon can bring as much as 80 percent or more of the region's yearly rainfall. In the winter, dry winds blow outward from the Asian continent to the ocean. Little rain falls during these months. Areas of Southeast Asia that are closest to the Equator have warm temperatures year-round. Rain falls more evenly there throughout the year.

Ocean currents also affect climate, especially on islands such as those of Japan. A warm-water current flows north along southeastern Japan. It adds moisture to the winter monsoon as it warms the land. A cold current flows southwest along Japan's Pacific coasts. It brings harsh, cold winters to Japan's northernmost areas.

The warm waters of the Tropics help form strong, hurricane-like storms called typhoons. Typhoons that arise in the Pacific can blow across coastal East Asia. High winds, large waves, and heavy rains during typhoons can cause much damage.

Determining Cause and Effect How does Siberia's landscape affect climate in China?

Climate Zones

Main Idea **East Asia generally has middle latitude climates like those in the United States, while Southeast Asia has mostly tropical climates.**

Geography and You Think about what winters are like where you live. Read to find out how much winter climates in East Asia and Southeast Asia differ from place to place.

A person traveling across East Asia and Southeast Asia would need clothes to suit almost every possible climate. China, the region's largest country, has the greatest range of latitudes and landforms. It also has the greatest variety of climates.

Dry Continental Climates

The climate is dry in the northern and far western parts of East Asia. In summer, this is caused by dry continental air blowing across Asia from the west. In winter, cold Arctic air from the north also carries little moisture. In addition, monsoon winds from the Pacific have released all their moisture by the time they reach inland areas.

Most of Mongolia and northern China receive only enough rain to create a steppe climate. Extensive grasslands grow in this mostly dry climate. The grasses support cattle, sheep, goats, and camels. Some years, however, a small amount of rain falls, but summers are dry. The weather pattern of a dry summer followed by a harsh winter is called a ***dzud*** by the people of Mongolia.

NATIONAL GEOGRAPHIC **Maps In MOtion** See StudentWorks™ Plus or glencoe.com.

Figure 1 **Climates of East Asia and Southeast Asia**

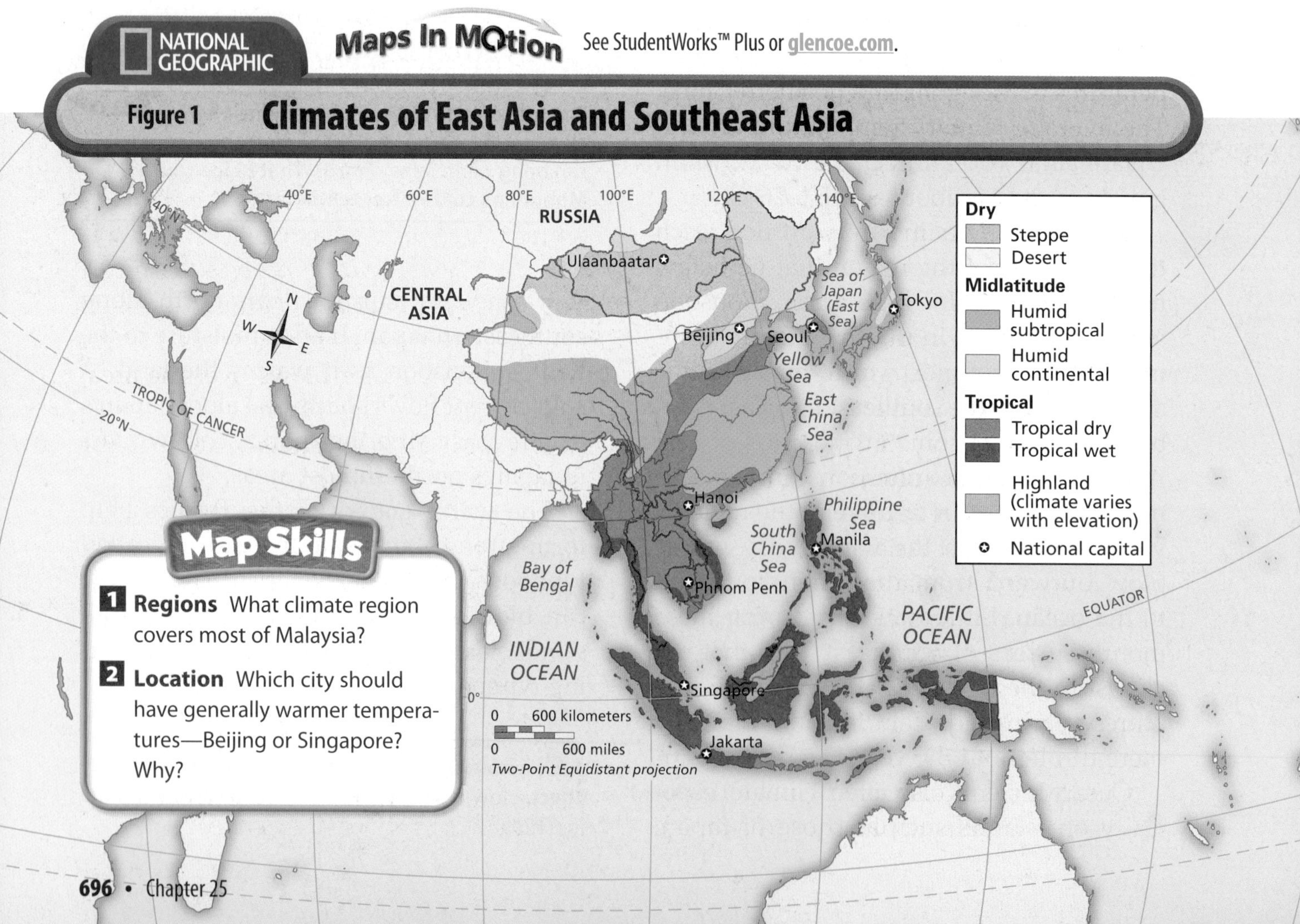

Map Skills

1 Regions What climate region covers most of Malaysia?

2 Location Which city should have generally warmer temperatures—Beijing or Singapore? Why?

Mongolians fear *dzud* conditions. This is because dry summers decrease the available food for their herds, which may then be too weak to survive a harsh winter. In recent years, Mongolia had a **series** of *dzud* conditions for three consecutive years. About a third of all the livestock in the country died—about 10 million animals.

Some areas in East Asia are extremely dry because they receive little rain throughout the year. Moisture that might reach these areas is blocked by surrounding mountains. The Gobi of southern Mongolia and northern China is a vast desert region. The name *Gobi* is a Mongolian word meaning "place without water." Even less rain falls in the Taklimakan (TAH·kluh·muh·KAHN), a smaller desert in western China. The eastern part of this desert averages less than 0.5 inch (1.27 cm) of rain per year. Not surprisingly, few people live in the Taklimakan.

Wet Continental Climates

Northeastern China, the northern part of the Korean Peninsula, and northern Japan have humid continental climates. This zone is marked by great differences in temperature during the year. Summers are warm, and winters are cold. On Hokkaido (hoh·KY·doh), the northernmost island in Japan, the average August temperature is 70°F (21°C), and the average January temperature falls to only 16°F (–9°C). Winters can be snowy. Snow often covers the island between November and April.

The rest of East Asia and the northern part of Southeast Asia have a humid subtropical climate. This climate is similar to that in the southeastern United States. Summer temperatures are slightly higher in the humid continental zone, and winters are much milder. The average July temperature in Hanoi, Vietnam, is 74°F (23°C)—not much more than in Hokkaido. The average January temperature in Hanoi, however, is 63°F (17°C). Thus, this climate is good for growing rice.

Winter in Northern Japan

NATIONAL GEOGRAPHIC

▲ These northern macaque monkeys grow thick coats of fur to protect them from the winter's cold in the northern islands of Japan. ***Place*** **What type of climate does northern Japan have?**

Tropical Climates

Much of Southeast Asia lies in the Tropics. This area sees little change in temperature during the year. It receives the direct rays of the sun in the summer, making temperatures very warm. In winter, warm air from the Equator blows over the area, keeping temperatures warm.

Though **sites,** or locations, near the Equator are generally warmer, sea breezes help keep coastal temperatures more moderate. Singapore, for example, has never had a temperature above 97°F (36°C) even though it is almost on the Equator. Altitude also keeps temperatures low in tropical areas.

Social Studies ONLINE

Student Web Activity Visit glencoe.com and complete the Chapter 25 Web Activity about Southeast Asian rain forests.

In the mountains that cross the islands of Borneo and New Guinea, temperatures can be quite cold. The Maoke (MOW·kay) Mountains in New Guinea, for example, are sometimes covered with snow, even though they are near the Equator.

Rains can be heavy in the Tropics. Areas of Indonesia receive about 120 inches (305 cm) a year. As much as 28 inches (71 cm) can fall in a day. Cyclones also affect Southeast Asia. In 2008 a cyclone killed more than 80,000 people in Myanmar.

Abundant rains support the growth of tropical rain forests. Rain forests are home to a tremendous variety of plants and animals. A small area can have as many as 100 different kinds of trees and countless other plants, including teak or other valuable woods. Malaysia's rain forests alone contain more than 14,000 species of flowering plants. The region's rain forests, however, are being cut down at a rapid rate. Thailand has lost nearly half its rain forests in less than 40 years. Other countries have undergone similar deforestation.

Deforestation has contributed to natural disasters in the region. For example, heavy rains soak treeless hillsides, leading to **landslides,** situations in which soil is washed down the hills. Landslides have buried villages in mud and killed villagers.

Highland Climates

Various areas of East Asia and Southeast Asia have highland climates, including mountain areas in Indonesia. Another highland climate zone occurs in southwestern China, which includes the Himalaya and the Plateau of Tibet. Temperatures in these areas tend to be cool. Temperatures drop even more in the mountains. Because these areas receive dry continental air, they tend to have dry landscapes.

Reading Check **Explaining** Why are some sites near the Equator cool?

Social Studies ONLINE
Study Central™ To review this section, go to glencoe.com.

Vocabulary

1. **Explain** the meaning of dzud and *landslide* by using them in a paragraph.

Main Ideas

2. **Explaining** How do ocean currents affect climates in Japan?
3. **Identifying** Use a diagram like the one below to identify climate zones found in China. Also identify characteristics of each climate.

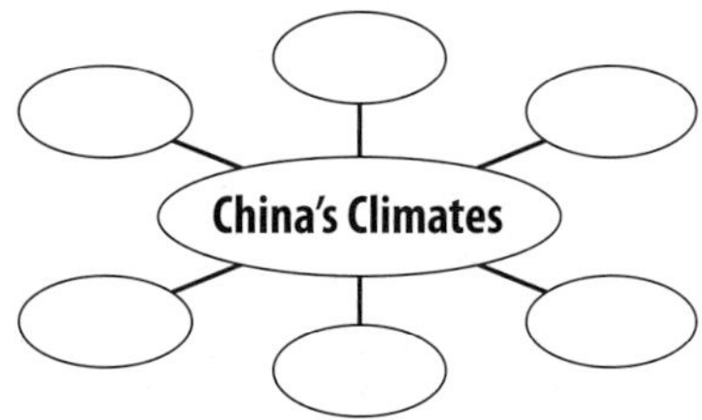

Critical Thinking

4. BIG Idea Why do so few people live in the Taklimakan?
5. **Making Connections** How does deforestation contribute to the problem of landslides in tropical rain forest regions?
6. **Challenge** Which climate zones in China are similar to climate zones in the United States?

Writing About Geography

7. **Using Your FOLDABLES** Use your Foldable to describe how elevation affects climate in East Asia and Southeast Asia.

Visual Summary

Rebuilding after a tsunami, Indonesia

Landforms

- The region's mountains lie in the western parts of East Asia.
- Most East Asians live in coastal lowland plains.
- Southeast Asia includes a mainland of peninsulas and thousands of islands.
- Areas of East Asia and Southeast Asia lie along the Ring of Fire and can experience volcanoes and earthquakes.

Han River, Vietnam

Seas and Waterways

- Oceans and seas provide important shipping routes as well as supplies of fish for food.
- China's major rivers—the Huang He (Yellow) and the Chiang Jiang (Yangtze)—support farming but also produce floods.
- The Chinese have built the Three Gorges Dam on the Chang Jiang to prevent flooding and to provide hydroelectric power.

Maoke Mountains, Indonesia

Natural Resources

- Oil deposits are found in China, Malaysia, and Indonesia.
- China has iron ore and coal deposits that support its steel industry.
- Southeast Asia's tropical rain forests provide valuable woods, such as teak and mahogany.

Coal miners, China

Climates

- Three dominant air masses affect climate in much of East Asia and Southeast Asia.
- Monsoons bring seasonal rainy or dry conditions to some parts of the region.
- Ocean currents affect climate, especially on islands such as those of Japan.

Study anywhere, anytime! Download quizzes and flash cards to your PDA from **glencoe.com**.

CHAPTER 25 ASSESSMENT

STANDARDIZED TEST PRACTICE

When you first get a test, do not just start to answer questions. Instead, first look through the test. How many questions are there? How many different sections? Are some questions worth more points than others? Pace yourself accordingly.

Reviewing Vocabulary

Directions: Choose the word(s) that best completes the sentence.

1. Japan and Taiwan are part of a chain of islands known as ______.

A a cordillera

B an archipelago

C a tsunami

D a gorge

2. Mainland Southeast Asia is crossed by ______, or several mountain ranges that run side by side.

A plateaus

B archipelagos

C cordilleras

D gorges

3. In Mongolia, a ______ is a weather pattern in which a harsh winter follows a dry summer.

A *dzud*

B loess

C yak

D gobi

4. Heavy rains hitting barren hillsides can often lead to ______.

A swamps

B typhoons

C loess

D landslides

Reviewing Main Ideas

Directions: Choose the best answer for each question.

Section 1 *(pp. 688–693)*

5. The islands of East Asia are part of an area known as the Ring of Fire. It is called this because ______.

A there are many forest fires in the area

B the area has very high temperatures year round

C the area has many earthquakes and volcanoes

D the area is surrounded by very high mountains

6. What are the two major oceans of East Asia and Southeast Asia?

A Indian and Atlantic

B Indian and Pacific

C Atlantic and Pacific

D Pacific and South China

Section 2 *(pp. 694–698)*

7. The climate in Mongolia and northern China is mostly ______.

A tropical

B temperate

C polar

D steppe

8. Most of Southeast Asia has a ______ climate.

A tropical

B desert

C cold

D temperate

GO ON

Critical Thinking

Directions: Choose the best answer for each question.

The Ring of Fire, also called the Circum-Pacific belt, is the zone surrounding the Pacific Ocean where about 90% of the world's earthquakes occur. Study the table below and then choose the best answers to the following questions.

Number of Earthquakes Worldwide, 2002–2006					
Magnitude	2002	2003	2004	2005	2006
8.0 to 9.9	0	1	2	1	1
7.0 to 7.9	13	14	14	10	9
6.0 to 6.9	130	140	141	144	126
5.0 to 5.9	1,218	1,203	1,515	1,699	1,368
4.0 to 4.9	8,584	8,462	10,888	13,917	11,030
3.0 to 3.9	7,005	7,624	7,932	9,173	8,455
2.0 to 2.9	6,419	7,727	6,316	4,638	3,412
1.0 to 1.9	1,137	2,506	1,344	26	16
0.1 to 0.9	10	134	103	0	3
No Magnitude	2,937	3,608	2,939	867	635
Total	27,454	31,419	31,194	30,475	25,055*
Estimated Deaths	1,685	33,819	284,010	89,354	6,595
* as of December 6, 2006					

Source: U.S. Geological Survey, 2006.

9. In what year between 2002 and 2006 did the most deaths from earthquakes occur?

A 2004
B 2002
C 2006
D 2005

10. How many earthquakes with a magnitude greater than 6.0 occurred in 2004?

A 1
B 103
C 157
D 2,939

Document-Based Questions

Directions: Analyze the document and answer the short-answer questions that follow.

> *An International Tsunami Survey Team (ITST) studying the effects of the December 26 tsunami on Indonesia's island of Sumatra documented wave heights of 20 to 30 m (65 to 100 ft) at the island's northwest end and found evidence suggesting that wave heights may have ranged from 15 to 30 m (50 to 100 ft) along a 100-km (60-mi) stretch of northwest coast. These wave heights are higher than those predicted by computer models made soon after the earthquake that triggered the tsunami. . . .*
>
> *The survey was conducted . . . in the province of Aceh, which lies only 100 km (60 mi) from the epicenter of the earthquake and sustained what many consider the worst tsunami damage of all affected areas.*
>
> —"Astonishing Wave Heights Among the Findings of an International Tsunami Survey Team on Sumatra," *Sound Waves Monthly Newsletter,* U.S. Geological Survey, 2005

11. Why may the province of Aceh have received the worst tsunami damage?

12. What did the survey team discover about the wave heights of the tsunami at Aceh? Why were those findings surprising?

Extended Response

13. In an essay, consider the positive and negative consequences of building the Three Gorges Dam. Then take a side and argue in favor of its viewpoint.

STOP

For additional test practice, use Self-Check Quizzes—Chapter 25 at glencoe.com.

Need Extra Help?

If you missed question...	1	2	3	4	5	6	7	8	9	10	11	12	13
Go to page...	689	690	696	698	689	690	696	697	689	689	690	690	691

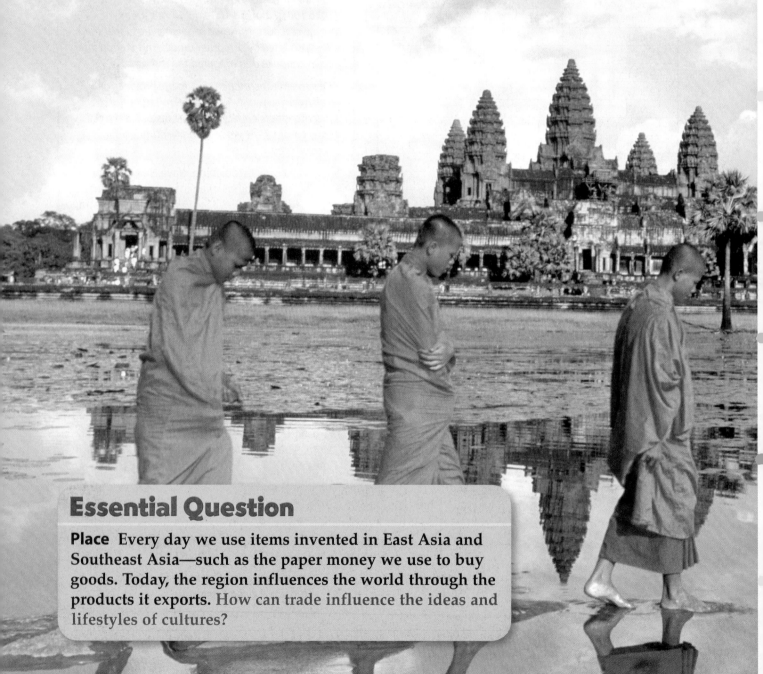

History and Cultures of East Asia and Southeast Asia

Essential Question

Place Every day we use items invented in East Asia and Southeast Asia—such as the paper money we use to buy goods. Today, the region influences the world through the products it exports. How can trade influence the ideas and lifestyles of cultures?

Angkor Wat, Cambodia

Section 1: History and Governments

BIG IDEA Patterns of economic activities result in global interdependence. Powerful local empires ruled early East Asia and Southeast Asia. Europeans seeking the region's resources gained control of much of the area beginning in the 1500s. By the mid-1900s, most Asian countries had gained independence. Today, several nations in East Asia and Southeast Asia have developed strong economies.

Section 2: Cultures and Lifestyles

BIG IDEA Culture influences people's perceptions about places and regions. Rapid population growth has created challenges for many countries in the region. Traditional beliefs and practices have influenced daily life, but the region has also been affected by modern technology and Western culture.

FOLDABLES™ Study Organizer

Organizing Information Make this Foldable to help you organize information about the history, peoples, and cultures of East Asia and Southeast Asia.

Step 1 Fold an 11 x 17 piece of paper lengthwise to create four equal columns.

Step 2 Label each column on your Foldable as shown.

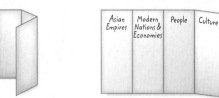

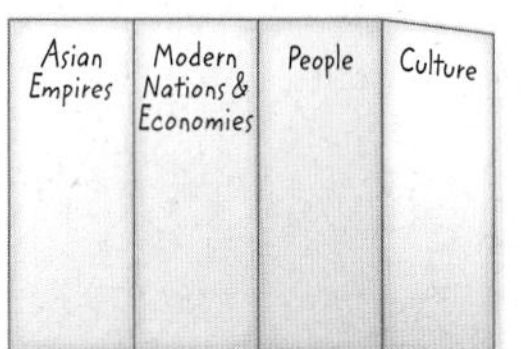

Reading and Writing As you read the chapter, take notes under the appropriate heading on your Foldable. Use your notes to write five quiz questions for each section of the chapter.

Visit glencoe.com and enter QuickPass™ code EOW3109c26 for Chapter 26 resources.

Guide to Reading

BIG Idea

Patterns of economic activities result in global interdependence.

Content Vocabulary

- dynasty *(p. 705)*
- porcelain *(p. 705)*
- census *(p. 706)*
- novel *(p. 706)*
- shogun *(p. 707)*
- samurai *(p. 707)*
- sphere of influence *(p. 708)*
- free port *(p. 712)*

Academic Vocabulary

- emerge *(p. 709)*
- dominate *(p. 709)*

Reading Strategy

Summarizing Use a diagram like the one below to list the key events and dates in China's history.

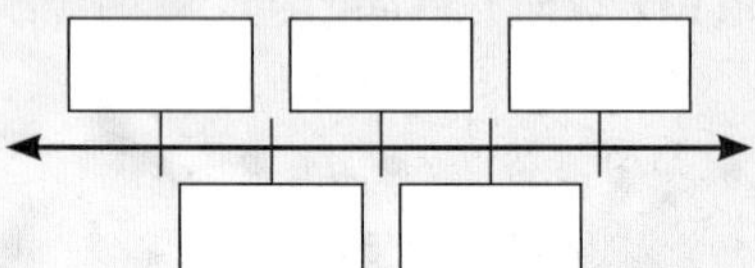

History and Governments

 Section Audio  Spotlight Video

Picture This A geisha may be Japan's most recognized symbol. *Geisha* means "artful person." For over 200 years, geisha have kept traditional Japanese culture alive. Geisha wear formal Japanese clothing. They learn to play the *shamisen,* a traditional three-stringed instrument that is similar to the banjo. Geisha also study calligraphy, dance, traditional poetry, and songs. You can learn more about the history of Japan by reading Section 1.

▼ Strolling in a Japanese garden

Asian Empires

Main Idea **Dynasties ruled large empires in East Asia and Southeast Asia.**

Geography and You Do you enjoy fireworks? The Chinese invented them many centuries ago. Read this chapter to learn about some of the inventions of East Asia and Southeast Asia that influence your life today.

The region of East Asia and Southeast Asia is home to some of the world's oldest civilizations. The earliest of these civilizations developed in China.

Early China

China's civilization began more than 4,000 years ago in the Huang He valley. Like the peoples of ancient Egypt, Mesopotamia, and South Asia, the early Chinese grew crops on rich soil left by river floods. Over time, they built cities and developed a writing system that is still in use today. The Chinese writing system uses characters to stand for entire words. This system differs from an alphabet, which has letters that stand for sounds.

For many centuries, until the early 1900s, emperors governed China. A **dynasty,** or line of rulers from a single family, would hold power until it was overthrown. Then a new leader would start a new dynasty.

During the Zhou (JOH) dynasty, China's best-known thinker was Confucius (also called Kongfuzi). He taught that people should be virtuous, or morally good, as well as loyal to their families. His way of thinking became known as Confucianism. Another thinker called Laozi (LOW·DZUH) helped found Daoism, a way of living in harmony with nature. Later, another important belief system called Buddhism was introduced to China from India. It became very popular in China. Confucianism, Daoism, and Buddhism have been major influences on East Asian life for centuries.

Sculpture of Laozi

Very little is known about the life of Laozi. Legend says that Laozi met a young Confucius, but scholars cannot confirm their meeting. ***Regions*** **What three religions have been major influences in East Asia?**

Under the Han dynasty, paper was first made in China about A.D. 100. Han government officials used paper to keep records. Han rulers also encouraged overland trade on the Silk Road. This large trading network stretched from China to Southwest Asia. The Chinese sent silk, tea, spices, paper, and a fine clayware known as **porcelain** (POHR·suh·luhn) over the Silk Road as far as the Mediterranean world. The Chinese received products, such as gold, silver, precious stones, and fine woods, from other lands.

Later, China's Tang and Song rulers built roads and waterways that made travel and trade within China easier. Farmers improved irrigation and were able to grow more rice. The greater amounts of food could support more people. As a result, China's population greatly increased.

Dynasties of China			
Dynasty	**Time Span**	**Dynasty**	**Time Span**
Xia	2200–1700 B.C.	**Tang**	A.D. 618–907
Shang	1766–1080 B.C.		
Zhou	1027–221 B.C.	**Song**	A.D. 960–1279
Qin	221–206 B.C.	**Yuan (Mongol)**	A.D. 1279–1368
Han	206 B.C.–A.D. 221	**Ming**	A.D. 1368–1644
Sui	A.D. 581–618	**Qing**	A.D. 1644–1911

New inventions changed Chinese life during the Tang and Song periods. The Chinese mixed iron with coal carbon to produce steel. They also developed a printing process using blocks of wood carved with characters. The invention of printing helped spread ideas more rapidly. Another invention was gunpowder, which they used in explosives and fireworks. The Chinese also invented the compass, which sailors used to help them find their way and to sail ships farther distances. In time, China's inventions spread to other parts of the world.

Later Chinese Dynasties

Later dynasties also contributed to China's growth. However, many wars also were fought. In 1211 well-trained Mongol warriors from Central Asia invaded and conquered most of China. Mongol rulers kept their own language, laws, and customs, yet they relied on Chinese officials to run the government.

Mongol rule unified China and eventually brought peace to the area. These stable conditions encouraged trade. One of the most famous European travelers to reach China was Marco Polo, who came from the city of Venice in Italy. The Mongol ruler was fascinated by Polo and sent him on many fact-finding trips. When Polo finally returned to Europe, he wrote a book about his adventures. It taught many Europeans about China.

In the late 1300s, the Chinese drove out the Mongol invaders, and the Ming dynasty arose. Ming rulers improved the examination system used to hire government officials. They also carried out a **census,** or a count of the number of people, so they could collect taxes more accurately. Finally, Ming rulers expanded roads and canals so that rice and other goods could be sent throughout the country.

In the capital of Beijing, the Ming built the Forbidden City. It was a large area of palaces and gardens that was off-limits to all except top officials. The emperors and their courts lived there for more than 500 years.

Under the Ming, the Chinese staged dramas and wrote **novels,** or long fictional stories. Ming rulers also sent voyages of exploration to South Asia and East Africa. Despite trade benefits, Chinese officials viewed the trips as costly and dangerous to China's traditional culture. They persuaded the Ming emperor to stop the voyages. As a result, China drew inward, and its overseas trade declined.

Korea and Japan

Chinese culture influenced other parts of East Asia. About 1200 B.C., Chinese settlers brought their culture to neighboring Korea. Later, Buddhism and Confucianism spread from China to Korea.

At the same time, Chinese and Korean culture spread from Korea to Japan. As a result, China and Korea both influenced Japan's civilization. In the A.D. 1400s, Japan, once ruled by clans, united under the Yamato dynasty. Yamato rulers adopted China's philosophy, writing system, art, sciences, and form of government.

Samurai Armor

NATIONAL GEOGRAPHIC

Samurai warriors often wore armor made of small pieces of brightly painted metal or leather laced together with silk or leather. ***Place*** **What was the samurai code of conduct?**

While the Japanese borrowed elements of foreign cultures, they also shaped those elements to suit their own culture. Wealthy nobles in the court of Japan's Yamato emperor developed new forms of literature written in Japanese. Buddhism mixed with Japan's local religion, Shinto.

By the A.D. 1100s, the armies of local nobles had begun fighting for control of Japan. Minamoto Yoritomo (mee·nah·moh·toh yoh·ree·toh·moh) became Japan's first **shogun,** or military leader. Land-owning warriors called **samurai** (SA·muh·RY) supported the shogun. The samurai lived by a strict code of conduct. This code demanded that a samurai be loyal to his master as well as brave and honorable. Although an emperor kept his title, the shoguns held the real power. The samurai helped shoguns govern Japan until the late 1800s.

Southeast Asia

Early peoples in Southeast Asia grew rice, raised cattle and pigs, and made metal goods. From the 100s B.C. to the A.D. 900s, the Chinese ruled much of what is now Vietnam. Hindu traders from India reached western parts of Southeast Asia by the A.D. 100s. Southeast Asians blended Hindu and Chinese ways with their own traditions.

During the A.D. 1100s, the Khmer people founded an empire in mainland Southeast Asia. They became wealthy from growing rice. They also built a large temple, the Angkor Wat, based on Indian and local designs. Angkor Wat served as a Hindu temple and a royal tomb.

Muslim Arab traders and missionaries settled coastal areas of Southeast Asia during the A.D. 800s. Eventually many people in these places converted to Islam.

Reading Check **Identifying Central Issues** How did China influence early Southeast Asia?

Modern Nations

Main Idea **After a period of European dominance, countries in the region regained their independence during the 1900s.**

Geography and You Do you want to be independent? What does the word mean to you? Read to find out how countries in the region gained independence during the 1900s.

By the 1600s, Europeans had arrived in East Asia and Southeast Asia. Their goals were to trade, to spread Christianity, and to claim territory.

NATIONAL GEOGRAPHIC

Uprising in China

In 1900 the Boxers, members of a secret Chinese society, attacked foreigners living in China. European, American, and Japanese forces landed and crushed the uprising. ***Place*** **Why did many Chinese oppose foreigners?**

Changes in China

By 1500, Asian goods such as silk, porcelain, and tea had become highly valued in Europe. European merchants began traveling to the region to trade for these goods. China did not welcome these traders. The Chinese tried to isolate their country from European influences. Europeans, however, wanted greater trade and used powerful warships to force China to open its ports. By the 1890s, European governments and Japan had claimed large areas of China as spheres of influence. A **sphere of influence** is an area of a country where a single foreign power has been granted exclusive trading rights.

Foreign interference in their country angered many Chinese. In 1911 this anger led to a revolution that ended the weak Qing dynasty and established a republic. The new government, however, could not establish control over the entire country. By 1927, a military leader named Chiang Kai-shek (JY·AHNG KY·SHEHK) had formed the Nationalist government. Meanwhile, Chiang's Communist opponent, Mao Zedong (MOW DZUH·DUNG), gained support from China's farmers. After years of civil war, the Communists won power in 1949. They set up the People's Republic of China on the Chinese mainland. The Nationalists fled to the offshore island of Taiwan. There, they set up a government called the Republic of China.

Modern Japan

Like China, Japan at first tried to isolate itself from Western influences. Then, in 1854 U.S. naval officer Matthew C. Perry pressured the Japanese to end their isolation and open their country to foreign trade. Not long afterward, rebel samurai forced shoguns to return full power to the emperor. Japan's new government quickly adopted western technology. Japan soon became an industrial and military power.

The Japanese islands, however, had few resources to support the country's new industry. Japan invaded neighboring countries to get those resources. By 1940, Japanese forces had gained control of Taiwan,

Korea, other parts of mainland Asia, and various Pacific islands. This expansion was one factor that led Japan to fight the United States and its allies in World War II. After its defeat in 1945, Japan became a democracy. Japan was stripped of its overseas territories and military might. However, Japan soon rebuilt its shattered economy and society. By the late 1900s, it had **emerged** as a global economic power.

A Divided Korea

After World War II ended, Korea was divided into American-backed South Korea and Communist-ruled North Korea. North Korea wanted to unite the two Koreas, so it invaded South Korea in 1950. United Nations forces led by the United States rushed to support South Korea. China's Communist leaders eventually sent troops to help North Korea. The Korean War ended in 1953 in a truce, but without a peace treaty or a victory for either side. Korea remains divided today.

Southeast Asia

During the 1800s and 1900s, European nations **dominated** Southeast Asia. They controlled the production and trade of the area's goods, such as sugar, coffee, tea, rubber, and oil. The United States gained power in the region after winning the Philippines from Spain in war. Only Siam, known today as Thailand, remained independent.

Resistance to colonial rule increased throughout Southeast Asia in the 1900s.

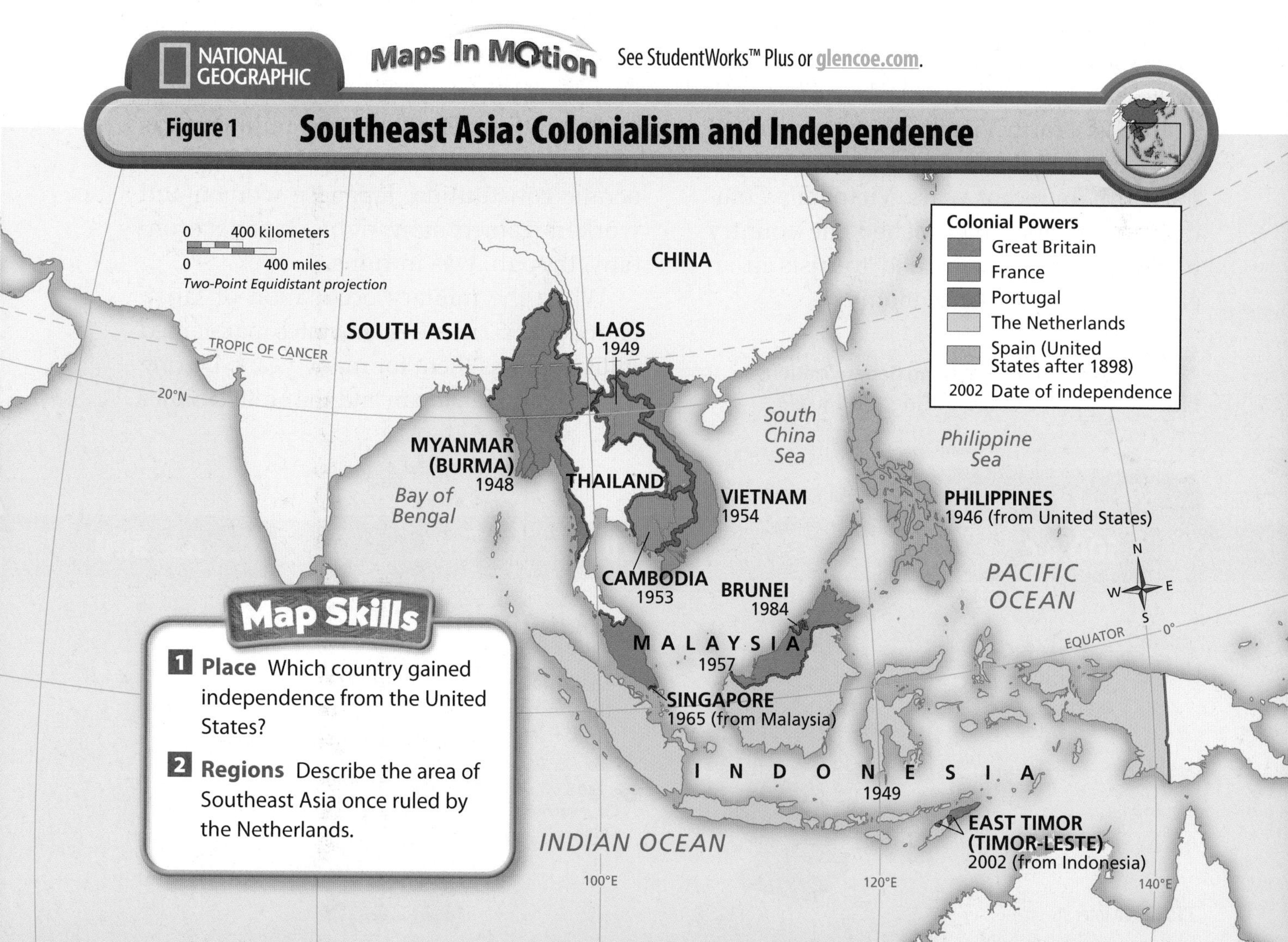

Figure 1 Southeast Asia: Colonialism and Independence

Map Skills

1. **Place** Which country gained independence from the United States?
2. **Regions** Describe the area of Southeast Asia once ruled by the Netherlands.

After World War II weakened European countries, they gave independence to colonies in the area. **Figure 1** on the previous page shows Southeast Asian countries that were once under colonial rule and their dates of independence.

After independence, political conflicts and wars raged throughout Southeast Asia. In Vietnam, Communist forces defeated the French in 1954. The Communists ruled North Vietnam, and an American-supported government ruled South Vietnam.

In the 1960s, fighting between these two groups led to the Vietnam War. During this conflict, American forces helped fight against the Communists. By the late 1960s, many Americans opposed the war. The United States eventually withdrew its troops in 1973. At least 2 million people, including 58,000 Americans, died in the Vietnam War. About 10 million South Vietnamese became refugees.

In 1975 North Vietnam's army reunited Vietnam and imposed Communist rule on the south. In recent years, Vietnam's Communist leaders have opened the country to Western businesses and tourists in an effort to improve the economy.

Reading Check **Making Generalizations** How did Japan change during the 1800s?

Economic Powers

Main Idea **Since 1945, many countries in the region have become great economic powers.**

Geography and You Do you think people who own their own businesses work harder than people who work for someone else? Why or why not? Read to learn how private ownership of business changed China and other countries in the region.

Following World War II, many countries in East Asia and Southeast Asia had weak economies. Within several decades, however, several countries developed into major economic powers.

Japan

At the end of World War II, Japan was occupied by troops of the United States and its allies. The Japanese military was reduced in size, and Japan adopted a democratic constitution. Japanese women and workers gained more rights. Japan's economy, though, was in ruins.

While the military occupation of Japan ended in 1952, the United States maintained military bases there for many years. During the Korean War, Japan and the United States

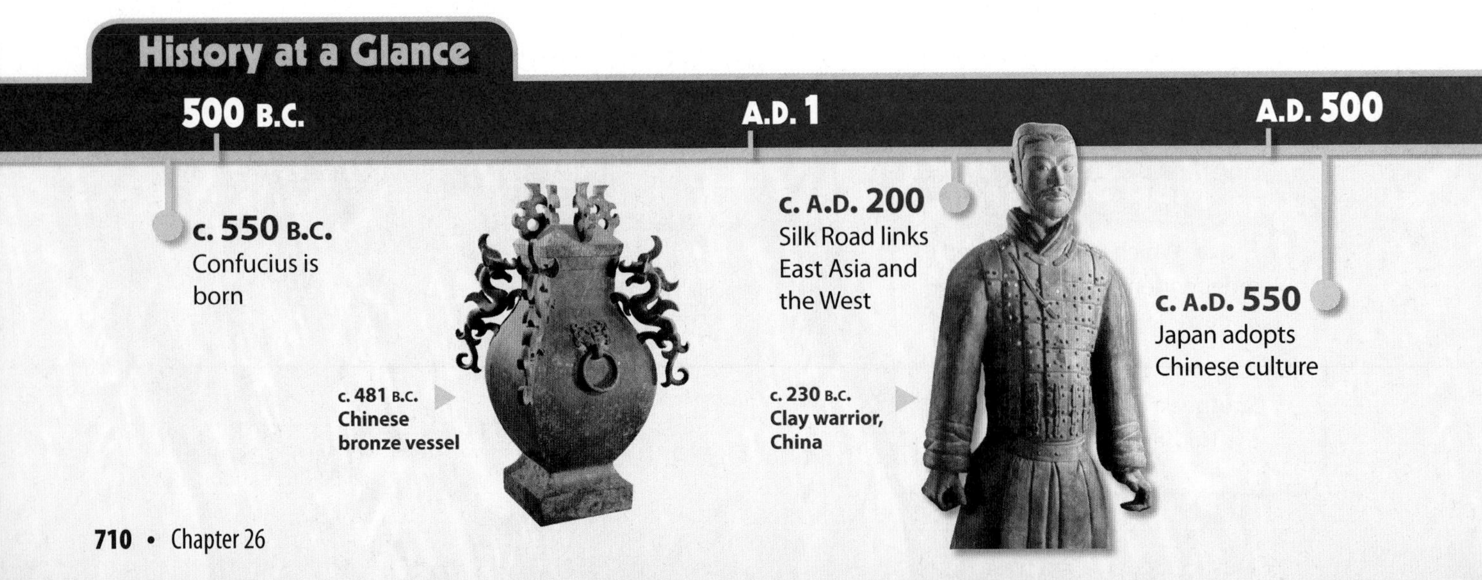
History at a Glance
500 B.C.
A.D. 1
A.D. 500
c. 550 B.C.
Confucius is born
c. A.D. 200
Silk Road links East Asia and the West
c. A.D. 550
Japan adopts Chinese culture
c. 481 B.C. Chinese bronze vessel
c. 230 B.C. Clay warrior, China

kept close ties. American troops needed war supplies, from medicines to trucks. To have a source of supplies nearby, the United States poured $3.5 billion into Japan's factories. Japanese shipbuilders, manufacturers, and electronics industries all benefited from American assistance.

This aid created an economic boom in Japan. The Japanese government worked closely with business leaders to plan the country's economic growth. For example, in the late 1950s, government and industry agreed to invest heavily in the research and development of electronics products for the home. Today, Japan has the world's second-largest economy after the United States. It is a major exporter of automobiles, cameras, and electronic goods such as computers, televisions, and sound systems.

The "Asian Tigers"

Since the 1960s, many economic changes have occurred in other countries in East Asia and Southeast Asia. Today, a group of countries and territories in East Asia and Southeast Asia has been nicknamed the "Asian Tigers." They include South Korea, Taiwan, Singapore, and the Chinese port of Hong Kong. Following Japan's example, these four "Tigers" have built strong, modern economies.

In the 1990s, South Korea changed from military rule to democracy. Its high technology and manufacturing industries have grown tremendously. South Korea today is a major exporter of ships, cars, computers, and electronic appliances.

The island of Taiwan lies about 100 miles (161 km) off China's coast. Taiwan's position as a nation is unusual. Following the Communist takeover of mainland China in 1949, the Nationalist government fled to Taiwan. From there, it claimed to be the rightful ruler of all of China. Today, Taiwan functions as an independent, democratic nation. China, however, claims Taiwan as a province that should be under Chinese control. For many years, relations between the two governments have been tense. Under its democratic government, Taiwan has developed a booming economy that produces computers, radios, televisions, and telephones.

The port of Hong Kong is also part of China. Hong Kong was ruled by the United Kingdom from the 1840s until 1997, when it passed to China as a special administrative region. Despite Chinese Communist rule, Hong Kong has been allowed to keep its strong, free market economy.

Another major port is Singapore, at the tip of Southeast Asia's Malay Peninsula.

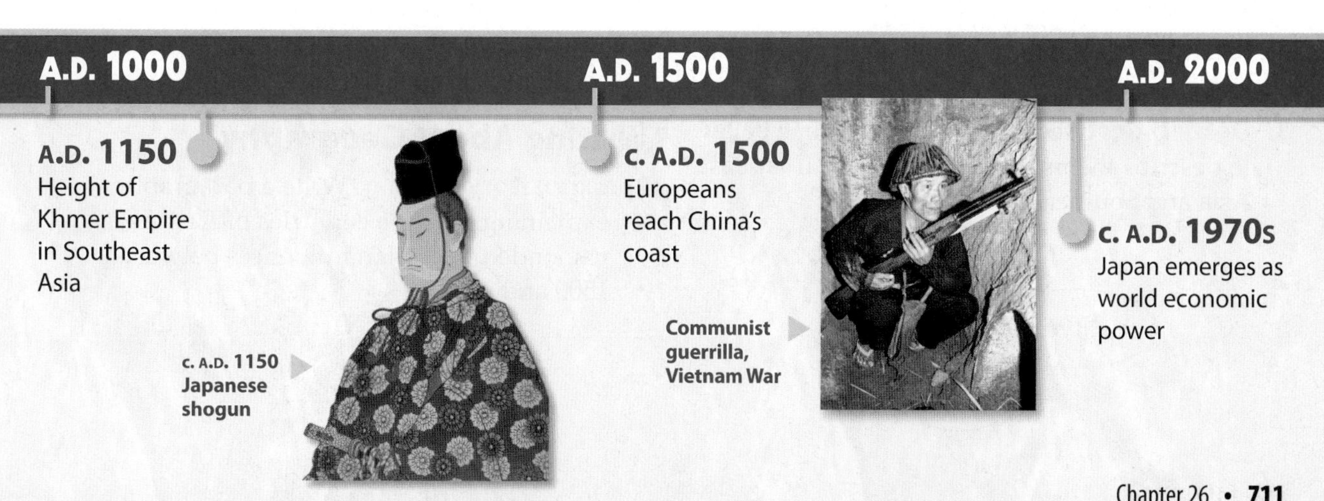
A.D. 1000
A.D. 1500
A.D. 2000
A.D. 1150
Height of Khmer Empire in Southeast Asia
c. A.D. 1500
Europeans reach China's coast
c. A.D. 1970s
Japan emerges as world economic power

c. A.D. 1150 Japanese shogun

Communist guerrilla, Vietnam War

Although small in size, Singapore has a highly productive economy. As a **free port,** Singapore is a place where goods can be unloaded, stored, and shipped again without payment of import taxes. These are taxes companies must pay to ship goods into a country. As a result, huge amounts of goods pass through Singapore.

China

China has also recently become a major economic power. China's economy, however, developed somewhat later than those of the "Asian Tigers." It took longer because, under China's Communist leaders, most businesses in the country were owned and run by the government. This system did not work well for the economy in China. China's farms and industries failed to produce enough of the kinds of goods that the economy needed in order to grow.

To spark growth, China began a number of free market reforms in 1979. The government relaxed its control over factories and farms. Factory managers could decide what goods to produce and what prices to charge. Farmers were allowed to sell their crops for a profit. China's Communist leaders also encouraged foreigners to set up businesses in parts of China.

These reforms made China's economy grow. Many Chinese began to enjoy a better standard of living. For the first time, China's people were able to buy consumer goods such as televisions and appliances. China's economic growth, however, has led to air and water pollution. Because China's growth has been so rapid, the country's pollution levels are among the highest in the world.

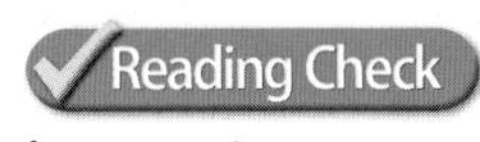

Explaining What is a free port?

Social Studies ONLINE
Study Central™ To review this section, go to glencoe.com.

Vocabulary

1. **Explain** the significance of:
   - a. dynasty
   - b. porcelain
   - c. census
   - d. novel
   - e. shogun
   - f. samurai
   - g. sphere of influence
   - h. free port

Main Ideas

2. **Identifying** What role did the samurai play in Japanese history?
3. **Describing** Use a diagram like the one below to describe the impact of World War II on East Asia and Southeast Asia.

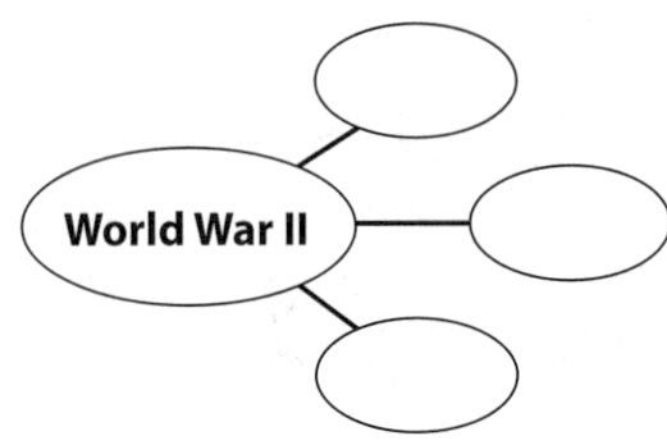

4. **Explaining** What changes led to economic growth in China starting in the late 1970s?

Critical Thinking

5. **BIG Idea** Explain how the "Asian Tigers" built strong economies.
6. **Challenge** Does a country have to be a democracy to have a prosperous free market economy? Why or why not?

Writing About Geography

7. **Expository Writing** Write a paragraph explaining how the countries of East Asia responded to foreign influences between 1500 and 1900.

Geography & History

Location, Location, Location

How can a small community avoid being dominated by a larger neighbor? Physical features can help.

Separating Two Lands Seas separate Japan from China, but they do not completely isolate Japan from its Asian neighbor. This fact has shaped Japan's history and culture. The Japanese were close enough to be influenced by the Chinese. Yet there was also enough distance that the Japanese developed a unique culture.

Since ancient times, people have used ships to cross between Japan and mainland Asia. Beginning in the A.D. 100s, regular travel between Japan and China opened Japan to Chinese influences.

During the A.D. 600s, the Japanese prince Shotoku wanted to learn from China's brilliant civilization. He sent officials and students to China to study. The Japanese learned about Buddhist and Confucian teachings. They also adopted the Chinese system of writing and artistic styles. Chinese artists came to Japan to decorate Buddhist temples. Japanese rulers patterned the design of their capital city and their government administration on that of China.

▲ **Mongol ships attacking Japan**

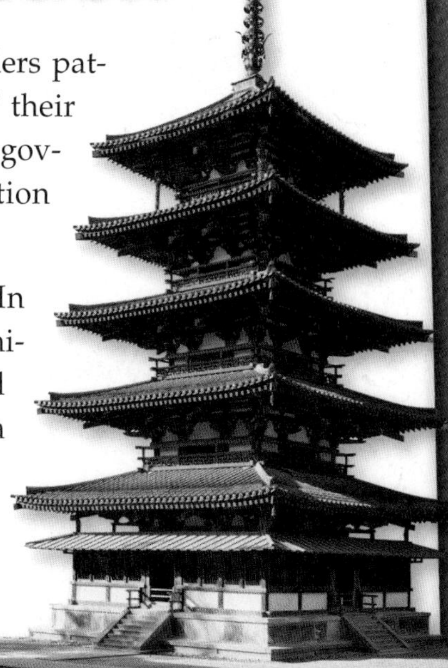

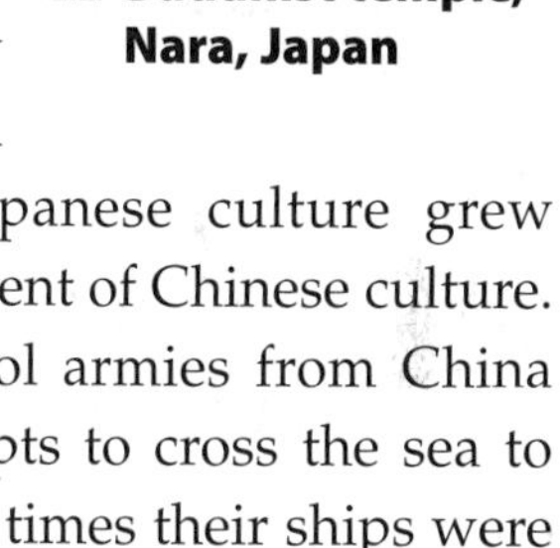

▲ **Buddhist temple, Nara, Japan**

Stepping Away In the A.D. 800s, the Chinese entered a period of turmoil. Partly as a result, the Japanese began to distance their culture from Chinese culture.

For the next several hundred years, Japan and China had relatively little contact with each other, and Japanese culture grew even more independent of Chinese culture. In the 1200s, Mongol armies from China made several attempts to cross the sea to conquer Japan. Both times their ships were destroyed by powerful typhoons.

These storms saved Japan from conquest. The Japanese, however, remained concerned about future invasions, and Japan isolated itself from China even further.

Today, Japan and China are close trading partners. They disagree, however, over many political issues.

Think About It

1. **Movement** How did the Chinese influence Japan?
2. **Human-Environment Interaction** What saved Japan from the Mongol invasions?

Guide to Reading

BIG Idea

Culture influences people's perceptions about places and regions.

Content Vocabulary

- megalopolis *(p. 716)*
- calligraphy *(p. 719)*
- pagoda *(p. 719)*
- haiku *(p. 719)*
- yurt *(p. 720)*

Academic Vocabulary

- overseas *(p. 716)*
- crucial *(p. 720)*

Reading Strategy

Identifying Central Issues Use a diagram like the one below to list key facts about population patterns in East Asia and Southeast Asia.

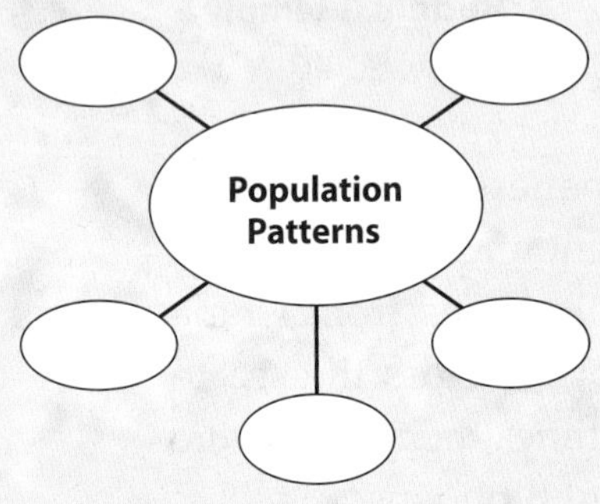

Cultures and Lifestyles

Picture This The Lion Dance is a popular form of entertainment during Chinese New Year celebrations and other special occasions. These four-legged creatures, made of colorful papier-mâché and fabric, are controlled by two people. One person holds the lion's head while another crouches under the tail. The lion dance is said to drive away the evil spirits and bring good luck and happiness. Read Section 2 to learn more about the cultures of East Asia and Southeast Asia.

Performing the Lion Dance in Beijing, China

Population Patterns

Main Idea **East Asia and Southeast Asia is one of the world's most densely populated regions.**

Geography and You Do you like having privacy? Do you live in a crowded neighborhood where privacy is hard to find? Read to learn about where people live in East Asia and Southeast Asia.

Countries in East Asia and Southeast Asia have rapidly growing populations. Today, more than two billion people live in the region—about one-third of the world's population. In many areas, the expanding population has placed a burden on available resources. Several countries have tried to resolve this problem by slowing the rate of population growth.

Population Growth

To control population growth, China's government enacted a policy in 1979 that encouraged families to have no more than one child. Although not followed by all Chinese, the "one child" policy has helped slow China's growth rate. Still, China continues to grow by millions of people each year. In 2005 China's population passed 1.3 billion. The greatest challenge is finding jobs for the many young people who enter the workforce each year. Despite economic growth, China has not been able to provide enough opportunities for new workers.

Other countries in East Asia and Southeast Asia also have seen rapid population increases due to high birthrates. Poorer countries in which agriculture is a major economic activity, such as Cambodia and Laos, have higher birthrates than wealthier countries, such as Japan and South Korea. The population growth rates of Japan, South Korea, and Taiwan are actually shrinking. These countries are expected to have fewer people by 2050.

Family in China

Most Chinese families follow the government policy that encourages one child per family. Because China's population is so large, however, annual population growth is still high.
Regions **How do population growth rates vary across the region?**

Where People Live

In East Asia, most people live crowded together in river valleys and basins or on coastal plains. There, the land and climate are highly favorable for agriculture and industry. These areas are among the most densely populated places on Earth. For example, the fertile Huang He basin in north central China has a population density of more than 1,000 people per square mile (400 per sq. km). By contrast, few people live in East Asia's interior, which is mountainous and has little vegetation. The large, rugged country of Mongolia, for example, has a population density of only 4 people per square mile (1.7 per sq. km).

Market Day in Seoul

Seoul, the capital of South Korea, is one of the most densely populated cities in the world. About 10 million people live in an area of 234 square miles (605 sq. km).
Movement **Why have people emigrated from the region?**

In Southeast Asia, populations are also unevenly distributed. About 60 percent of Indonesia's people live on Java, although this island is only 7 percent of Indonesia's land area. Singapore, the region's smallest country in land area, has the greatest population density—nearly 18,000 people per square mile (6,947 per sq. km).

Migrations

Nearly 60 percent of people in East Asia and Southeast Asia live in rural areas. In recent decades, however, many rural people in the region have moved to cities to find better-paying jobs and a higher standard of living. Cities in East Asia and Southeast Asia are centers of industry and commerce. Skyscrapers tower over busy streets, and bright neon signs advertise cars, electronics, and watches.

The region's cities are among the world's largest. In China, large cities, such as Shanghai, Beijing, and Guangzhou, have populations ranging from 6 million to more than 13 million. In Japan, several large coastal cities—Tokyo, Osaka, Nagoya, and Yokohama—lie so near each other that they form a **megalopolis,** or supersized urban area. Their combined population totals about 50 million people. Jakarta, the capital of Indonesia in Southeast Asia, has more than 11 million people. Another large Southeast Asian city is Manila in the Philippines, with a population of more than 10 million.

In addition to urban migration, many people in the region have left their homelands in recent decades to settle **overseas.** Between 1975 and 1990, hundreds of thousands of people left Vietnam and Laos to escape war and economic hardship. Many of the emigrants settled in the United States. One effect of overseas migrations is that some Southeast Asian countries have lost educated workers who would have contributed to their countries' economic growth.

Reading Check **Describing** Where do most people in East Asia live?

Analects

Sayings of Confucius recorded by his followers

23. Confucius said: "By nature men are pretty much alike; it is learning and practice that set them apart." [XVII:2]

28. Confucius said: "Learning without thinking is labor lost; thinking without learning is **perilous.**" [II:15]

40. Tzu Kung asked: "Is there any one word that can serve as a principle for the conduct of life?" Confucius said: "Perhaps the word 'reciprocity': Do not do to others what you would not want others to do to you." [XV:23]

56. Confucius said: "Nowadays a **filial** son is just a man who keeps his parents in food. But even dogs or horses are given food. If there is no feeling of reverence, wherein lies the difference?" [II:7]

87. Confucius said: "The gentleman understands what is right; the inferior man understands what is profitable." [IV:16]

88. Confucius said: "The gentleman cherishes virtue; the inferior man cherishes possessions. The gentleman thinks of **sanctions;** the inferior man thinks of personal favors." [IV:11]

90. Confucius said: "The gentleman seeks to enable people to succeed in what is good but does not help them in what is evil. The inferior man does the contrary." [XII:16]

97. Confucius said: "Lead the people by laws and regulate them by penalties, and the people will try to keep out of jail, but will have no sense of shame. Lead the people by virtue and restrain them by the rules of **decorum,** and the people will have a sense of shame, and moreover will become good." [II:3]

Analyzing Literature

1. **Analyzing Information** What did Confucius believe were qualities of a gentleman?
2. **Read to Write** Write a letter to a Chinese ruler advising him about how Confucius would have him govern his people.

Confucius

(551–479 B.C.)

Chinese philosopher Confucius taught about government, law, family relations, religion, and education. His ideas have deeply influenced life and thought in China for more than 2,000 years.

Background Information

Confucius lived when the Zhou dynasty was losing control of China. Rival kings fought each other for power. Disturbed by widespread suffering, Confucius criticized the evil ways and misrule of these kings. He believed virtue was the key to returning China to an era of good government and social harmony.

Reader's Dictionary

perilous: dangerous

filial: honoring one's parents

sanctions: limits on behavior

decorum: proper behavior

People and Cultures

Main Idea **Southeast Asia has greater ethnic diversity than East Asia.**

Geography and You Think about the different languages, music, and customs people have in the United States. Diversity is even greater in East Asia and Southeast Asia. Read to find out about the variety of cultures in the region.

The peoples of East Asia and Southeast Asia have a rich cultural heritage. Since ancient times, different local religions have greatly influenced the region. In the modern era, the rise of a global culture also has brought change to the region.

Ethnic Groups

Within each East Asian country, people tend to be ethnically similar. In Japan, about 99 percent of the population is ethnic Japanese and speaks the Japanese language. Koreans make up the largest ethnic group in their country and speak the Korean language. The people of Mongolia are ethnic Mongolian. About 90 percent of them speak the Khalkha Mongolian language.

In China, the Han ethnic group makes up about 92 percent of the population. The other 8 percent, however, belong to about 55 different ethnic groups. Most of these groups live mainly in western and northern China. Han Chinese, the most widely spoken language of China, has many dialects. Mandarin, the northern dialect, is China's official language. Cantonese, another major dialect, is spoken in southeastern China.

Southeast Asia has more ethnic diversity than East Asia. Among its many ethnic groups are Indonesians, Malays, Burmans, Vietnamese, Laotians, and Thais. Ethnic Chinese form an important minority in many of these countries. They are descendants of people who moved to this region from China. Other minority groups exist as well. Indonesia, with its many islands, has about 300 ethnic groups.

In addition to ethnic variety, hundreds of languages and dialects are spoken in Southeast Asia. Many of the languages were spoken by migrants or colonizers. In the Philippines, for example, Filipino, English, and Spanish are the major languages. Filipino, an official language of the Philippines, developed from the speech of the islands' early inhabitants. Spanish was brought to the Philippines when the islands were colonized by Spain. English, the second official language, was spoken after the United States gained control of the islands.

Religions

In addition to ethnic diversity, nearly all of the world's major religions are practiced in East Asia and Southeast Asia. Buddhism

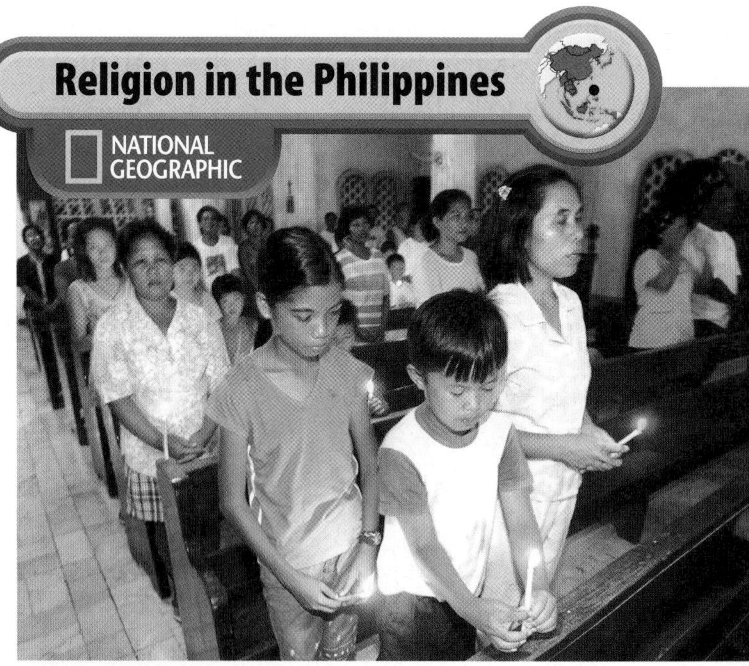

▲ Spanish rule brought the Roman Catholic religion to the Philippines. Today about 80 percent of Filipinos are Catholic. ***Regions*** **What parts of the region are mainly Muslim?**

spread to the region from South Asia about 2,000 years ago. Today, Buddhism is the major religion in Myanmar, Thailand, Laos, Cambodia, and Vietnam. It has a large following in China, North Korea, South Korea, and Japan.

In Japan, many people combine Buddhism with Shinto, the country's traditional religion. Shinto stresses that all parts of nature—humans, animals, plants, rocks, and rivers—have their own spirits. Communist governments have limited religious practice in China and North Korea, but Buddhist, Confucian, and other traditions still survive in both countries.

Other important religions in the region include Islam, Hinduism, Daoism, and Christianity. Arab traders brought Islam to East Asia and Southeast Asia many centuries ago. Today, most people in Indonesia, Malaysia, and western China are Muslims. Christianity became the dominant religion in the Philippines after that country came under Spanish rule in the 1500s.

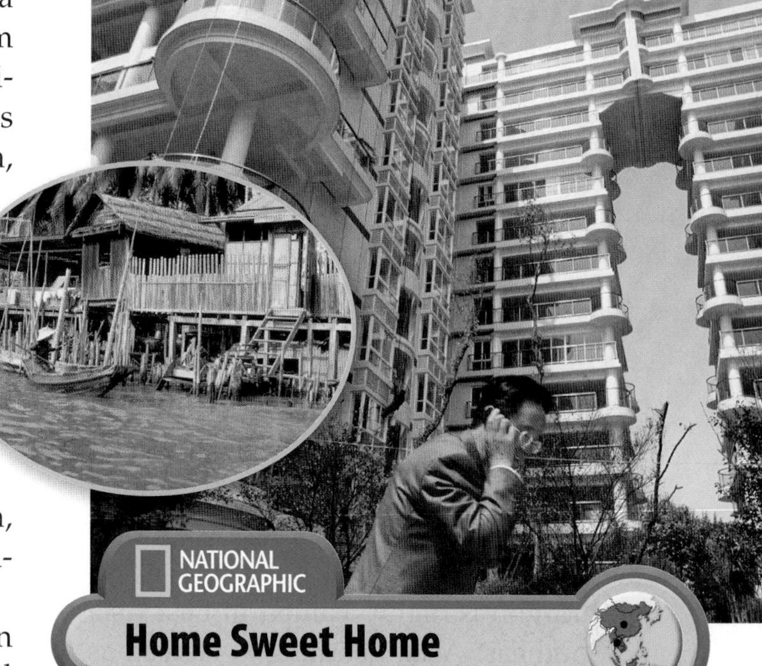

NATIONAL GEOGRAPHIC

Home Sweet Home

In the crowded cities of East Asia, like Chengdu, China (above), many people live in tall apartment buildings. In Thailand, some homes are built on stilts to protect against flooding (inset).
Place **What is a pagoda?**

The Arts

A number of art forms have long been popular in East Asia and Southeast Asia. In China, Korea, and Japan, artists have painted the rugged landscapes of their countries. Their works reflect the strong reverence for nature that is part of both Daoism and Shinto. Many paintings include poems written in an elegant brush stroke called **calligraphy,** which means "the art of beautiful writing." Craftspeople in East Asia and Southeast Asia are also skilled at weaving, carving, and making pottery.

The artistic skills of the region's people also appear in their architecture. Traditional temples, palaces, and houses are quite decorative. Many structures have several stories with tiled roofs that curve up at the edges. In East Asia, these buildings are called **pagodas** and serve as temples.

East Asians and Southeast Asians also have strong literary and theatrical traditions. Japanese poets often write **haiku,** which are brief poems that follow a specific structure, or organization. Many haiku reflect the deep feelings for nature that the Shinto religion teaches. Japan is famous for its different forms of theater, and Indonesia is known for its beautiful dances. Puppet plays are popular in many parts of Southeast Asia. Throughout East Asia and Southeast Asia, musicians play flutes, drums, gongs, and stringed instruments at various performances.

Social Studies ONLINE

Student Web Activity Visit glencoe.com and complete the Chapter 26 Web Activity about Asian art.

Daily Life

In East Asia and Southeast Asia, the family is the center of social life. This lifestyle largely reflects the ideas of Confucius, especially in East Asia. Young people are taught to show respect to older relatives. Confucius also taught women to obey their husbands. Some of these attitudes are changing, however, partly because of Western influences. In the past, marriages usually were arranged by parents, but today most people choose their own partners. Also, women in the region have gained more equality, and many now work outside the home.

Education is highly valued in East Asia and Southeast Asia. Learning is viewed as **crucial,** or extremely important, to the positive development of children and society. The program of study is rigorous. Some children attend school six days a week. The highly educated workforce has helped several countries build productive economies.

Housing varies widely across East Asia and Southeast Asia. Typical homes in crowded cities are modern but have limited space. In China, apartments in cities may house more than one generation. Houses in rural areas tend to be larger but simpler.

Traditional forms of housing fit the local environment. Rural Mongolians live in **yurts.** These large, circular structures made of animal skins can be packed up and moved from place to place. In Southeast Asia, houses are often built on poles, or stilts, to avoid flooding.

Rice makes up a major part of people's diets. It is often mixed with steamed or quick-fried vegetables and strips of meat or fish. Some areas are known for their hot and spicy foods. In recent years, Western foods such as hamburgers and fried potatoes have gained popularity in the region.

Reading Check **Explaining** Why is religious activity limited in China and North Korea?

Section 2 Review

Social Studies ONLINE
Study Central™ To review this section, go to glencoe.com.

Vocabulary

1. **Explain** the meaning of *megalopolis, calligraphy, pagoda, haiku,* and *yurt* by writing a sentence for each term.

Main Ideas

2. **Explaining** Why are population growth rates in South Korea and Taiwan different from those in Cambodia and Laos?
3. **Comparing and Contrasting** Use a Venn diagram like the one below to compare and contrast culture in Japan and Indonesia.

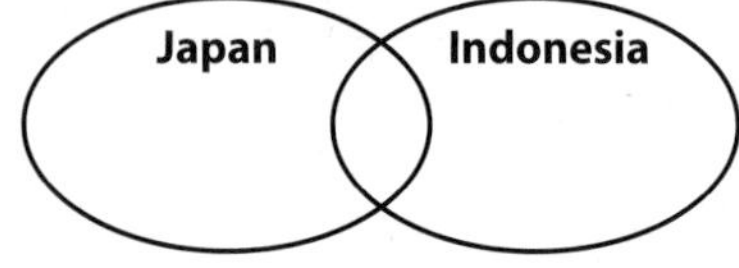

Critical Thinking

4. **Identifying Central Issues** Why are some countries in the region more ethnically diverse than others?
5. **BIG Idea** What is an example of an artistic practice in the region that reflects religious values?
6. **Challenge** How do you think traditional beliefs have influenced lifestyles in East Asia and Southeast Asia?

Writing About Geography

7. **Using Your FOLDABLES** Use your Foldable to write a paragraph describing the variety of religions in East Asia and Southeast Asia.

Visual Summary

Asian Empires

- A series of dynasties ruled China from ancient times to the early 1900s.
- In Japan, shoguns led armies of samurai warriors and governed the country.
- Southeast Asia, a major trading center, was ruled by several empires and kingdoms.

Modern Nations

- Europeans influenced China and Japan and controlled much of Southeast Asia in the 1800s.
- Japan adopted Western technology to become a modern military power.
- China has recently begun free market reforms and boosted its economy.
- Japan recovered after World War II and built one of the world's major economies.
- South Korea, Singapore, Taiwan, and Hong Kong, known as the "Asian Tigers," became economic powers in the late 1900s.

Commodore Matthew C. Perry's ship arrives in Japan, 1854

Horse and rider, China, Han dynasty

People

- Populations in the region grew rapidly in the 1900s. China's and other countries' population growth rates have since slowed.
- The region's river valleys and basins, coastal areas, and cities are very densely populated.
- Most countries in the region have many ethnic groups. Japan and the Koreas are exceptions.

Dancers, Bali, Indonesia

Culture

- Confucianism, Buddhism, and Daoism are major belief systems in East Asia and Southeast Asia. Islam, Shinto, Hinduism, and Christianity are also important in the region.
- Arts include landscape painting, weaving, carving, pottery, poetry, theater, and dance.
- People's lifestyles reflect traditional and modern influences.

Disabled Taiwanese artist

Study anywhere, anytime! Download quizzes and flash cards to your PDA from glencoe.com.

STANDARDIZED TEST PRACTICE

Do not keep changing your answers. Your first choice is usually the right one, unless you misread the question.

Reviewing Vocabulary

Directions: Choose the word(s) that best completes the sentence.

1. A line of rulers from a single family is called a ________.

A shogun
B samurai
C dynasty
D emperor

2. In order to collect taxes more accurately, officials from the Ming dynasty in China conducted a ________ to count the number of people.

A dynasty
B census
C sphere
D geisha

3. Many East Asian and Southeast Asian paintings include poems written in elegant brush strokes called ________.

A calligraphy
B haiku
C porcelain
D samurai

4. Brief Japanese poems called ________ follow a specific organization or structure.

A pagodas
B calligraphy
C yurts
D haiku

Reviewing Main Ideas

Directions: Choose the best answer for each question.

Section 1 *(pp. 704–712)*

5. In 1211 the Chinese were conquered by the ________, which resulted in a period of peace that helped boost trade.

A Japanese
B Mongols
C Koreans
D Khmer

6. In 1949 after years of civil war in China, the ________ won control of the government.

A Nationalists
B imperialists
C Communists
D capitalists

Section 2 *(pp. 714–720)*

7. In some wealthier countries of East Asia and Southeast Asia, such as ________, the birthrate is declining.

A Cambodia
B Laos
C Indonesia
D Japan

8. Which of the following countries in the region has the least amount of ethnic diversity?

A Japan
B Vietnam
C Laos
D Thailand

GO ON

Critical Thinking

Directions: Use the chart below and your knowledge of Chapter 26 to help you choose the best answer for each question.

Population Density in East Asia and Southeast Asia (people per square kilometer)

Country	1990	2000	2003
China	120	132	135
Hong Kong	5,230	6,200	6,300
Japan	332	344	342
South Korea	432	473	481
Singapore	4,814	5,885	6,134
Thailand	109	120	123

Source: United Nations Economic and Social Commission for Asia and the Pacific www.unescap.org/STAT/data/statind/pdf/t5_dec04.pdf

9. According to the chart above, which statement is accurate?

A Population density in these countries is decreasing.

B Population density is decreasing in some of these countries and increasing in others.

C Population density in these countries is increasing.

D There is virtually no change in population density from 1990 to 2003.

10. The most densely populated country in the region in both 2000 and 2003 was ______.

A Singapore

B Hong Kong

C Japan

D South Korea

Document-Based Questions

Directions: Analyze the document and answer the questions that follow.

Daoist practices were meant for people of all backgrounds. Below is an excerpt from an ancient Daoist text called the *Dao De Jing* (The Way of the Dao).

The more regulations there are,
The poorer people become.
The more people own lethal weapons,
The more darkened are the country and clans.
The more clever the people are,
The more extraordinary actions they take.
The more picky the laws are,
The more thieves and gangsters there are.

11. According to the reading, ______ will cause people to become poorer.

A too many regulations

B owning weapons

C fewer thieves and gangsters

D more clever people

12. According to the reading, there are more thieves and gangsters when ______.

A there are fewer lethal weapons

B there are more picky laws

C there are more countries and clans

D there are more clever people

Extended Response

13. The Chinese were known for inventions. Choose one Chinese invention and explain how it changed the world.

STOP

For additional test practice, use Self-Check Quizzes—Chapter 26 at glencoe.com.

Need Extra Help?

If you missed question...	1	2	3	4	5	6	7	8	9	10	11	12	13
Go to page...	705	706	719	719	706	708	715	718	715	715	705	705	706

TIME JOURNAL

"Hello! My name is Ferdian.

I am 13 years old and in 8th grade at Diponegoro 1 High School in Jakarta, Indonesia. Indonesia is a chain of islands in the Indian and Pacific Oceans. My city is on the island of Java. Read about my day."

6:00 a.m. Time to get up! I put on pants, a long white shirt, and a cap. This is traditional dress for Muslim boys here. My female classmates wear white headscarves, blouses, and blue skirts.

6:30 a.m. My brother and I watch Japanese cartoons on TV while eating our breakfast of *bubur ayam*, a thick rice porridge with pieces of chicken.

6:45 a.m. I kiss my mother's hand and say good-bye. (In Indonesia, we kiss our elders' hands as a sign of respect.) Then I hop on my bike and head to school.

7:00 a.m. I arrive just in time! I kiss my teacher's hand and greet my friends. Then we begin the first class of the day, where we learn about Indonesia's laws and government. It is interesting to me because I would like to be a member of the Indonesian Parliament someday.

8:30 a.m. In geometry class, I struggle to understand the lesson. Geometry is my least favorite subject! My teacher, Ms. Dwi, helps me.

10:00 a.m. It is time for morning prayers. I go with the other students to a mosque on the school property. First, I perform ablutions, or the washing of my hands, face, and feet. Then I say my prayers.

10:40 a.m. In religion class, I study the laws of Islam and learn Arabic. Arabic is the language of the Quran, the holy book of Islam.

11:30 a.m. School is dismissed. On most days, school ends at noon, but today is Friday. We get out early so that Muslim males can prepare for Friday prayers.

11:40 a.m. On my way home, I share the road with many cars and three-wheeled vehicles called *bajajs*. These loud, motorized vehicles are everywhere in Jakarta!

12:15 p.m. My best friend, Achmad, and I pray at a nearby mosque.

1:15 p.m. Achmad and I go to my house, where my mother serves us chicken soup over rice. During lunch, I tell my mom about my trouble in geometry. She tries to explain the lesson while we eat.

2:00 p.m. We play some games on my family's computer.

4:30 p.m. Achmad and I see some other boys from the neighborhood. We play soccer, my all-time favorite sport. Then we hang out and talk.

6:00 p.m. Oops! I almost forgot. It is time to feed my pet fish.

6:45 p.m. I perform my evening prayers. Then I join my mother and brother for dinner. (My father is a customs officer who works in another city. He is rarely home.) We have fried rice, and for dessert, pieces of banana that are dipped in batter and fried.

8:00 p.m. I do my homework. I think I finally understand the geometry lesson!

10:00 p.m. I go to sleep.

ILLUSTRATIONS BY BOOKMAPMAN

SEE YA! Ferdian says good-bye to his mother in a traditional Indonesian manner.

CIVICS LESSON Ferdian studies the constitution of Indonesia. Girls must cover their heads at school.

AT A MOSQUE As a devout Muslim, Ferdian prays several times a day.

What's Popular in Indonesia

JERRY ALEXANDER/GETTY IMAGES

Spices Often called the spice capital of the world, Indonesia produces cinnamon, vanilla, chili peppers, and many other spices.

Sepak takraw This sport is a cross between volleyball and soccer. Players on one team spike a ball over a net to the other team. They can use their feet, knees, shoulders, or heads—but no hands.

Shadow puppets The Indonesian art of *wayang kulit* is 1,000 years old. In it, a puppet master moves flat leather puppets behind an illuminated white screen. The audience sees only the puppets' shadows.

PETR SVARC/ ALAMY

Say It in Bahasa Indonesia

Most of Indonesia's inhabited islands have their own dialects. But almost all Indonesians also speak the national language, Bahasa Indonesia. Here are some phrases in Bahasa Indonesia.

What's new?
(often used as a greeting)
Apa kabar?
(AH·pah KAH·bahr)

Good-bye
Selamat tinggal
(suh·LAH·maht TING·ahl)

My name is ______.
Nama saya ______.
(NAH·mah SAH·yah)

OFF TO SCHOOL Ferdian Adhie Putranto (FUR·dee·an AD·hee poo·TRAHN·toe) sets off to school on a busy Jakarta street. The orange vehicle behind him has three wheels. It is an inexpensive way to get around Jakarta, the capital of Indonesia. The hat Ferdian is wearing is a traditional Indonesian style.

KEMAL JUFRI / POLARIS (4)

WENDY CHAN/GETTY IMAGES

CHAPTER 27

East Asia and Southeast Asia Today

Essential Question

Regions The region of East Asia and Southeast Asia has experienced amazing economic growth since the late 1900s. Many of the region's countries have become important trading partners of the United States and other nations around the world. What impact does rapid economic change have on the lives of people?

Hanoi, Vietnam

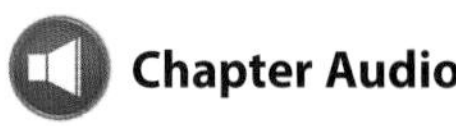

Section 1: China

BIG IDEA The characteristics and movement of people impact physical and human systems. In recent years, China's Communist government has moved toward freer economic policies.

Section 2: Japan

BIG IDEA People's actions can change the physical environment. In the past, Japan's tremendous economic growth hurt the environment.

Section 3: The Koreas

BIG IDEA Culture groups shape human systems. Opposing political views have made the two Koreas very different countries.

Section 4: Southeast Asia

BIG IDEA The physical environment affects how people live. Location, landforms, and distribution of resources make areas of Southeast Asia different from one another.

FOLDABLES™ Study Organizer

Categorizing Information Make this Foldable to help you collect information about the region of East Asia and Southeast Asia today.

Step 1 Fold the sides of a piece of paper into the middle to make a shutterfold.

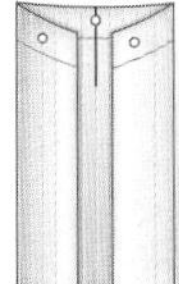

Step 2 Cut each flap at the midpoint to form four tabs.

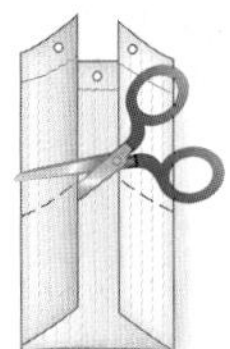

Step 3 Label the tabs as shown.

Reading and Writing After you have completed your Foldable, choose one of the subregions in the chapter and write a summary for that area today.

Visit glencoe.com and enter ***QuickPass***™ code EOW3109c27 for Chapter 27 resources.

Guide to Reading

BIG Idea

The characteristics and movement of people impact physical and human systems.

Content Vocabulary

- human rights *(p. 729)*
- exile *(p. 729)*

Academic Vocabulary

- reject *(p. 729)*
- nonetheless *(p. 732)*

Reading Strategy

Identifying Central Issues Use a chart like the one below to describe China in the past and today.

	Past	Today
Economy		
Hong Kong and Macao		
Environment		

China

Section Audio

Spotlight Video

Picture This This gallery of panda artwork is actually a workroom at a leather and fur factory in China. China is the world's leader in leather processing. As economies become more interdependent, though, China faces competition from leather manufacturers elsewhere. Now China is trying to focus on the quality of the leather it produces rather than the quantity. Read the section to learn more about issues facing China today.

▼ Leather processing in Jiaozuo, China

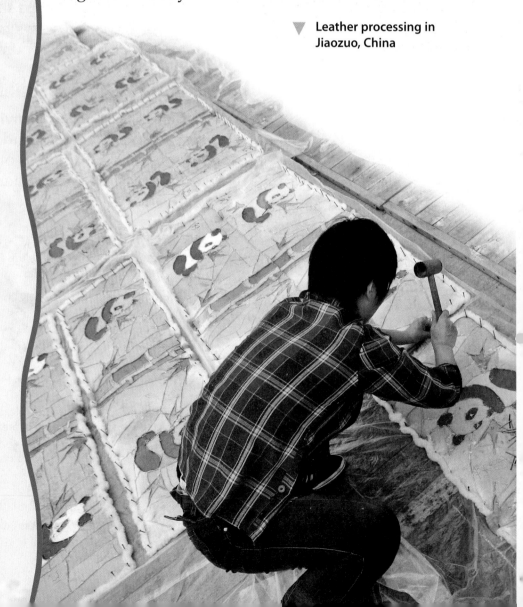

China's Government and Society

Main Idea **China's government keeps tight control over its citizens' political activities.**

Geography and You Think about how you might feel if you could not state your opinions freely. Read to find out how China's government treats citizens who oppose its policies.

Since China's Communist government took power in 1949, it has maintained strict political control. Even as Chinese society grows and changes, the country has seen little political change.

China's Government

China's government is controlled by the Chinese Communist Party, which greatly restricts other political groups. Government leaders are not freely elected by the people. They gain power through promotion in and loyalty to the Communist Party.

In recent years, China's government has allowed people more economic freedom, but it still has kept tight control over all political activities. It denies individual freedoms and acts harshly against Chinese citizens who criticize its actions. For example, in 1989 about 100,000 students and workers gathered in Beijing's Tiananmen (TEE·EHN·AHN·MEHN) Square to call for democracy. The government sent troops to break up the peaceful protest. Thousands of people were killed or injured.

Countries around the world have protested the Chinese government's harsh treatment of people who criticize it. They say that Chinese leaders have no respect for **human rights,** or basic freedoms such as freedom of speech and religion. China's leaders have also been criticized for their actions in Tibet, an area in southwestern China. Tibet was once a separate Buddhist kingdom. China took control of Tibet in 1950 and crushed a revolt there nine years later. The Tibetan people have demanded independence since then. The Dalai Lama (DAH·ly LAH·muh), Tibet's Buddhist leader, now lives in exile in India. To be in **exile** means to be forced to live somewhere other than your own country. The Dalai Lama has asked world leaders for support. China has **rejected,** or opposed, all demands to free Tibet.

NATIONAL GEOGRAPHIC

Political Protests

Some political protests by the Chinese people have been violently put down by government forces. ***Place*** **What occurred at Tiananmen Square?**

Chinese Society

About 65 percent of China's people live in rural areas. Most Chinese are crowded into the river valleys of eastern China.

Rural families here still use hand tools because machinery is too expensive. In recent years, village life has improved. For example, electricity is now available in many places. Most villagers live in larger houses than in the past—houses with three or four rooms. In addition, motorbikes, radios, and televisions are becoming more common.

China's cities have experienced even greater change. They are growing rapidly as people leave villages hoping to find better-paying jobs. Many city workers now earn enough money to buy extra clothes and televisions. In general, people in China's cities have a higher standard of living than do people in rural areas.

As a result of this movement, China has some of the world's largest cities. Nearly 160 Chinese cities have more than 1 million people. New office buildings, shopping malls, and apartments are being built rapidly in China's cities today.

Reading Check **Explaining** Why have other countries criticized China's government?

Economic Changes in China

Main Idea **China's government has introduced economic reforms in recent years.**

Geography and You Do you own anything that was made in China? Read to find out how changes have brought growth to China's economy.

As you read earlier, under Communist rulers, China's economy grew slowly at first. Beginning in the late 1970s, China's leaders began permitting some free market reforms in an attempt to improve the economy. Today people can choose the jobs they want and where to start their own businesses. Workers can keep the profits they make, and farmers have some control over the crops they grow and sell.

Economic Growth

Because of China's economic changes, the country now has one of the world's

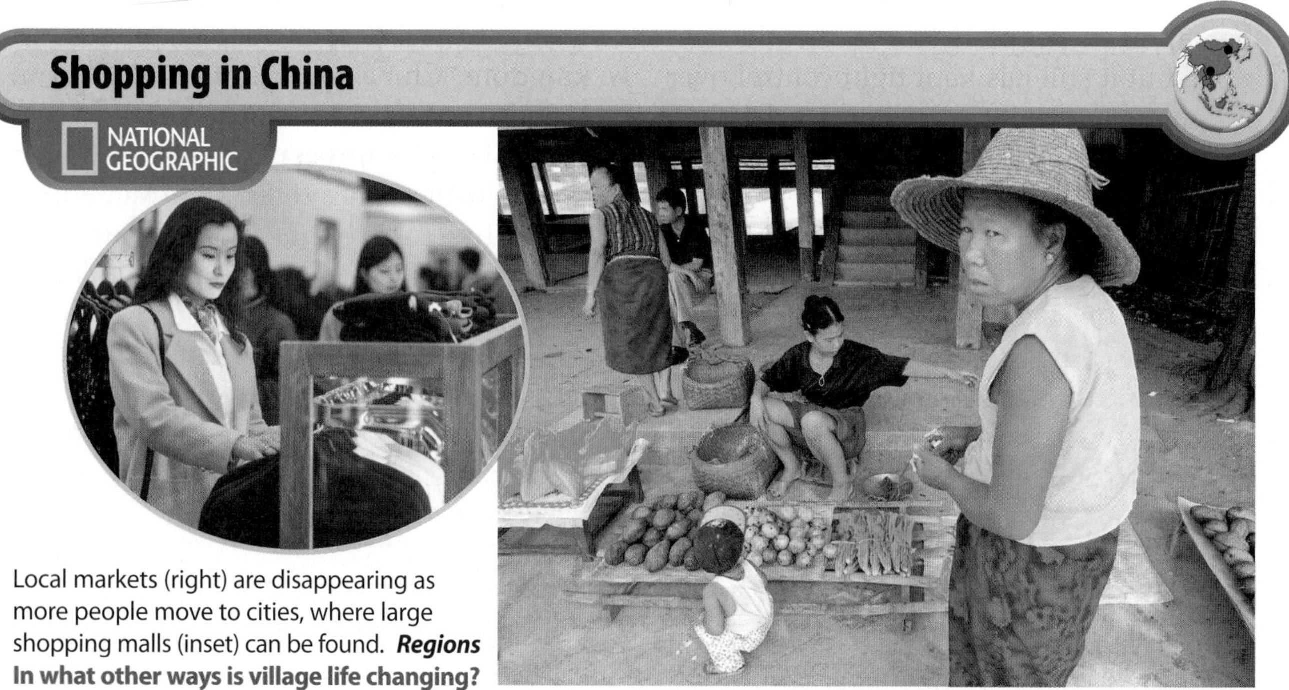

Local markets (right) are disappearing as more people move to cities, where large shopping malls (inset) can be found. ***Regions*** **In what other ways is village life changing?**

fastest-growing economies. Farm output has risen rapidly despite the fact that only 10 percent of China's land can be farmed. China has become a world leader in producing various agricultural products, including rice, tea, wheat, and potatoes.

Industry also has boomed. Chinese factories produce textiles, chemicals, electronic equipment, airplanes, ships, and machinery. Eager to learn new business methods, the Chinese have asked other countries to invest in Chinese businesses. Many companies in China are now jointly owned by Chinese and foreign businesspeople.

China's fast-growing economy has greatly increased the global demand for energy. China is the world's largest oil user after the United States, and the world's largest producer and user of coal.

Hong Kong and Macao

The territories of Hong Kong and Macao play an important role in the economic changes taking place in China. Both are leading centers of manufacturing, trade, and finance. Chinese leaders believe that successful businesses in these territories will help boost economic growth in the rest of the country.

Hong Kong and Macao were once controlled by European countries—Hong Kong by the United Kingdom and Macao by Portugal. China regained control of the two territories in the late 1990s. At that time, China promised to allow Western freedoms and markets to continue in these territories. Some people in the two territories, however, claim that China has broken this promise by interfering with citizens' rights.

Results of Growth

Because of economic growth, more of China's people are able to get better jobs and wages. They are also able to obtain more consumer goods. Not everyone has benefited equally from the country's economic boom, however. China's eastern coastal region produces about 60 percent of the country's goods and services. People who live in this area enjoy a more comfortable standard of living than those in other parts of China. Also, many Chinese find that prices have risen faster than their incomes. Some Chinese have become rich, while others remain poor.

Protection From Pollution

Beijing's poor air quality often forces residents to cover their noses and mouths to breathe.
Human-Environment Interaction **What are China's main energy sources?**

Social Studies ONLINE

Student Web Activity Visit glencoe.com and complete the Chapter 27 Web Activity about Hong Kong.

China's Environment

China's economic growth has also harmed the environment. China burns much coal for fuel. As a result, many cities have polluted air. Factory wastes, sewage, and fertilizers and pesticides also have poisoned water.

In 1998 China's government was forced to act to protect the environment. That year, the flooding of the Chang Jiang killed thousands of people. The cutting of trees and the draining of wetlands for farming had kept land upriver from soaking up the floodwaters. In response, the government began to turn farmland back into forests and wetlands. In another decision, the government also placed limits on burning coal for fuel. **Nonetheless,** pollution still threatens much of China.

Reading Check **Drawing Conclusions** How have recent economic changes affected Chinese society?

NATIONAL GEOGRAPHIC

Taiwan's Economy

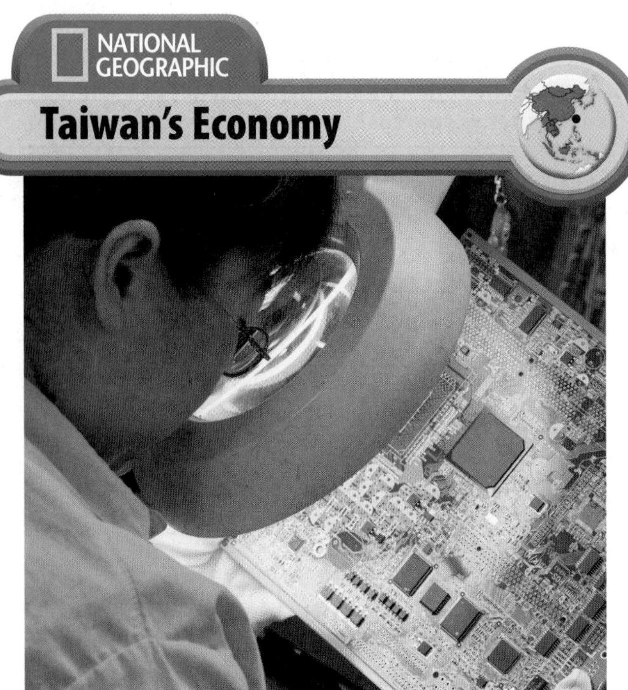

Taiwan's industries produce electronic goods, including computers and computer components. ***Regions*** **How is Taiwan's economic relationship with China changing?**

China's Neighbors

Main Idea **Taiwan and Mongolia have been influenced by Chinese ways and traditions.**

Geography and You Do you think the United States has influenced its neighboring countries? Read to find out how Taiwan and Mongolia, two very different places, have been connected to their neighbor, China.

Taiwan is an island close to China's mainland, and Mongolia is located along China's northern border. Throughout their histories, Taiwan and Mongolia have had close ties to their larger neighbor.

Taiwan

About 100 miles (161 km) off China's southeastern coast lies the island of Taiwan. A ridge of steep, forested mountains runs through Taiwan's center. On the east, the mountains descend to a rocky coastline. On the west, they fall away to a narrow, fertile plain, where most of Taiwan's people live.

Taiwan was a province of China for several hundred years. After the Communists defeated the Nationalists in 1949, Nationalist leaders and their followers fled to Taiwan and set up their own government. Since then, the governments of mainland China and Taiwan have both claimed the right to rule all of China. The government on the mainland considers Taiwan subject to its authority. Tensions between the two governments flare up when leaders in Taiwan discuss declaring independence.

Since 1949, Taiwan has developed a strong industrial economy. Taiwan's companies produce textiles, ships, and electronic products, such as computers and telephones. With a highly educated workforce, Taiwan has become a center for developing new products and improving old ones. In recent years, Taiwan's businesses have even invested in mainland China, helping to fuel economic growth there.

When the Nationalist government was set up in 1949, Taiwan was a dictatorship. In the late 1980s, however, the Nationalists' grip on the government loosened. Other political parties were allowed to form, and Taiwan moved toward democracy.

Mongolia

North of China is the landlocked country of Mongolia. Mongolia receives little rain, and steppe grasses and desert cover its vast landscape. Because of its harsh terrain, the country has few people compared to its large size.

For centuries, Mongolia's people were nomads who herded animals. During the 1200s, they used their horse-riding and military skills to create the Mongol Empire. This empire stretched from China to eastern Europe. Even today, many Mongolians raise horses, sheep, goats, cattle, and camels.

With its vast grasslands and herds of grazing animals, Mongolia has been called the "Texas of Asia." Important industries in Mongolia use products from these animals. Some factories use wool to make textiles and clothing, while others use the hides of cattle to make leather and shoes.

In 1990 Mongolia abandoned its 70-year-old Communist system and embraced political and economic reforms. In addition to these changes, many Mongolians have moved from rural areas to cities. Today, about one-fifth of Mongolia's people live in Ulaanbaatar, the capital.

Contrasting How is Taiwan's government different from that of mainland China?

Social Studies ONLINE
Study Central™ To review this section, go to glencoe.com.

Vocabulary

1. **Explain** how *exile* and *human rights* are related by using the terms in a paragraph.

Main Ideas

2. **Describing** How are China's rural and urban areas changing?
3. **Identifying** Use a diagram like the one below to identify China's products, both agricultural and industrial.

4. **Comparing and Contrasting** Compare and contrast Taiwan and Mongolia.

Critical Thinking

5. **BIG Idea** How have China's new economic policies affected the Chinese people?
6. **Challenge** What do you think might happen to Taiwan in the future? Why?

Writing About Geography

7. **Expository Writing** Write a paragraph describing China's economic growth and how it has both helped and harmed the country.

The News Media in China: Should the Government Regulate It?

The ways that people can get information in China are strictly regulated by the Chinese government. For example, government workers search for Web sites, e-mails, or blogs that contain words such as *democracy* and *freedom* and then block or delete them. Chinese leaders maintain that the regulations are to protect its citizens from harmful content. Others in China think that such monitoring is dangerous and violates China's constitution, which guarantees free speech.

For Regulation

The Internet is playing an important role in accelerating the development of China's national economy and science and technology. . . . At the same time, the question of how to ensure the secure operation of the Internet and the security of information . . . has aroused widespread concern throughout the community. [These laws are in place] to promote the good and eliminate the bad, encourage the healthy development of the Internet, safeguard the security of the State and the public interest, as well as protect the lawful interests of individuals, legal persons and other organizations.

—From the Decision of the Standing Committee of National People's Congress

Against Regulation

At the turning point in our history from a totalitarian to a constitutional system, depriving [denying] the public of freedom of speech will bring disaster for our social and political transition [change] and give rise to group confrontation and social unrest. . . . Experience has proved that allowing a free flow of ideas can improve stability and alleviate [ease] social problems.

—From a Public Letter Signed by Former Communist Party Officials and Scholars in Beijing

You Be the Geographer

1. **Identifying** What words and phrases in the Standing Committee's decision explain why the government regulates the Internet in China?
2. **Critical Thinking** How might the free flow of ideas help ease social problems?
3. **Read to Write** Write a paragraph that describes your opinion about regulating media sources like the Internet. Is there ever a time when government monitoring is necessary? Explain.

Guide to Reading

BIG Idea

People's actions can change the physical environment.

Content Vocabulary

- intensive agriculture *(p. 737)*
- trade deficit *(p. 738)*
- tatami *(p. 739)*
- kimono *(p. 740)*
- *anime (p. 741)*

Academic Vocabulary

- prohibit *(p. 737)*
- focus *(p. 740)*

Reading Strategy

Determining Cause and Effect Use a diagram like the one below to identify reasons behind Japan's successful economy.

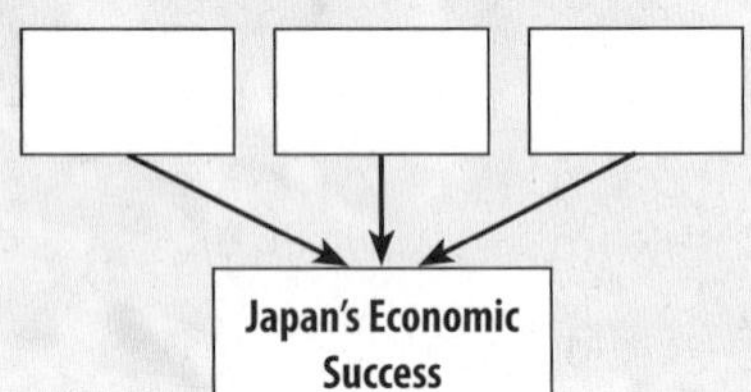

Japan

 Section Audio  Spotlight Video

Picture This Whether you are a tired tourist or a businessperson who missed a train and has nowhere to sleep, you can choose to rest in a Japanese capsule hotel. Ranging from $9 to $30 per night, capsule hotels offer people an inexpensive alternative to a standard hotel. Each capsule measures 6 feet long, 3 feet wide, and 3 feet (1 m) high. A capsule contains a mattress, a television, and an alarm clock. To learn more about Japan today, read Section 2.

▼ Overnight in Osaka

Government and Economy

Main Idea In Japan, government and business leaders work together to build the economy.

Geography and You Do you enjoy listening to music on compact discs? The Japanese helped invent the compact disc player. Read to find out how Japan became an economic power.

Japan lies off the coast of East Asia. The country consists of four main islands and thousands of smaller ones. The largest islands are Hokkaido (hoh·KY·doh), Honshu, Shikoku (shee·KOH·koo), and Kyushu (KYOO·SHOO). Tokyo, Japan's capital, is located on Honshu.

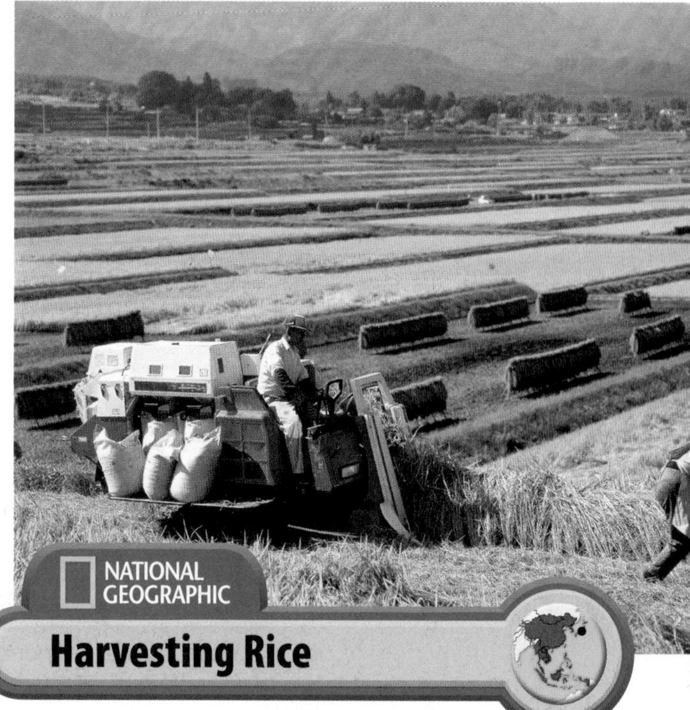

NATIONAL GEOGRAPHIC

Harvesting Rice

Rice production in Japan today depends heavily on machinery. ***Place*** **About how much of its food must Japan import?**

Japan's Government

Japan is a constitutional monarchy, which is a form of democracy. The emperor is the head of state, but elected officials run the government. Voters elect representatives to the Diet, or national legislature. The political party with the most members chooses a prime minister to lead the government.

At the urging of the United States, Japan's military forces have remained small. In addition, Japan's constitution **prohibits,** or forbids, Japan from being a military power. The United States is a close ally of Japan and keeps military bases there to defend Japan from attack.

Japan's Economy

Japan lacks military might, but it is a strong world economic power. This is an amazing accomplishment since Japan has few mineral resources. Japan has to import the raw materials it needs—iron ore, coal, and oil—to produce most manufactured goods. Farmland also is limited, so Japan's farmers practice **intensive agriculture.** This means they grow crops on every available piece of land. By using this method, Japan can produce enough rice to meet its needs. Still, the country must import about 60 percent of its food.

To pay for these imports, Japan has developed industries that produce goods to sell to other countries. Today, Japan is an industrial giant known around the world for the variety and quality of its manufactured products. Japan's modern factories use new technologies to make their products quickly and carefully. As a result, Japan is among the world's top producers of steel, cars, ships, and cameras and consumer electronics.

Japan's government and business leaders work together to advance the economy. Government-owned banks lend money to businesses so they can grow.

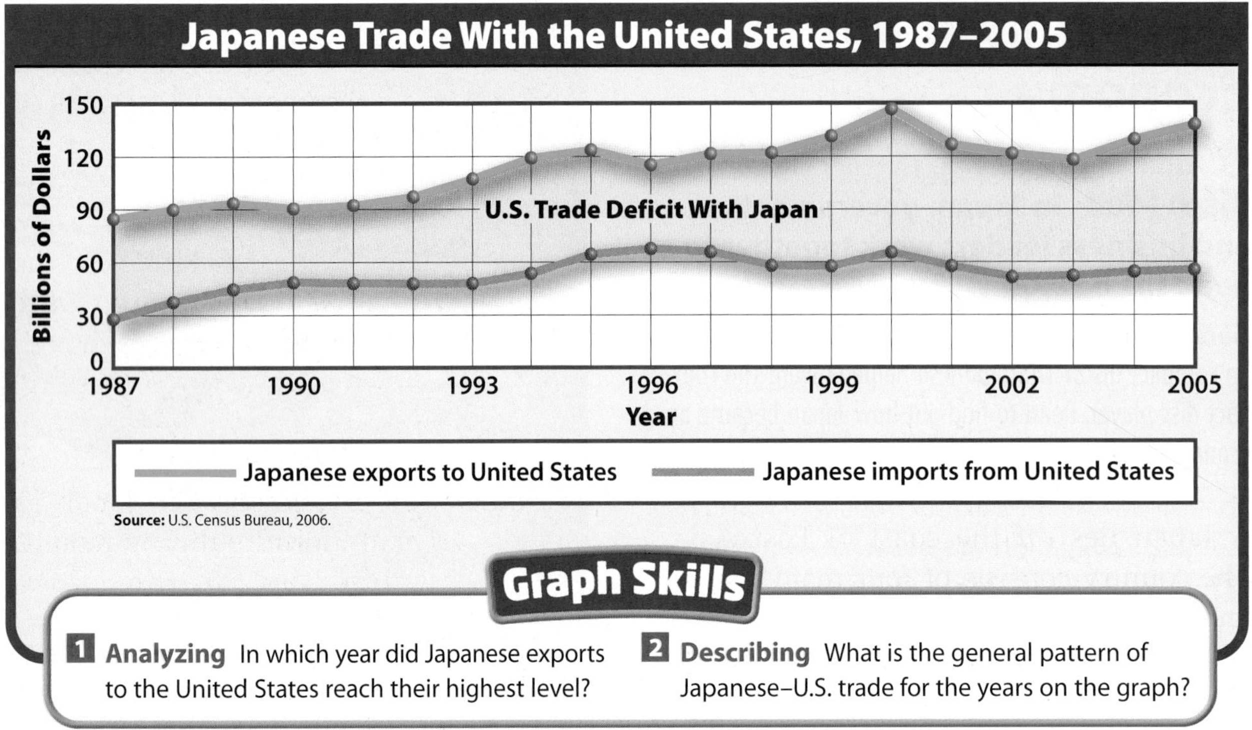

Graph Skills

1. **Analyzing** In which year did Japanese exports to the United States reach their highest level?
2. **Describing** What is the general pattern of Japanese–U.S. trade for the years on the graph?

The government also helps Japanese businesses by passing laws that make it difficult for other countries to sell their goods in Japan. These laws protect Japanese companies from the effects of foreign competition.

Trade restrictions, however, have led to protests from other industrial countries that want to sell more goods to Japan. As the graph above shows, the United States has long had a huge trade deficit with Japan. A **trade deficit** occurs when one country buys more goods from another country than it sells to that country. The United States has encouraged Japan to buy more American goods. In recent years, Japan has lowered trade barriers, but imports from the United States have not risen significantly.

Challenges for Japan

Despite its economic success, Japan faces four main challenges. First, Japan faces growing economic competition from South Korea, Taiwan, and China. This competition threatens Japan's position in world markets.

Japan also has the highest life expectancy in the world and a low birthrate. As a result, the country has an aging population and may soon face a shortage of workers, which will strain the economy. Taxes will need to rise to support health care benefits for the elderly.

Environmental pollution is also a major issue. Polluted air from power plants has produced acid rain. Although Japan has reduced environmental pollution and repaired some damage, more work needs to be done.

Finally, Japan faces a constant threat of destructive earthquakes. The Japanese islands lie on the Pacific Ring of Fire, an area of frequent tectonic activity. Earthquakes in this region are common. Because Japan's cities are so densely populated, a major earthquake would bring disaster to millions of people.

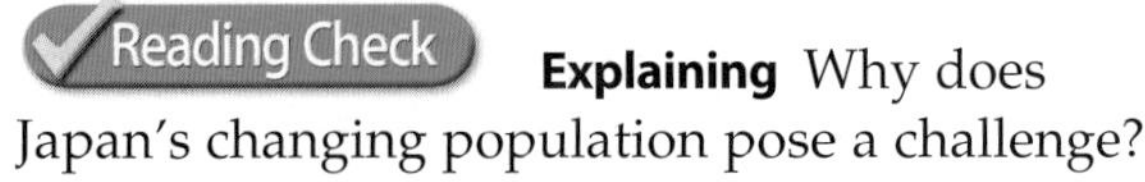

Explaining Why does Japan's changing population pose a challenge?

Life in Japan

Main Idea **Life in Japan is influenced by tradition and modern ideas.**

Geography and You Think about your family's traditions. How might they be different from those of families in other countries? Read to discover what traditions the Japanese people value.

Japanese culture is a mix of traditional and modern ways. Many features of daily life follow traditional customs and patterns of living. Many others show the influence of modern culture and new technology.

Cities and Countryside

Japan is the world's tenth-most-populous country and has a high population density. Mountains and highlands cover much of the country. Because of these landforms, most Japanese are crowded into urban areas on the coastal plains. Tokyo and its surroundings are home to more than 35 million people, about the same as the population of California. Japan's cities have tall office buildings and busy streets. Many city workers crowd into subway trains to get to work.

Because there is so little space, the Japanese have increased the land available to them. For example, they have built large islands using earth taken from high ground and deposited near a shoreline. One such island in Osaka Bay holds a major airport.

The Japanese have also adapted to crowded conditions by building small homes. Homes have an average of about four small rooms per family. In the cities, tall apartment buildings provide housing for many families.

In rural areas, people live in farming villages. These communities are often linked to nearby cities. Japan's rail system includes high-speed trains, local trains, and subways. City workers sometimes live in the villages and commute, or travel back and forth regularly, to jobs in the cities.

Changes in Daily Life

Living space is one example of the blending of traditional and modern ways of life in Japan. Traditional Japanese homes generally have wooden floors covered by tatami (tah·TAH·mee). **Tatami** are straw mats, each about 18 square feet (1.7 sq. m). The mats are very practical in Japan as they do not hold moisture during months of high humidity, as carpeting does.

Tokyo Driving Range

NATIONAL GEOGRAPHIC

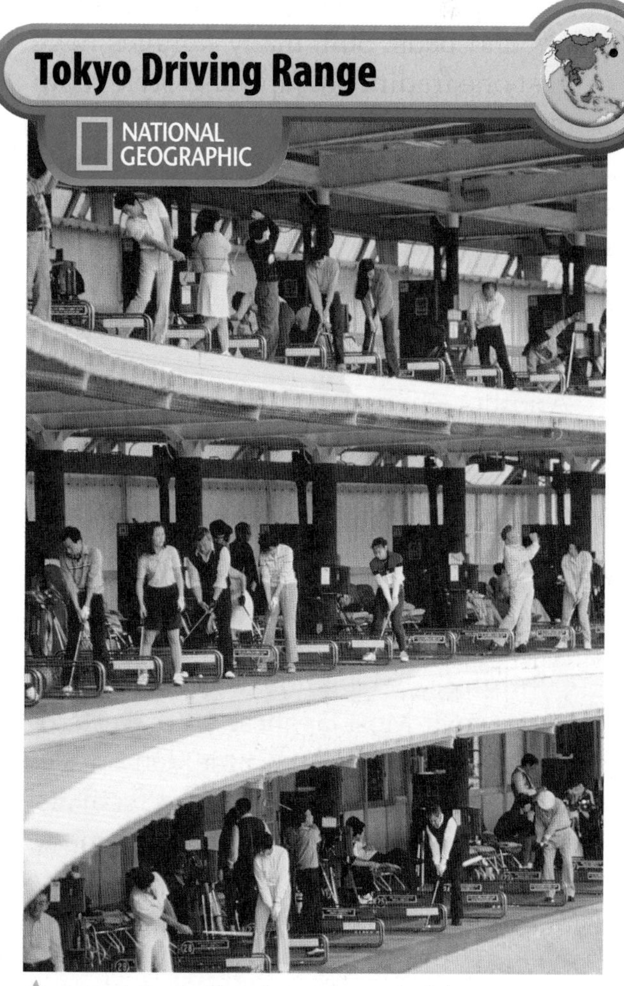

Japan's limited land area forces builders to come up with creative ideas to fit the country's people and their activities. ***Location*** **Where are most Japanese cities located?**

To keep the tatami matting clean, the Japanese usually take off their shoes before entering a room.

Traditional Japanese rooms are used as a living space during the day and as sleeping areas at night. This double use works well because rooms have little furniture. People sit on cushions on the floor during the day and sleep on portable mattresses, called futons, at night. These cushions and futons can be rolled up and stored easily when not in use.

Although some Japanese still maintain these traditions, modern housing styles have become common throughout Japan. In many modern homes, rooms are carpeted and equipped with furniture such as chairs and beds. Still, many families have at least one traditional Japanese-style room in their homes.

Living patterns in Japan also continue to change. In the past, grandparents, parents, and children often lived together in the same home. Today, it is more common for parents and children to live apart from the grandparents.

Another change is in the choices that women make. In the past, Japanese women married early and stayed home to raise children. Today, more women delay marriage in order to **focus** on their careers. When women do marry, they tend to have fewer children.

Clothing also blends traditional and modern ways. Most Japanese wear Western-style clothes most of the time. On special occasions, such as weddings and festivals, they wear the traditional Japanese garment, the kimono (kuh·MOH·NOH). A **kimono** is a long robe, usually made of silk, with an open neck and large sleeves. It has no fasteners, but is held in place by a wide sash. Kimonos are often dyed in bright colors with beautiful designs.

Family Ties

Enjoying a meal of sushi is a favorite pastime for Japanese families. ***Place*** **How are families changing in Japan?**

The blending of old and new styles has influenced cooking as well. People still eat the traditional Japanese meal of rice served with several other foods, including meat or fish, vegetables, and soup. However, many also enjoy food from other countries. For example, American hamburgers and fried chicken have become especially popular with Japanese children.

Religion

Japan's main religions are Shinto and Buddhism. Shinto began in Japan many centuries ago. Unlike most major world religions, Shinto has no organized set of teachings, no known historical founder, and no moral code. Instead, it focuses on respect for nature, love of simple things, and concern for cleanliness and good manners.

Buddhism came to Japan by way of China and Korea in the A.D. 500s. Buddhism teaches respect for nature and stresses the need to achieve inner peace.

For centuries, Japanese people have blended Buddhism and Shinto. They attend Shinto shrines for some events, such as New Year's Day and wedding ceremonies.

They attend Buddhist temples for other occasions, such as funerals and the midsummer festival that honors ancestors.

Japanese Pastimes

The people of Japan enjoy both traditional arts and modern pastimes. Many Japanese still enjoy traditional forms of theater, such as No and Kabuki (kuh·BOO·kee). The actors in No plays tell stories only through precise movements. By contrast, lively Kabuki theater uses brilliantly colored costumes, songs, and dances. Another popular kind of theater, called Bunraku (bun·RAH·koo), uses puppets to entertain audiences.

The Japanese also have embraced a more modern form of storytelling—movies. Several Japanese directors have won worldwide fame for their work. Even newer is ***anime*** (A·nuh·MAY), the Japanese style of animation that arose in the late 1900s. Comic books and cartoons using this style have become popular among young people not only in Japan but in other countries.

Large numbers of Japanese still practice traditional martial arts such as judo and kendo. Sumo wrestling, another traditional sport, also remains popular. In this sport, two opponents can use only their hands to try to force each other onto the ground or out of the fighting ring.

The Japanese enjoy more modern sports as well. The Japanese adopted baseball from the United States in the 1870s. Since then, it has become widely popular in Japan. Many children play on Little League teams, and Japan has its own baseball major leagues.

Reading Check **Describing** Describe the different forms of traditional Japanese theater.

Section 2 Review

Social Studies ONLINE
Study Central™ To review this section, go to glencoe.com.

Vocabulary

1. **Explain** the significance of:
   - **a.** intensive agriculture
   - **b.** trade deficit
   - **c.** tatami
   - **d.** kimono
   - **e.** *anime*

Main Ideas

2. **Explaining** How have Japanese banks contributed to the country's economic growth?
3. **Comparing and Contrasting** Use a Venn diagram like the one below to compare and contrast the traditional and the modern in Japanese daily life.

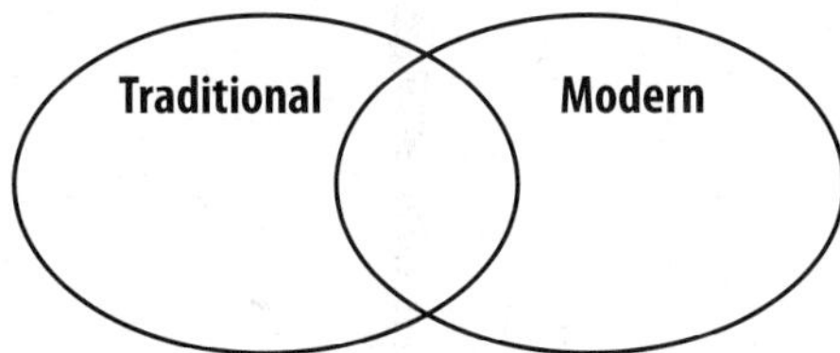

Critical Thinking

4. **Determining Cause and Effect** Why are Japanese goods in such high demand?
5. **BIG Idea** Explain how the Japanese use intensive agriculture and other methods to adapt to and change their physical environment.
6. **Challenge** Choose one of the four challenges facing Japan, and suggest ways that the country can meet the challenge.

Writing About Geography

7. **Descriptive Writing** Imagine you are visiting a family in Japan. Write a letter home describing how daily life in Japan differs from daily life at home.

The Koreas

Section Audio

Spotlight Video

Guide to Reading

BIG Idea

Culture groups shape human systems.

Content Vocabulary

- celadon *(p. 743)*
- hangul *(p. 743)*
- demilitarized zone (DMZ) *(p. 744)*
- land reform *(p. 744)*

Academic Vocabulary

- adapt *(p. 743)*
- controversy *(p. 745)*

Reading Strategy

Comparing and Contrasting Use a Venn diagram like the one below to compare and contrast the economies of South Korea and North Korea.

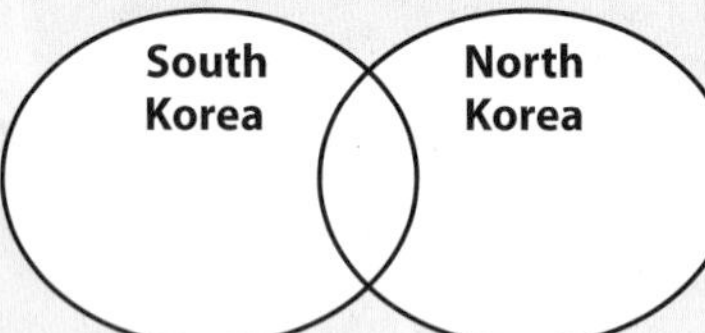

Picture This Jars of ginseng line the shelves of this market in Seoul, South Korea. Ginseng is a prized herb in East Asia. Some people believe that ginseng can lower blood sugar and cholesterol levels, protect against stress, and increase muscular strength. Roots of ginseng can live for 100 years. Some people claim that eating these old ginseng roots will also give one a long life. Read Section 3 to learn more about North and South Korea.

▼ Ginseng for sale

South Korea

Main Idea **South Koreans have built a strong economy and a modern culture.**

Geography and You Have you ever made changes in your life because of challenges you faced? Read to find out how South Korea's leaders made changes to improve their struggling economy.

The Korean Peninsula juts out from northern China between the Sea of Japan (East Sea) and the Yellow Sea. For centuries, this peninsula was the united country of Korea. Today the peninsula is divided into two nations. In the north is the Communist country officially titled the Democratic People's Republic of Korea. It is more generally referred to as North Korea. In the south is the Republic of Korea, or South Korea.

Korean Culture

Korean culture has been strongly influenced by China, its large neighbor. Both Confucianism and Buddhism spread to Korea from China. Korea's early kingdoms adopted the ideas of Confucius. Buddhism later affected Korean ideas and ways of life.

The Koreans also **adapted** Chinese culture to their own tastes. Koreans modified a type of Chinese pottery called **celadon** (SEH·luh·DAHN) by changing the color and cutting designs into the pots. In addition, Korea originally began using the Chinese writing system. In the 1400s, however, King Sejong ordered scholars to develop a different system. They produced **hangul** (HAHN·GOOL), a writing system that uses only 28 symbols—far fewer than the thousands of characters needed to write Chinese. This makes the Korean system much easier to learn.

NATIONAL GEOGRAPHIC

Patrolling the DMZ

A North Korean soldier adjusts a surveillance camera that looks over South Korean territory.
Place **How did the Korean War end?**

A Divided Korea

For centuries, Korea was independent. Then, in 1910 the Japanese conquered Korea and made it part of their empire. They governed the peninsula until the end of World War II.

After World War II, Korea became a divided country. Troops from the Communist Soviet Union took over northern Korea, and American troops occupied the southern half of the country. Eventually Korea was divided along the 38th parallel, or line of latitude.

In 1950 the armies of North Korea attacked South Korea in an effort to unite all of Korea under Communist rule. United Nations countries led by the United States rushed to South Korea's defense. This began the Korean War, which finally ended in 1953 without a victory for either side.

After the war, North Korea and South Korea were separated by a 2.5-mile (4-km) wide **demilitarized zone (DMZ)**. Both sides agree not to place any soldiers or weapons in the DMZ. Although the two Koreas have tried to settle their differences, relations between them remain tense.

South Korea's Economy

After the Korean War, South Korea's leaders set out to rebuild the country's economy. To help farmers, the government introduced **land reform.** This policy broke up large estates into smaller family farms. The government also supplied fertilizer to help farmers make the land more productive. As a result, South Korea was able to produce enough food to feed its people.

The government also set out to make South Korea an industrial country. It borrowed money from foreign banks to create industries that produced textiles, iron and steel, cars, ships, and electronic goods. The products were then sold abroad, earning enough money to repay the loans. As a result, South Korea's economy strengthened.

For many years, the government exported most goods made in South Korea. It also kept workers' wages low. Beginning in the 1980s, wages increased, and more goods were made available in South Korea. As a result, South Koreans now enjoy a higher standard of living than they did after the Korean War. Because it is an economic leader, South Korea is now taking on a more important role in world affairs. In 2006 Ban Ki-moon, a South Korean diplomat, became Secretary General of the United Nations.

The People

For centuries, most Koreans lived in the countryside and farmed. This way of life changed as South Korea industrialized. Today 80 percent of South Koreans live in cities. The largest city is Seoul, the capital. Seoul has about 10 million people—nearly one-fifth of the country's population. Other large cities are Taegu (TA·goo), a manufacturing center, and the ports of Inch'ôn (IHN·CHAHN) and Pusan (POO·SAHN).

Like Japan, South Korea's cities have a mix of traditional and modern architecture. These cities bustle with life and reflect the changes brought by modern technology. One of the most striking examples is the rise of cell phone usage. About 80 to 90 percent of South Korean adults and teens own cell phones. This is one of the world's highest rates of cell phone ownership.

South Korea's population is generally young. Although life expectancy is in the seventies, only 9 percent of the people are 65 years old or older. More than one-third of the country's population is younger than 30.

As in Japan, the population of South Korea is growing slowly. People tend to marry at a later age and generally have fewer children. In the past, families used to arrange marriages for their children, but today people typically choose their own marriage partners.

Korean Technology

South Korea today is a leading manufacturer of electronics. ***Place*** **How did South Korea's government create an industrial economy?**

Despite modernization, South Koreans remain devoted to their traditions. Buddhism, Confucianism, and Christianity are South Korea's major religions. The Koreans have developed their own culture, but Chinese religion and culture influenced the traditional arts of Korea. Like Japan, Korea has a tradition of martial arts. The martial art of tae kwon do (TY KWAHN DOH) began in Korea. Tae kwon do emphasizes mental discipline as well as self-defense.

South Koreans also celebrate traditional holidays. One of the most popular is Chuseok, the fall harvest festival. During this festival, Koreans leave rice cakes and other foods at cemeteries to honor their ancestors.

Reading Check **Making Generalizations** What have relations been like between the two Koreas?

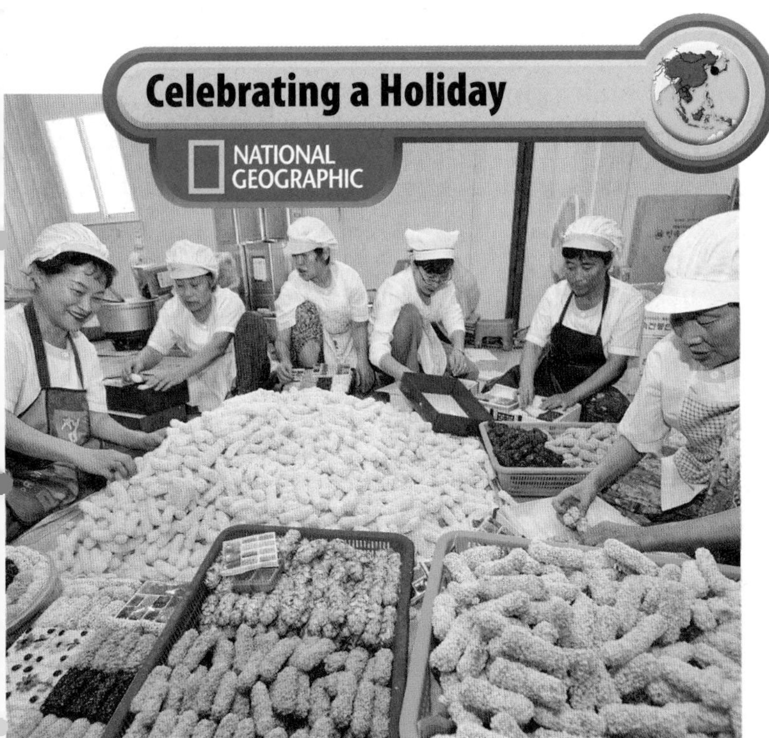

Food, like these rice cakes, is an important part of celebrating Chuseok, a holiday similar to Thanksgiving in the United States. ***Place*** **What are the major religions of South Korea?**

North Korea

Main Idea **North Korea is an isolated country whose people are very poor.**

Geography and You Imagine not being able to speak your opinion or choose a career for yourself. Read to learn about ways of life in Communist North Korea.

North Korea is very different from South Korea. North Korean rulers have isolated this country from the rest of the world. North Koreans face many hardships because of a struggling economy.

North Korea's Government

Since it was formed in 1948, North Korea has been ruled by Communist dictators. Kim Il Sung became North Korea's first ruler in the late 1940s. After Kim's death in 1994, his son Kim Jong Il became the ruler. Throughout North Korea, monuments have been built to honor these Communist leaders.

The North Korean government places the needs of the Communist system over the needs of its citizens. It controls all areas of life in North Korea. People have few freedoms. In addition, the government makes travel into and out of the country difficult.

In 2006, as a test, North Korea exploded a nuclear weapon. Concerned about this test, the United States, South Korea, and other countries urged North Korea to give up its goal of becoming a nuclear power. In 2007, this **controversy,** or disputed issue, seemed to be resolved when North Korea agreed to gradually shut down its nuclear reactors in return for fuel and aid.

The Economy

Unlike prosperous South Korea, North Korea has long been economically poor.

Coal and iron ore are plentiful, but the country's industries suffer from old equipment and power outages.

One reason for North Korea's poverty is that its Communist rulers devote many resources to the military. For many years, North Korea traded with the Soviet Union and Eastern Europe. After the fall of communism in Europe, North Korea's foreign trade declined. In addition, North Korea's government has done little to bring in investment from other countries.

Poverty is widespread throughout North Korea. The infant mortality rate in this country is four times higher than in South Korea. Because of harsh conditions, tens of thousands of people have left the country, mainly to go to China.

Another cause of hardship is a failed farm policy. North Korea's leaders ended private ownership of land and created large, government-owned farms. Instead of growing food for themselves, farmers must turn their harvests over to the government, which then distributes the food. Only a small area of North Korean land can be farmed, and the collective farms are unproductive. As a result, North Korea must import food for its people to survive. In some years, the government refused to accept food aid from world relief organizations, and many people died.

North Korea's government also controls industry, telling managers and workers what to produce. The country's chief products are iron and steel, chemicals, and textiles. North Korea also makes military weapons, some of which it sells to other countries.

Analyzing Information What factors contribute to North Korea's poverty?

Section 3 Review

Social Studies ONLINE
Study Central™ To review this section, go to glencoe.com.

Vocabulary

1. **Explain** the meaning of the following terms.
   a. celadon
   b. hangul
   c. demilitarized zone (DMZ)
   d. land reform

Main Ideas

2. **Identifying** Use a diagram like the one below to list key facts about the people of South Korea. Add as many ovals as you need.

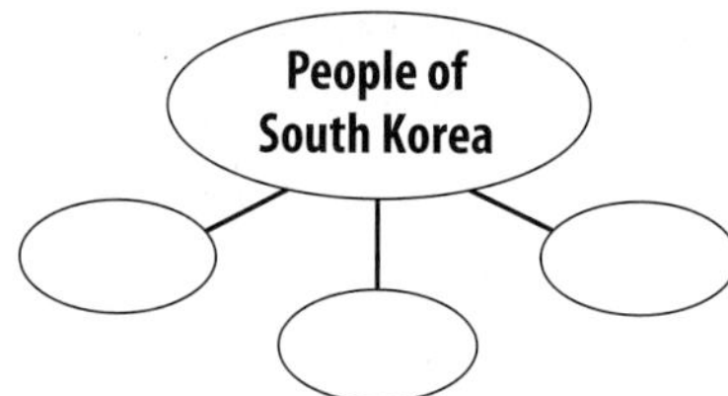

3. **Explaining** Why has agriculture been unsuccessful in North Korea?

Critical Thinking

4. **Analyzing Information** How did the policies of North Korea's government contribute to its economic problems?
5. **BIG Idea** How did Chinese culture shape life in Korea?
6. **Challenge** How do you think China might benefit if North Korea's economy improved?

Writing About Geography

7. **Persuasive Writing** Would you rather live in South Korea or North Korea? Write a paragraph stating your preference and explaining your answer.

TIME PERSPECTIVES

EXPLORING WORLD ISSUES

EAST ASIA'S RISING TIGERS

East Asian nations are soaring to new economic heights and having a major impact on our world.

Shanghai, China, is an Asian boomtown.

The countries of East Asia are changing like never before. In China, economic reforms have changed the Communist nation into a global powerhouse. Chinese businesses sell billions of dollars of goods on world markets, including ones in the United States. And in recent years, Asian democracies, such as South Korea, have roared forward with new political and economic freedoms.

Experts predict that Asian nations will continue to change in the new century. Many believe China will enjoy a bigger role in world affairs. Others think that relations between North Korea and South Korea—longtime enemies—will continue to improve. As the global economy becomes more connected, the links that bind these countries to each other—and to the world—will become more important.

MICHAEL PRINCE/CORBIS

China's Hu Jintao and President George W. Bush discuss trade.

CHINA'S NEW REVOLUTION

Liu Li's life began to change in 2003. That year, the rice farmer's daughter moved to the southern Chinese city of Kaiping. Liu found a job in a garment factory, where she sewed jackets that would be sold in the United States.

Surrounded by sewing machines, Liu tried to imagine the Americans who would wear the jackets. "They must be very tall and very rich," she said. "Beyond that, I can't really picture what their lives are like."

Most Americans would find it just as hard to picture Liu's life—or the amazing changes that are transforming her country. Over the past 17 years, China has become an economic giant. **Incomes**, or money people earn for work, have soared, and millions of people have prospered. Chinese shoppers spent some $600 billion in 2004 on goods such as clothes, cars, and cell phones. Even Wal-Mart, the U.S. retail company, has 51 stores in China.

A Bumpy Road

China's journey to prosperity has been a long and troubled one. The country began opening to the outside world in the 1850s. British gunships forced China's **imperial** government, which was ruled by an emperor, to trade with foreigners. For the next 100 years, foreign powers invaded Chinese **territory** and divided it among themselves.

In 1949, after a violent revolution, the Chinese Communist Party founded the People's Republic of China. For the next 30 years, the country had very little to do with the outside world.

Bursting With Shoppers

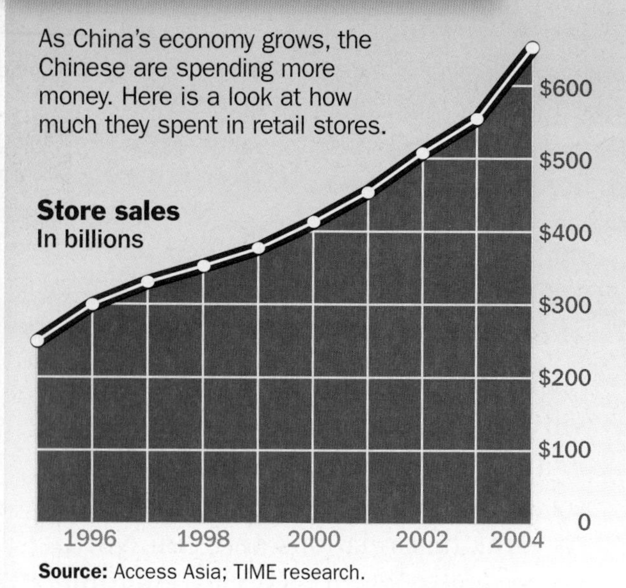

Source: Access Asia; TIME research.

INTERPRETING GRAPHS

Identifying Cause and Effect How much more did the Chinese spend in retail stores in 2004 than in 2000? How might the increase affect the country's economy?

More Chinese own cars than ever before.

Shops and restaurants line a busy outdoor mall in Nanjing, China.

Chinese workers make cloth at a textile factory.

Shoppers browse the goods sold at a Wal-Mart in Beijing, China's capital.

The Chinese Communist government controlled many aspects of everyday life. The state ran most of the nation's businesses, from farms to factories, and people were told where to live and work. During the 1960s and 1970s, anyone suspected of being interested in non-Chinese culture or politics was persecuted. Millions were sent to work in China's countryside, far from their loved ones.

These policies, however, produced a weak economy. China faced shortages of all kinds, including food and energy. In the late 1970s, the government began to reform the economy. Chinese citizens were allowed to own businesses and farm their own land. The chance to earn good pay in private business gave the Chinese a reason to work harder, and the economy began to grow. Since 1989, China's leaders have expanded these economic reforms, opening the country's economy to trade with other nations. The Chinese economy has grown ever since.

Getting Down to Business

During the recent reforms, China's relationship with the United States has changed. The two nations' economies have become more closely linked than ever before. In 2004 Americans bought $185 billion worth of Chinese goods. Many of the clothes, toys, and shoes available in the U.S. are made in China. China has invested billions of dollars in the U.S., making it an important economic partner.

But China is also an economic rival. Chinese companies often compete with American businesses for important **resources** like oil, steel, and iron. Some U.S. businesses complain that Chinese companies illegally reproduce American products like CD-ROMs, DVDs, and computer software.

Friend or Foe?

In April 2006, President Hu Jintao of China visited the U.S. He and President George W. Bush discussed ways to make sure trade between China and the U.S. benefits both countries. "The United States and China are two nations separated by a vast ocean, yet connected through a global economy that has created opportunity for both our people," said Bush. For each country, finding common ground will require a clearer understanding of each other if they are to remain partners in the future.

EXPLORING THE ISSUE

1. **Making Inferences** Wal-Mart, the American retail chain, operates 51 stores in China. How might that impact U.S.-China relations?

2. **Identifying Cause and Effect** In the 1970s, Chinese citizens were given the right to own private businesses and farms. Why did that help China's economy grow?

TENSION ACROSS THE BORDER

For nearly 60 years, bitterness and distrust have separated North Korea and South Korea. In 1945 Japan's defeat in World War II freed Korea from 36 years of Japanese colonial rule. But Korea was soon divided. The former Soviet Union supported the Communist government of the north, and the United States supported the democratic government in the south.

Relations between the two countries have been troubled. Both sides would like to reunite the Korean Peninsula under their form of government. In June 1950, the North Korean army invaded South Korea. After three years of bloody fighting that killed more than 2 million people, a cease-fire agreement was signed.

The war left the peninsula devastated and still divided. Millions of families were separated. Brothers and sisters, along with parents and children, were forced to live in different—and battling—countries.

A United Future?

Today, North Korea remains a Communist dictatorship, and its citizens have very few rights. The nation's government controls all parts of the economy. There are many shortages, including fuel, electricity, and food.

In recent years, South Korea has tried to encourage better relations by exchanging cultural groups and giving economic aid to its neighbor to the north. In 2000 the leaders of the two nations met for the first time since the Korean War. The meeting sparked hope for more peaceful relations and even reunification someday.

But North Korea has continued to build its military and its nuclear weapons program. The relationship between the two Koreas has been improving, but the U.S. and other countries fear that North Korea's nuclear weapons program poses a threat to peace. South Koreans hope that more talks and openness will preserve peace on the peninsula.

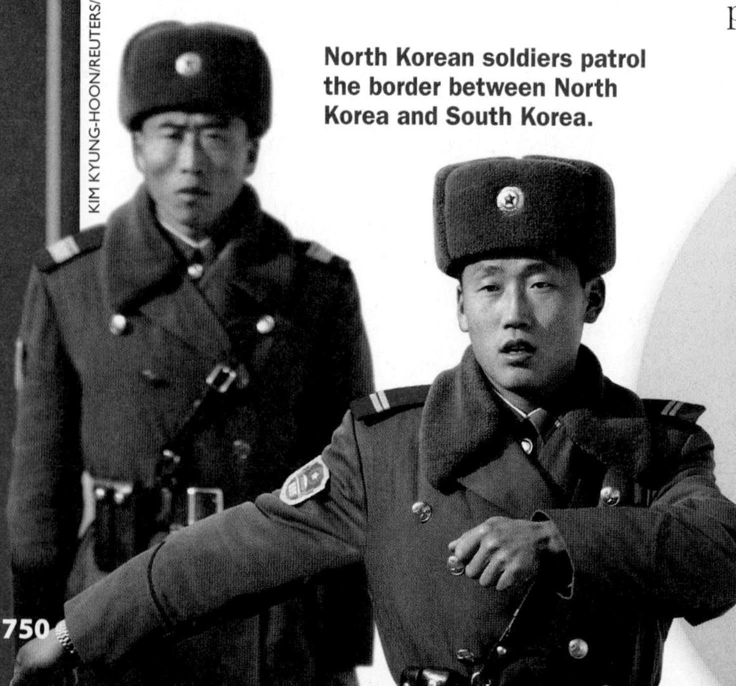

KIM KYUNG-HOON/REUTERS/CORBIS

North Korean soldiers patrol the border between North Korea and South Korea.

EXPLORING THE ISSUE

1. **Explaining** How might cultural exchanges and economic aid improve relations between the two Koreas?

2. **Analyzing Information** Does a nation have a right to strengthen its military power without interference from other nations? Why or why not?

People surf the Web at an Internet club in Beijing.

REVIEW AND ASSESS

UNDERSTANDING THE ISSUE

1 **Making Connections** How have economic reforms affected the incomes of Chinese citizens? What happened to China's territory when its imperial government was forced to trade with foreign powers?

2 **Writing to Inform** What freedoms do you enjoy as a U.S. citizen? Write a letter to a person in North Korea explaining the freedoms you have.

3 **Writing to Persuade** What is the most important fact Americans should know about China? Write your answer in a short essay.

INTERNET RESEARCH ACTIVITIES

4 The Chinese government limits information that can be posted on the Internet. Type in the key words "China," "freedom of speech," and "search engines" on an Internet search engine. Write a brief essay that discusses how the government controls what its citizens can see on the Internet.

5 Navigate to the Web site of the National Committee on U.S.–China Relations, www.ncuscr.org. Click on the "Student Leaders Exchange" link. Look at the photo gallery showing student visits to China. What do the pictures tell you about student life in China? Share your answers with your classmates.

BEYOND THE CLASSROOM

6 **Visit your school or local library** to learn more about China's recent history. Working in groups, learn what it was like to live in China before the economic reforms of the 1980s. Discuss your findings with your classmates.

7 **Research the reunification** of East and West Germany in the 1990s. What problems did the two countries face while reuniting political and economic systems? Can North Korea and South Korea learn from Germany's example? Write your findings in a short report.

High Speed at Home

South Korea ranks second in the world in percentage of population with access to high-speed, or broadband, Internet. This line graph shows high-speed home connections in South Korea and its neighbor Japan.

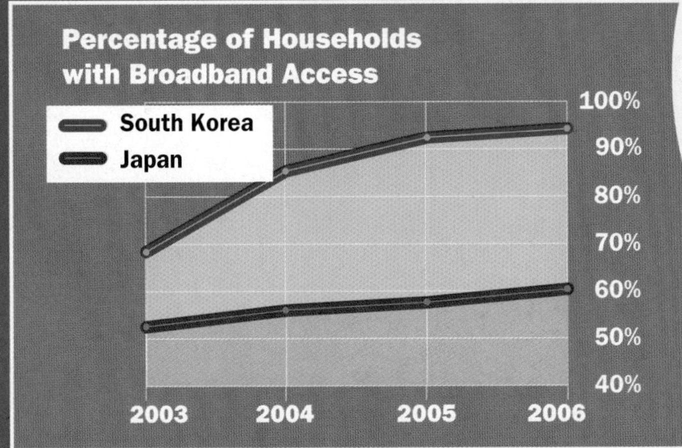

Source: Organization for Economic Cooperation and Development.

Building Graph Reading Skills

1. **Analyzing Data** In which year did household connections to the Internet in South Korea grow the most?
2. **Making Inferences** In 2006 about 94 percent of South Korean households had access to the Internet. How might that help South Korea compete economically with Asian countries?

Guide to Reading

BIG Idea

The physical environment affects how people live.

Content Vocabulary

- precious gem *(p. 753)*
- mangrove *(p. 757)*
- terraced field *(p. 758)*

Academic Vocabulary

- portion *(p. 754)*
- category *(p. 756)*

Reading Strategy

Identifying Central Issues Use a chart like the one below to describe the economic activities of four Southeast Asian countries.

Country	Economic Activities
1.	1.
2.	2.
3.	3.
4.	4.

Southeast Asia

 Section Audio  Spotlight Video

Picture This This mother uses baskets to carry her young son and a load of firewood. Her son's face is painted with a golden yellow paste from the *thanaka* tree. The sweet-smelling paint is used to protect the skin from the sun's harmful rays and to heal minor skin irritations. People also use the paint to decorate their skin. This woman and her son live in Bagan, a part of Myanmar that contains important historical sites, such as ancient Buddhist statues and temples. Read the section to learn about Southeast Asia today.

▼ Performing daily chores

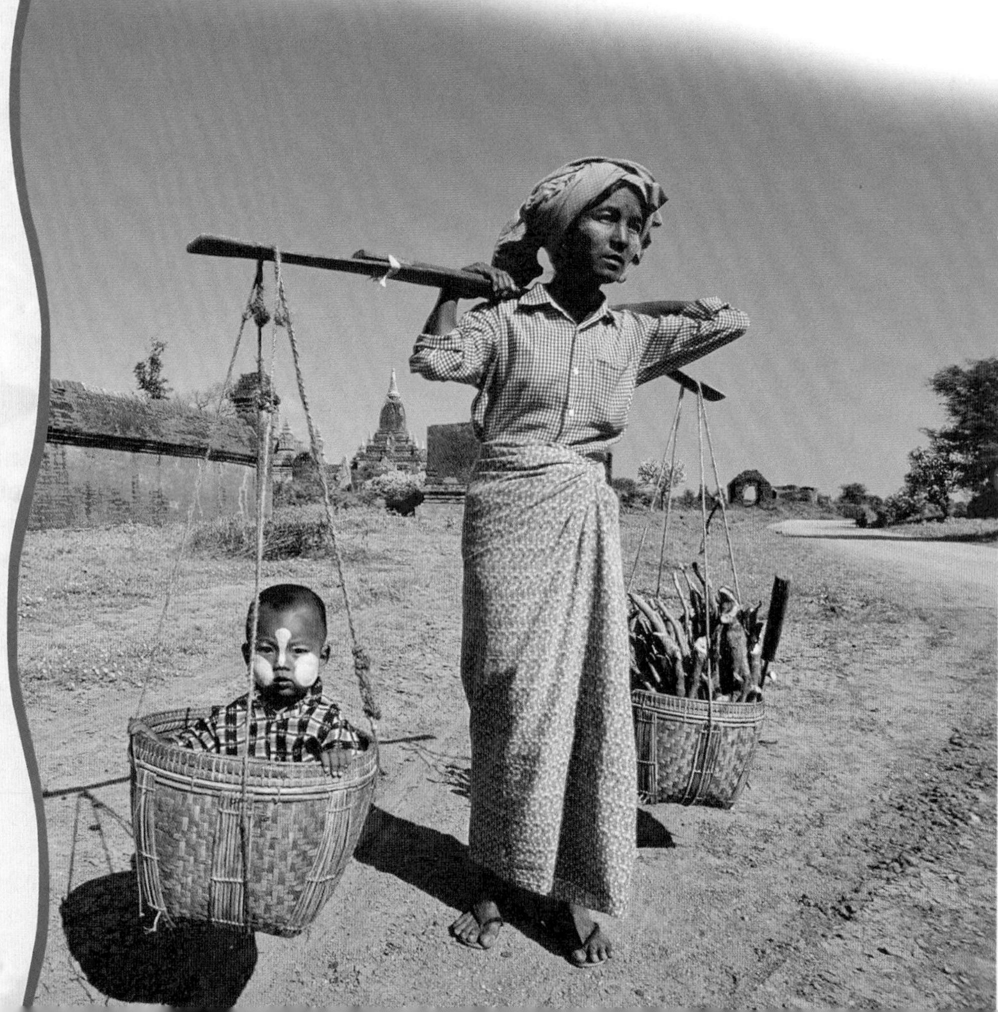

Mainland Southeast Asia

Main Idea **Mainland Southeast Asia has a mix of strong and weak economies.**

Geography and You Do you like rice? Rice is a staple food throughout Southeast Asia. Read to find out about the important resources of this region.

Some countries of mainland Southeast Asia have enjoyed economic growth by developing industries. Others have been less successful because of a lack of resources, years of conflict, or failed government policies.

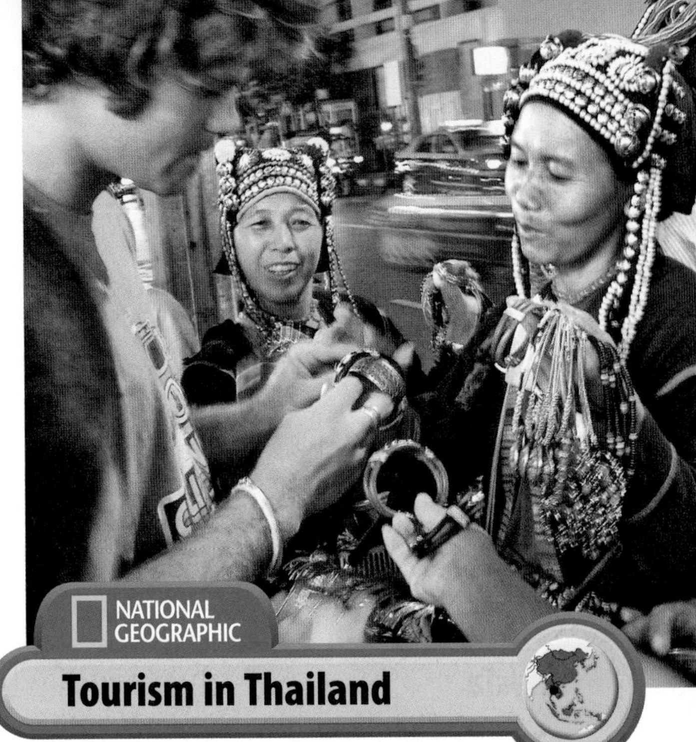

NATIONAL GEOGRAPHIC

Tourism in Thailand

Tourists contribute more than $5 billion to the Thai economy every year. ***Place*** **What kind of government does Myanmar have?**

Myanmar

Myanmar, formerly called Burma, is about the size of Texas. Approximately two-thirds of the country's people farm. Some farmers work their fields with tractors, but most rely on plows pulled by water buffalo. The main crops are rice, sugarcane, beans, and peanuts.

Myanmar exports wood products, gas, and foods such as beans and rice. The country provides most of the world's teakwood, which is used in shipbuilding and for making fine furniture. Myanmar's prized forests suffer from deforestation, however, because they are overharvested. The country also exports **precious gems,** or valuable stones such as rubies, sapphires, and jade.

About 70 percent of Myanmar's people live in rural areas. The most densely populated area of the country is the Irrawaddy River valley. Myanmar's largest city is Yangon, a city once known as Rangoon. It lies in the Irrawaddy's delta. Yangon is famous for its many gold-covered Buddhist temples. Buddhism is the main religion in Myanmar.

Myanmar was part of British India for many years, but it became an independent republic in 1948. Since then, military leaders have turned Myanmar into a socialist country in which the government runs the economy. No criticism of government policies is allowed. Some people, however, have tried to bring democracy to Myanmar. A woman named Aung San Suu Kyi (AWNG SAN SOO CHEE) has become a leader in this struggle. In 1991 she received the Nobel Peace Prize for her efforts. Despite her work, she still faces opposition from the government today.

Thailand

Thailand was known as Siam until the 1900s. It is the only Southeast Asian country that has never been a European colony. Thailand's people trace their independence as a kingdom back to the A.D. 1200s. Most people in the country practice Buddhism. Hundreds of Buddhist temples dot the cities and countryside.

Most Thais live in rural areas, although many look for jobs in Bangkok, the capital.

Bangkok has beautiful temples and royal palaces that are surrounded by modern skyscrapers, busy stores, and streets crowded with traffic.

Thailand's economy benefits from tourism and the export of tin, tungsten, precious gems, and rubber. In recent years, Thailand has attracted foreign investors to build industry. Thai workers now produce textiles, clothing, cars, and computer parts.

Malaysia

Malaysia lies on the southern end of the Malay Peninsula and also on part of the island of Borneo. Malaysia has long exported raw materials, such as palm oil, rubber, tin, and valuable woods. Recently, textiles, electronic goods, and cars have made up a growing **portion,** or part, of its exports. Malaysia's cities, such as the capital, Kuala Lumpur (KWAH·luh LUM·PUR), are now important centers of trade and industry for Southeast Asia.

After years of British rule, the mainland and island areas united to form Malaysia in 1963. Most of Malaysia's people belong to the Malay ethnic group and practice Islam. Chinese and South Asians of different religious backgrounds also make up the country's population.

Skyscrapers in Malaysia

NATIONAL GEOGRAPHIC

▲ The Petronas Towers are among the tallest buildings in the world and can be seen from anywhere in Kuala Lumpur. ***Place*** **Who controlled the Malay Peninsula before 1963?**

Singapore

Singapore lies off the southern tip of the Malay Peninsula. It is one of the world's smallest countries, yet it has one of the world's most productive economies. It was once covered by rain forests. Today, however, Singapore is filled with highways, factories, office buildings, and docks.

Trade is an extremely important economic activity in Singapore. Because it is a free port, Singapore is one of the world's busiest harbors. Huge amounts of goods pass through this port. Singapore's many factories make electronic goods, machinery, chemicals, and paper products. Because of their productive economy, Singapore's people enjoy a high standard of living.

Laos and Cambodia

Laos, located on Southeast Asia's mainland, is an economically poor country. Most people in Laos live in rural areas and grow rice and other food products along the fertile banks of the Mekong River. Industry is largely undeveloped because of isolation and years of civil war. Laos's Communist

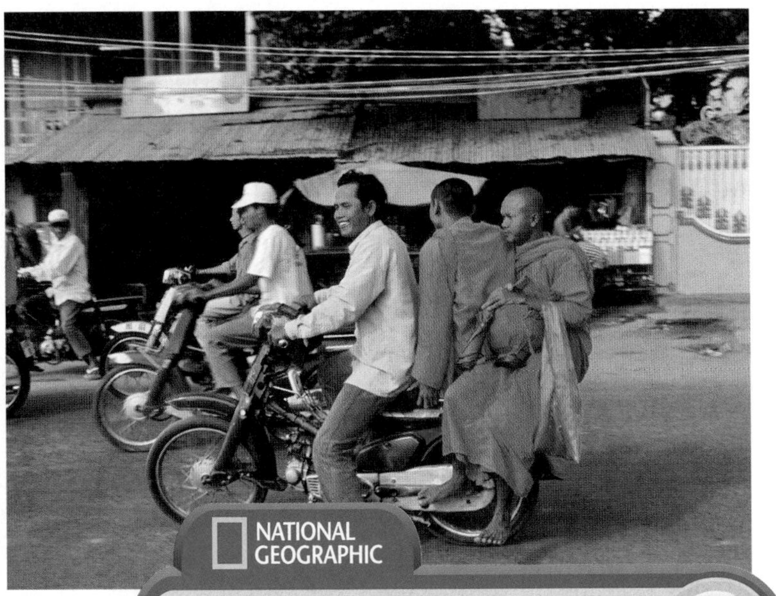

Transportation in Phnom Penh

Motos are small motorcycles that are a popular form of transportation in Cambodia's capital. ***Location*** **What type of industry is Cambodia's government trying to develop?**

government discourages religion, but most of Laos's people remain Buddhists.

South of Laos, the Mekong River divides the country of Cambodia in two. Cambodia's capital, Phnom Penh (puh·NAWM PEHN), lies on that river, as does the ancient site of Angkor, the capital of the Khmer (kuh·MEHR) Empire centuries ago. Not far from Angkor sits Angkor Wat, a complex of temples that is an architectural treasure.

In 1975 Communist rebels took control of Cambodia. Their harsh policies resulted in the deaths of as many as 2 million people and left the country in ruins. Although the Communists were later overthrown, Cambodia has still not recovered from their rule.

As in Laos, most people in Cambodia are Buddhists and live by growing rice. The country has few resources and is very poor. To rebuild the economy, Cambodia's leaders hope to develop a tourist industry, **focusing,** or centering, on Angkor Wat.

Vietnam

Vietnam's rapidly growing population is the largest in mainland Southeast Asia. Most Vietnamese live in rural villages, but Vietnam's cities are growing. Vietnam's largest city is Ho Chi Minh City, named for the country's first Communist leader. Located in the south, it used to be called Saigon. Vietnam's capital, Hanoi, is located in the north.

Vietnam has a number of natural resources, including coal, petroleum, and several metals. These resources are important exports. The Red River in the north and the Mekong River in the south both have fertile deltas. Farmers in these areas grow more than enough rice to feed Vietnam and to ship abroad. Other export products include rubber, tobacco, tea, coffee, shrimp, and fish. Vietnam's warm climate and natural beauty are attracting a growing number of tourists.

With its resources, thriving industry, and productive agriculture, Vietnam has enjoyed some economic growth. Its economy, however, has been held back by government policies and by wars. Vietnam's Communist rulers have loosened some controls, but they still must combat inflation and attract more foreign investors. In addition, Vietnam lacks modern transportation systems. For these reasons, although its economy is slowly growing, Vietnam remains a poor country.

Reading Check **Explaining** What advantages might help Vietnam's economy grow in the future?

Island Southeast Asia

Main Idea **The economies of island Southeast Asia depend on natural resources and agriculture.**

Geography and You Have you ever felt an earthquake? Read to find out how earthquakes have influenced Southeast Asia.

The island countries of Southeast Asia are Indonesia, East Timor, Brunei, and the Philippines. Indonesia is Southeast Asia's largest country in land area and population. The islands that make up Indonesia stretch about 3,200 miles (5,100 km) from east to west. By contrast, Brunei is the area's smallest country in both **categories,** or divisions.

Indonesia

Indonesia is an archipelago of thousands of islands. Sumatra, Java, and Celebes (SEH·luh·BEEZ) are the major islands. Indonesia also shares the islands of Borneo and Timor with other countries.

Indonesia lies where two of the Earth's tectonic plates meet. As a result, Indonesia has many active volcanoes and experiences earthquakes. An undersea earthquake off the shore of Sumatra in late 2004 launched a huge tsunami that struck Indonesia and other countries bordering the Indian Ocean. **Figure 1** shows the path of the tsunami's waves and its effects on the countries of the region. The disaster left as many as 200,000 dead in Indonesia alone.

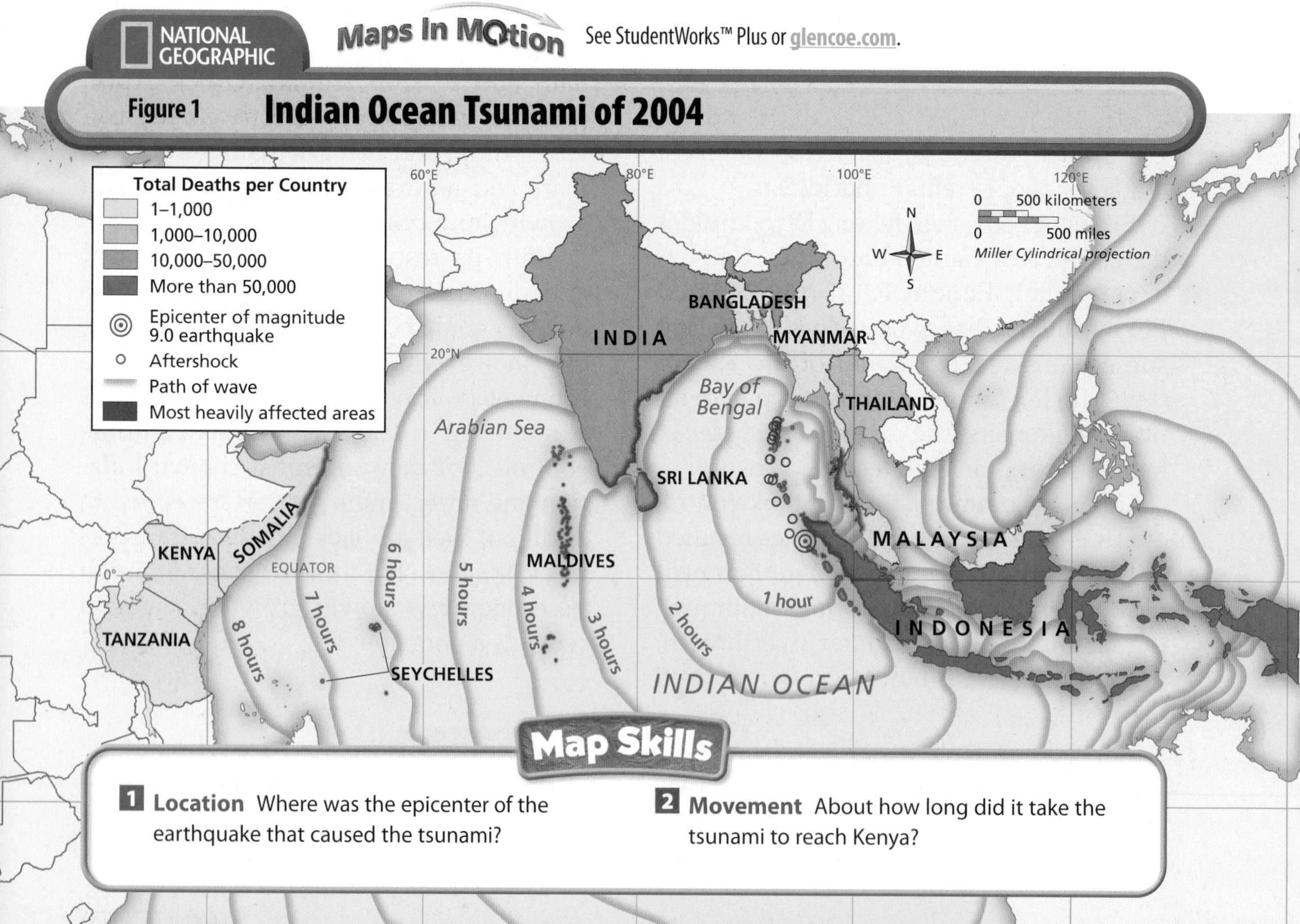

Figure 1 **Indian Ocean Tsunami of 2004**

Map Skills

1. **Location** Where was the epicenter of the earthquake that caused the tsunami?
2. **Movement** About how long did it take the tsunami to reach Kenya?

Such disasters take a heavy toll because Indonesia is one of the world's most densely populated countries. It has more than 220 million people—the fourth-largest population in the world. More than half the people live on the island of Java. Jakarta (juh·KAHR·tuh), Indonesia's capital and largest city, is located on Java. It has modern buildings and streets that are crowded with cars and bicycles.

Indonesia has large reserves of oil and natural gas. Its mines yield tin, silver, nickel, copper, bauxite, and gold. Dense rain forests provide valuable woods. Trees, however, are being cut down quickly, and the environment has suffered. Indonesia's extensive mangrove forests also are threatened by overharvesting. **Mangroves** are tropical trees that grow along coasts and help maintain the health of coastal environments. In Indonesia, people cut them down for timber and to clear the land for fish farming and other activities.

Many Indonesians still make a living by farming. They grow rice, cassava, sweet potatoes, and corn. Rubber trees, oil palms, tea, and coffee provide exports. For years, Indonesia produced raw materials. Recently, the government has tried to develop tourism, which has become an important industry, especially in Bali. This small island east of Java is famous for its beautiful beaches and elaborate Hindu temples. Unfortunately, recent terrorist attacks have occurred at popular tourist sites.

Indonesia became independent in 1949 after years of Dutch rule. Today the country has a democratic government. However, Indonesia's government has difficulty uniting the country. Most Indonesians are Muslims, but they live on scattered islands and belong to different ethnic groups. The government worries about calls for independence from some of these groups.

Muslim Indonesia

Muslim girls in Jakarta line up to give money to help the poor during the Islamic holy month of Ramadan. ***Place*** **Why is it difficult for Indonesia's government to unite the country?**

Keeping the country stable and united will be a key to its future.

Brunei and East Timor

On the northern coast of Borneo lies the small nation of Brunei (bru·NY). It is only slightly larger than Delaware in size and has only about 400,000 people. Brunei has grown wealthy by building its economy on its large oil and gas reserves. Because the country depends almost entirely on these resources, Brunei's economy often suffers when world prices fall. In addition, Brunei depends on imports to provide almost all of its food and manufactured goods.

Another small country—East Timor—lies on the eastern half of the island of Timor. About 800,000 people live in this country. East Timor is the newest country in Southeast Asia, having gained its independence from Indonesia in 2002. Most of East Timor's people live by farming. The country has oil and natural gas deposits that the government hopes to develop.

The Philippines

The Philippines is made up of about 7,000 islands in the South China Sea. The island group stretches about 1,150 miles (1,850 km) from north to south. Volcanic mountains dominate many of these islands. Rice, sugarcane, coffee, pineapples, bananas, and coconuts are grown for export. Farmers build terraces on steep mountain slopes. **Terraced fields** are strips of land cut out of a hillside like stair steps. These fields increase the amount of land that can be farmed and prevent soil from washing down the mountainside when it rains.

The Philippines has rich deposits of gold, iron ore, copper, lead, and zinc. Forests once covered much of the land, but deforestation has removed most of the valuable trees.

Cities in the Philippines are busy and modern. Manila, the capital, is a large commercial center. Factory workers here produce electronic goods, food products, chemicals, and clothing.

The Philippines was ruled by Spain for more than 300 years beginning in the 1500s. The country is named after a Spanish king during that time, Philip I. About 90 percent of Filipinos are Roman Catholic. The religion was brought to the islands by Spanish missionaries who marched with Spain's soldiers during the conquest of the islands. The Catholic Church remained powerful there for many years.

As a result of the Spanish-American War, the United States controlled the islands from 1898 until World War II. In 1946 the Philippines became an independent republic. Today the culture of the Philippines blends Malay, Spanish, and American ways.

Reading Check **Comparing** Compare the agricultural products of Indonesia and the Philippines.

Section 4 Review

Social Studies ONLINE
Study Central™ To review this section, go to glencoe.com.

Vocabulary

1. **Explain** the meaning of *precious gem, mangrove,* and *terraced field* by using each term in a sentence.

Main Ideas

2. **Recalling** What makes Thailand different from other countries of Southeast Asia?
3. **Analyzing** Use a diagram like the one below to show the effects of government policies on the economies of three countries in the region.

Government Policy	Effect

Critical Thinking

4. **Comparing and Contrasting** How do Thailand and Vietnam compare in the way they manage their economies?
5. BIG Idea How does the physical environment affect the people of Southeast Asia?
6. **Analyzing Information** Why is Brunei dependent on world markets?
7. **Challenge** Why are some countries of Southeast Asia struggling economically while others are economically strong?

Writing About Geography

8. **Using Your FOLDABLES** Use your Foldable to write a paragraph explaining what steps countries like Cambodia might take to develop a successful tourist industry.

Visual Summary

China and Its Neighbors

- China's economy has grown rapidly as a result of free market reforms.
- China's Communist leaders tightly control political activities.
- Taiwan has moved toward democracy and has a strong industrial economy.
- Herding remains an important economic activity in Mongolia.

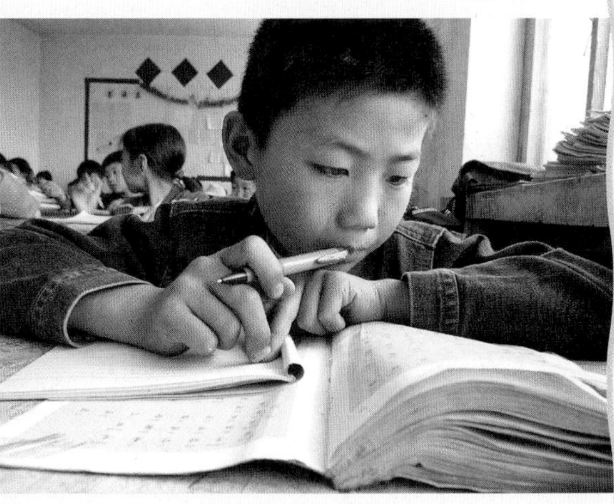

Rural school, China

Japan

- Japan has a large population but a small amount of livable space.
- Japan's powerful economy is based on the export of manufactured goods.
- The culture of Japan is a blend of traditional and modern ways.

Auto manufacturing, South Korea

The Two Koreas

- Korean culture is based on Chinese and local traditions.
- Communist North Korea and democratic South Korea have had a tense relationship.
- South Korea has a prosperous economy and modern society.
- North Korea is ruled by a Communist dictator. Its isolated economy is relatively poor.

Shuheki-en Garden, Japan

Southeast Asia

- Several countries have prosperous economies based on manufacturing and trade.
- Civil war and government policies prevent economic growth in some countries.
- Much of Southeast Asia's forests have been lost to deforestation.

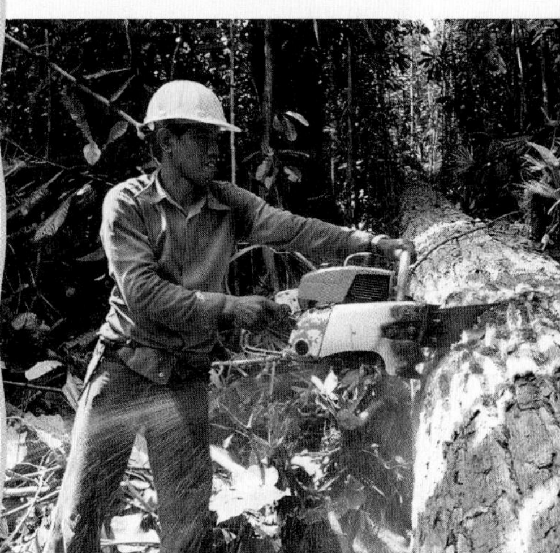

Logging in the Philippines

- In 2004 an undersea earthquake near Indonesia launched a huge tsunami that struck Indian Ocean countries.

Study anywhere, anytime! Download quizzes and flash cards to your PDA from glencoe.com.

CHAPTER 27 ASSESSMENT

STANDARDIZED TEST PRACTICE

When answering extended response questions, organize your thoughts before you begin writing. You will reduce the time you need to revise.

Reviewing Vocabulary

Directions: Choose the word(s) that best completes the sentence.

1. The Dalai Lama was forced to leave Tibet and live in ______.

A poverty
B disgrace
C isolation
D exile

2. In traditional Japanese homes, straw mats called ______ are used to cover floors.

A kimono
B tatami
C *anime*
D sushi

3. A writing system called hangul was developed in the 1400s to write the ______ language.

A Korean
B Chinese
C Japanese
D Mongolian

4. Tropical trees called ______ grow along coasts and help maintain the health of coastal environments.

A tatamis
B sushis
C mangroves
D landfills

Reviewing Main Ideas

Directions: Choose the best answer for each question.

Section 1 ***(pp. 728–733)***

5. China uses ______ to meet most of its energy needs.

A nuclear power
B coal
C solar power
D hydroelectric power

Section 2 ***(pp. 736–741)***

6. The government of Japan is a ______.

A Communist dictatorship
B absolute monarchy
C constitutional monarchy
D direct democracy

Section 3 ***(pp. 742–746)***

7. Since 1948, North Korea has been ruled as a ______ by Kim Il Sung and then by his son, Kim Jong Il.

A Communist dictatorship
B representative democracy
C constitutional monarchy
D absolute monarchy

Section 4 ***(pp. 752–758)***

8. Vietnam's economic growth has been slowed by government policies, wars, and ______.

A a scarcity of resources
B a poor transportation system
C declining tourism
D falling oil prices

GO ON

Critical Thinking

Directions: Use the chart below and your knowledge of Chapter 27 to help you choose the best answer for each question.

Population Growth Rates for Selected East Asian and Southeast Asian Nations

Nation	Total Population	Annual Population Growth Rate (percent)
Brunei	400,000	1.9
Cambodia	13,300,000	2.2
China	1,303,700,000	0.6
East Timor	900,000	2.7
Indonesia	221,900,000	1.6
Japan	127,700,000	0.1
Laos	5,900,000	2.3
Malaysia	26,100,000	2.1
Myanmar	50,500,000	1.2
Philippines	84,800,000	2.3
Singapore	4,300,000	0.6
Thailand	65,000,000	0.7
Vietnam	83,300,000	1.3

Source: *World Population Data Sheet,* 2005.

9. By far, the largest country in the region in terms of total population is ______.

A Indonesia

B Japan

C China

D Vietnam

10. The nation with the highest annual population growth rate over the past five years is ______.

A Brunei

B East Timor

C Laos

D Malaysia

Document-Based Questions

Directions: Analyze the document and answer the short-answer questions that follow.

> ***A Great Wall of Waste***
>
> *Plugging a cigarette into his mouth, He Shouming runs a nicotine-stained fingernail down a list of registered deaths in Shangba, dubbed "cancer village" by the locals. The Communist Party official in this cluster of tiny hamlets of 3,300 people in northern Guangdong province, he concludes that almost half the 11 deaths among his neighbours this year, and 14 of the 31 last year, were due to cancer.*
>
> *Mr. He blames . . . a nearby mineral mine . . . and . . . smaller private mines for spewing toxic waste into the local rivers. . . . [T]he water in the streams is indeed an alarming rust-red. A rice farmer complains of itchy legs from the paddies, and his wife needs a new kettle each month because the water corrodes metal. "Put a duck in this water and it would die in two days," declares Mr. He.*
>
> —The Economist, *August 19, 2004*

11. Why is Shangba called "cancer village"?

12. How do we know that something is wrong with the village's water supply?

Extended Response

13. If you were a resident of Shangba, what hazards would you likely face each day? Why?

STOP

For additional test practice, use Self-Check Quizzes—Chapter 27 at glencoe.com.

Need Extra Help?

If you missed question...	1	2	3	4	5	6	7	8	9	10	11	12	13
Go to page...	729	739	743	757	731	737	745	755	729	729	732	732	729

UNIT 10

Australia, Oceania, and Antarctica

CHAPTER 28
Physical Geography of Australia, Oceania, and Antarctica

CHAPTER 29
History and Cultures of Australia, Oceania, and Antarctica

CHAPTER 30
Australia, Oceania, and Antarctica Today

Sydney, Australia ▶

NATIONAL GEOGRAPHIC

NGS ONLINE For more information about the region, see www.nationalgeographic.com/education.

Regional Atlas

Australia, Oceania, and Antarctica

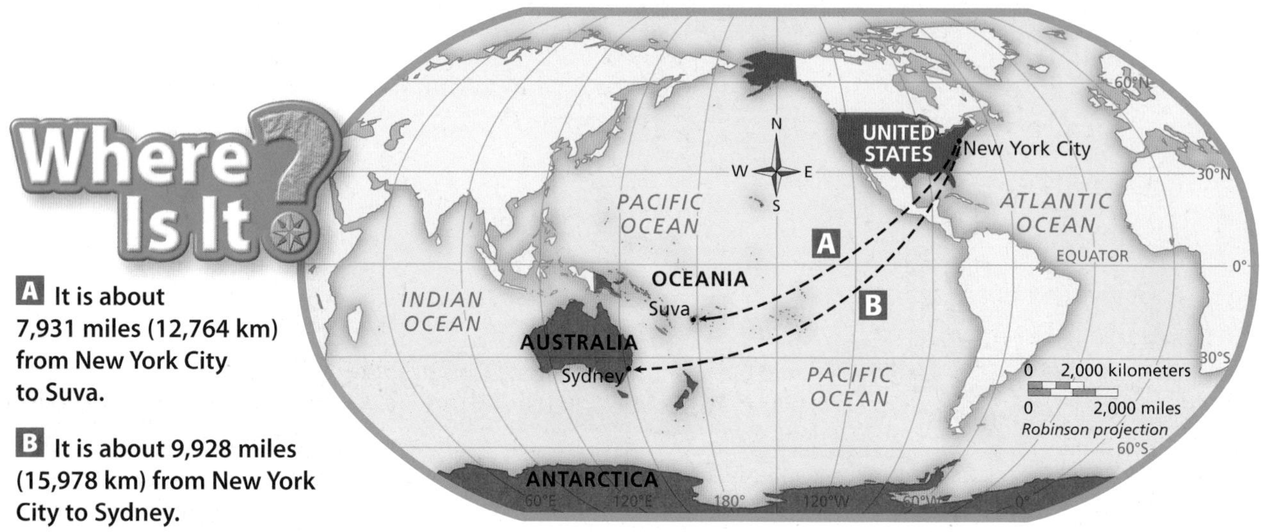

Where Is It?

A It is about 7,931 miles (12,764 km) from New York City to Suva.

B It is about 9,928 miles (15,978 km) from New York City to Sydney.

The land area of the region of Australia, Oceania, and Antarctica is about 8.8 million square miles (22.8 million sq. km), and it includes the continents of Australia and Antarctica. With an area of 3.0 million square miles (7.8 million sq. km), Australia is slightly smaller than the continental United States, whereas Antarctica is almost twice as large. The islands of Oceania make up a very small part of the region's land.

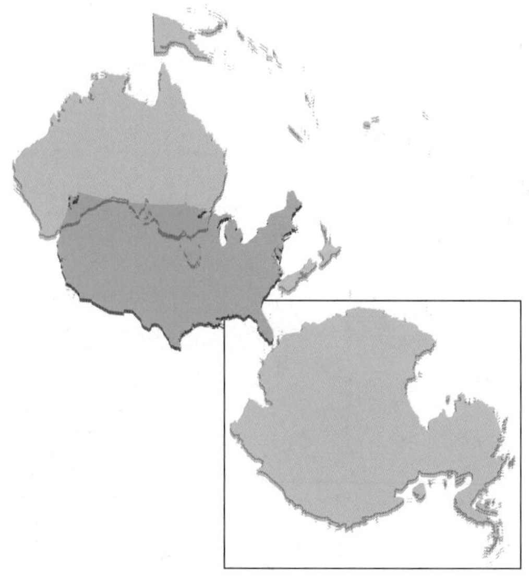

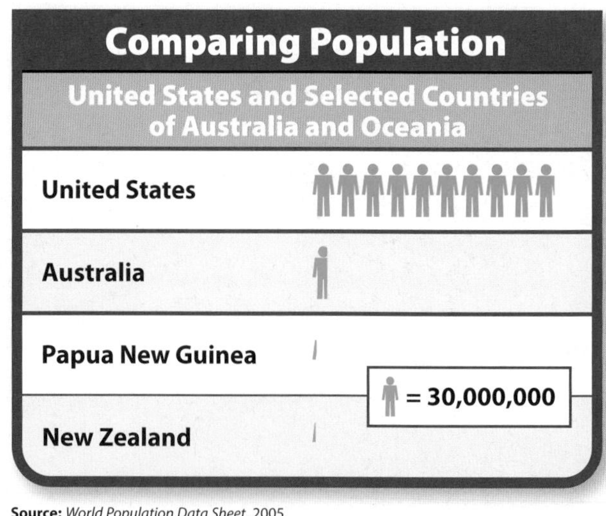

Comparing Population

United States and Selected Countries of Australia and Oceania

United States	10 figures
Australia	1 figure (partial)
Papua New Guinea	less than 1 figure
New Zealand	less than 1 figure

Figure = 30,000,000

Source: *World Population Data Sheet*, 2005.

GEO Fast Facts

Largest Coral Reef

Great Barrier Reef (Australia)
1,250 mi. (2,011 km) long

Longest River

Darling River (Australia)
1,702 mi. (2,739 km) long

Largest Lake

Lake Eyre (Australia)
3,600 sq. mi.
(9,324 sq. km)

Highest Point

Vinson Massif
(Antarctica)
16,066 ft.
(4,897 m) high

UNIT 10 Regional Atlas

Australia, Oceania, and Antarctica

PHYSICAL

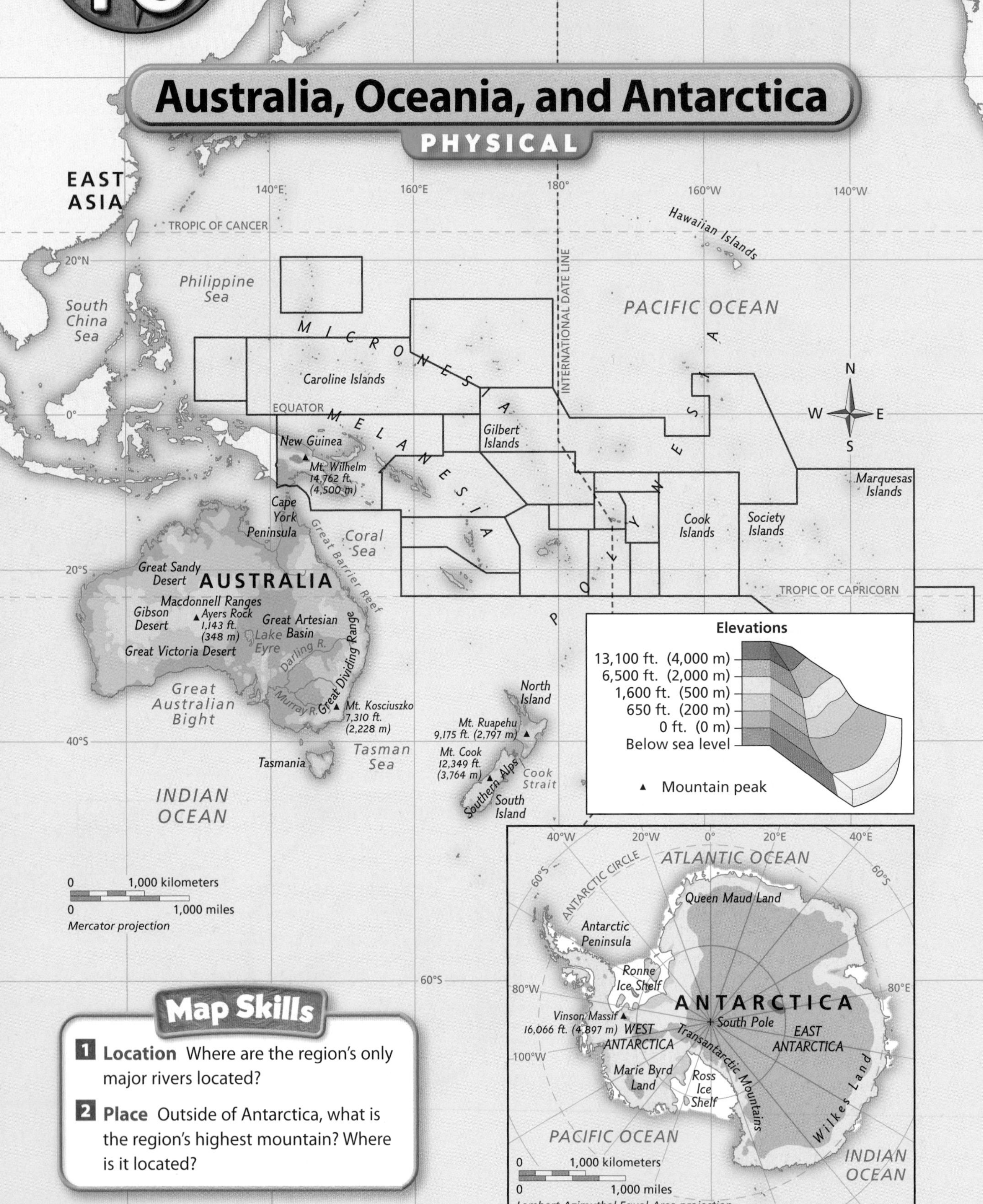

Map Skills

1. **Location** Where are the region's only major rivers located?
2. **Place** Outside of Antarctica, what is the region's highest mountain? Where is it located?

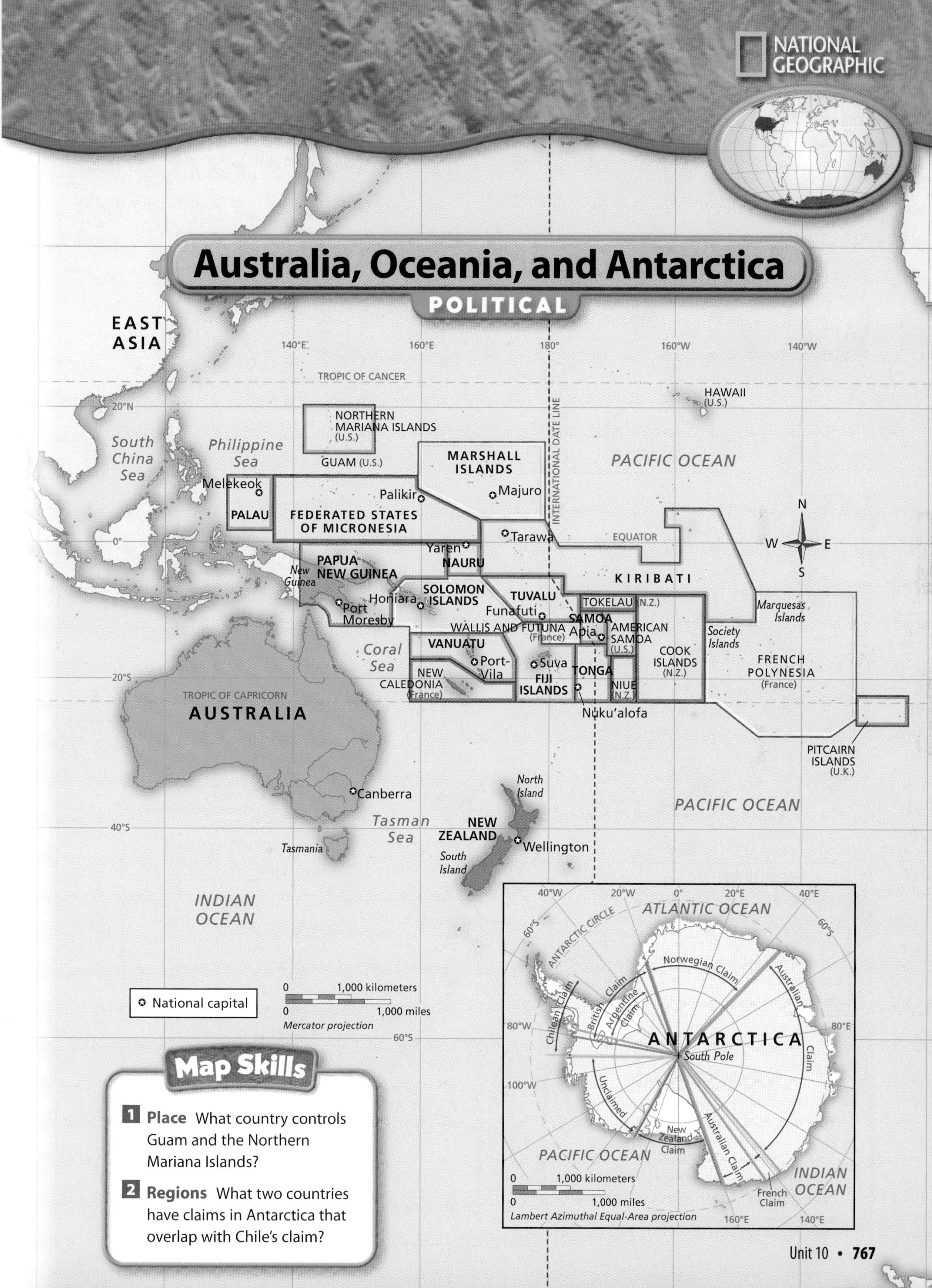

Map Skills

1 Place What country controls Guam and the Northern Mariana Islands?

2 Regions What two countries have claims in Antarctica that overlap with Chile's claim?

Regional Atlas

Australia, Oceania, and Antarctica

POPULATION DENSITY

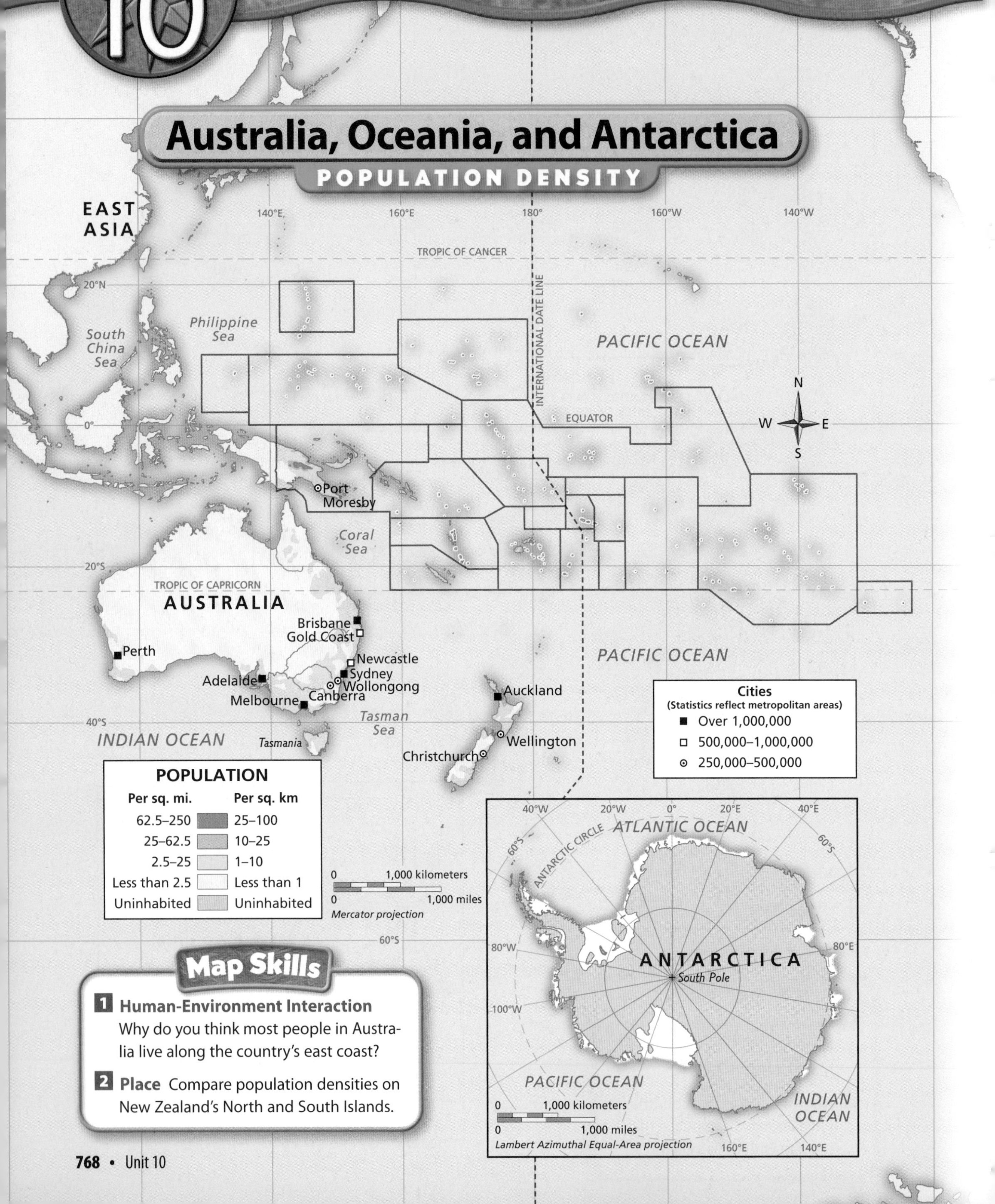

Map Skills

1 Human-Environment Interaction Why do you think most people in Australia live along the country's east coast?

2 Place Compare population densities on New Zealand's North and South Islands.

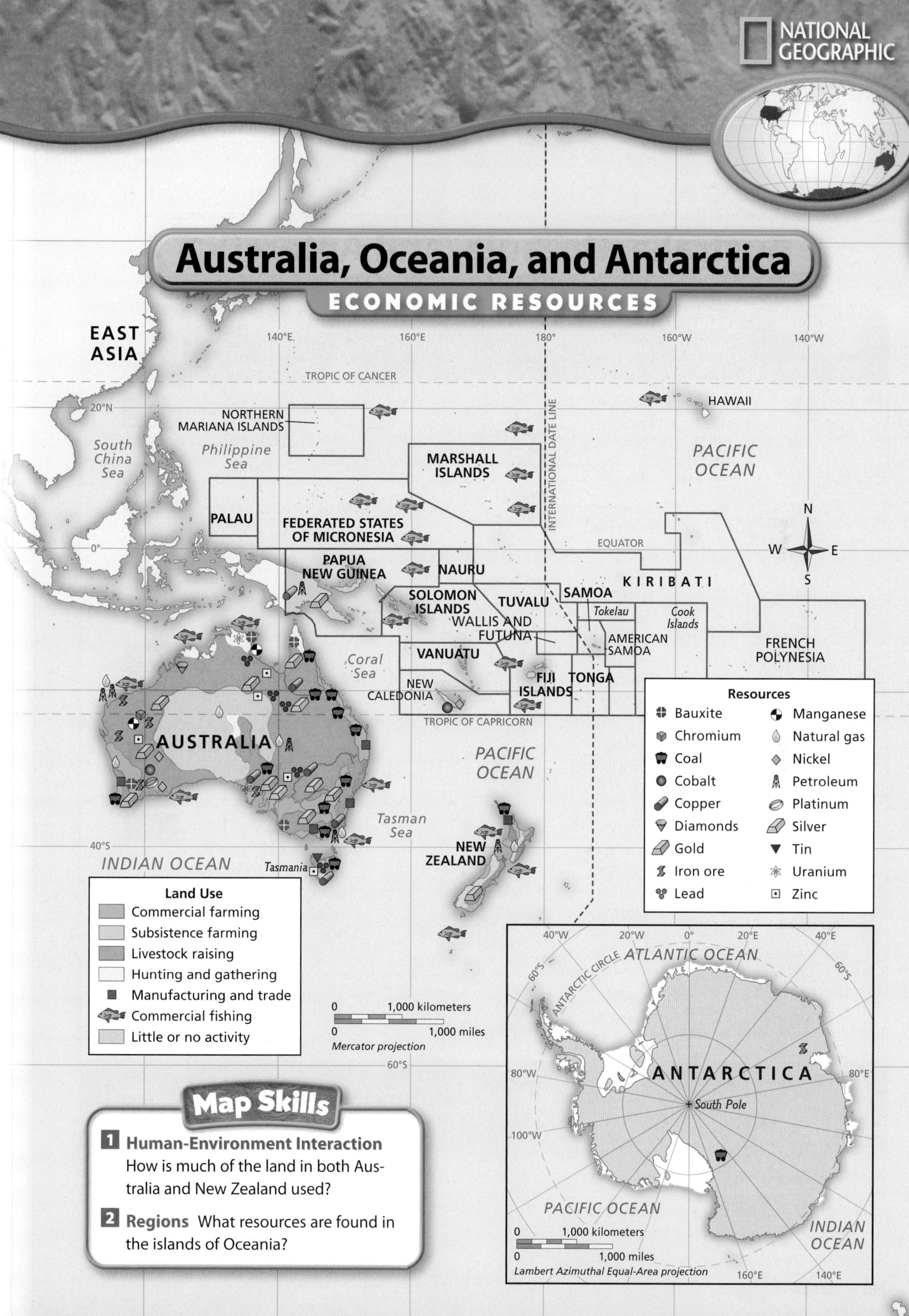

Map Skills

1 Human-Environment Interaction How is much of the land in both Australia and New Zealand used?

2 Regions What resources are found in the islands of Oceania?

Regional Atlas

Australia, Oceania, and Antarctica

Country and Capital	Literacy Rate	Population and Density	Land Area	Life Expectancy (Years)	GDP* Per Capita (U.S. dollars)	Television Sets (per 1,000 people)	Flag and Language
AUSTRALIA Canberra	100%	20,400,000 7 per sq. mi. 3 per sq. km	2,988,888 sq. mi. 7,741,184 sq. km	80	$30,700	716	English
FEDERATED STATES OF MICRONESIA Palikir	89.1%	100,000 370 per sq. mi. 143 per sq. km	270 sq. mi. 699 sq. km	67	$2,000	20	English
FIJI Suva	93.7%	800,000 113 per sq. mi. 44 per sq. km	7,054 sq. mi. 18,270 sq. km	68	$5,900	110	English
KIRIBATI Tarawa (Bairiki)	NA	100,000 355 per sq. mi. 136 per sq. km	282 sq. mi. 730 sq. km	63	$800	23	I-Kiribati, English
MARSHALL ISLANDS Majuro	93.7%	100,000 1,449 per sq. mi. 559 per sq. km	69 sq. mi. 179 sq. km	68	$1,600	information not available	Marshallese, English
NAURU Yaren	NA	10,000 1,111 per sq. mi. 435 per sq. km	9 sq. mi. 23 sq. km	61	$5,000	1	Nauruan, English
NEW ZEALAND Wellington	99%	4,100,000 39 per sq. mi. 15 per sq. km	104,452 sq. mi. 270,529 sq. km	79	$23,200	516	English, Maori
PALAU Melekeok	92%	20,000 112 per sq. mi. 43 per sq. km	178 sq. mi. 461 sq. km	70	$9,000	98	Palauan, English
UNITED STATES Washington, D.C.	97%	296,500,000 80 per sq. mi. 31 per sq. km	3,717,796 sq. mi. 9,629,047 sq. km	78	$40,100	844	English

*Gross Domestic Product

Countries and flags not drawn to scale

Australia, Oceania, and Antarctica

Country and Capital	Literacy Rate	Population and Density	Land Area	Life Expectancy (Years)	GDP* Per Capita (U.S. dollars)	Television Sets (per 1,000 people)	Flag and Language
PAPUA NEW GUINEA Port Moresby	64.6%	5,900,000 33 per sq. mi. 13 per sq. km	178,703 sq. mi. 462,839 sq. km	55	$2,200	13	Melanesian Pidgin
SAMOA Apia	99.7%	200,000 182 per sq. mi. 70 per sq. km	1,097 sq. mi. 2,841 sq. km	73	$5,600	56	Samoan
SOLOMON ISLANDS Honiara	NA	500,000 45 per sq. mi. 17 per sq. km	11,158 sq. mi. 28,889 sq. km	62	$1,700	16	Melanesian Pidgin, English
TONGA Nuku'alofa	98.5%	100,000 345 per sq. mi. 133 per sq. km	290 sq. mi. 751 sq. km	71	$2,300	61	Tongan, English
TUVALU Funafuti	NA	10,000 1,000 per sq. mi. 385 per sq. km	10 sq. mi. 26 sq. km	64	$1,100	9	Tuvaluan
VANUATU Port-Vila	53%	200,000 42 per sq. mi. 16 per sq. km	4,707 sq. mi. 12,191 sq. km	67	$2,900	12	Local Languages, Bislama
UNITED STATES Washington, D.C.	97%	296,500,000 80 per sq. mi. 31 per sq. km	3,717,796 sq. mi. 9,629,047 sq. km	78	$40,100	844	English

Sources: *CIA World Factbook*, 2005; Population Reference Bureau, *World Population Data Sheet*, 2005.

For more country facts, go to the **Nations of the World Databank** at glencoe.com.

Reading Social Studies

Monitoring and Clarifying

Learn It!

Monitoring is the ability to know whether you understand what you have read. For example, you might read a sentence that you do not understand. Monitoring helps you recognize that you do not understand the information.

After you recognize that you do not understand the information, then you must try to identify why you do not understand it. You must *clarify,* or clear up, the parts that are confusing.

- Read the following paragraph.
- Answer the questions in the "Monitor" column.
- Use the "Ways to Clarify" column for tips about how you can clarify your reading.

> Antarctica is covered by ice, but it receives little precipitation. Without the direct rays of the sun, Antarctica's air is so cold that it holds little moisture. This lack of humidity adds to the coldness, since humid air is needed to trap the sun's warmth. Whatever precipitation falls only adds to the ice covering the continent's surface. In fact, rainfall is so low that Antarctica is actually a desert.
>
> —*from page 786*

Monitor	Ways to Clarify
Are there any words that are unfamiliar?	Use the glossary, a dictionary, or context clues.
What is the main idea of the paragraph?	Try to summarize the information into one sentence to create a main idea.
Is it helpful to use what I already know about deserts to understand the text?	Think about what you have already learned about deserts. Then think about how this relates to Antarctica.
What questions do I still have after reading this text?	Write the questions and continue to read to find the answers.

Reading Tip

When text is difficult to understand, slow down your reading. When the text is easy to understand, you can speed up your reading.

2 Practice It!

- Read the following paragraph.
- Draw a chart like the one below to help you monitor and clarify your reading.
- As you monitor your reading, write questions or phrases in the first column that need clarifying.
- In the second column, write ways that you can clarify your reading.

> Christianity is the most widely practiced religion in Australia and Oceania today. It was brought by Europeans during the 1700s and 1800s and attracted many followers among local peoples. However, traditional religions are also still practiced in some areas. In Australia, for example, the Aborigines believe in the idea of Dreamtime, the time long ago when they say wandering spirits created the world. They believe that all natural things—rocks, trees, plants, animals, and humans—have a spirit and are related to one another.
>
> —*from page 803*

Monitor	Ways to Clarify
1.	1.
2.	2.
3.	3.

Read to Write Activity

Read the World Literature selection *The Bamboo Flute* in Chapter 29. As you read, write at least two questions about the selection. Your questions might include something you would like to ask the author so that you can clarify your reading. Share your questions with a partner.

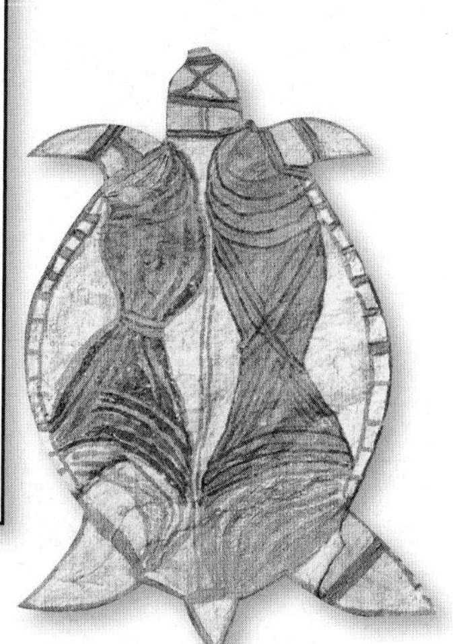

Aborigine art

3 Apply It!

Create a chart like the one above for each chapter in this unit. Stop often to record any part of the text that is especially difficult to understand. Be sure to clarify text that is confusing.

CHAPTER 28

Physical Geography of Australia, Oceania, and Antarctica

Essential Question

Human-Environment Interaction The region of Australia, Oceania, and Antarctica has great variety in landforms and climate. It includes high mountains, low plains, and tropical islands, as well as hot and cold deserts. How might the remoteness of a region make it different from other places?

Whitewater rafting, New Zealand

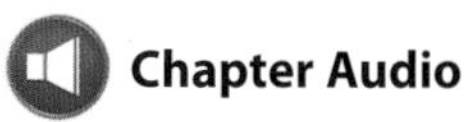

Chapter Audio

Section 1: Physical Features

BIG IDEA Physical processes shape Earth's surface. This region contains an amazing variety of landforms. Australia has mountains as well as vast plains. Most of Oceania's islands were formed from volcanic activity or from coral, the skeletons of hundreds of millions of small sea creatures. Some islands were formed by the rising and folding of rock on the ocean floor. Antarctica's mountains are surrounded by glaciers that are many feet thick.

Section 2: Climate Regions

BIG IDEA Places reflect the relationship between humans and the physical environment. The region's climates vary from tropical to polar. These different climate regions influence patterns of settlement and ways of life.

FOLDABLES™ Study Organizer

Summarizing Information Make this Foldable to help you summarize information about the landforms, resources, and climates of Australia, Oceania, and Antarctica.

Step 1 Fold an 11 x 17 piece of paper lengthwise to create 4 equal sections.

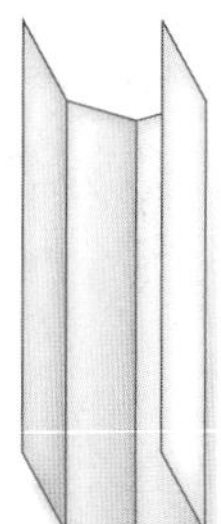

Step 2 Then fold it to form 5 columns.

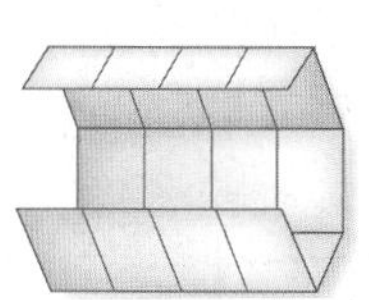

Step 3 Label your Foldable as shown.

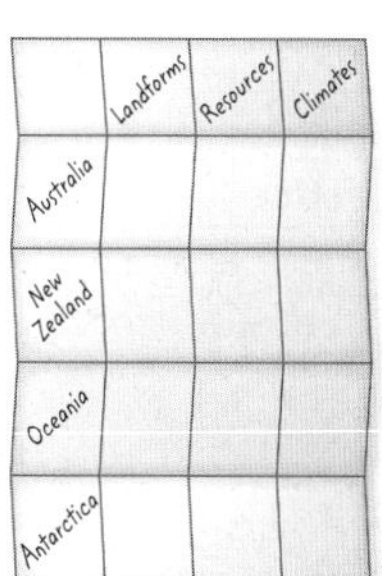

Reading and Writing After you have filled in your Foldable, use your notes to write an essay explaining why this region is unique compared to the other regions of the world that you have studied.

Visit glencoe.com and enter *QuickPass*™ code EOW3109c28 for Chapter 28 resources.

Guide to Reading

BIG Idea

Physical processes shape Earth's surface.

Content Vocabulary

- outback *(p. 777)*
- coral reef *(p. 777)*
- geyser *(p. 778)*
- high island *(p. 778)*
- low island *(p. 778)*
- atoll *(p. 778)*
- continental island *(p. 778)*
- ice shelf *(p. 778)*
- iceberg *(p. 778)*
- marsupial *(p. 779)*

Academic Vocabulary

- adjacent *(p. 777)*
- accurate *(p. 777)*

Reading Strategy

Identifying Use a chart like the one below to list examples of each type of landform in the areas within the region, and list a key fact about each.

	Highlands	Lowlands
Australia		
New Zealand		
Oceania		
Antarctica		

SECTION 1

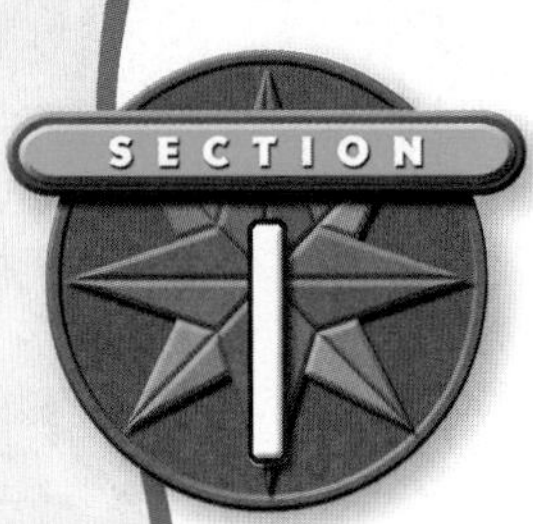

Physical Features

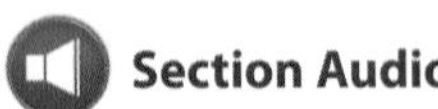

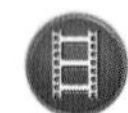

Section Audio   Spotlight Video

Picture This Catch the wave! There's no need to rush, though—this wave has been here for more than 2.7 billion years, and it is not going anywhere soon. The "wave" is actually a granite cliff face near Hyden, Australia. Weathering and erosion have undercut the cliff base, leaving a rounded overhang. As you read this section, you will learn more about the spectacular landforms of Australia, Oceania, and Antarctica.

▼ Wave Rock

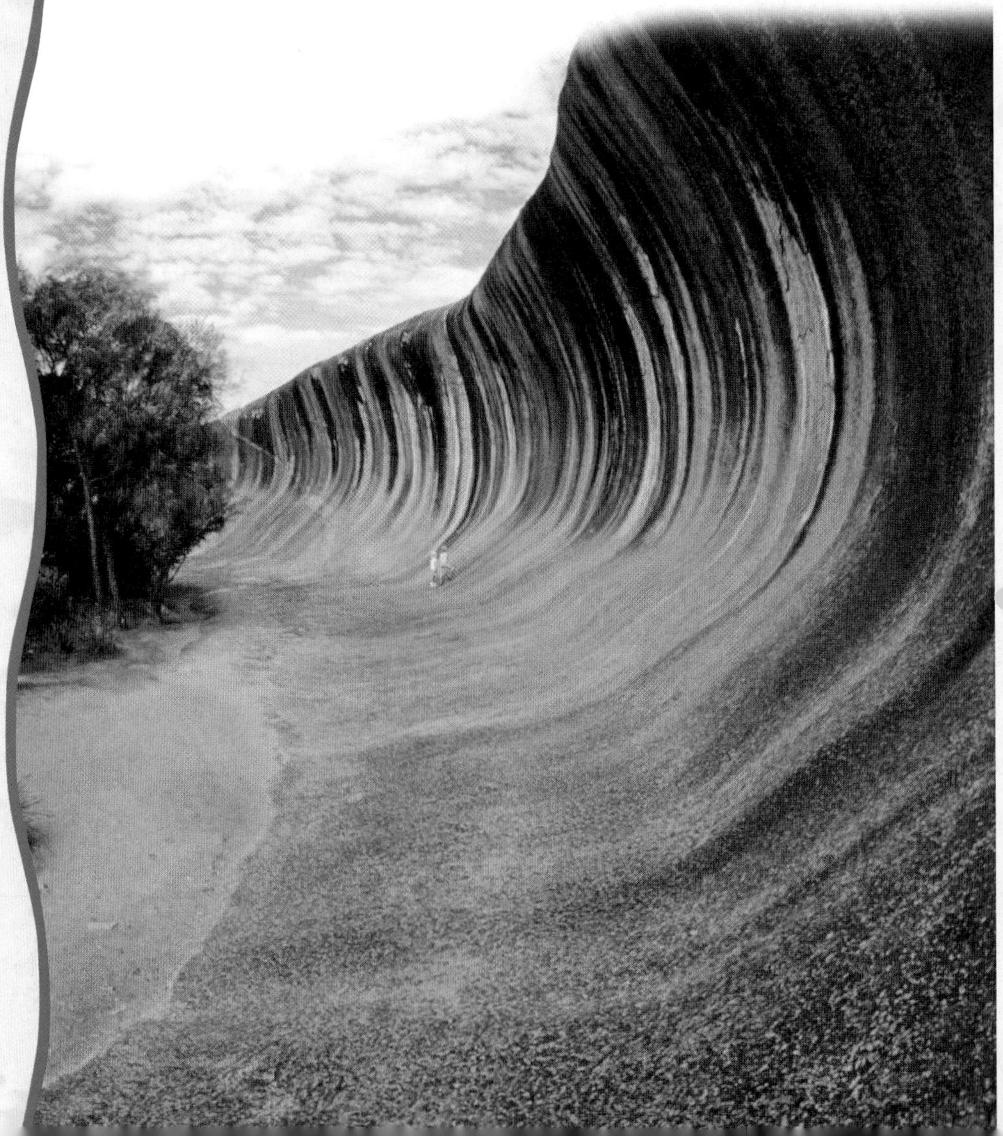

Landforms of the Region

Main Idea **Plate tectonics, erosion, and biological processes have shaped this region's landforms.**

Geography and You Have you ever seen beautiful tropical fish swimming on a coral reef? The world's largest coral reef lies off Australia's northeast coast. Read to learn more about it and other features in the region.

Australia, Oceania, and Antarctica lie almost entirely in the Southern Hemisphere. These areas form a huge region that reaches from north of the Equator to the South Pole.

Landforms of Australia

Australia, the sixth-largest country in the world, is also a continent. Despite its vast size, Australia is mostly flat. It also has low relief, or few differences in the elevations of **adjacent,** or neighboring, areas. Narrow plains run along the south and southeast of Australia. This land is Australia's best farmland. Much of the country's population lives here. Two major rivers, the Murray and the Darling, drain these areas.

A chain of mountains known as the Great Dividing Range stretches along Australia's eastern coast from the Cape York Peninsula to the island of Tasmania. Tasmania is also part of Australia. Although called a mountain chain, the Great Dividing Range is more **accurately,** or correctly, an escarpment. It is the rocky face of a plateau that plunges to lowland below.

To the west, the range blends into the outback. The **outback** is a vast area of plains and plateaus that is largely flat and dry. It also is dotted with isolated, heavily eroded masses of rock that stand above the lowland. Ayers Rock is one of these massive rocks. Also called Uluru, this rock is sacred to the Aborigines (A·buh·RIHJ·neez), who were the first people to settle Australia.

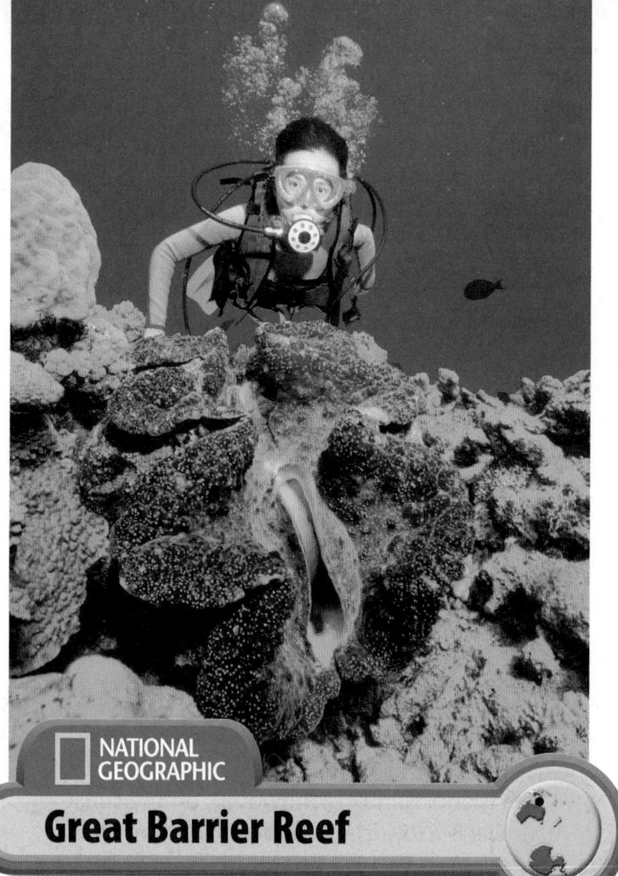

NATIONAL GEOGRAPHIC

Great Barrier Reef

The Great Barrier Reef is home to many creatures like this giant clam. Giant clams can live for 70 years and grow to be more than 3 feet (1 m) long.
Location **Where is the Great Barrier Reef located?**

Off Australia's northeastern coast lies the Great Barrier Reef. This natural wonder is the world's largest **coral reef,** a structure formed by the skeletons of small sea animals. Its colorful formations stretch about 1,250 miles (2,012 km).

Landforms of New Zealand

The country of New Zealand includes two main islands—North Island and South Island—as well as many smaller islands. The Cook Strait separates the two main islands.

Social Studies ONLINE

Student Web Activity Visit glencoe.com and complete the Chapter 28 Web Activity about the Great Barrier Reef.

The mountains of Bora Bora (left) are typical of a high island. Fairfax Island (inset), off the coast of Australia, is a low island. ***Place*** **What are the characteristics of a high island?**

New Zealand lies along a fault line where two tectonic plates meet. As a result, the large central plateau of North Island has active volcanoes as well as **geysers.** These hot springs carry steam and heated water to the Earth's surface, where they erupt as high as 60 feet (18 m) into the air.

Along the South Island's western coast are the Southern Alps. Snowcapped Mount Cook, New Zealand's highest peak, soars 12,349 feet (3,764 m) high. Glaciers lie on mountain slopes above green forests and sparkling blue lakes. Long ago, these glaciers cut deep fjords, or steep-sided valleys, into the mountains. The sea has filled the fjords with crystal-blue waters.

East of the Southern Alps stretch the Canterbury Plains. They form New Zealand's largest area of nearly flat land. The fertile Canterbury Plains are New Zealand's best farming area.

Landforms of Oceania

New Zealand is part of Oceania (OH·shee·A·nee·uh), a grouping of thousands of islands in the Pacific Ocean. These islands consist of three types: high, low, and continental.

Volcanic activity formed the mountainous **high islands** centuries ago. High islands, such as Tahiti and the Fiji Islands, feature mountain ranges split by valleys that fan out into coastal plains. The islands hold bodies of freshwater, and volcanic activity has provided fertile soil.

Coral, or skeletons of millions of tiny sea animals, formed the **low islands.** Many low islands, such as the Marshall Islands, are **atolls,** or low-lying, ring-shaped islands that surround shallow pools of water. Low islands have little soil.

Continental islands were formed centuries ago by the rising and folding of rock from the ocean floor. This movement was caused by tectonic activity. Inland from the coast, these islands hold rugged mountains, plateaus, and valleys. New Guinea and the Solomon Islands are continental islands.

Landforms of Antarctica

Antarctica lies at Earth's southern polar region. A long mountain range called the Transantarctic Mountains divides Antarctica into two distinct areas. To the east is a high plateau where the South Pole, the Earth's southernmost point, is located. To the west is a group of islands that are linked by ice.

A huge ice cap covers much of Antarctica. In some places, the ice is 2 miles (3 km) thick. At the coast, the ice cap spreads into the ocean. This layer of ice above water is called an **ice shelf.** Huge chunks of ice occasionally break off, forming **icebergs,** which float freely in the icy waters.

Reading Check **Explaining** How was the Great Barrier Reef formed?

Natural Resources

Main Idea **Australia, New Zealand, and Antarctica have many resources, but the islands of Oceania have relatively few.**

Geography and You You have probably seen photographs of the Australian outback or Antarctica's ice cap. Read to discover the kinds of unusual wildlife in the region.

Natural resources vary greatly throughout Australia, Oceania, and Antarctica. Australia mines rich mineral resources, including bauxite, copper, nickel, and gold. New Zealand has some deposits of gold, coal, and natural gas. Its rivers and dams supply hydroelectric power, and its hot springs provide geothermal energy.

The smaller islands of Oceania generally have few resources. Some larger islands have deposits of oil, gold, nickel, and copper.

Geologists have discovered that Antarctica is rich in mineral resources such as coal and iron ore. Tapping these minerals in the rugged, frozen landscape, however, would be very difficult and costly. Also, many nations have agreed not to mine this mineral wealth in order to protect Antarctica's environment.

Because the region is mostly islands that are far away from other land masses, it has long been isolated. As a result, some native plants and animals here are not found anywhere else in the world. Two well-known Australian animals are kangaroos and koalas. Both are **marsupials**, or mammals that carry their young in a pouch. New Zealand is home to the kiwi, a flightless bird that has become the country's national symbol.

Reading Check **Identifying** What energy resources are found in New Zealand?

Section Review

Social Studies ONLINE
Study Central™ To review this section, go to glencoe.com.

Vocabulary

1. **Explain** the significance of:
   - **a.** outback
   - **b.** coral reef
   - **c.** geyser
   - **d.** high island
   - **e.** low island
   - **f.** atoll
   - **g.** continental island
   - **h.** ice shelf
   - **i.** iceberg
   - **j.** marsupial

Main Ideas

2. **Summarizing** Use a diagram like the one below to identify and describe the three types of islands in Oceania.

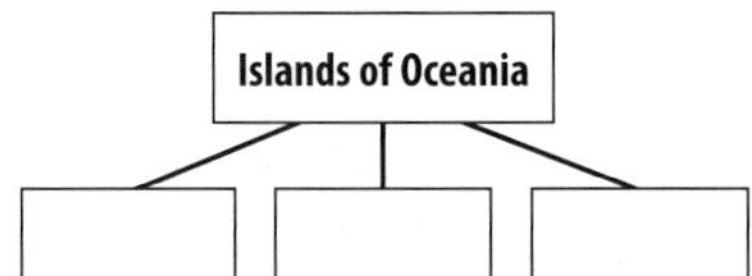

3. **Identifying** Identify the resources found in Australia, New Zealand, Oceania, and Antarctica.

Critical Thinking

4. **BIG Idea** Why are there geysers and volcanoes in New Zealand?
5. **Determining Cause and Effect** Why does this region have unique types of animals?
6. **Challenge** Considering Antarctica's physical geography, why would it be difficult to collect the area's resources?

Writing About Geography

7. **Descriptive Writing** Write a travel brochure describing one of the areas in this region in a way that encourages tourists to visit it.

Visiting Antarctica: Is Tourism Good for the Continent?

Antarctica is an increasingly popular tourist destination. Travel agencies organize ship cruises so that tourists can travel to Antarctica to observe the continent's unique environment. Some people believe that tourism helps visitors appreciate the region, which in turn encourages them to preserve it. Others, however, think that tourism is driven by companies that want to make a profit, with little concern for protecting Antarctica's fragile ecosystem.

Tourism

Education is an important component of any Antarctic expedition. Ship-based expeditions provide an opportunity for visitors to experience a wide range of areas of interest, including wildlife sites, historic sites, active research stations, and sites of exceptional wilderness and aesthetic [artistic] value. . . .

The benefits derived [gotten] from responsible tourism, such as better knowledge and appreciation of the region, are substantial.

—International Association of Antarctica Tour Operators

Against Tourism

Within a relatively short time, as the numbers of tourists continue to increase,...we may see the emergence of air-supported mass tourism in Antarctica....The problems of tourism, familiar everywhere else, have arrived in Antarctica....

What makes Antarctica a particular concern is that there is no regulation of tourism at present....[T]here is essentially no constraint [limit] on where you can go, what you can do, and how many of you can do it.

The practical consequence of this is that tourism is already exerting [placing] pressures on the Antarctic environment, and the increasing commercial interest is changing the nature of...Antarctic [politics]. Increasingly, commercial benefit, rather than concern for the environment, science, or international cooperation, is driving...Antarctic [politics].

—Antarctic and Southern Ocean Coalition

You Be the Geographer

1. **Analyzing** According to the International Association of Antarctica Tour Operators, why is tourism important?
2. **Critical Thinking** What does the Antarctic and Southern Ocean Coalition believe are the problems with tourism in Antarctica?
3. **Read to Write** Write a paragraph describing your opinion about whether tourists should be able to visit fragile ecosystems such as Antarctica.

Guide to Reading

BIG Idea

Places reflect the relationship between humans and the physical environment.

Content Vocabulary

- eucalyptus *(p. 784)*
- pasture *(p. 785)*
- breadfruit *(p. 785)*
- lichen *(p. 786)*

Academic Vocabulary

- distort *(p. 784)*
- duration *(p. 784)*
- sufficient *(p. 784)*

Reading Strategy

Determining Cause and Effect
Use a diagram like the one below to explain the effects of climate on life in each area.

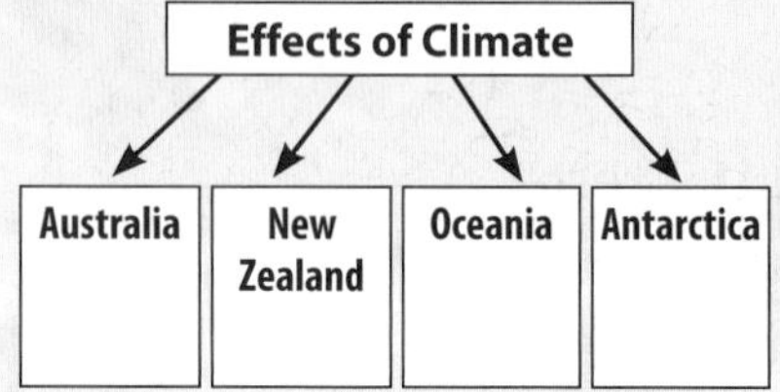

Climate Regions

 Section Audio  Spotlight Video

Picture This An ecologist uses his mountain-climbing skills to scale one of the world's tallest hardwood trees, the mountain ash. The mountain ash is a type of eucalyptus tree that can grow more than 300 feet (91 m) tall. It thrives in the wet forests in southern and eastern Australia and Tasmania. To learn more about climate and vegetation in Australia, Oceania, and Antarctica, read Section 2.

▼ Climbing a mountain ash

Climates of Australia

Main Idea **Australia has several climate regions, but much of the country is dry.**

Geography and You Think about what the landscape might look like in your community if it rarely rained. Read to find out about Australia's dry climate.

Australia is, in general, a dry continent. About one-third of its area is desert, and another third is partly dry steppe. These arid regions are mainly in the country's vast interior. Only the northern, eastern, and southwestern coastal areas receive plentiful rainfall.

The northern third of Australia lies in the Tropics. As a result, it is warm or hot year-round. The rest of the country lies south of the Tropics and has warm summers and cool winters.

Australia's Dry Areas

Figure 1 shows Australia's climate zones. Deserts cover much of the outback. Moist winds drop most rain on the coasts, and little rainfall reaches the interior. Deserts in south and central parts of Australia receive no more than 8 inches (20 cm) of rain a year.

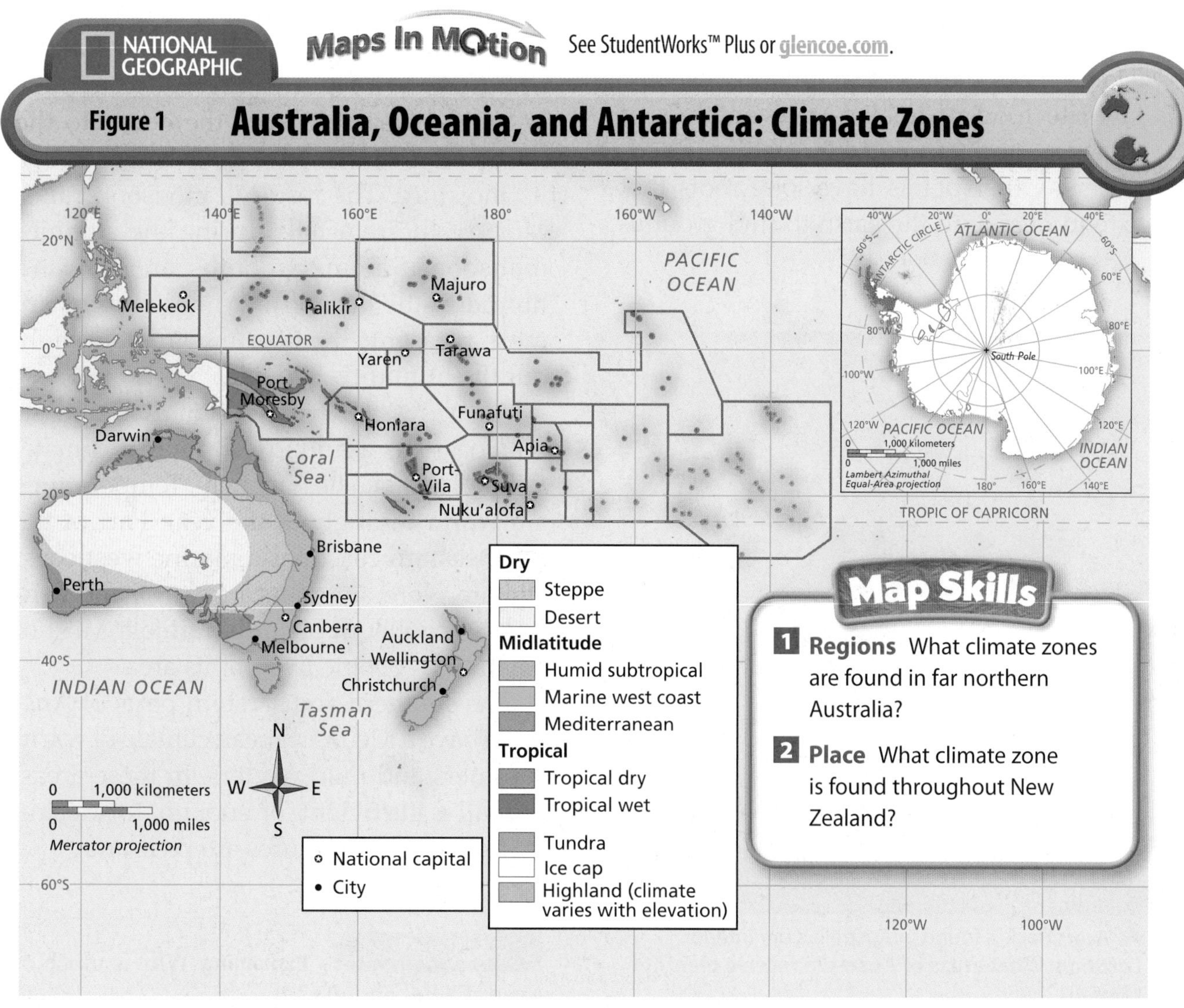

Figure 1 Australia, Oceania, and Antarctica: Climate Zones

Map Skills

1. **Regions** What climate zones are found in far northern Australia?
2. **Place** What climate zone is found throughout New Zealand?

However, yearly averages such as this may **distort,** or present a misleading impression of, normal climate patterns. Rainfall in any year can fall well short of the average, and rain may not fall for long periods of time.

A zone of milder steppe climate encircles Australia's desert region. Although rainfall is quite low, farming can still take place. When rains do reach desert and steppe areas, they often come in heavy bursts and can cause flash floods. High temperatures, however, cause any amount of rain that falls to quickly evaporate.

The plants that grow in central Australia are well suited for such dry conditions. **Eucalyptus** (YOO·kuh·LIHP·tuhs) trees, for example, are native to Australia and nearby islands. Their thick, leathery leaves—the favorite food of koalas—prevent loss of moisture and can survive rushing floodwaters. Other plants have long roots that extend deep into the earth to find groundwater for the **duration,** or length, of the long dry season.

Few people inhabit Australia's desert regions. In the lowlands west of the Great Dividing Range, however, lies a vast underground reservoir of water known as the Great Artesian Basin. The drilling of wells brings water to the surface and allows people to live in this region even though it is extremely dry.

▲ Ayers Rock is found in Australia's dry interior.
Location **What areas of Australia receive plentiful rainfall?**

Australia's Other Climate Zones

Although Australia's climates are mainly dry, the country also has other climate zones. The far north has a tropical savanna climate. This area is affected by the same kinds of monsoon winds that shape the climate in nearby areas of Asia. Dry, hot air over the interior of Australia rises and pulls in moist, warm air from the ocean to the north. As that air rises, it cools and drops its moisture. The result is monsoon rains. Most of the rain falls during the summer monsoon. Summer months are hot and humid in the far north. Winter months are more pleasant. Temperatures remain high, but the humidity is lower.

A narrow stretch of Australia's northeastern coast has a humid subtropical climate. Rainfall is much heavier here. Temperatures are warm throughout the year. Part of the southeastern coast has a marine west coast climate. Here, summers are warm, and winters are cool with plentiful rainfall. Most of Australia's people live in this area.

The southern and western parts of Australia have a Mediterranean climate of warm summers and mild winters. In these areas, rainfall is **sufficient,** or enough, for raising crops, and temperatures are pleasant.

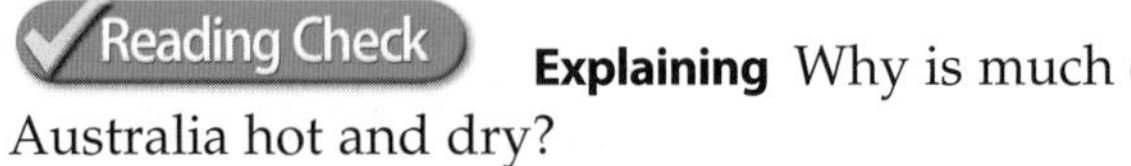

Explaining Why is much of Australia hot and dry?

Climates of Oceania

Main Idea **New Zealand has a mild climate, while the smaller islands of Oceania are mainly tropical.**

Geography and You Would you rather live in an area that has four seasons but mild weather, or one that is warm year-round? Read to find out about the climates of Oceania.

Oceania does not have as many climate zones as Australia. All of the countries of Oceania are islands, and the sea has a major effect on their climates.

New Zealand's Climate

Much of New Zealand has a marine west coast climate. Ocean winds here warm the land in winter and cool it in summer. Rainfall is generally plentiful, ranging from 25 to 60 inches (64 to 152 cm) during the year.

Temperatures in New Zealand show little change during the year. High temperatures in summer tend to be around 70°F (21°C), while highs in winter are 50°F (10°C) or above. The mild temperatures and plentiful rain promote the year-round growth of **pasture,** the grasses and other plants that are feed for grazing animals. As a result, many New Zealanders make their living by raising livestock.

Pacific Tropical Climates

Almost all of Oceania's smaller islands lie in the Tropics. Temperatures are generally warm, averaging around 80°F (27°C) throughout the year.

Rainfall in much of Oceania is seasonal. In some areas, heavy rains come in the spring and summer, while in other areas, heavy rainfall comes in the summer and fall. Strong typhoons can occur in this area.

Rainfall in Oceania is affected by an island's elevation. High islands tend to have more rain. They also have lower temperatures in mountainous areas. In contrast, low islands tend to be drier and warmer.

Vegetation in Oceania also depends on the type of island. High islands, with their higher rainfall and more fertile soil, tend to have a greater variety of plant life and support more farming. Low islands support only a few kinds of plants, such as coconut palms and breadfruit trees. **Breadfruit** is a starchy pod that can be cooked in several ways and is a food staple in Oceania.

Reading Check **Explaining** How does New Zealand's climate promote raising livestock?

Preparing Breadfruit

NATIONAL GEOGRAPHIC

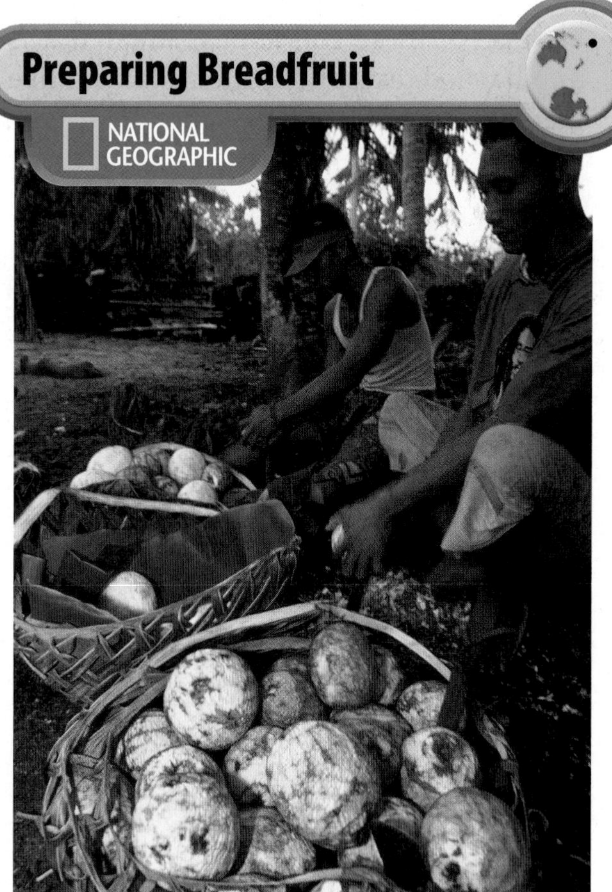

▲ In Samoa, breadfruit is seldom eaten raw. It is often peeled and baked in an *umu,* or Samoan oven. ***Place*** **What other type of tree is found on low islands?**

The Climate of Antarctica

Main Idea **Antarctica is a cold desert where no humans live permanently.**

Geography and You Have you ever heard of a place where it is too cold to snow? Read to find out why Antarctica is considered a cold desert.

Antarctica never receives the direct rays of the sun. As a result, temperatures there are bitterly cold. They reach no higher than –4° to –31°F (–20° to –35°C) in the summer. In the winter, they may plunge as low as –129°F (–89°C). Fierce winds that can reach 155 miles per hour (250 km per hr.) also whip across the surface of Antarctica. Because of these conditions, Antarctica is uninhabited, except for scientists who live there for brief periods.

Antarctica is covered by ice, but it receives little precipitation. Without the direct rays of the sun, Antarctica's air is so cold that it holds little moisture. This lack of humidity adds to the coldness, since humid air is needed to trap the sun's warmth. Whatever precipitation falls only adds to the ice covering the continent's surface. In fact, rainfall is so low that Antarctica is actually a desert.

Despite Antarctica's coldness, the continent does have life. Many different kinds of penguins live there, along with numerous marine mammals. All feed off the rich sea life in the surrounding waters. In rocky areas along the coasts, tiny, sturdy plants called **lichens** (LY·kuhnz) grow.

Reading Check **Explaining** Why is Antarctica considered a desert?

Section 2 Review

Social Studies ONLINE
Study Central™ To review this section, go to glencoe.com.

Vocabulary

1. **Explain** the meaning of *eucalyptus, pasture, breadfruit,* and *lichen* by using each word in a sentence.

Main Ideas

2. **Explaining** How have some plants adapted to Australia's dry climate?
3. **Comparing and Contrasting** Use a Venn diagram like the one below to compare and contrast rainfall, temperatures, and vegetation on Oceania's high and low islands.

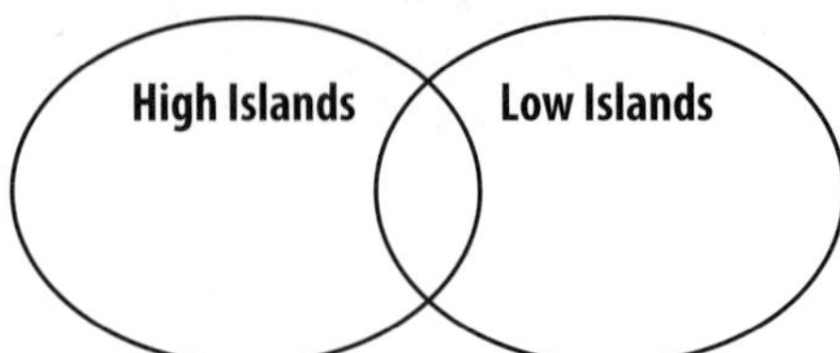

4. **Identifying** What sorts of plants and animals live in Antarctica?

Critical Thinking

5. **BIG Idea** Why is Antarctica uninhabited?
6. **Challenge** How do you think climate has affected settlement within the region?

Writing About Geography

7. **Using Your FOLDABLES** Using your Foldable, write a series of fictional journal entries describing the climates and vegetation you experience while traveling by ship in the Pacific Ocean, with stops in Oceania, Australia, and Antarctica.

Visual Summary

Landforms

- Australia, Oceania, and Antarctica form a huge region that reaches from north of the Equator to the South Pole.
- Australia is mainly flat with low relief. Erosion has worn down highland areas.
- Australia's Great Barrier Reef stretches 1,250 miles (2,012 km). It is the world's largest coral reef.
- New Zealand has high mountains and coastal lowlands.

Sea kayaking, Antarctica

- Volcanic activity formed many of Oceania's high islands; coral buildups created the low islands.
- Thick ice covers Antarctica's highlands and plains.

Kalgoorlie gold mine, Australia

Resources

- Australia and New Zealand are rich in mineral and energy resources.
- Islands in Oceania have few natural resources.
- Antarctica's natural resources are untapped.
- Because Australia and New Zealand remained isolated for a long period, they have many unique plants and animals.

Climates

- Australia has mainly warm, dry climates. The country's coasts have more moderate temperatures and receive more rainfall than inland areas.
- New Zealand, close to the sea, has moderate temperatures and ample rain.
- Oceania has tropical climates with warm temperatures year-round.
- Antarctica is a bitterly cold desert.

Beach in Fiji

Study anywhere, anytime! Download quizzes and flash cards to your PDA from **glencoe.com**.

CHAPTER 28 ASSESSMENT

STANDARDIZED TEST PRACTICE

Do not wait until the last minute to prepare for a test. Make sure you ask questions in class after you have studied, but before you take the test.

Reviewing Vocabulary

Directions: Choose the word(s) that best completes the sentence.

1. Australians call the vast interior of their country the ______.

A highlands
B escarpment
C outback
D geyser

2. Kangaroos and koalas are both ______, or mammals that carry their young in a pouch.

A marsupials
B invertebrates
C amphibians
D aviaries

3. ______ grows in the dry conditions of central Australia and is eaten by koalas.

A Breadfruit
B Lichen
C Pastures
D Eucalyptus

4. Despite Antarctica's cold temperatures, some plants, including ______, grow there.

A eucalyptus
B lichens
C breadfruit
D marsupials

Reviewing Main Ideas

Directions: Choose the best answer for each question.

Section 1 *(pp. 776–779)*

5. Off Australia's northeastern coast lies a huge structure called ______, which was formed by the skeletons of small sea animals.

A Uluru
B Ayers Rock
C the Great Barrier Reef
D Tasmania

6. New Zealand is part of a grouping of thousands of islands known as ______.

A Tasmania
B Oceania
C Micronesia
D Papua New Guinea

Section 2 *(pp. 782–786)*

7. Australia has several climate regions, but much of the country is ______.

A swampy
B mountainous
C fertile farmland
D dry

8. Much of New Zealand's climate is ______.

A marine west coast
B humid subtropical
C desert
D Mediterranean

GO ON

Critical Thinking

Directions: Base your answers to questions 9 and 10 on the circle graph below.

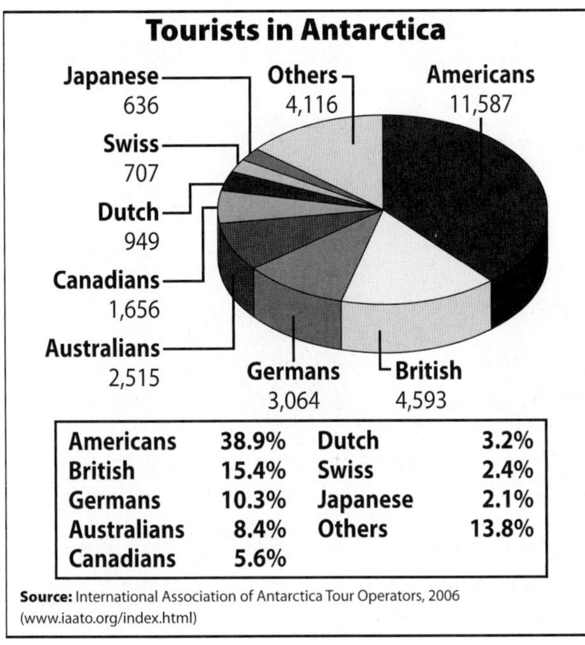

Americans	38.9%	Dutch	3.2%
British	15.4%	Swiss	2.4%
Germans	10.3%	Japanese	2.1%
Australians	8.4%	Others	13.8%
Canadians	5.6%		

Source: International Association of Antarctica Tour Operators, 2006 (www.iaato.org/index.html)

9. The graph shows that most tourists come from ________.

A Australia

B Canada

C the United States

D the United Kingdom

10. Which of the following statements is accurate?

A Most tourists to Antarctica were Japanese.

B About 30,000 tourists visited Antarctica.

C Most tourists to Antarctica were from Australia because it is closest to Antarctica.

D More than half the tourists to Antarctica in 2005–2006 were from the United Kingdom, Germany, and Australia combined.

Document Based Questions

Directions: Analyze the document and answer the questions that follow.

Reef Water Quality Protection Plan

The Great Barrier Reef is facing an increasing threat from a decline in the water quality . . . draining into the Reef lagoon. . . .

The Great Barrier Reef is a nationally and internationally significant area with outstanding natural values. It makes a major contribution to the local, regional, and national economy . . . .

Along with the largest system of coral reefs in the world, the Reef is home to extensive seagrass beds, mangrove forests, and sponge gardens. Many of the Reef's marine species rely on coastal freshwater wetlands and estuaries as breeding and nursery areas.

Extensive land development in the [river basins] adjacent to the Reef . . . has led to increased pollution of these rivers.

—Australia's Department of the Environment and Heritage, 2006.

11. According to the writer, why is the Great Barrier Reef important?

12. Describe the threats that currently exist to the Great Barrier Reef and the marine species that live there.

Extended Response

13. If you were a geologist, where in Australia, Oceania, and Antarctica would you suggest that a company search for minerals? Explain your reasons.

For additional test practice, use Self-Check Quizzes—Chapter 28 at glencoe.com.

Need Extra Help?

If you missed question...	1	2	3	4	5	6	7	8	9	10	11	12	13
Go to page...	777	779	784	786	777	778	783	785	780	780	777	777	779

CHAPTER 29

History and Cultures of Australia, Oceania, and Antarctica

Essential Question

Regions Australia, Oceania, and Antarctica are grouped together more because of their nearness to one another than because of any similarities among their peoples. Even though their cultures are different, many of the people share a similar history of colonization. Today, after achieving independence, this region's people are creating a new identity for themselves that blends traditional beliefs with modern ideas. How does a people's past influence its present and future?

Carving showing Maori canoe, New Zealand

BIG Ideas

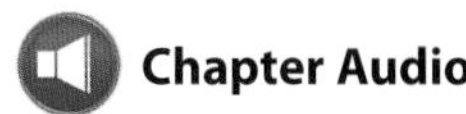

Chapter Audio

Section 1: History and Governments

BIG IDEA Geographic factors influence where people settle. Asian and Pacific peoples settled Australia, New Zealand, and Oceania thousands of years ago. Europeans later migrated to the region because of its rich resources.

Section 2: Cultures and Lifestyles

BIG IDEA Culture groups shape human systems. Peoples from different parts of the world have helped shape the cultures of Australia, New Zealand, and Oceania. For example, people of European descent make up the primary ethnic groups in Australia and New Zealand. However, the populations of native groups are growing. As a result of this population mix, cultures are changing.

FOLDABLES™ Study Organizer

Organizing Information Make this Foldable to help you organize information about the history, people, and daily life of Australia, Oceania, and Antarctica.

Step 1 Fold the bottom edge of a piece of paper up 2 inches to create a flap.

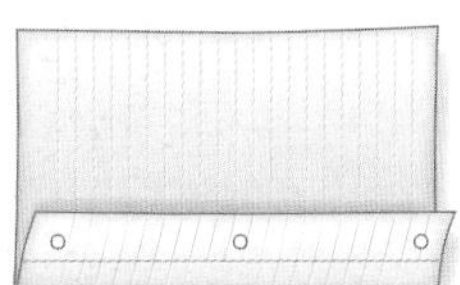

Step 2 Fold into thirds.

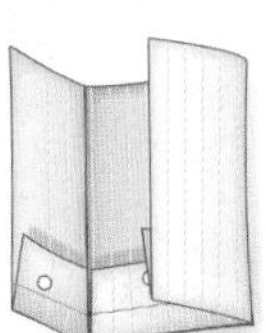

Step 3 Glue to form pockets and label as shown. Use pockets to hold notes taken on slips of paper.

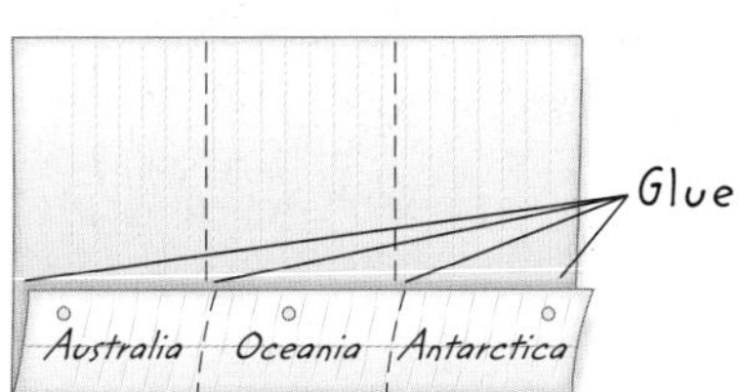

Reading and Writing As you read the chapter, take notes on slips of paper. Place each note in the correct pocket of your Foldable. Use your notes to write a short summary of the history and culture of either Australia or Oceania.

Social Studies ONLINE

Visit glencoe.com and enter *QuickPass*™ code EOW3109c29 for Chapter 29 resources.

History and Governments

 Section Audio  Spotlight Video

Guide to Reading

BIG Idea

Geographic factors influence where people settle.

Content Vocabulary

- boomerang *(p. 793)*
- trust territory *(p. 797)*

Academic Vocabulary

- acquire *(p. 794)*
- prime *(p. 795)*

Reading Strategy

Summarizing Use a chart like the one below to organize key facts about the first peoples to settle each area.

First Settlers		
Australia	New Zealand	Oceania

Picture This Future rodeo cowboys in Australia practice their skills using an empty steel drum suspended on a rope. A rodeo rider must ride an angry bull or a bucking horse for at least eight seconds. Falling off before the eight seconds is up results in a score of zero. During outback rodeos, Aborigines and Australians of European descent gather to compete in a variety of events. Read this section to learn more about the history of the people of Australia, New Zealand, and Oceania.

Rodeo practice in western Australia

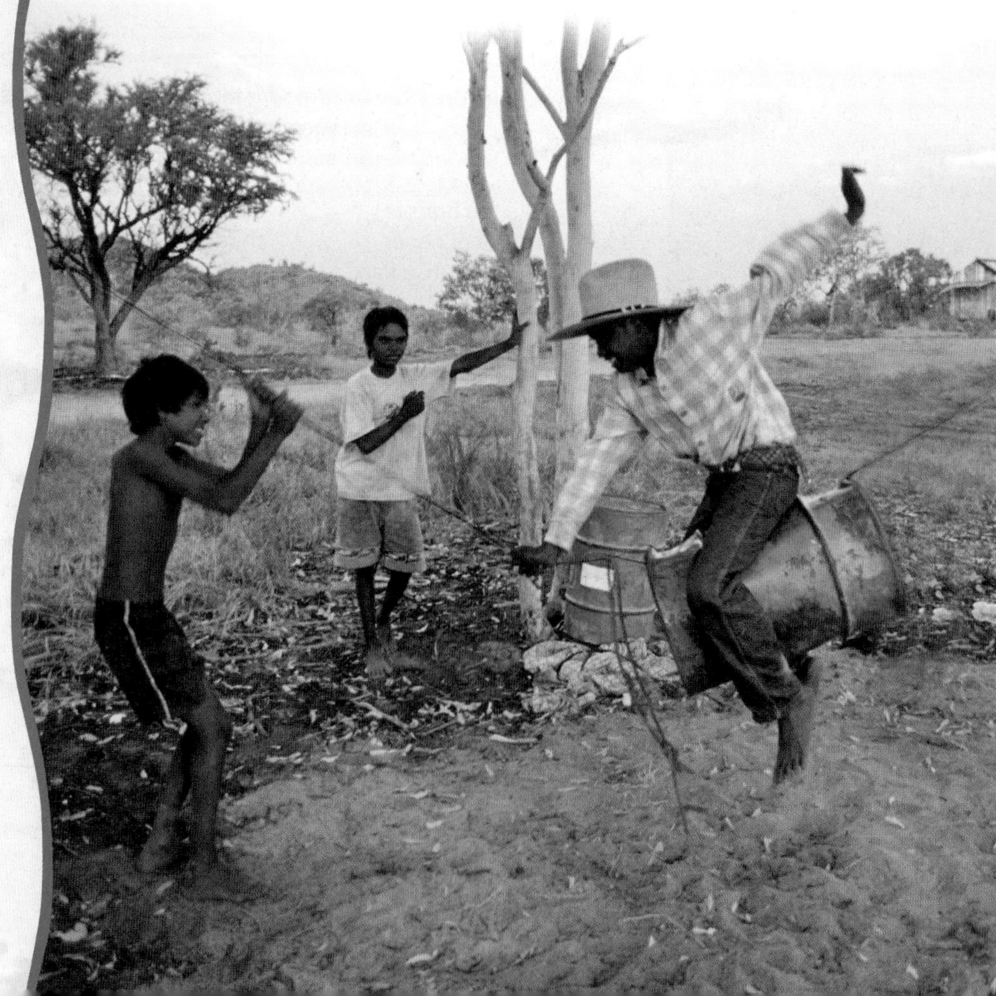

First Settlers

Main Idea **The region's first settlers came from Asia and islands in the Pacific Ocean.**

Geography and You Which groups first settled the area where you live? Read to find out about the Aborigines, the descendants of the first people who lived in Australia.

Some 40,000 years ago, the Earth was undergoing a period known as an Ice Age. During this period, temperatures were lower and much more of Earth's water was frozen in ice caps. Ocean levels were not as high as they are today. At this time, people from Southeast Asia traveled to the islands of Oceania and to Australia either by land or by canoe. Several thousand years later, the Ice Age ended and ocean levels rose. The people who had migrated to Oceania and Australia were cut off from the rest of the world.

Australia's Aborigines

The people now called Aborigines are the descendants of these first Australians. Early Aborigines traveled in small family groups around Australia. They hunted, gathered plants, and searched for water. To hunt for small animals, they developed a special weapon called a **boomerang.** It is a flat, bent, wooden tool that hunters throw to stun prey. If the boomerang misses its target, it sails back to the hunter.

The religion of the Aborigines focuses on the relationship of people to nature. The Aborigines believe that powerful spirits created the land and that their role as a people is to care for it. Aborigines had no large settlements, but ancient rock paintings and stories tell much about their early history.

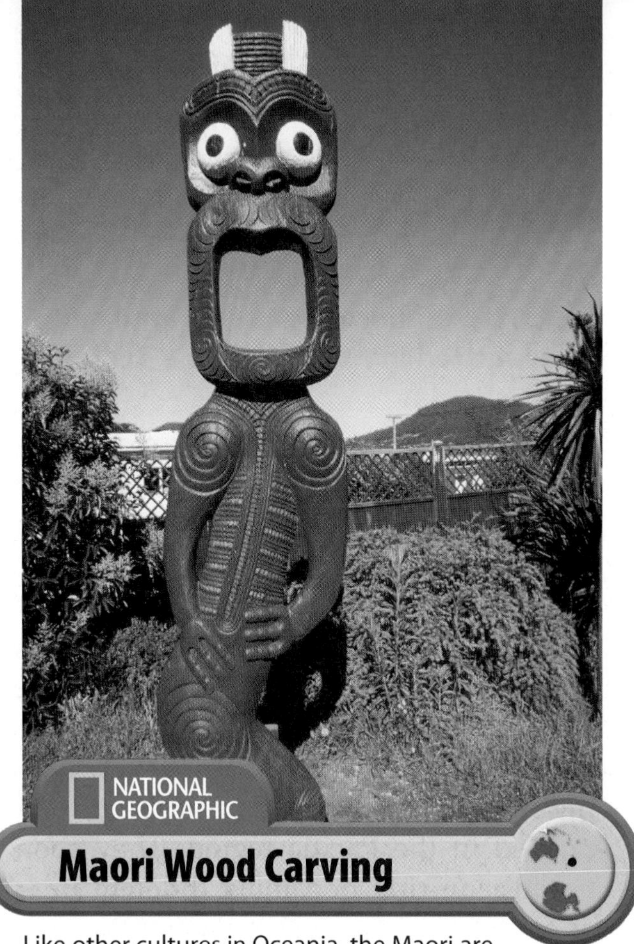

NATIONAL GEOGRAPHIC

Maori Wood Carving

Like other cultures in Oceania, the Maori are skilled wood-carvers. ***Location*** **From where did the Maori originally migrate?**

Pacific Migrations

About the time Australia was first inhabited, other people from Southeast Asia settled New Guinea and nearby islands. By 1500 B.C., they had developed large canoes that could travel long distances across the ocean. In time, settlers reached other remote islands, such as Fiji, Tonga, and Hawaii.

Between A.D. 950 and A.D. 1150, the Maori (MOWR·ee) people left Polynesia, a south central Pacific island area, and settled the islands of New Zealand. On New Zealand's North Island and South Island, they set up villages, hunted, fished, and farmed the land. Maori farmers, like other Pacific Islanders, grew root crops that they brought from Polynesia. They also developed skills in wood carving.

Reading Check **Describing** Describe the lifestyle of Australia's Aborigines.

The European Era

Main Idea **Europeans explored and later settled in Australia, New Zealand, and Oceania.**

Geography and You What role did European explorers and settlers play in the history of the Americas? Read to find out how Europeans also explored and settled Australia and Oceania.

The first settlers and their descendants developed thriving cultures in Australia, New Zealand, and Oceania. Life for these Pacific peoples changed, however, during the 1500s. At that time, Europeans first arrived in the Pacific region. They soon began **acquiring,** or gaining, different areas as colonies.

Arrival of Europeans

From the 1500s to the 1800s, Europeans from various countries explored the South Pacific region. Perhaps the most well-known explorer was the British sailor Captain James Cook. He made three voyages to the region between 1768 and 1779. Cook claimed eastern Australia for Great Britain, visited several Pacific islands, and circled Antarctica. He produced amazingly accurate records and maps of these areas.

When Cook landed in Australia, about 300,000 Aborigines lived there. The British, though, viewed the continent as uninhabited. The British government at first established a colony in Australia for convicts from overcrowded British prisons. Between 1788 and 1868, about 160,000 convicts were taken to Australia to serve their sentences. Once they had done so, most stayed in the new land.

By the mid-1800s, the British government stopped sending convicts to Australia. Many free British settlers began migrating to Australia's shores, however, in hopes of making a living or growing wealthy. They settled along the coasts and then moved farther inland. By 1861, the European population had passed 1 million.

Many of the settlers were farmers who grew wheat. Ranchers soon realized that conditions in Australia were perfect for raising sheep that produced a fine wool. Exports of wool became a major part of the economy. The discovery of several minerals also benefited the economy. Copper was the first metal found. The discovery of gold in 1851 led to a new rush of settlers and greatly increased Australia's population.

At first, relations between the Aborigines and the British were relatively peaceful. As the Europeans took more and more

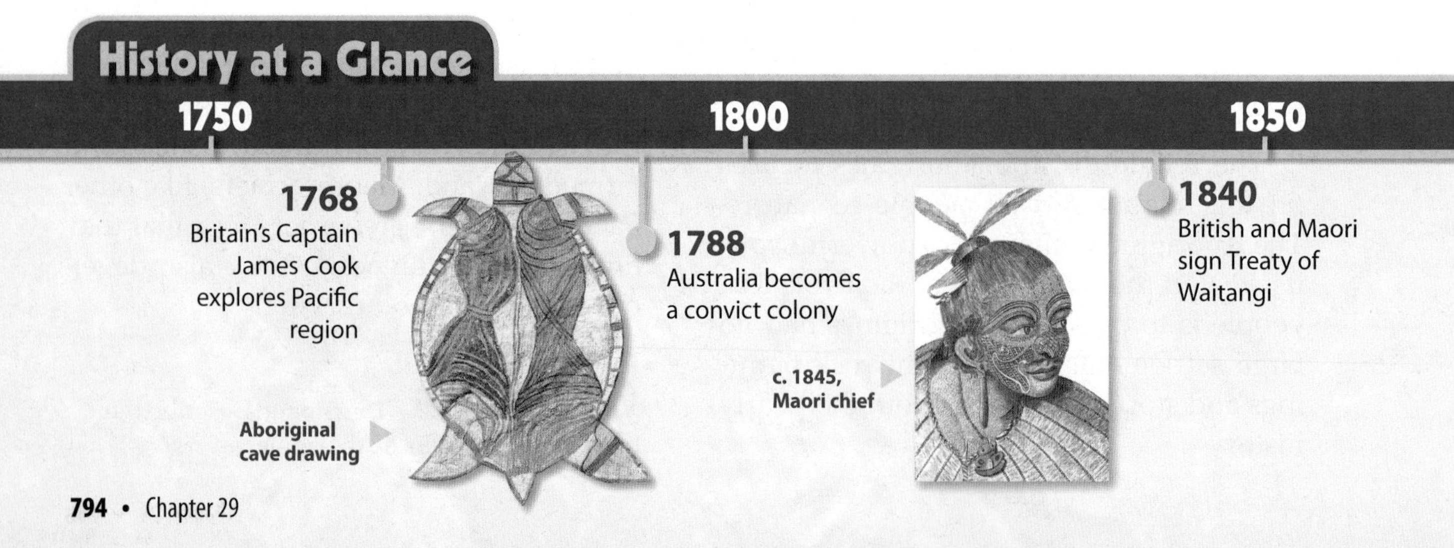
History at a Glance
1750
1800
1850
1768
Britain's Captain James Cook explores Pacific region
1788
Australia becomes a convict colony
1840
British and Maori sign Treaty of Waitangi
Aboriginal cave drawing

c. 1845, Maori chief

land for ranching and farming, however, the Aborigines were forced to defend their traditional hunting grounds. The Aborigines fought with spears and were easily defeated by the Europeans with rifles. To make matters worse, the Aborigines had no resistance to European diseases. By the late 1800s, warfare and illness had reduced the number of Australia's Aborigines to about 80,000.

Building Empires

The British government divided Australia into five separate colonies. Each colony had its own legislature that made laws for the people within its boundaries. Unlike other parts of the world, where voting was almost always limited to property owners, Australian colonies allowed all men to vote. Voting rights, as well as the right to local self-government, eventually led to a democratic government for all of Australia.

The first European settlers in New Zealand may have been shipwrecked sailors and escaped Australian convicts. In time, Australians set up small whaling settlements along New Zealand's coasts. British settlers arrived in New Zealand in the 1820s and 1830s. The Maori were able to hold off the newcomers until European diseases took a heavy toll. By 1840, the Maori population had been cut in half to 100,000. The same year, Maori chiefs signed the Treaty of Waitangi (WY·TAHNG·gee) with the British. Under this treaty, the Maori agreed to accept British rule in return for the right to keep their land. More British settlers moved onto Maori land, however, and war broke out in the 1860s. The Maori lost the war and much of their land. British colonists then built an economy based on dairy products and sheep.

As global trade grew, Europeans and Americans turned their attention to Oceania. The Pacific Islands offered **prime,** or very attractive, locations for trading ports and refueling stations for ships. As a result, Western countries began to colonize the Pacific Islands. As you can see in **Figure 1** on the next page, most of the Pacific area had come under the rule of foreign powers.

Foreign control increased trade and missionary activity. Over time, many Pacific Islanders accepted Christianity and some Western ways. The spread of Western diseases, however, continued to reduce the populations of the Pacific Islands.

Reading Check **Summarizing** What part did James Cook play in making Australia a British colony?

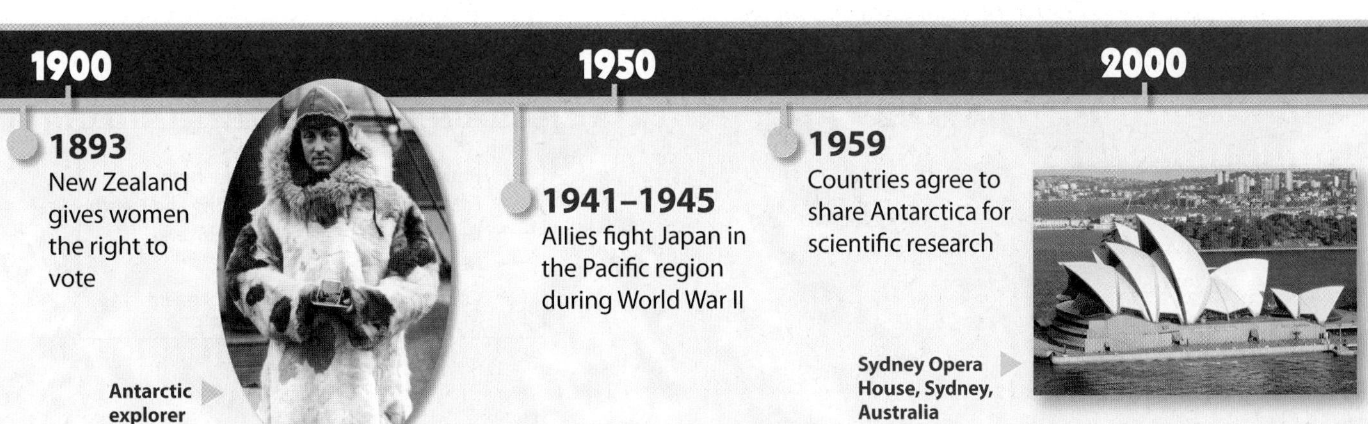
1900
1950
2000
1893
New Zealand gives women the right to vote
Antarctic explorer
1941–1945
Allies fight Japan in the Pacific region during World War II
1959
Countries agree to share Antarctica for scientific research
Sydney Opera House, Sydney, Australia

Independent Nations

Main Idea **Australia, New Zealand, and many islands in Oceania gained independence in the 1900s.**

Geography and You What does independence mean to you? Read to find out how countries in the region gained their independence.

During the early 1900s, the British colonies of Australia and New Zealand became independent countries. By 2000, most of the other Pacific Islands gained their independence.

Australia and New Zealand

Australia and New Zealand both won freedom from British rule peacefully. In 1901 the Australian colonies became an independent country known as the Commonwealth of Australia. New Zealand gained independence in 1907. Today, each country is a parliamentary democracy in which elected representatives choose a prime minister to head the government. Australia also has a federal government. Power to make and enforce laws is divided between the national government and the states and territories. In smaller New Zealand, the central government holds all major powers.

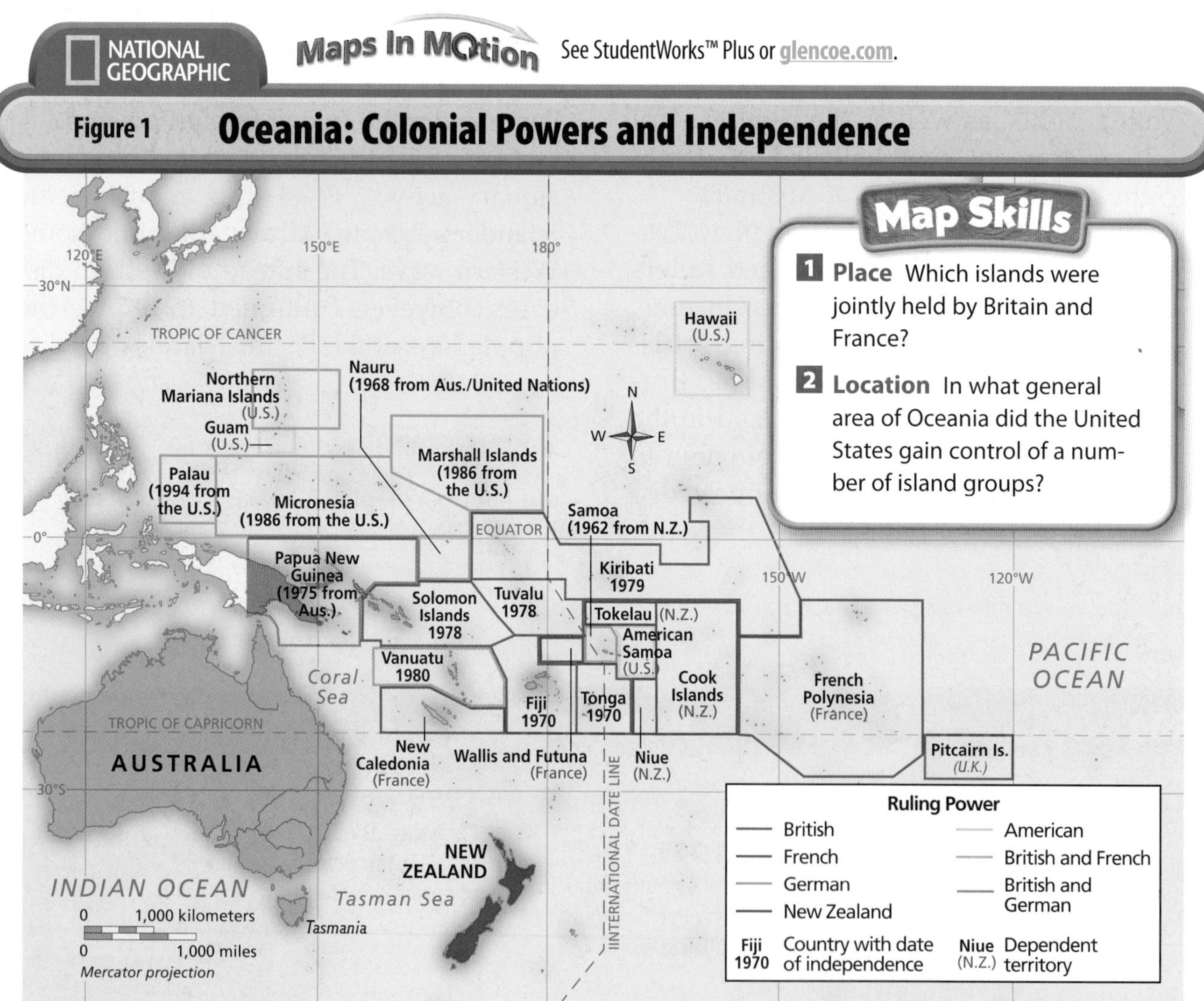

Figure 1 **Oceania: Colonial Powers and Independence**

Map Skills

1. **Place** Which islands were jointly held by Britain and France?
2. **Location** In what general area of Oceania did the United States gain control of a number of island groups?

In 1893 New Zealand became the first country in the world to give women the right to vote. Australia followed in 1902. New Zealand also was among the first countries to provide government help to the elderly, the sick, and the jobless. Some groups, however, such as the Aborigines in Australia and the Maori in New Zealand, have suffered discrimination. In addition, until the 1970s, Australia and New Zealand had laws that banned or limited certain immigrants, especially Asians.

Oceania and Antarctica

During the early 1900s, most of Oceania was still under foreign control. The two World Wars, however, changed the course of its history. After World War I, Germany's Pacific colonies came under Japan's rule. Then, in December 1941, Japan attacked the American fleet at Pearl Harbor, Hawaii. This attack brought the United States into World War II. During the conflict, the United States and Japan fought fierce battles on many Pacific islands. After Japan's defeat, Japan's Pacific territories—such as the islands of Micronesia—were turned over to the United States as trust territories. A **trust territory** is an area temporarily placed under the control of another country.

Since the 1960s, most Pacific trust territories and colonies have become independent. This change has not been easy for every country. For example, Fiji and the Solomon Islands have been torn by ethnic conflict since gaining their independence.

Antarctica's status also changed during the 1900s. Several countries had at first claimed land in Antarctica. Then, in 1959, many countries signed the Antarctic Treaty, agreeing to share the continent for peaceful scientific research. Since then, scientists from several different countries have worked in Antarctica.

Reading Check **Explaining** How did World Wars I and II affect Oceania?

Section Review

Social Studies ONLINE
Study Central™ To review this section, go to glencoe.com.

Vocabulary

1. **Explain** the meaning of *boomerang* and *trust territory* by using each word in a sentence.

Main Ideas

2. **Explaining** How did the first people arrive in the Pacific Islands?
3. **Identifying** Use a diagram like the one below to identify key events in the historical development of Australia.

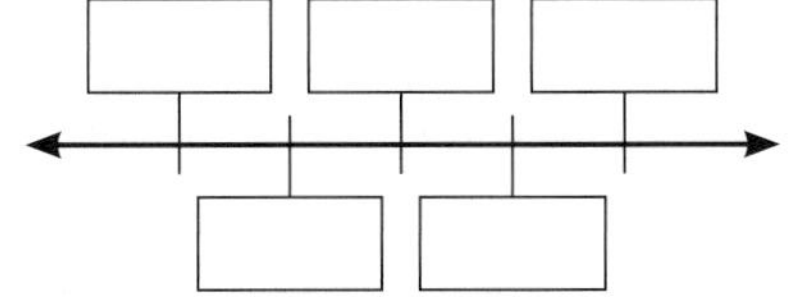

4. **Comparing** Compare the governments of Australia and New Zealand.

Critical Thinking

5. **BIG Idea** Why did it take thousands of years for people to settle all of Oceania?
6. **Challenge** Compare and contrast European colonialism in Australia and Oceania with that in another part of the world, such as Asia or Africa.

Writing About Geography

7. **Using Your FOLDABLES™** Use your Foldable to create an outline of the history of one of the subregions discussed in this section.

Populating Oceania

Centuries before Europeans left their continent to sail the oceans, Pacific Islanders traveled thousands of miles to settle new islands.

Far Travelers The Pacific Islands were first settled by people sailing from Southeast Asia. These peoples arrived at what are today the islands of Melanesia. About 1300 B.C., Pacific migrants made the longest journey when they reached Fiji, a group of islands midway between present-day Melanesia and Polynesia.

Early European explorers in Oceania refused to believe that the people there could travel the long distances between islands. The islanders had no large ships and still used stone tools. The Europeans, however, greatly underestimated the skills and technology of the Pacific peoples.

Canoes and the Stars Pacific Islanders traveled by canoes powered by the wind. The hull, or body, of each canoe was made by felling a tree and then hollowing it out. To travel long distances, they used double canoes. Having two hulls made the vessels more stable in the ocean. A platform that connected the two hulls carried people plus the plants and animals they brought along. The canoes held a mast with a sail made from leaves joined together.

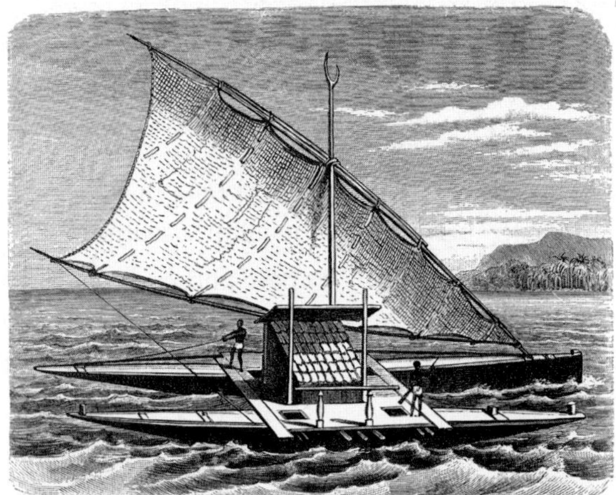

Polynesian double canoe

Polynesian shore

Islanders steered by carefully noting—and remembering—the positions of the stars when they rose and set each day. By comparing changes in the stars' positions, they knew how much progress they were making. They also saw whether they needed to change course.

Sailing skills were needed to avoid the dangers of the ocean crossing. Storms and strong winds could damage the ship. If the trip lasted too long, food might run out.

Despite these dangers, Pacific Islanders achieved one of the greatest migrations in human history. They crossed hundreds of miles of empty ocean to move among and populate the region's many island groups.

Think About It

1. **Movement** Why did Europeans not believe that the Pacific Islanders could travel long distances?
2. **Human-Environment Interaction** What skills and technology did the islanders use to make their journeys?

Guide to Reading

BIG Idea

Culture groups shape human systems.

Content Vocabulary

- bush *(p. 801)*
- station *(p. 801)*
- pidgin language *(p. 801)*
- action song *(p. 803)*
- *fale (p. 804)*
- poi *(p. 804)*

Academic Vocabulary

- sustain *(p. 800)*
- integral *(p. 803)*
- generation *(p. 804)*

Reading Strategy

Comparing and Contrasting Use a Venn diagram like the one below to compare and contrast the people of Australia and New Zealand with the people of Oceania.

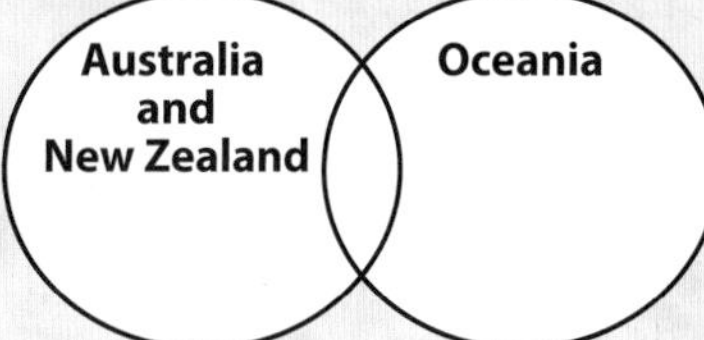

Cultures and Lifestyles

 Section Audio  Spotlight Video

Picture This It is so hot on an Australian beach, you could fry an egg! Well, a fiberglass egg, that is. The artist who created this sculpture was inspired by people who lounge on Australian beaches, hoping for a bronze tan. The sculpture is part of a popular event in Sydney that celebrates the summer lifestyles of Australians. Sculptors from around the world, as well as from Australia, contribute more than 100 works of art to the beach display. Other sculptures have included a pair of oversized sunglasses. Read this section to learn more about the culture and lifestyles of people living in Australia, New Zealand, and Oceania.

▼ Celebrating Australian summers

The People

Main Idea **The people of this region have varied ethnic backgrounds.**

Geography and You Have you ever traveled a great distance without seeing any signs of human life? Read to learn about an area of Australia where one can travel 100 miles (161 km) without seeing another human being.

The region of Australia, Oceania, and Antarctica spans a vast area. Much of its land is too dry, icy, or remote to **sustain,** or support, human settlement. Thus, few people live in this region compared to other world regions. At the same time, the region's population comes from diverse ethnic backgrounds.

Population Growth and Density

Because of vast differences in climate, landforms, and cultures, populations vary greatly throughout Australia, Oceania, and Antarctica. Australia is the region's most populous country, with more than 20 million people. New Zealand has about 4 million people. The population growth rate in both countries has slowed in recent years because of low birthrates. On the other hand, immigration has increased.

Population density differs throughout Australia and New Zealand. Very few people live in Australia's dry plateaus and deserts of the outback. Most Australians live in coastal areas that have a mild climate, fertile soil, and access to the ocean. Most of New Zealand's people live along coasts rather than in the rugged inland areas.

On the other islands of Oceania, the overall number of people is growing rapidly. This is because these islands have relatively young populations. Population density, however, varies greatly among the island countries. Oceania's most populous country is Papua New Guinea, with about 5.9 million people. Papua New Guinea has a large area—178,703 square miles (462,839 sq. km)—so its population density is only 33 people per square mile (13 per sq. km). By contrast, tiny Nauru has only 10,000 people and only 9 square miles (23 sq. km). As a result, Nauru has one of Oceania's highest population densities—about 1,111 people per square mile (435 per sq. km). Because many Pacific islands like Nauru are small in land area, overcrowding is a problem. As a result, some Pacific Islanders have begun migrating to other parts of the world.

NATIONAL GEOGRAPHIC

Schooling in the Outback

Because ranches in Australia's outback are large and far apart, children often attend school in small classrooms on the ranch. ***Location*** **Where do most of Australia's people live?**

Social Studies ONLINE

Student Web Activity Visit glencoe.com and complete the Chapter 29 Web Activity about the Aborigines.

Antarctica's forbidding icy landscape and polar climate do not support permanent human settlement. Research scientists and tourists make short-term visits.

Urban and Rural Life

More than 85 percent of the people in Australia and New Zealand live in coastal urban areas. Cities, such as Sydney and Melbourne in Australia, and Wellington and Auckland in New Zealand, are thriving commercial centers with modern buildings and transportation systems. Other countries in the region have few large urban areas. The largest is Port Moresby, the capital of Papua New Guinea. Most people in the Pacific Islands live in small rural villages.

A small number of Australians live in rural areas known as the **bush.** Some people who live in the bush work on cattle and sheep ranches, or **stations.** Others farm or work in mining camps. Because Australia's outback is so large, rural settlements are often far apart.

Ethnic Groups and Languages

People of European descent, mostly British and Irish, make up more than 90 percent of Australia's population. The population, though, is becoming increasingly diverse. Australia now receives more immigrants from Asia than from Europe. In addition, the Aborigine population has grown to 400,000. English is Australia's official language, but many Aboriginal languages are still spoken. Australian English is known for its unique words such as *barbie,* which means barbecue, and *G'day,* which is used as a greeting.

New Zealand's people are not as diverse as those of Australia. About 75 percent of the people are of European, mostly British, heritage. The Maori make up 13 percent of the population, and Asians and Pacific Islanders make up another 10 percent. Diversity is increasing as the Pacific Islander population grows. Nearly all New Zealanders speak English, although Maori is recognized as a second official language.

Melanesians, Micronesians, and Polynesians are the three large ethnic groups of Oceania. All three consist of many smaller groups. Altogether, the diverse peoples of Oceania speak more than 1,200 languages—as many as 700 languages in Papua New Guinea alone. Many Papuans speak a **pidgin** (PIH·juhn) **language** formed by combining parts of several different languages.

Most islands in Oceania also have small populations of European descent. The largest of the groups is in French Polynesia, where Europeans make up more than one-third of the population.

Reading Check **Describing** Describe Australia's urban/rural population pattern.

Melbourne, Australia

NATIONAL GEOGRAPHIC

Residents in Australia's large cities enjoy many conveniences, such as Melbourne's library, visible behind the sculpture in the foreground. ***Location*** **What is the largest city in Oceania outside of New Zealand?**

Garry Disher

(1949–)

Australian author Garry Disher grew up on a wheat and sheep farm that his family owned for generations. His stories capture the experience of living and working in north central Australia. Disher has won several awards for his work, including Australia's Children's Book of the Year Award in 1993 for *The Bamboo Flute.*

Background Information

The Bamboo Flute takes place in 1932. At that time, Australia was struggling with economic problems and a drought. Life was especially difficult for farmers and ranchers.

Reader's Dictionary

quarreling: arguing

paddock: an enclosed area for livestock

sodden: soaked through

THE BAMBOO FLUTE

By Garry Disher

There was once music in our lives, but I can feel it slipping away. Men are tramping the dusty roads, asking for work, a sandwich, a cup of tea. My father is bitter and my mother is sad. I have no brothers, no sisters, no after-school friends. The days are long. No one has time for music.

That's why I dream it.

I'm dreaming it now.

I'm dreaming a violin note, threading it through the **quarreling** cries of the dawn birds outside my window. . . .

The soft light of dawn is leaking through the gaps in the curtain. Five o'clock. I swing my feet on to the floor, drag on my clothes and boots, and leave the house.

My father is waiting for me at the **paddock** gate, his forearms on the top rail. . . .

He opens the gate and I follow him into the paddock. The grass is dewy this morning. Soon my boots feel lumpy and **sodden.**

The cows are in the farthest corner, of course. I watch my father tip back his throat and let out a whistle. . . .

Sometimes the cows respond, sometimes they don't. This time they don't, so we set out to fetch them. . . .

Milking is just one of the tasks in my endless day. After breakfast, there's a one-hour walk to school. Lessons, lunch, lessons—and rapped [tapped] knuckles when I fall asleep at my desk. Then the long walk home. Chop the firewood. Collect the eggs. Weed the vegetable patch. Homework. Teatime. Bedtime.

So I dream. Who wouldn't?

Analyzing Literature

1. **Making Inferences** Why does the boy say he can feel the music "slipping away"?
2. **Read to Write** Rewrite the passage above from the perspective of the father. Describe his thoughts and feelings.

Culture and Daily Life

Main Idea **Lifestyles in Australia and Oceania primarily reflect European and Pacific cultures.**

Geography and You Do the people in your community have similar backgrounds, or are they different? Read to find out about the culture and daily life in Australia and Oceania.

For hundreds of years, European and Pacific traditions have influenced the cultures of Australia and Oceania. Recently, American and Asian influences have grown. All of these groups have had an impact on the religion, arts, and daily life in Australia and Oceania.

Religions

Christianity is the most widely practiced religion in Australia and Oceania today. It was brought by Europeans during the 1700s and 1800s and attracted many followers among local peoples. However, traditional religions are also still practiced in some areas. In Australia, for example, the Aborigines believe in the idea of Dreamtime, the time long ago when they say wandering spirits created the world. They believe that all natural things—rocks, trees, plants, animals, and humans—have a spirit and are related to one another.

The Arts

The arts in Australia draw from the past as well as the present. The Aborigines created paintings on rocks to tell about the relationship of humans to nature. Australian painters of European descent looked to the Australian landscape for inspiration. Australia's writers and filmmakers have used local themes in many of their works.

Much of New Zealand's art is based on Maori culture. Maori artisans are skilled in canoe making, weaving, and wood carving. Although the Maori language is now written, storytellers still pass on the history and myths of long ago. The Maori also use songs and chants to tell stories. In the 1900s, they developed a new type of music called **action songs.** This art form blends traditional dance with modern music.

The spirited and graceful dances of Oceania are an **integral,** or necessary, part of important events. Pacific Islanders also use storytelling to pass on knowledge of their cultures. The stories often are told through the movements of dancers.

Daily Life

Modern and traditional ways both influence daily life in Australia and Oceania. People having European backgrounds typically live in nuclear families.

▲ Women from one of New Guinea's many ethnic groups perform a traditional ceremonial dance.
Place **Why is dancing important in Pacific Island cultures?**

Aborigines, Maori, and Pacific Islanders stress the extended family. Maori households often include relatives from three or four **generations,** or groups of people about the same age, living together. Males have traditionally been the head of the family in most societies, but women also head Maori families and some island groups.

One-floor brick or wood houses with tiled roofs are common in Australia. Many New Zealanders live in timber houses with porches or in stone cottages. City residents in both countries typically live in Western-style apartments or small houses. Traditional homes in Oceania have thatched or tin roofs that are held up by posts. Many homes, like the Samoan ***fale,*** have open sides that allow cooling ocean breezes to circulate. Blinds made of coconut palm leaves are lowered for privacy.

Meat is a major part of the Australian and New Zealand diet. Typical meals include lamb, beef, fish, or pork served with vegetables, bread, and fruit. Many people in Australia and New Zealand like to cook outdoors. In recent decades, American-style fast foods have gained popularity. People in Oceania eat a variety of foods including fish, pork, yams, taro, breadfruit, and fruit. Taro is a plant that grows a tuber, or fleshy bulb, that Pacific Islanders mash into a paste called **poi.**

Because of the generally pleasant weather, outdoor sports are popular. People swim and surf in the ocean, and scuba divers explore the area's many colorful coral reefs. Boat racing is a favorite sport in Oceania, and skiing and mountain climbing are popular in New Zealand and Australia. Rugby and cricket, brought by British settlers, are also popular in Australia and New Zealand.

Reading Check **Comparing** Compare family life among the region's people.

Section 2 Review

Social Studies ONLINE
Study Central™ To review this section, go to glencoe.com.

Vocabulary

1. **Explain** the significance of:
   - **a.** bush
   - **b.** station
   - **c.** pidgin language
   - **d.** action song
   - **e.** *fale*
   - **f.** poi

Main Ideas

2. **Identifying** What are the major languages of Australia and New Zealand?
3. **Categorizing** Use a chart like the one below to identify and describe the arts in each area.

Country	Arts
Australia	
New Zealand	
Oceania	

Critical Thinking

4. **Comparing** Compare the population growth of Australia with that of Oceania.
5. BIG Idea Why is Christianity the main religion in Australia, New Zealand, and Oceania?
6. **Challenge** How are European influences reflected in the lifestyles of the people in Australia and New Zealand?

Writing About Geography

7. **Persuasive Writing** Write an advertisement that the government of Australia could use to attract immigrants. Be sure to describe the benefits of moving to Australia.

Visual Summary

First Settlers

- Hunters from Southeast Asia settled Australia about 40,000 years ago.
- Pacific Islanders developed sailing skills that helped them travel to faraway islands.
- The Aborigines of Australia and the Maori of New Zealand developed complex cultures.

The European Era

- Europeans explored and settled Australia, New Zealand, and Oceania from the 1500s to the 1800s.
- Disease and warfare caused Aborigine and Maori populations to decline.
- Western nations colonized Pacific Ocean islands.

Maoris signing the Treaty of Waitangi

Independent Nations

- Australia and New Zealand gained independence in the early 1900s.
- Australia and New Zealand are parliamentary democracies.
- Most territories in Oceania gained independence after World War II.
- A number of countries signed an agreement to share Antarctica for scientific research.

Scientist, Antarctica

People

- Most people in Australia and New Zealand live in urban areas.
- Oceania has few large urban areas; most people live in small villages.
- Since World War II, the populations of Australia and New Zealand have become more ethnically diverse.
- Populations in Oceania are growing faster than in Australia and New Zealand.

Sydney, Australia

Culture

- Western culture has influenced the religion and art of the region.
- Homes, clothing, and activities reflect the region's generally pleasant climate.
- Modern and traditional ways influence daily life in Australia and Oceania.

Family, Papua New Guinea

Study anywhere, anytime! Download quizzes and flash cards to your PDA from **glencoe.com**.

CHAPTER 29 ASSESSMENT

STANDARDIZED TEST PRACTICE

To understand questions better, rewrite them in your own words if you have time. Be careful not to change the meaning.

Reviewing Vocabulary

Directions: Choose the word(s) that best completes the sentence.

1. Australian Aborigines developed a hunting weapon called a ________.

A *fale*
B Maori
C boomerang
D bamboo

2. A ________ is an area temporarily placed under control of another country.

A colony
B trust territory
C commonwealth
D sovereignty

3. Sparsely populated rural areas in Australia are referred to as the ________.

A bush
B barbie
C *fale*
D station

4. Samoan homes called ________ have open sides to allow cooling ocean breezes to circulate.

A stations
B pois
C pidgins
D *fales*

Reviewing Main Ideas

Directions: Choose the best answer for each question.

Section 1 ***(pp. 792–797)***

5. The first people to settle Australia came from ________ over 40,000 years ago.

A New Zealand
B Southeast Asia
C South America
D Europe

6. Australia and New Zealand both won their independence peacefully from ________.

A China
B Japan
C Britain
D Germany

Section 2 ***(pp. 799–804)***

7. The greatest diversity of languages in the region is found in ________.

A Antarctica
B New Zealand
C Australia
D Oceania

8. Traditional Maori and Pacific Islander societies used ________ to tell stories.

A Dreamtime
B food
C songs
D religion

GO ON

Critical Thinking

Directions: Base your answers to questions 9 and 10 on the chart below and your knowledge of Chapter 29.

Per capita GDP in Australia, United States, and Argentina (1990 international dollars)			
	Australia	United States	Argentina
1870	3,641	2,457	1,311
1890	4,433	3,396	2,152
1950	7,493	9,561	4,987
1998	20,390	27,331	9,219
Source: http://eh.net/encyclopedia/article/attard.australia			

9. When did per capita GDP in the United States first exceed that of Australia?

A between 1870 and 1890

B by 1950

C prior to 1890

D after 1998

10. Based on the table, which of the following statements is accurate?

A Argentina has had a weak economy since 1870.

B Australia had many more workers than the United States.

C In 1870 and 1890, Australians produced more per person than did Americans.

D Americans have produced more than both Australians and Argentinians since 1870.

Document-Based Questions

Directions: Analyze the document and answer the short-answer questions that follow.

When Captain Cook landed in Tahiti in 1769, he established some rules for contact with the people of the island. Three of the rules were:

> *1. To [attempt] by every fair means to Cultivate a Friendship with the Natives, and to treat them with all imaginable humanity.*
>
> *2. A Proper Person or Persons will be appointed to Trade with the Natives for all manner of Provisions, Fruits, and other Productions of the Earth; and no Officer or Seaman or other person belonging to the Ship, excepting such as are so appointed, shall Trade or offer to Trade for any sort of Provisions, Fruit or other Productions of the Earth, unless they have my leave so to do.*
>
> *5. No sort of Iron or anything that is made of Iron, or any sort of Cloth or other useful or necessary Articles, are to be given in Exchange for anything but Provisions.*
>
> —Captain Cook's Journal, First Voyage

11. Judging by the rules, what kind of relationship did Captain Cook want to establish with the people of Tahiti?

12. What kind of goods did Captain Cook want to acquire from the Tahitians in trade?

Extended Response

13. Imagine that you are a European traveling to Australia, New Zealand, or Oceania in the mid-1800s to start a new life. Where would you decide to settle? Why? What difficulties might you face?

STOP

Social Studies ONLINE

For additional test practice, use Self-Check Quizzes—Chapter 29 at glencoe.com.

Need Extra Help?

If you missed question...	1	2	3	4	5	6	7	8	9	10	11	12	13
Go to page...	793	797	801	804	793	796	801	803	796	796	794	794	794

"Hello! My name is Caitlin.

I'm 16 years old, and I live in the village of Agana Heights, Guam. The island of Guam is a U.S. territory, so people here are U.S. citizens. But we also have our own history and culture. Here's how I spend a typical day."

5:00 a.m. I wake up and shower. Then I dress in jeans and a light T-shirt (it is warm and humid here all year-round).

5:30 a.m. I eat breakfast with my mother, stepfather, and half sister, Tiyana. We have eggs and a kind of meat that is similar to ham. As we eat, we talk about our plans and schedules for the day.

6:15 a.m. I take a 15-minute walk to the bus stop and board the school bus. My school is only 10 miles from my house, but the bus makes a lot of stops. The trip seems to take forever!

7:15 a.m. I arrive at George Washington High School. It is one of Guam's four public high schools and has more than 2,700 students. Our school mascot is a gecko, which is a cool lizard that lives on the island.

7:50 a.m. I start the school day with art, my favorite class. After that, I move on to algebra and Spanish.

10:20 a.m. I go to my Chamorro class. The Chamorros were the original inhabitants of Guam, and many people here are of Chamorro descent. (Not me! My family moved to Guam from another Pacific island called Palau.) In this class, we study the Chamorro culture and language. We even learn Chamorro dances.

12:00 p.m. It is time for lunch. Today, a student rock band is playing outside. I grab some chips and enjoy the concert with my friends.

12:50 p.m. Classes resume. This afternoon, I have social studies and science. In science, we are learning how the principle of lift helps airplanes fly. I am fascinated because I want to be a flight attendant when I am older.

2:40 p.m. I ride the bus back to my stop and walk home. On my walk, I pass a fort that was built in the 1700s, when Spain ruled Guam. Antique cannons still point to the ocean.

4:15 p.m. It's time for chores. Usually I clean and vacuum the house, but today I have been asked to help my Aunt Larie rake her yard. Tomorrow she is having the whole family over for a traditional Palauan first birth ceremony. (My cousin Crystal just had her first baby!) I head to Larie's house, which is nearby.

6:00 p.m. I return home. I had a good time at my aunt's house. When we finished the yard work, she taught me a Palauan dance. It is very different from the Chamorro dances I learned at school.

6:15 p.m. Tiyana, my stepfather, and I have dinner together. We all help prepare *kadu*, a chicken soup with cabbage and other vegetables. My mom works as a cook in a restaurant, so she isn't home.

6:45 p.m. I study and do my homework. I take a short break to go to the local ball field and watch my 8-year-old cousin Asri play baseball.

9:00 p.m. I get ready for bed. Tomorrow will be another early day!

ILLUSTRATIONS BY BOOKMAPMAN

LUNCH BREAK Caitlin meets up with her friends during lunch period.

LET'S DANCE Caitlin and some of her family celebrate a cousin's birth with a Palauan dance.

FISH STORY This wood carving, called a storyboard, tells a myth about the sea.

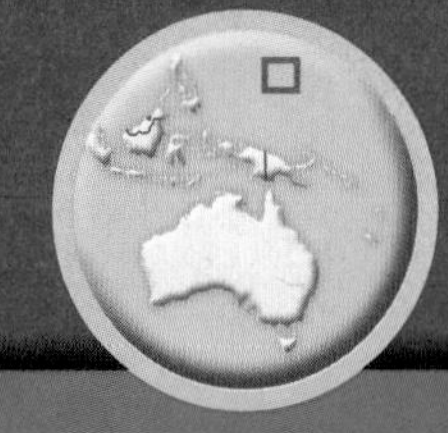

FLOWERS IN HER HAIR Caitlin Kesewaol (KATE·lin KEZ·wuhl) wears a tropical flower in her hair as she gets ready to take part in a family celebration. Like most teens on Guam, Caitlin usually dresses in jeans or shorts.

JEREMY NICHOLL / POLARIS (4)

What's Popular in Guam

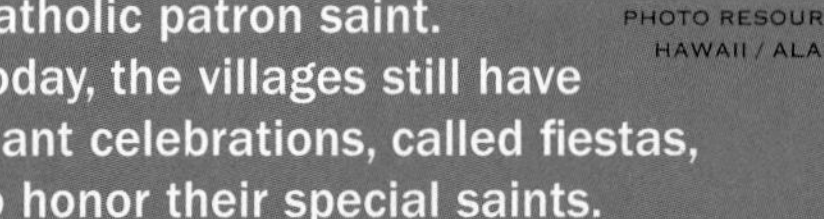

PHOTO RESOURCE HAWAII / ALAMY

Fiestas During Spanish rule, many Guamanians became Catholic, and each village adopted a Catholic patron saint. Today, the villages still have giant celebrations, called fiestas, to honor their special saints.

Voting Guam has a higher voter turnout rate than most other U.S. territories and states. Citizens here vote for their own delegate to the U.S. House of Representatives, but they cannot vote for president.

Betel nuts Many Guamanians chew these hard red nuts instead of bubble gum. The nuts are often sprinkled with lime. The tradition is passed from grandparents to grandchildren.

SUSAN LIEBOLD

Say It in Chamorro

Guam has two official languages, English and Chamorro. Chamorro has been spoken on Guam for thousands of years. Because the island was ruled by Spain for three centuries, many Spanish words made their way into the Chamorro language. Try these everyday Chamorro expressions.

Hello ***Hafa adai***
(HAH·fuh day)

Goodbye ***Adios***
(ah·dee·OHS)

My name is _____.
Si _____ yu.
(see _____ dzu.)

RUBBERBALL/PUNCHSTOCK

CHAPTER 30

Australia, Oceania, and Antarctica Today

Essential Question

Human-Environment Interaction The lands of Australia, Oceania, and Antarctica range from tiny islands to massive continents. Some places in this region have environments too harsh for people to live there permanently. Others have attractive climates but few resources. How might people survive in a land with limited resources?

Victoria, Australia

Chapter Audio

Section 1: Australia and New Zealand

BIG IDEA People's actions can change the physical environment. Extensive farming and ranching, along with other agricultural and economic practices, have affected Australia and New Zealand.

Section 2: Oceania

BIG IDEA Patterns of economic activities result in global interdependence. Many of Oceania's islands have limited resources and depend on tourism or aid from other countries to support their economies.

Section 3: Antarctica

BIG IDEA All living things are dependent upon one another and their surroundings for survival. Scientists fear that human activity may be harming plant and animal life in Antarctica.

FOLDABLES™ Study Organizer

Summarizing Information Make this Foldable to help you summarize information about Australia, Oceania, and Antarctica today.

Step 1 Place three sheets of paper on top of one another about 1 inch apart.

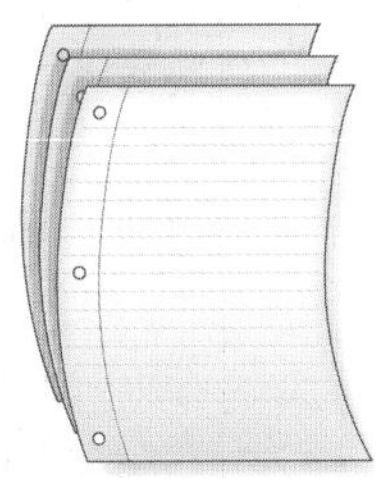

Step 2 Fold the papers to form 6 equal tabs.

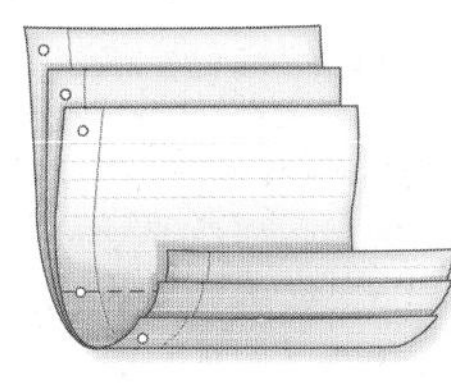

Step 3 Staple the sheets, and label each tab as shown.

Antarctica
Micronesia and Polynesia
Melanesia
New Zealand
Australia
Australia, Oceania, and Antarctica Today

Reading and Writing After you have finished taking notes in your Foldable, write a short essay describing the challenges faced by each area of the region.

Social Studies ONLINE

Visit glencoe.com and enter QuickPass™ code EOW3109c30 for Chapter 30 resources.

Guide to Reading

BIG Idea

People's actions can change the physical environment.

Content Vocabulary

- lawsuit *(p. 813)*
- merino *(p. 814)*
- kiwifruit *(p. 816)*

Academic Vocabulary

- consist *(p. 815)*
- acknowledge *(p. 815)*

Reading Strategy

Comparing and Contrasting

Use a Venn diagram like the one below to compare the economies of Australia and New Zealand.

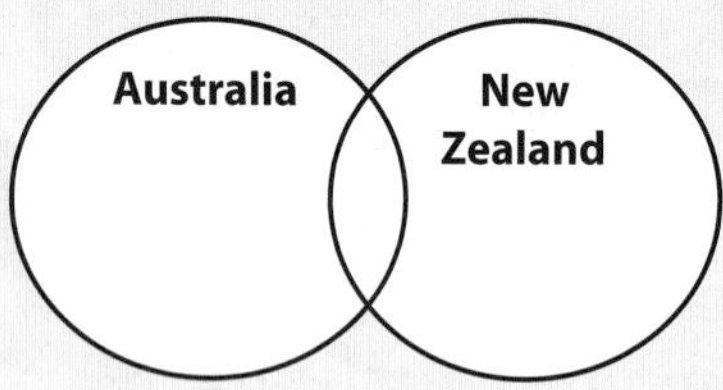

Australia and New Zealand

 Section Audio  Spotlight Video

Picture This 2,224 . . . 2,225 . . . 2,226 . . . People count the sheep running down the main street of Te Kuiti, New Zealand. The "Running of the Sheep" celebrates the New Zealand Shearing Championships that are held in Te Kuiti. Each year, the competition draws hundreds of sheep shearers to Te Kuiti, which is known as the "Shearing Capital of the World." You can learn more about Australia and New Zealand today as you read the following section.

▼ Celebrating the sheep in Te Kuiti, New Zealand

Australia

Main Idea **Australia has a strong economy, but economic growth has created serious challenges for its environment.**

Geography and You Do you think introducing a new species of toad to an area could harm the environment there? Read to find out about some of the environmental challenges Australians face.

In land area, Australia is the largest country in the region. It also is the richest in mineral resources. As a result of this wealth, Australia's people have built one of the most productive economies in the world. At the same time, Australians struggle with issues such as Aborigine land rights and protecting their environment.

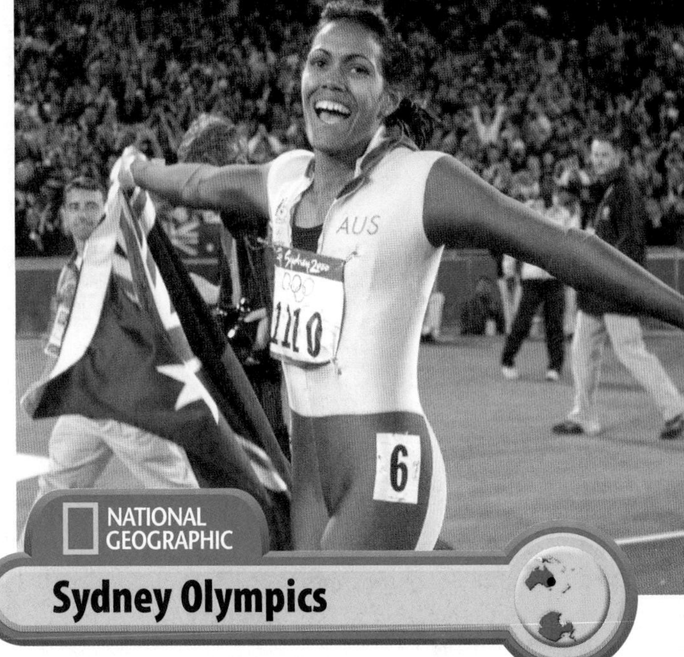

Sydney Olympics

Cathy Freeman, an Aborigine, competed in track for the Australian team at the 2000 Olympics, which were held in Sydney. ***Place*** **How have conditions for Australia's Aborigines changed?**

Australia's People

Despite its huge area, Australia has only 20.6 million people. The country has long needed skilled workers to develop resources and build its economy. Thus, the government has encouraged immigration.

At first, most immigrants came from European countries, especially the British Isles. Since the 1970s, the government has enacted programs to attract people from other regions. Today, immigrants come to Australia from various parts of Asia, South Africa, Latin America, and Oce–ania. Still, most Australians are of European descent.

One of the major challenges facing Australians today involves the Aborigines, the first people to settle Australia. For years, Aborigines suffered discrimination from white Australians. Recently the government has tried to improve conditions for the Aborigines, but problems still exist. For example, Aborigines tend to receive less education than white Australians. They also tend to work in lower-paying jobs and suffer more from poverty and poor health care.

In recent years, Aborigines have pushed more forcefully for their rights. In the late 1980s, a group of Aborigines tried to block mining on land they said belonged to their people. To do this, they filed a **lawsuit,** or legal action in court intended to address a problem. In 1992 a court ruled that the Aborigines controlled the land and had the right to request that mining be stopped. Later decisions in other cases extended the Aborigines' control over land that was being used for sheep ranches and other economic activities.

These legal decisions opened the possibility that Aborigines could claim much of the country's land. Other groups of Australians are worried that they might lose land to such Aborigine claims.

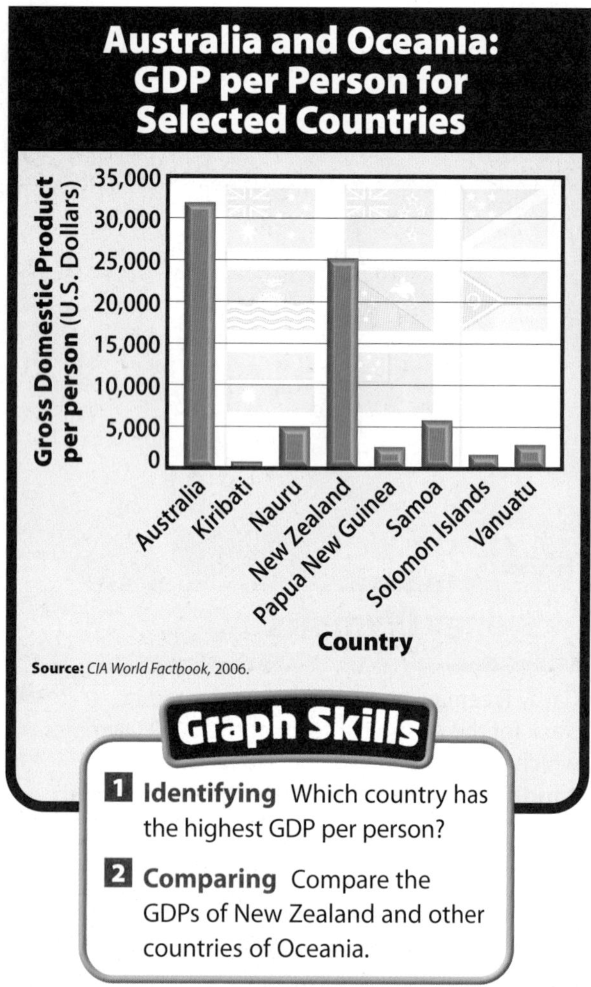

The government is trying to find a way to balance the claims of Aborigines and other landowners in Australia.

Australia's Economy

Australia's prosperous economy is partly based on the export of mineral and energy resources. These riches include iron ore, nickel, zinc, bauxite, gold, diamonds, coal, oil, and natural gas. China and Japan purchase large amounts of these Australian resources. The mining industry holds great promise for Australia's future growth.

Australia's dry climate and poor soils limit farming. Irrigation, however, allows farmers to grow grains, sugarcane, cotton, fruits, and vegetables. The main agricultural activity, though, is raising cattle and sheep. Australia is a major exporter of wool, lamb, beef, and cattle hides. Many sheep raised in the country are **merinos,** a breed of sheep known for its fine wool.

In the mid-1900s, manufacturing became an important part of Australia's economy. Many factories there produce processed foods, transportation equipment, cloth, and chemicals. High-technology industries, service industries, and tourism also play a large role in the economy. Most of Australia's industries are located near the cities of Sydney and Melbourne.

Environmental Challenges

Economic activities have damaged Australia's lands. Many trees have been cut down to provide more grazing land for sheep. These actions, along with overgrazing by large herds of animals, have removed plants and grasses that hold the soil in place. As a result, winds have blown away much of the topsoil.

In addition, Australia's unusual wildlife has been threatened by animals that settlers have brought from other areas. For example, Hawaiian cane toads were introduced to eat insects that damaged sugarcane crops. Unfortunately, the toads did not eat the insects. Their skin, however, contains poisons that kill other animals that eat the toads.

During the 1980s, more Australians grew concerned about their country's environment. As a result, Australians are now taking steps to try to preserve their land. Some people fear that these efforts are too extreme and will hurt the economy. Australians continue to debate how to solve their environmental problems.

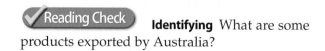

Reading Check **Identifying** What are some products exported by Australia?

New Zealand

Main Idea **New Zealand is a small country with a growing economy that is based on trade.**

Geography and You Do you like the green, fuzzy-skinned kiwifruit? Read to find out more about other products from New Zealand.

The islands of New Zealand support a growing economy. Recent immigrants who are attracted by the country's wealth have enriched the diversity of New Zealand's society.

The People of New Zealand

The population of New Zealand, like Australia, **consists** largely of the descendants of European immigrants. People of British and Irish descent make up the largest European groups. There are also people of German, Scandinavian, Croatian, and Dutch backgrounds.

The Maori, the first people to settle New Zealand, are the largest non-European group. They form about 15 percent of New Zealand's population. In 1840 Maori leaders signed the Treaty of Waitangi with Great Britain. This treaty **acknowledged,** or recognized, British rule over the islands. At the same time, the British promised to protect Maori land rights.

The Treaty of Waitangi has now become the basis for Maori claims to land in New Zealand. Some Maori have charged that, since 1840, Europeans have unfairly taken land from them. The Maori have won a number of lawsuits recognizing their right to land. As in Australia, some people of European descent fear that these lawsuits will cause them to lose their land and livelihoods.

Today New Zealand's population is changing. Many Pacific Islanders are moving to New Zealand to find work. In addition, New Zealand is attracting people from East Asia and Southeast Asia. While the population growth rate among these groups and the Maori is high, the growth rate among whites is low. It appears likely that in the future, the ethnic balance of the country will change.

The Economy of New Zealand

As in Australia, agriculture is important to New Zealand. New Zealand has millions of sheep—far more than it has people. The export of wool and meat has long been a major factor in the country's economy. New Zealand's cattle industry produces butter, cheese, and meat exports.

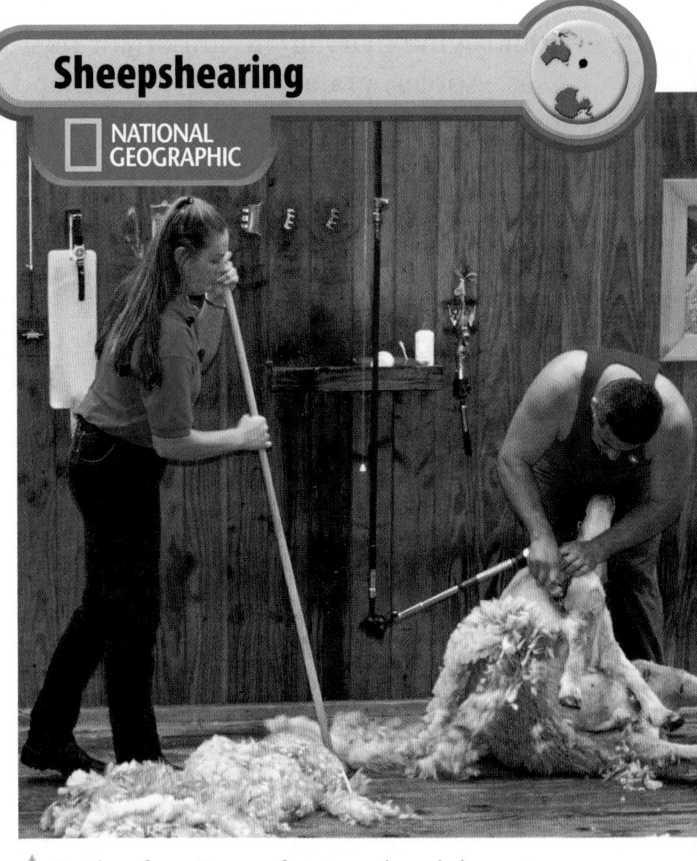

▲ Settlers from Europe first introduced sheep to New Zealand in the 1700s. Today New Zealand is the second-largest exporter of wool in the world. ***Place*** **What exports does New Zealand's cattle industry produce?**

Kiwifruit came to New Zealand from China in the early 1900s. Kiwi are high in vitamins and nutrients. ***Human-Environment Interaction*** **What other crops are grown in New Zealand?**

New Zealand recently has begun expanding its economy. Forests along the country's mountains are now being used to produce wood and paper products. Farming and wine making are important businesses. Apples, grapes, barley, wheat, and corn are the major crops. **Kiwifruit,** a small oval fruit with a brownish-green fuzzy skin, is another major agricultural product. You may have seen kiwi in your grocery store.

New Zealand's resources include gold, coal, and natural gas. New Zealand also uses hydroelectric and geothermal energy to produce electricity. Its main manufactured items include wood products, fertilizer, wool products, shoes, machinery, and vehicles. Service industries and tourism also are important.

Because New Zealand is a relatively small country, trade with other countries is a major part of its economy. In the past, New Zealand traded mostly with the United Kingdom and nearby Australia. Australia remains an important trading partner. The United Kingdom, however, has become a lesser partner as trade with the United States and East Asia has risen.

Explaining How is New Zealand's population changing?

Social Studies ONLINE
Study Central™ To review this section, go to glencoe.com.

Vocabulary

1. **Explain** the meaning of *lawsuit, merino,* and *kiwifruit* by using each term in a sentence.

Main Ideas

2. **Identifying** Use a diagram like the one below to identify Australia's mineral and energy resources.

3. **Describing** Describe how New Zealand's trading partners have changed over the years.

Critical Thinking

4. **Analyzing** How are Australia's and New Zealand's economies suited to their environments?
5. **BIG Idea** What economic activities pose challenges to the environment in Australia?
6. **Challenge** Do you think Australia and New Zealand can resolve the Aborigine and Maori land claims? Explain.

Writing About Geography

7. **Persuasive Writing** Imagine that you work for an Australian government agency charged with protecting the environment. Write a memo to the head of the agency outlining how to solve the country's environmental challenges.

TIME PERSPECTIVES

EXPLORING WORLD ISSUES

SAVING THEIR TREASURES

After years of neglect, New Zealand and Australia are working to preserve important cultures and natural resources.

A Maori and a New Zealander of European ancestry exchange a traditional Maori greeting.

New Zealand and Australia are nations with different ethnic groups—each group with its own rich culture and traditions. For nearly two centuries, European settlers and their descendants and local people—the Maori in New Zealand and the Aborigines in Australia—have lived side by side. But the groups have often struggled to accept and respect their differences.

Today, the two countries are working to preserve the cultures of each ethnic group and to work out their differences. Schools are teaching the language and traditions of local cultures. Governments are creating jobs for people. In addition to safeguarding cultural resources, Australia is working to save natural resources, such as the Great Barrier Reef. All threatened resources are treasures worth preserving.

TIME PERSPECTIVES

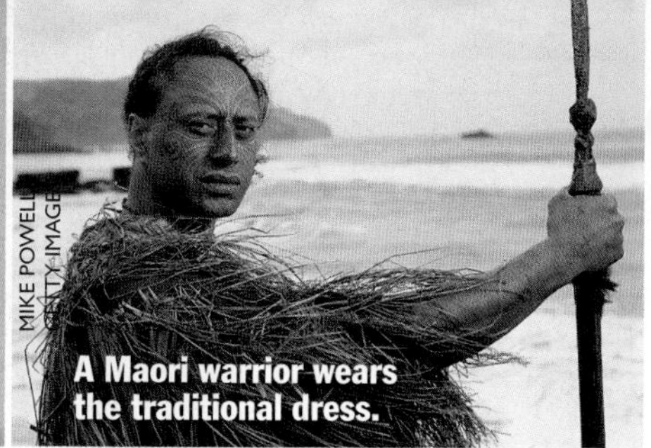

A Maori warrior wears the traditional dress.

A LONG STRUGGLE FOR RESPECT

For Angeline Greensill, the town of Raglan on New Zealand's North Island isn't just the place where she was born. The coastal town is also a link to her ancestors. She believes it is the home of Maori spirits who live on its green hills and sandy beaches. Greensill is a **Maori**, a native New Zealander. She and her family are members of the Tainui tribe, one of dozens of Maori ***iwis***, or tribes.

The Maori were the first people to reach New Zealand. Beginning around A.D. 600, they arrived in big canoes from islands in the Pacific Ocean and built a rich culture. They were primarily warriors who expressed themselves through songs, woodcarving, and tattooing. The Maori felt a deep connection to nature, their ancestors, and the land. "The earth is our mother," Greensill explained. "When we die, we go back to it."

A Battle to Survive

For nearly two hundred years, the Maori have been struggling to hold on to their culture and sacred land. In the 1820s, settlers from Great Britain began arriving in New Zealand in large numbers. Most settlers did not value Maori culture or their land rights. They moved onto Maori land and paid little or nothing for it. The Maori who tried to protect their land were often forced off it.

Most British were **ethnocentric**, or convinced their way of life was better than that of any other group. They believed the Maori would be better off if they gave up their traditional ways.

Maori Tribal Lands

About 200 years ago, New Zealand was home to dozens of *iwis*, or tribes. The map shows where the largest tribes were located.

INTERPRETING MAPS

Identifying Which tribe settled the farthest north on North Island?

Children learn the Maori language at a school in New Zealand.

Protestors at a rally demand fair treatment for the Maori.

Keisha Castle-Hughes stars in *Whale Rider*, a movie about a Maori girl.

PAULA SOUDA/CORBIS NEIL RABINOWITZ/CORBIS NEW ZEALAND FILM COMM./THE KOBAL COLLECTION

Broken Promises

In 1840 the colonial British and Maori tribal chiefs signed the **Treaty of Waitangi**. The treaty became New Zealand's founding document, much like the U.S. Declaration of Independence. In return for the right to rule New Zealand, the British promised to protect Maori land rights.

Over the next 150 years, however, the Maori lost control of most of their territory. Maori tribal chiefs sold some of it, but much of the land was taken by the government or illegally fenced off by British farmers. "The Maori started with 66 million acres," Greensill said. "By 1975, we had about 4 million."

Australia's Similar Mistakes

Australia's treatment of its original people, the **Aborigines**, resembles the treatment of the Maori in New Zealand. For 40,000 years the Aborigines lived throughout Australia. In 1788 British settlers arrived and immediately began to drive the Aborigines off their sacred tribal land. Many Aborigines who resisted the British were killed.

Righting Wrongs

How do you solve problems that began centuries ago? In recent years, New Zealand's government has been working to "close the gap" between its Maori citizens and those of European ancestry. To help keep Maori culture alive, schools are teaching the Maori language and traditions. The government is also working to provide the Maori with adequate jobs, health care, and housing. Since 2004, the Maori political party has represented the tribes in New Zealand's parliament.

The land ownership issue, however, has been difficult to resolve. The government cannot return land to the Maori without affecting the people who currently live on it. To address the land issues, the Waitangi Tribunal was established in 1975 to hear Maori land rights grievances. By June 2005, more than 1,200 Maori claims of land ownership had been registered with the tribunal. Some *iwis* have had their land returned, whereas others have been paid for land that they were forced to give up.

There is still a lot of work to be done, but by 2006, the gap between New Zealand's ethnic groups seemed a little smaller. "The future is looking good," Greensill said. "Maori people have a sense of awakening."

EXPLORING THE ISSUE

1. **Summarizing** What happened after the Treaty of Waitangi was signed?
2. **Making Inferences** How might the Waitangi Tribunal and the Maori political party help New Zealand's government create fair land rights policies?

SAVING AUSTRALIAN TREASURES

Ancient cultures are not the only endangered treasures in the South Pacific. In Australia, global warming is threatening a fragile ecosystem. The country's Great Barrier Reef is a chain of 2,900 coral reefs that stretches 1,240 miles (1,995 km) along the east coast. The giant ocean reef is home to thousands of fish, plants, and other marine life.

In recent years, some scientists believe that global warming is threatening to destroy the reef. As global warming heats the ocean's surface, a deadly situation called coral bleaching occurs. Coral contains tiny algae, or water plants, that give coral its vibrant color. Coral also uses the algae to create its food. At high temperatures, though, coral releases the algae. Without it, coral loses its beautiful color and dies. When the coral dies, so do many of the animals and plants that live in the reef. To save the aquatic ecosystem, Australia's government is studying ways to help the reef adapt to warmer waters.

REUTERS/WILL BURGESS

An animal rights activist protests the killing of kangaroos.

Human activity and pollution from fishermen and tourists have also done enormous damage to the reef. To decrease their impact, "no-take zones" were set up in 2004. No fishing or coral collecting is allowed in the zones.

Kangaroos Everywhere

Another **icon**, or symbol, of Australia is in danger of being damaged by human activity: the kangaroo. With 50 million of the pouched marsupials bouncing around, Australia's kangaroo population is immense. As they search for food, kangaroos often destroy farmland and cause car crashes. To control the kangaroo population, Australia's government permits a certain amount to be hunted.

Animal rights groups, however, are working to protect the kangaroos. They propose establishing safe places where tourists can view kangaroos in their natural habitat. "We want to promote kangaroos as part of the tourism industry," said one activist.

EXPLORING THE ISSUE

1. **Determining Cause and Effect** How might establishing "no-take zones" help preserve Australia's Great Barrier Reef?
2. **Explaining** Why might the creation of "safe places" for kangaroos help increase tourism in parts of Australia?

REVIEW AND ASSESS

Tourists wade in the waters at the Great Barrier Reef.

UNDERSTANDING THE ISSUE

1 **Making Connections** How did the Treaty of Waitangi fail to protect the rights of New Zealand's Maori tribes? How did British settlers' ethnocentric attitudes affect their relations with the Maori?

2 **Writing to Inform** Write a short article describing how Maori tribes lost their land to British settlers. Be sure to include the history of the Treaty of Waitangi.

3 **Writing to Persuade** Write a paragraph that starts with this sentence: Preserving the Great Barrier Reef is important because . . .

INTERNET RESEARCH ACTIVITIES

4 With your teacher's help, use Internet resources to learn more about Maori culture. Read about the history of the Maori language and the purpose of the Maori Language Commission. How important is language to a culture's survival? Write a short essay answering that question, using facts you find in your search.

5 Go to the Web site of the Great Barrier Reef Marine Park, www.gbrmpa.gov.au/. Click on the "Conservation, Heritage and Indigenous Partnerships" link. Read about the work being done to protect the many species and habitats found on the reef. Present your findings to your classmates.

BEYOND THE CLASSROOM

6 **Work in groups** to develop ways young people can learn about the culture and traditions of ethnic groups in your community. Write your suggestions on a poster. Include the names of social organizations that teach groups how to overcome the differences that exist between them. Bring the poster to class and display it.

7 **Visit your school or local library** to find books and articles on the impact of global warming on the Earth. Besides the Great Barrier Reef, research other ecosystems and species that are threatened by climate change. Discuss your findings with your friends and classmates.

A Natural Tourist Attraction

Human activity has had a huge effect on Australia's Great Barrier Reef. Here is a look at the number of tourists who have visited the fragile ecosystem in recent years.

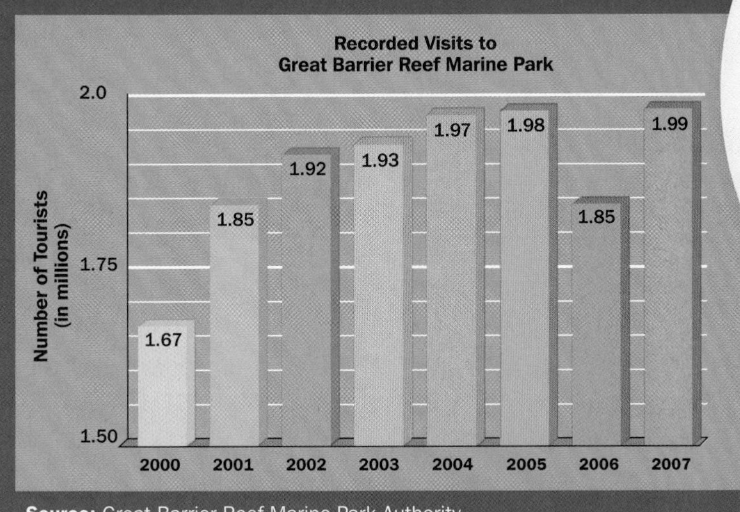

Source: Great Barrier Reef Marine Park Authority.

Building Chart Reading Skills

1. **Analyzing Information** How many more tourists visited the Great Barrier Reef Marine Park in 2007 than in 2000?
2. **Predicting** How might the increase in tourism to the Great Barrier Reef impact this fragile ecosystem?

Guide to Reading

BIG Idea

Patterns of economic activities result in global interdependence.

Content Vocabulary

- copra *(p. 823)*
- lingua franca *(p. 824)*
- *fa'a Samoa (p. 825)*
- habitat *(p. 826)*

Academic Vocabulary

- extract *(p. 823)*
- establish *(p. 825)*

Reading Strategy

Making Generalizations Use a diagram like the one below to write three important facts about the economies of Oceania in the smaller boxes. Then, in the larger box, write a generalization that you can draw from those facts.

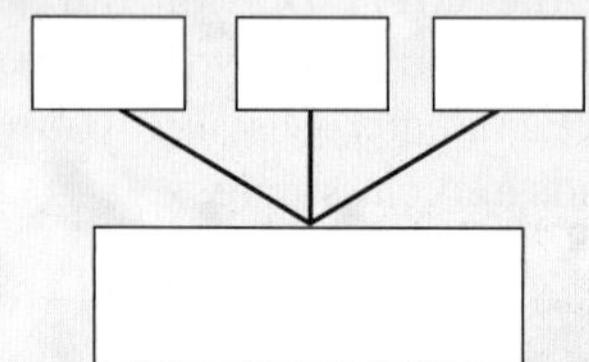

Oceania

 Section Audio  Spotlight Video

Picture This These ocean farmers use boats instead of tractors to harvest their crop of pearls in French Polynesia. Workers at the oyster laboratories select the best young oysters and prepare them to produce pearls. During "seeding," the workers place a tiny, round piece of mussel shell inside an oyster and then put it back into the water. The oyster eventually produces a material called "mother of pearl," which coats the seed and becomes a pearl. Read this section to learn more about islands in the Pacific Ocean.

▼ Pearl farm in French Polynesia

Melanesia

Main Idea **Although small in population, Melanesia includes diverse groups of people.**

Geography and You Have you ever eaten fresh coconut? It might have come from Papua New Guinea. Read to find out about Papua New Guinea and other island countries in the part of Oceania known as Melanesia.

Geographers group Oceania into three main island regions—Melanesia, Micronesia, and Polynesia. The islands of Melanesia lie across the Coral Sea from Australia. They stretch from New Guinea in the west to the Fiji Islands in the east.

Papua New Guinea

The largest and most populous country in Melanesia is Papua New Guinea (PA·pyu·wuh noo GIH·nee). It occupies the eastern half of the island of New Guinea and several hundred smaller islands. Nearly all of Papua New Guinea's people belong to different Papuan or Melanesian ethnic groups, which are closely related. More than 700 languages are spoken in the country.

Many people in Papua New Guinea live by subsistence farming. Others work on plantations that grow coffee, oil palm trees, cacao trees, and coconut palms. Coconut oil from **copra,** the meat from dried coconuts, is used to make margarine, soap, and other products.

Copra and other plantation products are produced for export. As a result, food must be imported for city dwellers. Papua New Guinea also supports its economy by **extracting,** or removing, oil, gold, copper, silver, iron, and zinc from deposits in the land and ocean floors.

Social Studies ONLINE

Student Web Activity Visit glencoe.com and complete the Chapter 30 Web Activity about the islands of Melanesia.

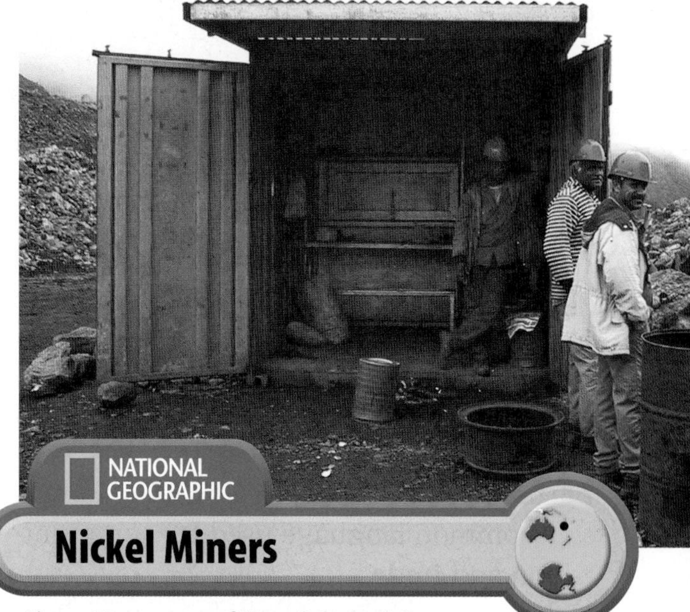

NATIONAL GEOGRAPHIC

Nickel Miners

About 20 percent of New Caledonia's population of 212,000 people work in industries such as mining. ***Place*** **What European country governs New Caledonia?**

Other Island Groups

On the other islands of Melanesia, most of the people belong to different Melanesian ethnic groups. In the Fiji Islands, however, the population is about evenly divided between Melanesians and South Asians. The ancestors of these South Asians were brought from British India in the late 1800s and early 1900s to work on sugarcane plantations. Melanesians and South Asians have struggled for control of the government. Fiji's economy has suffered from this conflict because foreign companies have been afraid to invest money there. The conflict has also hurt tourism by keeping travelers away from Fiji.

Most people in the Solomon Islands are ethnic Melanesians. They live by subsistence farming and fishing. They also tend to follow traditional ways. For example, some groups still use items such as shells or feathers as money.

Tradition is also strong in the volcanic island country of Vanuatu (VAN·WAH·TOO).

People there believe that the volcanoes hold spirits, which they honor in their religious ceremonies. Most people in Vanuatu are farmers. Tourism, however, is becoming increasingly important to the economy. There are more than 100 Melanesian languages spoken in the country. In order to communicate with each other, many people use Bislama as Vanuatu's **lingua franca**, or a common language used for communication and trade.

New Caledonia is a French-owned island territory. Rich nickel deposits provide the country's chief export. About one-third of the people are of French descent, and they control the economy. Some of New Caledonia's Melanesians want independence from France.

Reading Check **Explaining** How has ethnic conflict hurt Fiji's economy?

Drying Coconuts

NATIONAL GEOGRAPHIC

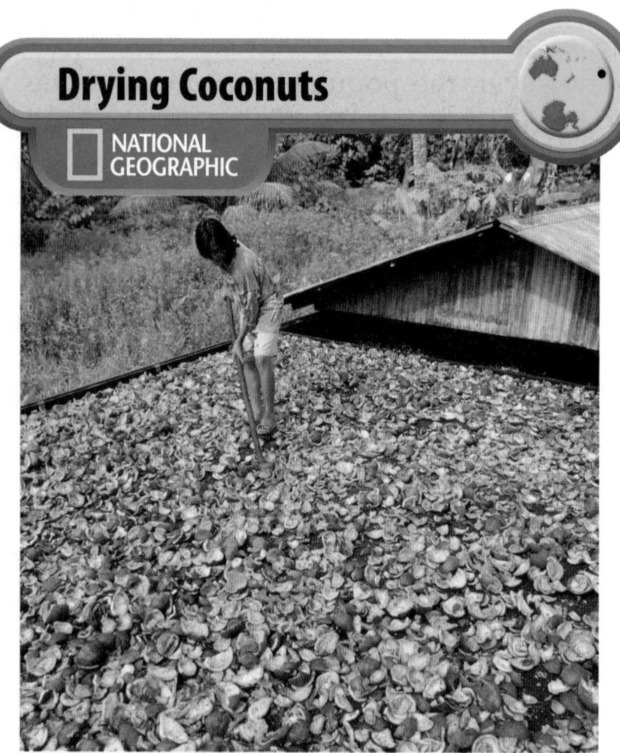

One method of drying coconut meat, or copra, is to spread it out and allow it to dry in the sun for up to a week. ***Location*** **Where is Melanesia located?**

Micronesia and Polynesia

Main Idea **Many people in Micronesia and Polynesia practice subsistence farming.**

Geography and You Have you ever eaten a juicy mango? The United States imports mangos from the Pacific region. Read to find out about other links between the Pacific Islands and the United States.

The island groups of Micronesia and Polynesia are scattered over a vast area of the Pacific Ocean. Both Micronesia and Polynesia are made up of high volcanic islands and low, ring-shaped atolls.

Micronesia

Despite their remote location in the Pacific Ocean, the islands of Micronesia have historic links to the United States. During World War II, the United States and Japan fought a number of bloody battles on the Micronesian islands. After World War II, most of Micronesia was temporarily turned over to the United States as territories.

Since the 1970s, most Micronesian islands have become independent. They include the Federated States of Micronesia, the Marshall Islands, Palau (puh·LOW), Nauru (nah·OO·roo), and Kiribati (KEE·ree·buhs). The Northern Mariana Islands and Guam are still territories of the United States.

American influence remains strong in Micronesia. Some islands benefit from having American military bases. The United States pays a fee to keep the bases, and the bases provide jobs to islanders. In addition, these tiny countries rely on aid from the American government because they have no major resources or industries.

Most people in Micronesia follow traditional lifestyles. Those on the volcanic and fertile high islands practice subsistence farming. They grow yams, sweet potatoes, and cassava. People on the low islands fish and grow breadfruit, taro, and bananas. Poor soil limits farm production, so most food is imported.

Several Micronesian islands have deposits of phosphate, a mineral salt that is used to make fertilizer. The Federal States of Micronesia and the Marshall Islands have phosphate deposits, but they lack the money to mine the resource. Kiribati and Nauru once had large amounts of phosphate. The phosphate deposits on Kiribati are gone now, and they are almost gone on Nauru. With the loss of its phosphate industry, Kiribati remains heavily dependent on foreign aid, mainly from Japan, the European Union, and Australia. As Nauru's phosphate supplies lessen, the Nauruan government is investing abroad and trying to develop service industries.

NATIONAL GEOGRAPHIC

Samoan Traditions

Samoans highly value community and hospitality. Traditional Samoan tattoo designs (inset) are often complicated patterns. ***Place*** **What are several important economic activities on Samoa?**

Polynesia

Polynesia is a vast island area that lies southeast of Micronesia. Today, after a period of European rule, some Polynesian islands, such as Samoa and Tonga, are independent. Others, such as French Polynesia, are still controlled by European countries.

Many people in Polynesia practice subsistence farming. Several island economies are so poor that they depend on foreign aid to help their people. Samoa and Tonga have built strong tourist industries. Both also earn money by exporting timber. Samoa has tried to prevent deforestation by **establishing,** or setting up, a program to replant trees as they are cut down. Tonga grows vanilla beans and coconuts as cash crops. Other important industries in Polynesia include canning tuna and issuing colorful postage stamps for collectors.

The people of Samoa proudly call their island "the Cradle of Polynesia." One of its islands, Savai'i, is thought to be the original home of the Polynesian people. Samoans call their way of life the ***fa'a Samoa.*** This way of life emphasizes living in harmony with the community and the land. The people of Samoa are well-known for their music, dance, and handicrafts. Their tattoos are also famous. In fact, the word *tattoo* comes from the Samoan language.

Environmental Issues

Some human activities have harmed people and environments in parts of Oceania. Nuclear weapons testing had a disastrous effect on people living in the region.

In the late 1940s, the United States and other countries carried out testing of nuclear weapons in the Pacific area. The dangers of the tests were not completely understood at the time. As a result, residents of nearby islands were exposed to radiation that caused deaths and illnesses. The radiation also poisoned the land, water, and vegetation. The testing was stopped, but the tests' effects on people and the environment continue years later.

In recent years, governments have taken more responsibility in dealing with nuclear issues. The United States, for example, has provided millions of dollars to help Marshall Islanders and their families who were affected by the atomic tests. United States aid has also been used to clean up the environment in testing areas. Still, by the late 1900s, islanders could not return to Bikini Atoll, where the United States began nuclear testing in 1946, because of radioactivity levels.

Concern about nuclear weapons has affected the actions of other governments in Oceania. In 1987 New Zealand forbade nuclear-powered and nuclear-armed ships from entering its waters. In the 1990s, France planned nuclear tests on an atoll in French Polynesia but cancelled those tests as a result of international protests.

In addition, phosphate mining has caused environmental damage. About 80 percent of Nauru cannot support human life, and native birds are threatened by the loss of their **habitats,** or living areas. Nauru is now seeking international aid to restore its land.

Reading Check **Explaining** What is phosphate and why is it important to some Micronesian economies?

Section 2 Review

Social Studies ONLINE
Study Central™ To review this section, go to glencoe.com.

Vocabulary

1. **Explain** the significance of:
   - **a.** copra
   - **b.** lingua franca
   - **c.** *fa'a Samoa*
   - **d.** habitat

Main Ideas

2. **Summarizing** Use a diagram like the one below to identify key facts about Papua New Guinea.

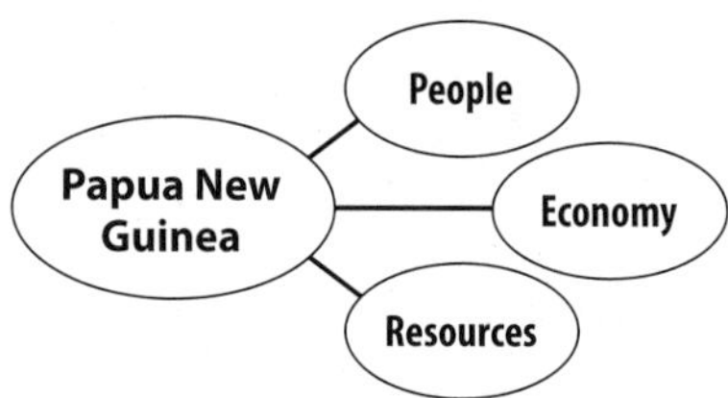

3. **Describing** Describe the lifestyles and economic activities of the peoples of Micronesia.

Critical Thinking

4. **Analyzing** Why is having a lingua franca important to a country such as Vanuatu?
5. **BIG Idea** How does Nauru benefit from trade?
6. **Challenge** Why are some economies in Oceania more successful than others?

Writing About Geography

7. **Expository Writing** Write a paragraph evaluating how the island nations of Oceania have tried to overcome their relative lack of resources.

Guide to Reading

Antarctica

BIG Idea

All living things are dependent upon one another and their surroundings for survival.

Content Vocabulary

- extinction *(p. 829)*
- krill *(p. 829)*
- ozone *(p. 830)*

Academic Vocabulary

- research *(p. 828)*
- specify *(p. 828)*

Reading Strategy

Organizing Use a format like the one below to make an outline of the section. Write each main heading on a line with a Roman numeral, and list important facts below it.

I. First Main Heading
   A. Key Fact 1
   B. Key Fact 2
II. Second Main Heading
   A. Key Fact 1
   B. Key Fact 2

 Section Audio  Spotlight Video

Picture This If you visit Antarctica, look for a red-and-white striped pole near the Amundsen-Scott South Pole Station. The actual geographic South Pole is located some distance away from the striped pole—buried under an ice sheet more than 1.5 miles (2.4 km) thick. The ice sheet moves more than 30 feet (9 m) per year, so every January 1st, a new brass marker is placed at 90°S to indicate the true geographic South Pole. Read this section to learn about Antarctica.

Amundsen-Scott South Pole Station, Antarctica

International Cooperation

Main Idea **Antarctica is a center of scientific research.**

Geography and You What rules do you and your classmates have for sharing spaces such as the lunchroom? Who makes the rules? Read to learn how the world's nations have established rules for Antarctica.

Antarctica was first sighted in the 1820s. Scientists and seal hunters began visiting parts of Antarctica's coasts, but the interior remained unexplored until the early 1900s. Then, in 1911, explorers reached the South Pole. This achievement opened the rest of the icy continent for exploration.

International Agreements

Hoping to find mineral resources, several countries claimed territory in Antarctica. Many other countries, including the United States, opposed the claims. During the 1950s, several countries began to cooperate, or work together, on scientific **research** in Antarctica.

To prevent any future conflicts, 12 countries signed the Antarctic Treaty in 1959. This agreement stated that Antarctica should be used only for peaceful, scientific purposes. It **specified,** or made clear, that Antarctica could not be used for weapons testing or any other military use.

Since 1959, forty-five countries have signed the Antarctic Treaty. These countries have agreed to forbid mining in Antarctica and to protect its environment.

GLOBAL CITIZENS

NAME: NIGEL WATSON **HOME COUNTRY:** New Zealand

ACHIEVEMENT: "Antarctica is home to a little known but immensely important part of the world's cultural heritage," says Nigel Watson, director of the Antarctic Heritage Trust. Watson works to protect structures left in Antarctica by famous explorers such as Sir Ernest Shackleton and Robert F. Scott. Damage from neglect, extreme weather, careless visitors, and penguin excrement are just a few of the challenges Watson faces in preserving the historic structures. One of the buildings in danger is a hut built in 1899. The hut was used by the first expedition that spent the winter in Antarctica.

QUOTE: ❝Our work will hopefully save the last link to Antarctica's first explorers—individuals of adventure and endurance—whose legacy can inspire future generations.❞

Watson shows Britain's Princess Anne the hut built by British explorer Robert F. Scott.

COURTESY NIGEL WATSON; (INSET) MARK BAKER/AGENCE FRANCE PRESSE/NEWSCOM

CITIZENS IN ACTION Why do many people think it is important to preserve links to our past?

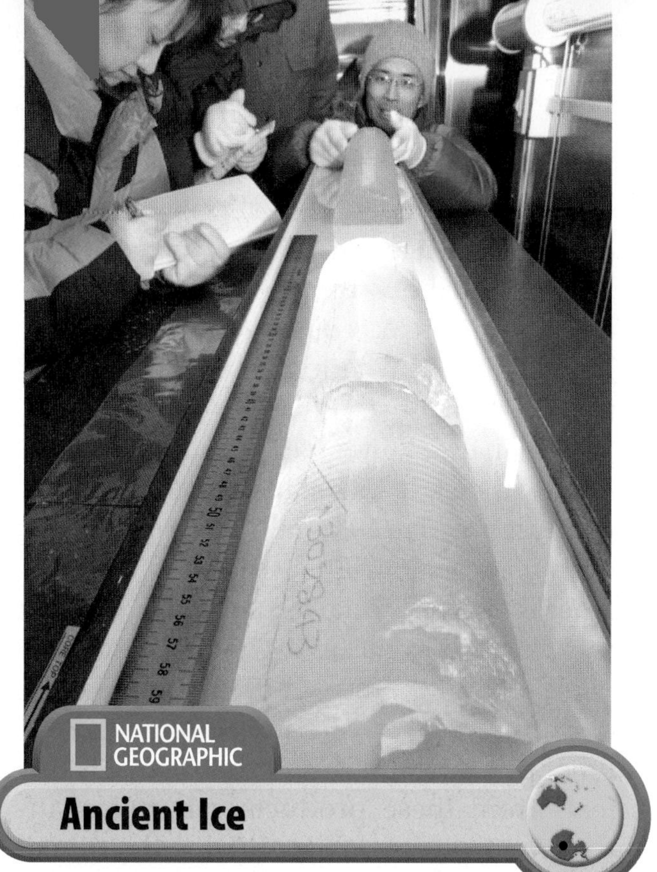

Ancient Ice

Scientists believe this block of Antarctic ice, drilled by a team of Japanese scientists, is more than 1 million years old. ***Human-Environment Interaction*** **In what year did explorers reach the South Pole?**

Scientific Research

Many countries today have scientific research stations in Antarctica. In January—summer in Antarctica—several thousand scientists from various countries come to study Antarctica's land, plants, animals, and ice. About 1,000 scientists stay even during the harsh polar winter.

Scientists carry out different kinds of research in Antarctica. Geologists have found the remains of trees from millions of years ago. They believe these findings show that Antarctica was once joined to Africa and South America. Climatologists study samples of ice from deep beneath the surface of the ice layer. These samples can reveal much about the climate from thousands of years ago.

Reading Check **Explaining** What is the purpose of the Antarctic Treaty?

Antarctica's Environment

Main Idea **Climate changes are affecting Antarctica's environment.**

Geography and You Could your daily actions impact Antarctica? Read to learn how efforts are being made to take care of Antarctica's fragile environment.

Antarctica has a harsh but fragile environment. Changes to that environment might affect people in other parts of the world.

Wildlife of Antarctica

Penguins, seals, fish, whales, and many kinds of flying birds live in or near the seas surrounding Antarctica. Whales and seals were once hunted nearly to **extinction,** or disappearance from the Earth. Countries around the world have since agreed to protect many of these creatures.

Environmental Challenges

Despite efforts to protect Antarctica's environment, the continent may face dangers due to human activity elsewhere in the world. As you have read, human activity may be contributing to global warming. Higher temperatures could lead to the loss of ice in and near Antarctica. The loss of ice means the loss of plants that live on that ice. These plants form the diet of **krill,** a tiny shrimp-like creature. Krill is the main food source for many larger species of Antarctic animals. Less plant life means less krill, and less krill threatens the survival of other animals.

In addition, an Antarctic ice melt could disastrously affect areas beyond Antarctica. Many scientists warn that sea levels will rise around the world, possibly flooding low islands in Oceania and highly populated coastal cities.

Maps In MOtion See StudentWorks™ Plus or glencoe.com.

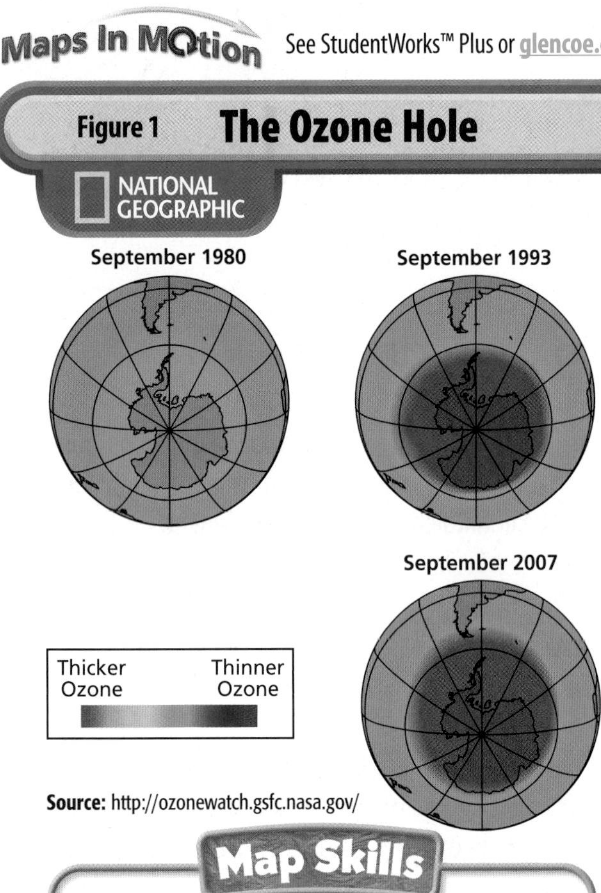

Map Skills

1. **Place** How has the ozone layer over Antarctica changed since 1980?
2. **Human-Environment Interaction** How might the loss of the ozone layer affect Antarctica?

Research in Antarctica also has revealed another challenge. A gas called **ozone** forms a layer around the Earth in the atmosphere. This ozone layer protects all living things on the Earth from certain harmful rays of the sun. In the 1980s, scientists noticed a weakening, or "hole," in the ozone layer above Antarctica. At its largest, in the early 2000s, this hole measured about 10.8 million square miles (28 million sq. km). Some scientists fear that the loss of ozone could lead to higher rates of skin cancer or contribute to global warming.

As a result, many countries have reduced or banned the use of aerosol sprays and other products linked to ozone loss. Chemicals from these products can collect in the atmosphere. When hit by the sun's rays, they form new chemicals that destroy the ozone. Scientists continue to watch the ozone layer over Antarctica.

Reading Check **Analyzing** How might human activities throughout the world affect Antarctica?

Section 3 Review

Social Studies ONLINE
Study Central™ To review this section, go to glencoe.com.

Vocabulary

1. **Explain** the meaning of *extinction, krill,* and *ozone* by using each term in a sentence.

Main Ideas

2. **Describing** What are some types of scientific research carried out in Antarctica?
3. **Illustrating** Use a diagram like the one below to illustrate the environmental challenges facing Antarctica.

Critical Thinking

4. **Determining Cause and Effect** Why did some nations make claims on Antarctica despite its harsh climate?
5. **BIG Idea** What effect might global warming have on Antarctica? How could that affect people in Oceania?
6. **Challenge** Do you think international cooperation in Antarctica would help nations work together on other issues? Why or why not?

Writing About Geography

7. **Using Your FOLDABLES™** Use your Foldable to write a paragraph explaining if you think the Antarctic Treaty should be continued or allowed to expire.

Visual Summary

Australia

- Australia's largely European population is becoming more diverse.
- The Aborigines still face problems in Australian society.
- Australia has rich minerals and productive farms and ranches.

Sheep, Australia

New Zealand

- New Zealand's population is mostly of European background.
- The Maori have laid claims to lands in New Zealand.
- New Zealand's agricultural economy depends on trade.

Christchurch, New Zealand

Family preparing meal, Fiji

Melanesia

- Papua New Guinea is Oceania's largest and most populous country.
- Most people in Melanesia practice subsistence farming.
- People in many areas of Melanesia follow traditional lifestyles.

Micronesia and Polynesia

- Many islands in Micronesia have close ties to the United States.
- Low-lying islands in Micronesia have to import food.
- Polynesian countries have built strong tourist industries.

Antarctica

- Many nations have agreed to set aside Antarctica for peaceful purposes.
- Antarctica is a major center of scientific research.
- Small animals and plants live in Antarctica. Larger animals thrive in nearby coastal waters.
- A number of problems threaten Antarctica's fragile environment.

South Georgia Island, Antarctica

Study anywhere, anytime! Download quizzes and flash cards to your PDA from glencoe.com.

CHAPTER 30 ASSESSMENT

STANDARDIZED TEST PRACTICE

TEST-TAKING TIP

On a multiple-choice test, remember that you are looking for the best answer, which might not necessarily be the only answer that applies.

Reviewing Vocabulary

Directions: Choose the word(s) that best completes the sentence.

1. Australian sheep called _______ are known for their fine wool.

A kiwis

B copras

C merinos

D krills

2. A small oval fruit with a brownish-green fuzzy skin known as a _______ is a major crop in New Zealand.

A krill

B kiwifruit

C copra

D merino

3. The meat of dried coconuts is called _______.

A kiwifruit

B krill

C merino

D copra

4. _______ is a common language used for communication and trade.

A Lingua franca

B Indigenous language

C Local dialect

D Pidgin dialect

Reviewing Main Ideas

Directions: Choose the best answer for each question.

Section 1 *(pp. 812–816)*

5. For years, white Australians discriminated against the _______, who were the first people to settle Australia.

A Merinos

B Aborigines

C Micronesians

D Maoris

6. In the Treaty of Waitangi, the British promised to

A provide jobs for all Maori.

B build houses for all Maori.

C protect Maori land rights.

D give the Maori the same rights as the Aborigines.

Section 2 *(pp. 822–826)*

7. _______ is the largest and most populous country in Melanesia.

A Polynesia

B Vanuatu

C Marianas

D Papua New Guinea

Section 3 *(pp. 827–830)*

8. The Antarctic Treaty of 1959

A stated that Antarctica should be used only for peaceful, scientific purposes.

B banned fishing in the waters around Antarctica.

C established colonies for all the countries that signed.

D divided Antarctica into sections for mining.

GO ON

Critical Thinking

Directions: Study the graph, and then choose the best answer for each question.

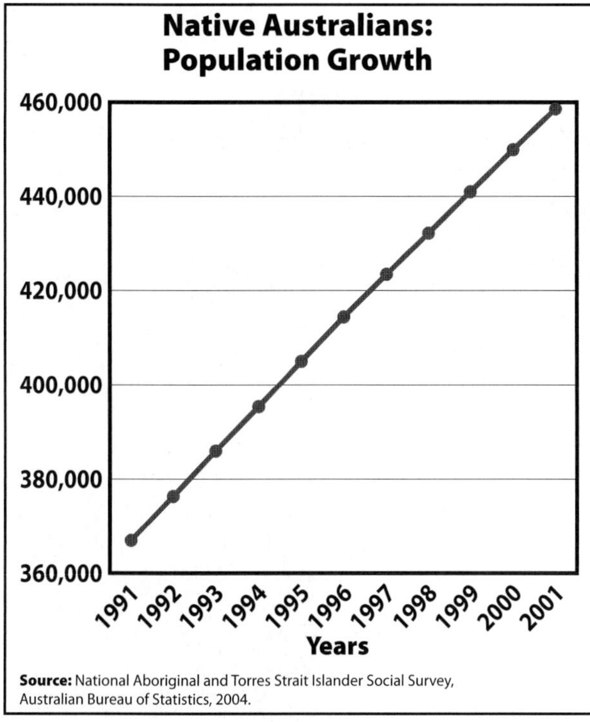

Source: National Aboriginal and Torres Strait Islander Social Survey, Australian Bureau of Statistics, 2004.

9. What was the trend in the Native Australian population from 1991 to 2001?

A increasing

B decreasing

C stable over the years

D decreasing then increasing

10. About how many more Native Australians were there in 2001 compared to 1991?

A 50,000

B 70,000

C 80,000

D 100,000

Document-Based Questions

Directions: Analyze the document and answer the short-answer questions that follow.

The Health and Welfare of Australia's Aboriginal and Torres Strait Islander Peoples, 2003

The diseases and conditions examined . . . include circulatory system diseases, diabetes, . . . kidney disease, cancer, respiratory diseases, communicable diseases, injury and poisoning, vision and hearing problems, oral health and mental health. For most of these conditions Indigenous peoples [Aborigines] had higher prevalence [occurrence] rates, higher hospitalisation rates and higher death rates than non-Indigenous Australians. Moreover, some of the chronic [long-lasting] diseases described here are diagnosed at a younger age in Indigenous persons than non-Indigenous persons, resulting in a lower quality of life at younger ages and premature mortality [death].

—Australian Institute of Health and Welfare

11. How would you characterize the diseases referred to in the document, and how do they affect Australian Aborigines?

12. How does the age at which these chronic diseases are diagnosed affect Australian Aborigines?

Extended Response

13. A member of Congress has proposed a bill to encourage tourism to Antarctica. Write a letter to your representative or senator in which you support or oppose this bill. Be sure to include your reasons.

For additional test practice, use Self-Check Quizzes—Chapter 30 at glencoe.com.

Need Extra Help?

If you missed question...	1	2	3	4	5	6	7	8	9	10	11	12	13
Go to page...	814	816	823	824	813	815	823	828	813	813	813	813	828–830

Appendix

Contents

What Is an Appendix?

What Is an Appendix?

An appendix is the additional material you often find at the end of books. The following information will help you learn how to use the Appendix in *Exploring Our World: People, Places, and Cultures.*

Skills Handbook

The Skills Handbook offers you information and practice using critical thinking and social studies skills. Mastering these skills will help you in all your courses.

Gazetteer

The Gazetteer (GA•zuh•TIHR) is a geographical dictionary. It lists many of the world's largest countries, cities, and important geographic features. Each entry also includes a page number telling where the place is shown on a map in the textbook.

English-Spanish Glossary

A glossary is a list of important or difficult terms found in a textbook. The glossary gives a definition of each term as it is used in the textbook. The glossary also includes page numbers telling you where in the textbook the term is used. Since words may have additional meanings, you may wish to use a dictionary to find other uses for them.

In *Exploring Our World: People, Places, and Cultures,* the Spanish glossary is included with the English glossary. The Spanish term is located directly across from the English term. A Spanish glossary is especially important to bilingual students, or those Spanish-speaking students who are learning the English language.

Index

The Index is an alphabetical listing that includes the subjects of the book and the page numbers where those subjects can be found. The index in this book also lets you know that certain pages contain maps, graphs, photos, or paintings about the subject.

Acknowledgments

This section lists photo credits and literary credits for the book. You can look at this section to find out where the publisher obtained the permission to use a photograph or to use excerpts from other books.

Test Yourself

Find the answers to these questions by using the Appendix on the following pages.

1. What does *famine* mean?
2. Where did you find what the word *famine* means?
3. What is the Spanish word for *availability?*
4. What skill is discussed on page 846?
5. What are the latitude and longitude of Moscow?
6. On what pages can you find information about the government of the United Kingdom?

Skills Handbook

Contents

Interpreting Political Cartoons

Why Learn This Skill?

Political cartoons express opinions through art. The cartoons appear in newspapers, magazines, books, and on the Internet. Political cartoons usually focus on public figures, political events, or economic or social conditions. This type of art can give you a summary of an event or circumstance, along with the artist's opinion, in an entertaining way.

1 Learn It!

Follow these steps to interpret political cartoons:

- Read the title, caption, or conversation balloons. They help you identify the subject of the cartoon.
- Identify the characters or people in the cartoon. They may be caricatures, or unrealistic drawings that exaggerate the characters' physical features.
- Identify any symbols. Symbols are objects that stand for something else. An example is the American flag, which is a symbol of our country. Commonly recognized symbols may not be labeled. Unusual symbols might be labeled.
- Examine the actions in the cartoon—what is happening and why?
- Identify the cartoonist's purpose. What statement or idea is he or she trying to express? Decide if the cartoonist wants to persuade, criticize, or just make people think.

2 Practice It!

On a separate sheet of paper, answer these questions about the political cartoon below.

1. What is the subject of the cartoon?
2. What words give clues to the meaning of the cartoon?
3. What item seems out of place?
4. What message do you think the cartoonist is trying to send?

3 Apply It!

Bring a newsmagazine to class. With a partner, analyze the message in each political cartoon you find in the magazine.

Predicting

Why Learn This Skill?

You have probably read about people making difficult decisions based on something they think *might* happen. You will have a better understanding of why people make certain choices when you consider the factors that influenced their decisions.

1 Learn It!

As you read a paragraph or section in your book, think about what might happen next. What you think will happen is your *prediction*. A prediction does not have a correct or incorrect answer. A prediction is an educated guess of what might happen next based on facts.

To make a prediction, ask yourself:

- What happened in this paragraph or section?
- What prior knowledge do I have about the information in the text?
- What similar circumstances do I know of?
- What do I think might happen next?
- Test your prediction: read further to see if you were correct.

▲ Aztec shield

2 Practice It!

To practice the skill, read the following paragraphs about the Aztec Empire. Then answer the questions.

In the late 1400s and early 1500s, Spanish explorers arrived in the Americas. They were greatly impressed by the magnificent cities and the great riches of the Native Americans.

In 1519 a Spanish army led by Hernán Cortés landed on Mexico's Gulf coast. He and about 600 soldiers marched to Tenochtitlán, which they had heard was filled with gold.

1. Choose the outcome that is most likely to occur between the Native Americans and the Spaniards.
   a. The Spaniards will conquer the Native Americans.
   b. The Native Americans will conquer the Spaniards.
   c. The two groups will become friends.
2. What clues in the text help you make your prediction?

3 Apply It!

Watch a television show or a movie. Halfway through the show, write down your prediction of how it will end. At the end of the show, check your prediction. Were you correct? What clues did you use to make your prediction?

Analyzing Library and Research Resources

Why Learn This Skill?

Imagine that your teacher asked you to write a report about the physical geography of Australia using library or Internet resources. Knowing how to choose sources that contain accurate information will help you save time in the library or on the Internet. You will also be able to write a better report.

1 Learn It!

Not all sources will be useful for your report on Australia's physical geography. Even some sources that involve topics about Australia will not always provide the information you want. In analyzing sources for your research project, choose items that are nonfiction and that contain the most information about your topic.

When choosing research resources, ask these questions:

- Is the information up-to-date?
- Does a book's index have several page references listed for the topic?
- Is the research written in a way that is easy to understand?
- Are there helpful illustrations and photos?

2 Practice It!

Look at the following list of sources. Which would be most helpful in writing a report on the physical geography of Australia? Explain your choices.

(1) *A current travel guide to Australia*

(2) *A book about Australia's landforms and climates*

(3) *A children's storybook about an Australian kangaroo*

(4) *A student's notes on the Internet about a family trip to Australia*

(5) *A study of the rise and fall of the British Empire*

(6) *A Web site with physical maps of Australia*

(7) *A book about Australian government*

(8) *A geographical dictionary*

3 Apply It!

Go to your local library or use the Internet to create a bibliography of sources you might use to write a report on the physical geography of Australia. Explain why you chose each source.

Uluru (Ayers Rock) in central Australia

Interpreting a Chart

Why Learn This Skill?

To make learning easier, you can organize information into groups of related facts and ideas. One way to organize information is with a chart. A chart presents written or numerical information in columns and rows. It helps you to remember and compare information more easily.

Learn It!

To organize information in a chart, follow these steps:

- Decide what information you must organize.
- Identify several major categories of ideas or facts about the topic, and use these categories as column headings.
- Find information that fits into each category, and write those facts or ideas under the appropriate column heading.

2 Practice It!

On a separate sheet of paper, answer the following questions using the chart at the bottom of this page.

1. What type of information does the chart contain?
2. What other related information appears in the chart?
3. Canada also exports clothing and beverages to the United States. Is it necessary to create a new chart to show this information?

3 Apply It!

Create a chart to track your school assignments. Work with five areas of information: Subject, Assignment, Description, Due Date, and Date Completed. Be sure to keep your chart up-to-date.

U.S. International Trade

	Japan	United Kingdom	Canada
Exports to U.S.	Engines, rubber goods, cars, trucks, buses	Dairy products, beverages, petroleum products	Wheat, minerals, paper, mining machines
Value of Exports to U.S.	$138 billion	$51.1 billion	$287.9 billion
Imports from U.S.	Meat, fish, sugar, tobacco, coffee	Fruit, tobacco, electrical equipment	Fish, sugar, metals, clothing
Value of Imports from U.S.	$55.4 billion	$38.6 billion	$211.3 billion

Source: *CIA World Factbook*, 2006; *United States Census Bureau, Foreign Trade Statistics*, 2005.

Making Comparisons

Why Learn This Skill?

Suppose you want to buy a portable CD player, and you must choose among three models. To make this decision, you would probably compare various features of the three models, such as price, sound quality, size, and so on. After you compare the models, you will choose the one that is best for you. In your studies of world geography, you must often compare countries of the world to identify patterns, make predictions, or make generalizations about regions.

Learn It!

When making comparisons, you identify and examine two or more places, peoples, economies, or forms of government. Then you identify any similarities between two topics, or ways the two topics are alike.

When making comparisons, apply the following steps:

- Decide what topics to compare. Clue words such as *also, as well as, like, same as,* and *similar to* can help you identify when topics are being compared.
- Read the information about each topic carefully.
- Identify what information is similar for both topics.

2 Practice It!

To practice the skill, analyze the information in the chart at the bottom of this page. Then answer these questions.

1. What countries are being compared?
2. What categories for each country are being compared?
3. In what ways are the United States and the United Kingdom similar?
4. Suppose you wanted to compare the two countries in more detail. What other categories might you use?

Apply It!

Think about two sports that are played at your school. Make a chart comparing categories such as where the games are played, who plays them, what equipment is used, and so on.

The United States and the United Kingdom

	United States	United Kingdom
Location	North America	Europe
Language	English	English
Form of Government	Federal republic	Constitutional monarchy
Popular Sports	Baseball, football, basketball	Soccer, rugby, cricket
Popular Foods	Hamburgers, hot dogs	Fish and chips, roast beef

Analyzing Primary Sources

Why Learn This Skill?

People who study history examine pieces of evidence to reconstruct events. These types of evidence—both written and illustrated—are called *primary sources*. Examining primary sources can help you understand the history of a place.

1 Learn It!

Primary sources are firsthand accounts that describe a historical event or time period. They can include letters, diaries, photographs and pictures, news articles, legal documents, stories, literature, and artwork.

Ask yourself the following questions when analyzing primary sources:

- What is the primary source?
- Who created it?
- Where is it from?
- When was it created?
- What does it reveal about the topic I am studying?

2 Practice It!

The following primary source is from *The Log of Christopher Columbus.* Christopher Columbus reached the new world on October 12, 1492. Columbus's entry explains what occurred when he and his shipmates encountered Native Americans. Read the entry, and then answer the questions that follow.

> *October 12:*
>
> *The people here called this island* Guanahani *in their language, and their speech is very fluent [easily flowing], although I do not understand any of it. They are friendly . . . people who [bear] no arms except for small spears, and they have no iron. I showed one my sword, and through ignorance he grabbed it by the blade and cut himself. . . .*
>
> *. . . They traded and gave everything they had with good will, but it seems to me that they have very little and are poor in everything. . . .*
>
> *This afternoon the people of San Salvador came swimming to our ships and in boats made from one log. They brought us parrots, balls of cotton thread, spears, and many other things. . . . For these items we swapped them little glass beads and hawks' bells.*
>
> —*The Log of Christopher Columbus*

1. Why did Columbus believe that the Native Americans had no knowledge about weapons?
2. Does Columbus fear the Native Americans? Explain.
3. What items did Columbus and his crew exchange with the Native Americans?
4. Why is this reading a primary source?

Apply It!

Find a primary source from your past, such as a photo, newspaper clipping, or diary entry. Explain to the class what it shows about that time in your life.

Recognizing Bias

Why Learn This Skill?

If you say, "Cats make better pets than dogs," you are stating a bias. A *bias* is an attitude that favors one way of thinking over another. It can prevent you from looking at a situation in a reasonable or truthful way.

1 Learn It!

Most people have feelings and ideas that affect their point of view on a subject. Their viewpoint, or *bias*, influences the way they interpret events. For this reason, an idea that is stated as a fact may really be only an opinion. Recognizing bias will help you judge the accuracy of what you read.

To recognize bias, follow these steps:

- Identify the speaker or writer and examine his or her views. Why did he or she speak or write about a particular issue?
- Look for language that shows emotion or opinion. Look for words such as *all*, *never*, *best*, *worst*, *might*, or *should*.
- Examine the information for imbalances. Is it written from one point of view? Does it take into consideration other points of view?
- Identify statements of fact. Factual statements usually answer the *who*, *what*, *where*, and *when* questions.
- Does the writer use facts to support his or her point of view?

2 Practice It!

Read the following statement about wildlife in Africa, and answer the questions below.

Mountain gorillas live in the misty mountain forests of East Africa. Logging and mining, however, are destroying the forests. Unless the forests are protected, the gorillas will lose their homes and disappear forever. As a concerned African naturalist, I must emphasize that this will be the worst event in Africa's history.

1. What problem is the speaker addressing?
2. What reasons does the speaker give for the loss of the forests?
3. What is the speaker's point of view, or bias?
4. What words give clues as to the speaker's bias?

3 Apply It!

Choose a letter from the editorial page of a newspaper. Summarize the issue being discussed and the writer's bias about the issue. Describe a possible opposing opinion and who might have it and why.

Mountain gorilla

Interpreting a Circle Graph

Why Learn This Skill?

Have you ever watched someone serve pieces of pie? When the pie is cut evenly, everyone's slice is the same size. If one slice is cut a little larger, however, someone else gets a smaller piece. A *circle graph* is like a sliced pie. In fact, a circle graph is also called a pie chart. In a circle graph, the complete circle represents a whole group—or 100 percent. The circle is divided into "slices," or wedge-shaped sections representing parts of the whole.

1 Learn It!

To read and interpret a circle graph, follow these steps:

- Read the title of the circle graph to find the subject.
- Study the labels or the key to see what each "slice" represents.
- Compare the sizes of the circle slices.

2 Practice It!

Study the circle graph on this page, and answer the following questions.

1. What is the subject of the circle graph?
2. On what do Americans spend most of their incomes?
3. On what do Americans spend the least portion of their incomes?
4. What is the total percentage of income spent on transportation and food?

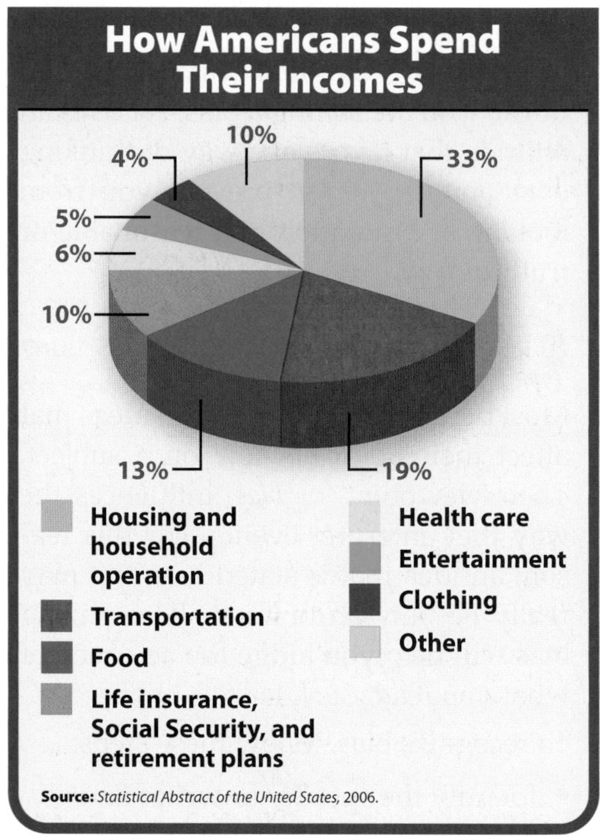

3 Apply It!

Quiz 10 friends about the capitals of India, Pakistan, and Bangladesh. Create a circle graph showing what percentage knew (a) all three capitals; (b) two capitals; (c) one capital; or (d) no capitals.

Sequencing Events

Why Learn This Skill?

Have you ever had to remember events and their dates in the order in which they happened? *Sequencing* means listing facts in the correct order that they occurred. A time line helps you do this. A time line is a diagram that shows how dates and events relate to one another. The years are evenly spaced along most time lines. Events on time lines are described beside the date on which they occurred.

Learn It!

To understand how to sequence events, follow these steps:

- As you read, look for dates or clue words that hint at chronological order, such as *in 2006, the late 1900s, first, then, finally*, and *after*.
- To read a time line, find the dates on the opposite ends of the time line. These dates show the range of time that is covered.
- Note the equal spacing between dates on the time line.
- Study the order of events.
- Look to see how the events relate to one another.

Practice It!

Examine the time line on this page and answer the following questions.

1. When does the time line begin? When does it end?
2. What major event happened in the late 1700s?
3. Did the Civil War begin before or after the United States entered World War I?
4. During what decade did the Cold War end?

Apply It!

List key events from one of the chapters in your textbook that covers the history of a region. Create a time line that lists these events in the order they occurred.

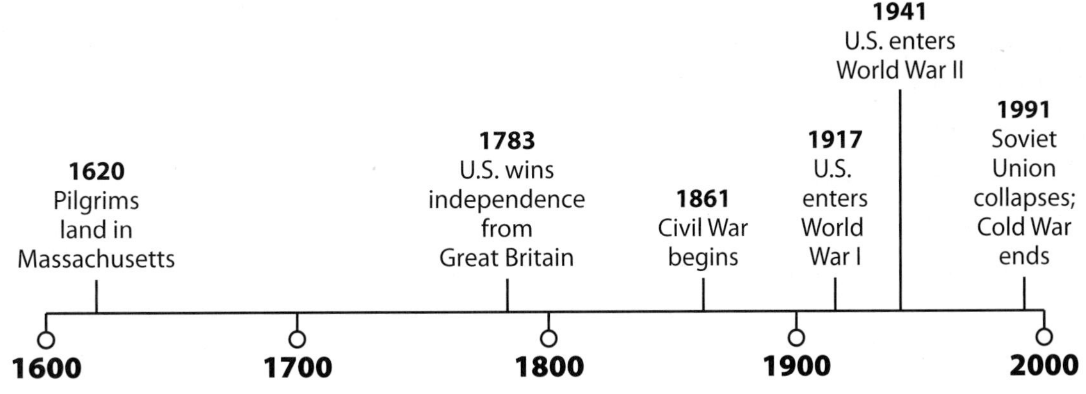
1620
Pilgrims land in Massachusetts
1783
U.S. wins independence from Great Britain
1861
Civil War begins
1917
U.S. enters World War I
1941
U.S. enters World War II
1991
Soviet Union collapses; Cold War ends
1600
1700
1800
1900
2000

Interpreting a Population Pyramid

Why Learn This Skill?

A population pyramid shows a country's population by age and gender. Geographers use population pyramids to plan for a country's future needs.

1 Learn It!

A population pyramid is two bar graphs. These bar graphs show the number of males and females living in a region. The number of males and females is given as a percentage along the bottom of the graph. The age range for each group is listed along the left side of the graph.

To interpret population pyramids, follow these steps:

- Look at the bar graphs for the male and female groups.
- Identify, for each group, the bars that indicate the largest percentage and the smallest percentage.
- Find the age range for these groups.
- If a country's population is *growing*, the pyramid will be large at the bottom. This shows that the country's population has a large number of children and young people.
- If a country's population is *declining*, the pyramid will be narrow at the bottom and wider at the top. This means that the country's population has a large number of elderly people.
- If a country's population is *stable*, the pyramid will have bars with similar lengths over several age ranges.

2 Practice It!

Study the 2007 population pyramid for Spain shown below, then answer the following questions.

1. Which age group makes up the largest portion of Spain's population?
2. Does it appear that Spanish men or Spanish women live longer? Explain.
3. What does the shape of the pyramid tell you about Spain's population?

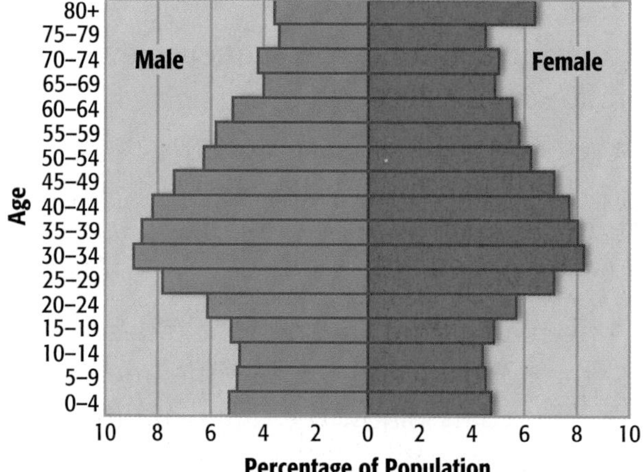

Source: *U.S. Census Bureau, International Data Base.*

3 Apply It!

Find the population pyramids for two countries at www.census.gov/ipc/www/idbpyr.html. Then write a paragraph to describe the similarities and differences between their populations.

Gazetteer

Gazetteer

A gazetteer (GA•zuh•TIHR) is a geographic index or dictionary. It shows latitude and longitude for cities and certain other places. Latitude and longitude are shown in this way: 48°N 2°E, or 48 degrees north latitude and two degrees east longitude. This Gazetteer lists many important geographic features and most of the world's largest independent countries and their capitals. The page numbers tell where each entry can be found on a map in this book. As an aid to pronunciation, most entries are spelled phonetically.

Abidjan [AH•bee•JAHN] Capital of Côte d'Ivoire. 5°N 4°W (p. RA22)
Abu Dhabi [AH•boo DAH•bee] Capital of the United Arab Emirates. 24°N 54°E (p. RA24)
Abuja [ah•BOO•jah] Capital of Nigeria. 8°N 9°E (p. RA22)
Accra [ah•KRUH] Capital of Ghana. 6°N 0° longitude (p. RA22)
Addis Ababa [AHD•dihs AH•bah•BAH] Capital of Ethiopia. 9°N 39°E (p. RA22)
Adriatic [AY•dree•A•tihk] **Sea** Arm of the Mediterranean Sea between the Balkan Peninsula and Italy. (p. RA20)
Afghanistan [af•GA•nuh•STAN] Central Asian country west of Pakistan. (p. RA25)
Albania [al•BAY•nee•uh] Country on the Adriatic Sea, south of Serbia. (p. RA18)
Algeria [al•JIHR•ee•uh] North African country east of Morocco. (p. RA22)
Algiers [al•JIHRZ] Capital of Algeria. 37°N 3°E (p. RA22)
Alps [ALPS] Mountain ranges extending through central Europe. (p. RA20)
Amazon [A•muh•ZAHN] **River** Largest river in the world by volume and second-largest in length. (p. RA17)
Amman [a•MAHN] Capital of Jordan. 32°N 36°E (p. RA24)
Amsterdam [AHM•stuhr•DAHM] Capital of the Netherlands. 52°N 5°E (p. RA18)
Andes [AN•DEEZ] Mountain system extending north and south along the western side of South America. (p. RA17)
Andorra [an•DAWR•uh] Small country in southern Europe between France and Spain. 43°N 2°E (p. RA18)
Angola [ang•GOH•luh] Southern African country north of Namibia. (p. RA22)
Ankara [AHNG•kuh•ruh] Capital of Turkey. 40°N 33°E (p. RA24)
Antananarivo [AHN•tah•NAH•nah•REE•voh] Capital of Madagascar. 19°S 48°E (p. RA22)
Arabian [uh•RAY•bee•uhn] **Peninsula** Large peninsula extending into the Arabian Sea. (p. RA25)
Argentina [AHR•juhn•TEE•nuh] South American country east of Chile. (p. RA16)
Armenia [ahr•MEE•nee•uh] European-Asian country between the Black and Caspian Seas. 40°N 45°E (p. RA26)
Ashkhabad [AHSH•gah•BAHD] Capital of Turkmenistan. 38°N 58°E (p. RA25)
Asmara [az•MAHR•uh] Capital of Eritrea. 16°N 39°E (p. RA22)
Astana Capital of Kazakhstan. 51°N 72°E (p. RA26)
Asunción [ah•SOON•see•OHN] Capital of Paraguay. 25°S 58°W (p. RA16)
Athens Capital of Greece. 38°N 24°E (p. RA19)
Atlas [AT•luhs] **Mountains** Mountain range on the northern edge of the Sahara. (p. RA23)
Australia [aw•STRAYL•yuh] Country and continent in Southern Hemisphere. (p. RA30)
Austria [AWS•tree•uh] Western European country east of Switzerland and south of Germany and the Czech Republic. (p. RA18)
Azerbaijan [A•zuhr•BY•JAHN] European-Asian country on the Caspian Sea. (p. RA25)

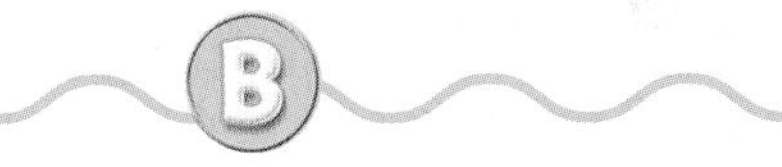

Baghdad Capital of Iraq. 33°N 44°E (p. RA25)
Bahamas [buh•HAH•muhz] Country made up of many islands between Cuba and the United States. (p. RA15)
Bahrain [bah•RAYN] Country located on the Persian Gulf. 26°N 51°E (p. RA25)
Baku [bah•KOO] Capital of Azerbaijan. 40°N 50°E (p. RA25)
Balkan [BAWL•kuhn] **Peninsula** Peninsula in southeastern Europe. (p. RA21)
Baltic [BAWL•tihk] **Sea** Sea in northern Europe that is connected to the North Sea. (p. RA20)
Bamako [BAH•mah•KOH] Capital of Mali. 13°N 8°W (p. RA22)
Bangkok [BANG•KAHK] Capital of Thailand. 14°N 100°E (p. RA27)
Bangladesh [BAHNG•gluh•DEHSH] South Asian country bordered by India and Myanmar. (p. RA27)
Bangui [BAHNG•GEE] Capital of the Central African Republic. 4°N 19°E (p. RA22)
Banjul [BAHN•JOOL] Capital of Gambia. 13°N 17°W (p. RA22)

Barbados [bahr•BAY•duhs] Island country between the Atlantic Ocean and the Caribbean Sea. 14°N 59°W (p. RA15)
Beijing [BAY•JIHNG] Capital of China. 40°N 116°E (p. RA27)
Beirut [bay•ROOT] Capital of Lebanon. 34°N 36°E (p. RA24)
Belarus [BEE•luh•ROOS] Eastern European country west of Russia. 54°N 28°E (p. RA19)
Belgium [BEHL•juhm] Western European country south of the Netherlands. (p. RA18)
Belgrade [BEHL•GRAYD] Capital of Serbia. 45°N 21°E (p. RA19)
Belize [buh•LEEZ] Central American country east of Guatemala. (p. RA14)
Belmopan [BEHL•moh•PAHN] Capital of Belize. 17°N 89°W (p. RA14)
Benin [buh•NEEN] West African country west of Nigeria. (p. RA22)
Berlin [behr•LEEN] Capital of Germany. 53°N 13°E (p. RA18)
Bern Capital of Switzerland. 47°N 7°E (p. RA18)
Bhutan [boo•TAHN] South Asian country northeast of India. (p. RA27)
Bishkek [bihsh•KEHK] Capital of Kyrgyzstan. 43°N 75°E (p. RA26)
Bissau [bihs•SOW] Capital of Guinea-Bissau. 12°N 16°W (p. RA22)
Black Sea Large sea between Europe and Asia. (p. RA21)
Bloemfontein [BLOOM•FAHN•TAYN] Judicial capital of South Africa. 26°E 29°S (p. RA22)
Bogotá [BOH•goh•TAH] Capital of Colombia. 5°N 74°W (p. RA16)
Bolivia [buh•LIHV•ee•uh] Country in the central part of South America, north of Argentina. (p. RA16)
Bosnia and Herzegovina [BAHZ•nee•uh HEHRT•seh•GAW•vee•nuh] Southeastern European country bordered by Croatia, Serbia, and Montenegro. (p. RA18)
Botswana [bawt•SWAH•nah] Southern African country north of the Republic of South Africa. (p. RA22)
Brasília [brah•ZEEL•yuh] Capital of Brazil. 16°S 48°W (p. RA16)
Bratislava [BRAH•tih•SLAH•vuh] Capital of Slovakia. 48°N 17°E (p. RA18)
Brazil [bruh•ZIHL] Largest country in South America. (p. RA16)
Brazzaville [BRAH•zuh•VEEL] Capital of Congo. 4°S 15°E (p. RA22)
Brunei [bru•NY] Southeast Asian country on northern coast of the island of Borneo. (p. RA27)
Brussels [BRUH•suhlz] Capital of Belgium. 51°N 4°E (p. RA18)
Bucharest [BOO•kuh•REHST] Capital of Romania. 44°N 26°E (p. RA19)
Budapest [BOO•duh•PEHST] Capital of Hungary. 48°N 19°E (p. RA18)
Buenos Aires [BWAY•nuhs AR•eez] Capital of Argentina. 34°S 58°W (p. RA16)
Bujumbura [BOO•juhm•BUR•uh] Capital of Burundi. 3°S 29°E (p. RA22)
Bulgaria [BUHL•GAR•ee•uh] Southeastern European country south of Romania. (p. RA19)
Burkina Faso [bur•KEE•nuh FAH•soh] West African country south of Mali. (p. RA22)
Burundi [bu•ROON•dee] East African country at the northern end of Lake Tanganyika. 3°S 30°E (p. RA22)

Cairo [KY•ROH] Capital of Egypt. 31°N 32°E (p. RA24)
Cambodia [kam•BOH•dee•uh] Southeast Asian country south of Thailand and Laos. (p. RA27)
Cameroon [KA•muh•ROON] Central African country on the northeast shore of the Gulf of Guinea. (p. RA22)
Canada [KA•nuh•duh] Northernmost country in North America. (p. RA6)
Canberra [KAN•BEHR•uh] Capital of Australia. 35°S 149°E (p. RA30)
Cape Town Legislative capital of the Republic of South Africa. 34°S 18°E (p. RA22)
Cape Verde [VUHRD] Island country off the coast of western Africa in the Atlantic Ocean. 15°N 24°W (p. RA22)
Caracas [kah•RAH•kahs] Capital of Venezuela. 11°N 67°W (p. RA16)
Caribbean [KAR•uh•BEE•uhn] **Islands** Islands in the Caribbean Sea between North America and South America, also known as West Indies. (p. RA15)
Caribbean Sea Part of the Atlantic Ocean bordered by the West Indies, South America, and Central America. (p. RA15)
Caspian [KAS•pee•uhn] **Sea** Salt lake between Europe and Asia that is the world's largest inland body of water. (p. RA21)
Caucasus [KAW•kuh•suhs] **Mountains** Mountain range between the Black and Caspian Seas. (p. RA21)
Central African Republic Central African country south of Chad. (p. RA22)
Chad [CHAD] Country west of Sudan in the African Sahel. (p. RA22)
Chang Jiang [CHAHNG jee•AHNG] Principal river of China that begins in Tibet and flows into the East China Sea near Shanghai; also known as the Yangtze River. (p. RA29)
Chile [CHEE•lay] South American country west of Argentina. (p. RA16)
China [CHY•nuh] Country in eastern and central Asia, known officially as the People's Republic of China. (p. RA27)
Chişinău [KEE•shee•NOW] Capital of Moldova. 47°N 29°E (p. RA19)
Colombia [kuh•LUHM•bee•uh] South American country west of Venezuela. (p. RA16)

Colombo [kuh•LUHM•boh] Capital of Sri Lanka. 7°N 80°E (p. RA26)
Comoros [KAH•muh•ROHZ] Small island country in Indian Ocean between the island of Madagascar and the southeast African mainland. 13°S 43°E (p. RA22)
Conakry [KAH•nuh•kree] Capital of Guinea. 10°N 14°W (p. RA22)
Congo [KAHNG•goh] Central African country east of the Democratic Republic of the Congo. 3°S 14°E (p. RA22)
Congo, Democratic Republic of the Central African country north of Zambia and Angola. 1°S 22°E (p. RA22)
Copenhagen [KOH•puhn•HAY•guhn] Capital of Denmark. 56°N 12°E (p. RA18)
Costa Rica [KAWS•tah REE•kah] Central American country south of Nicaragua. (p. RA15)
Côte d'Ivoire [KOHT dee•VWAHR] West African country south of Mali. (p. RA22)
Croatia [kroh•AY•shuh] Southeastern European country on the Adriatic Sea. (p. RA18)
Cuba [KYOO•buh] Island country in the Caribbean Sea. (p. RA15)
Cyprus [SY•pruhs] Island country in the eastern Mediterranean Sea, south of Turkey. (p. RA19)
Czech [CHEHK] **Republic** Eastern European country north of Austria. (p. RA18)

Dakar [dah•KAHR] Capital of Senegal. 15°N 17°W (p. RA22)
Damascus [duh•MAS•kuhs] Capital of Syria. 34°N 36°E (p. RA24)
Dar es Salaam [DAHR EHS sah•LAHM] Commercial capital of Tanzania. 7°S 39°E (p. RA22)
Denmark Northern European country between the Baltic and North Seas. (p. RA18)
Dhaka [DA•kuh] Capital of Bangladesh. 24°N 90°E (p. RA27)
Djibouti [jih•BOO•tee] East African country on the Gulf of Aden. 12°N 43°E (p. RA22)
Dodoma [doh•DOH•mah] Political capital of Tanzania. 6°S 36°E (p. RA22)
Doha [DOH•huh] Capital of Qatar. 25°N 51°E (p. RA25)
Dominican [duh•MIH•nih•kuhn] **Republic** Country in the Caribbean Sea on the eastern part of the island of Hispaniola. (p. RA15)
Dublin [DUH•blihn] Capital of Ireland. 53°N 6°W (p. RA18)
Dushanbe [doo•SHAM•buh] Capital of Tajikistan. 39°N 69°E (p. RA25)

East Timor [TEE•MOHR] Previous province of Indonesia, now under UN administration. 10°S 127°E (p. RA27)
Ecuador [EH•kwuh•dawr] South American country southwest of Colombia. (p. RA16)
Egypt [EE•jihpt] North African country on the Mediterranean Sea. (p. RA24)
El Salvador [ehl SAL•vuh•dawr] Central American country southwest of Honduras. (p. RA14)
Equatorial Guinea [EE•kwuh•TOHR•ee•uhl GIH•nee] Central African country south of Cameroon. (p. RA22)
Eritrea [EHR•uh•TREE•uh] East African country north of Ethiopia. (p. RA22)
Estonia [eh•STOH•nee•uh] Eastern European country on the Baltic Sea. (p. RA19)
Ethiopia [EE•thee•OH•pee•uh] East African country north of Somalia and Kenya. (p. RA22)
Euphrates [yu•FRAY•teez] **River** River in southwestern Asia that flows through Syria and Iraq and joins the Tigris River. (p. RA25)

Fiji [FEE•jee] **Islands** Country comprised of an island group in the southwest Pacific Ocean. 19°S 175°E (p. RA30)
Finland [FIHN•luhnd] Northern European country east of Sweden. (p. RA19)
France [FRANS] Western European country south of the United Kingdom. (p. RA18)
Freetown Capital of Sierra Leone. (p. RA22)
French Guiana [gee•A•nuh] French-owned territory in northern South America. (p. RA16)

Gabon [ga•BOHN] Central African country on the Atlantic Ocean. (p. RA22)
Gaborone [GAH•boh•ROH•nay] Capital of Botswana. (p. RA22)
Gambia [GAM•bee•uh] West African country along the Gambia River. (p. RA22)
Georgetown [JAWRJ•TOWN] Capital of Guyana. 8°N 58°W (p. RA16)
Georgia [JAWR•juh] European-Asian country bordering the Black Sea south of Russia. (p. RA26)
Germany [JUHR•muh•nee] Western European country south of Denmark, officially called the Federal Republic of Germany. (p. RA18)
Ghana [GAH•nuh] West African country on the Gulf of Guinea. (p. RA22)
Great Plains The continental slope extending through the United States and Canada. (p. RA7)
Greece [GREES] Southern European country on the Balkan Peninsula. (p. RA19)
Greenland [GREEN•luhnd] Island in northwestern Atlantic Ocean and the largest island in the world. (p. RA6)
Guatemala [GWAH•tay•MAH•lah] Central American country south of Mexico. (p. RA14)

Guatemala Capital of Guatemala. 15°N 91°W (p. RA14)
Guinea [GIH•nee] West African country on the Atlantic coast. (p. RA22)
Guinea-Bissau [GIH•nee bih•SOW] West African country on the Atlantic coast. (p. RA22)
Gulf of Mexico Gulf on part of the southern coast of North America. (p. RA7)
Guyana [gy•AH•nuh] South American country between Venezuela and Suriname. (p. RA16)

Haiti [HAY•tee] Country in the Caribbean Sea on the western part of the island of Hispaniola. (p. RA15)
Hanoi [ha•NOY] Capital of Vietnam. 21°N 106°E (p. RA27)
Harare [hah•RAH•RAY] Capital of Zimbabwe. 18°S 31°E (p. RA22)
Havana [huh•VA•nuh] Capital of Cuba. 23°N 82°W (p. RA15)
Helsinki [HEHL•SIHNG•kee] Capital of Finland. 60°N 24°E (p. RA19)
Himalaya [HI•muh•LAY•uh] Mountain ranges in southern Asia, bordering the Indian subcontinent on the north. (p. RA28)
Honduras [hahn•DUR•uhs] Central American country on the Caribbean Sea. (p. RA14)
Hong Kong Port and industrial center in southern China. 22°N 115°E (p. RA27)
Huang He [HWAHNG HUH] River in northern and eastern China, also known as the Yellow River. (p. RA29)
Hungary [HUHNG•guh•ree] Eastern European country south of Slovakia. (p. RA18)

Iberian [eye•BIHR•ee•uhn] **Peninsula** Peninsula in southwest Europe, occupied by Spain and Portugal. (p. RA20)
Iceland Island country between the North Atlantic and Arctic Oceans. (p. RA18)
India [IHN•dee•uh] South Asian country south of China and Nepal. (p. RA26)
Indonesia [IHN•duh•NEE•zhuh] Southeast Asian island country known as the Republic of Indonesia. (p. RA27)
Indus [IHN•duhs] **River** River in Asia that begins in Tibet and flows through Pakistan to the Arabian Sea. (p. RA28)
Iran [ih•RAN] Southwest Asian country that was formerly named Persia. (p. RA25)
Iraq [ih•RAHK] Southwest Asian country west of Iran. (p. RA25)
Ireland [EYER•luhnd] Island west of Great Britain occupied by the Republic of Ireland and Northern Ireland. (p. RA18)
Islamabad [ihs•LAH•muh•BAHD] Capital of Pakistan. 34°N 73°E (p. RA26)
Israel [IHZ•ree•uhl] Southwest Asian country south of Lebanon. (p. RA24)
Italy [IHT•uhl•ee] Southern European country south of Switzerland and east of France. (p. RA18)

Jakarta [juh•KAHR•tuh] Capital of Indonesia. 6°S 107°E (p. RA27)
Jamaica [juh•MAY•kuh] Island country in the Caribbean Sea. (p. RA15)
Japan [juh•PAN] East Asian country consisting of the four large islands of Hokkaido, Honshu, Shikoku, and Kyushu, plus thousands of small islands. (p. RA27)
Jerusalem [juh•ROO•suh•luhm] Capital of Israel and a holy city for Christians, Jews, and Muslims. 32°N 35°E (p. RA24)
Jordan [JAWRD•uhn] Southwest Asian country south of Syria. (p. RA24)

Kabul [KAH•buhl] Capital of Afghanistan. 35°N 69°E (p. RA25)
Kampala [kahm•PAH•lah] Capital of Uganda. 0° latitude 32°E (p. RA22)
Kathmandu [KAT•MAN•DOO] Capital of Nepal. 28°N 85°E (p. RA26)
Kazakhstan [kuh•ZAHK•STAHN] Large Asian country south of Russia and bordering the Caspian Sea. (p. RA26)
Kenya [KEHN•yuh] East African country south of Ethiopia. (p. RA22)
Khartoum [kahr•TOOM] Capital of Sudan. 16°N 33°E (p. RA22)
Kigali [kee•GAH•lee] Capital of Rwanda. 2°S 30°E (p. RA22)
Kingston [KIHNG•stuhn] Capital of Jamaica. 18°N 77°W (p. RA15)
Kinshasa [kihn•SHAH•suh] Capital of the Democratic Republic of the Congo. 4°S 15°E (p. RA22)
Kuala Lumpur [KWAH•luh LUM•PUR] Capital of Malaysia. 3°N 102°E (p. RA27)
Kuwait [ku•WAYT] Country on the Persian Gulf between Saudi Arabia and Iraq. (p. RA25)
Kyiv [KEE•ihf] Capital of Ukraine. 50°N 31°E (p. RA19)
Kyrgyzstan [KIHR•gih•STAN] Central Asian country on China's western border. (p. RA26)

Laos [LOWS] Southeast Asian country south of China and west of Vietnam. (p. RA27)

La Paz [lah PAHS] Administrative capital of Bolivia, and the highest capital in the world. 17°S 68°W (p. RA16)
Latvia [LAT•vee•uh] Eastern European country west of Russia on the Baltic Sea. (p. RA19)
Lebanon [LEH•buh•nuhn] Country south of Syria on the Mediterranean Sea. (p. RA24)
Lesotho [luh•SOH•TOH] Southern African country within the borders of the Republic of South Africa. (p. RA22)
Liberia [ly•BIHR•ee•uh] West African country south of Guinea. (p. RA22)
Libreville [LEE•bruh•VIHL] Capital of Gabon. 1°N 9°E (p. RA22)
Libya [LIH•bee•uh] North African country west of Egypt on the Mediterranean Sea. (p. RA22)
Liechtenstein [LIHKT•uhn•SHTYN] Small country in central Europe between Switzerland and Austria. 47°N 10°E (p. RA18)
Lilongwe [lih•LAWNG•GWAY] Capital of Malawi. 14°S 34°E (p. RA22)
Lima [LEE•mah] Capital of Peru. 12°S 77°W (p. RA16)
Lisbon [LIHZ•buhn] Capital of Portugal. 39°N 9°W (p. RA18)
Lithuania [LIH•thuh•WAY•nee•uh] Eastern European country northwest of Belarus on the Baltic Sea. (p. RA21)
Ljubljana [lee•oo•blee•AH•nuh] Capital of Slovenia. 46°N 14°E (p. RA18)
Lomé [loh•MAY] Capital of Togo. 6°N 1°E (p. RA22)
London Capital of the United Kingdom, on the Thames River. 52°N 0° longitude (p. RA18)
Luanda [lu•AHN•duh] Capital of Angola. 9°S 13°E (p. RA22)
Lusaka [loo•SAH•kah] Capital of Zambia. 15°S 28°E (p. RA22)
Luxembourg [LUHK•suhm•BUHRG] Small European country bordered by France, Belgium, and Germany. 50°N 7°E (p. RA18)

Macao [muh•KOW] Port in southern China. 22°N 113°E (p. RA27)
Macedonia [MA•suh•DOH•nee•uh] Southeastern European country north of Greece. (p. RA19). Macedonia also refers to a geographic region covering northern Greece, the country Macedonia, and part of Bulgaria.
Madagascar [MA•duh•GAS•kuhr] Island in the Indian Ocean off the southeastern coast of Africa. (p. RA22)
Madrid Capital of Spain. 41°N 4°W (p. RA18)
Malabo [mah•LAH•boh] Capital of Equatorial Guinea. 4°N 9°E (p. RA22)
Malawi [mah•LAH•wee] Southern African country south of Tanzania and east of Zambia. (p. RA22)
Malaysia [muh•LAY•zhuh] Southeast Asian country with land on the Malay Peninsula and on the island of Borneo. (p. RA27)
Maldives [MAWL•DEEVZ] Island country southwest of India in the Indian Ocean. (p. RA26)
Mali [MAH•lee] West African country east of Mauritania. (p. RA22)
Managua [mah•NAH•gwah] Capital of Nicaragua. (p. RA15)
Manila [muh•NIH•luh] Capital of the Philippines. 15°N 121°E (p. RA27)
Maputo [mah•POO•toh] Capital of Mozambique. 26°S 33°E (p. RA22)
Maseru [MA•zuh•ROO] Capital of Lesotho. 29°S 27°E (p. RA22)
Masqat [MUHS•KAHT] Capital of Oman. 23°N 59°E (p. RA25)
Mauritania [MAWR•uh•TAY•nee•uh] West African country north of Senegal. (p. RA22)
Mauritius [maw•RIH•shuhs] Island country in the Indian Ocean east of Madagascar. 21°S 58°E (p. RA3)
Mbabane [uhm•bah•BAH•nay] Capital of Swaziland. 26°S 31°E (p. RA22)
Mediterranean [MEH•duh•tuh•RAY•nee•uhn] **Sea** Large inland sea surrounded by Europe, Asia, and Africa. (p. RA20)
Mekong [MAY•KAWNG] **River** River in southeastern Asia that begins in Tibet and empties into the South China Sea. (p. RA29)
Mexico [MEHK•sih•KOH] North American country south of the United States. (p. RA14)
Mexico City Capital of Mexico. 19°N 99°W (p. RA14)
Minsk [MIHNSK] Capital of Belarus. 54°N 28°E (p. RA19)
Mississippi [MIH•suh•SIH•pee] **River** Large river system in the central United States that flows southward into the Gulf of Mexico. (p. RA11)
Mogadishu [MOH•guh•DEE•SHOO] Capital of Somalia. 2°N 45°E (p. RA22)
Moldova [mawl•DAW•vuh] Small European country between Ukraine and Romania. (p. RA19)
Monaco [MAH•nuh•KOH] Small country in southern Europe on the French Mediterranean coast. 44°N 8°E (p. RA18)
Mongolia [mahn•GOHL•yuh] Country in Asia between Russia and China. (p. RA23)
Monrovia [muhn•ROH•vee•uh] Capital of Liberia. 6°N 11°W (p. RA22)
Monetenegro [MAHN•tuh•NEE•groh] Eastern European country. (p. RA18)
Montevideo [MAHN•tuh•vuh•DAY•OH] Capital of Uruguay. 35°S 56°W (p. RA16)
Morocco [muh•RAH•KOH] North African country on the Mediterranean Sea and the Atlantic Ocean. (p. RA22)
Moscow [MAHS•KOW] Capital of Russia. 56°N 38°E (p. RA19)

Mount Everest [EHV•ruhst] Highest mountain in the world, in the Himalaya between Nepal and Tibet. (p. RA28)
Mozambique [MOH•zahm•BEEK] Southern African country south of Tanzania. (p. RA22)
Myanmar [MYAHN•MAHR] Southeast Asian country south of China and India, formerly called Burma. (p. RA27)

Nairobi [ny•ROH•bee] Capital of Kenya. 1°S 37°E (p. RA22)
Namibia [nuh•MIH•bee•uh] Southern African country south of Angola on the Atlantic Ocean. 20°S 16°E (p. RA22)
Nassau [NA•SAW] Capital of the Bahamas. 25°N 77°W (p. RA15)
N'Djamena [uhn•jah•MAY•nah] Capital of Chad. 12°N 15°E (p. RA22)
Nepal [NAY•PAHL] Mountain country between India and China. (p. RA26)
Netherlands [NEH•thuhr•lundz] Western European country north of Belgium. (p. RA18)
New Delhi [NOO DEH•lee] Capital of India. 29°N 77°E (p. RA26)
New Zealand [NOO ZEE•luhnd] Major island country southeast of Australia in the South Pacific. (p. RA30)
Niamey [nee•AHM•ay] Capital of Niger. 14°N 2°E (p. RA22)
Nicaragua [NIH•kuh•RAH•gwuh] Central American country south of Honduras. (p. RA15)
Nicosia [NIH•kuh•SEE•uh] Capital of Cyprus. 35°N 33°E (p. RA19)
Niger [NY•juhr] West African country north of Nigeria. (p. RA22)
Nigeria [ny•JIHR•ee•uh] West African country along the Gulf of Guinea. (p. RA22)
Nile [NYL] **River** Longest river in the world, flowing north through eastern Africa. (p. RA23)
North Korea [kuh•REE•uh] East Asian country in the northernmost part of the Korean Peninsula. (p. RA27)
Norway [NAWR•WAY] Northern European country on the Scandinavian peninsula. (p. RA18)
Nouakchott [nu•AHK•SHAHT] Capital of Mauritania. 18°N 16°W (p. RA22)

Oman [oh•MAHN] Country on the Arabian Sea and the Gulf of Oman. (p. RA25)
Oslo [AHZ•loh] Capital of Norway. 60°N 11°E (p. RA18)
Ottawa [AH•tuh•wuh] Capital of Canada. 45°N 76°W (p. RA13)
Ouagadougou [WAH•gah•DOO•goo] Capital of Burkina Faso. 12°N 2°W (p. RA22)

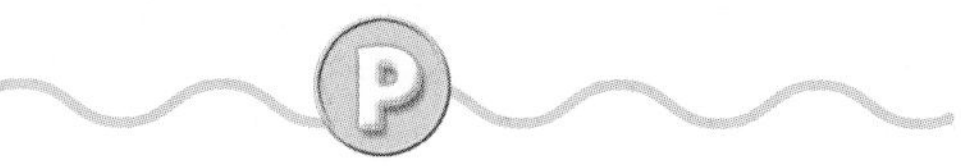

Pakistan [PA•kih•STAN] South Asian country northwest of India on the Arabian Sea. (p. RA26)
Palau [puh•LOW) Island country in the Pacific Ocean. 7°N 135°E (p. RA30)
Panama [PA•nuh•MAH] Central American country on the Isthmus of Panama. (p. RA15)
Panama Capital of Panama. 9°N 79°W (p. RA15)
Papua New Guinea [PA•pyu•wuh NOO GIH•nee] Island country in the Pacific Ocean north of Australia. 7°S 142°E (p. RA30)
Paraguay [PAR•uh•GWY] South American country northeast of Argentina. (p. RA16)
Paramaribo [PAH•rah•MAH•ree•boh] Capital of Suriname. 6°N 55°W (p. RA16)
Paris Capital of France. 49°N 2°E (p. RA18)
Persian [PUHR•zhuhn] **Gulf** Arm of the Arabian Sea between Iran and Saudi Arabia. (p. RA25)
Peru [puh•ROO] South American country south of Ecuador and Colombia. (p. RA16)
Philippines [FIH•luh•PEENZ] Island country in the Pacific Ocean southeast of China. (p. RA27)
Phnom Penh [puh•NAWM PEHN] Capital of Cambodia. 12°N 106°E (p. RA27)
Poland [POH•luhnd] Eastern European country on the Baltic Sea. (p. RA18)
Port-au-Prince [POHRT•oh•PRIHNS] Capital of Haiti. 19°N 72°W (p. RA15)
Port Moresby [MOHRZ•bee] Capital of Papua New Guinea. 10°S 147°E (p. RA30)
Port-of-Spain [SPAYN] Capital of Trinidad and Tobago. 11°N 62°W (p. RA15)
Porto-Novo [POHR•toh•NOH•voh] Capital of Benin. 7°N 3°E (p. RA22)
Portugal [POHR•chih•guhl] Country west of Spain on the Iberian Peninsula. (p. RA18)
Prague [PRAHG] Capital of the Czech Republic. 51°N 15°E (p. RA18)
Puerto Rico [PWEHR•toh REE•koh] Island in the Caribbean Sea; U.S. Commonwealth. (p. RA15)
P'yŏngyang [pee•AWNG•YAHNG] Capital of North Korea. 39°N 126°E (p. RA27)

Qatar [KAH•tuhr] Country on the southwestern shore of the Persian Gulf. (p. RA25)
Quito [KEE•toh] Capital of Ecuador. 0° latitude 79°W (p. RA16)

Rabat [ruh•BAHT] Capital of Morocco. 34°N 7°W (p. RA22)
Reykjavík [RAY•kyah•VEEK] Capital of Iceland. 64°N 22°W (p. RA18)

Rhine [RYN] **River** River in western Europe that flows into the North Sea. (p. RA20)
Riga [REE•guh] Capital of Latvia. 57°N 24°E (p. RA19)
Rio Grande [REE•oh GRAND] River that forms part of the boundary between the United States and Mexico. (p. RA10)
Riyadh [ree•YAHD] Capital of Saudi Arabia. 25°N 47°E (p. RA25)
Rocky Mountains Mountain system in western North America. (p. RA7)
Romania [ru•MAY•nee•uh] Eastern European country east of Hungary. (p. RA19)
Rome Capital of Italy. 42°N 13°E (p. RA18)
Russia [RUH•shuh] Largest country in the world, covering parts of Europe and Asia. (pp. RA19, RA27)
Rwanda [ruh•WAHN•duh] East African country south of Uganda. 2°S 30°E (p. RA22)

Sahara [suh•HAR•uh] Desert region in northern Africa that is the largest hot desert in the world. (p. RA23)
Saint Lawrence [LAWR•uhns] **River** River that flows from Lake Ontario to the Atlantic Ocean and forms part of the boundary between the United States and Canada. (p. RA13)
Sanaa [sahn•AH] Capital of Yemen. 15°N 44°E (p. RA25)
San José [SAN hoh•ZAY] Capital of Costa Rica. 10°N 84°W (p. RA15)
San Marino [SAN muh•REE•noh] Small European country located on the Italian Peninsula. 44°N 13°E (p. RA18)
San Salvador [SAN SAL•vuh•DAWR] Capital of El Salvador. 14°N 89°W (p. RA14)
Santiago [SAN•tee•AH•goh] Capital of Chile. 33°S 71°W (p. RA16)
Santo Domingo [SAN•toh duh•MIHNG•goh] Capital of the Dominican Republic. 19°N 70°W (p. RA15)
São Tomé and Príncipe [sow too•MAY PREEN•see•pee] Small island country in the Gulf of Guinea off the coast of central Africa. 1°N 7°E (p. RA22)
Sarajevo [SAR•uh•YAY•voh] Capital of Bosnia and Herzegovina. 43°N 18°E (p. RA18)
Saudi Arabia [SOW•dee uh•RAY•bee•uh] Country on the Arabian Peninsula. (p. RA25)
Senegal [SEH•nih•GAWL] West African country on the Atlantic coast. (p. RA22)
Seoul [SOHL] Capital of South Korea. 38°N 127°E (p. RA27)
Serbia [SUHR•bee•uh] Eastern European country south of Hungary. (p. RA18)
Seychelles [say•SHEHL] Small island country in the Indian Ocean off eastern Africa. 6°S 56°E (p. RA22)
Sierra Leone [see•EHR•uh lee•OHN] West African country south of Guinea. (p. RA22)
Singapore [SIHNG•uh•POHR] Southeast Asian island country near tip of the Malay Peninsula. (p. RA27)
Skopje [SKAW•PYAY] Capital of the country of Macedonia. 42°N 21°E (p. RA19)
Slovakia [sloh•VAH•kee•uh] Eastern European country south of Poland. (p. RA18)
Slovenia [sloh•VEE•nee•uh] Southeastern European country south of Austria on the Adriatic Sea. (p. RA18)
Sofia [SOH•fee•uh] Capital of Bulgaria. 43°N 23°E (p. RA19)
Solomon [SAH•luh•muhn] **Islands** Island country in the Pacific Ocean northeast of Australia. (p. RA30)
Somalia [soh•MAH•lee•uh] East African country on the Gulf of Aden and the Indian Ocean. (p. RA22)
South Africa [A•frih•kuh] Country at the southern tip of Africa, officially the Republic of South Africa. (p. RA22)
South Korea [kuh•REE•uh] East Asian country on the Korean Peninsula between the Yellow Sea and the Sea of Japan. (p. RA27)
Spain [SPAYN] Southern European country on the Iberian Peninsula. (p. RA18)
Sri Lanka [SREE LAHNG•kuh] Country in the Indian Ocean south of India, formerly called Ceylon. (p. RA26)
Stockholm [STAHK•HOHLM] Capital of Sweden. 59°N 18°E (p. RA18)
Sucre [SOO•kray] Constitutional capital of Bolivia. 19°S 65°W (p. RA16)
Sudan [soo•DAN] East African country south of Egypt. (p. RA22)
Suriname [SUR•uh•NAH•muh] South American country between Guyana and French Guiana. (p. RA16)
Suva [SOO•vah] Capital of the Fiji Islands. 18°S 177°E (p. RA30)
Swaziland [SWAH•zee•land] Southern African country west of Mozambique, almost entirely within the Republic of South Africa. (p. RA22)
Sweden Northern European country on the eastern side of the Scandinavian peninsula. (p. RA18)
Switzerland [SWIHT•suhr•luhnd] European country in the Alps south of Germany. (p. RA18)
Syria [SIHR•ee•uh] Southwest Asian country on the east side of the Mediterranean Sea. (p. RA24)

Taipei [TY•PAY] Capital of Taiwan. 25°N 122°E (p. RA27)
Taiwan [TY•WAHN] Island country off the southeast coast of China; the seat of the Chinese Nationalist government. (p. RA27)
Tajikistan [tah•JIH•kih•STAN] Central Asian country east of Turkmenistan. (p. RA26)

Gazetteer

Gazetteer

Tallinn [TA•luhn] Capital of Estonia. 59°N 25°E (p. RA19)
Tanzania [TAN•zuh•NEE•uh] East African country south of Kenya. (p. RA22)
Tashkent [tash•KEHNT] Capital of Uzbekistan. 41°N 69°E (p. RA26)
Tbilisi [tuh•bih•LEE•see] Capital of the Republic of Georgia. 42°N 45°E (p. RA26)
Tegucigalpa [tay•GOO•see•GAHL•pah] Capital of Honduras. 14°N 87°W (p. RA14)
Tehran [TAY•uh•RAN] Capital of Iran. 36°N 52°E (p. RA25)
Thailand [TY•LAND] Southeast Asian country east of Myanmar. 17°N 101°E (p. RA27)
Thimphu [thihm•POO] Capital of Bhutan. 28°N 90°E (p. RA27)
Tigris [TY•gruhs] **River** River in southeastern Turkey and Iraq that merges with the Euphrates River. (p. RA25)
Tiranë [tih•RAH•nuh] Capital of Albania. 42°N 20°E (p. RA18)
Togo [TOH•goh] West African country between Benin and Ghana on the Gulf of Guinea. (p. RA22)
Tokyo [TOH•kee•OH] Capital of Japan. 36°N 140°E (p. RA27)
Trinidad and Tobago [TRIH•nuh•DAD tuh•BAY•goh] Island country near Venezuela between the Atlantic Ocean and the Caribbean Sea. (p. RA15)
Tripoli [TRIH•puh•lee] Capital of Libya. 33°N 13°E (p. RA22)
Tshwane [ch•WAH•nay] Executive capital of South Africa. 26°S 28°E (p. RA22)
Tunis [TOO•nuhs] Capital of Tunisia. 37°N 10°E (p. RA22)
Tunisia [too•NEE•zhuh] North African country on the Mediterranean Sea between Libya and Algeria. (p. RA22)
Turkey [TUHR•kee] Country in southeastern Europe and western Asia. (p. RA24)
Turkmenistan [tuhrk•MEH•nuh•STAN] Central Asian country on the Caspian Sea. (p. RA25)

Uganda [yoo•GAHN•dah] East African country south of Sudan. (p. RA22)
Ukraine [yoo•KRAYN] Eastern European country west of Russia on the Black Sea. (p. RA25)
Ulaanbaatar [OO•LAHN•BAH•TAWR] Capital of Mongolia. 48°N 107°E (p. RA27)
United Arab Emirates [EH•muh•ruhts] Country made up of seven states on the eastern side of the Arabian Peninsula. (p. RA25)
United Kingdom Western European island country made up of England, Scotland, Wales, and Northern Ireland. (p. RA18)
United States of America Country in North America made up of 50 states, mostly between Canada and Mexico. (p. RA8)
Uruguay [YUR•uh•GWAY] South American country south of Brazil on the Atlantic Ocean. (p. RA16)
Uzbekistan [UZ•BEH•kih•STAN] Central Asian country south of Kazakhstan. (p. RA25)

Vanuatu [VAN•WAH•TOO] Country made up of islands in the Pacific Ocean east of Australia. (p. RA30)
Vatican [VA•tih•kuhn] **City** Headquarters of the Roman Catholic Church, located in the city of Rome in Italy. 42°N 13°E (p. RA18)
Venezuela [VEH•nuh•ZWAY•luh] South American country on the Caribbean Sea between Colombia and Guyana. (p. RA16)
Vienna [vee•EH•nuh] Capital of Austria. 48°N 16°E (p. RA18)
Vientiane [vyehn•TYAHN] Capital of Laos. 18°N 103°E (p. RA27)
Vietnam [vee•EHT•NAHM] Southeast Asian country east of Laos and Cambodia. (p. RA27)
Vilnius [VIL•nee•uhs] Capital of Lithuania. 55°N 25°E (p. RA19)

Warsaw Capital of Poland. 52°N 21°E (p. RA19)
Washington, D.C. Capital of the United States, in the District of Columbia. 39°N 77°W (p. RA8)
Wellington [WEH•lihng•tuhn] Capital of New Zealand. 41°S 175°E (p. RA30)
West Indies Caribbean islands between North America and South America. (p. RA15)
Windhoek [VIHNT•HUK] Capital of Namibia. 22°S 17°E (p. RA22)

Yamoussoukro [YAH•moo•SOO•kroh] Second capital of Côte d'Ivoire. 7°N 6°W (p. RA22)
Yangon [YAHNG•GOHN] City in Myanmar; formerly called Rangoon. 17°N 96°E (p. RA27)
Yaoundé [yown•DAY] Capital of Cameroon. 4°N 12°E (p. RA22)
Yemen [YEH•muhn] Country south of Saudi Arabia on the Arabian Peninsula. (p. RA25)
Yerevan [YEHR•uh•VAHN] Capital of Armenia. 40°N 44°E (p. RA25)

Zagreb [ZAH•GREHB] Capital of Croatia. 46°N 16°E (p. RA18)
Zambia [ZAM•bee•uh] Southern African country north of Zimbabwe. (p. RA22)
Zimbabwe [zihm•BAH•bway] Southern African country northeast of Botswana. (p. RA22)

Glossary/Glosario

- Content vocabulary are words that relate to geography content. They are **boldfaced** and highlighted yellow in your text.
- Words below that have an asterisk (*) are academic vocabulary. They help you understand your school subjects and are **boldfaced** in your text.

English

absolute location exact spot where a place is found (p. 15)

***access** a way or means of approach (p. 275)

***accumulate** to increase in amount (p. 48)

***accurate** exact (p. 777)

acid rain chemicals from air pollution that combine with precipitation (pp. 64, 171)

***acknowledge** recognize (p. 815)

***acquire** get (p. 794)

action song art form that arose in New Zealand in the 1900s and blends traditional dance with modern music (p. 803)

***adapt** change (pp. 125, 743)

***adequate** enough to satisfy a particular requirement (p. 450)

***adjacent** next to or near (p. 777)

***administrator** a person who manages or directs (p. 550)

***affect** to influence, or produce an effect upon (p. 275)

alluvial plain area built up by rich fertile soil left by river floods (pp. 444, 496)

***alter** to change (p. 57)

altitude height above sea level (p. 201)

amendment an addition to a legal document or law (p. 141)

anime Japanese style of animation that arose in the late 1900s and appears in comic books and cartoons (p. 741)

annex declare ownership of an area (p. 136)

***annual** occurring once a year (p. 539)

apartheid system of laws in South Africa aimed at separating the races (p. 555)

Español

ubicación absoluta punto exacto donde se encuentra un lugar (pág. 15)

***acceso** manera o medio de acercamiento (pág. 275)

***acumular** aumentar en cantidad (pág. 48)

***preciso** exacto (pág. 777)

lluvia ácida sustancias químicas producto de la contaminación ambiental que se mezclan con las precipitaciones (págs. 64, 171)

***reconocer** examinar con cuidado a algo o alguien (pág. 815)

***adquirir** conseguir (pág. 794)

canción activa forma de arte que surgió en Nueva Zelanda a principios del siglo XIX y que combina la danza tradicional con música moderna (pág. 803)

***adaptar** cambiar (págs. 125, 743)

***adecuado** suficiente para satisfacer una necesidad específica (pág. 450)

***adyacente** junto a o cerca de (pág. 777)

***administrador** persona que administra o dirige (pág. 550)

***afectar** influir o producir un efecto en algo o alguien (pág. 275)

planicie aluvial zona creada por el sedimento fértil depositado por las inundaciones de los ríos (págs. 444, 496)

***alterar** cambiar (pág. 57)

altitud altura sobre el nivel del mar (pág. 201)

enmienda adición a un documento legal o ley (pág. 141)

anime estilo de animación japonesa que surgió a finales del siglo XIX y que aparece en libros de cómics y dibujos animados (pág. 741)

anexar declarar posesión de un área (pág. 136)

***anual** que ocurre una vez por año (pág. 539)

apartheid sistema de leyes en Sudáfrica que tenía como fin la separación de las razas (pág. 555)

aquifer underground layer of rock through which water flows (pp. 52, 450)

acuífero capa de roca subterránea a través de la que fluye agua (págs. 52, 450)

archipelago group of islands (pp. 193, 689)

archipiélago grupo de islas (págs. 193, 689)

***assemble** put together (p. 235)

***reunir** juntar (pág. 235)

atmosphere layer of oxygen and other gases that surrounds Earth (p. 36)

atmósfera capa de oxígeno y otros gases que rodean la Tierra (pág. 36)

atoll circular shaped islands made of coral (pp. 613, 778)

atolón islas de forma circular formadas por corales (págs. 613, 778)

***attitudes** a particular feeling or way of thinking about something (p. 310)

***actitudes** sentimientos o maneras de pensar particulares respecto de algo (pág. 310)

***authority** power or influence over others (p. 300)

***autoridad** poder o influencia sobre los demás (pág. 300)

autonomy having independence from another country (pp. 339, 399)

autonomía ser independiente de otro país (págs. 339, 399)

***availability** state of being easy or possible to get or use (p. 52)

***disponibilidad** fácil o posible de obtener o usar (pág. 52)

axis imaginary line that passes through the center of Earth from the North Pole to the South Pole (p. 36)

eje línea imaginaria que atraviesa el centro de la Tierra desde el Polo Norte al Polo Sur (pág. 36)

B

ban legally block (p. 145)

prohibir impedir legalmente (pág. 145)

bazaar local marketplace in North Africa and Southwest Asia (p. 474)

bazar mercado local en África Septentrional y en el Sudoeste Asiático (pág. 474)

bedouin nomadic desert people of Southwest Asia who follow a traditional way of life (p. 492)

beduino habitantes nómades del desierto del sudoeste de Asia que tienen un estilo de vida tradicional (pág. 492)

***benefit** something that does good to a person or thing (pp. 373, 573)

***beneficio** bien que se hace a una persona o cosa (págs. 373, 573)

bilingual accepting two official languages; able to speak two languages (pp. 149, 332)

bilingüe que acepta dos idiomas oficiales, que puede hablar dos idiomas (págs. 149, 332)

biodiversity variety of plants and animals living on the planet (p. 66)

biodiversidad variedad de plantas y animales que viven en el planeta (pág. 66)

biome area that includes particular kinds of plants and animals adapted to conditions there (p. 60)

bioma área que incluye clases particulares de plantas y animales adaptadas a las condiciones del área (pág. 60)

biotechnology study of cells to find ways of improving health (p. 159)

biotecnología estudio de las células para descubrir maneras de mejorar la salud (pág. 159)

birthrate number of children born each year for every 1,000 people (p. 73)

índice de natalidad cantidad de niños nacidos por año cada 1,000 personas (pág. 73)

blizzard severe winter storm that lasts several hours and combines high winds with heavy snow (p. 128)

ventisca tormenta de invierno intensa que dura varias horas y combina vientos y nevadas fuertes (pág. 128)

bog low swampy area (p. 325)

ciénaga zona pantanosa y baja (pág. 325)

***bonds** a uniting or binding force or influence (p. 307)

***lazos** fuerza influyente que une o vincula (pág. 307)

boomerang flat, bent wooden tool of the Australian Aborigines that is thrown to stun prey when it strikes them and that sails back to the hunter if it misses its target (p. 793)

bumerán arma de madera curva y plana de los aborígenes australianos, que se arroja para golpear y aturdir a la presa y que regresa al cazador si no da en el blanco (pág. 793)

***boycott** to refuse to buy items from a particular country (p. 631)

breadfruit fruit from a tree of the same name that is a basic food in Oceania (p. 785)

***brief** not very long (p. 591)

brownfield sites that have been abandoned and may contain dangerous chemicals (p. 172)

bush rural areas in Australia (p. 801)

***boicot** negarse a comprar artículos de un país determinado (pág. 631)

árbol del pan fruta de un árbol del mismo nombre que es alimento básico en Oceanía (pág. 785)

***breve** no muy largo (pág. 591)

zona industrial abandonada sitios abandonados que pueden contener sustancias químicas peligrosas (pág. 172)

monte zonas rurales de Australia (pág. 801)

cacao tropical tree whose seeds are used to make chocolate and cocoa (p. 573)

caliph successor to Muhammad (p. 460)

calligraphy art of beautiful writing (pp. 473, 719)

canopy umbrella-like covering formed by the tops of trees in a rain forest (pp. 200, 540)

canyon deep valleys with steep sides (p. 119)

***capable** able to do one's job well (p. 629)

carnival large festival held each spring in countries in Latin America on the last day before the Christian holy period called Lent (p. 224)

casbah older section of Algerian cities (p. 487)

cash crop farm product grown for export (pp. 211, 505)

cassava a plant whose roots are ground to make porridge (p. 579)

caste social class a person is born into and cannot change (p. 628)

***category** division or grouping (p. 756)

caudillo Latin American ruler, often a military officer or wealthy individual ruling as a dictator (p. 213)

celadon highly prized style of pottery that has a greenish tint (p. 743)

census a count of the number of people living in an area or country (p. 706)

century a period of 100 years (p. 16)

city-state independent political unit that includes a city and the surrounding area (pp. 295, 457)

civil disobedience use of nonviolent protests to challenge a government or its laws (p. 631)

civilization highly developed culture (p. 86)

cacao árbol tropical cuyas semillas se utilizan para hacer chocolate y productos de cacao (pág. 573)

califa sucesor de Mahoma (pág. 460)

caligrafía arte de escribir con letra bella (págs. 473, 719)

bóveda de follaje cubierta en forma de sombrilla, formada por las copas de los árboles en una selva tropical (págs. 200, 540)

cañón valles profundos con laterales de pendientes agudas (pág. 119)

***capaz** capacidad de hacer bien el trabajo (pág. 629)

carnaval gran festival que se realiza cada primavera en países de América Latina el día antes del período sagrado cristiano llamado Cuaresma (pág. 224)

casbah parte más antigua de las ciudades argelinas (pág. 487)

cultivo comercial producto agrícola cultivado para la exportación (págs. 211, 505)

mandioca planta cuyas raíces se muelen para hacer tapioca (pág. 579)

casta clase social en la que nace una persona y que no se puede modificar (pág. 628)

***categoría** división o grupo (pág. 756)

caudillo gobernante latinoamericano, a menudo un oficial militar o un hombre rico que gobierna como dictador (pág. 213)

celadón estilo de cerámica muy valorado que posee un color verdoso (pág. 743)

censo recuento de la cantidad de personas que habitan en una zona o país (pág. 706)

siglo período de cien años (pág. 16)

ciudad-estado unidad política independiente que incluye una ciudad y el área circundante (págs. 295, 457)

desobediencia civil uso de protestas no violentas para desafiar a un gobierno o a sus leyes (pág. 631)

civilización cultura con un alto desarrollo (pág. 86)

civil war fight between opposing groups for control of a country's government (p. 488)

guerra civil lucha entre grupos opuestos para obtener el control del gobierno de un país (pág. 488)

clan large group of people who have a common ancestor in the far past (pp. 495, 564)

clan grupo extenso de personas que tienen un ancestro en común en el pasado lejano (págs. 495, 564)

classical referring to the civilizations of ancient Greece and Rome (p. 295)

clásico referente a las civilizaciones de la Grecia y Roma antiguas (pág. 295)

climate pattern of weather that takes place in an area over many years (p. 56)

clima conjunto de condiciones atmosféricas que ocurren en una zona durante muchos años (pág. 56)

climate zone areas that have similar patterns of temperature and rainfall, and may have similar vegetation (p. 59)

zona climática áreas que tienen patrones similares de temperatura y precipitaciones, y pueden tener vegetación similar (pág. 59)

Cold War period from about 1947 until 1991 when the United States and the Soviet Union engaged in a political struggle for world influence but did not fight each other (p. 393)

Guerra Fría período desde 1947 hasta 1991, cuando los Estados Unidos y la Unión Soviética se involucraron en una lucha política para influir en el mundo pero sin combatir entre sí (pág. 393)

collection part of the water cycle; process by which streams and rivers carry water that has fallen to the earth back to the oceans (p. 54)

escurrimiento parte del ciclo del agua; proceso en el que los ríos y arroyos llevan el agua que cayó a la tierra, de regreso a los océanos (pág. 54)

collectivization a system in which small farms were combined into huge state-run enterprises with work done by mechanized techniques in the hopes of making farming more efficient and reducing the need for farmworkers (p. 392)

colectivización sistema en el que pequeñas granjas se combinaron en grandes emprendimientos controlados por el estado; el trabajo se realizaba con técnicas mecanizadas con la esperanza de que la agricultura fuera más eficiente y que disminuyera la necesidad de empleados (pág. 392)

colony overseas settlement tied to a parent country (p. 136)

colonia asentamiento en el extranjero unido a un país madre (pág. 136)

command economy economic system in which the government decides how resources are used and what goods and services are produced (pp. 240, 349)

economía de mando sistema económico en el que el gobierno decide cómo se usan los recursos y qué bienes y servicios se producen (págs. 240, 349)

***comment** talk about (p. 224)

***comentar** hablar sobre algo (pág. 224)

commonwealth self-governing territory (p. 240)

mancomunidad territorio que se autogobierna (pág. 240)

communism system of government in which the government controls the ways of producing goods (p. 303)

comunismo sistema de gobierno en el que el estado controla los modos de producción de los bienes (pág. 303)

communist state country whose government has strong control over the economy and society as a whole (pp. 215, 391)

estado comunista país cuyo gobierno tiene un fuerte control de la economía y la sociedad como un todo (págs. 215, 391)

community neighborhood (p. 172)

comunidad vecindario (pág. 172)

***complex** highly developed (p. 210)

***complejo** altamente desarrollado (pág. 210)

compound group of houses surrounded by walls; a pattern of rural housing typical in parts of Africa south of the Sahara (p. 563)

complejo habitacional grupo de casas rodeadas por paredes; modelo de viviendas rurales típico de las zonas de África al sur del Sahara (pág. 563)

***comprise** to be made up of (p. 409)

***componer** estar compuesto de (pág. 409)

concentration large amount in one area (p. 614)

concentración cantidad grande en una zona (pág. 614)

condensation part of the water cycle; process by which water changes from gas to liquid (p. 54)

condensación parte del ciclo del agua; proceso en el que el agua cambia de la forma gaseosa a la forma líquida (pág. 54)

***conduct** carry out (p. 421)

***conducir** llevar a cabo (pág. 421)

Glossary/Glosario

coniferous referring to evergreen trees that have their seeds in cones (p. 285)

***consequence** result (p. 628)

conservation careful use of resources to avoid wasting them (p. 66)

***considerable** much (p. 200)

***consist** made up of (p. 815)

***constant** happening a lot or all the time (p. 47)

constitution document that describes the structure and powers of a government and the rights of people in a country (p. 589)

constitutional monarchy form of government in which a monarch is the head of state but elected officials run the government (pp. 322, 488)

***constrain** limit (p. 117)

***construct** to build (p. 691)

consumer goods products people buy for personal use (p. 666)

***contemporary** of the present time, modern (p. 643)

contiguous joined together inside a common boundary (p. 117)

continent large landmass that rises above an ocean (p. 45)

continental island island formed centuries ago by the rising and folding of the ocean floor due to tectonic activity (p. 778)

continental shelf plateau off of a continent that lies under the ocean and stretches for several miles (p. 50)

***contrast** showing the difference between two things when they are compared (p. 619)

***controversy** disputed issue (p. 745)

***convert** change from one to another (p. 389)

***cooperate** to work together (p. 662)

copra dried coconut meat (p. 823)

coral reef long undersea structure formed by the tiny skeletons of coral, a kind of sea life (p. 777)

cordillera region of parallel mountain chains (pp. 118, 690)

core area at the center of the Earth, which includes a solid inner core and a hot liquid outer core (p. 45)

***core** basic or fundamental (p. 141)

cottage industry home- or village-based industry in which people make simple goods using their own equipment (p. 656)

coníferas referente a árboles perennes que guardan sus semillas en conos (pág. 285)

***consecuencia** resultado (pág. 628)

conservación uso cuidadoso de los recursos para evitar su derroche (pág. 66)

***considerable** mucho (pág. 200)

***consistir** compuesto por (pág. 815)

***constante** que sucede la mayor parte del o todo el tiempo (p. 47)

constitución documento que establece la estructura y los poderes de un gobierno y los derechos de las personas de un país (pág. 589)

monarquía constitucional forma de gobierno en la que un monarca es la cabeza del estado pero en el que funcionarios electos controlan el gobierno (págs. 322, 488)

***obligar** limitar (pág. 117)

***construir** edificar (pág. 691)

bienes de consumo productos que las personas compran para uso personal (pág. 666)

***contemporáneo** del momento actual, moderno (pág. 643)

contiguo unidos juntos dentro de una misma frontera (pág. 117)

continente gran masa continental que se alza por encima de un océano (pág. 45)

isla continental isla formada hace siglos por el alzamiento y plegamiento del piso oceánico debido a la actividad tectónica (pág. 778)

plataforma continental meseta saliente de un continente que se encuentra bajo el océano y se extiende por varias millas (pág. 50)

***contrastar** mostrar la diferencia entre dos cosas cuando se las compara (pág. 619)

***controversia** tema en discusión (pág. 745)

***convertir** cambiar de una cosa a otra (pág. 389)

***cooperar** trabajar juntos (pág. 662)

copra pulpa seca del coco de la palma (pág. 823)

arrecife de coral gran estructura submarina formada por los diminutos esqueletos del coral, un tipo de vida marina (pág. 777)

cordillera región de cadenas montañosas paralelas (págs. 118, 690)

núcleo área del centro de la Tierra que incluye un núcleo interno sólido y un núcleo externo de líquido caliente (pág. 45)

***esencial** básico o fundamental (pág. 141)

industria casera industria basada en el hogar o en la aldea, donde la gente fabrica bienes sencillos con sus propios equipos (pág. 656)

coup action in which a group of individuals seizes control of a government (p. 394)

covenant agreement (p. 459)

crop rotation changing what crops farmers plant in a field from year to year (p. 65)

***crucial** very important (p. 720)

crust uppermost layer of the Earth (p. 45)

cultural diffusion process of spreading ideas, languages, and customs from one culture to another (p. 87)

culture way of life of a group of people who share similar beliefs and customs (p. 83)

culture region area that includes different countries that share similar cultural traits (p. 88)

cuneiform form of writing from ancient Mesopotamia that consisted of wedge-shaped markings pressed into clay tablets (p. 457)

***currency** money (p. 303)

current steadily flowing stream of water in the ocean (p. 57)

cyclone a storm with high winds and heavy rains (p. 619)

czar title given to the emperors of Russia's past (p. 390)

golpe acción en que un grupo de individuos toma el control de un gobierno (pág. 394)

pacto acuerdo (pág. 459)

rotación de cultivos cambio en los cultivos que los agricultores plantan en un campo entre un año y el siguiente (pág. 65)

***crucial** muy importante (pág. 720)

corteza capa superior de la Tierra (pág. 45)

difusión cultural proceso de divulgación de ideas, idiomas y costumbres de una cultura a otra (pág. 87)

cultura estilo de vida de un grupo de personas que comparten creencias y costumbres similares (pág. 83)

región cultural zona que incluye diferentes países que comparten características culturales similares (pág. 88)

cuneiforme forma de escritura de la antigua Mesopotamia, que consistía de marcas en forma de cuña prensadas en tabletas de arcilla blanda (pág. 457)

***moneda** dinero (pág. 303)

corriente curso de agua que fluye constantemente en el océano (pág. 57)

ciclón tormenta con fuertes vientos y abundantes lluvias (pág. 619)

zar título que recibían los emperadores de Rusia en el pasado (pág. 390)

death rate number of deaths per year out of every 1,000 people (p. 73)

decade a period of 10 years (p. 16)

deciduous trees that lose their leaves in the fall (p. 285)

***decline** a change to a lower state or level (p. 382)

decree order issued by a leader that has the force of law (p. 419)

default failure to make debt payments that are due to a lender (p. 250)

***define** to describe or establish (p. 50)

deforestation cutting down of forests without replanting new trees (pp. 65, 540)

delta area formed by soil deposits at the mouth of a river (p. 612)

demilitarized zone (DMZ) area in which neighboring countries agree not to place any soldiers or weapons (p. 744)

índice de mortalidad número de muertes por año cada 1,000 personas (pág. 73)

década período de diez años (pág. 16)

caducifolios árboles que pierden sus hojas en otoño (pág. 285)

***descenso** cambio a un nivel o estado inferior (pág. 382)

decreto orden emitida por un líder que tiene el poder de ley (pág. 419)

incumplimiento imposibilidad de realizar los pagos de deuda que se deben a un prestamista (pág. 250)

***definir** describir o establecer (pág. 50)

deforestación destrucción de bosques sin plantar nuevos árboles (págs. 65, 540)

delta área formada por depósitos de tierra en la desembocadura de un río (pág. 612)

zona desmilitarizada (DMZ) área en la que países vecinos acuerdan no apostar soldados ni armas (pág. 744)

democracy form of limited government in which power rests with the people, and all citizens share in running the government (pp. 85, 295)

democracia forma de gobierno limitado en el que el poder reside en la gente, y en el que todos los ciudadanos comparten la gestión del gobierno (págs. 85, 295)

deposit insurance government plan that promises to repay people who deposit their money in a bank if the bank should go out of business (p. 421)

seguros de depósito plan del gobierno que promete devolver el dinero a la gente que lo deposita en un banco si éste cierra sus puertas (pág. 421)

desalinization process of treating seawater to remove salts and minerals and make it drinkable (p. 450)

desalinización proceso para quitar las sales y minerales del agua marina y volverla potable (pág. 450)

***despite** in spite of (p. 690)

***a pesar de** aunque (pág. 690)

desertification process by which dry areas turn into desert (p. 541)

desertificación proceso en el que áreas secas se convierten en desiertos (pág. 541)

developed country country with an economy that has a mix of agriculture, a great deal of manufacturing, and service industries and that is very productive and provides its people with a high standard of living (p. 94)

país desarrollado país con una economía que combina agricultura, manufacturas e industrias de servicio; es muy productivo y proporciona a sus habitantes un nivel de vida alto (pág. 94)

developing country country that has limited industry, where agriculture remains important, incomes are generally low (p. 94)

país en vías de desarrollo país que posee una industria limitada, en el que la agricultura sigue siendo importante y los ingresos son, en general, bajos (pág. 94)

dharma duty in Hinduism (p. 628)

dharma el deber en el hinduismo (pág. 628)

dialect local form of a language that may have a distinct vocabulary and pronunciation (p. 84)

dialecto forma local de un idioma que puede tener un vocabulario y una pronunciación diferentes (pág. 84)

dictatorship form of government in which a leader rules by force and typically limits citizens' freedoms (pp. 86, 487)

dictadura forma de gobierno en el que un líder gobierna por la fuerza y, por lo general, limita las libertades de los ciudadanos (págs. 86, 487)

dietary law rules in certain religions that detail which foods people can and cannot eat and how food should be prepared and handled (p. 472)

ley de alimentación reglas en ciertas religiones que detallan los alimentos que la gente puede y los que no puede comer, y la manera en que deben prepararse y manipularse (pág. 472)

***differentiate** to make or become different in some way (p. 321)

***distinguir, diferenciar** hacer o volver diferente de algún modo (pág. 321)

***discriminate** to treat members of a particular group unfairly (p. 667)

***discriminar** tratar injustamente a los miembros de un determinado grupo (pág. 667)

discrimination unfair treatment of members of a particular group (p. 553)

discriminación trato injusto de los miembros de un determinado grupo (pág. 553)

***distinct** clearly different from one another (p. 618)

***distinto** claramente diferentes entre sí (pág. 618)

***distinctive** clearly different, unique (p. 473)

***distintivo** claramente diferente, único (pág. 473)

***distort** present in a manner that is misleading (p. 784)

***distorsionar** presentar de forma engañosa (pág. 784)

***distribute** to spread out (p. 56)

***distribuir** dispersar (pág. 56)

***diverse** varied (p. 125)

***diverso** variado (pág. 125)

divide the high point in a landmass that determines the direction rivers flow (p. 120)

divisoria de aguas punto alto de una masa continental que determina la dirección en la que fluye un río (pág. 120)

***document** an important paper (p. 322)

***documento** papel importante (pág. 322)

***dominant** having controlling influence over others (p. 296)

***dominante** que posee influencia controladora sobre los demás (pág. 296)

***dominate** control (p. 709)

***dominar** controlar (pág. 709)

dominion self-governing country in the British Empire (p. 139)

drought long period of time without rainfall (pp. 126, 539)

dry farming agriculture that conserves water and uses crops and growing methods suited to semiarid environments (pp. 339, 449)

***duration** length of time something lasts (p. 784)

dynasty line of rulers from a single family that holds power for a long time (p. 705)

dzong Buddhist center of prayer and study (p. 642)

dzud weather pattern in Mongolia in which a harsh winter follows a dry summer (p. 696)

dominio país con gobierno propio en el Imperio Británico (pág. 139)

sequía período de tiempo largo sin lluvia (págs. 126, 539)

agricultura de secano agricultura que conserva el agua y usa cultivos y métodos de crecimiento apropiados para medios semiáridos (págs. 339, 449)

***duración** lapso de tiempo en el que algo perdura (pág. 784)

dinastía línea de gobernantes de una única familia que mantiene el poder por un período largo (pág. 705)

dzong centro budista de oración y estudio (pág. 642)

dzud patrón climático en Mongolia en el cual un invierno severo sigue a un verano seco (pág. 696)

earthquake sudden and violent movement of the Earth's crust that shakes the land, and can cause great damage (p. 47)

economic system system that sets rules for deciding what goods and services to produce, how to produce them, and who will receive them (p. 94)

***economy** way of producing goods (p. 136)

ecosystem place shared by plants and animals that depend on one another for survival (p. 66)

ecotourism type of tourism in which people visit a country to enjoy its natural wonders (p. 540)

***element** part of something larger (p. 219)

***eliminate** to remove or get rid of (p. 392)

El Niño weather phenomenon marked by very heavy rains in western South America, often causing flooding; reduced rainfall in Southern Asia, Australia, and Africa; and severe storms in North America; opposite of **La Niña** (p. 58)

embargo ban on trade with a particular country (p. 497)

***emerge** to become known (p. 709)

emigrate to leave a country and move to another (p. 75)

emperor all-powerful ruler (p. 297)

***emphasis** special attention or importance given to something (p. 505)

empire collection of different territories united under the rule of one government (p. 210)

terremoto movimiento sorpresivo y violento de la corteza terrestre que sacude la tierra y puede ocasionar grandes daños (pág. 47)

sistema económico sistema que establece las reglas que deciden qué bienes y servicios se producen, cómo producirlos y quién los recibirá (pág. 94)

***economía** modo de producción de bienes (pág. 136)

ecosistema lugar compartido por plantas y animales que dependen unos de otros para sobrevivir (pág. 66)

ecoturismo tipo de turismo en el que la gente visita un país para disfrutar sus maravillas naturales (pág. 540)

***elemento** parte de algo más grande (pág. 219)

***eliminar** quitar o deshacerse de (pág. 392)

El Niño fenómeno meteorológico caracterizado por lluvias muy fuertes en la parte occidental de América del Sur, que suelen provocar inundaciones; lluvias reducidas en el sur de Asia, Australia y África, y grandes tormentas en América del Norte; opuesto de **La Niña** (pág. 58)

embargo prohibición de comercio con un país en particular (pág. 497)

***surgir** hacerse conocido (pág. 709)

emigrar dejar un país y mudarse a otro (pág. 75)

emperador gobernante todopoderoso (pág. 297)

***énfasis** atención o importancia especial que se da a algo (pág. 505)

imperio conjunto de diferentes territorios unidos bajo el control de un gobierno (pág. 210)

enclave small territory entirely surrounded by larger territory (pp. 508, 591)

***enormous** very big (p. 540)

environment natural surroundings of people (p. 15)

epic tales or poems about heroes or heroines (p. 472)

equinox either of the days in spring and fall in which the noon sun is overhead at the Equator and day and night are of equal length in both the Northern and Southern Hemispheres (p. 38)

erg large areas of soft sands and dunes in the Sahara (p. 448)

erosion process by which weathered bits of rock are moved elsewhere by water, wind, or ice (p. 48)

escarpment steep cliff at the edge of a plateau with a lowland area below (pp. 194, 531)

***establish** set up (p. 825)

estuary an area where river currents and ocean tide meet (p. 194)

ethnic cleansing forcing people from one ethnic or religious group to leave an area so that it can be used by another group (p. 356)

ethnic group people with a common language, history, religion, and some physical traits (pp. 84, 307)

eucalyptus tree found only in Australia and nearby islands that is well suited to dry conditions with leathery leaves, deep roots, and ability to survive when rivers flood (p. 784)

evaporation part of the water cycle; process by which water changes from liquid to gas (p. 53)

***eventual** happening at a later time (p. 613)

***evolve** develop (p. 147)

exile being forced to live outside one's country for political reasons (p. 729)

***exploit** use (p. 400)

export to sell goods or resources to other countries (p. 95)

***expose** to put on display; to leave without shelter or protection (p. 444)

extended family household made up of several generations, including grandparents, parents, and children (p. 564)

extinction complete disappearance from the Earth of a particular kind of plant or animal (p. 829)

***extract** remove (p. 823)

enclave territorio pequeño totalmente rodeado por un territorio más grande (págs. 508, 591)

***enorme** muy grande (pág. 540)

medio ambiente entorno natural de las personas (pág. 15)

épica relatos o poemas sobre héroes o heroínas (pág. 472)

equinoccio cualquier de los días en primavera y otoño cuando el sol del mediodía está sobre el Ecuador y el día y la noche tienen igual duración en los hemisferios Sur y Norte (pág. 38)

erg áreas grandes de arenas suaves y dunas en el Sahara (pág. 448)

erosión proceso por el que trozos de roca expuestos a la intemperie se mueven a otros sitios con el agua, el viento o el hielo (pág. 48)

escarpadura acantilado en gran pendiente en el borde de una meseta con una área de tierra baja debajo (págs. 194, 531)

***establecer** formar (pág. 825)

estuario área donde las corrientes del río y la marea del océano se encuentran (pág. 194)

limpieza étnica forzar a la gente de un grupo étnico o religioso a abandonar un área para que la use otro grupo (pág. 356)

grupo étnico personas con idioma, historia, religión y ciertas características físicas en común (págs. 84, 307)

eucalipto árbol que se encuentra solamente en Australia y las islas cercanas y que está bien adaptado a condiciones secas, con hojas coriáceas, raíces profundas y capacidad de sobrevivir con las inundaciones de los ríos (pág. 784)

evaporación parte del ciclo del agua; proceso en el que el agua cambia de la forma líquida a la forma gaseosa (pág. 53)

***final** que ocurre en un momento posterior (pág. 613)

***evolucionar** desarrollar (pág. 147)

exilio estar forzado a vivir fuera del país propio debido a razones políticas (pág. 729)

***explotar** usar (pág. 400)

exportar vender bienes o recursos a otros países (pág. 95)

***exponer** colocar en exhibición; dejar sin refugio o protección (pág. 444)

familia extendida familia formada por varias generaciones, incluidos los abuelos, los padres y los hijos (pág. 564)

extinción desaparición total de la Tierra de una clase particular de planta o animal (pág. 829)

***extraer** quitar (pág. 823)

fa'a Samoa Samoan way of life, which puts a heavy emphasis on living in harmony with the community and the land (p. 825)

***facilitate** make possible (p. 199)

fale traditional Samoan home that has no walls, opening the inside to cooling ocean breezes (p. 804)

famine severe lack of food (p. 73)

fault crack in the Earth's crust where two tectonic plates meet; prone to earthquakes (pp. 47, 507)

favela an overcrowded city slum in Brazil (p. 247)

***feature** a part or detail that stands out (p. 284)

federalism form of government in which power is divided between the federal, or national, government and the state governments (p. 140)

***fee** payment (p. 239)

fellahin peasant farmers of Egypt who rent small plots of land (p. 485)

fertility rate average number of children born to each woman (p. 307)

feudalism political and social system in which kings gave land to nobles in exchange for the nobles' promise to serve them; those nobles provided military service as knights for the king (p. 298)

***finance** provide funds or capital (p. 95)

***finite** limited in supply (p. 93)

fjord narrow, U-shaped coastal valley with steep sides formed by the action of glaciers (p. 327)

***focus** concentrate (p. 740)

fossil fuel oil, natural gas, or coal, which are an important part of the world's energy supply (p. 375)

free market type of economy in which people are free to buy, produce, and sell with limited government involvement (p. 159)

free port place where goods can be unloaded, stored, and shipped again without payment of import taxes (p. 712)

free trade removal of trade restrictions so that goods flow freely among countries (p. 96)

***fundamental** of central importance (p. 654)

fa'a Samoa estilo de vida de Samoa, que enfatiza la vida en armonía con la comunidad y la tierra (pág. 825)

***facilitar** hacer posible (pág. 199)

fale hogar tradicional de Samoa que no tiene paredes y se abre a las brisas refrescantes del océano (pág. 804)

hambruna falta grave de alimentos (pág. 73)

falla fractura en la corteza terrestre donde se unen dos placas tectónicas, propensa a los terremotos (págs. 47, 507)

favela barrio pobre y superpoblado de una ciudad en Brasil (pág. 247)

***característica** parte o detalle que se destaca (pág. 284)

federalismo forma de gobierno en el que el poder está dividido entre el gobierno federal, o nacional, y los estados (pág. 140)

***cargo** pagos (pág. 239)

labriego árabe campesinos agricultores de Egipto que arriendan pequeñas parcelas de tierra (pág. 485)

tasa de fertilidad cantidad de niños nacidos de cada mujer (pág. 307)

feudalismo sistema político y social en el que los reyes entregaban tierras a los nobles a cambio de su promesa de servirlos; dichos nobles brindaban servicio militar como caballeros del rey (pág. 298)

***financiar** proporcionar fondos o capital (pág. 95)

***finito** limitado en el suministro (pág. 93)

fiordo valle costero angosto y en forma de U con laderas abruptas que se formó por la acción de los glaciares (pág. 327)

***enfocar** concentrar (pág. 740)

combustible fósil petróleo, gas natural o carbón, que son una parte importante del suministro de la energía del mundo (pág. 375)

mercado libre tipo de economía en el que las personas son libres para comprar, producir y vender con participación limitada del gobierno (pág. 159)

puerto libre lugar donde se pueden descargar, almacenar y volver a embarcar mercancías sin pagar impuestos de importación (pág. 712)

libre comercio eliminación de restricciones comerciales de modo que los bienes circulen libremente entre los países (pág. 96)

***fundamental** de importancia central (pág. 654)

gaucho cowhand in Argentina (p. 249)

gasohol human-made fuel produced from mixing gasoline and alcohol made from sugarcane (p. 195)

generate make (p. 147)

generation groups of people about the same age (p. 804)

genocide mass murder of people from a particular ethnic group (pp. 508, 580)

Geographic Information System (GIS) combination of computer hardware and software used to gather, store, and analyze geographic information and then display it on a screen (p. 17)

geography study of the earth and its people (p. 15)

geothermal energy electricity produced by natural underground sources of steam (p. 327)

geyser spring of water heated by molten rock inside the earth that, from time to time, shoots hot water into the air (pp. 327, 778)

glacier giant sheets of ice (p. 119)

glasnost policy of political openness in the Soviet Union that allowed people to speak freely about the country's problems (p. 393)

globalization development of a worldwide culture with an interdependent economy (p. 89)

Global Positioning System (GPS) group of satellites that uses radio signals to determine the exact location of places on Earth (p. 17)

gorge like a canyon, a steep-sided valley formed when a river cuts through land that is being lifted upward (pp. 534, 691)

greenhouse effect buildup of certain gases in the Earth's atmosphere that, like a greenhouse, retain the sun's warmth (p. 64)

green revolution effort to use modern techniques and science to increase food production in poorer countries (p. 656)

griot storyteller from West Africa who relates oral traditions of the people (p. 562)

groundwater water that filters through the soil into the ground (p. 52)

***guarantee** promise (p. 159)

gaucho vaqueros de Argentina (pág. 249)

gasohol combustible fabricado por el ser humano, producto de la mezcla de gasolina y alcohol hecho de caña de azúcar (pág. 195)

generar hacer (pág. 147)

generación grupos de personas de aproximadamente la misma edad (pág. 804)

genocidio asesinato en masa de un grupo étnico particular (págs. 508, 580)

Sistema de Información Geográfica (GIS) combinación de hardware y software para obtener, almacenar y analizar información geográfica y luego exhibirla en una pantalla (pág. 17)

geografía estudio de la Tierra y sus habitantes (pág. 15)

energía geotérmica electricidad producida por fuentes de vapor subterráneas naturales (pág. 327)

géiser fuente de agua calentada por roca derretida dentro de la Tierra que de vez en cuando lanza agua caliente al aire (págs. 327, 778)

glaciar capas gigantes de hielo (pág. 119)

glásnost plan de acción de apertura política en la Unión soviética, que permitió que las personas hablaran libremente sobre los problemas del país (pág. 393)

globalización desarrollo de una cultura amplia mundial con una economía interdependiente (pág. 89)

Sistema de Posicionamiento Global (GPS) grupos de satélites que usan señales de radio para determinar la ubicación exacta de lugares en la Tierra (pág. 17)

desfiladero valle de laderas empinadas parecido a un cañón, que se forma cuando un río corta a través elterreno que se eleva (págs. 534, 691)

efecto invernadero acumulación de ciertos gases en la atmósfera de la Tierra que, al igual que un invernadero, retienen el calor del Sol (pág. 64)

revolución verde esfuerzo por usar las técnicas modernas y la ciencia para aumentar la producción de alimentos en los países más pobres (pág. 656)

griot narrador del África Occidental que relata tradiciones orales del pueblo (pág. 562)

agua subterránea agua que se filtra a través del suelo hacia las profundidades (pág. 52)

***garantizar** prometer (pág. 159)

habitat type of environment in which a particular animal species lives (pp. 578, 826)

hábitat tipo de medio ambiente en donde vive una especie particular de animal (págs. 578, 826)

haiku form of Japanese poetry known for being short and following a specific structure (p. 719)

haiku tipo de poesía japonesa conocida por su brevedad y su estructura específica (pág. 719)

hajj pilgrimage to the Muslim holy city of Makkah, the completion of which at least once in a lifetime is one of the Five Pillars of Islam (p. 471)

hajj peregrinaje hasta La Meca, ciudad sagrada de los musulmanes, cuya finalización al menos una vez en la vida es uno de los Cinco Pilares del Islam (pág. 471)

hangul writing system developed in the 1400s to write the Korean language in which characters represent sounds rather than ideas (p. 743)

hangul sistema de escritura desarrollado en los años 1400 para escribir el idioma coreano, en el que los caracteres representan sonidos en lugar de ideas (pág. 743)

heavy industry manufacture of goods such as machinery, mining equipment, and steel (p. 411)

industria pesada fabricación de bienes como maquinarias, equipo de minería y acero (pág. 411)

hieroglyphics system of writing that uses small pictures to represent sounds or words (pp. 209, 458)

jeroglíficos sistema de escritura que usa pequeños dibujos para representar sonidos o palabras (págs. 209, 458)

high island mountainous island in the Pacific Ocean formed by volcanic activity (p. 778)

isla volcánica isla montañosa en el Océano Pacífico formada por actividad volcánica (pág. 778)

high-technology industry areas of business that include making computers and other products with sophisticated engineering (p. 330)

industria de alta tecnología áreas de negocios que incluyen la creación de computadoras y otros productos con ingeniería sofisticada (pág. 330)

Holocaust mass killing of 6 million European Jews by Germany's Nazi leaders during World War II (p. 302)

Holocausto asesinato masivo de 6 millones de judíos europeos por parte de los líderes nazis de Alemania durante la Segunda Guerra Mundial (pág. 302)

human rights basic freedoms and rights, such as freedom of speech, that all people should enjoy (p. 729)

derechos humanos libertades y derechos fundamentales, como la libertad de expresión, que todas las personas deberían poseer (pág. 729)

hunter-gatherer person who moves from place to place to hunt and gather food (p. 549)

cazadores recolectores persona que se mueve de un lugar a otro para cazar y recolectar alimentos (pág. 549)

hurricane wind system that forms over the ocean in tropical areas and brings violent storms with heavy rains (p. 127)

huracán sistema de vientos que se forma sobre el océano en áreas tropicales y provoca tormentas violentas y fuertes lluvias (pág. 127)

iceberg huge piece of floating ice that broke off from an ice shelf or glacier and fell into the sea (p. 778)

iceberg masa enorme de hielo flotante que se separó de una plataforma de hielo o glaciar y cayó al mar (pág. 778)

ice shelf thick layer of ice that extends above the water (p. 778)

plataforma de hielo capa gruesa de hielo que se extiende por encima del agua (pág. 778)

***identical** exactly the same (p. 38)

***idéntico** exactamente lo mismo (pág. 38)

***impact** effect (p. 279)

***impacto** efecto (pág. 279)

import to buy resources or goods from other countries (p. 95)

importar comprar bienes o recursos a otros países (pág. 95)

***income** earned money (p. 350)

***ingreso** dinero ganado (pág. 350)

indigenous people descended from an area's first inhabitants (p. 149)

indígenas pueblo descendiente de los primeros habitantes de un área (pág. 149)

industrial diamond diamond used to make drills, saws, or grinding tools to process other materials rather than as a gemstone (p. 535)

diamante industrial diamante usado para hacer taladros, sierras o herramientas para moler; se usa para procesar otros materiales y no como piedra preciosa (pág. 535)

infrastructure system of roads and railroads that allows the transport of materials (pp. 375, 486)

infraestructura sistema de carreteras y vías férreas que permiten el transporte de materiales (págs. 375, 486)

***inhibit** limit (p. 375)

***inhibir** limitar (pág. 375)

***integral** necessary (p. 803)

***integral** necesario (pág. 803)

***intense** existing in an extreme degree (p. 444)

***intenso** que existe en un grado extremo (pág. 444)

intensive agriculture kind of farming that involves growing crops on every available piece of usable land (p. 737)

agricultura intensiva tipo de agricultura que implica el cultivo en toda porción de tierra disponible (pág. 737)

interdependence condition that exists when countries rely on each other for ideas, goods, services, and markets (p. 96)

interdependencia condición que existe cuando los países dependen uno del otro para obtener ideas, bienes, servicios y mercados (pág. 96)

***internal** existing or taking place within (p. 75)

***interno** que existe o sucede dentro de (pág. 75)

***invest** to lay out money so as to return a profit (pp. 335, 410)

***invertir** colocar dinero de manera de obtener una ganancia (págs. 335, 410)

irrigation process of collecting water and distributing it to crops (pp. 66, 457)

irrigación proceso de recolección de agua y su distribución en los cultivos (págs. 66, 457)

***isolate** to set or keep apart from others (p. 549)

***aislar** separar o mantener alejado de los otros (pág. 549)

***issue** problem (p. 248)

***asunto** problema (pág. 248)

isthmus narrow stretch of land connecting two larger land areas (p. 193)

istmo extensión estrecha de tierra que conecta dos masas de tierra más grandes (pág. 193)

jade shiny stone that comes in many shades of green (p. 209)

jade piedra brillante de varios tonos de verde (pág. 209)

jute plant with a tough fiber that is used for making rope, burlap bags, and carpet backing (p. 656)

yute planta con una fibra resistente que se usa para hacer sogas, bolsas de arpillera y bases para alfombras (pág. 656)

karma belief in Hinduism that one's actions in past lives determine the spiritual level into which one is reborn (p. 628)

karma creencia del hinduismo que sostiene que las acciones en las vidas pasadas determinan el nivel espiritual en el que se renace (pág. 628)

kibbutz settlement in Israel where settlers share all their property and make goods as well as carry out farming (p. 493)

kibutz asentamiento en Israel donde los colonos comparten su propiedad, producen bienes y se dedican a la agricultura (pág. 493)

kimono traditional Japanese clothing; a long robe, usually made of silk, with an open neck and large sleeves held in place by a wide sash (p. 740)

kimono atuendo tradicional japonés; túnica larga, generalmente de seda, con un escote abierto y amplias mangas que se sujeta a la cintura con una banda ancha (pág. 740)

kiwifruit small, fuzzy, brownish-colored fruit with bright green flesh (p. 816)

kiwi fruta pequeña de color marrón, cubierta de pelusa y con pulpa verde brillante (pág. 816)

krill tiny shrimplike sea creatures that provide food to whales and many other sea animals (p. 829)

krill diminutas criaturas marinas similares al camarón que son el alimento de las ballenas y muchos otros animales marinos (pág. 829)

lagoon shallow body of water in the center of an atoll (p. 613)

landlocked having no border with ocean or sea (pp. 275, 574)

land reform policy of South Korean government that broke large estates into smaller family farms and provided fertilizer for farmers (p. 744)

landslide disaster that occurs when soil is washed down steep hillsides due to an earthquake or heavy rain (p. 698)

La Niña weather phenomenon marked by unusually cool waters in the eastern Pacific and low amounts of rainfall there and heavier rains—and a greater chance of typhoons—in the western Pacific; opposite of **El Niño** (p. 58)

lawsuit legal action in which people ask for relief from some damage done to them by someone else (p. 813)

***layer** a thickness or fold (p. 64)

leap year year with 366 days, which happens every fourth year to make calendars match Earth's movement around the sun (p. 36)

***legal** based on laws (p. 458)

lichen tiny sturdy plants that grow in rocky areas (p. 786)

life expectancy number of years an average person is expected to live (p. 560)

light industry manufacture of consumer goods such as clothing, shoes, furniture, and household products (p. 411)

lineage larger family group with close blood ties, such as connection to a common grandmother or grandfather (p. 564)

lingua franca common language used for communication and trade (p. 824)

***link** connect (p. 666)

literacy rate percentage of people who can read and write (p. 239)

Llanos tropical grasslands that stretch through eastern Colombia and Venezuela (p. 194)

local wind wind pattern typical of a small area (p. 59)

loess fine-grained fertile soil deposited by the wind (p. 690)

low island type of island in the Pacific Ocean formed by the buildup of coral (p. 778)

laguna masa acuática poco profunda en el centro de un atolón (pág. 613)

sin salida al mar que no posee fronteras al mar o al océano (págs. 275, 574)

reforma agraria política del gobierno de Corea del Sur que separó grandes propiedades convirtiéndolas en granjas familiares más pequeñas y proveía fertilizante a los campesinos (pág. 744)

desplazamiento de tierra desastre que ocurre cuando la tierra se desliza pendiente abajo debido a un terremoto o fuertes lluvias (pág. 698)

La Niña fenómeno meteorológico caracterizado por aguas inusualmente frías y bajas cantidades de lluvia en el este del Pacífico, y fuertes lluvias (y una mayor posibilidad de tifones) en el oeste del Pacífico; opuesto de **El Niño** (pág. 58)

pleito acción legal en la que las personas exigen pago por algún daño que alguien les hizo (pág. 813)

***capa** grosor o doblez (pág. 64)

año bisiesto año con 366 días, que ocurre cada cuatro años para hacer que los calendarios coincidan con el movimiento de la Tierra alrededor del Sol (pág. 36)

***legal** basado en leyes (pág. 458)

liquen plantas diminutas y resistentes que crecen en áreas rocosas (pág. 786)

expectativa de vida cantidad de años promedio que se espera que viva una persona (pág. 560)

industria liviana fabricación de bienes de consumo como ropas, calzado, muebles y productos para la casa (pág. 411)

linaje grupo familiar amplio con lazos de sangre cercanos, como la relación con una abuela o abuelo en común (pág. 564)

lingua franca idioma en común usado para la comunicación y el comercio (pág. 824)

***unir** conectar (pág. 666)

tasa de alfabetización porcentaje de personas que saben leer y escribir (pág. 239)

llanos praderas tropicales que se extienden a través del este de Colombia y Venezuela (pág. 194)

viento local patrón de viento típico de un área pequeña (pág. 59)

loes tierra fina y fértil depositada por el viento (pág. 690)

isla de coral tipo de isla en el Océano Pacífico formada por la acumulación de coral (pág. 778)

magma hot melted rock inside the Earth that flows to the surface when a volcano erupts (p. 45)

***maintain** keep up (p. 248)

maize corn (p. 209)

***major** to be great in size or impact (p. 283)

malnutrition condition that results from people not getting enough nutrients because of not eating enough food or not eating a variety of foods (p. 559)

mangrove tree that grows along coastal areas in tropical regions; mangrove forests help maintain the health of coastal environments (p. 757)

mantle Earth's thickest layer, found between the core and the crust (p. 45)

maquiladora a foreign-owned factory in Mexico where workers assemble parts made in other countries (p. 235)

market economy economic system in which individuals make the decisions about how resources are used and what goods and services to provide (p. 350)

marsupial mammals that carry their young in a pouch (p. 779)

***media** types of communication such as the Internet, television, and radio (p. 159)

***medical** relating to the science or practice of medicine (p. 356)

megalopolis huge urban area made up of several large cities and nearby communities (pp. 117, 716)

merino breed of sheep known for especially fine wool (p. 814)

mestizo in Latin America, a person of mixed Native American and European heritage (p. 221)

middle class part of society that is neither very rich nor poor but has enough money to buy cars, new clothing, electronics, and luxury items (p. 410)

migrant worker person who earns a living by temporarily moving to a place separate from his or her home in order to work (pp. 236, 591)

migration movement of people (p. 219)

***militant** person who uses war or violence to accomplish goals (p. 340)

millennium a period of 1,000 years (p. 16)

magma roca caliente y derretida dentro de la Tierra que fluye a la superficie durante la erupción de un volcán (pág. 45)

***mantener** conservar (pág. 248)

maíz mazorca y grano (pág. 209)

***principal** grande en tamaño o impacto (pág. 283)

desnutrición situación que se da cuando las personas no obtienen los nutrientes necesarios porque no ingieren alimentos suficientes o no ingieren una variedad de alimentos (pág. 559)

mangle árboles que crecen a lo largo de las áreas costeras en las regiones tropicales; los bosques de mangles ayudan a mantener la salud de los medio ambientes costeros (pág. 757)

manto capa más gruesa de la Tierra que se encuentra entre el núcleo y la corteza (pág. 45)

maquiladora fábrica de propiedad extranjera en México donde los trabajadores montan piezas hechas en otros países (pág. 235)

economía de mercado sistema económico en el que las personas toman las decisiones sobre cómo se usan los recursos y qué bienes y servicios se proveen (pág. 350)

marsupial mamíferos que llevan a sus crías en una bolsa (pág. 779)

***medios de comunicación** tipos de comunicación como Internet, la televisión y la radio (pág. 159)

***médico** relativo a la ciencia o la práctica de la medicina (pág. 356)

megalópolis área urbanizada gigantesca compuesta por varias ciudades grandes y comunidades cercanas (págs. 117, 716)

merino raza de ovejas conocidas por su lana particularmente fina (pág. 814)

mestizo en América Latina, persona de herencia mixta de nativos americanos y europeos (pág. 221)

clase media parte de la sociedad que no es ni muy rica ni muy pobre, pero que posee suficiente dinero para comprar autos, ropas nuevas, y artículos electrónicos y de lujo (pág. 410)

trabajador migratorio persona que se gana la vida mudándose temporalmente a un sitio alejado de su hogar para poder trabajar (págs. 236, 591)

migración movimiento de personas (pág. 219)

***militante** persona que usa la guerra o la violencia para lograr objetivos (pág. 340)

milenio período de 1000 años (pág. 16)

missionary person who moves to another area to spread his or her religion (p. 389)

mistral cold, dry winter wind from the north that strikes southern France (p. 287)

monarchy government led by king or queen who inherited power by being born into ruling family (p. 86)

monotheism belief in one god (p. 459)

monsoon seasonal winds that blow steadily from the same direction for several months at a time but change directions at other times of the year (p. 618)

moshav settlement in Israel in which people share in farming, production, and selling, but each person is allowed to own some private property as well (p. 493)

mosque Islamic house of worship (p. 473)

mural large painting on a wall (p. 224)

multinational company company that has locations in more than one country (p. 332)

misionero persona que se muda a otra zona para difundir su religión (pág. 389)

mistral viento invernal, frío y seco, que llega al sur de Francia desde el norte (pág. 287)

monarquía gobierno conducido por un rey o una reina que heredaron el poder al nacer dentro de la familia reinante (pág. 86)

monoteísmo creencia en un dios (pág. 459)

monzón vientos estacionales que soplan sin cesar en la misma dirección durante varios meses y que luego cambian de dirección en otros momentos del año (pág. 618)

moshav asentamiento en Israel donde la gente comparte la agricultura, la producción y la venta, pero donde también cada persona puede poseer propiedad privada (pág. 493)

mezquita casa de culto islámica (pág. 473)

mural pintura grande sobre una pared (pág. 224)

compañía multinacional empresa que tiene ubicaciones en más de un país (pág. 332)

national debt money owed by the government (p. 250)

nationalism feelings of affection and loyalty towards one's country (pp. 397, 552)

nationalize taking control of an industry or company by a government (p. 661)

nation-state country formed of people who share a common culture and history (p. 299)

natural resource material from the Earth that people use to meet their needs (p. 93)

navigable referring to a body of water wide and deep enough for ships to use (pp. 119, 277)

neutrality refusal to take sides in a war between other countries (p. 336)

newly industrialized country country that is creating new manufacturing and business (p. 94)

newsprint type of paper used for printing newspapers (p. 162)

nirvana state of perfect peace and an end to the cycle of rebirth; the goal of following the Eightfold Path in Buddhism (p. 629)

nomad person who lives by moving from place to place to follow herds of migrating animals that they hunt or to lead herds of grazing animals to fresh pasture (p. 449)

***nonetheless** in spite of (p. 732)

deuda nacional dinero que el gobierno debe (pág. 250)

nacionalismo sentimiento de afecto y lealtad hacia el propio país (págs. 397, 552)

nacionalizar toma del control de una industria o empresa por parte del gobierno (pág. 661)

estado-nación país formado por personas que comparten una historia y una cultura en común (pág. 299)

recurso natural material del planeta Tierra que la gente usa para cubrir sus necesidades (pág. 93)

navegable referente a un cuerpo de agua con anchura y profundidad suficientes para que lo usen barcos (págs. 119, 277)

neutralidad negativa a tomar posición en una guerra entre otros países (pág. 336)

país recientemente industrializado país que está creando nuevos negocios y manufacturas (pág. 94)

papel de periódico tipo de papel usado para imprimir periódicos (pág. 162)

nirvana estado de paz perfecta y finalización del ciclo del renacimiento; meta de la Óctuple Senda en el budismo (pág. 629)

nómada persona que se traslada de un lugar al otro para perseguir rebaños de animales migratorios que caza o para guiar su rebaños a pasturas nuevas (pág. 449)

***sin embargo** a pesar de (pág. 732)

nonrenewable resource natural resource such as a mineral that cannot be replaced (p. 93)

novel long fictional story (p. 706)

nuclear family family group that includes only parents and their children (p. 564)

recurso no renovable recurso natural, como un mineral, que no se puede reemplazar (pág. 93)

novela historia larga de ficción (pág. 706)

familia nuclear grupo familiar que sólo incluye a los padres y los hijos (pág. 564)

oasis fertile area that rises in a desert wherever water is regularly available (p. 448)

obsidian hard, black, volcanic glass useful for making weapons (p. 209)

***occur** to be found in (p. 50)

oligarch member of a small ruling group that holds great power (p. 420)

***ongoing** in progress (p. 639)

oral tradition passing stories by word of mouth from generation to generation (p. 397)

orbit specific path each planet follows around the sun (p. 35)

outback inland areas of Australia west of the Great Dividing Range (p. 777)

***output** results of agricultural or industrial production (p. 506)

outsourcing hiring workers in other countries to do a set of jobs (p. 657)

overgraze problem that occurs when grazing animals like cattle and sheep eat too much of the sparse vegetation in a semiarid region (p. 575)

***overlap** to have in common (p. 653)

***overseas** beyond or across the seas (p. 716)

ozone gas that forms a layer around the Earth in the atmosphere; it blocks out many of the most harmful rays from the sun (p. 830)

oasis área fértil que se alza en el desierto donde hay agua regularmente disponible (pág. 448)

obsidiana vidrio volcánico de color negro y resistente, útil para fabricar armas (pág. 209)

***ocurrir** que se encuentra en algo (pág. 50)

oligarca miembro de un grupo pequeño de gobierno que posee gran poder (pág. 420)

***permanente** en progreso (pág. 639)

tradición oral pasar historias de boca en boca de generación en generación (pág. 397)

órbita trayectoria específica que cada planeta sigue alrededor del Sol (pág. 35)

outback áreas del interior de Australia al oeste de la Gran Cordillera Divisoria (pág. 777)

***productos** resultados de la producción industrial o agrícola (pág. 506)

tercerización contrato de trabajadores en otros países para hacer una serie de trabajos (pág. 657)

sobrepastoreo problema que ocurre cuando animales de pastoreo como el ganado y las ovejas comen demasiado de la escasa vegetación de una región semiárida (pág. 575)

***imbricar** tener en común (pág. 653)

***ultramar** más allá o a través de los mares (pág. 716)

ozono gas que forma una capa alrededor de la Tierra en la atmósfera; bloquea la mayoría de los rayos solares más dañinos (pág. 830)

pagoda tower with many stories built as a temple or memorial (p. 719)

Pampas treeless grassland of Argentina and Uruguay (p. 194)

parliamentary democracy form of government in which voters elect representatives to a lawmaking body called Parliament, and members of Parliament vote for an official called the prime minister to head the government (pp. 141, 323)

***participate** take part in (p. 147)

pagoda torre con varios pisos construida como templo o monumento conmemorativo (pág. 719)

pampas pradera desprovista de árboles en Argentina y Uruguay (pág. 194)

democracia parlamentaria forma de gobierno en la que los votantes eligen representantes para un cuerpo que crea las leyes denominado Parlamento, sus miembros votan a un funcionario llamado primer ministro como jefe del gobierno (págs. 141, 323)

***participar** tomar parte en (pág. 147)

pass space people can use to travel through a mountain range (p. 277)

pasture grasses and other plants that are ideal feed for grazing animals (p. 785)

peat plants partly decayed in water which can be dried and burned for fuel (p. 325)

pensioner person who receives regular payments from the government because he or she is too old or sick to work (p. 410)

perestroika policy of economic restructuring in the Soviet Union that called for less government control of the economy (p. 393)

***period** a portion of time (p. 379)

permafrost permanently frozen lower layers of soil found in the tundra and subarctic climate zones (p. 380)

pesticide powerful chemicals that kill crop-destroying insects (p. 66)

pharaoh name for powerful ruler in ancient Egypt (p. 458)

phosphate chemical salt used to make fertilizer (pp. 445, 485)

***physical** related to natural science (p. 15)

pidgin language language formed by combining parts of several different languages (pp. 221, 801)

plantation large farm (pp. 235, 552)

plate tectonics scientific theory that explains how processes within the Earth form continents and cause their movement (p. 46)

plaza public square in Latin American city around which government buildings and major churches were built (p. 233)

poaching illegal fishing or hunting (p. 445)

poi paste made in Oceania from the mashed tubers of the taro plant (p. 804)

polder reclaimed wetlands that use a system of dikes and pumps to keep out the sea's waters (p. 332)

***policy** a plan or course of action (p. 487)

pollutant chemical and smoke particles that cause pollution (p. 380)

polytheism belief in more than one god (p. 457)

pope head of the Roman Catholic Church (p. 298)

population density average number of people living in a square mile or square kilometer (p. 74)

porcelain high-quality kind of pottery that can be very thin and is covered with a shiny coating (p. 705)

paso lugar que la gente puede usar para viajar a través de una cordillera montañosa (pág. 277)

pastura pastos y otras plantas que son el alimento ideal para los animales de pastoreo (pág. 785)

turba plantas parcialmente descompuestas en agua que pueden secarse y usarse como combustible (pág. 325)

pensionado persona que recibe pagos regulares del gobierno dado que está demasiado anciana o enferma para trabajar (pág. 410)

perestroika política de reestructuración económica en la Unión Soviética que exigía menos control gubernamental de la economía (pág. 393)

***período** lapso de tiempo (pág. 379)

permafrost capas de suelo inferiores que están permanentemente congeladas en la tundra y en las zonas de clima ártico (pág. 380)

pesticida fuerte sustancia química que matan los insectos que destruyen los cultivos (pág. 66)

faraón nombre dado a los poderosos gobernantes del antiguo Egipto (pág. 458)

fosfato sal química usada para hacer fertilizante (págs. 445, 485)

***físico** relativo a las sciencias naturales (pág. 15)

lengua mixta idioma formado por la combinación de partes de varios idiomas diferentes (págs. 221, 801)

plantación granja grande (págs. 235, 552)

tectónica de placas teoría científica que explica cómo los procesos dentro de la Tierra forman los continentes y causan su movimiento (pág. 46)

plaza lugar público en las ciudades de América Latina alrededor del que se construyeron las iglesias principales y los edificios gubernamentales (pág. 233)

caza furtiva pesca o caza ilegal (pág. 445)

poi pasta comestible hecha con los tubérculos machacados del taro, en Oceanía (pág. 804)

pólder pantanos recuperados que usan un sistema de diques y bombas para mantener fuera a las aguas marinas (pág. 332)

***política** un plan o curso de acción (pág. 487)

contaminante sustancia química y partículas de humo que provocan contaminación (pág. 380)

politeísmo creencia en más de un dios (pág. 457)

papa líder de la Iglesia Católica Romana (pág. 298)

densidad de población cantidad promedio de personas que viven en una milla cuadrada o un kilómetro cuadrado (pág. 74)

porcelana tipo de cerámica de alta calidad que puede ser muy delgada y está cubierta con un revestimiento brillante (pág. 705)

***portion** part (p. 754)

potash mineral salt used in making fertilizer (p. 350)

prairie rolling inland grassland region with fertile soil (p. 118)

precious gem valuable stone (p. 753)

precipitation part of the water cycle; process by which water falls to the Earth as, for example, rain or snow (p. 54)

prevailing winds wind patterns that are similar over time (p. 57)

***primary** main or most important (pp. 400, 640)

***prime** very attractive (p. 795)

***principal** main or primary (p. 534)

***principle** rule or guideline (p. 140)

***prior** earlier in time or order (p. 419)

privatization transfer of ownership of businesses from the government to individuals (p. 409)

productivity measure of how much work a person produces in a set amount of time (p. 325)

***professional** person who works in a job that requires education and training (p. 657)

profit money a business earns after all its expenses are met (p. 159)

***prohibit** forbid (p. 737)

***promote** to put forward (p. 398)

prophet messenger of God (p. 459)

***porción** parte (pág. 754)

potasa sal mineral usada en la fabricación de fertilizantes (pág. 350)

pradera región interna de las llanuras, con pastizales ondulantes y tierras fértiles (pág. 118)

gema piedra valiosa (pág. 753)

precipitación parte del ciclo del agua; proceso mediante el que el agua cae a la Tierra, por ejemplo, como lluvia o nieve (pág. 54)

vientos predominantes patrones de viento que se mantienen similares con el paso del tiempo (pág. 57)

***primario** principal o más importante (págs. 400, 640)

***excelente** muy atractivo (pág. 795)

***principal** esencial o primario (pág. 534)

***principio** regla o pauta (pág. 140)

***precedente** anterior en tiempo u orden (pág. 419)

privatización transferencia de propiedad de los negocios del gobierno a particulares (pág. 409)

productividad medición de cuánto trabajo produce una persona en una cantidad fija de tiempo (pág. 325)

***profesional** persona que trabaja en un empleo que requiere educación y capacitación (pág. 657)

ganancia dinero que se gana en un negocio luego de cubrir todos los gastos (pág. 159)

***prohibir** vedar (pág. 737)

***promover** impulsar (pág. 398)

profeta mensajero de Dios (pág. 459)

Q

quota number limit on how many items of a particular product can be imported from a certain nation (p. 95)

cupo límite de la cantidad de artículos de un producto determinado que puede importarse de cierto país (pág. 95)

rain forest dense stand of trees and other growth that receives high amounts of precipitation each year (pp. 199, 540)

rain shadow effect of mountains that blocks rain from reaching interior regions (p. 59)

rationing making a resource available in limited amounts (p. 450)

refinery facility that turns petroleum into gasoline and other products (p. 446)

selva tropical agrupación densa de árboles y otras plantas que reciben grandes cantidades de precipitación cada año (págs. 199, 540)

sombra de lluvia efecto de las montañas que impiden que la lluvia alcance regiones interiores (pág. 59)

racionamiento acción de disponer de un recurso en cantidades limitadas (pág. 450)

refinería instalación que convierte el petróleo en gasolina y en otros productos (pág. 446)

refugee person who flees to another country to escape persecution or disaster (pp. 76, 553)

refugiado persona que huye a otro país para escapar de la persecución o el desastre (págs. 76, 553)

***regime** government (pp. 137, 492)

***régimen** gobierno (págs. 137, 492)

reincarnation belief that after a person dies, his or her soul is reborn into another body (p. 628)

reencarnación creencia que sostiene que el alma de una persona, después de muerta, renace en otro cuerpo (pág. 628)

***reject** oppose (p. 729)

***rechazar** oponer (pág. 729)

relative location description of where a place is in relation to the features around it (p. 15)

ubicación relativa descripción de dónde está un lugar en relación con las características a su alrededor (pág. 15)

***release** to relieve pressure (p. 45); to set free (p. 391)

***liberar** aliviar la presión (pág. 45); dar la libertad (pág. 391)

***reluctant** hesitant (p. 162)

***reacio** indeciso (pág. 162)

***rely** to depend on (pp. 330, 458)

***confiar** depender de (págs. 330, 458)

remittance money sent back home by workers who leave their home country to work in other nations (p. 240)

remesa dinero que envían a su hogar los trabajadores que abandonan su país para trabajar en otras naciones (pág. 240)

renewable resource natural resource that can be replaced naturally or grown again (p. 93)

recurso renovable recurso natural que puede reemplazarse o crecer nuevamente de manera natural (pág. 93)

republic government in which people choose their leaders (p. 296)

república gobierno en el que las personas eligen a sus líderes (pág. 296)

representative democracy form of government in which voters choose leaders who make and enforce the laws (p. 140)

democracia representativa forma de gobierno en el que los votantes eligen a sus líderes para que hagan cumplir las leyes (pág. 140)

***require** to have a need for (p. 564)

***requerir** tener una necesidad (pág. 564)

***research** work done by scientists or scholars (p. 828)

***investigación** trabajo realizado por científicos y estudiosos (pág. 828)

***reside** to live (p. 194)

***residir** vivir (pág. 194)

***resolve** to reach a decision about (p. 664)

***resolver** llegar a una decisión acerca de algo (pág. 664)

***restore** return; to put or bring back into existence or use (pp. 126, 582)

***restituir** devolver; restablecer al estado o uso (págs. 126, 582)

***restrict** to limit (p. 169)

***restringir** limitar (pág. 169)

reunification the act of being brought back together (p. 336)

reunificación el acto de ser vuelto a reunir (pág. 336)

***reveal** make known (p. 234)

***revelar** dar a conocer (pág. 234)

***revenue** incoming money (p. 474)

***ingreso** dinero obtenido por recaudación (pág. 474)

revolution one complete circuit around the sun (p. 36); sweeping change (pp. 215, 301)

revolución giro completo alrededor del Sol (pág. 36); cambio radical (págs. 215, 301)

***reverse** opposite (p. 38)

***reverso** opuesto (pág. 38)

rift valley area of low relief flanked by highland regions that is formed by the moving apart of two of the Earth's tectonic plates (p. 532)

valle de falla área de relieve bajo rodeada por regiones montañosas, formada por la separación de dos placas tectónicas de la Tierra (pág. 532)

rite of passage special ceremony that marks a particular stage in life, such as when young boys or girls reach adulthood (p. 562)

ritos de paso ceremonia especial que marca una etapa particular en la vida, por ejemplo, cuando los niños y las niñas alcanzan la edad adulta (pág. 562)

rotate to spin on an axis (p. 36)

rotar girar sobre su eje (pág. 36)

***route** a journey (p. 119)

***travesía** viaje (pág. 119)

samurai powerful land-owning warrior in Japan (p. 707)

saint Christian holy person (p. 472)

sanitation removal of waste products (p. 559)

sari traditional clothing for women in India; a long rectangular piece of cloth that can be draped around the body in several different ways (p. 644)

savanna broad grassland in the tropics with few trees (p. 541)

secular nonreligious (pp. 310, 491)

sedimentary rock type of rock formed when layers of sediment, or dirt from the ocean floor, are compressed together and harden (p. 444)

selva Brazilian name for the Amazonian rain forest (p. 247)

separatist movement campaign by members of an ethnic group to break away from the national government and form an independent state (p. 421)

serf farm laborer who could be bought and sold along with the land (p. 390)

***series** arranged in an order and alike in some way (pp. 531, 697)

***shift** to change from one to another (pp. 239, 577)

shipbreaking practice of bringing ashore large, oceangoing ships that have been retired from service; they are cut into pieces, and then sold (p. 664)

shogun military leader who ruled Japan in early times (p. 707)

***significant** important (p. 37)

silt small particles of rich soil (p. 444)

***similar** having qualities in common (p. 339)

sirocco hot winds from Africa that blow across southern Europe (p. 287)

sisal plant fiber used to make rope and twine (p. 578)

sitar long-necked instrument with seven strings on the outside and ten inside the neck that provides Indian music with a distinctive sound (p. 643)

***site** location (p. 697)

smog thick haze of smoke and chemicals (pp. 64, 236, 380)

samurai guerrero terrateniente y poderoso de Japón (pág. 707)

santo persona sagrada cristiana (pág. 472)

saneamiento eliminación de productos de desecho (pág. 559)

sari atuendo tradicional de las mujeres en la India; pieza de tejido grande y rectangular que se usa para envolver el cuerpo de distintas maneras (pág. 644)

sabana amplia pradera de los trópicos con pocos árboles (pág. 541)

secular no religioso (págs. 310, 491)

roca sedimentaria tipo de roca que se forma cuando las capas de sedimento o residuos del fondo oceánico se comprimen y endurecen (pág. 444)

selva nombre brasilero para los bosques tropicales del Amazonas (pág. 247)

movimiento separatista campaña realizada por los miembros de un grupo étnico para separarse del gobierno nacional y formar un estado independiente (pág. 421)

siervo labriego que podía comprarse y venderse junto con la tierra (pág. 390)

***serie** organizado en un orden y similar de alguna manera (págs. 531, 697)

***alternar** cambiar de una cosa a otra (págs. 239, 577)

desguace práctica que consiste en traer a tierra grandes barcos que ya no cumplen servicios, y desarmarlos en piezas que luego se venden (pág. 664)

shogun líder militar que gobernaba Japón en tiempos antiguos (pág. 707)

***significativo** importante (pág. 37)

lama pequeñas partículas de tierra fértil (pág. 444)

***similar** que posee cualidades en común (pág. 339)

siroco vientos calientes de África que soplan atravesando el sur de Europa (pág. 287)

sisal fibra vegetal que se usa para fabricar cuerdas y bramante (pág. 578)

sitar instrumento de mástil largo con siete cuerdas afuera y diez adentro del mástil, que le proporciona a la música india un sonido particular (pág. 643)

***sitio** ubicación (pág. 697)

smog neblina espesa, resultado de la combinación de humo y sustancias químicas (págs. 64, 236, 380)

social status person's position in the community (p. 562)

sodium nitrate mineral used in fertilizer and explosives (p. 252)

softwood wood of evergreen trees, often used in buildings or making furniture (p. 375)

solar system planets, along with their moons, asteroids and other bodies, and the sun (p. 35)

***sole** being the only one (p. 494)

***source** a point where something begins (p. 577)

***sparse** few or scattered (p. 450)

specialization focusing on certain economic activities to make the best use of resources (p. 330)

***specify** make clear (p. 828)

sphere of influence an area of a country where a single foreign power has been granted exclusive trading right (p. 708)

***stable** firmly established; not likely to change suddenly or greatly (pp. 213, 574)

station cattle or sheep ranch in rural Australia (p. 801)

***status** position or rank in relation to others (p. 627)

steppe partly dry grassland often found on the edges of a desert (p. 448)

stock part ownership in a company (p. 159)

***structure** an arrangement of parts (p. 589)

***style** form (p. 224)

subcontinent large landmass that is part of a continent (p. 611)

subregion smaller area of a region (p. 193)

subsidy special payment made by a government to support a particular group or industry (p. 340)

subsistence farm small plot of land on which a farmer grows only enough food to feed his or her family (pp. 235, 573)

suburb smaller community just outside a large city (p. 147)

succulent type of plant that has thick, fleshy leaves that can conserve moisture (p. 541)

***sufficient** enough (p. 784)

suffrage right to vote (p. 589)

estatus social posición de una persona en la comunidad (pág. 562)

nitrato de sodio mineral que se usa en fertilizantes y explosivos (pág. 252)

madera blanda madera de árboles siempre verdes, que se utiliza comúnmente en edificios o muebles (pág. 375)

sistema solar los planetas y sus lunas, los asteroides y otros cuerpos celestes además del Sol (pág. 35)

***único** ser uno solo (pág. 494)

***procedencia** punto en el que algo comienza (pág. 577)

***escaso** poco o disperso (pág. 450)

especialización concentrarse en ciertas actividades económicas para hacer el mejor uso de recursos (pág. 330)

***especificar** aclarar (pág. 828)

esfera de influencia área de un país donde un solo poder extranjero ha recibido el derecho exclusivo de comercio (pág. 708)

***estable** establecido con firmeza; sin probabilidades de cambiar repentina o ampliamente (págs. 213, 574)

estación rancho de ovejas o vacas en la zona rural de Australia (pág. 801)

***estatus** posición o rango en relación con otros (pág. 627)

estepa pastura parcialmente seca que suele encontrarse en los bordes de un desierto (pág. 448)

acción parte propietaria en una empresa (pág. 159)

***estructura** cierto tipo de arreglo de partes (pág. 589)

***estilo** forma (pág. 224)

subcontinente gran masa continental que es parte de un continente (pág. 611)

subregión zona más pequeña de una región (pág. 193)

subsidio pago especial hecho por un gobierno para apoyar a un grupo o industria particular (pág. 340)

agricultura de subsistencia parcela de tierra en la cual un granjero cultiva sólo los alimentos suficientes para alimentar a su familia (págs. 235, 573)

suburbio comunidad justo en las afueras de una ciudad grande (pág. 147)

suculenta tipo de planta que posee hojas gruesas y carnosas que conservan la humedad (pág. 541)

***suficiente** bastante (pág. 784)

sufragio derecho al voto (pág. 589)

summer solstice day that has the most daylight hours and the fewest hours of darkness (p. 37)

***survive** to remain alive (p. 620)

***sustain** support (p. 800)

solsticio de verano día que tiene la mayor cantidad de horas de luz diurna y la menor cantidad de horas de oscuridad (pág. 37)

***sobrevivir** permanecer vivo (pág. 620)

***sostener** soportar (pág. 800)

taiga large coniferous forests (p. 380)

tariff tax added to the price of goods that are imported (pp. 95, 170)

tatami straw mats that are traditionally used to cover floors in Japan (p. 739)

teak high-quality wood used to make buildings and ships because it is strong and durable (p. 693)

***technique** a method of accomplishing something (p. 65)

***technology** the application of scientific discoveries to practical use (p. 73)

***temporary** lasting for a limited time, not permanent (p. 661)

terraced fields strips of land cut out of a hillside like stair steps on which crops are planted (p. 758)

terrorism violence used against the people or government in the hopes of winning political goals (pp. 137, 464)

***theme** topic (p. 15)

theocracy form of government in which the leader claims to rule on behalf of a god (p. 457)

tornado severe windstorm that takes the form of a funnel-shaped cloud and often touches the ground (p. 127)

trade deficit situation that occurs when the value of a country's imports is higher than the value of its exports (pp. 169, 738)

trade sanction step taken to cut off trade with a country to show opposition to its government's actions (p. 487)

trade surplus situation that occurs when the value of a country's exports is higher than the value of its imports (p. 170)

***tradition** belief or custom handed down through generations (p. 562)

***transform** greatly change (p. 211)

***transport** move (p. 193)

trench deep cut in the ocean floor (p. 50)

taiga grandes bosques coníferos (pág. 380)

tarifa impuesto agregado al precio de los bienes que se importan (págs. 95, 170)

tatami esteras de paja que se usan tradicionalmente para cubrir los suelos en Japón (pág. 739)

teca madera de alta calidad que se usa para construir edificios y barcos debido a su resistencia y durabilidad (pág. 693)

***técnica** un método para lograr algo (pág. 65)

***tecnología** la aplicación de los descubrimientos científicos a un uso práctico (pág. 73)

***temporario** que dura un tiempo limitado, no permanente (pág. 661)

campos en terraza franjas de tierra que se excavan en la ladera de una colina como escalones, y en las que se plantan cultivos (pág. 758)

terrorismo violencia usada en contra del pueblo o del gobierno con la esperanza de alcanzar metas políticas (págs. 137, 464)

***tema** tópico (pág. 15)

teocracia forma de gobierno en la que el líder afirma reinar en nombre de un dios (pág. 457)

tornado intensa tormenta de viento que toma forma de embudo y suele tocar el suelo (pág. 127)

déficit comercial situación que ocurre cuando el valor de las importaciones de un país es mayor que el valor de sus exportaciones (págs. 169, 738)

sanción comercial acción realizada para bloquear el comercio con un país y así mostrar oposición a las acciones gubernamentales (pág. 487)

excedente comercial situación que ocurre cuando el valor de las exportaciones de un país es mayor que el valor de sus importaciones (pág. 170)

***tradición** creencia o costumbre que se transmite a través de las generaciones (pág. 562)

***transformar** cambiar mucho (pág. 211)

***transportar** mover (pág. 193)

fosa marina corte profundo en el suelo marino (pág. 50)

***trend** general tendency (p. 559)

tributary small river that flows into a larger river (p. 194)

Tropics area between the Tropic of Cancer and the Tropic of Capricorn, which has generally warm temperatures because it receives the direct rays of the sun for much of the year (pp. 38, 199)

trust territory area temporarily placed under control of another country (p. 797)

tsunami huge ocean wave caused by an earthquake on the ocean floor (p. 668)

tungsten metal combined with iron to make steel, also used to make the filaments in electric lightbulbs and parts of rockets that must resist high amounts of heat (p. 693)

***tendencia** predisposición general (pág. 559)

tributario río pequeño que fluye dentro de un río más grande (pág. 194)

trópicos área entre el Trópico de Cáncer y el Trópico de Capricornio con temperaturas generalmente cálidas, ya que recibe los rayos directos del Sol durante gran parte del año (págs. 38, 199)

territorio confiado área que se coloca temporalmente bajo el control de otro país (pág. 797)

tsunami enorme ola oceánica provocada por un terremoto en el fondo oceánico (pág. 668)

tungsteno metal combinado con hierro para crear acero, que se usa en los filamentos de las bombillas eléctricas y piezas de cohetes que deben resistir grandes cantidades de calor (pág. 693)

underemployment situation that arises when a worker must take a job that requires lesser skills than he or she possesses (p. 410)

***unify** to unite or bring together (p. 421)

***unique** being the only one of its kind (p. 89)

urban climate weather patterns in cities, including higher temperatures and distinct wind patterns, as compared to nearby rural areas (p. 61)

urban sprawl spread of human settlement into natural areas (p. 172)

urbanization growth of cities (pp. 75, 308)

subempleo situación provocada cuando un trabajador debe aceptar un trabajo que requiere menos habilidades que las que él posee (pág. 410)

***unificar** unir o juntar (pág. 421)

***exclusivo** ser el único en su tipo (pág. 89)

clima urbano patrones climáticos de las ciudades, incluyendo temperaturas más altas y patrones de vientos distintos, al compararse con áreas rurales cercanas (pág. 61)

expansión urbana expansión de asentamientos humanos en áreas naturales (pág. 172)

urbanización crecimiento de las ciudades (págs. 75, 308)

Glossary/Glosario

***vary** to be different (pp. 326, 619)

vaquero Mexican cowhand (p. 234)

***volume** amount (p. 411)

varna one of the four broad classes of human society under Hinduism (p. 627)

***variar** ser diferente (págs. 326, 619)

vaquero empleado que trabaja con el ganado en México (pág. 234)

***volumen** cantidad (pág. 411)

varna una de las cuatro clases más amplias de sociedad humana, según el hinduismo (pág. 627)

wadi dry riverbed that fills with water when rare rains fall in a desert (p. 448)

water cycle system in which water moves from the Earth to the air and back to the Earth (p. 53)

wadi (uadi) lecho de río seco que se llena de agua cuando las escasas lluvias caen en el desierto (pág. 448)

ciclo del agua sistema en el que el agua se mueve de la Tierra hacia el aire y luego de vuelta hacia la Tierra (pág. 53)

weather changes in temperature, wind speed and direction, and air moisture that take place over a short period of time (p. 56)

weathering process in which rock is broken into smaller pieces by water and ice, chemicals, or even plants (p. 47)

welfare state country where the government is the main provider of support for the sick, needy, and the retired (p. 307)

***widespread** scattered or found in a wide area (pp. 86, 591)

winter solstice day of the year that has the fewest hours of sunlight and the most hours of darkness (p. 38)

clima cambios en la temperatura, velocidad y dirección del viento, y humedad en el aire que duran un período breve (pág. 56)

deterioro por exposición proceso por el cual se rompen las rocas en pedazos más pequeños ocasionado por el agua y el hielo, los químicos o hasta los vegetales (pág. 47)

estado de bienestar país en el que el gobierno es el proveedor principal de ayuda para los enfermos, necesitados y jubilados (pág. 307)

***generalizado** que se distribuye o encuentra en un área amplia (págs. 86, 591)

solsticio de invierno día del año que tiene la menor cantidad de horas de luz diurna y la mayor cantidad de horas de oscuridad (pág. 38)

yurt large circular structure made of animal skins that can be packed up and moved from place to place; used as a home in Mongolia (p. 720)

yurt estructura circular amplia hecha con pieles de animales que puede empacarse y moverse de un lugar a otro; se usa como casa en Mongolia (pág. 720)

Index

c=chart m=map
d=diagram p=photo
g=graph q=quote

Index

Index

Index

Index

Index

Index

Index

Index

Index

Index

Index

Index

X

Text Acknowledgments

142—"Should the Electoral College Be Abolished?" by Kay J. Maxwell and Robert Hardaway. Published in *The New York Times Upfront,* October 11, 2004. Copyright © 2004 by Scholastic Inc. Reprinted by permission.; **143—**"The Votes that Count: Experts Debate the U.S. Election System" *TIME For Kids,* October 29, 2004. Copyright © 2004 TIME For Kids. Reprinted by permission.; **175—**From "Debate still rages on 10th anniversary of Quebec's sovereignty referendum" CBC News, October 30, 2005. Reprinted by permission of CBC Radio Canada.; **304—**From "Language is the key to integration" an interview with Armin Laschet, by Anna Reimann. Spiegel Online, January 27, 2006. Reprinted by permission.; **305—**From "Speaking up for multilingualism" by David Gordon Smith. Expatica Germany, January 26, 2006. Reprinted by permission of Expatica Communications BV, www.expatica.com; **376—**From "Boreal: The great northern forest," by Fen Montaigne, *National Geographic,* June 2002. Copyright © 2002 National Geographic Society. Reprinted by permission.; **377—**From *The Last of the Last: Old Growth Forests of Boreal,* edited by Sarah Lloyd. Copyright © 1999, Taiga Rescue Network. Reprinted by permission.; **453—**From "Aral Sea's reveal retreat" by Norman Hammond. (Times Online, October 23, 2006). Copyright © Norman Hammond, NI Syndication Limited, 2006. Reprinted by permission.; **467—**From "Returning Ramesses: An Egyptian patriarch goes home" by Peter Lacovara. Reprinted by permission of the author.; **536—**From "Where Do we Stand Today with Private Infrastructure?" by Michael U. Klein, from *Development Outreach,* March 2003. Reprinted by permission of World Bank Institute, World Bank, Washington, D.C. www1.worldbank.org/devoutreach/; **537—** From "The Rains Do Not Fall on One Person's Roof: An Interview with Rudolf Amenga-Etego" Pambazuka News, August 26, 2004. Reprinted by permission Fahamu.; **545—**Excerpt from *Dark Star Safari: Overland from Cairo to Cape Town* by Paul Theroux. Copyright © 2003 by Paul Theroux. Reprinted

Index

Acknowledgments

Text (continued from p. 898)

by permission of Houghton Mifflin Company. All rights reserved.; **623—**From "Unanswered Questions," *Nova*, May 25, 1999. From WGBH Educational Foundation. Copyright © 2000 WGBH/Boston. Reprinted by permission.; **659—**From "Cyber Coolies" by Harish Trivedi, *Little India*, October 2004. Reprinted by permission.; **701—**From "Astonishing Wave Heights Among the Findings of an International Tsunami Survey Team on Sumatra" by Helen Gibbons and Guy Gelfenbaum (Sound Waves, March 2005, soundwaves.usgs.gov). Reprinted by permission of the U.S. Geological Survey.; **780—**From "Tourism Overview" from www.iaato.org/tourism_overview.html. Reprinted by permission of the International Association of Antarctic Tour Operators.; **781—**From "ASOC's Antarctic Tourism Campaign" (www.asoc.org/what_tourism.htm). Copyright © 2007 ASOC. Reprinted by permission of The Antarctic and Southern Ocean Coalition (ASOC).; **842—**From *The Log of Christopher Columbus*, translation © 1987 by Robert H. Fuson. Reprinted by permission of Amelia Fuson.

Photographs

Cover-(tl) Keith Macgregor/Getty Images, (tc) Andrea Pistolesi/Getty Images, (c) Yann Arthus-Bertrand/CORBIS, (r) David Turnley/CORBIS, (bl) Chris Cheadle/Getty Images; **ii-1-**NASA/Photo Researchers; **ii-103-**Alan Schein Photography/CORBIS; **ix-**Markus Scholz/Peter Arnold, Inc.; **v-**Walter Bibikow/Getty Images; **vi-**Peter Guttman/CORBIS; **vii-**Remi Benali/CORBIS; **x-**AP Photo; **xxi-**PhotoDisc; **1-**(bl) photolibrary.com/Index Open, (bkgd) PhotoDisc/Getty Images; **11-**Cancan Chu/Getty Images; **12–13-**Keren Su/Getty Images; **14-**John Van Hasselt/CORBIS; **15-**age fotostock/SuperStock; **16-**Michael S. Yamashita/CORBIS; **18-**(t to b) ThinkStock/SuperStock, (1) Janet Foster/Masterfile, (2) Mark Tomalty/Masterfile, (3) age fotostock/SuperStock, (4) Jurgen Freund/Nature Picture Library; **20-**(t) David Young-Wolff/PhotoEdit, (tc) NOAA/CORBIS, (b) The Photolibrary Wales/Alamy Images, (bc) Martin Harvey/Getty Images; **22-**(t) Dorling Kindersley/Getty Images, (b) Steve Skjold/Alamy Images; **34-**NASA/Roger Ressmeyer/CORBIS; **36-**(l) Andrew Parker/Alamy Images, (r) World Perspectives/Getty Images; **39-**(t) Tony West/CORBIS, (cl cr) age fotostock/SuperStock, (b) StockTrek/Getty Images; **42–43-**Bill Hatcher/National Geographic Image Collection; **44-**David Parker/Photo Researchers; **46-**Anthony West/CORBIS; **48-**Rafiqur Rahman/Reuters/CORBIS; **49-**Paul Bigland/Lonely Planet Images; **50-**Jimmy Chin/National Geographic Image Collection; **51-**(t) Library of Congress, (b) John Lemker/Animals Animals; **52-**(l) Tom Bean/Getty Images, (r) James P. Blair/NGS/Getty Images; **54-**Yves Marcoux/Getty Images; **55-**China Newsphoto/Reuters/CORBIS; **57-**John Maier Jr/Argus Fotoarchiv/CORBIS SYGMA; **61-**AFP/Getty Images; **62-**Peter Arnold, Inc./Alamy Images; **63-**Vince Streano/age fotostock; **64-**Norbert Rosing/National Geographic Image Collection; **67-**(tc) age fotostock, (tr) Tom Uhlman/Alamy Images, (bl) Remi Benali/CORBIS, (br) Greg Stott/Masterfile; **70–71-**ML Sinibaldi/CORBIS; **72-**Keren Su/Getty Images; **74-**(l) Cancan Chu/Getty Images, (r) Greg Elms/Lonely Planet Images; **77-**Wayne R. Bilenduke/Getty Images; **78-**Reuters/Jagadeesh NV; **79-**(tl) Reuters/Bobby Yip, (tr) Danita Delimont/Alamy Images, (b) AP Photo; **80-**Reuters/Gary Hershorn; **81-**(t) Baldev/CORBIS, (b) Blend/Punchstock; **82-**Erich Schlegel/Dallas Morning News/CORBIS; **84-**Hideo Haga/HAGA/The Image Works; **86-**North Wind Picture Archives/Alamy Images; **87-**Kevin Lee/Getty Images; **89-**age fotostock/SuperStock; **90-**Charles O. Cecil/Alamy Images; **90–91-**W. Cody/CORBIS; **91-**Sean Sprague/The Image Works; **92-**Frederic J. Brown/AFP/Getty Images; **95-**(l) AP World Wide, (r) George Pimentel/wireimage.com; **97-**(tl) Jamal Said/Reuters/Landov, (cl) Richard I'Anson/Lonely Planet Images, (cr) Tom Hanson/Canadian Press/AP Images, (b) Jerry Alexander/Lonely Planet Images; **100-**(t to b) Creatas/SuperStock, AP Images, Jim Zuckerman/CORBIS, Kurt Scholz/SuperStock, Lisa Englebrecht/Danitadelimont.com, Gary Cook/Alamy Images; **100–101-**(bkgd) NASA; **101-**(t to b) Jose Azel/Getty Images, Macduff Everton/CORBIS, AP Images, ITAR-TASS/Vitaly Belousov/Newscom; **105-**(tl) Timothy O'Keefe/Index Stock Imagery, (tr) Christian Heeb/Aurora Photos, (bl) Jim Wark/Lonely Planet Images, (br) Kevin Horan/Getty Images; **113-**Tim Smith/Getty Images; **114–115-**age fotostock/SuperStock; **116-**Ralph Lee Hopkins/Lonely Planet Images; **117-**aerialarchives.com/Alamy Images; **118-**(l) Paul A. Souders/CORBIS, (r) Sarah Leen/National Geographic Image Collection; **119-**SuperStock; **121-**Richard Olsenius/National Geographic Image Collection; **123-**Mona Reeder/Dallas Morning News/CORBIS; **124-**Steve Terrill/CORBIS; **126-**Liz Condo/AP Images; **127-**Franz Marc Frei/CORBIS; **129-**(t) Paul A. Souders/CORBIS, (cl) Walter Bibikow/Getty Images, (cr) Eric Nguyen/Jim Reed Photography/CORBIS, (b) Robert Harding World Imagery/CORBIS; **132–133-**Gunter Marx Photography/CORBIS; **134-**Mark E. Gibson/CORBIS; **136-**(l) The British Museum/HIP/The Image Works, (r) FPG/Getty Images; **137-**(l) Swim Ink 2, LLC/CORBIS, (r) Tim Brakemeier/dpa/CORBIS; **138-**Joseph Barrak/AFP/Getty Images; **139-**Mary Evans Picture Library; **140-**Najlah Feanny/CORBIS; **142-**Ed Andrieski/AP Images; **142–143-**Photodisc/Getty Images; **143-**Max Whittaker/CORBIS; **144-**Oliver Strewe/Lonely Planet Images; **145-**Bob Daemmrich/The Image Works; **146-**Tony Vaccaro/Getty Images; **148-**(tl) CORBIS, (tr) Lucien Aigner/CORBIS, (b) Bettmann/CORBIS; **149-**Alan Marsh/Getty Images; **151-**(t) Roy Rainford/Robert Harding, (bl) Bettmann/CORBIS, (bc) Hulton Archive/Getty Images, (br) Tim Smith/Getty Images; **154–155-**Aaron Huey/Polaris; **155-**(tcr) Melanie Acevedo, (tr) Ryan Remiorz/AP Images, (bcr) Roy Morsch/age fotostock, (br) Purestock/Alamy Images; **156–157-**Charles O'Rear/CORBIS; **158-**Ed Kashi/CORBIS; **159-**Julia Malakie/AP Images; **161-**Ted Soqui/CORBIS; **163-**AP Photo; **164-**(t) Beth Dixson/Alamy Images, AP Photo; **165-**AP Photo; **166-**AP Photo; **167-**AP Photo; **168-**Bill Brooks/Alamy Images; **169-**Jason Kryk/AP Images; **170-**The Bill and Melinda Gates Foundation; **171-**Evan Vucci/AP Images; **173-**(t) Adam Pretty/Getty Images, (cl) David Leahy/Getty Images, (bl) John and Lisa Merrill/CORBIS, (br) Mark Elias/Bloomberg News/Landov; **176–177-**Stone/Getty Images; **179-**(tr) Jack Novak/SuperStock, (cl) Ken Fisher/Getty Images, (bl) Jeff Rotman/Getty Images, (br) Hubert Stadler/CORBIS; **189-**Tom Cockrem/Lonely Planet Images; **190–191-**Tony Savino/The Image Works; **192-**Jeremy Horner/CORBIS; **193-**Jon Arnold Images/SuperStock; **194-**Kit Houghton/CORBIS; **195-**Pete Oxford/Nature Picture Library; **197-**(t) Media Bakery, (b) The Bridgeman Art Library; **198-**David Lyons/National Geographic Image Collection; **200-**(l) Tui De Roy/Minden Pictures, (r) Brent Winebrenner/LPI; **203-**(tl) SuperStock, (tr) Trevor Smithers ARPS/Alamy Images, (c) Rodrigo Arangua/AFP/Getty Images, (b) Galen Rowell/Odyssey Productions; **206–207-**Danny Lehman/CORBIS; **208-**Carlos Lopez-Barillas/CORBIS; **209-**Robert Frerck and Odyssey Productions; **211-**Werner Forman/CORBIS; **212-**(l) Charles & Josette Lenars/CORBIS, (r) Brooklyn Museum/CORBIS, **213-**(l) Index, Museo de America, Madrid, Spain/The Bridgeman Art Library; (r) Angelo Cavalli/Getty Images; **214-**(l) H.N. Rudd/CORBIS, (r) Jose Fuste Raga/CORBIS; **216-**Tim Page/CORBIS; **216-217-**Kevin Schafer/Photographer's Choice RF/Getty Images; **217-**Alfredo Maiquez/Getty Images; **218-**Blaine Harrington III/CORBIS; **219-**Tom Cockrem/Lonely Planet Images; **220-**David Dudenhoefer/Odyssey Productions; **221-**(l) Chris Brandis/AP Images, (r) AM Corporation/Alamy Images; **222-**Robert Holmes/CORBIS; **223-**GM Photo Images/Alamy Images; **225-**(tl) Danny Lehman/CORBIS, (tr) Mario Algaze/The Image Works, (bl) Alfredo Dagli Orti/The Art Archive/CORBIS, (br) Adriano Machado/Reuters/CORBIS; **228–229-**Adriana Zehbrauskas/Polaris; **229-**(tr) Danita Delimont/Alamy Images, (cr) Ken Welsh/age fotostock, (br) Luc Novovitch/Alamy Images; **230–231-**Chad Ehlers/Getty Images; **232-**Lynsey Addario/CORBIS; **233-**World Pictures/Alamy Images; **234-**Danita Delimont/Alamy Images; **235-**Danny Lehman/CORBIS; **237-**Oswaldo Rivas/Reuters/CORBIS; **238-**(l) Mark Godfrey c. 2004 The Nature Conservancy, (r) Courtesy Marie Claire Paiz; **239-**Alejandro Ernesto/epa/CORBIS; **241-**AP Photo/Dolores Ochoa; **242-**Peter Arnold,Inc./Alamy Images; **243-**(tl) World Picture Library/Alamy Images, (tc) AP Photo/Alberto Cesar-Greenpeace/HO, (tr) Sue Cunningham Photographic/Alamy Images, (b) Reuters/Carlos Barria; **244-**Reuters/Jamil Bittar; **245-**Reuters/Paulo Whitaker; **246-**Roger Ressmeyer/CORBIS; **247-**Paulo Whitaker/Reuters/Landov; **249-**Vanderlei Almeida/AFP/Getty Images; **250-**Eduardo De Baia/AP Images; **251-**Eliseo Fernandez/Reuters/CORBIS; **253-**(t) Marcelo Sayao/epa/CORBIS, (cl) Marco Ugarte/AP Images, (cr) Juan Barreto/AFP/Getty Images, (b) Jeff Greenberg/age fotostock; **256–257-**Roberto Gerometta/Lonely Planet Images; **259-**(tl) Anthony West/CORBIS, (tr) Jon Arnold Images/Alamy Images, (bl) Owen Franken/CORBIS, (br) Walter Bibikow/Getty Images; **269-**Nicole Duplaix/NGS/Getty Images; **271-**Tim Graham/Getty Images; **272–273-**Richard Nebesky/Lonely Planet Images; **274-**Emile Luider/Getty Images; **275-**age fotostock/SuperStock; **276-**(l) Chase Jarvis/Getty Images, (r) Jon Arnold/SuperStock; **277-**Walter Geiersperger/CORBIS; **278-**age fotostock/SuperStock; **279-**(inset) Andra Maslennikov/Peter Arnold, Inc., Ben Osborne/Getty Images; **281-**Sergei Supinsky/AFP/Getty Images; **282-**Fernand Ivaldi/Getty Images; **286-**Raymond Gehman/CORBIS; **287-**George Simhoni/Masterfile; **289-**(tl) John Garrett/CORBIS, (cl cr) age fotostock/SuperStock, (br) Asgeir Helgestad/Nature Picture Library; **292–293-**Hubert Stadler/CORBIS; **294-**David Tomlinson/Lonely Planet Images; **295-**Ted Spiegel/CORBIS; **297-**(l) Matt Houston/AP Images, (r) Royalty-Free/CORBIS; **298-**C. Steve Vidler/eStock Photo; **299-**National Gallery Collection; By kind permission of the Trustees of the National Gallery, London/CORBIS; **300-**(l) Scala/Art Resource, (r) Royalty-Free/Getty Images; **301-**(l) Ashmolean Museum, University of Oxford, UK/Bridgeman Art Library, (r) Rèunion des Musèes Nationaux/Art Resource; **304-**Homer Sykes/Alamy Images; **304–305-**Royalty-Free/CORBIS; **305-**Robert Fried Photography; **306-**Jeff Morgan/Alamy Images; **307-**Gideon Mendel/CORBIS; **308-**Michel Euler/AP Images; **309-**Simeone Huber/Getty Images; **310-**Daniel Mihailescu/AFP/Getty Images; **313-**(t) Art Resource, (cl) The Art Archive, (cr) Jonathan Smith/Lonely Planet Images, (b) Masterfile; **316–317-**Richard Harbus/Polaris; **317-**(tr) Steven Mark Needham/Picturearts/Newscom, (cr) Franck Fife/AFP/Newscom, (br) Patrick Sheandell O'Carroll/Getty Images; **318–319-**Elliot Daniel/Lonely Planet Images; **320-**The Image Bank/Getty Images; **322-**Manfred Gottschalk/Lonely Planet Images; **323-**Jayanta Shaw/Reuters/CORBIS; **324-**(l) General Photographic Agency/Getty Images, (r) Mary Evans Picture Library/The Image Works; **325-**Gideon Mendel/CORBIS; **326-**scenicireland.com/Christopher Hill Photographic/Alamy Images; **327-**Palmi Gudmundsson/Getty Images; **329-**George F. Mobley/National Geographic Image Collection; **330-**Jack Dabaghian/Reuters/CORBIS; **331-**(l) Eddie Koegh/Reuters Photo Archive/Newscom, (r) Getty Images for Nike; **332-**Reuters/Luis D'Orey; **333-**K.M. Westermann/CORBIS; **334-**Winfried Rothermel/AP Images; **335-**Christian Charisius/Reuters/CORBIS; **336-**Martin Ruetschi/epa/CORBIS; **338-**Getty Images; **339-**age fotostock/SuperStock; **340-**(inset) Charles O'Rear/CORBIS, Jose Manuel Ribeiro/Reuters/Landov; **341-**Peter Adams/Getty Images; **343-**Brian Atkinson/Alamy Images; **344-**Jack Naegelen/Reuters/CORBIS; **345–346-**AP Images; **348-**Sergei Supinsky/AFP/Getty Images; **349-**Reuters/Fabrizio Bensch; **350-**Katarina Stoltz/Reuters/CORBIS; **352-**Sean Gallup/Newsmakers/Getty Images; **353-**Peter Turnley/CORBIS; **355-**Robb Kendrick/Getty Images; **357-**(t) Alain Nogues/CORBIS, (cl) Tim Graham/Getty Images, (cr) Bob Stern/The Image Works, (bl) Goddard Space Flight Center Scientific Visualization Studio/NASA, (br) Craig Pershouse/Lonely Planet Images; **360–361-**Viktor Korotayev/Reuters/CORBIS; **363-**(tl) Tkachev Andrei/ITAR-TASS/Landov, (tr) age fotostock/SuperStock, (bl) Konstantin Mikhailov/Nature Picture Library, (br) Roshanak.B/CORBIS; **369-**Novosti/Topham/The Image Works; **370–371-**Maria Stenzel/National Geographic Image Collection; **372-**NGS/Getty Images; **373-**Lee Foster/Lonely Planet Images; **376-**Osetrov Yury/ITAR-TASS/Landov; **376–377-**Karl Weatherly/Getty Images; **377-**Klaus Nigge/NGS/Getty Images; **378-**Rashid Salikhov/EPA/epa/CORBIS; **380-**(l) Maria Stenzel/National Geographic Image Collection, (r) SIME s.a.s/

eStock Photo; **381-**(r) Marc Garanger/CORBIS, (b) Wolfgang Kaehler/CORBIS; **382-**(inset) Konstantin Mikhailov/naturepl.com, Chris Niedenthal/Time Life Pictures/Getty Images; **383-**(l) Peter Guttman/CORBIS, (r) Peter Turnley/CORBIS, (b) Wade Eakle/Lonely Planet Images; **386–387-**Mark Sykes/Alamy Images; **388-**Stringer/AFP/Getty Images; **390-**(l) Martin Gray/National Geographic Image Collection, (r) Kremlin Museums, Moscow, Russia/The Bridgeman Art Library; **391-**(l) National Geographic Image Collection, (r) Douglas Kirkland/CORBIS; **392-**Yuri Kozyrev for Time; **393-**Central Press/Getty Images; **395-**Markus Scholz/Peter Arnold, Inc.; **396-**Ilya Naymushin/Reuters/CORBIS; **397-**Steve Vidler/SuperStock; **398-**(l) Igor Akimov/ITAR-TASS/Landov, (r) Topham/The Image Works; **399-**REZA/National Geographic Society Image Collection; **401-**(t) Hulton-Deutsch Collection/CORBIS, (cl) Topham/The Image Works, (cr) Marc Garanger/CORBIS, (b) Peter Turnley/CORBIS; **404–405-**Jeremy Nicholl/Polaris; **405-**(tr) Bill Aron/Photo Edit, (cr) Anatoly Rukhadze/ITAR-TASS Photos/Newscom, (br) Creatas/Punchstock; **406–407-**Belinsky Yuri/ITAR-TASS/CORBIS; **408-**Reuters/CORBIS; **409-**Ivan Sekretarev/AP Images; **410-**AP Images; **413-**AP Photo; **414-**Wolfgang Kaehler/CORBIS; **415-**(l c) AP Photo, (r) Peter Blakely/CORBIS SABA; **416-**Smolsky Sergei/ITAR-TASS/CORBIS; **417-**AP Photo; **418-**Richard Nowitz/National Geographic Image Collection; **419-**Reuters/CORBIS; **421-**Khasan Kaziyev/AFP/Getty Images; **423-**(t) Buddy Mays/CORBIS, (cl) Alexey Danichev/AFP/Getty Images, (bl) transit/Peter Arnold, Inc., (br) AP Images; **426–427-**Tim de Waele/CORBIS; **429-**(tl) Hanan Isachar/CORBIS, (tr) PCL/Alamy Images, (bl) Frank Krahmer/Masterfile, (br) The Photolibrary Wales/Alamy Images; **439-**Vittoriano Rastelli/CORBIS; **440–441-**Ed Kashi/CORBIS; **442-**Jonathan Blair/CORBIS; **443-**Look GMBH/eStock Photo; **444-**Bob Turner/Alamy Images; **447-**Kazuyoshi Nomachi/CORBIS; **448-**(inset) age fotostock/SuperStock, Sara-Jane Cleland/Lonely Planet Images; **451-**(t) Reza/National Geographic Image Collection, (cl) Peter Turnley/CORBIS, (cr) Janet Wishnetsky/CORBIS, (b) Eoin Clarke/Lonely Planet Images; **454–455-**Jean du Boisberranger/Getty Images; **456-**Archivo Iconografico, S.A./CORBIS; **458-**Werner Forman/CORBIS; **459-**SIME s.a.s/eStock Photo; **460-**Scala/Art Resource; **462-**(l) Rèunion des Musèes Nationaux/Art Resource, (r) Bettmann/CORBIS; **463-**(l) Hulton-Deutsch Collection/CORBIS, (r) Mohammed Adnan/AP Images; **464-**(l) Vittoriano Rastelli/CORBIS, (r) Ron Edmunds/AP Images; **466-**Sandro Vannini/CORBIS; **466–467** Royalty-Free/CORBIS; **467-**Jose Luis Pelaez, Inc./CORBIS; **468-**Kazuyoshi Nomachi/CORBIS; **469-**Mark Daffey/Lonely Planet Images; **470-**Bethune Carmichael/Lonely Planet Images; **471-**Awad Awad/AFP/Getty Images; **472-**David Silverman/Getty Images; **473-**Nigel Reed/Alamy Images Images; **474-**Zylberman Laurent/CORBIS; **475-**(l) Ana E. Fuentes/Zuma Press/Newscom, (r) Micheline Pelletier/CORBIS; **477-**(tl) Gianni Dagli Orti/CORBIS, (tr) Prisma/SuperStock, (bl) Anthony Plummer/Lonely Planet Images, (br) Peter Sanderrs/HAGA/The Image Works; **480–481-**David Rochkind/Polaris; **481-**(t) Jimin Lai/AFP Photo/Newscom, (c) Reuters/Bazuki Muhammad/Newscom, (b) Emmanuel Dunand/AFP/Getty Images/Newscom; **482–483-**Jose Fuste Raga/CORBIS; **484-**Ray Roberts/Alamy Images; **485-**Nasser Nasser/AP Images; **486-**AP Images; **487-**age fotostock/SuperStock; **489-**Ed Kashi/CORBIS; **490-**Nakheel Development/AP Images; **491-**Harvey Lloyd/Getty Images; **492-**Karel Prinsloo/AP Images; **494-**Paula Bronstein/Getty Images; **495-**AFP/Getty Images; **497-**Lynsey Addario/CORBIS; **499-**AP Photo; **500-**AP Photo; **501-**AP Photo; **502-**AP Photo; **503-**AP Photo; **504-**Stringer/AP Images; **505-**Guenter Fischer/imagebroker/Alamy Images; **506-**imagebroker/Alamy Images; **507-**Scott Peterson/Getty Images; **509-**(t) AFP/Getty Images, (c) age fotostock/SuperStock, (b) Ali Haider/epa/CORBIS; **512–513-**Jon Arnold Images/Alamy Images; **515-**(tl) Michael Freeman/CORBIS, (tr) Network Photographers/Alamy Images, (cl) Images of Africa Photobank/Alamy Images, (bl) David Keaton/CORBIS; **527-**Malcolm Linton/Liaison/Getty Images; **528–529-**SIME s.a.s/eStock Photo; **530-**Royalty-Free/CORBIS; **532-**(l) Winfried Wisniewski/Getty Images, (r) Jon Arnold Images; **533-**age fotostock; **534-**Michael Nichols/National Geographic Image Collection; **536-**Jacob Silberberg/Panos Pictures; **537-**Giacomo Pirozzi/Panos Pictures; **538-**George Steinmetz/CORBIS; **540-**Wolfgang Kaehler/CORBIS; **541-**AAI Fotostock; **543-**(t) Royalty-Free/CORBIS, (cl) Daryl Balfour/Getty Images, (cr) Peter Andrews/Reuters, (b) George Steinmetz/CORBIS; **546–547-**Ariadne Van Zandbergen/Lonely Planet Images; **548-**Jon Hicks/CORBIS; **549-**David Reed/CORBIS; **551-**AKG Berlin/SuperStock; **552-**(l) Timothy Kendall, (r) Diego Lezama Orezzoli/CORBIS; **553-**(l) Lauros/Giraudon, Bibliotheque Nationale, Paris, France/The Bridgeman Art Library, (c) Held Collection, National Museum, Lagos, Nigeria/The Bridgeman Art Library, (r) STR/epa/CORBIS; **556-**(l) PhotoDisc, (r) James L. Stanfield/National Geographic Society Image Collection; **557-**Carol Polich/Lonely Planet Images; **558-**Jean-Lèo Dugast/Peter Arnold, Inc.; **559-**Richard Lord/The Image Works; **560-**Anna Zieminski/Getty Images; **561-**age fotostock/SuperStock; **562-**Christy Gavitt/IPN/Woodfin Camp & Assoc.; **563-**Ann Johansson/CORBIS; **565-**(t) David Turnley/CORBIS, (cl) Bruce Coleman Inc./Alamy Images, (cr) Patrick Olear/PhotoEdit, (b) Frances Linzee Gordon/Lonely Planet Images; **568-**(l) AP Images, (c) Abayomi Adeshida; (r) Chris Hondros/Getty Images; **569-**(l) Abayomi Adeshida, (tr) Parick Olear/PhotoEdit, (cr) Abayomi Adeshida, (br) Jane Sweeney/Lonely Planet Images; **570–571-**Peter Horree/Alamy Images; **572-**James Marshall/CORBIS; **573-**James Marshall/The Image Works; **576-**Michael Lewis/National Geographic Image Collection; **577-**Andanson James/CORBIS SYGMA; **578-**Picture Contact/Alamy Images; **579-**(l) Don Boroughs/The Image Works, (r) Tony Karumba/AFP/Getty Images; **580-**Wolfgang Langenstrassen/dpa/CORBIS; **581-**Peter Turnley/CORBIS; **583-**Marcus Prior/WFP/EPA/CORBIS; **584-**AP Photo; **585-**AP Photo, (c) Reuters/Lud Gnago, (r) AP Photo; **586-**AP Photo; **587-**Patrick Robert/Sygma/CORBIS; **588-**Jeff Rotman/jeffrotman.com/AGPix; **589-**CORBIS SYGMA; **590-**(l) Irene Staunton, (b) AFP/Getty Images; **591-**Ron Giling/Peter Arnold, Inc.; **593-**(t) Marco Cristofori/Peter Arnold, Inc., (cl) Jack Sullivan/Alamy Images, (cr) Mike Hutchings/Reuters/CORBIS, (b) G P Bowater/Alamy Images; **596–597-**Keren Su/CORBIS; **599-**(t) Nature Picture Library/Alamy Images, (cl) DPA/AJI/The Image Works, (bl) Hemis/Alamy Images Images, (br) David H. Wells/CORBIS; **605-**Eye Ubiquitous/CORBIS; **607-**Wolfgang Langenstrassen/dpa/Landov; **608–609-**Robb Kendrick/Aurora Photos; **610-**Torleif Svensson/CORBIS; **611-**Jake Norton/Aurora Outdoor Collection/Getty Images; **612-**Prashant Panjiar; **613-**Michael S. Yamashita/CORBIS; **614-**Deshakalyan Chowdhury/AFP/Getty Images; **616-**Michelle Burgess/Visuals Unlimited, f1 online/Alamy Images; **617-**Andrew Errington/Getty Images; **619–621-**Reuters/CORBIS; **621-**(tr) Picture Finders Ltd./eStock Photo, (c) Thomas Mangelsen/Minden Pictures, (b) Eugen/zefa/CORBIS; **624–625-**Suzanne & Nick Geary/Getty Images; **626-**Rafiqur Rahman/Reuters/CORBIS; **627-**Harappa Archaeological Research Project, Courtesy Dept. of Archaeology and Museums, Govt. of Pakistan; **629-**Chester Beatty Library, Dublin/Bridgeman Art Library; **630-**(l) The Art Archive/CORBIS, (r) Martin Schoyen/The Schoyen Collection; **631-**(l) Yoshio Tomii/SuperStock, (r) AP Images; **632-**Mian Khursheed/Reuters/Landov; **633-**Medioimages/Getty Images; **634-**Robert Nickelsberg/Getty Images; **635-**(l) Reuters/Kimal Kishore, (c r) AP Photo; **636-**Kapoor Baldev/Sygma/CORBIS; **637-**Charles Sturge/Alamy Images; **638-**Michael Freeman/CORBIS; **639-**Shafiq Alam/AFP/Getty Images; **640-**Don Smith/Alamy Images; **642-**Neil Cooper/Alamy Images; **643-**(l) Wolfgang Langenstrassen/dpa/Landov, (r) Dinodia/The Image Works; **645-**(tl) Stapleton Collection/CORBIS, (cl) Abir Abdullah/epa/CORBIS, (cr) AP Images, (b) Dinodia/The Image Works; **648–649-**Sam Hollenshead/Polaris; **649-**(tr) Susan Liebold, (cr) AP Images, (br) DPA/The Image Works; **650–651-**Tom Cockrem/Lonely Planet Images; **652-**Amit Dave/Reuters/CORBIS; **654-**Ami Vitale/Reportage/Getty Images; **655-**(t) Jerry Bauer Photography, (b) Sucheta Das/Reuters/CORBIS; **656-**Joerg Boethling/Peter Arnold, Inc.; **658-**Sherwin Crasto/Reuters/CORBIS; **658–659-**Royalty-Free/CORBIS; **659-**Fredrik Renander/Alamy Images; **660-**Rafiqur Rahman/Reuters/CORBIS; **661-**Syed Zargham/Getty Images; **662-**(l) Anjum Naveed/Pool/Reuters/CORBIS, (r) Shakil Adil/AP Images; **663-**Rafiqur Rahman/Reuters/CORBIS; **665-**Jeremy Horner/CORBIS; **666-**Paul Dymond/Lonely Planet Images; **667-**Michael Melford/National Geographic Image Collection; **669-**(tr) Topham/The Image Works, (c) Howard Davies/CORBIS, (bl) Markus Kirchgessner/Bilderberg/Aurora Photos, (br) Devendra Man Singh/AFP/Getty Images; **672–673-**Pixtal/age fotostock; **675-**(t) Dean Conger/National Geographic Image Collection, (c) Keren Su/Lonely Planet Images, (bl) Jacques Langevin/CORBIS SYGMA; **683-**Eye Ubiquitous/CORBIS; **685-**AP Images; **686–687-**Raymond Gehman/National Geographic Image Collection; **688-**John Elk III/Lonely Planet Images; **689-**Josè Fuste Raga/zefa/CORBIS; **690-**WEDA/epa/CORBIS; **691-**NASA; **692-**Palani Mohan for Time; **693-**Michael S. Yamashita/CORBIS; **694-**Frans Lanting/Minden Pictures; **695-**(l) AFP/Getty Images, (r) Hamid Sardar/CORBIS; **697-**Roy Toft/National Geographic Image Collection; **699-**(tl) Tarmizy Harva/Reuters/CORBIS, (tc) age fotostock/SuperStock, (tr) Michael Reynolds/epa/CORBIS, (b) George Steinmetz/CORBIS; **702–703-**Gavin Hellier; **704-**Justin Guariglia/National Geographic Image Collection; **705-**Liu Liqun/CORBIS; **706-**(t) Rèunion des Musèes Nationaux/Art Resource, (cl) Asian Art & Archeology Inc./CORBIS, (cr) ChinaStock, (b) Royal Ontario Museum/CORBIS; **707-**Visual Arts Library, London/Alamy Images; **708-**Archivo Iconografico, S.A./CORBIS; **710-**(l) Asian Art & Archaeology, Inc./CORBIS, (r) O Louis Mazzatenta/NGS/Getty Images; **711-**(l) Asian Art & Archaeology, Inc./CORBIS, (r) Landov; **713-**(t) Archivo Iconografico, S.A./CORBIS, (b) Mary Evans Picture Library/Alamy Images; **714-**Keren Su/CORBIS; **715-**AP Images; **716-**Mark Henley; **717-**(t) Photolibrary, (b) The Art Archive; **718-**AP Images; **719-**(l) Carl & Ann Purcell/CORBIS, (r) Bob Sacha/CORBIS; **721-**(tl) Kimbell Art Museum/CORBIS, (tr) Steve Vidler/eStock Photo, (bl) The Art Archive, Patrick Lin/AFP/Getty Images; **724–725-**Kemal Jufri/Polaris; **725-**(tr) Jerry Alexander/Getty Images, (cr) Petr Svarc/Alamy Images, (br) Wendy Chan/Getty Images; **726–727-**K Eriksson/SV-Bilderdienst/The Image Works; **728-**Guang Niu/Getty Images; **729-**Reinhard Krause/Reuters/CORBIS; **730-**(l) Peter Turnley/CORBIS, (r) AP Images; **731-**Bobby Yip/Reuters/CORBIS; **732-**Brent Winebrenner/Lonely Planet Images; **734-**China Photo ASW/RCS/Reuters; **734–735-**Gregor Schuster/zefa/CORBIS; **735-**Lou Linwei/Alamy Images; **736-**Paul Chesley/Getty Images; **737-**Steve Vidler/SuperStock; **739-**Anna Clopet/CORBIS; **740-**Ryan McVay/Getty Images; **742-**Neil Beer/CORBIS; **743–744-**AP Images; **745-**epa/CORBIS; **747-**Michael Prince/CORBIS; **748-**(t) Brooks Kraft/CORBIS, (b) AP Images; **749-**(l) Paul Souders/CORBIS, (c r) AP Images; **750-**Kim Yung Hoon/Reuters/CORBIS; **751-**Liu Lioun/CORBIS; **752-**Sergio Pitamitz/CORBIS; **753-**AP Images; **754-**Free Agents Limited/CORBIS; **755-**Juliet Coombe/Lonely Planet Images; **757-**Choo Youn-Kong/AFP/Getty Images; **759-**(t) Yun Suk Bong/Reuters, (cl) AP Images, (cr) Digital Vision/Getty Images, (b) Frank Carter/Lonely Planet Images; **762–763-**R. Wallace/Stock Photos/zefa/CORBIS; **765-**(t) Fred Bavendam/Minden Pictures, (cr) Michael & Patricia Fogden/CORBIS, (bl) R. Ian Lloyd/Masterfile, (br) Gordon Wiltsie/Getty Images; **773-**SuperStock; **774–775-**age fotostock/SuperStock; **776-**Chris Mellor/Lonely Planet Images; **777-**Ken Usami/Photodisc Green/Getty Images; **778-**(l) Morrison/AUSCAPE/Minden Pictures, (r) Darrell Gulin/CORBIS; **780-**Wolfgang Kaehler/CORBIS; **780–781-**Royalty-Free/CORBIS; **781-**Panoramic Images/Getty Images; **782-**Bill Hatcher/NGS/Getty Images; **784-**Carl D. Walsh/Aurora Photos; **785-**Anders Ryman/Alamy Images; **787-**(t) age fotostock/SuperStock, (bl) Juliet Coombe/Lonely Planet Images, (br) Jon Arnold Images/SuperStock; **790–791-**David Wall/Lonely Planet Images; **792-**Richard I'Anson/Lonely Planet Images; **793-**Ken Gillham/Robert Harding World Imagery/CORBIS; **794-**(l) SuperStock, (r) Gottfried Lindauer/Getty Images; **795-**(l) N G Thwaites/Getty Images, (r) age fotostock/SuperStock; **798-**(t) Frans Lanting/CORBIS, (b) Private Collection/The Bridgeman Art Library; **799-**Will Burgess/Reuters/CORBIS; **800-**Bill Bachman/Alamy Images; **801-**Richard Nebesky/Lonely Planet Images; **802-**(t) Wallace-Crabbe/National Library of Australia, (b) CORBIS; **803-**Klaus Nigge/National Geographic Image Collection; **805-**(tr) Chris McGrath/Getty Images, (c) Grant Dixon/Lonely Planet Images, (bl) Mary Evans Picture Library, (br) Mrs Holdsworth/Robert Harding World Imagery/Getty Images; **808–809-**Jeremy Nicholl/Polaris; **809-**(tr) Photo Resource Hawaii/Alamy Images, (cr) Susan Liebold, (br) Rubberball/Punchstock; **810–811-**Bill Bachman/Wild Light; **812-**Michael Bradley/Getty Images; **813-**Jerry Lampen/Reuters/CORBIS; **815-**C. Dani/Peter Arnold, Inc.; **816-**Kevin Fleming/CORBIS; **817–818-**Mike Powell/Getty Images; **819-**(l) Paul A. Souders/CORBIS, (c) Neil Rabinowitz/CORBIS, (r) New Zealand Film Comm./The Kobal Collection; **820-**Reuters/Will Burgess; **821-**Dave G. Houser/Post-Houser Stock/CORBIS; **822-**Louie Psihoyos/CORBIS; **823-**Reuters; **824-**Giraud Philippe/CORBIS SYGMA; **825-**(l) Catherine Karnow/CORBIS, (r) LOOK Die Bildagentur der Fotografen GmbH/Alamy Images; **827-**Gordon Wiltsie/National Geographic Image Collection; **828-**(l) Courtesy Nigel Watson, (r) Mark Baker/Agence France Presse/Newscom; **829-**Kyodo/Landov; **831-**(t) Robyn Jones/Lonely Planet Images, (cl) Australian Picture Library/CORBIS, (bl) Christopher Groenhout/Lonely Planet Images, (br) Kurt Scholz/SuperStock; **834-**Wade Eakle/Lonely Planet Images; **837-**Jerry Barnett; **838-**Kunsthistorisches Museum, Wien oder KHM, Wien; **839-**Larry Williams/CORBIS; **843-**Michael Nichols/National Geographic Image Collection